COHERENCE AND QUANTUM OPTICS VIII

COHERENCE AND QUANTUM OPTICS VIII

Proceedings of the Eighth Rochester Conference on
Coherence and Quantum Optics, held at the
University of Rochester, June 13–16, 2001

Edited by

N. P. Bigelow, J. H. Eberly,
C. R. Stroud, and I. A. Walmsley

University of Rochester
Rochester, New York

KLUWER ACADEMIC / PLENUM PUBLISHERS
NEW YORK, BOSTON, DORDRECHT, LONDON, MOSCOW

Proceedings of the Eighth Rochester Conference on Coherence and Quantum Optics, held June 13–16, 2001, at the University of Rochester, Rochester, New York.

ISBN 0-306-48116-2

233 Spring Street, New York, New York 10013

http://www.kluweronline.com

10 9 8 7 6 5 4 3 2 1

A C.I.P. record for this book is available from the Library of Congress

Printed in the United States of America

PREFACE

The Eighth Rochester Conference on Coherence and Quantum Optics was held on the campus of the University of Rochester during the period June 13-16, 2001. This volume contains the proceedings of the meeting. The meeting was preceded by an affiliated conference, the International Conference on Quantum Information, with some overlapping sessions on June 13. The proceedings of the affiliated conference will be published separately by the Optical Society of America. A few papers that were presented in common plenary sessions of the two conferences will be published in both proceedings volumes. More than 268 scientists from 28 countries participated in the week long discussions and presentations.

This Conference differed from the previous seven in the CQO series in several ways, the most important of which was the absence of Leonard Mandel. Professor Mandel died a few months before the conference. A special memorial symposium in his honor was held at the end of the conference. The presentations from that symposium are included in this proceedings volume. An innovation, that we believe made an important contribution to the conference, was the inclusion of a series of invited lectures chaired by CQO founder Emil Wolf, reviewing the history of the fields of coherence and quantum optics before about 1970. These were given by three prominent participants in the development of the field, C. Cohen-Tannoudji, J.F. Clauser, and R.J. Glauber. Their lectures are included in the proceedings and should provide a valuable resource for historians of science. The historical sessions were also recorded for preservation in the History Center of the American Institute of Physics, and recordings are available through that office. A photographic record of the conference is also available in an extensive archive that is available electronically at http://www.optics.rochester.edu/~stroud/conference.

Production was managed with the important assistance of the Optical Society of America, through Deborah Herrin's office of New Product Development, which supplied staff backup for the accumulation and transmission to Rochester of the submitted papers before the meeting. OSA staff and its Internet-based submission-download procedures were used to regularize, to a much greater extent than could be achieved in previous CQO volumes, a uniform type face, page format and reference style. However, authors can subvert uniformity in a variety of ways, sometimes accidentally, and in any event strict uniformity is almost impossible to enforce when different Text Editors from the same commercial source are subtly incompatible with each other. The variations among the articles in the book thus serve, in a non-scientific sense, as indicators of the state of the art of automated production of conference proceedings at the turn of the 21st century. Interestingly, there is no 21st century method by which production of a useful Index can be automated, so the 766 entries, covering 8 pages themselves, were accumulated the old-fashioned way by hand and eye, page by page through the proofs.

The organizing committee for the conference included: G. S. Agarwal, N. P. Bigelow, J. H. Eberly, S. Haroche, P. L. Knight, S. Stenholm, C. R. Stroud, Jr., I. A. Walmsley, and H. Walther.

N. P. Bigelow, J. H. Eberly, C. R. Stroud, Jr., and I. A. Walmsley, joint secretaries.

CONTENTS

COHERENCE AND QUANTUM OPTICS VIII

The Early Days of Quantum Optics in France

C. Cohen-Tannoudji
Collège de France and Laboratoire Kastler-Brossel, département de physique de l'Ecole Normale Supérieure, 24 rue Lhomond, 75231 Paris cedex 05, France
cct@lkb.ens.fr

Abstract.

We review a few early studies on atom-photon interactions which took place in France after world war II and before the seventies. Several examples are discussed, including double resonance and optical pumping, radiofrequency mutiphoton processes, quantum interference effects in radiative processes, master equation description of atom-photon interactions, optical resonators and laser media, etc. We try to point out the continuity which exists between these early studies and more recent investigations in new research fields, such as laser cooling and trapping. Some general conclusions about science policy will be finally drawn from such a historical review.

Before starting, I would like first to express my warmest thanks to the organizers for their very kind invitation to speak in this session devoted to the history of quantum optics. I think that it is a very good idea to look back from time to time to the early stages of a research field. That can give a better perception of the present state of the field and provide some inspiration for future developments. There is a colleague and friend whom we miss very much today. Leonard Mandel is no longer with us. His contributions to the development of quantum optics are very deep and outstanding. I am sure that he would have liked very much the idea of an historical session.

1. Introduction

Quantum optics can be considered as a quantum theory of light and of its interactions with matter. To be more specific, let me focus on atom-photon interactions. This is a very broad field, so broad that we need a few guiding ideas to give an underlying frame to the discussion. Here are a few :

Basic conservations laws : they explain the exchanges of energy, linear and angular momentum between atoms and photons and they are at the root of important methods such as optical pumping and, more recently, laser cooling and trapping.

Higher order effects : they appear when the field intensity is so high that the first order perturbative treatment is no longer sufficient. This covers the whole field of multiphoton processes and non linear optics.

Linear superpositions of states : they are at the heart of quantum mechanics and they give rise to a wealth of quantum interference effects observable on the light emitted or absorbed by atoms. Linear superpositions of product states of two systems give rise also to entangled states which are so important in the new field of quantum information.

Master equations : they provide a quantum description of the evolution of the system. They can take different forms, depending on the coherent properties of the field: rate equations, optical Bloch equations. More recent developments include stochastic wave functions and Monte-Carlo simulations.

Laser development raised also several interesting questions such as the design of good optical resonators or the identification of efficient laser media.

When Joe invited me to talk in this session, he asked me to focus on the early contributions of France to quantum optics before the seventies. This is not an easy job because Science is above all international and it may be artificial and even somewhat pretentious to focus on a specific scientific community. Nevertheless, I will try to do it and to point out the connections and the continuity between these early studies and more recent developments of the field such as cavity QED, laser cooling and trapping or matter waves. By looking at the evolution of a scientific community in a given place and for a long period of time, I think also that one can draw a few conclusions on science policy.

2. Conservation of angular momentum

2.1. Polarization selection rules

Let us consider the three simple states of polarization, σ_+-polarization corresponding to a right circular polarization around the z-axis, σ_--polarization corresponding to a left circular polarization around the z-axis and π-polarization corresponding to a linear polarization parallel to the z-axis.

The polarization selection rules express that these three polarizations induce transitions between Zeeman sublevels m and m' such that :

$$\begin{aligned} \sigma_+ - \text{polarization} &\leftrightarrow \Delta \text{m} = \text{m}' - \text{m} = +1 \\ \sigma_- - \text{polarization} &\leftrightarrow \Delta \text{m} = \text{m}' - \text{m} = -1 \\ \pi - \text{polarization} &\leftrightarrow \Delta \text{m} = \text{m}' - \text{m} = 0 \end{aligned}$$

The interpretation of these selection rules is straightforward. They express that the angular momentum of the absorbed photon along the z-axis, which is equal to $+\hbar$ for σ_+ light, $-\hbar$ for σ_- light, 0 for π light, is transferred to the atom when the atom absorbs the photon so that the atomic magnetic quantum number *m* changes by +1, -1 or 0 . This connection between polarization selection rules and conservation of angular momentum was pointed out by A. Rubinowicz whose work inspired very much A. Kastler at the beginning of his investigation of resonance fluorescence. In fact the basic idea is that, by exciting atoms with polarized light, one can prepare them in well defined Zeeman sublevels. In other words, one can polarize atoms by light. The second important point is that, by detecting the light emitted or absorbed by an atom with a given polarization, one can determine the Zeeman sublevel occupied by the atom before the emission or absorption process. In other words, one can monitor the angular momentum of atoms by light.

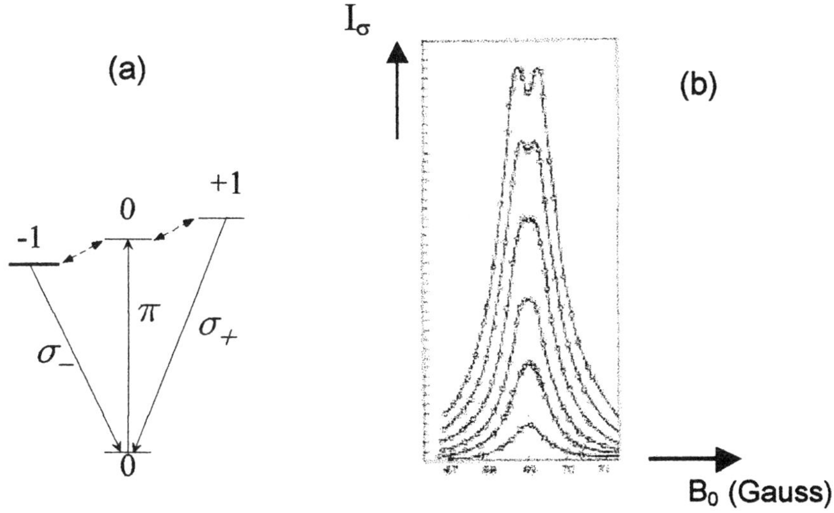

Fig. 1. (a) : Zeeman components of a transition 0⇔1. The excitation is π-polarized and one observes the σ-polarized reemitted light. (b) : Variations of the σ-polarized reemitted light versus the magnetic field which splits the 3 excited Zeeman sublevels. The various curves correspond to different values of the amplitude of the RF field which induces resonant transitions between the excited Zeeman sublevels. The amplitude of the RF field increases by a factor 7 between the smallest curve and the largest one.

2.2. Double resonance

The idea of double resonance was suggested by A. Kastler and J. Brossel in 1949 [1]. Let us consider the simple case of a transition connecting a lower state with angular momentum 0 to an excited state with angular momentum 1 with three Zeeman sublevels −1, 0, +1 (Fig.1-a). If one excites such an atom with π-polarized light, one excites only the sublevel 0 of the excited state. Then, during the time spent by the atom in the excited state, one can perform magnetic resonance and induce resonant transitions between this sublevel 0 of the excited state and the other sublevels −1 and +1. These transitions are easily detected by looking at the light emitted with a σ_+ or a σ_- polarization, or with a so called σ-polarization, that is to say, a linear polarization perpendicular to the magnetic field (Fig.1-a). Figure 1-b gives an example of experimental results which have been obtained by J. Brossel and F. Bitter on the even isotopes of mercury [2]. It represents the magnetic resonance curves in the excited state of the mercury atom detected by observing the σ-polarized fluorescence intensity as a function of the static magnetic field which produces the Zeeman splitting between the sublevels. These curves allow one to measure the magnetic moment of the excited state. By extrapolating the width of the curves at the limit of zero intensity of the RF field, one can also measure the natural width of the excited state.

2.3. Optical pumping

The double resonance method allowing the investigation of excited atomic states has been extended by A. Kastler to ground states, giving birth to the so called optical pumping method [3]. To explain optical pumping, let us consider the simple case of a transition connecting a ground state with an angular momentum ½ to an excited state with an angular momentum ½, so that there are two ground state Zeeman sublevels $g_{-1/2}$ and $g_{+1/2}$ and two excited Zeeman sublevels $e_{-1/2}$ and $e_{+1/2}$. If one excites such an atom with σ_+-polarized light, one excites only the transition $g_{-1/2}$ to $e_{+1/2}$ because this is the only transition with $\Delta m = +1$. Once the atom has been excited in $e_{+1/2}$, it can fall back by spontaneous emission either in $g_{-1/2}$, in which case it can repeat the same cycle, or in $g_{+1/2}$ by emission of a π-polarized photon. In the last case, the atom remains trapped in $g_{+1/2}$ because there is no σ_+ transition starting from $g_{+1/2}$. This gives rise to what is called an optical pumping cycle transferring atoms from $g_{-1/2}$ to $g_{+1/2}$ through $e_{+1/2}$. How can one detect such a transfer ? It clearly appears in figure 2 that atoms absorb σ_+ light only if they are in $g_{-1/2}$. If they are in $g_{+1/2}$, they cannot absorb σ_+ light because there is no σ_+ transition starting from $g_{+1/2}$. This means that any transfer from $g_{+1/2}$ to $g_{-1/2}$ which would be induced by a resonant RF field or by a relaxation process can be detected by monitoring the amount of light absorbed with σ_+-polarization.

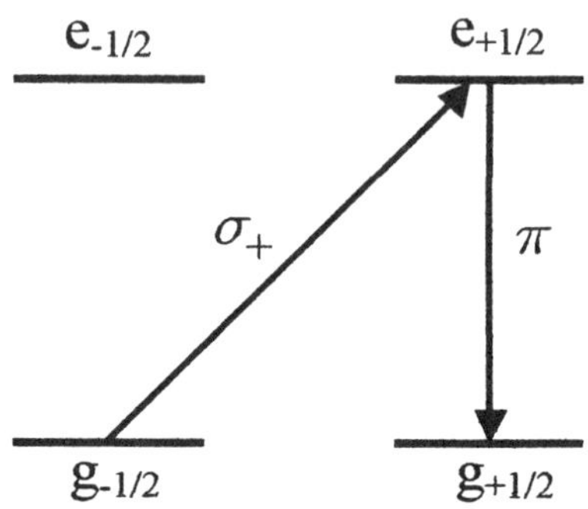

Fig. 2. Principle of optical pumping for a ½ ⇔ ½ transition. Atoms are transferred from $g_{-1/2}$ to $g_{+1/2}$ by an "optical pumping cycle" which consists of an absorption of a σ_+-polarized photon followed by the spontaneous emission of a π–polarized photon

2.4. Interesting features of these optical methods

A first interesting feature of these optical methods is that they provide a very efficient scheme for polarizing atoms at room temperatures and in low magnetic fields. Secondly, they have a very high sensitivity. The absorption of a single RF photon is detected by the subsequent optical absorption or emission of an optical photon and it is much easier to detect an optical photon than a RF photon because it has a much higher energy. It thus becomes possible to study very dilute systems where interactions are weak. The third significant point is that the Doppler effect of the RF transitions induced between Zeeman sublevels is negligible. In a sense, this can be considered as a Doppler free spectroscopy. At last, these optical methods allow one to study and to investigate non-equilibrium situations. Atoms are removed from their thermodynamic equilibrium and by observing the temporal variations of the absorbed or emitted light, one can study how the system returns to equilibrium using what can be called a pump-probe spectroscopy.

2.5. More recent developments

Optical pumping methods still find interesting applications. For example, gaseous samples of optically pumped He_3 atoms are used for studying the scattering of proton beams by polarized targets or for obtaining magnetic resonance images of the cavities of the human lung [4]. Another example is the field of spin polarized quantum systems where one studies the modifications of the transport properties of a spin polarized gas due to the postulate of symmetrisation which forbids certain collision channels [5].

3. Radiofrequency multiphoton resonances

The sensitivity of optical methods has allowed physicists in the beginning of the fifties to observe RF multiphoton resonances between Zeeman sublevels. This section is devoted to the description of a few of these experiments

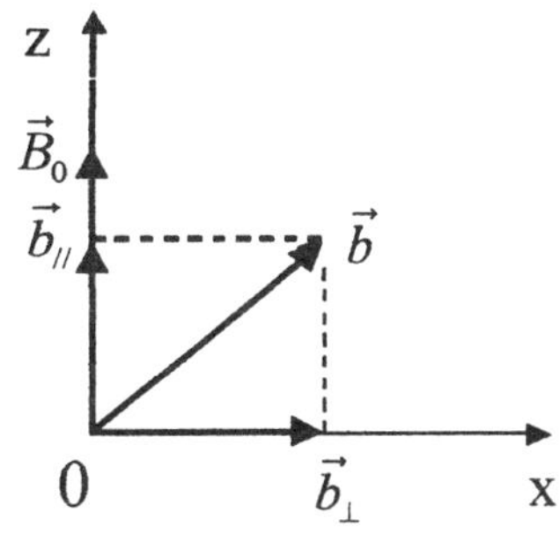

Fig. 3. The radiofrequency field $\vec{b}$ has a component, $\vec{b}_{//}$, parallel to the static field $\vec{B}_0$, which is associated with π-polarized photons with an angular momentum $J_z = 0$, and a component, $\vec{b}_\perp$, perpendicular to $\vec{B}_0$, which is associated with σ-polarized photons with an angular momentum $J_z = +\hbar$ or $-\hbar$.

3.1. Conservation laws for a multiphoton process

RF multiphoton resonances correspond to transitions between two Zeeman sublevels *m* and *m'* by absorption of a number *n* of photons with $n > 1$. These transitions can be easily understood in terms of two conservation laws. The first one is the conservation of total energy, $E_{m'} - E_m = n\hbar\omega$, where ω is the frequency of the RF field. The second one is the conservation of the total angular momentum $(m' - m)\,\hbar = \sum$ angular momenta of the absorbed photons. In a given experiment, if one takes the static field $\vec{B}_0$ aligned along the z-axis, one has to consider two components for the radiofrequency field $\vec{b}$: $\vec{b}_{//}$ along $\vec{B}_0$ and $\vec{b}_\perp$ perpendicular to $\vec{B}_0$ (Fig.3). The $\vec{b}_{//}$ component corresponds to the π-polarized photons with an angular momentum $J_z = 0$ and the $\vec{b}_\perp$ component to σ-polarized photons which have an angular momentum $J_z = +\hbar$ or $-\hbar$.

3.2. Two-photon transitions between unequally spaced Zeeman sublevels

Let us now consider the Zeeman hyperfine sublevels of an atom in a non-zero magnetic field. Because of the Back-Goudsmit decoupling, the Zeeman sublevels are generally not equidistant. Figure 4 pictures three such sublevels m_F, m_F+1, m_F+2. The two-photon

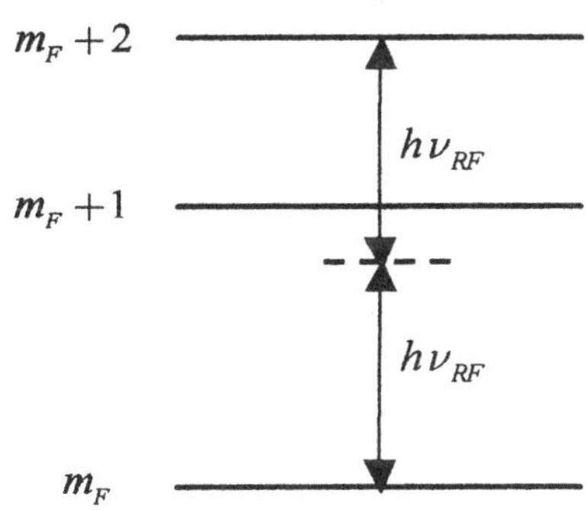

Fig. 4. Two-photon transition between unequally spaced Zeeman sublevels

transition between m_F and $m_F +2$ corresponds to a resonant process where the atom goes from m_F to $m_F +2$ by absorbing two σ_+-polarized RF photons. The conservation of total energy and angular momentum is clearly fulfilled if $E(m_F+2)-E(m_F)=2h\,\nu_{RF}$. It is clear also that such a higher order process goes through a non resonant intermediate state $m_F +1$. Finally, Fig. 4 shows that ν_{RF} is located between the resonant frequencies corresponding to a single σ_+ photon transition between m_F and $m_F +1$ and between m_F $+1$ and $m_F +2$. In fact, in the experiment, one keeps the frequency of the RF field fixed and the static field is scanned. In these conditions, the two-photon transition $m_F \rightarrow m_F+2$ appears between the two single-photon transitions $m_F \rightarrow m_F+1$ and $m_F+1 \rightarrow m_F+2$.

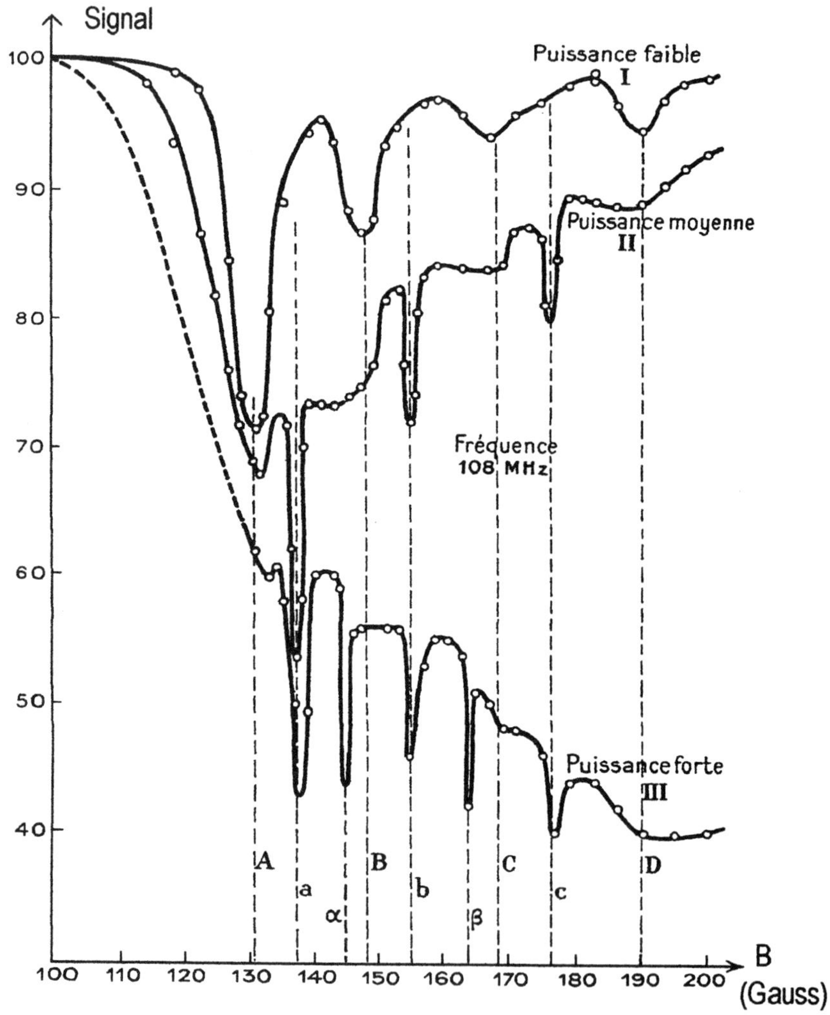

Fig. 5. Spectrum of multiphoton RF resonances observed in the ground state of optically pumped sodium atoms. The curves I, II, III correspond to increasing powers of the RF field. Resonances A, B, C, D correspond to single photon transitions; resonances a, b, c to two-photon transitions; resonances α, β to three-photon transitions (Figure taken from Ref. [7]).

3.3. Experimental observation with sodium atoms

Multiphoton resonances have been experimentally observed by A. Kastler, B. Cagnac and J. Brossel in 1953 [6-8]. Let us consider, for example, the F=2 hyperfine levels of sodium with five Zeeman sublevels m_F = +2, -1, 0, -1, -2 . The five Zeeman sublevels are not equidistant thus giving rise to four single-photon transitions at four different values of the magnetic field (curve I of Fig.5 corresponding to a weak RF power): resonance A corresponding to a transition between –2 and –1, resonance B corresponding to a transition between –1 and 0, resonance C corresponding to a transition between 0 and +1 and resonance D corresponding to a transition between +1 and +2. If the RF power of the field is increased (curve II of Fig.5), three two-photon transitions $\Delta m_F = 2$ appear : resonance a between -2 and 0, b between –1 and +1, c between 0 and +2. At last, if the RF power is further increased (curve III of Fig.5), two three-photon transitions α and β also appear : resonance α between -2 and +1 and β between –1 and 2.

3.4. Multiphoton resonances at subharmonics of the Larmor frequency

Multiphoton resonances can also take place between two adjacent Zeeman sublevels m_F and $m_F + 1$ with the absorption of one, two, three, four or more RF photons as represented in figure 6. In that case, it clearly appears that the Larmor frequency which gives the energy splitting between two Zeeman sublevels is equal to an integer multiple of the frequency ν_{RF} of the RF field .

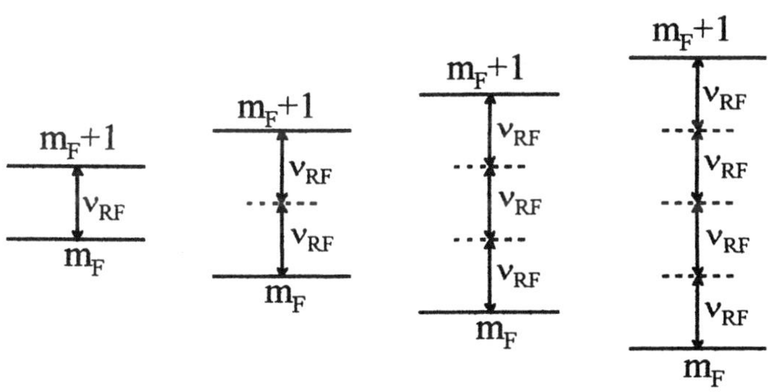

Fig. 6. Multiphoton RF resonances between 2 Zeeman sublevels m_F and m_F+1. For a n-photon resonance, the frequency ν_{RF} of the RF field is a subharmonic ν_L/n of the Larmor frequency ν_L which determines the splitting between the 2 Zeeman sublevels.

These resonances have been also observed in the ground state of sodium atoms. Figure 7 gives an example of such multiphoton resonances obtained by J. Margerie, J. Brossel and J. Winter in 1955 [9,10]. As in figure 5, three curves are represented for three increasing values of the RF field. For low values, one can see only two-photon transitions whereas, for higher values of the RF amplitude, one can also see three and four-photon transitions. Note that the conservation of angular momentum requires that one of the two photons absorbed in the resonance $\nu_L = 2\nu_{RF}$ must be π-polarized whereas the other one is σ_+-polarized. Similar arguments apply to the resonance $\nu_L = 4\nu_{RF}$ which must involve π-polarized RF photons as well as $\sigma_\pm$ ones. Coming back to Fig. 3, we see that the RF field must have both parallel and perpendicular components with respect to the static field. Note finally that Fig.7 clearly shows that these multiphoton resonances are broadened and shifted when the RF power is increased. The theory of this power broadening and power shift has been extensively developed by J. Winter [10].

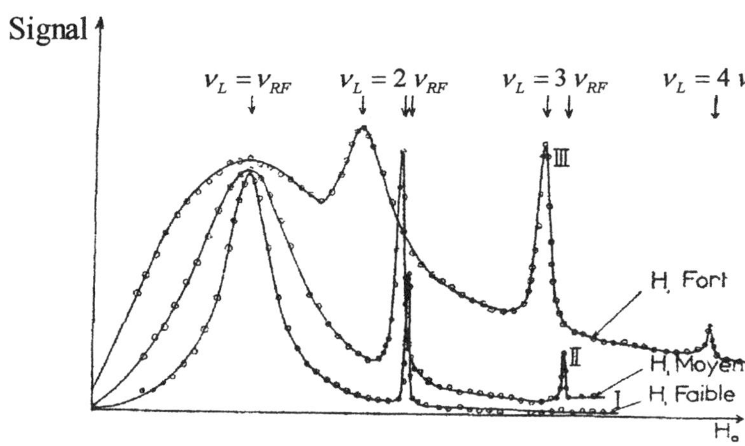

Fig. 7. Spectrum of multiphoton RF resonances $m_F \to m_F+1$ observed in the ground state of optically pumped sodium atoms. The curves I, II, III correspond to increasing powers of the RF field (Figure taken from Ref.[9]).

3.5. Earlier works on two-photon processes

The first theoretical treatment of two-photon processes was given by M. Goppert Mayer in 1931. It seems also that V. Hughes and L. Grabner observed experimentally in 1950 two-photon processes while studying the RF spectrum of RbF molecules using a molecular beam electric resonance method [11]. They observed an unpredicted group of lines appearing at one half the frequency of another line group. One possible explanation for this unpredicted group of lines was the existence of two-photon transitions, but this was not clearly demonstrated.

3.6. Subsequent works on higher order effects

The development of laser sources has made it possible to extend the observation of multiphoton processes to other frequency ranges, in particular to the optical range. I have selected some examples without attempting to be exhaustive.

Optical multiphoton processes involving multiphoton ionization and harmonic generation have been extensively studied by the Saclay group in France, in particular by C. Manus and G. Mainfray [12].

Another example that is worth mentioning is the new field of non linear optics which gave rise to a lot of investigations in several laboratories. Several French physicists were offered the opportunity to make a doctoral or postdoctoral stay in the laboratory of N. Bloembergen in Harvard : J. Ducuing, G. Bret, C. Flytzanis, N. Ostrowski and P. Lallemand. I will mention also the work developed at the CNET (Centre National d'Etudes des Télecommunications) involving molecular engineering for non linear optical materials. A third example of subsequent work involving multiphoton processes is Doppler free spectroscopy. Consider a two-photon transition where the two absorbed photons are propagating in opposite directions. The transfer of linear momentum from the field to the atom is then equal to zero because the momenta of the two absorbed photons cancel out. As a consequence, there is no Doppler effect and no recoil. This new Doppler-free method has been investigated by several groups : the group of V. Chebotev in Soviet Union [13], of B. Cagnac, G. Grynberg and F. Biraben in France [14], of N. Bloembergen, M. Levenson [15] and T. Hänsch [16] in the United States.

Finally, I will mention the possibility of using high intensity laser pulses for producing laser intertial confinement fusion. In 1964, A. Kastler published a "Note aux Comptes Rendus de l'Académie des Sciences"[17] where he mentioned the high temperature which could be obtained by concentrating the light from a high power laser source on a target. This note stimulated other groups, in particular a French group at the CEA (Commissariat à l'Energie Atomique) who investigated, in close collaboration with an industrial company (CGE- Marcoussis), the nuclear reactions of fusion produced by a a nanosecond laser pulse [18]. This was achieved one year after the first demonstration of this effect in Soviet Union by N. G. Basov's group with picosecond laser pulses [19]. Laser induced fusion has now become a very broad field which is outside the scope of this paper.

4. Quantum interference effects in radiative processes

Later on, the development of optical methods of radiofrequency spectroscopy made it possible to observe new interesting quantum interference effects between different emission or absorption amplitudes in radiative processes.

4.1 Coherent multiple scattering

We describe first an experiment which played an important role in the investigation of these problems. In 1966, M.-A. Guiochon-Bouchiat, J.-E. Blamont and J. Brossel observed a very intriguing effect represented in Fig. 8.[20,21] This figure shows double resonance curves observed on mercury atoms, for increasing values of the temperature of the reservoir of atoms, i.e. for increasing values of the vapour density. The surprising result is that the width of the curves decreases when the number of atoms increases. In such a situation, one would normally expect a broadening of these curves: when the density increases, the number of collisions increases, producing a broadening of the levels. In fact, the narrowing of the curves was rapidly understood as being due to the following effect. When a mercury atom undergoes magnetic resonance in the excited state, it is put in a linear superposition of the three excited Zeeman sublevels. This atom emits then a photon which can be reabsorbed by another atom if the vapour pressure is high enough. It turns out that this second atom is excited in a linear superposition of its three excited Zeeman sublevels which keeps a certain memory of the relative phases of

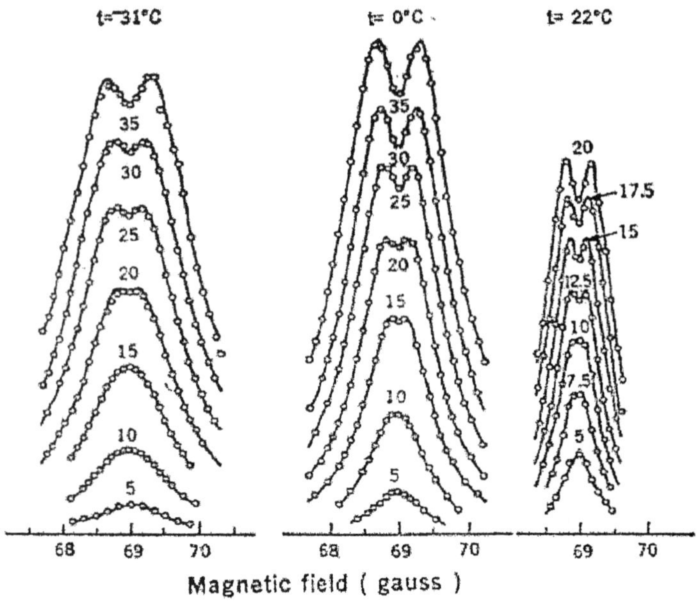

Fig. 8. Three sets of double resonance curves analogous to the set shown in Fig. 1b. These three sets are taken at different temperatures of the reservoir of atoms, increasing from the left to the right and indicated at the top of the figure. One clearly sees that the width of the curves decreases when the vapour pressure increases (Refs [20,21]).

the linear superposition in which the first atom was prepared before emitting his photon. In other words the "Zeeman coherence" in the excited state which determines the width of the magnetic resonance curve is partially conserved in a multiple scattering process and thus appears to have a longer lifetime.

4.2. Linear superpositions of excited Zeeman sublevels

The above described experiment demonstrates the importance of linear superpositions of excited Zeeman sublevels and the first question which can be asked is how they can be prepared. It is clear that a first possibility is the action of a RF field which transforms the state of an atom initially prepared in a well defined Zeeman sublevel into a linear superposition including the other sublevels to which the initial state is coupled by the RF field. Another possibility for producing linear superpositions of excited Zeeman sublevels is to excite the atom with a light whose polarization is a linear superposition of σ_+, σ_- and π polarizations, so that, starting from a well defined ground state sublevel, several excited Zeeman sublevels can be coherently excited by the incoming light field.

Another important question which can be asked is how do these linear superpositions of states show up in the emission process. The answer is that the light emitted by an atom in a linear superposition of excited Zeeman sublevels is not the same as the light emitted by a statistical mixture of these sublevels. Different quantum interference effects between different scattering amplitudes can be observed. Here are a few examples among others of such physical effects (for a review of the field, see for example Ref.[22] and references therein):

- Coherent multiple scattering, which has been described in § 4.1.

- Hanle effect [23].
- Light beats such as those observed by the group of G. Series revealing that the light emitted by an atom undergoing magnetic resonance in the excited state is modulated at the frequency of the RF field [24]. It looks like as if the various emission amplitudes were beating giving rise to modulations of the emitted light.
- Quantum beats effects resulting from a sudden excitation of the atom by a pulse of light. The pulse of light prepares the atom in a superposition of two excited Zeeman sublevels. The light subsequently emitted by the atom exhibits modulations resulting from an interference between the two emission amplitudes starting from the two excited Zeeman sublevels[25].
- Level crossing resonances which appear in the light emitted by the atom near the value of the field where two excited Zeeman sublevels cross [26].
- Detection of the first photon emitted in a radiative cascade which puts the atom in a certain linear superposition of the sublevels of the intermediate atomic state of the cascade.

4.3. Master equation for the atomic density matrix

To interpret the various quantum interference effects described above, a description of the atomic state in terms of the populations of the various energy sublevels is no longer sufficient. It is then necessary to use a density matrix description. The diagonal elements σ_{MM} of the density matrix σ give the populations of the energy sublevels, but σ has also off-diagonal elements, such as $\sigma_{MM'}$. These off-diagonal elements are very important because they describe the interference effects between the emission amplitudes starting from M and M'. It turned out subsequently that off-diagonal elements of the ground state density matrix are also important for describing the quantum interference effects between different absorption amplitudes starting from two different ground state sublevels.

It was therefore very important to establish the equation of the evolution of σ, the so called master equation . Such master equations were derived first for an excited state [27], and then for the whole optical pumping cycle [28]. This work can be now considered as the precursor of the master equation description of modern quantum optics.

One should also mention that the Bloch equations of Nuclear Magnetic Resonance (see for example [29]) were a great source of inspiration for describing the evolution of an ensemble of atoms coupled to optical fields.

5. Light broadening and light shifts of ground state sublevels

5.1. Theoretical predictions

One of the most important predictions of the master equation describing the evolution of the atomic density matrix during the optical pumping cycle is the fact that Zeeman atomic sublevels in the ground state are broadened and shifted by light. On the one hand, the light broadening Γ_M' of atomic sublevel g_M can be interpreted in terms of real absorptions of photons from the sublevel g_M, which shorten the lifetime of g_M. On the other hand, the light shift of g_M, denoted Δ_M' can be interpreted in terms of virtual absorptions or reemissions of photons. The theory shows that Γ_M' and Δ_M' are proportional to the light intensity and vary with the detuning δ of the light excitation as Lorentz absorption and dispersion curves, respectively. There is a certain analogy between Γ_M' and Δ_M', on the one hand, and the natural width Γ_M and the Lamb shift Δ_M

of an excited sublevel e_M, on the other hand. Γ_M' and Δ_M' can be considered as radiative corrections due to absorption and stimulated emission processes in the same way as Γ_M and Δ_M can be interpreted in terms of radiative corrections due to the coupling with the empty modes of the radiation field. Another analogy can also be drawn between Γ_M' and Δ_M' on the one hand and the imaginary and real part of the index of refraction of a light beam passing through the vapour, on the other hand.

5.2 Experimental observation

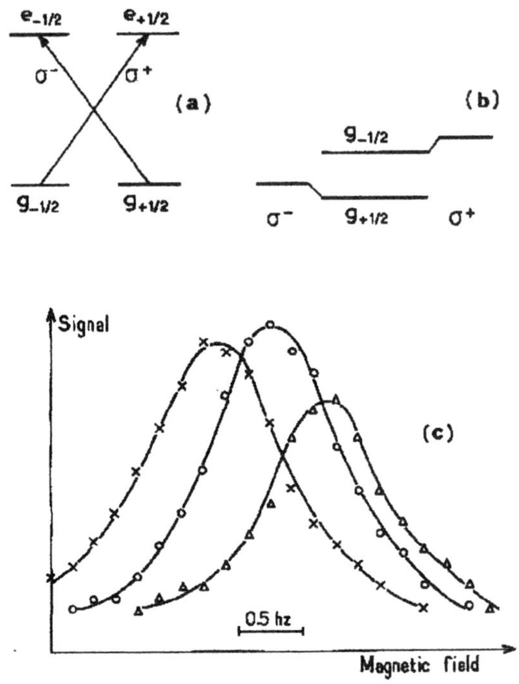

Fig. 9. First experimental observation of light shifts on the transition ½ → ½ of ^{199}Hg (from Ref. [30]). (a): For a σ_+ excitation, only sublevel $g_{-1/2}$ is light shifted, whereas for a σ_- excitation, only sublevel $g_{+1/2}$ is light shifted. (b): The splitting between the 2 Zeeman sublevels is increased by a (blue detuned) σ_+ excitation, whereas it is decreased by a blue detuned σ_- excitation. (c): Magnetic resonance signal vs magnetic field. The central curve (circles) is taken without the light-shifting beam. When this beam is introduced, either σ_+ - polarized (crosses) or σ_- - polarized (triangles), the magnetic resonance curve is light shifted in opposite directions.

Light shift and light broadening of ground state sublevels have been observed in optical pumping experiments performed on the odd isotopes of mercury [30,31]. These experiments done in 1961 were not using laser sources since these sources were not available at that time. In fact, light shifts could be easily observed because, in general, the two ground state sublevels undergo different light shifts depending on the light polarization. Consequently, the splitting between two Zeeman sublevels is modified by the light irradiation. And, since the lifetime of the atomic ground state is very long, giving rise to very narrow magnetic resonance curves, it is possible to detect such a modification of the splitting between two Zeeman sublevels by a shift of the magnetic resonance curves. In fact, the first experimental observation of light shifts gave evidence of light shifts which were only of the order of one Hertz. These studies also demonstrated that the broadening Γ_M' and the light shift Δ_M' of the ground state were varying with the detuning δ of the light beam with respect to the atomic transition frequency as absorption and dispersion curves, respectively [31].

5.3. Continuity with more recent works on radiative forces

About twenty years later, light shifts and light broadening of ground state sublevels were found to play an important role in the physics of radiative forces [32]. There are in fact two great categories of radiative forces : dissipative forces, which are proportional to Γ' and which are due to real absorption and spontaneous emission of photons; and reactive, or dipole forces, which are proportional to Δ', and which are due to virtual absorption and stimulated emission processes. A laser trap which can be achieved at the focus of a red detuned laser beam is a good illustrative example. Since the light shift Δ' due to the laser beam is negative for red detuning and maximum in absolute value at the focus of the laser beam, a potential well is formed where sufficiently cold atoms can be trapped. Another example is the atomic mirror formed by a blue detuned evanescent wave produced at the surface of a piece of glass. For blue detuning, the light shift Δ' is positive and decreases with the distance to the surface producing a potential barrier upon which atoms can bounce if they are cold enough. Finally, one can mention one important laser cooling mechanism, called "Sisyphus cooling", which is due to spatially modulated light shifts and optical pumping rates in a laser standing wave. A situation is achieved where a moving atom runs up potential hills more frequently than down, losing its kinetic energy (Fig.10). This provides a very efficient cooling mechanism which allows one to reach the microKelvin range [33].

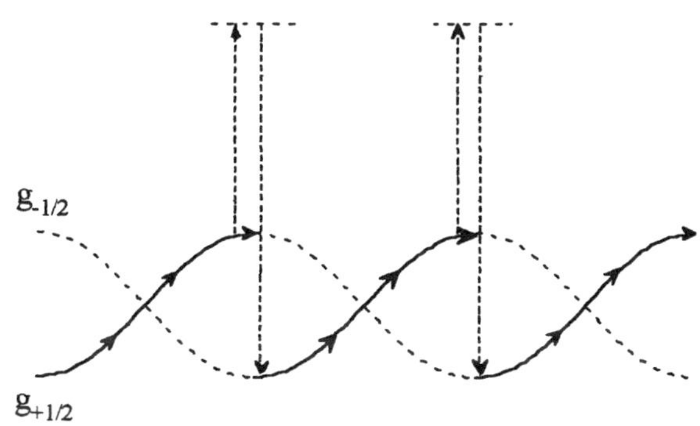

Fig. 10. Sisyphus cooling. In a laser standing wave, the intensity and the polarization of the laser electric field are spatially modulated, giving rise to correlated spatial modulations of the light shifts of two ground state sublevels and of the optical pumping rates between them. As a result of these correlations, a moving atom can run up potential hills more frequently than down.

5.4. Other physical effects connected with off diagonal elements of the ground state density matrix

The existence of Zeeman or hyperfine coherences in the atomic ground state manifests itself by several new physical effects which have been studied in detail. Let us mention a few of them:

- Zero field level crossing resonances which are the equivalent for the ground state of the Hanle effect in the excited state [34,35]. However, because the lifetime of the ground state is much longer than the lifetime of the excited state, these resonances are extremely narrow, with a width which can be as small as 10^{-6} Gauss. Such resonances have been used to build magnetometers with a sensitivity of 5×10^{-10} Gauss [35].

- Modulation of the absorption of a crossed beam [31] which are the extension to atomic ground states of the light beats and quantum beats observed on the light emitted from excited states.
- One can also mention the discovery of coherent population trapping [36] which consists of the optical pumping of atoms in a linear superposition of two atomic states which no longer absorbs light (the so called "dark states"). More recently, coherent population trapping has found new applications such as the achievement of subrecoil cooling by velocity selective dark states [37], self-induced transparency [38] and dramatic decrease of the light speed in an atomic medium. [39].

6. New investigations stimulated by the development of lasers in the sixties

6.1. Good optical resonators

When lasers started to be built in the early sixties, some physicists soon realized that the confocal Fabry-Perot, developed by P. Connes a few years before [40,41] was very useful. Such a scheme uses a cavity formed by two spherical mirrors having equal radii of curvature and coincident foci. It turned out to be a very convenient optical resonator and it played an important role in the development of laser source.

Another example of development which occurred in the sixties in France was the pulse compression scheme and the Gires-Tournois interferometer[42].

6.2. Atoms inside a Fabry-Perot

During the same period, in 1962, A. Kastler published a paper on the modification of the emission properties of atoms inside a Fabry-Perot [43]. He showed how, instead of having an isotropic diagram of emission, the emission of atoms should be concentrated in narrow rings corresponding to the modes of the Fabry-Perot. This paper can be considered as a precursor of cavity QED.

6.3. Laser action in semiconductors

Finally, let us mention another interesting development dealing with the possibility of having laser action in semiconductors. This question was first raised by Pierre Aigrain as early as 1958 in an unpublished lecture at a conference on solid state physics in electronics and telecommunications in Bruxelles. A few years later, Bernard and Duraffourg published a paper where they gave the condition for laser action, a condition which is now known as "Bernard-Duraffourg condition", connecting the differences between the quasi-Fermi energies of electrons (holes) in the conduction (valence) band to the energy of the emitted photons [44].

6.4. More recent developments

Laser sources introduced a spectacular improvement for spectroscopy and they stimulated the development of several new research fields. We mention here a few examples where French groups were very active.

- Investigation of forbidden atomic transitions and use of these transitions for getting evidence for parity violation in atomic physics. This research field was pioneered by M-A. and C. Bouchiat in France [45]
- Laser spectroscopy of series of radioactive isotopes which can bring interesting information on nuclear structures. P. Jacquinot and S. Liberman, from Laboratoire Aimé Cotton in Orsay, were engaged in international collaborations at C.E.R.N. (Geneva) for these problems [46]

- Development of the method of saturated absorption for high resolution spectroscopy by C. Bordé [47]
- Spectroscopy of highly excited Rydberg states and investigation, by the group of S. Haroche, of the radiative properties of these states [48]
- Investigation, by the group of M. Ducloy, of atom-surface interactions by laser spectroscopy [49]

7. Interaction atom – single mode laser field

When lasers appeared in the sixties, one important theoretical problem which was raised was the description of an atom interacting with a single mode coherent field. Two models were then developed in connection with this problem.

7.1. Jaynes-Cummings model

In this model [50], a two-level atom is put in a single mode cavity and interacts with a single-mode electromagnetic field. Formally, we have a spin ½ coupled to a one-dimensional harmonic oscillator. This model has proved to be extremely useful a few years later when experiments started to be achieved in what is called now cavity QED.

7.2. Dressed-atom approach. Atoms dressed by RF photons

Another model developed in the sixties was the so called dressed-atom approach where a two-level atom is interacting with a single-mode field in free space [51]. In this model, one considers a fictitious single-mode cavity containing N photons in a volume V (at the end of the calculation, one lets N and V tend to infinity with the ratio N/V maintained constant). This approach has provided a synthetic description and a better understanding of the various physical effects which can be observed in RF spectroscopy, in particular the shift and the broadening of multiphoton RF resonances. New effects were also predicted by this approach and observed experimentally, in particular the fact that the magnetic moment (or equivalently the g-factor) of an atom can be modified and even completely cancelled when this atom interacts with a high frequency RF field [52]. This modification of the g-factor is somewhat analogous to the electron spin anomaly which is actually a spontaneous radiative correction. An interesting question was then raised : is it possible to understand the spin anomaly $g - 2$ as a stimulated radiative correction due to vacuum fluctuations, which could be obtained by summing the effects, calculated here for the coupling between a magnetic moment and a RF field, over all modes of the electromagnetic field (with an energy $\hbar\omega / 2$ per mode)? The difficulty is that, in this way, one gets the wrong sign for $g - 2$! Understanding the origin of this minus sign stimulated several investigations and it turned out that the interpretation of $g - 2$ requires one to consider, not only the effects of vacuum fluctuations, but also those of radiation reaction [53].

7.3. Subsequent developments. Atoms dressed by optical photons

A few years later, in the seventies, the dressed-atom approach was extended to the optical domain where spontaneous emission plays an important role. This approach provided a very simple interpretation of many physical effects [51].

- Simple interpretation of the "Mollow resonance fluorescence triplet" in intense laser field in terms of transitions between dressed states.
- Simple interpretation of photon correlations and photon antibunching in single atom resonance fluorescence in terms of a radiative cascade of the dressed atom.

- Simple calculation of the so called "delay function" or "waiting time distribution" giving the distribution of the time intervals between two successive spontaneous jumps. This delay function has been extremely useful for developing Monte-Carlo simulations of the quantum evolution and for giving simple accounts of intermittent fluorescence and coherent population trapping.
- Interpretation of dipole forces in terms of spatial gradients of the dressed state energies when the atom is moving in a spatially inhomogeneous laser field.

Let us finally point out that photon correlations have been extremely important for studying and achieving entangled states of photon pairs emitted in an atomic radiative cascade. In this respect, one can mention the important work done by A. Aspect in France for testing Bell's inequalities [54-55].

Conclusion

I have tried in this lecture to review a few early developments of quantum optics which took place in France before the seventies. It seems clear that this research field has known a rather good start in this country and one can try to understand why. A first reason was probably the existence of a long standing tradition in optics in France, going back to Fresnel and continuing with Fabry, Pérot and many others. A second important point was the role played by a few strong personalities who where able to attract good students and to build creative schools around them. This was the case of A. Kastler and J. Brossel at Ecole Normale Supérieure in Paris, of A. Abragam in Saclay, of P. Jacquinot in Bellevue and Orsay. I should also mention the great efforts made in France after world war II for introducing young generations to modern physics. I still remember the very stimulating lectures on quantum mechanics which were given every week by A. Messiah in Saclay. Les Houches summer school which started in 1951 was also a unique place where every year about fifty young graduate or PhD students were spending two months following lectures given by the best physicists of the world. Finally, this period was a golden period for the expansion of research and it was relatively easy at that time to get long term supports from the national research institutions like Centre National de la Recherche Scientifique. Unfortunately, this combination of favorable circumstances is not always easy to achieve. More and more often, politicians now prefer to put the priority on specific short terms programs with the hope that scientific solutions will be rapidly found for the problems of health, energy, environment that we have to face. I think that it is not the good way. The history of science clearly shows that all important applications in physics such as lasers, transistors, magnetic resonance imaging, were not planned in advance. They came out after a high quality basic research essentially motivated by the desire to understand the physical mechanisms. Research is a long term investment for which perseverance and high quality standards are essential. I hope that an historical session like this one will help us to convey this message.

Acknowledgement

The author thanks Nadine Beaucourt for her help in the preparation of this manuscript, and Jean Brossel, Jean Dalibard and Franck Laloë for stimulating discussions.

References

[1] A. Kastler, J. Brossel, C. R. Acad Sci. **229**, 1213 (1949).
[2] J. Brossel, F. Bitter, Phys. Rev. **86**, 308 (1952).
[3] A. Kastler, J. de Physique **11**, 255 (1950).
[4] For a review of this field, see for example : E. Otten in *Atomic Physics 15*, H. B. van Linden van den Heuvell, J. T. M. Walraven, M. W. Reynolds eds (World Scientific, 1997).
[5] See for example F. Laloë and J. Freed, Scientific American **256**, April 1988, p. 94
[6] J. Brossel, B. Cagnac, A. Kastler, C. R. Acad. Sci. **237**, 984 (1953).

[7] J. Brossel, B. Cagnac, A. Kastler, J. Phys. Rad. **15**, 6 (1954).
[8] B. Cagnac, Thèse, Paris, Ann. de Phys .**6**, 467 (1960).
[9] J. Margerie, J. Brossel, C.R. Acad. Sci. **241**, 373 (1955).
[10] J. Winter, C.R. Acad Sci. **241**, 375 (1955).
[11] V. Hughes, L. Grabner, Phys. Rev. **79**, 314 and 829 (1950).
[12] See for example the review paper of G. Mainfray and C. Manus in "Multiphoton Ionization of Atoms" edited by S. Chin and P. Lambropoulos, Academic Press 1984.
[13] L. Vasilenko, V. Chebotayev, A. Shishaev, J.E.T.P. Lett **12**, 161 (1970).
[14] B. Cagnac, G. Grynberg, F. Biraben, J. Physique **34**, 845 (1973).
[15] M. Levenson, N. Bloembergen, Phys. Rev. Lett. **32**, 645 (1974)
[16] T. Hänsch, S. Lee, R. Wallenstein, C. Wieman, Phys. Rev. Lett. **34**, 307 (1975).
[17] A. Kastler, C. R. Acad. Sci. **258**, 489 (1964)
[18] F. Floux, D. Cognard, J-L. Bobin, F. Delobeau, C. Fauquignon, C. R. Acad. Sci. **B 269**, 697 (1969)
[19] N. Basov, S. Zakharov, P. Kryukov, Yu. Senatskii, V. Chekalin, J.E.T.P. Lett. **8**, 14 (1968)
[20] M.-A. Guiochon-Bouchiat, J.-E. Blamont, J. Brossel, C. R. Acad. Sci **243**, 1859 (1956)
[21] M.-A. Guiochon-Bouchiat, J.-E. Blamont, J. Brossel, J. Phys. Rad. **18**, 99 (1957)
[22] C. Cohen-Tannoudji, Physica Scripta, **T70**, 79 (1997).
[23] W. Hanle, Z. Phys. **30**, 93 (1924) ; **35**, 346 (1926).
[24] A. Corney, G. Series, Proc. Phys. Soc. **83**, 207 (1964).
[25] J. Dodd, R. Kaul, D. Warrington, Proc. Phys. Soc. **84**, 176 (1964).
[26] F. Colegrove, P. Franken, R. Lewis, R. Sands, Phys. Rev. Lett. **3**, 420 (1959).
[27] J.-P. Barrat, Proc. Roy. Society A **263**, 371 (1961).
[28] J.-P. Barrat, C. Cohen-Tannoudji, J. de Phys. **22**, 329 and 443 (1961).
[29] A. Abragam, "The principles of nuclear magnetism" (Clarendon Press, Oxford, 1961).
[30] C. Cohen-Tannoudji, C. R. Acad. Sci. **252**, 394 (1961).
[31] C. Cohen-Tannoudji, Thèse, Paris (1962), Ann. de Phys. 7, 423-495 (1962).
[32] For a review on radiative forces, see C. Cohen-Tannoudji, "Atomic motion in laser light", in *Fundamental systems in quantum optics*, J. Dalibard, J.-M. Raimond and J. Zinn-Justin eds, Les Houches session LIII (1990), 1-183 (North Holland 1992) and references therein.
[33] J. Dalibard, C. Cohen-Tannoudji, J. Opt. Soc. Am. **B6**, 2023 (1989).
[34] J.-C. Lehmann, C. Cohen-Tannoudji, C. R. Acad. Sci. **258**, 4463 (1964).
[35] J. Dupont-Roc, S. Haroche, C. Cohen-Tannoudji, Phys. Lett. **28 A**, 638 (1969).
[36] G. Alzetta, A. Gozzini, L. Moi, G. Orriols, Nuovo Cemento **B36**, 5 (1976).
For a review on coherent population trapping, see E. Arimondo, in Progess in Optics XXXV, edited by E. Wolf, (Elsevier,1996).
[37] A. Aspect, E. Arimondo, R. Kaiser, N. Vansteenkiste, C. Cohen-Tannoudji, Phys. Rev. Lett. **61**, 826 (1988)
[38] See the review paper of S. Harris in Physics Today, July 1997, p. 36 and references therein.
[39] See the paper of L. Hau in this volume.
[40] P. Connes, Rev. d'Opt. **35**, 37 (1956).
[41] P. Connes, J. de Phys. Rad. **19**, 262 (1956).
[42] F. Gires, P. Tournois, C. R. Acad Sci. **258**, 6112 (1964).
[43] A. Kastler, Applied Optics **1**, 17 (1962).
[44] M. Bernard, G. Duraffourg, Physica Statu Solidi **1**, 699 (1961).
[45] M.-A. Bouchiat, C. Bouchiat, Phys. Lett. **B46**, 111 (1974), J. de Phys. **35**, 899 (1974). For a recent review of the field, see " Parity Violation in Atoms and Electron Scattering", B. Frois and M.-A. Bouchiat eds (World Scientific, 1999).
[46] See S. Liberman in Atomic Physics 11, S. Haroche,J.-C. Gay, G. Grynberg eds (World Scientific, 1989).
[47] C. Bordé, C. R. Acad. Sci. **271 B**, 371 (1970).
[48] For a review, see S. Haroche in "New Trends in Atomic Physics", G. Grynberg and R. Stora eds (North-Holland, 1984).
[49] M. Ducloy in "Nanoscale Science and Technology", N. Garcia et al eds (Kluwer, 1998).
[50] E. Jaynes, F. Cummings, Proc. I.E.E.E. **51**, 89 (1963).
[51] For a review on the dressed atom approach, see C. Cohen-Tannoudji, J. Dupont-Roc, G. Grynberg, *Processus d'interaction entre photons et atomes* (InterEditions et Editions du CNRS, 1988). English version : *Atom-Photon Interactions. Basic processes and Applications* (Wiley-Interscience, 1992), Chapter VI.
[52] C. Cohen-Tannoudji, S. Haroche, C. R. Acad. Sci. **262**, 37 and 268 (1966).
[53] J. Dupont-Roc, C. Fabre, C. Cohen-Tannoudji, J. Phys. **B11**, 563 (1978).
[54] A. Aspect, P. Grangier, G. Roger, Phys. Rev. Lett. **49**, 91 (1982).
[55] A. Aspect, J. Dalibard, G. Roger, Phys. Rev. Lett. **49**, 1804 (1982).

Early History of Bell's Theorem

John F. Clauser
J.F. Clauser & Assoc., 817 Hawthorne Drive, Walnut Creek, CA 94596, USA †

Bell's Theorem is one of the most profound results in physics of the twentieth century. Not only does it have a significant impact on natural philosophy and on the true meaning of quantum mechanics, it also has stimulated important and practical new research in quantum optics. In 1972 at the CQ03 Rochester conference, in response to a number of disturbing issues and challenges then raised by Ed Jaynes concerning the foundations of quantum electrodynamics, I introduced the quantum optics community to Bell's Theorem and a few of the associated mysteries manifest in quantum entanglement. Given the widespread belief that the foundations of quantum mechanics were then well understood, needless to say, my 1972 talks were then met with considerable skepticism. Eventually, however, the importance of Bell's Theorem gained full acceptance, so that now it is discussed in many recent quantum mechanics and quantum optics textbooks. Similarly, my first experimental test (with Stuart Freedman in 1972) of the Bell-Clauser-Horne-Shimony-Holt prediction has since then been repeated and confirmed literally dozens of times, and that prediction now provides a standard quantitative measure of entanglement. Moreover, the fundamental ideas underlying Bell's Theorem have been found to be sufficiently useful and important, that it is doubtful that the parallel conference ICQI-2001 would have occurred without them. This article recounts the important historical events behind the development of Bell's Theorem.

† Formerly at Department of Physics, Univ. of California – Berkeley (retired).

Bell's Theorem provides us with a somewhat rare opportunity for experimental physics to answer questions in natural philosophy. Correspondingly, Abner Shimony and I open our 1978 review article, *Bell's theorem: experimental tests and implications*[1], with the sentence,

> *"Realism is a philosophical view in which external reality is assumed to exist and have definite properties, whether or not they are observed by someone."*

In that review we describe how the experimental evidence in hand in 1978 allowed us reasonably to assert that

> *"The conclusions* [from Bell's Theorem] *are philosophically startling; either one must totally abandon the realistic philosophy of most working scientists, or dramatically revise our concept of space-time."*

Despite numerous repetitions and extensions of the reviewed experiments and theory, along with considerable scrutiny of both since then, our 1978 conclusion appears to stand.

Bell's Theorem explicitly reveals highly peculiar and remarkable properties of quantum-mechanically "entangled" states that were previously not appreciated. Improved understanding of entanglement, in turn, has allowed it to be exploited on a wide variety of physical systems. As a result, Bell's Theorem has directly or indirectly fostered many spin-off practical applications and created whole new fields of study, including quantum cryptography, quantum communication, quantum computing, entanglement sharing, quantum cloning, etc. Its concepts are now common in undergraduate physics textbooks, as well as in a large body of scientific literature for the layman. One colossal recent spin-off proposal is by the US National Security Agency, who now envisions a quantum Internet that is absolutely secure against eavesdropping, and that is linked together by an international network of quantum computers, wherein quantum entanglement is then shared world-wide via a network of satellites. Thus, not only has Bell's Theorem deepened our understanding of quantum mechanics and the foundations of physics and natural philosophy, its full impact far exceeds the realm of natural philosophy, and has become an important heuristic tool for promoting revolutionary useful technology.

This note provides an overview of the history (through the early 80's) of the development of Bell's Theorem, along with its relation to the foundations of quantum mechanics. Prior to 1972, that development had proceeded quite independently from the study of quantum optics and quantum electrodynamics. However, the residual problems in quantum mechanics spilled over into quantum electrodynamics. In the late 60's and early 70's, Ed Jaynes had raised a number of related and disturbing parallel issues. Unlike many of quantum theory's founder's, however, Jaynes challenged the quantum optics community to provide hard experimental evidence to resolve these problems. In 1972 at the CQ03 Rochester conference, in response to Ed Jaynes' challenge, I brought these independent areas of study together, and introduced the quantum optics community to Bell's Theorem, its experimental tests, and a few of the profound associated mysteries manifest in quantum entanglement.

Acceptance of the importance of Bell's Theorem did not come easily. From the post-war years well into the 70's there was a widespread belief that the foundations of both quantum mechanics and quantum electrodynamics were then well understood. Needless to say, my CQ03 talks were met with considerable skepticism. That belief was so prevalent that it provided what then amounted to stigma against any challenges to or criticism of those foundations, and especially against any associated discussion of the notion of hidden variables in quantum mechanics. Nonetheless, quantum theory's founders had, in fact, left many significant gaps in the theory's conceptual foundations. More importantly, they also had promoted many incorrect, unfounded, and/or misleading conclusions. Much of the history behind Bell's Theorem involved filling in these gaps and correcting these errors, and furthermore, doing so in an unfriendly climate for a politically unreceptive audience. As a result of this stigma, much of John Bell's early work was published only as preprints and in

[1] J. F. Clauser and A. Shimony, Rep. Prog. Phys. (1978) **41** 1881-1927.

an "underground" newspaper[2], whose circulation was limited to members of a "quantum-subculture", and that probably cannot be found in most physics libraries. Only much later, it was collected to form his book *Speakable and unspeakable in quantum mechanics*[3].

Eventually, however, the importance of Bell's Theorem gained full acceptance, so that now it is discussed in many standard quantum mechanics and quantum optics textbooks. Similarly, my first experimental test (with Stuart Freedman in 1972) of the Bell-Clauser-Horne-Shimony-Holt prediction has since been repeated and confirmed literally dozens of times, and that prediction now provides a standard quantitative measure of entanglement.

An unresolved issue left by the founding fathers

While quantum mechanics was being initially formulated, there were many discussions about its meaning, its self-consistency, and about whether or not it provides a complete description of physical reality. Albert Einstein's famous debate with Niels Bohr at the 1927 Solvay Conference was primarily focused on the self-consistency of quantum mechanics *vis-à-vis* Heisenberg's uncertainty principle and the Copenhagen interpretation. At the end of the debate, Einstein conceded that quantum mechanics is indeed self-consistent. However, he remained adamantly undeterred in his belief that quantum mechanics, as a framework for physics, is incomplete. In his view, additional "hidden parameters" are needed to supplement quantum mechanics, and thereby to provide a full and complete description of physical reality.

If quantum mechanics is an incomplete theory, then the probabilities that one calculates from it are similar to those defined by Bernoulli and Pascal, i.e. they are quantitative measures of ignorance. Via classical probability theory, probabilities are modeled classically as sampling frequencies, in a conceptual parallel to the sampling of colored balls with varying weights from an urn. That is, given a blind choice of a ball from an urn, one estimates, for example, the probability (i.e. the sampling frequency) for choosing a ball with say a prescribed color. Alternatively, since the balls have weight as a second distinguishing characteristic among them, one may also estimate a probability for obtaining a prescribed weight. If the ball's weights are somehow correlated with their colors, one may also estimate a joint probability distribution of weight and color for these balls. In classical probability theory, however, no one ever doubts the existence of some actual set of balls within the urn, immediately prior to a sampling. That set correspondingly still has some preexisting color distribution, weight distribution, and joint color and weight distribution, although these distributions' properties may or may not be known. Thus, even when, "weight" and "color" are not quantum-mechanically "commuting observables", Einstein still expected that probabilities arise within quantum mechanics in exactly the same way that they do in classical probability theory, i.e. through ignorance of details about the sampling process, and/or ignorance of details of the parent distribution.

The measurement of some characteristic feature of a quantum-mechanical system is, however, not quite as simple as simply weighing a ball and/or observing its color. During Einstein's initial discussions with Bohr, Bohr talked about a "physical disturbance" that occurs during a measuring operation that is caused by the measuring apparatus, itself. Given only a limited set of then-known apparatus choices, Bohr argued that this disturbance is unavoidable in a quantum-mechanical measurement. In effect, Bohr was saying that one can measure a ball's weight, or its color, but not both, and somehow a measurement of a ball's weight can change its color. Nonetheless, the existence of such a physical disturbance mechanism is fully consistent with Einstein's view, and does not contradict any classical

---

[2] *Epistemological Letters* (Association Ferdinand Gonseth, Institut de la Methode, Case Postale 1081, CH-2501, Bienne.) This newspaper was somewhat unique for its time, in that it openly proclaimed that the usual stigma against hidden-variable theories, and the like, was to be absent for publications within it.

[3] J. S. Bell, *Speakable and unspeakable in quantum mechanics*, (Cambridge Univ. Press, Cambridge, 1987).

probability concepts. In Einstein's view, random results occur here because one is ignorant about both the details of the physical disturbance mechanism and the details of the *à priori* probability distribution of any physical variable or set of variables being measured. If one somehow were given a hypothetical "improved" apparatus that produces only a negligible disturbance, then the preexisting value of a variable could be measured in a single trial, and the *à priori* parent probability distribution could be determined with a large set of trials. In effect, such a hypothetical apparatus can still weigh the ball without thereby changing its color. Equivalently, given such a hypothetical apparatus, one can then measure simultaneously the preexisting values of two non-commuting variables, along with their *à priori* joint probability distribution. Bohr's initial arguments thus offered nothing to Einstein to counter his notion about the origins of randomness within quantum theory, and Einstein continued to believe that the uncertainty principle sets only practical limits to measurement precision, and that it is not an inherent fundamental limitation to the precision of quantum measurements.

Bohr, however, also maintained that quantum mechanics is a complete theory. If so, then the randomness observed in experimental outcomes is something very different from the probabilistic behavior of events envisaged by Bernoulli and Pascal, and that the non-existence of said hypothetical apparatus is highly relevant. In Bohr's view, probabilities arise within quantum mechanics, not through ignorance of the internal workings of nature, but in a fundamentally new and different way. In a complete theory, there is no preexisting *à priori* joint distribution of actual values that one may attempt to measure, since that complete distribution is not prescribed by quantum mechanics[4]. The prohibition against peeking inside the urn prior to sampling is now one of principle, and is not due to a limitation of the availability of suitable measuring devices. Bohr thus insisted that there is no such distribution, nor are there any details available in principle about the disturbance mechanism. In such case, one may not ask questions about details of said distributions, nor may one ask questions about details of the measurement process for any given trial.

A moment's reflection will convince the reader that Bohr's insistence on the nonexistence of an *à priori* joint distribution of actual values, is also a denial of the preexistence of any one value in any one trial. In essence, Bohr is denying the whole realistic philosophy, as defined above in the opening paragraph[5]. The random results observed in a quantum-mechanical measurement of a specified parameter are somehow a manifestation of the lack of a preexistence of a value for that parameter. Einstein evidently recognized that Bohr was denying realism. At one point in their discussions, Einstein reportedly asked Bohr "Do you really believe that the moon doesn't exist when no one is looking at it?" A recognition that the basic difference between their opposite views is simply acceptance versus denial of realism, then clarifies both of their logically following conclusions. Einstein is often quoted as saying that "God does not play at dice!" A more cogent metaphoric response that describes Bohr's view is then "Of course not; God has no dice to play with!" The apparent nonexistence of the abovementioned hypothetical "improved" apparatus is a <u>necessary</u> condition for Bohr's view to hold. However, from Einstein's view, issues about the existence or nonexistence of such an apparatus are all red herrings, since the apparent nonexistence of said hypothetical apparatus is, by no means, a <u>sufficient</u> condition for a denial of realism. Neither man, however, expected that the whole realistic philosophy would become amenable to experimental testing. It is thus probable that if Bell's Theorem had been available to them, their discussions and debates would have been remarkably different, and that both men would have taken keen interest in Bell's Theorem!

---

[4] The Wigner distribution is obviously not such a distribution, since it is not always non-negative.

[5] Max Jammer, in Sect. 4.2 of his book, *The Conceptual Development of Quantum Mechancs*, (McGraw-Hill, New York, 1966) argues that Bohr's denial of realism was in response to his being strongly influenced by the contemporary philosophers, Kierkegaard and Høffding.

Whether or not quantum mechanics is a complete theory is a nontrivial question, and its answer significantly determines the appropriate direction of future research in physics. Clearly, if quantum mechanics is incomplete, then a complete theory presumably exists that should be sought. On the other hand, if it is complete, then one may ask Einstein's questions, but one can never hope to find their answers, nor can one even hope to understand why it is vain to ask these questions.

Untidy legacy left by the founders of quantum mechanics

Soon after the Solvay Conference debates had ended, two heuristically important works were published. Both were written for the purpose of settling the completeness issue, once and for all. Unfortunately, the opposite result occurred, since these two works reached exactly opposite conclusions. In turn, these two were immediately followed by additional works that further sharpened their content.

One of these two important works is by John von Neumann (a former student of the mathematician David Hilbert). In his book, *Mathematische Grundlagen der Quantenmechanic*[6], von Neumann offers a "proof" of the non-existence of "hidden variables" within quantum mechanics. As a basis for this mathematical proof, he assumes that a linear operator (on a Hilbert space) is associated with each measurement apparatus, and that any measured result from this apparatus is an eigenvalues of this operator. He then proves that, within the mathematical structure of quantum mechanics, there cannot exist any additional (hidden) variables that then may be used to distinguish any further statistical differences between particles in the same "pure" state. Thereby, he shows that quantum mechanics is a complete theory. First, he concludes that

> "... *The only formal theory existing at the present time which orders and summarizes our experiences in this area in a half-way satisfactory manner, i.e. quantum mechanics, is in compelling logical contradiction with causality. Of course it would be an exaggeration to maintain that causality has thereby been done away with: quantum mechanics has, in its present form, several serious lacunae, and it may be even that it is false, although this latter possibility is highly unlikely, in the face of its startling capacity in the qualitative explanation of general problems, and in the quantitative calculation of special ones. ...* "

Further on, he strengthens this conclusion by stating that

> "... *we may say that there is at present no occasion and no reason to speak of causality in nature – because no experiment indicates its presence, since the macroscopic are unsuitable in principle, and the only known theory which is compatible with our experiences relative to elementary processes, quantum mechanics, contradicts it. ... Under such circumstances, is it sensible to sacrifice a reasonable physical theory for its sake?* "

Given the careful wording of these passages it should be noted in all fairness to von Neumann, that he does not prove, nor does he claim to prove the nonexistence of some other alternative theory that does not use his linear operator calculus, and whose predictions are equivalent to those of quantum mechanics. He just states that he believes that the existence of such an alternative theory is "highly unlikely". Unfortunately, many subsequent workers misinterpreted his work as a "proof" of the nonexistence of any such alternative theory.

The other one or the two important works is by Einstein, Podolsky and Rosen[7] (EPR). In their paper, they construct a hypothetical configuration that involves a pair of spatially separated but quantum-mechanically correlated particles. They further specify that these two particles are sufficiently far apart that it is impossible for the particles to physically interact with each other in any way. In this manner, EPR evade Bohr's "physical-disturbance"

---

[6] J. von Neumann, *Mathematische Grundlagen der Quantenmechanic*, (Springer-Verlag, 1932). English translation: *Mathematical Foundations of Quantum Mechanics*, (Princeton University Press, 1955).

[7] A. Einstein, B. Podolsky, and N. Rosen (1935), Phys. Rev. **47**, 777-80.

argument. EPR contend that a measurement at one location on one particle of the pair cannot "physically disturb" the other particle, nor can it disturb the other particle's measurement outcome. In the above ball and urn analogy, for a given trial there are now two urns, each containing one ball. At one urn, the ball's weight is measured, and at the other, the ball's color is measured. Knowing say the *à priori* color correlation between the two balls, one can then simultaneously determine each ball's weight and color. Quantum mechanics, however, does not simultaneously specify these two parameters, and therefore it must be considered incomplete. Correspondingly, it then must be supplemented with "hidden parameters" that provide such a specification for it to become complete.

Bohr then offered an important rebutal[8] to EPR that further complicated the saga. It significantly diminished the elegance of EPR's careful reasoning by its providing an important counter-argument. Bohr therein contends that, depending on how one words the "completeness" question, EPR's configuration can be viewed in a manner that escapes their conclusion. While many physicists view Bohr's arguments as a "refutation" of EPR's arguments, in fact, Bohr simply shows that his own view may be held consistently, even for EPR's configuration. First he concedes a lack of full generality to his own earlier arguments regarding a (presumably local) physical disturbance of either particle by its associated apparatus. Thus, he generalizes his earlier idea of a local physical disturbance now to include a possible global disturbance to the two-particle composite system by the two-component composite apparatus. If either of the component apparatuses that comprise the global apparatus is changed, then the net global disturbance by the composite apparatus is also changed. Since this global disturbance interacts with the composite two-particle system, then the predicted outcome for the joint measurement is also changed.

To anyone intent on obtaining a microscopic view of details of this interaction process, then action-at-a-distance clearly seems to be necessary for Bohr's global disturbance to occur. Bohr's argument is more readily understood, once that one recognizes that it is based on his denial of realism. That is, it is impossible to physically disturb something that doesn't exist! Thus, if there is no physical-space description of either the quantum mechanically entangled individual systems or of the actual disturbance process, then the required associated existence of non-local action-at-a-distance is a non-issue. The objectionable aspect of Bohr's description (action-at-a-distance) is then, consistently with Bohr's assertion of completeness, simply nonexistent, by definition!

To an avowed realist (such as Einstein), Bohr's "refutation" of the EPR argument seems like simply playing with words. Thus, Einstein remained adamantly unconvinced by Bohr's argument, saying,

> *"To believe this is logically possible without contradiction, but it is so very contrary to my scientific instinct that I cannot forego the search for a more complete conception."*

Reportedly, Einstein was also unconvinced by von Neumann's argument, but never published a critique of it.[9] Thus, despite the aims of von Neumann (and Bohr) and of EPR, their antipodal conclusions left the completeness issue in a very untidy state.

---

[8] N. Bohr, Phys. Rev. **48**, 696 (1935); Nature **136**, 65 (1935).

[9] Unfortunately, there does not appear to be any evidence that von Neumann and Einstein (whose offices at Princeton were just down the hall from each other) ever directly discussed the validity of von Neumann's "completeness proof". However, Abner Shimony relates a story, told to him by Peter Bergmann, about a time when he (Bergmann) asked Einstein for an opinion of this proof. Reportedly, Einstein was quite familiar with it. In response to Bergmann, he fetched and opened von Neumann's book, and then pointed to one of von Neumann's assumptions, upon which the proof is based. He then said that he saw absolutely no reason to believe that this assumption should hold in general for all alternative theories. Sadly, Einstein never published his criticism of von Neumann's proof. Evidently, the assumption that he then fingered is the same one (linearity) that is later criticized by Bell in his 1965 paper, and found there to be untenable in general.

Additional related comments about EPR were then offered by Erwin Schrödinger[10] and by Wendell Furry[11]. They both considered what would happen if the quantum-mechanical correlation between the particles somehow ceased to exist, once the two particles became sufficiently separated. Both men arrive at exactly the same conclusion. If this were to happen, then the specific experimental predictions by quantum mechanics would be altered significantly. Schrödinger's paper is also noteworthy in that it introduces the term, "entanglement", into our vocabulary for describing the two-particle correlation. Interestingly, these two authors also offer exactly opposite personal impressions of entanglement. Schrödinger calls entanglement "sinister", and views its presence as indicating a fatal flaw in quantum mechanics. Furry, on the other hand, views these result as indicating that any modified theory that omits entanglement must be immediately viewed as wrong, since quantum mechanics is "obviously" a correct theory.

The neglected experimental foundations of quantum mechanics

The untidy legacy left by the above authors was particularly acute, since neither side attempted to justify its position with hard experimental data. It was automatically assumed by them that, without any doubt, quantum mechanics is "obviously" the correct theory[12]. Indeed, none of the above authors ever cite any experimental evidence at all to back up their arguments. Instead, hard experimental evidence is replaced in the above arguments by a simple reliance on the self-consistency of the theory, whereupon only *Gedankenexperiments* are then needed for these arguments to proceed. Thus, it seems that all of these authors believed in what I will call here the "standard religion", that may be summarized as follows:

> *If a theory is self-consistent and elegant, and if it explains a significant body of experimental data that are inconsistent with previously held theories, then the theory must be accepted as gospel. Correspondingly, the theory needs no further testing, even in areas where its predictions may seem to be surprising and/or paradoxical.*

In later years, the phrase "it needs no further testing" became reinterpreted to mean, *it is sacrilege to even suggest that further tests should be performed.*

Said "religion" follows, in turn, directly from the above-quoted passages by von Neumann. Unfortunately, von Neumann's work evidences a total lack of any direct contact with any actual experiments. His "proof" is based only on the self-consistency of the operator calculus used by quantum mechanics. Throughout his book, he cites only one experiment, that by Compton and Simon[13], and then only to refute the short-lived Bohr- Kramers-Slater[14] predecessor theory. But after all, von Neumann was a mathematician. Unfortunately and correspondingly, his totally mathematical approach leaves an experimental physicist in the very awkward position of having no specifically defined experiment (or set of experiments) on which quantum mechanics is purportedly based! Needless to say, von Neumann also does

---

[10] E. Schrödinger, Proc. Camb. Phil. Soc. **31**, 555 (1935).

[11] W. H. Furry, Phys. Rev. **49**, 393 (1936); Phys. Rev. **49**, 476 (1936). The second paper emphasizes the differences between his and Schrödinger's views of this result.

[12] Whenever I asked anyone during the '60s about the experimental justification for using quantum mechanics to describe entangled-state systems, I was told that its accurate prediction of the spectrum of helium proved it to be universally correct for all such systems. However, not much sophistication is required (especially now in hindsight) to recognize that the energy levels exhibited by two electrons bound in a helium atom do not provide, by themselves, a very unconvincing example of the full range of remarkable phenomenology exhibited by an EPR system.

[13] A. H. Compton and A. W. Simon, Phys. Rev. **26**, 289-299 (1925). This experiment is a more precise repetition of an earlier experiment by W. Bothe and H. Geiger, Zeits. für Physik, **26**, 44 (1924). Schrödinger, however, subsequently found a semi-classical explanation of this experiment, and went on to propose an important relevant experiment. See "Splitting photons?".

[14] N. Bohr, H. A. Kramers, and J. C. Slater, Phil. Mag. [6] **47**, 785-802 (1924).

not specify any allowed error bounds for these unspecified experiments, which, if violated, might overturn or help confirm the far reaching conclusions that he draws.

Von Neumann is not alone in this regard. Bohr, EPR, Furry and Schrödinger also cite no actual experiments. Perhaps these authors all tacitly accepted and promoted the aforementioned religious belief because they were all theorists and felt that correspondingly they should act as clergymen. At the very least, they clearly show negligible respect for experimental physics as the final arbiter in physics. To their discredit, none of these authors even suggests in the above-cited works that experimental testing of the predictions for their *Gedankenexperiments* is in order. Perhaps their tacit reliance on and promotion of this religion was due to the fact that in the thirties, experimental physics was still in a rather crude state of development, and theorists then had no choice but to base their arguments on *Gedankenexperiments*, and then to stop at that point, because realizations of these *Gedankenexperiments* were well beyond the technology of the day.

Whatever justification these men may have had in this regard is perhaps unimportant now. Given their ecumenical leadership, and especially given Bohr's strong leadership, the net legacy of their arguments is that the overwhelming majority of the physics community accepted Bohr's "Copenhagen" interpretation as gospel, and totally rejected Einstein's viewpoint. Also, given von Neumann's daunting intellect, his proof was held to be sacrosanct, even to those who had never even perused its details. Moreover, since its details were rarely perused, over time, von Neumann's proof became commonly (and incorrectly) misinterpreted to imply that

> *No theory based on hidden variables is possible that gives the same experimental predictions as quantum mechanics!*

Correspondingly, the standard religion became hardened to hold that

> *Quantum theory is so successful, it is simply inconceivable that its foundations and/or predictions can be unsettled, misunderstood, or, God forbid, in error. Therefore, it warrants no further testing!*

as a nearly direct rephrasing of von Neumann's above claims. Unfortunately, this religious belief was held dogmatically by most physicists well into the '70's.

Challenging the common wisdom

Given the effectively religious impact of Bohr and von Neumann's teaching, little progress was made during the subsequent two to three decades towards any further development of our fundamental understanding of the meaning of quantum mechanics, or especially towards a resolution of the completeness problem. Given that the religion warrants against further direct testing of quantum theory's foundations, negligible effort was made in this area also, even after the state-of-the-art of experimental physics had advanced to the point where many new direct tests had become possible.

This neglect of the theory's foundations is quite unfortunate, in my opinion, because (as will be documented below) a large number of "killer" details have frequently slipped through cracks that exist between a theorist's view of a fundamental experiment and an experimentalist's view of that same experiment. Moreover, such "killer" details are frequently sufficiently important so as to totally invalidate many of the conclusions that are then commonly drawn from these experiments, even though those conclusions have become entrenched as an inseparable part of the physics community's common wisdom. John Bell's article, *Against Measurement*[15], is likewise critical of our community's blind acceptance of this religion, and is quite outspoken in suggesting that outstanding problems remain in our understanding of quantum theory, and that further careful study of the theory's foundations is warranted. Fortunately, within our community there were a few physicists who actually enjoy(ed) challenging this common wisdom. These are the iconoclasts of our profession.

[15] J. S. Bell, *Phys. World*, Aug. 33-40 (1990).

The next important step in the saga occurred in 1950, when David Bohm (a former student of Oppenheimer) published a new textbook, *Quantum Theory*[16]. The book's treatment of the underlying conceptual foundations of quantum mechanics is quite similar to other textbooks of the era. In it, Bohm follows von Neumann and expresses strong doubts about the existence of hidden parameters within quantum mechanics. However, he stops short of asserting that their existence is impossible. While he does cite von Neumann's book as a reference, he does not mention its "impossibility proof". Instead, he discusses the completeness of quantum mechanics via reference to the EPR configuration, and argues for completeness by defending Bohr's criticism of EPR's argument. However, in the process, Bohm adds one very important new idea to the discussion that significantly sharpens both the EPR configuration and its associated argument.

Originally, EPR had discussed entanglement in terms of the continuous variables, momentum and position. In this book, Bohm reformulates the configuration in terms of entangled discrete-state systems, i.e. in terms of the spin-component correlations of two spin-1/2 atoms in a spin-singlet entangled state. In a historical context, this sharpening is crucial for the work that follows.

One problem with writing a textbook is that it makes you think carefully about the book's subject matter. Following publication of this book, Bohm broke with tradition in two important ways. First, with his new student Yakir Aharonov, he published a paper[17] where, for the first time, they sought real experimental evidence for the peculiar properties of entanglement that are exhibited by the EPR configuration. Using a two-photon analogy of the two-spin configuration outlined in Bohm's book, they noted that the polarization correlation of gamma-ray pairs emitted during positronium annihilation is a real-life example of EPR's configuration. That correlation's magnitude had, in fact, recently been measured by Wu and Shaknov[18] at Columbia Univ. Bohm and Aharonov then point out that the correlation's measured magnitude agrees with quantum theory's prediction (with entanglement present), and significantly disagrees with the prediction by the Schrödinger-Furry hypothesis (with entanglement absent, once the particles become significantly separated). By publishing this paper, Bohm and Aharonov transcend the "standard religion", and actually consult an experiment! They also prove here what Furry could only assume but could not prove, i.e. that the Schrödinger-Furry hypothesis cannot be held as a correct description for nature.

Bohm, however, was just beginning his iconoclastic endeavors. Despite the fact that in his book he expresses strong doubts that hidden variables might exist, he now adds extreme untidiness to the existing situation by formulating a new "causal" theory that actually incorporates hidden parameters within it[19]. Given the stigma against such work, few people took Bohm's paper seriously. One who did take it seriously was Louis deBroglie. Inspired by this work, deBroglie reentered the arena and published a whole series of books[20] and papers promoting a similar theory. (Bohm and deBroglie's theories are both examples of what are commonly referred to as "pilot-wave theories".) The state of untidiness surrounding the issue of quantum theory's completeness had now blossomed to become a real mess, let alone a now seriously important skeleton in quantum theory's closet. Von Neumann's "proof" purportedly asserts that these theories cannot exist. But they obviously now do! Something clearly must be wrong, somewhere. Aware of von Neumann's "impossibility proof" Bohm attempts to

---

[16] D. Bohm, *Quantum Theory*, (Prentis Hall, Englewood Cliffs, NJ, 1950).

[17] D. Bohm and Y. Aharonov, Phys. Rev. **108**, 1070 (1957).

[18] C. S. Wu and I. Shaknov, Phys. Rev. **77**, 136 (1950).

[19] D. Bohm, Phys. Rev. **85**, 169 (1952). See also, D. Bohm and J. P. Vigier, Phys. Rev. **96**, 208 (1954). Fortunately, this work was published before the APS policy had been formulated. Presumably, it would have not passed muster under that policy.

[20] L. deBroglie, *Nonlinear Wave Mechanics*, (Elsevier, Amsterdam, 1960); *Ondes Electromagnetiques et Photons,* (Gauthier-Villars, Paris, 1968); and *Introduction to the Vigier Theory of Elementary Particles,* (Elsevier, Amsterdam, 1963).

clean things up a little, and offers an argument as to how von Neumann's proof may not apply to his theory.

John Bell unravels the confusion

The stage was finally set for John Bell to enter the picture to begin unraveling much of the accrued (but largely ignored) mess. While both von Neumann's and EPR's arguments are directed towards settling the issue about the possibility (or impossibility) of hidden variables within quantum mechanics, ironically Bell now shows, in a sense, that both arguments are effectively "wrong", at least in terms of their effective conclusions.

Bell first critically looked at von Neumann's argument. Given the existence of Bohm and deBroglie's theories, von Neumann's conclusion (or at least his perceived conclusion) obviously must be wrong! Bohm had given one argument as to how that might be so. However, Bell is neither very complimentary towards nor impressed by Bohm's counter to von Neumann's argument. In a masterfully written article that was published in a 1965[21], Bell scrutinizes von Neumann's requirement for a linear superposability of eigenvalues along with the linear superposability their associated operators. He then points out that this assumed requirement has no *à priori* reason to hold, in general, for a hidden-variables theory, especially when different operators correspond to physically different measuring apparatuses, and when these eigenvalues are also to be taken as the possible measurement outcomes for the associated apparatus. Furthermore, without this linearity assumption, von Neumann's argument neither obtains, nor constrains hidden-variable theories. Correspondingly, hidden variables are then no longer found to be "impossible" for alternative theories (to quantum mechanics) by von Neumann's argument.

At that time, many physicists who had looked carefully at these issues generally put much more reliance on von Neumann's rigorous mathematical argument than on Bohr's somewhat vague philosophical argument in casting their votes against hidden variables. However, given that von Neumann's argument can no longer be relied upon, Bell next wondered if EPR's precisely worded argument might be relied upon instead. Given that Bohm and deBroglie's theories had provided good test cases for von Neumann's argument, Bell (in his 1965 paper) offers a generalization of these theories to allow them to include the two-particle case needed by the EPR configuration. Bell's generalization does not follow exactly along the lines of what Bohm and deBroglie had wanted.[22] In fact, Bell found that no matter how hard he tried to formulate a two-particle generalization of these theories while simultaneously maintaining agreement with quantum theory's predictions, he could only succeed by incorporating non-local interactions between the separated particles into the generalization. But it is just these non-local interactions that EPR reasonably had insisted cannot and do not exist, given the arbitrarily wide separation of the two particles. Given Bell's difficulties in generalizing these theories in a purely local way, he then began to speculate that this difficulty might be generic to any possible hidden-variables generalization operating in the EPR configuration. Following this line of thinking further, Bell discovered an astonishing new result. This result (along with its many subsequent generalizations by others) is (are collectively) now referred to as Bell's Theorem. Bell evidently considered his new result so important that he published it in 1964[23], before publishing his 1965 review of von Neumann's impossibility proof.

Recall that EPR had argued earlier that considerations of their configuration show that hidden variables must exist. With supreme irony, Bell shows in his 1964 paper that further considerations of EPR's configuration indicate that exactly the opposite is true!

---

[21] J. S. Bell, Revs. Mod. Phys. **38**, 447 (1965).

[22] DeBroglie had earlier admitted to having encountered serious difficulties in generalizing his theory to describe more than one particle, and in affecting a transition from the configuration space in which quantum theory is formulated, to physical space, in which his and Bohm's theories are formulated.

[23] J. S. Bell, Physics, **1**, 195 (1964).

Therein he proves that for Bohm's two-spin configuration, the prediction by any deterministic local hidden variable theory for the polarization correlation coincidence rate's normalized angular dependence will necessarily be different from that rate's angular dependence predicted by quantum theory. Quantum mechanics predicts that the dependence will be precisely sinusoidal with unit amplitude. Bell's paper shows that for any deterministic local theory based on hidden variables, this dependence must be strongly non-sinusoidal if it has at least one point with the maximal correlation value that is required by quantum theory.

Theoretical truth versus experimental truth

Initially, Bell expected that his new result would be adopted as a straightforward replacement for von Neumann's argument. With this expectation, however, he initially fell into the trap of accepting the "standard religion". His 1964 paper cites no experimental data as to what the observed dependence actually is. Of course, he couldn't cite any such data, because none had yet been measured. Still, he appears to assume tacitly in that paper that experimental evidence for this rather dramatic prediction is already in hand. After concluding that quantum mechanics and a theory with local hidden variables give different statistical predictions, he further states

> *"Of course, the situation is different if the quantum mechanical predictions are of limited validity. Conceivably they might apply only to experiments in which the settings of the instruments are made sufficiently in advance to allow them to reach some mutual rapport by exchanging signals with velocity less than the speed of light. In this connection, experiments of the type proposed by Bohm and Aharonov, in which the settings are changed during the flight of the particles, are crucial."*

Bell's wording of this quoted passage has been very carefully chosen. Its converse conclusion is that it is correspondingly inconceivable that the quantum-mechanical predictions do not correctly predict experiments with static instrument settings, i.e. that the quantum-mechanical predictions for static instrument settings are indeed correct, and have been verified experimentally to be correct. While the passage does not actually assert that this converse conclusion holds true, it strongly suggests that it does. Thus, the passage tacitly implies and gives the misimpression that the existence of hidden variables (at least in situations with static analyzers) is, *ipso-facto*, refuted by his proof.

Nonetheless, Bell hedges his bets. Given that his arguments only consider the predictions for non-relativistic quantum theory, he appears to suggest that new keys to the hidden-variables mystery will be revealed only via considerations of relativistic corrections to quantum theory. Correspondingly, he goes on to promote Bohm and Aharonov's second-generation experiment with rapidly changed analyzers. However, such a proposal is clearly silly, unless one acknowledges the existence of first-generation experimental results with static analyzers. He was clearly bluffing, and did not really know, at this point, what the experimental status of his result really was. I personally will claim much of the credit for calling his bluff.

In the late sixties I was a graduate student at Columbia Univ., and was struggling, trying to understand quantum mechanics. I had read EPR's paper, and also had read Bohm's and deBroglie's work. While I had difficulty understanding the Copenhagen interpretation, the arguments by its critics seemed far more reasonable to me at that time. Since quantum mechanics is formulated in a configuration space, it then provides no physical-space model to consult, whereupon it is then difficult to visualize what might actually be happening, especially in a two- (or more-) particle system. I found Bohm's and deBroglie's works refreshing, since they do give real physical-space models of what is happening. I also found EPR's arguments much more persuasive than Bohr's, and was only vaguely aware of von Neumann's "proof" at this time. Hidden variables thus seemed (to me then) to be a perfectly logical solution to the problem[24]. By holding that politically incorrect opinion, however, I was then certainly branded as a heretic by many, and undoubtedly as a quack by others.

Then, I read Bell's 1965 and 1964 papers. Given my familiarity with Bohm and deBroglie's work, I was not particularly surprised by Bell's 1965 result. On the other hand, once I comprehended what the 1964 paper actually said, I was astounded by its result. However, I was not yet willing to accept the paper's far-reaching implications until I finally saw some experimental evidence that decided between its two significantly different predictions. Since Bell's paper is conspicuously vague on the experimental status of its prediction (but crystal clear on everything else), I suspected that Bell was bluffing and started to search through the literature for experimental results. I found none, other than that by Wu and Shaknov.

With regard to experimental testing, I saw at least one serious problem with Bell's 1964 result. That problem is that Bell's form for his result applies <u>only</u> to his idealized configuration, and thus cannot be directly tested experimentally. Bell (being a theorist) assumes an ideal apparatus and an ideal two-particle entangled-state preparation. However, his proof then relies on this idealization, and further uses it to show that a perfect correlation between the two particle's polarizations occurs for at least one setting of the polarization analyzers. Unfortunately, even an infinitesimal departure by the considered system from this idealized behavior then invalidates the applicability of Bell's proof to it, since then, the required single point with a maximal correlation no longer exists. Correspondingly, his result then applies only to ideal systems and not to real ones! (Damn theorists!) Obviously, in any real experiment, the required perfect correlation cannot and will not obtain.

Does my observation here then imply that Bell's exciting new result is no more applicable to the completeness problem in real-life situations than is von Neumann's result? I thus asked myself whether or not Bell's assumption of an idealized apparatus and system is, in fact, necessary for an alternative proof of his Theorem to obtain. Looking further, I found that this assumption can indeed be avoided. Thus, I succeeded in deriving a new more general form for his result that then <u>can</u> be tested experimentally. Thus holding an experimentally testable prediction, I then returned to the question about the result's experimental status. It became immediately clear that no definitive conclusion then obtains when this testable prediction is compared with the data available from the Wu-Shaknov experiment. Moreover, I found that it is straightforward to build a simple *ad-hoc* hidden-variables counter-example that yields predictions that are exactly the same as those from the quantum-mechanical prediction for the Wu-Shaknov experiment. Thus, I concluded that something beyond a simple extension of the Wu-Shaknov experiment is needed for a valid experimental test. What can it be?

While I was still searching for an appropriate experimental test of these new predictions, I gave a seminar on Bell's Theorem to Dan Kleppner's group at MIT. Dave Pritchard's thesis work in this group involved doing crossed-beam scattering experiments with alkali metal atoms. I suggested to Pritchard and Kleppner that 90° - COM scattering is a possible method for producing Bohm's two-particle spin-singlet entangled state, and that their experiments might be reconfigured thusly to allow measurement of the polarization correlation of the scattered atoms as a test of the Bell's Theorem prediction. A newly arrived postdoc in the group, Carl Kocher, attended the seminar. For his PhD thesis at UC Berkeley, he had just finished performing an experiment with Gene Commins, wherein they had measured the polarization correlation of photon pairs emitted in an atomic cascade[25]. My attention was called to this experiment during questions that followed my talk, whereupon I then realized that an experiment similar to it could provide a far easier and much better test. Unfortunately, Kocher and Commins had only measured two-photon coincidences at 0° and 90° relative analyzer orientation. As per Bell's assumption, these data points are located

---

[24] John Bell confesses in the preface to his book, *Speakable and unspeakale in quantum mechanics*, to being similarly enamored with these features of pilot-wave theories.

[25] C. A. Kocher and E. D. Commins, Phys. Rev. Lett., **18**, 575 (1969).

exactly where the two theories can give the same prediction. Also, as per my own generalization of Bell's result to provide an actual experimental prediction, I noted that their polarizer efficiencies were nowhere high enough for the experiment to be conclusive, in any case. Thus, their experiment provides no data at all for a test of hidden-variable theories. What is clearly needed at that point are data taken at angles intermediate between their measured points with an experiment that is similar to theirs, but that uses much better polarizers.

Finally, I decided to call Bell's bluff. I wrote letters to Bell, Bohm and deBroglie, asking all three of them (a) did they know of any experiments testing the result, and (b) did they agree that a repeat of the Kocher-Commins experiment with improved polarizers at intermediate angles would be convincing, and (c) how did they view the importance of such tests[26]. All three courteously responded and replied NO to (a) and YES to (b). Bell was particularly enthusiastic about the idea, and simultaneously also revealed his own agnostic religious beliefs about quantum theory. In response to (c), he said

> *"In view of the general success of quantum mechanics, it is very hard for me to doubt the outcome of such experiments. However, I would prefer these experiments, in which the crucial concepts are very directly tested, to have been done and the results on record. Moreover, there is always the slim chance of an unexpected result, which would shake the world!"*

The Vietnam War dominated the political thoughts of my generation. Being a young student living in this era of revolutionary thinking, I naturally wanted to "shake the world". Since I already believed that hidden variables may indeed exist, I figured that this was obviously the crucial experiment for finally revealing their existence. But if they do exist, then quantum mechanics must be verifiably wrong here, with its error having gone undiscovered heretofore. However, since there was virtually no experimental evidence then available for this configuration (plus or minus), that "slim" possibility could not be discounted. To me, the possibility of actually experimentally discovering a flaw in quantum mechanics was mind-boggling. Thus, I drafted an abstract for the Washington DC Spring Meeting of the American Physical Society proposing this experiment[27]. Immediately as it appeared in print in the APS Bulletin, I received a phone call from Abner Shimony (a former student of Eugene Wigner). He said that he and his student Mike Horne had come to the same conclusions[28]. Their pursuit of experimental evidence had led them to Frank Pipkin at Harvard, who had just performed an experiment that was similar to the Kocher-Commins experiment, in that it observed two-photon coincidences from an atomic cascade, but that did not include any polarization analysis. Pipkin's PhD student, Dick Holt, was now setting up to perform the experiment that I had just proposed. After comparing notes, we all agreed to coauthor a Phys. Rev. Letter formally presenting our conclusions[29]. Our collaborative effort is now commonly referred to by the initials, CHSH, with the paper's experimental prediction contained in what is now correspondingly known as the CHSH inequality. In the process of writing this paper, Abner, Mike and I forged a long lasting friendship that was to spawn many subsequent collaborations.

---

[26] According to David Wick [*The Infamous Boundary – Seven Decades of Controversy in Quantum Physics*, Birkhäuser, Boston, 1995], my 1969 letter was the first response to his 1964 paper that Bell had received

[27] J. F. Clauser, Bull. Amer. Phys. Soc. **14**, 578 (1969).

[28] Shimony's name was already familiar to me when he called, because I had earlier read some of his work on the "measurement problem" in quantum mechanics. This problem then leads to the Schrödinger's-Cat and Wigner's-Friend paradoxes, that are another untidy aspect of the foundations of quantum mechanics. Both paradoxes result from the peculiar properties of quantum entanglements.

[29] J, F, Clauser, M. A. Horne, A. Shimony, and R. A. Holt, Phys. Rev. Lett., **23**, 880 (1969). This paper first coined the term "Bell's Theorem".

Beware of the experimentalists lurking about

Upon receiving my PhD from Columbia, I moved to UC Berkeley to work in radio astronomy as a postdoc for Charlie Townes. Since Gene Commins was also at UC Berkeley, I suggested to both Townes and Commins that they allow me to resurrect the Kocher-Commins experimental apparatus, and to use it to test the CHSH-Bell prediction. Townes was intrigued by the idea, took it seriously, and offered half of my time to this project, while the other half was to be spent on radio astronomy. Commins offered Stuart Freedman, his new graduate student, to work with me on the project. Said "resurrection" actually involved building a whole new mammoth-size apparatus. Fortunately, Townes was very tolerant as my radio astronomy projects languished. Stu and I published our results in 1972[30], and Stu got his PhD with this experiment.

Our results agree exactly with the quantum-mechanical prediction, but exclude theories based on local hidden variables by about five standard deviations. This experiment is noteworthy in that it is the first to actually be able to draw such a conclusion! Its count rate was so low that the results are effectively unchanged, even if "accidental" (background) coincidences are not subtracted from the totals used in a comparison with the CHSH inequality.

Meanwhile, in competition with our experiment at Berkeley, Dick Holt and Frank Pipkin at Harvard had been pursuing their parallel experiment. It was similar to ours, except that it was much smaller and used a different atomic species and a different kind of polarizers. While our experiment was still in progress, they actually got the first results. Unfortunately, their results disagreed with the quantum-mechanical prediction, but did agree with the local hidden-variables prediction by the CHSH inequality. They decided to be cautious, however, and not to publish their results until they saw what our experiment at Berkeley revealed. Once we announced our results, they then opted not to publish theirs formally. Instead, they distributed an informal preprint that outlined what they had done and observed. In the end, they concluded that their result was due to an underlying systematic error.

By declining publication of their experimental results, they left ours as the only published results available. Given the importance of the issues at hand, our own results obviously needed confirmation. Obviously, I was no longer actively working in radio astronomy at this point. Fortunately, Howard Shugart allowed me to continue my work at UC Berkeley as a member of his atomic-beams group. Following my photon-splitting experiment (described below) I proceeded to repeat Holt and Pipkin's experiment with only a few minor modifications. Again, my new result[31] agrees with quantum theory's prediction, but significantly disagrees with the CHSH-Bell prediction. I also went further and tested the predictions of quantum theory for the circular polarization correlation[32]. These results also agree with quantum theory's prediction.

It should be noted that when we began our experiments at Berkeley, tunable lasers were not yet commercially available. Thus, all of the Berkeley experiments were done without their benefit. As a result, the count rates were very low, and the experiments were very slow and tedious. Finally, Ed Fry and his student Randal Thompson[33] performed a test of the CHSH prediction using one of the first commercially available dye lasers. Given the short counting periods afforded by their use of this laser, they could more readily search for possible systematic errors. Their experiment confirmed our "Berkeley" results, but disagreed with Harvard's results. The score was now three to one, in favor of quantum mechanics. That score was sufficiently convincing to allow Abner Shimony and me to draw the above-described conclusions in our review paper [Ref. 1].

---

[30] S. J. Freedman and J. F. Clauser, Phys. Rev. Lett. **28**, 938 (1972).

[31] J. F. Clauser, Phys. Rev. Lett., **36**, 1223, (1976).

[32] J. F. Clauser, Il Nuovo Cimento, **33B**, 740 (1976).

[33] E. S. Fry and R. C. Thompson, Phys. Rev. Lett. **37**, 465 (1976).

Generalization of the Bell and CHSH results to constrain local realism and space-time

While I was actively pursing my experimental efforts at UC-Berkeley, in parallel with that work, I also pondered some remaining details that still had not yet been addressed by CHSH and Bell. There were then at least four important unanswered questions:

(1) What are the fundamental assumptions underlying Bell's Theorem?

(2) What is the affect of the large number of unobserved particles (especially when the particles are optical photons) associated with a finite detector solid-angles and finite detector efficiencies? Also, given unobserved particles, what are the requirements for an experiment's design that then still allow its results to provide conclusive arguments to be made regarding the truth or falsity of the fundamental assumptions that, in fact, underlie Bell's Theorem?

(3) If these experimental requirements cannot readily be met in practice, then what is the least objectionable supplementary assumption that one can make to allow reasonably convincing inferences to be drawn from experiments that, for example, can be performed with optical photons?

(4) Also, if these requirements are not met in practice, is there at least one counterexample consistent with the fundamental assumptions underlying Bell's Theorem for experiments performed with say optical photons? Unfortunately, if such a counterexample exists, then only an experiment performed within these limits can be considered to be fully conclusive as a test of the fundamental assumptions underlying Bell's Theorem.

Mike Horne and I addressed and answered all four of these questions in our 1974 paper[34]. That paper is now commonly referred to by the initials, CH.

Regarding question (1), it may seem surprising and strange, at this point in the development, that the fundamental assumptions underlying Bell's Theorem were still not yet well defined. As a parallel problem to the gap between Bell's 1964 result and the CHSH result, this lack of definition again arises from differences between the idealized configuration considered by Bell, and the realizable configuration considered by CHSH. Here and elsewhere, it represents an example of crucial details falling through a gap that exists between a theorist's view of an experiment and an experimentalist's view of the same experiment.

Following EPR's lead, both Bell and CHSH initially had assumed that determinism is the basic underlying assumption for Bell's Theorem. This notion follows from EPR's definition of an "element of reality", as per

> *"If, without in any way disturbing a system, we can predict with certainty (i.e. with probability equal to unity) the value of a physical quantity, then there exists an element of physical reality corresponding to this physical quantity."*

While predictions with certainty are possible for idealized systems, unfortunately, no such predictions can be made for realizable systems (except, of course, for death and taxes). Following EPR, Bell effectively derives determinism in his 1964 paper for the EPR particles, via a use of the at least one data point with a perfect correlation. Unfortunately and conversely, determinism cannot be derived via this reasoning for any real system, which, of course, has no such point with said perfect correlation, and thus no predictions with probability equal to unity. Given this fact but still following EPR's ideas, CHSH simply and explicitly assumed determinism to hold for the purposes of their derivation[35]. However, Mike

---

[34] J. F. Clauser and M. A. Horne, Phys. Rev. D **10**, 526-535 (1974).

[35] Shortly following CHSH, Bell did extend the range of applicability of the CHSH result further to include the effects of additional hidden variables in each apparatus. [J. S. Bell, *Introduction to the hidden-variables question*, in *Foundations of Quantum Mechanics, Proceedings of the International School of Physics, "Enrico Fermi"*, B. d'Espagnat, ed. (Academic Press, New York, 1971), pp. 170-194]. However, in doing so he adds no new premises to Bell's Theorem in this new derivation, and similarly explicitly assumes determinism to hold.

Horne and I noticed the fact that the CHSH prediction also appears to hold for models in which determinism is neither assumed, nor in fact, even holds. This fact suggested to us that an assumption of determinism is not really necessary for Bell's Theorem to apply. Independently, John Bell had also noticed this fact and mentions it in a book review[36], wherein he comments that a hidden-variables model cannot give the probabilistic quantum-mechanical prediction for the two-spin system, even if a random-noise generator at every point in space-time[37] is assumed to influence the space-time evolution of the hidden variables. Such a system, by definition, does not have a deterministic evolution.

Mike and I thus asked ourselves the following questions: If determinism is not required of a system for Bell's Theorem's constraints to hold for it, what then is the fundamental characteristic of a physical system, such that it is constrained by Bell's Theorem. Also, how does one go about specifying said system's assumed nature within the Theorem's derivation? Our 1974 paper was the first publication that specifically answers these questions.

Historically, the CH results can be seen as a logical continuation of a sequence of quandaries faced by Einstein during his development of special relativity. In formulating special relativity, Einstein noted that one cannot ask the universal questions "When" and "Where" in a precise fashion, without first defining the associated primitive entities, "time" and "distance". He found that, at best, he could only provide purely operational definitions for these entities. That is, he defines "time" to be the stuff you measure with a clock, and "distance" to be the stuff you measure with a ruler. Similarly, to allow Einstein to ask the universal question "What" in a similarly precise fashion, EPR's definition of an "element of reality" effectively provides his initial attempt to help one to define the associated primitive entity, "object". But, as noted above, that definition does not apply to realizable systems. CH thus found it necessary to offer an improved and explicit operational definition for the notion of an "object". The historical development of Bell's Theorem is then seen to have a direct parallel to the development of special relativity. Once that a suitably precise definition is given for some fundamental "stuff" of nature, then and only then can new testable physics emerge.

In their 1974 paper, CH operationally define an "object" (within a local realistic theory) as stuff with measurable properties that also can be put in a box, i.e. stuff (along with an associated measuring apparatus) that can be spatially surrounded by a space-time Gaussian surface. CH supplement this definition with an associated definition of an "objective local theory" (subsequently also called a "local realistic theory" and/or a "theory of local realism" by Clauser and Shimony in Ref. [1]), as a theory that describes such objects, wherein action-at-a-distance is precluded, and wherein a measurement reveals local properties of an object (i.e. of said stuff within said box)[38]. Moreover, which properties of an object are to be measured may be arbitrarily chosen at the free will of the experimenter who operates the apparatus. Thus, for objects within such a theory, if the correlated properties of two objects in two disjoint boxes are measured at space-like separated measurement events, then absence of action-at-a-distance prevents the experimenter's choice of property measured in the first box (e.g. his choice of color vs. weight, in the earlier examples) from affecting the results of a measurement made on/in the second arbitrarily distant box. CH found that these very simple and naïve premises and definitions for <u>any</u> such local realistic theory of nature, then lead to a very general formulation of Bell's Theorem, that, in turn, then constrains the experimental predictions for said theory. The CH specific experimental predictions are contained in a new inequality that then must be satisfied by any such theory within local realism. That inequality is now commonly referred to as the CH inequality.

---

[36] J. S. Bell, Science, **177**, 880 (1972).

[37] Presumably, Bell may have had vacuum-fluctuations in mind for his random noise source, as a parallel to some of John Wheeler's space-time models,

[38] Under the CH definition, a "deterministic local hidden variables theory" then becomes a special case of the more general class of "local realistic theories".

Under CH's definition of an object, Bell's Theorem then not only provides hard new experimental predictions for a local realistic theory, but it also provides surprising and profound implications for quantum mechanics as well. One first surprise is that these two different theories (local realism and quantum mechanics) give measurably different predictions for realizable experiments. A second surprise is that Bell's Theorem shows that quantum mechanics does not describe objects that fall within the scope of CH's naïve definition. Thus, it appears that within quantum mechanics, a "quantum" object (however else one may choose to define it) is not something that can be put in a box! Unfortunately, it does not appear that Bohr (or anyone else) has ever provided a clear concise definition for an "object" within quantum mechanics. On the contrary (as is often noted by Bell in his book), the distinction between and definitions of the terms "observer", "apparatus", and "system" (where "system" presumably means "object"), all remain quite vague under the Copenhagen interpretation.

Question (2) naturally arises when one tries to test the CHSH experimental predictions using optical photons, as Stu Freedman and I had just done. In such an experiment, only about one emitted photon pair in a million is actually detected in coincidence. Most photon pairs emitted by the source are not detected at all. For most of the observed events from our cascade photon source, only one photon of the pair is actually detected. This situation is clearly quite far removed from that of the idealized configuration originally considered by Bell in his 1964 paper, wherein both particles of each and every emitted pair are subsequently detected, and where the pair is known to have been emitted (as per the discussion of "event-ready" detectors in Ref. [1]). Mike Horne and I realized that the associated enormous unobserved volume of phase space in such experiments leaves a lot of room for significant sampling biases to occur. Unlike the CHSH inequality, however, the CH inequality's prediction is carefully derived so as to render it unaffected by these biases. The experimental requirements for testing the CH inequality are also derived in the CH paper. Unfortunately, the consequence of using the CH inequality (instead of the CHSH inequality) for experimental tests is that an experiment performed with optical photons can no longer, by itself, provide a direct test of local realism[39]. Moreover, designing and carrying out a practical experiment that actually meets the CH inequality's experimental requirements is not an easy task. Fry, Walther, and Li[40] have recently offered such a design. As of this writing, no direct tests of the full CH inequality have yet been performed. Experimental tests of the CHSH inequality, rather than the CH inequality, are now commonly said to leave open the "detector-efficiency loophole".

Question (3) concerns the relationship between the experimental predictions made by the CHSH and CH inequalities. Since experiments performed with optical photons, as a general rule, do not violate the CH inequality, but can and do violate the CHSH inequality, it is natural to ask what additional assumptions can (or must) be added to allow the CHSH inequality and experiments using photons to still give reasonably persuasive (but perhaps not totally compelling) conclusions about the truth or falsity of local realism. That is, what is the best we can do with contemporary technology? To a similar end, CHSH had earlier derived their inequality by making an auxiliary assumption regarding the behavior of a photon at a polarizer that then allows an experimentally testable prediction to obtain. Thus, CHSH assume that whenever a photon encounters a polarizer, then the photon is either transmitted by the polarizer or not transmitted by the polarizer. Furthermore, given that two photons have emerged from their respective polarizers, their detection probability is assumed by CHSH to be then independent of the analyzer-pair settings. Here, CHSH have taken the concepts of "passage" and "non-passage" to be primitive, and thereby effectively also assume that photons are particle-like, for the purposes of the experiment. The CHSH auxiliary assumption obviously does not apply to a semi-classical radiation theory (such as Jaynes' "neo-classical

[39] This fact is still true, even for photon pairs that are generated by a parametric down-conversion source, as with most recent experimental tests of Bell's Theorem.

[40] E. S. Fry, T. Walther, and S. Li, Phys. Rev. A, **52**, 4381-4395 (1995).

radiation theory", see below) that is necessarily an example of the Schrödinger-Furry hypothesis[41]. In such a theory, each of the individual "photons" that comprise a pair of "coincident photons", must then be represented as a short pulse (wave packet) of classical electromagnetic radiation. Since each pulse is now classical, each "photon" is then partially passed and partially transmitted by a polarizer. Nonetheless, "photons" in such a theory are clearly "objects", under the CH definition.

The CHSH auxiliary assumption is thus deficient, in that it does not apply to this important class of local hidden-variable theories. CH noted, however, that even though the CHSH assumption does not apply to these theories, the CHSH inequality apparently does still apply, and Bell's Theorem clearly does constrain semi-classical radiation theories. To remedy this deficiency, CH provide a significantly improved auxiliary assumption that is also much weaker than the CHSH auxiliary assumption. It is called the "no-enhancement" assumption. It simply requires that the probability of an object's detection, after it has passed through (or partially passed through) an analyzer, can then be no greater than the probability would have been if the particle had not passed through the analyzer, i.e. than if the analyzer had been absent. This assumption clearly holds, for example, for semi-classical radiation theories when the objects are classical electromagnetic pulses and the analyzers are optical polarizers. Furthermore, it seems to be highly plausible and very difficult to avoid for almost any local realistic theory. Then, for all local realistic theories for which the CH "no-enhancement" assumption also holds, the CH inequality's prediction reduces to the CHSH inequality's prediction. In such cases, (except as noted below) experiments such as that by Freedman and Clauser then convincingly refute all such theories.

Question (4) arises in a quest for finding both necessary and sufficient conditions for the CH inequality to constrain local realism. Towards this end, CH provide an *ad hoc* counterexample, that is consistent with local realism, that does not violate the CH inequality, and that does violate the CHSH inequality. While the existence of this counterexample does not prove that the "no-enhancement" assumption is "necessary", it does show that at least some additional assumption is required for one to use the CHSH inequality to test local realism. Use of the CHSH inequality (as opposed to the CH inequality) is now commonly said to leave open the "detection-efficiency loophole".

Common confusion about count-rate normalization

For testing either the CH or CHSH inequalities, a normalization is required for the coincidence count rates as a function of relative polarizer orientation. A valid test of the CHSH inequality's prediction (as obtained from either the original CHSH auxiliary assumption and/or the CH no-enhancement assumption) calls for coincidence rate measurements to be taken, both with polarizers in place and with polarizers absent. The latter measurements then provide a normalization for the former. Here one calculates ratios of coincidence rates to coincidence rates, and since both rates have the same order of magnitude in photon counting experiments, a valid test is straightforward to obtain. The primary reason that direct tests of the CH inequality are much more difficult is that here, the polarizers are not removed. Instead, this normalization is done via the detector singles count rates (with the polarizers in place). For low detector efficiencies and/or a high particle loss rate via the broad angular correlation associated with a three-body-decay emission, the singles rates are generally much larger than the coincidence rates, the associated ratios are very small, and no violation of the CH inequality, by itself, can then occur. For this reason, a two-body-decay is required for such a test.

---

[41] The Schrödinger-Furry hypothesis represents probably the simplest form of a hidden-variable theory. It predicts a reduced-amplitude sinusoidal correlation. Bell's 1964 proof does not directly apply to it, since Bell's required perfect correlation (for at least one point) never obtains for it. This hypothesis, however, is constrained both by the CH inequality and by the CHSH inequality with the no-enhancement assumption.

Thus, obtaining a correct normalization for the coincidence count rates is always crucial for a proper and valid test to be performed. Unfortunately, this requirement is often either overlooked or significantly downplayed in its importance in many recent experiments. Furthermore, unless "event-ready detectors" are used, testing of either the CH or CHSH inequality is most readily done with only one detector following each analyzer. Indeed, the superiority of two-channel experiments (relative to four-channel experiments) is discussed, at length, in both the CH paper (Appendix B), and in the Clauser-Shimony review article [Ref. 1].

Note that various other more recent workers have been tempted to use an alternative scheme that is more along the lines of Bohm's original *Gedankenexperiment*, and thus also along the lines of the idealized experiment considered by Bell in his original 1964 paper. In these four-channel schemes, two counting channels follow each polarizer. Workers who succumb to this temptation generally ignore the need for event ready detectors, and commonly reason that an alternative and easier to implement normalization method will suffice. In such schemes, the polarizers are never removed. Instead, the observed coincidence rates are normalized to the sum of the four possible coincidence rates in all such channels. This alternative normalization method is adamantly not the CHSH (nor the CH) normalization method, although it is frequently (and incorrectly) attributed as such[42].

A four channel experiment was first used by Aspect, Grangier and Roger[43]. For that experiment, Aspect *et al.* necessarily rely on a very different set of auxiliary assumptions to allow a comparison of their results to Bell's Theorem. However, their set of assumptions is clearly much stronger than the CH no-enhancement assumption, and thus, allow a much larger "detection-efficiency loophole". The relative strength of their alternative auxiliary assumptions with the no-enhancement assumption is readily seen by noting that it is straightforward to build a rather simple counterexample that violates the CHSH inequality (configured as per the no-enhancement assumption), but that does not violate the count-rate inequality used by Aspect *et al.* For example, consider a simple local hidden-variable theory that is similar to the CH counterexample, but that does not include the enhancement used by that counterexample. This theory then predicts an angular dependence that has a reduced coincidence-rate sinusoidal amplitude along with a proportionately reduced constant offset from zero. An experiment that uses Aspect *et al.*'s normalization will then incorrectly conclude that this hidden-variable theory is refuted, and that quantum mechanics is correct. However, an experiment performed using the CHSH/CH normalization method will reach the opposite (but correct) conclusion.

Bell's response to CH and "local Beables"

Prompted by the CH paper's results, Bell's own fuller discussion of the fundamental basis for Bell's Theorem appeared two years later, but then only as a preprint[44]. It is titled *The theory of local beables*, and is based on the CH paper's results[45], which Bell therein attempts to extend. In this extension, Bell follows a line of reasoning that is essentially the same as that of an early rejected draft of the CH paper. In preparing the CH manuscript, Mike and I had been convinced by our discussions with Abner Shimony that such an extension does not

---

42 In the future, authors are hereby requested to please stop misquoting us!

43 A. Aspect, P. Grangier and G. Roger, Phys. Rev. Lett., **49**, 91 (1982).

44 J. S. Bell, *Communication at the 6th Gift Conf. Jaca, June 1975*, Res Th 2053-CERN. Perhaps, Bell's reason for not publishing this result in 1975 is that he may have been afraid that doing so might warrant his being branded a quack, given the then existing stigma against such work.

45 Bell therein acknowledges our prior work, by stating
"As regards the literature on the subject, I am particularly conscious of having profited from the paper of Clauser, Horne Holt, and Shimony which gave the prototype of (16), and from that of Clauser and Horne. As well as a general analysis of this topic this last paper contains a valuable discussion of how best to apply the inequality in practice."

work, and correspondingly we did not use it. Thus, Abner, Mike and I were together confident that Bell had pushed his arguments too far. Whereas the CH paper maintains the observer as free-willed and as a separate entity from the "objects" being observed, in the "Beables" paper, Bell further includes the observer as part of the objective reality being constrained. So doing, he is then able to hang the whole argument on locality considerations alone, whereupon considerations of objective reality are then no longer needed. But in doing so, the observer's free will to choose analyzer orientations is lost. Given that the observer unwittingly may be a preprogrammed automaton, his programming then defeats the locality considerations, and the arguments then leading to Bell's Theorem no longer follow. With Abner Shimony now taking the initiative, we three proceeded with a public dialog with Bell over this point in the abovementioned "quantum subculture" newspaper, "Epistemological Letters".[46] In this dialog, Bell effectively concedes our point. The dialog is also noteworthy in that many other important related issues are also discussed in detail therein, including the important concepts of "parameter independence" and "outcome independence".[47]

The quantum-optics community encounters related problems.

Despite the grand successes of quantum electrodynamics (QED) in accurately predicting the Lamb Shift, g-2 for an electron, etc. it was quite common in the 60's and 70's for atomic physicists to use semi-classical radiation theory (SCT) to explain their results, since doing so was much easier than using QED. However, any author who did so, also generally included an apology, conceding, "of course a correct treatment of this problem requires QED". Nonetheless, atomic physicists found that, with very few exceptions, SCT accurately predicts observed experimental results. As time progressed, the number of exceptions kept shrinking. It then became sort of a game among these workers to see just how often QED effects can be ignored. For example, as part of this game, Willis Lamb and Marlan Scully[48], and Peter Franken [49] showed that important observable aspects of the photoelectric effect can be derived without quantizing the radiation field, and Franken made a similar demonstration for resonance fluorescence. In fact, the use of SCT became so common[50] and gave results in agreement with atomic physics experiments so often, that various workers such as Franken began to jest that QED may be totally unnecessary! The game began to get serious, however, when iconoclast Ed Jaynes said that he was not joking, and formulated a new semi-classical theory as a serious alternative to QED. He referred to it as neoclassical radiation theory (NCT). Furthermore, he formally challenged the quantum-optics community to refute NCT with direct experimental evidence. Talk about a break from the "standard religion" (regarding acceptance of a theory without compelling experimental evidence) – Wow!

Jaynes' NCT assumes that (1) Schrödinger's equation governs the evolution of an atom's wave function, (2) the absolute square of an atom's wave function provides a

---

[46] See, J. S. Bell, A. Shimony, M. A. Horne, and J. F. Clauser, Dialectica, **39**, pp.85-110, (1985), for a republication of this interchange that also includes Bell's "Beables" paper.

[47] See also, A. Shimony, *Search for a Naturalistic World*, (Cambridge Univ. Press, Cambridge, 1993). More recently, the importance of these concepts was independently rediscovered by J. Jarrett [Noûs, **18**, 569 (1984)].

[48] W. E. Lamb and Scully, in *Polarization: Matière et Rayonnement*, (Presses Universitaires de France, Paris, 1969), edited by Société Française de de Physique.

[49] P. A. Franken, in *Atomic Physics: Pooceedings of the 1st International Conference on Atomic Physics,* (Plenum, New York, 1968) V. Hughes *et al.* eds., p.377.

[50] Strictly speaking a semi-classical treatment of such systems is a "hidden-variables" treatment of these systems, and, of course, such a treatment represents "forbidden thinking" under the canons of the "standard religion". However, it is well known in matters of both politics and religion, that issues of practicality commonly override such canons, via the use of "creative ambiguity", especially when appropriate obsequious apologies are also offered. Indeed, "creative ambiguity" allows the present existence only one China in the world.

physically real electric charge density, (3) Maxwell's equations (without field quantization) describe the physically real electromagnetic radiation field[51], and (4) radiation reaction effects are included, so that radiation emitted by an atom affects the atom's own evolution. Using only these elements, Jaynes and his students found that (without QED) they could predict absorption of radiation, spontaneous and stimulated emission of radiation, the Lamb shift, and the black body radiation spectrum[52].

Jaynes' work, however, had crossed a magic line in the sand and obviously had to be viewed as heresy. Heretofore, it was a firmly held belief that all of these latter effects do require a quantization of the radiation field, as is done in QED. But Jayne's' work now cast serious doubts on these cherished beliefs. Nonetheless, it would seem that if quantum theory rests on a solid foundation, then Jaynes' challenge should be easy and straightforward to meet. But Jaynes' challenge went unanswered for several years. A small crisis was thus at hand within the quantum-optics community.

In a historical context, Jaynes was iconoclastically following a path similar to that taken by Bohm, deBroglie, and Schrödinger. While Jaynes did not refer to NCT as a hidden-variables theory (perhaps because of the stigma against such theories), he did recognize that he was indeed tinkering with very fundamental issues within quantum theory. Unlike the theory's founders, however, Jaynes based his work on a very broad range of experimental evidence, most of which postdated the work by the theory's founders. Unfortunately, he was not yet familiar with the power of Bell's Theorem and the associated peculiarities of entanglement.

Upon reading Jaynes' work, I quickly realized that NCT is clearly a local realistic theory. The electromagnetic field and the charge density in NCT are, in principle, all measurable objectively real quantities. Photons (or whatever the stuff is that gives rise to the coincidences in the Freedman-Clauser experiment) within NCT thus qualify as objects, as per the above discussion. Moreover, to refute NCT, the full power of Bell's Theorem is not even needed, and a much simpler argument is sufficient to do so. I thus accepted Jayne's challenge[53].

The 1972 CQ03 - Third Rochester Conference on Coherence and Quantum Optics became a watershed for discussions of this issue, and many of the papers and discussions at this conference were directed toward resolving the issues that Jaynes had raised. At this conference, I pointed out to the participants that if NCT is applied to the EPR configuration, it essentially becomes an example of the Schrödinger-Furry hypothesis. Given the above-mentioned argument by Bohm and Aharonov, it is clear that the Wu-Shacknov experiment already refutes this hypothesis. I also pointed out that the Kocher-Commins experiment provides additional refutation, and offered a slightly more general proof of this fact than that given by Bohm and Aharonov. In his own written contribution to the Conference's Proceedings[54], Jaynes' publicly concedes that my arguments do indeed refute NCT. He soon dropped any further pursuit of this theory, and in a personal letter to me, both praised my work and admitted that his own efforts now "lay in ruins". In defense of Jaynes' valiant efforts, I contend that his efforts provide a heuristically valuable and therefore worthwhile exercise that is highly beneficial to our overall basic understanding of physics.

---

[51] Elements (1) - (3) will be recognized as together constituting the so-called "old Schrödinger interpretation". It is noteworthy that these elements are still essential primitives for x-ray crystallography.

[52] E, T. Jaynes and Cummings, Proc. IEEE **51**, 89 (1963); C. R. Stroud, Jr. and E. T. Jaynes, Phys. Rev. A **1**, 106 (1970); M. D. Crisp and E. T. Jaynes, Phys. Rev. **179**, 1253 (1969); **185**, 2046 (1969). See also, P. A. Nesbet, Phys. Rev. A **4**, 259 (1971).

[53] J. F. Clauser, Phys. Rev. A, **6**, 49 (1972).

[54] The proceedings of this conference are in *Coherence and Quantum Optics*, (Plenum Press, New York, 1973) L. Mandel and E. Wolf, eds.

Splitting photons?

To obtain an experimentally testable prediction, CHSH originally assumed that whenever a photon (or at least its detectable component part[55]) encounters a polarizer, then it is either transmitted by the polarizer or not transmitted by the polarizer. That is, CHSH, with their auxiliary assumption require that photons are always particle-like for the purposes of the experiment. While the CH no-enhancement assumption now obviates the CHSH auxiliary assumption, it was then still an open and relevant question as to whether or not there was any compelling *à priori* experimental evidence that justifies the CHSH assumption. That is, do photons clearly demonstrate an unambiguous particle-like behavior, at any time? If so, then under the CH definition of an object, this behavior may be used to constrain further the dimensions of the box that is needed to surround them, and thereby to limit further the number of viable counterexamples to the Freedman-Clauser experiment[56]. It is clear from the simple fact that one observes temporal coincidences between a sequentially emitted pair of cascade photons, that at least the second photon of a cascade must be temporally localized during its propagation. However, this observation, by itself, does not imply that the same photon is also spatially localized (i.e. confined within a box that does not always also still include the emitting atom), and especially spatially localized in directions perpendicular to its propagation direction.

Josef Jauch, who had earlier coauthored a standard textbook on QED[57], expresses a view that photons do indeed exhibit an unambiguous particle-like character[58]. He bases this view on an experiment that was originally proposed by Schrödinger. Earlier, Schrödinger had provided a causal interpretation of his own equation -- the so-called "old" Schrödinger interpretation – wherein a propagating electron therein has objectively real charge density waves associated with its propagating wave function. For the Compton scattering process, Schrödinger[59] shows that these waves can act as moving matter-wave gratings that Bragg-scatter classical electromagnetic wave packets in a manner that exactly predicts the results observed by Compton and Simon. Schrödinger was thus not convinced of the pivotal importance of the Compton-Simon experiment (as von Neumann apparently was) in compelling an acceptance of the Copenhagen interpretation. As an original heretic to the standard religion, he thus persuaded Ádám, Jánnosy and Varga (ÁJV)[60] to actually perform his proposed experiment. In their experiment, two independent photo-detectors are placed respectively in the transmitted and reflected beams of a half-silvered mirror. If photons have a particle-like character, i.e. if their detectable components are always spatially bounded and well localized, then photons impinging on the half-silvered mirror will not be split in two at this mirror. On the other hand, if they are purely wave-like in nature, (as with classical waves) then they can and will be split into two independent classical wave packets at this mirror. This fact then implies that if photons are purely wave-like (in this classical sense), then the two detectors will show coincidences when a single temporally localized photon is directed at said

[55] In a "pilot-wave" theory, separate component parts of a quantum particle exist that then respectively account for the particle-like and the wave-like aspects of its propagation.

[56] The huge detector efficiency loss that is due to the small solid-angles subtended by photon detectors in typical cascade photon experiments, can then be rendered effectively harmless by said significantly reduced box volume. This fact is used explicitly (but without citing my experimental justification for it) by Bell in the *Gedankenexperiment* that he depicts on p.107 of his book, *Speakable and unspeakable in quantum mechanics*.

[57] J. M. Jauch and F. Rohrlich, *The Theory of Photons and Electrons* (Addison-Wesley, Reading, MA, 1955).

[58] J. M. Jauch, *Dialogue on the Question ARE QUANTA REAL?* (Univ. of Geneva, preprint, 1971).

[59] E. Schrödinger, Physikalische Zeits. **23**, 301-303 (1922); Die Naturwissenschaften **12**, 720-724 (1924); Il Nuovo Cimento **9**, 162-170 (1958).

[60] A. Ádám, L. Jánnosy and P. Varga, Acta Phys. Hung., **4**, 301 (1955); Ann. Physik, **16**, 408, (1956).

mirror. One of these independent wave packets will be transmitted to illuminate the first detector, and the other will be reflected to illuminate the second detector, and both detectors will then have a finite probability of detecting the same photon (classical wave packet). This latter possibility, however, violates the predictions by QED, which prohibits such coincidences. ÁJV thus searched for anomalous coincidences between photomultiplier tubes that viewed the reflected and transmitted beams behind a half-silvered mirror. Unfortunately, their mirror was illuminated by a temporally continuous beam of light, and not by "tagged" single photons[61]. They found no such anomalous coincidences, and thereby concluded that, consistently with QED, photons do not split at the mirror and thus do exhibit a particle-like character.

A critical parameter in the ÁJV experiment is their photon detection efficiency. Its value then determines whether or not the sought-after anomalous coincidences can be distinguished from the large background of omnipresent "accidental" coincidences. Upon reading their paper, I was astonished by their claim to have achieved 10% photon detection efficiency. Stu Freedman and I had struggled with much better equipment to get an efficiency of about 0.1%. In fact, we actually measured this efficiency using the abovementioned photon-tagging scheme, by calculating the ratio of the coincidence count rate to the singles count rate for first photons of the cascade. How could ÁJV possibly claim to have achieved this absurdly high efficiency? To do so, they had ignored the solid-angle loss, and had quoted only the so-called photomultiplier quantum efficiency. But it is clear that for a worst-case spherically expanding wave, it is the overall photon detection efficiency that matters to the argument, and <u>not</u> the photomultiplier's quantum efficiency. Further study of their result and of similar repetitions of their experiment convinced me that none of these experiments convincingly demonstrated a particle-like behavior for photons. With low detection efficiencies, all of these experiments fell far short of having the required precision to do so. At best, they could only observe coincidences between two photons that accidentally were temporally overlapped. Viewed in this light, the ÁJV experiment was effectively a null experiment that, at best, sets an upper limit to the anomalous coincidence rate that might be present. Moreover, the upper limit that they obtain does not then even usefully constrain semi-classical radiation theories.

I thus decided to perform an experiment that has convincingly high photon detection efficiency, and also that is configured, not as a null experiment, but instead as one that gives a definitive answer. Its results were published in 1974[62]. This experiment, unlike that by ÁJV, exploits the temporal localization already exhibited by "tagged" photons. Also, unlike the ÁJV experiment, the experiment employs four photomultiplier tubes, operating pair-wise in coincidence, and two half-silvered mirrors. By a judicious application of the Cauchy-Schwarz inequality to four different measured coincidence rates, it is then possible to obtain the desired definitive (non-null) prediction for this configuration. The experiment's results show that both quantum mechanics and quantum electrodynamics hold true, and photons do not split at a half-silvered mirror. This experiment is further noteworthy, in that it represents the first observed violation of a Cauchy-Schwarz inequality, as derived from classical optics principles, and thus also represents the first observation of sub-Poissonian statistics for light.

Curiously, the results of this experiment also resolve another untidy issue left behind by quantum theory's founders. That issue is whether or not energy conservation in the photoelectric effect is point-wise exact, or only holds as a statistical average, as suggested earlier by Bohr, Kramers and Slater. (Recall that von Neumann cited the Compton-Simon

---

[61] A "tagged" single photon is readily produced by a cascade photon source, by using a coincidence gate to only look for a sequentially emitted second single photon from an atom that is undergoing a cascade decay, that immediately follows the detection of the first photon of the decay. Given a detection of the former, one then knows with a very high probability that the associated second photon has been emitted by the atom, within roughly an intermediate-state lifetime, following the first photon's emission.

[62] J. F. Clauser, Phys. Rev. D, **9**, 853 (1974).

experiment as a refutation of this theory.) Schrödinger, in his above-cited works, is critical of this conclusion, and reopens the question. As a logical follow-on, Schrödinger then worried about the particle-like character of photons and, as noted above, went on to propose the ÁJV experiment as a crucial test. Schrödinger (as a theorist), however, never carefully scrutinized ÁJV's experimental assumptions and parameters to note the difference between AJV's incorrect use of the photomultiplier quantum efficiency in place of the (correct) photon detection efficiency. As a result, no prior experiments had ever carefully and conclusively resolved this issue, that instead, heretofore had been simply dismissed by others on purely theoretical grounds. (Here we see more "killer" details falling through a critical gap between theorists and experimentalists.)

An ambiguity concerning energy conservation arises from the fact that there are really two independent definitions of electromagnetic energy – one from quantum mechanics (hv) and one from classical mechanics (integral over all space of the electromagnetic field energy). Quantum electrodynamics is formally based on an assumed strict equality of these two different independently defined quantities, and thus requires a strict point-wise conservation of energy. But, prior to my experimental result, this crucial fundamental assumption had not yet been directly experimentally demonstrated. My improved version of the ÁJV experiment now finally does just that. Grangier, Roger, and Aspect[63] have since performed a somewhat similar experiment that also observes sub-Poissonian statistics and thereby confirms my results.

Remaining locality loopholes

It should be noted that an experiment that is shown to violate the CH inequality then rules out all local realistic theories, except for one remaining special class of theories. These are theories for which Bell (following Bohm and Aharonov) noted in his original 1964 paper that an experiment performed with static analyzers is, itself, logically flawed, since it does not strictly enforce locality. Indeed, it is logically possible that in such an experiment the two different particle detectors, located on opposites sides of one's laboratory, are conspiratorially communicating with each other, with the specific motive of defeating the experimental test. To close this so-called "locality loophole", Bohm and Aharonov proposed that one should rapidly reorient the analyzers while the photons are in flight, and Bell echoes and promotes this suggestion in his 1964 paper. However, Bohm and Aharonov's artifice clearly does not close this loophole! Instead, it simply shifts the burden of perpetrating said conspiracy from one apparatus component to another.

Although logically, it must be admitted that such theories are indeed possible, in a sense they seem to be rather improbable, indeed almost "pathological". To accept the locality loophole as a reasonable possibility, it is first necessary to believe that a pair of detectors and analyzers that are several meters apart are somehow conspiring with each other, so as to defeat the experimenter, and thus somehow are converting a non-sinusoidal coincidence rate dependence on relative polarizer orientation into a sinusoidal dependence. However, once that one is willing to concede that these apparatus components are indeed capable of perpetrating such a conspiracy, then it seems to be equally reasonable (and similarly logically possible) to assume further that the electronics that control a pair of rapidly reorienting analyzers are (perhaps unwittingly) also equally participating in the same conspiracy, also then to defeat the experimenter's efforts with similarly devious tricks. To be sure, it may seem that a belief in such logically possible conspiracies requires a certain degree of paranoia. A logical consequence of admitting to this paranoia is then an associated recognition that every piece of equipment that has been in your laboratory for more than a few nanoseconds might possibly participate in such a grand conspiracy against you, whereupon no effort on your part can rule out this logical possibility! However, if all experimentalists were similarly paranoid, then experimental physics, in general, would seem to be a pointless endeavor.

[63] P. Grangier, G. Roger, and A. Aspect, Europhys. Lett . **20**, 1061 (1986).

participate in such a grand conspiracy against you, whereupon no effort on your part can rule out this logical possibility! However, if all experimentalists were similarly paranoid, then experimental physics, in general, would seem to be a pointless endeavor.

Without attributing paranoid tendencies for these workers, it is noteworthy that nearly a decade after the publication of the Freedman-Clauser results, Aspect *et al.*[64] took the possible existence of such conspiracies seriously, and following Bohm and Aharonov's original suggestion, performed a second-generation experiment with dynamic (time-varying) analyzers. Their experiment then rules out detector-analyzer conspiracies. Other workers are currently considering performing experiments that might rule out even more insidious higher-level conspiracies.

Conclusions

Clearly, quantum theory's predictive power and Bohr's Complementarity have both proven to be exceedingly robust while under attack by what is probably the most powerful challenge they have ever had to face – Bell's Theorem. History now shows that Bohr's views are probably the correct ones to hold. However, in retrospect, those who held to Einstein's view can, with reasonable justification, claim that Bohr, himself, was just lucky, in that he then had no real compelling (experimentally justifiable) reason to hold his view. Similarly, those who smugly held to Bohr's view now can say, also with a similarly reasonable justification, "See, we told you so!" Nonetheless, it appears reasonable to believe that neither camp can justifiably argue that the resolution, via Bell's Theorem, of the issues separating them was a waste of time. A lot of new physics was discovered in the process.

Given historical hindsight, I assert that our basic understanding of quantum theory has been significantly improved via Bell's Theorem and via its associated experimental testing, long after it was confidently asserted by many textbooks to be well understood. It is truly amazing that so many "killer" details slipped through cracks that existed between experimentalists and theorists. It is clearly of continuing importance for experimentalists and theorists to scrutinize each other's work with great care to try to eliminate such cracks. Given such hindsight, I also assert that it is clearly counterproductive to scientific progress for one camp smugly to hold to a belief that all problems are solved in any given area. It is even more counterproductive for this camp then further to rely on this belief to formulate a religious stigma against others who do not share their cherished belief. Indeed, history also shows that a prohibition against open discussion and experimental testing of the foundations of quantum theory, in turn, led to a significant delay of the discovery of important new applications of these foundations. Quantum cryptography, distributed entanglement, etc. undoubtedly would never have been envisioned without the intellectual challenges posed by Bell's Theorem.

---

[64] A. Aspect, J. Dalibard and G. Roger, Phys. Rev. Lett., **49**, 1804 (1982).

Cavity QED with Strong Coupling – Toward the Deterministic Control of Quantum Dynamics

H. J. Kimble and T. W. Lynn

Norman Bridge Laboratory of Physics 12-33, California Institute of Technology, Pasadena, CA 91125, USA

Abstract:

Many of the current efforts to control the dynamics of individual quantum systems take place within the setting of cavity quantum electrodynamics (QED). The coupling of an atomic dipole to the mode of an optical resonator has historically produced important quantum effects in the regime of weak coupling between dipole and cavity mode; more recent experiments access the regime of strong coupling and begin to enable control of the quantum states of single atoms and single-photon fields through a coherent coupling that exceeds dissipative rates in the system. We briefly review the historicl path to strong coupling and the variety of experiments involving single-quantum cavity QED. Current achievements and future challenges are illustrated through further discussion of two ongoing experiments in out group: one pursuing quantum feedback to trap single atoms in a cavity mode with single photons, the other building capability for quantum logic by using a FORT to hold atoms within a cavity mode. [Note that the presentation on which this paper is based can be accessed at http://www.its.caltech.edu/˜qoptics/cqed.html]

1 Brief Overview of Cavity QED

Cavity quantum electrodynamics (QED) provides the setting for a wide variety of techniques for the control of individual quantum systems. Numerous protocols exploit the coherent interaction between an atomic dipole and the mode of an optical resonator to achieve deterministic control of both atomic states and states of the light field[1]. Such observation and control of individual quanta are crucial in the developing field of quantum information science; strongly-coupled cavity QED systems are uniquely suited for the implementation of many proposals, for example in the areas of quantum networks and communication[2].

Contemporary experiments in cavity QED build on a rich history of work on the coupling of atomic dipoles to the electromagnetic field in the presence of boundaries or materials which alter the structure of the quantized field. A diverse body of research has investigated changes in atomic radiative processes caused by the presence of a boundary[3]; for example, boundary-induced atomic level shifts form the basis of the Casimir effect[4] and numerous other phenomena[5]. Examples abound of experiments in which these effects occur, either as the primary observation or as important corrections to be taken into account[6].

In the special case of an atom placed inside the mode volume of an optical resonator or cavity, the cavity geometry defines single isolated modes of the radiation field that can interact coherently with the atomic dipole. The dipole-field coupling is given by $H_{int} = \hbar g(\sigma_+ a + \sigma_- a^\dagger)$, where $\sigma_\pm$ are dipole raising and lowering operators, $(a, a^\dagger)$ are field annihilation and creation operators for the cavity mode, and g is one half of the single-photon Rabi frequency. This interaction gives rise to the well-known Jaynes-Cummings ladder of eigenstates for the coupled atom-cavity system[7], and correspondingly to the vacuum Rabi splitting for the system's resonant frequencies[8].

While the Jaynes-Cummings ladder describes the coherent atom-cavity interaction, the physics of cavity QED is ultimately that of an open quantum system; the atom couples to continuum modes not associated with the cavity, and in any finite-Q cavity the cavity modes themselves are imperfectly isolated from the environment. In this case, the system's behavior changes qualitatively depending on the ratio of coherent evolution to dissipative coupling with the environment[9, 10]. For cavity QED the coherent evolution is characterized by g, with dissipation occurring through cavity decay at rate κ and atomic spontaneous emission at rate γ. Depending on the relative sizes of these rates, cavity QED systems exhibit a range of phenomena well known within the setting of the dynamics of open quantum systems, including diverse laser transitions[11], quantum coherence even in the presence of dissipation[12], and optical bistability[13].

Cavity QED phenomena can be classified in the regimes of weak or strong coupling according to the balance between coherent and incoherent dynamics[10, 14]. For this purpose it is useful to define two dimensionless parameters, the critical atom and photon numbers $N_0 \sim \kappa\gamma/g^2$ and $m_0 \sim \gamma^2/g^2$; these describe the number of quanta of each kind required to saturate the system's response. For $(N_0, m_0) \gg 1$ dissipative processes dominate and individual quanta do not significantly alter the system dynamics. This regime of weak coupling includes traditional optical bistability phenomena as well as cavity-modified atomic spontaneous emission and atom-modified cavity decay linewidths. Conversely, for $(N_0, m_0) \ll 1$ coherent dynamics overtake dissipation and a single atom or photon completely alters the system's response. In this regime of strong coupling, cavity QED can be exploited for control of individual quantum systems, deterministic preparation of quantum states of light, and numerous protocols in the area of quantum logic.

We consider a typical cavity QED experiment, in which an atom traverses the mode of an optical cavity in time T as the system undergoes dynamics as prescribe by rates$g$, κ, and γ. A principle requirement for strong coupling is the realization of a small cavity mode volume V_m, since the single-photon Rbi frequency $2g$ scales as $1/\sqrt{V_m}$. This requirement drives a tren toward smaller cavity lengths l and cross-sectional areas for the cavit mode. However, cavity decay scales inversely with the length, $\kappa \sim 1/l$ so as cavity size decreases the mirror finesse must increase concomitantly to allow for true single-photon, single-atom interactions in a strongly coupled system. Finally, within the regime of $(m_0, N_0) \ll 1$ the atom-cavity interaction time T limits the total number of cycles of coherent interaction, as well as the amount of information extracted from the system. Thus, for the purpose of many quantum control and quantum logic protocols it is also critical to achieve long T, specifically $1/T \ll (g, \gamma, \kappa)$.

The pursuit of strong coupling encompasses a variety of current experiments based on diverse atomic systems. These include Rydberg atoms in microwave cavities[15, 16, 17], alkali atoms in Fabry-Perot cavities[18, 19, 20, 21, 22, 23], and more recently trapped ions within an optical cavity[24, 25]. Several developing techniques promise yet more improvements in critical parameters – for example, lengthening T via micromagnetic traps for neutral atoms[26], or decreasing (m_0, N_0) by the use of solid-state microcavities[27].

2 Experiments in Optical Cavity QED

For the remainder of this paper, we focus on progress in optical cavity QED in Fabry-Perot cavities, and principally in the quantum optics group at Caltech. This system is the setting for an experimentally demonstrated quantum phase gate[18], and for numerous other protocols in quantum logic and quantum communication[2, 28, 29, 30]. The major challenge in the implementation of these protocols is the realization of long interaction times T for an atom in the cavity; in particular, accurate quantum state control requires precise knowledge of the coupling g and thus of the atom's location within the cavity during a control operation. Several groups have worked toward this end by combining the techniques of laser cooling and trapping with cavity QED[20, 21, 31, 32], beginning with the work of Mabuchi et al. in 1996[19]. These techniques have enabled real-time sensing of atomic motion via the atom-cavity coupling [33, 34]. Two current experiments in our group have used different approaches to achieve longer interaction times and better control of the atom-cavity coupling.

We first discuss an experiment that employs a very short (10-μm) cavity and exploits the resulting high signal-to-noise and large forces to trap an atom using the quantized cavity field at the level of a single photon. The atom's position within the cavity can be tracked in real time with this "atom-cavity microscope"[34], leading to current efforts using active feedback to further cool the atomic motion. This trapping and cooling strategy may lead to an experimental testing ground for quantum feedback theory as the precision of the atomic position measurement comes close to the standard quantum limit.

In a second experiment, atoms are held within the cavity mode volume by a far-off-resonance trap (FORT) which provides a nearly classical trapping potential[35]. The light fields generating the FORT and the cavity QED interaction are both resonant with the cavity, but as different longitudinal modes, generated and detected independently from one another. Thus, with this approach we can envision trapping an atom with the FORT while the cavity QED field is free to switch as required for a specific quantum logic operation

While the two experiments display quite different phenomena due to the different properties of the two cavities as well as the distinct strageties employed, they share a common basis in cold-atom cavity QED preparation and detection techniques Figure 1 shows a schematic diagram of the experimental procedure, in which Cesium atoms (atomic resonance frequency ω_a) are collected in a magneto-optical trap (MOT) directly above the cavity mirrors, cooled to temperatures of 3-10 μK, and dropped through the cavity. The geometry of the mirror substrates cuts off most of the atomic flux so that one atom at a time transits the cavity mode of length l and Gaussian waist w_0. In both experiments the cavity resonance ω_c is tuned near but slightly below the atomic resonance frequency so that $\omega_c - \omega_a = \Delta_{ac} < 0$. The cavity is continuously driven by a probe laser at frequency $\omega_p = \omega_a + \Delta_p$, and the transmission of this beam through the cavity is monitored via balanced heterodyne detection. For a probe resonant with the empty cavity, $\omega_p = \omega_c$, transmission falls as an atom enters the cavity mode and becomes strongly coupled to the cavity; for a probe red-detuned from both atom and cavity (ω_p near the lower vacuum Rabi sideband[8]), transmission is low for the empty cavity and is highest when an atom is in the regions of strongest coupling. Thus the monitored photocurrent carries real-time information about th atomic position, with high signal-to-noise even for probe strengths corresponding to < 1 intracavity photons.

2.1 *The Atom-Cavity Microscope: Trapping and Feedback in a Manifestly Quantum Regime*

In the first experiment we trap a single atom using an average of 1 photon in the cavity QED field itself. [34, 36] The first excited state of the coupled atom-cavity system is a strong-field-seeking state, in which the atom experiences an attractive effective potential in the cavity mode. A deep ($\sim$ 2.5mK) potential associated with this single-quantum interaction arises from the use of a short cavity with very small mode volume

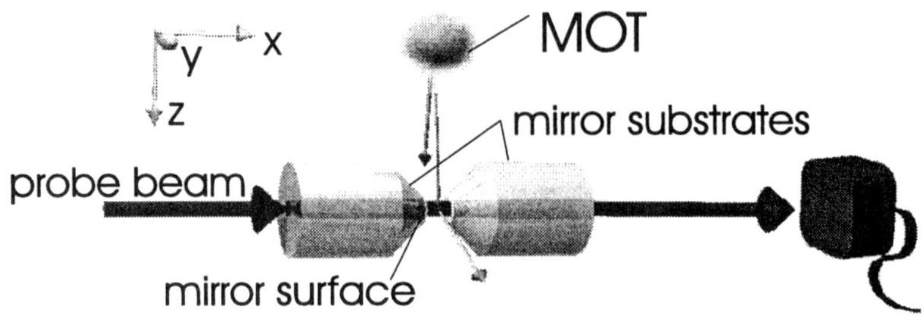

Fig. 1. In both experiments described, Cesium atoms are dropped from a MOT and a small fraction of these fall one by one through the mode of an optical cavity. The cavity mode is a Gaussian in the radial (y,z) directions and has a standing-wave structure in the axial (x) direction. Transmission of a probe beam through the cavity is monitored to sense an atom's motion through the cavity mode; the atom is then trapped via intracavity light forces. The figure shows one falling atom which misses the cavity mode altogether and a second atom which enters the cavity, is trapped in the mode, and eventually escapes.

(l = 10.9μm, w_0 = 14.1μm). For this cavity we have mirror finesse $\mathcal{F} = 480,000$ and $(g, \kappa, \gamma)/2\pi = (110, 14.2, 2.6)$MHz, giving critical photon and atom numbers $m_0 = 2.7 \cdot 10^{-4}$, $N_0 = 6 \cdot 10^{-3}$. The coherent interaction energy exceeds other relevant energies in the problem, in particular the atomic kinetic energy $E_k \sim k_B \cdot 0.5mK$ acquired in the 3-mm fall from the MOT to the cavity mode. Thus an atom can be caught within the cavity if the system can be driven from its ground state to the trapping state when the atom is at a maximum of the cavity field[37].

To implement this triggering strategy, we drop atoms through the cavity mode while monitoring cavity transmission with a weak probe beam of detuning $\Delta_p < \Delta_{ac} < 0$ and 0.05 intracavity photons. Such a weak probe allows high signal-to-noise for observation of an atom entering the cavity mode, but does not significantly populate the excited states of the atom-cavity system. Once the probe transmission rises above a predetermined threshold, indicating that an atom is in a region of strong coupling to the cavity mode, we switch the probe power up to a level of typically 0.3 intracavity photons to create a confining potential around the atom.

Figure 2(a) shows the resulting cavity transmission for an atom trapped in this way. Note that while the "high" probe level was set at 0.3 intracavity photons, the level rises to $|\langle \hat{a} \rangle|^2 \sim 1$ (where $\hat{a}$ is the cavity field annihilation operator) during parts of the measurement record due to the atom's motion within the cavity field. Oscillations in transmission arise from atomic motion toward and away from the cavity axis, with the level falling back to 0.3 photons when the atom eventually heats out of the trap and escapes from the cavity. Using this protocol we have observed mean atom dwell times in the cavity of 340μs, with some rare events lasting up to several milliseconds, as compared with the $\sim$ 75μs free-fall time for an atom to traverse the cavity mode. Lifetimes are limited by heating associated with the many decays and re-excitations the atom-cavity system experiences during a single atomic dwell time. It is important to note that, while atoms are trapped via their dipole interaction with a red-detuned light field in a manner reminiscent of the more familiar free-space situation, the dynamics associated with the atom-cavity system are quantitatively and qualitatively different from that case[38]. In particular, a free-space potential of equal intensity would exhibit much greater diffusive heating and thus would largely fail to trap the atoms we see in our experiment; furthermore, the single-atom sensing which enables our trapping strategy is only available through the use of the cavity QED system. A detailed treatment of trapping in cavity QED in the single-photon limit is given in Ref.[38], including an analysis of our experiment as well as the work of Pinkse et al.[21].

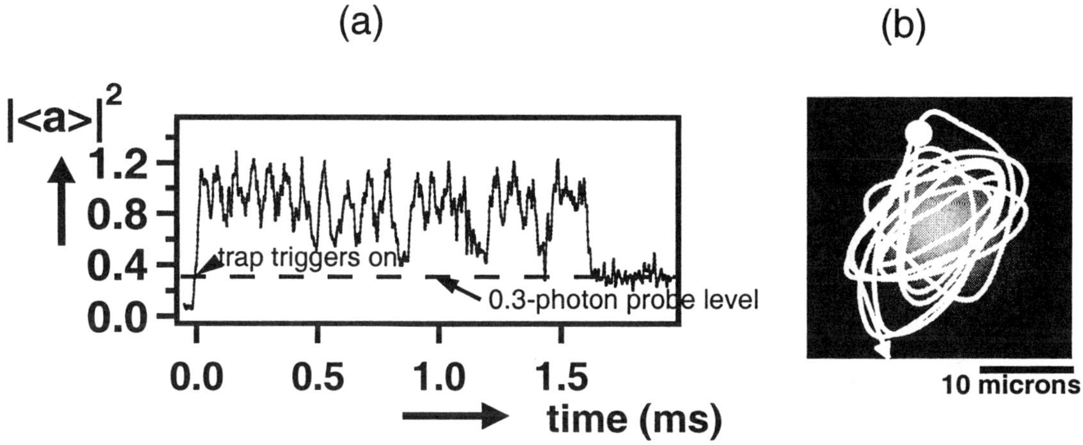

Fig. 2. (a) Cavity transmission record for an atom trapped with parameters $\Delta_{ac} = -47$ MHz, $\Delta_p = -125$ MHz, and 0.3 photons in the empty cavity. Motion of the trapped atom within the cavity mode can be tracked via the oscillations in cavity transmission. (b) Atomic trajectory in the radial (y,z) plane as reconstructed from the transmission data of (a). The size of the white ball is an estimate of error in the reconstructed position throughout the trajectory.

The most interesting feature of the measurement record for a trapped atom is undoubtedly the oscillation in transmission. The period and amplitude of these oscillations agrees quantitatively with atomic orbital motion in the Gaussian mode transverse to the cavity axis. Trapped atoms in our system are tightly confined close to a single antinode of the field and we do not observe detectable signatures from axial motion[34, 38], leading to bounds on the size of any small-amplitude axial excursion. Thus transmission provides a direct, real-time record of an atom's radial distance from the cavity axis; this record can also be used along with the known effective potential to reconstruct two-dimensional trajectories as shown in Figure 2(b) and discussed in Ref.[34]. Such reconstructions, as tested by applying the method to simulated atomic trajectories, represent an atomic position measurement which is about a factor of five above the standard quantum limit (SQL)[34, 39].

Using this real-time position information to feed back to the trapping potential and cool an atom in the cavity is the object of current work on this experiment. An initial step involves modulating the intensity of the probe beam to create a steep potential when the atom is moving away from the cavity axis, and a shallow potential when the atom is falling into the cavity axis. This form of feedback should circularize atomic orbits but cannot cool them further, as the effective potential is cylindrically symmetric about the cavity axis. In order to remove atomic angular momentum, this technique must be combined with a symmetry-breaking force; for example, light coupled into a transverse mode of the cavity can be used to "kick" an atom towards lower angular momentum until it is cooled to the cavity axis.

Further extensions of this active feedback-based cooling include a variety of different feedback protocols. Full detection of the amplitude and phase of the cavity output field would offer increased flexibility for design of the feedback loop. Another avenue is the extension of this technique to detect and control axial motion; under current conditions, axial motion large enough to detect also induces rapid heating which causes the atom to escape completely in just a few axial periods. However, with this complication overcome, axial motion could be differentiated from radial motion by virtue of the two-order-of-magnitude separation of timescales for the two directions, with typical radial oscillation frequencies of $\sim 10-20$ kHz contrasting with typical axial oscillation frequencies of $\sim 1-2$ MHz. Finally, many of these efforts will gain added significance as the measurement accuracy is improved to bring the atomic position measurement closer to the SQL and thus to bring the feedback loop into the realm of quantum servo theory.

2.2 *FORT in the Cavity: The Challenge of Trapping Atoms for Quantum Logic*

Quantum logic and quantum state preparation in the atom-cavity system require freedom to manipulate the quantized cavity field while an atom is independently held in a region of strong coupling. A second cavity QED experiment in our group pursues this goal of independent trapping using a FORT inside the cavity.[35] As in the experiment discussed above, Cesium atoms fall several millimeters from a MOT and are detected as they transit the mode of a Fabry-Perot cavity; single atoms are trapped one-at-a-time in the cavity mode. However, they are trapped not by the coherent atom-cavity coupling but by the approximately classical potential of a separately generated far-off-resonant trap.

The cavity used in this experiment has slightly larger dimensions ($l = 44.6\mu$m, $w_0 = 24\mu$m) and lower mirror finesse $\mathcal{F} = 420,000$, giving $(g, \kappa, \gamma)/2\pi = (32, 4, 2.6)$MHz. While the cavity QED interaction involves a cavity mode nearly resonant with the Cesium atomic transition at 852 nm, the FORT potential derives from a separate laser beam at 869 nm coupled into another longitudinal mode of the cavity. In this far-detuned limit the scattering of FORT photons by the atom is strongly suppressed, and the effect on the atom is nearly that of a classical potential arising from spatially dependent light shifts of the atomic ground and excited levels.

These light shifts interact with the eigenvalue structure of the strongly coupled atom-cavity system, so that the resultant spectrum for the composite system is given by the so-called "well-dressed" states[40]. Because the FORT and cavity QED light are of significantly different wavelengths they are not perfectly overlapped over the length of the cavity axis. Thus in the presence of both FORT and cavity QED light the eigenvalue structure of the atom-cavity system is different at each trapping well (i.e. antinode of the FORT) along the length of the cavity. Without a priori knowledge of which well an atom occupies, it is not straightforward to interpret the transmission record of a cavity QED probe field throughout the time an atom is being trapped by the FORT. Rather, measurements of the FORT lifetime are more easily performed by a trapping cycle in which the cavity QED field is used for initial sensing, this field is then switched off when the FORT is triggered on to trap an atom, and finally the FORT is switched off after a given time interval and the cavity QED field is turned on to probe for the remaining presence of an atom in the cavity. In our initial demonstration, a FORT lifetime of 28 ms was measured by this technique[35].

Operating the FORT in a resonant mode of the cavity has advantages as well as disadvantages, such as the technical complications of the system's eigenvalue structure as discussed above. While a simpler option might seem to be coupling a FORT beam through the cavity from the side, this geometry does not lend itself to achieving atomic localization in the Lamb-Dicke regime. Furthermore, the resonant cavity enhancement factor is very useful for obtaining FORT potentials deep enough to overcome not just the initial atomic kinetic energy, but also cavity QED-induced level shifts of order $\hbar g$. Our cavity has a finesse of approximately 350,000 at the FORT wavelength, so that only $\sim 7\mu$W of injected power is required to obtain a peak ground state shift of ~ 75 MHz.

In addition, the imperfect spatial overlap between FORT and cavity QED fields, while it complicates detection strategies, may ultimately provide a useful tool for identifying which antinode an atom occupies and even for cooling atoms in the combined FORT and cavity QED fields[41].

The high cavity finesse at FORT wavelengths, while providing a useful buildup of intensity in the trapping beam, also introduces a sensitivity to phase and frequency noise which appears to dominate current FORT lifetime limits. Measured lifetimes fall well below the $\sim$ 100s limits from background gas collisions at the chamber pressure of 10^{-10} Torr. Rather, the observed lifetimes are consistent with heating due to intensity noise on the intracavity FORT beam[42]. This intensity noise arises from relative frequency jitter between the FORT laser and cavity resonance; the cavity's high finesse converts this otherwise minor frequency noise into intracavity FORT amplitude noise which heats trapped atoms[43].

Current technical improvements to the FORT laser's frequency locking serve to reduce this source of noise. However, investigations of the noise sources indicate an interesting origin for this noise above about 500 kHz, in the frequency range most relevant to heating atomic motion along the cavity axis. Measured noise in this frequency range can largely be attributed to thermal vibrations of the mirror substrates, which change the effective cavity length and thus its resonant frequency. Figure 3 shows a measured spectrum of noise on the FORT intensity transmitted through the cavity. For comparison, we show a theoretical spectrum calculated from mirror motion with each elastic mode of the mirror substrates excited by $k_B T$ of thermal energy. The theoretical spectrum comes from numerical computation of the normal modes of a right-circular cylinder; the effect of each elastic mode on the cavity length as seen by the intracavity FORT field is accounted for with a weighting by effective mass[44]. Differences between measured noise and this initial calculation may stem from the actual cone shape of the mirrors, which we are currently treating in a finite element analysis in collaboration with Dennis Coyne of the LIGO project.

This noise source may be addressed by shifting the FORT wavelength yet further from resonance, to a longitudinal mode at a wavelength of around 940 nm. Here the cavity finesse falls to approximately 1,000-3,000, thus significantly reducing the conversion of incident laser frequency jitter and thermally induced mirror motion to FORT intensity noise. Certain choices of wavelength in this range also allow a FORT which traps both ground and excited atomic states with the same position-dependent level shift[45], thus mitigating the complicated eigenvalue structure and detection issues discussed above. In addition to these solutions for cavity QED with well-localized atoms, the FORT sensitivity to vibrations of the mirror substrates presents scientific opportunities in its own right[46, 47]. One can imagine selective cooling of a single mode of vibration using feedback, or using this probe to reach the standard quantum limit for observation of the motion of a test mass.

3 Conclusions and Future Directions

Our current experiments aim to cool atoms in a cavity via quantum feedback mechanisms and cavity-QED assisted cooling in a FORT[41]. With greater control of the atomic motional state within the cavity, quantization of the atomic center of mass motion will become an important effect; diverse new dynamical effects involve the interplay of atomic wavepackets with single-photon cavity fields[48, 40].

In the realm of cavity design and advances in critical parameters, (m_0, N_0), the limits of Fabry-Perot technology[49] are rapidly approaching as cavity lengths shrink towards the shortest possible value, $\lambda/2$. Photonic bandgap and other solid-state microcavities

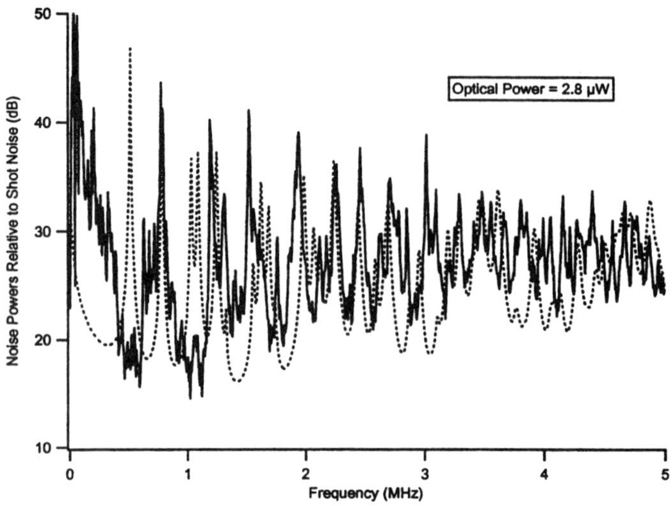

Fig. 3. Measured FORT intensity noise power spectrum (solid trace) compared to the spectrum predicted by calculation of thermally excited vibrational motion of the mirror substrates (dashed trace). Note that the theoretical trace is for a right circular cylinder, whereas the actual mirror substrates are of a more complex shape. Apart from a choice of the quality factor for the mode at 800kHz, the comparison is absolute with no adjustable parameters.

offer a direction for new cavity designs to achieve future advances in critical parameters (m_0, N_0).

These and other efforts promise to bring cavity QED yet further into the realm of strong coupling and coherent quantum state control. Additional benefits should result from the combination of cavity QED with other advances, such as ion-trapping, motional control of atoms in optical lattices, and even quantum teleportation of atomic wavepackets[50]. A common thread throughout these endeavors is the use of strong coupling to enable sensing and control, for quantum logic and other applications of fundamental interest in the study of individual quantum systems.

4 Acknowledgements

We gratefully acknowledge the contributions of the many people who have been involved in the work described above, including Kevin Birnbaum, Joseph Buck, Andrew Doherty, Christina Hood, Ron Legere, Jason McKeever, Christoph Naegerl, Dan Stamper-Kurn, David Vernooy, and Jun Ye. We also thank Dennis Coyne and Kent Blackburn of LIGO for assistance with analysis of thermal noise and mirror vibrations. This work has been funded by the National Science Foundation, by the Caltech MURI on Quantum Networks administered by the US Army Research Office, and by the Office of Naval Research.

References and links

1. For a review of experiments in the field, see *Cavity Quantum Electrodynamics*, ed. P. Berman, Academic Press, San Diego, 1994.
2. J. I. Cirac, S. J. Van Enk, P. Zoller, H. J. Kimble, and H. Mabuchi, Phys. Scripta **T76**, 223 (1998).
3. E. M. Purcell, Phys. Rev. **69**, 6 81 (1946); K. H. Drexhage, in Progress in Optics (ed. E. Wolf), vol. XII, pp. 163-232, New York, Elsevier(1974).

4. H. B. G. Casimir and D. Polder, Phys. Rev. **73**, 360 (1948); P. Goy et al., Phys. Rev. Lett. **50**, 1903 (1983); L. S. Brown and G. Gabrielse, Rev. Mod. Phys. **58**, 233 (1986); D. J. Heinzen and M. S. Feld, Phys. Rev. Lett. **59**, 2623 (1987); Y. Zhu et al., Phys. Rev. Lett. **61**, 1946 (1988); C. I. Sukenik et al., Phys. Rev. Lett. **70**, 560(1995); S. K. Lamoreaux, Phys. Rev. Lett. **78**, 5 (1997); U. Mohideen and A. Roy, Phys. Rev. Lett. **81**, 4549 (1998).
5. E. A. Hinds, Adv. Atom. Mol. Opt. Phy. **28**, 237 (1990).
6. G. Gabrielse and H. Dehmelt, Phys. Rev. Lett. **55**, 67 (1985); R. Hulet, E. Hilfer, D. Kleppner, Phys. Rev. Lett. **55**, 2137 (1985); F. diMartini et al., Phys. Rev. Lett. **59**, 2955 (1987); D. Klepper and S. Haroche, Physics Today (January, 1989); D. Meschede, Phys. Reps. **211**, 201 (1992); S. Peil and G. Gabrielse, Phys. Rev. Lett. **83**, 1287 (1999).
7. E. T. Jaynes and F. W. Cummings, P. IEEE **51**, 89 (1963).
8. J. J. Sanchez-Mondragon, N. B. Narozhny, and J. H. Eberly, J. Opt. Soc. Am. B **1**, 518 (1984).
9. For a textbook development of cavity QED basics, see S. Haroche, in ***Fundamental Systems in Quantum Optics, Les Houches Session LIII***, eds. J. Dalibard, J.-M. Raimond, and J. Zinn-Justin, North-Holland, Amsterdam, 1992.
10. H. J. Kimble, Phys. Scripta **T76**, 127 (1998).
11. See e.g. I. R. Senitzky, Phys. Rev. **128**, 2864 (1962); H. Haken, Z. Phys. **181**, 96 (1964); M. Lax, Phys. Rev. **145**, 110 (1966); M. Scully and W. E. Lamb, Jr., Phys. Rev. **159**, 208 (1967).
12. E. C. G. Sudarshan, Phys. Rev. Lett. **10**, 277 (1963); R. J. Glauber, in Quantum Optics and Electronics (ed. C. de Witt, A. Blandin, and C. Cohen-Tannoudji), pp. 63-185, New York, Gordon and Breach (1965).
13. See e.g. the review article by L. A. Lugiato, Progress in Optics **21**, 69 (1984).
14. See H. J. Kimble in ***Cavity Quantum Electrodynamics***, ed. P. Berman, Academic Press, San Diego, 1994.
15. For a scientific tutorial on microwave cavity QED see Pierre Meystre; "Cavity Quantum Optics and the Quantum Measurement Process," Progress in Optics XXX, ed. E. Wolf (Elsevier Science Publishing, 1992).
16. S. Brattke, B. T. H. Varcoe, and H. Walther, Phys. Rev. Lett **86**, 3534 (2001).
17. G. Nogues et al., Nature **400**, 239 (1999); P. Bertet et al., Nature **411**, 166 (2001).
18. Q. A. Turchette et al., Phys. Rev. Lett. **75**, 4710 (1995).
19. H. Mabuchi, Q. A. Turchette, M. S. Chapman, and H. J. Kimble, Opt. Lett. **21**, 1393 (1996).
20. Work in progress in the group of M. S. Chapman, described at http://www.physics.gatech.edu/ultracool/
21. P. Munstermann et al., Phys. Rev. Lett. **82**, 3791 (1999); P. W. H. Pinkse et al., Nature **404**, 365 (2000).
22. S. L. Mielke, G. T. Foster, and L. A. Orozco, Phys. Rev. Lett. **80** 3948 (1998); G. T. Foster, S. L. Mielke, and L. A. Orozco, Phys. Rev. A **61** 053821 (2000); G. T. Foster et al., Phys. Rev. Lett. **85**, 3149 (2000).
23. An et al., PRL **73**, 3375(1995); An and Feld, PRA **56**, 1662(1997).
24. Work in progress in the group of R. Blatt, described at http://heart-c704.uibk.ac.at/
25. Work in progress in the group of H. Walther, described at http://www.mpq.mpg.de/laserphysics.html; see also their article in these proceedings.
26. H. Mabuchi, M. Armen, B. Lev, M. Loncar, J. Vuckovic, H. J. Kimble, J. Preskill, M. Roukes, and A. Scherer, "Quantum networks based on cavity QED," to appear in Proceedings of the First International Conference on Experimental Implementations of Quantum Computation (2001).
27. For several reviews of the field see "Optical Processes in Microcavities," Y. Yamamoto and R. Slusher, Physics Today (June, 1993); ***Optical Processes in Microcavities***, ed. R. K. Chang and A. J. Campillo (World Scientific, 1996); G. Khitrova et al., Rev. Mod. Phys. **71**, 1591 (1999).
28. T. Pellizzari, S. Gardiner, I. Cirac, and P. Zoller, Phys. Rev. Lett. **75**, 3788 (1995).
29. J. I. Cirac, P. Zoller, H. J. Kimble, and H. Mabuchi, Phys. Rev. Lett. **78**, 3221 (1997).
30. S. J. Van Enk, J. I. Cirac, and P. Zoller, Science **279**, 205 (1998).
31. C. J. Hood et al., Phys. Rev. Lett. **80**, 4157 (1998).
32. Work in progress in the group of T. Kuga, described at http://photon.c.u-tokyo.ac.jp/
33. H. Mabuchi, J. Ye, and H. J. Kimble, Applied Phys. **B68**, 1095 (1999).
34. C. J. Hood, T. W. Lynn, A. Doherty. A. S. Parkins, and H. J. Kimble, Science **287**, 1447 (2000).
35. J. Ye, D. W. Vernooy, and H. J. Kimble, Phys. Rev. Lett. **83**, 4987 (1999).
36. J. Ye, C. J. Hood, T. Lynn, H. Mabuchi, D. W. Vernooy, H. J. Kimble, IEEE Trans. on Inst. and Meas. **48**, 608 (1999).
37. S. Haroche, M. Brune, and J. M. Raimond, Europhys. Lett. **14**, 19 (1991); B. G. Englert et al., Europhys. Lett. **14**, 25 (1991); M. O. Scully et al., Phys. Rev. Lett. **76**, 4144 (1996); A. C. Doherty et al., Phys. Rev. A **56**, 833 (1997).
38. A. C. Doherty, T. W. Lynn, C. J. Hood, and H. J. Kimble, Phys. Rev. A **63**, 013401 (2001); available as quant-ph/0006015.
39. H. Mabuchi, Phys. Rev. A **58**, 123 (1998).

40. D. W. Vernooy and H. J. Kimble, Phys. Rev. A **56**, 4287 (1997).
41. S. J. van Enk, J. McKeever, H. J. Kimble, and J. Ye, Phys. Rev. A **64**, 013407 (2001); available as quant-ph/0005133.
42. T. A. Savard, K. M. O'Hara and J. E. Thomas, Phys. Rev. A **56**, R1095 (1997).
43. C. W. Gardiner, J. Ye, C. Nägerl and H. J. Kimble, Phys. Rev. A **61**, 045801 (2000).
44. A. Gillespie and F. Raab, Phys. Lett. A **178**, 357 (1993).
45. H. J. Kimble, C. J. Hood, T. W. Lynn, H. Mabuchi, D. W. Vernooy, and J. Ye in *Laser Spectroscopy: XIV International Conference*, eds. Rainer Blatt, et al. (World Scientific, Singapore, 1999).
46. Y. Hadjar Y, P. F. Cohadon, C. G. Aminoff, M. Pinard, and A. Heidmann, Europhys. Lett. **47**, 545 (1999); F. Cohadon, A. Heidmann, and M. Pinard, Phys. Rev. Lett. **83**, 3174 (1999).
47. I. Tittonen, G. Breitenbach, T. Kalkbrenner, T. Muller, R. Conradt, S. Schiller, E. Steinsland, N. Blanc, and N. F. de Rooij, Phys. Rev. A **59**, 1038 (1999).
48. P. Meystre, E. Schumacher, and S. Stenholm, Opt. Comm. **73**, 443 (1989); W. Ren and H. J. Carmichael, Phys. Rev. A **51** 752 (1995); M. O. Scully, G. M. Meyer, and H. Walther, Phys. Rev. Lett. **76** 4144 (1996); A. M. Herkommer, H. J. Carmichael, and W. P. Schleich, Quant. Semiclass. Opt. **8** 189 (1996); A. C. Doherty et al., Phys. Rev. A **57**, 4804 (1998).
49. C. J. Hood, H. J. Kimble, and J. Ye, Phys. Rev. A **64** 033804 (2001); available as quant-ph/0101103.
50. A. S. Parkins and H. J. Kimble, J. Opt. B - Quant. Semiclass. Opt. **1**, 496 (1999).

Fast and slow light

Peter W. Milonni
Los Alamos National Laboratory, Los Alamos, New Mexico 87545 USA

Abstract: Experimental demonstrations of superluminal as well as highly subluminal group velocities have rekindled interest in an old subject and might also lead to useful applications. Some basic physical considerations relating to these developments are revisited.

There is much interest in group velocities v_g that are either very large ($v_g > c$, the speed of light in vacuum) or very slow ($v_g << c$, or even $v_g = 0$). The first case has a long history, and was studied intensively by Sommerfeld, Brillouin and others, who showed among other things that superluminal group velocities are not in conflict with the special theory of relativity. The second case has received less attention, as $v_g < c$ is generally deemed "normal." Attention here is focused first on some physical considerations relating to the impossibility of superluminal transmission of information.

1 Causality

The most primitive notion of causality is that "the effect should not precede the cause." This leads to the requirement that no signal can travel faster than c. For suppose an event at (x, t) causes an event at $(x + \Delta x, t + \Delta t)$ via some signal with velocity u. In some other reference frame with relative velocity v ($< c$) the time interval between the events is

$$\Delta t' = \frac{\Delta t - (v/c^2)\Delta x}{\sqrt{1 - v^2/c^2}} = \frac{\Delta t(1 - uv/c^2)}{\sqrt{1 - v^2/c^2}} . \tag{1}$$

Then a superluminal signal velocity ($u > c$) implies there are v's for which Δt and $\Delta t'$ have opposite signs: the order of "cause" and "effect" would be different for different observers. Special relativity and causality thus forbid superluminal *signal* velocities. But it is instructive to consider examples where this fact is not immediately obvious.

Fermi's problem and photodetection theory

Fermi [1] addressed the question of signal propagation velocity in QED using a simple model: there are two two-level atoms separated by a distance r in free space, in a state at $t = 0$ in which atom A is in its upper-energy state, atom B is in its lower state, and there are no photons. Fermi wanted to show that the probability for atom B to be excited remains zero until the time $t = r/c$ it takes for light to propagate from A to B. However, he made an approximation that, as a matter of principle, rendered his conclusion invalid [2],[3].

Coherence and Quantum Optics VIII
Edited by Bigelow et al., Kluwer Academic/Plenum Publishers, 2003

In the model considered by Fermi each component of the electric field operator associated with either atom in the Heisenberg picture has the form (carets denote operators) [3]

$$\hat{E}(\mathbf{r},t) \propto \int_0^t dt' \hat{\sigma}_x(t') \int_{-\infty}^{\infty} d\omega \omega^2 e^{i\omega(t'-t+r/c)} \propto \theta(t-r/c)\frac{\partial^2}{\partial t^2}\hat{\sigma}_x(t-r/c) \ , \quad (2)$$

where for simplicity we consider only the far field. $\hat{\sigma}_x$ is the two-state Pauli operator in the standard notation, and $\theta(t)$ is the unit step function. The field operator has the same form as its classical counterpart, and in particular it depends on atom operators at the retarded time $t-r/c$ and vanishes for $t < r/c$. Atom B therefore cannot be excited before $t = r/c$.

Fermi, however, restricted himself at the outset to energy-conserving "essential states," and his neglect of other states is equivalent to a rotating-wave approximation (RWA). When formulated in the Heisenberg picture, the RWA implies the effective electric field operator $\hat{E}(\mathbf{r},t) = \hat{E}^{(+)}(\mathbf{r},t)+\hat{E}^{(-)}(\mathbf{r},t)$, where $\hat{E}^{(-)}(\mathbf{r},t) = \hat{E}^{(+)}(\mathbf{r},t)^\dagger$ and

$$\hat{E}^{(+)}(\mathbf{r},t) \propto \int_0^t dt' \hat{\sigma}(t') \int_0^{\infty} d\omega \omega^2 e^{i\omega(t'-t+r/c)} \ , \quad (3)$$

where $\hat{\sigma}$ is the two-state lowering operator. In contrast to the exact expression for the electric field operator [equation (2)], *the positive- and negative-frequency parts of the field are not retarded.* Fermi circumvented the problem by extending the lower limit of integration over ω to $-\infty$, in which case $\hat{E}^{(+)}(\mathbf{r},t) \propto \theta(t - r/c)(\partial^2/\partial t^2)\hat{\sigma}(t-r/c)$ and it follows that atom B cannot be excited before the time $t = r/c$. His extension of the integration range is a good approximation, and it leads to the correct probability for atom B to be excited [3]. But by making this approximation Fermi did not rigorously prove that atom B cannot be excited before $t = r/c$.

A very similar situation occurs in the standard theory of photodetection. There one assumes (energy-conserving) detection processes in which electrons gain energy as photons are annihilated. This RWA leads to field correlation functions involving normally ordered positive- and negative-frequency field operators which, as already noted, are not retarded. For this reason the theory has been criticized for being "acausal." However, causality is easily restored, as in Fermi's problem, by working with the *full* (causal) electric field operator $\hat{E}(\mathbf{r},t)$ and making the RWA at a later stage of the formulation rather than at the beginning [3].

No perfect amplifier

It is well known that the "spooky action at a distance" suggested by Einstein-Podolsky-Rosen (EPR) correlations cannot be used for superluminal communication. But suppose we have two photons in a correlated state of polarization such that a measurement of the polarization of one allows us to predict with certainty the polarization of the other, and that we send the other into an amplifier that results in $N >> 1$ identical photons. Letting the N photons pass through polarization-dependent beam splitters, we could determine with a high degree of accuracy whether a measurement of linear or circular polarization was made on the unamplified photon of the original EPR pair. In other words, since an observation of the polarization of the unamplified photon determines the polarization of its EPR partner, and therefore that of the amplified field with a macroscopically large number of photons, it appears possible to communicate superluminally: an observer A can send a 0 or a 1 to observer B by choosing to measure linear or circular polarization, and the information as to which choice was made is then immediately available to B when he determines whether the field at his end has linear or circular polarization [4].

This superluminal communication scheme fails because photons cannot be perfectly copied: the amplifier will produce photons with polarization different from the incident photon by spontaneous emission—we cannot have stimulated emission without spontaneous emission, and therefore we cannot amplify an arbitrary polarization [5]. More generally, quantum theory does not permit the cloning of a single quantum [6].

It is worth noting that this conclusion does not require that an optical amplifier be polarization-dependent. It is possible for an amplifier to produce a final state that is independent of the polarization of the incident photon, but such an amplifier does not violate the no-cloning theorem: "the essential element that prevents cloning is here seen to be spontaneous emission, rather than any dependence of amplifier gain on polarization" [7].

No local determination of a single-particle eigenstate

Another interesting idea is a "superluminal quantum Morse telegraph" involving EPR polarization-correlated photon pairs and a Michelson interferometer in which one of the mirrors is replaced by a phase-conjugating mirror (PCM) [8]. In contrast to an ordinary mirror, where circular (but not linear) polarization is reversed by reflection, polarization does not change upon "reflection" from a PCM. It would seem therefore that, when one of the ordinary mirrors of a Michelson interferometer is replaced by a PCM, there will be interference of the two propagation paths when the incident light is linearly polarized but not when it is circularly polarized. (In the latter case the fields from the PCM and the ordinary mirror are orthogonally polarized and so do not interfere.) With a steady stream of polarization-correlated EPR pairs, therefore, an observer A would be able to superluminally communicate to an observer B, equipped with the PCM-modified Michelson, whether he is measuring linear or circular polarization.

The interference properties assumed for the PCM-modified Michelson interferometer have in fact been verified experimentally for coherent laser fields. However, a simple calculation shows that for a single incident photon there is no interference between the two paths, regardless of the photon's polarization [9]. This is related to the fact that, like an amplifier, a PCM must have quantum noise—it can spontaneously emit a photon whose polarization is uncorrelated with that of the incident photon [10].

It must be impossible to have an apparatus that could *locally* determine the polarization state of a photon, without making measurements on a (nonlocal) EPR partner. Similarly it must be impossible to have an apparatus that faithfully preserves the polarization of a photon entering with (arbitrary) linear polarization while reversing the helicity of a photon entering with circular polarization [11]. Otherwise, as examples such as these suggest, superluminal communication would be possible.

Causality thus puts strong constraints on possible physical processes. It might be argued that all such constraints follow "simply" from the fact that field commutators must vanish for spacelike intervals; if we ignore quantum noise in an amplifier, for example, we are in effect ignoring part of the electric field operator and so we should not expect to maintain the correct commutation relations.

No perfect filter

But causality imposes requirements on physical processes even in classical theory. Consider a "black box" that produces an output $F_{out}(t)$ that (a) depends linearly on the input $F_{in}(t)$ and (b) is time-independent in the sense that a shift in time of the input produces the same shift in time of the output. Thus we write

$$F_{out}(t) = \int_{-\infty}^{\infty} dt' G(t-t') F_{in}(t') \ , \tag{4}$$

or, in terms of the corresponding Fourier transforms, $f_{out}(\omega) = g(\omega) f_{in}(\omega)$.

Suppose there is no input until the time $t = 0$, so that for $t < 0$ there is destructive interference of the Fourier components of F_{in}. Causality requires that there be no output before $t = 0$. Now suppose that our black box does nothing but absorb the frequency component ν. Then its output would be $F_{out} = F_{in} - F_{in,\nu}$, which does not vanish for $t < 0$—there would be an output before there were any input. It must therefore be impossible to have such a perfect filter that absorbs one frequency without affecting any other frequency. A real filter must also produce phase shifts in all the other Fourier components, in such a way that they interfere destructively for all $t < 0$, so that indeed there is no output before any input [12].

The mathematical expression of this requirement, of course, is a dispersion relation between the absorptive and dispersive parts of a response function. No output before any input means that $G(t-t') = 0$ for $t < t'$, so that

$$g(\omega) = \int_{-\infty}^{\infty} d\tau G(\tau) e^{i\omega\tau} = \int_{0}^{\infty} d\tau G(\tau) e^{i\omega\tau} \ . \tag{5}$$

The important point is that causality requires that the integral extends over only half the τ axis. Then $g(\omega)$ has an analytic continuation in the upper half of the complex ω plane, and it is this analyticity that leads to the Hilbert-transform relations (i.e., dispersion relations) between the real and imaginary parts of $g(\omega)$ [13].

2 Causality in electromagnetic propagation

The causal relation $\mathbf{P}(\omega) = (1/4\pi)[\epsilon(\omega)-1]\mathbf{E}(\omega)$ between the polarization $\mathbf{P}$ and the electric field $\mathbf{E}$ at frequency ω has the input-output form discussed above, and it follows that the dielectric constant $\epsilon(\omega)$ has an analytic continuation in the upper half of the complex ω plane. It doesn't necessarily follow from this alone, however, that the refractive index $n(\omega) = \sqrt{\epsilon(\omega)}$ has the same property; there can be branch points in $n(\omega)$.

Consider the propagation of a scalar plane wave incident from vacuum onto a dielectric medium occupying the half-space $z > 0$. Let the incident field be $E(0,t) = \int_{-\infty}^{\infty} d\omega A(\omega) e^{-i\omega t}$. Propagation over the distance z multiplies each Fourier component of the field by $\exp(ikz) = \exp[i\omega n(\omega) z/c]$, so that $E(z,t) = \int_{-\infty}^{\infty} d\omega A(\omega) e^{-i\omega[t-n(\omega)z/c]}$ or

$$E(z,t) = \int_{-\infty}^{\infty} dt' G(z, t-t') E(0,t') \ , \tag{6}$$

where

$$G(z,\tau) = \frac{1}{2\pi}\int_{-\infty}^{\infty} d\omega e^{-i\omega[\tau - n(\omega)z/c]} . \tag{7}$$

Actually this equation should be modified to include the Fresnel transmission coefficient at the interface at $z = 0$, but this does not affect any fundamental results relating to causality.

Sommerfeld and Brillouin [14] assumed a particular form for the complex refractive index [they were concerned with regions of anomalous dispersion] and showed that signal velocities must be less than c in a medium with such a refractive index. Kramers (1927) showed that this was true in any medium satisfying the dispersion relation, and later Krönig (1942) proved that the dispersion relation is a necessary and sufficient condition for causality.

If $n(\omega)$ is analytic in the upper half of the complex plane and if $n(\omega) \to 1$ as $\omega \to \infty$, then a simple application of Cauchy's theorem to equation (7) implies the causality condition $G(z,\tau) = 0$ for $\tau < z/c$. It is of course physically reasonable that $n(\omega) \to 1$ as $\omega \to \infty$, and the refractive index for virtually any system one imagines has an analytic continuation in the upper half plane.

[An exception is the case where the damping mechanism is radiative reaction as described by the Abraham-Lorentz theory. $n(\omega)$ in this case has a pole in the upper half-plane resulting from an ω^3 appearing in a denominator. In fact this model is well known to be acausal. But the famous "preacceleration" problem occurs on such a short time scale that quantum effects must be considered, i.e., the classical model breaks down.]

If the incident field $E(0,t) = 0$ for $t < 0$, and $G(z,\tau) = 0$ for $\tau < z/c$, then $E(z,t) = 0$ for $t < z/c$. Sommerfeld and Brillouin defined a *signal* as a train of oscillations that starts from zero at some instant. Then, for a physically admissible refractive index, *no signal can propagate faster than c.*

What happens after $t = z/c$ depends on the particular form of $n(\omega)$ and $E(0,t)$. Because the medium cannot respond instantaneously to the applied field, the wave *front* velocity is always equal to c. After $t = z/c$ there are precursors to the main part of the propagating field, the first precursor arising from the high-frequency parts of the field. A detailed discussion of precursors is given by Oughstun and Sherman [15].

3 Superluminal group velocities

If $E(0,t) = A(t)\exp(-i\omega_o t)$, where $A(t)$ varies sufficiently slowly on times scales $\sim \omega_o^{-1}$, it is often useful to make the *group velocity approximation* in which we replace $k(\omega)$ by $k(\omega_o) + (dk/d\omega_o)_{\omega_o}(\omega - \omega_o) \equiv k_o + (\omega - \omega_o)/v_g$. It is straightforward to show from equations (6) and (7) that in this approximation [17]

$$E(z,t) = e^{-i(\omega_o t - k_o z)} e^{[c^{-1} - v_g^{-1}]z\partial/\partial t} A(t - z/c) = e^{-i(\omega_o t - k_o z)} A(t - z/v_g) \tag{8}$$

if $A(t)$ is differentiable. The group velocity

$$v_g = c/[n + \omega dn/d\omega]_{\omega_o} . \tag{9}$$

In a region of "normal" dispersion $[(dn/d\omega)_{\omega_o} > 0]$, the group velocity is less than the phase velcity $c/n(\omega_o)$. In a region of anomalous dispersion $[(dn/d\omega)_{\omega_o} < 0]$, however, $v_g > c/n(\omega_o)$. Because v_g can exceed c in a region of anomalous dispersion, and because v_g was generally thought to be the velocity of energy propagation, $v_g > c$ was at one time thought to be in conflict with the special theory of relativity. This conflict was resolved when Sommerfeld and Brillouin proved that the *signal* velocity must be less than c even in a region of anomalous dispersion.

Regarding the velocity v_E of energy transport, simple models based on the Lorentz model of a dielectric give the result that v_E, defined as the ratio of the (time-averaged) Poynting vector and the total energy density (including that of the medium), is $\leq c$ for a monochromatic wave [15],[16]. In the absence of absorption $v_E = v_g$, and $v_E \cong v_g$ if there is absorption but the field frequency is far from an absorption resonance. Because the v_E used in these models is based on the assumption of a monochromatic field, it cannot be associated with a signal velocity. There is some controversy surrounding the definition of the velocity of energy transport [18], [19], all stemming from the fact that part of the total energy density is stored for a finite time in the medium.

Confusion surrounding the meaning of group velocity did not end with the work of Sommerfeld and Brillouin. It is asserted that the concept of group velocity lacks physical significance when $v_g > c$ (or has other interesting values—see below) because, in contradiction to equation (8), the pulse will become highly distorted as a consequence of strong dispersion implied by $v_g > c$. This is a sensible but incorrect conclusion. Worse, it is often implied that there would be a conflict with special relativity if the group velocity were larger than c. This is false because, to repeat, group velocity is not in general a signal velocity.

There are related discussions that can be criticized for the same reason if one is sufficiently pedantic. For instance, it is argued that the propagation of x-rays in glass is consistent with special relativity because the group velocity (though not the phase velocity) is less than c [20]. However, the argument is incomplete without some justification for assuming that the group velocity is actually a signal velocity—which it effectively is *in this case* because x-rays are far from absorption resonances. Otherwise, what would be the conclusion if for a different medium the group velocity were negative?

Observations of superluminal group velocities in optics

Basov *et al.* (1966) demonstrated that the group velocity of a laser pulse in an amplifier could exceed c due to a pulse reshaping in which the front part of the pulse leaves less gain available for the back part, resulting in an advancment of the peak of the pulse. Icsevgi and Lamb argued that Basov *et al.* in their analysis assumed "an unphysical input pulse extending to infinity at both ends." The measurement of pulse velocities greater than c does not contradict special relativity because "an experimental apparatus can only trace the bulk of the pulse" [21], which does not constitute a signal.

Garrett and McCumber [22] showed that superluminal group velocities can also appear when a Gaussian pulse propagates in an *absorbing* medium, provided that the pulse bandwidth is much smaller than the width of the absorption line and the medium is sufficiently short. This results from a pulse reshaping and advancement process in which the back part of the pulse is more strongly absorbed than the front part. Equation (9) indicates that the group velocity can be infinite or even negative if anomalous dispersion is sufficiently strong ($dn/d\omega$ sufficiently negative). An infinite group velocity means that the peak of the pulse emerging from the medium occurs at the same instant as the peak of the pulse entering the medium. A negative group velocity means that the peak of the emerging pulse occurs at an earlier time than the peak of the incident pulse.

Furthermore this can occur without significant pulse distortion, albeit with significant absorption. Experimental observations of these remarkable effects were reported in 1982 [23].

More recent interest in superluminal group velocities is due in considerable part to the work of Chiao's group [24]. They performed experiments in which the central frequency of a single-photon wave packet was that for minimum transmission through a multilayered dielectric, in which case the exponentially decaying, evanescent behavior of the transmitted wave corresponds to quantum tunnelling. The tunnelling of individual photons through the barrier was observed using a coincidence counting technique, which revealed that a single-photon wave packet crossed the barrier with a velocity $1.7c \pm 0.2c$ and that the transmitted wave packet, though highly attenuated, had approximately the same Gaussian shape as the incident wave packet. Other groups have confirmed in other optical tunnelling situations that the group velocity associated with the tunnelling is superluminal. These developments are reviewed by Chiao and Steinberg [24],[25].

Chiao *et al.* suggested that it should also be possible to observe superluminal group velocities of *unattenuated* pulses propagating off resonance in amplifying media. This superluminal propagation would be more dramatic than in the case of tunnelling not only because it does not involve strong attenuation, but also because of the related fact that the propagation paths are much longer [24].

The observation of distortionless pulses propagating with superluminal group velocity *and* relatively small change in amplitude has recently been reported by Wang's group [26]. In their experiment anomalous dispersion occurred in a Raman gain medium (cesium vapor) in the spectral region between two gain lines. They reported a 62-ns pulse advancement compared to the 0.2-ns time for propagation in vacuum.

Signals and analytic continuation

Equation (8) states that propagation over a distance z corresponds to an analytic continuation over the time $z/c - z/v_g$ of the vacuum-propagated pulse envelope $A(t - z/c)$. Thus a superluminal group velocity does not imply a superluminal propagation of *information*, since there is no information in $A(t - z/v_g)$ that is not already contained in $A(t - z/c)$.

New information is therefore propagated only if $A(t - z/c)$ does not have an analytic continuation. In this case the second equality in (8) is invalid, while the first equality holds up to a time at which $A(t - z/c)$ or one of its derivatives has a discontinuity. Thereafter the pulse evolution becomes much more complicated than a simple undistorted propagation at the superluminal group velocity. The point of singularity behaves like a Sommerfeld-Brillouin front, which propagates at c. In other words, a true signal requires non-analyticity, and cannot transmit information at a velocity $> c$. The experimental observations of superluminal group velocities do not contradict Einstein causality!

The case of a differentiable (e.g., Gaussian) wave form propagating without distortion at a superluminal group velocity is nevertheless remarkable when one considers that (1) points at time t on the transmitted pulse are causally determined by points at $t < z/c$ on the incident pulse, and yet (2) the transmitted pulse advances at the velocity $v_g > c$. This means, as indicated in Figure 1, that if $(z/c - z/v_g)$ is much larger than the pulse duration, the peak of the transmitted signal is reconstructed entirely from a small tail of the incident pulse; if

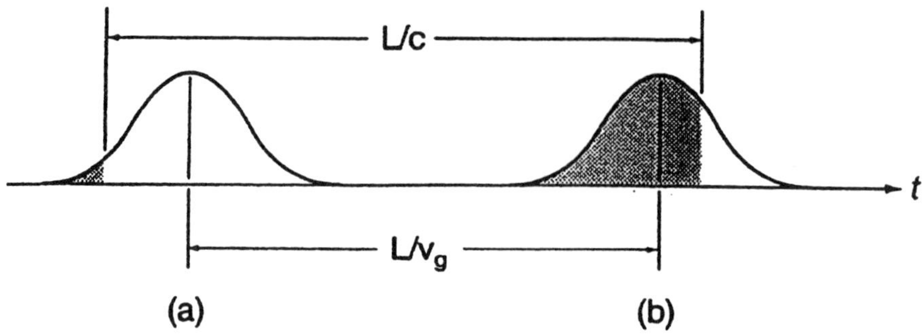

Figure 1: Incident (a) and transmitted (b) pulses for a propagation length z and group velocity $v_g > c$. The shaded portion of (b) is completely determined by the shaded portion of (a).

$(z/c - z/v_g)$ is large enough, nearly the entire transmitted pulse is reconstructed by analytic continuation of a tiny portion of the incident pulse.

Note that the peak of the transmitted pulse is not causally connected to the peak of the incident pulse, so that in particular the observation that the pulse peak moves superluminally does not contradict Einstein causality. A simple analogy would be the motion of a spot of light made by shining a rotating flashlight onto a distant wall. The spot can in principle move superluminally, but there is no violation of causality because the spot at one instant is not the source of the spot at a later instant. The recent experimental observations of superluminal group velocities [24], [26] involve a linear response of the medium to the field, and the exchange of energy between the medium and the front and back parts of the pulse, leading to the pulse advancement, can be interpreted using classical spectral arguments [27].

Linearity also makes it straightforward to perform a quantum-mechanical calculation using a simple model in which an absorbing dielectric is described as a collection of identical two-level atoms and the source of light is a single excited atom outside the dielectric [28]. It is shown that, if the source atom is suddenly excited at time $t = 0$, then the probability of detecting a photon at a point inside the dielectric and at a distance z from the source atom is zero before the time z/c. This is the analogue of the classical result that a sharp wavefront cannot propagate faster than c. However, if the probability $P(t)$ that the source atom is excited varies smoothly in time, then the photon counting rate $R(t)$ at an ideal detector located at z is proportional to $\exp(-2\omega_o n_I(\omega_o) z/c)P(t - z/v_g)$, where n_I is the imaginary part of the refractive index. Thus, if $v_g > c$, the peak probability of producing a "click" at the detector can occur earlier than is possible when there is no medium between the detector and the source atom. This is analogous to what has been observed in the optical tunnelling experiments of Chiao *et al.*

Quantum limitations

Chiao *et al.* [29] (CKK) have suggested that superluminal group velocities are possible in an optical amplifier whose relaxation times are long compared with the pulse duration, and that this can occur even at the single-photon level. They refer to a single-photon wave packet propagating with superluminal group velocity as an "optical tachyon." To avoid spontaneous emission noise they required that the radiative lifetime be large compared to the propagation time through the amplifier, which in turn should be large compared to the pulse duration. In order to avoid superfluorescence (SF), the time delay before the peak of the SF pulse should be larger than all these other times.

Aharonov *et al.* [30] (ARS) have argued from general considerations and a model involving coupled inverted pendula that "unstable systems such as media with inverted atoms" have unstable modes whose quantum fluctuations will overcome a superluminal pulse at the single-photon level. This conclusion is based on the reconstruction of the superluminal pulse from a small tail as discussed earlier (Figure 1). If the incident field is very weak, the small tail whose analytic continuation determines the propagated pulse can have a very small probability of containing even a single photon, and consequently the superluminal pulse will have a very small signal-to-noise ratio (SNR). One of the conditions imposed by ARS for the observation of superluminality is that $(v_g/c-1)L/c >> \tau_p$, where L is the length of the amplifier and τ_p is the pulse duration: the separation in time of pulses propagating with velocities c and v_g must be large compared with the pulse duration in order to strongly distinguish experimentally between luminal and superluminal propagation.

It has been shown [31] that the quantum noise in the ARS model corresponds to superfluorescence in a medium of inverted two-level atoms. Even at times short compared with the delay time for the buildup of the peak of the SF pulse, the initial onset of SF is sufficient to make the SNR very small at the one- or few-photon level when the ARS condition of strong distinguishability is imposed. However, the results are not inconsistent with CKK: even at the one-photon level it is possible to have SNR > 1 if the temporal separation $(v_g/c-1)L/c$ is smaller than the pulse duration, i.e., if we do not impose the ARS condition but instead allow for a weaker degree of experimental "distinguishability." $(v_g/c-1)L/c < \tau_p$ does not, of course, preclude the observation of superluminal propagation, and in fact this was the case in the tunnelling experiments of Chiao *et al.* [24].

Quantum noise associated with spontaneous emission also acts to effectively retard the measured pulse by producing a background level of irradiation that must be exceeded before it can be asserted that the pulse has arrived. We can define a "signal" arrival time operationally as the time at which the integrated photocurrent exceeds the background level by some predetermined factor. More precisely, the observable corresponding to this definition of the arrival time is [32]

$$\hat{S}(L.t) = \eta \int_{t_o-T/2}^{t} dt' \hat{E}^{(-)}(L,t)\hat{E}^{(+)}(L,t') \tag{10}$$

at the exit point ($z = L$) of the medium, and we define the arrival time as the time at which the optical signal-to-noise ratio [33]

$$SNR(L,t) = \frac{(\langle \hat{S}_1(L,t)\rangle - \langle \hat{S}_0(L,t)\rangle)^2}{\langle \Delta^2 \hat{S}(L,t)\rangle} \tag{11}$$

reaches a prescribed threshold level. Here $\hat{S}_1$ and $\hat{S}_0$ are defined by equation (10) with and without the input pulse, respectively. $t_o = T_c + L/c$, where T_c is the time corresponding to the peak of the pulse, T is a time window of typically a few times the pulse width, and η is an efficiency factor that is unimportant for the present discussion. This definition of a "signal" arrival time is obviously

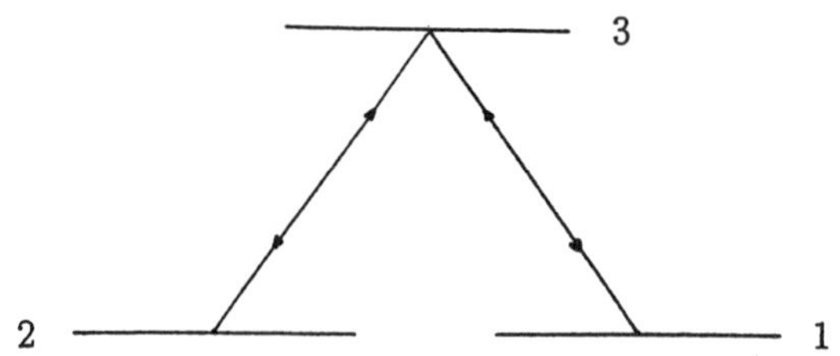

Figure 2: Three-state atom in which the two allowed transitions $1 \leftrightarrow 3$ and $2 \leftrightarrow 3$ have the same transition frequency and electric dipole moment.

somewhat arbitrary, but not unreasonable. When computations are performed using parameters appropriate to the gain medium in the experiments of Wang *et al.* [26], it is concluded that the corresponding velocity of the operationally defined "signal" is less than c [32].

Arrival times and velocities defined classically are also subjective, and also depend on the detection process. Sommerfeld and Brillouin defined a "signal velocity" as the velocity of the point of half-maximum of a wave packet. Brillouin noted that [34]

> ... the definition of the signal velocity is somewhat arbitrary ... The signal does not arrive suddenly; there is a quick but still continuous transition from the very weak intensity of the [precursors] to that corresponding to the signal. A detector set to detect an intensity equal to 1/4 the final intensity will detect the arrival of the signal in agreement with the above arbitrary definition; if the detector is more or less sensitive, then it will detect the arrival of the signal a little earlier or later.

Superluminal propagation in an absorber would not be affected by quantum noise associated with spontaneous emission. Even in this case, however, the "arrival time" cannot be inferred before the detection of the first photon. The photon wave packet (probability amplitude) in any case propagates causally, even though a photon detector can be expected to "click" earlier than it could if the wave packet were propagating in vacuum.

Recent studies of "metamaterials" may lead to superluminal pulses propagating with neither spontaneous emission noise nor significant attenuation [35].

4 Halting and storage of light

Recent observations of extremely *small* group velocities are based on quantum interference effects that cause greatly reduced absorption and very rapid variation with frequency of the refractive index.

The interference effect here can be understood using a three-state model with equal transition dipole moments (μ) and equal applied electric field amplitudes ($\mathcal{E}$) at two equal transition frequencies (Figure 2). The effect of the applied field on the atom in this model is described by the interaction Hamiltonian

$$H_{int} = -\mu\mathcal{E}[|3\rangle\langle 1| + |3\rangle\langle 2|] + \text{h.c.} \tag{12}$$

Define two superposition states $|C\rangle = |1\rangle + |2\rangle$ and $|NC\rangle = |1\rangle - |2\rangle$. Obviously $\langle 3|H_{int}|C\rangle = -2\mu\mathcal{E}$ and $\langle 3|H_{int}|NC\rangle = 0$.

The "noncoupled" state $|NC\rangle$ is a *dark state*: an atom in such a superposition state does not interact with the applied field. When the applied field is

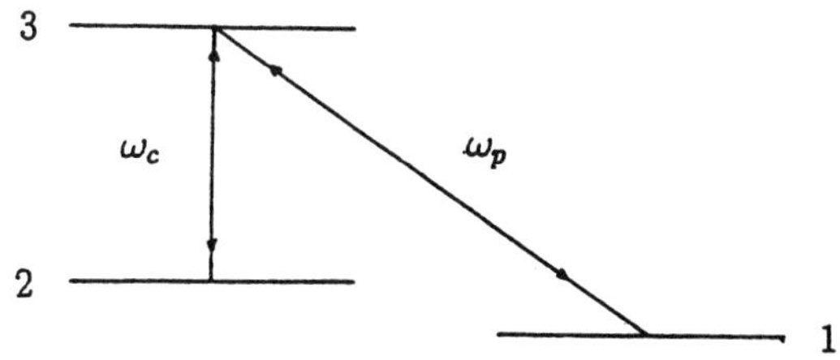

Figure 3: As a consequence of coherent population trapping in the presence of the coupling field (ω_c), a probe field at frequency ω_p can propagate without absorption.

turned on our three-state atom can be pumped into state 3, from which it can go by spontaneous emission into either the coupled state $|C\rangle$ or the noncoupled state $|NC\rangle$. Once in the latter state it is trapped. Eventually the atom will find itself in the (nonabsorbing) dark state by this process of "coherent population trapping" [36].

Coherent population trapping in the presence of a coupling field can lead to electromagnetically induced transparency (EIT) [37] for a probe field—that is, the probe can propagate without absorption while the atoms remain unexcited (Figure 3). Associated with EIT is a rapidly varying refractive index, and consequently a very small group velocity, and furthermore the refractive index is unity and the group velocity dispersion is zero at line center [38].

Hau *et al.* [39] first observed such extremely small group velocities: $v_g = 17$ m/s in a Bose-Einstein condensate of sodium atoms. The vanishing of the group velocity dispersion allows a pulse to propagate with essentially no distortion. Among the interesting observations was the spatial compression of the EIT pulse by the factor c/v_g, without a change in the peak amplitude of the electric field. A pulse that would be about 750 m long in free space is observed to be compressed to about 42 μm along the direction of propagation in the ultracold gas. This results from energy being transferred from the front part of the pulse to the atoms and the coupling field, and then being returned to the pulse at its back part.

Ultracold temperatures are not required for ultraslow group velocities. Kash *et al.* [40] observed group velocities $\sim$ 90 m/s in ^{87}Rb gas at 360 K, and pointed out that Doppler averaging in EIT and related phenomena is mitigated by a two-photon Doppler-free effect that can occur for copropagating coupling and probe fields. Budker *et al.* [41] have reported an 8 m/s group velocity in room-temperature ^{85}Rb vapor. The ultraslow group velocity in this case arises from an extremely narrow nonlinear magneto-optic resonance.

Dark-state polaritons

Let $\Omega(t)$ be the Rabi frequency associated with the coupling field at the frequency ω_c in Figure 3, and denote the operator for the slowly varying envelope of the field at the probe frequency ω_p by $\hat{E}(z,t)$. From the Heisenberg equations of motion it follows that

$$\left(\frac{\partial}{\partial t} + c\frac{\partial}{\partial z}\right)\hat{E}(z,t) = igN\hat{\sigma}_{13}(z,t)\ . \tag{13}$$

Here $\hat{\sigma}_{13}$ is the lowering operator for the $1 \leftrightarrow 3$ transition, N is the number density of atoms, and g is the atom-probe coupling constant proportional to the transition dipole matrix element μ_{13}. $\Omega(t)$ is a c-number, i.e., the coupling field is treated as a prescribed, classical field.

The right-hand side of equation (13) is determined by the Heisenberg equations of motion for the operators $\hat{\sigma}_{ij}$. If the probe field is sufficiently weak, so that $\hat{\sigma}_{11} \approx 1$, then

$$\hat{\sigma}_{13} \approx -\frac{i}{\Omega(t)} \frac{\partial}{\partial t} \hat{\sigma}_{12} \ . \tag{14}$$

$\hat{\sigma}_{12}(z,t)$ can in turn be approximated by $-g\hat{E}(z,t)/\Omega(t)$ in the adiabatic approximation in which Ω varies slowly. Then

$$\left(\frac{\partial}{\partial t} + c\frac{\partial}{\partial z}\right) \hat{E}(z,t) \approx -\frac{g^2 N}{\Omega(t)} \frac{\partial}{\partial t} \left[\frac{\hat{E}(z,t)}{\Omega(t)}\right] . \tag{15}$$

In the limiting case in which Ω is constant, this implies the group velocity $v_g = c/(1 + g^2 N/\Omega^2)$ [42]. This approximation describes rather well the group velocity observed in the experiment of Hau *et al.* [39].

The operator [43]

$$\hat{\Psi}(z,t) = \cos\theta(t)\hat{E}(z,t) - \sin\theta(t)\sqrt{N}\hat{\sigma}_{12}(z,t) \ , \tag{16}$$

where

$$\cos\theta(t) = \frac{\Omega(t)}{\sqrt{\Omega^2(t) + g^2 N}} \ , \qquad \sin\theta(t) = \frac{g\sqrt{N}}{\sqrt{\Omega^2(t) + g^2 N}} \ , \tag{17}$$

satisfies the equation of motion

$$\left[\frac{\partial}{\partial t} + c\cos^2\theta(t)\frac{\partial}{\partial z}\right] \hat{\Psi}(z,t) = 0 \ . \tag{18}$$

This equation has a shape-preserving solution with time-dependent group velocity $v_g(t) = c\cos^2\theta(t)$:

$$\hat{\Psi}(z,t) = \hat{\Psi}(z - c\int_0^t dt' \cos^2\theta(t'), t = 0) \ . \tag{19}$$

This shape-preserving solution, based on the adiabatic approximation, is closely related to the "adiabatons" discussed by Grobe *et al.* [42]. Note that $\hat{\Psi}$ is neither a pure electric field operator nor a pure atomic operator, but a combination of the two. It is easily shown to have bosonic commutation relations, and Fleischhauer and Lukin [43] refer to its bosonic quasiparticles (number states) as *dark-state polaritons.* These polaritons involve the operator $\hat{\sigma}_{12}$ associated with a *non-allowed* (Raman) transition, thus distinguishing them from the more familiar polaritons defined in terms of allowed transitions. They are called "dark-state" polaritons because they do not contain the excited atomic state 3.

Coherent control of light propagation

The group velocity of the dark-state polaritons depends on the mixing angle θ between their atomic and photonic components. The fact that this mixing angle depends on $\Omega(t)$ means that the group velocity can be controlled by the coupling field. Thus, if θ is changed from 0 to $\pi/2$, the group velocity changes from c

to 0. In this case the dark-state polariton goes from being purely photonic to being purely atomic; the light pulse is effectively halted and coherently stored in the medium. Changing θ back to 0 should result in a "re-acceleration" of the light pulse as the polariton changes from purely atomic to purely photonic [43]. Equation (16) shows that this exchange between atomic and photonic components of the dark-state polariton occurs via the atomic coherence of the Raman transition $1 \leftrightarrow 2$. The storage time of the light when the coupling field is adiabatically turned off is therefore limited not by excited-state relaxation but by the $1 \leftrightarrow 2$ coherence lifetime, which is typically very large.

This halting, storage, and re-acceleration of light pulses was very recently reported by Liu *et al.* [44], who used sodium vapor at a temperature ~ 1 μK, a density $\sim 10^{13}$ cm^{-3}, and a propagation length of 339 μm. Phillips *et al.* [45] reported similar observations using rubidium at 70-90 ^{o}C, $10^{11} - 10^{12}$ atoms/cm^3, and a 4-cm path length. Both groups observed light storage for a period ~ 1 ms, approximately the coherence lifetime of their Raman transitions.

It should by now be evident that light pulses can be made to propagate with essentially any group velocity, including $v_g < 0$, $v_g = 0$, and $v_g = \infty$, and that the concept of group velocity does not necessarily lose any meaning or utility when it is dramatically different from c.

Acknowledgement

I thank especially R. Y. Chiao, J. Peatross, A. M. Steinberg, and L. J. Wang for enlightening discussions.

References

[1] E. Fermi, "Quantum theory of radiation," Rev. Mod. Phys. **4**, 87-132 (1932).

[2] "I am reminded how much trouble Fermi had ... in his 1932 RMP article on QED with ... the velocity of light problem" — W. E. Lamb, Jr., private communication.

[3] P. W. Milonni, D. F. V. James, and H. Fearn, "Photodetection and causality in quantum optics," Phys. Rev. A**52**, 1525-1537 (1995) and references therein.

[4] N. Herbert, "FLASH—A superluminal communicator based upon a new type of quantum measurement," Found. Phys. **12**, 1171-1179 (1982).

[5] P. W. Milonni and M. L. Hardies, "Photons cannot always be replicated," Phys. Lett. A**92**, 321-322 (1982).

[6] W. K. Wooters and W. H. Zurek, "A single quantum cannot be cloned," Nature **299**, 802-803 (1982) and also references therein.

[7] L. Mandel, "Is a photon amplifier always polarization dependent?," Nature **304**, 188-188 (1983).

[8] A. Garuccio, "On the contradiction between quantum mechanics and relativity: a superluminal quantum Morse telegraph," in *The Present Status of the Quantum Theory of Light*, ed. S. Jeffers *et al.* (Kluwer, Dordrecht, 1997).

[9] K. Furuya, P. W. Milonni, A. M. Steinberg, and M. Wolinsky, Phys. Lett. A**251**, "Failure of a proposed superluminal scheme," 294-296 (1999).

[10] A. L. Gaeta and R. W. Boyd, "Quantum noise in phase conjugation," Phys. Rev. Lett. **60**, 2618-2621 (1988); P. W. Milonni, E. J. Bochove, and R. J. Cook, "Fluorescence near a phase-conjugating mirror: amplification of quantum noise," Phys. Rev. A**40**, 4100-4102 (1989).

[11] A. Peres, private communication.

[12] J. S. Toll, "Causality and the dispersion relation: logical foundations," Phys. Rev. **104**, 1760-1770 (1956).

[13] H. M. Nussenzveig, *Causality and Dispersion Relations* (Academic Press, N. Y., 1972).

[14] L. Brilluoin, *Wave Propagation and Group Velocity* (Academic Press, N. Y., 1960).

[15] K. E. Oughstun and G. C. Sherman, *Electromagnetic Pulse Propagation in Causal Dielectrics* (Springer-Verlag, Berlin, 1994).

[16] R. Loudon, "The propagation of electromagnetic energy through an absorbing dielectric," J. Phys. A**3**, 233-245 (1970).

[17] G. Diener, "Superluminal group velocities and information transfer," Phys. Lett. A**223**, 327-331 (1996).

[18] G. Diener, "Energy transport in dispersive media and superluminal group velocities," Phys. Lett. A**235**, 118-124 (1997).

[19] J. Peatross, S. A. Glasgow, and M. Ware, "Average energy flow of optical pulses in dispersive media," Phys. Rev. Lett. **84**, 2370-2373 (2000).

[20] R. P. Feynman, R. B. Leighton, and M. Sands, *The Feynman Lectures on Physics* (Addison-Wesley, Reading, Mass., 1964), p. 48-7.

[21] A. Icsevgi and W. E. Lamb, Jr., "Propagation of light pulses in a laser amplifier," Phys. Rev. **185**, 517-545 (1969).

[22] C. G. B. Garrett and D. E. McCumber, "Propagation of a gaussian light pulse through an anomalous dispersion medium," Phys. Rev. A**1**, 305-313 (1970).

[23] S. Chu and S. Wong, "Linear pulse propagation in an absorbing medium," Phys. Rev. Lett. **48**, 738-741 (1982).

[24] See the following reviews and references therein: R. Y. Chiao, "Population inversion and superluminality," in *Amazing Light*, ed. R. Y. Chiao (Springer-Verlag, N. Y., 1996); R. Y. Chiao and A. M. Steinberg, "Tunneling times and superluminality," in *Progress in Optics XXXVII*, ed. E. Wolf (Elsevier, Amsterdam, 1997).

[25] See also the recent work of M. Mojahedi, E. Schamiloglu, F. Hegeler, and K. J. Malloy, "Time-domain detection of superluminal group velocity for single microwave pulses," Phys. Rev. E**62**, 5758-5766 (2000).

[26] L. J. Wang, A. Kuzmich, and A. Dogariu, "Gain-assisted superluminal light propagation," Nature **406**, 277-279 (2000).

[27] J. Peatross, M. Ware, and S. A. Glasgow, "Role of the instantaneous spectrum on pulse propagation in causal linear dielectrics," J. Opt. Soc. Am. A (in press).

[28] P. W. Milonni, K. Furuya, and R. Y. Chiao, "Quantum theory of superluminal pulse propagation," Optics Express **8**, 59-65 (2001).

[29] R. Y. Chiao, A. E. Kozhekin, and G. Kurizki, "Tachyonlike excitations in inverted two-level media," Phys. Rev. Lett. **77**, 1254-1257 (1996).

[30] Y. Aharonov, B. Reznik, and A. Stern, "Quantum limitations on superluminal propagation," Phys. Rev. Lett. **81**, 2190-2193 (1998).

[31] B. Segev, P. W. Milonni, J. F. Babb, and R. Y. Chiao, "Quantum noise and superluminal propagation," Phys. Rev. A**62**, 022114-1-15 (2000).

[32] A. Kuzmich, A. Dogariu, L. J. Wang, P. W. Milonni, and R. Y. Chiao, "Signal velocity, causality, and quantum noise in superluminal light pulse propagation," Phys. Rev. Lett. **86**, 3925-3929 (2001).

[33] E. Desurvire, *Erbium-Doped Fiber Amplifiers: Principles and Applications* (Wiley, N. Y., 1994), Chapter 2; B. E. A. Saleh and M. C. Teich, *Fundamentals of Photonics* (Wiley, N. Y., 1991), Chapter 16.

[34] Reference [14], p. 79.

[35] R. W. Ziolkowski, "Superluminal transmission of information through an electromagnetic metamaterial," Phys. Rev. E**63**, 046604-1-13 (2001).

[36] See, for instance, P. L. Knight, M. A. Lauder, and B. J. Dalton, "Laser-induced continuum structure," Phys. Rep. **190**, 1-61 (1990).

[37] S. E. Harris, "Electromagnetically induced transparency," Phys. Today **50**(7), 36-42 (1997); M. O. Scully and M. S. Zubairy, *Quantum Optics* (Cambridge University Press, Cambridge, UK, 1997).

[38] S. E. Harris, "Dispersive properties of electromagnetically induced transparency," Phys. Rev. A**46**, R29-32 (1992).

[39] L. V. Hau, S. E. Harris, Z. Dutton, and C. H. Behroozi, "Light speed reduction to 17 metres per second in an ultracold atomic gas," Nature **297**, 594-598 (1999).

[40] M. M. Kash, V. A. Sautenkov, A. S. Zibrov, L. Hollberg, G. R. Welch, M. D. Lukin, Y. Rostovtsev, E. S. Fry, and M. O. Scully, "Ultraslow group velocity and enhanced nonlinear optical effects in a coherently driven hot atomic gas," Phys. Rev. Lett. **82**, 5229-5232 (1999).

[41] D. Budker, D. F. Kimball, S. M. Rochester, and V. V. Yashchuk, "Nonlinear magneto-optics and reduced group velocity of light in atomic vapor with slow ground state relaxation," Phys. Rev. Lett. **83**, 1767-1770 (1999).

[42] R. Grobe, F. T. Hioe, and J. H. Eberly, "Formation of shape-preserving pulses in a nonlinear adiabatically integrable system," Phys. Rev. Lett. **73**, 3183-3186 (1994).

[43] M. Fleischhauer and M. D. Lukin, "Dark-state polaritons in electromagnetically induced transparency," Phys. Rev. Lett. **84**, 5094-5097 (2000).

[44] C. Liu, Z. Dutton, C. H. Behroozi, and L. V. Hau, "Observation of coherent optical information storage in an atomic medium using halted light pulses," Nature **409**, 490-493 (2001).

[45] D. F. Phillips, A. Fleischhauer, A. Mair, R. L. Walsworth, and M. D. Lukin, "Storage of light in atomic vapor," Phys. Rev. Lett. **86**, 783-786 (2001).

Controlling Quantum Rotation With Light

I. Sh. Averbukh*, R. Arvieu†, and M. Leibscher*

* *Department of Chemical Physics, The Weizmann Institute of Science, Rehovot 76100, Israel*
† *Institut des Sciences Nucléaires, F 38026 Grenoble Cedex, France*
e-mail: Ilya.Averbukh@weizmann.ac.il

Abstract: Semiclassical catastrophes in the dynamics of a quantum rotor (molecule) driven by a strong time-varying field are considered. We show that for strong enough fields, a sharp peak in the rotor angular distribution can be achieved via time-domain focusing phenomenon, followed by the formation of angular rainbows and glory-like angular structures. Several scenarios leading to the enhanced angular squeezing are proposed that use specially designed and optimized sequences of pulses. The predicted effects can be observed in many processes, ranging from molecular alignment (orientation) by laser fields to heavy-ion collisions, and the squeezing of cold atoms in a pulsed optical lattice.

1. Introduction

Driven rotor is a standard model in classical and quantum nonlinear dynamics studies[1]. Increasing interest in the problem has arisen because of recent atom optics realization of the quantum δ-kicked rotor [2, 3], and novel experiments on molecule orientation (alignment) by strong laser fields[4, 5, 6, 7]. A strong enough laser field creates the so-called pendular molecular states [8, 9, 10] that are hybrids of field-free rotor eigenstates. By adiabatically turning on the laser field, it is possible to trap a molecule in the ground pendular state, thus leading to molecular alignment. The only way to reach a considerable degree of alignment in this approach is by increasing the intensity of the field. However, many applications may require only a transient molecular alignment (orientation), where the molecular angular distribution becomes extremely squeezed at some predetermined moment of time. It is well known that a physically related problem of squeezed states generation in a harmonic system may be solved by a proper time-modulation of the driving force (parametric resonance excitation). Behavior of a rotor in general, strong, time-varying fields is a much less-studied problem, although it is understood that the long-persisting beats in the molecular angular distribution may be induced by short laser pulses [11, 12, 13, 14, 15, 16, 17].

In the present paper, we analyze generic features in the dynamics of a quantum rotor driven by strong pulses, and present a strategy for efficient squeezing of the rotor angular distribution by a sequence of pulses of moderate intensity. The results of our research are related to a number of physical processes, ranging from molecular alignment (orientation) by laser fields to heavy-ion collisions, and the trapping of cold atoms by a standing light wave. The paper is organized as following. Section II discusses semiclassical catastrophes in the dynamics of a driven quantum rotor with the help of a simple two-dimensional model [18]. Section III applies these generic results to a thermal system of cold atoms driven by a pulsed optical lattice [19]. Section IV focuses on 3D effects important in the processes of molecular alignment and orientation by strong laser pulses [20].

2. Two-dimensional rotor

For the sake of clarity, we start with the simplest model of a two-dimensional rotor described by the time-dependent Hamiltonian

$$\widehat{H} = \frac{\widehat{L}^2}{2I} + V(\theta, t) \tag{1}$$

where $\widehat{L}$ is the angular momentum operator, and I is the momentum of inertia of the rotor. This model contains most of the physics we want to present here. Some important additional effects that appear in 3D will be discussed below in Section IV. For a "linear 2D molecule" having a permanent dipole moment μ, and driven by a linearly polarized field, the interaction potential is

$$V(\theta, t) = -\mu\mathcal{E}(t)\cos(\theta) \tag{2}$$

where $\mathcal{E}(t)$ is the field amplitude (i. e., of a half-cycle pulse), and θ is the polar angle between the molecular axis and the field direction. In the absence of interaction with a permanent dipole moment, the external field may couple with induced anisotropic molecular polarization. This interaction (being averaged over fast optical oscillations) may result in the interaction potential proportional to $\cos^2(\theta)$ (see Eq.(17) in Section IV) [21]. Although these two forms of $V(\theta, t)$ may lead to different physical consequences (i. e., orientation vs alignment), the effects we will present are more or less insensitive to the choice of interaction. Therefore, we prefer to start with the more simple situation described by Eq.(2).

By introducing dimensionless time $\tau = t\hbar/I$, and interaction strength $\varepsilon = \mu\mathcal{E}(t)I/\hbar^2$, the Hamiltonian can be written as

$$\widehat{H} = -\frac{1}{2}\frac{\partial^2}{\partial\theta^2} - \varepsilon(\tau)\cos(\theta)$$

The wave function of the system can be expanded in the eigenfunctions of a free rotor

$$\Psi(\theta, \tau) = \frac{1}{\sqrt{2\pi}}\sum_{n=-\infty}^{+\infty} c_n(\tau)e^{in\theta}$$

In the absence of the field, the wave function takes the form

$$\Psi(\theta, \tau) = \frac{1}{\sqrt{2\pi}}\sum_{n=-\infty}^{+\infty} c_n(0)e^{-in^2\tau/2+in\theta} \tag{3}$$

Despite a simple form of Eq.(3), the wave function exhibits extremely rich space-time dynamics. In particular, it shows periodic behavior in time with the period $T_{rev} = 4\pi$ (full revival) and a number of fractional revivals at $\tau = p/sT_{rev}$ (p and s are mutually prime numbers) [22]. An analytical solution valid for a general time-dependent field is unknown even for this simplest model. Much effort has been devoted to the case of extremely short field pulses ($\delta-$kicks) (see, e. g. [1], and references therein). In general, as a result of a single kick applied to the rotor at $\tau = \tau_k$, the coefficients c_n transform as

$$c_n(\tau_k + 0) = \sum_{m=-\infty}^{+\infty} i^{n-m}J_{n-m}(P)c_m(\tau_k - 0), \tag{4}$$

where

$$P = \int_{-\infty}^{+\infty} \varepsilon(\tau) d\tau,$$

and $J_n(P)$ is the Bessel function of the nth order. The result of multiple kicks applied at different times can be obtained by combining transformations (4) after each kick with a free evolution according to Eq.(3) between the kicks. If the kicks are applied periodically to the system with the period T_{rev}, the system does not show chaotic behavior, and the energy accumulates quadratically with time (the so-called "quantum resonance" [23, 24]). It is, therefore, quite natural to examine potential accumulation of angular squeezing of the rotor wave function under the "quantum resonance" excitation. In this case, because of the exact quantum revivals at the free-evolution stages, the effect of N kicks of a magnitude P is equivalent to the action of a single strong pulse of strength NP (see, e. g. [23]). In Figure 1, we show numerically calculated time evolution of the probability density $|\Psi(\theta, \tau)|^2$ after a relatively strong kick of a magnitude $P = 85$ applied at $\tau = 0$. Initially the rotor was in the ground s-state ($c_n(0) = \delta_{n0}$). For the chosen values of τ, several distinct phenomena can be seen in these plots. First of all, the wave function shows an extreme narrowing in the region of small θ after some delay following the kick [Fig. 1 (b)]. The physics of this effect may be understood with the help of the following semiclassical arguments. Consider an ensemble of randomly oriented classical rotors subject to a kick. The angular velocity of a rotor located at the angle θ is

$$\omega(\theta) = -P \sin(\theta) \tag{5}$$

just after the kick, assuming negligible initial velocity. For rotors from the region of small $\theta << 1$, the acquired velocity is linearly proportional to their initial angle, so that all of them arrive at the focal point $\theta = 0$ at the same time

$$\tau_f = 1/P. \tag{6}$$

This phenomenon is quite similar to the focusing of light rays by a thin optical lens. For $P >> 1$, the shape of the distribution at the focusing time τ_f is dictated by the aberration mechanism (deviation of the $\cos(\theta)$ potential from the parabolic one), and it is P-independent. We consider the orientation factor $O =< 1 - \cos(\theta) >$ (where angular brackets mean averaging over the state of rotor) as a measure of the rotor orientation. For large enough P, the time-dependent orientation factor (for the initial s-state) may be easily estimated by averaging over the initially uniform classical ensemble of rotors having the velocity distribution of Eq. (5): $O(\tau) = 1 - J_1(P\tau)$. Here $J_1(x)$ is Bessel function of the first order. The minimal value $O \approx 0.418$ of the orientation factor is, in fact, achieved in the post-focusing regime, at $\tau \approx 1.84\tau_f$. As seen in Fig. 1 (c), a new phenomenon can be observed in the angular probability distribution just after the focusing. Sharp singular-like features are formed in the distribution, which are moving with time. Each of these features has a typical asymmetric shape, with pronounced oscillations on one side and an abrupt drop down on the other side. Again, the origin of this effect can be traced in the time evolution of a classical ensemble of initially motionless rotors.

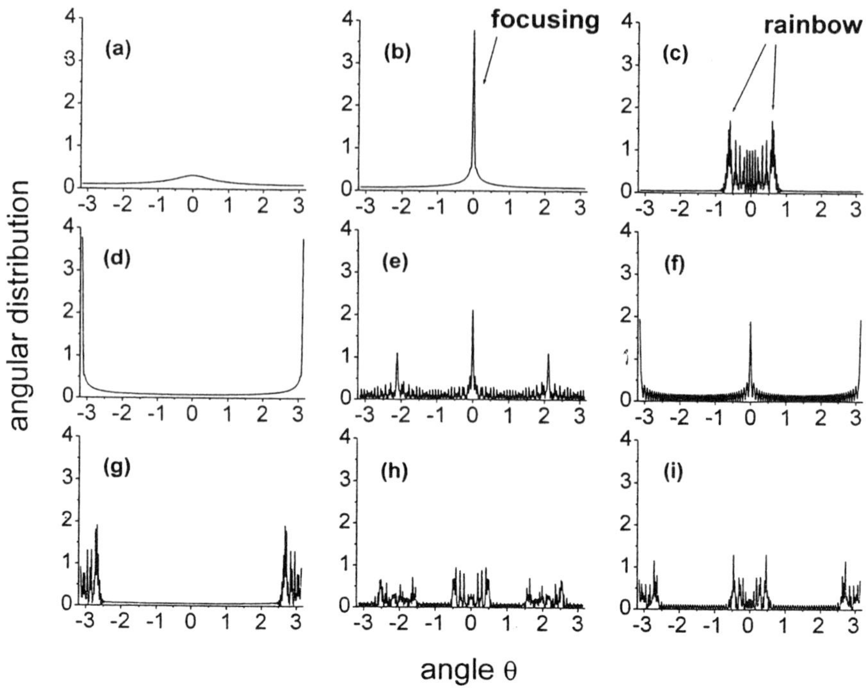

Fig. 1. Angular distribution of a quantum rotor excited by a strong δ-kick ($P = 85$). The graphs correspond to (a) $\tau = 0.5\tau_f$, (b) $\tau = \tau_f$, (c) $\tau = 2\tau_f$, (d) $\tau = \tau_f + T_{rev}/2$, (e) $\tau = \tau_f + T_{rev}/3$, (f) $\tau = \tau_f + T_{rev}/4$, (g) $\tau = 1.8\tau_f + T_{rev}/2$, (h) $\tau = 1.8\tau_f + T_{rev}/3$, and (i) $\tau = 1.8\tau_f + T_{rev}/4$, respectively.

After a kick applied at $\tau = 0$, the motion of the rotors is described by

$$\theta = \theta_0 - P\sin(\theta_0)\tau \ (mod 2\pi) \tag{7}$$

where θ_0 is the initial angle. For $\tau < \tau_f$ Eq.(7) represents a one-to-one mapping $\theta(\theta_0)$ [see Fig. 2 (a)]. At $\tau = \tau_f$ the curve $\theta(\theta_0)$ touches the horizontal axis [Fig. 2 (b)]. At $\tau > \tau_f$ the angle θ_0 becomes a multi-valued function of θ [Figs. 2 (c), (d)]. The classical time-dependent angular distribution function of the ensemble is given by

$$f(\theta, \tau) = \sum_a \frac{f(\theta_0^a, \tau = 0)}{|d\theta/d\theta_0^a|} \tag{8}$$

The summation in Eq. (8) is performed over all possible branches of the function $\theta_0(\theta)$ defined by Eq.(7). It follows immediately from Eq.(8) that even for a smooth initial distribution, $f(\theta, \tau)$ may exhibit a singular behavior near the angles where $d\theta/d\theta_0^a \to 0$. The quantum nature of the rotor motion replaces the classical singularities by sharp maxima in the probability distribution with the Airy-like shape typical to rainbow phenomena. Indeed, this effect is similar to the formation of caustics in the wave optics [25], and rainbow-type scattering in optics and quantum mechanics [26, 27, 28]. We should stress, however, that the long-time asymptotic regimes are radically different for the classical and truly quantum motion of the rotor. Thus, contrary to the classical limit, in which the caustics exist forever, they gradually disappear in the quantum case because of the overall decay of the initial rotational wave packet. On even longer time scale, another quantum phenomenon can be seen, namely revivals and fractional revivals

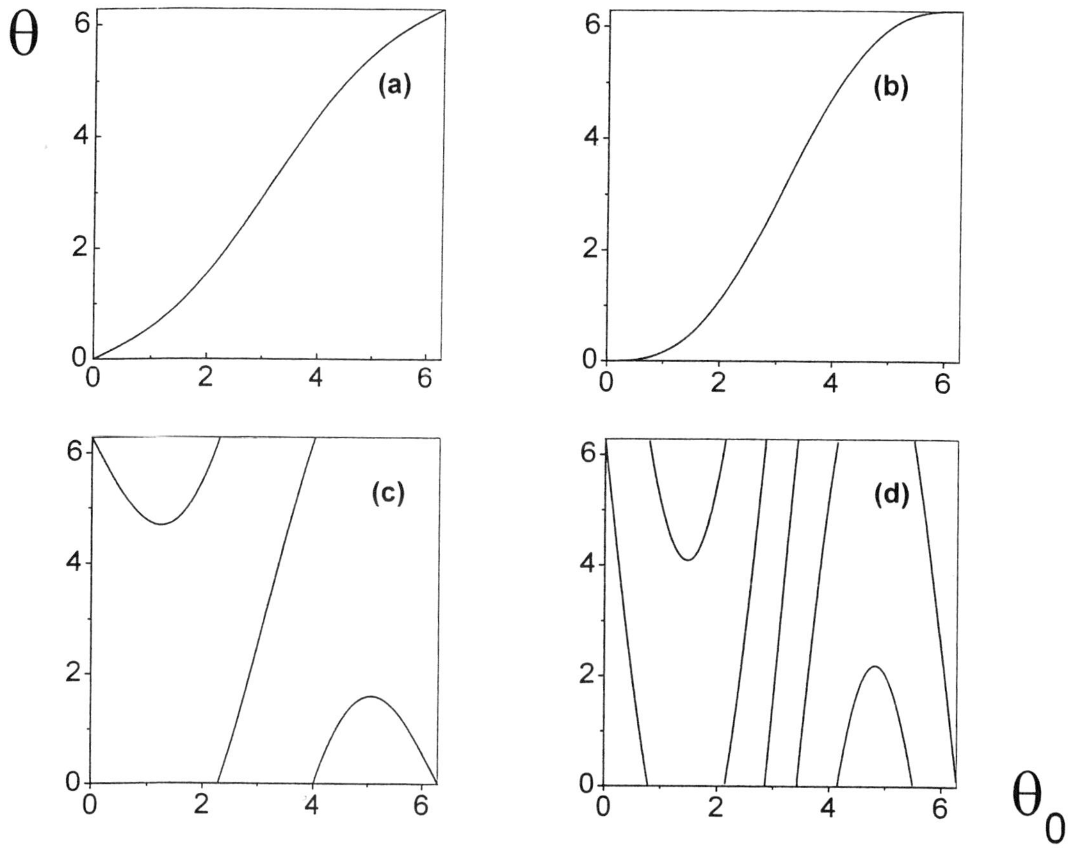

Fig. 2. Classical map representing the final angle θ as a function of initial angle θ_0 for (a) $\tau = 0.5\tau_f$, (b)$\tau = \tau_f$, (c) $\tau = 3\tau_f$, and (d) $\tau = 10\tau_f$.

As we have demonstrated, a mere application of $\delta-$kicks at the condition of "quantum resonance" does not lead to accumulated angular squeezing, and the orientation is saturated at some finite asymptotic level. Here we suggest an excitation scheme that exhibits the desired accumulation property. As previously mentioned, the wave function of the rotor reaches the state of the maximal orientation (i. e., minimal O value) after a certain delay $\Delta\tau_1$ following the application of the first kick at $\tau = \tau_1 = 0$. We suggest to apply the second kick at $\tau_2 = \Delta\tau_1$. Immediately after the second kick, the system will keep the same probability density distribution. On the other hand, $\tau = \tau_2$ will no longer be a stationary point for $O(\tau) =< 1-\cos(\theta) > (\tau)$. The orientation factor $O(\tau)$, and its derivative are continuous and periodic functions of time in the course of a free evolution. Therefore, $O(\tau)$ will reach a new minimum at some point $\tau_2 + \Delta\tau_2$ in the interval $[\tau_2, \tau_2 + T_{rev}]$. Clearly, the new minimal value of the orientation factor is smaller than the previous one. By continuing this way, we will apply short kicks at iterative time instants $\tau_{k+1} = \tau_k + \Delta\tau_k$. By construction of this pulse sequence, the squeezing effect will accumulate with time, in contrast to the "quantum resonance" excitation. This is demonstrated by Fig. 3, which shows calculated sequences $\{\Delta\tau_k\}$ and $\{< 1-\cos(\theta) > (\tau_k)\}$ for a rotor initially in the s-state and being kicked by pulses with $P = 3$. The logarithm of the orientation factor gradually decreases, without any sign of saturation.

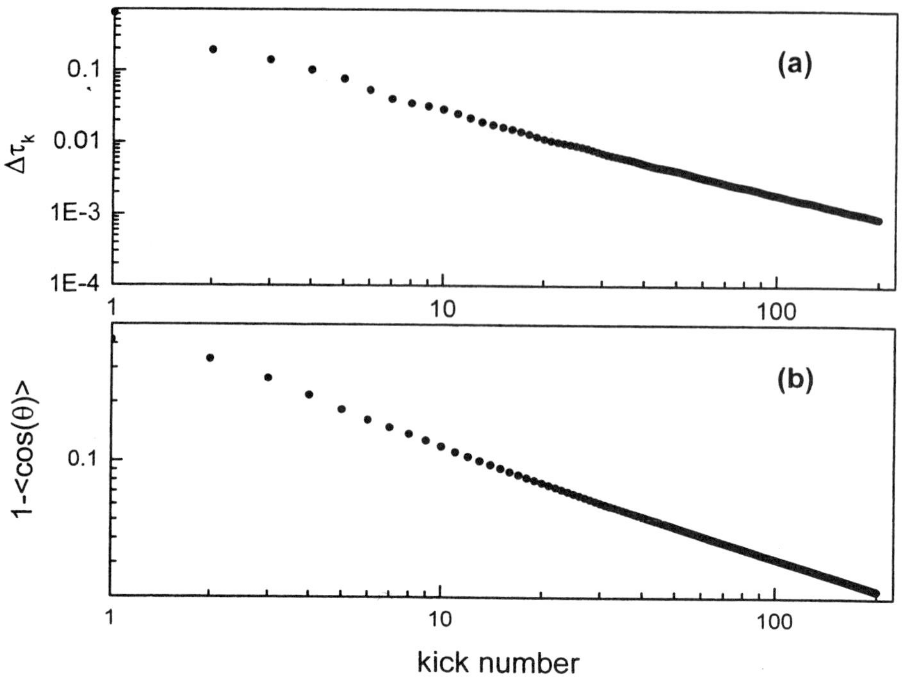

Fig. 3. Accumulative angular squeezing. Graphs are shown in double logarithmic scale.

At the stage of a well-developed squeezing ($O << 1$), the $\cos(\theta)$ -potential may be approximated by a parabolic one. It can be easily shown (both classically and quantum mechanically) that in this limit our strategy provides the following recurrent relationships for the time intervals $\Delta\tau_k$, successive values of the angular variance $u_k =< \theta^2 >_k \approx 2O_k$ and normalized variance of the angular velocity $w_k =< (-id/d\theta)^2 >_k /P^2$:

$$\begin{aligned} \Delta\tau_k &= \frac{u_k}{u_k + w_k} \\ u_{k+1} &= u_k - \frac{u_k^2}{u_k + w_k} \\ w_{k+1} &= w_k + u_k \end{aligned} \tag{9}$$

For large k, the last two finite-difference equations may be replaced by a system of coupled differential equations. The latter has an exact solution providing the following asymptotics: $< \theta^2 >_k \propto 1/\sqrt{k}$ and $\Delta\tau_k \propto 1/k$. This result is in good agreement with the numerically observed power-laws behavior of the graphs 3 (a) and 3 (b), and it describes correctly their slopes at $k >> 1$. We note, that in contrast to the wave optics (in which the size of the focal spot is diffraction limited), our system may be, in principle, "unlimitedly" squeezed in angle. We also note that a quasi-periodic sequence of kicks applied at $\tau_{k+1} = \tau_k + \Delta\tau_k + T_{rev}$ provides the same squeezing scenario for a quantum rotor. The introduction of the T_{rev}-shift between pulses may be useful in the practical realizations of the scenario to avoid the overlap between short excitation pulses of a finite duration.

3. Squeezing of atoms in a pulsed optical lattice

In the present Section we apply the above results to another well-known system : cold atoms interacting with a pulsed optical lattice [19]. Optical lattices are periodic potentials for neutral atoms induced by standing light waves formed by counter-propagating laser beams. When these waves are detuned from any atomic resonance, the ac Stark shift of the ground atomic state leads to a conservative periodic potential with spatial period $\lambda/2$, half the laser wavelength (for a review, see, e. g. [29]). Such lattices present a convenient model systems for solid state physics and nonlinear dynamics studies. In contrast to traditional solid state objects, the parameters of optical lattices (i. e. lattice constant, potential well depth, etc.) are easily controllable. Many fine phenomena that were long discussed in solid state physics, have been recently observed in corresponding atom optics systems. Bloch oscillations [30], or the Wannier-Stark ladder [31, 32] are only few examples to mention. Time-modulation of the frequency and intensity of the constituent laser beams provide tools for effective modeling of numerous time-dependent nonlinear phenomena. Since the initial proposal [33], and first pioneering experiments on atom optics realization of the δ-kicked quantum rotor [2], cold atoms in optical lattices provide also new grounds for experiments on quantum chaos.

We describe atoms as two-level systems with transition frequency ω_0, interacting with a standing optical wave that is linearly polarized and has the frequency of ω_l. If the detuning $\Delta_l = \omega_0 - \omega_l$ is large compared to the relaxation rate of the excited atomic state, the internal structure of the atoms can be neglected and they can be regarded as point-like particles. In this approximation, the Hamiltonian for the atomic motion is

$$H(x,p,t) = \frac{p^2}{2m} - V(t)\cos(2k_l x)\,, \tag{10}$$

where m is the atomic mass, $k_l = \omega_l/c$ is the wave number of the standing wave. The depth of the potential produced by the standing wave is $V(t) = \hbar\Omega(t)^2/8\Delta_l$, where $\Omega(t) = 2|d_z E(t)|/\hbar$ is the Rabi frequency, $\vec{d}$ is the atomic dipole moment and $E(t)$ is the time-dependent amplitude of the light field. In the case of rather short laser pulses, this Hamiltonian corresponds to that of the two-dimensional δ-kicked rotor considered in the previous Section, with $\theta = 2k_l x$. In accordance with the previous discussion, many aspects of the dynamics of this system can be explained with semiclassical arguments. Here we provide results for a classical description of atoms in a pulsed optical lattice, with thermal effects taken into account.

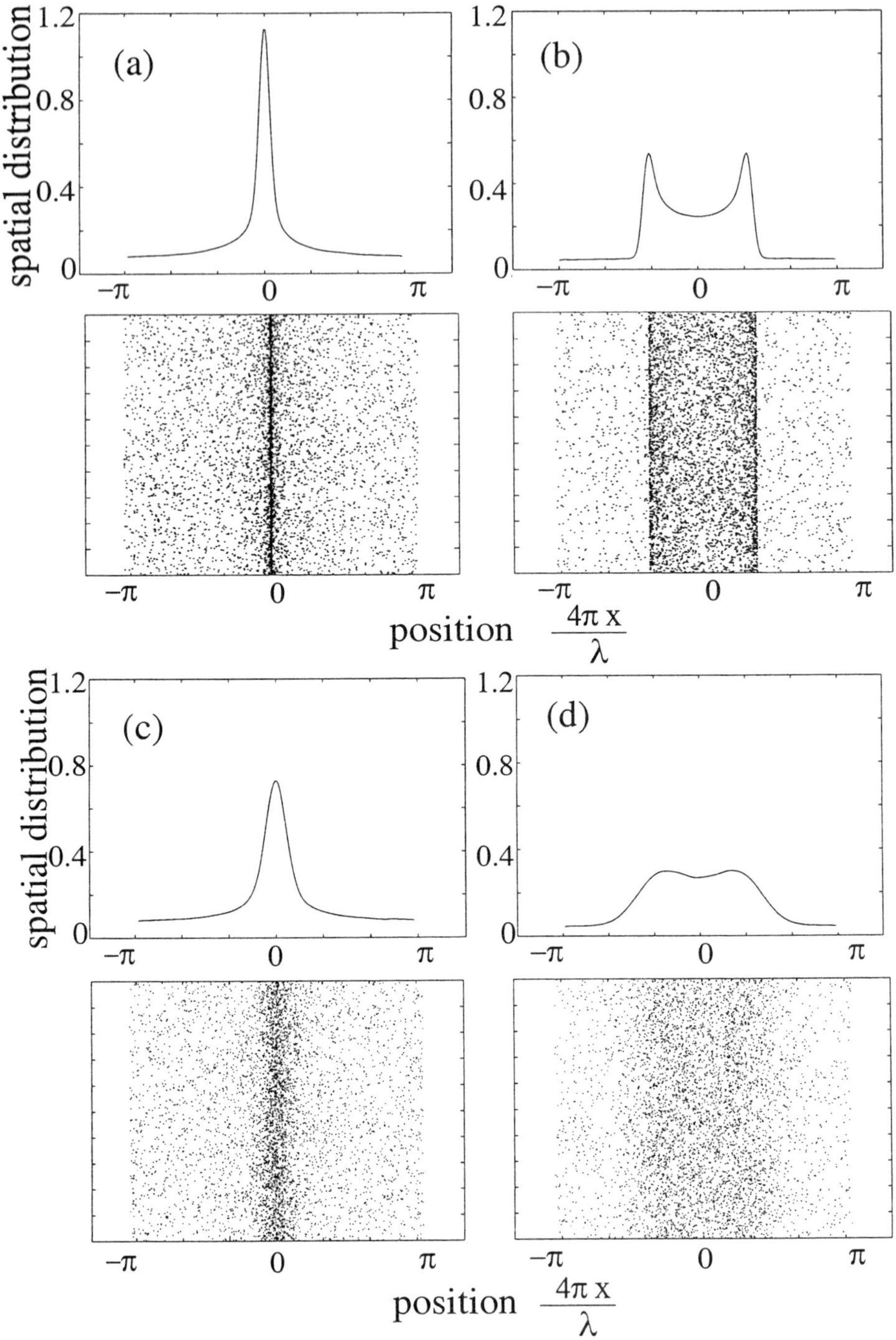

Fig. 4. Spatial distribution of a classical ensemble of atoms after a single δ-kick. In Figs. (a) and (b), the initial temperature, T of the ensemble is zero. In (c) and (d), $k_B T = (1/9)k_l^2[\int_{-\infty}^{\infty} V(t)dt]^2/m$, where $V(t)$ is the depth of the potential produced by the standing wave, and k_B is Boltzmann constant. Figures (a) and (c) show the spatial distribution at focal time $\tau = \tau_f$, while in figure (b) and (c) $\tau = 2.5\tau_f$.

We performed a Monte-Carlo simulation of the dynamics of an ensemble of δ-kicked particles which were initially uniformly distributed in space and had a thermal momentum distribution. Figure 4 shows the spatial distribution of atoms at different times after a single kick. In the upper row, the initial temperature of the ensemble is zero, and we can observe focusing [Fig. 4(a)] and formation of caustics [Fig. 4(b)] as it was described in Section II. In Fig. 4(c) and (d), the average thermal energy of the initial ensemble is comparable with the energy supplied by a kick. As a result, instead of sharp peaks in the spatial distribution (like in Figure 4 (a) and (b)), we observe some broader spatial structures that are still reminiscent of the focusing phenomenon and the rainbow effect.

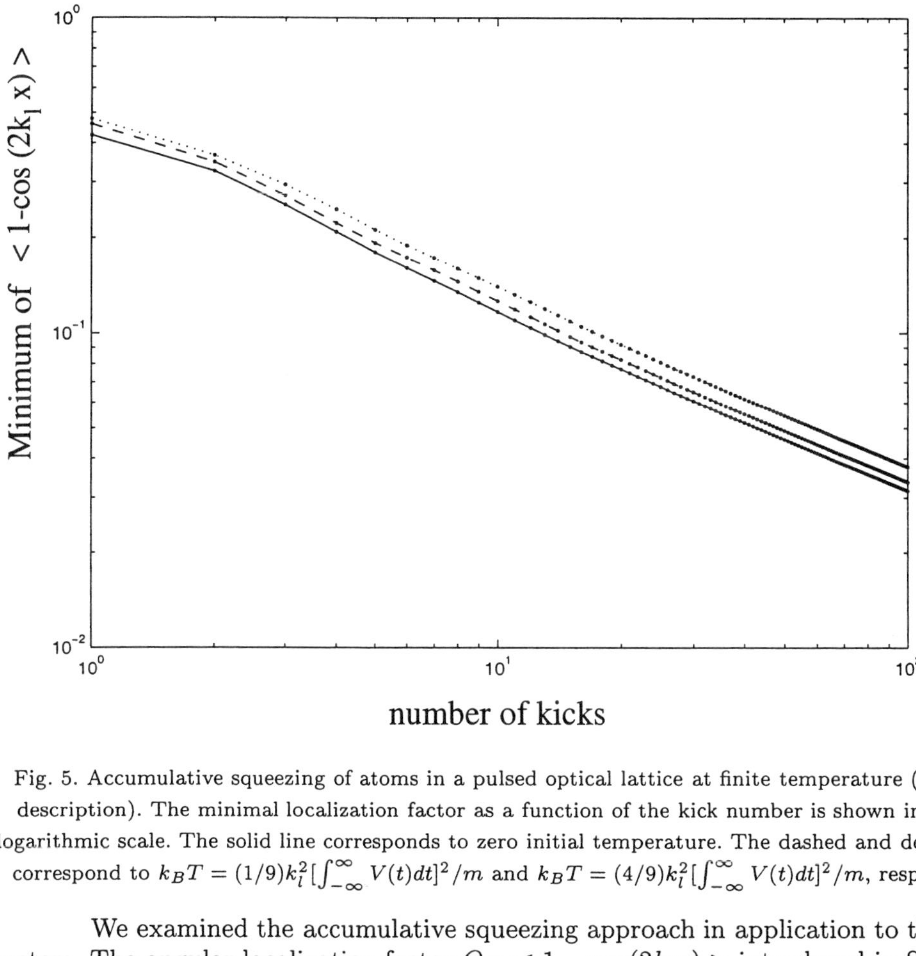

Fig. 5. Accumulative squeezing of atoms in a pulsed optical lattice at finite temperature (classical description). The minimal localization factor as a function of the kick number is shown in double logarithmic scale. The solid line corresponds to zero initial temperature. The dashed and dotted lines correspond to $k_BT = (1/9)k_l^2[\int_{-\infty}^{\infty} V(t)dt]^2/m$ and $k_BT = (4/9)k_l^2[\int_{-\infty}^{\infty} V(t)dt]^2/m$, respectively.

We examined the accumulative squeezing approach in application to the above system. The angular localization factor $O =< 1 - \cos(2k_lx) >$ introduced in Section II describes now the spatial width of atomic groups localized in the minima of the light-induced potential. In the classical limit, the localization factor after applying n kicks is given by

$$O(t_n) = 1 - \frac{1}{2\pi}\int_{-\infty}^{\infty} d\omega_0 \int_{-\pi}^{\pi} d\theta_0 \rho(\omega_0)\cos\theta_n, \tag{11}$$

where θ_n is determined by

$$\theta_n = \theta_{n-1} + \Delta\tau_n \left(\omega_0 - \sum_{i=0}^{n-1}\sin\theta_i\right). \tag{12}$$

Here, $\omega_0 = 2k_lv_0$, where v_0 is the initial velocity of an atom, and $\rho(\omega_0)$ describes the thermal distribution of the velocities. Figure 5 displays the amount of spatial squeezing for the series of 100 kicks for different initial temperatures. It can be seen, that after the first few kicks the localization factor demonstrates a negative power dependence as a function of the kick number (a straight line in the double logarithmic scale). Although, the system demonstrates a reduced squeezing for higher initial temperatures, the slope of all of the curves in Figure 5 is the same after the first several kicks, in full agreement with the general arguments of the previous Section (see Eqs.(9). Therefore, the accumulative squeezing scenario may be an effective and regular strategy for atomic localization even at finite temperatures.

However, this does not mean that the accumulative squeezing is the only one (or the most effective) squeezing strategy. We have studied the best localization that can be achieved with *a given number* of identical δ-kicks, by minimizing the localization factor Eq.(11) with respect to the delay times $\Delta\tau_n$ between the kicks. For clarity, we present here only the results for zero initial temperature of the atoms. Table 1 shows the best values of the localization factor found for up to five kicks, and compares them with the results of the accumulative squeezing strategy with the same number of kicks. While the maximal atomic localization that can be achieved with two pulses is almost the same for accumulative squeezing and for the optimal sequence of two pulses, the optimized results for three and more pulses are much better.

For illustration, we choose the sequence of four optimal pulses to visualize the dynamics behind the localization process. Figure 6 (a) shows the spatial distribution of the ensemble of atoms at the time of arrival of the second pulse. Note that the second pulse is not applied at the time of the maximal localization. On the contrary, the optimized procedure finds it favorable to wait after the focusing event, until the distribution becomes rather broad. The optimal four-pulse sequence we found requires applying the third and the forth pulses simultaneously, thus producing an effective "double pulse". The spatial distribution at the time of the combined third and fourth pulses can be seen in Figure 6 (b). The distribution is only slightly more localized than in Fig. 6(a), and only the last pulse (with the double strength) squeezes the ensemble at the time of the maximal localization, thus bringing most of the atoms to the optical lattice minima [see Fig. 6 (c)]. For illustration, we choose the sequence of four optimal pulses to visualize the dynamics behind the localization process. Figure 6 (a) shows the spatial distribution of the ensemble of atoms at the time of arrival of the second pulse. Note that the second pulse is not applied at the time of the maximal localization. On the contrary, the optimized procedure finds it favorable to wait after the focusing event, until the distribution becomes rather broad. The optimal four-pulse sequence we found requires applying the third and the forth pulses simultaneously, thus producing an effective "double pulse". The spatial distribution at the time of the combined third and fourth pulses can be seen in Figure 6 (b). The distribution is only slightly more localized than in Fig. 6(a), and only the last pulse (with the double strength) squeezes the ensemble at the time of the maximal localization, thus bringing most of the atoms to the optical lattice minima [see Fig. 6 (c)].

No of kicks	O_{acc}	O_{opt}
2	0.33	0.31
3	0.26	0.20
4	0.21	0.11
5	0.18	0.07

Table 1. The table shows the minimal value of the localization factor that can be achieved with a fixed number of kicks using accumulative squeezing scenario (O_{acc}) and optimized sequence of pulses (O_{opt}).

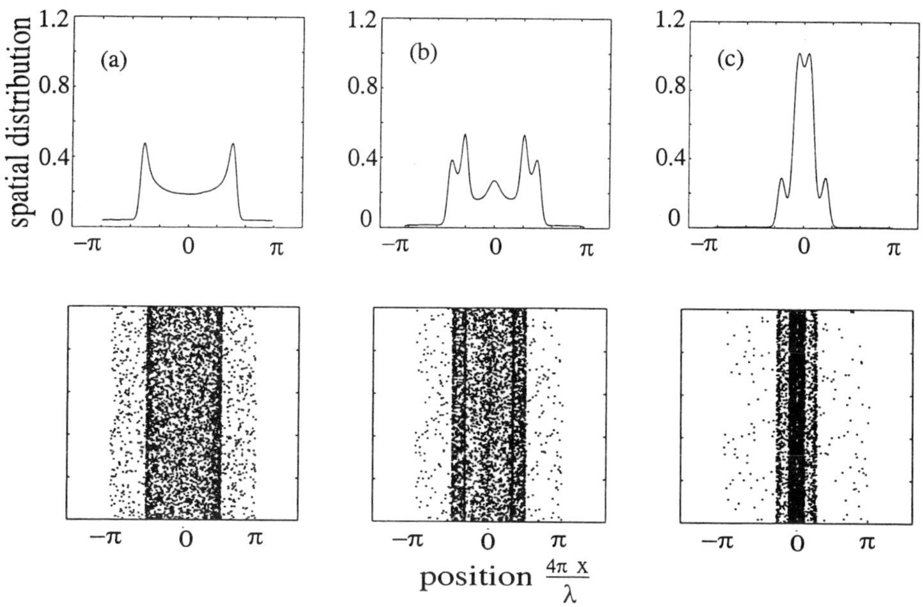

Fig. 6. Spatial distribution for the optimized sequence of four δ-pulses. The upper row shows the spatial distribution averaged over 100 atomic ensembles (each containing 5000 atoms), the lower figures show the distribution of atoms in one of the ensembles. In (a), the distribution is plotted at the time of the second pulse, that is delayed by $\tau = 3.02\tau_f$ with respect to the first pulse, in (b) - at the time of the (combined) third and fourth pulses delayed by $\tau = 1.35\tau_f$ after the second pulse. Figure (c) shows the distribution of atoms at the time of the maximal squeezing.

4. Orientation and alignment of a 3D rotor

Under certain conditions, the process of molecular orientation (or alignment) by laser fields can be described by a strongly driven three-dimensional rigid rotor model. Although many features in the dynamics of 3D rotors are similar to those already discussed for the two-dimensional case, there are two principal differences that we want to emphasize. The first one may be traced in the evolution of a classical ensemble of 3D-rotors being initially at zero temperature. The probability of finding a rotor (driven by a linearly polarized field) at a certain solid angle element $\sin\theta d\theta$ is determined by the initial distribution function, $f_0(\theta_0)$ as follows:

$$f(\theta)\sin\theta d\theta = f_0(\theta_0)\sin\theta_0 d\theta_0. \tag{13}$$

Therefore, the probability density of finding a rotor at the angle θ at some latter time is

$$f(\theta) = \sum_a \frac{f_0(\theta_0^a)\sin\theta_0^a}{|d\theta/d\theta_0^a|\sin\theta}, \tag{14}$$

where the summation is done over all branches of the $\theta_0(\theta)$ function (see Section II). The probability density $f(\theta)$ has a singularity if one of the factors in the denominator is zero. As in the 2D case, the zeroes of the fraction $|d\theta/d\theta_0^a|$ give rise to the formation of the rainbow-like structures. But in addition, the geometrical factor $\sin\theta$ can get zero too, which leads to the additional singularities at $\theta = 0$ and $\theta = \pi$. This kind of singularities are responsible for the formation of the *corona* and *glory* effects in the optical and quantum-mechanical scattering [26, 27].

The second difference appears at finite initial temperature of the ensemble of 3D rotors. Because of the conservation of the angular momentum projection onto the field polarization direction, an effective repulsive centrifugal force appears that prevents the rotors from reaching the exact $\theta = 0$ and $\theta = \pi$ orientations. As a result, two *holes* in the angular distribution of a driven 3D thermal ensemble should be always present at $\theta = 0$ and $\theta = \pi$ [34].

We present below the results of Monte-Carlo simulation of the dynamics of a classical ensemble of three-dimensional rotors (linear molecules) driven by a linearly polarized time-dependent field. The corresponding Hamiltonian is

$$H = \frac{1}{2mr^2}\left(p_\theta^2 + \frac{p_\phi^2}{\sin^2\theta}\right) - \mu\mathcal{E}(t) r\cos\theta, \tag{15}$$

where m is the reduced mass of the molecule, r is the (fixed) distance between the atoms, μ is permanent dipole moment, θ and ϕ are Euler angles, and p_θ and p_ϕ are related canonical momenta. For the driving field described by a $\delta-$ pulse, the equations of motion can be easily integrated (see [35]).

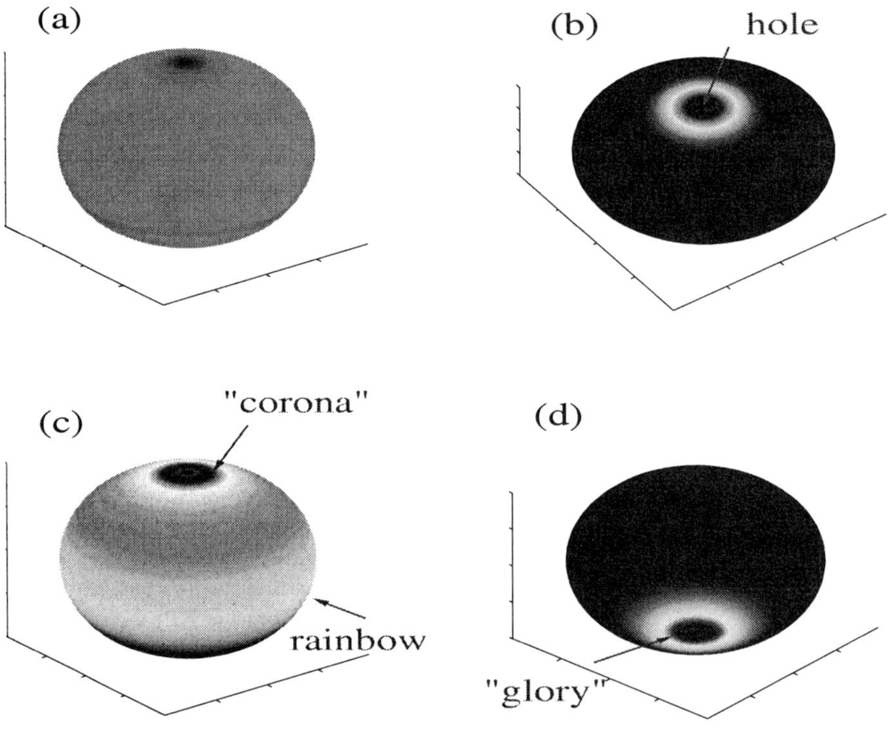

Fig. 7. Time evolution of the angular distribution of a (classical) ensemble of δ-kicked 3D rotors at finite temperature. The probability of finding a rotor inside the solid angle element $\sin\theta d\theta d\phi$ is plotted on a sphere. Picture (a) shows the distribution at $\tau = 0$ (time of the kick). Figures (b), (c), and (d) correspond to $\tau = \tau_f$, $\tau = 3.3\tau_f$ and $\tau = 5\tau_f$, respectively. The initial temperature corresponds to $k_B T = (1/100)\mu^2 r^2 [\int_{-\infty}^{\infty} \mathcal{E}(t)dt]^2/I^2$.

At zero initial temperature, $p_{\theta_0} = 0$ and $p_{\phi_0} = 0$, and the angle $\phi(t)$ is an additional constant of motion. In this case, the dynamics of the system is reduced to that of a two-dimensional rotor, beside the effect of the geometrical factor described above. Figure 7 demonstrates the time evolution of the angular distribution of an ensemble with finite initial temperature. Here, the probability of finding a rotor inside the solid angle element $\sin\theta d\theta d\phi$ is plotted on a sphere. Red color corresponds to the high probability density while the blue color presents a low probability. Initial isotropic ensemble corresponds to an uniform solid angle distribution [see Fig. 7 (a)]. In Figure 7 (b), the distribution is plotted at the "focal time" defined according to Eq.(6). In the two-dimensional case considered in the previous Sections, the distribution at the focal time is characterized by a sharp peak at $\theta = 0$ and a broad background for larger values of θ. In the three-dimensional case, the distribution has an additional sharp dip at $\theta = 0$, as discussed above. Figures 7 (c) and (d) show the angular distribution at $t > t_f$. In Fig. 7 (c) the formation of the "*corona*" around $\theta = 0$ can be seen. In addition, we can observe a ring of relatively high probability moving from the north pole to the south pole of the sphere. This ring is a three-dimensional analog of the rainbow structure modified by the thermal effects. After the ring arrives at the south pole, a singular feature appears around $\theta = \pi$, with a hole in the center caused the repulsive centrifugal force [Fig. 7 (d)]. The formation of this sharp and robust structure in the angular distribution is analogous to the glory effect in the wave optics.

All the described phenomena, namely, focusing, caustics creation, and accumulating squeezing are not specific for the simplest models considered above, but are rather common features that can be observed under general conditions of a strong excitation of the quantum rotor. Indeed, for strong enough driving field one may neglect the initial rotational energy stored in the rotor, and use the above quasiclassical ideas to describe its dynamics. The role of the initial state of the rotor (even thermally averaged) is reduced to the formation of a frozen classical-like initial angular distribution of the rotational ensemble. For example, for the dipole-type interaction, Eq. (2), the angular focusing may be analyzed by considering classical dynamics at small angles ($\theta \ll 1$):

$$\left[\frac{d^2}{d\tau^2} + \epsilon(\tau)\right]\theta = 0. \tag{16}$$

Here, $\tau = t\hbar/I$, and $\epsilon = \mu\mathcal{E}(t)I/\hbar^2$, and the initial condition is $\theta(\tau_0) = \theta_0$, $d\theta(\tau_0)/d\tau = 0$ (where τ_0 is any moment in the past before the beginning of the pulse). Focusing time τ_f is defined by $\theta(\tau_f) = 0$. Because of the linearity of Eq.(16), the position of the focusing times, and number of focusing events do not depend on θ_0, but are determined only by the properties of the pulse $\epsilon(\tau)$. This boundary problem may be solved analytically in several special cases only (for δ-pulse considered above, or for a step-like $\epsilon(\tau)$ dependence). In general, determination of focusing times requires numerical solution of Eq.(16). Based on the previous analysis, we expect that a new angular rainbow appears immediately after each focusing event.

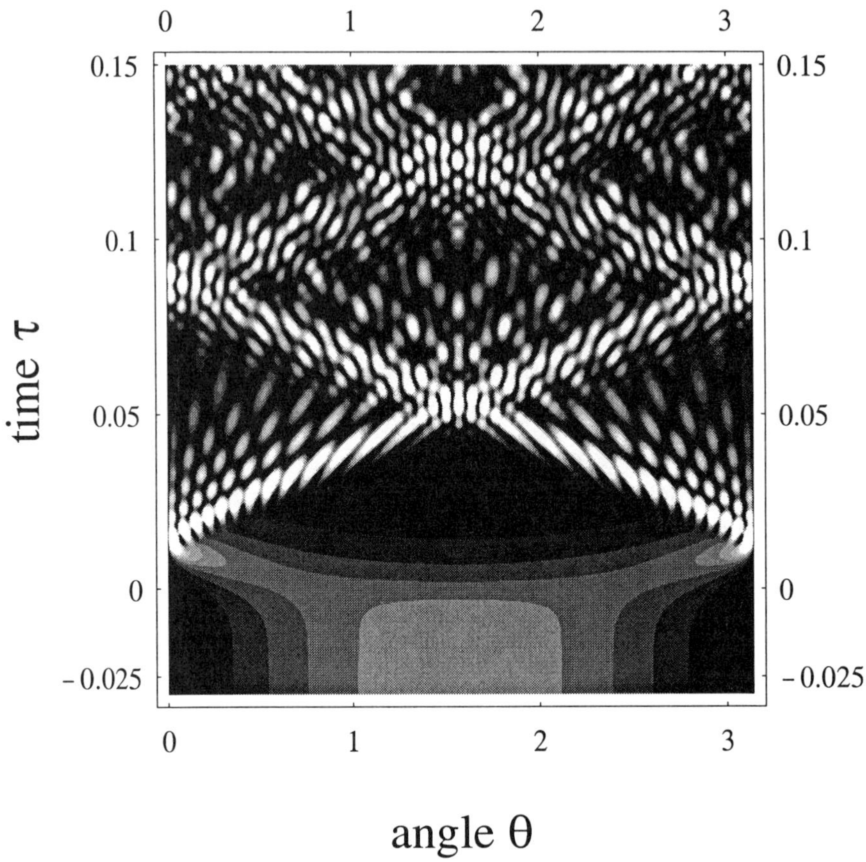

Fig. 8. Contour plot for the time-dependent angular distribution function $2\pi\sin(\theta)|\Psi(\theta,\tau)|^2$ of a three-dimensional rotor (molecule) subject to a strong "polarization-type" interaction, Eq.(17) with a Gaussian pulsed laser field: $\mathcal{E}^2(t)(\alpha_{\|}-\alpha_{\perp})I/4\hbar^2 = 3\times 10^3\exp[-(\tau/0.01)^2]$. Here $\tau = t\hbar/I$. The molecule resides initially in the isotropic ground angular state $(J=0, m=0)$. Angular focusing and rainbows emerging from each of the focal regions can be seen. The angular rainbows are made of moving rings located symmetrically with respect to the equatorial plane.

Figure 8 shows numerically calculated time-evolution of a three-dimensional *quantum* rotor (linear molecule) coupled to the external laser field via the anisotropic polarization interaction. The time averaged value of this interaction is (see, e. g.[21, 9])

$$V(\theta,t) = -1/4\mathcal{E}^2(t)[(\alpha_{\|}-\alpha_{\perp})\cos^2(\theta)+\alpha_{\perp}] \tag{17}$$

Here $\alpha_{\|}$ and $\alpha_{\perp}$ are the components of the polarizability, parallel and perpendicular to the molecular axis, and $\mathcal{E}(t)$ is the *envelope* of the laser pulse. In this case there are two opposite equilibrium directions $\theta = 0, \pi$, so the field aligns the molecules but not orients them. The applied pulse has a Gaussian shape of a finite duration (i. e. it is not a δ-kick). As seen from Fig. 8, both focusing and caustics formation occur in the system, similar to the previous consideration, with the differences attributed to another symmetry of the problem. For instance, the angular rainbows are made of moving rings located symmetrically with respect to the equatorial plane. Note, that the glory and corona are not seen because of the $\sin(\theta)$ factor incorporated into the distribution function shown in Fig. 8.

5. Conclusions

The predicted effects may be observed in a wide range of systems with strongly driven rotational degrees of freedom. Possible examples range from heavy-ion collisions (when highly excited wave packets of nuclear rotational states are produced [12]) to molecules subject to strong laser pulses, and cold atoms trapped by standing light waves. The spectacular features described in this paper may be observed in the spatial distribution of an atomic ensemble driven by pulsed optical lattices. Recently, the accumulated squeezing scenario [18] has been realized experimentally in this system [36]. Moreover, the related squeezing approaches may find application in atom lithography of ultra-high resolution [37, 38]. In the case of molecules, the considered effects may reveal themselves in the angular distribution of fragments produced by intense laser-field molecular interaction. The most direct evidence can be achieved in a two-pulse experiment, in which the first strong non-resonant pulse attempts to orient (align) the molecular ensemble, while the second short delayed pulse creates fragment ions.

References

[1] F. Haake, *"Quantum Signatures of Chaos"*, Springer-Verlag, (Berlin, 1991)

[2] F. L. Moore, J. C. Robinson, C. F. Bharucha, B. Sundaram, and M. G. Raizen, Phys. Rev. Lett. **75**, 4598 (1995)

[3] H. Ammann, R, Gray, I. Shvarchuck, and N. Christensen, Phys. Rev. Lett. **80**, 4111 (1998)

[4] D. Normand, L. A. Lompre, and C. Cornaggia, J. Phys. **B** 25, L497 (1992)

[5] P. Dietrich, D. T. Strickland, M. Laberge, and P. B. Corkum, Phys. Rev. A **47**, 2305 (1993)

[6] G. R. Kumar *et al.*, Phys. Rev. A **53**, 3098 (1996)

[7] J. Karczmarek, J. Wright, P. Corkum, and M. Ivanov, *Phys. Rev. Lett.* **82**, 3420 (1999).

[8] B. A. Zon and B. G. Katsnelson, Sov. Phys. JETP 42, 595 (1976)

[9] B. Friedrich and D. Herschbach, Phys. Rev. Lett. **74**, 4623 (1995); J. Phys. Chem. **99**, 15686 (1995).

[10] W. Kim and P. M. Felker, J. Chem. Phys. **104**, 1147 (1996)

[11] C. H. Lin, J. P. Heritage, and T. K. Gustafson, Appl. Phys. Lett. **19**, 397 (1971); J. P. Heritage, T. K. Gustafson, and C. H. Lin, Phys. Rev. Lett. **34**, 1299 (1975).

[12] L. Fonda, N. Mankoč-Borštnik, and M. Rosina, Phys. Rep. **158**, 159 (1988)

[13] P. M. Felker, J. Phys. Chem. **96**, 7844 (1992)

[14] T. Seideman, J. Chem. Phys. **103**, 7887 (1995); *ibid.* **106**, 2881 (1997)

[15] T. Seideman, Phys. Rev. Lett. **83**, 4971 (1999)

[16] J. Ortigoso *et al.*, J. Chem. Phys. **110**, 3870 (1999)

[17] L. Cai, J. Marango and B. Friedrich, Phys. Rev. Lett. **86**, 775 (2001)

[18] I. Sh. Averbukh and R. Arvieu, Phys. Rev. Lett. **87**, 163601 (2001)

[19] M. Leibscher and I. Sh. Averbukh, submitted, 2001

[20] I. Sh. Averbukh, R. Arvieu, and M. Leibscher, to be published

[21] R. W. Boyd, *Nonlinear Optics*, Academic Press, (Boston,1992)

[22] I. Sh. Averbukh and N. F. Perelman, Phys. Lett. A **139**, 449 (1989); Sov. Phys. JETP **69**, 464 (1989)

[23] F. M. Izrailev *et al.*, Proc. Conf. on Stochastic Behavior in Classical and Quantum Hamiltonian Systems, Como, Italy (Springer-Verlag, Berlin, 1977), p.334

[24] F. M. Izrailev and D. L. Shepelyansky, Dokl. Akad. Nauk SSSR **249**, 1103 (1979); Sov. Phys. Dokl. **24**, 996 (1979)

[25] Yu. A. Kravtsov and Yu. I. Orlov, *"Caustics, Catastrophes and Wave Fields"* (Springer Series on Wave Phenomena, 15), 2nd edition, (Springer-Verlag, Berlin, 1999).

[26] K. W. Ford and J. A. Wheeler, Ann. Phys. **7**, 259 (1959)

[27] M. V. Berry, Adv. Phys. **25**, 1 (1976)

[28] S. D. Bosanac, J. Chem. Phys. **95**, 5732 (1991)

[29] P. S. Jessen and I .H. Deutsch, Adv. Atm Mol. Opt. Phys. **37**, 95 (1996)

[30] M. B. Dahan, E. Peik, J. Reichl, Y. Castin, and C. Salomon, Phys. Rev. Lett. **76**, 4508 (1996)

[31] S. R. Wilkinson, C. F. Bharucha, K. W. Madison, Qian Niu, and M. G. Raizen, Phys. Rev. Lett. **76**, 4512 (1996)

[32] Qian Niu, Xian-Geng Zhao, G. A. Gerogakis, and M. G. Raizen, Phys. Rev. Lett. **76**, 4504 (1996)

[33] R. Graham, M. Schlauutmann, and P. Zoller, Phys. Rev. Lett. **45**, R19 (1992)

[34] B. A. Zon, Eur. Phys. J. **D8**, 377 (2000)

[35] H. Goldstein, *Classical Mechanics* (Addison-Wesley, Reading, 1980)

[36] M. Raizen (private communication)

[37] I. Sh. Averbukh, patent pending

[38] for a recent review of atom lithography see, e. g. a special issue on nanomanipulation of atoms, Appl. Phys. B **70**, issue 5 (2000) (D. Meschede and J. Mlynek eds)

Retrodictive Quantum Optics

Stephen M. Barnett[(1)], John Jeffers[(1)] and David T. Pegg[(2)]

(1)Department of Physics and Applied Physics, University of Strathclyde, Glasgow G4 0NG, UK
(2)Faculty of Science, Griffith University, Brisbane Q4111, Australia
steve@phys.strath.ac.uk

Abstract: Retrodiction is making statements about the past based on current knowledge. In quantum theory this means assigning the state on the basis of a *later* measurement event as opposed to the more usual *earlier* preparation event. Quantum retrodiction provides new insights into familiar phenomena within quantum optics and is the natural way to analyze quantum communication systems.

1 Introduction

Prediction concerns statements about the future based on presently avaliable knowledge. There are many familiar examples of predictive disciplines including meteorology, share dealing and most of physics. Retrodiction [1, 2] (or postdiction [3]) makes statements about the past based on present knowledge. Although less familiar, there are many retrodictive pursuits including archaeology, forensic science and (because of the finite speed of light) cosmology. It is usually the case that we are unable to make absolute statements and so must settle for probabilistic statements. This is true for both prediction and retrodiction.

It is helpful to formulate retrodiction in terms of a communications problem that can be formulated as follows. Alice wishes to send a message to Bob. She selects one of an agreed set of possible messages M_i to send to Bob who detects a signal S_j. Alice knows the message and can make a *prediction* of the signal detected by Bob in the form of the conditional probability $P(S_j|M_i)$. Bob's problem is to determine the message selected by Alice given his detected signal. This means making a *retrodiction* of the message in the form of the conditional probability $P(M_i|S_j)$. These conditional probabilities are related by Bayes' theorem [4]:

$$P(S_j, M_i) = P(S_j|M_i)P(M_i) = P(M_i|S_j)P(S_j). \tag{1}$$

Here $P(S_j, M_i)$ is the probability that message M_i is selected and that signal S_j is detected, while $P(M_i)$ and $P(S_j)$ are respectively the *a priori* probabilities that message M_i is selected and that signal S_j is detected.

Quantum mechanics is usually formulated as predictive theory in that the state of a system is assigned on the basis of a preparation event. This state evolves in time and is used to make predictions about the outcome of a later measurement. For a quantum communications system this means calculating the predictive conditional probability $P(S_j|M_i)$. This approach can be combined with Bayes' theorem to obtain the retrodictive conditional probability $P(M_i|S_j)$. An alternative approach is to employ retrodictive quantum mechanics in which the state between the events of preparation and measurement is assigned on the basis of the measurement outcome. As an example, consider a Stern-Gerlach experiment in which a spin is prepared with $S_z = +\hbar/2$ at time t_0 and a subsequent measurement at time t_1 finds the result $S_x = +\hbar/2$. The predictive

Coherence and Quantum Optics VIII
Edited by Bigelow et al., Kluwer Academic/Plenum Publishers, 2003

formalism assigns the state $|\uparrow\rangle$ to the spin in the time between the preparation and measurement events. The retrodictive formalism, however, assigns the state $|\rightarrow\rangle$ to the spin.

Retrodictive quantum mechanics can be applied to problems with time evolution of the state by means of time-reversed Schrödinger equation. Consider first the situation in which Alice selects, with equal prior probabilities, one of the eigenstates $|i\rangle$ of some non-degenerate observable and sends, at time t_0 a system prepared in this state to Bob. At some later time t_1, Bob measures an observable and finds the result f associated with the non-degenerate eigenstate $|f\rangle$ of the measured observable. At the time t between preparation and measurement, the state of the system in the predictive formalism is $\hat{U}(t,t_0)|i\rangle$, where $\hat{U}(t,t_0) = \exp[-i\hat{H}(t-t_0)/\hbar]$ is the unitary time-evolution operator. The probability that Bob will find the result f, given that Alice selected the state $|i\rangle$ is given by the familiar expression:

$$P(f|i) = \left|\langle f|\hat{U}(t_1,t_0)|i\rangle\right|^2 . \tag{2}$$

Bob knows his measurement result, f, and can use retrodictive quantum mechanics to calculate the probability that Alice selected i. He assigns the state $\hat{U}^\dagger(t,t_1)|f\rangle$ to the system at the time t between preparation and measurement. The probability that Alice selected the state $|i\rangle$ given that Bob found the result f is

$$P(i|f) = \left|\langle i|\hat{U}^\dagger(t_1,t_0)|f\rangle\right|^2 . \tag{3}$$

Alternatively, Bob could calculate this probability using Bayes' theorem and Eq. (2).

It is not necessary to restrict ourselves to situations in which Alice selects randomly from a complete orthonormal basis set. Suppose that Alice selects from a known but arbitrary (not necessarily orthogonal) set of states $\{|i\rangle\}$ with prior probabilities $P_p(i)$. In this case, if Bob's measurement finds the result f he still assigns the state $\hat{U}^\dagger(t,t_1)|f\rangle$ to the system prior to the measurement and now calculates the probability that Alice selected i using

$$P(i|f) = \frac{\left|\langle i|\hat{U}^\dagger(t_1,t_0)|f\rangle\right|^2 P_p(i)}{\sum_{i'}\left|\langle i'|\hat{U}^\dagger(t_1,t_0)|f\rangle\right|^2 P_p(i')} . \tag{4}$$

This result is fully consistent with Bayes' theorem. Indeed, it may be shown that the retrodictive formalism can be *derived* from predictive quantum mechanics plus Bayes' theorem [5].

2 Quantum optical examples

Assigning states on the basis of measurement results leads to a physical picture that is very different from that associated with conventional predictive quantum theory. We illustrate this point with three simple examples involving the interaction of a single atom with the electromagnetic field. A more complete discussion of these can be found in [2].

2.1 *Spontaneous emission*

Consider a single two-level atom prepared at time t_0 in its excited state $|e\rangle$ with the surrounding electromagnetic field in its vacuum state. At a later time, t_1, a photon is detected at a distance d from the atom. Naturally, causality requires that $t_1 > t_0 + d/c$. Conventional predictive quantum theory assigns to the combined atom-field system, at times t ($t_0 < t < t_1$), an entangled superposition of the states $|e\rangle|\text{vac}\rangle$ and $|g\rangle|1_{\mathbf{k},\lambda}\rangle$ where $|g\rangle$ is the atomic ground state and $|1_{\mathbf{k},\lambda}\rangle$ denotes a field state with just one photon in the mode labeled $(\mathbf{k}, \lambda)$. The atomic state "collapses" to $|g\rangle$ at the time t_1 when the photon is detected in the distant photodetectector. In the retrodictive formalism, however, the detection of the photon at time t_1 leads us to put the field in a one photon state immediately *before* t_1. From this we retrodict that the field was in the single photon state immediately after time $t_1 - d/c$. At this time the one-photon field state is projected onto the entangled atom-field state. In contrast to the predictive description, the collapse of the atomic state occurs at the time we normally associate with the emission of the photon.

2.2 *Photon antibunching*

We can easily extend the above discussion to the fluorescence from a driven two-level atom, which leads to photon antibunching [6]. This is the intrinsically quantum phenomenon in which there is a time delay between successive photodetection events. We find that, following a detection we have to wait for a time Δt (determined by the atomic absorption rate) before we have a significant probability for detecting a second photon. In the retrodictive formalism the mechanism is clear. If a photon is detected at time t_1 then the atom state collapses to $|g\rangle$ at time $t_1 - d/c$. It will not have a significant probability of emitting again until time $t_1 - d/c + \Delta t$ after which there is a significant probability that the atom has made the transition to its excited state. Thus the detector will generally not detect another photon until time $t_1 + \Delta t$. This explanation makes explicit use of the retrodictive assignment of the state $|g\rangle$ to the atom between the times $t_1 - d/c$ and $t_1 - d/c + \Delta t$. It is also remarkably similar to the physical interpretation given to this phenomenon [6]: "The interpretation of this result is simple. It is due to the quantum nature of the scattering. Knowing that any susequent emission must begin with an excited atom, a delay corresponding to the time taken for the atom to be re-excited is naturally expected." This interpretation makes implicit use of the retrodictive formalism. We should also note that the interpretation of Monte Carlo [7, 8] simulations is similarly retrodictive in that the jumps associated with distant measurements interrupt the evolution of the system at times preceding the measurements.

2.3 *Kocher-Commins experiment*

We can also apply the retrodictive formalism to more complicated systems including the cascade emission of two photons [9, 10]. In this experiment, an atom decays from a $J = 0$ excited state to a $J = 0$ ground state via a triply degenerate $J = 1$ intermediate state. Two detectors, b and c, placed on opposite sides of the atom (on the z-axis) detect the photons emitted in the first and second transitions respectively. The detectors incorporate adjustable polarizers at angles ϕ_b and ϕ_c to the x-axis. This arrangement means that only transitions involving superpositions of the intermediate states with $m = \pm 1$ are detectable.

In the predictive formalism we associate an entangled polarization state with the two photons. In the linear polarization basis this state is $2^{-1/2}(|x_b\rangle|x_c\rangle + |y_b\rangle|y_c\rangle$, where x and y denote linear polarization in these directions. Detection of a photon at b constitutes a measurement of its polarization and instantaneously collapses the distant c-photon onto the state $\cos\phi_b|x_c\rangle + \sin\phi_b|y_c\rangle$. The probability amplitude for detection of the c-photon is determined by projection onto the measurement state $\cos\phi_c|x_c\rangle + \sin\phi_c|y_c\rangle$ giving

$$\cos\phi_b\cos\phi_c + \sin\phi_b\sin\phi_c = \cos(\phi_b - \phi_c). \tag{5}$$

We can also describe this in terms of a part retrodictive and part predictive picture. Detection of the first (b) photon leads us to assign the state $\cos\phi_b|x_b\rangle + \sin\phi_b|y_b\rangle$ to the field prior to detection. We can follow the retrodictive evolution back to the time of emission. At this time, projection of the backward evolved field state onto the entangled atom-field state leaves the atom in the intermediate state:

$$|k\rangle = 2^{-1/2}(-\exp(-i\phi_b)|m = 1\rangle + \exp(i\phi_b)|m = -1\rangle). \tag{6}$$

The knowledge of this intermediate state allows to assign (predictively) the state $\cos\phi_b|x_c\rangle + \sin\phi_b|y_c\rangle$ to the photon emitted towards detector c. The probability that this photon will be detected is then determined by the familiar Malus law.

3 Retrodiction for coupled systems

Retrodiction for coupled systems introduces the new element that we might have knowledge of the constituent parts at different times. We require a sufficiently flexible formulation to be able to account for such situations. Consider a pair of quantum systems A and B. Each of these is prepared at time t_0 and measured at a later time t_1. In between these times, the two systems are coupled together and the effect of this coupling is expressed in terms of a unitary operator $\hat{U}$.

3.1 *Predictive state*

In conventional predictive quantum theory, the initial states of both systems are determined and given by the preparation event. These lead us to assign at time t_0 to the systems the initial state $|A_0\rangle \otimes |B_0\rangle$. The unitary evolution produces, at the measurement time t_1, the state $\hat{U}|A_0\rangle \otimes |B_0\rangle$. If we also know that a measurement has been performed on B with the result associated state $|B_1\rangle$ then the *predictive* state of system A at this time, conditioned on all this information is

$$|\psi_A^{\text{pred}}\rangle = N\langle B_1|\hat{U}|A_0\rangle \otimes |B_0\rangle, \tag{7}$$

where N is a normalization factor. If no measurement is performed on system B then we can only give a mixed state description of system A in the form of the density operator

$$\hat{\rho}_A^{\text{pred}} = \text{Tr}_B\left[\hat{U}|A_0\rangle \otimes |B_0\rangle\langle B_0| \otimes \langle A_0|\hat{U}^\dagger\right]. \tag{8}$$

This state can be used to predict the outcome of a measurement performed on system A. The probability that a measurement of the non-degenerate observable $\hat{A}$ will give the results corresponding to its eigenstate $|a_i\rangle$ is $\langle a_i|\hat{\rho}_A^{\text{pred}}|a_i\rangle$.

3.2 *Retrodictive state*

An analogous procedure can be applied in retrodictive quantum theory. If we know the results of measurements on systems A and B at time t_1 then we can assign to the systems the final state $|A_1\rangle \otimes |B_1\rangle$. The time-reversed unitary evolution then produces, at the preparation time t_1, the state $\hat{U}^\dagger|A_1\rangle \otimes |B_1\rangle$. If we also know that system B was prepared in the state $|B_0\rangle$ then the *retrodictive* state of system at time t_0, conditioned on all this information is

$$|\psi_A^{\text{retr}}\rangle = N'\langle B_0|\hat{U}^\dagger|A_1\rangle \otimes |B_1\rangle, \tag{9}$$

where N' is a normalization factor. If, as above, no measurement has been performed on system B, but its initial state is known, then we can only assign to system A a retrodictive density operator at time t_0:

$$\hat{\rho}_A^{\text{retr}} = \kappa\langle B_0\hat{U}^\dagger|A_1\rangle\langle A_1| \otimes \hat{\mathbf{1}}_B\hat{U}|B_0\rangle, \tag{10}$$

where κ is a normalization constant and $\hat{\mathbf{1}}_B$ is the identity operator for system B. This may be used to make retrodictions about the initially selected state. If system A was prepared in one of the set of states $\{|i\rangle\}$ occuring with *a priori* probabilities $\{P_p(i)\}$ then the probability that the state $|i\rangle$ was prepared, given the result of the later measurement is

$$P(i|A_1, B_0) = \frac{\langle i|\hat{\rho}_A^{\text{retr}}|i\rangle P_p(i)}{\sum_{i'} \langle i'|\hat{\rho}_A^{\text{retr}}|i'\rangle P_p(i')} \tag{11}$$

This is the generalization of the earlier expresion (4) for uncoupled systems. A more detailed discussion of the retrodictive density operator can be found in [11]. We should note that Eq. (9) has general vlidity and can be applied to study strongly coupled systems [12]. For situations involving weak coupling to a large environment we can introduce a retrodictive master equation [13].

3.3 *Open systems*

In quantum optics we are often required to model systems coupled to a much larger environment. This entails a loss of information to the environment and consequently, the use of density operators. The evolution of such density operators can often be described by means of a master equation [14]. The general Markovian master equation for the predictive density operator for the system S can be witten in the standard Lindblad form

$$\dot{\hat{\rho}}^{\text{pred}} = -i\hbar^{-1}\left[\hat{H}_S, \hat{\rho}^{\text{pred}}\right] + \sum_q \left[2\hat{A}_q\hat{\rho}^{\text{pred}}\hat{A}_q^\dagger - \hat{A}_q^\dagger\hat{A}_q\hat{\rho}^{\text{pred}} - \hat{\rho}^{\text{pred}}\hat{A}_q^\dagger\hat{A}_q\right], \tag{12}$$

where $\hat{H}_S$ is the Hamiltonian for the system alone and $\hat{A}_q$ is a system operator. The form of this equation is such as to ensure that probability is conserved so that the density operator remains Hermitian and positive and that its trace is always unity.

We can apply the results of the preceding subsection to derive a retrodictive master equation for the system [13]. This allows us to make retrodictions about the preparation of the system of interest given the result of a later measurement on the system and knowledge of the *initial* state of its environment. The retrodictive master equation corresponding to the predictive equation (12) is

$$\begin{aligned} \dot{\hat{\rho}}^{\text{retr}} = -i\hbar^{-1}\left[\hat{H}_S, \hat{\rho}^{\text{retr}}\right] + \sum_q \left[2\hat{A}_q^\dagger\hat{\rho}^{\text{pred}}\hat{A}_q - \hat{A}_q^\dagger\hat{A}_q\hat{\rho}^{\text{pred}} - \hat{\rho}^{\text{pred}}\hat{A}_q^\dagger\hat{A}_q\right] \\ -2\hat{\rho}^{\text{retr}}\text{Tr}_S\left\{\hat{\rho}^{\text{retr}}\sum_q\left[A_q^\dagger, \hat{A}_q\right]\right\}, \end{aligned} \tag{13}$$

where the "dot" denotes differentiation with respect to the premeasurement time.

We have applied this retrodictive master equation to study retrodictive phenomena for two-level atoms [11] and for a single mode of the electromagnetic field [15]. As an example, consider a single field mode coupled to an absorbing environment at zero temperature. The evolution of this system can be described by the familiar master equation [14]

$$\dot{\hat{\rho}}^{\text{pred}} = \Gamma\left[2\hat{a}\hat{\rho}^{\text{pred}}\hat{a}^\dagger - \hat{a}^\dagger\hat{a}\hat{\rho}^{\text{pred}} - \hat{\rho}^{\text{pred}}\hat{a}^\dagger\hat{a}\right], \tag{14}$$

where $\hat{a}$ is the usual annihilation operator. The retrodictive counterpart to this master equation is then

$$\dot{\hat{\rho}}^{\text{retr}} = \Gamma\left[2\hat{a}^\dagger\hat{\rho}^{\text{retr}}\hat{a} - \hat{a}\hat{a}^\dagger\hat{\rho}^{\text{retr}} - \hat{\rho}^{\text{retr}}\hat{a}\hat{a}^\dagger\right], \tag{15}$$

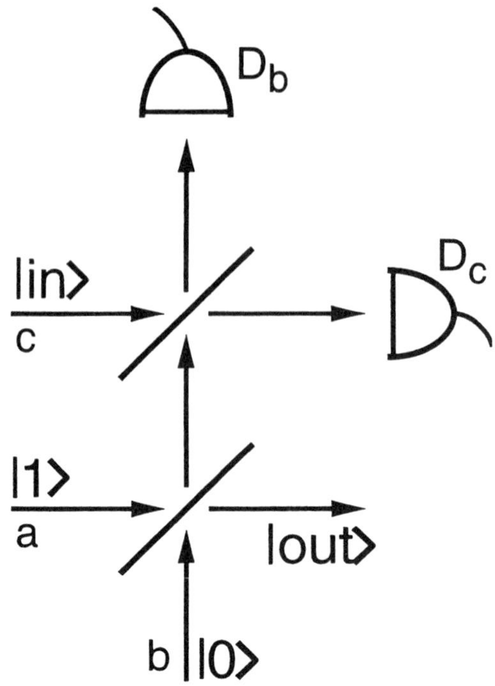

Fig. 1. The quantum scissors device (after [2]).

which we recognise as the master equation for a linear *amplifier*. This means that retrodicting the prepared input state for an attenuating (amplifying) channel corresponds to predicting the measured output signal for an amplifying (attenuating channel). An example of the application of this idea is given in [15].

4 Quantum scissors

The retrodictive formalism is also useful in prompting new ideas. As an example, we present a brief account of the "quantum scissors" device for producing a controlled superposition of the vacuum and one-photon states [16, 17]. The device comprises two 50:50 beam splitters and two photodetectors as depicted in the figure. The field state to be truncated is introduced into mode c:

$$|\mathrm{in}\rangle = \gamma_0|0\rangle_c + \gamma_1|1\rangle_c + \gamma_2|2\rangle_c + ..., \tag{16}$$

where $|n\rangle_c$ denotes the state with n photons in mode c. The other input modes a and b are prepared, in their respective one-photon and vacuum states $|1\rangle_a$ and $|0\rangle_b$. The quantum scissors works on those occasions that a single photon is detected in D_b and none are detected in D_c.

A retrodictive analysis begins with the detection events and assigns on the basis of these the premeasurement retrodictive state $|1\rangle_b \otimes |0\rangle_c$. This then is the retrodictive state at the output of the upper beam splitter. Applying the transformation for a symmetric beam splitter allows us to retrodict the state prior to this beam splitter to be

$$2^{-1/2}(|1\rangle_b \otimes |0\rangle_c - i|0\rangle_b \otimes |1\rangle_c). \tag{17}$$

We know that mode c was prepared in the state (17). Projecting this state onto (18) and renormalizing gives the retrodictive state

$$ic_0^*|1\rangle_b + c_1^*|0\rangle_b, \tag{18}$$

where $c_n = \gamma_n(|\gamma_0|^2 + |\gamma_1|^2)^{-1/2}$, for the mode between the two beam splitters.

The next step is to use predictive quantum theory for the lower beam splitter. This beam splitter transforms the input state $|1\rangle_a \otimes |0\rangle_b$ into the entangled state [18]

$$2^{-1/2}(|1\rangle_a \otimes |0\rangle_b + i|0\rangle_a \otimes |1\rangle_b). \tag{19}$$

If we neglect the unimportant phase shift associated with distance between the two beam splitters, then we retrodict that the state emerging in mode a is found by projecting the retrodictive state (19) onto (20) to give the state

$$c_0|0\rangle_a + c_1|1\rangle_a. \tag{20}$$

This is a truncated and renormalized version of the original state $|\text{in}\rangle$ with only contributions from the vacuum and one-photon states retained. The analysis presented here is somewhat idealized. A more complete analysis, however, shows that the technique is highly robust against effects such as finite detector efficiency and should provide a practical technique for creating superpositions of the vacuum and one-photon states [17].

We note that none of the light input into the device reaches the output mode. In this sense, the device operates by means of a nonlocal effect. More surprising is that the output state leaves device *before* the input light enters it. We might ask if this allows us to send information into the past. The answer, of course, is no and the reason is that we have no control over the detection events at D_b and D_c. The device will only work as described on those occasions that the right number of counts are recorded. It may be possible to use this device to provide input into a quantum computer before the required qubit state is available. The quantum computer will then work correctly on those occasions when the detection events confirm that the earlier input state was correct.

5 Conclusion

Quantum mechanics is usually formulated as a predictive theory. This means assigning a state to the system of interest on the basis of a preparation event. The state then evolves as described by the Schrödinger equation or a suitable master equation. The evolved state is then used to make predictions about the outcome of some measurement performed at a later time.

Some problems, including that of communications, require the ability to make retrodictions, that is statements about an earlier preparation event on the basis of the result of a later measurement. Such problems can be solved using predictive quantum mechanics together with Bayes' theorem. A more natural, although equivalent, approach is to use the retrodictive formalism. This assigns the state between the preparation and measurement times on the basis of the measurement result. We have outlined how the retrodictive formalism works and illustrated the very different physical picture it offers of elementary quantum optical phenomena. The retrodictive formalism is relatively new, but some of the tools for applying it have been developed. These include a retrodictive master equation to enable the study of realistic open systems.

Finally, it is worth emphasizing that retrodiction is not simply time-reversal. The retrodictive state at the preparation time will not, in general, be the same as the state selected in the preparation event. This is because the retrodictive state contains information about the measurement result. The predictive state, however, contains information about the preparation event. The predictive state is used to make predictions about the measurement outcome, while the retrodictive state is used to make retrodictions about the preparation event.

We are grateful to our collaborators, Rodney Loudon, Ottvia Jedrkiewicz and Lee Phillips for many interesting discussions. This work was supported by the UK Engineering and Physical Sciences Research Council, the Australian Research Council, the Royal Society of Edinburgh and the Scottish Executive Education and Lifelong Learning Department.

References and links

1. Y. Aharonov, P. G. Bergman and J. L. Lebowitz, "Time symmetry in the quantum process of measurement," Phys. Rev. **134**, B1410-1416 (1964).
2. D. T. Pegg and S. M. Barnett, "Retrodiction in quantum optics," J. Opt. B: Quantum Semiclass. Opt **1**, 442-445 (1999).
3. F. J. Belinfante, *Measurements and time reversal in objective quantum theory* (Pergamon Press, 1975).
4. G. E. P. Box and G. C. Tiao, *Bayesian inference in statistical analysis* (Addison-Wesley, 1973) p. 10.
5. S. M. Barnett, D. T. Pegg and J. Jeffers, "Bayes' theorem and quantum retrodiction," J. Mod. Opt. **47**, 1779-1789 (2000).
6. D. F. Walls and G. J. Milburn, *Quantum optics* (Springer-Verlag, 1994) p. 221 and references therein.
7. J. Dalibard, Y. Castin and K. Mølmer, "Wave-function approach to dissipative processes in quantum optics," Phys. Rev. Lett. **68**, 580-583.
8. H. Carmichael *An open systems approach to quantum optics* (Springer-Verlag, 1993).
9. C. A. Kocher and E. D. Commins, "Polarization correlation of photons emitted in an atomic cascade," Phys. Rev. Lett. **18**, 575-577 (1967).
10. A. Aspect, P. Grangier and G. Roger, "Experimental tests of realistic local theories via Bell's theorem," Phys. Rev. Lett. **47**, 460-463 (1981).
11. S. M. Barnett, D. T. Pegg, J. Jeffers and O. Jedrkiewicz, "Atomic retrodiction," J. Phys. B: At. Mol. Opt. **33**, 3047-3065 (2000).
12. J. Jeffers, S. M. Barnett and D. T. Pegg, (article in this volume).
13. S. M. Barnett, D. T. Pegg, J. Jeffers and O. Jedrkiewicz "Master equation for retrodiction of quantum communication signals," Phys. Rev. Lett. **86**, 2455-2458 (2001).
14. See for example S. M. Barnett and P. M. Radmore, *Methods in theoretical quantum optics* (Oxford University Press, 1997) p. 161.
15. S. M. Barnett, D. T. Pegg, J. Jeffers, O. Jedrkiewicz and R. Loudon, "Retrodiction for quantum optical communications," Phys. Rev. A **62**, 022313 (2000).
16. D. T. Pegg, L. S. Phillips and S. M. Barnett, "Optical state truncation by projection synthesis," Phys. Rev. Lett. **81**, 1604-1606 (1998).
17. S. M. Barnett and D. T. Pegg, "Optical state truncation," Phys. Rev. A **60**, 4965-4973 (1999).
18. This has been verified in the classic experiment by P. Grangier, G. Roger and A. Aspect, "Experimental evidence for a photon anticorrelation effect on a beam splitter: a new light on single-photon interference," Europhys. Lett. **1**, 173-179 (1986).

Spontaneous Intrinsic Decoherence in Two and Three Slits Young Interference Experiments

Rodolfo Bonifacio and Stefano Olivares

Dipartimento di Fisica, Università degli Studi di Milano, INFM and INFN Sezione di Milano, via Celoria 16, I - 20133 Milano, Italy

Abstract: We discuss the strict relation between the Scrhödinger spread and the single particle Young interference experiment with two and three slits. Moreover, while in the two slits case there is always a maximum at the center of the screen, we demonstrate that, with three slits, one can have zero intensity (at the center) for certain *discrete* values of the spacing between the slits and of the distance of the screen. Finally, we show the possibility that spontaneous intrinsic decoherence (SID) destroys interference pattern and that decoherence becomes stronger at the macroscopic limit.

In standard text books, single particle Young interference is described solving the Schrödinger equation in momentum space, deriving a time independent angular distribution. However, the real interference pattern must be described in the position space and time dependence explicitly appears. We show that interference is strictly connected to the Schrödinger spread, feature that does not appear in the momentum representation. The optical analogy is described putting at the end of the calculations $t = L/v$, where L is the distance from the screen and v is the particles velocity. The usual Young interference pattern appears only for time long enough so that the two wave packets coming from the slits completely overlap. Adding a third slit, we find a very puzzling aspect of quantum interference, which, to our knowledge, has not been described before. The central peak vanishes and becomes a zero intensity point for *discrete values of the spacing between the slits and of the distance of the screen.* Up to now, it is not obvious to us if this effect has a classical analogy.

Since interference pattern is time dependent, fluctuations of the evolution time $t = L/v$ induced by velocity spread give rise to decoherence[1, 2], i.e. to the reduction of fringe visibility. This is a *spontaneous decoherence* because it is not induced by the interaction with the environment[3]. In the paper [2] we have described spontaneous decoherence in cavity QED. Here velocity fluctuations can have experimental reasons due to thermal fluctuations or intrinsic ones due the Heisenberg uncertainty principle (HUP). In the last case one has *spontaneous intrinsic decoherence* (SID). Furthermore, in reference [1], using the Tam Mandelstam inequality[4], it has been suggested that in SID these time fluctuations have an intrinsic inferior limit given by the *energy-time*

$$\tau_E \equiv \frac{\hbar}{2\,\Delta\mathcal{H}} \tag{1}$$

where $\Delta\mathcal{H}$ is the variance of the system Hamiltonian $\mathcal{H}$. In this paper we describe

Coherence and Quantum Optics VIII
Edited by Bigelow et al., Kluwer Academic/Plenum Publishers, 2003

spontaneous decoherence in two and three slits Young interference experiments. The conditions to observe spontaneous decoherence due to classical (thermal) fluctuations of the particles velocity and SID, coming from the velocity spread due to HUP, are specified.

1. Two Slits Young Experiment

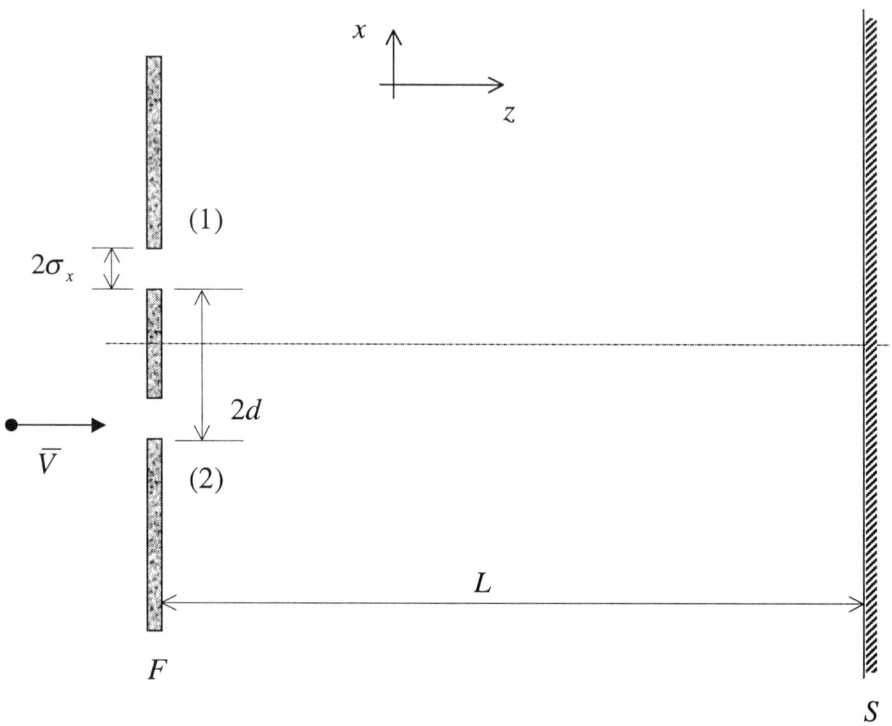

Fig. 1. Setup of the Young experiment described in the text.

In a previous work[5] we have discussed the time dependence of the Young two slits interference pattern for single particle, showing that interference is closely related to the free particle Schrödinger spread. We consider a beam of particles moving along $\hat{z}$ axis with mean velocity $\overline{V}$ and with so low intensity that one particle at a time arrives at a screen F (Fig. 1). The initial wave function, when particles pass through the slits, is chosen to be a superposition of two Gaussian wave packets, so that at time t one has

$$\Psi(x,t) \propto \Psi_1(x,t) + \Psi_2(x,t) \tag{2}$$

where[6]

$$\Psi_j(x,t) \propto \exp\left\{-\frac{(x-x_j)^2}{4\sigma(t)^2}\left(1 - i\frac{\sigma_v}{\sigma_x}t\right)\right\}, \quad j = 1,2 \tag{3}$$

with $x_1 = -x_2 = d$ and with $\sigma(t) = \sqrt{\sigma_x^2 + \sigma_v^2 t^2}$, where $\sigma_x \sigma_v = \hbar/2m$. From now on, it is useful to introduce the dimensionless quantities

$$\overline{x} = \frac{x}{\sigma_x}, \quad \overline{d} = \frac{d}{\sigma_x}, \quad \overline{t} = \frac{\sigma_v t}{\sigma_x} = \frac{\hbar}{2m\sigma_x^2}t. \tag{4}$$

In Fig. 2 it is shown what happens when the system evolves: interference appears as soon as the exponential wings of the two wave packets overlap. The probability density

$P(\bar{x},\bar{t})$ of finding a particle in $\bar{x}$ at time $\bar{t}$ is

$$P(\bar{x},\bar{t}) \equiv |\Psi(\bar{x},\bar{t})|^2 = |c(\bar{t})|^2 \left(G_+ + G_- + 2\sqrt{G_+G_-}\cos\omega\bar{t}\right) \tag{5}$$

where $|c(\bar{t})|^2$ is a normalization factor and

$$G_\pm = \exp\left\{-\frac{(\bar{x}\pm\bar{d})^2}{2(1+\bar{t}^2)}\right\}, \qquad \omega = \frac{\bar{x}\bar{d}}{1+\bar{t}^2}. \tag{6}$$

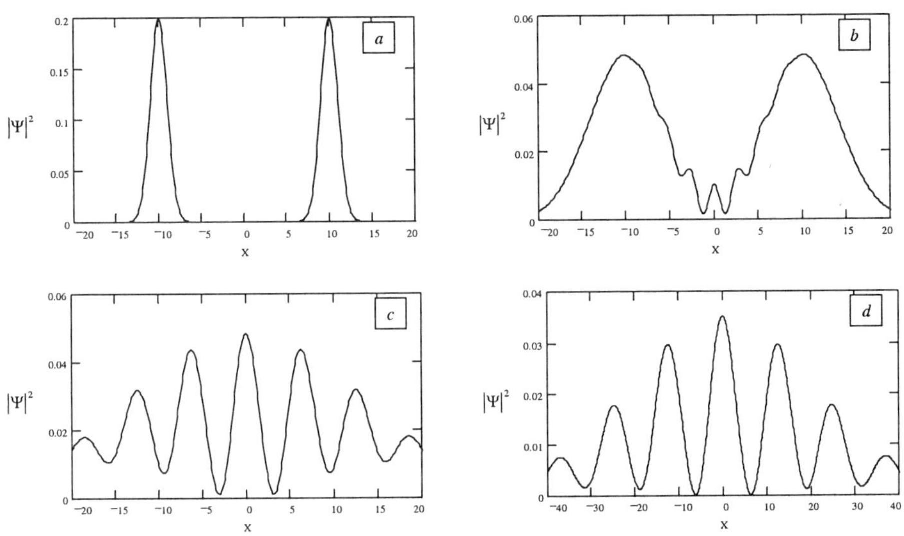

Fig. 2. Time evolution of the two slits interference pattern with $d/\sigma_x = 10$ for different values of the dimensionless time $\bar{t}$: *a*) $\bar{t}=0$; *b*) $\bar{t}=4$; *c*) $\bar{t}=10$; *d*) $\bar{t}=20$. Here $\bar{t}=\sigma_v t/\sigma_x = L/R$, where $R = 2\sigma_x^2/\lambda\!\!\!^{-}$ is the Rayleigh range.

Note that the time dependent interference term $\cos\omega\bar{t}$ is modulated by

$$\sqrt{G_+G_-} = \exp\left\{-\frac{\bar{x}^2}{2(1+\bar{t}^2)}\right\}\exp\left\{-\frac{\bar{d}^2}{2(1+\bar{t}^2)}\right\} \tag{7}$$

and, under the condition

$$\bar{t} \gg \bar{d} \gg 1, \tag{8}$$

Eq. (7) can be approximated by

$$\sqrt{G_+G_-} \approx \exp\left\{-\frac{\bar{x}^2}{2\bar{t}^2}\right\}, \tag{9}$$

i.e one has a Gaussian modulation. Hence the interference term is relevant only for $\bar{x} \ll \bar{t}$ ($x \ll \sigma_v t$). Furthermore, using the scaling (4), Eqs. (8) become

$$\sigma_v t \gg d \gg \sigma_x \tag{10}$$

which imply complete overlap of the wave functions: the interference pattern provides a direct evidence of the Schrödinger spread. The maxima occur at $\omega\bar{t} = 2n\pi$ (see Eq. (5)) so that by Eqs. (6) and (4) we can write (in the limit of Eq. (8))

$$\bar{x}_n = \frac{2n\pi\bar{t}}{\bar{d}}. \tag{11}$$

Hence, from Eq. (9), these maxima are modulated by

$$\exp\left\{-\frac{\bar{x}_n^2}{2\bar{t}^2}\right\} = \exp\left\{-\frac{1}{2}\left(\frac{2n\pi}{\bar{d}}\right)^2\right\} \tag{12}$$

so that the number of visible fringes is proportional to $\bar{d} = d/\sigma_x$, as in the optical case.

An optical analogy with the "classical" optical Young interference experiment can be obtained interpreting $P(x,t)$ (coming from (5) using (4)) as the probability density of finding a particle on a screen S at distance L such that $t = L/\overline{V}$, where $\overline{V}$ is the mean velocity perpendicular to the plane $\hat{x}$ of the slits (see Fig. 1). In this way Eq. (11) gives

$$\frac{x_n}{L} = \frac{\lambda}{2d}n \quad \left(\lambda = \frac{h}{m\overline{V}} = 2\pi\lambda\!\!\!^{-}\right) \tag{13}$$

which represents the usual condition for constructive interference assuming $\lambda \ll d \ll L$. All the previous equations can be translated in optical terms identifying $t = L/\overline{V}$ and $\lambda\!\!\!^{-} = \hbar/m\overline{V}$. In particular the conditions $\bar{d} \ll \bar{t}$ and $\bar{x} \ll \bar{t}$ becomes

$$\frac{d}{L} \ll \vartheta_{diff}, \quad \vartheta = \frac{x}{L} \ll \vartheta_{diff} \quad \left(\vartheta_{diff} \equiv \frac{\lambda\!\!\!^{-}}{2\sigma_x} \ll 1\right), \tag{14}$$

i.e. the angle d/L and the observation angle $\vartheta = x/L$ must be smaller than the diffraction angle ϑ_{diff} . Furthermore $\bar{t} \gg 1$ is equivalent to

$$\bar{t} = \frac{L}{R} \gg 1 \tag{15}$$

where $R = 2\sigma_x^2/\lambda\!\!\!^{-}$ is the Rayleigh range, which is a measure of the divergence of a Gaussian beam whose transverse width is σ_x. The first of conditions (14) guarantees that particles arrives at the center of the screen S. The translation in optical terms can be similarly done for the other equations. For example, Eq. (9), using the previous definitions, can be written under the form of the well-known diffraction envelope:

$$\sqrt{G_+G_-} \approx \exp\left\{-\frac{x^2(2\sigma_x)^2}{2L^2\lambda\!\!\!^{-2}}\right\} = \exp\left\{-\frac{\vartheta^2}{2\vartheta_{diff}^2}\right\}. \tag{16}$$

Finally, we emphasizes that our simple one dimensional model practically gives the same results of a more rigorous three dimensional calculation by B. G. Englert[7].

2. Three Slits Young Interference

The three slits interference experiment we are going to describe presents some really puzzling aspects. The two slits interference pattern shows a central maximum, i.e. particles have a high probability to reach the center of the screen. Suppose, now, to open a new slit between the two slits as in Fig. 3. One would expect that the intensity of the central maximum is further increased. On the contrary, we will show that for discrete values of the spacing between the slits and of the distance of the screen one can have zero intensity at the center: no particle goes there! This puzzle can be exposed in a different way. Let us consider a single hole: one sees a well defined diffraction pattern, i.e. all the particles practically go in front of the slit. If we open two slits symmetrically apart the central one, for discrete values no particle arrives at the center. To our knowledge this effect was not previously described.

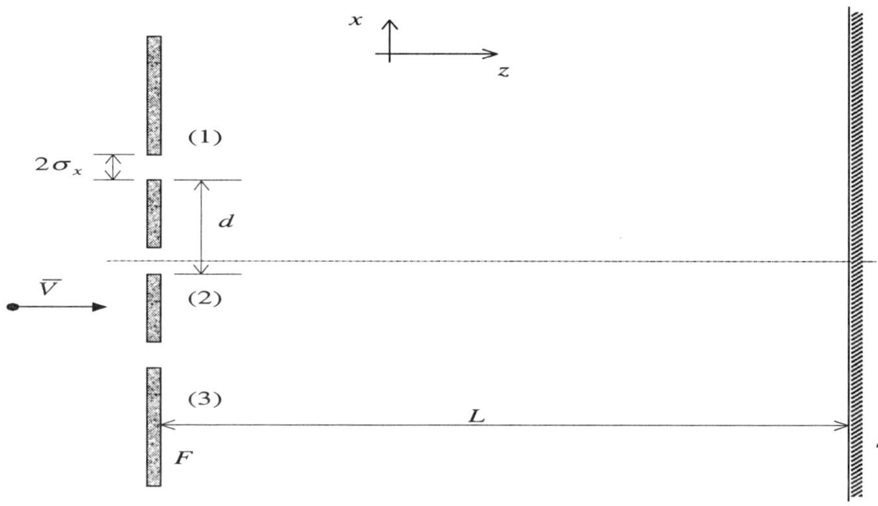

Fig. 3. Setup of the three slits Young experiment described in the text.

We consider a particles beam moving along $\hat{z}$ axis with velocity $\overline{V}$ (see Fig. 3). Using the one dimensional time dependent Schrödinger equation and assuming that at time $t = 0$, when particles pass through the slits, the state is a superposition of three Gaussian wave packets, at time t one has

$$\Psi(x,t) \propto \Psi_0(x,t) + \Psi_1(x,t) + \Psi_2(x,t) \tag{17}$$

where $\Psi_j(x)$, $j = 0,1,2$, are given by Eq. (3) with $x_1 = -x_2 = d$ and $x_0 = 0$. Hence the probability density of finding a particle on the screen S at position $\overline{x}$ is

$$\begin{aligned} P(\overline{x},\overline{t}) &\equiv |\Psi(\overline{x},\overline{t})|^2 \\ &= |\hat{c}(\overline{t})|^2 \Big(G_+ + G_- + G_0 + 2\sqrt{G_+G_-}\cos\omega_1\overline{t} + \\ &\quad +2\sqrt{G_0G_-}\cos\omega_2\overline{t} + 2\sqrt{G_+G_0}\cos\omega_3\overline{t}\Big) \end{aligned} \tag{18}$$

where $|\hat{c}(\overline{t})|^2$ is a suitable normalization factor and

$$G_\pm = \exp\left\{ -\frac{(\overline{x}\pm\overline{d})^2}{2(1+\overline{t}^2)} \right\}, \quad G_0 = \exp\left\{ -\frac{\overline{x}^2}{2(1+\overline{t}^2)} \right\}, \tag{19}$$

$$\omega_1 = \frac{\overline{x}\overline{d}}{1+\overline{t}^2}, \quad \omega_2 = \frac{(\overline{x}-\overline{d}/2)(\overline{d}/2)}{1+\overline{t}^2}, \quad \omega_2 = \frac{(\overline{x}+\overline{d}/2)(\overline{d}/2)}{1+\overline{t}^2}. \tag{20}$$

Specializing Eq. (18) for $\overline{x} = 0$, one has

$$|\Psi(0,\overline{t})| = |\hat{c}(\overline{t})|^2 \left\{ 1 + 4G + 4\sqrt{G}\cos\left(\frac{\overline{d}^2}{4(1+\overline{t}^2)}\overline{t} \right) \right\} \tag{21}$$

with

$$G = \exp\left\{-\frac{\overline{d}^2}{2(1+\overline{t}^2)}\right\}. \tag{22}$$

Maxima and minima in $\overline{x} = 0$ are obtained imposing

$$\frac{\overline{d}^2}{4(1+\overline{t}^2)}\overline{t} = n\pi, \tag{23}$$

so one has maxima or minima for even or odd n, respectively. Moreover the minimum of intensity is exactly equal to zero if we require $G = 1/4$, i.e.

$$\frac{\overline{d}^2}{2(1+\overline{t}^2)} = 2\ln(2). \tag{24}$$

Eq. (23) and Eq. (24) bring to the final conditions

$$\overline{t} = \frac{L}{R} = \overline{t}_n = \frac{\pi}{\ln(2)}n \approx 4.5n, \tag{25}$$

$$\overline{d} = \overline{d}_n = 2\sqrt{\frac{(\ln(2))^2 + \pi^2 n^2}{\ln(2)}} \approx 7.7n. \tag{26}$$

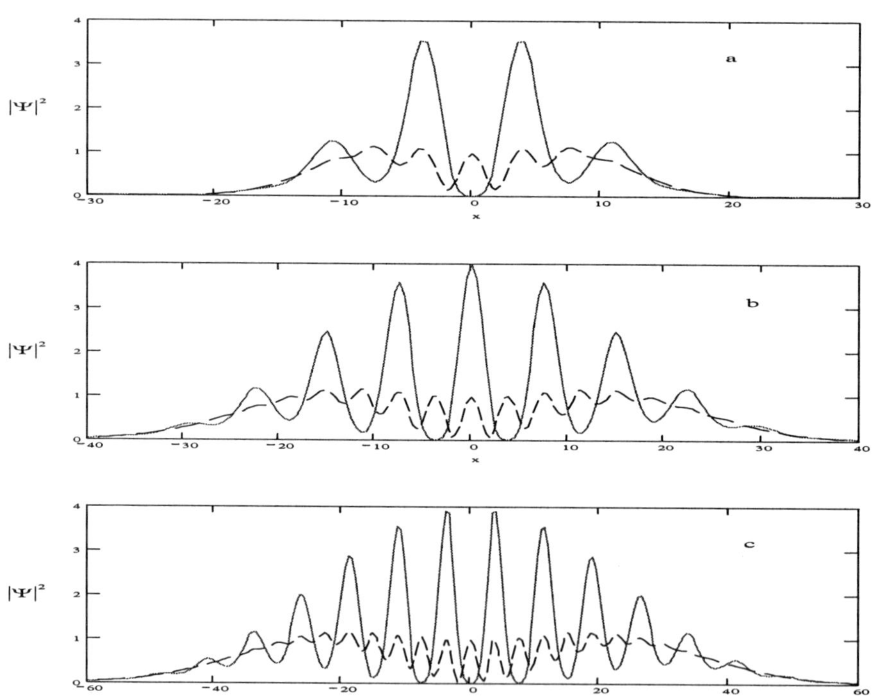

Fig. 4. Comparison between three slits (solid line) and two slits (dash line) interference pattern for different values of n, $\overline{t}_n = L_n/R$ and $\overline{d}_n$ (see the text): a) $n = 1$, $\overline{t}_1 \approx 4.5$ and $\overline{d}_1 \approx 7.7$; b) $n = 2$, $\overline{t}_2 \approx 9.1$ and $\overline{d}_2 \approx 15.2$; c) $n = 3$, $\overline{t}_3 \approx 13.6$ and $\overline{d}_3 \approx 22.7$. Notice that, with three slits, for odd n there is zero intensity at the center, whereas for even n there is a maximum.

In Fig. 4 we plot the comparison between two and three slits cases for $n = 1, 2, 3$. This Figure, for $n = 1$ and $n = 3$, explicitly shows that a new slit between the two slits of a Young experiment can inhibit particles to reach that place where they had high probability to go, so no particle arrives at that point! The corresponding physical optical values of L and d directly come from relations (4), (25) and (26), i.e.

$$L_n \approx 4.5\, R n \tag{27}$$

$$d_n \approx 7.7\, \sigma_x\, n \tag{28}$$

where we remember that $\lambda\!\!\!^{-} = \hbar/m\overline{V}$ and $R = 2\sigma_x^2/\lambda\!\!\!^{-}$.

3. Decoherence and Spontaneous Intrinsic Decoherence in Interference Experiments

First of all we briefly summarize the model independent formalism introduced in previous papers[2, 8, 9] to describe decoherence due to fluctuations in the evolution time. The density operator which takes into account these fluctuations is the time averaged of the usual density operator, i.e.

$$\overline{\rho}(t) = \int_0^\infty \mathrm{d}t'\, \mathcal{P}(t, t', \tau)\, \rho(t') \tag{29}$$

where the weight function is given by the Γ-distribution function

$$\mathcal{P}(t, t', \tau) = \frac{\exp(-t'/\tau)}{\tau} \frac{(t'/\tau)^{t/\tau - 1}}{\Gamma(t/\tau)}. \tag{30}$$

Here τ is a characteristic time which rules time fluctuations and can also be a function of t [2, 8, 9] and can have experimental or intrinsic origin. In general if τ is limited by τ_E, one has always a contribution given by intrinsic decoherence.

In the interference case, since $t = L/v$, time fluctuations are induced by velocity spread as in the case of cavity QED described in references [2, 9]. If the system is initially in a pure state, as assumed above, from Eq. (29) one obtains

$$\overline{P}(\overline{x}, \overline{t}) \equiv \langle \overline{x} | \overline{\rho}(\overline{t}) | \overline{x} \rangle = \int_0^\infty \mathrm{d}\overline{t}'\, \mathcal{P}(\overline{t}, \overline{t}', \overline{\tau})\, P(\overline{x}, \overline{t}') \tag{31}$$

where the expression of $P(\overline{x}, \overline{t}')$ is given by (5) and (18) in the case of two and three slits, respectively, and $\overline{\tau}$ is given by

$$\overline{\tau} \equiv \frac{\sigma_v}{\sigma_x} \tau, \tag{32}$$

where the scaling (4) has been used for all variables.

We assume τ to be the uncertainty in arrival time of particles at the screen, so that[9, 10]

$$\tau = \frac{\Delta_z(t)}{\overline{\mathcal{V}}} \quad \text{where} \quad \Delta_z(t) = \sqrt{\sigma_z^2 + \sigma_{\overline{V}}^2 t^2 + \Delta_{\overline{\mathcal{V}}}^2 t^2}, t = \frac{L}{\overline{\mathcal{V}}}. \tag{33}$$

Here $\Delta_z(t)$ is the total uncertainty in position along $\hat{z}$ axis[5], which is obtained summing up the square of two errors: an intrinsic one, $\sigma_z^2 + \sigma_{\overline{V}}^2 t^2$ (with $\sigma_z \sigma_{\overline{V}} = \hbar/2m$), i.e. the quantum Schrödinger free particle spread along $\hat{z}$ axis, and an "extrinsic" one, $\Delta_{\overline{\mathcal{V}}}^2 t^2$, i.e. the spread due to classical uncertainty of velocity distribution, e.g. thermal distribution. A more rigorous derivation of Eq. (33) is given in reference [9]. Note that

τ is always greater than $\tau_E = \sigma_z/\overline{\mathcal{V}} = \hbar/2\sigma(\mathcal{H}_\ddagger)$, where $\mathcal{H}_\ddagger$ is the kinetic energy along $\hat{z}$ axis and $\sigma(\mathcal{H}_\ddagger)$ its standard deviation, in agreement with Eq. (1). Assuming t such that $(\sigma_{\overline{V}}^2 + \Delta_{\overline{\mathcal{V}}}^2)t^2 \gg \sigma_z^2$, from Eqs. (32) and (33) one directly obtains

$$\overline{\tau} = a\overline{t} \quad \text{with} \quad a = \frac{\sqrt{\sigma_{\overline{V}}^2 + \Delta_{\overline{\mathcal{V}}}^2}}{\overline{\mathcal{V}}}. \tag{34}$$

If we take $\overline{\tau}/\overline{t} = \tau/t = a \ll 1$, $\mathcal{P}(\overline{t}, \overline{t}', \overline{\tau})$ can be approximated with the Gaussian

$$\mathcal{P}(\overline{t}, \overline{t}', \overline{\tau}) \approx \frac{1}{\sqrt{2\pi\overline{t}\overline{\tau}}} \exp\left\{-\frac{(\overline{t}-\overline{t}')^2}{2\overline{t}\overline{\tau}}\right\}. \tag{35}$$

In this limit we can perform the average (31) obtaining the analytical expression for the two slits case:

$$\overline{P}(\overline{x}, \overline{t}) = b(\overline{t})\left\{G_+ + G_- + 2\sqrt{G_+G_-}e^{-D}\cos\omega\overline{t}\right\} \tag{36}$$

where $b(\overline{t})$ is a normalization factor and

$$D \approx \frac{\overline{x}^2\overline{d}^2}{2}\frac{\overline{\tau}}{\overline{t}^3} = \frac{1}{2}\frac{x^2(2d)^2}{L^2\lambda^2}a \tag{37}$$

where we used Eqs. (4) and (34). The behavior of $\overline{P}(\overline{x}, \overline{t})$, as a function of $\overline{x}$ at a given $\overline{t} = L/R$, is shown in Fig. 5 for different values $\overline{d} = d/\sigma_x$ and a. The term e^{-D} is another Gaussian *decoherence envelope not due to diffraction*, as the one of Eq. (16), *but due to velocity spread*. It is very important and significative to note that it goes as $(2d)^2$, i.e. the square of the distance between slits, as it is typical of decoherence phenomena[3]. In general, diffraction envelope and decoherence one are both present in experiments, but the latter becomes visible as soon as $D \geq 1$ and it dominates diffraction when

$$a > \left(\frac{\sigma_x}{d}\right)^2 \tag{38}$$

which gives a lower limit to the relative velocity spread a.

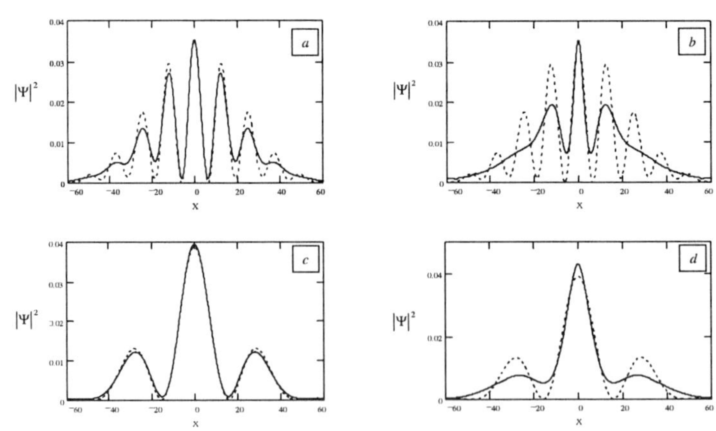

Fig. 5. Behaviour of the two slits interference pattern with decoherence (solid line) and without decoherence (dotted line) at time $\overline{t} = L/R = 20$ for different values of d/σ_x and a: a) $d/\sigma_x = 10$, $a = 0.01$; b) $d/\sigma_x = 10$, $a = 0.1$; c) $d/\sigma_x = 4$, $a = 0.01$; d) $d/\sigma_x = 4$, $a = 0.1$.

We can also write Eq. (37) as

$$D = \frac{1}{2}\left(\frac{\vartheta}{\vartheta_{int}}\right)^2 a \tag{39}$$

where $\vartheta_{int} = \lambda\!\!\!^{-}/2d$ (see Eq. (13)) and $\vartheta \approx x/L$. If one evaluates D in the interference pattern maxima (13), one obtains

$$D = 2\pi^2 n^2\, a. \tag{40}$$

Hence decoherence becomes more and more efficient as n and a increase (see Fig. 5).

Now, thanks to Eq. (34) we are able to study the classical and the quantum limit: 1) in the *classical limit* $\Delta^2_{\overline{V}} \gg \sigma^2_{\overline{V}}$, then Eq. (39) is ruled by the ratio $a = \Delta_{\overline{V}}/\overline{V}$, so that decoherence has a classical origin; 2) in the *quantum limit* , defined as $\Delta^2_{\overline{V}} \ll \sigma^2_{\overline{V}}$, one has *spontaneous intrinsic decoherence* (SID) due to the quantum velocity spread $\sigma_{\overline{V}} \geq \hbar/2m\sigma_z$. In this limit $a = \sigma_{\overline{V}}/\overline{V}$ and decoherence has an intrinsic, Quantum Mechanical origin related to HUP.

Moreover, at the macroscopic limit $\lambda\!\!\!^{-}$ becomes very small. However if one also decreases σ_x, so that $\lambda\!\!\!^{-}/\sigma_x$ remains constant, hence the diffraction envelope does not suppress the interference pattern; on the contrary, the decoherence envelope kills interference keeping constant the distance between slits (see Eq. (37)).

Finally, when one applies our theory of spontaneous decoherence to the three slits setup of the interference experiment, one proceeds as the two slits case, performing the time average of Eq. (18) and obtaining

$$\begin{aligned} \overline{P}(\overline{x},\overline{t}) \;\; = \;\; & |\tilde{c}(\overline{t})|^2 \Big(G_+ + G_- + G_0 + 2\sqrt{G_+G_-}e^{-(D_{+,-})}\cos\omega_1\overline{t} + \\ & +2\sqrt{G_0G_-}e^{-(D_{0,-})}\cos\omega_2\overline{t} + 2\sqrt{G_+G_0}e^{-(D_{0,+})}\cos\omega_3\overline{t}\Big) \end{aligned} \tag{41}$$

where $|\tilde{c}(\overline{t})|^2$ is a normalization factor and

$$D_{+,-} \approx \frac{1}{2}\frac{x^2(2d)^2}{L^2\lambda\!\!\!^{-\,2}}a, \tag{42}$$

$$D_{0,\pm} \approx \frac{1}{2}\frac{(x \pm d/2)^2 d^2}{L^2\lambda\!\!\!^{-\,2}}a, \tag{43}$$

with a given by Eq. (34). Notice that decoherence still goes as the square of the distance between slits as in the previous case.

Therefore, decoherence becomes more efficient increasing a and *spontaneous intrinsic decoherence* becomes observable when the longitudinal quantum velocity spread is dominant as discussed in the two slits case. In Fig. 6 we plotted the comparison of three slits interference pattern with and without decoherence for different values of the velocity spread a.

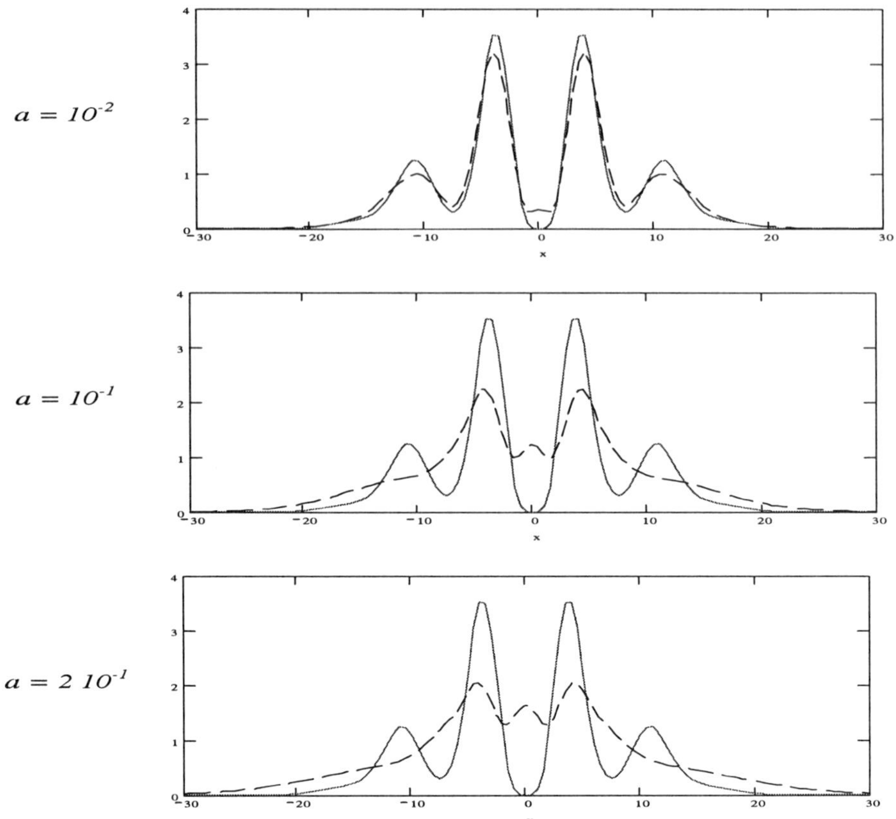

Fig. 6. Behaviour of three slits interference pattern with decoherence (dash line) and without decoherence (solid line) at time $\bar{t} = L/R \approx 4.5$ and $d/\sigma_x \approx 7.7$ for different values of a (looking down: $a = 10^{-2}$, $a = 10^{-1}$and $a = 2\,10^{-1}$).

4. Conclusions

We showed that interference pattern of a two and three slits experiment provides a direct evidence of the free particle Schrödinger spread. Furthermore, in the three slits experiment, we demonstrated that for *discrete values of the distances between the slits and of the distance of the screen* it is possible to inhibit particles to reach the center of the screen itself, while in the two slits case there is a maximum of the probability density. Moreover, we studied the fringes visibility, that now can be reduced by two different envelopes: the usual *diffraction envelope* and a new *decoherence envelope*, which is a straightforward consequence of the formalism presented in reference [1] to describe decoherence due to fluctuations in the evolution time. In this paper these fluctuations are induced by particle velocity spread as described in cavity QED[2, 9]. This spread *can* have a classical contribution (thermal spread and so on) which can be made arbitrarily small with a very monochromatic beam, but it *always* has a finite quantum contribution due to the Heisenberg uncertainty principle. When this contribution is dominant, one should observe *spontaneous intrinsic decoherence*, i.e. decoherence not due to interaction with environment or to experimental fluctuation of some parameter. In general, we find that decoherence is proportional to the *square of the distance of the slits*, to the velocity spread parameter a and becomes very "efficient" at the macroscopic limit.

One can think that our results can be easily reproduced just performing an average with the classical velocity distribution along $\hat{z}$ axis. However this would give an expression of the decoherence exponent D similar to Eqs. (39), (42) and (43), but with two basic differences: a^2 instead of a and a will be only a function of the classical contribution $\Delta_{\overline{V}}$. Furthermore, if $a \ll 1$, one will obtain a much smaller decoherence than the one of our approach.

References

[1] R. Bonifacio, "Time as a statistical variable and intrinsic decoherence", Nuovo Cimento **114B**, 473–488 (1999).

[2] R. Bonifacio, S. Olivares, P. Tombesi, and D. Vitali, "Model independent approach to nondissipative decoherence", Phys. Rev. A **61**, 053802-1–053802-8 (2000).

[3] W. H. Zurek, "Decoherence and the Transition from Quantum to Classical", Phys. Today **44**, 36–44 (October 1991), and references therein.

[4] A. Messiah, *Quantum Mechanics* (North-Holland Publishing Company, Amsterdam, 1961), Vol. I, Chap. VIII § 13.

[5] R. Bonifacio, and S. Olivares "Young's experiment, Schrödinger's spread and spontaneous intrinsic decoherence", Z. Naturforsch. **56a**, 41–47 (2001).

[6] C. Cohen–Tannoudji et al., *Quantum Mechanics* (John Wiley & Sons, New York, 1977), Vol. 1.

[7] B. G. Englert, "Lecture notes on: Complementarity, Wave-Particle Duality, And All That" (1998).

[8] R. Bonifacio, S. Olivares, P. Tombesi, and D. Vitali, "Non–dissipative decoherence in Rabi oscillation experiments", J. Mod. Opt. **47**, 2199–2211 (2000).

[9] R. Bonifacio, and S. Olivares "Spontaneous Intrinsic Decoherence in Rabi Oscillations Experiments" in *Quantum Communication, Computing, and measurement 3*, P. Tombesi, O. Hirota, eds. (Kluwer Academic/Plenum Publishers, 2001).

[10] A. Messiah, *Quantum Mechanics* (North-Holland Publishing Company, Amsterdam, 1961), Vol. I, Chap. IV § 10.

All optical formation of a Bose Einstein condensate

M. Chapman
School of Physics, Georgia Institute of Technology, Atlanta, GA 30332-0430
michael.chapman@physics.gatech.edu

Abstract: We have created a Bose-Einstein condensate of 87Rb atoms directly in an optical trap. We employ a quasi-electrostatic dipole force trap formed by two crossed CO_2 laser beams. Loading directly from a sub-Doppler laser-cooled cloud of atoms results in initial phase space densities of ~1/200. Evaporatively cooling through the BEC transition is achieved by lowering the power in the trapping beams over ~2 s. The resulting condensates are F=1 spinors with 3.5 x 10^4 atoms distributed between the m_F=(-1,0,1) states.

Faster-than-light propagations and their applications

Raymond Y. Chiao, Claus Ropers, Daniel Solli, and Jandir M. Hickmann
Department of Physics, University of California
Berkeley, California 94720-7300
phone: (510)642-7166, fax: (510)643-8497
e-mail: chiao@physics.berkeley.edu

Abstract: Recent experiments have confirmed our predictions that the group velocity of light pulses propagating in transparent optical media can exceed c. In electronics, the causality principle does not forbid negative group delays of analytic signals in electronic circuits, in which the peak of an output pulse leaves the exit port of a circuit *before* the peak of the input pulse enters the input port. In condensed matter physics, negative transmission times of atoms are possible through superfluid helium slabs and through atomic BECs. Relativity is not violated by these phenomena.

1 INTRODUCTION

In this paper, we discuss some recent experimental and theoretical developments concerning faster-than-light propagation phenomena in optics, electronics, and condensed matter physics. For an earlier critical review of this subject, see [1], in which the phenomena of superluminal (i.e., faster-than-c) tunneling times of photons, and the theoretical predictions of superluminal light pulse propagation in transparent media, were reviewed. We hope here to be able to help and correct the commonly held, but mistaken, belief that only the phase velocity in a medium can exceed c, but not the group velocity. We discuss here several situations in which the group velocity in fact exceeds c, and point out some applications of these phenomena, such as to speed-up of computer circuits by using negative group delays to cancel out deleterious positive group delays. We also apply these ideas to novel situations in condensed matter physics, such as to the transmission of atoms through a slab of superfluid helium, or through an atomic Bose-Einstein condensate. The transmission times of such atoms, under the appropriate experimental conditions, can be superluminal, or even negative.

2 PHASOR DESCRIPTION OF SUPERLUMINAL PROPAGATION

A simple phasor picture (see Figure 1) helps explain how such superluminal group velocities can occur for all kinds of waves (e.g., for light and for matter waves), in complete generality. The peak of a Gaussian wave packet, indicated by point B in Figure 1, is the moment when all the phasors which represent the various Fourier components of the wave packet, line up in a straight line, so that the resultant phasor has a maximum amplitude. Therefore, the peak of the wave packet represents a point of maximum constructive interference. By the same token, a point in the early part of the wave packet where the amplitude is small, represented by point A in Figure 1, is a moment of almost total destructive interference, in which the same phasors curl up to form a polygon which almost closes in on itself. Hence, the resultant amplitude at point A in the early tail of the Gaussian wave packet, is small.

After propagation through a dispersive medium, there is no physical reason why the phasors at the early point A cannot precess due to dispersion, in such a way that the

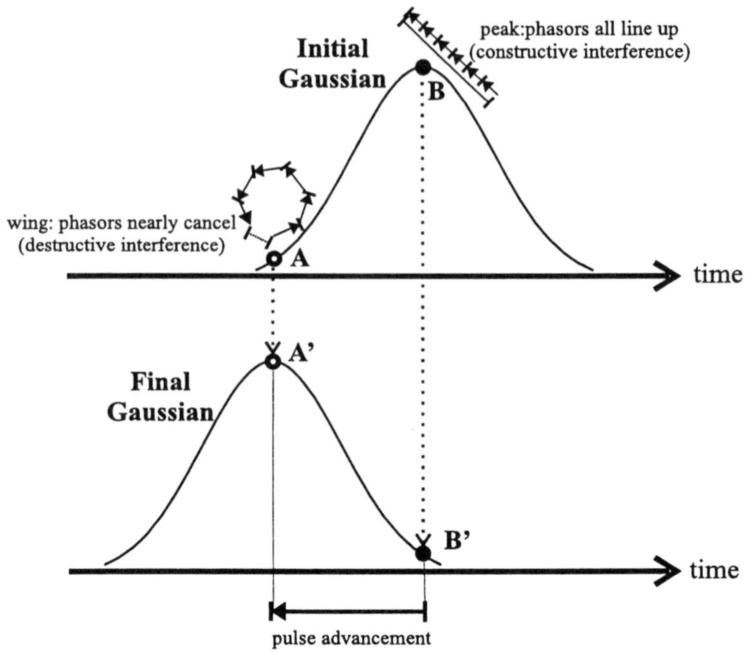

Fig. 1. Phasor description of superluminal propagation.

curled-up polygon becomes uncurled, forming a large resultant phasor at point A'. The phasors at this point are now aligned in constructive interference, and the peak occurs *earlier* in time in the wave packet compared with what would have happened in the vacuum. By the same reasoning, the aligned phasors at point B now curl up to form a polygon which almost closes in on itself, leading to almost total destructive interference at point B' in the trailing tail of the wave packet. The result is an *advancement* of the entire Gaussian wave packet relative to vacuum propagation; this occurs whenever the spectrum of this wave packet lies in a region of negative wave-number dispersion relative to that of the vacuum. In the case of optics, this can occur in a spectral region of anomalous dispersion.

Furthermore, if the dispersion in the group velocity vanishes at the carrier frequency, the entire wave packet can propagate superluminally with negligible change in shape. Moreover, by the superposition principle, one can add up many such wave packets to form an arbitrary, analytic waveform, which can propagate faster than c with negligible distortion.

Much more commonly, the medium possesses a *positive* dispersion, and the phasors precess in the opposite sense from the above case of *negative* dispersion. This results in a *retardation* of the Gaussian wave packet relative to vacuum propagation, rather than an *advancement*. For example, in optical media such as ordinary glass, one normally observes *subluminal* pulse propagation in spectral regions of normal dispersion. Nevertheless, the underlying physical mechanisms for both the retardation and the advancement of the wave packet are identical, apart from the *sign* of the precession of the phasors, and both are possible in principle.

However, it should be emphasized that there are conditions on the above phasor-precession processes, which must be fulfilled before they can be physically significant. In general, the group advancements or retardations of an incident wave packet will be destroyed by the *decoherence* of the system. For example, in the optical case it is necessary to maintain the phase coherence of a light pulse during the phasor-precession

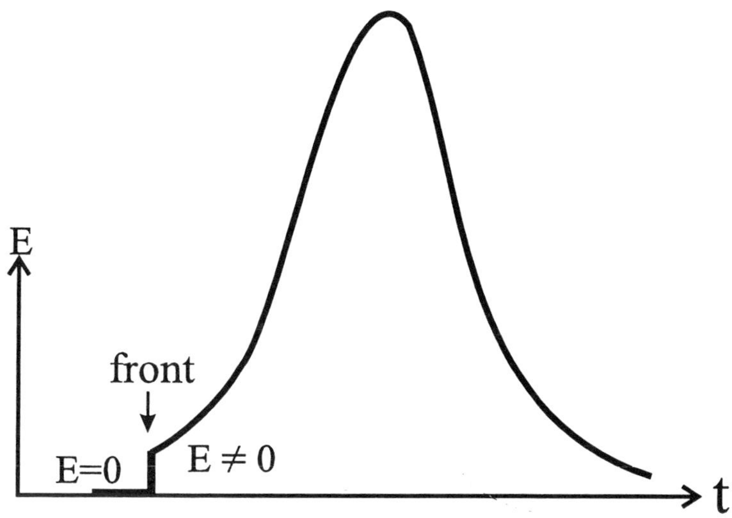

Fig. 2. Front defined by discontinuity.

process which produces the advanced peak; otherwise, the constructive interference will be destroyed. In particular, for an optical medium with gain, stimulated emission must always be accompanied by spontaneous emission. Hence, spontaneous emission will lead to phasor decoherence, and this will limit the maximum possible amount of superluminal advancement [2]. One possible definition of a signal velocity in terms of a signal-to-noise ratio which is limited by amplified spontaneous emission, leads to a velocity which is less than c [3]. Thus, signals so defined cannot propagate faster than c.

Why do such superluminal pulse propagation phenomena not violate relativity? The answer can be seen in Figure 2, where the Gaussian wave packet is multiplied by a step function, such that there results a *front* in its early tail, ahead of which the electromagnetic field is strictly zero. Here, all phasors representing the field ahead of the front have strictly zero length. Since the discontinuity at the front contains Fourier components of infinite frequency, where the index of refraction of any dispersive medium approaches unity, it follows that the front velocity is exactly c, as was pointed out by Sommerfeld and Brillouin [4]. Although the peak of the Gaussian wave packet moves at the group velocity, which may exceed c, this peak can never overtake the front. The reason is that there is no way that phasors of zero length ahead of the front can ever produce a resultant phasor of finite length. Thus, no signal can overtake the front, and this sets a fundamental limit on superluminal pulse propagation, in agreement with the principle of causality in special relativity.

It is important to distinguish between two kinds of signal waveforms in physics: *analytic* waveforms, such that of a Gaussian wave packet, and *nonanalytic* wave forms, such as that of a step function. Any small but finite piece of the the early part of an analytic wave form can be extrapolated into the future by means of a Taylor series; therefore, there is no surprise associated with the arrival of any of the features of the subsequent waveform, including its peak. Conversely, the same extrapolation of the signal ahead of the Sommerfeld front would predict that the waveform would remain identically zero for all time. When the front arrives, there is a genuine surprise associated with its arrival [5]. This suggests another definition of signal velocity in terms of discontinuities [1]. Again, signals so defined cannot travel faster than c.

An important special case of superluminal propagation occurs when the group velocity becomes negative. Recent optical experiments at Princeton NEC [6] have verified the

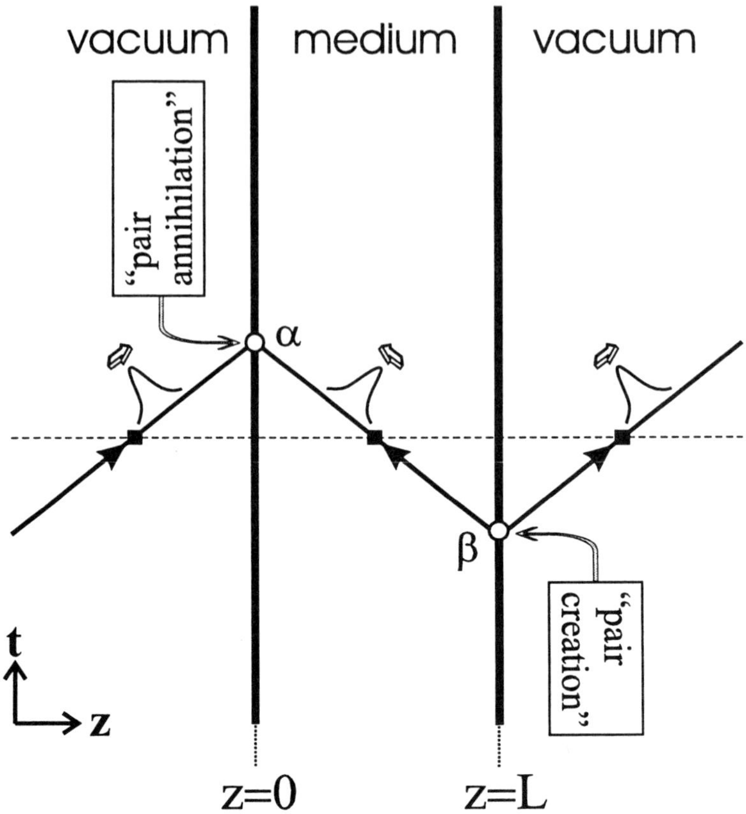

Fig. 3. zigzag world-line diagram.

prediction by one of the authors and his co-workers that superluminal pulse propagation, in particular, propagation with a negative group velocity, can occur in transparent media with optical gain [7]. These experiments have shown that a laser pulse can indeed propagate with little distortion in an optically pumped cesium vapor cell with a *negative* group velocity: The peak of the output laser pulse left the output face of the cell *before* the peak of the input laser pulse entered the input face of the cell. This implies a *negative time of flight* of the pulse through the cell. A negative-time-of-flight process is represented in Figure 3 by a zigzag world-line for the motion of the peak of the wave packets. Note that a time slice through the middle of the zigzag world-line shows that at that moment, three wave packets are simultaneously present in the system: the input wave packet propagating to the right, whose peak is about to enter the input face of the cell, a wave packet propagating to the left, which is in the middle of the cell, and an output wave packet propagating to the right, whose peak has already left the output face of the cell. The question immediately arises as to how energy can be conserved in such a process. One answer is that there can be optical gain, for example, through the population inversion of an atomic system: the atoms can then possess enough extra energy which can then be "loaned" to the light in order to produce the two extra pulses at point β at the output face of the medium, in a process reminiscent of "pair creation" (in the sense of the creation of a pair of wave packets). The borrowed energy is later restored to the inverted atomic system at point α at the input face of the medium, in the process reminiscent of "pair annihilation" (in the sense of the annihilation of a pair of wave packets). Note that for this zigzag process, if one were to *double* the cell length, the output laser pulse would have to come out *twice* as early.

This process is reminiscent of the zigzag world line introduced by Feynman in quantum electrodynamics shown in Figure 4(b). In this diagram, Feynman interpreted the backwards-in-time propagation of a particle (an electron) as the forwards-in-time propagation of an antiparticle (a positron). Point b, which corresponds to electron-positron pair creation in the vicinity of a nucleus of charge +Ze, occurs *earlier* in time than point a, which corresponds to pair annihilation in the vicinity of the nucleus. The net result of this Feynman process is that the peak of the scattered electron wave packet comes out *earlier* from the scattering region than when it entered [8]. This is in contrast to the more intuitive process shown in Figure 4(a), where the electron wave packet comes out *later* from the scattering region than when it entered. Both processes are possible, and must be added coherently when the final states are indistinguishable. However, there is a fundamental limit on the magnitude of the wave packet advancement associated with the counter-intuitive zigzag diagram. Note that the intermediate state consisting of the positron portion of the zigzag is a *virtual state.* The energy-time uncertainty principle

$$\Delta E \Delta t \sim \hbar, \tag{1}$$

allows the temporary borrowing of the energy ΔE from the nuclear Coulomb field at point b, as long as the borrowed energy is paid back later at point a after a short time Δt, where Δt is the duration of the intermediate state during which the positron exists. The resulting negative group delay for the scattered electron wave packet is plotted schematically in Figure 5. Note that there results a "negative Hartman effect," in which a negative group delay saturates at a value given by the above uncertainty principle. This behavior is analogous to the positive Hartman effect seen for wave packets which tunnel through a barrier. In this effect, the tunneling particle experiences a positive group delay, which saturates for thick barriers at a positive value given by the uncertainty principle. The Hartman effect [9] represents the thick-barrier limit of the Wigner tunneling time [10], which is also plotted in Figure 5, for the sake of comparison.

We performed some early experiments on the speed of the quantum tunneling process [11]. We found that a photon tunneled through a barrier at an effective group velocity which was faster than c. In these experiments, spontaneous parametric down-conversion was used as a light source which emitted randomly, but simultaneously, two photons at a time (i.e., photon "twins"). These photons were detected by means of two equidistant Geiger counters (silicon avalanche photodiodes), so that the time at which a "click" was registered was interpreted as the time of arrival of the photon. Coincidence detection was used to detect these photon twins. One photon twin traverses a tunnel barrier, whilst the other traverses an equal distance in the vacuum.

The idea of the experiment was to measure the time of arrival of the tunneling photon with respect to its twin, by measuring the time difference between the two "clicks" of their respective Geiger counters. (We employed a two-photon interference effect in the Hong-Ou-Mandel interferometer [12] in order to achieve sufficient time resolution.) The net result was surprising: On the average, the Geiger counter registering the arrival of the photon which tunneled through the barrier clicked *earlier* than the Geiger counter registering the arrival of the photon which traversed the vacuum. This indicates that the process of tunneling in quantum physics is superluminal.

The earliest experiment to demonstrate the existence of faster-than-c group velocities was performed by Chu and Wong at Bell Labs. They showed that picosecond laser pulses propagated superluminally through an absorbing medium in the region of anomalous dispersion inside the strong optical absorption line [13]. This experiment was reproduced in the millimeter range of the electromagnetic spectrum by Segard and Macke [14]. These experiments verified the prediction of Garrett and McCumber [15] that Gaussian-shaped pulses of electromagnetic radiation could propagate with faster-

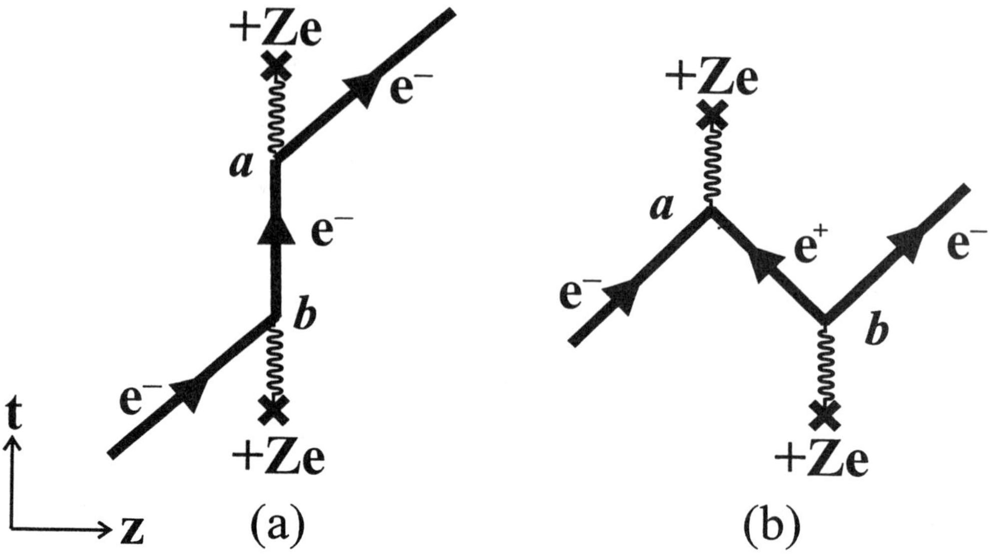

Fig. 4. Feynman processes in QED for multiple scattering of an electron from a nuclear (+Ze) Coulomb field. In the zigzag process (b), pair creation occurs at point b before pair annihilation occurs at point a. Lorentz transformations leave the interval between a and b space-like, and hence "superluminal."

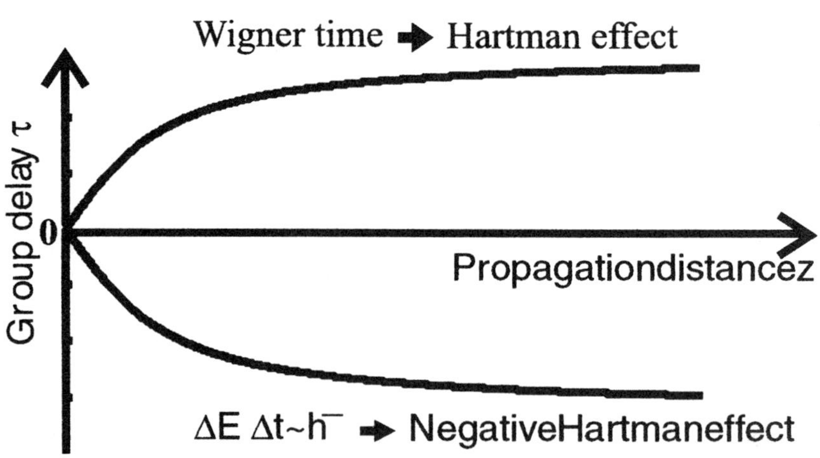

Fig. 5. Positive group delays in Wigner tunneling time lead to a positive Hartman effect; negative group delays (see Figure 4(b)) lead to a negative Hartman effect.

than-c group velocities in regions of anomalous dispersion associated with an absorption line. Negative group velocities were also observed to occur in these early experiments. However, these kinds of superluminal pulse propagation phenomena were not known to occur in *transparent* optical media at the time.

Subsequently, we observed these counter-intuitive pulse sequences in experiments on electronic circuits [16]. In the first of these experiments, we used an electronic circuit which consisted of an operational amplifier with a negative feedback circuit containing a passive RLC network. This circuit produced a negative group delay similar to that observed in the recent optical pumping experiment [6]: The peak of the output voltage pulse left the output port of the circuit *before* the peak of the input voltage pulse entered the input port of the circuit. Such a seemingly anti-causal phenomenon does not in fact violate the principle of causality, since there is sufficient information in the early portion of any analytic voltage waveform to reproduce the entire waveform earlier in time. We showed that causality is solely connected with the occurrence of discontinuities, such as "fronts" and "backs" of signals, and not with the peaks of voltage waveforms, and, therefore, that causal loop paradoxes could never arise [17]. Since there was gain in these electronic circuits, the output signal was not strongly attenuated, in contrast to the earlier optical experiments.

We believe that these counter-intuitive ideas can be applied to the design of microelectronic devices, in particular, computer chips [18]. This is timely, since it is widely believed that Moore's law for microprocessor performance will fail to hold in the next decade due to a "brick wall" arising from fundamental physical limitations [19]. Therefore, there have been many proposals for new transistor technologies to try to solve this problem [20][21]. At the present time, the "transistor latency" problem is one of the main factors limiting computer performance, although the "propagation delays" due to the RC time constants in the interconnects between individual transistors on a computer chip are beginning to be another serious limiting factor. As the scale of microprocessor circuits fabricated on a silicon wafer is reduced to become ever smaller in size, the transistor switching time becomes increasingly faster, but the propagation delay from transistor to neighboring transistor becomes increasingly longer [22]. This will still be true even after new technologies to replace MOSFETS with faster devices are implemented.

3 GENERAL PRINCIPLES FOR GENERATING NEGATIVE GROUP DELAYS IN ELECTRONICS

We now begin our discussion of superluminal effects in electronic circuits, using the concept of negative group delays as the starting point. Electronic circuits are usually very small in size compared with the wavelengths corresponding to the typical frequencies of operation of these circuits; thus, the retardation due to the speed of light across these circuits is usually negligible. Nevertheless, a concatenation of such negative-delay circuits interspersed periodically along a transmission line, could lead to superluminal propagation of pulses with a negative group velocity. Hence, we focus here only on how a negative group delay can be generated in general.

3.1 *Negative group delays necessitated by the "Golden Rule" for operational amplifier circuits with negative feedback*

In Figure 6, we show an operational amplifier with a signal entering the noninverting (+) port of the amplifier. The output port of the amplifier is fed back to the inverting (−) port of the amplifier by means of a black box, which represents a passive linear circuit with an arbitrary complex transfer function $\widetilde{F}(\omega)$ for a signal at frequency ω. We thus

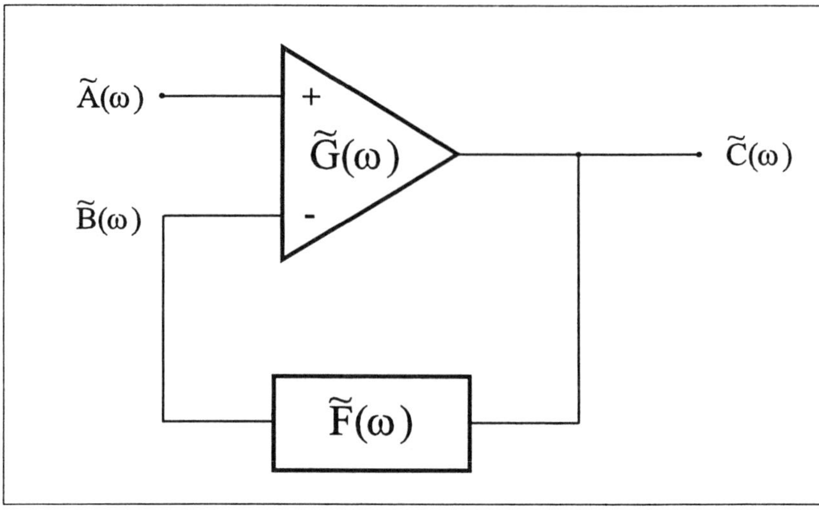

Fig. 6. Operational amplifier circuit with negative feedback.

have a linear amplifier circuit with a negative feedback loop containing a passive filter. In general, the transfer function of any passive linear circuit, such as a RC low-pass filter, will always lead to a *positive* propagation delay through the circuit.

However, for operational amplifiers with a sufficiently high gain-feedback product, the voltage difference between the two input signals arriving at the inverting and noninverting inputs of the amplifier must remain small at all times. The operational amplifier must therefore supply a signal with a *negative* group delay at its output, such that the *positive* delay from the passive filter is exactly canceled out by this negative delay at the inverting (−) input port. The signal at the inverting (−) input port will then be nearly identical to that at the noninverting (+) port, thus satisfying the "Golden Rule" which demands small voltage differences at all times. The net result is that this negative feedback circuit will produce an output pulse whose peak leaves the output port of the circuit *before* the peak of the input pulse arrives at the input port of this circuit.

In Figure 7, we show experimental evidence for this counter-intuitive behavior for the special case of an RLC tuned bandpass circuit in the negative feedback loop [16]. The peak of an output pulse is *advanced* approximately by 12 milliseconds relative to the input pulse. The output pulse has obviously not been significantly distorted with respect to the input pulse by this linear circuit, apart from a slight amplification factor. Also, note that the size of the advance of the output pulse is comparable in magnitude to the width of the input pulse. At these very low frequencies, the role of spontaneous emission is entirely negligible, and the pulse advance can obviously satisfy Rayleigh's criterion for pulse resolution, as can be seen by inspection of the data shown in Figure 7.

That causality is not violated is demonstrated in a second experiment, in which the input signal voltage is very suddenly shorted to zero the moment it reaches its maximum. The result is shown in Figure 8. By inspection, we see that the output signal is also very suddenly reduced to zero voltage at essentially the same instant in time that the input signal has been shorted to zero. This demonstrates that the circuit cannot advance in time truly *discontinuous* changes in voltages: These are the only points on the signal waveform which are directly connected by the principle of causality [17]. However, for the *analytic* changes of the input signal waveform, such as those in the early part of the Gaussian input pulse which we used, the circuit evidently has the ability to extrapolate the input waveform into the future, in such a way as to reproduce

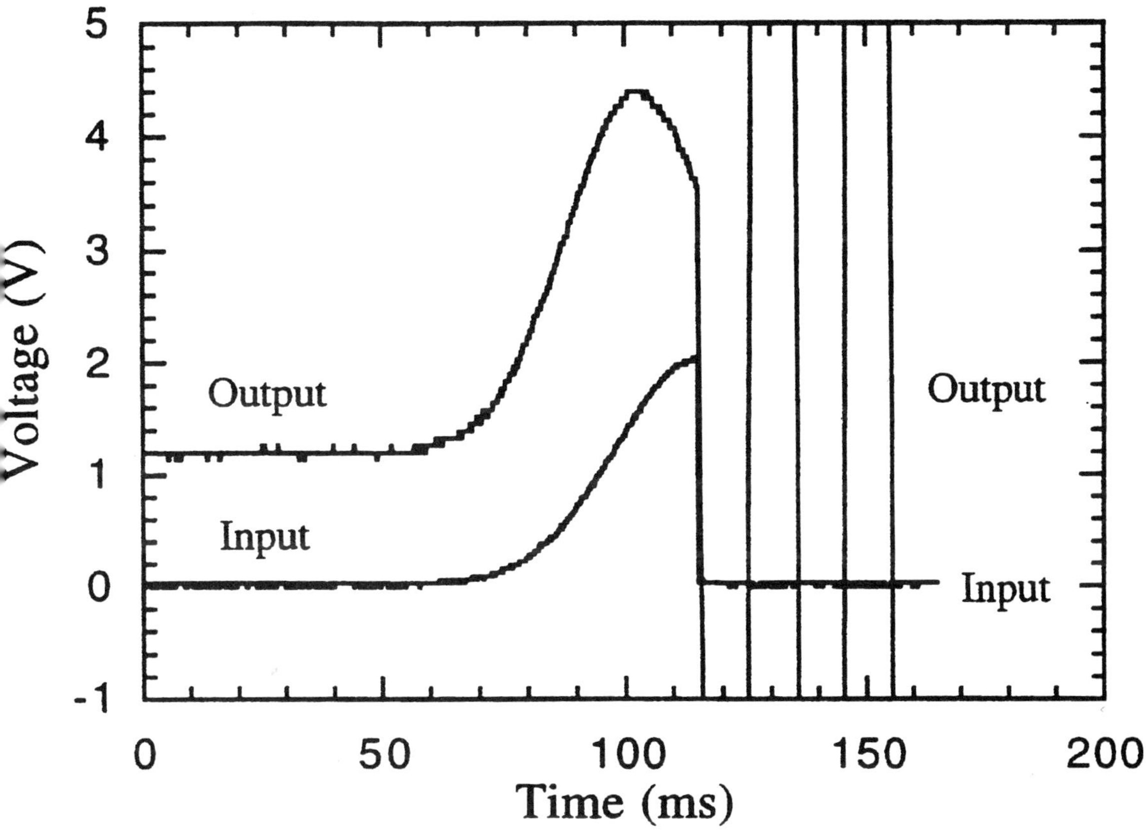

Fig. 8. Experimental results showing that discontinuities cannot be advanced.

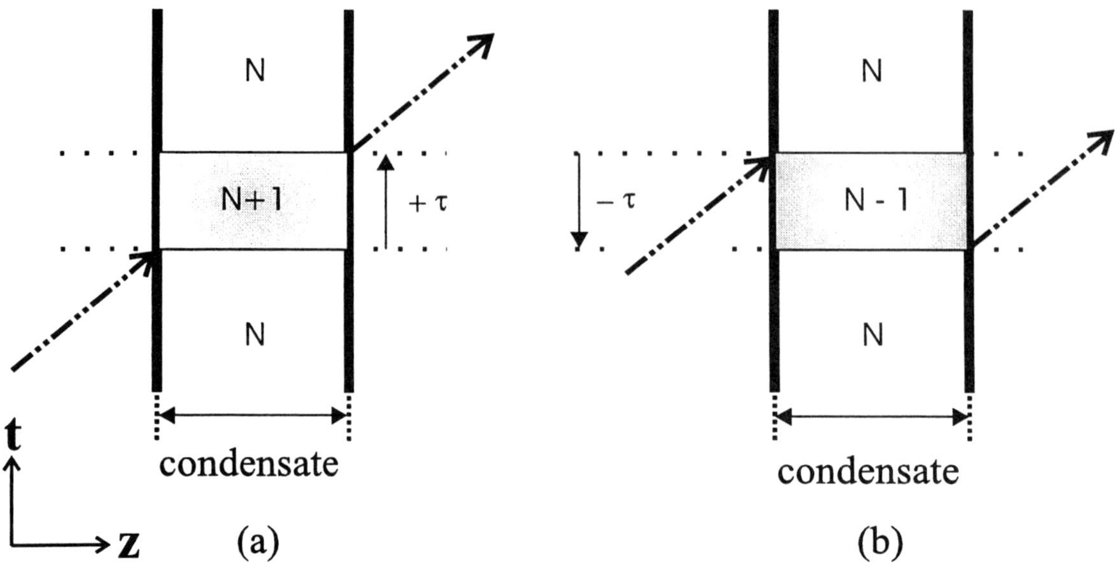

Fig. 11. Condensate-mediated processes associated with the coherent transmission of an incoming atom.

necessity of a self-consistent many-body solution for a system of strongly interacting bosons.

Circumventing all these difficulties, we give here some general arguments for why there can in principle be anomalous transmission times in the transmission of a helium atom through a superfluid helium slab, if measurements with sufficiently high time resolution are performed.

Halley et al. [34] proposed the possibility of the transmission of helium atoms through a superfluid helium slab due to a condensate-mediated process with transmission time delays independent of the slab thickness. Consider a superfluid helium slab in an N particle ground state. In the Halley et al. process, the transmission of an incoming helium atom through the slab occurs via a virtual transition of the N particle ground state plus the incident particle, to the $N+1$ particle ground state, followed by the coherent reemission of one particle on the opposite side of the superfluid (see Figure 11(a)). Following the steps of the paper first suggesting this process [34], we consider the transfer Hamiltonian

$$H_T = \sum_{\mathbf{k}'} (T^L_{\mathbf{k}'} b_{\mathbf{k}',L} a_0^\dagger + T^R_{\mathbf{k}'} b^\dagger_{\mathbf{k}',R} a_0 + h.c.) \tag{10}$$

The $b^\dagger$ and b operators are creation and annihilation operators of particles with momentum $\mathbf{k}'$ at the left (L) and right (R) surfaces; the $a_0^\dagger$ and a_0 operators create and annihilate the particles in the condensate part of the fluid. The coefficients T depend on the microscopic structure of the fluid surface and must be determined by experiment.

We assume that $|left\rangle = |N, N_0\rangle|\mathbf{k}\rangle_L|vac\rangle_R$ is the initial state of the system with a total number of N atoms in the fluid, out of which N_0 particles are in the condensate; one helium atom with momentum $\mathbf{k}$ is incident from the left. A transmission of an atom from the left to the right side of the slab can occur in the second-order process shown in Figure 11(a), for which the transition matrix element is the following:

$$\frac{\langle right|H_T|i\rangle\langle i|H_T|left\rangle}{E - E_i}. \tag{11}$$

Here $|i\rangle = |N+1, N_0+1\rangle|vac\rangle_L|vac\rangle_R$ is the intermediate state, E_i its energy, and $|right\rangle = |N, N_0\rangle|vac\rangle_L|\mathbf{k}\rangle_R$ the final state.

Using the standard definitions for the creation and annihilation operators, it is easy to show that the numerator of Eq. (11) is

$$(N_0+1)T_{\mathbf{k}}^L T_{\mathbf{k}}^R. \tag{12}$$

In this scattering process, the energy in the initial and the final state is given by the sum of the kinetic energy of the incoming particle plus the energy of the N particle ground state

$$E = \epsilon_{\mathbf{k}} + E_N \tag{13}$$

The energy in the intermediate state is

$$E_i = E_{N+1}. \tag{14}$$

This leads to an energy denominator in second-order perturbation theory of

$$\Delta E = \epsilon_{\mathbf{k}} + E_N - E_{N+1} = \epsilon_{\mathbf{k}} + |\mu|. \tag{15}$$

Here, μ is the chemical potential (−7.16 K in the case of superfluid helium). Thus, the transition matrix element for this process is

$$\frac{(N_0+1)T_{\mathbf{k}}^L T_{\mathbf{k}}^R}{\epsilon_{\mathbf{k}} + |\mu|}. \tag{16}$$

If the intermediate state used above were the only one coupling the initial and final states, a transition rate in the form of Fermi's golden rule could be given by squaring the transition amplitude [34]. However, a *coherent* sum of the transition amplitudes over all possible intermediate states has to be performed, and there is another process which was not taken into account in Ref. [34]. This process is equally important and leads to *negative* time delays in transmission. It is pictured in Figure 11(b), and we will refer to it as the primed process. A helium atom approaches the slab, and, before the atom reaches the superfluid, the condensate coherently emits another atom on the other side of the slab. In this intermediate state there are two atoms outside of an $N-1$ particle ground state. Finally, the atom on the left side of the superfluid gets absorbed into the condensate, and the fluid is brought back into the N particle ground state.

The energy for the intermediate state in the primed process is

$$E_i' = 2\epsilon_{\mathbf{k}} + E_{N-1}. \tag{17}$$

The energy difference between the initial and the intermediate state is

$$\Delta E' = \epsilon_{\mathbf{k}} + E_N - (2\epsilon_{\mathbf{k}} + E_{N-1}) = -\epsilon_{\mathbf{k}} - |\mu| = -\Delta E. \tag{18}$$

The numerator of Eq. (11) is now

$$N_0 T_{\mathbf{k}}^R T_{\mathbf{k}}^L, \tag{19}$$

so that the transition matrix element for the primed process is

$$-\frac{N_0 T_{\mathbf{k}}^R T_{\mathbf{k}}^L}{\epsilon_{\mathbf{k}} + |\mu|}. \tag{20}$$

For plane waves, the final states of these two condensate-mediated processes are indistinguishable. Therefore the two transition matrix elements must be added up coherently. Comparison of Eq. (16) with Eq. (20) shows that the two processes destructively interfere with each other, and that they almost cancel out in the large N_0 limit. This implies that the transmission probability for very long, single-particle wave packets via a condensate-mediated process becomes vanishingly small.

The energy-time uncertainty relation restricts the durations of the intermediate states, so that the transmission times will be of the order of $\pm\hbar/|\Delta E|$ for the unprimed and primed processes, respectively. For low energy incident atoms the transmission times are approximately given by $\pm\hbar/|\mu|$, which is of the order of a picosecond in the system considered. In order to detect either one of these processes experimentally, the final states have to be made distinguishable. This could be achieved by temporal resolution, i.e., the formation of atomic wave packets with a duration comparable to the lifetimes of the respective intermediate states, so that the transmitted wave packets would be Rayleigh resolved, and an actual measurement of the time sequence of events could be performed. A wave packet this short in time will have an energy uncertainty comparable to the size of the chemical potential, so that Rayleigh resolution will only barely be possible.

It should be noted that both of the processes considered here depend in the same way on the macroscopic coherence and on the off diagonal long range order of this quantum system. Both processes will lead to faster-than-light effects. In the primed process this is due to the intrinsic negativity of the transmission time. In the unprimed process faster-than-light effects can be achieved by the choice of a sufficiently thick slab, since the transmission time, although positive, is independent of the slab thickness, similar to the case of the Hartman effect in tunneling discussed earlier.

4.2 *Positive and negative transmission times of atoms through an atomic BEC*

In connection with condensate-mediated transmission of atoms, it is simpler to consider the recently observed atomic BECs [35], which are weakly-interacting Bose gases, than superfluid helium, which is a strongly-interacting Bose liquid. The microscopic theory of a dilute, weakly-interacting Bose gas is well characterized in terms the Bogoliubov theory [36], which has been verified by experiment. In the case of the atomic BECs, the chemical potential sets the energy scale for the uncertainty-principle lifetime or duration of the intermediate state, and hence for the transmission times (either positive or negative) through the condensate of low-energy incident atoms identical to the atoms of the BEC. The chemical potential μ in the condensate can be calculated in the Bogoliubov approximation, and is given by

$$\mu \approx N_0 \frac{2\pi\hbar^2 a}{mV}, \tag{21}$$

where N_0 is the number of atoms in the condensate, a is the S-wave scattering length, m is the mass of the atom, and V is the volume of the condensate. The typical experimental parameters for a BEC consisting of sodium atoms are $a = 2.75$ nm, $N_0/V = 1.5 \times 10^{14}$ cm^{-3}, and $D = 10$ μm for the typical size of the condensate [37]. The speed of sound for such a condensate has been observed to be $v_s \approx 1.0$ mm/s, which is close to that predicted by the Bogoliubov theory. From these numbers, we infer that the chemical potential for an atom in the typical sodium BEC is $\mu \approx 7.4 \times 10^{-24}$ erg ≈ 54 nK. The typical time scale associated with condensate-mediated processes is therefore $\hbar/\mu$, which is two orders of magnitude shorter than the transmission time due to sound wave propagation across the condensate. Hence it should be easy to distinguish the condensate-mediated processes from sound-wave mediated processes. Note that the time scale $\hbar/\mu$ is independent of the size D of the condensate, and leads to a Hartman-like effect, which resembles the Hartman effect in tunneling. The positive and negative condensate-mediated transmission times, or group delays, which correspond to the two processes depicted in Figure 11(a) and (b), respectively, are

$$\tau_{\pm} \approx \pm \frac{\hbar}{\varepsilon_{\mathbf{k}} + \mu} \pm \frac{\hbar}{\mu} \approx \pm 0.14 ms. \tag{22}$$

These times are much larger than those observed in the photon tunneling experiments. Again, as in the case of superfluid helium, it should barely be possible to distinguish, by Rayleigh's resolution criterion, a negative group delay from a positive group delay, for cold atoms whose incident kinetic energy is comparable to the chemical potential. Nevertheless, this effect should be experimentally observable with a sufficiently high signal-to-noise ratio, as has already be done in the case in the measurement of the tunneling times for photons. Furthermore, each atomic transmission event detected in an experiment would result in a definite *sign* for the transmission time for that event. The post-selection of rare, condensate-mediated transmission events leads to a "weak measurement" of this post-selected subensemble in the sense of Aharonov [38], and results, as in the case of tunneling [39], in surprising "weak values," such as *negative* transmission times. Thus, not only can light pulses be slowed to a stop in atomic BECs [40], but atoms can also be transmitted superluminally through such BECs.

5 CONCLUSIONS

There is a widespread view among electrical engineers and physicists that although the phase velocity can exceed the vacuum speed of light, the group velocity can never do so. Otherwise, signals would be able to propagate faster than light, since conventional wisdom equates the group velocity with the signal velocity. From this conventional point of view, the group velocity is essentially the same as the energy velocity in transparent media, and the latter could never exceed c. Several generations of students have been taught this. Many of the standard textbooks and handbooks state this conventional viewpoint, some, however, with qualifications which unfortunately are not strong enough, so that the net result is still misleading. For example, *The Electrical Engineering Handbook* in its discussion concerning the group velocity states the following [41]:

> "When traveling in a medium, the velocity of energy transmission (e.g. a light pulse) is less than c, and is given by [the group velocity]."

This statement, and other similar statements in many of the early standard textbooks in optics and classical electrodynamics, are misleading. As a result, we have been blinded by our misconceptions, and thereby have been prevented from exploring and discovering many new, interesting, and possibly important, phenomena, which could have been discovered long ago. Some of these are only now being uncovered, and some of these phenomena may in fact lead to important applications, such as the speed-up of computers.

The effects reported in this paper do not violate relativity. The front velocity of Sommerfeld and Brillouin, which is strictly c, is the only velocity which is relevant to relativity. However, the group, the energy, and Brillouin's "signal," velocity can all exceed c, without violating the principle of relativistic causality [1]

6 ACKNOWLEDGMENTS

This work was supported by the ONR, NSF, and NASA. We thank S. Coen, M. Haakestad, E. R. Krotcheck, C. McCormick, K. B. Whaley and M. Mojahedi for helpful discussions. J. M. H. is on a sabbatical leave from the Department of Physics, Federal University of Alagoas, Maceió, AL, Brazil with financial support from CNPq.

References and links

1. R. Y. Chiao and A. M. Steinberg, in *Progress in Optics*, edited by E. Wolf, Vol. XXXVII, (Elsevier, Amsterdam, 1997), p. 347.
2. B. Segev, P. W. Milonni, J. F. Baab, and R. Y. Chiao, Phys. Rev. A **62**, 022114 (2000).
3. A. Kuzmich, A. Dogariu, L. J. Wang, P. W. Milonni, and R. Y. Chiao, Phys. Rev. Lett. **86**, 3925 (2001).
4. *Wave Propagation and Group Velocity* by L. Brillouin (Academic Press, New York, 1960).
5. We thank J. C. Garrison for pointing out the importance of the idea of "surprise" in discontinuous signals.
6. L. J. Wang, A. Kuzmich, and A. Dogariu, Nature **406**, 277 (2000); A. Dogariu, A. Kuzmich, and L. J. Wang, Phys. Rev. A **63**, 053806 (2001).
7. R. Y. Chiao, Phys. Rev. A**48**, R34 (1993); A. M. Steinberg and R. Y. Chiao, Phys. Rev. A **49**, 2071 (1994); E. L. Bolda, J. C. Garrison, and R. Y. Chiao, Phys. Rev. A 49, 2938 (1994); R. Y. Chiao in *Amazing Light: A Volume Dedicated to Charles Hard Townes on his 80th Birthday* edited by R. Y. Chiao (Springer Verlag, New York, 1996), p. 91; R. Y. Chiao and A. M. Steinberg, in the *Proceedings of the Nobel Symposium No. 104* held in Gimo, Sweden on June 13-17, 1997 on 'Modern Studies of Basic Quantum Concepts and Phenomena,' edited by E. B. Karlsson and E. Braendas, Physica Scripta T**76**, 61 (1998).
8. *Quantum Electrodynamics* by R. P. Feynman (W. A. Benjamin, New York, 1961), p. 75. The zigzag process of Fig. 4(b) involves a fermionic system in a *microscopic* effect. Bosonic systems allow *macroscopic* effects, which can arise because they can exhibit off diagonal long range order with little decoherence. In Compton scattering also, there is a process in which a photon leaves the scattering region *before* it enters. We thank C. McCormick for pointing this out to us.
9. T. E. Hartman, J. Appl. Phys. **33**, 3427 (1962).
10. E. P. Wigner, Phys. Rev. **98**, 145 (1955).
11. A. M. Steinberg, P. G. Kwiat, and R. Y. Chiao, Phys. Rev. Lett. **71**, 708 (1993); A. M. Steinberg and R. Y. Chiao, Phys. Rev. A **51**, 3525 (1995).
12. C. K. Hong, Z. Y. Ou, and L. Mandel, Phys. Rev. Lett. **59**, 2044 (1987).
13. S. Chu and S. Wong, Phys. Rev. Lett. **48**, 738 (1982); **49**, 1293.
14. B. Segard and B. Macke, Phys. Lett. A **109**, 213 (1985).
15. C. G. B. Garrett and D. E. McCumber, Phys. Rev. A **1**, 305 (1970).
16. M. W. Mitchell and R. Y. Chiao, Phys. Lett. A **230**, 133 (1997); M. W. Mitchell and R. Y. Chiao, Am. J. Phys. **66**, 14 (1998).
17. J. C. Garrison, M. W. Mitchell, R. Y. Chiao, E. L. Bolda, Phys. Lett. A**245**, 19 (1998).
18. *Microelectronic circuits* by A. S. Sedra and K. C. Smith (4th edition, Oxford University Press, Oxford, 1998).
19. P. A. Packan, Science **285**, 2079 (1999).
20. Y. Taur, IEEE Spectrum **36**, 25 (1999).
21. L. Geppert, IEEE Spectrum **37**, 46 (2000).
22. D. A. B. Miller, Proceedings of the IEEE **88**, 728 (2000).
23. *Electrodynamics of Continuous Media* by L. D. Landau and E. M. Lifshitz (Pergamon Press, Oxford, 1960), p. 256.
24. *Fields and Waves in Communication Electronics* by S. Ramo, J. R. Whinnery, and T. van Duzer (John Wiley & Sons, New York, 1965), p. 623.
25. E. L. Bolda, R. Y. Chiao, and J. C. Garrison, Phys. Rev. A **48**, 3890 (1993).
26. R. Loudon, J. Phys. A **3**, 233 (1970).
27. E. O. Schulz-Dubois, Proc. IEEE **57**, 1748 (1969).
28. K. E. Oughstun and S. Shen, J. Opt. Soc. Am. B **5**, 2395 (1988).
29. J. Peatross, S. A. Glasgow, and M. Ware, Phys. Rev. Lett. **84**, 2370 (2000).
30. M. A. H. Tucker and A. F. G Wyatt, Science **283**, 1150 (1999).
31. P. A. Mulheran and J. C. Inkson, Phys. Rev. B **46**, 5454 (1992).
32. F. Dalfovo , A. Fracchetti, A. Lastri, L. Pitaevskii , S. Stringari, Phys. Rev. Lett. **75**, 2510 (1995).
33. C. E. Campbell, E. Krotscheck, and M. Saarela, Phys. Rev. Lett. **80**, 2169 (1998).
34. J. W. Halley, C. E. Campbell, C. F. Giese, and K. Goetz, Phys. Rev. Lett. **71**, 2429 (1993); A. K. Setty, J. W. Halley, and C. E. Campbell, Phys. Rev. Lett. **79**, 3930 (1997).
35. M. H. Anderson, J. R. Ensher, M. R. Matthews, C. E. Wieman, and E. A. Cornell, Science **269**, 198 (1995); C. C. Bradley, C. A. Sackett. J. J. Tollett, and R. G. Hulet, Phys. Rev. Lett. **75**, 1687 (1993); K. B. Davis, M.-O. Mewses, M. R. Andrews, N. J. van Druten, D. S. Durfee, D. M. Kurn, and W. Ketterle, Phys. Rev. Lett. **75**, 3969 (1995); D. Kleppner, Physics Today **49**, 11 (1996); A. S. Parkins and D. F. Walls, Physics Reports **303**, 1 (1998).
36. N. Bogoliubov, J. Phys. (USSR) **11**, 23 (1947); R. Y. Chiao, Opt. Comm. **179**, 157 (2000).

37. W. Ketterle and S. Inouye, arXiv:cond-mat/0101424 29 Jan 2001.
38. Y. Aharonov, D. Z. Albert, and L. Vaidman, Phys. Rev. Lett. **60**, 1351 (1988); Y. Aharonov and L. Vaidman, Phys. Rev. A **41**, 11 (1990).
39. A. M. Steinberg, Phys. Rev. A **52**, 32 (1995); A. M. Steinberg, Phys. Rev. Lett. **74**, 2405 (1995).
40. C. Liu, Z. Dutton, C. H. Behroozi, L. V. Hau, Nature **409**, 490 (2001).
41. *The Electrical Engineering Handbook* edited by R. C. Dorf (2nd edition, CRC Press/IEEE Press, Boca Raton, 1997), p. 812.

Beyond diffraction limit in optical lithography via quantum entanglement

J. P. Dowling

NASA-JPL, California Institute of Technology
Mail Stop 126-347, 4800 Oak Grove Drive, Pasadena, California 91109

ABSTRACT: The ever-increasing power of computers entirely rely on the fine technique of lithography which writes the circuit images onto semiconductor chips. The resolution in optical lithography is limited by so-called Rayleigh diffraction limit of of $\lambda/2$, where λ is the optical wavelength. Hence the conventional way to achieve better resolution is to use shorter wavelength light. But the imaging technique, however, easily becomes problematic for UV or X-ray region. By using entangled photon pairs we may overcome such difficulties.

"Stopping" of light and quantum memories for photons

Michael Fleischhauer

Fachbereich Physik, Univ. Kaiserslautern, D-67663 Kaiserslautern, Germany
mfleisch@physik.uni-kl.de

Abstract: The theory and first experimental results for a controlled and reversible transfer of the coherence and the quantum state of a photonic wavepacket to collective atomic spin excitations are presented. Potential applications to quantum memories for photons are discussed and the role of decoherence processes in systems with collective excitations are analyzed.

Quantum aspects of imaging

A. Gatti
Istituto Nazionale per la Fisica della Materia and Dipartimento di Scienze dell'Università dell'Insubria, Via Valleggio 11, 22100 Como, Italy
alessandra.gatti@mi.infn.it

Abstract: Quantum imaging is a recently born research field, which studies how quantum entanglement properties of light beams in the spatial domain can be exploited in order to improve the quality of processing of images and of parallel signals. This talk will first address some general concepts as spatially multimode squeezing and spatial entanglement, and describe some optical devices that are able to generate light beams with these features; then some interesting approaches that are presently pursued in this field will be illustrated.

Coherence and Quantum Optics VIII
Edited by Bigelow et al., Kluwer Academic/Plenum Publishers, 2003

Polarization dynamics of a two-photon laser

Daniel J. Gauthier, Olivier Pfister,[†] William J. Brown,[‡] and Michael D. Stenner

Duke Univesity, Department of Physics and
The Fitzpatrick Center for Photonics and Communication Systems,
Durham, North Carolina 27708, USA

The high degree of temporal coherence of laser light arises from a complex interplay between the fundamental light-matter interactions of absorption, spontaneous emission, and stimulated emission. The coherence properties of the generated light can be altered significantly, and often in a surprising manner, by modifying the type of light-matter interaction on which the laser is based. As an example, the two-photon laser [1, 2] is based on the higher-order two-photon stimulated emission process, whereby two incident photons stimulate an excited atom to a lower energy state and four photons are scattered, as shown schematically in Fig. 1. While replacing the standard one-photon stimulated emission process by a high-order one might be expected to give rise to subtle differences observable only at the quantum level, it has been predicted that there will be dramatic changes in both the microscopic and macroscopic laser behavior even when many atoms participate in the lasing process. The reason for these differences is that the two-photon stimulated emission rate depends quadratically on the incident photon flux, resulting in an inherently nonlinear light-matter interaction.

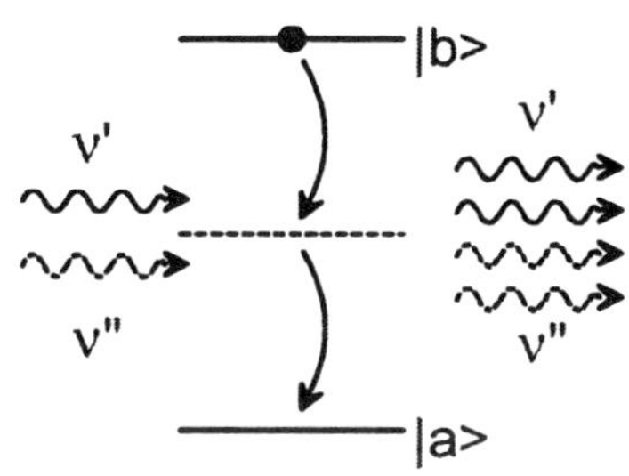

Fig. 1. The idealized two-photon stimulated emission process.

We have recently achieved degenerate two-photon optical lasing [3] using laser-driven potassium atoms contained in a very high finesse optical resonator. This new type of quantum oscillator displays the expected turn-on behavior whereby it can only turn on when an injected coherent field brings the intercavity light above a critical value. Once the laser is above threshold, we observe polarization instabilities and chaos even though the total intensity remains nearly constant. This observation is not encompassed within the framework of current theoretical models of two-photon lasers and results from the atomic Zeeman degeneracy of the states participating in the two-photon stimulated emission process. We suggest that such a two-photon laser could emit polarization-entangled twin beams of light if this degeneracy is lifted, which may have quantum information applications.

To achieve two-photon optical lasing, we use a two-photon gain medium consists of a dense, collimated, effusive beam of laser-driven ^{39}K atoms that passes through the center of the optical resonator. We use the D_1 transition of ^{39}K (6 MHz natural linewidth) whose atomic states are denoted by $|\alpha F_\alpha M_\alpha\rangle$, with $\alpha = g$ for $4^2S_{1/2}$ and $\alpha = e$ for $4^2\mathrm{P}_{1/2}$ levels, and F and M are the quantum numbers for the total angular momentum and its projection on the quantization axis $\hat{z}$. See the energy level diagrams shown in Fig. 2. The ground-state hyperfine splitting is equal to 462 MHz and is denoted by Δ_g. To create a population inversion for multiphoton transitions starting from the state $|g22\rangle$, the atoms are continuously optically pumped into this state by two $\hat{\sigma}_+$-polarized fields whose frequencies are set close to ω_{21} and ω_{22} (Bohr frequencies are denoted by $\omega_{F_e F_g}$). Note that the atoms are decoupled from the optical pumping fields when they are in this state. The maximum atomic number density in the interaction region is $\sim 2 \times 10^{11}$ atoms/cm^3, the atomic beam diameter is 2.5 mm, and the measured residual Doppler absorption width is 30 MHz.

Two-photon stimulated emission into the laser cavity mode (frequency ω_l) occurs when the optically pumped atoms are illuminated by an intense $\hat{\sigma}_-$-polarized Raman pump beam of frequency ω_d, and ω_l is adjusted to the two-photon transition frequency ($\simeq \omega_d + \Delta_g/2$). Photons are added two at a time to the cavity field via stimulated emission when two photons are annihilated from the $\hat{\sigma}_-$-polarized Raman pump field and the atom undergoes a multi-photon transition from the state $|g22\rangle$ to any one of the states $|g1M_g\rangle$, as shown schematically in Fig 2. Note that such a hyper-Raman process can be mapped onto the idealized two-photon stimulated emission process shown in Fig. 1 by working in a dressed-state basis. The atomic, Raman pump beam, and optical cavity axes are mutually orthogonal, thereby suppressing competing phase-matched processes that might give rise to parametric oscillation.

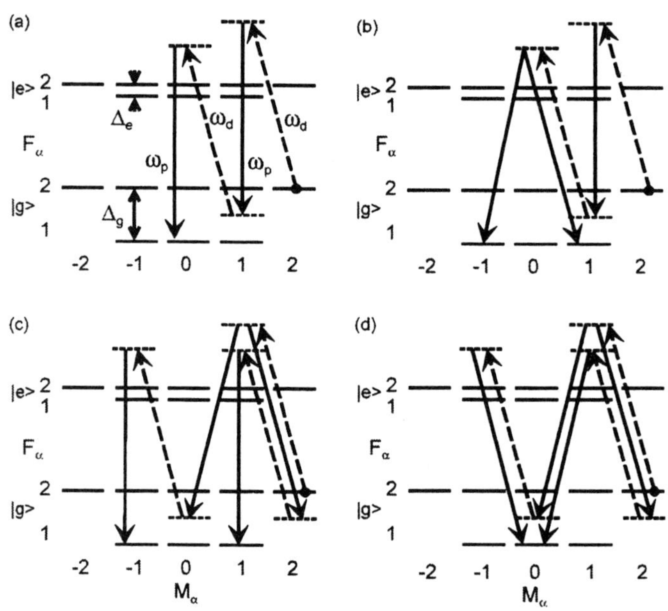

Fig. 2. Two-photon Raman scattering diagrams showing the quantum pathways for the possible states of polarization of the two-photon laser field (solid line) and the circularly polarized Raman pump beam (dashed line).

Two-photon amplification

Before discussing the two-photon laser properties, we demonstrate that this system can give rise to two-photon amplification. We characterize the gain medium by measuring the amplification of an intense $\hat{z}$-polarized probe beam (frequency ω_p) in the absence of the resonator [4]. A typical probe-beam transmission spectrum is displayed in Fig. 3 where the probe- and Raman pump-beam intensities have been made strong enough to induce the multi-photon process shown in Fig. 2(a). The Raman pump-beam intensity is ~ 25 W/cm^2, the atomic number density is $\sim 10^{11}$ atoms/cm^3, and $\omega_d - \omega_{22} = 522$ MHz. We observe several stimulated emission features that disappear when the optical pumping beams are blocked. We identify the mechanisms responsible for each gain and absorption feature by their location and probe-power dependence. The lowest-order contributions to these processes are diagramed as insets to Fig. 3. The two strongest gain peaks arise from one-photon Raman scattering: (α) at $\omega_p = \omega_d$ and (δ) at $\omega_p = \omega_d + \Delta_g$. The two-photon gain feature (γ) is located about halfway between (α) and (δ), $(\omega_p = \omega_d + \Delta_g/2$ in lowest order). The doublet (ϵ) is "pump-dressed" one-photon gain. The weaker peaks are one-photon-emitting transitions originating from ground states much less populated than $|g22\rangle$, as well as gain and absorption lines arising from ^{41}K (6.7% abundance).

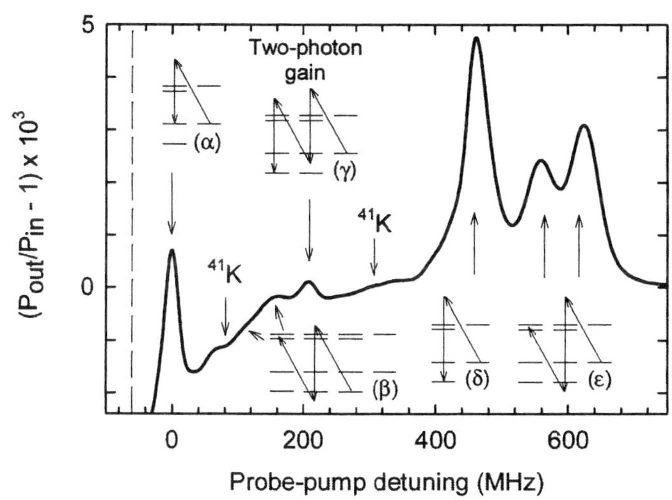

Fig. 3. Intense probe beam transmission spectrum of the Raman-pumped atomic beam.

To demonstrate that the peak denoted by (γ) is a genuine two-photon process, we measure the dependence of its size on the intensity of the probe beam. This intensity dependence is different for one- and two-photon amplification processes. This can be understood simply from an analysis of the amplification process, considering a plane wave propagating through a homogeneously broadened thin medium of length ℓ. The gain experienced by a probe beam and its resonant unsaturated limit are given by

$$G_2 = \frac{I}{I_o} - 1 = \frac{g_2 \ell I_o}{1 + \Delta_2^2/\Gamma_2^2 + I_o^2/I_{2s}^2} \rightarrow g_2 \ell I_o, \tag{1}$$

where I (I_o) are the output (input) probe intensities, g_2 is the unsaturated two-photon gain coefficient, I_{2s} is the saturation intensity, Γ_2 is the homogeneous width, and Δ_2 is the detuning from resonance that include the light shifts of the transitions induced by the probe beam. Equation 1 show that the unsaturated two-photon gain is *proportional* to I_o.

Figure 4 displays the height of peak (γ) versus the input probe power. A clear linear increase of the gain is observed for powers up to about 3 mW (24 W/cm^2 intensity), for which a saturation plateau occurs at a maximum peak height of 4×10^{-4}. This observed linear dependence below saturation is a clear demonstration that photons are emitted two at a time. The solid line shows the best fit of Eq. 1 to the data with $g_2 = 0.90(6)10^{-4}$ cm/W, and $I_{2s} = 36(3)$ W/cm^2.

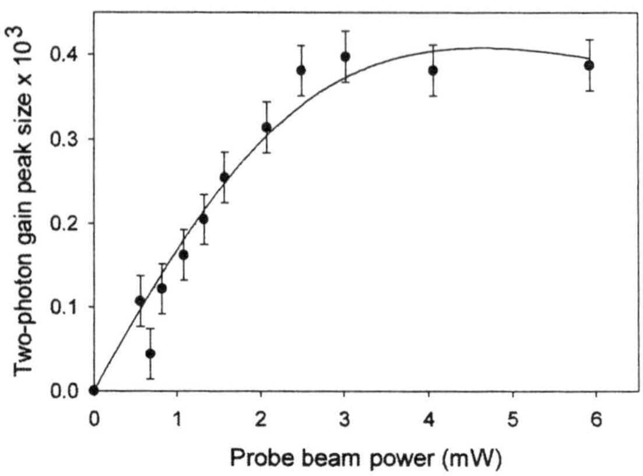

Fig. 4. Size of the two-photon gain peak as a function of the input probe power.

Two-photon lasing

Reaching the threshold for two-photon lasing requires the use of high-quality laser mirrors (developed for cavity quantum electrodynamics experiments) because of the small size of the two-photon gain. The laser resonator is placed in a vacuum chamber (see Fig. 5) and consists of two high-reflectivity ultra-low-loss mirrors of radius of curvature R=5 cm set close together at a distance of $L = 1.464$ cm in a linear (standing-wave) transverse-mode-degenerate sub-confocal configuration. This configuration suppresses normal one-photon lasing that can arise from spectrally nearby gain features (one-photon lasing can prevent or obscure two-photon lasing). The laser mirrors are mounted on a super-invar structure to minimize thermal drift of L, with one mirror attached to an electronically controlled mount to permit cavity alignment in vacuum. The other mirror is mounted on a piezoelectric ceramic for fine scanning of the cavity length. The mirrors have a transmissivity of $T = 2{\times}10^{-4}$ as measured by the manufacturer (Research Electro-Optics). We find that the cavity finesse $\mathcal{F} = 15,400 \pm 150 \simeq \pi/(T + A)$ using the ring-down technique, yielding a loss per mirror $A = (4 \pm 2) \times 10^{-6}$.

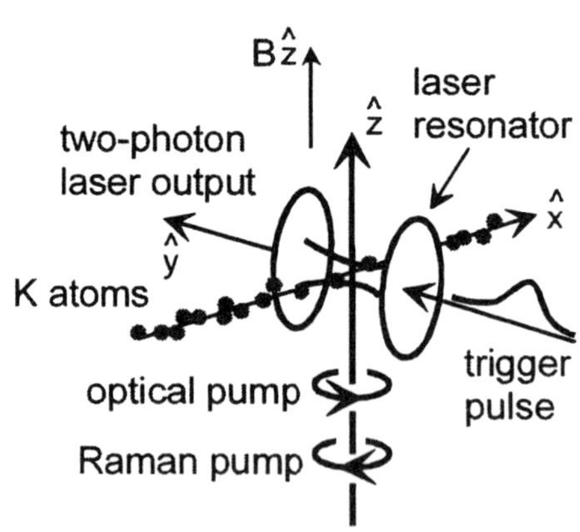

Fig. 5. Two-photon laser experimental setup.

We use a short cavity length so that the free spectral range ($FSR = c/2L$ =10.24

GHz) is larger than the frequency difference between the two-photon gain feature and all other nearby one-photon gain features, which are within 400 MHz of the two-photon frequency. We set the cavity length precisely to a sub-confocal condition $L = R[1 \pm \cos(\pi/p)]$ with $p = 4$ so that the transverse modes group into degenerate clusters, spaced by $c/2pL$=2.56 GHz and centered on the longitudinal mode frequencies. The spacing between these clusters is also larger than the frequency difference between the two- and one-photon gain features, thus ensuring that one-photon lasing will not occur on a moderate-order transverse mode. Finally, to suppress one-photon lasing on high-order transverse modes (which are no longer frequency degenerate with the lower-order modes due to spherical aberrations), we accurately position under vacuum 400 μm pinholes in front of each cavity mirror and along the cavity axis. With the pinholes in place, $\mathcal{F}$=15,140, resulting in a longitudinal mode linewidth $FSR/\mathcal{F}$ = 680 kHz. Within the cavity mode volume, we estimate that there are a maximum of 7×10^6 atoms. Throughout the experiments, the pump beam frequency is adjusted so that it is tuned to the blue side of the ω_{22} transition by 512 MHz. The mutually orthogonal geometry of our experiment suppresses competing phase-matched processes, ensuring that we are constructing a true two-photon laser and not a parametric oscillator.

Our experimental system is very rich because there are four quantum pathways, depicted in Fig. 2, that connect the initial state $|g22\rangle$ and the final states $|g1M_g\rangle$, differ only by the state of polarization of the cavity field, and are frequency degenerate at zero magnetic field. The two-photon resonator can support any state of polarization (it is measured to be highly isotropic) hence all pathways of Fig. 2 must be considered.

Conclusive evidence that we have achieved two-photon lasing is indicated by the threshold behavior of the laser. Briefly, the threshold condition for all lasers is that the round-trip gain must equal the round-trip loss. For one-photon lasers this yields the well known result that lasing will commence when a uniquely defined minimum inversion density (proportional to the gain) is attained via sufficient pumping. The situation is more complicated for the two-photon laser because the unsaturated gain increases with increasing inversion density *and* with increasing cavity photon number. This results in a threshold condition specified by a uniquely defined minimum inversion density and cavity photon number [5]. If the laser is pumped strongly but with no photons in the cavity initially, it cannot turn on unless quantum fluctuations [6] or an injected field brings the intracavity light above the critical value. In addition, once the minimum photon number exists in the cavity, the photon number undergoes a run-away process, growing rapidly until the two-photon transition is saturated [5]. Therefore, the two-photon laser operates in the saturated regime (a source of optical nonlinearity) even at the laser threshold, giving rise to the possibility that the laser will display dynamical instabilities.

To observe the threshold behavior, we measure the power of the beam emitted from the two-photon laser resonator and inject pulses of light into the cavity to trigger the laser to the 'on' state and shown in Fig. 6(a) for $\sim3.3\times10^5$ injected photons. The duration of the injected pulse is indicated by the dashed vertical lines. It is seen that the power emitted by the cavity rises to approximately 0.2 μW and remains at this value, corresponding to an average cavity photon number of 2.2×10^6 and an intracavity intensity of 7.3 W/cm^2 at the cavity waist.

While the total emitted power is seen to remain nearly constant, we find that the two-photon laser displays polarization instabilities. By placing a linear polarization before the detector, it is seen in Fig. 6(b) that the state of polarization undergoes very regular oscillations of period 0.11 μs, with a 50% depth of modulation. Similar dynamical behavior is observed for all polarizer orientations, suggesting that the state of polarization is elliptical with an ellipticity of 0.5 and a rotating major axis. We find

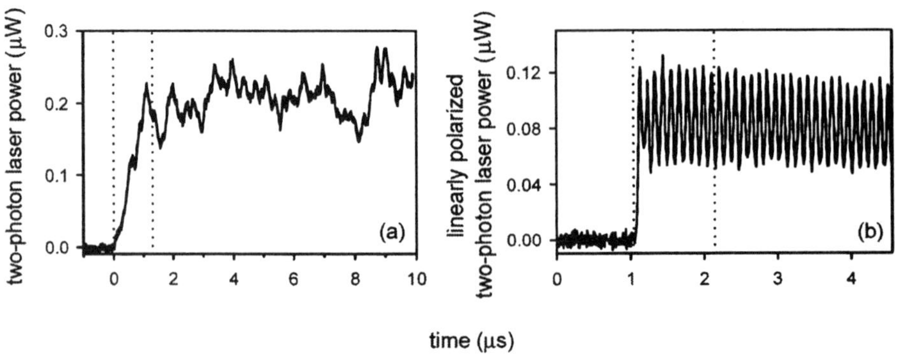

Fig. 6. (a) Turn on behavior of the two-photon laser. (b) Polarization dynamics of the two-photon laser.

that the laser dynamics is quite sensitive to the applied magnetic field: an increase by as little as 0.5 Gauss is sufficient to generate a much more complicated pattern, turning into a possibly chaotic pattern for a field strength of 2 Gauss. We believe that it may be possible to suppress the instabilities using a stronger magnetic field to lift the degeneracy of the different quantum pathways.

Lifting the degeneracy between the pathways also has the promise of yielding polarization entanglement. Consider, for simplicity, a single emission event. When the Zeeman levels are degenerate, the superposition of all possible amplitudes gives a photon pair (1,2) in the state $[\alpha_1|\hat{z}\rangle_1 + \beta_1|\hat{x}\rangle_1][\alpha_2|\hat{z}\rangle_2 + \beta_2|\hat{x}\rangle_2]$ with $|\alpha_i^2| + |\beta_i^2| = 1$, where $|\hat{\varepsilon}\rangle_i$ ($\varepsilon = x, y$) is the polarization-labeled Fock state of the ith photon of the emitted pair. In a strong magnetic field, the $\hat{x}\hat{z}$ and $\hat{z}\hat{x}$ pairs become off-resonant and should stop lasing, creating the polarization-entangled state $\alpha_1\alpha_2|\hat{z}\rangle_1|\hat{z}\rangle_2 + \beta_1\beta_2|\hat{x}\rangle_1|\hat{x}\rangle_2$.

In conclusion, we have described the design and characteristics of a Raman two-photon amplifier and laser. The multilevel nature of the two-photon transition allows interesting effects such as polarization instabilities and possible photon entanglement. This opens up exciting lines of study in nonlinear laser dynamics and quantum optics.

This work is supported by NSF grant PHY-9876988.

†University of Virginia, Dept. of Physics, 382 McCormick Rd, Charlottesville, VA 22904.
‡Corvis Corp., 7015 Albert Einstein Drive, Columbia, MD 21046.

References and links

1. A.M. Prokhorov, Science **10**, 828 (1965); P.P. Sorokin and N. Braslau, IBM J. Res. Dev. **8**, 177 (1964); R.Z. Garwin, *ibid.* **8**, 338 (1964).
2. D.J. Gauthier, Q. Wu, S.E. Morin, and T.W. Mossberg, Phys. Rev. Lett. **68**, 464 (1992).
3. O. Pfister, W.J. Brown, M.D. Stenner, and D.J. Gauthier, Phys. Rev. Lett. **86**, 4512 (2001).
4. O. Pfister, W.J. Brown, M.D. Stenner, and D.J. Gauthier, Phys. Rev. A **60**, R4249 (1999).
5. H.M. Concannon and D.J. Gauthier, Opt. Lett. **19**, 472 (1994); D.J. Gauthier and H.M. Concannon, in *Nonlinear Spectroscopy of Solids: Advances and Applications*, B. Di Bartolo and B. Bowlby (Plenum Press, New York, 1994), pp. 365-384; C.Z. Ning and H. Haken, Z. Phys. B - Condensed Matter **77**, 157 (1989).
6. M. Brune, J.M. Raimond, P. Goy, L. Davidovich, and S. Haroche Phys. Rev. Lett. **59**, 1899 (1987).

Atomic Clocks and Cold Atom Scattering

Chad Fertig, Ronald Legere,* J. Irfon Rees, and Kurt Gibble†
Yale University, P. O. Box 208120, New Haven, CT 06520-8120
(203) 432-6365, Fax: (203) 432-9710, kurt.gibble@yale.edu

Servaas Kokkelmans‡ and Boudewijn J. Verhaar
Eindhoven Technological University, Eindhoven, The Netherlands
31 40 247 4877, B.J.Verhaar@phys.tue.nl

Abstract: Laser cooling has dramatically shifted the relative importance of various systematic errors of fountain clocks relative to beam clocks. Optimizing the performance of fountain clocks will demand shot-noise limited detection of a large number of atoms and possibly spin squeezed atomic states. Fountain clocks also enable a novel quantum atom-optics precision scattering measurement.

1. Introduction

Laser cooling has dramatically shifted the relative importance of various systematic errors of fountain clocks relative to beam clocks since there is a very large frequency shift due to the collisions of cold atoms. This requires operating with a small atomic density so that optimizing the performance of fountain clocks will demand shot-noise limited detection of a large number of atoms, and possibly spin squeezed atomic states.

We demonstrate a prototype of a laser-cooled ^{87}Rb fountain clock and measure the frequency shift due to cold collisions. The shift is fractionally 30 times smaller than that in a laser-cooled Cs clock, allowing a Rb clock to use higher densities for the same clock accuracy. We observe a density dependent pulling by the microwave cavity and use it to cancel the collision shift.

To reach the atom shot noise limit, we are developing a new atomic state detection scheme that directly measures the population difference of the two clock states using FM absorption spectroscopy. We also demonstrate a juggling Rb fountain clock that we will use to study the collisions of juggled balls of atoms. Juggling can significantly improve the clock's short-term stability without requiring greater signal-to-noise or a larger cold collision frequency shift. Finally, a juggling fountain clocks also enables a novel quantum atom-optics precision scattering measurement.

2. Laser-cooled ^{87}Rb fountain clock

The most serious systematic error in laser-cooled clocks is the frequency shift due to cold collisions.[1] Tiesinga *et al.* first calculated this shift[2] and Gibble and Chu[3] measured the shift for laser-cooled Cs clocks to be $\delta\nu/\nu = -1.7\times10^{-12}$ at a density of 10^9 cm^{-3}. Due to Cs_2 molecular resonances, the frequency shift cross section has nearly the maximal value of $\lambda_{dB}{}^2/2\pi$, where λ_{dB} is the de Broglie wavelength. This large cross section led us to examine clocks based on other atoms, for which the cold collision shift might be smaller.[4,5]

A schematic of our ^{87}Rb fountain clock is shown in Figure 1. Using 0.9W of light from a Ti:Sapphire laser delivered by an optical fiber, atoms are collected from the room temperature Rb vapor in the vapor-cell MOT. The atoms are launched upwards and cooled to 1.8 μK in the moving frame. The atoms then pass through 2 microwave cavities that are normally used to prepare half of the atoms in the 5S_ $|F=1,m=0\rangle$ state.[6] Here, to achieve high density so we can best measure the small collision shift, the selection cavities are not used. Instead, we optically pump the atoms into $|1,-1\rangle$ just below the clock cavity. We use successive pulses of

---

* Present address: MIT Lincoln Laboratory, 244 Wood Street, Lexington MA 02420.
† Present address: Penn State University, 104 Davey Lab, University Park, PA 16802.
‡ Present address: JILA, University of Colorado Boulder, Box 440, Boulder, CO 80303.

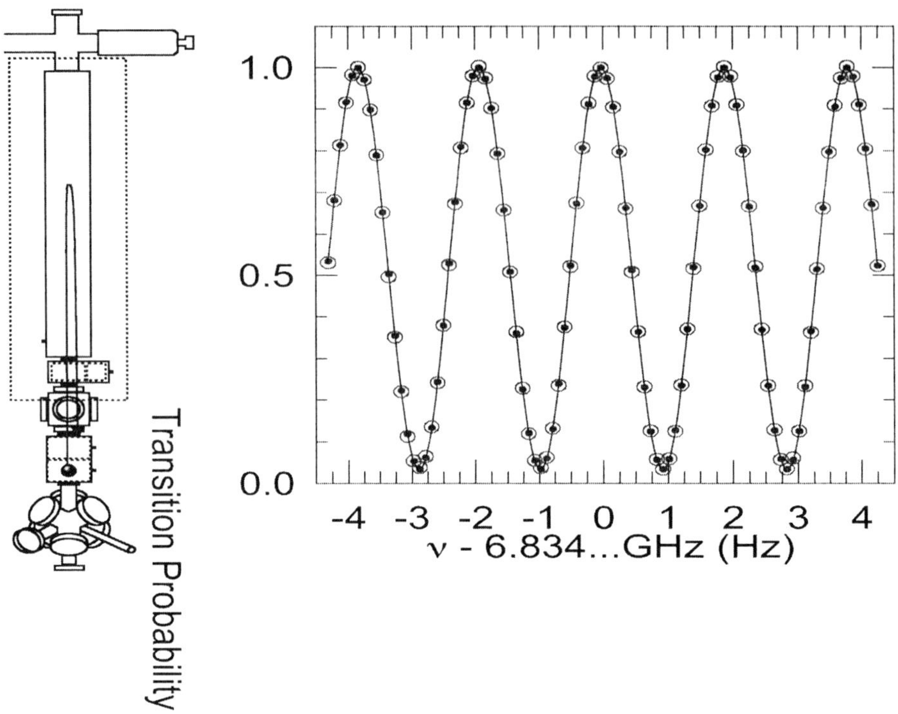

Figure 1. Schematic of ^{87}Rb fountain clock and Ramsey fringes at 6.834 GHz. The large circles are the data and the small are a fit to the data. The linewidth is 0.95 Hz and the S/N=200.

circularly polarized light from diode lasers tuned to the 5S_ F=2 → $5P_{3/2}$ F′=1′ and 1 → 1′ transitions. Microwaves from a horn then transfer atoms from $|1,-1\rangle$ state to the $|2,0\rangle$ state and then a variable number of atoms from $|2,0\rangle$ to $|1,0\rangle$. Any atoms remaining in F=2 are cleared with light tuned to the 2→3′ transition. In this way, we can vary the density preparing a maximum of 70% of the atoms in $|1,0\rangle$ at a temperature of 5 μK, with fewer than 1% in $|1,\pm1\rangle$. To prepare the atoms in $|2,0\rangle$, we add another microwave pulse to transfer the atoms in $|1,0\rangle$ to $|2,0\rangle$ and then repump any atoms in F=1 with light tuned to 1→2′.

The state prepared atoms enter the magnetic shielding and experience a 6.8 GHz microwave pulse in the rectangular TE_{102} clock cavity, creating a coherent superposition of $|1,0\rangle$ and $|2,0\rangle$. This coherence precesses as the atoms are slowed by gravity and return through the clock cavity. The second microwave pulse converts the phase difference between the atomic coherence and the microwave field into a population difference. We detect the transition probability using a laser tuned to 2→3′ producing Ramsey fringes as in Figure 1. To normalize the fringes, we also detect the total number of atoms by repumping the population in $|1,0\rangle$ using a laser beam tuned 1→2′, followed by a second detection laser pulse. In Figure 1, the interrogation time is T=0.526 s which gives a linewidth of Δν=0.950 Hz. With our detection signal-to-noise S/N=500 on a single launch, the atomic transition frequency of 6,834,682,610.9 Hz[7] can be determined in 1 s with a precision of $\delta\nu/\nu=\Delta\nu/\pi\nu S/N=9\times10^{-14}$. However, the short-term instability of our local oscillator limits the S/N to 200 or $\delta\nu/\nu=2.1\times10^{-13}$ for 1s of averaging. The magnetic bias field is 710 μG in the cavity and flight region which produces a fractional quadratic Zeeman shift[1] for this field of $\delta\nu/\nu = 4\times10^{-14}$.

During the precession time above the clock cavity, collisions between cold atoms shift the phase of their coherence producing a frequency shift of the clock. To measure this frequency shift, we vary the atomic density on successive fountain launches and look for a relative shift of the Ramsey fringes. In Figure 2 we show a series of measurements of the frequency as a

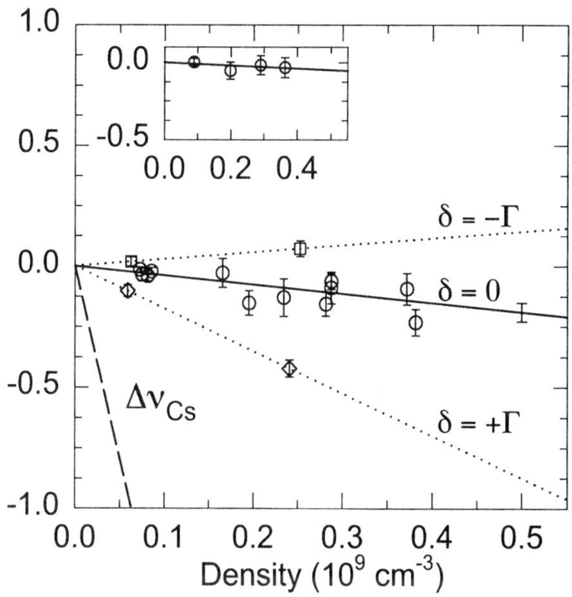

Figure 2. Measured ^{87}Rb cold collision shift (circles and solid line). The shift is -0.38(8) mHz for a density of $1.0(6)\times10^9$ cm^{-3}. The dashed line is the measured shift for Cs.[3] The dotted lines (diamonds and squares) show the density dependence for clock cavity detunings of $\delta=\pm\Gamma$. The inset shows a cancelled collision shift for δ = □30 kHz and atoms prepared in $|1,0\rangle$.

function of density(circles) for atoms prepared in $|1,0\rangle$ and in $|2,0\rangle$. The extrapolated shift for a density of $1.0(6)\times10^9$ cm^{-3} is $-0.38(8)$ mHz.[8,9] We also show the shift for Cs which is fractionally 30 times larger.[3] The measured shift agrees with that calculated in Ref. [5] and also a recent reanalysis of the ^{87}Rb interactions.[10,11]

The measured frequency differences in Figure 2 have a precision of $\pm2\times10^{-15}$. At the 10^{-15} level, there are several potential error sources. The only source that is explicitly density dependent is the pulling of the transition frequency by the coupling of the atoms to the microwave cavity. In NMR, the effect is known as radiation damping where the field radiated by the magnetization of the sample builds up in the microwave cavity and causes the Bloch vector to decay.[12] In hydrogen masers, it is called cavity pulling and is used to cancel the collisional frequency shift.[13] Here, more apparent than the decay of the atomic coherence is a small phase shift. When the cavity is detuned from the atomic transition frequency, the field radiated by the atoms is phase shifted relative to the applied field. The Bloch vector precesses about the total field leading to a phase error proportional to $\mu_0\hbar\mu_B^2N\omega\delta/(\delta^2+\Gamma^2)V_{cav}$.[14] Here, N is the number of atoms, ω is the transition frequency, δ is the cavity detuning, Γ is the cavity HWHM, V_{cav} is the effective cavity volume, and μ_B is the Bohr magneton.

In Figure 2 we also show the measured density dependent frequency shift when we detune the clock cavity by $\pm\Gamma$ for atoms prepared in $|1,0\rangle$ (diamonds and squares). The cavity detuning can significantly influence the density dependence. An intentional detuning of the cavity cancels the cold collision shift as shown in the inset. This spin-exchange tuning[13] has advantages for insuring immunity to long-term variations in the number of trapped atoms. Moreover, density extrapolations can be more accurate since the extrapolation does not depend on accurate density ratios.[4]

The cavity pulling has several unique dependences that we observe. For example, the cavity pulling not only reverses with the cavity detuning δ but also with the initial population

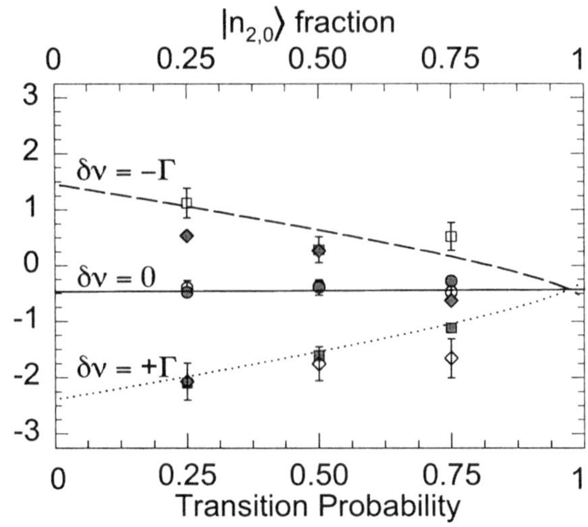

Figure 3. Density dependent shift versus transition probability on a single clock cavity passage for clock cavity detunings of $\delta=+\Gamma$ (diamonds and dotted) and $-\Gamma$ (squares and dashed line). Open (filled) data points are for atoms prepared in $|1,0\rangle$ ($|2,0\rangle$). For $\delta=0$, the circles and solid line are plotted versus the fraction of atoms in $|2,0\rangle$ in the fountain, giving the individual shift contributions from $|1,0\rangle$ and $|2,0\rangle$ state populations. Typical error bars are shown for open points.

inversion of $|1,0\rangle$ and $|2,0\rangle$. In addition, the cavity pulling depends on the transition probability during each cavity passage. For a transition probability of 0.5 on the first interaction($\pi/2$ pulse), a positive cavity detuning δ pulls the frequency higher (lower) for atoms prepared in $|2,0\rangle$ ($|1,0\rangle$). On the second $\pi/2$ pulse, if the microwave frequency is tuned to the side of a Ramsey fringe, the microwave field is phase shifted by $\pi/2$ so that it is parallel or anti-parallel to the Bloch vector. Therefore, the Bloch vector does not precess and the cavity pulling has no first order effect during the second interaction. For a transition probability of 0.25 in the first passage (0.33π pulse), the cavity pulls in the same direction during both cavity passages since the Bloch vector precesses during both. However, the effect of the second passage is small because the number of atoms is 8 times less due to their ballistic expansion. For a transition probability of 0.75 (0.67π pulse), the first cavity passage has a small effect as the Bloch vector is first perturbed in one direction as it precesses to $\theta=\pi/2$ and then the direction of the perturbation reverses once there is a population inversion. On the second passage, the perturbation remains reversed so that the total cavity pulling is small.

We show the cavity pulling effects in Figure 3. We plot the density dependent shift versus the atomic transition probability during the first cavity passage for $n=10^9$ cm^{-3}. For a cavity tuned below the atomic resonance, $\delta=-\Gamma$, and atoms prepared in $|1,0\rangle$, the frequency shift is positive and large for a small transition probability (open squares and dashed line). When the atoms are prepared in $|2,0\rangle$, the filled squares show a large negative density dependent shift for a small transition probability. The dotted line and open (filled) diamonds show the opposite density dependence for $\delta=+\Gamma$, and $|1,0\rangle$ ($|2,0\rangle$) state preparation.

To model the data in Figure 3, we account for the different cold-collision frequency shifts λ_{10} and λ_{20} for unequal $|1,0\rangle$ and $|2,0\rangle$ populations, and the cavity pulling. For no cavity pulling ($\delta=0$), we see essentially no shift as a function of the population in $|2,0\rangle$ during the precession time. We model the 18 measurements in Figure 3 accounting for the partial frequency shift cross sections and the cavity pulling by integrating the time evolution of the Bloch vector as the atom passes through the microwave field profile of the clock cavity. The model agrees and we find $2(\lambda_{10}-\lambda_{20})/(\lambda_{10}+\lambda_{20})=0.1(6)$. This constrains a determination of the ^{87}Rb interactions that suggests either -1.63 or 0.08.[10]

We tune the clock cavity by $\delta=\pm\Gamma$ by changing its temperature by $\mp$2.5K. To measure the frequency response of the cavity, we detect the AC Zeeman shift of the clock states due to a strong microwave sideband that is injected into the clock cavity.[6] The loaded Q of our TE_{102} copper clock cavity is 13,000 and its resonance is measured with an accuracy of 5 kHz.

3. Direct detection of population differences

Because the cold collision frequency shift in ^{87}Rb fountain clocks is fractionally at least 30 times smaller than that for Cs,[8,9] ^{87}Rb clocks can operate at higher densities and with consequently smaller projection noise. Nonetheless, the ^{87}Rb cold collision frequency shift cross-section is sufficiently large to suggest using as few atoms as possible. To retain sufficient short-term stability to reach the clock's potential accuracy, the detection of the transition probability in the clock should be at least nearly shot-noise limited. Accuracies exceeding 10^{-16} will likely require a short-term stability approaching 10^{-14} per launch. This stability of 10^{-14} per launch requires a detection signal-to-noise ratio S/N of nearly 5,000:1 for ^{87}Rb.

The detections schemes currently used in Cs fountain clocks detect the atomic population in F=4 and then either the population in F=3 or the total number of atoms.[8, 15-17] In this way, the transition probability can be normalized to the total number of atoms launched, which is typically stable to not better than 1%. The temporal separation of these two detection pulses aliases instabilities in the detection system, such as the detection laser frequency noise, into the detected signal, limiting the S/N. To date, the best achieved is S/N = 2300:1.[15] We are currently exploring detection schemes that simultaneously probe both atomic populations to avoid this difficult source of noise.

We have demonstrated a new state detection scheme that directly detects a population difference. We use FM absorption spectroscopy using 2 lasers tuned near the 5S_ F=2 $\rightarrow$ $5P_{3/2}$ F′=3′ and the 1$\rightarrow$0′ transitions as shown in Figure 4. Because the lasers tuned near 2$\rightarrow$3′ & 1$\rightarrow$0′ are on opposite sides of the lines, the partial absorption of each sideband by the atoms creates AM that is phase shifted by 180 degrees as shown in Figure 5(a). Thus, when the AM's produced by the populations in each clock state have equal magnitudes, the laser light impinging upon the detection photodiode has no AM and the output of the demodulation mixer is zero. Because the detection of both populations is simultaneous, phase locking the lasers probing the two transitions can eliminate the sensitivity to laser frequency noise. McGuirk *et al.* have demonstrated another realization of this idea by stopping the atoms projected into the upper hyperfine state, allowing the atoms projected into the lower state to fall for a short time (spatially separating the atoms), repumping the lower state atoms, and then using differential absorption of the 2 laser beams derived from the same laser (to avoid a sensitivity to laser frequency noise).[18]

The most difficult aspect of our scheme is efficiently detecting the F=1 population. While the 1$\rightarrow$0′ transition is a closed system, the atoms can quickly optically pump into a non-

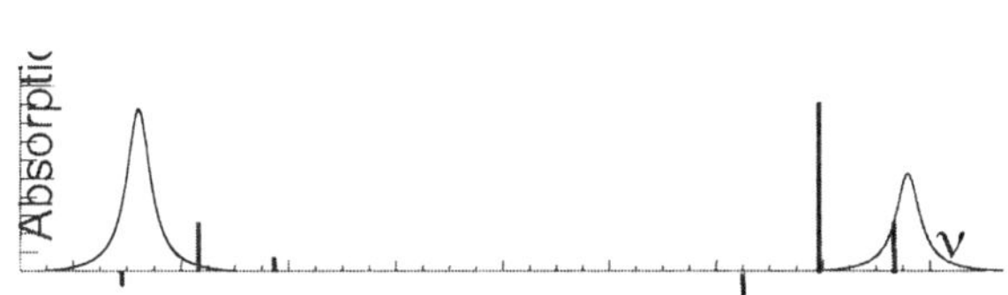

Figure 4. Schematic of ^{87}Rb clock state absorption and the FM laser spectrum. The frequency of the FM is near 50 MHz.

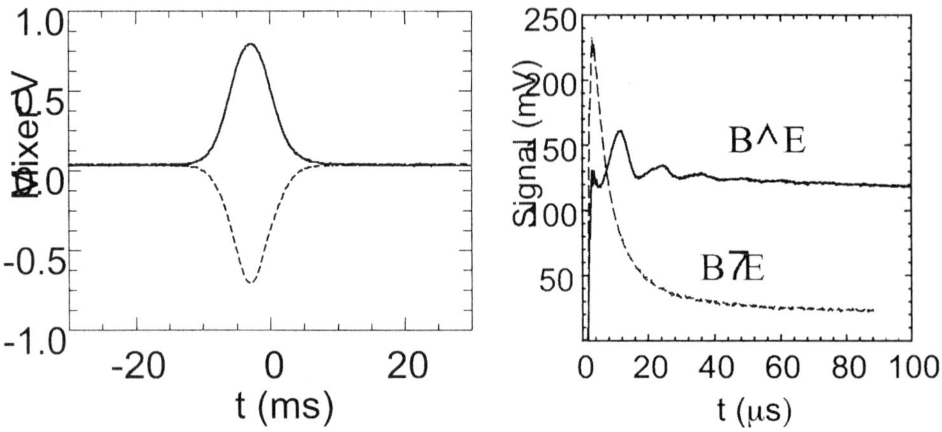

Figure 5. a) Demodulated time-of-flight FM absorption signal from population in the F=1 state (solid) and F=2 state (dashed). With equal populations, the demodulating mixer output is 0. b) Absorption for the $1 \rightarrow 0'$ transition as a function of time for B perpendicular to the laser polarization (solid) and parallel (dashed).

absorbing m_F state. By applying a magnetic field perpendicular to the linear polarization of the laser, the atomic absorption is much higher (Figure 5(b)). In Figure 4, we adjust the laser probing $2 \rightarrow 3'$ to be much less intense than the laser probing $1 \rightarrow 0'$. Our work to date has almost entirely focused on reducing the ratio of the background detection noise, when there are no atoms, relative to the size of the $1 \rightarrow 0'$ signal with atoms present.

The most troublesome background noise is residual AM from the EOM phase modulator. We choose the modulation frequency to be not close to piezo-acoustic resonances in the EOM, which can cause a very large and spurious AM. Spatial inhomogeneities in EOM also lead to AM noise. We control the inhomogeneity by launching the laser beam through a single-mode optical fiber after it passes through the EOM. We then servo the AM by detecting the residual AM after the fiber and feeding back with a DC voltage applied across the EOM.[19] Currently this leads to a noise level of 0.26 mV. There is however a quadrature noise that can be much larger. Using balanced detection or different modulation schemes may solve this problem.

4. A juggling Rb clock

One can achieve higher stabilities and eliminate the dead time (reducing the requirements for the local reference oscillator) by *juggling* atoms in the fountain[20] as depicted in Figure 6. With a S/N = 4600 on a single launch, the frequency uncertainty would be $\delta\nu/\nu = \Delta\nu/(\pi \sqrt{S/N}) = 1\times10^{-14}$. If the cycle time is 1 s, then the fractional instability of the clock after 1s of averaging is $\sigma_y(\tau{=}1s) = 1\times10^{-14}\ \tau^{-}$. By launching balls of atoms at a rate of 25 s^{-1}, the dead time is eliminated and gives a short-term instability of $\sigma_y(\tau) = 2\times10^{-15}\ \tau^{-}$. This improvement is achieved without the technical difficulty of higher S/N or higher atomic densities. There are 2 important problems: 1) shutters, as shown in Figure 6, must be used to block the trapping and cooling light from the interrogation region of the clock; 2) collisions between juggled balls of atoms will shift the frequency of the clock.

We have demonstrated a first version of a juggling Rb fountain to study the frequency shift due to collisions between 2 balls of atoms. The schematic of our single magneto-optic trap (MOT) Rb fountain[6] is shown in Figure 7. Here we rapidly launch 2 balls of atoms at the same velocity from a single trapped ball of atoms. After trapping and launching one ball from the MOT, approximately half of the atoms are depumped into the F=1 ground state. The trapping light is turned back on, recapturing the atoms remaining in F=2. A carefully masked repumping beam keeps these atoms trapped in F=2 until they are launched, 10 to 30 ms later.

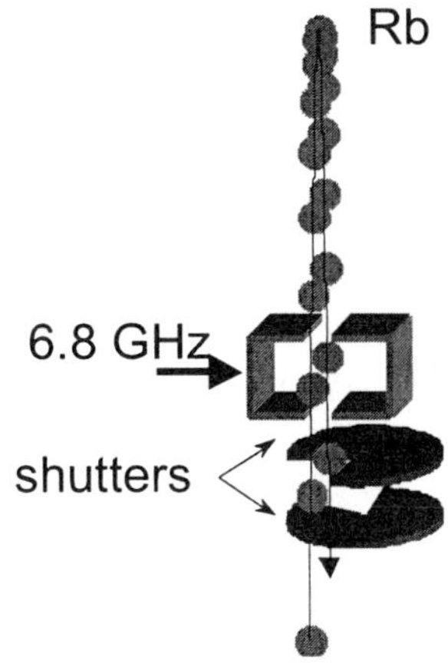

Figure 6. Schematic for a juggling fountain clock. Balls of laser-cooled atoms are launched with a delay shorter than the flight time above the microwave cavity.

This second ball is also depumped after launch and then the F=1 $m_F=0$ atoms in each ball are transferred to the F=2 $m_F=0$ state with microwaves. The atoms remaining in F=1 are then cleared with a laser beam tuned to the $1 \rightarrow 0'$ transition.

In Figure 7 we also show the Allan variance for this juggling clock versus the microwaves supplied to the clock that are generated from a quartz oscillator. The triangles show the stability of a "clock" based on each of the two juggled balls of atoms as compared to the quartz oscillator, in agreement with the previously measured stability of the oscillator. We also compare the relative stability of the "clock" based on the first juggled ball against the "clock" based on the second juggled ball, which removes the long-term drift of the quartz oscillator.

In Figure 8 we show a calculation of the s-wave juggling frequency shift for ^{87}Rb. The first Ramsauer-Townsend[21] frequency shift null occurs at 0.12 mK, corresponding to a time delay of 22 ms. Unfortunately, the peak of the s-wave juggling shift is very nearly at a time delay of 44 ms. Therefore, for a juggling rate of 1/(22 ms), the energy for collisions between

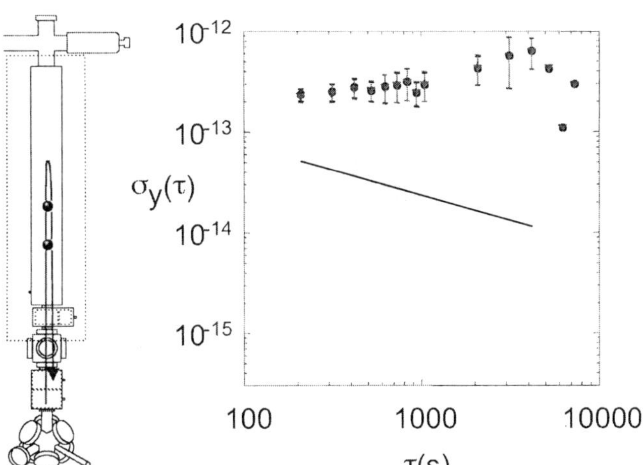

Figure 7. Juggling ^{87}Rb clock. The triangles are the Allan variance for the ^{87}Rb clock versus a quartz oscillator showing the expected deviations of the oscillator. Considering each juggled ball as an individual clock, the squares show the relative stability of these two "clocks" without the oscillator instability.

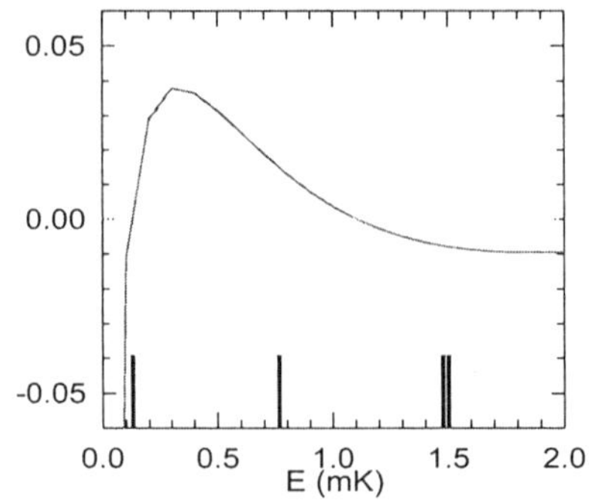

Figure 8. Juggling frequency shift for s-waves. The juggling shift is zero for time delays of 22ms (0.12mK) and 66 ms (1.1mK). The energy spectrum shown corresponds to juggling pattern for alternate launch delays of 22 and 55 ms.

every other ball would be near the peak of the juggling shift. To cancel the juggling shift, one can launch with a more sophisticated pattern. A pattern that cancels the s-wave shift launches balls with alternate delays of 22 and 55 ms. The energy spectrum for this pattern is also shown in Figure 8. Each ball collides with 1 ball with a 22 ms delay for which there is no shift. Each ball also collides with 1 ball at 55 ms delay, for which the shift is positive, and with 2 balls at 77 ms, where the shift is negative and half as large as the shift at 55 ms, so that there is no net frequency shift. There is no p-wave shift for identical atoms due to Bose symmetry. Higher partial waves will generally have random signs and all of the shifts decrease at higher energies. It is likely that a pattern similar to this will allow a high juggling rate and cancel the total juggling collision shift.

5. Direct measurement of scattering phase shifts

We are also currently building a juggling atomic fountain to perform a new type of scattering experiment that accurately measures atomic scattering phase shifts using atomic clock techniques. The first juggled ball will be prepared in one of the clock states and the microwaves excite and probe the atoms during successive passages through the cavity. The second ball is prepared in an internal energy eigenstate. When the two balls collide, the atoms in the first ball are in a coherent superposition of the two clock states and each of these states scatters off of the second ball, experiencing an s-wave phase shift. In a clock, the unscattered as well as the scattered components of each atom are detected, resulting in a density dependent frequency shift.[22,23] Here, however, we exclude the unscattered part of each (clock) atom in the first ball and detect only the scattered fraction after it returns through the microwave cavity. Because the scattering differentially shifts the phases of the two internal states of the clock atoms, the microwaves convert this phase difference into a population difference on the return passage. By detecting the populations, the difference in the s-wave scattering phase shifts can be directly measured as a phase shift of a Ramsey fringe pattern as shown in Figure 9. This phase shift does not depend on the atomic density (in the single collision limit); the number of scattered atoms, the amplitude of the Ramsey fringe, is proportional to the density.

This technique will enable extremely precise measurements of the differences of scattering phase shifts for a variety of spin states in a range of well-defined collision energies and can dramatically enhance our understanding of low energy atomic scattering. In other areas of physics, density independent scattering phase shifts are also directly measured. These include correlation measurements,[24] ratios of cross sections and generally any measurement of the angular dependence of low energy scattering. Here, cold-atom clock techniques allow us to

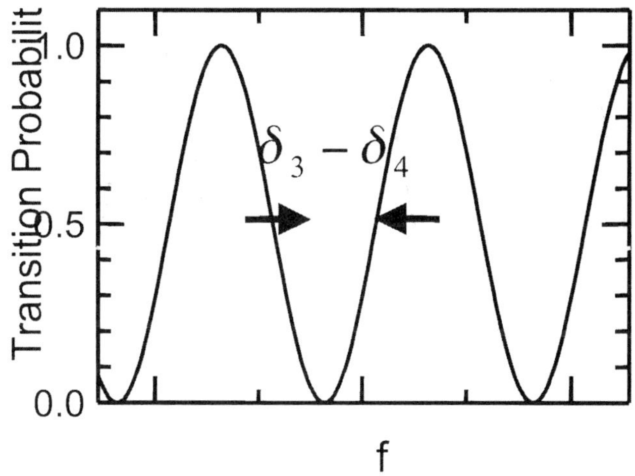

Figure 9. The phase shift (solid) of the Ramsey fringes when detecting only the scattered part of each atom in a juggling atomic clock, is δ_3□δ_4, where δ_3 & δ_4 are the s-wave phase shifts of the two clock states in the juggling collision. For reference, the Ramsey fringe when the entire atom is detected is also shown (dashed).

perform a phase measurement as a frequency measurement; in essence, this enables the high potential accuracy.

6. Acknowledgments

We acknowledge financial support from the NASA Microgravity program, the NSF, a NIST Precision Measurement Grant, and the Stichting voor Fundamenteel Onderzoek der Materie. C.F. acknowledges support from a NASA GSRP fellowship. We also acknowledge stimulating discussions with Mark Kasevich on atom shot-noise limited detection.

1. See K. Gibble and S. Chu, Metrologia **29**, 201 (1993).
2. E. Tiesinga, B. J. Verhaar, H. T. C. Stoof, and D. van Bragt, Phys. Rev. A **45**, 2671 (1992).
3. K. Gibble and S. Chu, Phys. Rev. Lett. **70**, 1771 (1993).
4. K.Gibble and B. J. Verhaar, Phys. Rev. A **52**, 3370 (1995).
5. S. J. J. M. F. Kokkelmans, B. J. Verhaar, K. Gibble, and D. J. Heinzen, Phys. Rev. A. **56**, 4389 (1997).
6. C. Fertig and K. Gibble, IEEE Trans. Instr. Meas. **48**, 520 (1999).
7. S. Bize, Y. Sortais, M. S. Santos, C. Mandache, A. Clairon, and C. Salomon, Europhys. Lett, **45**, 558 (1999).
8. C. Fertig and K. Gibble, Phys. Rev. Lett. **85**, 1622 (2000).
9. Y. Sortais, S. Bize, C. Nicolas, A. Clairon, C. Salomon, and C. Williams, Phys. Rev. Lett. **85**, 3117 (2000).
10. S. Kokkelmans and B. J. Verhaar, private communication (2000).
11. Carl Williams, private communication (2000).
12. S. Bloom, J. Appl. Phys. **28**, 800 (1957).
13. S. B. Crampton, Phys. Rev. **158**, 57 (1967).
14. M. E. Hayden and W. N. Hardy, J. Low Temp. Phys. **99**, 787 (1995).
15. G. Santarelli, Ph. Laurent, P. Lemonde, A. Clairon, A. G. Mann, S. Chang, A. N. Luiten, and C. Salomon, Phys. Rev. Lett. **82**, 4619 (1999).
16. S.R. Jefferts, D.M. Meekhof, J. Shirley, and T.E. Parker, submitted to Metrologia (2000).
17. S. Weyers, A. Bauch, U. Hubner, R. Schroeder, and Ch. Tamm, IEEE Tran. Ultrason. Ferr. Freq. Contr. **47**, 432 (2000).
18 J. M. McGuirk, G. T. Foster, J. B. Fixler, and M. A. Kasevich, Optics Lett. **26**, 364-366 (2001).
19. N. C. Wang and J. L. Hall, J. Opt. Soc. Am. B **2**, 1527 (1985).
20. R. Legere and K. Gibble, Phys. Rev. Lett. **81**, 5780 (1998).
21. Ramsauer, Ann. der Physik **72**, 345 (1923); Townsend and Bailey, Phil. Mag. **43**, 593 (1922).
22. E. Tiesinga, B. J. Verhaar, H. T. C. Stoof, and D. van Bragt, Phys. Rev. A 45, 2671 (1992);
23. K. Gibble and S. Chu, Phys. Rev. Lett. 70, 1771 (1993).
24. See D.K.McDaniels et al., Phys. Lett. **1**, 295 (1962).

Manipulating entanglement with atoms and photons in a box

S. Haroche[1,2], M. Brune[2], J.M. Raimond[2]

[1] *Collège de France, 11 place M. Berthelot, F-75231 Paris Cedex 05*
[2] *Laboratoire Kastler Brossel, Département de Physique de l'Ecole Normale Supérieure,*
24 rue Lhomond, F-75231 Paris Cedex 05, France

Abstract: We summarize recent experiments on entanglement, quantum logic, quantum measurement, complementarity and decoherence performed with photons and atoms in cavities.

1 Introduction

Entanglement is an essential concept in quantum physics. Composite systems, after their different parts have interacted, generally end up in a non separable quantum state. This has numerous consequences in microscopic physics. Non-locality and the generation of Einstein-Podolsky-Rosen (EPR) pairs of particles, as well as the Schrödinger cat paradox in quantum measurement are deeply connected with entanglement. This concept is also intimately related to the notion of complementarity. Whether the wave or the corpuscular aspect of a particle reveals itself in an interferometry experiment is indeed dependent upon the entanglement building up between the particle and a "which-path" detector, which can be the interferometer itself. Entanglement is also deeply connected with the process of decoherence. The loss of quantum coherence in large systems prepared into states superpositions is due to the unavoidable entanglement of this system with its environment. Various experiments in quantum optics have recently checked different aspects of entanglement. These experiments have involved twin-photon beams, trapped ions or atoms and photons in cavities (cavity QED systems). Beyond their fundamental interest to check basic aspects of quantum physics and to realize text book experiments, these studies are interesting because they constitute a simple demonstration of the basic steps required to process quantum information. Two-level atoms, or single photons can indeed be considered as carriers of binary information. The logical values (0 and 1) of usual information theory can in these systems be prepared and analyzed in state superpositions. The interaction between two or more such "qubits" can lead to bit-entanglement and opens the way to new applications in cryptography, communication and computation. In this article, we will review the experiments we have performed on entanglement manipulation at Ecole Normale Supérieure , using a cavity QED set-up. We will only present the main features of the experiments. More details about the set-up design and about the experiments themselves can be found in a recent review article [1].

2 The apparatus

Our apparatus is designed to manipulate individual Rydberg atoms and microwave photons as they interact in a controlled and coherent way in a high Q superconducting cavity. The set-up, described in [1, 2, 3, 4], comprises an open geometry cavity made of two spherical superconducting mirrors facing each other, in a Fabry-Perot configuration. It stores microwave photons (51.1 GHz frequency, 6 mm wavelength) for up to 1 millisecond and let them interact with velocity selected circular Rydberg atoms crossing the cavity one by one, at mid-distance between the mirrors. The set-up is cooled

Coherence and Quantum Optics VIII
Edited by Bigelow et al., Kluwer Academic/Plenum Publishers, 2003

to about 1 K to minimize thermal radiation noise. The atoms are detected after they have crossed the cavity by field ionization. Some experiments are performed by directly detecting the atomic final energy state and gaining in this way information on the atom-cavity field energy exchange process. In other experiments, atomic state superpositions are used. Atoms are sent across the cavity in a superposition of energy levels and the states are mixed again after the cavity interaction. This is performed with the help of two auxiliary classical microwave field pulses sandwiching the cavity. The successive application of these pulses constitutes a Ramsey interferometer [5]. Interference fringes are obtained in the probability for finding the atom in a given final state, when the frequency of the Ramsey pulses is tuned across a transition between two Rydberg levels. By studying how the fringes phase and amplitude is affected by the presence of photons in the cavity, one gains information on the atom-photon interaction process [6]. This set-up bears strong similarities with the Rydberg atom micromaser [7, 8], which makes use of a different type of cavity, without Ramsey interferometer. Other cavity QED experiments are performed in the optical domain with microscopic sized cavities [9, 10].

3 Resonant atom-photon and atom-atom entanglement

The basic interaction process of an atom with the cavity field results in the so called quantum Rabi oscillation. It is observed when the cavity mode is exactly resonant with a transition linking two adjacent circular Rydberg states e and g. When an atom in level e is sent in an initially empty cavity, it falls down to level g and emits reversibly a photon. This photon can subsequently be absorbed, resulting in a reexcitation of the atom. This reversible process leads to a quantum mechanical oscillation between the initial and final combined atom-field states $|e,0\rangle$ and $|g,1\rangle$, at frequency $\Omega/2\pi$, equal in our experiment to $50kHz$.

The quantum Rabi oscillation provides a simple and flexible tool to entangle atoms and photons. If the Rabi oscillation in vacuum is interrupted mid-way between the upper and the lower atomic state (50% transition probability between the levels or $\pi/2$ Rabi pulse), the final state of the atom-cavity field system is a maximally entangled state [16]. The atom-field correlations survive after the atom leaves the cavity. Any measurement on the atom has an immediate effect, at a distance on the cavity field and experiments of the Einstein-Podolsky-Rosen (EPR) type [17] can be performed on this entangled pair.

If the Rabi oscillation lasts twice as long (100% transition probability between the levels or π Rabi pulse), the atom and the cavity field fully exchange their energy [18]. An atom initially in the lower state of the transition entering in a cavity containing one photon will end up with certainty in the upper state, leaving the cavity empty. This process can be used to transform an atom-field entanglement into an atom-atom one. The experiment involves two atoms crossing the cavity one after the other, the first initially in the upper level, the second in the lower one. The first atom undergoes a $\pi/2$ Rabi pulse and gets entangled with the cavity. The second is submitted to a π pulse, copying the cavity state and getting entangled in the process with the first atom. We have checked this entanglement by performing various measurements on the final states of the two atoms and studying their correlations [16]. This massive EPR pair of particles could be used for Bell's inequality tests. Note that pairs of entangled atoms have also been realized with ions in traps [19].

4 Direct atom-atom entanglement via controlled cavity assisted collision

The atom-atom entanglement process via real cavity photon exchange has a fidelity intrinsically limited by cavity losses and thermal photon background. These limitations can, at least in principle, be avoided by entangling directly the two atoms, without having to rely on the production of an intermediate photon state in the cavity [30]. To achieve this, we make the two atoms interact simultaneously with the cavity C [20]. This procedure is illustrated on figure (1a). The atoms A_1 and A_2, effusing from oven O, prepared in box B respectively in the circular states e and g at times differing by T, have different velocities. Atom A_1 overcomes A_2 at cavity center. Both atoms are later detected by the state-selective detector D. A space-time diagram showing the evolution of the two atoms is shown in figure (1b).

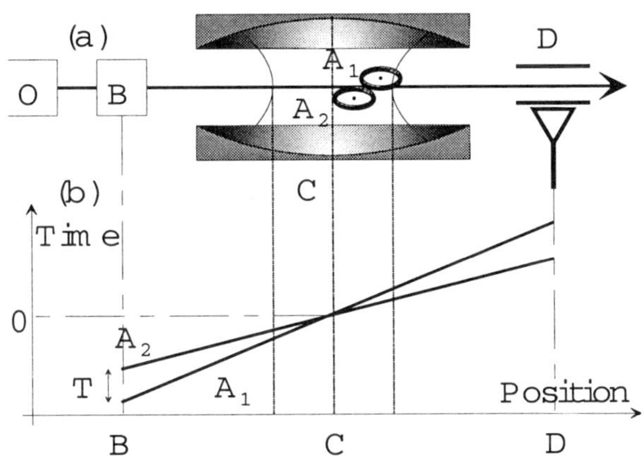

Fig. 1. (a) Scheme of the experimental apparatus. (b) Space-time diagram depicting the sequence of events. Atoms A_1 and A_2, sent at different times with different velocities, simultaneously cross the cavity axis. They are finally detected by field ionization in D.

The closest cavity mode is slightly detuned from the $e \to g$ transition at frequency ω by an amount $\delta > \Omega$. Due to energy non-conservation, no real photon emission can occur in this case. Atom A_1 can however virtually emit a photon immediately reabsorbed by A_2. This leads, as in the resonant case, to atom-atom entanglement. The system ends up in a superposition of the combined $|e, g\rangle$ and $|g, e\rangle$ states. The quantum mechanical amplitudes of this superposition are periodic functions of the duration of this cavity assisted collision process (this duration depending in turn on the controlled atomic velocities). The frequency associated to this second order collision process is $\Omega^2/4\delta$. By repeating the experiment many times, we have reconstructed the joint probabilities $P(e_1, g_2)$ and $P(g_1, e_2)$ for finding finally the atom pair in states e, g and g, e. We have plotted in figure (2) these probabilities versus a dimensionless parameter η, essentially equal to ω/δ [20]. The oscillations of $P(e_1, g_2)$ and $P(g_1, e_2)$ are well accounted for by theoretical models (solid and dashed lines in figure (2)). We have in particular realized a situation of maximum entanglement by adjusting δ (and thus η) to the value corresponding to $P(e_1, g_2) = P(g_1, e_2) = 0.5$. By performing measurements on the two atoms in a coherent basis admitting superposition of energy states as eigenstates, we have also checked the coherent nature of the two-atom state prepared in this way. This new way to entangle atoms might be used to demonstrate elementary steps of quantum logic, essentially immune from noise due to cavity decay or thermal field background.

5 Atom-photon phase gate

Let us now come back to the resonant atom-cavity case, leading to real atom photon exchange. After a full cycle of quantum Rabi oscillation (2π Rabi pulse), the combined atom-field system comes back to the initial state, but the phase of its wave function has undergone a π-shift. For example, if the atom is initially in the lower state of the

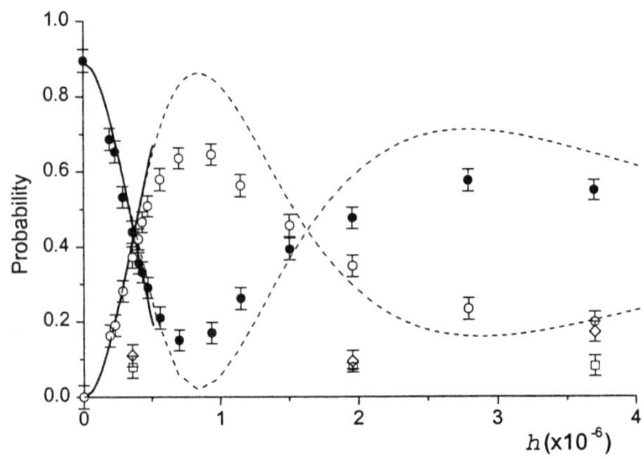

Fig. 2. Joint detection probabilities versus the detuning parameter η. $P(e_1, g_2)$ and $P(g_1, e_2)$ (solid and open circles) oscillate in a symmetrical way, reflecting the atom-atom energy exchange enhanced by the cavity. The solid line represents the predictions of a simple analytical model, in the $\eta < 5\,10^5$ range where it applies. The dashed lines present the result of a numerical integration of the system's evolution. The spurious channels probabilities, $P(e_1, e_2)$ and $P(g_1, g_2)$ (open squares and diamonds respectively) stay below the 10% level.

transition with one photon in the cavity, it emerges in the same state, leaving the photon in the cavity, but with the sign of its wave function changed. This evolution is reminiscent of the sign change of a spin state undergoing a full 2π rotation [21]. On the contrary, if the cavity is initially empty, the sign of the atomic state is unaltered. If we consider that the atom on one hand and the field in the cavity on the other hand are quantum bits, the 2π Rabi pulse couples them according to the conditional dynamics of a quantum phase gate. We have studied the operation of this gate and shown that it works in a fully coherent and reversible way [2]. Similar gates have been operated in ion trap experiments [22]. Multiple of 2π Rabi pulses are also essential for the generation of "trapping states" of the Rydberg atom micromaser [23].

6 Quantum Non-Destructive measurement of a single photon

We have applied our quantum phase gate to perform a non-destructive measurement of a single photon in the cavity [3]. The phase change of the atom's wave function when it undergoes a 2π Rabi pulse can be translated into an inversion of the phase of the fringe pattern of the Ramsey interferometer sandwiching the cavity (see section 2). By setting the interferometer at a fringe extremum, we correlate in this way the photon number (0 or 1) to the final state of the atom. The atom appears as a "meter" measuring, without destroying it, a single photon in the cavity. We have demonstrated the operation of this quantum non-destructive scheme and shown that the same photon can be measured repeatedly by successive atoms, without being absorbed. This is quite different from ordinary photon counting procedures, which absorb light quanta. Previous quantum non-demolition procedures based on non-linear optical processes were restricted to the detection of fields containing large photon numbers [24].

7 Multiparticle entanglement

By combining quantum Rabi pulses of various duration and auxiliary Ramsey pulses on successive atoms crossing the cavity, one can generate and analyse entangled states involving more than two particles. For example, by applying a $\pi/2$ Rabi pulse on a first atom, one entangles it to a 0/1 photon field. A second atom then undergoes a 2π

Rabi pulse combined to Ramsey interferometry, in order to measure this field. Before this atom is detected, a three-particle entanglement involving the two atoms and the photon field is generated. The field state is finally copied on a third atom, initially in the lower level of the atomic transition resonant with the cavity mode. This atom undergoes a π Rabi pulse, absorbing the photon and getting in this way correlated to the first two atoms. The characteristics of this three-particle entangled state are analysed by performing various measurements on the three atoms. These measurements involve the application of auxiliary Ramsey pulses after the atoms have interacted with the cavity field [4]. The procedure can be generalized to situations of increasing complexity, with larger numbers of atoms [25]. Entanglement involving more than two particles has also been studied by other quantum optics techniques [26, 27].

8 Complementarity experiments at the quantum classical boundary

We have also used our cavity QED apparatus to perform a complementarity test very closely related in its principle to the double slit Gedanken experiment described by Bohr at the 1927 Solvay conference. Discussing complementarity with Einstein, Bohr had analyzed a situation where photons or electrons are crossing a Young interferometer in which one slit is carried by a very light assembly, free to move independently of the other (see Fig. (3) adapted from [28]). In an ordinary Young apparatus with fixed slits, interference fringes reveal the wave nature of the particle. In the Bohr design, the momentum kick imparted to the moving slit by the deflected particle provides a which path information, suppressing the fringes and revealing the corpuscular aspect. As emphasised by Bohr, a quantum description of the slit is essential to understand the experimental conditions determining fringe visibility.

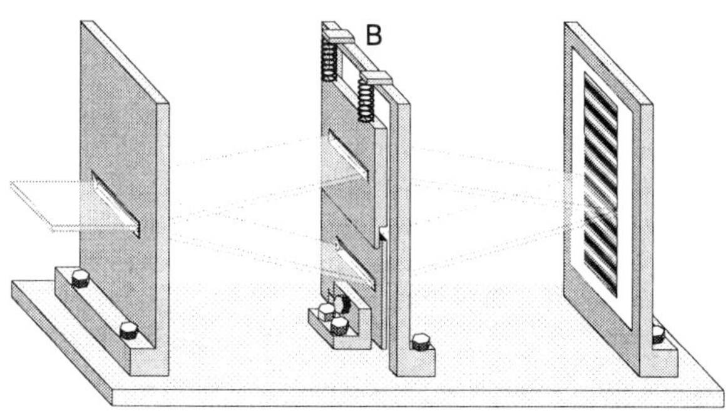

Fig. 3. Sketch of an interferometer with a recoiling slit, adapted from Bohr.

The concept of entanglement underlines an essential feature of this thought experiment. The particle and the quantum moving slit are in fact the two components of a correlated EPR system. The trajectory of the particle gets entangled to the motion of the slit. Observing (really or virtually) the slit motion lifts the ambiguity of the particle's trajectory and suppresses the interference. If the upper slit in Fig. (3) is fixed (classical limit), its final state is the same whether the particle crosses it or not. There is no particle-slit entanglement and fringes are visible. Intermediate situations can be considered, by varying for example the mass of the upper slit in Fig. (3). There can be a continuous transition from the quantum slit case (very small mass) to the classical one (infinite mass). In between, we expect fringes with a limited contrast, which would measure the degree of entanglement between the particle and the interferometer.

Designing an actual Young interferometer with ultra-light quantum slits would be extremely difficult. The Bohr-Einstein situation can however be translated into our cavity QED system, presenting the same conceptual features and experimentally easier to realise. Following previously published proposals [29, 30], we use, instead of a Young device, a double pulse oscillatory field Ramsey interferometer [31] operating with Rydberg atoms. The field pulses, resonant on a transition between the two atomic energy levels e and g, act as atomic energy beam splitters transforming these states into superpositions, and can give rise to interference effects in the probability for finding the atom in a given final state.

The first resonant field pulse is a small coherent field injected through a wave guide inside the long lived cavity mode. Depending upon the average photon number N in this field, it can be considered either as a microscopic (small N values) or as a macroscopic (large N's) "beam splitter". The second pulse is produced downstream in the Ramsey zone R_2. This latter pulse, produced by photons which are fed into a low Q field mode and recycled at a very fast rate, can always be considered as "macroscopic" and classical [32]. The first and second pulses in this experiment are thus respectively equivalent to the upper and lower slits in the Bohr's design. If the average photon number N in the first coherent field pulse is large, Ramsey fringes with maximum contrast are obsereved. When N gets smaller, the one photon change produced by the atomic transition from level e to g leads to atom-field entanglement, in the same way as the recoil motion of the upper slit in Figure (3) leads to slit–particle entanglement. The contrast of the resulting Ramsey fringe pattern is then reduced. At the limit where the coherent field in this first pulse is down to the vacuum field, the first quantum $\pi/2$ Rabi pulse results in maximum atom-"beam-splitter" entanglement and the fringes vanish altogether. Figure (4) shows the fringe contrast as a function of N. The points are experimental and the curve reproduces the theoretical variations of the degree of entanglement between the first pulse coherent field and the atom (corrected by a fixed factor accounting for our interferometer imperfections). This study can thus be considered as a realistic version of the Bohr gedanken experiment.

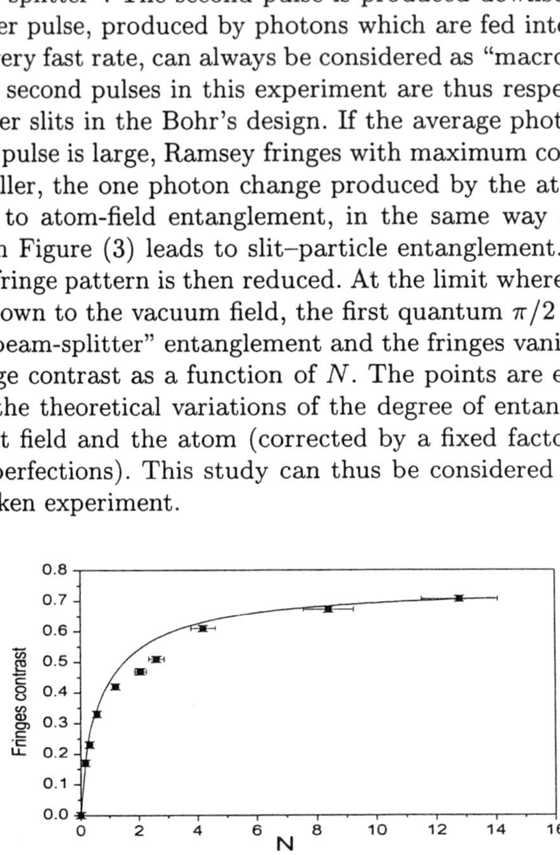

Fig. 4. Fringe contrast as a function of the mean photon number N. The points are experimental. The line represents the theoretical predictions corrected for the imperfections of our interferometer.

9 Schrödinger cats and decoherence

Entanglement based on dispersive effects can also be realized when the atoms and the cavity field are non-resonant. In this case, matter and radiation cannot exchange energy, but dispersive light-shift type effects [11] can be used to entangle the two systems. A non-resonant atomic medium corresponds to a real index of refraction shifting the field's frequency in the cavity. The Rydberg atom to cavity coupling is so strong that a single atom, crossing the cavity can shift its frequency by several kHz, a quite noticeable effect. Moreover, this shift has a sign depending upon the atom's energy state. If the atom is sent across the cavity in a superposition of states, the cavity ends up with two different frequencies at the same time. This leads to very simple entanglement situations. We have for example prepared a small coherent field, containing on average 3 to 10 photons in the cavity, by coupling it to an external source. We have then let this field interact with a single atom in a state superposition. The field frequency shift resulted in the appearance of two field components with different phases, correlated to the two atom's energy states [33]. We have studied in details this situation, strongly reminiscent of the famous Schrödinger's cat, suspended coherently between life and death [34]. By sending after a delay a second probe atom across the cavity, we have monitored the disappearance of the coherence between the two field components and studied how it occurred faster and faster as the separation between these components was increased [33]. This experiment provided a quantitative test of decoherence theories [35]. Similar Schrödinger cat and decoherence studies have been performed in ion trap systems [36, 37].

10 Perspectives

The experiments described above illustrate fundamental concepts of quantum theory and demonstrate the feasibility of various elementary steps in quantum information processing. Our set-up is however limited by several imperfections, making it difficult to implement more complex operations involving larger numbers of gates or qubits. One important problem is related to the preparation of our single atoms. We rely so far on a weak excitation into Rydberg levels of a velocity selected atomic beam. Due to Poissonian statistics, we can ensure that each preparation pulse does not contain more than one atom only by reducing the excitation to the point that only 0.1 to 0.2 Rydberg atoms are prepared on average. Most pulses then contain no atom at all and the corresponding sequences must be rejected. This results in data acquisition times which become prohibitive when one has to deal with more than three successive atoms in an experimental sequence. To solve this problem, we are planning to implement deterministic atom preparation schemes allowing us to get single atoms on demand, with essentially unit probability. One way to achieve this would be to start from a small tightly bound magneto-optical trap of cold atoms in their ground state and to excite the Rydberg atom by laser excitation in this trap. The dipole-dipole coupling between closely spaced Rydberg atoms should prevent then the excitation of more than one atom at a time (so-called dipole-dipole blockade [38]). Slow atoms, interacting for a long time with the cavity field could then be released one at a time across the set-up, making it possible to achieve new experiments (teleportation of atomic states [39], tests of Bell's inequalities with massive particles, implementation of simple quantum algorithms). Experiments in which a single slow atom would cross two cavities and entangle their fields are also considered.

Acknowledgements

Laboratoire Kastler Brossel is a Unité Mixte du Centre National de la Recherche Scientifique (CNRS) et de l'Université Pierre et Marie Curie. The ENS Cavity QED group is supported in part by the European Community and Japan Science and Technology Corporation (International Cooperation Research Project, Quantum Entanglement Project).

References and links

1. J.M. Raimond, M. Brune and S. Haroche, Rev. Mod. Phys. In press (2001).
2. A. Rauschenbeutel, G. Nogues, S. Osnaghi, P. Bertet, M. Brune, J.M. Raimond, and S. Haroche, Phys. Rev. Lett. **83** (1999) 5166.
3. G. Nogues, A. Rauschenbeutel, S. Osnaghi, M. Brune, J. M. Raimond, and S. Haroche, Nature **400** (1999) 239.

4. A. Rauschenbeutel, G. Nogues, S. Osnaghi, P. Bertet, M. Brune, J. M. Raimond, and S.Haroche, Science **288** (2000) 2024.
5. N. Ramsey, *Molecular Beams*, Oxford University Press, New York 1985.
6. M. Brune, P. Nussenzveig, F. Schmidt-Kaler, F. Bernardot, A. Maali, J. M. Raimond, and S.Haroche, Phys. Rev. Lett. **72** (1994) 3339.
7. G. Raithel, C. Wagner, H. Walther, L. M. Narducci, and M. O. Scully, in *Cavity Quantum Electrodynamics*, edited by P. Berman, Academic Press, New York 1994, p.57.
8. M. Brune, J. M. Raimond, P. Goy, L. Davidovich, and S. Haroche, Phys. Rev. Lett. **59** (1987) 1899.
9. P. Münstermann, *et al.*, Phys. Rev. Lett. **82** (1999) 3791.
10. C. J. Hoo, *et al.*, Science **287** (2000) 1447.
11. S. Haroche, and J. M. Raimond, in *Cavity Quantum Electrodynamics*, edited by P. Berman, Academic Press, New York 1994, p.123.
12. J. H. Eberly, N. B. Narozhny, and J. J. Sanchez-Mondragon, Phys. Rev. Lett. **44** (1980) 1323.
13. M. Brune, *et al.*, Phys. Rev. Lett. **76** (1996) 1800.
14. G. Rempe, H. Walther, and N. Klein, Phys. Rev. Lett. **58** (1887) 353.
15. B. Varcoe, S. Brattke, and H. Walther, Nature, **403** (2000) 743.
16. E. Hagley, *et al.*, Phys. Rev. Lett. **79** (1997) 1.
17. A. Einstein, B. Podolsky, and N. Rosen, Phys. Rev. **47** (1935) 777.
18. X. Maître, *et al.*, Phys. Rev. Lett. **79** (1997) 769.
19. Q. A. Turchette, *et al.*, Phys. Rev. Lett. **81** (1998) 3631.
20. S. Osnaghi *et al.*, Phys. Rev. Lett. **87**, 037902 (2001).
21. H. Rauch, A. Zeilinger, G. Badurek, and A. Wilfing, Phys. Lett. **54 A** (1975) 425.
22. C. Monroe, *et al.*, Phys. Rev. Lett. **75** (1995) 4714.
23. M. Weidinger, B. Varcoe, R. Heelein, and H. Walther, Phys. Rev. Lett. **82** (1999) 3795.
24. P. Grangier, A. L. Levenson, and J. P. Poizat, Nature, **396** (1998) 537.
25. N. D. Mermin, Phys. Rev. Lett. **65** (1990) 1838.
26. J.-W. Pan, *et al.*, Nature **403** (2000) 515.
27. C. A. Sackett, *et al.*, Nature **404** (2000) 256.
28. Bohr, N. in *Quantum Theory and Measurement*, (eds. J.A. Wheeler and W.H. Zurek) p9, Princeton University Press, (1983).
29. B.G. Englert, H. Walther and M. O. Scully, Appl. Phys. B **54**, 366 (1992).
30. S.B. Zheng, Opt. Comm. **173**, 265 (2000).
31. P. Bertet et al, Nature **411**, 166 (2001).
32. J.I. Kim et al., Phys. Rev. Lett. **82**, 4737 (1999).
33. M. Brune, *et al.*, Phys. Rev. Lett. **77** (1996) 4887.
34. E. Schrödinger, Naturwissenschaften **23** (1935) 807; translated in english in *Quantum Theory and Measurement*, edited by J.A.Wheeler and W.Zurek, Princeton University Press, Princeton 1983 p. 152.
35. W. H. Zurek, Physics Today, **44** 10 (1991) p.36.
36. C. Monroe, *et al.*, Science **272** (1996) 1131.
37. C. J. Myatt, *et al.*, Nature **403** (2000) 269.
38. M.D. Lukin *et al.*, Phys. Rev. Lett. **87**, 037901 (2001).
39. L. Davidovich, N. Zagury, M. Brune, J. M. Raimond, and S. Haroche, Phys. Rev. A **50** (1994) R895.

Taming light with cold atoms: light at bicycle speed .... and slower yet!

L. Hau
Lyman Laboratory, Harvard University, Cambridge MA 02138
hau@physics.harvard.edu

Abstract: The latest results from our Slow Light/Bose-Einstein condensation experiment will be presented. We are using a combination of laser and evaporative cooling to create Bose-Einstein condensates of sodium atoms in a '4D' magnetic bottle. We have succeeded in reducing the light speed in a Bose condensate to the speed of a bicycle by using the effect of electromagnetically induced transparency. Most recently we have brought light to a complete standstill. The results have implications for quantum information processing, for probing and manipulation of Bose-Einstein condensates, and for nonlinear optics.

Coherence and Quantum Optics VIII
Edited by Bigelow et al., Kluwer Academic/Plenum Publishers, 2003

Dark Resonances in Solids for Quantum Computing and Slow Light

Philip R. Hemmer
Air Force Research Laboratory, Sensors Directorate, Hanscom AFB, MA 01731
Tel: 781-377-5170, Fax: 781-377-2836, Philip.Hemmer@hanscom.af.mil

Abstract: Electromagnetically induced transparency and related optical dark-resonance techniques have found many potential applications such as quantum computing, slow light, optical memory, and nonlinear optics. For most of these applications, solid state implementations are preferred. To this end, I will describe our current research and recent experimental results toward the implementation of efficient dark resonance techniques in solids, with emphasis on quantum computing and slow light.

1. Introduction

Dark resonance techniques such as electromagnetically induced transparency [1] (EIT) and coherent population trapping [2] (CPT) have been explored for numerous potential applications. Examples include lasing without inversion [3], ultra-low threshold nonlinear optics, and novel optical memories [4]. Nonlinear optics applications include frequency upconversion [5], aberration correction [6], sensitive infrared image detection [7], and squeezed light generation [8]. Recently, dark resonance techniques have also been used for quantum computing and ultra-slow group-velocity light [9], where these two are related by the fact that slowed or stopped light [10] has the potential to be a high fidelity storage technique for quantum information.

For most of these applications, a solid state medium is preferred because it simplifies the eventual large-scale, practical implementation. To this end, much of our current research has been devoted to identifying materials and techniques for the realization of efficient dark resonances in solids. The goal is to transition the most promising proof-of-principle experiments, made in vapors, to the solid state. Materials considered so far include doped crystals, glasses, and semiconductors, organic molecules, and quantum well structures. So far, the most successful demonstrations have been accomplished in spectral hole burning materials [11]. This is because at low temperatures the dopants in these materials act much like the isolated atoms in a vapor.

2. Materials requirements

The key requirements for efficient dark resonance excitation in solids are a long two-photon coherence lifetime, strong single-photon optical transitions, but preferably lifetime limited, and of course allowed two-photon transitions. While this last requirement may seem obvious and can almost always be satisfied in vapors, it severely reduces the number of promising solid state materials.

For a long two-photon coherence lifetime, spin systems are by far the best. For nuclear spins, coherence lifetimes, T_2, are typically on the order of milliseconds at low temperatures and microseconds at room temperature, although coherence lifetimes as long as fractional seconds have been seen at room temperature in certain semi-solid materials [12]. Even more impressive are the long spin-state population lifetimes, T_1, that can be on the order of seconds to hours or longer. These long population lifetimes are significant because a number of experimental techniques have been devised, for nuclear magnetic resonance (NMR) applications, to increase the effective spin coherence lifetime. In principle, T_2 can be increased to T_1 using such techniques. Long coherence and population lifetimes are also seen for electron spin systems, although in general electron spin lifetimes tend to be from one to three orders of magnitude shorter. Aside from these shorter lifetimes, electron spin coherences are still long lived in comparison to other coherences, such as those due to phonons or optical transitions and have the advantage of being easier to manipulate than nuclear spins because of their stronger coupling to applied fields.

In atomic vapors, two-photon coherences involving electron or nuclear spins are easy to excite optically because of spin-orbit coupling. In solids, however, spin-orbit coupling is usually quenched because of low symmetry. Therefore, in general, solid materials whose dopants have a high degree of site symmetry are most likely to have allowed two-photon transitions involving spin flips, although there are notable exceptions to this rule as will be seen later.

The final requirement, for strong but lifetime limited (single-photon) optical transitions, is more difficult to satisfy. This is because strong optical transitions in solids are often broadened into bands or at least strongly coupled to phonons, giving broad and intense phonon sidebands and strong non-radiative decay channels. Nonetheless, narrow, long-lived optical transitions are possible in solids, especially those involving inner shell transitions, as in rare-earth doped materials. However, these transitions are forbidden in free atoms, and usually quite weak in solids.

3. Spectral hole burning materials

In spite of all these restrictions there is a class of solid materials, collectively known as spectral hole burning (SHB) materials, that satisfy most of the requirements for efficient dark resonance excitation. These materials are commonly used for high performance optical memories [13] and processors [14]. They are characterized by a strong, inhomogeneously broadened, zero-phonon line on one or more optical transitions, at least at low temperatures. A subset of these materials also have long-lived spin sublevels in the ground state.

In the following, we will concentrate on two particular spectral hole burning materials: Pr^{3+} doped Y_2SiO_5 (Pr:YSO) [15] and nitrogen-vacancy (N-V) [16] color centers in diamond.

4. Efficient dark resonance excitation in Pr:YSO

Among SHB materials, Pr:YSO has the advantage of a very long spin lifetime in the ground state. It also has an optical transition that is relatively strong compared to other rare-earth-based spectral holeburning materials, though still much weaker than a free alkali atom such as sodium. Finally, even though the dopant has very low site symmetry, two-photon spin-flip transitions are allowed due to a change in the quantization axis when the Pr ion is optically excited.

Experimentally, we have observed dark resonance excitation efficiencies approaching 100 % in Pr:YSO [17], as evidenced by almost complete suppression of the resonant optical absorption, shown in Figure 1(b). Figure 1(a) shows a simplified energy level diagram of Pr:YSO, where each of the nuclear-spin sublevels of the ground and excited states is actually a Kramer's doublet. This high degree of induced transparency disagrees with theoretical predictions, based on the ~ 4.4 GHz inhomogneously broadened zero-phonon line of the optical transition. This is because theory requires an optical Rabi frequency that is comparable to the ~ 4.4 GHz inhomogeneously broadened optical transition in order to achieve a transparency greater than about 50 %. The estimated optical Rabi frequencies in this experiment are many orders of magnitude smaller than this. In part, this discrepancy can be accounted for by considering the dynamics of persistent spectral holeburning as follows: When only the probe and coupling lasers needed for dark resonance excitation are present (see Figure 1(a)), all Pr atoms whose transition frequencies are near enough to interact are optically pumped into the third ground spin sublevel. Here they remain for a time on the order of the ground state population lifetime, T_1, that is greater than 100 seconds. This effectively causes the crystal to become transparent to the probe and coupling fields via persistent spectral hole burning, as illustrated in Figure 1(c). However, when the third, or repump, laser is applied some of these atoms are returned to the sublevels needed for dark resonance excitation. This repumping action produces what is known as an anti-hole, as illustrated in Figure 1(d). In our experiments, the width of this anti-hole is determined approximately by laser jitter, which is on the order of 1 MHz and is much narrower than the

4.4 GHz zero-phonon line. Since only these repumped atoms have a significant interaction with the coupling and probe lasers, the dark resonance can be saturated when the optical Rabi frequency becomes comparable to the anti-hole width. Thus, efficient dark resonance excitation becomes possible at the laser intensities used in the experiment.

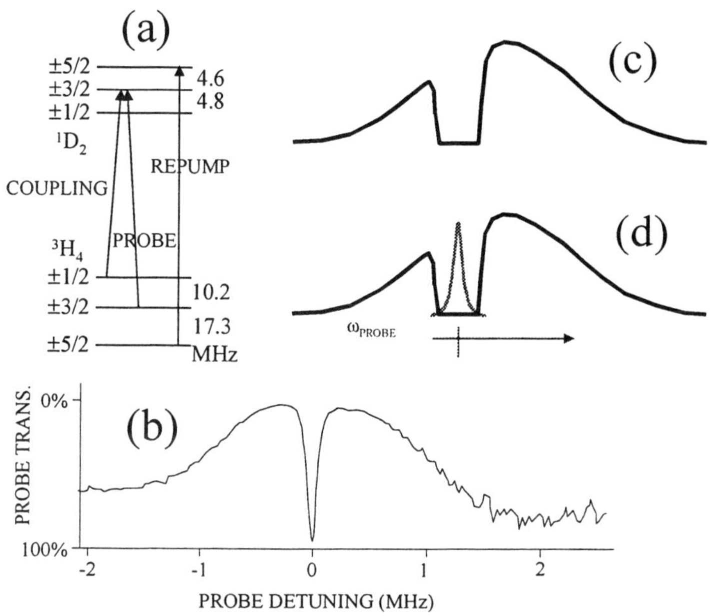

Figure 1. (a) Simplified energy level diagram of Pr:YSO. (b) Experimentally observed Raman dark resonance with near 100% efficiency. Peak optical absorption: ~ 95%. Crystal length: 3mm. Crystal temperature: 5K. (c) Schematic illustration of spectral hole burned by probe and coupling lasers in absence of repump. (d) Schematic illustration of anti-hole produced by repump laser.

5. Slow light in Pr:YSO

The high degree of induced transparency seen experimentally in Pr:YSO makes this material suitable for transitioning dark resonance applications that previously required either an intense short-pulse laser or an unbroadened system, such as an ultra cold vapor. One such application is ultra-slow light [18]. To this end, we measured pulse delays and deduced light speeds for a variety of coupling laser intensities, as shown in Figure 2(a). Light speeds as slow as 45 meters/seconds were observed, which is comparable to group velocities observed in hot and cold vapors [9]. For eventual application to quantum information storage, it is important to show that the slowed light pulse can be contained entirely in the medium. Experimentally, this means that the input and delayed pulses must be well separated. Figure 2(b) shows the observed delay of a probe pulse for various two-photon detunings. On resonance, complete separation of input and delayed pulses is seen. Note that in Figure 2(b), the input probe pulse is visible because the crystal only absorbs about 90 % of the input probe light. Moreover, the delayed pulse has smooth edges because the sharp edges of the input pulse represent high-frequency components that are significantly detuned from two-photon resonance and hence are not slowed.

For quantum information storage, the group velocity of the probe light must be reduced to zero by turning off the coupling laser. This storage technique, sometimes called "stopped light," effectively transfers the quantum information from the light to the spins in the medium [10]. A similar storage technique, Raman excited spin echoes, was demonstrated by our group several years ago [4]. Experimentally, the difference is that coupling and probe lasers are switched on at the same time for Raman excited spin echoes, whereas for stopped light the coupling laser is switched on first. Although this may seem like a minor difference, it completely changes the physics. By turning on the coupling laser first, the information in the probe pulse is stored by adiabatic passage as in EIT. In principle, this allows 100 % of the quantum information in the probe pulse to be stored. In contrast, Raman excited spin echoes have a maximum efficiency of less than 50 % since significant population is left in the optical excited state.

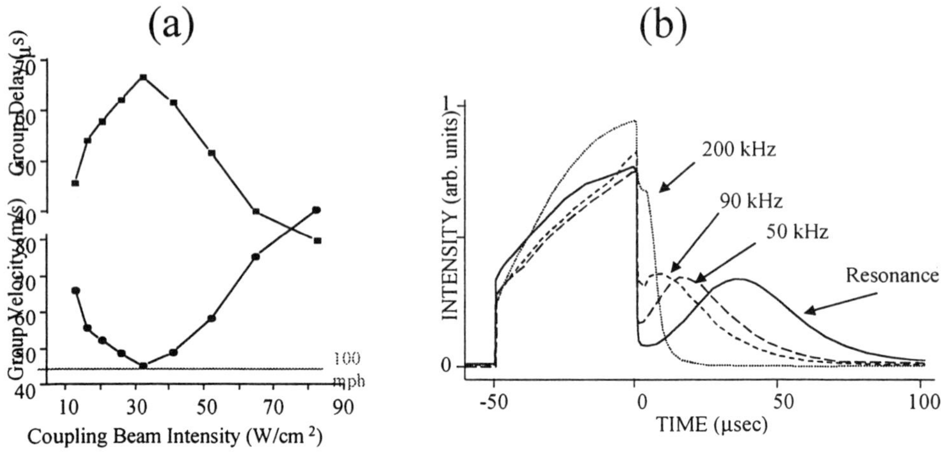

Figure 2. (a) Observed group delays and group light velocities as a function of coupling laser intensity. (b) Observed probe pulse delays for various two-photon detunings. The first pulse is due to probe light that is not absorbed by the crystal. The sharp edges of the probe pulse are not slowed.

6. Quantum computing using optically excited spins in solids

The long lifetime of spin coherences in atomic systems, along with the possibility of controlled excitation using Raman-resonant optical fields, make them well suited for quantum computing. Indeed, successful quantum computing demonstrations in trapped ions take full advantage of such techniques [19]. If similar techniques can be applied to optically active spin systems in solids, highly scalable quantum computers should be possible.

The key requirements for quantum computing using spins in solids are the same as for other systems: the ability to load information, perform single-qubit and two-qubit nonlinear operations, and to readout the results. In Pr:YSO, the ability to load information and perform single-qubit operations has been demonstrated, using dark-resonance techniques.

To perform nonlinear two-qubit operations, optical coupling techniques such as cavity QED [20] or near dipole-dipole [21] interactions can be used, in conjunction with optical Raman dark resonances. One advantage of using optical coupling to entangle spin-based qubits is that optical interactions are of longer range than spin-spin interactions. Another advantage of optical coupling, that is unique to spectral hole burning materials, is illustrated in Figure 3. By choosing dopant atoms whose optical transition frequencies differ by an amount equal to the ground state spin sublevel splitting, a chain of pairwise-coupled qubits can be constructed [22]. As shown, each pair of atoms has an optical transition frequency in common with its neighbor and can be used to couple their spins. Each atom also has an auxiliary transition that can be selectively excited by a laser to control this coupling, so as to perform the desired nonlinear quantum logic operations. It is known that a general-purpose quantum computer can be constructed using a long chain of such pairwise nonlinear interactions. The number of qubits in this chain is determined in principle by the ratio of the optical inhomogeneous width to the ground state spin splitting, or ~ 250 for Pr:YSO.

The final requirement for quantum computing, readout of the result, requires that the final state of a single atom be detected. For a weak optical transition, such as in Pr:YSO, this is problematic. For this reason, we have been exploring solid materials with strong, electric-dipole-allowed optical transitions.

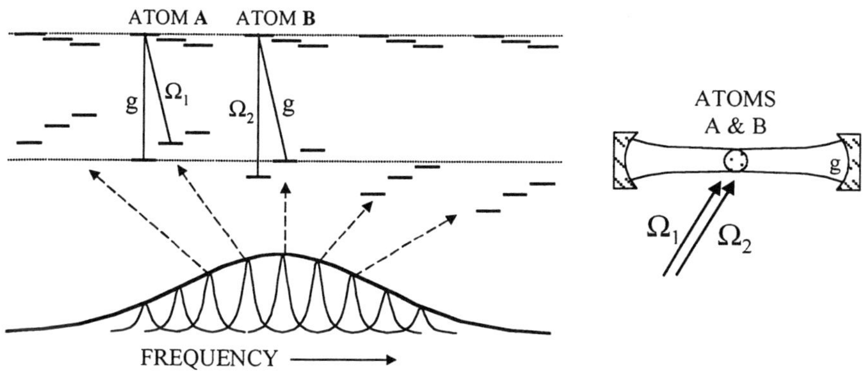

Figure 3. Schematic illustration of a chain of pairwise coupled qubits. The qubits are spin-based, but are coupled optically using either an optical cavity, as shown, or near dipole-dipole interactions. The vacuum Rabi frequency, g, drives transitions of both atoms. The quantum logic is controlled by selective excitation using lasers having Rabi frequencies Ω_1 and Ω_2. The maximum number of coupled qubits is given by the ratio of inhomogeneous width to ground state spin sublevel splitting: 250 for Pr:YSO and 6000 for N-V diamond.

7. Efficient dark resonance excitation in N-V diamond

The key advantage of N-V diamond is its strong optical transition [23], which has an oscillator strength 10^6 times larger than Pr:YSO, comparable to that of a free alkali atom such as sodium. Because of this strong optical transition, the detection of single N-V centers has been demonstrated experimentally, even at room temperature [24]. It also has a relatively long spin coherence lifetime compared to other electron-spin systems, on the order of 40 microseconds. Unfortunately, the spin population lifetime, T_1, is much shorter than in Pr:YSO. Another dis-advantage is that optical Raman transitions are only weakly allowed. Fortunately, by applying a ~ 1 kGauss magnetic field along the quantization axis, it is possible to induce strong mixing of the ground state spin sublevels, so that Raman transitions become more strongly allowed [25]. A simplified energy level diagram of N-V diamond is shown in Figure 4(a). As seen, the ground state is an electron-spin triplet, and the applied magnetic field mixes the $S_z = 0$ and $S_z = -1$ sublevels. Finally, N-V diamond exhibits persistent holeburning via site rearrangement with a quantum efficiency of less than 1 %. To control bleaching caused by this persistent holeburning, a repump laser is applied, that excites the phonon sideband, as shown in Figure 4(b).

Experimental results in Figure 4(c) demonstrate the excitation of a large-amplitude spin coherence by optical Raman resonant lasers [26]. This spin coherence is detected using a background-free, nondegenerate four-wave mixing technique. The frequency width of the observed lineshape is consistent with the ~ 5 MHz inhomogeneous width of the ground-state electron-spin transition, which is much narrower than the 50 MHz homogeneous width of the optical transition. The structure on either side of the main line is evidence of the hyperfine structure, caused by the interaction of electron spin with the nitrogen nuclear spin.

Experimental evidence of the efficient excitation of dark resonances in N-V diamond is shown in Figure 4(d). Induced transparencies as large as 17 % have been observed, not shown. This corresponds to about 70 % of the maximum theoretical transparency, based on the fact that at most only 1 in 4 of the N-V centers is oriented in the correct direction for a given applied magnetic field. The inability to reach 25 % transparency is due to the fact that the optical inhomogeneous width, 750 GHz, is much larger than the optical Rabi frequencies. Unlike the case of Pr:YSO, the repump laser in N-V diamond drives a broad phonon sideband, and a narrow anti-hole is not produced. Nonetheless, a theoretical fit to the lineshape in Figure 4(d) shows that the optically resonant N-V centers are rendered almost completely transparent by the dark resonance interaction. The fitted optical Rabi frequency is about 160 MHz. When compared to the known 40 microsecond spin coherence lifetime, this implies that 1000's of quantum logic operations can in principle be performed within the coherence lifetime.

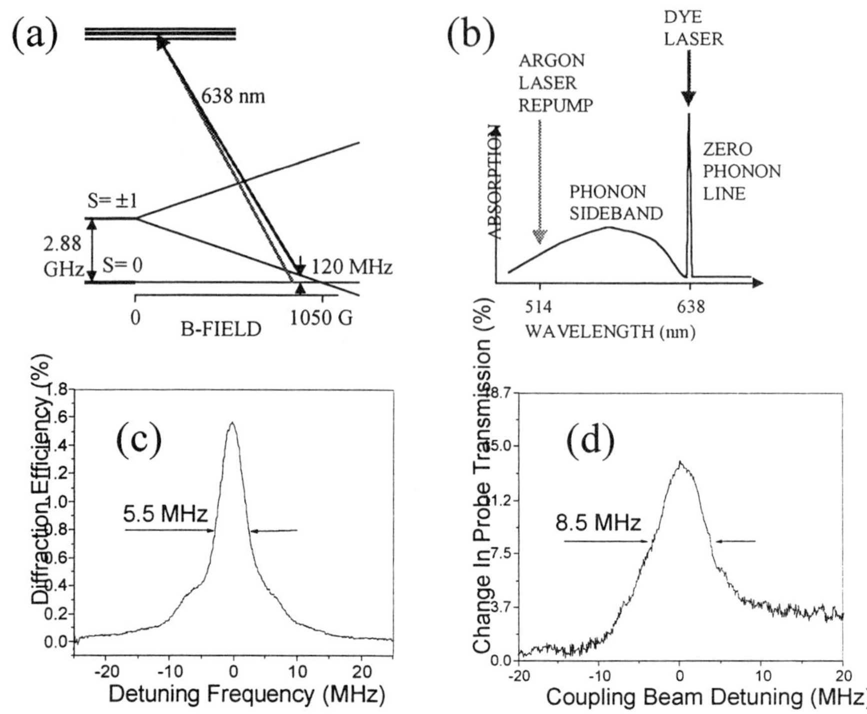

Figure 4. (a) The use of an applied magnetic field to mix the ground state spin sublevels and enhance the strength of the optical Raman transition. (b) Schematic illustration of the role of the repump laser tuned to the phonon sideband. (c) Dark-resonance excited spin coherence in N-V diamond detected by NDFWM. The width is comparable to the spin transition inhomogeneous width, which is narrower than the 50 MHz optical transition homogeneous linewidth. (d) Experimental Raman dark resonance in N-V diamond. The maximum induced transparency possible is 25%, due to the fact that only 1 in 4 N-V centers are orientated in the correct direction.

8. Summary

In summary, we have explored the requirements for efficient excitation of optical Raman dark resonances in solids. We have also demonstrated efficient dark resonance excitation in two materials: Pr:YSO and N-V diamond. In Pr:YSO, ultra-slow light has been observed. In N-V diamond, the ability to perform 1000's of quantum logic operations has been demonstrated in principle.

9. References

1. S. E. Harris, "Electromagnetically induced transparency," *Phys. Today* **50**(7) (1997) 36; S. E. Harris, L. V. Hau, *Phys. Rev. Lett.*, **82** (1999) 4611; O. Schmidt, R. Wynands, Z. Hussein, D. Meschede, *Phys. Rev. A.*, **53** (1996) R27; M. Xio, Y. Li, S. Jin, J. Gea-Banacloche, *Phys. Rev. Lett.*, **74** (1995) 666; A. Kasapi, M. Jain, G. Y. Yin, S. E. Harris, *Phys. Rev. Lett.*, **74** (1995) 2447; S. E. Harris, J. E. Field, A. Kasapi, *Phys. Rev. A*, **46** (1992) R29.

2. H.R. Gray, R.M. Whitley, and C.R. Stroud, "Coherent trapping of atomic populations," *Opt. Lett.* **3**, pg 218 (1978); G. Alzetta, A. Gozzini, L. Moi, and G. Orrilos, "An experimental method for the observation of rf transitions and laser beat resonances in oriented Na vapor," *Nuovo Cimento B* **36** (1976) 5.

3. S. Harris, "Lasers without inversion; interference of lifetime-broadened resonances," *Phys. Rev. Lett.* **62** (1989) 1033; A. S. Zibrov, M. D. Lukin, D. E. Nikonov, L. Hollberg, M. O. Scully, V. L. Velichansky, and H. G. Robinson, "Experimental demonstration of laser oscillation without population inversion via quantum interference in Rb," *Phys. Rev. Lett.* **75** (1995) 1499.

4. B.S. Ham, M.K. Kim, P. R. Hemmer, and M. S. Shahriar, "Frequency-selective Time-domain Optical Data Storage by Electromagnetically Induced Transparency in a Rare-Earth Doped Solid," *Opt. Lett.* **22**, 1849 (1997).

5. S. E. Harris, J. E. Field, and A. Imamoglu, "Nonlinear optical processes using electromagnetically induced transparency," *Phys. Rev. Lett.* **64** (1990) 1107; M. Jain, H. Xia, G. Y. Yin, A. J. Merriam, and S. E. Harris, "Efficient nonlinear frequency conversion with maximal atomic coherence," *Phys. Rev. Lett.* 77 (1996) 4326.

6. V. S. Sudarshanam, M. Cronin-Golomb, P. R. Hemmer, and M. S. Shahriar, "Turbulence Aberration Correction with High-Speed High-Gain Optical Phase Conjugation in Sodium Vapor," *Optics Letters* **22**, 1141 (1997).

7. S.F. Yelin and P.R. Hemmer, "Sensitive Detection based on Multiphoton Transitions in Atoms and Quantum Wells," submitted, preprint in: **quant-ph/0012136**.

8. M.S. Shahriar and P.R. Hemmer, "Generation of Squeezed States and Twin Beams via Non-Degenerate Four Wave Mixing in an Ideal Λ system," *Opt. Commun.* **158,** 273-286 (1998); "Distortion-free Gain and Noise Correlation in Sodium Vapor with Four Wave Mixing and Coherent Population Trapping," T. T. Grove, M. S. Shahriar, P. R. Hemmer, Prem Kumar, V. S. Sudarshanam, and M. Cronin-Golomb, *Optics Letters* **22**, 769 (1997).

9. L. V. Hau, S. E. Harris, Z. Dutton, C. Behroozi, *Nature*, **397** (1999) 594; M. Kash, V. Sautenkov, A. Zibrov, L. Hollberg, G. Welch, M. Lukin, Y. Rostovsev, E. Fry, M. Scully, *Phys. Rev. Lett.*, **82** (1999) 5229.

10. C. Liu, Z. Dutton, C.H. Behroozi, and L.V. Hau, *Nature* **409**, (2001) 490; D.F. Phillips, A. Fleischhauer, A. Mair, R.L. Walsworth, *Phys. Rev. Lett.*, **86** (2001) 783.

11. R.M. MacFarlane and R.M. Shelby, "Coherent Transients and Holeburning Spectroscopy of Rare Earth Solids" in *Spectroscopy of Solids Containing Rare Earth Ions*, ed by A.A Kaplyanskii and R.M. MacFarlane, Elsevier Science Publishers 1987, Chpt. 3.

12. S. Fernbach and W.G. Proctor, "Spin-echo memory device," *Jour. of Appl. Phys.* **26**, 170 (1955)

13. H. Lin, T. Wang, and T. W. Mossberg, *Opt. Lett.* **20** (1995) 1658.

14. X. A. Shen and R. Kachru, "Optical header recognition by spectroholographic filtering," *Opt. Lett.* **20** (1995) 2508.

15. R. W. Equall, R. L. Cone, and R. M. Macfarlane, "Homogeneous broadening and hyperfine structure of optical transitions in Pr^{3+}:Y_2SiO_5," *Phys. Rev. B* **52** (1995) 3963; K. Holliday, M. Croci, E. Vauthey, and U.P. Wild, *Phys. Rev. B* **47** (1993) 14741.

16. R.T. Harley, M.J. Henderson, and R.M. Macfarlane, "Persistent spectral hole burning of colour centres in diamond," *J. Phys.* C **17**, pg L233 (1984).

17. B.S. Ham, P. R. Hemmer, and M. S. Shahriar, "Efficient Electromagnetically Induced Transparency in a Rare-Earth Doped Crystal", *Optics Communications* **144**, 227 (1997).

18. A.V. Turukhin, V.S. Sudarshanam, M. S. Shahriar, J. A. Musser, and P.R. Hemmer, "Observation of Ultraslow Group Velocity of Light in a Solid," Submitted, preprint in: **quant-ph/0010009**.

19. C. Monroe, D. M. Meekhof, B. E. King, W. M. Itano, and D. J. Wineland, "Demonstration of a Fundamental Quantum Logic Gate," *Phys. Rev. Lett.* **75** (1995) 4714.

20. T. Pellizzari, S. A. Gardiner, J. I. Cirac, and P. Zoller, "Decoherence, continuous observation, and quantum computing: a cavity QED model," *Phys. Rev. Lett.* **75** (1995) 3788.

21. M. D. Lukin and P. R. Hemmer, "Quantum Entanglement via Optical Control of Atom-Atom Interactions," *Phys. Rev. Lett.* **84**, pp. 2818-2821 (2000).

22. M.S. Shahriar, P.R. Hemmer, S. Lloyd, J. Bowers, and A.E. Craig, "Solid State Quantum Computing Using Spectral Holes," **quant-ph/0007074**.

23. E. van Oort, N.B. Manson and M. Glasbeek, "Optically detected spin coherence of the N-V diamond centre in its triplet ground state," *J. Phys.* C **21**, pg 4385 (1988).

24. Gruber, D. Drabenstedt, C. Tietz, L. Fleury, J. Wrachtrup, C. von Borczyskowski, *Science* **276**, pg 2012 (1997).

25. Wei and N.B. Manson, "Observation of the dynamic Stark effect on electromagnetically induced transparency," *Phys. Rev. A* **60**, pg 2540 (1999).

26. P.R. Hemmer, A.V. Turukhin, M.S. Shahriar, and J.A. Musser, "Raman Excited Spin Coherences in N-V Diamond," Accepted in Optics Letters, preprint in **quant-ph/0007114**.

Quantum Dot Single Photon Source

A. Kiraz, P. Michler, C. Becher, B. Gayral, Lidong Zhang, E. Hu, W.V. Schoenfeld, and P.M. Petroff, and A. Imamoğlu

Department of Electrical and Computer Engineering, University of California, Santa Barbara, California 93106

Abstract: Single photon sources play a key role in quantum communication and computation schemes. Using pulsed laser excitation of a single quantum dot (QD), a single photon source that generates a train of single photon pulses is demonstrated. Generated photon pulses have a vanishing probability for detection of two or more photons. For a spectrally isolated self-assembled InAs QD, nearly 100% of the excitation pulses lead to emission of a single photon, yielding an ideal single photon source. The QD single-exciton (1X) resonance that generates the single-photon pulses can be resonantly coupled to a high quality factor microdisk whispering gallery mode (WGM), by varying the sample temperature. We demonstrate exciton lifetime reduction by a factor of 2 due to the Purcell effect, on resonance.

1 Introduction

A deterministic source of single-photons is an enabling technology for a number of quantum information protocols [1]. For example, an essential el ement of secure key distribution in quantum cryptography is an optical source emitting a train of pulses that contain one-and-only-one photon [2]. Since measurements unavoidably modify the state of a single quantum system, an eavesdropper cannot gather information about the secret key without being noticed, provided that the pulses used in transmission do not contain two or more photons. For applications in quantum cryptography, specific properties of single-photon pulses, such as precise photon energy, pulse duration and bandwidth are not of primary importance.

Recently, it has also been shown that availability of a single-photon source enables implemention of efficient quantum computation using only linear optical elements and efficient photo-detectors [3]. This remarkable proposal relies heavily on two-photon interference on a beam-splitter for realization of probabilistic quantum gates and state preparation. It is therefore crucial to have a source that generates identical transform-limited single-photon pulses with vanishing time-jitter.

We report the experimental demonstration of a triggered single-photon source which uses the anharmonicity of single quantum dot (QD) multi-exciton transitions and pulsed laser excitation to regulate the photon generation process [4]. Realization of such a source, termed single photon turnstile device [5], has been one of the holy grails of quantum electronics research, since it represents the ultimate limit in the quantum control of photon generation process. Unlike the room temperature proposals based on single molecules [6, 7] and nitrogen vacancies in diamond [8, 9], the QD based device can in principle generate transform-limited single-photon pulses. Since QDs can be coupled to photonic nanostructures that act as nano-cavities with a large Purcell factor, it is also possible to obtain a high single-photon collection efficiency [10].

2 Single Photon Generation from a Single Quantum Dot

It is known that a driven single anharmonic quantum system, such as an atom or a molecule, exhibits photon antibunching, i.e. a dead time between successive photon emission events [11]. Using Hanbury, Brown and Twiss (HBT) type photon correlation measurements [12], photon antibunching has been observed in a variety of single quantum emitters, e.g., an atom [13], a stored ion [14], a molecule [15], a semiconductor quantum dot [17], and a single nitrogen-vacancy center in diamond [8, 9]. Photon antibunching is a necessary but not sufficient condition for a single photon turnstile device; an additional mechanism for regulating the excitation process is required to realize single photon pulses.

The QD single photon source that we realized is based on a single QD embedded in a high quality factor (Q) microcavity structure [18]. To ensure single photon generation at the fundamental QD exciton transition (1X), we adjust the pump power so that two or more electron-hole pairs are captured by the QD during each excitation pulse. The energy of the photons emitted during relaxation depends significantly on the number of multiexcitons that exist in the QD, due to Coulomb interactions enhanced by strong carrier confinement [18]. If the total recombination time of the multiexciton QD state is longer than the recombination time of the free electron-hole pairs, each excitation pulse can lead to at most one photon emission event at the 1X transition. Therefore, regulation of photon emission process can be achieved due to a combination of Coulomb interactions creating an anharmonic multiexciton spectrum and slow relaxation of highly-excited QDs leading to vanishing re-excitation probability following the photon emission event at the 1X transition [18]. Since the QD 1X recombination has been shown to be predominantly radiative [17], every excitation pulse from the mode-locked laser will generate an ideal single photon pulse.

3 Sample Description

Fig. 1. The microdisk structure that consists of a 5 μm diameter disk and a 0.5 μm post. The GaAs disk area that supports high-quality factor whispering gallery modes is 200 nm thick and contains InAs quantum dots.

Our samples were grown by molecular beam epitaxy (MBE) on a semi-insulating GaAs substrate. Figure 1 shows the microdisk structure which consists of a 5 μm diameter disk and a 0.5 μm $Al_{0.65}Ga_{0.35}As$ post. The disk area consists of 100 nm GaAs, an InAs QD layer, and 100 nm GaAs. Details of the microdisk processing can be found in Ref. [19]. The QDs were grown using the partially covered island technique [20] with a gradient in the QD density reaching from $\leq 10^6$ cm^{-2} to $\sim 10^{10}$ cm^{-2} across the sample wafer. The QDs possess a diameter of $\sim$ 40 - 50 nm and a height of $\sim$ 3 nm, emitting in the wavelength range from 920 to 975 nm.

4 Experimental Setup

The experimental setup consists of a low-temperature diffraction-limited scanning optical microscope for spatially resolved photoluminescence (PL) spectroscopy and a HBT setup for photon correlation measurements. The system provides spectral resolution of 70 μeV, spatial resolution of 1.7 μm, and temporal resolution of 420 ps. The microdisks were mounted in a He gas flow cryostat. Optical pumping was performed with a mode-locked femtosecond ($\sim$ 250 fs) Ti:sapphire laser that operates at 750 nm and has a pulse repetition rate of 82 MHz. The electron-hole pairs were mainly generated in the GaAs barriers and subsequently captured by the QDs within a short timescale ($<$ 35 ps [21]). A microscope objective (with numerical aperture NA = 0.55) was used to focus the excitation laser onto the sample and to collect the emitted PL from the QDs. The collected light was spectrally filtered and then split with a 50/50 beamsplitter. The resulting two light beams were focused onto two single-photon-counting avalanche photodiodes (SPAD). The measured quantity is the number of photon pairs $n(\tau)$ with arrival time separation of $\tau = t_{start} - t_{stop}$, which is equivalent to the unnormalized correlation function $G^{(2)}(\tau)$ in the limit where the reciprocal of the average counting rate is much longer than the measured time separation τ between photon pairs [12].

5 Experimental Results

Photon correlation measurements carried out under pulsed excitation yield signatures for turnstile operation, by discriminating between one and two-photon (Fock-state) pulses as well as coherent-state pulses [6]. For a pulsed periodic coherent source which emits Poissonian light, the peak at $\tau = 0$ would be identical to the peaks at integer multiples of the repetition rate T_{rep}, for all values of the mean photon number. In contrast, for an ideal turnstile device the peak at $\tau = 0$ is absent [22].

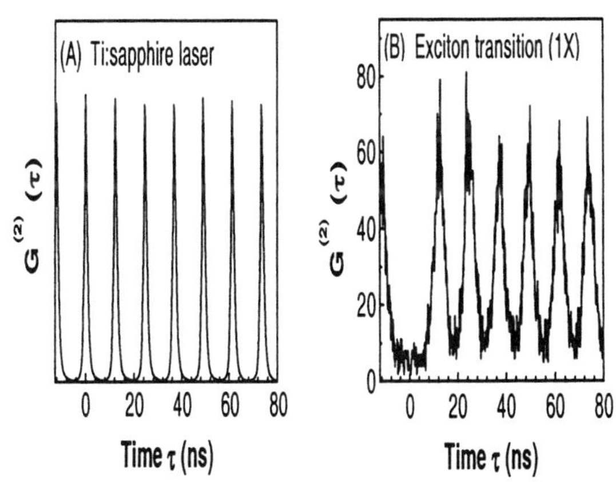

Fig. 2. Measured unnormalized correlation function $G^{(2)}(\tau)$ of (A) a mode-locked Ti:sapphire laser (FWHM = 250 fs), and (B) a single QD excitonic ground state (1X) emission under pulsed excitation conditions (82 MHz). The QD 1X transition was out of resonance with the microdisk modes.

Figure 2 shows the measured unnormalized correlation function $G^{(2)}(\tau)$ for (A) the pulsed Ti:sapphire laser, and (B) the 1X transition of a QD that is far detuned from all WGMs (T=4 K). The pump intensity in this experiment corresponds to an excitation of the QD where the 1X emission is well into the saturation regime [23]. As expected, the measured $G^{(2)}(\tau)$ of the pulsed Ti:sapphire l aser exhibits peaks at integer multiples of T_{rep} = 12.27 ns with negligible signal in between the peaks. The measured $G^{(2)}(\tau)$ of the QD 1X emission at T = 4 K (Fig. 2 B) also shows peaks at integer multiples of T_{rep}, indicating the locking of the photon emission to the pulsed excitation. But in contrast to the mode-locked laser, the peak at $\tau = 0$ is no longer present, i.e., the probability of finding a second photon following the detection of the first photon at $\tau = 0$ vanishes. This is the principal result of our work: absence of the peak at $\tau = 0$ provides strong evidence for an ideal single photon turnstile operation.

The lifetime of the single exciton ground-state transition (1X) was determined from cw antibunching experiments to be 2.2 ns, which is the shortest possible total recombination time for a multiply-excited QD. As the recombination time in the GaAs barrier and the wetting layer is considerable faster (100 - 200 ps), no free carriers are available to re-excite the QD after the 1X recombination process. As discussed earlier, only one 1X recombination process can occur per excitation pulse under these conditions. To ensure that a single photon is indeed emitted for each excitation pulse, the pump power of the excitation laser should be adjusted so that the probability of having no injected electron-hole pair in the QD is negligible. The fact that the photon correlation measurement depicted in Fig. 2 B was obtained well in the saturation regime ensures that QD is multiply-excited in our experiments. In addition, the quantum efficiency of the QD has to be high $\eta \sim 1$ to avoid nonradiative recombination processes. Recent experiments reported in Ref. [17] have shown that for our samples the dominant recombination mechanism is radiative. These facts allow us to conclude that the generated light at the excitonic ground-state transition energy 1X is a stream of single photons with a repetition rate of 82 MHz.

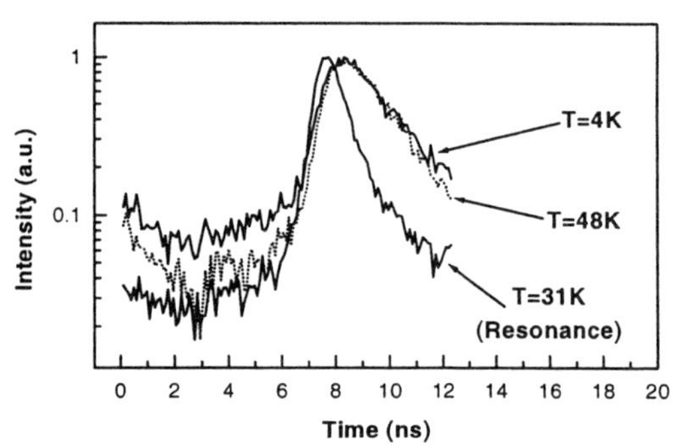

Fig. 3. Time correlated single photon counting measurements on the single-exciton (1X) transition of a QD: out of resonance with the WGM (4 K, 48 K), in resonance with the WGM (31 K).

As indicated earlier, the QDs that act as a source of single-photons are embedded in a microdisk structure that supports high quality WGMs. Even though the experimental results shown in Fig. 2 B were obtained when the QD 1X resonance was far away from all WGMs, it is in principle possible to bring the QD 1X recombination and cavity modes in resonance by changing the sample temperature. In the weak-coupling cavity-QED regime with fast cavity-decay, Purcell effect will then force the photons from 1X recombination to be emitted predominantly into the cavity mode. This will in turn ensure that the single-photon collection efficiency is enhanced and the time-jitter reduced. Purcell effect is therefore a crucial element for applications requiring a high-collection efficiency single-photon sources.

To demonstrate the Purcell effect, we have carried out time-correlated single photon counting (TCSPC) experiments [24] on a different QD, whose 1X resonance energy is only 1.5 meV away from a WGM with $Q \sim 6500$ at $T = 4K$. Figure 3 shows the experimental data at 3 different temperatures. These measurements have all been performed in low excitation regime where the decay time corresponds to the lifetime of the 1X transition [25]. Measured identical decay times of 1.8 ns at 4 K and 48 K show that when the 1X transition is not coupled to any WGMs its lifetime shows no dependence on temperature between 4-48 K. Hence, nonradiative recombination does not significantly effect the dynamics of the 1X transition within this temperature range. When the 1X emission is in resonance with the WGM at 31 K, the decay time is reduced to 850 ps, inferring a Purcell factor of 2.

6 Conclusion

We have demonstrated almost ideal single photon turnstile operation by pulsed laser excitation of a single InAs QD at 4 K with an overall detection efficiency of $5 \cdot 10^{-5}$. We envision that the operation temperature for the QD turnstile device can be extended to 77 K, and the overall detection efficiency can be increased to > 10% by using microcavity coupling. Furthermore, due to the Purcell effect, the QD turnstile device can generate single photon pulses with a repetition rate as high as 1 GHz. Since our single photon source is based on a semiconductor nanostructure, it can in principle be integrated into photonic circuits. All of these features make this QD turnstile device suitable for applications such as quantum cryptography and quantum information processing.

Acknowledgements

This work was supported by David and Lucile Packard Fellowship, and a grant from ARO. P.M. acknowledges support from the Max Kade Foundation; he is currently at the University of Bremen. C.B. is supported by Deutsche Forschungsgemeinschaft; he is currently at the University of Innsbruck.

References and links

1. D. Bouwmeester, A. Ekert, A. Zeilinger, *The Physics of Quantum Information* (Springer, Berlin, 2000).
2. C. H. Bennett, G. Brassard, "Quantum Cryptography: Public Key Distribution and Coin Tossing," Proc. IEEE Int. Conference on Computers, Systems and Signal Processing, 175 (IEEE, New York, 1984).
3. E. Knill, R. Laflamme, and G.J. Milburn,"A scheme for efficient quantum computation with linear optics," Nature **409**, 46 (2001).
4. P. Michler, A. Kiraz, C. Becher, W.V. Schoenfeld, P.M. Petroff, Lidong Zhang, E. Hu, and A. Imamoğlu, "A quantum dot single-photon turnstile device," Science **290**, 2282 (2000).
5. A. Imamoğlu, Y. Yamamoto, "Turnstile device for heralded single photons: coulomb blockade of electron and hole tunneling in quantum confined p-i-n heterojunctions," Phys. Rev. Lett. **72**, 210 (1994).
6. C. Brunel, B. Lounis, P. Tamarat, M. Orrit, "Triggered source of single photons based on single molecule fuorescence," Phys. Rev. Lett. **83**, 2722 (1999).
7. B. Lounis, W. E. Moerner, "Single photons on demand from a single molecule at room temperature," Nature (London) **407**, 491 (2000).
8. C. Kurtsiefer, S. Mayer, P. Zarda, H. Weinfurter, "Stable solid-state source of single photons," Phys. Rev. Lett. **85**, 290 (2000).
9. R. Brouri, A. Beveratos, J.-P. Poizat, P. Grangier, "Photon antibunching in the fluorescence of individual color centers in diamond," Opt. Lett. **25**, 1294 (2000).
10. A. Kiraz, P. Michler, C. Becher, B. Gayral, A. Imamoğlu, Lidong Zhang, E. Hu, W.V. Schoenfeld, and P.M. Petroff, "Cavity-quantum electrodynamics using a single InAs quantum dot in a microdisk structure," Appl. Phys. Lett. **78**, 3932 (2001).
11. D. F. Walls, G. J. Milburn, *Quantum Optics* (Springer, Berlin, 1994).
12. R. Hanbury Brown, R. Q. Twiss, "Question of correlation between photons in coherent light rays," Nature (London) **178**, 1447 (1956).
13. H. J. Kimble, M. Dagenais, L. Mandel, "Photon antibunching in resonance fluorescence," Phys. Rev. Lett. **39**, 691 (1977).
14. F. Diedrich, H. Walther, "Nonclassical radiation of a single stored ion," Phys. Rev. Lett. **58**, 203 (1987).

15. Th. Basché, W. E. Moerner, M. Orrit, H. Talon, "Photon antibunching in the fluorescence of a single dye molecule trapped in a solid," Phys. Rev. Lett. **69**, 1516 (1992).
16. P. Michler, A. Imamoğlu, M. D. Mason, P. J. Carson, G. F. Strouse, S. K. Buratto, "Quantum correlation among photons from a single quantum dot at room temperature," Nature (London) **406**, 968 (2000).
17. C. Becher, A. Kiraz, P. Michler, A. Imamoğlu, W. V. Schoenfeld, P. M. Petroff, Lidong Zhang, and E. Hu , "Nonclassical radiation from a single self-assembled InAs quantum dot," Phys. Rev. B **63**, 121312(R) (2001).
18. J.-M. Gérard, B. Gayral, "Strong Purcell effect for quantum boxes in three-dimensional solid-state microcavities," J. Lightwave Technol. **17**, 2089 (1999).
19. P. Michler, A. Kiraz, Lidong Zhang, C. Becher, E. Hu, A. Imamoğlu, "Laser emission from quantum dots in microdisk structures," Appl. Phys. Lett. **77**, 184 (2000).
20. J. M. Garcia, T. Mankad, P. O. Holtz, P. J. Wellman, P. M. Petroff, "Electronic states tuning of InAs self-assembled quantum dots," Appl. Phys. Lett. **72**, 3172 (1998).
21. S. Raymond, S. Fafard, P. J. Poole, A. Wojs, P. Hawrylak, S. Charbonneau, D. Leonard, R. Leon, P. M. Petroff, J. L. Merz, "State filling and time-resolved photoluminescence of excited states in InxGa1-xAs/GaAs self-assembled quantum dots," Phys. Rev. B **54**, 11548 (1996).
22. A. Imamoğlu, H. Schmidt, G. Woods, M. Deutsch, "Strongly interacting photons in a nonlinear cavity," Phys. Rev. Lett. **79**, 1467 (1997).
23. E. Dekel, D. V. Regelman, D. Gershoni, E. Ehrenfreund, W. V. Schoenfeld, P. M. Petroff, "Cascade evolution and radiative recombination of quantum dot multiexcitons studied by time-resolved spectroscopy," Phys. Rev. B **62**, 11038 (2000).
24. D.V. O'Connor and D. Phillips, *Time Correlated Single Photon Counting* (Academic Press, London, 1984).
25. G. Bacher, R. Weigand, J. Seufert, V.D. Kulakovskii, N.A. Gippius, A. Forchel, K. Leonardi, and D. Hommel, "Biexciton versus exciton lifetime in a single semiconductor quantum dot," Phys. Rev. Lett. **83**, 4417 (1999).

Coherence and collective behavior in Bose-Einstein condensates

W. Ketterle
Research Laboratory of Electronics, Department of Physics, and Center of Ultracold Atoms at MIT and Harvard
Massachusetts Institute of Technology, Cambridge, MA 02139, USA
ketterle@mit.edu

Abstract: Bose-Einstein condensed atomic gases are a new class of quantum fluids with many novel properties. This new subfield overlaps with condensed matter physics in the study of many-body physics, and with quantum optics since Bose condensates are sources for beams of coherent atoms - atom lasers. The talk will discuss examples from both areas. The study of quantized vortices and superfluidity illustrate the condensed matter physics aspect, and the study of coherence and matter-wave amplification are examples for the area of atom optics.

Sub-diffraction-limited resolution with squeezed light

Mikhail I. Kolobov[1], **Claude Fabre**[2], **Pierre Scotto**[3], **Pere Colet**[3], **and Maxi San Miguel**[3]

[1] *Laboratoire PhLAM, Université de Lille 1, F-59655 Villeneuve d'Ascq cedex, France*

[2] *Laboratoire Kastler Brossel, Université Pierre et Marie Curie, Case 74, F-75252 Paris cedex 05, France*

[3] *Instituto Mediterráneo de Estudios Avanzados (IMEDEA, CSIC-UIB), Universitat de les Illes Balears, E-07071 Palma de Mallorca, Spain*

Abstract: We discuss a possibility of achieving sub-diffraction-limited resolution in coherent optical imaging using multimode squeezed light. For this purpose, we formulate a quantum theory of superresolution for object reconstruction in terms of prolate spheroidal functions. We derive the expression for ultimate resolution limit in the reconstructed object using an illumination of the full object plane by a multimode squeezed vacuum. We show that the gain in superresolution using multimode squeezed light is maximum when the Shannon number S of the imaging system is smaller than unity. This situation is realized, for example, in scanning microscopy. We present some preliminary numerical simulations aiming to demonstrate the feasibility of significant superresolution for $S < 1$.

1 Introduction

A new area of quantum optics known as "quantum imaging" has emerged recently as a synthesis of the branch of quantum optics studying the spatial quantum fluctuations[1, 2, 3] with the classical optics of image formation, microscopy, resolution, et cetera. The aim of quantum imaging is to investigate the modifications introduced by the quantum-mechanical nature of light into the classical imaging. Recent publications have addressed such questions as quantum limits on measurement of small displacements[4] and quantum limits on optical resolution[5]. In particular, in these publications it was demonstrated theoretically that use of multimode squeezed light can significantly improve the resolution beyond the standard quantum limit in displacement measurements and can also beat the Rayleigh resolution limit in a diffraction-limited optical system. The theoretical prediction[4] was recently confirmed experimentally[6].

A well-known Rayleigh resolution criterion states that the resolution in an optical system is limited by diffraction on a system pupil[7]. This criterion was formulated for a simple visual observation and is based on a presumed resolving capabilities of a human eye. However, it is recognized now that modern CCD cameras allow us to achieve the performance very much exceeding that of a visual observation. For example, experimental measurements of displacements in the nanometer range has been performed to detect deflection of glass fibers[8, 9, 10], microscopic phase objects[11], movement of biological, subcellular vesicles[12], measurement of ultra-weak absorption using the mirage effect[13], or in atomic force microscopy[14]. The sensitivity of all these measurements

Coherence and Quantum Optics VIII
Edited by Bigelow et al., Kluwer Academic/Plenum Publishers, 2003

is ultimately limited not by diffraction, but by the quantum fluctuations of the light beam used in the experiment.

The question of the role of quantum fluctuations in optical resolution is very closely related to the so-called "superresolution" techniques. Recently in a series of papers[15, 16, 17, 18] it was shown both theoretically and experimentally that, when tightly focused beams are used for illumination with either coherent or incoherent light, the Rayleigh criterion underestimates the correct resolution limit. The authors of these papers have demonstrated that a suitably designed optical system can robustly achieve an increase in linear resolution that approaches twice the Rayleigh criterion Following these demonstrations the authors have proposed two new superresolution scanning microscopes in Refs. [19, 20]. The superresolution in these microscopes is achieved by means of a special optical mask placed in the image plane.

Since in these demonstrations and in the proposed superresolution scanning microscopy the ultimate resolution is limited not by diffraction but by the signal-to-noise ratio in the detection scheme, we believe that further improvement in superresolution can be achieved using multimode squeezed light. As shown in numerous previous publications, such light allows one to increase the signal-to-noise ratio in different imaging schemes.

In this paper we shall present in more details as compared to Ref. [5] the quantum theory of superresolution which takes into account the quantum fluctuations of light illuminating the object and the vacuum fluctuations outside the object. This theory allows us to consider the illumination of the object with multimode squeezed light. We shall estimate qualitatively the quantum limits of resolution in the case of illumination with light in coherent state and illumination with multimode squeezed light. Finally we shall present the preliminary numerical simulations of superresolution achievable for a one-dimensional two-slit object for the Shannon number smaller that unity.

2 Classical superresolution

For simplicity we shall consider here the case of a one-dimensional coherent optical imaging. However, our results can be easily generalized for the two-dimensional imaging and also for the case on incoherent illumination as will be shown elsewhere. The scheme of diffraction-limited coherent optical imaging is shown in Fig. 1. An object of finite size X is placed in the object plane y. The first lens L_1 performs the Fourier transform of this object into the Fourier plane ξ where a pupil of size d is located. The second lens L_2 performs the inverse Fourier transform and creates an image in the image plane x. The finite size of the pupil introduces a finite bandwidth in the transmission of the spatial frequencies by the system, so that the image is not an exact copy of the object.

"Superresolution" is an attempt to restore object details beyond the Rayleigh limit[21] or equivalently to recover the object spectrum outside the spatial-frequency band of the system. This amounts to an extrapolation of the object spatial-frequency spectrum outside the band. Such an extrapolation appears to be feasible under some assumptions, such as analyticity of the object spectrum, which holds in the case of an object of finite size X (see Fig. 1). In this case the analytic continuation of the object spectrum outside the band allows in principle for unlimited resolution. However such a superresolution is extremely sensitive to the noise. The ultimate limit of superresolution is therefore determined by the quantum fluctuations of light, that is the smallest amount of noise in the optical measurement allowed by the quantum theory.

Let us introduce dimensionless variables $s = 2x/X, s' = 2y/X$, and the space-bandwidth product $c = \dfrac{\pi d X}{2\lambda f}$. In terms of these variables the transformation L of the

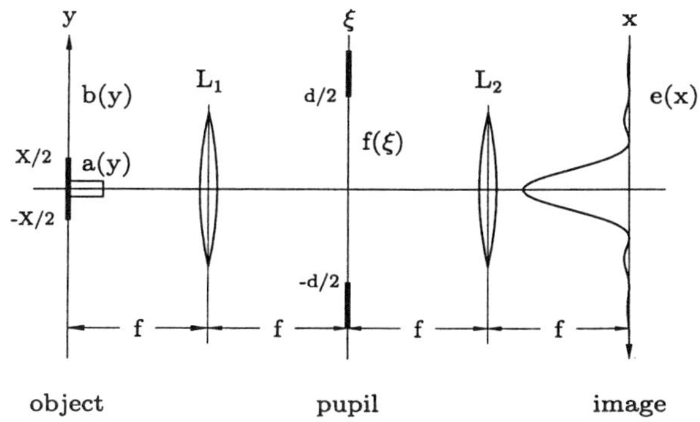

Fig. 1. Schematic of one-dimensional diffraction-limited coherent optical imaging.

classical object amplitude $a(s')$ into the image amplitude $e(s)$ reads:

$$\begin{aligned} e(s) &\equiv (La)(s) \\ &= \int_{-1}^{1} \frac{\sin[c(s-s')]}{\pi(s-s')} a(s')\, ds', \qquad -\infty < s < \infty. \end{aligned} \tag{1}$$

The problem of reconstruction of the object $a(s')$ from a detected image $e(s)$ in the absence of noise is equivalent to inversion of the integral operator L. The operator $L^\star$ adjoint to L is given by [15]

$$(L^\star f)(s') = \int_{-\infty}^{\infty} \frac{\sin[c(s'-s)]}{\pi(s'-s)} f(s)\, ds, \qquad |s'| \leq 1. \tag{2}$$

The product $A = L^\star L$ is the self-adjoint operator,

$$(Af)(s) = \int_{-1}^{1} \frac{\sin[c(s-s')]}{\pi(s-s')} f(s')\, ds', \qquad |s|, |s'| \leq 1, \tag{3}$$

studied by Slepian and Pollak[22]. The orthonormal system of eigenfunctions of A is given by

$$\varphi_k(s') = \frac{1}{\sqrt{\lambda_k}} \psi_k(s'), \qquad |s'| \leq 1, \tag{4}$$

where $\psi_k(s)$ are the linear prolate spheroidal functions [22, 23], and λ_k are the corresponding eigenvalues. The functions $\varphi_k(s)$ form a basis in $L^2(-1,1)$ and may be considered as "elements of information" of the input object. The eigenvalues λ_k are an infinite set of real, positive numbers obeying $1 \geq \lambda_0 > \lambda_1 > \ldots > 0$. For small k the λ_k fall off slowly with k until the index reaches the critical value, $k = S$, called the Shannon number,

$$S = \frac{2c}{\pi} = \frac{dX}{\lambda f}, \tag{5}$$

beyond which the λ_k rapidly approach zero.

Using the fundamental properties of the linear prolate spheroidal wave functions,

$$\int_{-1}^{1} \frac{\sin[c(s-s')]}{\pi(s-s')} \psi_k(s')\, ds' = \lambda_k \psi_k(s), \tag{6}$$

$$\int_{-\infty}^{\infty} \frac{\sin[c(s-s')]}{\pi(s-s')} \psi_k(s')\, ds' = \psi_k(s), \tag{7}$$

we obtain

$$L\varphi_k = \sqrt{\lambda_k}\psi_k, \qquad L^\star \psi_k = \sqrt{\lambda_k}\varphi_k. \tag{8}$$

Note that the functions $\psi_k(s)$ are defined on the real axis $-\infty < s < \infty$, and the functions $\varphi_k(s')$ on the interval $-1 \leq s' \leq 1$. Expanding the object amplitude over the functions $\varphi_k(s')$ and the image amplitude over $\psi_k(s)$, we can easily find the relation between the expansion coefficients of the object and the image. Indeed, since the functions $\varphi_k(s')$ form a complete orthonormal set in $[-1,1]$ we can write the object amplitude as

$$a(s') = \sum_{k=0}^{\infty} a_k \varphi_k(s'), \qquad |s'| \leq 1, \tag{9}$$

with the coefficients a_k given by

$$a_k = \int_{-1}^{1} a(s')\varphi_k(s')\, ds'. \tag{10}$$

A similar expansion can be written for the image amplitude in terms of functions $\psi_k(s)$

$$e(s) = \sum_{k=0}^{\infty} e_k \psi_k(s), \qquad -\infty < s < \infty. \tag{11}$$

Substituting these expansions into Eq. (1) and using first of Eqs. (8) we obtain the following relation between a_k and e_k:

$$e_k = \sqrt{\lambda_k} a_k. \tag{12}$$

Let us denote by $\tilde{a}_k$ the expansion coefficients of the object reconstructed from the measured image $e(s)$. In the absence of noise these coefficients are obtained by dividing the image coefficients e_k by $\sqrt{\lambda_k}$. Thus, for a noise-free image the object reconstruction can be performed exactly, $\tilde{a}_k = a_k$, i. e. without resolution limit. The ultimate accuracy in the determination of $\tilde{a}_k$ will therefore be determined by quantum fluctuations in the measurement of e_k.

3 Quantum theory of superresolution

In the quantum theory the object amplitude $a(s')$ and the image amplitude $e(s)$ become operators obeying the standard commutation relations (see, for example[1]). We can use Eqs. (9),(11) now treating the expansion coefficients a_k and e_k as photon annihilation operators. The operators a_k in the object plane obey the following commutation relations:

$$[a_k, a_l^\dagger] = \delta_{kl}, \qquad [a_k, a_l] = 0. \tag{13}$$

The same commutation relations must be satisfied by the image coefficients e_k. However, Eq. (12) does not preserve the commutation relations (13). The reason for this is that classical imaging equation (1) takes into account only nonzero field amplitude in the region $|s'| \leq 1$ of the object plane. The rest of this plane $|s'| > 1$ is ignored because there the classical field amplitude is zero. In the quantum theory this region must be taken into account to guarantee the conservation of the commutation relations.

To include the region $|s'| > 1$ in the object plane into the theory we introduce another set of functions:

$$\chi_k(s') = \frac{1}{\sqrt{1-\lambda_k}} \psi_k(s'), \qquad |s'| > 1. \tag{14}$$

Using the following properties of the prolate spheroidal wave functions,

$$\int_{-\infty}^{\infty} \psi_k(s)\psi_l(s)\, ds = \delta_{kl}, \tag{15}$$

$$\int_{-1}^{1} \psi_k(s)\psi_l(s)\, ds = \lambda_k\delta_{kl}, \tag{16}$$

it is easy to show that functions $\chi_k(s')$ form a complete orthonormal set in the region $|s'| > 1$. Therefore, we can use them as a basis for expansion of the field outside the object.

Now the operator of the total field in the object plane, $-\infty < s' < \infty$, can be written as

$$a(s') = \sum_{k=0}^{\infty} a_k\varphi_k(s') + \sum_{k=0}^{\infty} b_k\chi_k(s'), \tag{17}$$

where the operators b_k satisfy the commutation relations similar to Eq. (13). Substituting the expansion (17) into Eq. (1) we obtain the following relation between the coefficients in the object and the image plane,

$$e_k = \sqrt{\lambda_k}a_k + \sqrt{1-\lambda_k}b_k. \tag{18}$$

It is easy to verify that this transformation preserves the commutation relations of the operators, $[a_k, a_l^\dagger] = [b_k, b_l^\dagger] = [e_k, e_l^\dagger] = \delta_{kl}$.

Equation (18) is completely equivalent to the transformation performed by a beam splitter. Indeed, if we consider the operators a_k and b_k as the photon annihilation operators in the modes defined by prolate spheroidal waves incoming to the beam splitter with the amplitude transmission coefficient $\sqrt{\lambda_k}$ and the reflection coefficient $\sqrt{1-\lambda_k}$, then e_k is the photon annihilation operator in the kth mode of the transmitted wave.

From Fig. 1 one may think that the vacuum fluctuations coming from the region $|\xi| > d/2$ in the Fourier plane outside the pupil should be also taken into account. Indeed, when treating the field in the Fourier plane as an operator we must include the contribution from this region into the resulting field in the image plane. However, the advantage of expansion (11) is that the field from this region does not contribute to the expansion coefficients e_k of the image because it is orthogonal to the prolate spheroidal wave functions. This property was pointed out by Bertero and Pike in[15] for the out-of-band classical noise and remains valid in the quantum theory as we proceed to prove now.

For this proof we shall use the following property of the linear prolate spheroidal functions,

$$\int_{-1}^{1} \psi_k(s)e^{i\omega s}\, ds = i^k\left(\frac{2\pi\lambda_k}{c}\right)^{1/2}\psi_k(\omega/c), \tag{19}$$

which can be found, for example, in Ref. [23]. Taking Fourier transform of both sides of Eq. (19),

$$\int_{-\infty}^{\infty} e^{-i\omega s'}\int_{-1}^{1} \psi_k(s)e^{i\omega s}\, ds\, d\omega = 2\pi\int_{-1}^{1} \psi_k(s)\delta(s'-s)\, ds = i^k\left(\frac{2\pi\lambda_k}{c}\right)^{1/2}\int_{-\infty}^{\infty} \psi_k(\omega/c)e^{-i\omega s'}\, d\omega, \tag{20}$$

and using the property of the δ-function, we arrive at the following result,

$$\int_{-\infty}^{\infty} \psi_k(s)e^{i\omega s}\, ds = \begin{cases} i^k\left(\frac{2\pi}{\lambda_k c}\right)^{1/2}\psi_k(\omega/c), & \text{if } |\omega| \le c \\ 0, & \text{if } |\omega| > c. \end{cases} \tag{21}$$

From this equation it follows that $\psi_k(s)$ is a bandlimited function with vanishing Fourier spectrum outside the region of frequencies $|\omega| \leq c$.

Let us now consider the field operator $f(\omega)$ on the left side of the pupil plane of Fig. 1. This field is related to the total field operator in the object plane through the transformation performed by the lens L1,

$$f(\omega) = \frac{X}{2} \frac{1}{\sqrt{\lambda f}} \int_{-\infty}^{\infty} [a(s)p(s) + b(s)(1 - p(s))] e^{i\omega s} \, ds, \tag{22}$$

with $\omega = 2\xi c/d$ being the dimensionless spatial frequency. Here we have used the characteristic function $p(s)$ for the interval $[-1, 1]$ equal to unity inside this interval and zero outside it, which allows us to write the total field operator in the object plane in compact form. The field operator $e(s)$ in the image plane can be represented as the sum of two contributions coming from the area inside the pupil, $|\omega| \leq c$, and outside it, $|\omega| > c$,

$$e(s) = \frac{d}{2c} \frac{1}{\sqrt{\lambda f}} \Big[\int_{|\omega| \leq c} f(\omega) e^{-i\omega s} \, d\omega + \int_{|\omega| > c} \tilde{f}(\omega) e^{-i\omega s} \, d\omega \Big] \tag{23}$$

where $f(\omega)$ is just given by Eq. (22), whereas $\tilde{f}(\omega)$ is the operator associated with the unknown field present in the area of the pupil plane located outside the pupil.

From the first sight it seems that both contributions in Eq. (23) into $e(s)$ have to be taken into account. Let us however evaluate the contributions of these two areas into the coefficient e_k form the Eq. (11):

$$e_k = \int_{-\infty}^{\infty} e(s)\psi_k(s) \, ds = \int_{|\omega| \leq c} f(\omega) \int_{-\infty}^{\infty} \psi_k(s) e^{-i\omega s} \, ds \, d\omega + \int_{|\omega| > c} \tilde{f}(\omega) \int_{-\infty}^{\infty} \psi_k(s) e^{-i\omega s} \, ds \, d\omega \tag{24}$$

As follows from Eq. (21), the second term in the sum in Eq. (24) vanishes. Thus, there is no contribution from the quantum-mechanical field operator $\tilde{f}(\omega)$ outside the pupil into all the coefficients e_k and therefore into the field $e(s)$ in the image plane.

We will assume that we use a homodyne detection technique of the image that allows us to register any of the quadrature components of the field $e(s)$,

$$e(s) = e_1(s) + ie_2(s), \tag{25}$$

Using Eq. (18) we can express the variances of the expansion coefficients e_{1k} and e_{2k} of these quadrature components through the variances of coefficients a_{1k}, a_{2k} and b_{1k}, b_{2k} in the object plane,

$$\langle (\Delta e_{\mu k})^2 \rangle = \lambda_k \langle (\Delta a_{\mu k})^2 \rangle + (1 - \lambda_k) \langle (\Delta b_{\mu k})^2 \rangle, \tag{26}$$

with $\mu = 1, 2$ for corresponding quadratures. From Eq. (12) the fluctuations of the expansion coefficients of the reconstructed object are obtained as follows,

$$\langle (\Delta \tilde{a}_{\mu k})^2 \rangle = \frac{\langle (\Delta e_{\mu k})^2 \rangle}{\lambda_k}. \tag{27}$$

Let us assume that the field in the object plane is in a multimode coherent state at any point $|s'| \leq 1$. Since there is no light outside the region $[-1, 1]$ this corresponds to the vacuum state of all operators b_k. In this case the variances of the coefficients $a_{\mu k}$ and $b_{\mu k}$ in the object plane are equal to $\langle (\Delta a_{\mu k})^2 \rangle = \langle (\Delta b_{\mu k})^2 \rangle = \frac{1}{4}$, and the variances of the coefficients for the reconstructed object read

$$\langle (\Delta \tilde{a}_{\mu k})^2 \rangle = \frac{1}{4\lambda_k}. \tag{28}$$

As the eigenvalues λ_k become rapidly very small for $k > S$, the corresponding variance will become very large which will forbid a precise determination of the coefficient $\tilde{a}_{\mu k}$. Therefore, formula (28) sets the standard quantum limit in reconstruction techniques used in superresolution.

Since imaging equation in the form (18) is equivalent to the transformation of two fields by a beam splitter it gives us an idea on how to reduce the fluctuations in the reconstructed object below the standard quantum limit. It is well known that such an improvement can be achieved by illuminating the open port of the beam splitter by a squeezed vacuum with properly chosen squeezed quadrature. Since in our case the role of such an open port is played by the region $|s'| > 1$ outside the object, we expect that superresolution beyond the standard quantum limit can be achieved by illuminating this region by a multimode squeezed vacuum with a light spot much larger than the size of the object. Such a light can be produced by a traveling-wave optical parametric amplifier (OPA) or a degenerate optical parametric oscillator (OPO) below threshold in a confocal cavity [1].

We may expect that an even better result for the object reconstruction can be obtained when not only the area outside the object but the object itself is illuminated by multimode squeezed light with nonzero mean amplitude. However, to use the advantage of such illumination we have to make sure that squeezing in the incoming light is not destroyed by absorption in the object. Therefore, squeezed light illumination should be advantageous for pure phase or for weakly absorbing objects.

In practice, one can use a single source of multimode squeezed vacuum with a large transverse area and mix it with a coherent light wave in the central part illuminating the object using a weakly reflecting small size mirror. This coherent light wave is needed for producing a nonzero field component in the image plane and plays the role similar to the local oscillator in the homodyne detection technique. The quantum fluctuations of this coherent light wave can be eliminated by using the balanced homodyne detection of two images with two CCD cameras.

To evaluate the fluctuations of the reconstructed object in this case we assume for the sake of simplicity that squeezing has infinitely large spatial bandwidth, i. e. is the same for all coefficients a_k and b_k, $\langle(\Delta a_{\mu k})^2\rangle = \langle(\Delta b_{\mu k})^2\rangle = \frac{1}{4}e^{\pm 2r}$, where the "-" sign corresponds to the squeezed quadrature $\mu = 1$, and the "+" sign to the stretched one, $\mu = 2$, and r is the squeezing parameter. This gives

$$\langle(\Delta\tilde{a}_{\mu k})^2\rangle = \frac{1}{4\lambda_k}e^{\pm 2r}, \tag{29}$$

It follows from this equation that the variance of the reconstruction coefficients $\tilde{a}_{1k}$ in the squeezed quadrature component $\tilde{a}_1(s')$ can be significantly reduced below the standard quantum limit for large squeezing, $r \gg 1$. For $r = 0$ we recover the standard quantum limit of Eq. (28).

To estimate the resolution length D (the smallest object detail reconstructed from the image) from Eq. (28) and its improvement obtained using Eq. (29) we will use the arguments similar to Ref. [15]. For definiteness we shall consider an amplitude object, i. e. an object recorded in the amplitude quadrature component of the field, $\langle a^\dagger(s')a(s')\rangle = \langle a_1(s')a_1(s')\rangle$. Let us define the signal S and the noise B related to the reconstructed object as

$$S = \int_{-1}^{1} \langle\tilde{a}_1(s')\tilde{a}_1(s')\rangle\, ds', \quad B = \int_{-1}^{1} \langle\left(\Delta\tilde{a}_1(s')\right)^2\rangle\, ds'. \tag{30}$$

These quantities can be evaluated as $S = \langle N\rangle$, where $\langle N\rangle$ is the total number of photons in the object, and

$$B \simeq \frac{e^{-2r}}{4} \sum_{k=0}^{Q} \frac{1}{\lambda_k} \simeq \frac{e^{-2r}}{4} \frac{1}{\lambda_Q}. \tag{31}$$

Here Q is the index of the highest eigenfunction in the reconstructed object. In the second equality we have used the fact that λ_Q is the smallest eigenvalue and estimated the sum as $1/\lambda_Q$. Setting the signal-to-noise ratio equal to unity we can find the smallest eigenvalue λ_Q that can be recovered from the image measurement,

$$\lambda_Q \simeq \frac{e^{-2r}}{4\langle N \rangle}. \tag{32}$$

Using the tables of eigenvalues λ_k[23] or calculating them numerically we can evaluate the index Q of the highest eigenfunction $\varphi_Q(s)$ with a precisely known coefficient in the expansion of the reconstructed object. Knowing that $\varphi_Q(s)$ has exactly Q zeros on the interval $[-1, 1]$ we can estimate the resolution length D as

$$D \simeq X/(Q+1) = R\left(\frac{S}{Q+1}\right), \tag{33}$$

where $R = X/S = \lambda f/d$ is the Rayleigh resolution length. From this equation we can interpret the number $Q + 1 \equiv M$ as an effective number of degrees of freedom in the object which, as follows from Eq. (32), is a function of the total photon number $\langle N \rangle$ and the squeezing parameter r. The case of superresolution corresponds to $D < R$ or, equivalently, $M > S$. For classical noise the dependence of M on the classical signal-to-noise ratio was studied in[15, 21] with an important conclusion that significant superresolution can be achieved for small values of the Shannon number S. Since our Eq. (32) is an extension to the quantum domain of analogous equation studied in Refs. [15, 21] we conclude that a similar result holds true for quantum fluctuations.

A scanning optical microscope is operated in a small Shannon number configuration[24]. It is therefore a good candidate for a practical application of our analysis. We will consider it in more details in a forthcoming paper together with the influence of the finite pixel size in the image plane and the finite coherence area of squeezed light in the object plane.

4 Numerical simulations of superresolution

We now turn to a numerical study of the superresolution problem with the aim, on one side, to demonstrate practically the possibility of resolution beyond the Rayleigh limit and, on the other side, to describe quantitatively some of the qualitative results derived in the previous section.

Our ultimate goal is to investigate numerically the role of quantum fluctuations in the superresolution and the possibility of improving the superresolution using multimode squeezed light. However, before attacking the problem with quantum fluctuations in the detected object, we would like to verify that we are not limited by numerical errors in achieving superresolution *without* quantum fluctuations. These numerical errors in our simulations may be considered as technical noise in a real experiment and at this stage we would like to make sure that our "technical noise" is much smaller than the quantum noise that we intend to "measure".

To calculate the prolate spheroidal functions $\psi_k(s)$ and the corresponding eigenvalues λ_k we have decided to solve numerically the Fredholm integral equation (6). We followed the method proposed in [25]. Evaluating the integral in Eq. (6) by means of a Gauss-Legendre quadrature [26], we obtain the following discrete form of this equation,

$$\sum_{j=1}^{N} w_j \frac{\sin[c(s_i - s_j)]}{\pi(s_i - s_j)} \psi_k(s_j) = \lambda_k \psi_k(s_i), \tag{34}$$

which involves the Gauss-Legendre weights w_j and the zeros of the Legendre polynomials $s_j, j = 1, ..., N$ as specified by the Gaussian quadrature algorithm [26]. Those were evaluated using standard algorithms [27]. Eq. (34) can be directly solved with standard numerical methods for matrix diagonalization which give access to the eigenvalues λ_k and the corresponding eigenvectors, i. e. the values of the prolate functions $\psi_k(s_j)$ at the N Gauss-Legendre abscissas $-1 \leq s_j \leq 1$. To compute the value of the prolate functions $\psi_k(s)$ at any arbitrary point s and in particular outside the interval $[-1, 1]$, we use again Eq. (6), valid for any value of s, in its discretized form:

$$\psi_k(s) = \frac{1}{\lambda_k} \sum_{j=1}^{N} w_j \frac{\sin[c(s - s_j]}{\pi(s - s_j)} \psi_k(s_j), \tag{35}$$

which only involves on the right-hand-side the quantities previously determined from the diagonalization procedure. In Fig. 2 we illustrate two prolate spheroidal function calculated using this procedure for the typical value of the Shannon number $S = 0.6$ which will be used throughout our numerical simulations.

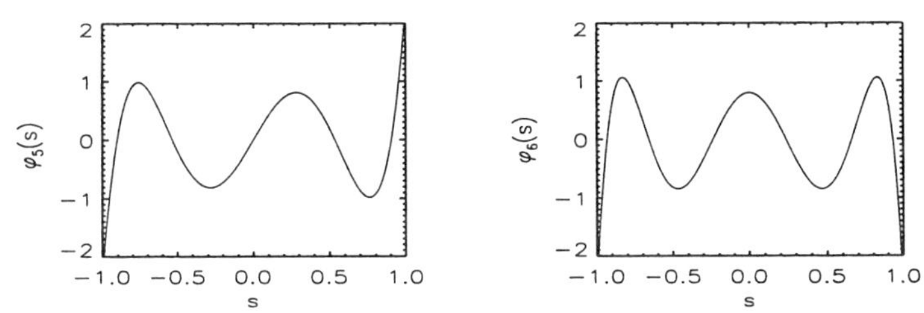

Fig. 2. Eigenfunctions $\varphi_5(s)$ and $\varphi_6(s)$ given by Eq. (4) for the Shannon number $S = 0.6$.

One would expect to access the higher and higher eigenvalues λ_k and the corresponding eigenfunction $\psi_k(s)$ by increasing the number of points N used for the Gauss-Legendre quadrature. Practically, however, only a finite number of eigenvalues and eigenvectors can be calculated with accuracy. This is essentially due to the fact that eigenvalues λ_k decrease extremely rapidly as a function of k for $k > S$. In a numerical calculation involving sums and subtractions, like typically a diagonalization routine, this is a fundamental problem. Roughly speaking, this will prevent from precise calculation of any eigenvalue smaller than the numerical precision of the machine, which is typically of the order of 10^{-16}, and a fortiori of the corresponding eigenvector. This numerical noise represents a strong limitation: for a typical value $S = 0.6$ of the Shannon number and using double precision computation we found that only seven first eigenvalues and eigenfunctions could be computed with reliable accuracy.

Going back to the reconstruction problem, the fact that we only have a finite number of functions $\psi_k(s)$ at disposition is equivalent to the truncation of the exact summation formula (9). As a consequence, the reconstruction method will be imperfect. But even truncated, this procedure represents a substantial improvement with respect to the Rayleigh resolution limit, as will be illustrated in the following. For this purpose, we considered an object consisting of two slits of finite width ϵ, separated by a given distance b. For a Shannon number $S = 0.6$, the intensity distribution in the image plane, plotted in Fig. 3, clearly shows that the Rayleigh resolution criterium does not allow us to resolve the slits: for this Shannon number, the width of the diffraction spot associated with each slit is much larger than the distance between the slits.

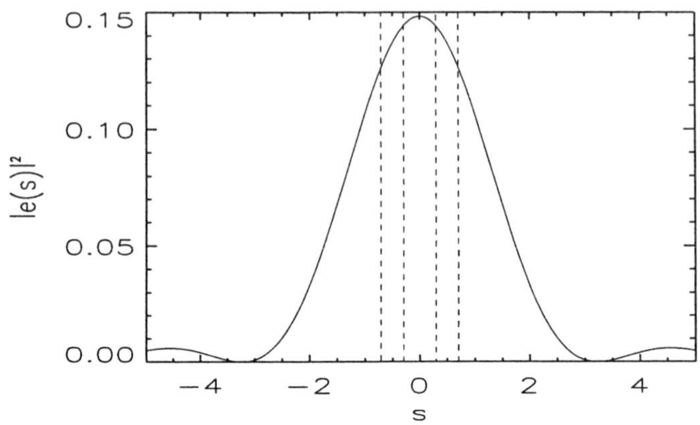

Fig. 3. Intensity distribution in the image plane for a two-slit object. For comparison, the position of the two slits in the object plane has been included as a dotted line. The distance between the slits is $b = 1$, and their width $\epsilon = 0.4$.

However, the result of the truncated reconstruction procedure

$$a_{\rm r}(s) = \sum_{k=0}^{6} a_k \varphi_k(s), \tag{36}$$

which is represented in Fig. 4, together with the exact object for comparison, allows for good resolution of the two slits. The difference between the object $a(s)$ and the reconstructed object $a_{\rm r}(s)$ is of course due to the fact that only seven first terms in the exact expansion (9) are taken into account.

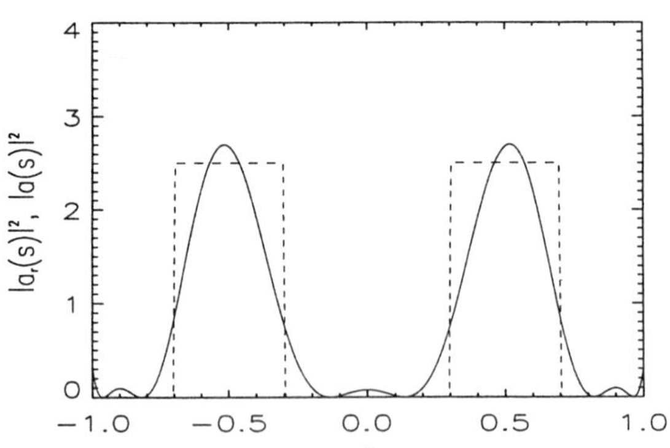

Fig. 4. Comparison of the object intensity distribution $|a(s)|^2$ (dotted line) with the intensity distribution $|a_{\rm r}(s)|^2$ (solid line). The parameters are the same as in Fig. 3.

This first result allows us to conclude that the reconstruction method with prolate spheroidal functions gives a significant improvement with respect to the Rayleigh criterium. In order to quantify this improvement, we can define the equivalent of the Rayleigh length associated with this method as the smallest distance between two-point objects which can be resolved. Here, we choose a somewhat different approach, which leads to quantitatively equivalent results: we first decreased the width of each slits, keeping constant the distance between them. For reasonably broad slits, the width of the reconstructed object follows the width of the object itself, as shown in Fig. 4. However, it turns out that in the limit of small widths, the width of the reconstructed object finally does not depend any more on the width of the object but saturates at a finite value. This limiting case is plotted in Fig. 5. In analogy to the original definition of the Rayleigh length, the width of each peak in the reconstructed object can be identified with the minimal length that can be resolved in $a_r(s)$. Roughly, it corresponds to the variation scale of the highest prolate function $\psi_6(s)$ used in the truncated sum.

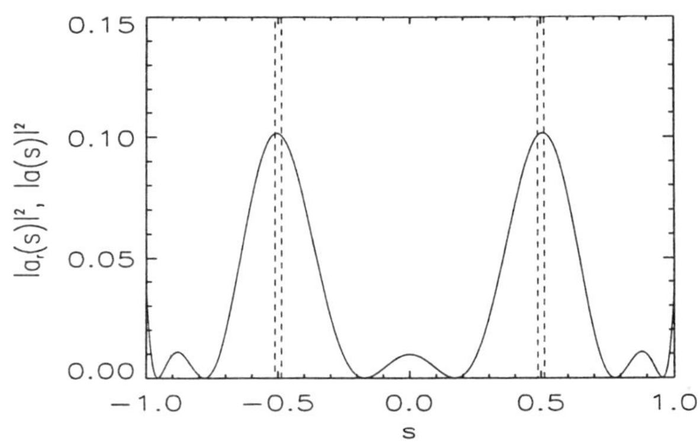

Fig. 5. Reconstructed object in the limit of very thin slits. $\epsilon = 0.01$. The distance between each maximum of the intensity distribution and the first zero represents the new resolution length allowed by the reconstruction procedure.

The improvement of resolution with respect to the classical Rayleigh limit can be quantified by simply comparing this ultimate resolution length with the Rayleigh length, which can be read out from Fig. 3 as the distance between the maximum and the first zero of the intensity distribution. In the case of $S = 0.6$ considered here, we find a substantial improvement by a factor of more than 10.

It should be noted that such an improvement is possible only for Shannon numbers S of the order or less that unity. Our simulations with larger Shannon numbers from $S = 5$ to $S = 7$ confirm that much lesser superresolution can be achieved in these cases. For example, for $S = 5$ we have found a superresolution factor of 2.2.

In conclusion, our preliminary numerical simulations of the two-point object reconstruction using prolate spheroidal functions for Shannon numbers smaller than unity demonstrate a possibility of superresolution of one order of magnitude over the Rayleigh limit. This is an encouraging result for further development of our project which aims to take into account numerically the quantum fluctuations of the illuminating light wave. We hope that simulations with multimode squeezed light as illuminating source will allow us to demonstrate an appreciable superresolution for $S < 1$.

Acknowledgments

This work was supported by the Network QUANTIM (IST-2000-26019) of the European Union. The Laboratoire Kastler Brossel of the Ecole Normale Supérieure and the Université Pierre et Marie Curie is associated with the Centre National de la Recherche

Scientifique.

References and links

1. M. I. Kolobov, Rev. Mod. Phys. **71**, 1539 (1999).
2. L. A. Lugiato, M. Brambilla, and A. Gatti, in *Advances in Atomic, Molecular, and Optical Physics*, Vol. 40, edited by B. Bederson and H. Walther (Academic, Boston, 1999), p. 229.
3. A. Gatti, E. Brambilla, M. I. Kolobov, L. A. Lugiato, J. Opt. B: Quant. Semiclass. Opt. **2**, 196 (2000).
4. C. Fabre, J. B. Fouet, and A. Maître, Opt. Lett. **25**, 76 (2000).
5. M. I. Kolobov and C. Fabre, Phys. Rev. Lett. **85**, 3789, (2000).
6. N. Treps, U. Andersen, B. Buchler, P. K. Lam, A. Maître, H. Bachor, C. Fabre, "Crossing the standard quantum limit for high-sensitivity measurements in optical images using non-classical light", submitted to Nature (2001).
7. J. Rayleigh, in *Collected Optics Papers of Lord Rayleigh*, part A, (Opt. Soc. Am., Washington,1994), p. 117.
8. A. Flock and D. Strelioff, Nature, **310**, 397 (1984).
9. J. J. Art, A. C. Craftford, and R. Fettiplace, J. Physiol. (London)**371**, 18P (1986).
10. S. Kamimura, Appl. Opt. **26**, 3425 (1987).
11. W. Denk and W. W. Webb, Appl. Opt. **29**, 2382 (1990).
12. J. Jelles, B. J. Schnapp, and M. P. Scheetz, Nature, **331**, 450 (1988).
13. D. Fournier, A. Boccara, N. Amer, and R. Gerlach, Appl. Phys. Lett. **37**, 519 (1980).
14. C. Putman, B. De Grooth, N. Van Hulst, and J. Greve, J. Appl. Phys. **72**, 6 (1992).
15. M. Bertero, and E. R. Pike, Opt. Acta **29**, 727 (1982).
16. M. Bertero, P. Boccacci, and E. R. Pike, Opt. Acta **29**, 1599 (1982).
17. M. Bertero, P. Brianzi, P. Parker, and E. R. Pike, Opt. Acta **31**, 181 (1984).
18. M. Bertero, C. De Mol, E. R. Pike, and J. G. Walker Opt. Acta **31**, 923 (1984).
19. J. G. Walker, E. R. Pike, and M. Bertero, British patent89/13129 (6 July 1989).
20. R. E. Davies, Ph. D. dissertation (Department of Physics, King's College, University of London,1990).
21. M. Bertero, and C. De Mol, in *Progress in Optics* Vol. XXXVI, edited by E. Wolf (North-Holland, Amsterdam, 1996), p. 129.
22. D. Slepian and H. O. Pollak, Bell System Tech. J. **40**, 43 (1961);
23. B. R. Frieden, in *Progress in Optics* Vol. IX, edited by E. Wolf (North-Holland, Amsterdam, 1971), p. 311.
24. T. Wilson and C. J. R. Sheppard, *Theory and Practice of Scanning Optical Microscopy* (Academic Press, London, 1984).
25. W. P. Latham and M. Tilton, Appl. Opt. 26, 2658 (1987).
26. B. Carnahan, H. A. Luther and J. O. Wilkes, *Applied Numerical Methods* (Wiley, New York, 1969), p. 100.
27. W. H. Press, B. P. Flannery, S. A. Teukolsky and W. T. Vetterling, *Numerical Recipes (Fortran Version)* (Cambridge University Press, Cambridge, 1989), p. 125.

Fiber generated quantum correlations for quantum-optical communications

P. Kumar
Center for Photonic Communication and Computing, ECE Department, Northwestern University ,2145 N. Sheridan Road, Evanston, IL 60208-3118, USA
kumarp@ece.northwestern.edu

Abstract: One of the main challenges facing the development of Quantum Information Technology is the difficulty associated with transmission of quantum information over lossy and noisy channels. Our approach to mitigating this problem relies on the use of fiber-based devices generating entanglement to be used as a shared quantum resource in the quantum communication network. This approach offers the advantage of technological synergy between the generation and propagation components of the overall quantum network. In this paper we present our initial work that is focused on the realization of pulsed, quantum-correlated systems that can be used as building blocks for fiber-based sources of quadrature and polarization entanglement. We have evaluated several different schemes in order to determine which one can satisfy the demanding specifications of the quantum network.

The authors acknowledge support from the U. S. Army Research Office through the MURI grant DAAD19-00-1-0177.

JC photons in frequency continua: Exact solution and quantum features

C. K. Law and P. T. Leung

Department of Physics, The Chinese University of Hong Kong, Shatin
Hong Kong, SAR, China

1. Introduction

In recent years, extensive studies of the Jaynes-Cummings (JC) model and its variations have provided many examples capturing quantum features of photon-atom interaction in a cavity [1]. An important situation of interest is the case where the cavity is leaky. At times long comparing with the relaxation times, the atom settles in its ground state and photons leave the cavity. The transmitted photons carry information about the intra-cavity photon-atom dynamics. In order to study the nonclassical properties of the field in detail, it is crucial to obtain the state vector of the system explicitly.

Recently we have reported an exact analytic solution of the JC model in leaky cavities [2]. The field quantization in our approach is based on the normal modes of the cavity plus its environment [3]. This allows us to address directly the state vector of the field without making an artificial separation between the system and bath. The state vector of the transmitted field is embedded in a vast Hilbert space characterized by continuous frequency modes. Such a pure-state approach has been applied to the subject of cavity modified spontaneous atomic decay [4, 5, 6, 7], which concerns single photon emission in vacuum. However, the rich structure of multi-photon states in the field continuum is just beginning to be investigated.

In this paper we first review the system and our solution. Then we describe two interesting features: (a) frequency entanglement and (b) quantum efficiency of stimulated emission observed in our solution. The former deals with quantum correlation between photons with different colors, and the latter quantifies the degeneracy of photons generated in open environment. Both (a) and (b) reveal quantum interference effects in the continuum.

2. The model Hamiltonian and solution

We consider a model Hamiltonian describing a two-level atom in a leaky cavity,

$$H = \omega_a |e\rangle\langle e| + \sum_k \omega_k a_k^\dagger a_k + \sum_k \Big(g_k a_k |e\rangle\langle g| + g_k^* a_k^\dagger |g\rangle\langle e| \Big). \tag{1}$$

Here $|e\rangle$ and $|g\rangle$ are the excited state and ground state of the atom with ω_a being the transition frequency. a_k and $a_k^\dagger$ are the annihilation and creation operators for the field modes, which are labeled by the continuous index k, and the field frequency of each mode is denoted by ω_k. The atom-field coupling g_k is obtained from a 1-D cavity model with high finesse mirrors so that $|g_k|^2$ is proportional to a Lorentzian [3], i.e.,

$$g_k = \frac{\eta}{\Delta_k + i\kappa}. \tag{2}$$

Here η is a coupling constant, κ is the leaky rate of a quasi-mode of the cavity with a frequency ω_c, and $\Delta_k = \omega_k - \omega_c$. Note that we have restricted our study to the Jaynes-Cummings situation where only one quasi-mode is involved. In this model, we also ignore the spontaneous emission to the side cavity modes.

Given an initial condition where the quasi-mode has $N > 0$ photons and the atom is in the excited state, we have found an exact solution of the system state vector using Laplace transform method. In terms of the cavity mode creation operator $a_c^\dagger = \lambda^{-1} \sum_k g_k^* a_k^\dagger$, where

$$\lambda^2 \equiv \sum_k |g_k|^2, \tag{3}$$

and the initial state is given by

$$|\Psi(0)\rangle = \frac{a_c^{\dagger N}}{\sqrt{N!}} |vacuum\rangle \otimes |g\rangle . \tag{4}$$

At a time $t \gg \kappa^{-1}$ when all the field leaks out of the cavity and the atom decays to the ground state, the state vector for the $N+1$-photon field can be decomposed in the form,

$$|\psi_F\rangle = |\phi_0\rangle + |\phi_1\rangle + |\phi_2\rangle + ... + |\phi_N\rangle \tag{5}$$

where

$$|\phi_j\rangle = \sum_{k_1 > k_2 > ... > k_{N+1}} f(k_1, k_2, ..., k_{N+1}) \sum_P X_j \, |k_1, k_2, ..., k_{N+1}\rangle , \tag{6}$$

and $j = 0, 1, ..., N$. Here the sign $\sum_P$ denotes the summation over all permutation of ks', and we have defined,

$$f(k_1, k_2, ..., k_{N+1}) \equiv i g_{k_1}^* g_{k_2}^* \cdots g_{k_{N+1}}^* e^{-i(\Delta_{k_1} + \Delta_{k_2} + ... + \Delta_{k_{N+1}})t} \tag{7}$$

$$\begin{aligned} X_j &= \sum_{m_0} \sum_{m_1} \cdots \sum_{m_j} D^{(j)}_{m_0 m_1 ... m_j} \left(\Delta_{k_1} + \Omega^{(0)}_{m_0}\right)^{-1} \left(\Delta_{k_1} + \Delta_{k_2} + \Omega^{(1)}_{m_1}\right)^{-1} \cdots \\ &\quad \cdots \times \left(\Delta_{k_1} + \Delta_{k_2} + ... + \Delta_{k_{j+1}} + \Omega^{(j)}_{m_j}\right)^{-1} \end{aligned} \tag{8}$$

where $m_i = \pm$. The complex Rabi frequencies $\Omega^{(q)}_\pm$ and the coefficients $D^{(q)}_{m_0 m_1 ... m_q}$ are given by (assuming $\omega_a = \omega_c$)

$$\Omega^{(q)}_\pm = \pm \sqrt{\lambda^2 (q+1) - \kappa^2/4} + i\,(q + 1/2)\,\kappa, \tag{9}$$

$$\begin{aligned} D^{(q)}_{m_0 m_1 ... m_q} &= -\frac{\lambda^{2q}}{2} \left[i m_q + \frac{\kappa}{\sqrt{4\lambda^2 (q+1) - \kappa^2}} \right] \\ &\quad \times \left[m_q \delta_{q0} + (1 - \delta_{q0})\, m_0 \prod_{j=1}^{q} \frac{m_j}{\sqrt{4\lambda^2 j - \kappa^2}} \right], \end{aligned} \tag{10}$$

where $q = 0, 1, 2, .., N$.

It is interesting that the decomposition form of $|\psi_F\rangle$ in Eq. (5) can be interpreted as a superposition of states classified by the *number of photon 'spectators'*. We see that $|\phi_0\rangle$ involves only the vacuum Rabi frequency $\Omega^{(0)}_\pm$. It seems that the atom executes vacuum Rabi oscillation and the N cavity photons do not participate the interaction. Similarly, the highest Rabi frequency in $|\phi_j\rangle$ is the $j-$photon Rabi frequency $\Omega^{(j)}_\pm$. We may interpret the $N - j$ photons as spectators and the remaining j photons interact with the excited atom.

3. Quantum features of JC photons

In this section, we study two aspects of photon state $|\psi_F\rangle$ which demonstrate the rich quantum structure in frequency continua.

3.1 Frequency entanglement

To illustrate photon correlation in frequency variables, we look at the case $N = 1$, which is the case corresponds to the decay of an atom in the presence of a single cavity photon. The final state is a two-photon state described by an amplitude $C(k_1, k_2)$. In Fig. 1 we show the absolute square of the two-photon amplitudes for two coupling strengths. In the case of weak coupling (Fig. 1a), $|C(k_1, k_2)|^2$ is localized at $\Delta_{k_1} = \Delta_{k_2} = 0$, i.e., photon frequencies mostly coincide with the cavity resonance frequency. The effect of Rabi-splitting becomes apparent in the strong coupling regime (Fig. 1b), where $|C(k_1, k_2)|^2$ is localized at two shifted frequencies.

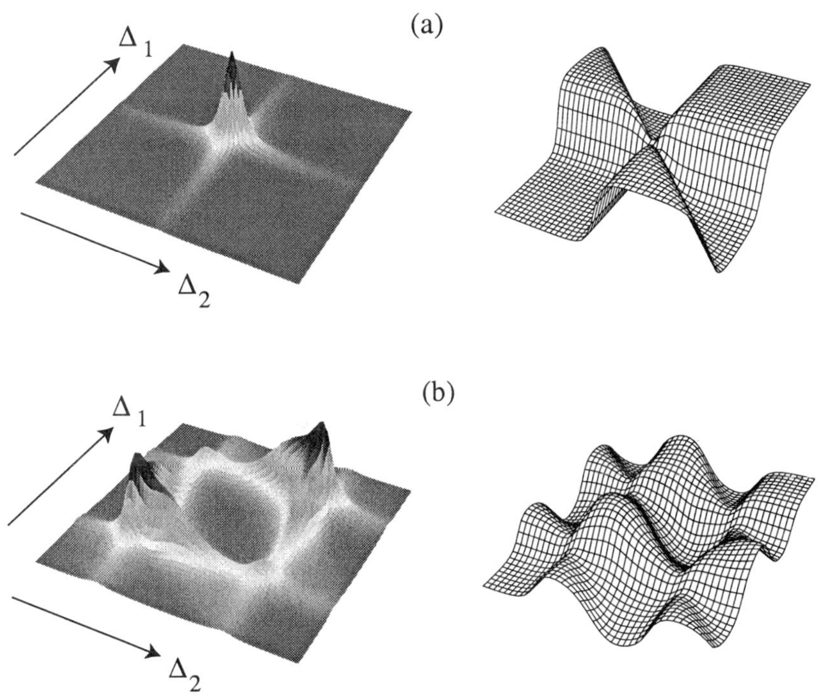

Fig. 1. A visualization of the two-photon states for: (a) $\lambda = 0.5\kappa$ and (b) $\lambda = 3\kappa$. The left figures are plots of $|C(k_1, k_2)|^2$ where the ranges of detuned frequencies Δ_1 and Δ_2 are both taken from -5κ to $+5\kappa$. The right figures are the corresponding phase of $C(k_1, k_2)$ in the same ranges of Δ_1 and Δ_2.

We see that both frequencies are shifted either to the blue side or the red side with respect to ω_c. The probability of finding a photon pair with shifts in opposite directions (i.e., blue-red) is relatively small. This correlation indicates a form of entanglement in the frequency variable. We may write the two-photon state symbolically as

$$|\psi_F\rangle \approx \sum |blue_1, blue_2\rangle + |red_1, red_2\rangle \tag{11}$$

which can be interpreted as a collection of Bell's states. A precise analysis of frequency entanglement can be performed via the Schmidt decomposition method [8]. For two indistinguishable particles as in our case, the Schmidt decomposition identifies the set of orthogonal modes that photons go as pairs. The corresponding set of eigenvalues provides a measure of entanglement. Our calculation shows that the entropy of entanglement is an increasing function of λ/κ in the strong coupling regime: $20 > \lambda/\kappa > 1.5$ [9].

We have also examined the case of $N = 2$, i.e., a three-photon final state. We find a similar frequency correlation between the three photons. That is, the three photons are more likely to have positive detunings (blue) or negative detunings (red), i.e., a GHZ type correlation.

3.2 Quantum efficiency of stimulated emission

For the $N+1$ photons described by $|\psi_F\rangle$, a fundamental question about the structure of the state is whether these photons occupy in the same mode or not. Although the N photons are initially in the same cavity mode, the photon-atom interaction can alter such a degeneracy as the system evolves. A particularly interesting issue is how to enhance stimulated effects so that the photon emitted by the atom can mostly occupy in the same mode of the stimulating photons.

To study the problem, let us consider the mode associated with the the freely evolving cavity field without interacting with the atom. The creation operator for such a mode is defined by

$$A_t^\dagger = \lambda^{-1} \sum_k g_k^* a_k^\dagger(0) e^{-i\Delta_k t}, \tag{12}$$

which defines a M-photon state in the mode by

$$|M\rangle_t = \frac{A_t^{\dagger M}}{\sqrt{M!}} |vacuum\rangle . \tag{13}$$

Therefore $|M\rangle_0$ represents the state initially having M photons inside the cavity. The subsequent free evolution of the M photons at a later time t is described by $|M\rangle_t$. We emphasize that these M photons are still in the same mode, but such a mode is parametrized by the time t corresponding to a form of field wavepacket transmitted out of the cavity.

Now we introduce the overlap probability at $\kappa t \gg 1$,

$$D = |\langle \psi_F \mid N+1\rangle_t|^2 \tag{14}$$

Such a quantity determines the *efficiency of stimulated emission*. For example, the case $D = 1$ means a 'perfect stimulated emission' because in the N photons stimulate the atom to emit a photon exactly into the *same* mode where the N photons reside.

In Fig. 2a we show D as a function of λ/κ for small values of N. Note that λ/κ is the ratio between the coherent photon-atom coupling rate and the cavity damping rate. For a given N, there exists a value λ/κ that maximizes D. We see that maximum value of D achievable is less than one, because there are always spontaneous processes competing with the stimulated emission process. Nevertheless, the optimal value of D increases as N increases. In the large N limit, D approaches one if $\lambda\sqrt{N} \approx 1.57\kappa$ is satisfied (Fig. 2b). This is a condition that indicates the required matching between the cavity decay rate and the photon-atom coupling in order to make the stimulated emission as the dominant process.

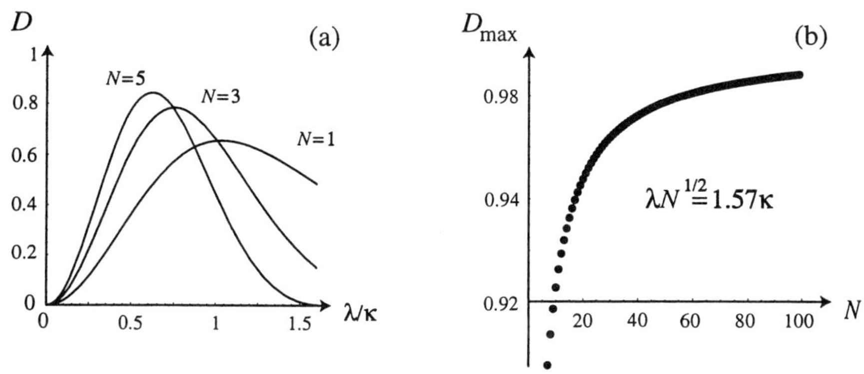

Fig. 2. (a)Dependence of D on λ/κ (b) The maximum value of D achievable as a function of N (occurring at $\lambda\sqrt{N} \approx 1.57\kappa$)

4. Conclusions

We have exactly solved and studied the multi-photon state resulting from the Jaynes-Cummings interaction. The quantum features of JC photons in continuous frequency space are rich as we have seen from the two features in the last section. Our solution is limited to situations in which photons are initially stored in the cavity. Therefore a more complete picture should include situations in which the state of the input photons can be arbitrary. This is a more challenging problem which requires the knowledge of the propagator of the system [9]. However, a judicious choice of input photons may provide more intriguing output photons via the JC interaction. We will address this issue elsewhere in the future.

This work is supported by a direct grant (Project ID: 2060150) from the Chinese University of Hong Kong, and a grant from the Hong Kong Research Grants Council (Grant No: CUHK 4282/009)

References

1. For reviews, see for example, H. I. Yoo and J. H. Eberly, Phys. Rep. **118**, 239 (1985); B. W. Shore and P. L. Knight, J. Mod. Opt. **40**, 1195 (1993).
2. C. K. Law, T. W. Chen, and P.T. Leung, Phys. Rev. A **61**, 02308 (2000).
3. M. Ley, and R. Loudon, J. Mod. Opt. **34**, 227 (1987).
4. H. M. Lai, P. T. Leung, and K. Young, Phys. Rev. A **37**, 1597 (1988).
5. M. Lewenstein, J. Zakrzewski, and T. W. Mossberg, Phys. Rev. A **38**, 808 (1988).
6. X. P. Feng and K. Ujihara, Phys. Rev. A **41**, 2668 (1990).
7. B. M. Garraway, Phys. Rev. A **55**, 2290 (1997); B. M. Garraway, Phys. Rev. A **55**, 4636 (1997).
8. C. K. Law, I. A. Walmsley, and J. H. Eberly, Phys. Rev. Lett. **84**, 5304 (2000).
9. T. W. Chen, C. K. Law, and P. T. Leung (in preparation).

A Bose Einstein condensate of metastable helium atoms

F. Pereira Dos Santos, J. Léonard, Junmin Wang[a], C. J. Barrelet,
F. Perales[b], E. Rasel[c], C. S. Unnikrishnan[d], M. Leduc, C. Cohen-Tannoudji
Collège de France,
Laboratoire Kastler Brossel*, Département de Physique,
Ecole Normale Supérieure,
24 rue Lhomond, 75231 Paris Cedex 05, France

Abstract: We recently observed a Bose-Einstein condensate in a dilute gas of ^{4}He in the 2^3S_1 metastable state. We give here a brief description of the successive experimental steps which led to the Bose-Einstein transition at 4.7 μK : loading of a large number of atoms in a MOT, efficient transfer into a magnetic Ioffé-Pritchard trap and optimization of the evaporative cooling ramp.

1 Introduction

Recently two groups in France reported the first observation of Bose Einstein condensation (BEC) of helium 4 atoms in the 2^3S_1 metastable state [1, 2], following the condensation of ^{87}Rb, ^{23}Na, and ^{7}Li in 1995 [3, 4, 5] and of atomic hydrogen in 1998 [6]. The method to reach BEC with metastable helium atoms uses similar routes as for alkali atoms. However the case of the metastable helium BEC is different because it carries a large internal energy (19.8 eV per atom), which can be exploited in several ways.

We here describe the experimental procedure used to reach metastable helium BEC by our group at ENS.

2 MOT

It is first necessary to trap a sample containing a large number of atoms, with a lifetime as long as possible. In this respect, metastable helium is more delicate in comparison with alkalis, first because one has to start by exciting the ground state helium beam and this process has a low efficiency, second because metastable helium atoms have very large inelastic collision rates due to Penning collisions enhanced by light, which severely limit the atomic density.

The loading of the MOT is explained in detail in reference [7]. The cold gas of helium atoms is trapped into a MOT at the center of a small quartz cell under ultra high vacuum conditions. The atoms are manipulated by an infra red laser at 1083 nm operating on the $2^3S_1 \rightarrow 2^3P_2$ transition (a laser diode oscillator followed by an Yb doped fiber amplifier [7]). The coils Q1 and Q2 of the MOT are the same as those used in the magnetostatic trap described in section 4 (see figure 2). For the MOT, the detuning of the laser beams δ is large, -45 MHz and the intensity of the beams is also large, 8 mW/cm^2 per beam. The half-life of the MOT is only 100 ms, because of the Penning collisions. The loading time of the MOT from the Zeeman slowed beam is thus short, of order 1 s. The total number of trapped atoms is measured, according to the procedure described in [8], to be 10^9 atoms, with a temperature of about 1 mK.

3 Optical detection of the cold cloud

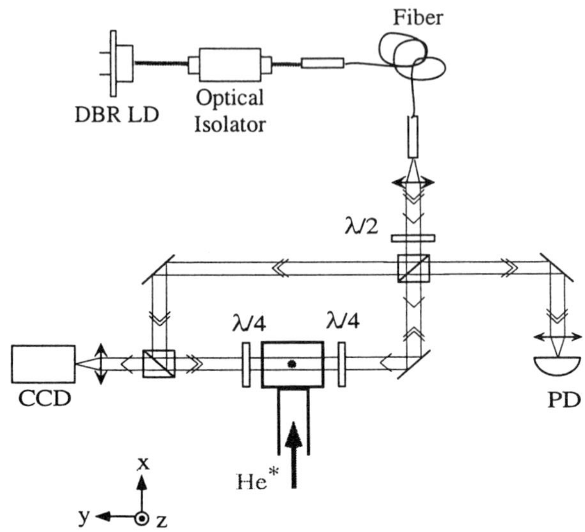

Fig. 1. Detection Setup. DBR LD is a Distributed Bragg Reflector Laser Diode, PD a photodiode and CCD a CCD camera.

The same optical detection scheme of the cold cloud is used through all the temperature range from the MOT down to the condensate. It is represented in figure 1. The source is a DBR laser diode (SDL-6702-H1), operating at 1083 nm at resonance with the atomic frequency. We image the shadow of the atomic cloud on a CCD camera (model Hamamatsu C4880) which has a low quantum efficiency of 1.5%. In order to obtain a sufficient signal to noise ratio on the pictures, it is required to operate at not too low an intensity (typically $I/I_{sat} = 0.2$), with large exposure times (typically 100 μs) and a moderate magnification. The resolution is thus limited to 24 μm by the pixel size of the camera. One uses a standing wave in order to avoid as much as possible any pushing effect due to the radiation pressure occurring if a travelling wave were used. From the absorption images taken by the camera we extract the size and the temperature of the cloud.

To measure the total number of atoms, we illuminate the cloud with the probe beam at resonance, but in a travelling wave configuration. One just rotates the $\lambda/2$ plate in 1. We measure the power absorbed in the probe beam on the photodiode PD (see figure 1). The power required to saturate the transition is about 10 mW. It allows us to measure the number of atoms in the MOT, where the absorption is about 1 mW, which corresponds to 10^9 atoms. When the cloud is evaporated in the magnetic trap, the power absorbed in the probe pulse becomes weaker, so that it is difficult to measure the total absorption on the photodiode PD for low temperature clouds. This problem could be overcome by making the probe beam size smaller.

The number of atoms in the evaporated clouds is deduced from the absorption images on the CCD camera, corrected in order to take into account various effects that tend to lower it [2].

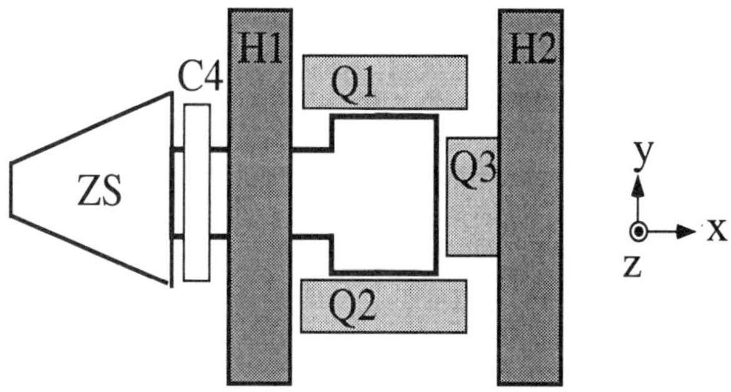

Fig. 2. Magnetostatic trap. ZS is the Zeeman slower coil, Q1, Q2 and Q3 the Ioffe-Prichard coils, H1 and H2 are the bias compensation coils. C4 is a coil used to compensate the field created by ZS at the center of the cell.

4 Loading the magnetic Ioffe Pritchard trap

The magnetostatic trap is sketched in figure 2. It is constituted by three coils Q1, Q2 and Q3 of relatively small size which realize a Ioffe Pritchard trap. A set of two larger Helmholtz coils H1 and H2 are added, which compensate the bias field down to about 5 G. Also, we can use an additional set of Helmholtz coils wound on top of coils H1 and H2, in order to finely adjust the bias and reduce it even more for further compression of the trap. The trap depth is then about 30 mK. The trap frequency along x is $\nu_x = 115$ Hz. The radial frequency is $\nu_\perp = 190$ Hz without the compensation of the bias, and 985 Hz with a 5.1 G bias field.

When the loading of the MOT is over, the center of the cloud has to be shifted by about 5 mm along the x direction in order to be at the central position of the magnetostatic trap. This displacement is produced by the magnetic field of the Zeeman Slower (ZS in figure 2). During this process, which lasts for 10 ms before reaching an equilibrium, the laser beams of the MOT are left on.

The spatial shift of the cloud is followed by a compression phase, in order to increase the density before loading the magnetic trap.

We then apply an optical molasses phase, in order to decrease the temperature of the gas before the loading into the magnetic trap. The molasses pulse duration is of 0.5 ms, the equilibrium temperature being typically 300 μK.

The next step is an optical pumping phase. It is used to put as many atoms as possible in the same low-field seeking magnetic substate of the 2^3S_1 state so as not to loose most of them when loaded into the magnetostatic trap, which only retains atoms in the $m_J = +1$ sublevel. The efficiency of the optical pumping pulse is measured by comparing the number of atoms transferred in the magnetic trap with and without optical pumping. Using the optical pumping pulse typically multiplies this number by a factor of 2.

In order not to loose on the phase-space density, a proper mode matching has to be made to adapt the atomic cloud in the MOT to the potential created by the magnetic trap. The best method is to keep the same aspect ratio (1.65) and temperature (300 μK) before and after transfer. We end up with 7×10^8 trapped atoms, compared to 10^9 atoms initially in the MOT.

Then, the cloud is compressed by ramping the current in the three coils to its final value of 46.6 A. This compression stage has to be adiabatic, which implies to increase the trap confinement on a time scale much longer than the oscillation period. We choose typically 1 s. After a final compression stage, the same current (46.6 A) runs through all the coils of the trap. The temperature, measured by time of flight, has increased up to 1.4 mK. The central density is calculated to be 8.4×10^{10} atoms/cm^3, using the measured temperature and number of atoms of the thermalized cloud.

Once the atomic cloud has been loaded into the magnetic trap, one can measure the number of atoms as a function of time. The result of this measurement is shown in figure 3.

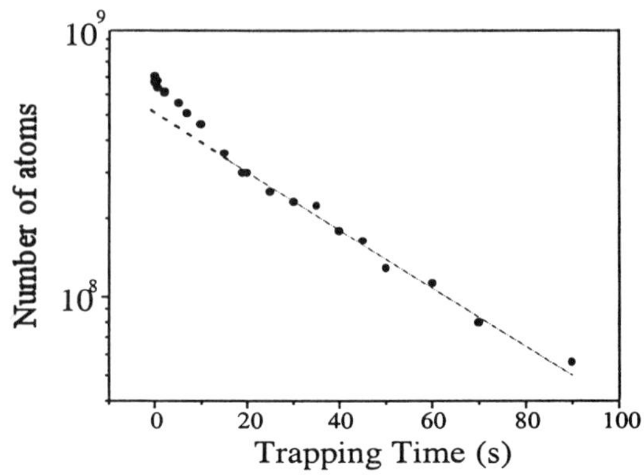

Fig. 3. Lifetime of the cloud in the magnetostatic trap.

The half-time is measured to be about 15 s. This rather long lifetime shows that the Penning ionization collisions between metastable helium atoms are inhibited (at least by one order of magnitude) by the spin polarization in the $m_J = +1$ sublevel, as predicted by theory [9, 10, 11, 12] and already observed to some extent in a MOT [13], and in a magnetic trap [14]. Indeed, if the collision rate had been equal to that in the non polarized case, namely $\beta \simeq 10^{-10}$ cm^3/s [15, 16, 17, 12, 18], the half-time of the cloud in the magnetostatic trap would not have exceeded 1 s. The decay is not purely exponential yet (see figure 3). One possible explanation is that the initial loss is due to the natural evaporation of the hottest atoms. If we fit the last part of the decay curve with an exponential law (straight line in figure 3), we extract a lifetime due to background collisions of 40 s.

The value of the lifetime was found quite dependent on the operation conditions of the atomic source, indicating that helium atoms in the ground state, not perfectly spatially filtered by the apparatus, contribute significantly to the residual vacuum. If after loading the magnetic trap one disconnects the atomic source from the rest of the set up, the lifetime goes up to 88 s.

5 Evaporative cooling

Before starting the evaporative cooling, it is useful to estimate the initial collisional rate, in order to determine whether the evaporative cooling can enter the so-called runaway regime, [19].

In order to estimate $\bar{\gamma}_{col}$, the mean rate of elastic collisions, we measure the time τ it takes for a cloud initially out of equilibrium to return to thermal equilibrium in the trap. This thermalization time is inversely proportional to the collisional rate [20, 21].

To prepare the cloud out of equilibrium, we compress the trap by lowering the bias field within 20 ms. At the end of the compression, the temperatures along the transverse and longitudinal directions are different. One can then monitor the evolution of the ellipticity of the cloud during the thermalization, which relaxes to its equilibrium value in the trap, as shown in figure 4.

We fit the ellipticity by an exponential law, which gives a thermalization time $\tau = 92$ ms. The mean collisional rate is thus about 30 s^{-1}, at a temperature of 1.4 mK, for a

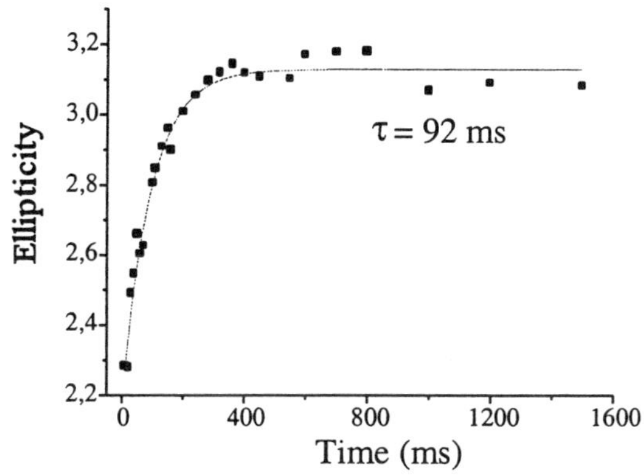

Fig. 4. Thermalization of the cloud after loading the magnetostatic trap. The ellipticity of the cloud relaxes to its value at thermal equilibrium with a time constant of 92 ms. The final temperature is 1.4 mK.

measurement performed with 5×10^8 atoms. The evaporation should thus be fast, as compared to the lifetime of the trap.

Evaporative cooling is performed by inducing RF spin-flips. The radio-frequency of the evaporation is lowered in a piecewise linear fashion in 6 steps. The duration and power of each segment is adjusted in order to increase the collisional rate, but minimizing the loss of atoms. The evaporation ramp starts at a frequency of 200 MHz and lasts for about 8 s. Figure 5 shows the temperature (measured by time of flight on the CCD camera at the end of each segment of the evaporation ramp) as a function of time. The parameter η, defined as the ratio between the cut-off energy and the thermal energy $k_B T$, remains larger than 6 during all the evaporation cooling.

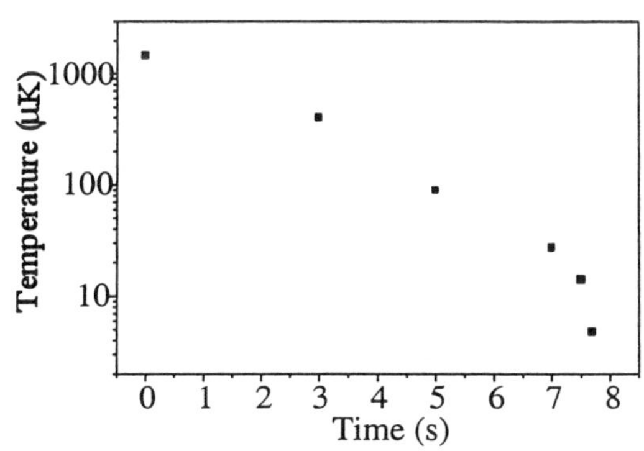

Fig. 5. Temperature during the evaporative cooling ramp. The temperature is measured by time of flight after each segment of the evaporation ramp.

Figure 6 shows the increase in phase space density as well as the decrease in number of atoms during the evaporation ramp. This cooling process is very efficient as the phase space density increases by 6 orders of magnitude, while only two orders of magnitude are lost on the number of atoms.

6 Bose Einstein condensation

When the final frequency of evaporation falls below about 14.5 MHz, one crosses the BEC transition.

On the time of flight pictures taken with the CCD camera, we observe a double structure in the density profile of the cloud, corresponding to a Bose-Einstein conden-

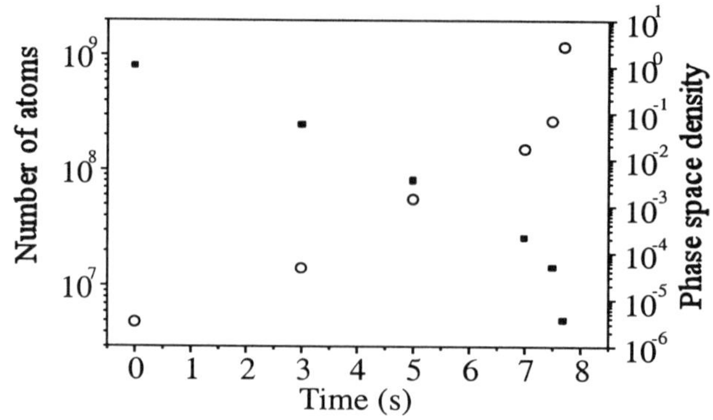

Fig. 6. Evolution of the number of atoms (squares) and the phase-space density (open circles) during the evaporative cooling.

Fig. 7. Three dimensional absorption image of the cloud after the BEC transition has been crossed.

sate and a non-condensed cloud (see figure 7). The pictures are fitted by the sum of a g_2 function, valid for the thermal cloud close to the transition, and an integrated paraboloidal distribution, that describes the density profile of the condensate in the Thomas-Fermi limit. From the pictures, we can extract the condensed fraction as a function of the normalized temperature which we show in figure 8. The experimental points deviate strongly from the theoretical curve for the non-interacting cloud, which could be explained by mean-field effects.

We measure a critical temperature $T_c = 4.7 \pm 0.5\mu$K. The number of atoms at the transition is inferred from the critical temperature, using the $N_c = 1.202(k_B T_c/\hbar\bar{\omega})^3$, where $\bar{\omega}$ is the geometrical average of the frequencies of the trap. We deduce $N_c = 8.2 \times 10^6$ atoms with an uncertainty of about 30%. We arbitrarily choose this value of N_c to estimate the scattering length a.

This allows us to calibrate the absorption images of the cloud in the region 5 to 1 μK, and extract the number of atoms in the condensate by comparison with the number at the transition (see figure 9). The number of atoms in the condensate is at most $N_0 = 6 \times 10^5$, and is roughly constant in the range from 1 to 3 μK.

Finally, we can give an estimate of the scattering length a using $a = \sigma/15N_0 \times (2\mu/\hbar\bar{\omega})^{5/2}$, where $\sigma = (\hbar/m\bar{\omega})^{1/2}$ is the characteristic size of the ground state of the trap, and μ is the chemical potential. Within the Thomas-Fermi approximation, μ can be deduced from the measurement of the size of the condensate [26], $\mu = 1.4 \times 10^{-29}$J. We finally find $a = 16 \pm 8$ nm, which is consistent with theoretical calculations [10, 18], as well as with the measurement of [1], which gives $a = 20 \pm 10$ nm.

The typical lifetime of the number of atoms in the condensate is about 2 s. In the present stage of the experiment, it is not possible to discriminate between 2-body or

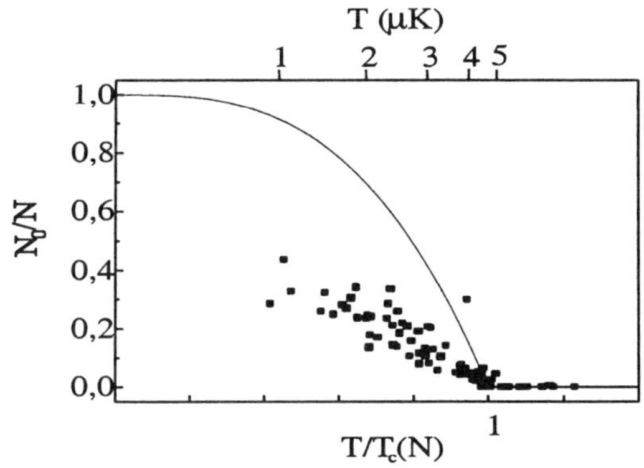

Fig. 8. Condensed fraction as a function of the reduced temperature, as well as the temperature. N_0 is the number of atoms in the condensate, and N the total number of atoms. The squares are the experimental data, whereas the line is the prediction for a non-interacting gas.

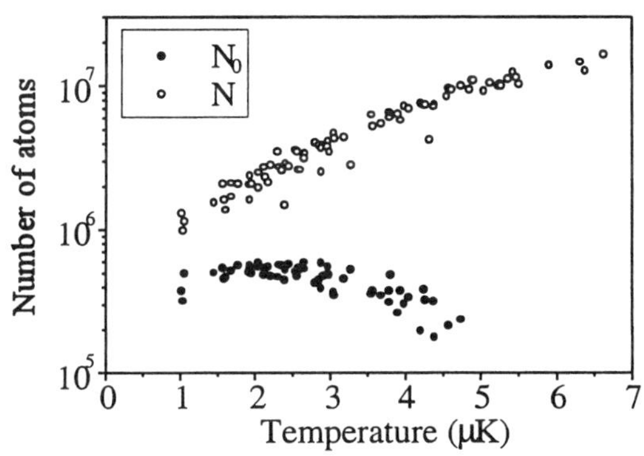

Fig. 9. Number of atoms N_0 in the condensate, and total number of atoms N, as a function of the temperature.

3-body decay processes. If one assumes now that 3-body collisions between atoms in the condensate are responsible for the decay of N_0 [27], one can give an upper bound for the rate constant L. Fitting our data gives $L \leq (2.8 \pm 0.2) \times 10^{-27}$ cm^6/s. This value can be compared to theoretical predictions [28, 29], which finds $L = 3.9\hbar a^4/2m = 2 \times 10^{-27}$ cm^6/s. A more precise measurement would require to improve the stability of the experiment in order to decrease the statistical errors, and include the presence of the atoms remaining in the thermal cloud in the analysis of the condensate decay.

7 Conclusion

We conclude that the procedure for BEC of a metastable helium gas is now well mastered. The most important key point to pay attention to is by far the achievement of a large enough initial density before starting the evaporation, because this is the condition for a fast enough process as compared to the intrinsic lifetime of the cloud. This implies that great care be given to the efficient loading of the MOT. The detection is purely optical, which has the advantage of allowing the measurement of the size of the atomic cloud. A crucial improvement will be to introduce a channelplate as in [1]. An ideal experiment would combine both the optical and the channelplate detection systems.

[a] Permanent address: Institute of Opto-Electronics, Shanxi University, 36 Wucheng Road, Taiyuan, Shanxi 030006, China.

[b] Permanent address: Laboratoire de Physique des Lasers, UMR 7538 du CNRS,

Université Paris Nord, Avenue J.B. Clément, 93430 Villetaneuse, France.

[c] Present address : Universität Hannover, Welfengarten 1, D-30167 Hannover, Germany.

[d] Permanent address : TIFR, Homi Bhabha Road, Mumbai 400005, India.

* Unité de Recherche de l'Ecole Normale Supérieure et de l'Université Pierre et Marie Curie, associée au CNRS (UMR 8552).

References and links

1. A. Robert, O. Sirjean, A. Browaeys, J. Poupard, S. Nowak, D. Boiron, C. I. Westbrook, and A. Aspect, Science Magazine **292**, 463 (2001)
2. F. Pereira Dos Santos, J. Léonard, Junmin Wang, C. J. Barrelet, F. Perales, E. Rasel, C. S. Unnikrishnan, M. Leduc, C. Cohen-Tannoudji, Phys. Rev. Lett **86**, 3459 (2001)
3. M. H. Anderson, J. Ensher, M. Matthews, C. Wieman, and E. Cornell, Science Magazine **269**, 198 (1995)
4. K. B. Davis, M.O. Mewes, N. Van Druten, D. Durfee, D. Kurn, and W. Ketterle, Phys. Rev. Lett. **75**, 3969 (1995)
5. C. C. Bradley, C. A. Sackett, J. J. Tollett, and R. G. Hulet, Phys. Rev. Lett. **75**, 1687 (1995); see also C. C. Bradley, C. A. Sackett, and R. G. Hulet, Phys. Rev. Lett. **78**, 985 (1997)
6. D. Fried, T. Killian, L. Willmann, D. Landhuis, S. Moss, D. Kleppner, and T. Greytak, Phys. Rev. Lett. **81**, 3811 (1998)
7. F. Pereira Dos Santos, F. Perales, J. Léonard, A. Sinatra, Junmin Wang, F. S. Pavone, E. Rasel, C. S. Unnikrishnan and M. Leduc, Eur. Phys. J. AP **14**, 69 (2001)
8. F. Pereira Dos Santos, F. Perales, J. Léonard, A. Sinatra, Junmin Wang, F. S. Pavone, E. Rasel, C. S. Unnikrishnan and M. Leduc, Eur. Phys. J. D **14**, 15 (2001)
9. G.V. Shlyapnikov, J.T.M. Walraven, U.M. Rahmanov, and M.W. Reynolds, Phys. Rev. Lett. **73**, 3247 (1994)
10. P. O. Fedichev, M. W. Reynolds, U. M. Rahmanov and G. V. Shlyapnikov, Phys. Rev. A **53**, 1447 (1996)
11. V. Venturi, I. B. Whittingham, P. J. Leo and G. Peach, Phys. Rev. A **60**, 4635 (1999)
12. V. Venturi and I. B. Whittingham, Phys. Rev. A **61**, 060703-1 (2000)
13. N. Herschbach, P.J.J. Tol, W. Hogervorst, W. Vassen, Phys. Rev. A **61**, 050702(R) (2000).
14. S. Nowak, A. Browaeys, A. Robert, D. Boiron, C. I. Westbrook, and A. Aspect, Appl. Phys. B **70**, 455 (2000)
15. H.C. Mastwijk, J.W. Thomsen, P. van der Straten and A. Niehaus, Phys. Rev. Lett. **80**, 5516 (1998)
16. M. Kumakura and N. Morita, Phys. Rev. Lett. **82**, 2848 (1999).
17. P.J.J. Tol, N. Herschbach, E.A. Hessels, W. Hogervorst, W. Vassen, Phys. Rev. A **60**, R761 (1999)
18. P.J. Leo, V. Venturi, I. B. Whittingham and J.F. Babb, Phys. Rev. A **64**, 042710 (2001)
19. J. Walraven, in *Quantum Dynamics of Simple Systems*, Vol. 44 of *Scottish Universities Summer Schools in Physics Proceedings*, edited by G.-L. Oppo, S. Barnett, E. Riis, and W. Wilinson (Institute of Physics, Bristol, 1996), p. 315
20. C. R. Monroe, E. A. Cornell, C. A. Sackett, C. J. Myatt, and C. E. Wieman, Phys. Rev. Lett. **70**, 414 (1993)
21. H. Wu and C. Foot, J. Phys. B **29**, L321 (1996)
22. A. Browaeys, A. Robert, O. Sirjean, J. Poupard, S. Nowak, D. Boiron, C. I. Westbrook, and A. Aspect, Phys. Rev. A **64**, 034703-1 (2001)
23. S. Giorgini, L. P. Pitaevskii and S. Stringari, Phys. Rev. A **54**, R4633 (1996)
24. G. Baym, J.-P. Blaizot, M. Holzmann, F. Laloë and D. Vautherin, Phys. Rev. Lett. **83**, 1703 (1999)
25. Effects due to mean field have been calculated for an harmonic trap [23] and effects due to quantum correlations have been calculated in a box [24]. The shift of the critical temperature T_c can be quite large ($\delta T_c/T_c \simeq 15\%$) for both of these effects, due to the large value of the scattering length a, but they have opposite signs. It would be interesting to study them both in more detail for an harmonic trap.
26. See for example, F. Dalfovo, S. Giorgini, L. P. Pitaevskii and S. Stringari, Rev. Mod. Phys. **71**, 463 (1999)
27. A. Aspect, private communication
28. P. O. Fedichev, M. W. Reynolds and G. V. Shlyapnikov, Phys. Rev. Lett. **77**, 2921 (1996)
29. B. P. Esry, C. H. Greene and J. P. Burke, Phys. Rev. Lett. **83**, 1751 (1999)

Quantum interferometry with fiber solitons - a direct experimental test of non-separability

N. Korolkova, Ch. Silberhorn, O. Glöckl, F. König and G. Leuchs

Zentrum für Moderne Optik an der Universität Erlangen-Nürnberg, Staudtstraße 7/B2, D-91058 Erlangen, Germany.

leuchs@physik.uni-erlangen.de
http://kerr.physik.uni-erlangen.de/

Abstract: We present a scheme for a direct experimental test of non-separability of a quantum state of two intense optical beams. On this basis, new entanglement-based quantum interferometry schemes using fiber solitons are suggested implying linear or nonlinear coupling of interacting pulses. The prospects of quantum interferometry with fiber solitons for quantum information processing are discussed.

Quantum interferometry refers to interference setups involving light beams with distinct quantum properties [1] which aim at overcoming the precision limit imposed by uncorrelated quantum uncertainties. Quantum interferometry with bright soliton pulses covers a wide range of applications from high precision measurements to quantum information processing using continuous variables.

Soliton pulses possess several features which are very promising for the new and rapidly growing field of quantum communication using quantum continuous variables. First, fiber solitons are intense ultra-short pulses taking full advantage of the optical non-linearity in a fiber to generate non-classical light. Second, the stability of a soliton during the propagation allows one to achieve high visibility in interferometric measurements. For specific applications one can exploit internal entanglement of soliton pulses which manifests itself in phase-photon number correlations and spectral photon number correlations. Interpulse correlations are introduced by interference and pulse collisions.

Continuous variables are quantum variables which take an infinite continuous spectrum of values, like position and momentum of a particle. One of the most attractive features of continuous variable quantum information is the possibility of information processing exclusively with linear inter-channel operations [2]. An additional advantage is the availability of efficient well-controllable sources and efficient and relatively simple detection systems. The basic example is the elegant way to generate continuous variable entanglement by overlapping two squeezed beams on a beam splitter [3, 4]. Nevertheless, also non-linear inter-channel operations are readily available in the regime of intense optical beams. With soliton pulses, the non-linear coupling for entanglement generation can be implemented using the cross Kerr interaction in a soliton collision.

Entanglement is well known to be a crucial resource for quantum information processing. Hence there is a strong interest not only in efficient and reliable sources but also in the experimental control and evaluation of entanglement. In this paper we report on an experimental protocol for verification of continuous variable entanglement in direct detection. We further extend it for the purposes of high-precision interfero-

Coherence and Quantum Optics VIII
Edited by Bigelow et al., Kluwer Academic/Plenum Publishers, 2003

metric measurements and for quantum communication applications. The experimental realization of these schemes using fiber optical solitons is presented. The prospects of quantum interferometry and information processing using non-linear coupling of soliton pulses are briefly discussed.

An entangled state is a non-separable quantum state of a system which cannot be represented as a convex sum of product states of two subsystems. For discrete variables, i.e. for the Hilbert space of 2×2 and 2×3 dimensions, a necessary and sufficient criterion for separability is the positive partial transpose $\hat{\rho}^T$ of the density matrix [5] known as the Peres-Horodecki criterion. Recently, the Peres-Horodecki criterion was extended to higher dimensions of the Hilbert space, i.e. to continuous variables, using two different approaches. The analysis of Simon [6] is accomplished using the formalism of the phase space Wigner function $W(x,p)$ [7]. It is based on the recognition that the partial transpose $\hat{\rho}^T$ is equivalent to a mirror reflection in phase space:

$$\hat{\rho} \longrightarrow \hat{\rho}^T \qquad \Longleftrightarrow \qquad W(x,p) \longrightarrow W(x,-p). \tag{1}$$

Here x, p are canonical variables position and momentum. Duan et al [8] are deriving the non-separabilty criterion in the spirit of two-mode squeezing [7]:

$$V(\hat{x}_1 + \hat{x}_2) \to 0, \qquad V(\hat{p}_1 - \hat{p}_2) \to 0 \tag{2}$$

where $V(\hat{A})$ denotes a variance $\langle \hat{A}^2 \rangle - \langle \hat{A} \rangle^2$ of an observable $\hat{A}$. They introduce the EPR-like operators in the form of joint variables:

$$\hat{u} = |a| \; \hat{x}_1 + \frac{1}{a} \; \hat{x}_2, \qquad \hat{v} = |a| \; \hat{p}_1 - \frac{1}{a} \; \hat{p}_2 \tag{3}$$

where a is an arbitrary non-zero real number. A separable state of such a joint system described by a density matrix $\hat{\rho}$ can be written as:

$$\hat{\rho} = \sum_i p_i \; \hat{\rho}_{i1} \otimes \hat{\rho}_{i2} \tag{4}$$

where $\hat{\rho}_{i1}$, $\hat{\rho}_{i2}$ denote density matrices of two subsystems 1 and 2. The total variance of $\hat{u}$ and $\hat{v}$ can be derived assuming that (4) holds. The lower bound for this total variance represents a separability condition. A sufficient criterion for an arbitrary two-mode state is then obtained in a form:

$$V(\hat{u})_\rho + V(\hat{v})_\rho \geq a^2 + \frac{1}{a^2}. \tag{5}$$

Furthermore, it provides a necessary and sufficient criterion for a certain class of Gaussian two-mode states [8].

We show that the criterion of Duan et al [8] and Simon [6] has a distinct advantage for experimental quantum communication as it can be expressed in terms of a single directly observable quantity. In Eqn. (5) the variances $V(\hat{u})$, $V(\hat{v})$ are presented without any additional normalization. For analytical analysis it is not needed: the reference level is a noise variance of the coherent beam or of the coherent vacuum which is equal to unity in these notations. However, in an experiment the respective reference is represented by the shot noise level and should be always explicitly determined in each measurement run. To account for this, we rewrite the Peres-Horodecki non-separability criterion (5) for continuous variables X_j, Y_j, $j = 1, 2$ of two optical modes 1 and 2 [6, 8] in the form:

$$V_{sq}^{\pm}(\hat{X}) + V_{sq}^{\mp}(\hat{Y}) < 2 \tag{6}$$

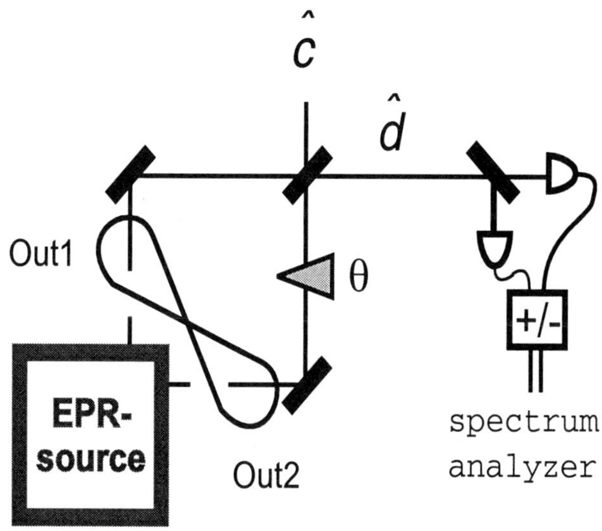

Fig. 1. Direct detection of the non-separability criterion for continuous variables.

where $V_{sq}^{\pm}(\hat{X})$, $V_{sq}^{\mp}(\hat{Y})$ are the normalized squeezing variances known from the context of two-mode squeezing [7]:

$$V_{sq}^{\pm}(\hat{X}) = \frac{V(\hat{X}_1 \pm g\,\hat{X}_2)}{V(\hat{X}_{1,\mathrm{SN}} + g\hat{X}_{2,\mathrm{SN}})}, \tag{7}$$

$$V_{sq}^{\mp}(\hat{Y}) = \frac{V(\hat{Y}_1 \mp g\,\hat{Y}_2)}{V(\hat{Y}_{1,\mathrm{SN}} + g\hat{Y}_{2,\mathrm{SN}})}. \tag{8}$$

The field modes are denoted by the respective subscripts, SN denotes the shot noise limit of the respective beam and g is the variable gain. The squeezing variances (7, 8) with the appropriate normalization correspond to the total variances (5) where the joint variables (3) for the optical beams are expressed in terms of amplitude and phase quadratures. For a particular choice of the $\hat{X}$, $\hat{Y}$ basis which ensures maximal available correlations and for the optimized gain g^{opt}, inequality (6) gives a necessary and sufficient condition for a state being entangled [8]. In general, it provides a sufficient criterion of non-separability of a Gaussian two-mode state [6, 8].

The Peres-Horodecki criterion in form (6) can be readily observed in an experiment. In Fig. 1 the source of the EPR-entanglement is treated as a black box with two output beams. It could be any source of continuous variable entanglement, an OPO setup or two interfering squeezed beams or any other generation process. The particular setup of Fig. 1 is designed for the intense output beams 1 and 2 of the same coherent amplitude α. The two relevant output modes are superimposed at a beam splitter (Fig. 1). The field operators of the beams after this interference are denoted $\hat{c}$, $\hat{d}$. To adjust the relative interference phase, a variable phase shift θ can be introduced in one of the arms. The balanced detection is performed in one of the output arms, $\hat{c}$ or, equivalently, $\hat{d}$. Difference and sum photocurrents are recorded and the relative phase θ is scanned. For the implementation with intense optical beams all the formulae should be rewritten in terms of the operators $\delta\hat{X}_j$, $\delta\hat{Y}_j$, $j = 1, 2$ with zero mean value. Here the linearized approach for bright beams is used: $\hat{X}_j = \langle X \rangle + \delta\hat{X}_j$, $\hat{Y}_j = \langle Y \rangle + \delta\hat{Y}_j$, reflecting that the entangled quantities in this case are quantum uncertainties of intense light fields. The

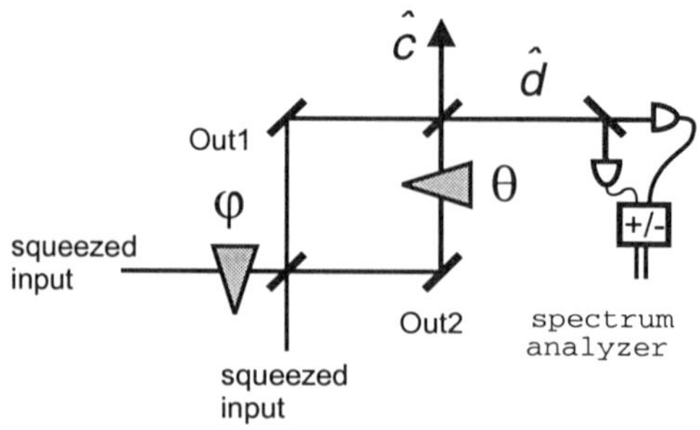

Fig. 2. Setup for entanglement-based quantum interferometry.

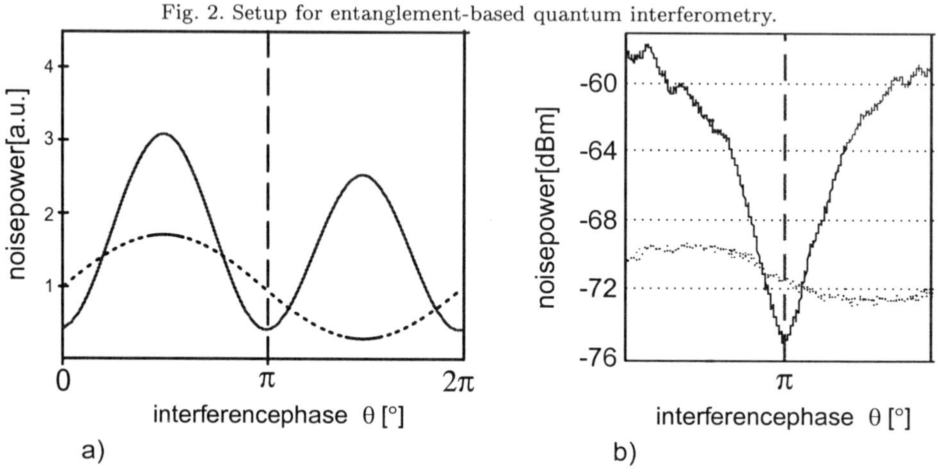

Fig. 3. a) Calculated intensity noise variances (solid line) and corresponding shot noise levels (dashed) at the output of the interferometer (Fig. 2) versus variable phase shift θ for $\varphi = \pi/4$. b) Experimental data: Amplitude noise variances for the indirect detection of the phase correlation. For these data the initial squeezed light beams forming the entangled pair interfere with a phase difference of $\phi = 30°$. For $\phi = 90°$ phase difference the noise variance detected in "+" channel will be below shot noise for all phase angles θ (see Fig. 4).

non-separability criterion of Eq. (6) then reads:

$$V_{sq}^{\pm}(\delta\hat{X}) + V_{sq}^{\mp}(\delta\hat{Y}) < 2. \tag{9}$$

We turn now to the description of the noise variance measurements using the setup of Fig. 1. The difference photocurrent provides the shot noise level reference:

$$V(\delta\hat{X}_{\mathrm{c,d,SN}}) = \alpha^2\,(1 \pm \cos\theta)\,. \tag{10}$$

The two signs refer to the two complementary outputs. The sum photocurrent delivers the amplitude noise variance of the signal in the output $\hat{c}$ or $\hat{d}$:

$$\langle\left(\delta\hat{X}_{\mathrm{c,d}}\right)^2\rangle = \frac{1}{2}\alpha^2\left[(1 \pm \cos\theta)^2\, V_{\mathrm{sq}}^{+}(\delta\hat{X}) + \sin^2\theta\; V_{\mathrm{sq}}^{-}(\delta\hat{Y})\right]. \tag{11}$$

From the measurements (10, 11) the normalized noise variance can be inferred:

$$V\left(\delta\hat{X}_{\mathrm{c,d}}\right) = \frac{\langle\left(\delta\hat{X}_{\mathrm{c,d}}\right)^2\rangle}{V(\delta\hat{X}_{\mathrm{c,d,SN}})} =$$

$$\frac{1}{2}\left[(1 \pm \cos\theta)\, V_{\text{sq}}^{+}(\delta\hat{X}) + \frac{\sin^2\theta}{(1 \pm \cos\theta)}\, V_{\text{sq}}^{-}(\delta\hat{Y})\right]. \tag{12}$$

It contains the information about both relevant squeezing variances $V_{\text{sq}}^{+}(\delta\hat{X}), V_{\text{sq}}^{-}(\delta\hat{Y})$ which characterize the quantum correlations between optical modes 1, 2 (Fig. 1). For $\theta = \frac{\pi}{2}$ the measured noise variance (12) is directly proportional to the Peres-Horodecki criterion for continuous variables:

$$V(\delta X_{\text{c,d}}) = \frac{1}{2}\left[V_{\text{sq}}^{+}(\delta X) + V_{\text{sq}}^{-}(\delta Y)\right]. \tag{13}$$

For the output entangled beams having symmetric circular uncertainty regions $V_{\text{sq}}^{+}(\delta\hat{X}) = V_{\text{sq}}^{-}(\delta\hat{Y}) = V_{\text{sq}}$ equation (12) takes the form:

$$V\left(\delta\hat{X}_{\text{c,d}}\right) = \frac{\langle\left(\delta\hat{X}_{\text{c,d}}\right)^2\rangle}{V(\delta\hat{X}_{\text{c,d,SN}})} = V_{\text{sq}} \tag{14}$$

and the non-separability condition reads: $V_{\text{sq}} < 1$ independent of the relative interference phase θ. Using (13) or (14), the non-separability of the state can be reliably verified experimentally using the interferometric scheme of Fig. 1.

In the experiment with fiber solitons quantum entanglement is generated via superimposing two amplitude squeezed beams on a beam splitter [4], provided the phase φ of this interference is adjusted properly. To prove the non-separability of the state it is sufficient to superimpose the output modes on a second beam splitter and to record the amplitude noise variance in addition to the shot noise of one of the resulting output beams $\hat{c}$ and $\hat{d}$ (Fig. 2). A value of $V_{sq}^{+}(\delta\hat{X}) + V_{sq}^{-}(\delta\hat{Y}) = 0.80 \pm 0.03$ was observed, demonstrating the high non-separability of the state [4]. Provided the variance of the sum of the amplitudes of beams 1 and 2 is already known, the quantum phase correlation can be inferred from the measured amplitude noise of $\hat{c}$ or $\hat{d}$ (Fig. 1, 2). A quantum phase correlation of up to 4.0 ± 0.4 dB was verified from the data of Fig. 3.

Note, that all indicated experimental values describe the noise statistics of the light without correction for linear losses. The photo-detectors showed a detection efficiency of around $92 \pm 5\%$ and the noise powers were recorded for a detection frequency of 10 MHz with a resolution bandwidth of 300 kHz. The measured data cited here are corrected for the electronic noise of the photo-detectors and the spectrum analysers. For quantum information applications the dark noise can limit the availability of the entanglement. If one corrects only the shot noise level for electronic noise to evaluate the quantum limit and for an uncorrected signal noise, the value of $1.12 \pm 0.06 < 2$ is observed in the experiment for the Peres-Horodecki criterion.

The phase φ of the first interference in Fig. 2 can also be controlled. That means that the degree of entanglement between the output beams 1 and 2 is tunable. The noise variances and mean value of the signal at the output of the interferometer are recorded by balanced detection in at least one of the output ports (see also Fig. 1). From these measurements the noise variance of the signal normalized to its shot noise level can be inferred: $V(\delta X_{c,d}) = \langle\left(\delta\hat{X}_{\text{c,d}}\right)^2\rangle / V(\delta\hat{X}_{\text{c,d,SN}})$. In this way the noise reduction at the output of the interferometer can be evaluated with respect to the interference phases φ and θ. This is illustrated in Figures 3 and 4.

For the phase of the first interference equal to zero, $\varphi = 0$, no entanglement is generated. In this case the setup (Fig. 2) represents just a trivial Mach-Zehnder interferometer with two squeezed inputs transformed to two squeezed outputs $\hat{c}, \hat{d}$, if the phase of the second interference is adjusted properly. The phase relation equal to $\varphi = \frac{\pi}{2}$

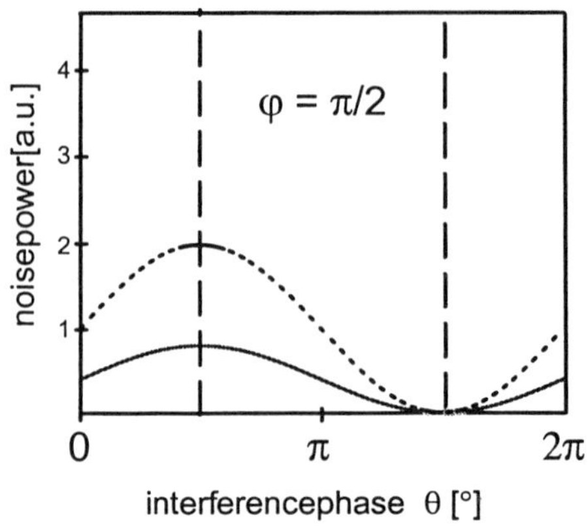

Fig. 4. Calculated intensity noise variances (solid line) and corresponding shot noise levels (dashed) at the output of the interferometer (Fig. 2) versus variable phase shift θ for $\varphi = \pi/2$ (compare Fig. 3).

corresponds to the maximal available entanglement between beams 1 and 2 (Fig. 4). The dependence of the second interference on the phase is given then by equation (12). Note that the noise reduction observed for $\varphi = \frac{\pi}{2}$ is insensitive to the phase θ: the ratio $V(\delta\hat{X}_{c,d}) = \langle\left(\delta\hat{X}_{\mathrm{c,d}}\right)^2\rangle/V(\delta X_{\mathrm{c,d,SN}}) < 1$ remains constant and less than unity for all values of θ (Fig. 3 (c)). Figure 3 (b) shows the intermediate situation with a close to maximal degree of entanglement (see also [4]).

An attractive feature of the setup for the direct detection of non-separability is the possibility to operate it as a setup for high-precision measurements of small phase modulation. The standard quantum limit for minimal resolvable phase modulations derived from the Heisenberg uncertainty principle reads:

$$\delta\theta_{\mathrm{min}} = \sqrt{\frac{1}{n}} \tag{15}$$

where n is a total photon number in the measured mode [7, 1]. We show that the quantum entanglement between beams 1 and 2 enables one to go beyond this resolution limit using the setup of Fig. 2.

The limit for minimal resolvable phase modulation can be derived in terms of signal-to-noise ratio [7]:

$$\mathrm{SNR} = \frac{\langle\hat{n}\rangle}{\sqrt{V(\hat{n})}} = \frac{\langle\hat{X}_{c,d}\rangle}{\sqrt{V(\delta\hat{X}_{c,d})}} \tag{16}$$

where $\hat{n} = \hat{c}^{\dagger}\hat{c}$ is the photon number operator in mode $\hat{c}$ with the mean value $\langle\hat{n}\rangle = n$ (analogously for $\hat{d}$). The smallest detectable phase shift is defined to be that for which $\mathrm{SNR} = 1$.

The analysis of the setup of Figure 2 on this basis yields the following results. For the phase relation $\varphi = 0$ there is no interferometric performance of the setup as the signal is zero. For Figure 3 minimal resolvable phase modulations around $\theta = 0 + k\pi$ are:

$$\delta\theta_{\mathrm{min}} = \sqrt{\frac{2V_{\mathrm{sq}}^{+}(\delta\hat{X})}{n}} \tag{17}$$

where $V_{\mathrm{sq}}^{+}(\delta\hat{X}) = V(\delta\hat{X})$ is the squeezing variance determining the maximum available quality of entanglement for the input squeezed beams. For the situation corresponding

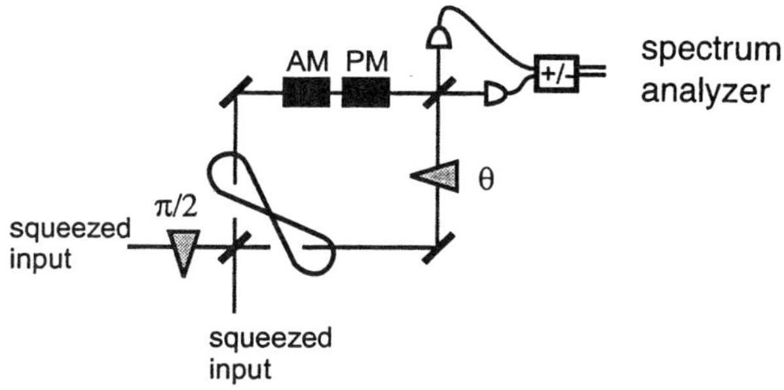

Fig. 5. Setup for quantum dense coding with bright entangled beams.

to maximal available entanglement between the beams (Fig. 4) the smallest resolvable phase modulations around $\theta = \frac{\pi}{2} + k\pi$ reduce to

$$\delta\theta_{\rm min} = \sqrt{\frac{V_{\rm sq}^{+}(\delta\hat{X})}{n}}. \tag{18}$$

The setup reaches the best phase resolution performance for the best entanglement.

One of the related quantum communication protocols is quantum dense coding [9, 10] which aims at increasing the channel information capacity as compared with classical coherent-state communication. Quantum dense coding was first suggested [9] and realized in an experiment [11] for discrete quantum variables like polarization states of entangled photon pairs. Recently this idea was extended to quadrature components of the electromagnetic field using entangled light beams as quantum channel [10]. For such quantum continuous variables, also quantum dense coding refers to the ability to read out amplitude and phase modulations with a precision below the limit given by the Heisenberg uncertainty relation. The first experimental implementation of quantum dense coding for continuous variables has been recently reported [12].

Figure 5 shows the interferometric setup of Figure 2 modified to implement dense coding using intense entangled beams. We would like to emphasize the role of the quantum correlations $V_{\rm sq}^{+}(\delta\hat{X})$, $V_{\rm sq}^{-}(\delta\hat{Y})$ on the scheme performance, which means on the enhanced resolution of small amplitude and phase modulations with respect to the quantum limit. We assume $V_{\rm sq}^{+}(\delta\hat{X}) = V_{\rm sq}^{-}(\delta\hat{Y}) = V_{\rm sq}$ which is required for the phase independent performance of the scheme. The balanced detection of the combined mode of both outputs delivers the information about small imposed amplitude $\delta X_{\rm m}$ and phase $\delta Y_{\rm m}$ modulations:

$$V^{+} = \alpha^2 \left(V_{\rm sq} + V(\delta\hat{X}_{\rm m})\right), \tag{19}$$

$$V^{-} = \alpha^2 \left(V_{\rm sq} + V(\delta\hat{Y}_{\rm m})\right) \tag{20}$$

where $V^{\pm}$ are the noise variances recorded in the sum and difference detector channels, respectively. Quantum EPR-like correlations between the bright beams in the two interferometer arms imply $V_{\rm sq} \ll 1$. Hence the use of quantum entanglement enhances the resolution of quadrature modulations beyond the quantum limit for all values of θ (19, 20). This enables quantum dense coding independent of the interference phase θ. The phase-insensitivity eliminates the need for extensive setup stabilization. It makes the scheme more reliable and easier to handle, in particular for the communication over long distances.

For fiber solitons, linear interference of non-classical beams is not the only way for the generation and exploitation of quantum correlations for quantum information purposes.

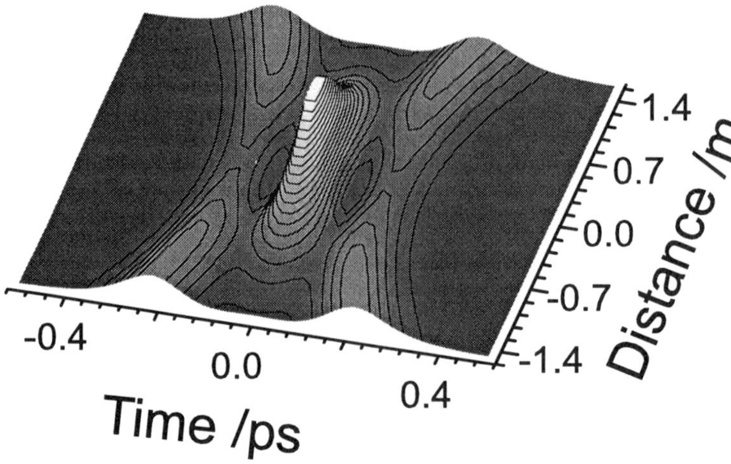

Fig. 6. The collision of two solitons in the time domain. In the interaction the pulse shapes are strongly distorted. The outgoing solitons, however, recover their initial energy, velocity, and shape.

The availability of ultra-short intense pulses allows for implementation of effective non-linear coupling due to the cross Kerr interaction. When two solitons collide in a fiber, they recover their initial energy, velocity and shape after the collision (Fig. 6). Due to the Kerr effect induced cross-phase modulation, however, they experience a permanent mutual phase shift which is proportional to the photon number of the other soliton. This effect was used to entangle the quantum variables of colliding solitons aiming at the optical implementation of a quantum non-demolition measurement of the photon number in a pulse [13]. The phase shift of one pulse (probe pulse) can be utilized as a readout for the photon number of the other (signal) pulse. In other words, the phase shift of the probe soliton due to the Kerr nonlinear interaction becomes entangled with the photon number of the signal soliton by the collision.

Alternatively, one can make use of the multimode quantum structure of each pulse [14] and its evolution in a Kerr non-linear interaction via collision [15]. The spectral quantum structure of colliding solitons was shown to contain transient inter-pulse photon number correlations [15]. During the collision, the spectral power distributions associated with each soliton are shifted apart (Fig. 7). This transient frequency shift is the reason for the phase shift mentioned above, in close analogy to phase changing collisions of atoms. It depends on the photon number and the relative velocity of the colliding solitons. The collision-induced transient changes in the spectrum produce therefore spectral inter-pulse correlations. These transient quantum correlations may be used as a nonlinear coupling mechanism for a back-action evading quantum measurement of the soliton photon number [15, 16]. In the experiment [16], the soliton exit the fiber at mid-collision and the transient frequency shift is thus frozen. In the next step the probe soliton is spectrally filtered and its photon number is directly detected. The measured photon number is quantum correlated with the photon number of the signal soliton [16].

The quantum information prospects of such non-linear coupling and of the QND interactions deserve further investigations. QND measurements were proposed for entanglement purification for continuous variable entanglement [17], for an eavesdropping attack in the context of quantum cryptography [18], for a quantum optical bus system and for monitoring in quantum computing [19]. The use of QND-generated entanglement for quantum teleportation was proposed [20] for schemes based on optical parametric processes. Many applications for a back-action evading measurement can be derived be-

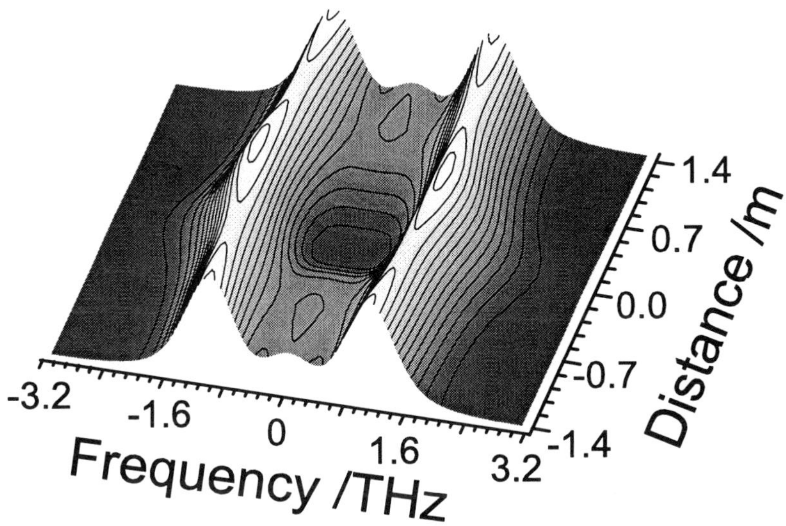

Fig. 7. The collision of two solitons in the frequency domain. During the collision at z=0 there is a transient shift of the spectral densities of the pulses away from each other. The frequency of a soliton at z=0 thus depends on the photon number of the other soliton.

cause it is a paradigm for a quantum-mechanical measurement. It seems worth exploring the possibility for the implementation of quantum logic operations using quantum entanglement generated in soliton collisions. The scheme of Figure 8 shows the conceptual possibility to exploit the described nonlinear soliton coupling in quantum information processing. The first (probe) soliton collides sequentially with two other (signal) solitons, which have eventually experienced additional processing, e.g processing needed for swapping of the quantum variable from photon number to phase. Through the measurement performed on the first soliton, the two other solitons become entangled. The degree and kind of emerging correlations depend on the particular implementation of the nonlinear interaction. The scheme might also allow for implementation of conditional operations, e.g. a conditional phase shift. Yet, there are no concrete experimental proposals with such systems so far. Experimental details and feasibility of the scheme of Figure 8, as well as the nature of the EPR-like correlations generated in QND interactions, still require further investigations.

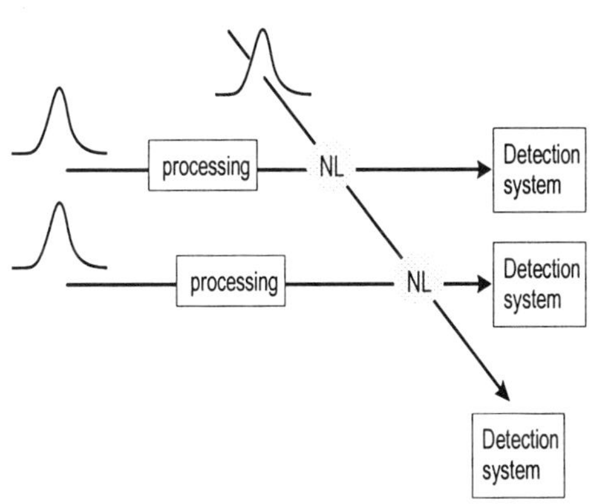

Fig. 8. Generation of quantum correlated solitons via nonlinear coupling due to cross Kerr interaction in a fiber.

In conclusion, quantum entanglement is a decisive resource for quantum information processing. Hence controlled manipulation of entanglement, its reliable generation and evaluation are the primary tasks of experimental quantum communication. We report on the interferometric scheme designed for the direct measurement of the continuous variable Peres-Horodecki criterion. It can be modified to implement a variety of quantum interferometric techniques with intense light pulses, also for continuous variable quantum communication. Quantum information processing using continuous variables has emerged as an alternative approach to quantum information with single photons and atoms described with discrete variables. The main motivation for this new approach is the availability of controlled sources, efficient detection systems and easy-to-handle processing using linear elements. Alternatively, non-linear coupling can be also efficiently implemented with intense soliton pulses.

References and links

1. F. De Martini, G. Denardo, and Y. Shih (Eds.), "Quantum Interferometry," Proceedings of an Adriatic Workshop, Trieste, March 1996, (VCH: Weinheim New York Baesl Cambridge Tokyo 1996).
2. S. Lloyd and S. L. Braunstein, "Quantum Computation over Continuous Variables," Phys. Rev. Lett. **82**, 1784 (1999).
3. A. Furusawa, J. L. Sørensen, S. L. Braunstein, C. A. Fuchs, H. J. Kimble, and E. S. Polzik, "Unconditional quantum teleportation," *Science*, **283**, 706 (1998).
4. Ch. Silberhorn, P. K. Lam, O. Weiß, F. König, N. Korolkova, and G. Leuchs, "Generation of Continuous Variable Einstein-Podolsky-Rosen Entanglement via the Kerr Nonlinearity in an Optical Fibre," Phys. Rev. Lett. **86**, 4267 (2001).
5. M. Horodecki, P. Horodecki, and R. Horodecki, "Separability of mixed states: necessary and sufficient conditions," Phys. Lett. , **A 223**, 1 (1996).
6. R. Simon, "Peres-Horodecki Separability Criterion for Continuous Variable Systems," Phys. Rev. Lett. **84**, 2726 (2000).
7. D. F. Walls and G. J. Milburn, "Quantum optics," (Springer Verlag: Berlin 1994).
8. L.-M. Duan, G. Giedke, J. I. Cirac, and P. Zoller, "Inseparability Criterion for Continuous Variable Systems," Phys. Rev. Lett. **84**, 2722 (2000).
9. C. H. Bennett and S. J. Wiesner, "Communication via one- and two-particle operators on Einstein-Podolsky-Rosen states," Phys. Rev. Lett. **69**, 2881 (1992).
10. S. L. Braunstein and H. J. Kimble, "Dense coding for continuous variables," Phys. Rev. A **61**, 042302 (2000).
11. K. Mattle, H. Weinfurter, P. G. Kwiat, and A. Zeilinger, "Dense Coding in Experimental Quantum Communication," Phys. Rev. Lett. **76**, 4656 (1996).
12. Y. Zhang, C. Xie, and K. Peng, "Quantum teleportation and dense coding by means of bright amplitude-squeezed light and direct measurement of a Bell state," *Quantum Semiclass. Opt.*, **3**, 293 (2001).
13. S. R. Friberg, S. Machida, and Y. Yamamoto, "Quantum-nondemoliton measurement of the photon number of an optical soliton," Phys. Rev. Lett. **69**, 3165 (1992).
14. S. Spälter, N. Korolkova, F. König, A. Sizmann and G. Leuchs, "Observation of multimode quantum correlation in fiber optical solitons," Phys. Rev. Lett. **81**, 786 (1998).
15. F. König, M. A. Zielonka, and A. Sizmann, "Transient quantum correlations of interacting solitons," Phys. Rev. Lett. , submitted.
16. F. König, B. Buchler, T. Rechtenwald, A. Sizmann, and G. Leuchs, "QND detection of the soliton photon number by spectral filtering," Conference on Laser and Electro-Optics / Europe - European Quantum Electornics Conference CLEO/Europe-EQEC Focus Meeting 2001, Munich, Germany, June 18-22, 2001, *CLEO/Europe-EQEC Focus Meeting Conference Digest*, 296.
17. L.-M. Duan, G. Giedke, J. I. Cirac, and P. Zoller, "Entanglement purification of Gaussian continuous variable quantum states," Phys. Rev. Lett. **84**, 4002 (2000).
18. N. Gisin, G. Ribordy, W. Tittel, and H. Zbinden, "Quantum Cryptography," quant-ph/0101098 (2001).
19. M. Ozawa, "Quantum nondemolition monitoring of universal quantum computers," Phys. Rev. Lett. **80**, 631 (1998).
20. D. B. Horoshko and S. Ya. Kilin, "Quantum teleportation using quantum nondemolition technique," Phys. Rev. A **61**, 032304 (2000).

Nonlinear atom optics: bosons vs fermions

P. Meystre
Optical Sciences Center, University of Arizona, Tucson, AZ 85721
pierre.meystre@optics.arizona.edu

Abstract: Nonlinear atom optics, the matter-wave analog of nonlinear optics, has recently been demonstrated in several experiments in quantum degenerate bosonic systems. Atomic four-wave mixing, as well as the nonlinear mixing between optical and matter waves, have been realized. In a parallel development, quantum degenerate atomic Fermi systems have become available. This leads one to ask whether nonlinear atom optical effects can occur in such systems, or whether they result from "Bose enhancement", and hence are inherent to bosonic fields. To answer this question we compare atomic four-wave mixing in quantum degenerate gases of bosonic and fermionic atoms. We find that in principle, matter-wave gratings formed from either bosonic or fermionic atoms can exhibit nearly identical Bragg scattering and four-wave mixing properties. This effect is due to constructive many-particle quantum interferences, which in the boson case are interpreted as "Bose enhancement." This implies that effects such as coherent matter-wave amplification and superradiance can occur in degenerate Fermi gases.

Coherence and Quantum Optics VIII
Edited by Bigelow et al., Kluwer Academic/Plenum Publishers, 2003

Instability, area quantization and area gaps in cavity-accelerated superradiant atom-cavity systems

C. Greiner, T. Wang, T. Loftus, B. Boggs and T. W. Mossberg
Oregon Center for Optics and Department of Physics, University of Oregon, Eugene, OR 97403

Abstract: We study the interaction of pulsed excitation fields with optically thin atomic ensembles subject to strong cavity-accelerated superradiance. Excitation of superradiant instabilities in homogeneously broadened atom-field systems is seen to lead to a quantization of the allowed intracavity pulse area. In inhomogeneously broadened atom-cavity systems superradiant instability causes area quasi-discretization and gaps (i.e. excluded regions) between allowed intracavity areas. The resulting area discontinuities are demonstrated in cryogenically coherence-stabilized Tm^{3+} ions.

The radiative decay of excited atoms proceeds via a variety of environment-dependent mechanisms. For isolated atoms, the coupling of the atomic transition to a continuum of incoherent vacuum-state modes results in exponential decay with a rate proportional to the ambient vacuum spectral energy density. [1] Very different decay dynamics are found in situations where atoms are strongly coupled to slowly decaying field modes, as present in high-finesse resonators, [2] when they are exposed to stimulating fields derived from external sources or other ensemble atoms, or when they are part of a highly correlated atomic ensemble, wherein correlations originate either spontaneously [3-5] or are impressed externally. Dicke explored the latter scenarios in his seminal 1954 paper on superradiance. [6] As pointed out by Dicke, correlated emission may lead to both sub- and superradiant behavior. It can occur in samples whose spatial extent is small compared to the relevant emission wavelength or much larger provided phased-array correlation is present. It has proven difficult to study many aspects of the correlated-emission scenario since the effect requires ensembles of substantial optical thickness in order to be dominant [7] and the preparation and decay of extended large-absorption samples is complicated by propagation effects.[8] Recently, a new correlated-emission scenario, cavity-accelerated superradiance, was shown to constitute an essentially ideal realization of the large-sample superradiance scenario originally envisioned by Dicke in his 1954 paper.[9] In this case, an optically thin atomic sample is coupled to a rapidly decaying mode of an optical resonator to produce strong superradiant emission unperturbed by cavity memory and field propagation effects.

In the present work, we consider the interaction of a pulsed optical field with an optically thin ensemble of atoms subject to strong cavity-accelerated superradiance. We investigate several scenarios comprising homogeneously and inhomogeneously broadened atomic ensembles and traveling as well as standing-wave cavities. Optical fields of controlled pulse area are injected along the cavity's axis creating a phased-array atomic excitation. Through cavity-accelerated superradiance, atomic emission strongly modifies the injected pulse areas. In homogeneously broadened atomic systems, superradiance is found to lead to a quantization of the total intracavity pulse area. Atomic emission clamps intracavity pulse areas to discrete values even when the injected area is continuously varied. In inhomogeneously broadened systems, atomic emission acts to produce sharp discontinuities in the variation of intracavity area as a function of injected area with intracavity area partially locked between discontinuities. In effect, certain bands of intracavity pulse areas become forbidden causing gaps between the allowed area values. Both effects can be seen to follow from a superradiant instability of the atom-field system wherein the atomic ensemble rapidly relaxes to stable states and in the process controls allowed intracavity pulse areas. In the following, we explore the cavity-accelerated superradiant system via numerical and analytic approaches and we report experimental results providing evidence for the pulse area discontinuities predicted.

Consider a system comprising an ensemble of N initially unexcited two-level atoms located inside a unidirectional four-mirror ring cavity (Figure 1). Two of the mirrors are partial with reflectivity R. The cavity has a photon lifetime, ϑ_c, which is short compared to all other system time scales, and atoms within the cavity experience a single-photon Rabi frequency, $\Sigma_0/2B$. The relative shortness of the cavity lifetime allows us to focus on a regime of atom-cavity dynamics, where the system behavior is dominated by the atomic evolution and unperturbed by cavity dynamics. The plane-wave excitation field has a rectangular temporal profile with a duration ϑ_{exc} and begins exciting the atoms at t = 0. Note that $\vartheta_{exc} >> \vartheta_c$. The center frequencies of the excitation field, atomic absorption line and cavity resonance are coincident. The pulse areas of the input, intracavity and output fields are denoted 2_i, 2_C and 2_{out}, respectively. We calculate $2_C(t)$ through a numerical integration of the Maxwell-Bloch equation. We define the excitation pulse area 2_{exc} to be the intracavity pulse area that accrues during $0 \# t \# \vartheta_{exc}$. The total intracavity pulse area $2_C^{tot} = 2_C(t6\infty)$. Post excitation superradiant emission is responsible for the difference between 2_C^{tot} and 2_{exc}. For all calculations, homogeneous atomic decoherence and cavity loss mechanisms other than mirror transmission are ignored.

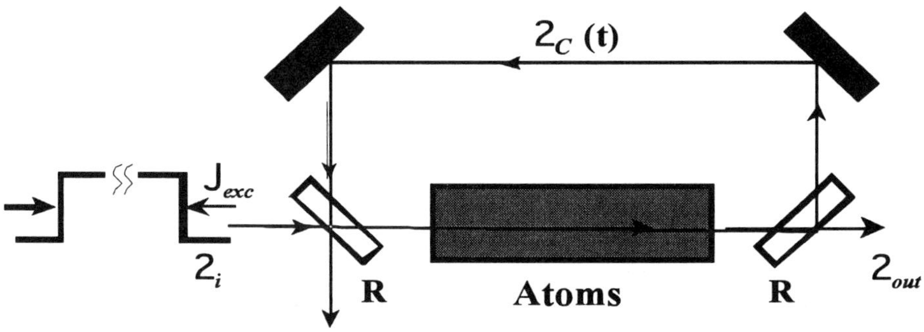

Figure 1. Model system. R, mirror reflectivity; 2_i, 2_c, 2_{out}, areas of the input, intracavity and output fields; ϑ_{exc}, excitation-pulse duration.

As a first model system, denoted the homogeneous system, we assume that the atoms all have the same resonance frequency. In Figure 2, as a solid line, we plot 2_C^{tot} as a function of 2_{exc} for N = 10^8. The dashed line corresponds to the case of an empty cavity (N = 0). Note how continuous ranges of excitation pulse areas 2_{exc} are mapped to discrete values of $2_C^{tot} = m$ _ 2B, where m = 0,1,2,J. Transitions between different values of 2_C^{tot} occur whenever 2_{exc} passes through an odd multiple of B. Since $2_{out} = (1-R)^{1/2}$ 2_C^{tot}, quantization of 2_C^{tot} implies quantization of 2_{out}.

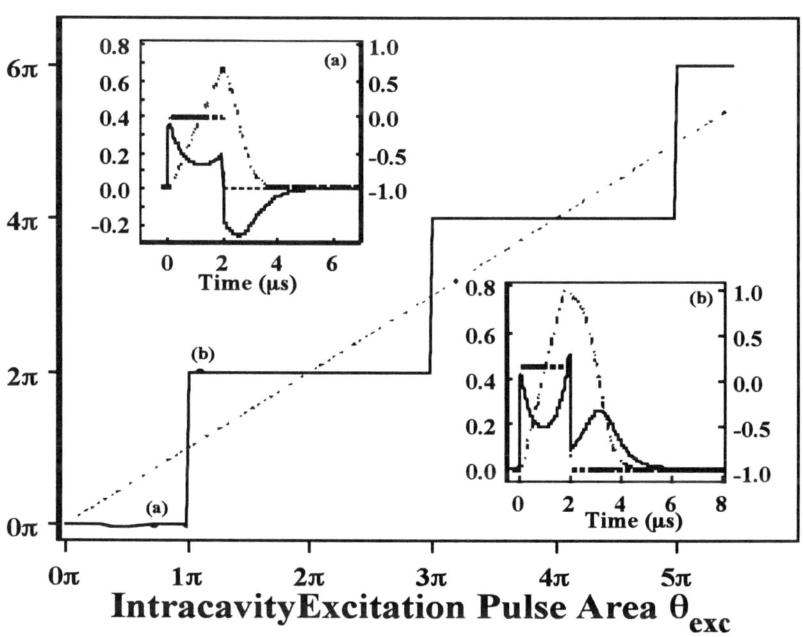

Figure 2. 2_C^{tot} versus 2_{exc} for the homogeneous system. The solid and dashed lines denote the cases with and without atoms, respectively. Insets: Solid (Dashed) lines, intracavity Rabi frequency Σ_C versus time with (without) atoms; dotted-dashed lines, atomic inversion w versus time. Upper (Lower) inset, 2_{exc} .0.73 B (2_{exc} . 1.1 B). For all results shown, R = .98, $\Sigma_0/2B$ = 100 Hz, ϑ_c = 12 ns, ϑ_{exc} = 2μs.

The insets to Fig 2 show the time evolution of the intracavity Rabi frequency (solid line) $\Sigma_C(t) = d2_C^{tot}(t)/dt$ and atomic inversion w (dotted-dashed line) for 2_{exc} . 0.73 B (a) and 2_{exc} . 1.1 B (b). The dashed lines in the insets show the intracavity Rabi frequency generated by the input field in the absence of atoms. After the excitation ends, sustained superradiant atomic emission is observed for several microseconds. Since the cavity decay time ϑ_c is much shorter than the superradiant emission time ϑ_R, the intracavity field is driven by and follows the long-lived atomic polarization once the input field terminates. In the upper inset, where 2_{exc} < B, the superradiant emission is anti-phased with respect to the excitation field, leading to zero total output pulse area. In the lower inset, where 2_{exc} > B, the sample radiates in phase with the excitation field yielding a net pulse area of 2B. The atom-field dynamics at higher thresholds are similar, except that the atoms undergo an integer number of full Rabi cycles prior to the subsequent superradiant decay. Significantly, the atomic inversion evolves back to -1 both above and below the discontinuity, indicating complete depletion of the atomic excitation energy by the superradiant process.

Note that the superradiance time scale, $\vartheta_R = 1/(\vartheta_c\_\Sigma_0^2\_N)$ [10], can be controlled through choice of cavity and atom parameters thereby providing a means to tailor the temporal properties of the intracavity and output fields. The area quantization effect depicted in Figure 2 should persist even to the level of single atoms and photons as long as $\vartheta_R << T_2$, where T_2 is the homogeneous atomic decoherence time.

Inhomogeneous dephasing, present in many atomic systems, frequently obscures phenomena relying on atomic coherence. To investigate its effect on the area quantization

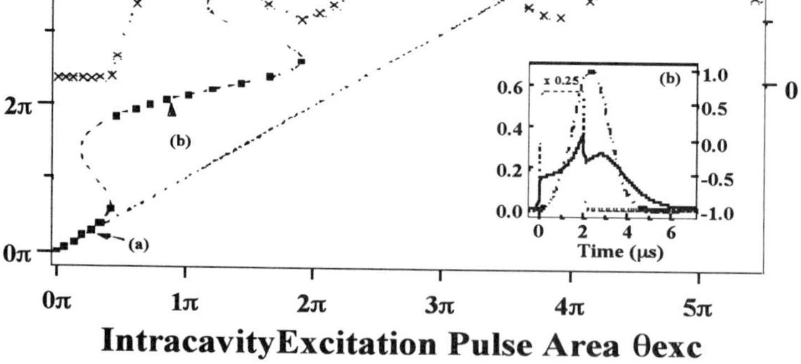

Figure 3. Squares, dashed and dotted -dashed lines, 2_C^{tot} versus 2_{exc} for the inhomogeneous system (see text), X's, fractional reradiated energy; Insets: Solid (dashed) lines, intracavity Rabi frequency S_C versus time for cavity with (without) atoms; dotte d-dashed lines, atomic inversion w at linecenter versus time. Upper (lower) inset, 2_{exc} = 0.26 B (2_{exc} = 0.86 B). See Figure 2 for other system parameters.

we modify the previous system by replacing the homogeneously broadened atomic ensemble with N . 3 x 10^{10} atoms that are distributed across a Gaussian frequency profile with a full-width-at-half-maximum $)<_{inhom}$ = 90 MHz. The inhomogeneous atomic ensemble exhibits a small signal single-pass optical thickness ∀L = 0.1 at linecenter. Approximately 1.6 _ 10^8 atoms are located within the excitation pulse bandwidth of

~1/ϑ_{exc}, making the number of excited atoms comparable to the number in the homogeneous system. A 100 ns linear rise/fall time is incorporated in the excitation pulse.

Figure 3 shows 2_C^{tot} as a function of 2_{exc} for the inhomogeneous system with (squares) and without (dashed line) intracavity atoms. Strict quantization of 2_C^{tot} is no longer observed. Instead, we find quasi-discretized regions of reduced slope in the 2_C^{tot} versus 2_{exc} plot, which are separated by abrupt discontinuous changes in 2_C^{tot}. The values of 2_{exc} at which the discontinuities in 2_C^{tot} occur are no longer uniformly spaced and differ from the homogeneous case.

Recall that in the homogeneous system 2_C^{tot}-discontinuities occur due to phase changes of the superradiant field when the excitation pulse drives the atom through its excited state at values of 2_{exc} = (2*m*+1) _ B, where *m* = 0,1,2,J. In contrast, in the inhomogeneous system post-excitation superradiance is always in phase with the excitation field, yielding 2_C^{tot} ∃ 2_{exc} and as a consequence more subtle dynamics must underlie the discontinuities. Note that the gaps in allowed 2_C^{tot} values are centered about values of 2_C^{tot} = (2*m*+1)_B. This suggests that the discontinuities occur when the combined effect of initial excitation and post-excitation superradiance drives linecenter atoms through total inversion, switching them from an absorptive to an emissive state. In contrast to the homogeneous system, post-excitation superradiance can act to increase linecenter atomic excitation by virtue of superradiant release of energy stored in detuned atoms. In the insets to Fig. 3, we show the temporal dynamics of the total intracavity field and the linecenter atomic inversion for values of 2_{exc} indicated. Above the 2_C^{tot}-discontinuity, the linecenter inversion continues to grow after the initial excitation has terminated, indicating the aforementioned energy transfer from detuned atoms. Subsequently, the line center atoms superradiantly damp to the ground state while below the discontinuity they do not. Interestingly, the fraction of the spectrally integrated energy that is absorbed from the excitation field and is superradiantly re-emitted (X's in Fig. 3) is always less than unity and exhibits oscillations correlated with the 2_C^{tot}-discontinuities. The reradiated energy fraction peaks when 2_C^{tot} . 2*m*_ B, not when area discontinuities occur.

Assuming that VL << 1, R . 1 and)<$_{inhom}$ >> 1/ϑ$_{exc}$ an analytic relationship between 2_i and 2_C^{tot} can be derived that provides additional insights into the inhomogeneously broadened system's dynamics:

$$\theta_i = \frac{1}{\sqrt{1-R}}\left[(1-R)\theta_C^{tot} + \frac{\alpha L}{2}\sin\theta_C^{tot}\right]. \tag{1}$$

Using numerical results to map 2_i onto 2_{exc}, Eq. 1 is plotted as the dotted-dashed line in Fig. 3. Note how the numerical calculations are duplicated except in the regions of area discontinuity where the analytic result is multivalued. Eq. 1 indicates that the 2_C^{tot}-discontinuities are observable in the inhomogeneous system only if VL/2(1-R) > 1. This condition may be compared to the premise that strong superradiant effects become appreciable only when VL/1 in the absence of an accelerating cavity.[7] An analytic result analogous to Eq. 1 for a two-mirror cavity has been reported previously.[11] In Ref. 11, the multi-valued regions were discussed in terms of the steady-state concept of optical bistability. It remains to be determined how this concept applies to the transient scenario under consideration, which evolves fully deterministically suggesting single-valued mapping of specific inputs to specific outputs as observed in our simulations.

In the experiment described below, a standing-wave cavity is

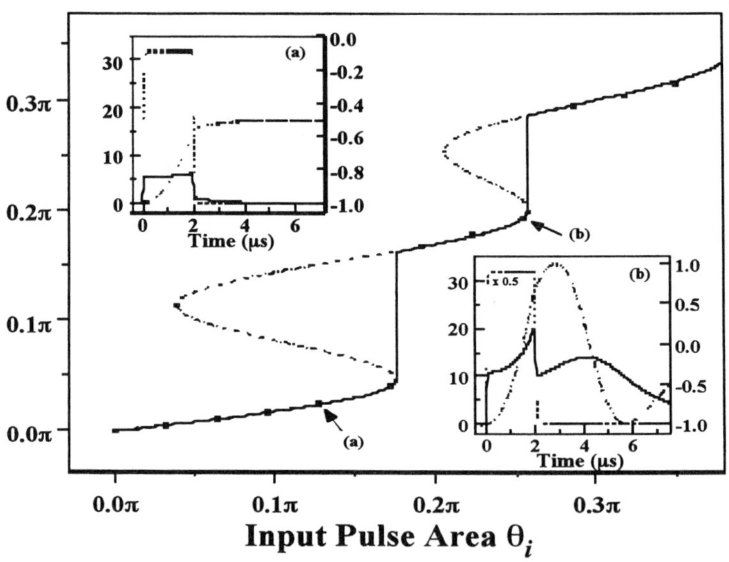

Figure 4. 2_{out} versus 2_i for inhomogeneously broadened atoms in a standing-wave cavity. Insets: Solid (dashed) lines, transmitted Rabi frequency Σout versus time for cavity with (without) atoms; dotted-dashed lines, linecenter atomic inversion w at a cavity antinode versus time. Left (Right) inset, 2_i = 0.13 B (2_i = 0.25 B).

employed and input and output (as opposed to intracavity) signals are measured. For comparison purposes, we present in Fig. 4, as solid squares, a relationship between 2_{out} and 2_{in} calculated for a standing-wave cavity with an inhomogeneously broadened atomic ensemble. The dotted-dashed line in Fig. 4, represents the multi-valued analytic relationship between 2_{out} and 2_{in} calculated for this system. Single-valued portions of this relationship are highlighted by the solid line. Significantly, quasi-discretization of output area, forbidden area regions and discontinuities survive the strong spatial inhomogeneity of a standing-wave cavity. The insets of Fig. 4 show the Rabi frequency (solid lines) of the transmitted field and linecenter inversion (dotted-dashed lines) at a cavity anti-node versus time both above and below the first area discontinuity. Qualitative behavior is similar to the inhomogeneously broadened but spatially homogeneous ring-cavity scenario.

To test the above predictions we have performed experiments on Tm^{3+} ions doped in a thin slice of YAG crystal and located within a symmetric two-mirror standing-wave cavity.[9] The cavity parameters are: R = 98%, ϑ_c . 1.5 ns and $\Sigma_0/2$ B . 170 Hz. The ions are excited on their $^3H_6(1)$ - $^3H_4(1)$ transition. The sample and cavity are cooled to –4.6 K, where the transition has an inhomogeneous linewidth of ~ 20 GHz, a homogeneous dephasing time T_2 .20 µs, an excited state ($^3H_4$) lifetime T_1 ~ 800 µs, and a wavelength 8_{Air} . 793.17 nm. The single-pass linecenter optical thickness, ∀L, is ~0.1. We estimate N .5_10^8, yielding ϑ_R.1 µs, comparable to the simulations.

The experimentally accessible quantities are cavity input and output optical power as a function of time. The input field is known to be of constant phase and the theory predicts that the inhomogeneously broadened Tm^{3+} system produces an output field of constant phase. We therefore assume that E % $P^{1/2}$ in calculating areas from observed powers. The cavity was axially excited by a 2-µs-long pulse (rise/fall time ~100 ns) of variable power (50 µW - 1.6 mW). Output signals corresponding to various input powers were recorded.

The results of our measurements are shown in Figure 5. Each solid square shown represents an average over 15 individual excitation cycles. The insets show measured output power versus time for the 2_i values indicated. The dashed line depicts the linear relationship between 2_{out} and 2_i expected in the absence of absorbers. Clear evidence of the discontinuity predicted in Fig. 4 is observed. The softness of the transition and the absence of higher order discontinuities may follow from the effects of nonuniform transverse intensity across the cavity mode involved, homogeneous dephasing, and the presence of orientationally non-equivalent Tm^{3+} sites within the YAG host that have

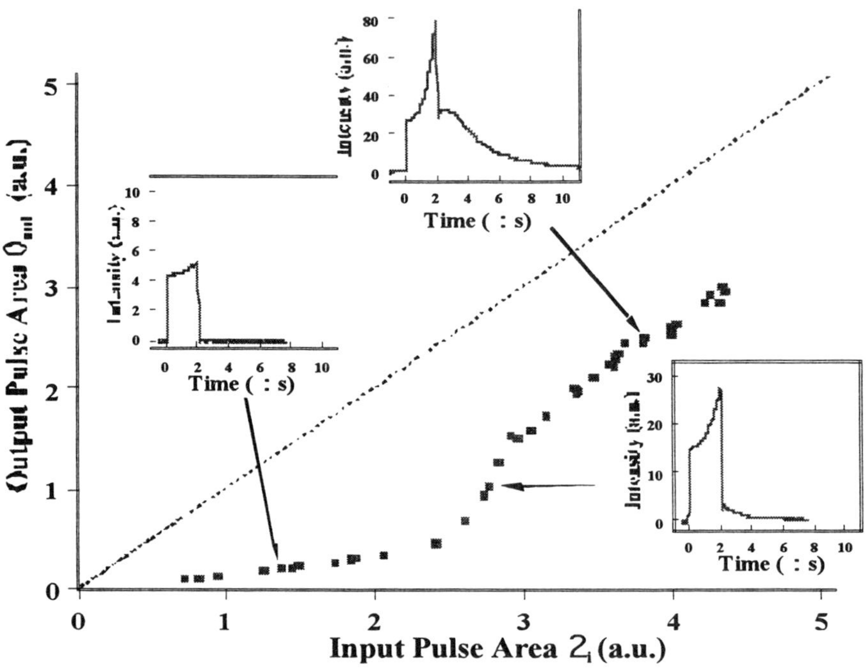

Figure 5. Experimental data; Squares (dashed line): 2_{out} versus 2_i with (without) atoms; Insets: cavity output power versus time; The excitation field is on for $0 < t < 2$ μs.

different atom-field coupling constants [12] and therefore experience different pulse areas.

In summary, we have explored one example of the new physics that can arise in an atomic system dominated by strong cavity-accelerated superradiance. The clamping of continuously varying input areas to discrete output areas found with homogeneous absorbers provides a powerful means of locking excitation areas to specific values regardless of fluctuations in input areas. Such a mechanism may be useful in quantum computing or other scenarios where precise and reliable atomic excitation is required. More generally, the output area discontinuities observed represent an extreme input-output nonlinearity, which may provide, for example, a means of producing novel devices such as "area" transistors. Application of very small input area perturbations can be used to switch very large area perturbations on the output.

We appreciate support from the U.S. Air Force Office of Scientific Research under contract No. F 49620-99-0214.

References

[1] V. F. Weisskopf and E. Wigner, Z. Phys. **63**, 54 (1930).

[2] E. T. Jaynes and F. W. Cummings, Proc. IEEE **51**, 89(1963).

[3] N. Skribanowitz, I. P. Herman, J. C. MacGillivray and M. S. Feld, Phys. Rev. Lett. **30**, 309 (1973); M. Gross, C. Fabre, P. Pillet and S. Haroche, Phys. Rev. Lett. **36**, 1035 (1976); A. Flusberg, T. Mossberg and S. R. Hartmann, Phys. Lett. **58A**, 373 (1976); H. M. Gibbs, Q. H. F. Vrehen and H. M. J. Hikspoors, Phys. Rev. Lett. **39**, 547 (1977).

[4] N. E. Rehler and J. E. Eberly, Phys. Rev. A **3**, 1735 (1971); R. Bonifacio, P. Schwendimann and F. Haake, Phys. Rev. **A** 4, 302, 854 (1971); R. Bonifacio and L. A. Lugiato, Phys. Rev. A **11**, 1507 (1975).

[5] Reviews on the subject can be found in: M. S. Feld and J. C. MacGillivray, in *Coherent Nonlinear Optics*, edited by M. S. Feld and V. S. Letokokv (Springer, Berlin, 1980),7; Q. H. F. Vrehen and H. M. Gibbs, in *Dissipative Systems in Quantum Optics* (Springer Topics in Current Physics 21), edited by R. Bonifacio (Springer, Berlin, 1982), 111; M. G. Benedict, A. M Ermoloaev, V. A. Malyshev, I. V. Sokolov and E. D. Trifonov, *Superradiance : Multiatomic Coherent Emission* (Institute of Physics Publishing, Bristol and Philadelphia, 1993).

[6] R. H. Dicke, Phys. Rev. **93**, 99 (1954).

[7] R. Friedberg and S. R. Hartmann, Phys. Lett. **37A**, 285 (1971).

[8] D. J. Heinzen, J. E. Thomas, and M. S. Feld, Phys. Rev. Lett. **54**, 677 (1985); J. C. MacGillivray and M. S. Feld, Phys. Rev. A **14**, 1169 (1976); D. C. Burnham and R. Y. Chiao, Phys. Rev. **188**, 667, (1969).

[9] C. Greiner, B. Boggs, and T.W. Mossberg, Phys. Rev. Lett. **85**, 3793 (2000).

[10] R. Bonifacio and L. A Lugiato, Opt. Commun. **47**, 79(1983), L. Moi, P. Goy, M. Gross, J. M. Raimond, C. Fabre, and S. Haroche, Phys. Rev. A 27, 2043 (1983).

[11] S. M. Zakharov, JETP **81**, 452 (1995).

[12] C. Greiner, B. Boggs, T. Loftus, T. Wang and T. W. Mossberg, Phys. Rev. A **60,** R2657 (1999); Y. Sun, G. M. Wang, R. L. Cone, R. W. Equall and M. J. M. Leask, Phys. Rev. B **62**, 155443 (2000).

Multitime coherent spectroscopies of complex molecules: from infrared to X-ray

S. Mukamel
Department of Chemistry, University of Rochester, Rochester. NY 14627-0216
mukamel@chem.rochester.edu

Abstract: Femtosecond visible and infrared analogues of multiple-pulse NMR techniques provide novel snapshot probes into the structure and electronic and vibrational dynamics of complex molecular assemblies such as photosynthetic antennae, proteins, and hydrogen-bonded liquids. The response of complex molecules to a series of femtosecond optical laser pulses provides a multidimensional view into molecular structure as well as a vibrational motions, interactions, and relaxations. In complete analogy with NMR techniques, spreading the spectroscopic information in more than one dimension helps resolve congested spectra, eliminates certain static broadening mechanisms and provides structural and dynamical information unavailable from one-dimensional measurements. The basic concepts underlying these generalized techniques will be surveyed.

A classical-oscillator description of these spectroscopies in terms of interacting quasiparticles (rather than transitions among global eigenstates) is developed and sets the stage for designing new pulse sequences and inverting the multidimensional signals to yield molecular structures. The folded and unfolded state of a helical heptapeptide following a T-jump and their signatures in 2D photon echo spectroscopy are simulated. A unified description of resonant multiple--pulse experiments in coupled spin-1/2 half systems in NMR spectroscopy and two-level systems in optical spectroscopy is presented.

The connection between the NMR product operator formalism and the Liouville space pathways in optical spectroscopy is established. We show how the information obtained in various strong-field two- and three-pulse NMR experiments can be extracted using weak fields with controlled phases. These results set the stage for the design of sequences of weak optical pulses that accomplish the same goals as multidimensional NMR spectroscopy. Time-resolved x-ray spectroscopies are formulated similarly in terms of nonlinear response functions and Liouville-space pathways. Simulations of frequency and wavevector resolved femtosecond resonant x-ray emission of molecular chains will be presented

Wave-particle correlations of non-classical light.

G. T. Foster†, L. A. Orozco, J. E. Reiner, W. P. Smith

Department of Physics and Astronomy, SUNYSB, Stony Brook, NY 11794-3800 USA.

H. J. Carmichael

Department of Physics, University of Oregon, Eugene, OR 97403-1274 USA.

P. R. Rice

Department of Physics, Miami University, Oxford, OH 45056 USA.

Abstract: We present our investigations of a third order correlation function of the electromagnetic field. The correlation function is subject to Schwartz inequalities and under conditions where third-order moments of the noise are negligible is the Fourier transform of the spectrum of squeezing. We present measurements and calculations of this correlation function in a strongly coupled system of cavity QED that produces non-classical light.

1 Introduction

The nature of light is a topic that has animated discussions in physics since the time of Isaac Newton. The argument of wave versus particle is resolved in quantum electrodynamics by a formalism that combines both of these aspects. The formalism is fundamentally statistical as is the case for quantum phenomena in general. It is through the statistical uncertainty, fluctuations, that the wave and particle natures of light sit self-consistently side by side. Two lines of experiments have been followed: those measuring the particle aspect of light (correlations between pairs of photon detection using the Hanbury Brown and Twiss technique [1]) and those studying the wave aspect of light (squeezing experiments that measure the fluctuation variance of the wave amplitude [2, 3]). No attempt has previously been made to draw on the particle and wave aspects together by correlating a photon detection with fluctuations of the electromagnetic wave amplitude. We have managed this, using as a source a cavity QED system. The source is known to emit nonclassical light; this has been demonstrated in both photon correlation [4, 5, 6] and squeezing measurements [7, 8, 9]. We now have a way to observe the fluctuation of the wave amplitude of light underlying these measurements by exploiting the conditioning on a photon detection to catch it as it occurs [10, 11].

2 Wave-particle correlation function

Correlation functions are an ideal tool for characterizing the electromagnetic field. Most correlation measurements emphasize even order functions of the field. (See for example Ref. [12].) We work with a third order correlation function [10, 11] of an electromagnetic field mode $\hat{b}$ (signal) that has a non-zero, steady state average $\langle\hat{b}\rangle = \lambda$. Carmichael *et al.* have generalized it to any source [10], but for the purpose of this contribution it is sufficient to limit the discussion to the detection scheme of Fig. 1. The apparatus measures the correlation between the amplitude and the intensity of the electromagnetic field. The particle produces a trigger "click" and the wave gets recorded in the photocurrent output of a balanced homodyne detector (BHD). The measured correlation is:

$$\langle : (\hat{b}^\dagger \hat{b})(0)\hat{Q}_\theta(\tau) : \rangle = \langle : (\hat{Q}_\theta(\tau) : \rangle_c, \tag{1}$$

where $\hat{Q}_\theta \equiv \frac{1}{2}(\hat{b}e^{-i\theta} + \hat{b}^\dagger e^{i\theta})$ is the quadrature amplitude selected by the local oscillator phase θ. The colons indicate normal ordering and the operator is time ordered. Eq. (1) is the measurement of the field after a photon has been detected.

Using the quantum regression theorem (see for example a similar treatment for the second order intensity correlation in Ref. [?]), we can write the correlation function, denoted by the subscript c, as the field evolution conditioned to detection of one photon.

Now normalize Eq. 1 by $\sqrt{\eta}\,|\langle \hat{b} \rangle| \langle \hat{b}^\dagger \hat{b} \rangle$, where η is the coupling efficiency into the BHD, and separate the fluctuations from the mean field by writing $\hat{b} = |\langle \hat{b} \rangle| e^{i\theta} + \Delta \hat{b}$. When the mean field has the same phase as that set for the local oscillator and for Gaussian fluctuations (third-order moments vanish) the correlation function is

$$h_\theta(\tau) = \cos\theta + 2\frac{\langle : \Delta\hat{Q}_\theta(0)\Delta\hat{Q}_\theta(\tau) : \rangle}{\lambda^2 + \langle \Delta\hat{b}^\dagger \Delta\hat{b} \rangle} + \xi(\tau), \tag{2}$$

where $\Delta\hat{Q}_\theta \equiv \frac{1}{2}(\Delta\hat{b}e^{-i\theta} + \Delta\hat{b}^\dagger e^{i\theta})$ and $\xi(\tau)$ is the residual shot noise, with correlation function

$$\overline{\xi(0)\xi(\tau)} = \frac{\Gamma}{16\eta N_s |\langle \hat{b} \rangle|^2} e^{-\Gamma\tau} \tag{3}$$

where Γ is the BHD bandwidth and N_s is the number of triggers the correlator has received from photon detections. The strength of the local oscillator is irrelevant, it cancels out of this correlation function and it only enters in the shot noise that should be averaged out to obtain a good signal to noise ratio.

A remarkable connection with non-classicality follows since the spectrum of squeezing is the cosine transform of the normalized correlation function,

$$S(\theta, \nu) = 4F \int_0^\infty d\tau \cos(2\pi\nu\tau)[\overline{h_\theta(\tau)} - 1], \tag{4}$$

where $F = 2\kappa\langle \hat{b}^\dagger \hat{b} \rangle$ is the photon flux into the correlator and $\overline{h_\theta(\tau)}$ is the average of $h_\theta(\tau)$ with respect to $\xi(\tau)$ taken with many realizations N_s of the measurement. This average is independent of the fraction of light, $1 - r$, sent to the BHD.

Notice that the measured correlation function Eq. (2) depends on the BHD efficiency, η, only in the residual shot noise term, which vanishes after averaging. The conditional measurement avoids the degradation of squeezing encountered by the standard squeezing measurement.

A related conditional detection scheme has been proposed and implemented for observing photon number states in parametric down conversion [13, 14]. That scheme does not display the efficiency independence observed here.

The signature of squeezing is a reduction in the shot noise level which is interpreted as a deamplification of the vacuum fluctuations. The wave-particle correlation function brings another interpretation. The focus is on the fluctuations of the light emitted by the source. The shot noise is irrelevant, ideally being eliminated through the conditional average. The fluctuations are nonclassical. The fluctuation intensity can be written in terms of the individual quadrature variances:

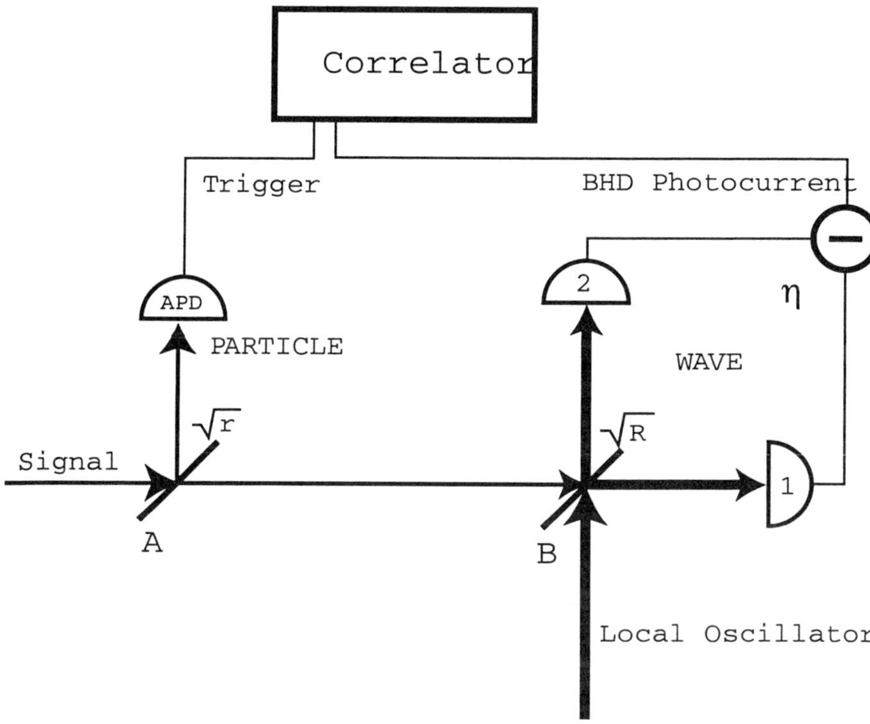

Fig. 1. Schematic of the wave-particle correlator.

$$\langle \Delta\hat{b}^\dagger \Delta\hat{b} \rangle = \langle :(\Delta\hat{Q}_\theta)^2: \rangle + \langle :(\Delta\hat{Q}_{\theta+\pi/2})^2: \rangle. \tag{5}$$

Classically, the quadrature variances are positive. It follows, from Eqs. (2) and (5) as Schwarz inequalities, that

$$0 \leq \overline{h_{0^\circ}(0)} - 1 \quad \leq \quad \frac{2}{1 + \lambda^2/\langle \Delta\hat{b}^\dagger \Delta\hat{b} \rangle}, \tag{6}$$

$$|\overline{h_{0^\circ}(\tau)} - 1| \quad \leq \quad |\overline{h_{0^\circ}(0)} - 1| \tag{7}$$

One quadrature variance is negative, however, if there is squeezing below the quantum limit, then the squeezed (unsqueezed) fluctuations violate the lower (upper) bound of (6), and fluctuations of either quadrature amplitude may violate (7).

Quantum mechanics accommodates this behavior through the reduction of the wavefunction, where triggering on a photon detection collapses the equilibrium state of the source into a conditional state which evolves back to equilibrium in a characteristic way [15]. The size of this collapse depends on the strength of the nonlinearity of the source. The source must be able to deliver more than one excitation during its characteristic fluctuations time to measure the wave-particle correlation.

3 The Cavity QED experiments

Cavity QED [16] is an area of quantum optics that has some of its origins in optical bistability [17]. Of the two components of a cavity QED system: Atoms and Electromagnetic Field, we focus our attention on the latter. Microwave Cavity QED systems have been used recently to detect a single photon nondestructively, prepare multiparticle entanglement [18, 19], and to produce photon number states and trapping states of

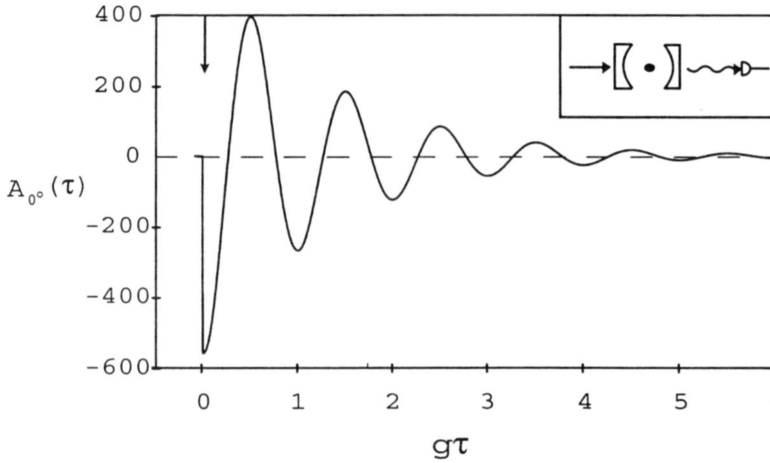

Fig. 2. Single event showing the regression to steady state of the normalized field after the detection of a single photon for cavity QED with a single atom and low excitation.

the electromagnetic field [20, 21]. Operated at optical frequencies, cavity QED systems can now trap single atoms in the electric field of the cavity when its average occupation is about one photon [22, 23, 24].

The theoretical basis of this system is the Jaynes-Cummings model [25], which treats a two-level atom at rest interacting with a single mode of the electromagnetic field. The experimental conditions that have to be treated by extensions of the theoretical model include atomic and cavity dissipation, interactions with more than one atom, an external field driving the cavity mode, atomic motion, and the spatial inhomogeneity of the light field in the cavity.

The system size and dynamics are characterized by two dimensionless numbers: The saturation photon number n_0 and the single atom cooperativity C_1. They scale the influence of a photon and the influence of an atom in cavity QED. These two numbers relate the reversible dipole coupling of a single atom with the cavity mode (g_0) with the irreversible coupling to the reservoirs through cavity (κ) and atomic polarization and inversion ($\gamma_\perp, \gamma_\|$) decays: $C_1 = g_0^2/2\kappa\gamma_\perp$ and $n_0 = 2\gamma_\perp\gamma_\|/3g_0^2$. We have cavities that operate in the strong coupling regime of cavity QED ($n_0 < 1$, and $C_1 > 1$). Strictly speaking, the coupling constant (g_0) is spatially dependent with a maximal value at the center of the Gaussian transverse mode and in an antinode of the standing wave. Moving atoms may be described by an effective coupling constant.

We study optical field dynamics that happens in time scales of the order of tens of nanoseconds. This is a distinctive feature of cavity QED; in other areas of Quantum Optics the dynamics evolve on a much shorter time scale.

The equilibrium state of the cavity QED system before a photodetection is:

$$|\psi\rangle = |G\rangle\{|0\rangle + \lambda|1\rangle + ...\} \tag{8}$$

where $|G\rangle$ represents all atoms in their ground state, $\lambda \ll 1$ and depends on the input driving field ϵ. After the photodetection occurs the new conditioned state is

$$|\psi\rangle = |G\rangle\{|0\rangle + \lambda(1 + \zeta f(t))|1\rangle + ...\} \tag{9}$$

This is different from the initial state because ζ is not zero, and $f(t)$ has to decay (oscillating at the vacuum Rabi frequency) to zero over time as the system re-equilibrates [15].

As an example of this conditional evolution Fig. 2 shows a calculation of the field amplitude A normalized to the steady state in a cavity QED system of similar parameters to our experiment. This is the regression to steady state following the detection of a photon escaping through the cavity mode.

Our cavity QED apparatus consists of four key components: the cavity, the atomic beam, an excitation laser, and the detector system. We drive the cavity system from one side, and observe the transmitted light out the other. Although we have implemented many cavities, an example of the three rates that characterize our cavity QED system are $(g_0, \kappa, \gamma_\perp)/2\pi = (12, 8, 3)$ MHz.

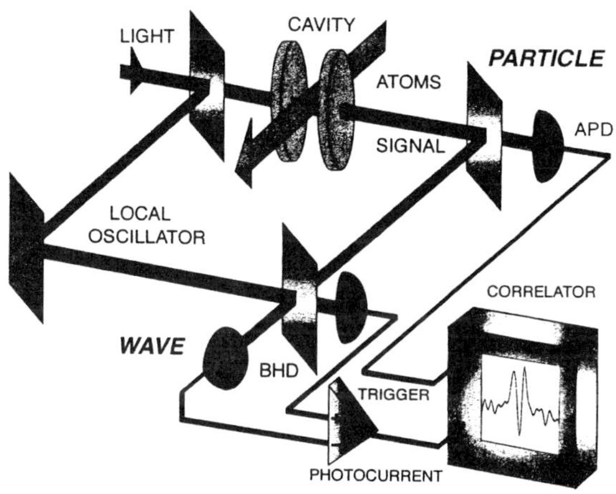

Fig. 3. Wave-particle correlator with cavity QED system.

The optical cavity is formed by two high reflectivity curved mirrors. The specific parameters for the cavities employed in these experiments are: waist of the TEM_{00} mode $w_0 = 21\mu m$, length $l = 410\mu m$, input transmission mirror 10 ppm, output transmission mirror 285 ppm. Piezo-electric transducers attached to the mirror holders allow us to control, through feedback, the mirror spacing.

An effusive oven produces a thermal beam of Rb atoms in a chamber pumped by a diffusion pump. Final collimation is provided by a 50 μm slit in the cavity holder. The oven is heated to $\approx$ 430 K. A computer controlled feedback system maintains the temperature of the oven within ± 0.1 K. The 1 mm wide oven opening and the aperture before the cavity form a beam with an angular spread of 3.4 mrads. A liquid nitrogen cooled Cu sleeve surrounds the cavity to reduce background atomic vapor.

A modified Coherent 899-01 titanium sapphire (Ti:Sapph) laser provides light for the signal beam and auxiliary beams for laser frequency locking, cavity locking, and optical pumping. The signal and lock beams are on resonance with the $5S_{1/2}, F = 3, m_F = 3 \rightarrow 5P_{3/2}, F = 4, m_F = 4$ transition of ^{85}Rb at 780 nm. A double passed acousto-optic modulator in the laser lock allows us to adjust its frequency around the Rb resonance.

An optical pumping beam of diameter 2 mm, parallel to the cavity mode, intersects the atomic beam 1 cm before the atoms enter the cavity. A 2.5 Gauss uniform magnetic field applied along the axis of the cavity provides a quantization axis such that the circular polarization of the optical pumping beam takes all the atoms into the ground state magnetic sublevel $F = 3, m_F = 3$.

An electro-optic modulator generates FM sidebands on the lock beam. The reflected lock beam is used to hold the cavity on resonance. During data collection, we send the beam through a chopper wheel which alternately passes the lock beam and opens the path from the cavity to the detectors at about 1.1 kHz. The lock and signal beams have orthogonal polarizations before the cavity. The circular polarization of the signal beam is in the same sense as the optical pumping beam. On the output side of the cavity, the beam passes through a $\lambda/4$ plate before a polarizing beam splitter sends the lock beam to a detector. The signal beam is directed to the wave-particle correlator.

Fig. 3a shows our wave-particle correlator. Light of wavelength 780 nm from theTi:Sapph laser enters a Mach-Zehnder interferometer, driving the cavity QED system in one arm and providing a local oscillator (LO) for the balanced homodyne detector (BHD) in the other. The BHD has strong common mode suppression and permits the phase sensitive measurement of the field quadrature to be made. Part of the signal escaping the cavity creates photodetections at the avalanche photodiode (APD), EG&G SPCM-AQ-151, which trigger a digital oscilloscope, LeCroy 9354A, to record in coincidence the BHD photocurrent. 15% or more of the signal is directed to the particle detector, represented in the figure by a single APD. The other 85% of the signal goes to the wave detector, provided by the BHD, where it is mixed with the LO and detected with fast photodiodes, EOT2030. We lock the relative phase between the signal and LO by feeding an electrical signal to a piezo electric transducer in the Mach Zehnder. To derive the error for this feedback loop, we use the interference fringes from an auxiliary diode laser locked with FM sidebands to an I_2 resonance around 640 nm. The photocurrent is AC amplified by 65 dB, 70 MHz low pass filtered, and sampled and averaged with the fast digital oscilloscope (LeCroy 9354A) at 2 Gs/s.

The averaging and processing is done in the oscilloscope. It takes about five minutes of data collection to acquire 60,000 averages. The maximum number of averages is limited by the 8 bit digitizer of the oscilloscope. If the processing of the information could be done faster we would be able to acquire a correlation much faster, since the rate of "clicks" is tens of kHz.

Operating at higher excitation we obtain the results of Fig. 4. There is still a strong oscillation of the conditional field. Inequality (7) is violated with no violation of the lower bound of (6). The amplitude of the latter is reduced due to spontaneous emission [26], and the background is likely caused by atomic beam fluctuations in combination with spontaneous emission. The spectrum of squeezing [Fig. 4(b)] separates the nonclassical fluctuations from the classical fluctuations. To obtain it, we symmetrize the scaled signal in Fig. 4(a) and take the Fourier transform as specified in Eq. (4). The spectrum is negative in the vicinity of the vacuum Rabi frequency, evidencing, in frequency space, the time inequality violations. We measure $\sim$ 5% reduction of the noise below the standard quantum limit. This determination is independent of the coupling efficiency into the BHD.

4 Conclusions

We have measured the wave-particle correlation function, $h(\tau)$, in our cavity QED system. It records the conditional time evolution of the field of a fraction of a photon [11]. Detection of a photon prepares a state of well-defined phase that evolves back to equilibrium via a damped vacuum Rabi oscillation. We record the regression of the field amplitude. The correlation function is nonclassical and provides an efficiency independent path to the spectrum of squeezing.

The notable feature of our measurement is that according to the reduction postulate, the detection of a photon (particle aspect of light) conditionally selects a field amplitude (wave aspect of light) oscillating in time with an anomalous phase. Our experiment catches the field fluctuations as they occur.

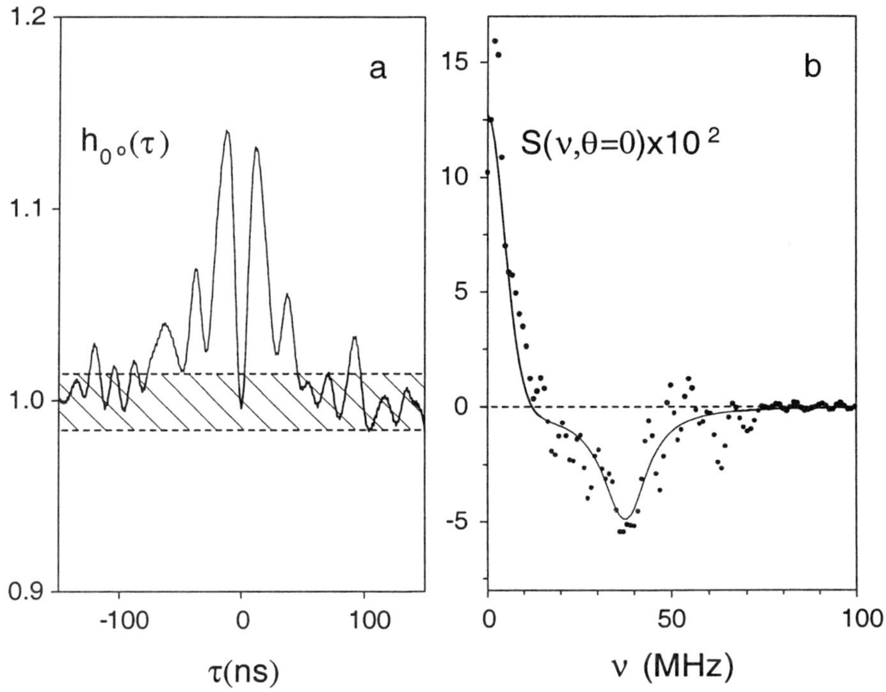

Fig. 4. a) $h_{0°}(\tau)$ shows *nonclassical* oscillatory behavior, and b) $S(\nu, \theta = 0°)$ showing squeezing below the standard quantum limit. The shaded region in a is classically allowed. The dashed line in b is the boundary between classical (above) and non-classical (below).

5 Acknowledgments

This work has been supported by the National Science Foundation through grants PHY 0099576 and PHY 0097691.

†Present address: Dept. Physics, Yale University, New Haven CT 06520-8120 USA.

References and links

1. R. Hanbury-Brown and R. Q. Twiss, "Correlation between photons in two coherent beams of light." Nature **177**, 27 (1956).
2. *Squeezed States of the Electromagnetic Field*, edited by H. J. Kimble and D. F. Walls, special issue of J. Opt. Soc. Am. B **4**, 1450 (1987).
3. *Squeezed Light*, edited by R. Loudon and P. L. Knight, special issue of J. Mod. Opt. **34**, 709 (1987).
4. G. Rempe, R. J. Thompson, R. J. Brecha, W. D. Lee, and H. J. Kimble, "Optical bistability and photon statistics in cavity quantum electrodynamics." Phys. Rev. Lett. **67**, 1727 (1991).
5. S. L. Mielke, G. T. Foster, and L. A. Orozco "Non-classical Intensity Correlations in Cavity QED." Phys. Rev. Lett. **80**, 3948 (1998).
6. G. T. Foster, S. L. Mielke, and L. A. Orozco "Intensity correlations in cavity QED." Phys. Rev. A **61**, 53821 (2000).
7. M. G. Raizen, L. A. Orozco, M. Xiao, M., T. L. Boyd, and H. J. Kimble, " Squeezed-state generation by the normal modes of a coupled system." Phys. Rev. Lett. **59**, 198 (1987).
8. L. A. Orozco, M. G. Raizen, Min Xiao, R. J. Brecha, and H. J. Kimble, "Squeezed State Generation in Optical Bistability." J. Opt. Soc. Am. B **4**, 1490 (1987).
9. D. M. Hope, H. A. Bachor, D. E. McClelland, and A. Stevenson, "The Atom-Cavity System as a Generator of Quadrature Squeezed States." Appl. Phys. B **55**, 210 (1992).

10. H. J. Carmichael, H. Castro Beltran, G. T. Foster, and L. A. Orozco, "Giant Violations of Classical Inequalities Through Conditional Homodyne Detection of the Quadrature Amplitudes of Light." Phys. Rev. Lett. **85**, 1855 (2000).
11. G. T. Foster, L. A. Orozco, H. M. Castro-Beltran, and H. J. Carmichael, "Quantum State Reduction and Conditional Time Evolution of Wave-Particle Correlations." Phys. Rev. Lett. **85**, 3149 (2000).
12. L. Mandel, E. Wolf, *Optical Coherence and Quantum Optics*, (Cambridge University Press, New York, 1995).
13. B. Yurke and D. Stoler, "Measurement of Amplitude Probability-Distributions for Photon Number Operator Eigenstates." Phys. Rev. A **36**, 1955 (1987).
14. A. I. Lvovsky, H. Hansen, T. Aichele, O. Benson, J. Mlynek, and S. Schiller, "Quantum State Reconstruction of the Single-Photon Fock State." Phys. Rev. Lett. **87**, 050402 (2001).
15. H. J. Carmichael, R. J. Brecha, and P. R. Rice, "Quantum interference and collapse of the wavefunction in cavity QED." Opt. Comm. **82**, 73 (1991).
16. P. Berman, ed., *Cavity Quantum Electrodynamics*, Supplement 2 of Advances in Atomic, Molecular and Optical Physics series (Academic Press, Boston, 1994).
17. L. A. Lugiato, "Theory of Optical Bistability", in *Progress in Optics*, E. Wolf, ed. (North-Holland, Amsterdam, 1984), Vol. XXI, p. 69-216.
18. G. Nogues, A. Rauschenbeutel, S. Osnaghi, M. Brune, J. M. Raimond, and S. Haroche "Seeing a single photon without destroying it." Nature **400**, 239 (1999).
19. A. Rauschenbeutel, G. Nogues, S. Osnaghi, P. Bertet, M. Brune, J. M. Raimond, and S. Haroche "Step-by-step engineered multiparticle entanglement." Science **288**, 2024 (2000).
20. B. T. H. Varcoe, S. Brattke, M. Weidinger, and H. Walther, "Preparing pure photon number states of the radiation field." Nature **403**, 743 (2000).
21. M. Weidinger, B. T. H. Varcoe, R. Heerlein, and H. Walther, "Trapping states in the micromaser." Phys. Rev. Lett. **82**, 3795 (1999).
22. C. J. Hood, R. W. Lynn, A. C. Doherty, A. S. Parkins, and H. J. Kimble, "The atom-cavity microscope: single atoms bound in orbit by single photons." Science **287**, 1447 (2000).
23. P. W. H. Pinkse, T. Fischer, P. Maunz, and G. Rempe, "Trapping an atom with single photons." Nature **404**, 365 (2000).
24. A. C. Doherty, T. W. Lynn, C. J. Hood, and H. J. Kimble, "Trapping of Single Atoms with Single Photons in Cavity QED." Phys. Rev. A **63**, 013401 (2001).
25. E. T. Jaynes, F. W. Cummings, "Comparison of Quantum and Semiclassical radiation Theories with Application to the Beam Maser." Proc. IEEE **51**, 89 (1963).
26. J. E. Reiner, W. P. Smith, L. A. Orozco, H. J. Carmichael, P. R. Rice, "Time evolution and squeezing of the field amplitude in cavity QED." To appear in J. Opt. Soc. Am. B (2001).

Real-Time Control of Atomic Motion Using Feedback

J. L. Hanssen, V. Milner, T. P. Meyrath, and M. G. Raizen

Center for Nonlinear Dynamics and Department of Physics, The University of Texas at Austin, Austin, Texas 78712-1081

raizen@physics.utexas.edu

Abstract: We demonstrate real-time feedback control of atomic center of mass motion in a cloud of laser cooled cesium atoms. Using the velocity measurement technique of recoil-induced resonances, the average velocity of the atomic cloud is obtained. This information is used to give a correction from a moving optical lattice. We are able to bring the cloud of atoms to a stop with a controllable amount of heating.

OCIS codes: (020.0020) Atomic and molecular physics; (070.4560) Optical data processing

1 Introduction

Manipulating ensembles of particles with feedback has evolved significantly since its use to increase the luminosity of antiproton beams [1]. This has led to proposals for cooling neutral atoms [2] and ions [3], as well as manipulation of wavepackets with ultrafast laser pulses [4]. The basic idea for feedback control of a physical system involves several steps. First, a measurement is made which probes the state of the system. Second, a correction to the system state is performed based on the result of the measurement. These together constitute feedback which may be used to control the system. The measurement as well as correction method vary among the various proposals and experiments in the field. In an experiment where the state of a system is to be controlled, it is desirable for the measurement to be as nondestructive as possible. Some experimental work [5] has involved the use of a high finesse optical cavity to monitor the position of an atom in real-time and to trigger on a far-off-resonance dipole force trap (FORT) to confine the atom to a cavity mode. Other recent work [6] has involved making position measurements of atoms in an optical lattice and feedback to alter the oscillations of the atoms within the lattice. In this demonstration of real-time feedback control, we use the method of recoil-induced resonances (RIR) [7, 8] to make a center-of-mass momentum measurement of a cloud of atoms and a moving optical lattice for changing the state of the system.

This experimental demonstration of real-time feedback control is a simplified single iteration model of stochastic cooling as discussed in Ref. [2]. A stochastic cooling setup entails a measurement and feedback determined correction as discussed here, but also includes a mixing process whereby the particles are allowed to rethermalize. In such a case, many iterations may be used to damp momentum fluctuations for instance. In this demonstration, a measurement and correction are applied but without subsequent state mixing. This is not stochastic cooling, but rather a single iteration of a possible feedback sequence. Instead of damping out the momentum fluctuations of the ensemble, the average momentum of the entire group is damped. A measurement of the velocity distribution is made in order to obtain the center of mass velocity of the ensemble of particles. The information is used in real-time to provide feedback correction to the system in the form of a time dependent optical potential. This potential is a moving 1D optical lattice which traps the atoms and slows them to rest in the laboratory frame.

The measurement-correction feedback iteration is done with relatively minor heating of the ensemble.

2 Experimental Setup

Our experiments are performed using cesium atoms from a standard vapor-cell magneto-optic trap (MOT). The MOT apparatus used is described in detail in Ref. [9]. We trap approximately 10^6 cesium atoms with Gaussian distributions in position and momentum with widths of $\sigma_y = 180\ \mu$m and $\sigma_p = 8.5\ \hbar k$, respectively. Here, $k = 2\pi/\lambda$ is the wavenumber, with wavelength $\lambda = 852$ nm. The experiment starts when the MOT is turned off and the atomic cloud is allowed to fall freely under the influence of gravity for 10 ms, during which it expands to a width $\sigma_y = 350\ \mu$m. After these 10 ms, the cloud is subjected to a velocity measurement through the method of recoil-induced resonances (RIR) [7, 8].

The RIR measurement involves two beams with frequency difference $2\delta\nu$ intersecting in the atomic cloud at angle θ. The beams are both detuned from atomic resonance by $\Delta_{\rm L}$ ($\cong$ 6 GHz in our case). Stimulated Raman transitions between motional states are driven which result in coherent scattering of photons from one beam to the other. The resonance condition [10] for these transitions is

$$2\delta\nu = \frac{2v_y}{\lambda}\sin(\theta/2) \tag{1}$$

where v_y is the velocity in the direction of momentum change between the beams which is vertical in the present work. By choosing a value of $2\delta\nu$, a certain velocity class of the ensemble undergoes stimulated Raman transitions making the scattering rate dependent on the population in the different motional states. Sweeping $2\delta\nu$, therefore, may be used to probe the system's velocity distribution. If done linearly at the appropriate rate, the power in the beams will change in proportion to the derivative of the atomic momentum distribution. This method provides a measure of the velocity distribution but is of limited sensitivity due to mechanical and electrical noise. The sensitivity of this measurement can be greatly enhanced by the application of frequency-modulation (FM) spectroscopy techniques [10].

The experimental setup used is shown in Fig. 1 and is similar to that which is described in Ref. [10]. The two RIR beams start as a single beam generated by a Ti:Sapphire laser which produces approximately 500 mW of light detuned 6 GHz red of the $6S_{\frac{1}{2}}(F = 4) \rightarrow 6P_{\frac{3}{2}}(F' = 5)$ transition near 852 nm. This beam is sent through an acousto-optic modulator (AOM) operating at 80 MHz. This modulator (AOM1) is used to split the original beam into a pair which have a frequency difference of 80 MHz. This pair is to become the pump and probe beams.

The zeroth order beam from AOM1 is double passed through AOM2 operating at 54 MHz+$\delta\nu$ where $\delta\nu$ is a variable frequency. The double pass allows the driving frequency of the AOM to be swept without steering the beam. It is then spatially filtered to clean the beam profile and aligned to hit the atomic cloud after a 10 ms free fall. The beam has a frequency that is shifted 108 MHz+$2\delta\nu$ relative to the original beam, and has a Gaussian intensity profile with a $1/e^2$ waist of 750 μm. This is the pump beam.

The first order beam from AOM1 is double passed through an electro-optic phase modulator (EOM) operating at 28 MHz. The EOM is used to put sidebands on the carrier at $\pm$28MHz . We drive the EOM to produce sidebands 2.5 dB below the carrier. The purpose of double passing the EOM is for reduction of residual amplitude modulation created by the EOM as the birefringence of the lithium niobate crystal drifts in time resulting in polarization modulation of the output beam at the driving frequency. Without compensation, the parasitic change in polarization results in a change

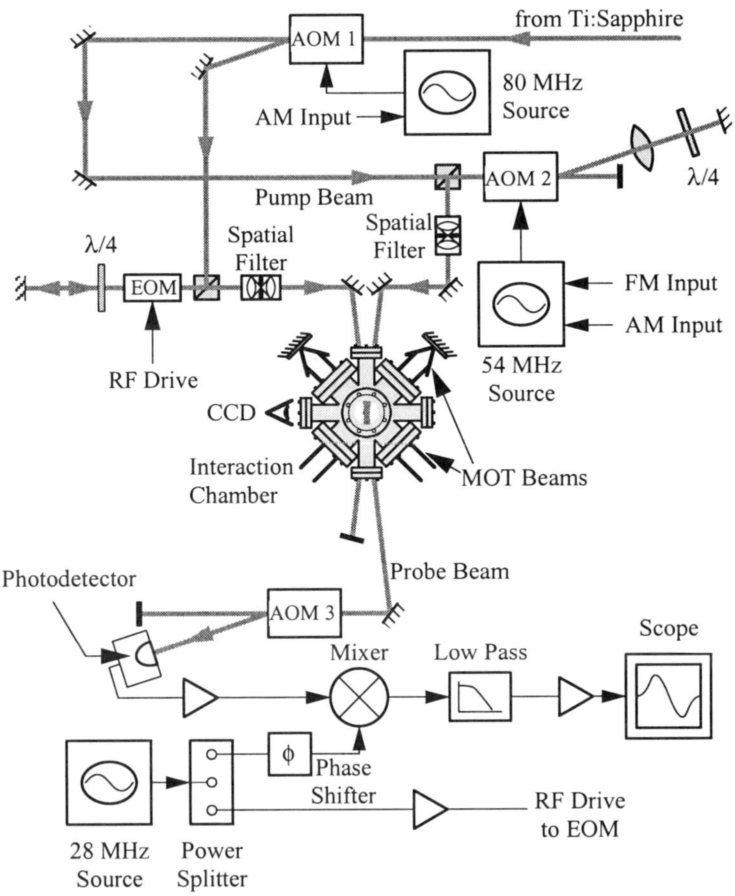

Fig. 1. Experimental setup

in beam amplitude upon passing through the polarizing beam splitter cube that follows the EOM. The double pass setup with the quarter-wave plate allows us to correct for the drifts of EOM birefringence and reduce this effect. Following the EOM double pass, the beam is spatially filtered and aligned to intersect the atomic cloud where it has a $1/e^2$ waist of 750 μm. Since the beam was originally derived as the first order of AOM1, the carrier is 80 MHz above the original Ti:Sapphire beam with sidebands at 53 MHz and 108 MHz. This is the probe beam.

The stimulated Raman transitions between motional states result from the interaction of the pump beam (108 MHz+$2\delta\nu$ relative to the original Ti:sapphire beam) and the upper sideband of the probe beam (at 108 MHz) which intersect at an angle $\theta = 4.5°$. The frequency difference here is $2\delta\nu$, which is experimentally controlled. Depending on the frequency difference, Raman transitions are driven as the resonance condition is met, as in Eq. (1). A frequency difference of $2\delta\nu = 0$ Hz, for example, corresponds to rest in the laboratory frame, whereas $2\delta\nu = \pm 320$ Hz corresponds to velocity classes at $\pm 1\ v_r$, where v_r is the recoil velocity for cesium. For a typical RIR measurement, $\delta\nu$ is linearly swept over a range that includes the entire velocity distribution. The best sweep rate for an optimal signal depends on the power in the beams as well as the detuning from atomic resonance [10]. Typically, the pump and probe beams have powers of 35 mW and 15 mW, respectively, with $\delta\nu$ swept at a rate of 74.3 MHz/s.

After the beams interact with the atoms, they exit the chamber where the pump beam is blocked. The probe beam passes through AOM3 after which the first order is detected by a photodiode. This additional AOM is used as a fast shutter to protect the photodetector from receiving intensities above the damage threshold during the correction phase which will be described shortly. Prior to interaction with the atoms, the FM sidebands on the probe beam were equal in intensity. The RIR interaction with the atoms results in the scattering of photons from the pump beam into the upper sideband of the probe beam. The imbalance between the FM sidebands results in the amplitude modulation measured by the photodetector. The output of the detector is, therefore, a signal at the sideband frequency (28 MHz) with an amplitude that is proportional to the intensity mismatch between the sidebands (i.e. to the scattering rate that is proportional to the derivative of the velocity distribution).

The output of the photodetector is amplified and mixed with the original 28 MHz signal appropriately phase shifted which is subsequently low-pass filtered to remove the high frequency component and is then amplified. The result is a DC level with the dispersive RIR signal on top of it. The DC level fluctuates to some degree due to fluctuations of the residual amplitude modulation. A typical RIR scan is shown as plot (a) in the timing diagram in Fig. 2.

The RIR measurement determines the average momentum of the cloud of atoms as a zero-crossing of the RIR curve, and this information is used in a feedback scheme to stop the atoms. The intersecting beams create a standing wave or one dimensional lattice that is moving at a velocity of

$$v_{\mathrm{sw}}^{\mathrm{lab}} = 2\delta\nu \frac{\lambda}{2\sin(\theta/2)} \tag{2}$$

with respect to the laboratory and

$$v_{\mathrm{sw}}^{\mathrm{atoms}} = 2\delta\nu \frac{\lambda}{2\sin(\theta/2)} - \frac{\overline{p}}{m} \tag{3}$$

with respect to the atoms. Assuming that the atoms are not initially at rest in the laboratory frame, a linear frequency sweep uniformly decelerates the lattice until it comes to a stop with respect to the atoms. Beyond which, the lattice accelerates in the opposite direction until it comes to a stop in the laboratory frame. The zero crossing of the RIR lineshape corresponds to when the lattice is stationary with respect to the atoms. By turning up the intensity of the beams at the zero crossing, deep potential wells are created that drag the atoms along with them as the standing wave comes to rest in the laboratory frame. This procedure of measurement and correction thereby stops the cloud of atoms in the laboratory frame.

To accomplish this, the RIR signal from the amplifier is fed into an analog electronic trigger circuit. The trigger circuit measures the DC level and subtracts it from the total signal. This leaves only the dispersive RIR lineshape which is compared to a reference voltage level. A trigger signal is generated once the signal passes the threshold level (see plot (a) in Fig.2). Note that the trigger pulse actually comes prior to the zero crossing of the dispersive signal. To compensate for this offset, appropriate delays are added to the electronics that are triggered off this pulse. These delays are typically 20μs and are not visible on the timing diagram in Fig. 2. The output of this comparator trigger is then used to initiate the sequence that stops the atoms. A timing diagram of this sequence is shown in Fig. 2.

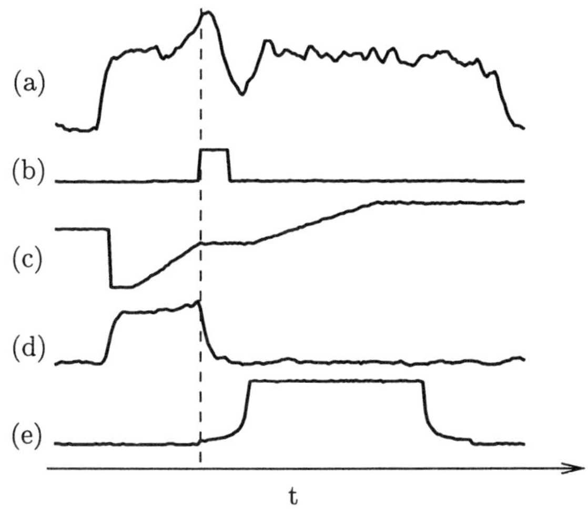

Fig. 2. Timing diagram of voltage signals involved in feedback scheme. The vertical axis is arbitrarily scaled for each signal.

The first curve (a) in Fig. 2 is the untriggered RIR signal shown for reference. It is proportional to the derivative of the momentum distribution. Curve (b) shows the generated trigger pulse, curve (c) shows the linear frequency sweep which changes slope after the trigger for the reason explained below, curve (d) is an RIR signal that was used in a trigger sequence, and curve (e) shows the control signal for the correction which is achieved by increasing the intensity of both RIR beams. The combination of signals (c) and (e) brings the atoms to a stop in the laboratory frame as described above. During the high intensity phase (after the trigger), AOM3 is turned off (via an RF switch) to protect the photodetector. This is seen in curve (d) which is the triggered RIR signal. At the point of trigger on the curve, the light to the photodetector is turned off and the signal vanishes (compare with curve (a)).

The increased intensity of the RIR beams provides the correction. The intensity in each beam increases according to $I(t) = I(0)/(1+\Gamma t)^2$ [11]. This is an adiabatic increase to avoid heating the atoms during this process due to a sudden momentum kick. In the experiment we used $\Gamma = 44.4$ kHz which disturbed the sample very little. Turning on the lattice more quickly was observed to heat the cloud significantly. On the other hand, size constraints limit the amount of time allowed for the adiabatic turn on. The beam size and cloud size are matched for an optimal RIR signal. Therefore, if the intensity is not raised quickly, the atoms will move out of the beam overlap region, and will not be caught in the moving lattice. The initial powers in the pump and probe beams were 35mW and 15mW, and were adiabatically increased to 75mW and 70mW, respectively. Curve (e) in Fig. 2 is the amplitude modulation signal to AOM1 and AOM2 (with different heights for each) that provides this increase to higher intensity.

The frequency sweep is also altered after the trigger pulse. At first, the linear frequency sweep is stopped during the adiabatic turn on. Ideally, the lattice velocity should follow the free fall velocity of the atoms during the adiabatic turn on, but that is difficult due to signal delays (especially due to the low pass filter), therefore we simply hold the lattice at its current speed. Secondly, the acceleration of the lattice after the adiabatic turn-on (proportional to the slope of the frequency) is changed. In an accelerating lattice, atoms will undergo Landau-Zener tunneling [12], escaping the lattice, and will not be stopped. The constant acceleration is decreased to avoid this tunneling, but kept high enough so that atoms would not exit the beam overlap before being stopped. The initial lattice acceleration for the RIR scan was 810 m/s^2 and was changed to 150 m/s^2 for slowing.

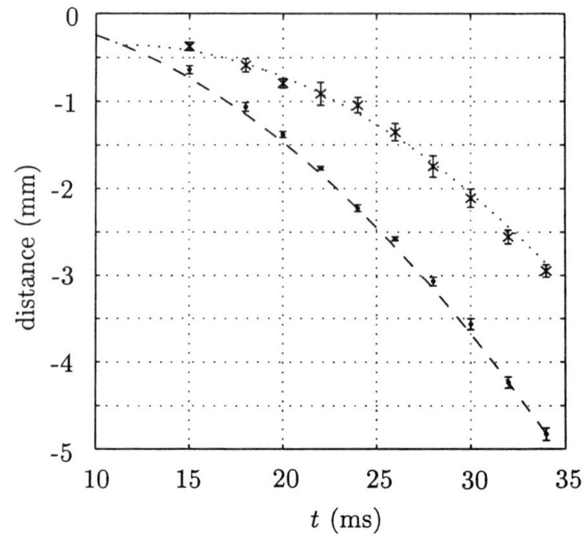

Fig. 3. Center of mass position of a cloud of atomic cesium falling. The data on the lower curve (•) is falling under gravity with no correction. The data on the upper curve (×) is falling with a corrective slow-down at10 ms.

3 Results

The primary result of this experiment is shown in Fig. 3. The data points on the lower curve (•) correspond to the initial cloud of atoms falling under gravity with no corrective slow-down. These are averaged over five free fall sequences. The dashed curve is a ballistic trajectory of the MOT falling under the influence of gravity.

The data points on the upper curve (×) correspond to atoms that were stopped and then allowed to fall again averaged over ten correction sequences. For the first 10 ms, the atoms follow the same free fall curve as described previously. The RIR and slow-down sequence starts at 10 ms after the initial release from the MOT. The sequence stops the falling atoms and then releases them to fall again approximately 1.5 ms later. The dotted curve is a ballistic trajectory with an initial momentum of 0 $\hbar k$ at 11.5 ms. There is good agreement between the data and the idealized trajectory.

Ideally, the trigger pulse turns on the deep well lattice at a time when both the average velocity of the atoms and the lattice are the same. In this case, we achieve stopping efficiency of approximately 80% of the atoms. Perfect stopping efficiency is not possible in our system due to the lower intensities in the wings of the beams creating smaller potential wells that can not slow the atoms. Another source of lower efficiency is noise on the RIR signal which causes the trigger time to vary. This leads to a velocity mismatch between the atoms and the lattice. Typically, the noise causes triggering within a window of 7 recoil velocities about the ideal trigger point. This problem is observed to result in loading efficiencies into the lattice as low as 50%.

To measure the heating effects of this method we plot the width of the cloud σ_y with time in Fig. 4. From this fit, we determine the momentum width in $\sigma_y^2(t) = \sigma_y^2(0) + t^2\sigma_p^2/M^2$. Before the correction sequence the momentum distribution has a width of $\sigma_p = 8.5\ \hbar k$ and after $\sigma_p = 12.2\ \hbar k$.

Heating results from various sources including spontaneous emission and coherent processes involved in the RIR measurement. The beams operate at a detuning of 6 GHz

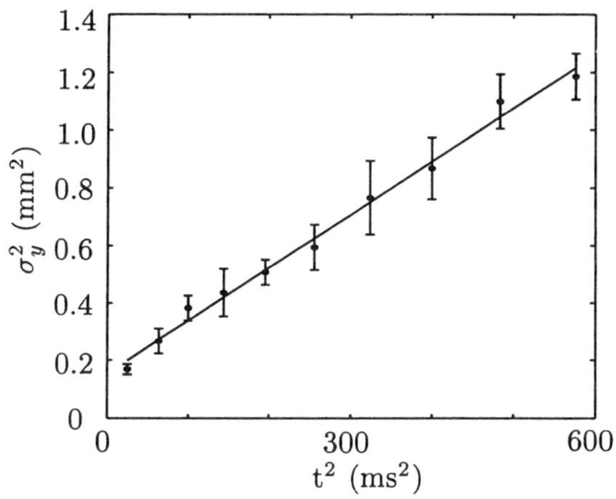

Fig. 4. Temperature measurement of the cloud after stopping showing the cloud's expansion as a function of time.

red from the atomic transition which allows for a reasonably small rate of spontaneous emission. We measure heating to about $\sigma_p = 9\ \hbar k$ from this effect alone. A larger detuning may be used to decrease the spontaneous emission at the expense of a lower well depth for the corrective slow-down.

The RIR measurement is only approximately nondestructive. There is an increase in the momentum width due to the coherent scattering of photons [10]. In our system, the momentum distribution after an RIR measurement when not followed by a corrective slow-down has a width of $\sigma_p = 16\ \hbar k$. In the correction scheme, this heating effect is not fully present since the RIR scan is halted after the trigger which takes place before the halfway point in the scan. The heating effect of the coherent scattering may be reduced by decreasing the scattering rate. This may be accomplished by reducing the power in the beams and/or increasing the detuning. Of course, this is at the expense of the signal for the obvious reason that the signal is smaller if fewer photons are scattered. Fundamentally, the measurement can not be fully nondestructive. Ultimately, the degree of nondestructiveness is determined by the threshold of the RIR measurement, which ideally is dictated by the shot-noise limit.

4 Conclusions

The basic result of this proof-of-principle experiment is that real-time measurement and correction feedback are possible with a controllable amount of perturbation. The center of mass motion of the atoms was controlled in real-time with minimal disturbance to the sample. As demonstrated, this method does not rely on any *a priori* information about atomic motion. As long as the average velocity of the cloud is within the measurement range of the RIR scan, the scheme will work on any unknown average velocity.

5 Acknowledgements

This work was supported by the National Science Foundation, the R. A. Welch Foundation, the U.S.-Israel Binational Foundation, the Sid W. Richardson Foundation, and the Texas Advanced Research Program. T.P.M. acknowledges the support from a National Science Foundation Graduate Research Fellowship. The authors would like to thank B. Gutierrez-Medina, W. H. Oskay, and D. A. Steck for helpful discussions and suggestions.

References and links

1. S. von der Meer "Stochastic Cooling and the Accumulation of Antiprotons," Rev. Mod. Phys. **57**, 689-697 (1985).
2. M. G. Raizen, J. Koga, B. Sundaram, Y. Kishimoto, H. Takuma, and T. Tajima "Stochastic Cooling of Atoms Using Lasers," Phys. Rev. A **58**, 4757-4760 (1998).
3. G. Zhang, J. Shen, J. Dai, and H. Zhang "Cooling Charged Particles in a Paul Trap by Feedback Control," Phys. Rev. A **60**, 704-707 (1999).
4. T. C. Weinacht, J. Ahn, and P. H. Bucksbaum "Controlling the Shape of a Quantum Wavefunction," Nature (London) **397**, 233-235 (1999).
5. J. Ye, D. W. Vernooy, and H. J. Kimble "Trapping of Single Atoms in Cavity QED," Phys. Rev. Lett. **83**. 4987-4990 (1999).
6. N. V. Morrow, S. K. Dutta, and G. Raithel "Feedback Control of Atomic Motion in an Optical Lattice," Submitted for publication (2001).
7. J. Guo, P. R. Berman, B. Dubetsky, and G. Grynberg "Recoil-Induced Resonances in Nonlinear Spectroscopy," Phys. Rev. A **46**, 1426-1437 (1992).
8. J. Guo and P. R. Berman "Recoil-Induced Resonances in Pump-Probe Spectroscopy Including Effects of Level Degeneracy," Phys. Rev. A **47**, 4128-4134 (1993).
9. D. A. Steck, V. Milner, W. H. Oskay, M. G. Raizen "Quantitative Study of Amplitude Noise Effects on Dynamical Localization," Phys. Rev. E **62**, 3461-3475 (2000).
10. M.C. Fischer, A. M. Dudarev, B. Gutierrez-Medina and M. G. Raizen "FM Spectroscopy in Recoil-Induced Resonances," J. Opt. B: Quantum Semiclass. Opt. **3**, 279-287 (2001).
11. A. Kastberg et. al. "Adiabatic Cooling of Cesium to 700nK in an Optical Lattice," Phys. Rev. Lett. **74**, 1542-1545 (1995).
12. C. F. Bharucha, K. W. Madison, P. R. Morrow, S. R. Wilkinson, Bala Sundaram, and M. G. Raizen "Observation of Atomic Tunneling from an Accelerating Optical Potential," Phys. Rev. A **55**, R857-R860 (1997).

Single atoms and single photons in cavity quantum electrodynamics

G. Rempe, T. Fischer, M. Hennrich, A. Kuhn,
T. Legero, P. Maunz, P.W.H. Pinkse, T. Puppe
Max-Planck-Institut für Quantenoptik, Hans-Kopfermann-Str. 1, D-85748 Garching, Germany
Tel: +49 (0)89 32905 701, Fax: +49 (0)89 32905 311, E-mail: gerhard.rempe@mpq.mpg.de

Abstract: We review some recent work performed with single moving atoms strongly coupled to high-finesse optical cavities, emphasizing cavity-mediated light forces and vacuum-stimulated generation of single photons.

1. Introduction

Recent years have witnessed remarkable progress in the experimental realization of intrinsically quantum mechanical systems. An example in cavity quantum electrodynamics is a single atom interacting with a single mode of an optical cavity. Experiments can now be performed in the so-called strong-coupling regime, where a single quantum of energy is exchanged between the atom and the cavity many times before escaping into the environment. Two conditions must be fulfilled to achieve such a novel situation: First, the characteristic (vacuum Rabi) oscillation frequency as determined by the atom-cavity coupling constant, g, must be large. This requires a large electric field per photon and, hence, a cavity with a small mode volume. Second, the energy loss rate must be small. Two decay channels have to be considered here: one is spontaneous emission of the atom into the free-space continuum of electromagnetic modes not covered by the cavity mode. It gives rise to a decay of the atomic dipole at rate γ, equal to half the spontaneous line width for a radiatively broadened atomic transition. This decay channel is hardly affected in a Fabry-Perot type cavity supporting a mode with a small solid angle. The other loss channel is the escape of a photon from the cavity. Both the mirror transmission and the mirror absorption lead to a decay of the cavity field at rate κ. To decrease this decay rate, low-loss mirrors must be used. As the mirror separation should be small, ultra-high reflectivity mirrors are an essential prerequisite for experimental research in the regime of strong coupling.

It is a remarkable feature of the strong-coupling regime, that the presence of an atom changes the cavity transmission dramatically. This effect allows one to detect a single atom inside the cavity with high temporal resolution [1, 2]. Moreover, the light force from a single photon in the cavity can significantly influence the motion of a cold atom, provided the kinetic energy of the atom is smaller than the energy associated with the oscillatory energy exchange between the atom and the cavity. Indeed, recent experiments have shown that it is possible to channel single atoms through the nodes and antinodes of a standing wave cavity field containing much less than one photon on average [3, 4]. A single atom could even be trapped in a light field containing about one photon on average [5, 6]. An interesting application of this technique is a high-resolution microscope for single atoms [6], where higher-order transverse modes of the cavity should allow one to obtain two-dimensional position information [7]. The dynamical interplay between the motion of the atom and the intensity of the light field in the cavity can also be used to cool the atom by means of cavity dissipation [8, 9]. In addition, the strongly coupled atom-photon system is interesting for the processing of quantum information. For example, it was proposed to use this system as a triggered source of single photons [10, 11]. In principle, such light pulses can be generated on demand with near-unity efficiency. A first experiment in this direction with single atoms falling through a single-mode cavity has recently been performed [12].

The following text summarizes very briefly some results on cavity-mediated light forces in section 2, before addressing the new light source in section 3. Experimental details can be found in the figure captions and, of course, in the referenced literature.

2. Single-photon light force

In a first series of experiments, we have investigated the light forces in a cavity containing at most a few photons on average. In our setup (see figure 1), laser-cooled rubidium atoms are collected in a magneto-optical trap. They are launched upwards in an atomic fountain and enter the high-finesse cavity (F≈430,000) which is placed close to the point where the atoms are expected to turn around under the influence of gravity. This makes possible to perform experiments in a regime, where the kinetic energy of the atom is smaller than the vacuum Rabi energy. Hence, it can be expected that forces of a light field containing about one photon on average induce significant motional effects. Another advantage of the setup is that individual atoms can be observed for a long time. Moreover, the ability to control the power of the laser pumping the cavity field allows us to manipulate in real time the motion of an isolated atom. Depending on the flux from the fountain we can study single atoms in transit, trap an atom or study many atoms in simultaneous transits.

2.1 Single-atom transits

If the light field is red or blue detuned relative to the atomic transition, the atoms are attracted towards or repelled, respectively, from the antinodes of the standing-wave light field in the cavity. This effect can be observed by counting the number of atoms detected close to an antinode, where the atom-cavity coupling constant is largest. These atoms lead to deep dips in the cavity transmission (see figure 2a)), when the laser is in resonance with the empty cavity. When scanning this frequency relative to the atomic frequency, it was found that the number of deep dips is much larger for red detuning than for blue detuning.

Other information about light forces is obtained from the intensity autocorrelation function of the transmitted light, which reveals a fast oscillation of the intracavity field. This is caused by atoms channeling through the standing light wave, thereby modulating the light intensity at twice the oscillation frequency. As the dependence of the cavity transmission on the atomic position is nonlinear, the amplitude of this oscillation is more pronounced for blue detuning, where the atoms channel through narrow nodes, than for red detuning, where the

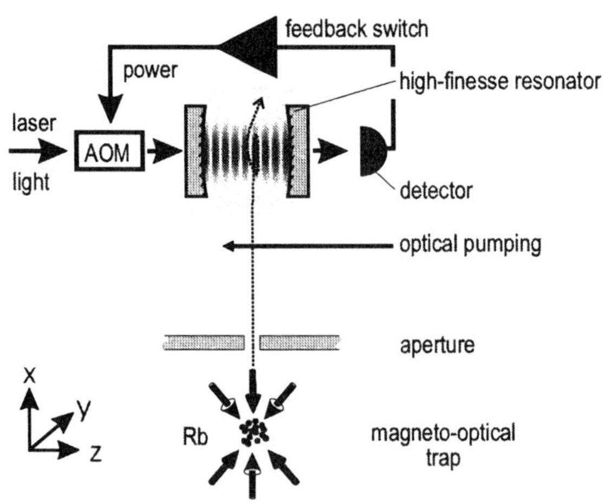

Fig. 1: Experimental setup [2]. ^{85}Rb atoms are collected from a background vapor, cooled by polarization gradient light forces and subsequently launched into a high-finesse optical cavity. On their way up, the atoms are pumped into the F=3, m_F=3 Zeeman level of the $5S_{1/2}$ ground state. For a mirror separation of 116 μm, a mode waist of 29 μm and right circularly polarized light resonant with the $5S_{1/2}$ F=3 – $5P_{3/2}$ F=4 transition at 780 nm, the atom-cavity coupling constant at an antinode of the standing-wave cavity mode is $g=2\pi \times 16$ MHz. The decay rates of the atomic dipole, γ, and the cavity field, κ, are $(\gamma,\kappa)=2\pi \times (3,1.5)$ MHz, respectively. The intensity of the laser pumping the cavity field can be switched to a higher value upon the detection of an atom in the cavity.

atoms channel through wide antinodes. When comparing the experimental data with the results of a numerical simulation, good agreement could only be obtained when the dipole force, momentum diffusion and the velocity-dependent force were taken into account [4]. Of these forces, the velocity-dependent force is particularly interesting. It has the novel feature that it allows to cool an atom without relying on spontaneous emission of the atom [8, 9]. In fact, the dissipation required for cooling can be provided by the decay of the cavity field. In principle, this allows to cool a two-level atom to a temperature well below the Doppler limit.

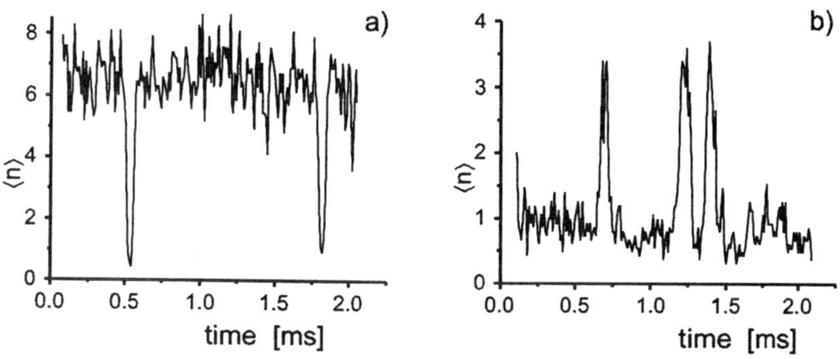

Fig. 2: Real-time detection of individual atoms [2]. The intracavity photon number is plotted while several atoms pass the cavity mode one after the other with a speed of about 80 cm/s. a) the laser is on-resonance with the cavity, but red detuned from the atom by $\Delta_A = -2\pi \times 26.5$ MHz. For these parameters, the two atoms passing the cavity decrease the transmission. b) for the same atom-laser detuning, but the cavity is blue shifted from the laser by $-\Delta_C = 2\pi \times 4$ MHz. This results in an increase of the transmission upon passage of three consecutive atoms. As the laser is detuned from the atom, dispersive effects dominate. Hence, the signals shown in a) and b) can be interpreted as being due to a shift of the cavity frequency by the atom's index of refraction.

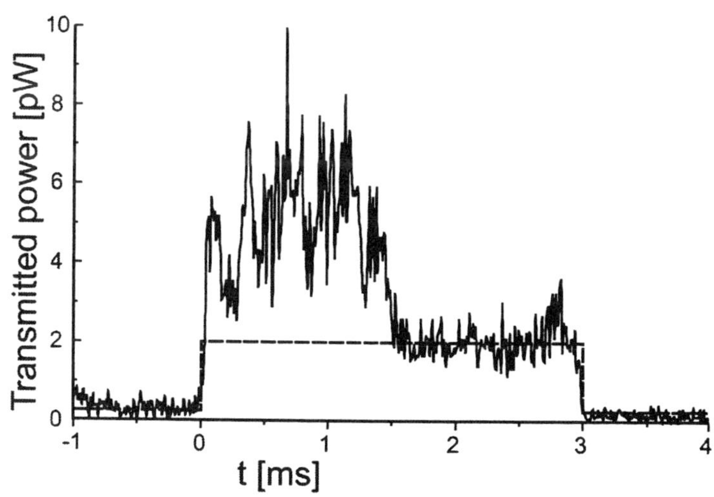

Fig. 3: Single-atom trapping [5]. The number of photons transmitted through the cavity is counted for time intervals of 10μs. At time t=0, an atom traversing the cavity triggers an 8-fold increase of the pump power (dashed line). The increase in the dipole potential causes the atom to remain in the cavity for about 1.5 ms. The pump power is switched back to its original value after 3 ms. The laser is red detuned from the atom and the cavity by $\Delta_A = -2\pi \times 35$ MHz and $\Delta_C = -2\pi \times 5$ MHz, respectively. The average trapping time of 280 μs is limited by transverse heating caused by spontaneous emission events. A single photon in the cavity corresponds to a transmitted power of 0.9 pW, leading to about 20 detected photons in 10 μs. Apart from shot noise, the changes of the transmitted power reflect the motion of the trapped atom.

2.2 Single-atom trapping

Mechanical binding of a single atom and a single photon was first envisioned about a decade ago for long-lived Rydberg atoms in a microwave cavity [13, 14]. It has recently been observed in the optical domain where light forces are much larger [5, 6]. In order to trap an atom in the conservative dipole potential in the cavity, the laser light coupled into the cavity is suddenly switched to a higher value when an atom is detected near an antinode. This increases the depth of the optical dipole potential and holds the atom in a cavity field containing about one photon on average. The subsequent motion of the trapped atom induces pronounced changes of the cavity transmission, which is monitored in real time. An example is displayed in figure 3. The observed signals are in good qualitative agreement with predictions based on a quantum jump Monte Carlo simulation of the trajectory of a point-like atom. Moreover, the fourth-order intensity correlation function displays a periodic structure, which we attribute to sudden flights of the atom between different antinodes of the standing wave. These flights are initiated by fluctuations of the dipole force heating the atom in the direction of the cavity axis. They are terminated by the velocity-dependent force cooling the atom for the parameters of the experiment. The cooled atom can be recaptured in one of the neighboring not-too-distant antinodes (instead of hitting one of the mirrors). Hence, the long trapping times observed in the experiment are further evidence for the existence of a cavity-mediated friction force.

2.3 Long range many-atom light force

If two or more atoms are placed inside the mode of the cavity, every atom can change the light field dramatically. This also changes the light forces on the other atoms. In other words, a light force between atoms is mediated by the cavity. This force is quite different from that in free-space, because the interaction strength is not a function of the inter-atomic distance, but rather of the position of the atoms in the cavity, which is macroscopic. We have found evidence for such a long-range dipole-dipole interaction in the spectra of a high-finesse cavity filled with ultra-cold rubidium atoms [15]. The forces between these strongly coupled atoms influence the spatial distribution of the atoms, which manifests itself in an asymmetric normal-mode spectrum, as displayed in figure 4. The asymmetry occurs already for photon numbers well below 0.05. Good agreement between the observed data and calculated spectra could only be achieved when including in the Monte-Carlo simulation the full motional dynamics of the many-atom system. For this purpose, we extended the existing work on the dipole force and the diffusion coefficient for a single atom [16] to that in the presence of many atoms [17].

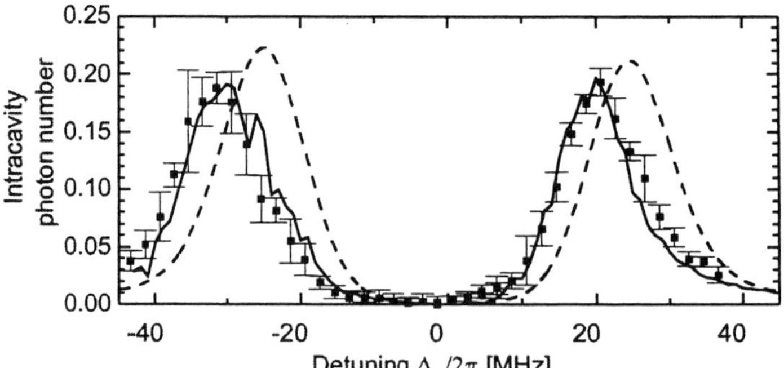

Fig. 4: Normal mode spectrum [15]. The number of photons in the cavity is measured as a function of the detuning, Δ_C, of the laser from the cavity, which is resonant with the free-space atoms. Symbols represent measured data. The dashed line comes from a semi-classical simulation ignoring light forces. The solid line is the result of a Monte-Carlo simulation including the motion of several strongly coupled atoms simultaneously.

2.4 Atomic kaleidoscope

An interesting application of the strongly coupled atom-cavity system is to observe a single moving atom with high spatial and temporal resolution. For example, an atom moving along the standing-wave induces a fast modulation of the intracavity light field [5, 18]. In principle, this allows one to measure the position of the atom in the direction of the cavity axis. In the transverse plane, however, position information is much more difficult to extract because of the rotational symmetry of the fundamental TEM_{00} cavity mode [6]. This disadvantage can be circumvented by using higher-order frequency-degenerate transverse modes [7]. Here, the atom couples different modes depending on its position. This produces a pattern of bright and dark regions at the cavity output, which changes when the atom moves, thereby resembling a toy kaleidoscope. A measurement of this intensity distribution with, e.g., a two-dimensional detector array allows one to reconstruct the atomic position and, hence, the atomic trajectory. An important feature of this scheme is that it does not require knowledge of the light forces to trace a trajectory. Instead, the position of the atom can directly be deduced from the measured intensity distribution.

3. Single-photon light source

In a second experiment performed with different parameters, the atom-cavity system is used as a novel source for single photons. A light source emitting single photons on demand into a well-defined mode of the radiation field would be a useful device for, e.g., quantum cryptography. Single-photon pulses would also allow one to process quantum information with purely linear optical elements [19]. However, most schemes used to generate single photons so far rely on spontaneous emission or on parametric down conversion and, hence, produce photons at more or less random times. Triggered photon sources were developed only during the last few years. Here, different schemes were investigated, like a turnstile device based on the Coulomb blockade mechanism in a quantum dot [20], the fluorescence from a single molecule [21, 22], a single (Nitrogen vacancy) color center in diamond [23, 24], or a single quantum dot [25]. All these new schemes allow one to emit a single photon upon an external trigger event. However, the photon is emitted into many modes of the radiation field and usually shows a broad energy distribution. Only recently, cavity-enhanced spontaneous emission techniques for single-photon generation have been proposed [26] and demonstrated [27] with quantum dots between mirrors.

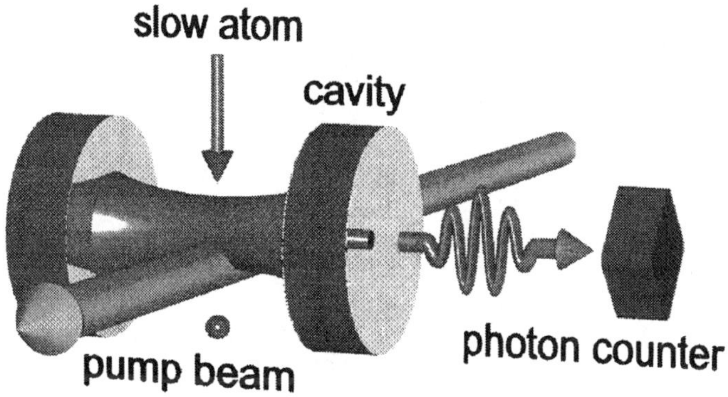

Fig. 5: Experimental setup. Laser-cooled ^{85}Rb atoms are released from a magneto-optical trap and fall with a speed of 2 m/s through the TEM_{00} mode of a high-finesse optical cavity. The two mirrors are separated by 1 mm, and their transmissions differ by a factor of 25. The pump beam crosses the cavity axis at right angle and is placed slightly downstream in the path of the atoms. The photons emitted from the cavity through the less reflecting mirror are registered by a single-photon counting avalanche diode with a quantum efficiency of 50%.

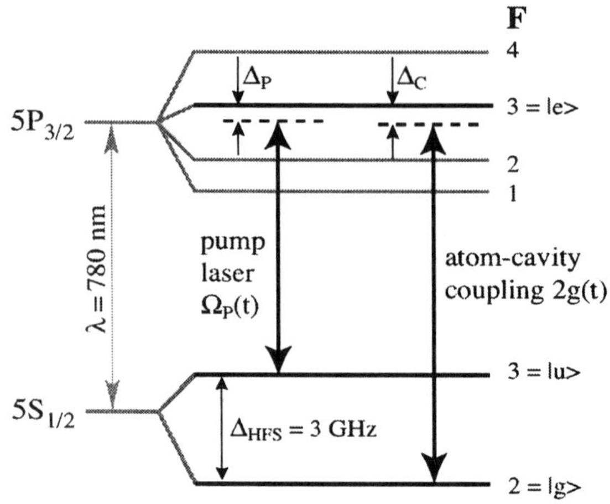

Fig. 6: Single-photon generation scheme. A Λ-type three-level scheme is realized by the two $5S_{1/2}$ hyperfine ground states and the F=3 hyperfine level of the electronically excited state $5P_{3/2}$. The atom interacts with the mode of an optical cavity, whose frequency is close to the atomic transition frequency between the excited state and the F=2 ground state. For this transition, the vacuum Rabi frequency of the strongly coupled atom-cavity system, 2g(t), is time dependent as the atom moves across the cavity mode. The atom is also exposed to a pump laser beam which is near resonant with the atomic transition between the excited state and the F=3 ground state which is the initial state of the atom. For a moving atom, the Rabi frequency of the pump beam, $\Omega_P(t)$, is also time dependent.

We have realized a new scheme where an atom emits exactly one photon into a single mode of a high-finesse (F≈61,000) optical cavity [9]. A sketch of the experimental setup is displayed in figure 5, and the relevant energy levels of the atom are shown in figure 6. In our scheme, a two-photon Raman transition is induced between two long-lived atomic ground states [8]. The Raman pump transition is excited by a laser beam crossing the cavity transverse to its axis. The Stokes transition is stimulated by the vacuum field of the cavity. In this way, the atom absorbs a photon from the pump laser and emits a photon into the initially empty cavity. To prevent excitation of the atom and, hence, avoid losses due to spontaneous emission into modes not covered by the solid angle of the cavity, the pump laser and the cavity field are arranged in a counter-intuitive interaction sequence, like in the well-known STIRAP process [28]. For an atom falling downwards, this is realized by shifting the Gaussian mode of the pump laser slightly below the fundamental mode of the cavity. Hence, the atom interacts first with the cavity vacuum, then with the pump laser. For an atom initially prepared in an internal state not coupled to the cavity mode ($|u\rangle$ in figure 6), and a slowly moving atom, the state vector of the combined atom-cavity system adiabatically follows a dark state throughout the interaction. This state has no contribution from the atom's excited state. The final state that is reached when the atom leaves the cavity mode ($|g\rangle$ in figure 6) is also a dark state, as it cannot be excited by the pump laser. Hence, the atom cannot emit a second photon. A further advantage of our stimulated rapid adiabatic passage technique is that a single photon is generated with, in principle, near-unity efficiency.

The interaction sequence realized in the experiment ensures that one atom passing through the cavity places exactly one photon into the empty cavity mode. This photon is detected as soon as it leaves the cavity through the less-reflecting mirror. Experimental results are shown in figure 7. The sub-natural line width of the observed spectrum clearly indicates that the excited state is not populated. Therefore, line broadening due to spontaneous emission does not occur and photons of well-defined frequency and direction are produced. We emphasize, that the characteristics of the light pulse are completely determined by the adjustable cavity parameters as well as the timing and the amplitude of the exciting pump laser.

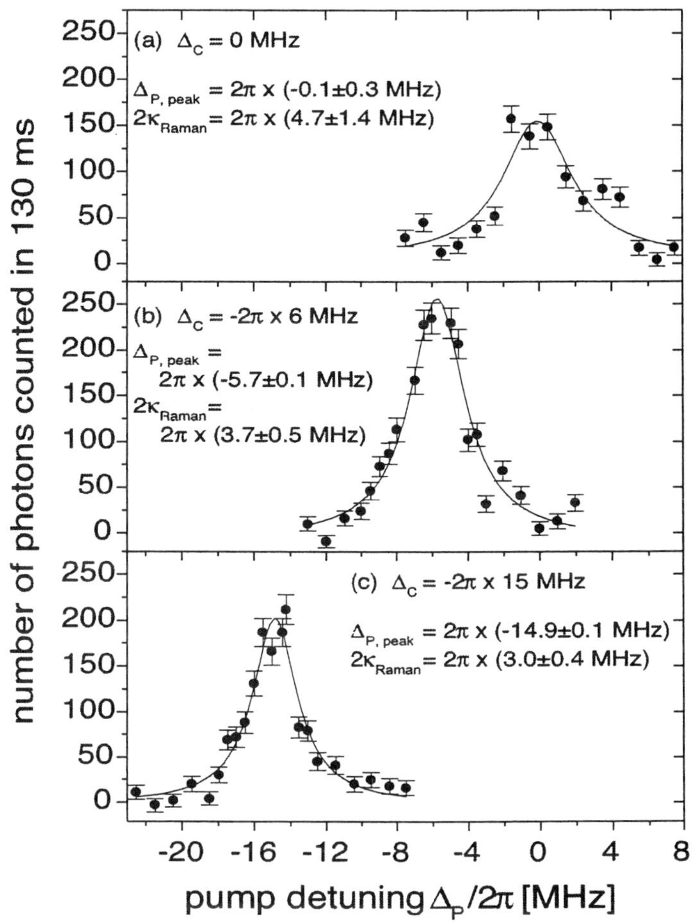

Fig. 7: A genuine cavity QED light source [12]. The number of photons emitted from the cavity is plotted as a function of the pump detuning, Δ_P, for three different cavity detunings, Δ_C (these detunings are defined in figure 6 and differ from the detunings used in figures 2-4 above). The solid lines are Lorentzian fits to the data. To register this data, atoms were loaded into the magneto-optical trap and dropped through the cavity 50 times. The total observation time was 130 ms. When the cavity is detuned from the atomic resonance, the emission peak follows the two-photon Raman resonance condition, $\Delta_P = \Delta_C$. This shift proves that the observed photons are produced by Raman scattering. Moreover, it shows that the light emission is not just the result of a pump transition followed by an enhanced spontaneous emission event into the cavity. The observed line width is significantly smaller than the natural line width, $2\gamma = 2\pi \times 6$ MHz, of the atom. In particular, for $\Delta_C = -2\pi \times 15$ MHz, the line width is 3 MHz and approaches the cavity line width, $2\kappa = 2\pi \times 2.5$ MHz. The decay of the cavity field limits the width of the Raman transition.

All observed features are in good agreement with theory. We therefore conclude that the scheme works as expected. Hence, it should also be possible to generate single, well-characterized photons on demand, provided the Raman excitation is performed in a controlled, triggered way. Moreover, our photon generation scheme opens up new prospects for quantum-state teleportation, since the photon generation depends on the initial state of the atom. For example, if the atom is prepared in a coherent superposition of the two ground states prior to the interaction, this state will be mapped onto the emitted photon. A second atom placed in another cavity could act as a receiver, and with the suitable pump pulse sequence applied to the emitting and the receiving atom, a quantum teleportation of the atom's internal state could be realized – even a quantum network of cavities seems at reach [29].

4. Acknowledgement

Some of the described experiments were performed at the University of Konstanz. Funding by the DFG focused research program, "Quantum Information Processing," the QUBITS and QUEST projects of the European Union is gratefully acknowledged.

5. References

1. H. Mabuchi, Q. A. Turchette, M. S. Chapman, and H. J. Kimble, "Real-time detection of individual atoms falling through a high-finesse optical cavity," Opt. Lett. **21**, 1393-1395 (1996).
2. P. Münstermann, T. Fischer, P. W. H. Pinkse, and G. Rempe, "Single slow atoms from an atomic fountain observed in a high-finesse optical cavity," Opt. Comm. **159**, 63-67 (1999).
3. C. J. Hood, M. S. Chapman, T. W. Lynn, H. J. Kimble, "Real-Time Cavity QED with Single Atoms," Phys. Rev. Lett. **80**, 4157-4160 (1998).
4. P. Münstermann, T. Fischer, P. Maunz, P. W. H. Pinkse, and G. Rempe, "Dynamics of Single-Atom Motion Observed in a High-Finesse Cavity," Phys. Rev. Lett. **82**, 3791-3794 (1999).
5. P. W. H. Pinkse, T. Fischer, P. Maunz, and G. Rempe, "Trapping an atom with single photons," Nature **404**, 365-368 (2000).
6. C. J. Hood, T. W. Lynn, A. C. Doherty, A. S. Parkins, and H. J. Kimble, "The Atom-Cavity Microscope: Single Atoms Bound in Orbit by Single Photons," Science **287**, 1447-1453 (2000).
7. P. Horak, H. Ritsch, T. Fischer, P. Maunz, T. Puppe, P. W. H. Pinkse, and G. Rempe, "An optical kaleidoscope using a single atom," quant-ph/0105048.
8. P. Horak, G. Hechenblaikner, K. M. Gheri, H. Stecher, and H. Ritsch, "Cavity-Induced Atom Cooling in the Strong Coupling Regime," Phys. Rev. Lett. **79**, 4974-4977 (1997).
9. V. Vuletić and S. Chu, "Laser Cooling of Atoms, Ions, or Molecules by Coherent Scattering," Phys. Rev. Lett. **84**, 3787-3790 (2000).
10. C. K. Law and H. J. Kimble, "Deterministic generation of a bit-stream of single-photon pulses," J. Mod. Opt. **44**, 2067-2074 (1997).
11. A. Kuhn, M. Hennrich, T. Bondo, and G. Rempe, "Controlled generation of single photons from a strongly coupled atom-cavity system," Appl. Phys. B **69**, 373-377 (1999).
12. M. Hennrich, T. Legero, A. Kuhn, and G. Rempe, "Vacuum-Stimulated Raman Scattering Based on Adiabatic Passage in a High-Finesse Optical Cavity," Phys. Rev. Lett. **85**, 4872-4875 (2000).
13. S. Haroche, M. Brune, and J. M. Raimond, "Trapping atoms by the vacuum field in a cavity," Europhys. Lett. **14**, 19-24 (1991).
14. B.-G. Englert, J. Schwinger, A. O. Barut, and M. O. Scully, "Reflecting slow atoms from a micromaser field," Europhys. Lett. **14**, 25-31 (1991).
15. P. Münstermann, T. Fischer, P. Maunz, P. W. H. Pinkse, and G. Rempe, "Observation of Cavity-Mediated Long-Range Light Forces between Strongly Coupled Atoms," Phys. Rev. Lett. **84**, 4068-4071 (2000).
16. G. Hechenblaikner, M. Gangl, P. Horak, and H. Ritsch, "Cooling an atom in a weakly driven high-Q cavity," Phys. Rev. A **58**, 3030-3042 (1998).
17. T. Fischer, P. Maunz, T. Puppe, P. W. H. Pinkse, and G. Rempe, "Collective light forces on atoms in a high-finesse cavity," New Journal of Physics **3**, 11.1-11.20 (2001).
18. P. W. H. Pinkse, T. Fischer, P. Maunz, T. Puppe, and G. Rempe, "How to catch an atom with single photons," J. Mod. Opt. **47**, 2769-2787 (2000).
19. E. Knill, R. Laflamme, and G. J. Milburn, "A scheme for efficient quantum computation with linear optics," Nature **409**, 46-52 (2001).
20. J. Kim, O. Benson, A. Kann, and Y. Yamamoto, "A single-photon turnstile device," Nature **397**, 500-503 (1999).
21. C. Brunel, B. Lounis, P. Tamarat, and M. Orrit, "Triggered Source of Single Photons based on Controlled Single Molecule Fluorescence," Phys. Rev. Lett. **83**, 2722-2725 (1999).
22. B. Lounis and W. E. Moerner, "Single photons on demand from a single molecule at room temperature," Nature **407**, 491-493 (2000).
23. C. Kurtsiefer, S. Mayer, P. Zarda, and H. Weinfurter, "Stable Solid-State Source of Single Photons," Phys. Rev. Lett. **85**, 290-293 (2000).
24. R. Brouri, A. Beveratos, J.-P. Poizat, and P. Grangier, "Photon antibunching in the fluorescence of individual color centers in diamond," Opt. Lett. **25**, 1294-1297 (2000).
25. C. Santori, M. Pelton, G. Solomon, Y. Dale, and Y. Yamamoto, "Triggered Single Photons from a Quantum Dot," Phys. Rev. Lett. **86**, 1502-1505 (2001).
26. O. Benson, C. Santori, M. Pelton, and Y. Yamamoto, "Regulated and Entangled Photons from a Single Quantum Dot," Phys. Rev. Lett. **84**, 2513-2516 (2000).
27. P. Michler, A. Kiraz, C. Becher, W. V. Schoenfeld, P. M. Petroff, Lidong Zhang, E. Hu, and A. Imamoglu, "A Quantum Dot Single-Photon Turnstile Device," Science **290**, 2282-2285 (2000).
28. For a review, see, e.g., K. Bergmann and B. W. Shore, in *Molecular Dynamics and Stimulated Emission Pumping*, H. L. Dai and R. W. Field, eds., (World Scientific, Singapore, 1995), 315-373.
29. J. I. Cirac and P. Zoller and H. J. Kimble, and H. Mabuchi, "Quantum State Transfer and Entanglement Distribution among Distant Nodes in a Quantum Network," Phys. Rev. Lett. **78**, 3221-3224 (1997).

Nontrivial analogies between quantum optics and Bose-Einstein condensation

K. Rzazewski
Center for Theoretical Physics, Polish Academy of Sciences, Al. Lotników 32/46, 02-668 Warsaw, Poland
kazik@cft.edu.pl

Abstract: There are close analogies between linear and nonlinear processes involving light and involving bosonic atoms. We are used to passing from classical to quantum description and back for the light fields. The same may be applied to atoms. I will present our recent results concerning:
1. Quantitative study of the interacting Bose gas near the critical temperature.
2. Multimode model of the photoassociation of ultracold molecules.
3. Quantitative estimate of condensate losses in the collision of two condensates.

Hyperentanglement in Parametric Down-Conversion

Alexander V. Sergienko,[1,2] Mete Atatüre,[1]
Giovanni Di Giuseppe, [2] Matthew D. Shaw,[2]
Bahaa E. A. Saleh,[2] and Malvin C. Teich[1,2]

Quantum Imaging Laboratory, [1] Department of Physics and [2] Department of Electrical and Computer Engineering, Boston University, 8 Saint Mary's Street, Boston, MA 02215

Abstract:

A theory of spontaneous parametric down-conversion, which gives rise to a quantum state that is simultaneously entangled in three-dimensional wavevector and polarization, allows us to understand the unusual characteristics of fourth-order quantum interference in many experiments, including ultrafast parametric down-conversion, the specific example illustrated in this paper. The comprehensive approach provided here permits the engineering of quantum states suitable for quantum information schemes and new quantum technologies.

OCIS codes: 42.50.Dv, 42.65.Re, 42.65.Ky, 03.67a

Entanglement [1] is, undoubtedly, one of the most fascinating features of quantum mechanics. Spontaneous parametric down-conversion (SPDC) [2], a nonlinear optical phenomenon, has been one of the most widely used sources of entangled quantum states. In this process, pairs of photons are generated in a state that can be entangled in frequency, momentum, and polarization when a laser beam illuminates a nonlinear optical crystal. The experimental arrangement for producing entangled photon pairs is simple both in conception and in execution.

Ironically, a significant number of experimental efforts designed to verify the nonseparability of entangled states are carried out in the context of models that fail to access the overall relevant Hilbert space, but rather are restricted to only a *single* kind of entanglement, such as entanglement in energy [3], momentum [4], or polarization [5]. Inconsistencies in the analysis of down-conversion quantum-interference experiments can emerge under such circumstances, as highlighted by the failure of the conventional theory [6] of ultrafast parametric down-conversion to characterize quantum-interference experiments [7].

In this paper we present a complete quantum-mechanical analysis of entangled-photon state generation via SPDC, considering *simultaneous* entanglement in three-dimensional wavevector (see Fig. 1) and polarization at the generation, propagation, and detection stages (see Fig. 2). As one specific example of the applicability of this approach, we use it to describe both new and previously obtained [7] results of SPDC experiments with a femtosecond pump. Our analysis confirms that the inconsistencies between existing theoretical models and the observed data in femtosecond down-conversion experiments can indeed be attributed to a failure of considering the full Hilbert space spanned by the simultaneously entangled quantum variables. Femtosecond SPDC models have heretofore ignored transverse wavevector components and have thereby not accounted for the previously demonstrated angular spread [8] of the down-converted light. The approach presented here is suitable for Type-I, as well as Type-II, spontaneous parametric down-conversion.

Our study leads to a deeper physical understanding of hyperentangled photon states and, concomitantly, provides a route for engineering these states for specific applications, including quantum information processing.

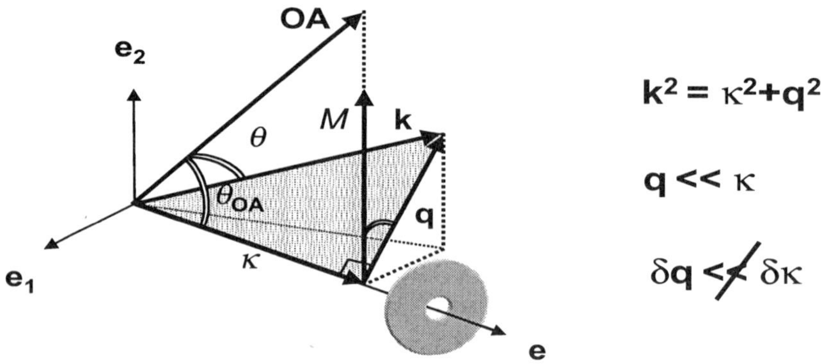

Fig. 1. Decomposition of a three-dimensional wavevector (**k**) into longitudinal (κ) and transverse (**q**) components. The angle between the optical axis of the nonlinear crystal (OA) and the wavevector **k** is θ. The angle between the optical axis and the longitudinal axis (e_3) is denoted θ_{OA}.

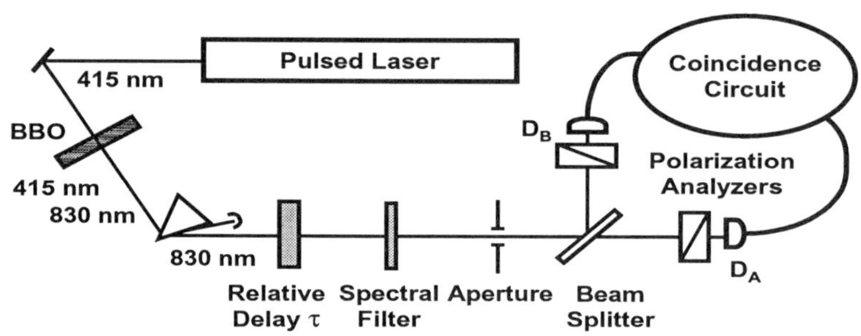

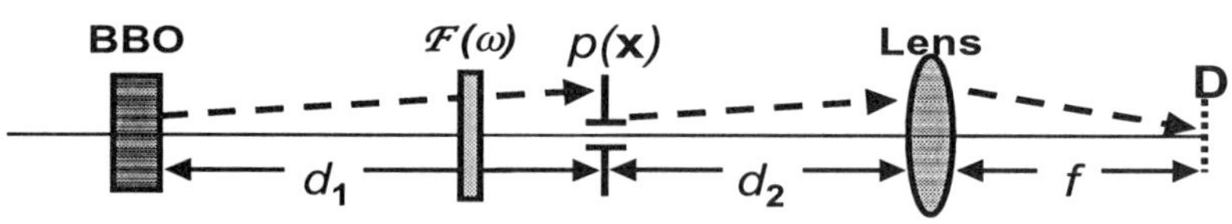

Fig. 2. (a) Schematic of the experimental setup for observation of quantum interference using femtosecond SPDC. (b) Detail of the path from the crystal output plane to the detector input plane.

Hyperentangled-State Generation.—With this motivation we present a complete multidimensional analysis of the entangled-photon state generated via SPDC. To admit a broad range of possible experimental schemes we consider, in turn, three general and fundamentally distinct stages in any experimental apparatus: the generation, propagation, and detection of the quantum state [9].

We begin with generation. By virtue of the weak nonlinear interaction, we consider the state generated within the confines of first-order time-dependent perturbation theory:

$$|\Psi^{(2)}\rangle \sim \frac{\mathrm{i}}{\hbar}\int_{t_0}^{t} dt'\ \hat{H}_{\mathrm{int}}(t')\ |0\rangle\,. \tag{1}$$

Here $\hat{H}_{\mathrm{int}}(t')$ is the interaction Hamiltonian, (t_0, t) is the duration of the interaction, and $|0\rangle$ is the initial vacuum state. The interaction Hamiltonian governing this phenomenon is [10]

$$\hat{H}_{\mathrm{int}}(t') \sim \chi^{(2)} \int_V d\mathbf{r}\ \hat{E}_p^{(+)}(t',\ \mathbf{r})\, \hat{E}_o^{(-)}(t',\ \mathbf{r})\, \hat{E}_e^{(-)}(t',\ \mathbf{r})\ +\ \mathrm{H.c.}\,, \tag{2}$$

where $\chi^{(2)}$ is the second-order susceptibility and V is the volume of the nonlinear medium in which the interaction takes place. The symbol $\hat{E}_j^{(\pm)}(t',\ \mathbf{r})$ represents the positive- (negative-) frequency portion of the jth electric-field operator, with the subscript j representing the pump (p), ordinary (o), and extraordinary (e) waves at time t' and position $\mathbf{r}$, and H.c. stands for Hermitian conjugate. Because of the high intensity of the pump field we take the coherent-state laser beam to be classical, with an arbitrary spatiotemporal profile given by

$$E_p(\mathbf{r}, t) = \int d\mathbf{k}_p\ \tilde{E}_p(\mathbf{k}_p) e^{\mathrm{i}\mathbf{k}_p\cdot\mathbf{r}} e^{-\mathrm{i}\omega_p(\mathbf{k}_p)t}\,, \tag{3}$$

where $\tilde{E}_p(\mathbf{k}_p)$ is the complex-amplitude profile of the field as a function of the wavevector $\mathbf{k}_p$.

In contrast with previous models we consider the wavevector to be three-dimensional, with a transverse wavevector $\mathbf{q}_p$ and frequency ω_p, so that Eq. (3) takes the form

$$E_p(\mathbf{r}, t) = \int d\mathbf{q}_p\, d\omega_p\ \tilde{E}_p(\mathbf{q}_p; \omega_p) e^{\mathrm{i}\kappa_p z} e^{\mathrm{i}\mathbf{q}_p\cdot\mathbf{x}} e^{-\mathrm{i}\omega_p t}\,, \tag{4}$$

where $\mathbf{x}$ spans the transverse plane perpendicular to the propagation direction z. In a similar way the signal and idler fields can be expressed in terms of the quantum-mechanical creation operators $\hat{a}^\dagger(\mathbf{q}, \omega)$ for the $(\mathbf{q}, \omega)$ modes as

$$\hat{E}_j^{(-)}(\mathbf{r}, t) = \int d\mathbf{q}_j\, d\omega_j\ e^{-\mathrm{i}\kappa_j z} e^{-\mathrm{i}\mathbf{q}_j\cdot\mathbf{x}} e^{\mathrm{i}\omega_j t}\, \hat{a}_j^\dagger(\mathbf{q}_j, \omega_j)\,, \tag{5}$$

where the subscript $j = o, e$. The longitudinal component of $\mathbf{k}$, denoted κ, can be written in terms of the $(\mathbf{q}, \omega)$ pair as [9]

$$\kappa = \sqrt{\left[\frac{n_e(\omega, \theta)\,\omega}{c}\right]^2 - |\mathbf{q}|^2}\,, \tag{6}$$

where θ is the angle between $\mathbf{k}$ and the optical axis of the nonlinear crystal, $n_e(\omega, \theta)$ is the extraordinary index of refraction in the nonlinear medium, and c is the speed of light in vacuum. Note that the extraordinary refractive index, $n_e(\omega, \theta)$, in Eq. (6) should be replaced by the ordinary refractive index, $n_o(\omega)$, when calculating κ for ordinary waves.

Substituting Eqs. (4) and (5) into Eqs. (1) and (2) yields the wavefunction at the output of the nonlinear crystal:

$$|\Psi^{(2)}\rangle \sim \int d\mathbf{q}_o d\mathbf{q}_e\, d\omega_o d\omega_e\; \Phi(\mathbf{q}_o, \mathbf{q}_e; \omega_o, \omega_e)\hat{a}_o^\dagger(\mathbf{q}_o, \omega_o)\hat{a}_e^\dagger(\mathbf{q}_e, \omega_e)|0\rangle\,, \tag{7}$$

with

$$\Phi(\mathbf{q}_o, \mathbf{q}_e; \omega_o, \omega_e) \;=\; \tilde{E}_p(\mathbf{q}_o + \mathbf{q}_e; \omega_o + \omega_e)\, L\, \mathrm{sinc}(\frac{L\Delta}{2})e^{-\mathrm{i}\frac{L\Delta}{2}}\,. \tag{8}$$

Here $\Delta = \kappa_p - \kappa_o - \kappa_e$ where κ_j ($j = p, o, e$) is related to the indices $(\mathbf{q}_j, \omega_j)$ via relations similar to Eq. (6). The nonseparability of the function $\Phi(\mathbf{q}_o, \mathbf{q}_e; \omega_o, \omega_e)$ in Eqs. (7) and (8), recalling (6), is the hallmark of *simultaneous* multi-parameter entanglement.

Hyperentangled-State Propagation.—Propagation between the planes of generation and detection is characterized by the classical transfer function of the optical system. The biphoton probability amplitude at the space-time coordinates $(\mathbf{x}_A, t_A)$ and $(\mathbf{x}_B, t_B)$, where detection will take place, is defined by [10],

$$A(\mathbf{x}_A, \mathbf{x}_B; t_A, t_B) = \langle 0|\hat{E}_A^{(+)}(\mathbf{x}_A, t_A)\hat{E}_B^{(+)}(\mathbf{x}_B, t_B)|\Psi^{(2)}\rangle\,. \tag{9}$$

The explicit forms of the quantum fieldspresent at the detection locations are represented by

$$\begin{aligned}
\hat{E}_A^{(+)}(\mathbf{x}_A, t_A) &= \int d\mathbf{q}\, d\omega\; e^{-\mathrm{i}\omega\, t_A}\,[\mathcal{H}_{Ae}(\mathbf{x}_A, \mathbf{q}; \omega)\hat{a}_e(\mathbf{q}, \omega) + \mathcal{H}_{Ao}(\mathbf{x}_A, \mathbf{q}; \omega)\hat{a}_o(\mathbf{q}, \omega)]\,, \\
\hat{E}_B^{(+)}(\mathbf{x}_B, t_B) &= \int d\mathbf{q}\, d\omega\; e^{-\mathrm{i}\omega t_B}\,[\mathcal{H}_{Be}(\mathbf{x}_B, \mathbf{q}; \omega)\hat{a}_e(\mathbf{q}, \omega) + \mathcal{H}_{Bo}(\mathbf{x}_B, \mathbf{q}; \omega)\hat{a}_o(\mathbf{q}, \omega)]\,,
\end{aligned} \tag{10}$$

where the transfer function $\mathcal{H}_{ij}$ ($i = A, B$ and $j = e, o$) describes the propagation of a $(\mathbf{q}, \omega)$ mode from the nonlinear-crystal output plane to the detection plane. Substituting Eqs. (7) and (10) into Eq. (9) yields a general form for the biphoton probability amplitude:

$$\begin{aligned}
A(\mathbf{x}_A, \mathbf{x}_B; t_A, t_B) \;= \int d\mathbf{q}_o d\mathbf{q}_e\, d\omega_o d\omega_e \quad & \Phi(\mathbf{q}_o, \mathbf{q}_e; \omega_o, \omega_e) \\
& \times \Big[\mathcal{H}_{Ae}(\mathbf{x}_A, \mathbf{q}_e; \omega_e)\mathcal{H}_{Bo}(\mathbf{x}_B, \mathbf{q}_o; \omega_o)\, e^{-\mathrm{i}(\omega_e t_A + \omega_o t_B)} \\
& + \mathcal{H}_{Ao}(\mathbf{x}_A, \mathbf{q}_o; \omega_o)\mathcal{H}_{Be}(\mathbf{x}_B, \mathbf{q}_e; \omega_e)\, e^{-\mathrm{i}(\omega_o t_A + \omega_e t_B)}\Big]
\end{aligned} \tag{.11}$$

This function can be separated into polarization-dependent and -independent terms, as necessary, for any particular configuration. By choosing explicit forms of the functions $\mathcal{H}_{Ae}$, $\mathcal{H}_{Ao}$, $\mathcal{H}_{Be}$, and $\mathcal{H}_{Bo}$, the overall biphoton probability amplitude can be sculpted as desired.

Hyperentangled-State Detection.—The formulation of the detection process depends on the scheme to be used. Slow detectors, for example, impart temporal integration while finite area detectors impart spatial integration. Quantum-interference experiments typically make use of just such detectors. Under these conditions, the coincidence count rate R is readily expressed in terms of the biphoton probability amplitude:

$$R = \int d\mathbf{x}_A d\mathbf{x}_B \, dt_A dt_B \, |A(\mathbf{x}_A, \mathbf{x}_B; t_A, t_B)|^2 . \tag{12}$$

Example: Quantum Interference in Ultrafast SPDC.—We now consider a particular example that demonstrates the validity of our analysis: an ultrafast polarization quantum-interference experiment of the form illustrated in Fig. 2(a). Details of the experimental arrangement and protocol can be found in an earlier work [7]; in the analysis offered there we made use of a phenomenological model that considered a collection of contributions from different regions in the nonlinear crystal that, in the absence of a full quantum-mechanical model, were conjectured to be independent and distinguishable. With the help of the general spatiotemporal quantum-mechanical approach developed here, we are now in a position to provide a complete analysis of those data along with new data in which filtering was used, presented in Fig. 3 and Fig. 4, respectively. For a finite-bandwidth pulsed plane-wave pump, such as that used in these experiments,

$$\tilde{E}_p(\mathbf{q}_p; \omega_p) = \delta(\mathbf{q}_p) \tilde{E}_p(\omega_p) , \tag{13}$$

where the delta function $\delta(\mathbf{q}_p)$ represents the absence of transverse wavevector components. For the polarization-interferometer arrangement illustrated in Fig. 2(a), in the presence of a polarization-dependent relative temporal delay τ, Eq. (10) can be conveniently separated into polarization-dependent and -independent terms via the relation

$$\mathcal{H}_{ij}(\mathbf{x}_i, \mathbf{q}; \omega) = (\mathbf{e}_i \cdot \mathbf{e}_j) \, e^{-\mathrm{i}\omega\tau\delta_{ej}} H(\mathbf{x}_i, \mathbf{q}; \omega) , \tag{14}$$

where $i = A, B$ and $j = e, o$. The symbol δ_{ej} is the Kronecker delta so that $\delta_{ee} = 1$ and $\delta_{eo} = 0$. The unit vector $\mathbf{e}_i$ describes the orientation of the polarization analyzers in the experimental apparatus [see Fig. 2(a)], while $\mathbf{e}_j$ is the unit vector that describes the polarization of the down-converted photons; the function $H(\mathbf{x}_i, \mathbf{q}; \omega)$ is the transfer function of the polarization-independent elements of the system such as free space, filters, apertures, and lenses, as illustrated in Fig. 2(b). The explicit form of H for Fig. 2(b) is calculated to be, in the Fresnel approximation,

$$H(\mathbf{x}, \mathbf{q}; \omega) = \left[e^{\mathrm{i}\frac{\omega}{c}(d_1 + d_2 + f)} e^{-\mathrm{i}\frac{\omega|\mathbf{x}|^2}{2cf}\left[\frac{d_2}{f} - 1\right]} e^{-\mathrm{i}\frac{d_1 c}{2\omega}|\mathbf{q}|^2} \tilde{P}\left(\frac{\omega}{cf}\mathbf{x} - \mathbf{q}\right) \right] \mathcal{F}(\omega) , \tag{15}$$

where d_1, d_2, and f (focal length of the lens) are indicated, $\tilde{P}$ is the aperture function $p(\mathbf{x})$ in the Fourier domain, and $\mathcal{F}(\omega)$ is the spectral filter function.

Using Eqs. (14) and (15) in Eq. (11), the biphoton probability amplitude for the arrangement shown in Fig. 2(a) therefore becomes

$$A(\mathbf{x}_A, \mathbf{x}_B; t_A, t_B) = \int d\mathbf{q}_o d\mathbf{q}_e \, d\omega_o d\omega_e \, \Phi(\mathbf{q}_o, \mathbf{q}_e; \omega_o, \omega_e) e^{-\mathrm{i}\omega_e\tau}$$

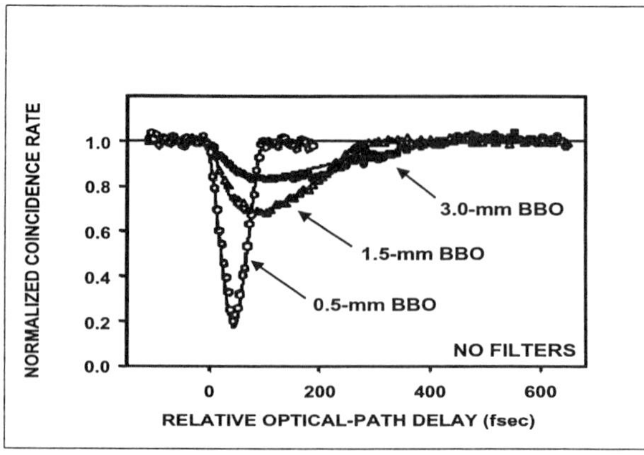

Fig. 3. Experimental (symbols) and theoretical (solid curves) results for the normalized coincidence rate for BBO crystals of three different lengths (hexagons: 0.5 mm; triangles: 1.5 mm; circles: 3.0 mm) as a function of the relative optical-path delay τ. As the crystal length increases the fringe visibility diminishes substantially and a dramatic asymmetry emerges. No free parameters are used to fit the data.

$$\times \Big[(\mathbf{e}_A \cdot \mathbf{e}_e)(\mathbf{e}_B \cdot \mathbf{e}_o)\ H(\mathbf{x}_A, \mathbf{q}_e; \omega_e) H(\mathbf{x}_B, \mathbf{q}_o; \omega_o) e^{-i(\omega_e t_A + \omega_o t_B)} + (\mathbf{e}_A \cdot \mathbf{e}_o)(\mathbf{e}_B \cdot \mathbf{e}_e)\ H(\mathbf{x}_A, \mathbf{q}_o; \omega_o) H(\mathbf{x}_B, \mathbf{q}_e; \omega_e) e^{-i(\omega_o t_A + \omega_e t_B)}\Big] . \quad (16)$$

Using this form for the biphoton probability amplitude in Eq. (12) yields the coincidence-count rate as a function of the polarization-dependent temporal delay τ.

Discussion.—Figure 3 displays the observed normalized coincidence rates (fourth-order quantum-interference patterns) for 0.5-, 1.5-, and 3.0-mm BBO crystals (symbols), in the absence of spectral filtering, along with the expected theoretical curves (solid), as a function of relative optical-path delay τ. The asymmetry of the observed interference pattern clearly increases with crystal thickness, as the $(\mathbf{q}, \omega)$ modes overlap less in space at the detection plane. This decreased overlap leads to increased distinguishability, which is manifested as loss of visibility and asymmetry in the interference pattern.

Figure 4 provides a set of data collected in a similar fashion, but this time observed in the presence of a narrowband (9-nm) spectral filter $\mathcal{F}(\omega)$, as illustrated in Fig. 2(b). The most dramatic effect of including the filter is the symmetrization of the quantum-interference patterns. Since $\mathbf{q}$ and ω are intrinsically linked by Eq. (8), the imposition of spectral filtering restricts the allowable transverse wavevector spread. Spectral and spatial filtering therefore have similar effects for non-cross-spectrally pure light, such as that generated in SPDC [8].

The physical origin of these observations resides in the spatial shift of the down-converted photons as they propagate through the birefringent crystal. A photon pair generated at the output plane of the crystal exits without relative temporal or spatial shifts. A pair generated at the input plane

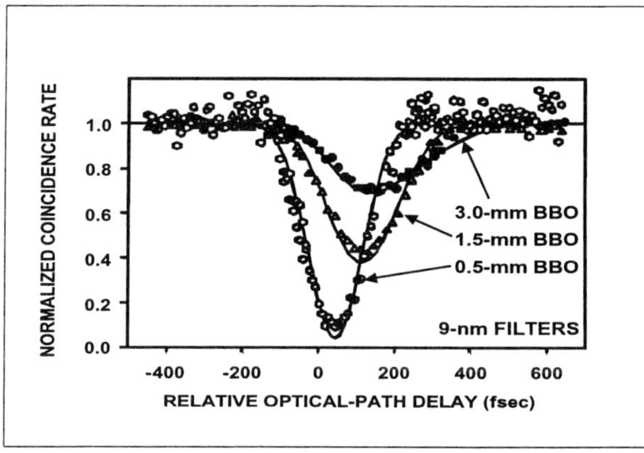

Fig. 4. Plots similar to those in Fig. 2 in the presence of an interference filter of 9-nm bandwidth. The patterns are symmetrized.

of the crystal, in contrast, must propagate through the entire crystal before exiting; the ensuing spatial and temporal shifts in principle tag the birthplace of every pair. Even when the experimental configuration is such that temporal indistinguishability is achieved, knowledge of the birthplace suppresses the quantum interference. The combination of free-space propagation and the small aperture lead to diffraction of the SPDC beams, which results in increased overlap and therefore reduced spatial shift. Indeed, as the aperture size becomes even smaller, the observed quantum-interference patterns ultimately revert to those calculated using the one-dimensional model that has traditionally been used.

In summary, we observe that the multi-parameter entangled nature of the two-photon state generated by SPDC allows transverse features, represented by their wavevectors, to play a role in polarization-based quantum interference experiments. The interference patterns generated in these experiments are, as a result, affected by the profiles of the apertures in the optical system which admit wavevectors in specified directions. The quantitative agreement between the experimental results and the theoretical results from the formalism presented in this paper, confirm this interplay. In contrast to the usual single-direction polarization entangled state, the wide-angle polarization entangled state offers a richness that can be exploited in a variety of applications involving quantum information processing.

Acknowledgments.— This work was supported by the National Science Foundation.

References

1. E. Schrödinger, Naturwissenschaften **23**, 807 (1935); **23**, 823 (1935); **23**, 844 (1935) [Translation in *Quantum Theory and Measurement*, edited by J. A. Wheeler and W. H. Zurek (Princeton University Press, Princeton, 1983)].
2. S. E. Harris, M. K. Oshman, and R. L. Byer, Phys. Rev. Lett. **18**, 732 (1967); D. Magde and H. Mahr, Phys. Rev. Lett. **18**, 905 (1967).

3. C. K. Hong, Z. Y. Ou, and L. Mandel, Phys. Rev. Lett. **59**, 2044 (1987); P. G. Kwiat, A. M. Steinberg, and R. Y. Chiao, Phys. Rev. A **47**, R2472 (1993).
4. J. G. Rarity and P. R. Tapster, Phys. Rev. Lett. **64**, 2495 (1990).
5. Z. Y. Ou and L. Mandel, Phys. Rev. Lett. **61**, 50 (1988); Y. H. Shih and C. O. Alley, Phys. Rev. Lett. **61**, 2921 (1988); Y. H. Shih and A. V. Sergienko, Phys. Lett. A **191**, 201 (1994); P. G. Kwiat, H. Weinfurter, A. Zeilinger, A. V. Sergienko, and Y. H. Shih, Phys. Rev. Lett. **75**, 4337 (1995).
6. G. Di Giuseppe, L. Haiberger, F. De Martini, and A. V. Sergienko, Phys. Rev. A **56**, R21 (1997); E. Keller and M. H. Rubin, Phys. Rev. A **56**, 1534 (1997); W. P. Grice, R. Erdmann, I. A. Walmsley, and D. Branning, Phys. Rev. A **57**, R2289 (1998); J. Peřina, Jr., A. V. Sergienko, B. M. Jost, B. E. A. Saleh, and M. C. Teich, Phys. Rev. A **59**, 2359 (1999).
7. M. Atatüre, A. V. Sergienko, B. M. Jost, B. E. A. Saleh, and M. C. Teich, Phys. Rev. Lett. **83**, 1323 (1999).
8. A. Joobeur, B. E. A. Saleh, and M. C. Teich, Phys. Rev. A **50**, 3349 (1994); C. H. Monken, P. H. Souto Ribeiro, and S. Pádua, Phys. Rev. A **57**, 3123 (1998); B. E. A. Saleh, A. Joobeur, and M. C. Teich, Phys. Rev. A **57**, 3991 (1998).
9. B. E. A. Saleh, A. F. Abouraddy, A. V. Sergienko, and M. C. Teich, Phys. Rev. A **62**, 043816 (2000).
10. B. R. Mollow, Phys. Rev. A **8**, 2684 (1973); J. Peřina, Z. Hradil, and B. Jurčo, *Quantum Optics and Fundamentals of Physics* (Kluwer, Boston, 1994).

Coherence in an Optically Trapped Fermi Gas

K. M. O'Hara, M. E. Gehm, S. R. Granade, M.-S. Chang and J. E. Thomas

Physics Department, Duke University, Durham, NC 27708-0305

jet@phy.duke.edu

Abstract: We demonstrate forced evaporative cooling of two-state mixtures of lithium fermions in a stable optical trap. This system offers exciting prospects for attaining extremely long hyperfine coherence lifetimes at high density, and for studying multiple-particle coherence in Fermi superfluids.

1 Introduction

Trapping and cooling of neutral fermionic atoms offers exciting prospects for fundamental precision studies of single- and multi-particle coherence in Fermi gases for which the interaction strength, density, and temperature can be experimentally controlled. At low temperatures, where s-wave scattering is dominant, the Pauli exclusion principle forbids scattering in a single component Fermi gas, so that coherence degrading two- and three-body collisions are suppressed [1]. Hence, a dense one-component sample in a pure superposition state can exhibit extremely long coherence lifetimes. Two-component Fermi gases may permit new studies of superfluidity and of multiparticle coherence which result from Cooper pairing at low temperatures [2].

To enable trapping of arbitrary single- and two-state mixtures of fermions, we have developed an all-optical trap based on an ultrastable CO_2 laser. This trap has been used to confine ^{6}Li atoms in a nearly spin-state-independent potential. With a trap lifetime of 370 seconds, this is the first all-optical trap to achieve confinement times comparable to the estimated background gas collision time at 10^{-11} Torr [3]. The parameters for this trap are given in Table 1.

Measured Lifetime	370 sec
Well depth	0.33 mK
Background pressure	$< 10^{-11}$ Torr
Measured residual heating rate	< 6 nK/sec

Table 1. CO_2 Laser Trap Parameters

The extreme detuning from resonance and the infrared wavelength lead to optical scattering rates of only one photon per hour per atom and a corresponding recoil heating rate of 8 pK/sec. The inherent stability of the CO_2 laser suppresses exponential heating arising from laser intensity noise [4, 5]. The heating time constant estimated from the measured noise spectrum of the laser is 46,000 seconds, much longer than the trap lifetime. Hence, the fundamental limit to the residual heating rate is expected to arise from diffractive background gas collisions [6, 7]. For ^{6}Li in an optical trap with a depth of 0.3 mK , the diffractive heating rate is estimated to be $\simeq$ 2.6 nK/sec at 10^{-11} Torr, comparable to the net residual heating rate in our trap, which is measured to be $\simeq$ 6 nK/sec. As a result of the small residual heating rate, the stable CO_2 laser trap is well-suited for evaporative cooling to temperatures below the Fermi temperature. Since the trapping potential is nearly state-independent and the residual heating rate is very small, the CO_2 laser trap is ideal for exploring both pure superposition states in Fermi gases and interactions in two-component mixtures.

Evaporation to low temperatures will enable long coherence lifetimes to be achieved in a single component Fermi gas, where s-wave scattering will be suppressed. Long coherence lifetimes can be attained by employing magnetic field independent transitions between ground hyperfine states in an all-optical trap. For fermions, magnetic field independent transitions occur between hyperfine states with magnetic quantum numbers M and g-factors of opposite sign, while for bosons, both states have $M = 0$. Unfortunately, at high density, both fermions and bosons undergo inelastic spin-exchange collisions between two-particle states of the same total magnetic quantum number. This leads to rapid decay of the number of trapped atoms. For example, for fermions with upper(lower) hyperfine states $F_2(F_1)$, the two particle state $|F_2, M; F_1, -M\rangle$ can decay to $|F_1, M : F_1, -M\rangle$. Inelastic collisions, as well as collisional shifts, can be greatly suppressed by using fermionic atoms in pure coherent superposition states [1]. In this case, as long as the coherence does not dephase, collisions are suppressed by the Pauli exclusion principle.

For two-component mixtures of neutral fermions with strongly attractive interactions, i.e., large and negative scattering lengths, a transition to a superfluid is predicted to occur at experimentally accessible temperatures [2]. Certain stable two-state mixtures of ^{6}Li and ^{40}K which have very strong, magnetically tunable, attractive interactions are promising candidates for observing this transition. However, the required spin states are expelled from a magnetic trap, necessitating the use of an all-optical trap with a low residual heating rate. If temperatures well below the Fermi temperature can be obtained, optical trapping of two-state mixtures may enable the superfluid regime to be explored.

In the following sections, we discuss the basic physics of coherence in trapped fermionic superposition states, the superfluid state, and our experiments on evaporation in two-state mixtures.

2 Hyperfine Coherence in a Single Component Fermi Gas

In the CO_2 laser trap, there is very little perturbation of the ground state hyperfine transition frequency. However, there is a small difference in the confining potentials of the upper and lower hyperfine states arising from the small difference in the optical detuning for these states. This finite difference between the confining potentials of the upper and lower hyperfine states produces a difference between the vibrational frequencies of the upper and lower hyperfines states, $\Delta\nu_{vib} = (\nu_{HF}/2\nu_0)\nu_{vib}$, where ν_{HF} is the ground state hyperfine splitting. Here ν_0 is the resonance frequency of the first optical transition and ν_{vib} is the vibrational frequency.

The vibrational frequency difference leads to a spread in the hyperfine transition frequency which is proportional to the spread in vibrational energy. This causes the coherence to dephase unless the atoms are very cold. To minimize the dephasing rate, an ultracold gas of ^{6}Li can be confined in a shallow CO_2 laser trap, where the minimum trap depth is limited only by the gravitational force. In this case, hyperfine coherence kHz $\nu_y = 3.3$ kHz and $\nu_z = 180$ Hz, where the trap laser beam propagates along **z**. Using the measured total power as a constraint, we obtain the trap intensity $1/e^2$ radii, $w_x = 50\,\mu$m and $w_y = 67\,\mu$m, and the axial intensity $1/e^2$ length $z_f \simeq 1.13$ mm, where z_f is consistent with the expected Rayleigh length within 15%. In the single beam trap, up to 4×10^5 atoms are loaded. A retroreflecting, polarization flipping mirror is used to increase the initial well depth, increasing the maximum number of loaded atoms to 8×10^5.

After loading and optical pumping into the two lowest ^{6}Li hyperfine states, a magnetic field of 130 G is applied, yielding a scattering length of -130 a_0. Evaporative cooling proceeds rapidly, the retroreflected beam is extinguished, and the temperature reaches 0.1 of the single beam well depth, or 30 μK. This yields an initial phase-space density of 3×10^{-3} at a well depth of 300 μK.

Forced evaporative cooling of the two-component gas of ^{6}Li is accomplished by lowering the depth of the optical trap. When the well depth is adiabatically lowered during evaporation, the rate of energy loss from the trap obeys the approximate evolution equation,

$$\dot{E} = U\dot{N} + \frac{\dot{U}}{U}\frac{E}{2}. \tag{11}$$

Eq. 11 follows from the s-wave Boltzmann equation in the ergotic approximation, with a time-dependent harmonic potential, neglecting background gas collisions. Here, N is the number of atoms and U is the trap depth. In the classical harmonic oscillator limit, $E = 3Nk_BT$ is the total energy of the trapped gas. The first term denotes the energy loss due to evaporation, while the second term is the loss of energy of the trapped atoms as the confining potential is adiabatically lowered. If evaporation is carried out near stagnation, the average thermal energy $k_BT = U/\eta$ with $\eta >> 1$. For our optical trap, $\eta \simeq 10$. Solving Eq. 11 with a fixed value of $U/kT = \eta$, it is easy to show that the phase space density, neglecting effects of Fermi statistics, scales with number and trap depth as

$$\rho = \rho_0 \left(\frac{N_0}{N}\right)^{\eta-4} = \rho_0 \left(\frac{U_0}{U}\right)^{\frac{3}{2}\frac{\eta-4}{\eta-3}}. \tag{12}$$

An important feature of Eq. 12 is that the increase in phase space density with decreasing number is identical to that obtained using a radio-frequency-knife method with a fixed trap depth and the same value of η. This is a consequence of the adiabatic energy loss which ensures that the phase-space density does not change as the trap depth is lowered in the absence of evaporation. For an optical trap with $\eta = 10$, only a modest decrease in number is needed to greatly increase the phase space density. Unlike evaporation from a fixed well, however, the collision rate for a fixed cross section decreases as $(U/U_0)^{5/7}$ for $\eta = 10$, and runaway evaporation is not achieved. Nevertheless, for ^{6}Li, where the scattering length is anomalously large, evaporation is still rapid and background collision induced loss can be minimized despite the reduced collision rate. This is confirmed by detailed modeling using a Boltzmann equation which includes Fermi statistics and background gas collisions [15]. For our initial phase space density of 3×10^{-3} we expect that lowering the well depth by a factor of 100 will yield a degeneracy.

In our first experiments, the well depth is exponentially lowered by a factor of 100 over 90 seconds. The highest phase-space density occurs at 60 seconds and is 2×10^{-2}, at a temperature of 12.6 μK in the recompressed well, or 2.2 times the Fermi temperature, T_F. Further improvements to the loading and to the well lowering procedure (which was not optimized) are expected to produce a degenerate two-state mixture at a temperature below $0.5\,T_F$.

5 Acknowledgements

This research was supported by the National Aeronautics and Space Administration, the Army Research Office, and the National Science Foundation.

References and links

1. K. Gibble and B. J. Verhaar, "Eliminating Cold-Collision Frequency Shifts," Phys. Rev. A **52**, 3370 (1995).
2. H. T. C. Stoof, M. Houbiers, C. A. Sackett, and R. G. Hulet, "Superfluidity of Spin-Polarized ^{6}Li," Phys. Rev. Lett. **76**, 10 (1996); See also, M. Houbiers, R. Ferwerda, H. T. C. Stoof, W. I. McAlexander, C. A. Sackett, and R. G. Hulet, "Superfluid State of Atomic ^{6}Li in a Magnetic Trap," Phys. Rev. A **56**, 4864 (1998).
3. K. M. O'Hara, S. R. Granade, M. E. Gehm, T. A. Savard, S. Bali, C. Freed, and J. E. Thomas, "Ultrastable CO_2 Laser Trapping of Lithium Fermions," Phys. Rev. Lett. **82**, 4204 (1999).
4. T. A. Savard, K. M. O'Hara, and J. E. Thomas, "Laser-Noise-Induced Heating in Far-off Resonance Optical Traps," Phys. Rev. A **56**, R1095 (1997).
5. M. E. Gehm, K. M. O'Hara, T. A. Savard, and J. E. Thomas, "Dynamics of Noise Induced Heating in Atom Traps," Phys. Rev. A **58**, 3914 (1998).
6. S. Bali, K. M. O'Hara, M. E. Gehm, S. R. Granade, and J. E. Thomas, "Quantum-Diffractive Background Gas Collisions in Atom-Trap Heating and Loss," Phys. Rev. A Rapid Comm. **60**, R29 (1999).
7. For a comprehensive treatment of background gas collision induced heating including multiple scattering, see H. C. W. Beijerinck, "Heating Rates in Collisionally-Opaque Alkali-metal Traps: Role of Secondary Collisions," Phys. Rev. A (2000).
8. K. M. O'Hara, M. E. Gehm, S. R. Granade, S. Bali, and J. E. Thomas, "Stable, Strongly Attractive, Two-State Mixture of Lithium Fermions in an Optical Trap," Phys. Rev. Lett. **85**, 2092 (2000).
9. M. Tinkham, ***Introduction to Superconductivity*** (McGraw-Hill, New York, 1975), pp. 16-71.
10. G. M. Bruun and C. W. Clark, "Detection of the BCS Transition of a Trapped Fermi Gas," see http://xxx.lanl.gov/abs/cond-mat/9906392 (25 June, 1999).
11. W. Zhang, C. A. Sackett, and R. G. Hulet, "Optical Detection of a Bardeen-Cooper-Schrieffer Phase Transition in a Trapped Gas of Fermionic Atoms," Phys. Rev. A **60**, 504 (1999).
12. B. DeMarco, and D. S. Jin, "Onset of Fermi Degeneracy in a Trapped Atomic Gas," Science **285**, 1703 (1999).
13. A. G. Truscott, K. E. Strecker, W. I. McAlexander, G. B. Patridge, and R. G. Hulet, "Observation of Fermi Pressure in a Gas of Trapped Atoms" *Science*, **291**, 2570-2572, (2001).
14. F. Schreck, G. Ferrari, K. L Corwin, J. Cubizolles, L. Khaykovich, M.-O. Mewes, and C. Salomon, "Sympathetic Cooling of Bosonic and Fermionic Lithium Gases Toward Quantum Degneracy," Phys. Rev. A **64**, 011402 (R), (2001).
15. For a discussion of evaporative cooling in an optical trap, see K. M. O'Hara, Ph.D. Thesis, (Duke University, 2000), unpublished.
16. K. M. O'Hara, S. R. Granade, M. E. Gehm, and J. E. Thomas, "Loading Dynamics of CO_2 Laser Traps," Phys. Rev. A **63**, 043403 (2001).

Generation of photon number states by cavity quantum electrodynamics

S. Brattke, G.R. Guthöhrlein, M. Keller, W. Lange, B. Varcoe, H. Walther

Sektion Physik der Universität München and
Max-Planck-Institut für Quantenoptik
85748 Garching, Fed. Rep. of Germany

herbert.walther@mpq.mpg.de
http://mste.laser.physik.uni-muenchen.de/

Abstract: Recent widely discussed applications in quantum information and quantum cryptography require radiation sources able to produce a fixed number of photons. In this paper we review the work performed in our laboratory to produce such fields on demand. One setup described is based on the micromaser operated under the conditions of a trapping sate. The second device uses a single ion in an optical cavity.

1 Introduction

The many applications discussed in quantum communication and quantum cryptography require sources able to produce a preset number of photons. Single photons are, for example, a necessary requirement for secure quantum communication [1, 2, 3], for quantum cryptography [4] and in special cases also for quantum computing [5]. However, photon fields with fixed photon numbers are also interesting from the point of view of fundamental physics since they represent the ultimate non-classical limit of radiation. When the photon number state is generated by strong coupling of excited-state atoms, a corresponding number of ground-state atoms is simultaneously populated. Such a system therefore produces a fixed number of atoms in the lower state as well. This type of atom source is a long sought after *gedanken* device [6]. Single photons have been generated by several processes such as single-atom fluorescence [7], single-molecule fluorescence [8], two-photon down-conversion [9], Coulomb blockade of electrons [10], and one- and two-photon Fock states have been created in the micromaser [11] (see also [12]). As these sources do not produce the photons on demand, they are better described as "heralded" photon sources, because they are stochastic either in the emission direction or in the time of creation. A source of single photons or even more generally Fock states created on demand has not yet been demonstrated. Cavity quantum electrodynamics (QED) provides us with both the possibility of generating a photon at a particular time and localising its emission direction. To this end there have been several proposals making use of high Q cavities that are basically capable of serving as sources of single photons [3, 13, 14, 15]. The current paper reviews the work on a microwave source able to produce a preset number of photons and lower state atoms. The principle of the source and the first experimental demonstration will be described. It is based on the One-Atom Maser and allows the generation of a specified photon Fock state ($n \geq 1$) *on demand*, without the need for conditional measurements, being therefore independent of detector efficiencies.

In the second part of the paper the work towards a new single photon source in the visible spectral range is described. This source uses a single trapped ion which is placed into a cavity. The first progress towards the realization of this source will be reported.

Coherence and Quantum Optics VIII
Edited by Bigelow et al., Kluwer Academic/Plenum Publishers, 2003

Steady-state operation of the One-Atom Maser or micromaser has been studied extensively both theoretically [16] and experimentally, and has already been used to demonstrate many quantum phenomena of the radiation field such as sub-Poissonian statistics [17], the collapse and revival of Rabi oscillations [18], and entanglement between the atoms and cavity field [19]. More recently two experiments have demonstrated that Fock states (ie. states with a fixed photon number) can be readily created in the normal operation of the maser, by means of either state reduction [11] or steady state operation of the micromaser in a trapping state [12]. State reduction is possible owing to the entanglement between the state of the outgoing atoms and the cavity field; detection of a lower state atom means that a field originally in an n photon Fock state is projected onto the $n+1$ state [20]. As a source of single photons such a source can be compared to two photon down conversion, in which an idler beam is used to herald the creation of a photon in the signal beam. Both are subject to the same limitation in that the creation of the Fock state is unpredictable, and imperfect detectors further reduce the probability that a state, once created, is also detected. In contrast it is shown here that the micromaser can be used to prepare Fock states with small photon numbers in the cavity *on demand* and independent of detection efficiencies. Simultaneously the same number of ground state atoms are produced with an efficiency of up to 98%.

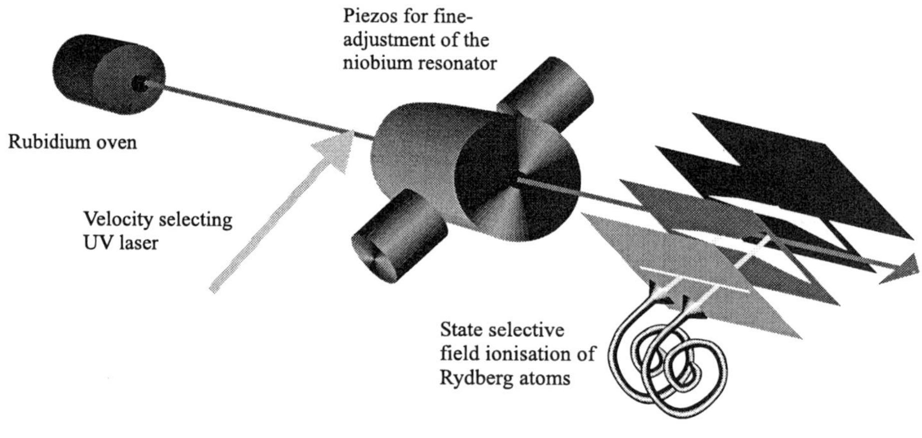

Fig. 1. The atoms leaving the rubidium oven are excited into the $63P_{3/2}$ Rydberg state by means of a UV laser at an angle of 11°. After the cavity the atoms are detected by state-selective field ionisation. Tuning of the cavity is performed with two piezo translators. An auxiliary atomic beam (not shown) is used to stabilise the laser frequency. The laser is locked to a Stark-shifted atomic resonance of the auxiliary beam, thus allowing the velocity subgroup selected by excitation to be continuously changed within the range of the velocity distribution of the atoms.

Trapping states are a feature of the low temperature operation of the micromaser, for which the steady state photon distribution closely approximates a Fock state under certain conditions. They are typical of strongly coupled systems. They occur when atoms perform an integer number, k, of Rabi cycles under the influence of a fixed photon number n:

$$\sqrt{n+1}gt_{\text{int}} = k\pi, \tag{1}$$

where g is the effective atom-field coupling constant and t_{int} is the interaction time. Trapping states are characterised by the number of photons n and the number of Rabi cycles k. The trapping state $(n, k)=(1, 1)$ therefore refers to the one photon, one Rabi oscillation trapped field state. In other words, trapping states occur when the interaction time is chosen such that the emission probability becomes zero for certain operating parameters of the maser. Therefore at some time during the steady state operation

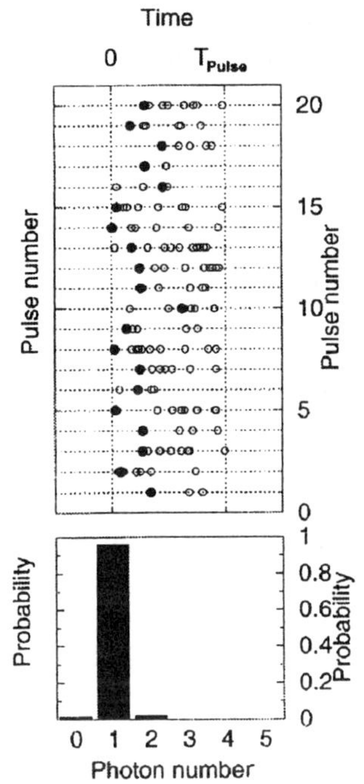

Fig. 2. A simulation of a subset of twenty subsequent atom bunches after the cavity and the associated probability distribution for photons or lower-state atom production (filled circles represent lower-state atoms and open circles represent excited-state atoms). The start and finish of each pulse is indicated by the vertical dotted lines marked 0 and τ_{pulse}, respectively. The operating conditions are the (1,1) trapping state ($gt_{int} = 2.2$) conditions. The size of the atoms in this figure is exaggerated for clarity. With the real atomic separation there are 0.06 atoms in the cavity on average (i.e., the system operates in the one-atom regime). The other parameters are $\tau_{pulse} = 9{,}92\ \tau_{cav}$, $n_{th} = 10^{-4}$ and $N_a = 7$ (see also Ref. [23]).

of the micromaser in a trapping state the field will enter a Fock state and become stabilised. The particular Fock state is known and is determined by the interaction time between atom and cavity as given by the trapping state formula (Eq. 1). The Fock state, once prepared is preserved owing to the trapping condition with a minimum probability of photon emission. Following the state preparation, the beam of pump atoms can be turned off and the Fock state remains in the cavity for the duration of the cavity decay time. For simplicity we will concentrate in the following on the preparation of a one photon Fock state, however, the method can also be generalised to the generation of fields of higher photon numbers.

The micromaser setup used for the experiments is shown in Fig. 1 and is operated in the same way as described in [12]. Briefly, a ^{3}He-^{4}He dilution refrigerator houses the closed superconducting microwave cavity. A rubidium oven provides two collimated atomic beams: the main beam passing directly into the cryostat and a second used to stabilise the laser frequency [12]. A frequency doubled dye laser (λ = 297 nm) was used to excite rubidium (^{85}Rb) atoms to the Rydberg $63P_{3/2}$ state from the $5S_{1/2}$(F = 3) state. The cavity is tuned to the 21.456 GHz transition from the $63P_{3/2}$ state to the $61D_{5/2}$ state, which is the lower or ground state of the maser transition. For this experiment a cavity with a Q-value of 4×10^{10} was used, this corresponds to a field decay time of 0.6 s or a photon lifetime of 0.3 s. This Q-value is the largest ever achieved in this type of experiment and the photon lifetime is more than two orders of magnitude higher than that achieved in related experiments [21]. This cavity is used to study micromaser operation in great detail. To realise the Fock state it is necessary to switch the excitation of the Rydberg atoms on and off in a predefined pulse sequence, this was achieved by means of an intensity modulating electro optic modulator triggered by control software. The pulse duration and pulse separation can both be tailored to the conditions required for the particular experiment.

To demonstrate the principle of this source, Fig. 2 shows a sequence of twenty subsequent pulses obtained via a Monte Carlo simulation [22] of the micromaser operating in the (1, 1) trapping state. In each pulse there is a single emission event, producing a single lower state atom and leaving a single photon in the cavity. In the case of the loss of a photon by dissipation, one of the next incoming excited state atoms will restore the single photon Fock state. This condition has been observed in [12] when sub-Poissonian atom statistics were measured with the maser operating in a trapping state. The influence of thermal photons and variations in interaction time or cavity tuning further complicates this picture, resulting in reduced visibility of steady state Fock states [24]. Pulsed excitation as discussed here, however, reduces these perturbations and the Fock state maintains a high visibility.

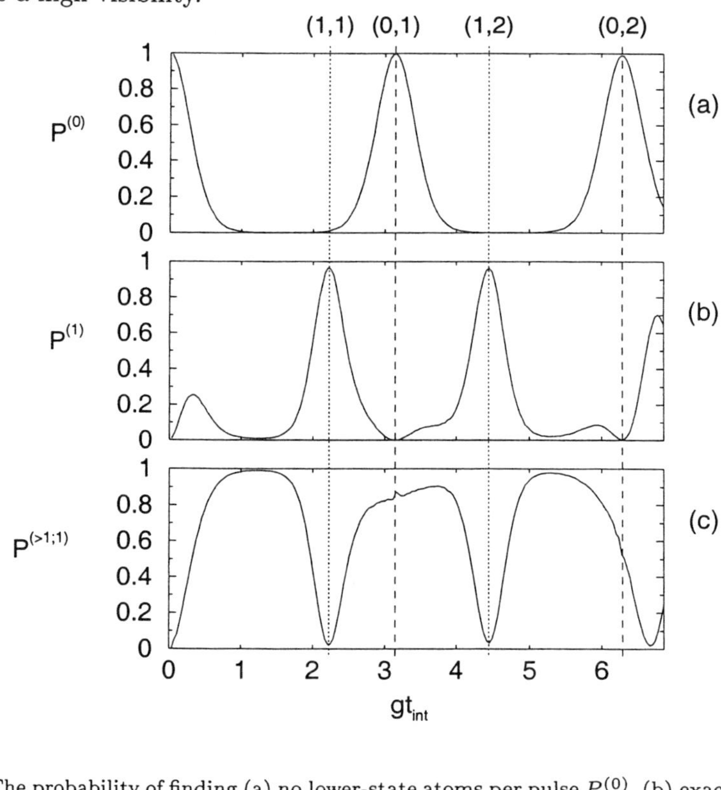

Fig. 3. The probability of finding (a) no lower-state atoms per pulse $P^{(0)}$, (b) exactly one lower-state atom per pulse $P^{(1)}$, and (c) of finding a second lower-state atom, if one has already been detected $P^{(>1;1)}$. The parameters are $\tau_{\text{pulse}}=0.02\tau_{\text{cav}}$, $N_{\text{a}}= 7$ atoms and $n_{\text{th}} = 10^{-4}$. The maximum value of $P^{(1)}$ is 98% for the (1, 1) trapping state.

Figure 3 shows three curves obtained again from a computer simulation, that illustrate the behaviour of the maser under pulsed excitation as a function of interaction time for more ideal (but achievable) experimental parameters. The simulations show the probability of finding; no ground state atom per pulse ($P^{(0)}$); finding exactly one ground state atom per pulse ($P^{(1)}$); and the conditional probability of finding a second ground state atom in a pulse already containing one ($P^{(>1;1)}$). It is shown below that the latter plot of the conditional probability, $P^{(>1;1)}$, has the advantage of being especially suitable for the comparison between theory and experiment as it is relatively insensitive to the detection probability for atoms in the upper and lower maser levels.

From the simulations it follows that with an interaction time corresponding to the (1, 1) trapping state, both one photon in the cavity and a single atom in the lower state are produced with a 98% probability. In order to maintain an experimentally verifiable quantity, most of the simulations presented relate to the production of lower state atoms rather than to the Fock state left in the cavity. Pulse lengths on the other hand are rather short ($0.01\tau_{\rm cav} \leq \tau_{\rm pulse} \leq 0.1\tau_{\rm cav}$) so there is little dissipation and the one photon state in the cavity following the pulse is very close to the probability of finding an atom in the lower state. Note that at no time in this process is a detector event required to project the field, the field evolves to the trapping state as a function of time, when the suitable interaction time has been chosen.

The variation of the time when an emission event occurs during an atom pulse in Fig. 2 is due to the variable time spacing between the atoms as a consequence of Poissonian statistics and the stochasticity of the quantum process. The atomic rate therefore has to be high enough that there will be a sufficient number of excited atoms per laser pulse, in order to maintain the 98% probability of an atom emitting.

To guarantee single-atom single-photon operation, the duration of the preparation pulses must be short in relation to the cavity decay time. For practical purposes, the pulse duration should be smaller than $0.1\tau_{\rm cav}$ for dissipative losses to be less than 10%. Apart from reducing the fidelity of the Fock state produced, losses increase the likelihood of a second emission event leading to larger number of lower state atoms than photons in the field; whereby the 1:1 correspondence between both would be lost. Shorter atom pulses reduce the dissipative loss, however the number of atoms per cavity decay time (usually labelled $N_{\rm ex}$) must be larger than ten times the threshold value of the atomic flux to realise the Fock source with a significant fidelity. Since a minimum atom number is required to produce the desired state, care must also be taken to avoid atom beam densities violating the one-atom-at-a-time condition.

For a large range of operating conditions, the production of Fock states of the field and single lower state atoms is remarkably robust against the influence of thermal photons, variations of the velocity of atoms and other influences such as mechanical vibrations of the cavity. Much more so than the steady state trapping states, for which highly stable conditions with low thermal photon numbers are required [12, 24].

An obvious side effect of the production of a single photon in the mode is, as mentioned already, that a single atom in the lower state is produced. This atom is in a different state when it leaves the cavity and is therefore distinguishable from the pump atoms, hence under this operation, the micromaser also serves as a source of single atoms in a particular state, a requirement for many proposed experiments [6, 25].

Although the distribution of lower state atoms leaving the cavity will be maximally sub-Poissonian, the arrival time of an atom within the pumping pulse still shows a small uncertainty, the upper limit of which is determined by the pump pulse duration in the range of $0.01 - 0.1\tau_{\rm cav}$ for the parameters used in this paper. The separation of the pulses is $\geq 3\tau_{\rm cav}$ leading to a small relative variation in the arrival times. If one would increase the pump rate still further, the pulse lengths could be further reduced and the arrival of an atom becomes even more predictable.

The present setup of the micromaser was specifically designed for steady state operation and is therefore not ideal for the parameter range presented here. However the current setup does permit a comparison between theory and experiment in a relatively small parameter range. The experimental test relies on the measurement of an absolute number of atoms and although the operation of the Fock source is independent of detector efficiencies, the experimental test is blurred by the fact that the state selective

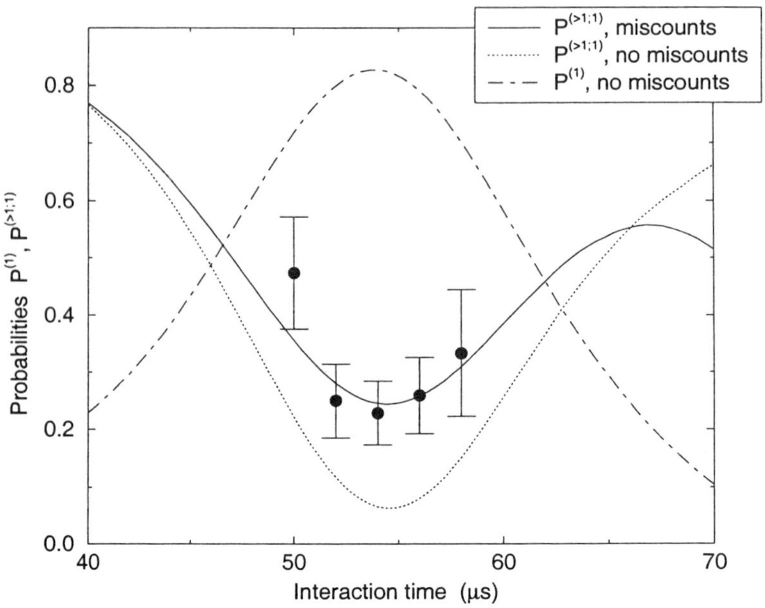

Fig. 4. Comparison between theory and experiment. Investigated here is the behaviour shown in Fig. 3. The experimental data is evaluated according to Eq. 2. This ratio is to the absolute detector efficiency and is dependent only on the relative detector efficiencies and the miscount probability which can be measured experimentally. The relative detector efficiencies can be considered approximately equal. On the other hand a calculation of the probability of finding one lower state atom per pulse is highly dependent on the absolute detector efficiency. The two theoretical curves presented are a theoretical prediction of Fock state creation and a theoretical result which takes into account the experimentally measured miscounts of 7% in the lower state detector and 2% in the excited state detector. Given the extreme conditions for the operation of the apparatus, there is an excellent agreement between theory and experiment. The single atom probability is evaluated to be 83.2%. The parameters for the experiment were, $\tau_{cav} = 300$ ms, $\tau_{\text{pulse}} = 0.066\tau_{cav}$, pulse spacing of 1 s, $n_{\text{th}} = 0.03$, $N_{\text{a}} = 4$.

field ionisation detectors for the Rydberg atoms do not reach an efficiency of 100%. Atoms in a particular state will therefore be missed leading to wrong or misleading results. In order to circumvent this disadvantage, it is useful to measure population correlations between subsequent atoms instead. Owing to the strong coupling between the atoms and cavity, the cavity field and the state of the pumping atom are entangled following the interaction. A subsequent pumping atom will thereby also become entangled with any previous one, thus the population correlations between subsequent atoms are determined by the particular dynamics of the atom-cavity interaction. The connection between population correlations and the micromaser dynamics, has been studied in detail in previous papers [19, 26]. It is important to note that even in the presence of lost counts, the correlations between subsequently detected atoms are maintained. Conditioning the experimentally measured parameter on the detection of atom pairs that contain at least one lower state atom, provides both a value appropriate to the existent correlation and -at the same time- directly related to the total probability of finding one atom per pulse.

By means of an extremely high cavity Q factor (4×10^{10}) an N_{ex} of approximately 60 was accessible for a short range of interaction times around the maximum in the Maxwell-Boltzmann velocity distribution, which happens to correspond to the interaction time for the (1, 1) trapping state. A pulse length of $\tau_{\text{pulse}} = 0.066\ \tau_{\text{cav}}$ leading to an average of 4 atoms per pulse was chosen as a compromise between the effects of dissipation and external influences, while still providing a pump rate above the threshold for single photon Fock state production. Figure 4 shows the results of the comparison of theory and experiment for a scan over the (1, 1) trapping state. Plotted here is the ratio of two atom events to the total number of two atom events detected that contain at least one lower state. That is,

$$P^{(>1;1)} = \frac{N_{\text{gg}}}{N_{\text{gg}} + N_{\text{eg}} + N_{\text{ge}}} \tag{2}$$

where for example N_{eg} is the probability of detecting a two atom event containing first an upper state atom (e) and then a lower state atom (g) in any given pulse. The number of three atom events detected is negligible and can be ignored as a contributing factor. Constructing the parameter in this way ensures that there is a one to one correspondence between the correlation parameter and the maximum probability of finding exactly one atom per pulse. That is, to a good approximation for pump rates and pulse durations employed in this experiment, the difference between the upper bound $P^{\max}$ and the measured correlation $P^{(>1;1)}$ gives the probability of finding exactly one atom per pulse ($P^{(1)}$). $P^{(>1;1)}$ in Eq. 2 is independent of the absolute detector efficiency and depends only on the relative detector efficiencies and the miscount probability (the probability that a given atomic level is detected in the wrong detector), each of which has been measured experimentally. The theoretical curves of Fig. 4 represent an evaluation of the probability of finding exactly one atom per pulse and the conditional probability introduced in Fig. 3. The curves are evaluated both for the ideal situation of no detector miscounts and for the measured detector miscounts of 7% in the lower state detector and 2% in the excited state detector. When the miscounts are incorporated into the data there is an excellent match between the experimental points and the theoretical curve.

As the present apparatus was designed for operating the micromaser in steady state conditions, the current results were obtained under non-ideal operating parameters of the apparatus. In order to overcome this disadvantage future developments will incorporate two improvements, to increase the atomic flux and to introduce a second pulse of atoms with variable velocity to act as a field probe. With these changes it will be easier to arrive at the optimum conditions for the Fock source and, in addition, will allow direct measurements of the cavity photon number by means of Rabi oscillations of the probe atom [11] and further studies of quantised field effects.

The source presented here has the significant advantage over our previous method of Fock state creation [11] of being unconditional and therefore significantly faster in preparing a target quantum state. Previously the state was prepared by the dynamics of the interaction of excited state atoms with the cavity field. State reduction occurred on detection of a lower state atom indicating the preparation of the desired cavity field. In the current experiment however the cavity field is correctly prepared in 83.2% of the pulses, independently of state reduction and hence atomic detection efficiency. Simply detecting the lower state atom as it emerges from the cavity increases the fidelity of the one photon Fock state to $\geq 95\%$ at the moment of detection (incorperating both dissipative losses and detector miscounts). Assuming 40% detector efficiency, detection of a lower state atom within any given preparation pulse will occur with a 36.8% probability. This should be compared to the 95% fidelity of the measured Fock state in [11], in which there was a $\leq 1\%$ probability of detecting the correctly prepared state. Thus along with the first observation of the operation of a single photon Fock state source, we also have an order of magnitude improvement of Fock state creation over our previous experiment.

In the second part of this paper we would like to report on the progress of our work on ions in optical cavities. The interaction of a single atom with a single field mode of a high finesse cavity has been the subject of a number of experiments in the field of cavity QED [15]. However, most of these investigations suffer from a lack of control over the position of the atom, which results in non-deterministic fluctuations of the coupling between atoms and field. In this context, the strong localisation and position control available when an ion trap is combined with an optical cavity would be a big step forward and would become a key technology for future progress in cavity QED in the optical range. We are presently implementing two experiments exploiting the localisation of an ion in a cavity. By pulsed excitation of a maximally coupled ion, single-photon wave packets may be emitted from the cavity on demand [14, 15, 27] (single-photon gun). Under conditions of strong coupling, a single calcium ion in the cavity provides sufficient gain to build up a laser field [13]. Analogous to a single ion in free space, which previously was shown to be an excellent source of antibunched light [28, 29], radiation from a single-ion laser has nonclassical photon statistics and correlations.

An equally attractive goal in the area of cavity QED is the simultaneous interaction of two or more ions with a single cavity mode. Due to the linear geometry of our trap several ions can be stored within the mode volume. As a first test, we have placed an array of two ions in the cavity field and observed the total fluorescence. We succeeded in matching the ion crystal to the two maxima of the TEM_{01} mode of the cavity. In such a configuration the cavity field may be used to entangle the two ions [30, 31]. This is a promising alternative to schemes involving the ions' motional degrees of freedom, since there is no need for cooling the vibrational modes of the string below the Doppler temperature. Using a cavity to perform quantum operations on adjacent pairs of ions in a long string is a viable route to a scalable quantum computer.

In the following we give a progress report of our experiment. We are using a linear trap with Ca ions (see Fig. 5). As an initial test of the setup for the above mentioned cavity QED experiments, we have used the trapped Ca ions to probe the optical field in the cavity. The Ca ion is sensitive to radiation close to the resonance line $4^2S_{1/2} - 4^2P_{1/2}$ at a wavelength of $\lambda = 397$ nm. The fluorescent light emitted by the ion is collected with a lens (numerical aperture = 0.17) and detected with a photomultiplier tube (overall detection efficiency $\eta \approx 10^{-4}$). The observed fluorescence rate R is proportional to the local intensity of the optical field at the position $\vec{r}$ of the ion, i.e. $R \propto I(\vec{r})$, provided there is no saturation of the atomic transition. By scanning the position of the ion in the field and detecting the fluorescence rate at each point, a high-resolution map of the optical intensity distribution is obtained. It should be noted that a single ion can also probe the amplitude distribution $\vec{E}(\vec{r})$ of the light field and hence measure its phase. To this end heterodyne detection of the fluorescent light must be used, with the exciting laser as a local oscillator [7, 29].

With the single ion as a probe, we have investigated the eigenmodes of a Fabry-Perot resonator formed by two mirrors (radius of curvature = 10 mm) at a distance of L = 6 mm (Fig. 5). The transverse mode pattern is described by Hermite-Gauss functions with a beam waist $w_0 \approx 24 \mu m$, while in the direction of the cavity axis a standing wave builds up. In the experiment, a particular cavity mode is excited by a laser beam with a power of a few hundred nanowatts at 397 nm. The length of the cavity is actively stabilised to this mode.

An ion is loaded in the trap after electron-impact ionisation of calcium atoms. Since the electron beam and the calcium beam would degrade the optical mirrors and make stable trapping difficult, we use a linear trap and load it in a region spatially separated from the observation zone, as shown in Fig. 5. Subsequently, DC electrodes along the

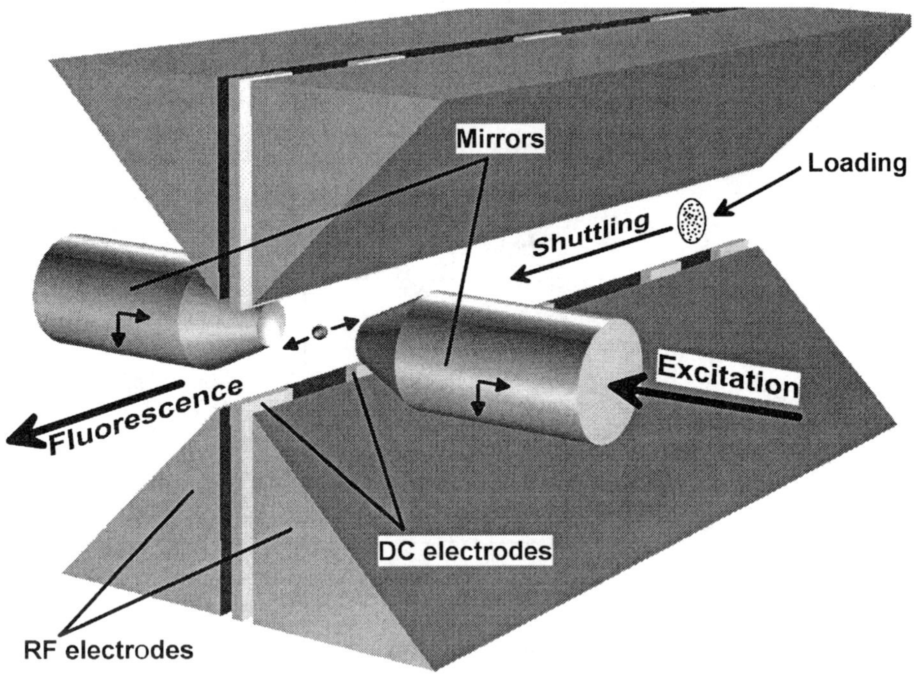

Fig. 5. Experimental arrangement of trap electrodes and cavity mirrors. The ion is loaded at the rear end of the trap and shuttled to the mirror region. Fluorescence is observed from the side of the cavity. For scans in the direction of the trap axis, the ion is moved with DC electrodes. In all other directions, the cavity is translated relative to the ion's position, as indicated by arrows.

axis are employed to shuttle the ion over a distance of 25 mm to the uncontaminated end of the trap, where the cavity is located, oriented at right angles to the trap axis. Residual DC fields in the radial direction must be carefully compensated with correctional DC voltages to place the ion precisely on the nodal line of the RF field (coinciding with the trap axis), to avoid the trapping field exciting the micromotion of the ion.

In the direction of the trap axis, the ion is confined in a DC potential well, which is approximately harmonic with an oscillation frequency of $\omega_z \approx 300$kHz. By applying asymmetric voltages, the minimum of the potential well and thus the equilibrium position of the ion is moved along the trap axis. By simultaneously monitoring the fluorescence, we have sampled one-dimensional cross sections of the cavity mode. The width of the ion's wave function in the axial potential well is a few hundred nanometres, which provides sufficient resolution to map the transverse mode pattern with an intensity distribution varying on a scale given by the cavity waist w_0.

Figure 6 shows scans of the first four TEM_{0n} modes of the cavity obtained in this way. The fluorescence data is not entirely symmetric because of a small displacement and rotation of the cavity eigenmodes with respect to the trap axis. In each plot, an inset indicates the path along which the ion is scanned. The solid curves in Fig. 6 are obtained from a fit using Hermite-Gauss functions and take into account saturation of the ion's transition. The influence of saturation is apparent in Fig. 6c, where a slightly higher intensity was injected into the cavity. In all cases, the correspondence with the measured fluorescence is excellent.

The ion's motion must be restrained to the trap axis, since off the axis the radio–frequency field of the trap would lead to micromotion. To scan other dimensions of the field, the sample must be moved. In our experiment this is done by piezoelectrically

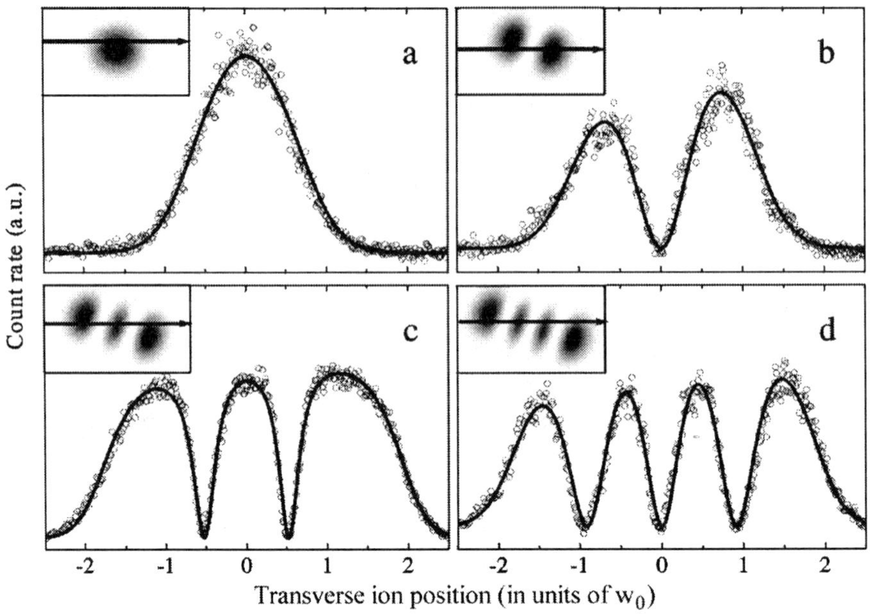

Fig. 6. Transverse profiles of the Hermite-Gauss modes of the cavity, obtained by monitoring the ion's fluorescence while scanning over a range of 120 μm. The solid line is a fit including saturation of the transition. The inset shows the calculated intensity distribution of the mode and indicates the scan path. The modes are a) TEM_{00}, b) TEM_{01}, c)TEM_{02}, d) TEM_{03}.

translating the entire cavity assembly perpendicular to the trap axis.

RF confinement of the ion perpendicular to the trap axis is also harmonic, but the corresponding oscillation frequency $\omega_r \approx 1.1 MHz$ is larger than the axial frequency so that field structures in the radial direction of the trap are better resolved. The resolution achieved with our method may be determined most accurately by probing the standing-wave field created between the cavity mirrors, which varies on a scale of $\lambda/2$. To this end the cavity was moved parallel to its axis while keeping the ion stationary and monitoring its fluorescence. Figure 7 shows the mapping of the cavity field obtained in this way. A pronounced standing-wave pattern is observed with a visibility of 40 %.

Taking advantage of the excellent localisation in ion traps, we have performed the most precise measurement to date of a three-dimensional spatial structure of an optical field over a range of up to 100 μm. As a demonstration we scanned modes of a low-loss optical cavity. The precise positioning we achieve implies a deterministic control of the coupling between ion and field. At the same time, the field and the internal states of the ion are not affected by the trapping potential. What we have realised, therefore, is an ideal system for cavity QED with a single particle.

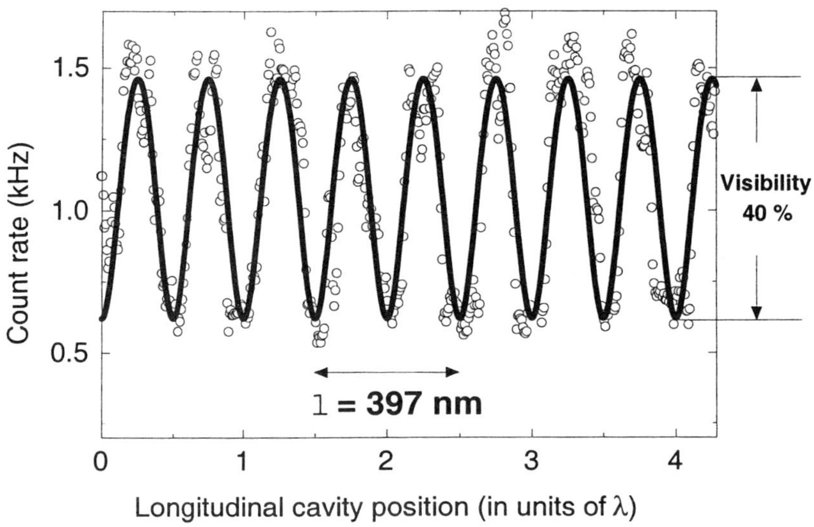

Fig. 7. Single-ion mapping of the longitudinal structure of the cavity field. The visibility is determined by the residual thermal motion of the Doppler-cooled ion. It corresponds to a resolution of 60 nm. The localisation of the ion's wave packet in this measurement is 16 nm.

2 Conclusion

In this paper we reviewed our work on the generation of photon number states on demand using the micromaser. In addition, we describe first experiments using a single trapped ion in connection with an optical cavity. The system described in the second part of the paper shows a lot of promise and will open a series of new interesting applications.

References and links

1. H. Zbinden, N. Gisin, B. Huttner, and W. Tittel, "Practical aspects of quantum cryptographic key distribution," J. Cryptol. **13**, 207–220 (2000); H.-K. Lo and H. F. Chau, "Unconditional security of quantum key distribution over arbitrarily long distances," Science **283**, 2050–2056 (1999).
2. K. M. Gheri, C. Saavedra, P. Törmä, J. I. Cirac, and P. Zoller, "Entanglement engineering of one-photon wave packets using a single-atom source," Phys. Rev. A **58** R2627—R2630 (1998); S. J. van Enk, J. I. Cirac, and P. Zoller, "Ideal quantum communication over noisy channels: a quantum optical implementation," Phys. Rev. Lett. **78**, 4293–4296 (1997); S. J. van Enk, J. I. Cirac, and P. Zoller, "Photonic channels for quantum communication," Science **279**, 205–208 (1998).
3. J. I. Cirac, P. Zoller, H. J. Kimble, and H. Mabuchi, "Quantum state transfer and entanglement distribution among distant nodes in a quantum network," Phys. Rev. Lett. **78**, 3221–3224 (1997); A. S. Parkins, P. Marte, P. Zoller, and H. J. Kimble, "Synthesis of arbitrary quantum states via adiabatic transfer of Zeeman coherence," Phys. Rev. Lett. **71**, 3095–3096 (1993); S. Parkins and H. J. Kimble, "Quantum state transfer between motion and light," J. Opt. B. **1**, 496 (1999).
4. T. Jennewein, C. Simon, G. Weihs, H. Weinfurter, and A. Zeilinger, "Quantum cryptography with entangled photons," Phys. Rev. Lett. **84**, 4729-4732 (2000); D. S. Naik, C. G. Peterson, A. G. White, A. J. Berglund, and P. G. Kwiat, "Entangled state quantum cryptography: eavesdropping on the Ekert protocol," Phys. Rev. Lett. **84**, 4733-4736 (2000); W. Tittel, J. Brendel, H. Zbinden, and N. Gisin, "Quantum cryptography using entangled photons in energy-time Bell states," Phys. Rev. Lett. **84**, 4737–4740 (2000).
5. The error tolerant quantum computing proposal by D. Gottesman and I. L. Chuang, "Demonstrating the viability of universal quantum computation using teleportation and single-qubit operations," Nature (London) **402**, 390–393 (1999) requires that a "quantum resource" be supplied *on demand* to facilitate computation. Such a source can be provided by the apparatus considered here. See also J. Preskill, "Plug-in quantum software," Nature (London) **402**, 357–358 (1999);

D. Johnathon and M. B. Plenio, "Entanglement-assisted local manipulation of pure quantum states," Phys. Rev. Lett. **83**, 3566-3569 (1999).

6. Sources of single atoms, such as the one described in this paper, are routinely employed for hypothetical tasks such as the creation of an atomic beam with arbitrary timing sequence or for the stabilisation of cavity states. For an example see, D. Vitali, P. Tombesi and G. Milburn, "Quantum-state protection in cavities," Phys. Rev. A **57**, 4930–4944 (1998).
7. J. T. Höffges, H. W. Baldauf, W. Lange, and H. Walther, "Heterodyne measurement of the resonance fluorescence of a single ion," J. Mod. Opt. **44**, 1999–2010 (1997).
8. C. Brunel, B. Lounis, P. Tamarat, and M. Orrit, "Triggered source of single photons based on controlled single molecule fluorescence," Phys. Rev. Lett. **83**, 2722–2725 (1999).
9. C. K. Hong and L. Mandel, "Experimental realization of a localized one-photon state," Phys. Rev. Lett. **56**, 58–60 (1986).
10. J. Kim, O. Benson, H. Kan, and Y. Yamamoto, "Single-photon turnstile device," Nature (London) **397**, 500–503 (1999).
11. B. T. H. Varcoe, S. Brattke, M. Weidinger, and H. Walther, "Preparing pure photon number states of the radiation field," Nature (London) **403**, 743–746 (2000).
12. M. Weidinger, B. T. H. Varcoe, R. Heerlein, and H. Walther, "Trapping states in the micromaser," Phys. Rev. Lett. **82**, 3795-3798 (1999).
13. G.M. Meyer, H.-J. Briegel, and H. Walther, "Ion-trap laser," Europhys. Lett. **37**, 317–322 (1997).
14. C. K. Law and J. H. Eberly, "Arbitrary control of a quantum electromagnetic field," Phys. Rev. Lett. **76**, 1055–1058 (1996); C. K. Law and H. J. Kimble, "Deterministic generation of a bit-steam of single-photon pulses," J. Mod. Opt. **44**, 2067–2074 (1997); P. Domokos, M. Brune, J. M. Raimond and S. Haroche, "Photon-number-state generation with a single two-level atom in a cavity: a proposal," Eur. Phys. J. D **1**, 1–4 (1998).
15. A. Kuhn, M. Hennrich, T. Bondo, and G. Rempe, "Controlled generation of single photons from a strongly coupled atom-cavity system," Appl. Phys. B **69**, 373–377 (1999); P. W. H. Pinkse, T. Fischer, P. Maunz, and G. Rempe, "Trapping an atom with single photons," Nature (London) **404**, 365–368 (2000); J. Ye, D. W. Vernooy, and H. J. Kimble, "Trapping of single atoms in cavity QED," Phys. Rev. Lett. **83**, 4987–4990 (2000); C. J. Hood, T. W. Lynn, A. C. Doherty, A. S. Parkins, and H. J. Kimble, "The atom-cavity microscope: single atoms bound in orbit by single photons," Science **287**, 1447–1453 (2000).
16. See, for example, M. O. Scully and M. S. Zubairy, *Quantum Optics* (Cambridge University Press, 1997).
17. G. Rempe, F. Schmidt-Kaler, and H. Walther, "Observation of sub-Poissonian photon statistics in a micromaser," Phys. Rev. Lett. **64**, 2783–2786 (1990).
18. G. Rempe, H. Walther, and N. Klein, "Observation of quantum collapse and revival in a one-atom maser," Phys. Rev. Lett. **58**, 353–356 (1987).
19. B. Englert, M. Löffler, O. Benson, M. Weidinger, B. Varcoe, and H. Walther, "Entangled atoms in micromaser physics," Fortschr. Phys. **46**, 897–926 (1998).
20. J. Krause, M. O. Scully, and H. Walther, "State reduction and $|\, n\rangle$-state preparation in a high-Q micromaser Phys. Rev. A **36**, 4547–4550 (1987).
21. G. Nogues, A. Rauschenbeutel, S. Osnaghi, M. Brune, J. M. Raimond and S. Haroche, "Seeing a single photon without destroying it," Nature (London) **400**, 239–242 (1999).
22. A detailed account of the simulations used in this paper and a comparison with ideal micromaser theory can be found in S. Brattke, B.-G. Englert, B. T. H. Varcoe, and H. Walther, "Fock states in a cyclically pumped one-atom maser," J. Mod. Opt. **47**, 2857–2867 (2000).
23. S. Brattke et al., "Preparing Fock states in the micromaser," Optics Express **8**, 131–144 (2001).
24. B. T. H. Varcoe, S. Brattke, and H. Walther, "Generation of Fock states in the micromaser," J. Opt. B. **2**, 154–157 (2000).
25. Proposals such as the teleportation of an atomic state using multiple atomic beams would be substantially enhanced when atoms arrive on demand rather than by chance. See for example: L. Davidovich, N. Zagury, M. Brune, J. M. Raimond, and S. Haroche, "Teleportation of an atomic state between two cavities using nonlocal microwave fields," Phys. Rev. A **50**, R895–R898 (1994); J. I. Cirac and A.S. Parkins, "Schemes for atomic-state teleportation," Phys. Rev. A **50**, R4441–R4444 (1994); M. H. Y. Moussa, "Teleportation with identity interchange of quantum states," Phys. Rev. A **55**, R3287–R3290(1997).
26. H.-J. Briegel, B.-G Englert, N. Sterpi, and H. Walther, "One-atom maser: statistics of detector clicks," Phys. Rev. A **49**, 2962–2985 (1994).
27. M. Hennrich, T. Legero, A. Kuhn, and G. Rempe, "Vacuum-stimulated Raman scattering based on adiabatic passage in a high-finesse optical cavity," Phys. Rev. Lett. **85**, 4872–4875 (2000).
28. F. Diedrich and H. Walther, "Nonclassical radiation of a single stored ion," Phys. Rev. Lett. **58**, 203–206 (1987).
29. J.T. Höffges, H. W. Baldauf, T. Eichler, S.R. Helmfrid, and H. Walther, "Heterodyne measurement of the fluorescent radiation of a single trapped ion," Opt. Comm. **133**, 170–174 (1997).

30. T. Pellizzari, S. A. Gardiner, J. I. Cirac, and P. Zoller, "Decoherence, continuous observation, and quantum computing, a cavity QED model," Phys. Rev. Lett. **75**, 3788–3791 (1995).
31. S. B. Zheng and G. C. Guo, "Efficient scheme for two-atom entanglement and quantum information processing in cavity QED," Phys. Rev. Lett. **85**, 2392–2395 (2000).

Superluminal Light Pulse Propagation at a Negative Group Velocity

L. J. Wang, A. Dogariu, and A. Kuzmich

NEC Research Institute, 4 Independence Way, Princeton, NJ 08540, USA

Lwan@research.nj.nec.com

http://www.neci.nj.nec.com/homepages/lwan

Abstract: Anomalous dispersion cannot occur in a transparent, passive medium where electromagnetic radiation is being absorbed at all frequencies, as pointed out by Landau and Lifshitz. Here we show, using gain-assisted linear anomalous dispersion, that superluminal pulse propagation can occur in a transparent medium when gain is present. Superluminal light pulse propagation is observed at a negative group velocity of $-c/315$ through an atomic medium with little pulse distortion. Experimentally, a pulse propagating through the atomic vapor cell has its peak reaching the exit side before entering it, resulting in a negative transit time. This counterintuitive effect is a direct result of the rephasing process owing to the wave nature of light and is not at odds with causality.

1 INTRODUCTION

It is well known that a number of velocities are involved in the propagation of a light pulse through a transparent medium. First, there is the "wave (phase) velocity," c/n, determined by the refractive index n. The envelope of the pulse, however, travels at the group velocity $v_g = c/n_g$, determined by the group index: $n_g = n + \nu dn/d\nu$. In a dispersive medium where the the wave velocity depends on the frequency, i.e., $dn/d\nu \neq 0$, the group velocity can be very different from the phase velocity. The topic has been extensively discussed in the past [1, 2]. Furthermore, Sommerfeld and Brillouin considered to define a "signal velocity," and has concluded that it is impossible to do so for a noiseless "classical" analytical signal. Hence, they concluded that the proper defination of a signal is the propagation of an abrupt discontinuity; the resulting "front velocity" should be considered the velocity of signal transfer. And finally, there is the problem of defining the velocity of energy transfer [1-4]. Of course, in vacuum, all five velocities are equal to c. In this paper, we will examine closely the superluminal group velocity in a transparent anomalously dispersive medium. The discussion of energy velocity is beyond the scope of this present paper.

A pulse of light propagates at the group velocity in a dispersive medium. In recent years, much attention has been paid to the case where an artificial steep normal dispersion is realized in an EIT medium. Consequently, very slow or vanishingly small group velocity can be achieved [5-8]. To the opposite, if a medium can be obtained to be both transparent and anomalously dispersive, i.e., $n + dn/d\nu < 1$, the group velocity $v_g = c/n_g$ will exceed c ("superluminal" group velocity), or even become negative. Of course,it is well known that inside an absorption line, the dispersion is anomalous, resulting in a superluminal group velocity [1,2,9-12]. However, when absorption is large, the medium becomes opaque.

Generally, for all passive, transparent matter such that it is absorptive at all frequencies in the electromagnetic spectrum, the medium's optical dispersion is normal.

Landau and Lifshitz [2] showed that under the condition

$$Im\,[\chi(\nu)] \geq 0, \quad \text{for any } \nu \tag{1}$$

and in the special case for media with a magnetic permeability $\mu(\nu) = 1$, two inequalities hold simultaneously:

$$\begin{aligned} n_g(\nu) &= \frac{d\,[n(\nu)\nu]}{d\nu} > n(\nu), \\ \text{and} \qquad n_g(\nu) &= \frac{d\,[n(\nu)\nu]}{d\nu} > \frac{1}{n(\nu)}. \end{aligned} \tag{2}$$

Here n_g is the group index. Obviously, we have $n_g > 1$.

However, for media with gain, the general assumption in Eq.(1) no longer holds. In a series of papers [13-18], Chiao and coworkers showed theoretically that anomalous dispersion can occur inside a transparent material. It was predicted that, by using a gain doublet [18], it is possible to obtain a transparent anomalous dispersion region where the group velocity of a light pulse exceeds c.

Here we use gain-assisted linear anomalous dispersion to demonstrate superluminal light pulse propagation with a negative group velocity through a transparent atomic medium [19-21]. We place two Raman gain peaks closely to obtain an essentially lossless anomalous dispersion region that results in a superluminal group velocity. The group velocity of a pulse in this region exceeds c and can even become negative [13-22], while the shape of the pulse is essentially preserved. We measured a negative group velocity index of $n_g = -315(\pm 5)$. Experimentally, a light pulse propagating through the atomic vapor cell exits from it earlier than propagating through the same distance in vacuum by a time difference that is 315 times of the vacuum light propagation time L/c.

2 TRANSPARENT ANOMALOUS DISPERSION

We consider a classical Lorentz oscillator model of the refractive index. It is easy to write the dielectric susceptibility of the medium:

$$\chi(\nu) = -\frac{Ne^2}{4\pi\varepsilon_0 m\omega_0} \times \frac{1}{\nu - \nu_0 + i\gamma} = -f \times \frac{M}{\nu - \nu_0 + i\gamma}, \tag{3}$$

where N is the atomic density and γ is the atomic transition linewidth. $M = \nu_P^2/\nu_0$ with ν_P being the effective plasma frequency and f being the oscillator strength. For gain resonance, a negative oscillator strength $f = -1$ is assigned [15]. Hence for two closely placed gain lines, the dielectric susceptibility becomes:

$$\chi(\nu) = \frac{M}{\nu - \nu_1 + i\gamma} + \frac{M}{\nu - \nu_2 + i\gamma}. \tag{4}$$

For a dilute gaseous medium, we obtain from Eq.(4) for the refractive index $n(\nu) = n'(\nu) + i\,n''(\nu) = 1 + \chi(\nu)/2$ and the real and imaginary parts of the refractive index are plotted in Fig.1(a). It is evident from Fig.1(a) that a steep anomalous dispersion region appears without the heavy absorption present. In fact, a residual gain persists. Furthermore, with the correct choice of experimental parameters, the steep drop of refractive index as a function of frequency can be made a mostly linear one in this region. Thus a light pulse with a frequency bandwidth within this narrow linear anomalous dispersion region will experience almost no change in pulse shape upon propagating through the medium.

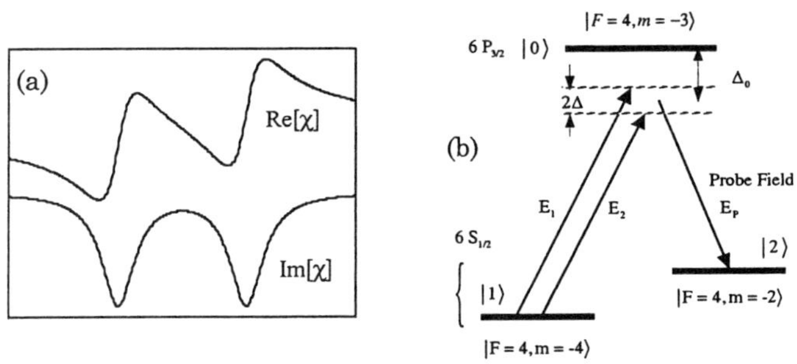

Fig. 1. Gain-assisted anomalous dispersion. (a) frequency-dependent gain coefficient and refractive index, (b) schematic atomic level diagram.

The principle of the experimental realization of a transparent, anomalously dispersive medium is illustrated in Fig.1(b). In a gaseous medium of atoms each of which has three levels: an excited state $|0\rangle$ and two ground states $|1\rangle$ and $|2\rangle$, we first prepare all atoms to be in a ground state $|1\rangle$ via optical pumping. For simplicity, let us first ignore the Doppler shift and assume that the atoms are at rest. We apply two strong continuous-wave (CW) Raman pump light beams E_1 and E_2 that propagate through the atomic medium. The frequencies of E_1 and E_2, ν_1 and ν_2, are different by a small amount 2Δ and both fields are detuned from the atomic transition frequency ν_{01} ($|1\rangle$ to $|0\rangle$) by a large average amount Δ_0. Since the Rabi frequencies associated with the fields E_1 and E_2 are small compared with the common detuning Δ_0, the atoms mostly remain in state $|1\rangle$. When a probe light beam E_P is introduced, a Raman transition can occur causing an atom to absorb a Raman pump photon from the fields E_1 or E_2 and emit a photon into the field E_P while in the mean time making a transition from $|1\rangle$ to $|2\rangle$. Obviously, there are two frequencies where the gain in the probe field is maximized. The maximum gain occurs when the probe field is resonant with the Raman transitions caused by either of the two pump fields E_1 and E_2. We hence obtain a medium whose optical susceptibility for the probe field is described by Eq.(4). Here we have

$$M_{1,2} = \frac{|\mu_{02}|^2}{4\pi\hbar\varepsilon_0}\frac{|\Omega_{1,2}|^2}{|\Delta_0|^2}N, \tag{5}$$

with ν_{02}, $\Omega_{1,2}$, and N being the dipole moment of the $|0\rangle$ to $|2\rangle$ atomic transition, the Rabi frequencies of the Raman pump fields E_1 and E_2, and the effective atomic density difference of states $|1\rangle$ to $|2\rangle$, respectively.

Of course, in a gaseous atomic medium, there is Doppler broadening that demands the expression in Eq.(5) being modified. Specifically, the common detuning in Eq.(5), Δ_0, needs to be replaced by the shifted detuning $\Delta_0 + \nu_0 V/c$, where ν_0 is the carrier frequency of the light field and V is the velocity of those atoms in a specific velocity group $G(V)$ for which the atoms move along the light propagation direction with a speed V. Owing to the co-linear propagation geometry, the Doppler shifts in both detunings Δ_1 and Δ_2 are essentially the same. Hence, the shift in frequencies in $\Delta_1 - \Delta_2$ or $\nu - \nu_0 \pm \Delta\nu$ are essential zero to within an error of $\Delta\nu \cdot V/c$ which is negligible. Therefore, Eq.(5) need only be modified by replacing the expression of the M-factor with:

$$M = N \cdot \frac{|\mu_{02}\cdot \hat{e_2}|^2}{4\pi\hbar\,\varepsilon_0}\int dV \frac{|\Omega_1|^2}{(\Delta_0 + \nu_0\cdot V/c)^2}G(V). \tag{6}$$

The effects of this modification are two fold. First, the quadratic dependence on $1/(\Delta_0 + \nu_0 \cdot V/c)^2$ in Eq.(6) is an even function which prevents cancellation of the effect

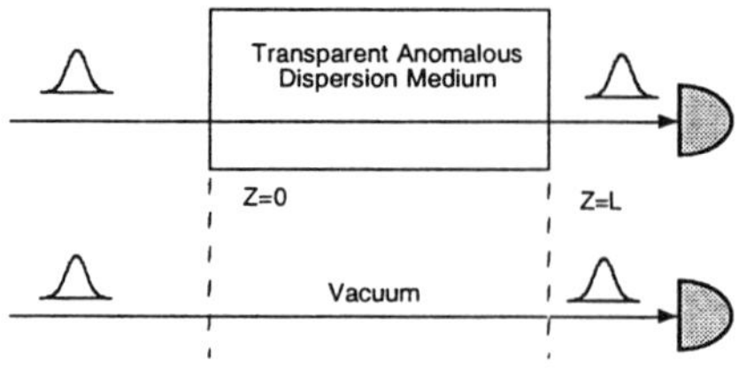

Fig. 2. Pulse propagation through a medium of a length L at a group velocity $v_g = c/(n + \nu\, dn/d\nu)$. and through vacuum for the same length.

due to Doppler broadening. This is a direct result of using Raman transitions. Second, for some velocity groups where the shifted detuning $\Delta_0 + \nu_0 \cdot V/c$ may vanish. However, for these velocity groups, the Raman pump beams act like reversed optical pumping which empty these velocity groups such that $G(V_0) = 0$. This is similar to the "spectral hole burning" effect commonly know in laser physics. Furthermore, the atoms reversely pumped away in these velocity groups act like a broadband weak absorber that helps to compensate the residual gain as shown in Fig.1(a). The spectral linewidths of the gain lines are approximately given by $\gamma \approx v_R/2\pi\, w_0$, due to transit broadening [23]. Here v_R is the mean atomic velocity in the radial direction and w_0 is the mean radius of the Gaussian beam.

3 Pulse Propagation in an Anomalously Dispersive Medium

Now, we consider the propagation of a light pulse of an arbitrary shape through a transparent anomalous dispersing medium of a length L as illustrated in Fig.2. For a scalar light pulse that is decomposed into its positive and negative frequency parts:

$$E(z,t) = E^{(+)}(z,t) + E^{(-)}(z,t), \tag{7}$$

we have for its Fourier decomposition:

$$E^{(+)}(z,t) = \frac{1}{\sqrt{2\pi}} e^{-i\,[\omega_0 t - k(\omega_0) z]} \int d\Omega\, E^{(+)}(\Omega)\, e^{-i\,\{\Omega\, t - [k(\omega_0+\Omega) - k(\omega_0)] z\}}, \tag{8}$$

where ω_0 is the carrier frequency of the light pulse. Inside the transparent anomalous dispersion medium, if over the narrow bandwidth of the incident light pulse $E(\omega - \omega_0)$, the gain is essentially constant, the propagation is governed by the wave vector $k(\omega)$. We can expand the wave vector:

$$k(\omega) = k(\omega_0) + \frac{1}{v_g}(\omega - \omega_0) + \frac{1}{2} \left.\frac{d^2k}{d\omega^2}\right|_{\omega_0} \cdot (\omega - \omega_0)^2 + \cdots, \tag{9}$$

in a Taylor series. When these nonlinear terms in Eq.(9) are negligible, i.e., the dispersion is essentially linear, from Eqs.(8) and (9) we obtain:

$$E^{(+)}(L,t) = g \cdot e^{-i\,(\omega_0 t - k_0 \cdot L)}\, E^{(+)}(0, t - L/v_g), \tag{10}$$

where $g \approx 1$ is a gain factor. Hence, the intensity of the light pulse as a function of time measured with a detector, $I(L,t)$, is related to the incident pulse's time-dependent intensity by:

$$I(L,t) = I(0, t - L/v_g). \tag{11}$$

Ordinarily, in a normal dispersion medium, the group velocity $v_g < c$. Hence, the output intensity of a pulse propagating through the medium is retarded by the propagation time L/v_g, resulting in a delay longer than the vacuum transit time L/c. In a transparent anomalous dispersion medium, the group velocity $v_g = c/[n + \nu\, dn/d\nu]$ can exceed c provided the anomalous dispersion is sufficiently strong such that: $n + \nu\, dn/d\nu < 1$. In this case, the group velocity becomes superluminal: $v_g > c$, resulting in a "*superluminal transit time*:" $L/v_g < L/c$.

Furthermore, when the transparent anomalous dispersion becomes stronger to yield: $n + \nu\, dn/d\nu = 0$, the group velocity $v_g = c/[n + \nu\, dn/d\nu]$ approaches infinity, resulting in a "*zero transit time*", such that Eq.(11) gives $I(L,t) = I(0,t)$. In this case, the output pulse and the input pulse vary the same way in time and there is no time delay experienced by the pulse propagating through the medium.

Finally, when the transparent anomalous dispersion becomes very steep, the dispersive term $\nu\, dn/d\nu$ which is negative becomes very large in its magnitude such that $|\nu\, dn/d\nu| \gg 1$, resulting in a negative group velocity $v_g = c/[n + \nu\, dn/d\nu] < 0$. In this case, Eq.(11) gives $I(L,t) = I(0, t + |L/v_g|)$, where the quantity $|L/v_g| = |n_g|\, L/c$ is positive and can becomes very large compared to the vacuum transit time L/c. This means that the intensity at the output of the medium of length L, $I(L,t)$, will vary in time *earlier* than that of the input pulse $I(0,t)$. Thus in this case, a "*negative transit time*" can be observed. The time difference between the output pulse and the input pulse in the form of a pulse advance, is $|n_g|$ fold of the vacuum transit time L/c. Practically, since the shape of the pulse is not changed, this results in a rather counterintuitive phenomenon where a certain part of the light pulse has already exited the medium before the corresponding part of the incident light pulse even enters, by a time difference that is $|n_g|$ times of the vacuum transit time L/c.

This rather counterintuitive effect is a result of the wave nature of light.

4 EXPERIMENTS

From an experimental point of view, one must satisfy the following requirements. First, a gain doublet must be obtained for which the anomalous dispersion between the gain lines can become linear to avoid any pulse distortion. In previous work, excited state population inversion was considered to obtain gain[16]. However, spontaneous emission and the short excited state life time would cause such gain doublets to be very difficult to sustain. It is important to have a steady state gain with a lifetime longer than the pulse duration τ to avoid transient effects and the associated complications. Second, the medium must be transparent since opaque anomalous dispersion has been long known and has resulted in controversies in terms of interpretations. While ideally the dispersion shown in Fig.1 is transparent, residual absorption and the associated loss are often present and cannot be simply discarded in experimental situations. Third, in order to show superluminal light pulse propagation in a linear regime, one must employ a very weak light pulse for which the photon number is far less than the atomic number in order to avoid Raman gain saturation. Fourth, in order to achieve a reasonable accuracy in the measurement, a system should be designed to demonstrate a negative group velocity. In this case, the pulse advancement under conventional experimental situation will be substantially large compared with commonly obtained accuracy (about 1ns). A number of other experimental conditions also have to be considered such as atomic density, polarization decay time, etc. and they will be discussed as well.

4.1 *Experimental Setup*

The experiment is performed using an atomic Cesium (Cs) vapor cell at 30°C and the main setup is shown in Fig.3. The cesium atoms are confined in a 6-cm-long Pyrex glass cell coated with paraffin for the purpose of maintaining atomic ground-state spin

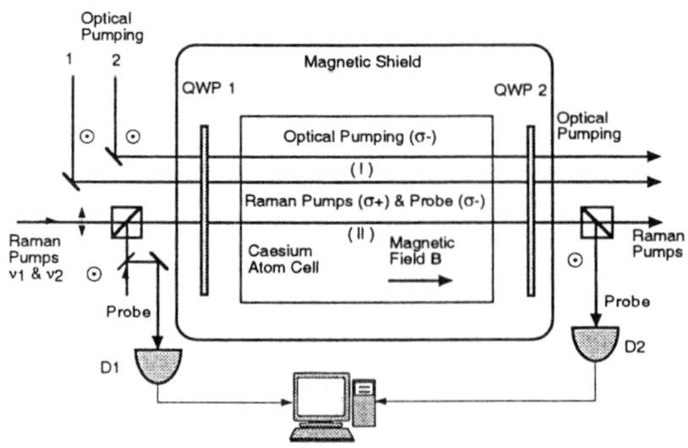

Fig. 3. Schematic experimental set up. Two optical pumping beams tuned to the Cesium (Cs) atomic D1 and D2 transitions prepare the atoms in its ground state hyperfine sublevel $|F = 4, m = -4\rangle$. Two Raman pump beams and a Raman probe beam derived from a common narrow linewidth diode laser propagate colinearly parallel to a small magnetic field B through the atomic cell. Two $\lambda/4$-plates (QWP1 and 2) are used to prepare the three light beams into the corresponding circular polarization states and then separate them for analysis.

polarization. The atomic cell is placed inside a magnetic shield made of a thin layer of high-μ metal material inside which the Earth magnetic field is reduced to sub-milli-Gauss level. A Helmholtz coil (not shown in Fig.3) produces a uniform magnetic field inside the magnetic shield parallel to the propagation direction of all optical beams. This uniform field is approximately 1 Gauss serving the purpose of defining a quantization axis for optical pumping. Inside the magnetic shield, the air temperature is controlled using a heater servo system in order to control the temperature of the Cesium cell. During data acquisition, this control system is turned off to avoid any stray magnetic field. Having good thermal insulation, the temperature of the atomic cell remains the same during the data acquisition time.

In region-I of Fig.3, two optical pumping laser beams prepare almost all Cesium atoms into the ground-state hyperfine magnetic sublevel $6S_{1/2}$, $|F = 4, m = -4\rangle$ that serves as the state $|1\rangle$ in Fig.1(b). Laser-1 is a narrow linewidth diode laser locked to the 852-nm D_2 transition of Cs using a Lamb-dip technique and empties the $6S_{1/2}$, $F = 3$ hyperfine ground states. Laser-2 is a broadband tunable Ti:sapphire laser tuned to the 894-nm D_1 transition of Cesium. The linewidth of laser-2 covers transitions from both the $6S_{1/2}$, $F = 4$ and $F = 3$ hyperfine ground states to the $6P_{1/2}$ excited state. Both laser beams are initially linear polarized and are turned into left-hand polarization ($\sigma-$) using a quarter-wave plate placed before the atomic cell. Inside the vapor cell, Cesium atoms collide with the paraffin-coated glass walls, the atoms change their velocities inside the Doppler profile. However, their ground state spin polarizations are not changed during collisions. Hence, all atoms inside the entire Doppler broadening profile are optically pumped into the ground state $|F = 4, m = -4\rangle$ quickly. The mean dephasing time of the ground state spin polarization of Cesium atoms in a paraffin coated cell is of order a fraction of a second.

In region-II, three light beams derived from the same narrow linewidth diode laser propagate colinearly through the cell. Two strong continuous-wave (CW) Raman pump beams are right-hand polarized ($\sigma+$) and a weak Raman probe beam is left-hand polarized ($\sigma-$). Using three acousto-optical modulators (AOM's), the frequency difference of the two Raman pump beams can be tuned continuously over a few MHz while the probe beam can also be tuned in frequency and can be operated in both CW or pulsed mode. The typical carrier frequency of the AOM's is 80 MHz and the linewidth is about 20 kHz. A residual optical beam that is shifted in frequency by 80 MHz generated from the

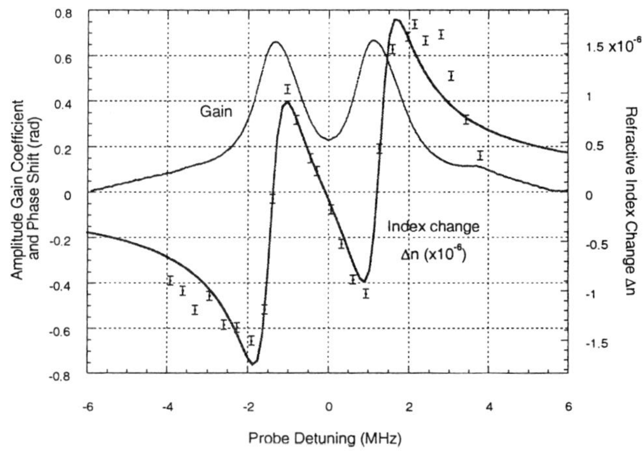

Fig. 4. Measured refractive index and gain coefficient. The superposed curve over the index data is obtained using Eq. (30) with parameters $\nu_0, \Delta\nu$, and γ obtained experimentally. The length of the cell is 6 cm.

same AOM which modulates the probe beam is also available for the refractive index measurement.

4.2 *Experimental Methods and Results*

First, we operate the Raman probe beam in a tunable CW mode to measure the gain and refractive index of the atomic system as a function of the probe frequency detuning. Fig.4 shows the measured gain coefficient and the refractive index. In order to obtain the gain coefficient, we first measure the intensity of the transmitted probe beam as a function of probe frequency. We then extract the gain coefficient. The refractive index is measured using a radio-frequency (rf) interferometric technique. The superimposed curve is obtained from Eq.(4) using parameters obtained from the gain measurement. From Fig.4, we readily see that a negative change of Δn= -1.8x10^{-6} in the refractive index occurs over a narrow probe frequency range of $\Delta\nu$= 1.9 MHz between the two gain lines. Using the expression of the group-velocity index, we obtain the result $n_g = -330(\pm 30)$ in that frequency region. The 10% error reflects the accuracy of the phase measurement.

Next, a pulsed Raman probe beam is employed to observe the superluminal propagation. A near Gaussian probe pulse with a 2.4 μsec FWHM is generated by applying a biased sinusoidal electronic signal to the probe beam A/O modulator. The repetition rate is 50 kHz. A portion of the pulsed probe beam is divided at a beam-splitter before the atomic cell and aligned onto photodiode D1 as a reference. Because the total number of atoms in the probe volume limits the maximum net energy gain of the probe pulse, we use a very weak probe beam ($\approx 1\mu$W) in order to avoid saturation and hence to optimize the anomalous dispersion. A high sensitivity avalanche photodiode, reverse-biased below breakdown, serves as detector D2 to measure the weak probe pulse that propagates through the atomic cell. The photoelectric current produced by detector D2 is converted to a voltage signal using a 500-Ω load resistor and recorded by a digitizing oscilloscope using a synchronized output signal from the pulse generator as the trigger. Pulses from detector D1 are also recorded.

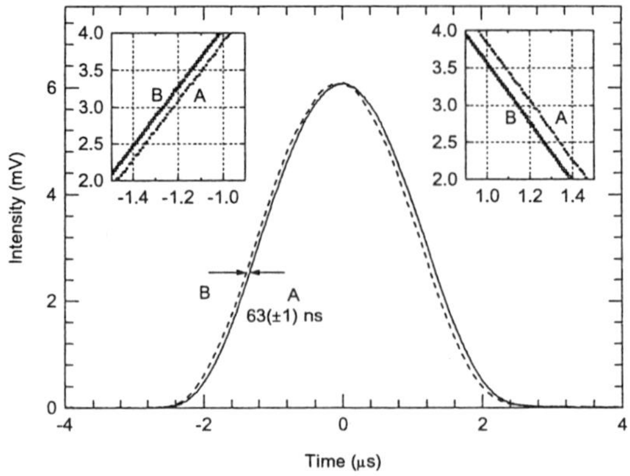

Fig. 5. Measured pulse advancement for a light pulse traversing through the Cesium vapor. (A), light pulse far off-resonance from the Cesium D2 transitions propagating at speed c through 6 cm of vacuum. (B), same light pulse propagating through the same Cs-cell near resonance with a negative group velocity $-c/315$. Insets show the front and trail parts of the pulses. A and B are both the average of 1000 pulses. Off-resonance pulse (A) is normalized to the magnitude of (B).

In order to measure the pulse propagation time, we first tune the diode laser that produces the Raman pump and probe beams far off-resonance from the 852 nm Cesium D2 lines (by 2.5 GHz) to measure the time-dependent probe-pulse intensity. When the laser is placed far off-resonance, the atoms have no effect and the probe pulse propagates at the speed c inside the cell. We then tune the diode laser back to within the Doppler absorption profile and lock it on its side. Using the same synchronized pulse generator output signal as the trigger, we record the time-dependent probe pulse intensity measured by detector D2. We verify that no systematic drift is present by tuning the laser off-resonance again by the same amount and record the probe pulse signal; the two off-resonance pulses are identical to within less than 1 ns. Probe pulses both on and off-resonance are shown in Fig.5 (average of approximately 1000 pulses). Curve A in Fig.5 shows the light pulses far off-resonance from the Cesium D2 transitions propagating at speed c through 6 cm of vacuum. Curve B shows the same light pulse propagating through the same Cs-cell near resonance at a negative group velocity $-c/315$. Probe pulses on resonance show a 40% transmittance and this is due to the broadband absorption of those atoms reverse pumped away from the state $|F = 4, m = -4\rangle$. It is evident that there is almost no change in the pulse shape. The front edges and the trailing edges of the pulses are shown in the insets; both edges are shifted forward by the same amount.

Using a least square fitting procedure, we obtain a pulse advancement shift of 63 ($\pm$1) nsec. Compared with the 0.2 nsec propagation time for light to traverse the 6-cm length of the atomic cell in vacuum, the 63 nsec advancement gives an effective group index of $n_g = 315(\pm 5)$. This is in good agreement with that inferred from the refractive index measurement. The pulses measured with detector D1 are also recorded in the sequence of the off-, on-, off-resonance pulse propagation measurements and are found to be identical to within 1.5 ns.

We note here that the measured superluminal pulse propagation inside the transparent anomalous dispersion medium is a linear effect. We further estimate the photon number per pulse and the interacting atomic number to show that there is no saturation effect present. In the experiment, the measured voltage signal peak strength is $V_p = \alpha\xi\, RG\, \dot{N}_{ph}\hbar\omega_0$, where $\alpha \approx 0.5A/W$ is the photo responsivity of the avalanche photo detector and $\xi = 0.2$ is an effective efficiency of the detection imaging system.

R=500 Ω is the load resistance. $G \approx 80$ is the avalanche gain. $\hbar\omega_0 = 1.5eV$ is the photon energy. Hence, we obtain the peak photon rate $\dot{N}_{ph} \approx 5 \times 10^{12}/sec$. Using the 2.4 μsec FWHM as the pulse duration, each probe pulse contains approximately 1.2×10^7 photons. On the other hand, inside the volume of the probe light beam $\pi w_0^2 L$, there are on average $N \pi w_0^2 L$ atoms at any given moment, where $N \approx 10^{11} cm^{-3}$ is the atomic density. The beam radius is approximately 90 μm and the Cesium cell is of a length 6 cm, the atoms inside the beam volume is approximately 1.4×10^8 at any given moment. However, since atoms are coming in and out of this volume within an average time of $2 w_0/V_R \approx 1\,\mu sec$, during the 2.4 μsec pulse duration, there are approximately 3.4×10^8 atoms inside the volume of the light pulse, much larger than the photon number per pulse. Hence, gain saturation effects are insignificant and the observed superluminal pulse propagation is a linear effect.

The anomalous dispersion and the resulting advance in pulse propagating through the medium depends on a few parameters. Particularly, the pulse advance $|\Delta T|$ is linear to the atomic density N and the Raman pumping Rabi frequency $|\Omega_{1,2}|^2$ which in turn is linear to the Raman pump power. These parameter dependence, as well as the dependence on detuning, can be tested and detailed experimental results and theoretical analyses are given in reference [20].

5 CONCLUSIONS

Here we note that the observed superluminal and negative group velocity is a result of the wave nature of light. The measured negative and superluminal group velocity of a light pulse propagating through a transparent anomalous dispersion medium is due to the physical effect of "rephasing" [20,22]. Specifically, inside a medium with refractive index n, the effective wavelength of a light ray is modified: $\lambda' = \lambda/n$, where λ is the vacuum wavelength. It is easy to derive:

$$\frac{d(\lambda/n)}{d\lambda} = \frac{n_g}{n^2}. \tag{12}$$

Under the condition $n_g < 0$, we have $d(\lambda/n)/d\lambda < 0$. Hence a longer wavelength (redder) component of the incident pulse becomes a shorter wavelength ray inside the medium, and vice versa. This results in an unusual situation where the phases of the different frequency components of a pulse become aligned at the exit surface of the medium earlier than even in the case of the same pulse propagating through the same distance in a vacuum.

Finally, we note that a superluminal group velocity is not at odds with causality or special relativity [1-4,9-22]. Simply put, causality only requires that the signal velocity, instead of the group velocity, be limited by c; the signal velocity is different from the group velocity, as first noted again by Sommerfeld and Brillouin [1]. More recently, we proposed to define signal operationally [21] according to common laboratory practice. In the experimental cases where the medium is transparent, we must consider the excess noise due to the gain lines. We found that, in these cases the excess spontaneous emission noise from the gain resonances retards the unset of the signal by retarding the time at which a prescribed signal-to-noise ratio (SNR) is reached [21]. Hence, it was concluded that the quantum noise associated with an amplifier (which is related to the "no-cloning" theorem and the linear nature of quantum mechanics) is associated with the basic requirement of causality that states that no "signal" can be transmitted faster than c.

References and links

1. L. Brillouin, *Wave Propagation and Group Velocity* (Academic Press, New York, 1960).
2. L. D. Landau and E. Lifshitz, *Electrodynamics of Continuous Media*, p.286 (Pergamon, Oxford, 1960).
3. G. Diener, Phys. Lett. A **223**, 327 (1996).
4. G. Diener, Phys. Lett. A **235**, 118 (1997).
5. L. V. Hau, S. E. Harris, Z. Dutton, and C. H. Behroozi, Nature (London) **397**, 594 (1999).
6. M. M. Kash, et al., Phys. Rev. Lett. **82**, 5229 (1999); Budker, et al *ibid* **83**, 1767 (1999).
7. C. Liu, Z. Dutton, C. H. Behroozi, L. V. Hau, Nature (London) **409**, 490 (2001).
8. D. F. Phillips, A. Fleischhauer, A. Mair, R. L. Walsworth, and M. D. Lukin, Phys. Rev. Lett. **86**, 783 (2001).
9. C. G. B. Garrett and D. E. McCumber, Phys. Rev. A **1**, 305 (1970).
10. S. Chu and S. Wong, Phys. Rev. Lett. **48**, 738 (1982).
11. B. Segard and B. Macke, Phys. Lett. A **109**, 213 (1985).
12. A. M. Akulshin, S. Barreiro, and A. Lezama, Phys. Rev. Lett. **83**, 4277 (1999).
13. R. Y. Chiao, in *Amazing Light, a Volume Dedicated to C. H. Townes on His Eightieth Birthday*, Ed. R. Y. Chiao, (Springer, New York, 1996), p.91.
14. R. Y. Chiao and A. M. Steinberg, in *Progress in Optics*, Ed. E. Wolf (Elsevier, Amsterdam, 1997), p.345.
15. R. Y. Chiao, Phys. Rev. A **48**, R34 (1993).
16. E. L. Bolda, R. Y. Chiao, and J. C. Garrison, Phys. Rev. A **48**, 3890 (1993).
17. R. Y. Chiao, J. Boyce, and J. C. Garrison, in *Fundamental Problems in Quantum Theory, in Honor of Professor John A. Wheeler*, D. M. Greenberg and A. Zeilinger, eds., Ann. N. Y. Acad. of Sci. **755**, 400 (1993).
18. A. M. Steinberg and R. Y. Chiao, Phys. Rev. A **49**, 2071 (1994).
19. L. J. Wang, A. Kuzmich, and A. Dogariu, Nature **406**, 277 (2000).
20. A. Dogariu, A. Kuzmich, and L. J. Wang, Phys. Rev. A **63**, 053806 (2001).
21. A. Kuzmich, A. Dogariu, L. J. Wang, P. W. Milonni, and R. Y. Chiao, Phys. Rev. Lett. **86**, 3925 (2001), and references therein.
22. A. Dogariu, A. Kuzmich, H. Cao, and L. J. Wang, Opt. Exp. **8**, 344 (2001).
23. J. E. Thomas and W. W. Quivers, Phys. Rev. A **22**, 2115 (1980).

Decoherence of trapped-atom motional state superpositions[1]

D. J. Wineland,[#] C. Monroe,[‡] C. J. Myatt,[†]
Q. A. Turchette,[§] C. A. Sackett,[*] D. Kielpinski,[¶]
B. E. King,[$] and W. M. Itano[#]

[#] *NIST, Boulder, CO, 80303*

[‡] *Dept. of Physics, Univ. Michigan, Ann Arbor, MI 48109-1120*

[†] *Precision Photonics, Boulder, CO, 80301*

[§] *Research Electro-Optics, Boulder, CO, 80301*

[*] *Dept. of Physics, Univ. Virginia, Charlottesville, VA 22904*

[¶] *Dept. of Physics, MIT, Cambridge, MA, 02139*

[$] *NIST, Gaithersburg, MD, 20899-8424*

david.wineland@boulder.nist.gov

Abstract: Experiments that investigate the decoherence of superpositions of motional states of trapped ions are described. Decoherence is characterized by the loss of contrast in Ramsey-type interferometer experiments involving superpositions of two motional coherent states or two motional Fock states that are subject to stochastically fluctuating electric fields.

1 Introduction

Decoherence in quantum systems [1] has been a subject of enduring interest because its fundamental causes are a subject of debate and more recently because it is *the* problem that must be overcome in quantum information processing [2]. Here, we briefly examine decoherence as it affects quantum superpositions of motional states of a single trapped atom. Since this problem is closely related to decoherence of superposition of states of a single mode of the electromagnetic field, we can rely on a rather large body of theoretical research in quantum optics to guide us; see for example Refs. [3–11].

For brevity, we only summarize the aspects of trapped atomic ion state-manipulation that are relevant for the experiments on $^9Be^+$ ions described below; further details can be found in Ref. [12] and references therein. To a good approximation, a trapped ion can be viewed as being confined in a three-dimensional harmonic well. We consider only motion along one direction; pure states of this motion can be written as superpositions of Fock states

$$\psi_{motion} = \sum_{n=0}^{\infty} C_n |n\rangle, \tag{1}$$

and with laser cooling we will only consider such superpositions for small values of n. The ion has two internal "spin" states that we label $|\downarrow\rangle$ and $|\uparrow\rangle$. The motional and spin states can be manipulated with coherent laser beams through stimulated-Raman

[1] Contribution of NIST; not subject to U. S. copyright

transitions. By this means we can perform operations of the form

$$|\downarrow\rangle|n\rangle \rightarrow cos(\theta)|\downarrow\rangle|n\rangle + e^{i\phi}sin(\theta)|\uparrow\rangle|n'\rangle \tag{2}$$

and, in principle, starting from the state $|\downarrow\rangle|n=0\rangle$, we can synthesize any desired entangled superposition state between the motion and spin [13]. After each experiment, we measure the internal state of the ion, and by repeating the experiment many times we can obtain the probability of finding the ion in the $|\downarrow\rangle$ (or $|\uparrow\rangle$) state.

Decoherence of the ion motion occurs naturally due to ambient fluctuating fields; however, in most cases, for convenience, we can "engineer" the decoherence in order to speed up the experiments.

2 Trapped-ion motional decoherence

We assume that we are interested in characterizing the decoherence of a superposition state $\alpha|\psi_1\rangle + c_2|\psi_2\rangle$ where $|\psi_1\rangle$ and $|\psi_2\rangle$ are states of a single mode of motion (here taken to be the z-mode). This mode of motion we call the quantum "system" under investigation. In a basic model of decoherence [1,3], it is assumed that this state couples to the outside world or "environment," whose initial state is $|\phi_e\rangle$. Therefore, the initial state of the motion and environment can be written $\psi_0 = (c_1|\psi_1\rangle + c_2|\psi_2\rangle) \otimes |\phi_e\rangle$. After the motional state couples to the environment, the system and environment evolve to

$$\psi_0 = (c_1|\psi_1\rangle + c_2|\psi_2\rangle) \otimes |\phi_e\rangle \rightarrow \psi_{final} = c_1|\psi_1'\rangle|\phi_{e1}\rangle + c_2|\psi_2'\rangle|\phi_{e2}\rangle. \tag{3}$$

For a specific type of system/environment coupling, we can assume $|\psi_1'\rangle = e^{i\zeta_1}|\psi_1\rangle$ and $|\psi_2'\rangle = e^{i\zeta_2}|\psi_2\rangle$; therefore, after the interaction, $\psi_{final} = c_1|\psi_1\rangle|\phi_{e1}\rangle + e^{i(\zeta_2-\zeta_1)}c_2|\psi_2\rangle|\phi_{e2}\rangle$ and we see how the initial states become correlated with different states of the environment. Now, if $\langle\phi_{e2}|\phi_{e1}\rangle \simeq 0$, and if the final environment states are uncontrolled and unmeasured (or unmeasurable), we must ignore these degrees of freedom (mathematically trace over them) such that the pure state ψ_{final} becomes a statistical mixture expressed by the density matrix $\rho_{final} = |c_1|^2|\psi_1\rangle\langle\psi_1| + |c_2|^2|\psi_2\rangle\langle\psi_2|$. Hence, the off-diagonal terms or "coherences" of the pure state density matrix $(c_1|\psi_1\rangle + c_2|\psi_2\rangle)(c_1^*\langle\psi_1| + c_2^*\langle\psi_2|)$ are lost to the environment.

It is sometimes useful to incorporate a quantum measuring device or quantum "meter" into the scheme above using a "von Neumann chain" [3]. Here we assume that the quantum system, initially given by $c_1|\psi_1\rangle + c_2|\psi_2\rangle$, is first coupled to a quantum meter and in general, this combination is then coupled to the environment. In the first stage of coupling, we have

$$(c_1|\psi_1\rangle + c_2|\psi_2\rangle) \otimes |\psi_M\rangle \otimes |\phi_e\rangle \rightarrow (c_1|\psi_1'\rangle|\psi_{M1}\rangle + c_2|\psi_2'\rangle|\psi_{M2}\rangle) \otimes |\phi_e\rangle. \tag{4}$$

Upon coupling to the environment, the evolution is

$$(c_1|\psi_1'\rangle|\psi_{M1}\rangle + c_2|\psi_2'\rangle|\psi_{M2}\rangle) \otimes |\phi_e\rangle \rightarrow c_1|\psi_1''\rangle|\psi_{M1}'\rangle|\phi_{e1}\rangle + c_2|\psi_2''\rangle|\psi_{M2}'\rangle|\phi_{e2}\rangle, \tag{5}$$

and, as before, if we assume $\langle\phi_{e2}|\phi_{e1}\rangle \simeq 0$ and that the final environment states are uncontrolled and unmeasured (or unmeasurable), then the final state of the system and meter is expressed by the density matrix

$$|c_1|^2|\psi_1''\rangle\langle\psi_1''||\psi_{M1}'\rangle\langle\psi_{M1}'| + |c_2|^2|\psi_2''\rangle\langle\psi_2''||\psi_{M2}'\rangle\langle\psi_{M2}'|. \tag{6}$$

Therefore the correlation between the system and meter states is established, yielding the expected classical result, and the coherences are lost to the environment. Including the quantum meter more closely describes the ion experiments discussed below because the "system" is the ion's motion, which is not directly measured, but can be coupled to the ion's spin which is the meter that is measured (by laser scattering). In the first experiment described below we will consider a case that is described by expression (5) but we have a situation where $|\phi_{e1}\rangle \simeq |\phi_{e2}\rangle \simeq |\phi_e\rangle$. However since, in practice, we can't measure $|\phi_e\rangle$, and because $|\phi_e\rangle$ changes from experiment to experiment, we must average over the distribution of $|\phi_e\rangle$ states, which leads to decoherence very similar to the case where $\langle\phi_{e2}|\phi_{e1}\rangle \simeq 0$.

2.1 *High-temperature amplitude reservoir*

Decoherence of harmonic oscillator superposition states, with coupling to a variety of reservoir environments, has been investigated extensively in theory; see for example Refs. [3–11]. The model in these studies is a system harmonic oscillator coupled to a bath of environment quantum oscillators. The interaction Hamiltonian for this situation can be modelled as

$$H_I = \hbar \sum_{i=0}^{\infty} \Gamma_i a b_i^\dagger + h.c., \tag{7}$$

where a is the lowering operator for the system oscillator, $b_i^\dagger$ is the raising operator for the ith environment oscillator, and Γ_i gives the strength of coupling between the system oscillator and the ith environment oscillator. The case where the system oscillator is a single mode of the electromagnetic field has been investigated extensively in the experiments of Haroche *et al.* [14]. This model and other similar ones in the specific context of trapped-ion experiments are discussed theoretically in Refs. [12, 15–18]. One model assumes that the motion of a trapped ion couples to the (noisy) uniform electric field $\mathbf{E}$ caused by the environment oscillators through the potential $U = -q\mathbf{x}\cdot\mathbf{E}$, where $\mathbf{x}$ is the displacement of the ion from its equilibrium position. Such a model corresponds to the case of a noisy electric field due to a resistor coupled between the trap electrodes [12, 19], and is described by the Hamiltonian in Eq. (7).

In the experiments, to speed up decoherence, we can simulate a hot resistor coupled between trap electrodes by applying random uniform electric fields that have a spectrum that overlaps the ion's axial-motion frequency ω_z [20, 21]. We generate these fields in the trap by applying voltages to one of the trap electrodes. A commercial function generator produces pseudo-random voltages that are applied through a band-pass filter centered near ω_z, defining the frequency spectrum of the reservoir. We measured the initial coherence and subsequent decoherence of the motion state superpositions with single-atom interferometry, as described in detail in Refs. [20, 21].

Briefly, we first created the entangled system/meter state shown in the left hand side of expression (5) with $|\psi_1'\rangle = |\alpha\rangle, |\psi_{M1}\rangle = |\uparrow\rangle, |\psi_2'\rangle = |\alpha'\rangle$, $|\psi_{M2}\rangle) = |\downarrow\rangle$, and $c_1 = c_2 = \frac{1}{\sqrt{2}}$ as indicated in part (2) of Fig. 1. Here, $|\alpha\rangle$ and $|\alpha'\rangle$ are coherent states of amplitude α and α'. This state is usually called a "Schrödinger-cat" state. The superposition was then coupled to the reservoir (the applied fluctuating fields) for a fixed time. Finally, the components of the motional-state superposition were overlapped by reversing the steps which initially created it, a final $\pi/2$ pulse was applied to the internal states, and the internal state was measured. The experiment was repeated many times; after each experiment, the internal state of the ion was measured as a function of the relative phase ϕ_R of the creation and reversal steps, and the contrast of the resulting interference fringes characterizes the amount of coherence remaining after coupling to the reservoir. The results are displayed in Fig. 2. We observe that the contrast of the interferometer fringes decays as $\exp(-\kappa|\alpha - \alpha'|^2\langle E_n^2\rangle)$, where $\sqrt{\langle E_n^2\rangle}$ is the r.m.s. value of the applied fluctuating fields and κ is a constant. This scaling was also observed for the ambient fluctuating fields in the experiment [20, 21]. This exponential dependence

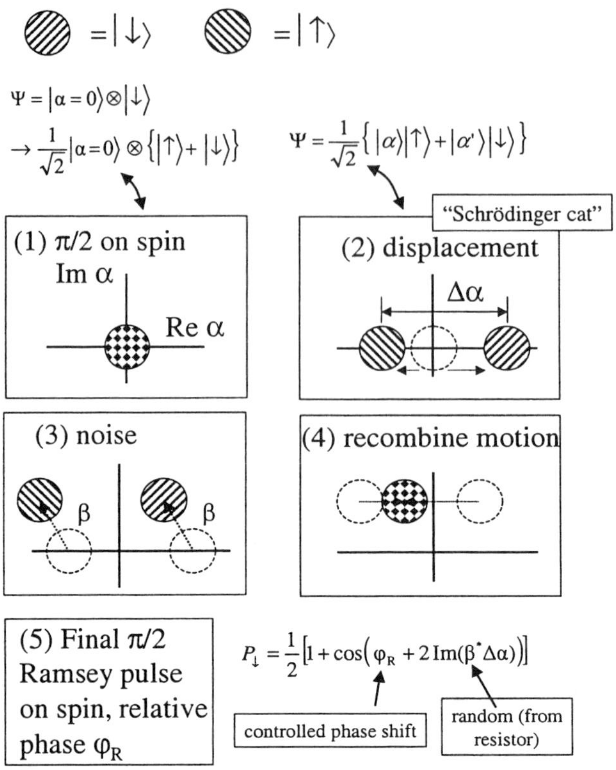

Fig. 1. Pictorial representation in phase space of a Ramsey interferometer [22] experiment designed to observe decoherence of coherent-state superpositions caused by coupling to an amplitude reservoir (from Refs. [20] and [21]). Starting from the initial system/meter state $|\alpha = 0\rangle|\downarrow\rangle$, we create the Schrödinger-cat state shown in panel (2) by applying a $\pi/2$ pulse on the spin followed by a spin-dependent optical-dipole-force displacement [23]. Noise is then applied to the motion simulating coupling to a hot resistor. A (random) displacement β in phase space results as shown in panel (3). Subsequently we reverse the steps used to create the cat state but with a phase shift ϕ_R on the final Ramsey $\pi/2$ pulse. Finally, we measure the probability $P_\downarrow$ of finding the ion in the $|\downarrow\rangle$ internal state. In the absence of noise ($\beta = 0$), we obtain the characteristic sinusoidal oscillations as a function of ϕ_R of Ramsey interferometry; with noise, the Ramsey fringe contrast is reduced because we must average β over a distribution of values characteristic of the applied amplitude noise.

of the decoherence on the "size" of the cat-state $\Delta\alpha \equiv |\alpha - \alpha'|$ agrees with theoretical predictions and indicates why it is so difficult to preserve superpositions except on a microscopic scale.

The decoherence observed here fits into the general scheme outlined above with however some notable differences. First, the environment acted on the system (motional state) and not the meter (spin), so that $|\psi'_{M1}\rangle = |\psi_{M1}\rangle$ and $|\psi'_{M2}\rangle = |\psi_{M2}\rangle$ in expression (5). (Of course, when we finally measured the spin state, we coupled the meter to the environment through laser scattering, but this is not the environment coupling we want to study here.) Also, even though we must have $|\phi_{e1}\rangle \neq |\phi_{e2}\rangle$ because of the difference in back-action from the states $|\alpha\rangle$ and $|\alpha'\rangle$ to the environment, to a very good approximation the back-action of the system on the environment is negligible so that for all practical purposes, $\langle\phi_{e2}|\phi_{e1}\rangle = 1$ and $|\phi_{e1}\rangle = |\phi_{e2}\rangle = |\phi_e\rangle$. Here we have a situation in which we could, in principle, measure $|\phi_e\rangle$ without disturbing it appreciably and apply an operation which undoes its effect. In the case of engineered noise, we could readily do this, but for the case of the ambient noise, this is, in practice, impossible and we must average over the ensemble of environment states $\{|\phi_e\rangle\}$ leading to the expected decoherence exhibited in Fig. 2.

In related experiments discussed elsewhere [20,21], we characterized the decoherence when we applied the same amplitude reservoir to superpositions of two Fock states and also when we apply a phase reservoir, characterized by the interaction Hamiltonian

$$H_I = \hbar a a^\dagger \sum_{i=0}^{\infty} \Gamma_i b_i^\dagger + h.c. \tag{8}$$

to the Schrödinger-cat state and to superpositions of two Fock states.

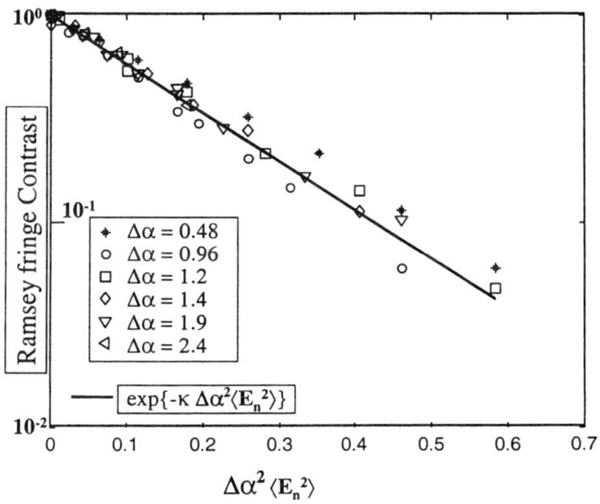

Fig. 2. Loss of coherence of a Schrödinger-cat state caused by coupling to an amplitude reservoir (from Ref. [20]). The amplitude noise was applied for a fixed time interval (3 μs) but with varying amplitude. The contrast for all experiments is normalized to the contrast observed in the absence of applied noise. The expected scaling was observed showing that the decoherence rate is proportional to the square of the phase-space separation $\Delta\alpha$ of the cat-state components.

2.2 *"Engineered" zero-temperature amplitude reservoir*

In the cavity-QED decoherence experiments of Ref. [14], where the system oscillator's frequency is around 50 GHz and the ambient temperature is 0.6 K, the equilibrium oscillator quantum number is around 0.05 and the amplitude reservoir's temperature could be assumed to be 0 K to a good approximation. In the ion experiments, the oscillator's frequency is around 5 MHz and the ambient temperature is near room temperature, implying the equilibrium oscillator number is very large. (Actually, the ambient fluctuating fields behave as if the reservoir were much hotter [24]). Nevertheless, we can simulate a $T = 0$ reservoir, as suggested by Poyatos *et al.* [10]. The basic process is laser cooling as indicated in Fig. 3. Coherent Raman beams couple the states $|\downarrow\rangle|n\rangle \leftrightarrow |\uparrow\rangle|n-1\rangle$ with Rabi rate Ω^*. At the same time, an optical pumping beam causes spontaneous Raman transitions from $|\uparrow\rangle|n\rangle$ to $|\downarrow\rangle|n\rangle$ through the excited P states at rate γ. From the diagram in Fig. 3, we see that all populations are driven towards the state $|\downarrow\rangle|0\rangle$ thereby simulating a $T = 0$ reservoir. By varying the strength of the Raman and optical pumping couplings and time, we have control over the reservoir parameters.

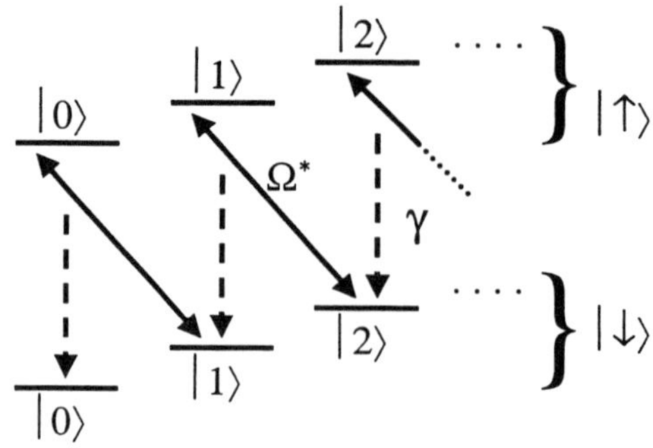

Fig. 3. Schematic diagram depicting an "engineered" T = 0 reservoir. A red sideband drives transitions [20] $|\downarrow\rangle|n\rangle \leftrightarrow |\uparrow\rangle|n-1\rangle$, while spontaneous-Raman scattering is used to make transitions $|\uparrow\rangle|n\rangle$ to $|\downarrow\rangle|n\rangle$ at rate γ. When $\gamma \gg \Omega^*$, the system acts as a $T = 0$ reservoir for motional states.

In the experiments [20, 21], we examined the time evolution of the coherence of the Fock state superposition $\psi = \frac{1}{\sqrt{2}}(|0\rangle + |2\rangle)|\downarrow\rangle$ for varying lengths of reservoir interaction time. To observe this coherence, we made an interferometer similar to one described in the previous section and illustrated in Fig. 4 . (Note that in this experiment, the spin serves multiple functions: it allows us to construct the motional-state interferometer, make measurements and, in the middle of any particular interferometer experiment, it also acts as part of the environment.) The resulting data are shown in Fig. 4. Each data point is the contrast of the interference fringes after the system interacts with the reservoir for a time τ. We show two cases, $\gamma \gg \Omega^*$ and $\gamma < \Omega^*$. In the first case, we observed the decay of the coherence due to coupling to the simulated T=0 reservoir. In this case, it is natural to assume that the reservoir is the spin plus the rest of the environment, which includes the spontaneously scattered photons. (Since we don't rigorously satisfy $\gamma \gg \Omega^*$, we also see some initial non-exponential decay characteristic of the Zeno effect [20, 21]). When $\gamma < \Omega^*$, the coherence between the $|0\rangle$ and $|2\rangle$ state disappears and reappears over time, with an overall decay of the fringe contrast. The underlying effect is population transfer back and forth (Rabi flopping) between the states $|\downarrow\rangle|2\rangle$ and $|\uparrow\rangle|1\rangle$, with a coupling of the $|\uparrow\rangle|1\rangle$ state to the outside environment through spontaneous Raman scattering. If we make $\gamma \to 0$, then we see near perfect revival of the fringe contrast, as expected [21]. For $\gamma \to 0$, we have restricted the environment to be just the internal spin states. In this case, we can reverse the effects of decoherence (of the $\frac{1}{\sqrt{2}}(|0\rangle + |2\rangle)$ state). (A scheme for observing a similar reversal in the context of cavity-QED is discussed in Ref. [25].)

In this experiment, when the atom scatters a photon through spontaneous Raman scattering, no measurement we do will allow us to restore the initial superposition. That is, the emission (and subsequent absorption by a measuring apparatus or the environment) of a spontaneous photon projects the atom into a definite state ($|\downarrow\rangle|1\rangle$ in the context of Fig. 3) and phase information is lost. This is the same situation as in the experiments of Brune *et al.* [14]. The decoherence in this second ion experiment is to be contrasted with the first, where we could in principle detect the environment and reverse its effects. Note that although the data for $\gamma < \Omega^*$ illustrate how coherence transferred to the environment can be recovered, an alternative explanation would say that by transferring the $|\downarrow\rangle|2\rangle$ component of the superposition to the $|\uparrow\rangle|1\rangle$ state, we can gain "which-path" information in our interferometer - the paths being the $|\downarrow\rangle|0\rangle$ and $|\downarrow\rangle|2\rangle$ parts of the superposition. The oscillation in which-path information is analogous to

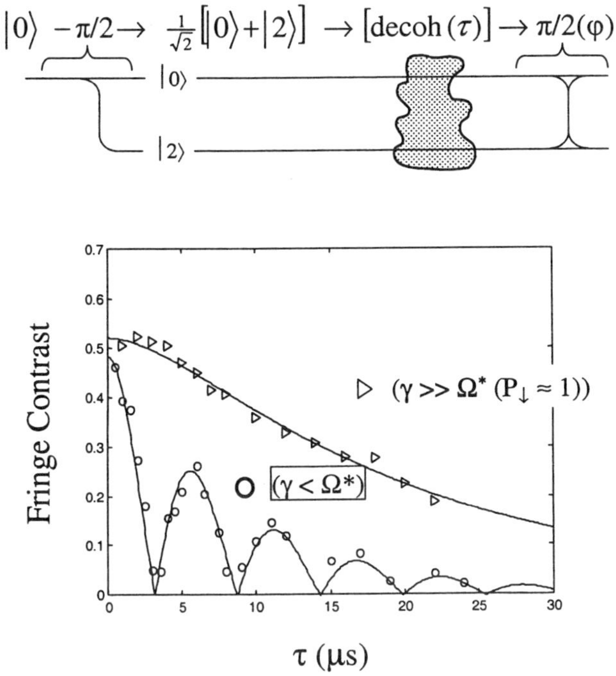

Fig. 4. Decoherence of the motional superposition state $\frac{1}{\sqrt{2}}(|0\rangle + |2\rangle)$ coupled to an engineered $T = 0$ reservoir (from Refs. [20] and [21]). In the upper part of the figure we show schematically how the motional state Ramsey-interferometer works, where the reservoir is applied during the time indicated by the shaded area. In the lower part of the figure, we show the interferometer contrast as a function of time for two cases of the relative values of Ω^* and γ. The initial contrast (without decoherence) is not unity due to imperfections in the motional-state Ramsey interferometer.

that illustrated by the experiments of Chapman *et al.* [26], Dürr *et al.* [27], Kokorowski *et al.* [28], and Bertet *et al.* [29].

This second experiment also illustrates a fundamental dilemma in explaining decoherence. If we restrict the size of the environment to be the spin ($\gamma \to 0$), then the coherence information is not lost and can be recovered, as indicated by the revival of the contrast. Even in the case of photon emission (spontaneous emission at rate γ in Fig. 3), the information need not be lost if the photon is emitted into a lossless cavity where it can be recovered, as illustrated in the cavity-QED experiments [30, 31]. Therefore it seems that decoherence is useful to describe situations, where for practical and technical reasons, we cannot retain the overall system information or the overall system cannot be regarded as being closed. However, these limitations seem only to be practical and not fundamental *unless* some mechanism, that is so far missing in quantum mechanics, can give rise to intrinsic decoherence . For a summary of alternatives, see for example, Ref. [32].

2.3 *Motional decoherence in practice*

The ambient motional decoherence observed in the NIST experiments can be characterized by the model described in the first case above; its characteristics are the same as that caused by thermal electronic noise in the resistance of the electrodes or resistance coupled to the electrodes. At the relatively low ion oscillation frequencies used, where the characteristic wavelength is much larger than the electrode spacing, this could be described by Johnson noise associated with lumped circuit elements attached to the electrodes. Typical heating rates (expressed as quanta per second from the motional ground state) are observed to be around $\langle dn/dt\rangle \simeq 10^3 - 10^4\ s^{-1}$ for $\nu_z \simeq 10$ MHz and the distance from the ion to the nearest electrode surface around 150 μm [24]. The problem is that given our best estimates of the electrode resistance [12] and attached circuit elements, the electrodes would have to be at a temperature of 10^6 K or higher to explain most of the heating results [24]. The principal cause of the anomalously large fluctuating fields and resultant heating is still not understood at this time, but some of its characteristics are known. It appears to be rather broad band (no sharp resonances observed from 2 to 20 MHz) and is likely caused by electric-field noise emanating from the electrodes. For example, heating from collisions due to background gas or an electron field emission source would tend to heat the internal modes of two ions at about the same rate as the center-of-mass (COM) modes. However, we observe that the internal mode heating is negligible compared to the heating of the COM modes [33] indicating that the fields at the site of the ions are approximately uniform spatially. (We note that the experiments of Rohde *et al.* [34] disagree with this finding.). A general trend is that the heating is proportional to $1/R^\kappa$ where $\kappa > 2$. For thermal electronic noise from circuit resistance, $\kappa = 2$ [12]. If the noise were due to fluctuating patch fields, $\kappa = 4$ [24]. The scatter of the data using $^9Be^+$ ions is rather large but is consistent with $\kappa = 4$. Data on other traps [34,35] appear also to be consistent with κ equal to 4 or more.

In the experiments of Ref. [24], there was some indication that beryllium deposition onto the electrodes caused a higher heating rate. (Beryllium ions are created in the trap by ionizing neutral beryllium atoms that pass near the center of the trap after being emitted from a wide angle source. Some of the beryllium atoms from the source are deposited on electrodes.) By masking the electrodes from direct beryllium deposition, preliminary evidence indicates a significant drop ($\simeq 100$) in heating rate (M. Rowe *et al.*, unpublished). In any case, we conjecture that for clean metal electrode surfaces, free from oxides or adsorbed gases (which could support mobile electrons), the heating should approach that predicted by thermal electronic noise $\langle dn/dt\rangle \simeq 1\ s^{-1}$.

3 Acknowledgements

This work was supported by NSA/ARDA, ONR, and ARO. We thank B. DeMarco, D. Leibfried, M. Lombardi, D. Smith, and D. Sullivan for helpful comments on the manuscript.

References and links

1. W. H. Zurek, "Decoherence, einselection, and the quantum origins of the classical," quant-ph/0105127 (2001).
2. D. P. DiVincenzo, in *Scalable Quantum Computers*, S. L. Braunstein and H. K. Lo, eds., (Wiley-VCH, Berlin, 2001), pp. 1–13.
3. W. H. Zurek, "Decoherence and the transition from quantum to classical," Physics Today **44,** 36–44 (October, 1991).
4. A. O. Caldeira and A. J. Leggett, "Influence of damping on quantum interference: an exactly soluble model," Phys. Rev. A **31,** 1059–1066 (1985).
5. D. F. Walls and G. J. Milburn, "Effect of dissipation on quantum coherence," Phys. Rev. A **31,** 2403–2408 (1985).
6. M. J. Collett, "Exact density-matrix calculations for simple open systems," Phys. Rev. A **38,** 2233–2247 (1988).
7. D. F. Walls and G. J. Milburn, *Quantum Optics*, 1st ed. (Springer, Berlin, 1994).
8. W. Vogel and D. G. Welsch, *Quantum Optics*, 1st ed. (Akademie Verlag, Berlin, 1994).
9. V. Bužek and P. L. Knight, "Quantum interference, superposition states of light, and nonclassical effects," Prog. Opt. **34,** 1–158 (1995).
10. J. F. Poyatos, J. I. Cirac, and P. Zoller, "Quantum reservoir engineering with laser cooled trapped ions," Phys. Rev. Lett. **77,** 4728–4731 (1996).
11. W. P. Schleich, *Quantum Optics in Phase Space*, 1st ed. (Wiley-VCH, Berlin, 2001).

12. D. J. Wineland, C. Monroe, W. M. Itano, D. Leibfried, B. E. King, and D. M. Meekhof, "Experimental issues in coherent quantum-state manipulation of trapped atomic ions," J. Res. Nat. Inst. Stand. Tech. **103,** 259–328 (1998).
13. C. K. Law and J. H. Eberly, "Arbitrary Control of a Quantum Electromagnetic Field," Phys. Rev. Lett. **76,** 1055–1058 (1996).
14. M. Brune, E. Hagley, J. Dreyer, X. Maître, A. Maali, C. Wunderlich, J. M. Raimond, and S. Haroche, "Observing the progressive decoherence of the "meter" in a quantum measurement," Phys. Rev. Lett. **77,** 4887–4890 (1996).
15. S. Schneider and G. J. Milburn, "Decoherence in ion traps due to laser intensity and phase fluctuations," Phys. Rev. A **57,** 3748–3752 (1998).
16. M. Murao and P. L. Knight, "Decoherence in nonclassical motional states of a trapped ion," Phys. Rev. A **58,** 663–669 (1998).
17. S. Schneider and G. J. Milburn, "Decoherence and fidelity in ion traps with fluctuating trap parameters," Phys. Rev. A **59,** 3766–3774 (1999).
18. R. Bonifacio, S. Olivares, P. Tombesi, and D. Vitali, "Model-independent approach to nondissipative decoherence," Phys. Rev. A **61,** 053802–1–8 (2000).
19. D. J. Wineland and H. G. Dehmelt, "Principles of the stored ion calorimeter," J. Appl. Phys. **46,** 919–930 (1975).
20. C. J. Myatt, B. E. King, Q. A. Turchette, C. A. Sackett, D. Kielpinski, W. M. Itano, C. Monroe, and D. J. Wineland, "Decoherence of quantum superpositions through coupling to engineered reservoirs," Nature **403,** 269–273 (2000).
21. Q. A. Turchette, C. J. Myatt, B. E. King, C. A. Sackett, D. Kielpinski, W. M. Itano, C. Monroe, and D. J. Wineland, "Decoherence and decay of motional quantum states of a trapped atom coupled to engineered reservoirs," Phys. Rev. A **62,** 053807–1–22 (2000).
22. N. F. Ramsey, *Molecular Beams* (Oxford University Press, London, 1963).
23. C. Monroe, D. M. Meekhof, B. E. King, and D. J. Wineland, "A "Schrödinger cat" superposition state of an atom," Science **272,** 1131–1136 (1996).
24. Q. A. Turchette *et al.*, "Heating of trapped ions from the quantum ground state," Phys. Rev. A **61,** 063418–1–8 (2000).
25. J. M. Raimond, M. Brune, and S. Haroche, "Reversible decoherence of a mesoscopic superposition of field states," Phys. Rev. Lett. **79,** 1964–1967 (1997).
26. M. S. Chapman, T. D. Hammond, A. Lenef, J. Schmiedmayer, R. A. Rubenstein, E. Smith, and D. E. Pritchard, "Photon scattering from atoms in an interferometer: coherence lost and regained," Phys. Rev. Lett. **75,** 3783–3787 (1995).
27. S. Dürr, T. Nonn, and G. Rempe, "Fringe visibility and which-way information in an atom interferometer," Phys. Rev. Lett. **81,** 5705–5709 (1998).
28. D. A. Kokorowski, A. D. Cronin, T. D. Roberts, and D. E. Pritchard, "From single- to multiple-photon decoherence in an atom interferometer," Phys. Rev. Lett. **86,** 2191–2195 (2001).
29. P. Bertet, S. Osnaghi, A. Rauschenbeutel, G. Nogues, A. Auffeves, M. Brune, J. Raimond, and S. Haroche, "A complementarity experiment with an interferometer at the quantum-classical boundary," Science **411,** 166–170 (2001).
30. X. Maître, E. Hagley, G. Nogues, C. Wunderlich, P. Goy, M. Brune, J. M. Raimond, and S. Haroche, "Quantum memory with a single photon in a cavity," Phys. Rev. Lett. **79,** 769–772 (1997).
31. B. T. H. Varcoe, S. Brattke, M. Weidinger, and H. Walther, "Preparing pure photon number states of the radiation field," Nature **403,** 743–746 (2000).
32. A. J. Leggett, "Quantum theory; weird and wonderful," Physics World **12,** 73–77 (December, 1999).
33. B. E. King, C. S. Wood, C. J. Myatt, Q. A. Turchette, D. Leibfried, W. M. Itano, C. Monroe, and D. J. Wineland, "Cooling the Collective Motion of Trapped Ions to Initialize a Quantum Register," Phys. Rev. Lett. **81,** 1525–1528 (1998).
34. H. Rohde, S. T. Gulde, C. F. Roos, P. A. Barton, D. Leibfried, J. Eschner, F. Schmidt-Kaler, and R. Blatt, "Sympathetic ground-state cooling and coherent manipulation with two-ion crystals," J. Opt. B: Quantum Semiclass. Opt. **3,** S34–S41 (2001).
35. F. Diedrich, J. C. Bergquist, W. M. Itano, , and D. J. Wineland, "Laser Cooling to the Zero Point Energy of Motion," Phys. Rev. Lett. **62,** 403–406 (1989).

Recent advances in teleportation of entanglement and in fullerene interference

A. Zeilinger
Institute of Experimental Physics, University of Vienna, Boltzmanngasse 5, A-1090 Vienna, Austria
anton.zeilinger@univie.ac.at

Abstract: Extending earlier results on photon teleportation and entanglement swapping it was recently possible to teleport one photon of an entangled pair such that a statistically significant violation of Bell inequality between the other member of the original pair and the teleported one was obtained. This constitutes a definite confirmation of the quantum nature of our photon entanglement procedure. It also implied that the teleported state not only was unknown, it was even undefined at the moment of teleportation. In a parallel development in fullerence interference a precision measurement of the Heisenberg uncertainty relation for these large molecules was achieved. Furthermore, more recently, we were able to realise experimentally diffraction of fullerene molecules at standing light waves.

Coherence and Quantum Optics VIII
Edited by Bigelow et al., Kluwer Academic/Plenum Publishers, 2003

Nonclassical, two-photon interferometry and lithography with high-gain optical parametric amplifiers

Elna M. Nagasako, Sean J. Bentley and Robert W. Boyd
Institute of Optics, University of Rochester, Rochester, New York 14627

Girish S. Agarwal
Physical Research Laboratory, Navrangpura, Ahmedabad-380 009, India

Two-photon entangled states have been utilized in many configurations that exploit the novel fourth-order interferometric properties of these states, most recently in a proposal for the achievement of sub-Rayleigh-limit resolution [1]. While the interferometric properties of biphotons have been extensively analyzed under a variety of arrangements, the impact of multiphoton states on these properties has not been extensively explored.

In this poster, the contributions to the output of four-port fourth-order interferometers are considered, with low- and high-gain terms distinguished. The relative contribution of these terms in the Hong-Ou-Mandel and quantum lithography arrangements for parametric amplifier and coherent state input are analyzed. While the presence of terms arising from multiphoton input has significant effects on interferometer visibility, the desired properties are preserved in the case of quantum lithography and lost in the case of the Hong-Ou-Mandel interferometer.

A four-port interferometer can generate coincidence counts at its exit via four paths. These paths can be divided into single-input paths, in which the detected photons arise from only one input arm, and dual-input paths, in which both input arms contribute one photon. The two dual-input paths differ only in the specific mapping of each input arm to an output arm and are the only paths present when the input is a biphoton. Single-input paths are present when multiple-photon inputs are used and differ only in whether it is the signal or the idler arm that contributes both detected photons. These paths are not present with a biphoton input and, when introduced by increasing gain, contribute to a degradation of interferometric visibility.

The joint detection probability for the Hong-Ou-Mandel interferometer is given by

$$\langle \hat{a}_1^\dagger \hat{a}_2^\dagger \hat{a}_2 \hat{a}_1 \rangle = \frac{1}{4}\left[\langle \hat{a}_s^\dagger \hat{a}_s^\dagger \hat{a}_s \hat{a}_s \rangle + \langle \hat{a}_s^\dagger \hat{a}_i^\dagger \hat{a}_i \hat{a}_s \rangle + \langle \hat{a}_i^\dagger \hat{a}_s^\dagger \hat{a}_s \hat{a}_i \rangle + \langle \hat{a}_i^\dagger \hat{a}_i^\dagger \hat{a}_i \hat{a}_i \rangle\right]$$
$$+\frac{1}{2}\langle \hat{a}_s^\dagger \hat{a}_s^\dagger \hat{a}_i \hat{a}_i \rangle - \frac{1}{2}\langle \hat{a}_s^\dagger \hat{a}_i^\dagger \hat{a}_s \hat{a}_i \rangle$$

and gives the following results for each of the states considered. For a biphoton input, $\langle \hat{a}_1^\dagger \hat{a}_2^\dagger \hat{a}_2 \hat{a}_1 \rangle = 0$. For a Fock states in both signal and idler arms, the joint detection probability is $\frac{1}{2}m(m-1)$ and for a parametric amplifier the probability is $\overline{m}^2$.Figure 1 shows this result with the contributions of the single- and dual-input terms displayed separately. The dual-input terms combine to give a net probability of zero; these are the only terms present for a biphoton input, leading to the coincidence count cancellation that is the hallmark of the Hong-Ou-Mandel interferometer. The single-input terms that become dominant as the photon number increases lead to a joint detection probability of $\overline{m}^2$ for the parametric amplifier. We can note that this has the same form as the probability for a coherent state input $\langle \hat{a}_1^\dagger \hat{a}_2^\dagger \hat{a}_2 \hat{a}_1 \rangle = 0$, with both being equal to the product of the single detector probabilities, however this apparent similarity arises through the contribution of different terms in each case.

Coherence and Quantum Optics VIII
Edited by Bigelow et al., Kluwer Academic/Plenum Publishers, 2003

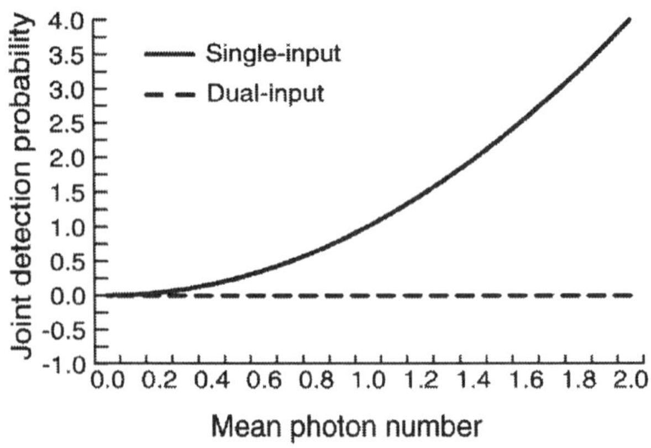

Fig. 1. Single-input and dual-input contributions to the output of a Hong-Ou-Mandel interferometer. The dual-input contribution vanishes at all values of the mean photon number.

While the interplay of the interferometer coefficients and the states' expectation values lead to similar results with coherent and parametric amplifier inputs for the HOMI, the different weightings for each state present in the quantum lithography case lead to qualitatively different results for the parametric amplifier and coherent output states, even at high gain. The joint detection probability for the quantum lithography configuration reduces to $\langle \hat{a}_1^\dagger \hat{a}_2^\dagger \hat{a}_2 \hat{a}_1 \rangle = 4(1-\sin^2\chi)$ for a biphoton input, $(1+\sin^2\chi)[2m(m-1)]+(1-\sin^2\chi)[4m^2]$ for a $|m\,m\rangle$ input, and $(1+\sin^2\chi)[4\overline{m}^2]+(1-\sin^2\chi)[8\overline{m}^2+4\overline{m}]$ for a parametric amplifier input, where the terms proportional to $1+\sin^2\chi$ arise from the single-input paths and the $1-\sin^2\chi$ terms arise from the dual-input paths. Figure 2 shows the parametric amplifier joint detection probability, with the relative size of the single-input versus the dual-input contribution plotted as a function of mean photon number. Recalling that the optimum visibility of unity is achieved with a biphoton input which has no single-input contribution, it can be seen that as the single-input contribution increases relative to the dual-input contribution the visibility decreases to its limiting value of 1/5.

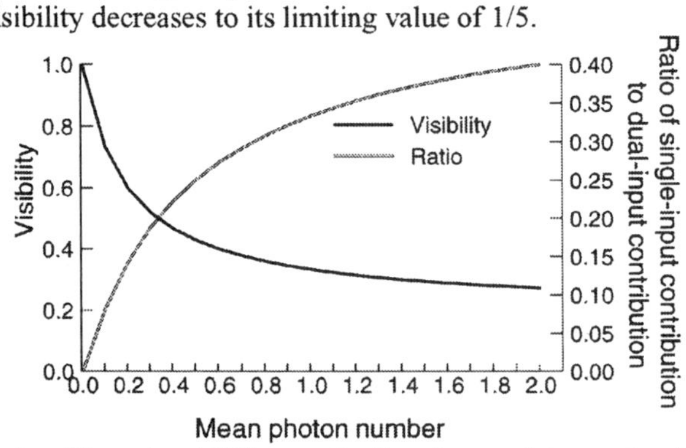

Fig. 2. Quantum lithography visibility as a function of mean photon number. The visibility decreases as the relative contribution of the single-input terms increases.

1. A.N. Boto, P. Kok, D.S. Abrams, S.L. Braunstein, C.P. Williams, and J.P. Dowling, Phys. Rev. Lett. **85**, 2773 (2000).

Simultaneous quantum state measurements using array detection

A.M. Dawes and M. Beck
Department of Physics, Whitman College, Walla Walla, Washington 99362
Ph: 509-527-5260, Fax: 509-527-5904,beckmk@whitman.edu

Here we present the results of the first experiment that uses array detection to measure quantum states of an optical beam [1]. We demonstrate that an array detector can be used to simultaneously measure the quantum states of many different spatial modes of the same beam. Furthermore, we show that array detectors can allow for an improvement in effective detection efficiency over standard detectors when using balanced homodyne detection. This improvement comes from the fact that the local oscillator (LO) and signal fields need not be mode-matched when using array detectors. In our experiment array detection is found to be over forty times more efficient than standard detection for measurements of a particular field mode. The technique we use for determining the state of our field modes is quantum state tomography (QST), and yields the density matrix in the Foch-state basis ρ_{mn} [2,3].

We wish to measure the quantum mechanical state of a light mode that is described by a real transverse spatial mode function $u_m(x)$. The positions at the center of the detector pixels are given by x_j, where j is an integer that labels the pixel. Light from the signal field of interest interferes with an LO field on a 50/50 beamsplitter. The LO is a plane wave in a large amplitude coherent state $\left|\beta e^{i\phi}\right\rangle$. The beams exiting the beamsplitter are detected with array detectors. Photoelectron numbers from corresponding pixels on each detector are subtracted, yielding a set of photoelectron difference numbers that are labeled by pixel: ΔN_j. The rotated quadrature amplitude of mode m, $q_{m\phi}$, is given in terms of the difference numbers by

$$q_{m\phi} = \frac{1}{\beta}\left(\frac{D_x}{2}\right)^{1/2} \sum_j \Delta N_{j\phi} u_m(x_j) , \tag{1}$$

where D_x is the width of the measured mode [1]. The subscript ϕ indicates that the measured difference numbers and the quadrature amplitude depend on the phase of the LO. Since we perform many measurements of $q_{m\phi}$ on an identically prepared ensemble, we are able to determine the quantum mechanical state of the field corresponding to mode m [2,3]. By choosing different mode functions, it is possible to use Eq. (1) to simultaneously determine the quadrature amplitudes (and hence the quantum state) of many different spatial modes for the same set of measurements. In our experiments we use the corrected difference number $\Delta N_{j\phi} - \left\langle \Delta N_j \right\rangle_{vac}$ in place of $\Delta N_{j\phi}$ in Eq. (1). Here $\left\langle \Delta N_j \right\rangle_{vac}$ is the average difference number when the signal field is in a vacuum state.

The signal field we use varies linearly with position and has a π phase shift in the middle. In Fig. 1 we plot the corrected difference number as a function of pixel number observed across the detector for a signal field in a coherent state having a mean of approximately 1 photon. Figure 1(a) depicts a single exposure, while Fig. 1(b) depicts an average over 200 exposures. These curves display the spatial variation of the amplitude of the electric field of the signal, consistent with a field that has a linear amplitude variation, with a π phase shift in the middle (the difference counts tend to be negative for half the beam, and positive for the other half). Changing the LO phase by π causes the slope of the curves in Fig. 1 (a) & (b) to invert, as expected. While these curves contain large noise (due to the shot-noise of the LO and imperfect subtraction of the vacuum difference level), they can clearly be seen to have

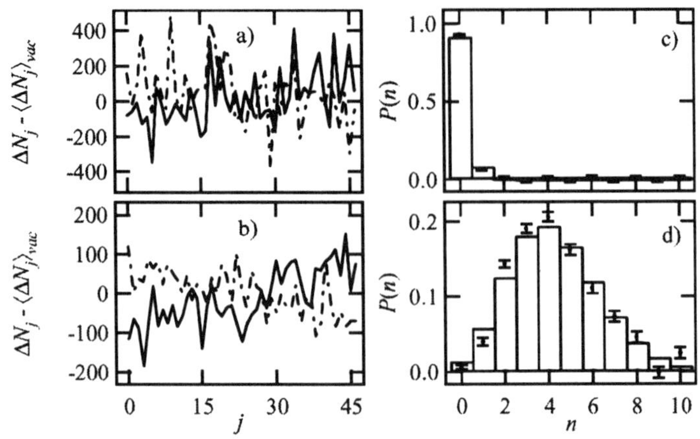

Fig. 1 (a) & (b): The corrected difference number for a signal mode in a coherent state with a mean of approximately 1 photon: (a) is for a single exposure, while (b) is for average of 200 exposures. The two curves in each figure correspond to two values of the local oscillator phase that differ by π. (c) & (d): Photon number distributions of the measured states corresponding to: (c) $u_{con}(x)$ and (d) $u_{lin}(x)$. The points represent measured values, while the bars correspond to theoretical coherent states having the same mean number of photons.

opposite slopes. An average of one photon in the signal beam, even on single shots as shown in Fig. 1(a), can lead to macroscopic differences in the detected signal across the array.

Since the signal is linear in position, we have measured the state of the field corresponding to the properly normalized mode $u_{lin}(x) = \left(12/D_x{}^3\right)^{1/2} x$, where we choose x=0 to be at the center of the range of pixels we are measuring. A homodyne detector using ordinary single detectors would not resolve spatial variation; this would correspond in our scheme to measuring a mode function that was constant across the array $u_{con}(x) = (1/D_x)^{1/2}$. In Fig. 1 (c) & (d) we show photon number distributions $P(n) = \rho_{nn}$ for the two different spatial modes. Using these distributions we find that the constant mode contains an average of 0.09 photons, while the linear mode contains an average of 4.3 photons. Thus, we see that standard single detectors would have an effective detection efficiency that is over 40 times smaller than our array detector for measuring a mode with a linear spatial variation. By comparing the experimental and theoretical curves it is seen that the measured photon number distributions are nearly those of coherent states.

In conclusion, we have experimentally demonstrated the use of an array detector to perform state measurements on an optical beam having a nontrivial spatial distribution. Array detection has an advantage over standard detection for this task because it allows for the simultaneous determination of the quantum states of multiple spatial modes in the beam, and it also allows for greater effective detection efficiency.

[1] M. Beck, "Quantum state tomography with array detectors," Phys. Rev. Lett. **84**, 5748-5751 (2000); A.M. Dawes and M. Beck, "Simultaneous quantum-state measurements...," Phys. Rev. A **63**, 040101(R) (2001).

[2] D.T. Smithey, M. Beck, M.G. Raymer and A. Faridani, "Measurement of the Wigner ...,"Phys. Rev. Lett. **70**, 1244-1245 (1993).

[3] U. Leonhardt, *Measuring the Quantum State of Light* (Cambridge University Press, Cambridge, England, 1997); G.M. D'Ariano, in *Quantum Optics and the Spectroscopy of Solids*, edited by T. Hakioglu and A.S. Shumovsky (Kluwer, Dordrecht, 1997), p. 139.

Quantum positioning system

V. Giovannetti, S. Lloyd*, and L. Maccone

Massachusetts Institute of Technology, Research Laboratory of Electronics

* *Department of Mechanical Engineering MIT 3-160, Cambridge, MA 02139, USA.*

Abstract: We show how quantum entanglement and squeezing can be employed in measuring a position in space, gaining an accuracy increase over classical localization procedures. We develop a positioning protocol based on frequency-entanglement of number-squeezed pulses, which may be used also for synchronizing the clocks of distant parties, and show how the accuracy increase scales with the system resources and how it compares with analogous, but classical, protocols.

OCIS codes: (030.1670) Coherent optical effects; (270.6570) Squeezed states

Quantum entanglement and squeezing have been exploited in the context of interferometry, frequency measurements, lithography, and algorithms (see for example Refs. [1]) in order to beat classical limitations. In this paper, we show that entanglement and squeezing can also be used to increase the accuracy of positioning measurements. Typically, such measurements can be performed by sending M pulses (each composed of N photons in average) from the party that is to be localized to M receiving stations and measuring their times of arrival. We will show that, by using entangled-squeezed light, it is possible to obtain an accuracy enhancement of $\sqrt{MN}$ over the classical case. A detailed derivation of the methods described here can be found in [2].

Consider the frequency-entangled state

$$|\Psi\rangle \equiv \int d\omega\ \phi_\omega\ |N_\omega\rangle_1 |N_\omega\rangle_2 \cdots |N_\omega\rangle_M \ , \tag{1}$$

where ϕ_ω is a spectral envelope function, and $|N_\omega\rangle_i$ is the N-photons Fock state (number squeezed) of frequency ω, directed towards the i-th receiving station. In the case of ideal detectors, the probability to detect all the photons is given by [3]

$$p(\{t_{i,k}\}) \propto \left\langle : \prod_{i=1}^{M} \prod_{k=1}^{N} E_i^{(-)}(t_{i,k}) E_i^{(+)}(t_{i,k}) : \right\rangle \ , \tag{2}$$

where $t_{i,k}$ is the detection time of the k-th photon at the i-th receiving station. In Eq. (2), the signal field at the i-th detector position is $E_i^{(-)}(t) \equiv \int d\omega\ a_i^\dagger(\omega)\ e^{i\omega t}$ and $E_i^{(+)} \equiv \left(E_i^{(-)}\right)^\dagger$, where $a_i(\omega)$ is the field annihilator of a quantum of frequency ω. The calculation of this probability with the state $|\Psi\rangle$ yields

$$p(\{t_{i,k}\}) \propto \left| g\left(\sum_{i=1}^{M} \sum_{k=1}^{N} t_{i,k}\right)\right|^2 \ , \tag{3}$$

where $g(t)$ is the Fourier transform of ϕ_ω. The uncertainty Δt in the mean time of arrival

$$\langle t \rangle = \frac{1}{MN} \sum_{i=1}^{M} \sum_{k=1}^{N} t_{i,k} \tag{4}$$

can be immediately obtained from Eq. (3), resulting in

$$\Delta t = \frac{\Delta\tau}{MN} , \tag{5}$$

where $\Delta\tau^2$ is the variance of the distribution $g(t)$. This result must be compared to what one would obtain in the classical case where one employs M unentangled pulses each in a coherent state with N average photons. For a fair comparison with the state $|\Psi\rangle$, we must consider the situation in which all the photons are described by the same bandwidth function ϕ_ω of Eq. (1). In this case, because of the Poissonian statistics, the times of arrival of each photon are independent. One can thus apply the central limit theorem, obtaining an accuracy

$$\Delta t \gtrsim \frac{\Delta\tau}{\sqrt{MN}} , \tag{6}$$

where, by using Eq. (2), it is possible to show that time of arrival uncertainty of the single photon $\Delta\tau$ is the same as in Eq. (5). The comparison between (5) and (6) shows that entanglement and number-squeezing can increase the accuracy of the mean time of arrival (and hence of positioning) by a factor $\sqrt{MN}$.

Using experimental techniques analogous as the ones described in [4], at least the simple case of $M = 2$ detectors and $N = 1$ photons per pulse can be implemented in practice.

Assuming that the distance between different parties is known, the same protocol can also be employed to enhance the accuracy of the synchronization of their clocks. Namely, they have to exchange entangled and squeezed pulses while measuring the pulses transit times.

In addition to the difficulty of creating the requisite entanglement, the primary drawback of this scheme is the sensitivity to loss. However, for this same reason, the procedure exhibits improved security: because the timing information resides in the correlations between pulses, an eavesdropper who intercepts some but not all of the photons obtains no timing information. Finally, the frequency entanglement allows similar schemes to be highly robust against pulse broadening due to transit through dispersive media as experimentally shown in [5]. Since dispersion is one of the main limiting factors in employing classical narrow pulses, this is a very attractive characteristic for practical applications.

This work was funded by the ARDA, NRO, and by ARO under a MURI program.

References and links

1. C. M. Caves, Phys. Rev. D **23**, 1693 (1981); R. S. Bondurant and J. H. Shapiro, Phys. Rev. D **30**, 2548 (1984); B. Yurke, S. L. McCall, and J. R. Klauder, Phys. Rev. A **33**, 4033 (1986); J. J. Bollinger, *et al.*, Phys. Rev. A **54**, R4649 (1996); A. N. Boto, *et al.*, Phys. Rev. Lett. **85**, 2733 (2000); L. K. Grover, Phys. Rev. Lett. **79**, 325 (1997).
2. V. Giovannetti, S. Lloyd, and L. Maccone, Eprint quant-ph/0103006.
3. R. J. Glauber, Phys. Rev. **130**, 2529 (1963).
4. C. K. Hong, Z. Y. Ou, and L. Mandel, Phys. Rev. Lett. **59**, 2044 (1987).
5. A. M. Steinberg, P. G. Kwiat, and R. Y. Chiao, Phys. Rev. Lett. **68**, 2421 (1992).

High-frequency, time-domain, optical homodyne tomography

M. Bellini, P.L. Ramazza
*Istituto Nazionale di Ottica Applicata, L.go E. Fermi, 6, I-50125 Florence. ITALY
bellini@ino.it*

A. Zavatta
Dept. of Systems and Informatics, University of Florence, Florence, ITALY

F. Marin, and F.T. Arecchi*
Dept. of Physics, University of Florence, Florence, ITALY

Abstract: We present the first time-domain measurements of pulsed, balanced homodyne detection at very high repetition rates (82 MHz). The technique is demonstrated by tomographic reconstruction of few-photon coherent states.

Homodyne detection has been extensively used to obtain tomographic reconstructions of CW optical fields, such as vacuum, coherent or squeezed states [1]. In order to access states that exhibit stronger non-classical features, one has to exploit the higher nonlinearities made available by the high peak intensities of pulsed lasers. A frequency-domain approach, with a spectral analysis of the difference current in the homodyne photodetectors, was originally developed to study the noise in the field quadratures at definite radio-frequencies [2]. Successively, a pulsed balanced homodyne detection at zero frequency (dc), consisting in the measure of integrated photocurrents on a pulse-by-pulse basis in the time-domain, has been adopted to achieve a full reconstruction of weak field states [3]. The task of performing this time-domain homodyne detection is however quite demanding, since it requires very low electronic noise and high subtraction efficiencies over very large frequency bandwidths (from dc to a few times the repetition rate of the laser). Mostly due to these technical limitations, all the existing applications of the latter technique have been limited to pulse repetition rates well below the megahertz.

Here we present an efficient pulsed optical homodyne apparatus that is able to work at the high repetition frequencies characteristic of the commonly used mode-locked laser systems. We have demonstrated its capabilities by tomographic reconstruction of coherent states with a low number of photons ($\langle n \rangle \sim 1$).

A commercial Ti:sapphire laser emitting 3 ps pulses at 786 nm and at a repetition rate of about 82 MHz, is used as the source for the local oscillator (LO) beam in the homodyne setup. The apparatus consists of a non-polarizing cube beam-splitter for the mixing with the field state under study, and of two 80-mm lenses that focus the beams onto the photodetectors. Two EG&G-FFD40 photodiodes are connected so that their difference photocurrent is directly fed to a home-made amplifier with a bandwidth of about 100 MHz. The output pulses have a FWHM duration of about 5 ns and successive pulses from the laser (separated by 12 ns) are thus well distinct from each other. The amplified difference signal is acquired by a fast (4 Gs/s) digital scope, able to store sequences of up to 4000 laser pulses with a vertical resolution of 8 bits and a temporal sampling of 250 ps. Each of these sequences, covering a time window of 50 μs, is then immediately analyzed to extract the areas of all the 4000 pulses and measure their variance.

We have tested the conditions for shot-noise-limited operation by verifying that the measured variances scale linearly with the power from the LO up to a mean input power level of about

3-4 mW, where some saturation of the detector response and inefficiencies in the subtraction start to take place. At the optimum LO power, the ratio between the shot noise and the electronic noise is about 5. The correctness of this method of acquisition and analysis has been further verified by comparing the results obtained in the time domain with those derived by an integration of the frequency-domain noise signal acquired by a spectrum analyzer.

In order to test the possibility of performing homodyne measurements and tomographic reconstructions, we have generated a few-photon coherent state from a strongly filtered portion of the same laser beam and we have spatially and temporally superimposed it to the LO beam on the beam-splitter. The relative phase between the two fields has been controlled by applying a voltage to a PZT holding one of the mirrors in the LO beam path.

We have performed measurements of coherent states with different average photon numbers, ranging from 0 to about 5. Each measurement consists of 8 sequences of 4,000 pulses for each of the 25 values of the relative phase in the $[0,\pi]$ interval, for a total of 800,000 points per coherent state (Fig.1).

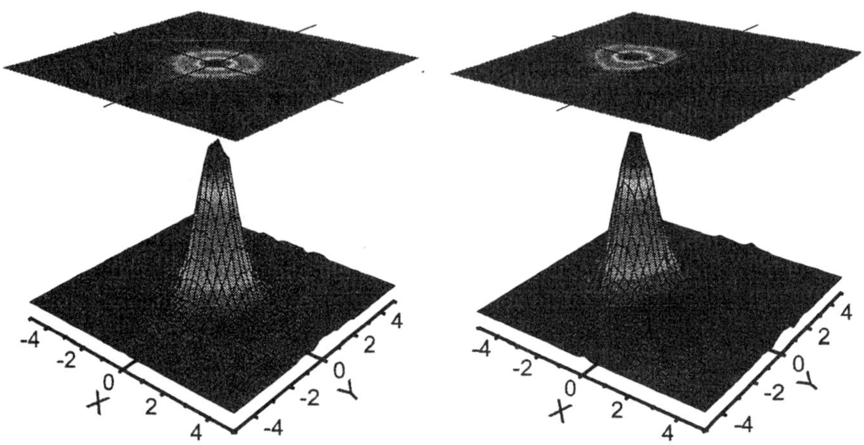

Fig.1. Reconstructed Wigner functions for a vacuum field and for a coherent state with <n>~1, acquired with the high-frequency technique described in the text.

The demonstration of reliable quantum tomographic reconstruction at high frequencies opens new possibilities in the investigation of highly non-classical states of the electromagnetic field. The use of high repetition rates has been proposed [4] as a way to circumvent the limitations connected with the finite efficiency of detectors. In a conditional homodyne measurement one has to make sure that n clicks in the detectors really correspond to n photons arriving to the measurement apparatus. This can be only assured if the gain in the generation is kept sufficiently low so as to have a negligible probability of generating $n+1$ photons. Unfortunately, low gains also mean low production rates and impractically long acquisition times unless much higher pump rates are used.

We are currently investigating the possibility to apply this technique to the tomographic reconstruction of Fock states (note that the high pump rate will make states with n>1 much more accessible) and of Schroedinger's cat-like states of the optical field.

References:

[1] G. Breitenbach, S. Schiller and J. Mlynek, Nature 387, 471 (1997)
[2] R.E. Slusher, P. Grangier, A. LaPorta, B. Yurke and M.J. Potasek, Phys. Rev. Lett. 59, 2566 (1987)
[3] D.T. Smithey, M. Beck, M.G. Raymer and A. Faridani, Phys. Rev. Lett. 70, 1244 (1993)
[4] A. Montina and F.T. Arecchi, Phys. Rev. A 58, 3472 (1998)

Coherent control of atoms in an optical lattice

H.L. Haroutyunyan and G. Nienhuis

Huygens Laboratorium, Universiteit Leiden
Postbus 9504, 2300 RA Leiden, The Netherlands

nienhuis@molphys.leidenuniv.nl

Abstract: An exactly solvable model is presented for the dynamics of a quantum particle in a periodic potential. The possibilities for coherent control of the state by an additional time-dependent uniform force are analyzed.

OCIS codes: (020.1670) Coherent optical effects; (140.3320) Laser cooling

The Wannier-Stark problem of a quantum particle in a periodic potential and suffering a uniform force has regained interest due to the emergence of optical lattices of atoms. Counterpropagating light beams provide the periodic potential, and a time-dependent frequency difference of the beams induces acceleration of the periodic potential, which is equivalent to a uniform force in the comoving frame. In a constant uniform force a wavepacket is expected to perform Bloch oscillations, provided that interband crossings are negligible. These Bloch oscillations have been observed in an optical lattice [1]. When the uniform force is modulated, dynamical localization of the particle position can occur [2, 3].

In the present paper we discuss the posibilities to prepare the state of particles in a periodic potential by tailoring the time-dependent force. We assume that the atoms are sufficiently cooled, so that only the lowest energy band is populated. For simplicity we adopt the tight-binding limit, where only the ground levels in neighboring wells are coupled. The state in well n located at $x = na$ is indicated as $|n\rangle$. These states play the role of the basis of localized Wannier states. In this basis the total Hamiltonian is

$$H = \frac{1}{2}\hbar\Omega(B_+ + B_-) + \hbar\Delta(t)B_0, \tag{1}$$

where $B_\pm$ and B_0 are displacement and position operators, defined by $B_\pm|n\rangle = |n \pm 1\rangle$, and $B_0|n\rangle = n|n\rangle$. The frequancy Ω measures the coupling between neighboring wells, and $\Delta(t)$ measures the time dependence of the uniform force. By using the commutation algebra of these operators, one can evaluate the full evolution operator, which we specify by its matrix

$$\langle n|U(t,0)|m\rangle = (ie^{-i\beta})^{-n+m}e^{-in\phi}J_{n-m}(\alpha)\ , \tag{2}$$

where the parameters are defined by the relations

$$\phi(t) = \int_0^t dt'\ \Delta(t')\ ,\ \alpha(t)e^{i\beta(t)} = \Omega\int_0^t dt' e^{i\phi(t')}\ . \tag{3}$$

The parameter α determines the spreading of an initally localized state. One conclusion of (2) is that average motion of a wavepacket can only occur when an initial phase coherence between neighboring states exists, corresponding to a non-zero expectation value $\langle B_+\rangle$.

The case of a constant force is the standard situation where Bloch oscillations occur. When Δ is constant, the Hamiltonian is time-independent and the energy eigenvalues

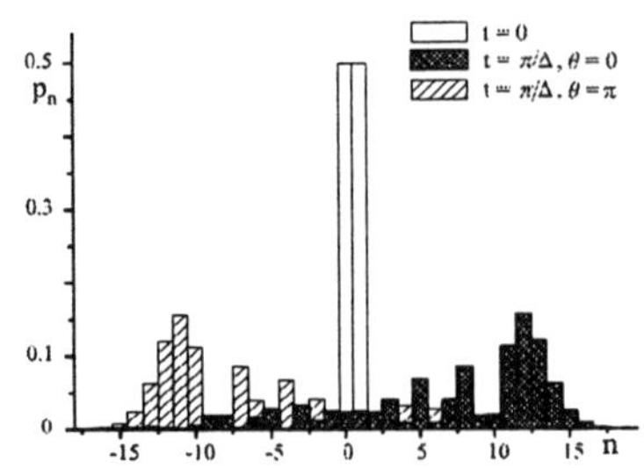

Fig. 1. Plot of oscillating population distribution over Wannier states, for $\Omega/\Delta = 6$.

form so called Wannier-stark ladder. They are integer multiples of Δ, so that each solution of the Schrödinger equation is periodic with the Bloch period $2\pi/\Delta$.

For an initial superposition state of only two neighboring states, such as $|\Psi(0)\rangle = (|0\rangle \pm e^{i\theta}|1\rangle)/\sqrt{2}$ the average position already exhibits an oscillation with an amplitude $\Omega/2\Delta$. This is illustrated in the figure for a constant uniform force.

A wider wavepacket is found to display Bloch oscillations with the amplitude Ω/Δ, which is in accordance with the effective velocity $v(t) = \Omega a \sin(\Delta t)$. This velocity can be understood as the group velocity defined by $\hbar v = dE/dk$, with $E = \hbar\Omega\cos(ka)$ the dispersion relation between the energy E and the quasimomentum. The external force makes the quasimomentum time-dependent, so that $k(t) = -\Delta t/a$. The dispersion relation follows from diagonalizing the Hamiltonian (1) with $\Delta = 0$, which produces the Bloch states $|k\rangle \propto \sum_n |n\rangle \exp(inka)$. In addition, the width is found to vary over the Bloch period. For Gaussian wavepackets, this time variation can be evaluated analytically.

Similar techniques can be applied when the uniform force is modulated according to the expression $\Delta(t) = \Delta_0 + \delta\cos(\omega t)$. Then the parameter $\alpha(T)$ after one oscillation period $T = 2\pi\omega$ takes the value

$$\alpha = 2\Omega\sin(\Delta_0 T/2)\sum_n J_n(\frac{\delta}{\omega})\frac{1}{\Delta_0 + n\omega}\,, \tag{4}$$

in terms of Bessel functions. This expression is valid provided that none of the denominators in (4) vanishes, and it indicates a bound on the spreading. On the other hand, at a resonant value $\Delta_0 + n_0\omega = 0$ of the force, the spreading of the packet becomes unbounded, as described by the expression $\alpha = \Omega t J_{n_0}(\delta/\omega)$. This spreading can be suppressed when the ratio δ/ω between the frequency and the amplitude of the modulation coincides with a zero of the Bessel function. This subtle transition between bounded (quasi)periodic behavior and unbounded spreading of the wavepacket of a quantum particle illustrates dynamical localization [2, 3]. It also shows the richness of our simple model system.

References and links

1. M. Ben Dahan, E. Peik, J. Reichel, Y. Castin and C. Salomon, "Bloch oscillations of atoms in an optical potential," Phys. Rev. Lett. **76**, 4508 -4511 (1996); E. Peik, M. Ben Dahan, I. Bouchoule, Y. Castin and C. Salomon, "Bloch oscillations of atoms, adiabatic rapid passage, and monokinetic atomic beams," Phys. Rev. A **55**, 2989-3001 (1997).
2. D.H. Dunlap and V.M. Kenkre, "Dynamic localization of a charged particle moving under the influence of an electric field," Phys. Rev. B **34**, 3625-3633 (1986); V.M. Kenkre and S. Rhagavan, "Dynamic localization and related resonance phenomena," J. Opt. B: Quantum Semiclass. Opt. **2**, 686-693 (2000).
3. C.F. Bharucha, J.C. Robinson, F.L. Moore, B. Sundaram, Q. Niu and M.G. Raizen, "Dynamical localization of ultracold sodium atoms," Phys. Rev. E **60**, 3881-3885 (1999).

Photoionization of Bose-Einstein condensates

D. Ciampini, F. Fuso, J. H. Müller, M. Anderlini, O. Morsch, E. Arimondo

Istituto Nazionale per la Fisica della Materia, Dipartimento di Fisica Università di Pisa, Via F. Buonarroti 2, I-56127 Pisa, Italy
ciampini@df.unipi.it

Abstract: Photoionization of Bose-Einstein condensates is an interesting topic at the interface between plasma physics and quantum statistics. We irradiated a BEC of rubidium atoms with ionizing laser pulses and studied losses and density profile modifications in the remaining trapped cloud of atoms.

The recent development of laser cooling and trapping techniques has made possible the controlled realization of dense and cold atomic samples, thus opening the way for spectroscopic investigations in the low and ultra-low temperature regimes not accessible with conventional techniques. In recent years there has been an increased interest in the study of ultra-cold plasmas [1] and of photoionization processes of cold atoms, providing the opportunity to perform accurate measurements of the photoionization cross section [2].

The realization of Bose-Einstein condensates (BEC) of alkali atom vapors have attracted much interest to new aspects of photon-matter interaction arising from the coherent nature of the atomic ensemble. Quantum collective effects give rise to a significant modification of the optical properties of atomic samples.

On the theoretical side, attention has been paid recently to the analysis of photoionization of a BEC by monochromatic laser light [3]. Due to the coherent nature of the initial atomic ensemble and narrow spectral width of the light source, the occupation numbers of the final states can become close to unity, especially for excitation close to threshold. The products of the photoionization process (electrons and ions) obey Fermi-Dirac statistics, thus the ionization rate should be slowed down and determined by a balance between the laser light acting on the atoms and the rate of escape of the products from the system.

A crucial point in every treatment of photoionization processes is the role of interactions between charged and neutral particles of the condensate and between the charged particles themselves. The first consideration is that just after an act of photoionization the fermions are very far from thermodynamical equilibrium, hence the well known formulae for transport processes derived for the case of a plasma deviating slightly from the equilibrium state cannot be used. Secondly, the wavepacket of a particular fermion is spread widely and overlaps significantly with wavepackets of many other fermions, beyond the standard binary-collision approach.

A single charged particle in a cloud of dense cold atoms will polarize the surrounding ground state atoms, influencing the density distribution of the cloud and thus the cloud-averaged interaction energy. This energy is also sensitive to the to the details of the 2-body correlation function for small distances, which before the ionization is governed by the van der Waals interaction between ground state atoms. This correlation is not included in a mean-field (Gross-Pitaevskii type) description of the condensate.

In our lab, an apparatus for laser cooling and Bose-Einstein condensation of rubidium atoms is in operation. Depending on the chosen sequence of cooling and trapping steps, neutral samples with temperatures ranging from a few hundreds of μK down to tens of nK can be obtained. Typical densities (10^{10} - 10^{12} atoms/cm^3) are large enough to perform reliable investigations by absorption or emission techniques. Changing the trap parameters, we can study photoionization of atomic samples with different phase space densities, reaching the condensed phase with a number of atoms ($\approx 10^4$) that allows a dynamical analysis of the cold atom sample (temperature, density distribution).

In a first stage, the photoionization of the atomic sample has been achieved by driving a two photon non-resonant transition from the ground state. The radiation was provided by an excimer-pumped dye laser used at a wavelength close to 594 nm, the two photon ionization threshold, with intensities in the tens of MW/cm^2 range and a pulse duration of 16 ns. To extract the photoionization cross-section we experimentally looked at the remaining cloud of atoms, since a charge detector was not present in the vacuum cell. We first irradiated the cold atoms of the Magneto Optical Trap (MOT) with the ionizing laser, studying the induced additional losses. During the decay time of the MOT the atoms received $\approx$500 ionizing pulses, the effect on the decay rate is of the order of 20%. In a second experiment we studied the effect of a single ionizing pulse on the BEC sample, measuring the loss of atoms from the magnetic trap and the density profile modifications of the condensed clouds (Fig.1).

The interpretation of these preliminary results suggests the occurrence of different processes, presumably initiated by multi-photon ionization, involving collisions between cold electrons and ultra-cold neutrals, recombination at low temperatures and collective processes in the weak and cold plasma.

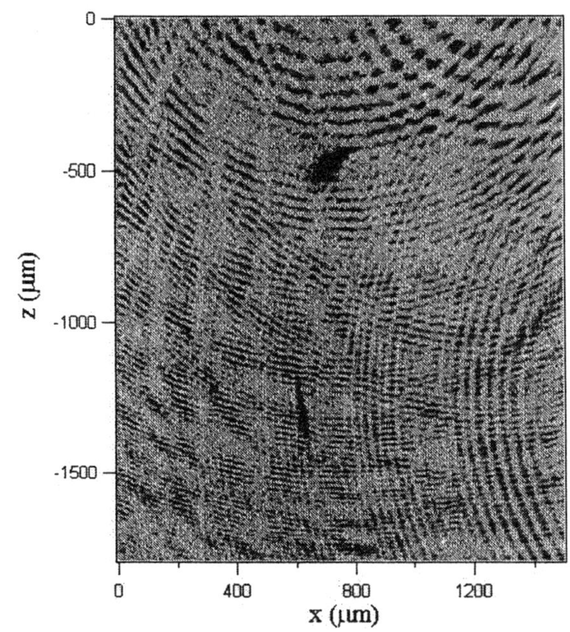

Fig. 1. Shadow image of a condensate after a single dye laser pulse. The BEC is deformed and a secondary cloud is present (The periodic circular fringe patterns are an artifact of the imaging technique).

[1] T.C. Killian et al., "Creation of an Ultracold Neutral Plasma", Phys. Rev. Lett. **83** 4776 (1999); S. Kulin et al., "Plasma Oscillations and Expansion of an Ultracold Neutral Plasma", Phys. Rev. Lett. **85** 318 (2000).

[2] previous experiments in our lab.: O. Maragò et al., "Photoionization cross sections for excited laser-cooled cesium atoms", Phys. Rev. A **57** R4110 (1998); E. Arimondo et al., "Laser Cooling and Photoionization of Alkali Atoms", Appl. Surf. Sci. **154-155** 527 (2000); F. Fuso et al., "Ion processes in the photoionization of laser cooled alkali atoms", Opt. Commun. **173** 223 (2000).

[3] I.E. Mazets, " Photoionization of neutral atoms in a Bose-Einstein condensate", Quantum Semiclass. Opt **10** 675-681 (1998).

Quantum noise in coupled atomic and molecular Bose-Einstein condensates

J.J. Hope, M.K. Olsen and C.M. Savage

Department of Physics and Theoretical Physics, Australian National University, ACT 0200, Australia

+61 2 61252780,+61 2 61250741(fax),Joseph.Hope@anu.edu.au

Most theoretical descriptions of multi-component Bose-Einstein condensates (BEC) have relied on semiclassical analysis such as the Gross-Pitaevskii equation (GPE). This semiclassical approximation allows the quantum field to be described by a single wave-function and it has been widely successful in describing the dynamical features of weakly interacting dilute gas Bose-Einstein condensates. It includes the effects of s-wave interactions, and can be readily generalised to include multi-component condensates with inter-species couplings [1, 2]. In contrast, we analyse a non-linear, experimentally accessible system which requires a full quantum field calculation [3].

There has been much interest recently in the production of a molecular Bose-Einstein condensate (MBEC) from the photoassociation of an atomic Bose-Einstein condensate (BEC) of a weakly interacting dilute gas [1,4-12] This process not only produces a new species of BEC, but does it through a non-linear coupling which allows the possibility of quantum statistical effects becoming evident in the dynamics. We use the functional positive-P representation to calculate the full quantum dynamics of the atomic and molecular fields, and show that in certain parameter regimes the Gross-Pitaevskii equation (GPE) can give incorrect results, even for the atomic and molecular populations.

The Bose-enhancement of the photoassociation of atoms from a trapped BEC leads to giant, collective oscillations between the atomic and molecular populations. This enhancement of a chemical process was dubbed "superchemistry" by Heinzen *et al.* when they modelled it using a two-component GPE [1]. A more recent model using the Hartree-Fock-Bogoliubov method includes pair correlations in the atomic field, and showed discrepancies with the results obtained with the GPE [10]. To investigate the full effects of the quantum nature of the fields, we develop a set of stochastic equations for the atomic and molecular fields based on the functional positive-P representation [13]. The GPE is simply the Itô form of the stochastic partial differential equations with the noise terms removed. The dynamical statistical properties of a BEC have been examined using stochastic differential equations in the positive-P and Wigner representations [14, 15, 16], with success limited by the tendency for the individual trajectories to diverge for many physically interesting parameter regimes.

We model an atomic BEC coupled to an excited molecular field by a far detuned laser, which is in turn coupled to a stable molecular level by a second laser. In order for two lasers to provide a two-photon transition without populating the excited MBEC, the single photon detuning must be made very large. We then map the equations of motion for the density matrix to an equation of motion for the functional positive-P representation. The resulting Fokker-Planck equation can be converted to a set of stochastic field equations. The stochastic integration was then performed numerically with the XMDS package [17], and the results are shown in Fig. 1.

Although the GPE solution exhibits a revival in atomic population after this regime, we can see that it is clearly inadequate for predictions over this timescale. The positive-P solution has a sampling error which can be reduced arbitrarily, given a sufficient number of trajectories. The GPE does not contain the spontaneous component of the process

Coherence and Quantum Optics VIII
Edited by Bigelow et al., Kluwer Academic/Plenum Publishers, 2003

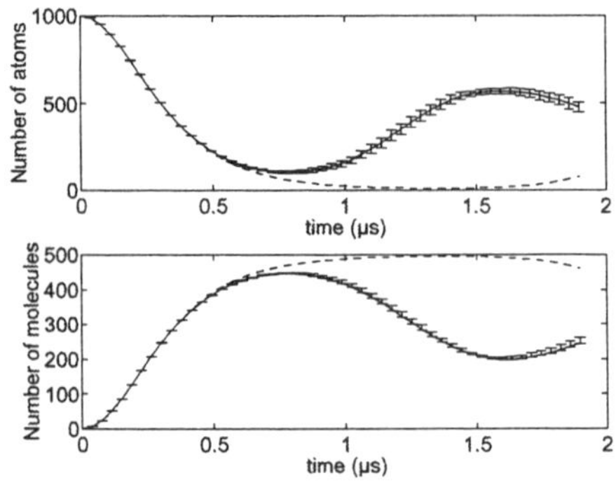

Fig. 1. Atomic and molecular populations for a fast coupling rate, assuming that the molecules have the same scattering length as the atoms. The solid line is the result of the positive-P calculation and the dashed line is the solution of the GPE. The error bars on the positive-P solution are due to the sampling error.

which is dominant when there are very few atoms.

The positive-P representation contains the full description of the quantum field, including all quantum statistical effects such as the pair-correlation effects as described by Holland *et al.* [10]. It is limited by the behaviour of the individual trajectories which do not have to behave in a physical fashion, and can become unstable over time in the absence of damping.

While it may seem reasonable to expect that the quantum statistics will tend to affect the multi-time correlations of the field rather than the mean field, our result here shows it is also important to include them when considering the equations of motion for moments of the mean field. The signature of the breakdown of the GPE occurs in the simplest experimental observable - the total atomic and molecular populations.

References and links

1. D.J. Heinzen, R. Wynar, P.D. Drummond and K.V. Kheruntsyan, "Superchemistry: Dynamics of coupled atomic and molecular Bose-Einstein condensates", Phys. Rev. Lett. **84**, 5029, (2000).
2. R.J. Ballagh, K. Burnett and T.F. Scott, "Theory of an output coupler for Bose-Einstein condensed atoms", Phys. Rev. Lett. **78**, 1607 (1997).
3. J.J. Hope and M.K. Olsen, "Quantum superchemistry: dynamical quantum effects in coupled atomic and molecular Bose-Einstein condensates", Phys. Rev. Lett. **86**,3220(2001).
4. B. Levi, "Hot prosects for ultracold molecules", Physics Today **53** No. 9, 46 (2000).
5. P.D.Drummond, K.V. Kheruntsyan and H. He, "Coherent molecular solitons in Bose-Einstein condensates", Phys. Rev. Lett. **81**, 3055 (1998).
6. R. Wynar, R.S. Freeland, D.J. Han, C. Ryu, D.J. Heinzen, "Molecules in a Bose-Einstein condensates", Science **287**, 1016 (2000).
7. P.S. Julienne, K. Burnett, Y. B. Band and W. C. Stwalley, "Stimulated Raman molecule production in Bose-Einstein condensates", Phys. Rev. A **58**, R797 (1998).
8. M. Mackie, R. Kowalski and J. Javanainen, "Bose-stimulated Raman adiabatic passage in photoassociation", Phys. Rev. Lett. **84**, 3803, (2000).
9. J.J. Hope, M.K. Olsen and L.I. Plimak, "Multimode model of formation of molecular Bose-Einstein condensates by Bose-stimulated Raman adiabatic passage", Phys. Rev. A **63**,043603(2001).
10. M. Holland, J. Park and R. Walser, cond-mat/0005062.
11. M.K. Olsen, J.J. Hope and L.I Plimak, "Photon blockade and quantum dynamics in intracavity coherent photoassociation of Bose-Einstein condensates", Phys. Rev. A *in press*.
12. K. Góral, M. Gajda and K. Rzążewski, cond-mat/0006192.
13. P.D. Drummond and C.W. Gardiner, J. Phys. A **13**, 2353 (1980).
14. M.J. Steel *et al*, "Dynamical quantum noise in trapped Bose-Einstein condensates", Phys. Rev. A **58**, 4824, (1998).
15. P.D. Drummond and J.F. Corney, "Quantum dynamics of evaporatively cooled Bose-Einstein condensates", Phys. Rev. A **60**, R2661 (1999).
16. U. Poulsen and K. Mølmer, cond-mat/0006030.
17. G.R. Collecutt, P.D. Drummond and H. He, http://www.physics.uq.edu.au/xmds/

Cavity-QED with a single ion

Gerhard R. Guthöhrlein, Matthias Keller, Wolfgang Lange, and Herbert Walther
Max-Planck-Institut für Quantenoptik, Hans-Kopfermann-Str. 1, 85748 Garching, Germany
Phone +49 (89) 32905-0, fax +49 (89) 32905-200, e-mail wfl@mpq.mpg.de

Kazuhiro Hayasaka
Kansai Advanced Research Center, Communications Research Lab., 588-2 Iwaoka, Nishi-ku, Kobe 651-24, Japan
Phone +81 (78) 969-2223, fax +81 (78) 969-2229, e-mail hayasaka@crl.go.jp

1. Introduction

Radio-frequency ion-traps provide ideal conditions for storing and manipulating single particles and localizing them with a precision far below their resonance wavelength. This property is applied in our experiment to investigate the interaction between ions and the field in an optical cavity, implementing a completely deterministic dynamical evolution of the system. This is in contrast to previous cavity QED experiments using atomic beams [1] or atomic fountains [2]. In these cases, deterministic control over the motion of the atoms is not possible, because the randomly varying arrival times and trajectories result in a fluctuating atom-field interaction strength. In the ion-based cavity-QED system we have set up, a continuous interaction of a single particle with the photon field at a pre-defined position is realized. Applications include the generation of single photon pulses as well as a single-ion laser.

2. Experimental Setup

We use a linear RF ion trap to store single $^{40}Ca^{+}$-ions. In order to avoid the coating and the charging of the mirrors in the ionization process, the trap is loaded in a region separated from the cavity by 25 mm (Fig. 1). Subsequently, the ion is shuttled along the trap axis, reaching the cavity region with an efficiency of 50 %. After careful compensation of offset voltages, an ion localization of $\lambda/10$ is achieved, with λ=397 nm being the wavelength of the $S_{1/2}$-$P_{1/2}$ resonance transition. The position of the cavity with respect to the ion is controlled with an equivalent precision by a three-axis piezo-electric stage upon which the cavity is mounted.

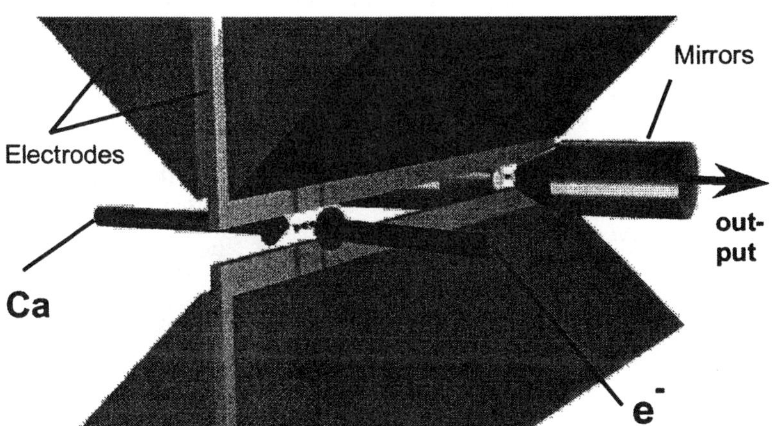

Fig. 1. Experimental setup of linear trap and optical cavity. The trap is loaded in the front region and a single ion is shuttled along the trap axis into the cavity region

For coupling to the optical field, we have chosen the $D_{3/2}$ to $P_{1/2}$ transition in $^{40}Ca^+$, because at a wavelength of λ=866 nm mirror losses can be minimized to the order of 10 ppm. This corresponds to a cavity decay rate as low as $\kappa/2\pi$=0.24 MHz. The $P_{1/2}$ level is pumped from the $S_{1/2}$ ground state, while the $D_{3/2}$ level is emptied via the $P_{3/2}$ state (Fig. 2). With a cavity length of 1 mm, the atom-cavity coherent coupling constant is $g/2\pi$=3.9 MHz, which is larger than the corresponding atomic decay rate $\gamma/2\pi$=1.7 MHz.

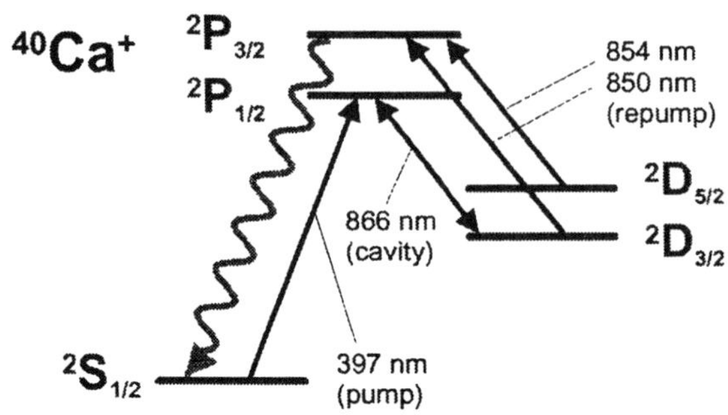

Fig. 2. Level scheme of $^{40}Ca^+$, showing the 866 nm cavity transition as well as pumping transitions at 397 nm, 850 nm and 854 nm.

To check the control we have over the atom-cavity coupling, we used the $^{40}Ca^+$-ion to probe the intensity distribution in the cavity. Scanning the relative position of ion and cavity, while at the same time monitoring the fluorescent light emitted by the ion, we are able to position the ion at the maximum of the transverse TEM_{00}-mode, as well as at an antinode of the standing wave pattern in the longitudinal direction. The resolution achieved in this way was $\lambda/10$.

3. Single-ion cavity QED

We will employ the setup described above to deterministically generate single photon pulses [3]. For this purpose we couple the $D_{3/2}$ to $P_{1/2}$ transition in $^{40}Ca^+$ at λ=866 nm to an optical cavity tuned to this wavelength, with mirror losses as low as 10 ppm (cavity decay rate $\kappa/2\pi$=0.24 MHz). Using the adiabatic passage technique with a pump pulse on the $S_{1/2}$-$P_{1/2}$ transition, a pulse containing exactly one photon is emitted from the cavity with an efficiency of 97%. The system will also be used as a source of cw radiation by pumping the $P_{1/2}$ level continuously and emptying the $D_{3/2}$ level via the $P_{3/2}$ state. In this case lasing is achieved with a single ion and radiation with non-classical properties is emitted from the cavity [4]. In the same regime, we can also probe the overlap between the ion's matter-wave and the optical standing wave in the cavity [5].

[1] C. J. Hood, T. W. Lynn, A. C. Doherty, A.S. Parkins, and H. J. Kimble, "The atom-cavity microscope," Science 287, 1447-1453 (2000).
[2] P. W. H. Pinkse, T. Fischer, P. Maunz, G. Rempe , "Trapping an atom with single photons," Nature 404, 365-368 (2000).
[3] C. K. Law and H. J. Kimble, "Deterministic generation of a bit-stream of single-photon pulses," J. Mod. Opt. 44, 2067-2074 (1997).
[4] G. M. Meyer, H. J. Briegel, and H. Walther, "Ion-trap laser," Europhys. Lett. 37, 317-322 (1997).
[5] M. Löffler, G. M. Meyer, H. Walther , "One-atom laser with quantized centre-of-mass motion," Europhys. Lett. 40, 263-268 (1997).

Superradiance in photonic crystals

C R Bennett and M. Babiker
Department of Physics, University of York, Heslington, York YO10 5DD, England
crb5@york.ac.uk

J. B. Kirk
Department of Physics, University of Essex, Colchester, Essex CO4 3SQ, England

Abstract: Electric field Green functions are derived for a photonic crystal in the form of an ordered array of metallic sheets. We outline how the Green functions permit the study of spontaneous emission as well as Dicke-type superradiance and explore screening and dipole orientation effects.

OCIS codes: (240.0310) Thin films; (270.6630) Superradiance, superfluorescence

An ordered array of thin metallic sheets separated by identical dielectric layers constitute a one-dimensional photonic crystal inside which electromagnetic fields interact with dipole emitters embedded in the structure. The decay rates, together with Dicke-type cooperative effects [1] can be evaluated once the electric field Green functions are derived. The required Green functions must satisfy the jump condition to be imposed on the magnetic field vector due to the areal conductivity of the metallic sheets and must also conform with Bloch's theorem due to the periodicity along the growth direction. The derivation leads first to mixed (real space, reciprocal space) Green functions and the all-space Green functions follow from these by Fourier transformation. The explicit forms of the Green functions will not be displayed here.

The emission rates for a dipole moment μ, of frequency ω_0, with a given orientation as either parallel ($\|$) or perpendicular (z) to the metallic sheet planes and which is positioned at a point $\mathbf{r} = (\mathbf{r}_\|, z)$ within the crystal can be obtained from the Green functions as follows [2]

$$\Gamma_\|(\mathbf{r}_\|, z) = \frac{2\mu^2}{\hbar^2}\mathrm{Im}\left[G_{xx}(\mathbf{r}_\|, z; \mathbf{r}_\|, z; \omega_0)\right]; \qquad \Gamma_z(\mathbf{r}_\|, z) = \frac{2\mu^2}{\hbar^2}\mathrm{Im}\left[G_{zz}(\mathbf{r}_\|, z; \mathbf{r}_\|, z; \omega_0)\right]. \quad (1)$$

Figure 1 exhibits these rates against dipole frequency for a dipole at position $z = d/2$ in a metallic sheets superlattice of periodicity length $d = 500\text{Å}$ and sheet charge density $n_s = 1.7\times10^{20}\mathrm{m}^{-2}$. The results are compared with the situations in which the dipole is positioned in front of a single layer and when the sheets are considered uncoupled. Note the symmetry in the exact superlattice curves in Figs. 1(a) and (b) and their relative agreement or otherwise with the single sheet and uncoupled sheets results.

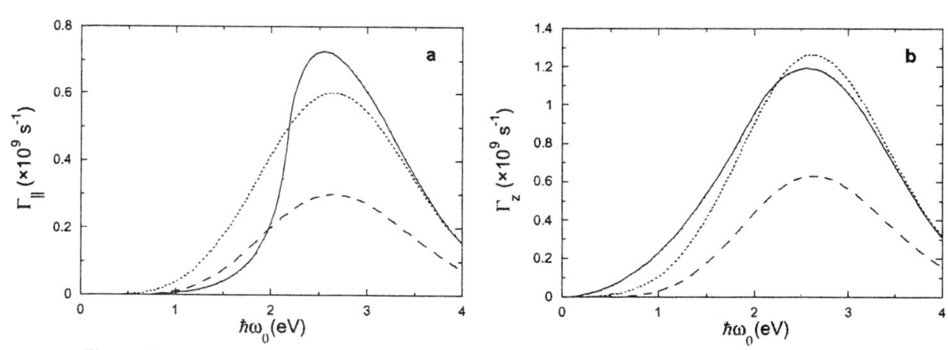

Fig. 1: Spontaneous emission rates for (a) a dipole parallel and (b) perpendicular to the layers using the exact superlattice result (solid), the single layer result (dashed) and the uncoupled sheets result (dotted).

It is well known that Dicke-type super-radiance and sub-radiance phenomena can be realised in free space for a system of identical two-level atoms [1]. These pair cooperative effects are currently of much interest, especially to the emerging field of quantum logic [3]. It is interesting to examine how such cooperative phenomena are affected by the environment when the dipoles are positioned at separate points within a superlattice of metal layers. There are two types of spontaneous emission rates of this system, one corresponding to the symmetric state (+) and another to the antisymmetric state (-). It is straightforward to show that the evaluation of these rates can be carried out in terms of the Green function formalism and we arrive at the required results in the form

$$\Gamma^{\pm}_{\hat{\mu}_1\hat{\mu}_2}\left(\mathbf{r}_{\|},z_1,z_2\right)=\frac{\mu^2}{\hbar^2}\,\mathrm{Im}\Big\{G_{\hat{\mu}_1\hat{\mu}_1}\left(\mathbf{r}_{\|},z_1;\mathbf{r}_{\|},z_1;\omega_0\right)+G_{\hat{\mu}_2\hat{\mu}_2}\left(\mathbf{r}_{\|},z_2;\mathbf{r}_{\|},z_2;\omega_0\right)$$
$$\pm\left[G_{\hat{\mu}_1\hat{\mu}_2}\left(\mathbf{r}_{\|},z_1;\mathbf{r}_{\|},z_2;\omega_0\right)+G_{\hat{\mu}_2\hat{\mu}_1}\left(\mathbf{r}_{\|},z_2;\mathbf{r}_{\|},z_1;\omega_0\right)\right]\Big\}, \quad (2)$$

where the subscripts $\hat{\mu}_1$ and $\hat{\mu}_2$ are unit vectors indicating directions of the dipoles 1 and 2, respectively. Figures 2(a) and (b), respectively, show the variations of $\Gamma^{\pm}_{xx}\left(\mathbf{r}_{\|},z_1,z_2\right)$ and $\Gamma^{\pm}_{zz}\left(\mathbf{r}_{\|},z_1,z_2\right)$ with the position z_2 of dipole 2, such that dipole 1 is fixed at $z_1 = d/2$. The dipole frequency is such that $\hbar\omega_0 = 2\text{eV}$ and the sheet density is $n_s = 1.7\times10^{20}\text{m}^{-2}$. It can be seen that there are superradiance and sub-radiance effects when the dipoles are positioned between two adjacent sheets. The most striking feature, however, is the discontinuity in $\Gamma^{\pm}_{zz}\left(\mathbf{r}_{\|},z_1,z_2\right)$ at the positions of the layers surrounding dipole 1. These are attributable to the discontinuities in the electric field vector due to the polarisation charges at the metallic sheets.

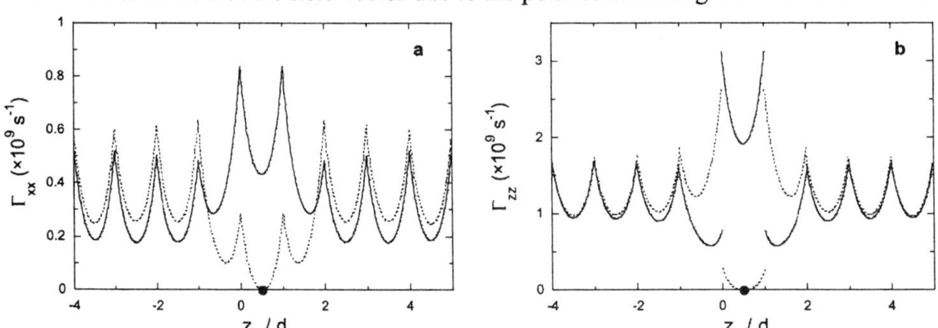

Fig. 2: The variation of the cooperative decay rate for two dipoles (a) parallel and (b) perpendicular to the layers where the solid (dotted) curves are for a symmetric (antisymmetric) initial state.

An experimental set-up to explore the phenomena discussed in this paper should not be too hard to realise. We propose adopting the monolayer assembly technique as used by Drexhage [4], which has recently been the basis of a number of experiments on related surface effects. By repeatedly producing periodically spaced layers of metal a superlattice may be built and by inserting molecular monolayers at certain positions during the growth process all the situations examined here could, in principle, be investigated. Spontaneous decay and cooperative effects should be evident in the flourescence output arising in the various situations.

1. R. H. Dicke, "Coherence in spontaneous radiation processes", Phys. Rev. **93** 99-110 (1954).
2. S. M. Barnett, B. Huttner, R. Loudon and R. Matloob, "Decay of excited atoms in absorbing dielectrics", J. Phys. B: At. Mol. Opt. Phys. **29** 3763-3781 (1996).
3. G. K. Brennan, I. H. Deutsch and P. S. Jensen, "Entangling dipole-dipole interactions for quantum logic with neutral atoms", Phys. Rev. A **61** 2309-2318 (2000).
4. K. H. Drexhage, "Interaction of light with monomolecular dye lasers" in *Progress in Optics XII*, ed. E. Wolf 163-232 (North Holland, Amsterdam 1974).

Temporal and spatial coherence properties of nonstationary polychromatic light sources in one-dimensional photonic band-gap structures.

A. Albertoni,
INFM at Dipartimento di Energetica, Università di Roma "La Sapienza", Via Scarpa 16, 00161 Roma, Italy And Dipartimento di Elettronica, Politecnico di Bari, Via Orabona 4, 70125 Bari, Italy.

C. Sibilia, M.Centini , G.D'Aguanno , M. Bertolotti
INFM at Dipartimento di Energetica, Università di Roma "La Sapienza". Via Scarpa 16, 00161 Roma, Italy +39 06 4991 6541 , +39 06 442 40 183 , concita.sibilia@uniroma1.it

M. Scalora, M. J. Bloemer, C. M. Bowden
U.S. Army Aviation and Missile Command, Weapon Sciences Directorate, AMSMI-RD-WS-ST Redstone Arsenal,Huntsville, Alabama 35898-5000.

1.Introduction

In recent years one-, two- and three-dimensional, periodic, dielectric and metallo-dielectric structures have attracted a great deal of attention[1, 2]. In such materials, electromagnetic field propagation is forbidden for a range of frequencies, and allowed for others. The nearly complete absence of some frequencies in the transmitted spectrum is referred to as a photonic band gap (PBG), in analogy to semiconductor band gaps We have investigated the temporal coherence and the spatial coherence over PBG by using the plane wave spectrum extension of the matrix method [3]. This method gives us the PBG's spatial temporal output field and hence the possibility of calculation of all the functions of coherence: in this work we have studied the two frequency spectral density, the cross correlation and the complex degree of coherence.

2. Coherence properties of gaussian shaped pulses over PBG structure

Under condition of fully coherent light sources and by using directly the statistical definition of degree of coherence in time and space domains , we have defined the time of coherence (t_c) and the transversal length of coherence (l_c) in the PBG .Two type of source are investigated: the time Gaussian shaped plane wave and a Gaussian pulsed laser beam ; the coherence time (t_c) in the first case and the transversal length of coherence (l_c) in the second have been studied for the light field disturbance at the PBG output. The degree of coherence $\gamma(t_1,t_2)$ in the temporal domain is used to calculate the t_c,(defining t_c as the time where γ is unitary in modulus) and $\gamma(\rho_1,\rho_2)$ in the space domain is used to calculate l_c, where (l_c) indicate the radius of the circle $\rho_c=l_c$ where $|\gamma|=1$

$$\gamma(\rho_1,\rho_2,t_1,t_2)=\frac{\Gamma(\rho_1,\rho_2,t_1,t_2)}{\{I(\rho_1,t_1)I(\rho_2,t_2)\}^{1/2}} \tag{1}$$

where

$\Gamma(\rho_1,\rho_2,t_1,t_2)=\langle V^*(\rho_1,t_1)V(\rho_2,t_2)\rangle=\int_{-\infty}^{+\infty}\int_{-\infty}^{+\infty}\exp[i(\omega_1 t_1-\omega_2 t_2)]s(\rho_1,\rho_2,\omega_1,\omega_2)d\omega_1 d\omega_2$ is

the cross-correlation function at two spatial points ρ_1 and ρ_2 ($\rho_1,\rho_2\in\mathbf{D}$) and at two istants t_1 and t_2; in space-frequency domain the two-frequency cross-spectral density is

$$s(\rho_1, \rho_2, \omega_1, \omega_2) = \langle u^*(\rho_1, \omega) u(\rho_2, \omega_2) \rangle \tag{2}$$

which is the correlation momentum between ω_1 and ω_2 of the spectral process at two points ρ_1 and ρ_2.

To put into evidence the modification in the coherence properties introduced by the propagation through the PBG, we have superimposed on the Gaussian input disturbance (fully coherent) a completely uncorrelated Gaussian white noise (0.1% additive in amplitude, incoherent and white in band $[-\Omega_0, +\Omega_0]$). To obtain the time of coherence we calculate, assuming the noise two-frequency cross-spectral density as $s^N(\omega_1, \omega_2) = N^2 rect_{\omega 1}(-\Omega_0, \Omega_0) \cdot rect_{\omega 2}(-\Omega_0, \Omega_0)$, the complex degree of coherence in temporal domain:

$$|\gamma_T(t_1, t_2)| = \left| \frac{\Gamma(t_1, t_2) + 2N^2 i[\sin(\Omega_0 t_1) - \sin(\Omega_0 t_2)]}{\Gamma(t_1, t_1)\Gamma(t_2, t_2)} \right| \cong \left| \frac{\Gamma(t_1, t_2)}{\Gamma(t_1, t_1)\Gamma(t_2, t_2)} \right| = |\gamma(t_1, t_2)| \tag{3}$$

where N is the ampluitude of the noise. For the input pulse, with noise, changes of the coerence time (increase) have been found for propation at the band edge.

The analysis on the spatial coherence properties has been performed: an increase of spatial coherence, for suitable spectral regions has been calculated.

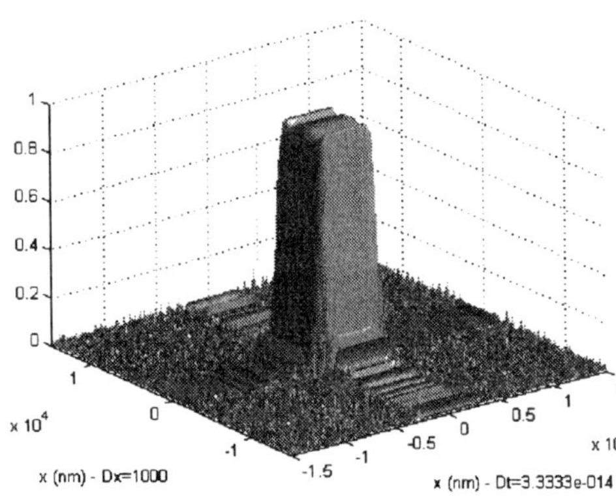

Fig. 1. The input modulus of the complex degree of coherence, in the spatial case; the pulse have a transversal dimension of 2Δt (2000*nm*).The pulse propates at the band edge.

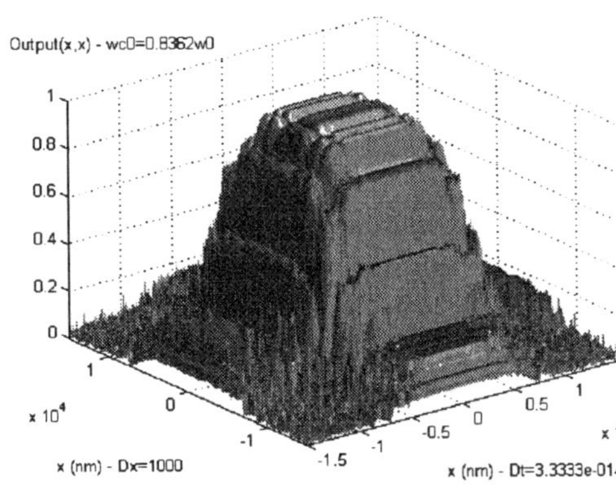

Fig. 2. The output modulus of the complex degree of coherence after propagation in PBG band edge, in the same conditions of Fig. 1.

3.References

[1] M. Scalora, J. P. Dowling, M. J. Bloemer, A. S. Manka, C. M. Bowden, J. Haus, "Pused propagation near reflective surfaces: Application to photonic band-gap structures and the question of superluminal tunneling times", Phisical Review A, **52** (1995) 726-731.
[2] J. M. Bendickson, J. P. Dowling, M. Scalora, " Analytic expression for the electromagnetic mode density in finite, one dimensional, photonic band gap structures " Phisical Review E, **53** (1996) 4107 - 4112.
[3] A .Albertoni et al. . to be published.

Effect of measurements of collective observables on a single atom

S. Wallentowitz

Fachbereich Physik, Universität Rostock, Universitätsplatz 3, D-18051 Rostock, Germany

wal@physik5.uni-rostock.de

Abstract: The effect of measurements of collective observables of N-atom systems on the single bosonic atom in the system is derived and an interpretation of the resulting projective dynamics of the single atom is given.

Consider a system of N atoms where one measures a collective (i.e., additive) observable of the system. An example would be to measure the center-of-mass position (or momentum) of a cloud of N atoms in free space or in a trapping potential. The question now arises as to how a single atom in the many-atom system is affected by such a measurement. The motivation for this question can be given by the fact that the treatment of the N-atom problem is in general impractical so that reduced single-atom descriptions are desired. Moreover, the single-atom description of collective measurements is particularly useful in the context of quantum trajectories, that describe the free evolution of a single atom, being frequently interrupted by measurements of collective variables and eventually subsequent applications of feedback.

Let us start from the N-atom density operator $\hat{\varrho}_N$ and consider the measurement of the observable $\hat{A} = \sum_{i=1}^{N} \hat{a}_i$, where $\hat{a}_i$ is the contribution of the ith atom. When measuring the collective observable $\hat{A}$ with measured outcome $A_{\rm m}$, the projected density operator after the measurement $\hat{\varrho}'_N(A_{\rm m})$, being the N-atom quantum state conditioned on the measurement outcome $A_{\rm m}$, reads [1, 2]

$$\hat{\varrho}'_N(A_{\rm m}) = \hat{P}_N(A_{\rm m})\, \hat{\varrho}_N\, \hat{P}^\dagger_N(A_{\rm m}), \tag{1}$$

where $\hat{P}_N(A) = \hat{P}^\dagger_N(A)$ is the measurement projector defined as $\hat{P}_N(A) = |A\rangle\langle A|$ with $|A\rangle$ being the N-atom eigenstate of the observable $\hat{A}$ with eigenvalue A.[1] The probability (density) to measure the value $A_{\rm m}$ is given by the trace of the (unnormalized) projected density operator: $\mathrm{prob}(A_{\rm m}) = \mathrm{Tr}\, \hat{\varrho}'_N(A_{\rm m})$.

Introducing the second-quantized bosonic field operators $\hat{\psi}(x)$ and $\hat{\psi}^\dagger(x)$ with $[\hat{\psi}(x), \hat{\psi}^\dagger(x')] = \delta(x - x')$ and performing a mode expansion with respect to the single-atom eigenmodes $\phi_a(x)$, that correspond to the eigenvalues a of the single-atom observable $\hat{a}$, the mode operators $\hat{\psi}_a$, $\hat{\psi}^\dagger_a$ are obtained with the commutator relation $[\hat{\psi}_a, \hat{\psi}^\dagger_{a'}] = \delta_{a,a'}$. The collective observable reads then $\hat{A} = \sum_a a\, \hat{\psi}^\dagger_a \hat{\psi}_a$ and for describing the effect of the measurement projection (1) on the single atom we may now introduce the reduced single-atom density matrix as

$$\rho_{a,a'} = \frac{1}{N} \left\langle \hat{\psi}^\dagger_{a'}\, \hat{\psi}_a \right\rangle, \tag{2}$$

where the expectation value is obtained by use of the N-atom density operator $\hat{\varrho}_N$.

[1] For simplicity a possibly existing degenerated auxiliary degree of freedom is omitted here.

Coherence and Quantum Optics VIII
Edited by Bigelow et al., Kluwer Academic/Plenum Publishers, 2003

The action of the collective measurement on a single atom can then be shown to be,

$$\rho'_{a,a'}(A_{\rm m}) = \frac{1}{N} \left\langle \hat{P}^\dagger_N(A_{\rm m})\, \hat{\psi}^\dagger_{a'}\, \hat{\psi}_a\, \hat{P}_N(A_{\rm m}) \right\rangle, \tag{3}$$

where $\rho'_{a,a'}(A_{\rm m})$ is the conditioned single-atom state after the measurement, where the probability (density) for measuring $A_{\rm m}$ is $\text{prob}(A_{\rm m}) = \sum_a \hat{\rho}'_{a,a}(A_{\rm m})$. In general, the right-hand side of Eq. (3) cannot be expressed in terms of single-atom density matrices since the projector $\hat{P}_N(A_{\rm m})$ introduces higher-order correlations. However, performing a cumulant expansion up to second order leads to the following approximation

$$\rho'_{a,a'}(A_{\rm m}) = \rho_{a,a}\, \frac{\delta(a-a')}{\sqrt{2\pi\, \mathcal{C}_2(a)}}\, \exp\left\{ -\frac{[a+\mathcal{C}_1(a)-A_{\rm m}]^2}{2\,\mathcal{C}_2(a)} \right\} dA_{\rm m}, \tag{4}$$

where for simplification the case of a continuous spectrum of eigenvalues is considered. The first two cumulants are defined via

$$N\rho_{a,a}\,\mathcal{C}_1(a) \;=\; \left\langle \hat{\psi}^\dagger_a \hat{A}\, \hat{\psi}_a \right\rangle, \tag{5}$$

$$N\rho_{a,a}\,\mathcal{C}_2(a) \;=\; \left\langle \hat{\psi}^\dagger_a \left[\hat{A}-\mathcal{C}_1(a)\right]^2 \hat{\psi}_a \right\rangle. \tag{6}$$

For a system with exactly $N=1$ atoms both cumulants naturally vanish and the single-atom measurement projection is recovered from Eq. (4). In general, however, they still contain higher-order atom-atom correlations that are generated by the measurement projection of the N-atom quantum state. In Hartree–Fock approximation these correlations can be expressed in terms of single-atom density matrices and a closed system of equations on the single-atom level is obtained.

The truncation at the second-order cumulant can be justified by means of the central-limit theorem: Since N atoms contribute to the measured value $A_{\rm m}$, according to the central-limit theorem the probability distribution becomes Gaussian for large N regardless of the shape of the single-atom probability distributions for the individual contributions a_i. The eigenstate $\phi_a(x)$ onto which the single-atom state is projected to on average is given by the condition $a+\mathcal{C}_1(a)=A_{\rm m}$. This means that the measured value is composed of the single-atom contribution plus $\mathcal{C}_1(a)$, which then follows to be the mean-field contribution of the remaining $N-1$ atoms. Furthermore, the uncertainty with which the single-atom state is projected is given by the second-order cumulant $\mathcal{C}_2(a)$. This uncertainty describes the fact that even for a precise measurement of the collective observable, apart from the case of condensed atoms $[\mathcal{C}_2(a)=0]$, the individual atoms do not perfectly behave identical and therefore are not projected onto identical states.

In conclusion the above described formulation of collective measurements may prove to be useful in the case of measurements on trapped atoms for instance, including the case where a large fraction of the atoms is condensed. It allows the generalization of measurement schemes known from single-atom experiments to the many-atom case, where additional many-atom effects occur.

References and links

1. J. Von Neumann, "Mathematical foundations of quantum mechanics," (Princeton University Press, Princeton, 1955).
2. P. Busch, P.J. Lahti, and P. Mittelstaedt, "The quantum theory of measurement," (Springer Verlag, Berlin, 1991).

Experimental demonstration of wave chaos in a conventional optical resonator

J.P. Woerdman, J. Dingjan and M.P. van Exter

Leiden University, Huygens Laboratory,
P.O. Box 9504, 2300 RA Leiden, The Netherlands
e-mail: qo@molphys.leidenuniv.nl, fax: ++(31)715275819

Abstract: We have observed wave chaos in an extremely simple system, namely a passive optical resonator consisting of 3 standard laser mirrors. Nonintegrability, introduced via nonparaxial aberrations, leads to chaos, as evidenced by a statistical analysis of the eigenfrequency spectrum.

Wave chaos (or quantum chaos) refers to the wave-mechanical behavior of systems that are chaotic in the classical limit. The most striking effects of wave chaos are found in the energy spectrum of chaotic systems, where nearest-neighbor level distributions and long-range correlations are drastically altered. This topic has developed into a large field; particularly fruitful has been the microwave implementation of wave chaos, using a stadium-type high-Q resonator in order to achieve nonintegrability [1].

Experimental realization of wave chaos in the *optical* domain has been hindered by the fact that it is very difficult to fabricate a stadium-type high-Q optical resonator. Transient chaotic lasing has been observed in deformed dye-doped liquid microspheres [2]; however, this approach is not likely to allow full study of the spectral and spatial signatures of chaos.

We present here a novel approach based upon an *open* passive resonator defined by standard high-reflectivity laser mirrors. Nonintegrability is introduced by operating the resonator in the *nonparaxial* regime; this destroys the separability of the intra-cavity wave field into longitudinal and transverse modes. The violation of paraxiality can be parametrized, in lowest order, as the 5 Seidel aberrations. Depending on the symmetry of the latter, one (or both) of the transverse mode indices loose their meaning.

For a 2-mirror cavity the numerical aperture (NA) must become rather large in order to achieve the required amount of nonintegrability. Therefore we use a 3-mirror cavity, with a curved folding mirror (Fig. 1). This introduces relatively large astigmatism and coma, so that a very modest NA (in our case ~ 0.02) is already sufficient. The eigenfrequency spectrum of the resonator of Fig. 1 is measured in transmission by injecting a HeNe beam and scanning the cavity length. The cavity has a reflection finesse of $\sim 2 \times 10^3$ and it supports $\sim 10^4$ low-loss transverse modes. Before injection, the HeNe beam passes through a weak diffusor in order to have reasonable overlap with the transverse-mode manifold.

The finesse of the individual resonances in the measured spectrum is not notably deteriorated due to the folding-mirror-induced aberrations. This result, which is *not* a priori obvious, shows that the mode manifold spanned by the mirrors is apparently a well-isolated part of the ray phase space, as in the case of a truly closed cavity. We thus have the necessary ingredients of chaos in phase space: the rays that bounce between the mirrors are subject to repeated stretching and folding since, apparently, the nonparaxial aberrations hardly couple to the extra-cavity modes.

To actually prove the occurrence of chaos we take the squared modulus of the Fourier

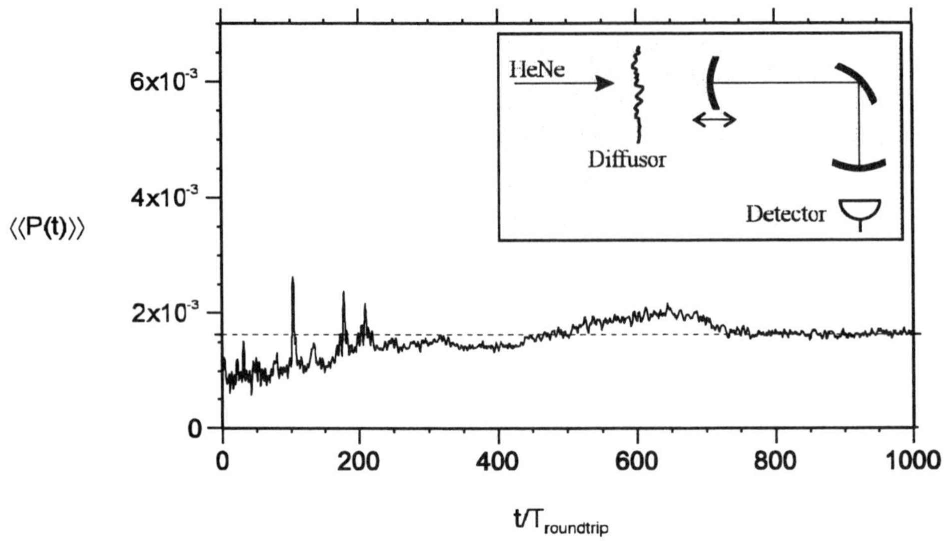

Fig. 1. Doubly averaged wave packet survival probability as a function of the number of round trip times; the inset shows the experimental set-up.

transform of the measured transmission spectrum, and perform a double averaging [3]. First we perform ensemble averaging by (slightly) varying the cavity length, and subsequently initial-state averaging, by varying the position of the diffusor. This leads to the so-called wave-packet survival probability $\langle\langle P(t)\rangle\rangle$; the inverse of the asymptotic value $\langle\langle P(\infty)\rangle\rangle$ directly gives the number of excited modes ($\sim 10^4$). For a regular system $\langle\langle P(t)\rangle\rangle \geq \langle\langle(P(\infty)\rangle\rangle$ for all t. For a chaotic system $\langle\langle P(t)\rangle\rangle$ *must* fall below its asymptote for sufficiently small times t [3]; fig. 1 shows that it does so indeed. This small-time behavior of $\langle\langle P(t)\rangle\rangle$ corresponds to "spectral rigidity" or "level repulsion": the number of levels, N, in a given spectral interval fluctuates much less (namely as $\sqrt{\ell nN}$) than in a nonchaotic spectrum ($\sqrt{N}$). The spikes on top of the curve in Fig. 1 represent scars; these have been discussed elsewhere [4].

Due to its extreme simplicity the nonparaxial optical resonator opens a whole range of exciting possibilities to study wave chaos in the optical domain. Apart from the spectral aspects we point in particular to the *spatial* aspects, which can be directly studied with a CCD camera. This should allow very well controlled experimental study of Anderson localization of light [5]. We may also study its precursor, *i.e.* coherent backscattering [5]. Finally, the open nature of the chaotic resonator easily allows to insert any desired nonlinear medium, *e.g.* a *cw* gain medium. In this way we may study the intriguing quantum fluctuations that have been predicted for a chaotic laser [6].

References and links

1. H.-J. Stöckmann, *Quantum Chaos* (Cambridge University Press, 1999).
2. S. Chang, R.K. Chang, A.D. Stone, and J.U. Nöckel, "Observation of emission from chaotic laser modes in deformed microspheres: displacement by the stable- orbit modes", J.Opt.Soc.Am.B **17**, 1828–1834 (2000).
3. J. Wilkie and P. Brumer, "Time-dependent manifestations of quantum chaos", Phys.Rev.Lett. **67**, 1185–1188 (1991).
4. J. Dingjan, E. Altewischer, M.P. van Exter, and J.P. Woerdman, submitted to Nature.
5. D.S. Wiersma, P. Bartolini, A. Lagendijk, and P. Righini, "Localization of light in a disordered medium", Nature **390**, 671–673 (1997).
6. C.W.J. Beenakker, "Thermal radiation and amplified spontaneous emission from a random medium", Phys.Rev.Lett. **81**, 1829–1831 (1998); G. Hackenbroich, C. Viviescas, B. Elattari, and F. Haake, "Photocount statistics of chaotic lasers", quant-phys/0102028.

Bell state preparation using pulsed non-degenerate two-photon entanglement

Yoon-Ho Kim, Sergei P. Kulik, and Yanhua Shih

Department of Physics University of Maryland, Baltimore County
Baltimore, Maryland 21250

In femtosecond pulse pumped SPDC, polarization Bell states have been realized only after post-selections of the two-photon amplitudes are made: especially, spectral post-selection by using narrow-band filters. Instead of the spectral post-selection, one can use a thin nonlinear crystal. However, both methods reduce the available flux of the entangled photon pair significantly.

In this paper, we report a Bell state preparation experiment in which we effectively eliminate *any post-selection* in femtosecond pulse pumped SPDC for the first time. Other features in our scheme include: (i) collinear SPDC makes the alignment much easier and the visibility as high as 92% is observed, (ii) Alice and Bob shares photons of different frequencies entangled in both space-time and polarization, (iii) thick crystals can be used to increase the intensity (without losing the visibility). These features make our scheme a good source of Bell states for quantum information experiments.

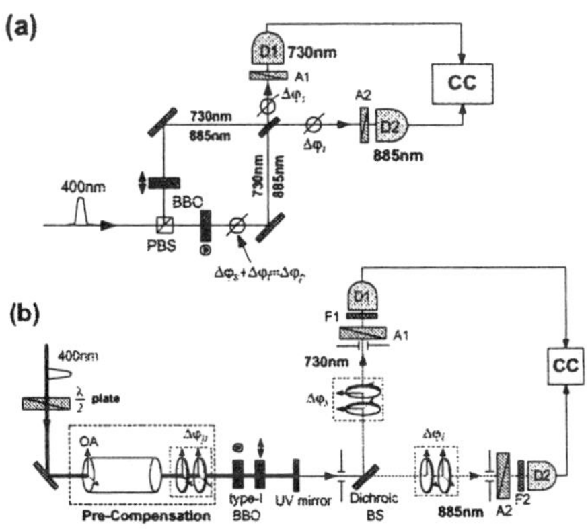

Fig. 1. (a) Principle schematic of the experiment. Non-degenerate collinear type-I SPDC occurs at the nonlinear crystal placed in each arm of the MZI. (b) Schematic of the experimental setup. Note that three different phases can be observed.

The basic idea of the experiment is illustrated in Fig.1(a). A 45° polarized femtosecond laser pulse (central wavelength $\lambda_p = 400$nm and pulse duration $\sigma = 80$fsec.) enters the Mach-Zehnder interferometer (MZI) which contains a type-I nonlinear crystal in each arm. One crystal has its optic axis oriented vertically ($\updownarrow$) and the other horizontally ($\odot$). Non-degenerate collinear type-I SPDC occurs, with equal probability, at each crystal (signal wavelength = 730nm and idler wavelength = 885nm) and they are mixed at the dichroic beamsplitter which directs 730nm photons to detector D_1 and 885nm to detector D_2. The quantum state generated from the vertically oriented crystal ($\updownarrow$) is $|H_{730}\rangle|H_{885}\rangle$ and from the horizontally oriented one ($\odot$) is $|V_{730}\rangle|V_{885}\rangle$. When the MZI is balanced, the quantum state after the MZI is (*without* throwing away any amplitudes)

$$|\Phi\rangle = |V_{730}\rangle_1|V_{885}\rangle_2 + e^{i\Delta\varphi}|H_{730}\rangle_1|H_{885}\rangle_2, \tag{1}$$

where $\Delta\varphi$ is the relative phase between the two amplitudes and it can easily be varied by scanning one of the mirror of the MZI.

In the experiment, we have used a collinear scheme where two type-I BBO crystals are placed collinearly in the pump beam path, see Fig1(b). Two type-I BBO crystals with thickness 3.4mm each (the first one is oriented horizontally and the second one is oriented vertically) are then pumped by a 45° polarized pump pulse. Due to the dispersion, $|V_{730}\rangle|V_{885}\rangle$ from the first BBO ($\odot$) and $|H_{730}\rangle|H_{885}\rangle$ from the second BBO ($\updownarrow$) are distinguishable in time. To make them indistinguishable in time, one needs to compensate the delay experienced by the SPDC photon pairs at each crystal. The compensator consists of a quartz rod and two quartz plates whose optic axes are oriented vertically. By tilting the two quartz plates in the opposite directions, the phase delay $\Delta\varphi_p$ can be varied to prepare a Bell state. This is demonstrated in Fig.2(a) where we study the space-time interference as a function of $\Delta\varphi_p$. Note that $\Delta\varphi_s$ and $\Delta\varphi_i$ are fixed. The observed modulation period is 400nm, see Fig.2(a), which agrees with the theory.

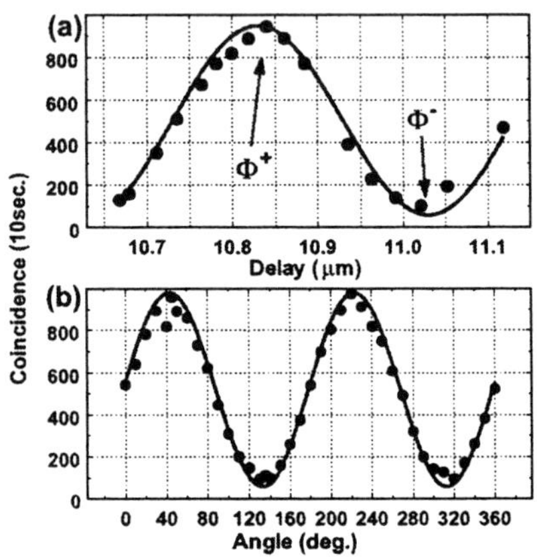

Fig. 2. Experimental data. (a) Space-time interference by varying the pump phase $\Delta\varphi_p$ when $\theta_1 = \theta_2 = 45°$. Transition from Φ^+ to Φ^- is clearly demonstrated. (b) Polarization interference at Φ^+. A_1 is fixed at $\theta_1 = 45°$ and A_2 is rotated, i.e., θ_2 is varied. Solid lines are the theoretical curves.

Unlike the usual degenerate two-photon sources, this source has one distinctive feature: one can vary three different phases independently. To demonstrate this interesting feature, we observed the space-time interference by varying the relative phases of the signal ($\Delta\varphi_s$) and the idler ($\Delta\varphi_i$) independently. When the signal phase $\Delta\varphi_s$ was varied, the signal wavelength (730nm) modulation was observed in coincidence rate, while varying the idler phase $\Delta\varphi_i$, the idler wavelength (885nm) modulation was observed. We have also tested the condition $\Delta\varphi_p = \Delta\varphi_s + \Delta\varphi_i$ by varying the signal phase $\Delta\varphi_s$ and the idler phase $\Delta\varphi_i$ at the same time. As expected, 400nm modulation was observed in coincidence. (Experimental data could not be included due to the page limit.)

In summary, we have demonstrated a novel scheme to prepare pulsed entangled photon pairs from which all four Bell states can be easily obtained. Amplitude and spectral post-selection are not necessary. The visibility and the photon flux is greatly enhanced by this method although a femtosecond pulse laser is used as a pump.

Efficient collection of polarization-entangled photons on type-II parametric down conversion

Ch. Kurtsiefer, M. Oberparleiter, J. Volz and H. Weinfurter

Sektion Physik der Universität München, Schellingstr. 4/III, 80799 München, Germany

email: christian.kurtsiefer@physik.uni-muenchen.de

Abstract: We report on efficient collection of photon pairs from type-II parametric down conversion. We obtained 10000 polarization-entangled photon pairs per second in single mode optical fibers with a frequency-doubled diode laser as a pump.

OCIS codes: (270.4180) Multiphoton processes, (260.2510) Fluorescence

Experiments with entangled photon pairs on various concepts of Quantum mechanics have been significantly advanced with the development of parametric down conversion sources, in particular by using a type-II parametric fluorescence process[1].

There, a pump photon is converted in a nonlinear optical crystal into two orthogonally polarized photons, signal and idler, obeying energy and momentum conservation. For a fixed pump photon momentum and a given idler frequency, there is a one-dimensional manifold of emission directions for the idler photon, and a corresponding one for the signal photon, forming two emission cones. The nonlinear crystal can be oriented such that the two cones intersect, defining two directions along which the polarization of each emitted photon is undefined, but perfectly anti-correlated with the polarization of the other one. Provided complete indistinguishability of which photon belongs to which emission cone, a polarization-entangled pair of photons is obtained[1].

To optimize collection efficiency, we consider the wavelength dependency of the opening angles of the emission cones. For a given spectral width, the signal and idler light emitted along the intersection directions is dispersed over a certain angular width. By matching the receiving field modes of a pair of single mode optical fibers to that angular dispersion, optimal collection of parametric fluorescence light can be obtained without the need of interference filters to determine the optical bandwidth. The divergence of the collected fluorescence distribution then defines a source region being mapped to the receiving fiber. Restricting the pump mode to that region then allows optimal usage of the pump light for the creation of parametric fluorescence photons.

In a first experiment [3], we used an Argon ion laser ($\lambda_p = 351.1$ nm) to pump a 2 mm thick BBO crystal (see figure 1(a)). The pump laser was focused to a waist of 90 μm under an angle of $\Theta_P = 49.7°$ with respect to the optical axis. The modes of collecting single mode optical fibers were matched to the down conversion region such that the acceptance angle was equal to the divergence angle for a wavelength range of 4.6 nm FWHM using Sellmeier equations. Figure 1(b) shows the coincidence count rates and the count rates for the individual detection events as a function of UV pump power P. We determined a coincidence/single count ratio (i.e., an overall efficiency) of $0.286 \pm .001$ for the whole range of pump power. At $P = 465$ mW, we observed a coincidence count rate of 360800 s^{-1}. For low powers, we found a slope of 900 coincidences per second and mW in the single mode optical fibers.

To verify the polarization entanglement of the photon pairs, we measured the polarization correlations in a HV- and ±45° basis for $P = 400$ mW, finding $98.2 \pm 0.1\%$ and $96.3 \pm 0.1\%$, respectively, after correction for accidentals. Using (raw) data for the evaluation of a CHSH-type Bell inequality, we obtain $S = -2.6979 \pm 0.0034$, i.e., a violation of 204 standard deviations for a measurement time of 1 s per angle setting.

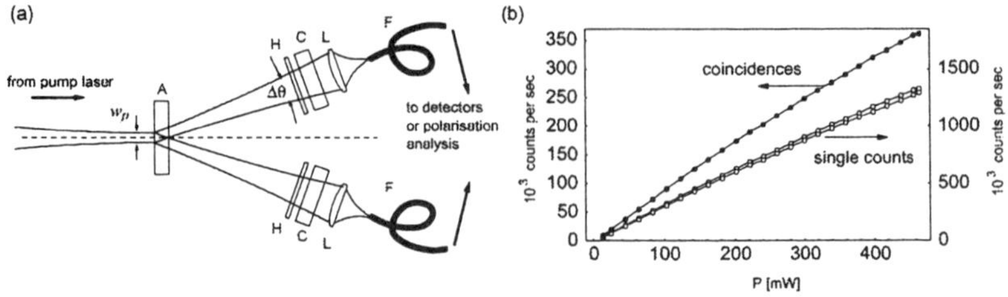

Fig. 1. (a) Setup for collecting polarization-entangled photon pairs from type-II parametric down conversion from a BBO crystal A. (b) Coincidence and individual count rates from photons in single mode optical fibers as a function of the pump power at λ=351 nm.

If this optimization strategy is combined with a resonant enhancement of the pump light[2], one can obtain high rates of entangled photon pairs already from diode-laser based sources. In the experiment shown in figure 2, we used a frequency-doubled laser diode at a wavelength of λ=856 nm to pump a second cavity containing the BBO crystal for parametric down conversion. We obtained 10000 polarization-entangled photon pairs per second in single-mode optical fibers from 7 mW of blue light. The entanglement was verified again by measuring polarization correlations both in the (H,V)-basis and in a ±45° basis to be better than 97% and 95%, respectively.

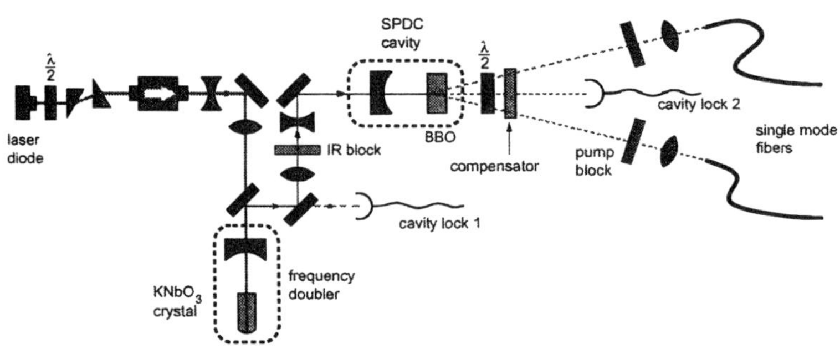

Fig. 2. A diode-laser based source for polarization-entangled photon pairs, consisting of a frequency-doubled laser diode, feeding a second cavity for pump enhancement of the parametric down-conversion process.

To summarize, we were able to significantly increase the intensity of a type-II parametric down conversion source for polarization-entangled photons with $\lambda \approx 800$ nm, so that diode-laser based systems can be used in possible cryptography applications.

References and links

1. P.G. Kwiat, K. Mattle, H. Weinfurter, A. Zeilinger, A.V. Sergienko, Y. Shih, "New High-Intensity Source of Polarisation-Entangled Photon Pairs", Phys. Rev. Lett. **75**, 4337-4340 (1995).
2. M. Oberparleiter and H. Weinfurter, "Cavity-enhanced generation of polarization-entangled photon pairs", Opt. Comm. **183**, 133-137 (2000).
3. Ch. Kurtsiefer, M. Oberparleiter, and H. Weinfurter, " High efficiency entangled photon pair collection in type II parametric fluorescence", quant-ph/0101074
4. J. Volz, Ch. Kurtsiefer and H. Weinfurter, "Compact All-Solid-State-Source of Polarisation-Entangled Photon Pairs", *in preparation*

A Stochastic Map for Qubit Decoherence

Sonja Daffer, Krzysztof Wódkiewicz, John K. McIver

Center for Advanced Studies, Department of Physics and Astronomy, University of New Mexico, Albuquerque, NM, USA

sonja@phys.unm.edu, wodkiew@phys.unm.edu, jmciver@unm.edu

Abstract: In the presence of noise, pure states become mixed states. This process may be viewed as a stochastic map, taking an intial density operator ρ into another density operator $\Phi(\rho)$. We explicitly calculate a non-unital stochastic map with unequal damping eigenvalues and provide a geometrical picture of the effect of stochastic noise on the set of pure state qubit density operators.

OCIS codes: (270.2500) Fluctuations, relaxations, and noise

Decoherence of a two-level atom is described by the dynamics of a Bloch vector inside a unit sphere governed by a set of Bloch equations. The decoherence is caused by the presence of noise and may be viewed as a stochastic operation acting on the Bloch vector in the form of a mapping:

$$\Phi : \vec{b} \rightarrow \vec{b'}. \tag{1}$$

Since there is a correpondence between the Bloch vector $\vec{b}$ and the density operator ρ, we see that the stochastic map is a superoperator which maps density operators into density operators. The image of the map, $\Phi(\rho)$, can be viewed as a sum of conjunctive transformations of the density matrix which is of the form $A_k^{\dagger}\rho A_k$. Alternatively, we can expand the density operator in the Pauli basis, $\{I, \sigma_x, \sigma_y, \sigma_z\}$, and consider how the components of ρ transform under the map. This latter transformation is characterized by a $4x4$ matrix representation of Φ. In terms of the Bloch vector, a general stochastic map may be written in the form

$$\Phi : \vec{b} \rightarrow \vec{b'} = \mathbf{\Lambda}\vec{b} + \vec{b_o} \tag{2}$$

where $\vec{b_o} = (t_{10}, t_{20}, t_{30})$ are translations. Due to the presence of translations, the transformation is affine.

To guarantee that the map Φ transforms the density operator into another density operator we see that the Bloch vector can only be transformed into a vector contained in the interior of the Bloch sphere. This is only possible if the eigenvalues Λ_i are contractions. In what follows, we provide an explicit construction of a stochastic map Φ from the Bloch equations by considering a two-level atom in a squeezed vacuum reservoir. In terms of the $4x4$ matrix, this map can also be expressed as

$$\mathcal{T}_{\mathcal{D}} = \begin{pmatrix} 1 & 0 & 0 & 0 \\ 0 & \Lambda_1 & 0 & 0 \\ 0 & 0 & \Lambda_2 & 0 \\ (1-\Lambda_3)w_{eq} & 0 & 0 & \Lambda_3 \end{pmatrix} \tag{3}$$

with

$$0 < \Lambda_3 < \Lambda_1 < \Lambda_2 < 1 \tag{4}$$

and affine shifts

$$t_{01} = 0, t_{02} = 0, t_{03} = w_{eq}(1-\Lambda_3), \qquad -1 < w_{eq} < 0. \tag{5}$$

The matrix (3) transforms the Bloch sphere $u^2 + v^2 + w^2 = 1$ into an ellipsoid inside the Bloch ball. That is, the image of the set of pure state density matrices under the stochastic map is given by the family of ellipsoids:

$$\left(\frac{u}{\Lambda_1}\right)^2 + \left(\frac{v}{\Lambda_2}\right)^2 + \left(\frac{w}{\Lambda_3} + w_{eq}\left(1 - \frac{1}{\Lambda_3}\right)\right)^2 = 1. \tag{6}$$

The shift, $w_{eq}(1 - \Lambda_3)$, determines the center of the ellipsoid while the eigenvalues, $\{\Lambda_1, \Lambda_2, \Lambda_3\}$, define the lengths of the axes. If we start with the set of pure state density matrices which lie on the Bloch sphere, then as time progresses, the sphere is deformed into a contracted ellipsoid. Because of the squeezed vacuum, the 2-component experiences very little damping while the 1-component is rapidly damped. The explicit expression for the shift in terms of w_{eq} indicates the ellipsoid is translated in the negative 3-direction over time and settles into an equilibrium or fixed point.

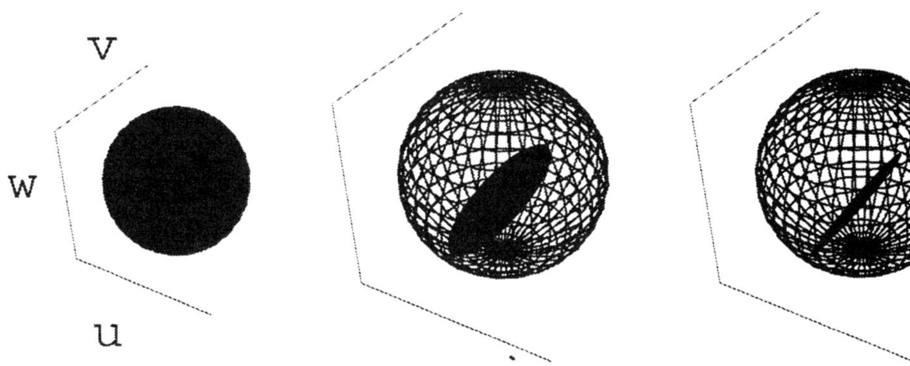

Fig. 1. The effect of noise on the set of qubit density operators for an atom in a squeezed vacuum. Parameters are N=1,M=$\sqrt{2}$,t=0,.5,1.

References and links

1. K. Wódkiewicz "Stochastic Decoherence of Qubits" Optics Express **8**, No. 2, 145 (2001).
2. M.B. Ruskai, E. Werner, S. Szarek, "A Characterization of Completely-Positive Trace-Preserving Maps on 2x2 Matrices," lanl preprint **quant-ph/0005004**.
3. C. King, M.B. Ruskai, "Minimal Entropy of States Emerging from Noisy Channels, lanl preprint **quant-ph/9911079**.
4. H.J. Briegel, B.-G. Englert, "Quantum Optical Master Equations: The Use of Damping Bases," Phys. Rev. A **47**, 3311-3328 (1993).
5. P. Meystre, M. Sargent III, "Elements of Quantum Optics," ,Third Edition, Springer-Verlag, 1999.
6. C.W. Gardiner, "Quantum Noise," Springer-Verlag (1991).
7. K. Kraus, "States, Effects and Operations: Fundamental Notions of Quantum Theory," Springer-Verlag (1983).
8. W.F. Stinespring, "Positive Functions of C*-Algebras," Proceedings of the American Mathematical Society, **Vol.6**, No. 2., 211-216 (Apr., 1955).

Fighting decoherence with pointer engineering

A.R.C. Carvalho, P. Milman, R.L. de Matos Filho, and L. Davidovich

Instituto de Fíisica, Universidade Federal do Rio de Janeiro, Cx. P. 68528, 21945-970 Rio de Janeiro, RJ, Brazil

Fighting decoherence has become a major chalenge in the last years, motivated by recent progress in the theory of quantum information processing, which relies on the possibility of preserving quantum coherence. It is also of interest to high-precision frequency measurements in ion traps. Several strategies have been devised, including quantum error correction schemes, feedback implementations, the realization of q-bits in symmetric subspaces decoupled from the environment, and dynamical decoupling techniques.

In linear ion traps, by far the most important decoherence effect is the one associated with the motional state [1]. In the present paper, we show this effect can be suppressed by generating, through the technique of "reservoir engineering" [2], artificial reservoirs associated with properly chosen pointer observables [3], which have the states to be preserved as their eigenstates, and which dominate over other dissipation processes.

Our strategy for the protection of a given quantum state against the environment consists in engineering, via external driving, a system-enviroment coupling, with the net effect of adding to the master equation of the system an extra term, thus getting

$$\frac{d\hat{\rho}}{dt} = \mathcal{L}\hat{\rho} + (\Gamma_{\text{eng}}/2)\left(2\,\hat{d}\,\hat{\rho}\,\hat{d}^{\dagger} - \hat{d}^{\dagger}\,\hat{d}\,\hat{\rho} - \hat{\rho}\,\hat{d}^{\dagger}\,\hat{d}\right). \tag{1}$$

where $\mathcal{L}$ is the usual Lindblad super-operator (corresponding to a damping constant γ). The operator $\hat{d}$ is chosen so that the state one wants to protect is the *only* steady state of Eq. (1) without the environment term $\mathcal{L}\hat{\rho}$. For $\Gamma_{\text{eng}} \gg \gamma$, the steady state of the new master equation (1) will be very close to the state to be protected (if the state is not unique, the term $\mathcal{L}\hat{\rho}$ could still induce transitions between the steady states). Besides, *any* state of the system will decay into the state chosen to be protected. Therefore, this is also a procedure for preparing quantum states in the presence of decoherence.

We apply this method to the one-dimensional motion of the center-of-mass of an ion confined in an electromagnetic trap. The reservoir engineering process will be implemented by letting the ion interact with several laser beams of adequate frequencies and intensities, which are quasi- resonant to an electronic transition $|1\rangle \leftrightarrow |2\rangle$ of frequency ω_{21}. We consider the ion to be in the regime of resolved sidebands, given by $\nu_u \gg \Gamma, |\Omega_n|$, where Γ is the electronic energy decay rate, ν_u is the ion vibrational frequency along axis $\mathbf{u}$, and Ω_n is the (complex) Rabi frequency corresponding to laser n, tuned to the k-th red vibrational sideband of the ion. Under these conditions the interaction term corresponding to laser n can be described, in the interaction picture, by [4]

$$\hat{\tilde{H}}_{\text{int}}(t) = \hbar g(\hat{A}_{21}\hat{d} + \hat{A}_{12}\hat{d}^{\dagger}) = \frac{1}{2}\hbar\Omega_n(i\eta_n)^k\hat{A}_{21}\hat{f}_k(\hat{\tilde{a}}^{\dagger}\hat{a})\,\hat{\tilde{a}}^{k} + H.c., \tag{2}$$

where g is taken to be real and $\hat{f}_k(\hat{a}^{\dagger}\hat{a}) = e^{-\eta_n^2/2}\sum_{l=0}^{\infty}[(-1)^l\eta_n^{2l}/l!(l+k)!](\hat{a}^{\dagger})^l\hat{a}^l$. Here, the operators $\hat{a}$ and $\hat{A}_{21}$ are the annihilation operator of a quantum of the ionic vibrational motion and the electronic flip operator, respectively. The quantity $\eta_n = \sqrt{\hbar(\mathbf{k}_n\cdot\mathbf{u})^2/2M\nu_u}$ is the Lamb–Dicke parameter with respect to the direction of

Coherence and Quantum Optics VIII
Edited by Bigelow et al., Kluwer Academic/Plenum Publishers, 2003

vibration, fixed by the unit vector $\mathbf{u}$. M is the ion's mass and $\mathbf{k}_n$ is the wave vector of laser n. It is assumed that the Lamb-Dicke parameters corresponding to directions orthogonal to $\mathbf{u}$ are much smaller than one, as it is the case in linear traps.

Assuming that the spontaneous decay rate Γ of the electronic state 2 is by far the largest rate in the system, we may eliminate adiabatically the population of level 2, as well as the coherence ρ_{12}, getting an approximate master equation for the reduced density operator corresponding to the vibrational mode in the form (1), with a decay constant $\Gamma_{\rm eng} = 4g^2/\Gamma$. The engineered reservoir will dominate as long as $\Gamma_{\rm eng} \gg \gamma$.

In order to protect a state $|\psi\rangle$, we look for $\hat{d}$ such that $\hat{d}|\psi\rangle = \lambda|\psi\rangle$ with $\lambda = 0$, and make sure that there is no further eigenstate $|\phi\rangle$ of $\hat{d}$ such that $[\hat{d}, \hat{d}^\dagger]|\phi\rangle = 0$. This implies that $|\psi\rangle$ is the only pure steady state of (1) without $\mathcal{L}\hat{\rho}$. Since $\Gamma_{\rm eng} \gg \gamma$, this yields a good approximation of the corresponding steady state of (1).

We consider the protection of the class of states $|\psi\rangle = \sum_{n=0}^{N} c_n|n\rangle$, where $|n\rangle$ is an energy eigenstate of the vibrational motion of the trapped ion (assumed to vibrate in the x-direction) and $c_n \neq 0$, and establish the relationship between the laser amplitudes so that this class of states is protected. In the special case of the "qubit" state $|\psi\rangle = c_0|0\rangle + c_1|1\rangle$, our method requires three lasers, two of them along the x-axis, with Rabi frequencies Ω_{x1} and Ω_{x2}, and a third one with Rabi frequency Ω_y, along an orthogonal axis y, satisfying the following conditions (asuming $\eta_y \gg \eta_{x1} = \eta_{x2}$): $\eta\Omega_{x1}/i\Omega_{x2} = -c_1/c_0$ and $\Omega_y/i\Omega_{x2} = \exp(-\eta^2/2)(c_1/c_0)(1-\eta^2)/\eta$. Typical values which allow the protection of this state against the environment are $\Gamma \approx 4$ MHz, $\Omega_1 \approx 2$ MHz, $\eta = 0.2$, $\nu \approx 20 - 30$ MHz, with $\gamma \ll \Gamma_{\rm eng} = \eta^2{\Omega_1}^2/\Gamma \approx 40$ kHz. Fig. 1a displays the fidelity $F(t) = \mathrm{Tr}\{\hat{\rho}(0)\hat{\rho}(t)\}$, with the ion initially in the vibrational state $|\psi\rangle = \frac{1}{2}(|0\rangle + |1\rangle)$. The system rapidly reaches a steady state with fidelity very close to unity ($\eta = 0.2$).

Fig. 1 (b) displays the results obtained for superpositions of coherent states [5].

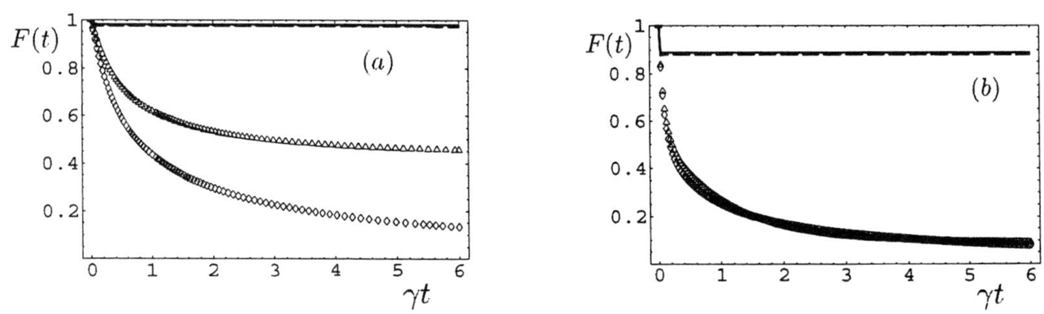

Fig. 1. Time evolution of the fidelity $F(t)$ for several initial nonclassical states, in the presence of a thermal reservoir with $\langle n_{\rm th}\rangle = 0.5$ (full curves and triangles) and a random field (dashed curves and losangles) . For the triangles and losangles, $\Gamma_{\rm eng} = 0$. The full and dashed curves correspond to $\Gamma_{\rm eng} = 40\gamma$ for (a) and $\Gamma_{\rm eng} = 150\gamma$ for (b). The nonclassical states are: (a) $|\psi\rangle = \frac{1}{\sqrt{2}}(|0\rangle + |1\rangle)$ and (b) $|\psi\rangle = \frac{1}{\sqrt{2}}(|\alpha\rangle + i|-\alpha\rangle)$, with $\alpha^2 = 3$.

References and links

1. D.J. Wineland *et al* , J. Res. Natl. Inst. Stand. Technol. **103**, 29 (1998).
2. J. F. Poyatos, J. I. Cirac, and P. Zoller, Phys. Rev. Lett. **77**, 4728 (1996).
3. W. H. Zurek, Phys. Rev. D **24**, 1516 (1981); **26**, 1862 (1982); Physics Today **44**, 36 (1991).
4. W. Vogel and R. L. de Matos Filho, Phys. Rev. A **52**, 4214 (1995).
5. A.R.R. Carvalho, P. Milman, R.L. de Matos Filho, and L. Davidovich, Phys. Rev. Lett. **86**, 4988 (2001).

Uncertainty Relations Associated with Correlations in Mixed Quantum States

Sergey A. Ponomarenko* **and Emil Wolf**

Department of Physics and Astronomy,
University of Rochester, Rochester, NY 14627

* *email: ponos@pas.rochester.edu*

The conventional uncertainty relation (UR) is a cornerstone of the modern quantum theory of measurement [1], and consequently is treated in most quantum-mechanics textbooks. However, contrary to the widespread view, the usual "textbook" UR pertains to separate rather than joint measurements [2] of observables $\hat{A}$ *or* $\hat{B}$ on a system in a state $|\psi\rangle$. In essence, the uncertainty relation indicates that due to intrinsic indeterminacy of the quantum state, the product of the variances of two noncommuting observables $\hat{A}$ and $\hat{B}$ cannot be less than a certain value, which is expressed mathematically as:

$$\langle(\Delta\hat{A})^2\rangle\,\langle(\Delta\hat{B})^2\rangle \geq \frac{1}{4}|\langle\psi|[\hat{A},\hat{B}]_-|\psi\rangle|^2. \tag{1}$$

Here $[\hat{A},\hat{B}]_- \equiv \hat{A}\hat{B}-\hat{B}\hat{A}$ is the commutator of a pair of operators, and $\langle(\Delta\hat{A})^2\rangle = \langle\hat{A}^2\rangle - \langle\hat{A}\rangle^2$ is the variance of the operator $\hat{A}$. It should be noted that the inequality (1) pertains to a *pure* state. Uncertainty relations of this type were extensively studied as early as in the 1930's [3]. There has also been a considerable interest in uncertainty relations associated with joint measurements of noncommuting observables [5-8] as well as in the generalized parameter-based UR's that do not explicitly depend on the expectation value of the commutator [8, 9]. Further, a generalization of the Heisenberg-type UR (1) to open quantum systems was obtained [10, 11], and the nature of the states that minimize such a generalized UR was examined. These studies have shown that, at least when the observables $\hat{A}$ and $\hat{B}$ are the coordinate and the momentum, the most general minimum-uncertainty state must be a pure state [10].

In the present paper, we consider separate measurements of either of the two *arbitrary* noncommuting Hermitian operators $\hat{A}$ and $\hat{B}$ on a quantum system in a mixed state. We introduce generalized measures for deviations of $\hat{A}$ and $\hat{B}$ from their expectation values. These novel measures fully incorporate *correlations* in the density operator, which are specified by its off-diagonal elements in the Hilbert space of the appropriate observables. We then utilize the generalized variances to derive the associated Heisenberg-type uncertainty relations, and we show that the new UR's reduce to the conventional one for pure states. Next, we demonstrate that the explicit incorporation of quantum coherences into the structure of the novel UR's relaxes the requirement that the minimum-uncertainty states be pure states. In particular, we find that in the important case of the coordinate and momentum operators, the most general minimum-uncertainty states are the *displaced squeezed thermal* states. Further, we compare our algebraic approach to that based on the phase-space formalism employed in Ref. [12].

In order to study correlations in an open quantum system , we introduce a pair of generalized variances of the observable $\hat{A}$ which (*i*) explicitly incorporate quantum coherences and (*ii*) reduce to the conventional measure of the mean-square deviation from the expectation value of $\hat{A}$ for the measured system in a pure state. Let

$$\langle(\Delta\hat{A})^2\rangle_- \equiv -\operatorname{Tr}([\Delta\hat{A},\hat{\rho}]_-^2) \tag{2}$$

Coherence and Quantum Optics VIII
Edited by Bigelow et al., Kluwer Academic/Plenum Publishers, 2003

and

$$\langle(\Delta\hat{A})^2\rangle_+ \equiv \mathrm{Tr}([\Delta\hat{A},\hat{\rho}]_+^2). \tag{3}$$

Here $[,]_+$ stands for the anticommutator of two operators, i. e. $[\hat{A},\hat{B}]_+ \equiv \hat{A}\hat{B}+\hat{B}\hat{A}$, and $\Delta\hat{A} = \hat{A} - \langle A\rangle$, where $\langle A\rangle = \mathrm{Tr}(\hat{A}\hat{\rho})$ is the expectation value of the operator $\hat{A}$. It should be noted that the minus sign in front of the right-hand side of Eq. (2) is necessary to ensure the nonnegativity of the variance. A detailed analysis reveals that the quantity given by equation (2) is the width of the correlations between any pair of eigenvalues of the observable $\hat{A}$ in the mixed state of the system, specified by the density operator $\hat{\rho}$. The other generalized variance corresponds to a mean square deviation in the extended Hilbert space of the observable $\hat{A}$. The generalized variances of *any* pair of noncommuting observables $\hat{A}$ and $\hat{B}$ are shown to satisfy the pair of Heisenberg-Robertson type uncertainty relations

$$\begin{aligned}\langle(\Delta\hat{A})^2\rangle_-\langle(\Delta\hat{B})^2\rangle_+ &\geq |\mathrm{Tr}([\hat{A},\hat{B}]_-\hat{\rho}^2)|^2,\\ \langle(\Delta\hat{B})^2\rangle_-\langle(\Delta\hat{A})^2\rangle_+ &\geq |\mathrm{Tr}([\hat{A},\hat{B}]_-\hat{\rho}^2)|^2.\end{aligned} \tag{4}$$

The equality sign in these two inequalities pertains to a mixed state with the smallest amount of correlation. In general, the density operator of such a state is shown to be determined from the equation

$$[\Delta\hat{B},\hat{\rho}]_+ + i\lambda[\Delta\hat{A},\hat{\rho}]_- = 0. \tag{5}$$

In the important particular case, when $\hat{A}$ and $\hat{B}$ are the coordinate and the momentum operators or the two quadratures of a single mode of the electromagnetic field, we found that the minimum uncertainty state (or the minimum correlation state) is a displaced squeezed thermal state. Such a state can be produced, for instance, by degenerate four-wave mixing in a cavity coupled to a thermal reservoir [13]. A full account of this research is scheduled to be published in the June issue of Phys. Rev. A [14].

References and links

1. V. B. Braginsky and F. Ya. Khallili, *Quantum Measurement* (Cambridge University Press, Cambridge, 1992).
2. For a recent review on this subject, see M. G. Raymer, Am. J. Phys. **62**, 986 (1994).
3. H. P. Robertson, Phys. Rev. **35**, 667A (1930); *ibid.*, **46**, 794 (1934).
4. E. Arthurs and J. L. Kelly, Jr., Bell Syst. Tech. J. **44**, 725 (1965).
5. J. Hilgevoord and J. Uffink, in *Sixty-Two Years of Uncertainty*, edited by A. I. Miller (Plenum, New York, 1990), p. 121.
6. Y. Yamamoto and H. A. Haus, Rev. Mod. Phys. **58**, 1001 (1986).
7. S. Steholm, Ann. Phys. **218**, 233 (1992).
8. C. W. Helstrom, *Quantum Detection and Estimation Theory* (Academic, New York, 1976).
9. S. L. Braunstein, C. M. Caves, and G. J. Milburn, Ann. Phys. (N.Y.), **247**, 135 (1996).
10. D. Stoler and S. Newman, Phys. Lett. **38A**, 433 (1971).
11. V. V. Dodonov, E. V. Kurmyshev, and V. I. Man'ko, Phys. Lett. **79A**, 150 (1980).
12. S. Chountasis and A. Vourdas, Phys. Rev. A, **58**, 1794 (1998).
13. G. S. Agarwal and G. Adam, Phys. Rev. A, **38**, 750 (1988).
14. S. A. Ponomarenko and E. Wolf, Phys. Rev. A, in press.

Observation of Gouy scars in optical wave chaos

J. Dingjan, M.P. van Exter, and J.P. Woerdman

Leiden University, Huygens Laboratory,
P.O. Box 9504, 2300 RA Leiden, The Netherlands

e-mail: dingjan@molphys.leidenuniv.nl, fax: ++(31) 71 527 5819

Abstract: We report experimental observation of scars in the wave chaos occurring in a 3-mirror optical resonator. The scars are associated with Gouy phases that are rational fractions of 2π; these lead to periodic ray paths which close after a certain number of round trips.

Wave chaos, also known as quantum chaos, refers to the behavior of wave-mechanical systems that are chaotic in the classical limit, i.e., for infinitely small wavelengths. This field has enjoyed considerable interest, both from a theoretical and an experimental point of view [1]. Because the concept of a ray trajectory, or a bouncing particle, loses its significance in a wave-mechanical system, the key manifestation of classical chaos, i.e. the exponential divergence of neighboring trajectories, is no longer valid. Instead, wave chaos manifests itself in the eigenvalue spectrum, and hence in the dynamical behavior, of a system. Nevertheless, the classical framework can still exert an influence, and can express itself as *scars*[2], where a wave function shows a tendency to concentrate around a trajectory that, in the classical limit, would be periodic and unstable.

Here we report the observation of scars in optical wave chaos; as system we use a 3-mirror resonator (with a spherical folding mirror) in the non-paraxial regime [3]. The transmission spectrum, or Fabry-Perot spectrum, of such a resonator shows a seemingly random or a quasi-periodic structure, depending on the length L of the resonator (Fig. 1b gives some examples).

Wilkie and Brumer have developed a general statistical method in order to check whether a random looking spectrum is indeed chaotic [4]. They transform the spectrum into a time-domain wave-packet survival probability $P(t)$, which is then doubly averaged, first over the initial conditions and then over the configuration ensemble. Translated to our system this means averaging over the in-coupling and over the length L of the 3-mirror resonator.

We have analyzed transmission spectra as in Fig. 1b with this method. The result is shown in Fig. 1a. First we notice that the curve $\langle\langle P(t)\rangle\rangle$ dips below its long-time asymptote, thus confirming the existence of overall wave chaos [3, 4]; the key point, however, is the occurrence of sharp spikes on the curve $\langle\langle P(t)\rangle\rangle$ at relatively short times. We conjecture that these spikes correspond to scars; in order to prove this we have to show, as we will do next, that the spikes correspond, in the classical limit, to trajectories that are both periodic and unstable.

Earlier experiments on *paraxial* multi-mode two-mirror resonators have shown that periodic structures are the result of resonator Gouy phases that are rational fractions of 2π, i.e. $\theta = 2\pi K/N$, leading to a high degree of frequency degeneracy [5]. Equivalently, cavity configurations corresponding to such Gouy phases support marginally stable periodic orbits, where the period, in units of the resonator round trip time, is given by N, and which make K transverse excursions before closing. *Outside* the paraxial limit, or, equivalently, in systems that have strong aberrations, this simple description breaks

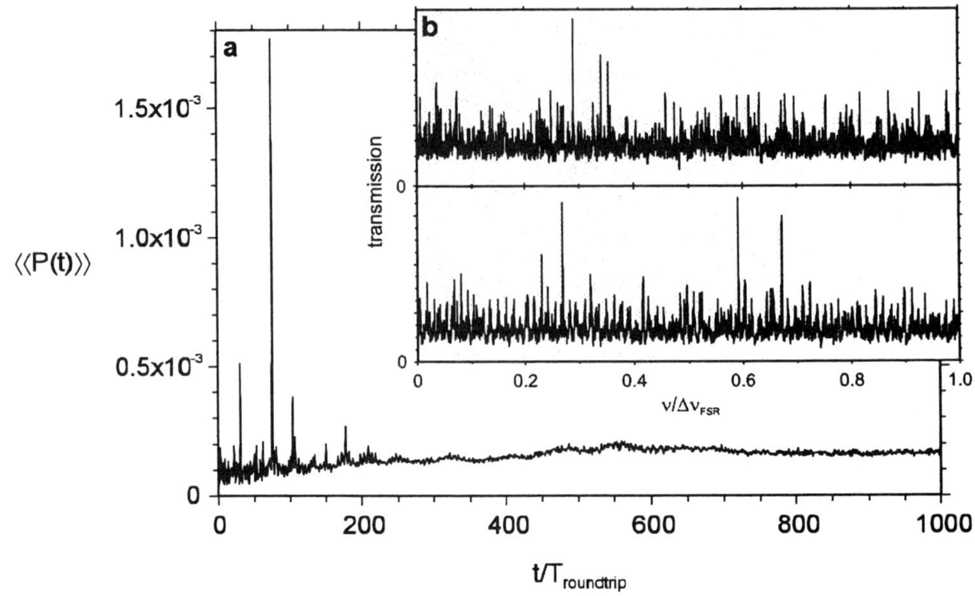

Fig. 1. a) Doubly-averaged wave-packet survival probability. b) Irregular Fabry-Perot spectrum (top), quasi-periodic Fabry-Perot spectrum (bottom).

down. However, we can still characterize periodic orbits using the "quantum numbers" K and N; these orbits are now localized. We have performed a stability analysis for a number of these non-paraxial periodic orbits, and found that, barring special cases, all orbits studied are unstable. Since every orbit has to be treated on an ad-hoc basis, a general proof of instability is presently out of reach. However, we suspect that periodic orbits in aberrated optical resonators are generally unstable.

We conclude that these fractional Gouy orbits leave a clear fingerprint as scars in the optical wave chaos occurring in a non-paraxial 3-mirror resonator.

References and links

1. M.C. Gutzwiller, *Chaos in Classical and Quantum Mechanics* (Springer-Verlag, New York, 1990); F. Haake, *Quantum Signatures of Chaos* (Springer-Verlag, Berlin, 1991); H.-J. Stöckmann, *Quantum Chaos* (Cambridge Universtiy Press, Cambridge, 1999);
2. E.J. Heller, "Bound-state eigenfunctions of classically chaotic Hamiltonian systems: scars of periodic orbits", Phys. Rev. Lett. **53**, 1515–1518 (1984)
3. J. Dingjan, E. Altewischer, M.P. van Exter, and J.P. Woerdman, submitted to Nature.
4. J. Wilkie, and P. Brumer, "Time-dependent manifestations of quantum chaos", Phys. Rev. Lett. **67**, 1185–1188 (1991)
5. J. Dingjan, M.P. van Exter, and J.P. Woerdman, "Geometric modes in a single-frequency Nd:YVO$_4$ laser", Opt. Comm. **188**, 345–351 (2001)

Adiabatic Wavepacket Propagation

Markku Jääskeläinen, Stig Stenholm

Royal Institute of Technology, Stockholm, Sweden

mrq@physics.kth.se

Abstract: The adiabatic approximation has been applied to wavepacket dynamics in order to reduce the numerical complexity and gain insight into quantum dynamics.

OCIS codes: (350)Other areas of optics, (350.5500) Propagation, (350) Other areas of optics, (350.7420) Waves

In quantum mechanics the adiabatic approximation [1] has been used in many different contexts. As adiabatic approximations often turn out to be more useful than expected from the limiting conditions, it seems natural to try to use them in numerical wave packet dynamics as well. Here an adiabatic basis of discrete eigenstates is used to reduce the complexity of the problem. When the physical parameters of the problem allow for adiabatic propagation the discrete basis states decouple making it possible to model the system as be as a number of independent one dimensional problems in different potentials.

For a quantum mechanical problem with time independent hamiltonian in a two dimensional configuration space the dynamics will evolve according to the time dependent Schrödinger equation:

$$i\hbar\frac{\partial\Psi(x,y,t)}{\partial t} = -\frac{\hbar^2}{2m}\nabla^2\Psi(x,y,t) + V(x,y)\Psi(x,y,t) \tag{1}$$

In many cases of interest the propagation will be occur along the minimal valleys of the potential energy surface. Assuming that these minima constitute a smooth, connected curve it becomes possible to make a coordinate change to a local Frenet frame, with the path length(s) and the transversal distance (ξ) to the path as coordinates. Changing coordinates from (x, y) to (s, ξ) and taking the minimal curve to turn sufficiently slowly, the Laplacian can be approximated by:

$$\nabla^2 \approx \frac{\partial^2}{\partial s^2} + \frac{\partial^2}{\partial \xi^2} + \kappa(s)\frac{\partial}{\partial \xi} \tag{2}$$

Where $\kappa(s)$ is the curvature of the path. If we now assume that the time independent Schrödinger equation can be solved along the transversal direction at each point of the minimal curve and make the following ansatz using the eigenfunctions as a basis set:

$$\Psi(s,\xi,t) = \sum_{n=0}^{\infty}\phi_n(s,t)\eta_n(s,\xi) \tag{3}$$

Where the set of transversal eigenfunctions satisfy a one dimensional time independent Schrödinger equation in the transversal coordinate:

$$E_n(s)\eta_n(s,\xi) = -\frac{\hbar^2}{2m}\frac{\partial^2\eta_n(s,\xi)}{\partial\xi^2} + U(s,\xi)\eta_n(s,\xi) \tag{4}$$

Where the potential is taken with respect to the minima of the potential along the transversal direction:

Coherence and Quantum Optics VIII
Edited by Bigelow et al., Kluwer Academic/Plenum Publishers, 2003

$$U(s,\xi) = V(s,\xi) - V(s,0) \tag{5}$$

Inserting Eqs.(2,3,4,5)into Eq.(1), multiplying by $\eta_m(s,\xi)$, followed by integration over the transversal coordinate ξ and using orthonormality of the basis set, we arrive at:

$$i\hbar\frac{\partial\phi_m(s,t)}{\partial t} = -\frac{\hbar^2}{2m}\frac{\partial^2\phi_m(s,t)}{\partial s^2} + \Big(V(s,0) + E_n(s)\Big)\phi_m(s,t) + \sum_{n=0}^{\infty} T_{nm}(s)\phi_n(s,\chi) \tag{6}$$

where T_{nm} are the matrix elements:

$$T_{n,m}(s) = -\frac{\hbar^2}{2m}\int_{-\infty}^{\infty}\eta_n(s,\xi)\Big(\frac{\partial^2}{\partial s^2} + \kappa(s)\frac{\partial}{\partial\xi}\Big)\eta_m(s,\xi)d\xi \tag{7}$$

The basic features of any potential with a minima is captured by the Harmonic Oscillator. Taylor expanding the potential energy surface along the transversal coordinate to second order we get:

$$V(s,\xi) = V(s,0) + \frac{1}{2}\frac{\partial^2 V(s,0)}{\partial\xi^2}\xi^2 + O(\xi^3) \tag{8}$$

Inserting the quadratic term of Eq.(8) into Eq.(4) as $U(s,\xi)$ makes it possible to use the Harmonic Oscillator eigenfunctions as basis:

$$\eta_n(s,\xi) = \Big(\frac{\alpha(s)}{\sqrt{\pi}2^n n!}\Big)^{1/2} e^{-\frac{1}{2}\alpha(s)^2x^2} H_n(\alpha(s)\xi) \tag{9}$$

Where the oscillator width:

$$\alpha(s) = \Big(\frac{d^2V}{d\xi^2}\Big)^{1/4} \tag{10}$$

is given by the transversal curvature of the potential energy surface at the bottom. If the potential energy surface changes sufficiently slowly over lengths comparable to the de Broglie wavelength of the wave packet, we expect the coupling between transversal states to be negligible and the corresponding one dimensional equations to decouple. In order to derive a condition for adiabaticity the timescales of internal dynamics of the transversal motion has to be compared to the time scale of change in the potential along the minimal curve as the wave-packet traverses it. When the coupling between the levels in the Harmonic Oscillator model is weak. The the full problem is expected to be well described by the set of equations:

$$i\hbar\frac{\partial\phi_n}{\partial t} = -\frac{\hbar^2}{2m}\frac{\partial^2\phi_n}{\partial^2 s} + \hbar\omega(s)(n+\frac{1}{2})\phi_n \tag{11}$$

where the potential energy surface along the minimal curve has been set equal to zero. In order to test for adiabaticity the solution given by eq.11 has to be compared with the full numerical solution of eq.1.

References and links

1. B.H. Bransden, and C.J. Joachain, "Introduction to Quantum Mechanics," Longman Scientific and Technical, (1989)

Squeezing light in linear dispersive dielectrics

Victor V. Kozlov

Abteilung für Quantenphysik, Universität Ulm, D-89069, Germany

Abstract: Correlations between frequency and position induced in a quantized wave packet propagating in a medium with non-zero group-velocity dispersion, e.g. an optical fiber, are used for generating squeezed states of light. The method does not require a nonlinearity.

OCIS codes: (270.5530) Pulse propagation and solitons, (270.6570) Squeezed states; (320.0320) Ultrafast optics.

A common feature of all known methods of squeezing is the use of the *nonlinear* character of a light-matter interaction. For a review see [1]. We propose an alternative way of generating squeezed states which has no need for a nonlinearity. The method relies on spreading a quantum wave packet while propagating through a medium with nonzero group-velocity dispersion (GVD). Initial quantum fluctuations in central frequency (or momentum) of the wave packet result in corresponding fluctuations of speed, and via the speed, cause a linearly increasing spread in the position. The GVD-induced quantum-mechanical correlations between the frequency and the position result in a squeezing effect.

For demonstrating squeezing effects, it is most natural to exploit the description where propagating wave packets of light appear as moving quantum particles. Well-known contradictions with localizing photons noted in [2] and reviewed later with emphasis on quantum optics in [1], vanish in the quasi-monochromatic approximation. Adopting the latter, we may introduce electromagnetic position and momentum operators. The position operator is defined as

$$\hat{X} \equiv \int t\, \hat{\phi}^\dagger(z,t)\, \hat{\phi}(z,t)\, dt\, \hat{N}^{-1}\,, \tag{1}$$

a direct analog of the classical definition of the temporal peak (more accurately, center of gravity) of a pulse with shape $\phi(z,t)$. The field operator $\hat{\phi}$ is related to the positive-frequency part of the electric field operator $\hat{E}$ as $\hat{E}(z,t) = \sqrt{\hbar k_0/4\pi\epsilon_0 S}\,\hat{\phi}(z,t)$ $\exp[i(k_0 z - \omega_0 t)]$, where k_0 and ω_0 are central wave number and frequency of the wave packet and S is the cross-sectional area of the beam. The normalization is chosen such that the integral $\int \hat{\phi}^\dagger \hat{\phi}\, dt$ gives the photon number operator $\hat{N}$. In the quasi-monochromatic approximation, the field operators obey equal-space commutation relation $[\hat{\phi}(z,t), \hat{\phi}^\dagger(z,t')] = \delta(t-t')$.

An operator is canonically conjugate to position if it obeys the commutation relation: $[\hat{X},\ \hat{P}] = i$, where we omit $\hbar$ for brevity. It is easy to check that such an operator $\hat{P}$ has the form

$$\hat{P} \equiv \frac{i}{2} \int \left(\frac{\partial \hat{\phi}^\dagger}{\partial t}\, \hat{\phi} - \hat{\phi}^\dagger\, \frac{\partial \hat{\phi}}{\partial t} \right) dt\,, \tag{2}$$

which as a consequence can be called (total) momentum.

We apply the fully canonical quantization scheme of the electromagnetic field in Hopfield's dielectric [3] to the case of lossless fibers with dispersion relation $k(\omega) = k_0 + k'(\omega - \omega_0) + (1/2)k''(\omega - \omega_0)^2$, and derive evolution equation for the field:

$$i\frac{\partial}{\partial z}\hat{\phi}(z,\tau) = -\frac{k''}{2}\frac{\partial^2}{\partial \tau^2}\hat{\phi}(z,\tau), \tag{3}$$

where $\tau = t - k'z$, k' is the inverse group velocity and k'' is the GVD.

For simplicity, we introduce a decomposition of the field operator into classical and quantum parts as $\hat{\phi} = \phi + \hat{v}$, and disregard all terms nonlinear in $\hat{v}$. Eq. (3) is then used for deriving the evolution equations for linearized versions of the position and momentum operators (1) and (2). The resultant equations are solved giving

$$\hat{P}(z) = \hat{P}_0 \quad \text{and} \quad \hat{X}(z) = \hat{X}_0 + (k''z/n_0)\,\hat{P}_0\,, \tag{4}$$

where label 0 marks quantities evaluated at $z = 0$. The reference frame is chosen such that mean position and mean momentum are zero: $P_0 \equiv i\int(\partial\phi^*/\partial\tau)\phi\, d\tau = 0$ and $X_0 \equiv n_0^{-1}\int \tau|\phi|^2\, d\tau = 0$. Here, $n_0 \equiv \int|\phi|^2 d\tau$ is the mean photon number.

As an example, we consider the propagation of an initially transform-limited Gaussian wave packet of duration τ_p in a coherent state with a mean shape $\phi(0,\tau) = \sqrt{n_0/\sqrt{\pi}\tau_p}\,\exp(-\tau^2/2\tau_p^2)$ through the dispersive fiber. The wave packet continuously changes with propagation:

$$\phi(z,\tau) = \sqrt{\frac{n_0}{\sqrt{\pi}T_p(z)}}\,\exp\left[\frac{\tau^2}{2T_p^2(z)}\right]\exp\left[i\frac{\tau^2}{2T_p^2(z)}\frac{z}{Z_\mathrm{D}} - \frac{i}{2}\mathrm{arctg}\frac{z}{Z_\mathrm{D}}\right], \tag{5}$$

with $Z_D \equiv \tau_p^2/k''$ and $T_p(z) = \tau_p\sqrt{1 + (z/Z_\mathrm{D})^2}$.

To demonstrate squeezing effects, we define the quadrature $\hat{q}$ as a rotation of the $(\hat{P},\hat{X})$ vector on angle ψ: $\hat{q} = r\,\hat{P}\,\cos\psi + r^{-1}\,\hat{X}\,\sin\psi$ with $r \equiv \tau_p/\sqrt{n_0}$. The conjugate quadrature $\hat{p}$ is constructed as $\hat{p} = \hat{q}(\psi + \pi/2)$. The pair of quadratures obeys commutation relation $[\hat{q},\hat{p}] = i$, and give the same product of dispersions in vacuum state as the original position and momentum variables. We then define a squeezed state by the condition that there exists an angle ψ for which the dispersion $\langle(\Delta\hat{q})^2\rangle(z)$ is smaller than in the vacuum state: $\langle(\Delta\hat{q})^2\rangle(z) < \langle(\Delta\hat{q})^2\rangle_0 = 1/2$. Using (4) and (5) we get

$$\langle(\Delta\hat{q})^2\rangle_\mathrm{min} = \frac{1}{2}\left[\sqrt{1 + \left(\frac{z}{2Z_\mathrm{D}}\right)^2} - \frac{z}{2Z_\mathrm{D}}\right]^2, \tag{6}$$

where $\psi = \psi_\mathrm{min}$ is chosen according to $\mathrm{ctg}(2\psi_\mathrm{min}) = -z/2Z_\mathrm{D}$ for minimizing the variance. The deeper the pulse propagates into the medium the larger degree of squeezing is reached (provided the linearization approximation is still valid). Thus, 10 dB squeezing is achieved after a 10-fs pulse propagated 10 cm in a conventional fiber with $k'' = 10\ \mathrm{ps}^2/\mathrm{km}$ ($Z_\mathrm{D} = 1$ cm). We shall also propose an experimental setup for detecting the momentum-position squeezed states.

This work was partially supported by a grant from Corning Inc. to the Rochester Theory Center, and the programme "QUIBITS" of the European Commission.

[1] L. Mandel and E. Wolf, *Optical Coherence and Quantum Optics* (Cambridge University Press, 1995).

[2] T. D. Newton and E. P. Wigner, Rev. Mod. Phys. **21**, 400 (1949).

[3] B. Huttner and S. M. Barnett, Phys. Rev. A **46**, 4306-4322 (1992).

Intensity–Field and Intensity–Intensity Correlations in a Strongly Coupled Cavity QED System.

W. P. Smith[1], J. E. Reiner[1], S. Kuhr[2], and L. A. Orozco[1]

[1] *Dept. of Physics and Astronomy, SUNYSB, Stony Brook NY 11794-3800, USA*

[2] *Institut für Angewandte Physik, Universität Bonn, 53115 Bonn, Germany*

Phone: (631) 632 8138, Fax: (631) 632 8176, Wade.Smith@SUNYSB.EDU

Abstract: Measurements of intensity–field and intensity–intensity correlations show different aspects of the non-classical states produced by an optical cavity QED system and reveal its conditional dynamics.

Cavity QED, given its specific components and dynamics, is a preferred system for understanding fundamental questions of quantum mechanics such as dissipation, control, and entanglement. The dynamics of the system are characterized by the ratios of the reversible dipole coupling, g_0, of a single atom with the cavity mode to the irreversible coupling to the reservoirs through cavity, κ, and atomic polarization and inversion decays, $\gamma_\perp, \gamma_\|$. The strong coupling regime of cavity QED, realized in our experiments, requires that $\gamma_\perp\gamma_\|/g_0^2 < 1$, and $g_0^2/\kappa\gamma_\perp > 1$. When operating in this regime the excitation present in the cavity QED system is shared by the coupled atom–cavity and its exchange occurs at a rate that is compatible with available detection technology.

We study the conditional dynamics of our cavity QED system with two different correlation techniques, the well established intensity–intensity correlation, $g^{(2)}(\tau)$, and a new intensity–field correlation, $h(\tau)$. Both correlation functions depict the evolution of the system conditioned on the detection of a first photon.

The intensity–intensity correlation shows that the state of the electromagnetic field emitted by the cavity QED system is non-classical. It presents antibunching and sub-Poissonian statistics. If the cavity QED system is driven more intensely the emitted light is bunched, with no indication of non-classicality nor of the dynamical processes that are happening in the two quadratures of the electromagnetic field.

The intensity–field correlation [1, 2] permits us to follow the evolution of the mean amplitude of the conditional electromagnetic field emitted by our cavity QED system. The information obtained from the field is not the same as that of the intensity since we can separate the quadratures of the field and analyze how they both evolve. Fig. 1 shows a schematic of the experimental apparatus for the field-intensity correlator. The detection of a photon is correlated with the photocurrent record from a balanced homodyne detector. The measured signal in the homodyne detector is proportional to a correlation between the intensity $\hat{a}^\dagger\hat{a}$ and the field $\hat{a}$ of the light,

$$\langle : (\hat{a}^\dagger\hat{a})(0)\hat{Q}_{\varphi+\theta}(\tau) : \rangle \tag{1}$$

where the quadrature of the electromagnetic field $\hat{Q}_{\varphi+\theta}(\tau) \equiv [\hat{a}e^{-i(\varphi+\theta)} + \hat{a}^\dagger e^{i(\varphi+\theta)}]/2$, and $\hat{a}$ is the annihilation operator for a cavity photon. The phase θ is set by homodyne detection, φ by the phase of the mean cavity field.

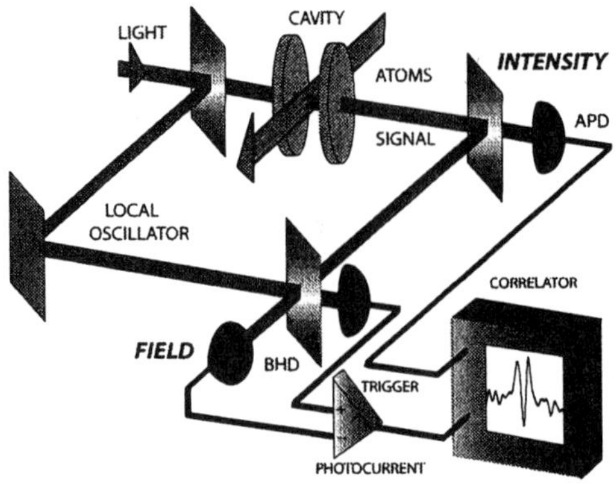

Fig. 1. Simplified diagram of the experimental setup. The fluctuations of the signal field are detected with a balanced homodyne detector (BHD) situated at the output beam splitter of a Mach-Zehnder interferometer, with one arm containing the cavity and the other a local oscillator (LO). The photocurrent is AC amplified, filtered and then sampled with a fast digital oscilloscope. The remaining signal is directed to an avalanche photodiode (APD). A photon detection at the APD triggers the digital oscilloscope to record the BHD photocurrent.

Figure 2 shows an example of the two correlation functions in the low intensity regime. In our experiment we explore the different ways the two measurements reveal the subtlety of the non-classicality of the system. These differences become more pronounced as the coupling constant g_0 increases and when the system is driven away from the low intensity limit, in particular when spontaneous emission begins to affect the conditional field dynamics.

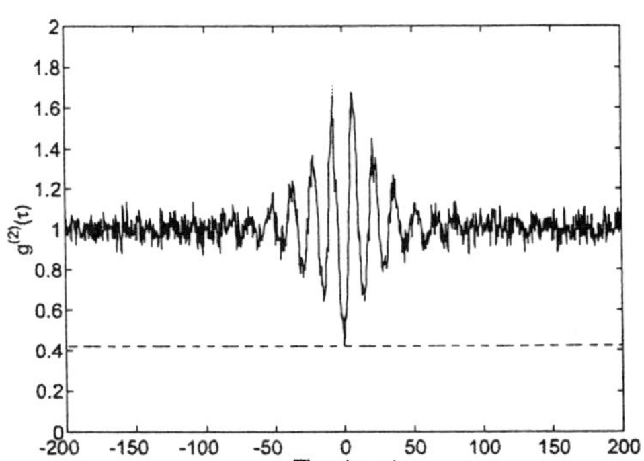

Fig. 2. The measured intensity–intensity correlation function ($g^{(2)}(\tau)$) in the low intensity limit. The cavity driving field is on atomic resonance, with parameters (g_0, κ, $\gamma_\perp$)$/2\pi$ = (8.8, 6.0, 3.0) MHz. $N \approx 23$ is the effective number of atoms coupled to the cavity mode.

Work supported by NSF. S. K. acknowledges support from DAAD.

References and links

1. G. T. Foster, L. A. Orozco, H. M. Castro-Beltran, H. J. Carmichael, "Quantum State Reduction and Conditional Time Evolution of Wave-Particle Correlations in Cavity QED," Phys. Rev. Lett. **85**, 3149 (2000).
2. H. J. Carmichael, H. M. Castro-Beltran, G. T. Foster, and L.A. Orozco, "Giant Violations of Classical Inequalities through Conditional Homodyne Detection of the Quadrature Amplitudes of Light," Phys. Rev. Lett. **85**, 1855 (2000).

Separation approach for Jaynes-Cummings model in multi-level atom

Ali Kamli and Smail Bougouffa
Department of Physics, Faculty of Science, King Khalid University, Abha P. O. Box 9003, Saudi Arabia
Fax + 966 7 231 32 96, aakamli@yahoo.com

Abstract : We investigate a new separation approach of coupled differential equations appropriate for three atom-field system. We deduce the separability condition for three level atom Jaynes-Cummings model and solve exactly for the probability amplitudes.
OCIS codes: (270.0270) Quantum optics

1- Introduction

Coupled Schrodinger equations are needed in the formulation of various physical problems but are usually not easy to handle and one must frequently have recourse to numerical techniques. It may, however, be worthwhile if the physical models can be constructed in such a manner that the coupled system can either be solved analytically or transformed into another system in which the equations are decoupled and solvable separately. A prime example in quantum optics is the case of three level atom interacting with quantized field. The problem is that one has to deal with a system of coupled differential equations and this does not always permit exact solutions. However, it may be possible under certain conditions to obtain closed analytic solutions. The coupled system of equations considered here may arise in three level atom interacting with a single mode quantized field. It is also applicable to the situation where the atom is interacting with multi mode field, after having employed the pseudo mode and factorization method as described in [1].

2- Mathematical Formulation

The three level V system under consideration comprises two upper levels 1 and 3 with amplitudes c_1 and c_3, and a ground state with amplitude c_2. The upper levels are coupled by dipole transitions to the ground state 2 while the upper levels are dipole forbidden. The coupled Schrodinger equations of motions for the amplitudes c's are:

$$\begin{aligned} i\dot{c}_1 &= \omega_1 c_1 + g_1 c_2 \\ i\dot{c}_3 &= \omega_3 c_3 + g_2 c_2 \\ i\dot{c}_2 &= \omega_\lambda c_2 + g_1^* c_1 + g_2^* c_3 \end{aligned} \qquad (1)$$

where ω_1 and ω_3 are the frequencies of the states 1 and 3, and ω_λ is the field mode frequency, and g_1 and g_2 are the frequency-dependent couplings of the atomic transitions 1-2 and 3-2 respectively, and the ground state frequency is assumed zero. To solve system (1) we assume $c_1=X_1c_2$ and $c_3=X_2c_2$ where X_1 and X_2 are continuous functions of time. Substituting in (1) we obtain 3 equations for the amplitude c_2 only, and from the compatibility of these equations we get coupled nonlinear Riccati equations for X_1 and X_2. This coupled system may be separated if X_1X_2=constant. This requirement along with the compatibility of equations for c_2, leads to value of 1/2 for the constant and $|g_1|^2=|g_2^*|^2=|g|^2$, and $\omega_\lambda=(\omega_1+\omega_3)/2$. With these conditions the Riccati coupled equations for X's can be separated and solved exactly.

3- Application to Jaynes-Cummings model

We shall apply the above approach to the case where the atom is interacting with a single mode quantized (vacuum) field, i.e., the Jaynes-Cummings model (JCM) in the three level atom. This model exhibits interesting nontrivial quantum features despite its conceptual simplicity [2-4], and some of these features have recently started to be observed experimentally in cavity quantum electrodynamics experiments (see[3]).

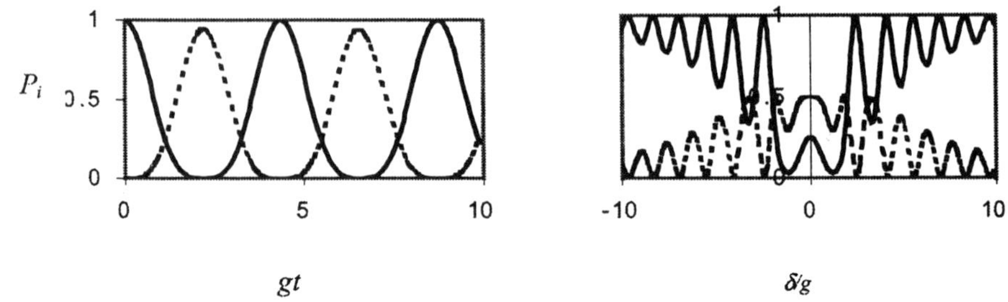

Fig.1. The variations of the populations P_1 (solid) and P_3 (dotted) with (a) time, gt and (b) frequency difference δ/g.

The three level system is described by Eq. (1) and we assume the initial condition $c_1(0)=1$, and all other probability amplitudes vanish, which simply means that the atom is initially in the upper level and there are no photons in the cavity mode at t=0. The level populations $P_1=|c_1|^2$ and $P_3=|c_3|^2$ are given as:

$$P_1 = [\cos^2(\mu t/2) + \frac{\delta^2}{4\mu^2}\sin^2(\mu t/2)]^2 \quad , \quad P_3 = 4|g|^4 \frac{\sin^4(\mu t/2)}{\mu^4} \qquad (2)$$

where $\mu^2=\delta^2/4+2|g|^2$, and $\delta=(\omega_1-\omega_3)$ measures the frequency difference of the two upper levels, and $P_1+P_2+P_3=1$. Fig.1a displays the probabilities P_1 and P_3 as functions of time gt for $\tilde{\delta}$g=0.5, and in Fig.1b we plot P_1 and P_3 as functions of $\tilde{\delta}$g for gt=10.0.

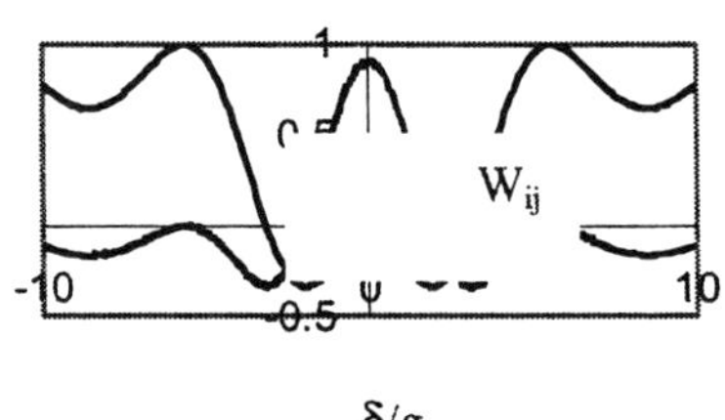

Fig.2. The three level atom inversions W_{12} (solid line) and W_{32} (dotted) as functions of δ/g.

We illustrate in Fig.2 the population inversions $W_{ij}=P_i-P_j$ as functions of $\tilde{\delta}$ g for gt=5.0. In this work we have presented an exact analytic solution for three level atom system.

4- References

[1] B.M.Garraway and P.L.Knight, "Cavity modified quantum beats", Phys. Rev. A54, 3592-3602(1996).

[2] H.I.Yoo and J.H.Eberly, " Dynamical theory of an atom with two or three levels interacting with quantized cavity fields" Phys. Rep., **118**,239(1985).

[3] B.M.Shore and P.L.Knight, " The Jaynes-Cummings model ", J. Mod. Opt., **7**,1195(1993).

[4] A.S.Shumovsky, E.I.Aliskenderov, and F.L.Kien, "exactly solvable three-level two-mode model with multiphoton transitions", J. Phys., A **18**, L 1031 (1985).

Ion Coulomb-crystals: A medium for storage of quantum states of light?

Anders Mortensen, Frank K. Jensen, Dirk Voigt, and Michael Drewsen
Institute of Physics and Astronomy, University of Aarhus, Ny Munkegade, DK-8000 Aarhus C, Denmark
drewsen@ifa.au.dk

Abstract:

Following the idea of a recent paper [M. D. Lukin, S. F. Yelin, and M. Fleischhauer, Phys. Rev. Lett. **84**, 4232 (2000)], a two-species ion Coulomb-crystal inside an optical cavity is proposed as a medium for storage of the quantum state of a light pulse.

1. Introduction

Recently, two independent experiments have demonstrated that a pulse of coherent light can be temporarily stored as a collective excitation of a cold atomic ensemble [1,2]. The idea behind these experiments suggests that not only coherent light pulses, but also light with non-classical characters can be stored. Both experiments relied on a very large reduction of the group velocity of the light pulse as well as on the atomic medium being optically thick in the absence of a control laser beam. It has, however, been shown that these requirements can be relaxed if the atomic ensemble is placed within a high-finesse optical cavity [3], where the light pulse, due to multiple reflections, will pass the atoms several times.

In the present contribution, we consider the prospect of implementing the idea of Ref. [3] using a two-species ion Coulomb-crystal inside an optical cavity as a medium for storage of quantum states of light.

2. Coulomb-crystals

When trapped ions are cooled below a certain critical temperature (typically about ten milli-Kelvin) they start to form spatially ordered structures sometimes referred to as ion Coulomb-crystals or just ion crystals. Such crystals of various sizes containing single atomic ion species have now for more than ten years been investigated in various types of traps. Recently, also multi-component crystals, and in particular two-component crystals (bi-crystals), have attracted attention. In quantum optics, short strings of several ion species have, e.g., been proposed applied in the implementation of a quantum computer [4], while larger bi-crystals have so far mainly been considered interesting objects either in non-neutral plasma physics [5] or in atomic and molecular physics [6]. In the latter case, the possibility of sympathetic translational cooling of various atomic or molecular ion species, through Coulomb interactions with ion species that are directly laser-cooled, is very attractive. It has been demonstrated that the translational temperature of the sympathetically cooled ions can easily be below 100 mK and that the two species at such low temperatures in, e.g., linear Paul traps generally separates radially with the lighter ions situated closest to the trap axis [5,6]. Figure 1 below, which shows a picture of a bi-crystal of laser cooled Ca^+ ions surrounding an invisible cylindrical shaped core of O_2^+ ions, illustrates this point.

Coherence and Quantum Optics VIII
Edited by Bigelow et al., Kluwer Academic/Plenum Publishers, 2003

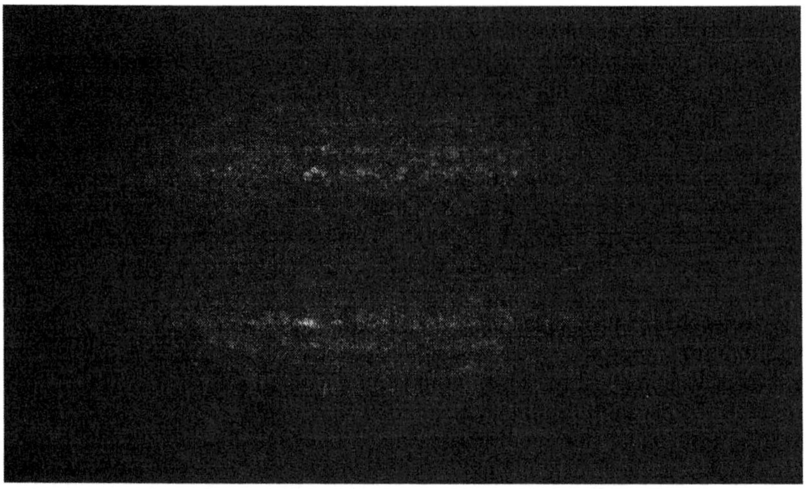

Fig. 1. A picture of an ion bi-crystal consisting of an invisible cylindrical core of O_2^+ ions surrounded by heavier fluorescing $^{40}Ca^+$ ions.

3. The ion crystal quantum memory

In order to implement the idea of Ref. [3], a relatively strong coupling between the atoms and the cavity mode is needed and the atoms should stay within the mode volume at least during the storage and read-out process. Ions in the nearly cylindrical shaped core of an ion bi-crystal situated within the TEM_{00} mode volume of a high-finesse optical cavity fulfill these criteria even for times up to seconds. Furthermore, in order to make the storage of quantum states of light possible, it is necessary that the storage medium consists of effective three-level Λ-atoms where one transition is nearly resonant with the quantum light. Since the wavelength of the transitions between the metastable 3d D-states and the 4p P-states of the Ca^+ isotopes are in the region of 850 nm, where tunable sources of non-classical light already exist, such ions seem to be a good choice for a storage medium. To realize the quantum memory, we envision a bi-crystal consisting of a cylindrical core of Ca^+ ions sympathetically cooled by surrounding laser-cooled heavier isotopes of Ca^+ or other ion species such as Sr^+, Yb^+, or Ba^+. In a linear Paul trap, the length of the core of the bi-crystals can be at least up to a few millimeters long, and the ratio of the length to the diameter can be matched to the cavity mode by changing the trap parameters. With typical ion densities of 10^8 cm^{-3}, a medium of about 1000 ions seems to be feasible. Since this is relatively few compared to the number of atoms in the experiments of Ref. [1,2], an ion Coulomb-crystal memory device will only be able to store light pulses consisting of up to tens of photons with high fidelity. However, for some quantum information applications this could be sufficient.

In the presentation, details on an actual realization of a memory device based on a linear Paul trap will be discussed.

[1] C. Liu, Z. Dutton, C. H. Behroozi, and L. V. Hau, "Observation of coherent optical information storage in an atomic medium using halted light pulses", Nature **409**, 490-493 (2001).
[2] D. F. Phillips, A. Fleischhauer, A. Mair, R. L. Walsworth, and M. D. Lukin, "Storage of light in atomic vapor", Phys. Rev. Lett. **86**, 783-786 (2001).
[3] M. D. Lukin, S. F. Yelin, and M. Fleischhauer, "Entanglement of atomic ensembles by trapping correlated photon states", Phys. Rev. Lett. **84**, 4232-4235 (2000).
[4] E. Peik, J. Abel, Th. Becker, J. von Zanthier, H. Walther, "Sideband cooling of ions in radio-frequency traps", Phys. Rev. A **60**, 439-449 (1999).
[5] L. Hornekær, N. Kjærgaard, A. M. Thommesen, and M. Drewsen, "Two-component Coulomb crystals", Phys. Rev. Lett. **86**, 1994-1997 (2001).
[6] K. Mølhave and M. Drewsen, "Formation of translationally cold MgH^+/MgD^+ molecules in an ion trap", Phys. Rev. A **62**, 0011401(R)1-4 (2000).
[7] E. S. Polzik, J. Carri, and H. J. Kimble, "Spectroscopy with squeezed light", Phys. Rev. Lett. **68**, 3020-3023 (1992).

Chaotic walking of a cold atom in an ideal cavity

S. V. Prants and L.E.Kon'kov

Laboratory of Nonlinear Dynamical Systems, V.I.Il'ichev Pacific Oceanological Institute of the Russian Academy of Sciences, 690041 Vladivostok, Russia

fax: 007(4232)312573, e-mail:prants@mail.ru

The interaction of cold atoms with coherent fields involves not only the internal atomic transitions and photon states, but also the center-of-mass motion of atoms. We consider the simple situation: a two-level atom moves inside a high-finesse Fabry-Perot cavity and interacts with a single-mode standing-wave electromagnetic field whose spatial variation creates an one-dimensional optical potential. The respective Hamiltonian

$$\hat{H} = \frac{\hat{p}^2}{2m} + \frac{1}{2}\hbar\omega_a\hat{\sigma}_z + \hbar\omega_f\left(\hat{a}^\dagger\hat{a} + \frac{1}{2}\right) + \hbar\Omega_0\sin(k\hat{x} + \varphi)(\hat{a}^\dagger\hat{\sigma}_- + \hat{a}\hat{\sigma}_+) \tag{1}$$

generates in the semiclassical approximation the equations of motion

$$\dot{\xi} = \alpha\rho, \quad \dot{\rho} = -u\sin\xi, \quad \dot{u} = \delta v, \quad \dot{v} = -\delta u + 2Nz\cos\xi, \quad \dot{z} = -2v\cos\xi \tag{2}$$

for normalized expectation values of the atomic population inversion $z =< \hat{\sigma}_z >$, the position $\xi = k_f < \hat{x} >$, the momentum $\rho =< \hat{p} > /\hbar k_f$, the collective atom-field variables $u =< \hat{a}^\dagger\hat{\sigma}_- + \hat{a}\hat{\sigma}_+ >$ and $v =< i(\hat{a}^\dagger\hat{\sigma}_- - \hat{a}\hat{\sigma}_+) >$, where k_f is the wave number of the standing wave. The operators are averaged over an initial quantum state of the full system. The set (2) possesses the integral of motion, $W = \alpha\rho^2/2 - u\cos\xi - \delta z/2$, and three control parameters, the normalized recoil frequency $\alpha = \hbar k_f^2/m\Omega_0 \equiv 2\omega_R/\Omega_0$, the normalized detuning $\delta = (\omega_f - \omega_a)/\Omega_0$, and the total number of excitations $N =< \hat{a}^\dagger\hat{a} + (\hat{\sigma}_z + 1)/2 >$.

The set of equations (2) may be considered as a Hamiltonian system with two degrees of freedom, the slow translational one (ξ and ρ) and the fast atom-field one (u

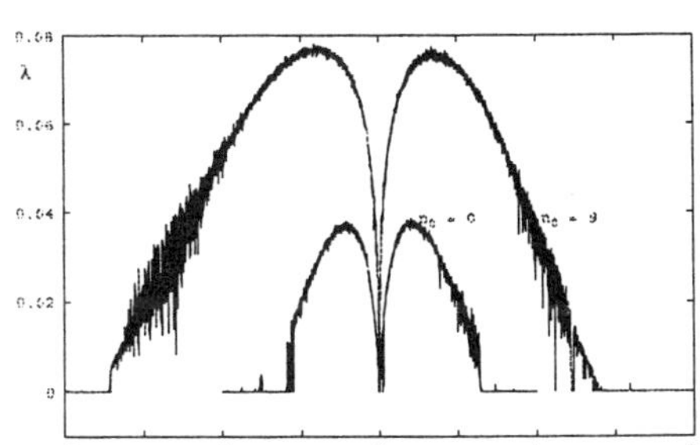

Fig. 1. The maximal Lyapunov exponent λ versus the normalized detuning $\delta = (\omega_f - \omega_a)/\Omega_0$

Coherence and Quantum Optics VIII
Edited by Bigelow et al., Kluwer Academic/Plenum Publishers, 2003

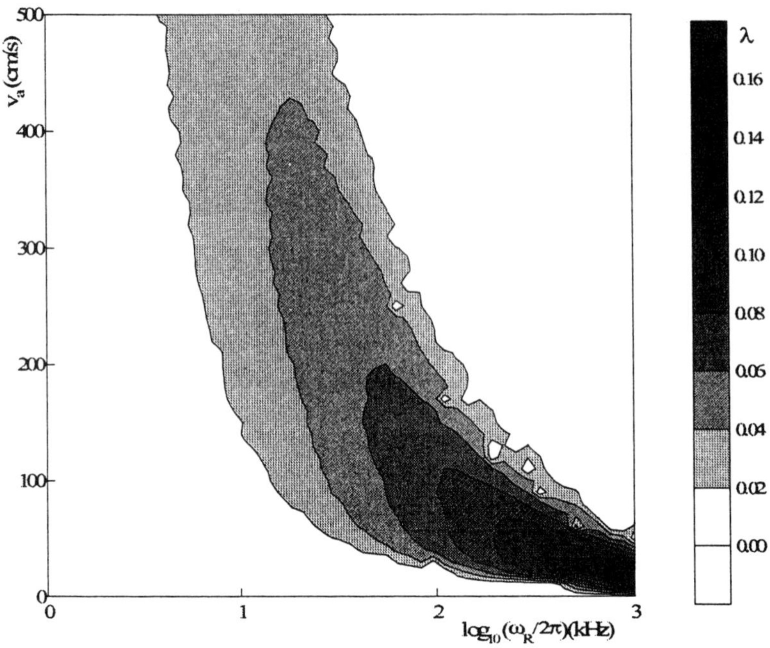

Fig. 2. The maximal Lyapunov exponent λ versus the initial atomic velocity v_a (cm/s) and the recoil frequency on a logarithmic scale $\log_{10}(\omega_R/2\pi)$ (KHz)

and v). At exact resonance, $\delta = 0$, the fast and slow variables are separated from each other resulting in regular translational and internal motion. What will happen out of resonance at $\delta \neq 0$? All the numerical experiments are carried out in the ranges of magnitudes of the control parameters in the strong-coupling regime. The normalized recoil frequency is chosen to be $\alpha = 10^{-3}$, and the total number of excitations is supposed to be $N = 10$ corresponding to the mean number of intracavity photons to be $n_0 = 9$. Extremal sensitivity to initial conditions, a hallmark of dynamical chaos, is characterized by the positive values of the maximal Lyapunov exponent λ. Fig. 1 demonstrates the dependence of λ on the detuning with initially vacuum cavity field ($n_0 = 0$) and with comparatively large mean number of the intracavity photons ($n_0 = 9$). It is evident from the figure that the motion is regular ($\lambda \simeq 0$) with very small detunings and with comparatively large values of δ in accordance with our analytical speculation. In order to estimate the range of the system's parameters where the motion is expected to be chaotic it is useful to compute the so-called topographical λ maps showing the values of λ with given initial conditions as a function of two of the system's parameters at fixed values of the other parameters. Such maps give a representative "portrait" of chaos. The λ map in Fig. 2 shows the values of the maximal Lyapunov exponent of the atom-field system (2) with $\delta = 0.5$ and $N = 10$ as a function of the initial atomic velocity v_a and of the common logarithm of the recoil frequency $\omega_R/2\pi$. Under the other equal conditions, the larger the recoil energy is, the stronger will be chaos. More heavy atoms are expected to move more regularly. It is a light He* atom may possess the normalized recoil frequency of the order of $\alpha \simeq 10^{-3}$ that corresponds to the recoil frequency $\omega_R/2\pi \simeq 10^3$ KHz.

Thus, we show in this report that a two-level atom may walk chaotically in the periodic single-mode standing-wave field in the absense of any sources of random fluctuations.

Feedback with conditional measurements in cavity QED

J. E. Reiner, W. P. Smith, L. A. Orozco
Department of Physics and Astronomy, SUNYSB, Stony Brook NY 11794-3800
Phone 1 (631) 632 7453, Fax 1 (631) 632 8176 Joseph.Reiner@sunysb.edu

Abstract:

Control of a quantum system with feedback is a fundamental problem in quantum coherence. We are studying the use of a passive feedback system triggered by photon detections from a cavity QED system to cancel the decay in a conditional intensity [$g^{(2)}(\tau)$] measurement.

OCIS codes: (270.0270) Quantum Optics; (270.2500) Fluctuations, relaxations, and noise; (999.9999) Quantum Feedback

Quantum feedback has been proposed as a way to fight both dissipation and decoherence in a quantum system [1]. The method for implementing quantum feedback depends upon both the system and the purpose of the feedback. Some feedback schemes try to stabilize a system by forcing it into a particular quantum state [2]. Other schemes wish to protect the system's evolution from environmental effects [3].

Correlation functions utilize conditional measurements of the intensity [4] or of the field [5] in quantum optical systems such as cavity QED. These measurements rely on the preparation of a well defined quantum state by the detection of a first photon, which then evolves back to steady state in a time set by the decay of the system. The state of the cavity QED system is non-classical because it demonstrates antibunching, sub-Poissonian statistics and squeezing.

The recent development of the wave-particle correlator has allowed us to measure the conditional time evolution of a cavity QED field mode [5]. This measurement is carried out by correlating the detection of a photon with the recorded output from the photocurrent of a balanced homodyne detector (BHD). This photocurrent, which is proportional to the cavity field, is averaged over many starts to remove the shot noise.

We have studied numerically the use of this wave-particle correlator to actively feedback information about the field into the driving intensity of the cavity QED system [6]. We found that feeding back the running average of the measured photocurrent removes the contribution of the BHD shot noise in the feedback. The experimental implementation requires answering a series of questions on the time response of the opto-electronics and the speed of signal processing for close-loop feedback.

To address part of those questions we have now turned our attention to the intensity correlation function. Its measurement does not require control or knowledge of a phase, but it still takes many realizations in order to build enough statistics to get the conditional evolution of the intensity.

Figure 1 shows a schematic of our experimental setup to measure the correlation function $g^{(2)}(\tau)$ of our cavity QED system [7]. A beam of optically pumped rubidium atoms intersects the mode of a high Finesse cavity at ninety degrees. The size of the cavity is such that the coherent dipole coupling between an atom and the field of a single photon is larger than the decay through spontaneous emission or the escape out of the cavity mode through the mirrors. This apparatus, under appropriate circumstances, measures the probability for the cavity to emit a photon at a time τ conditioned on a photon emission at time $\tau = 0$. The modification that we introduce is in the amplitude modulator before the cavity QED system.

We do not actively feedback the driving intensity in a $g^{(2)}(\tau)$ measurement; instead we rely on the measured dynamics of the system. The regression to steady state is dominated by the exchange of excitation at a rate given by the coupling between the atoms and the cavity, also called the vacuum Rabi frequency. We first measure $g^{(2)}(\tau)$ without any input modulation. This measurement gives the frequency and phase of the regression to steady state after the detection of the first photon. We next adjust the amplitude modulator to produce a sinusoidal modulation in phase with the Rabi oscillation triggered by the detection of the first photon. Finally we control the size of the modulation in order to achieve the proper oscillation for the $g^{(2)}(\tau)$ measurement. The scheme relies on the continuous preparation of the same state every time, although some of the experimental parameters may vary, for example there are fluctuations in the number of atoms interacting at any given time with the cavity mode.

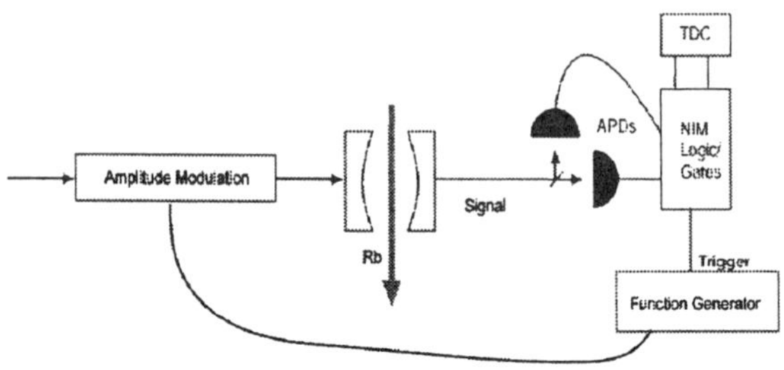

Fig. 1. A schematic diagram of the proposed setup for passively stabilizing the vacuum-Rabi oscillations from a cavity QED system. The light emitted by the cavity goes to a 50/50 beam splitter where avalanche photodiodes (APD) produce start and stop pulses that after shaping go to a time to digital converter (TDC) to measure the separation and histogram it. Once the histogram is complete, we generate a modulation signal that we feedback into the drive whenever we detect a start pulse.

The passive quantum feedback in the conditional intensity measurement works in the simulations. Its experimental implementation requires implementation of opto-electronics that are necessary for future active feedback schemes, both in the field and in the intensity. This scheme serves as a proof of principle for using any feedback in our cavity QED system.

Work supported by NSF.

References:

1. H. M. Wiseman, "Quantum theory of continuous feedback," Phys. Rev. A **49,** 2133 (1994).
2. J. Wang and H. M. Wiseman, "Feedback-stabilization of an arbitrary pure state of a two-level atom," quant-phy/0008003, Submitted to Phys. Rev. A.
3. P. Tombesi and D. Vitali, "Physical realization of an environment with squeezed quantum fluctuations via quantum-nondemolition-mediated feedback," Phys. Rev. A **50**, 4253 (1994).
4. S. L. Mielke, G. T. Foster, and L. A. Orozco, "Non-classical intensity correlations in cavity QED," Phys. Rev. Lett. **80**, 3948 (1998).
5. G. T. Foster, L. A. Orozco, H. M. Castro-Beltran, H. J. Carmichael, "Quantum State Reduction and Conditional Time Evolution of Wave-Particle Correlations in Cavity QED," Phys. Rev. Lett. **85**, 3149 (2000).
6. J. E. Reiner, W. P. Smith, J. Wang, H. M. Wiseman, and L. A. Orozco, "Using quantum feedback to preserve the vacuum Rabi oscillations in cavity QED," Presented at the QELS 2001 meeting held in Baltimore, MD. 5/6-5/11.
7. G. T. Foster, S. L. Mielke, L. A. Orozco, "Intensity correlations in cavity QED," Phys. Rev. A **61**, 053821 (2000)

Effects of decoherence on entangled atomic wave functions in microcavities

A. S. Majumdar[1] and N. Nayak[2]

[1]*S. N. Bose National Centre for Basic Sciences, Block JD, Sector III, Calcutta 700098, India*

archan@boson.bose.res.in

[2]*Department of Physics, Texas A&M University, College Station, Texas 77843-4242, USA*

nayak@atlantic.tamu.edu

Abstract: We show that the effects of cavity dissipation and atomic decay on entangled atomic wave functions can be monitored and tested in a real micromaser. A curious example of multiparticle enhancement of nonlocality is observed.

OCIS codes: (270.0270,999.9999) Photon statistics, Entanglement, Decoherence

Micromaser dynamics is a natural candidate for studying varied aspects and consequences of entanglement of atomic wave functions with cavity photons[1]. In this article we consider the violation of a Bell-type inequality using a realistic model for the micromaser[2]. Our purpose is to study the effect of both cavity dissipation and atomic decay towards diminishing atom-photon entanglement in an actual experiment[3].

The generation of correlations between the wave functions of two or more atoms is achieved in the following way. A two-level atom initially in its upper excited state $|e>$ traverses a high-Q single mode cavity. The cavity is in a steady state and tuned to a single mode resonant with the transition $|e>\rightarrow|g>$. The emerging single-atom wavefunction is a superposition of the upper $|e>$ and lower $|g>$ state, and it leaves an imprint on the photonic wavefunction in the cavity. After leaving the cavity, the atom passes through an electromagnetic field which gives it a $\pi/2$ pulse the phase of which can be varied for different atoms. The atom then reaches the detector, placed at a sufficient distance, capable of detecting the atom only in the upper or lower state. This process is repeated for a similar second atom. Though there is no spatial overlap between the two atoms, the entanglement of their wavefunctions with the cavity photons can be used to formulate a Bell-type inequality on the detection probabilities for the two atoms[1]:

$$\begin{aligned} B &\equiv |E^{\lambda}(\phi_1,\phi_2) - E^{\lambda}(\phi_1,\phi_3)| \\ &+ sign(E_0)[E^{\lambda}(\phi_2,\phi_3) - E_0] \end{aligned} \qquad (1)$$

where λ is the hidden variable, ϕ's are the phases of the $\pi/2$ pulse, and the quantum mechanical analog of E is given by

$$\begin{aligned} E(\phi_1,\phi_2) &= P_{ee}(\phi_1,\phi_2) + P_{gg}(\phi_1,\phi_2) \\ &- P_{eg}(\phi_1,\phi_2) - P_{ge}(\phi_1,\phi_2) \end{aligned} \qquad (2)$$

where $P_{eg}(\phi_1,\phi_2)$ stands for the probability that the first atom is found to be in state $|e>$ after traversing the $\pi/2$ pulse with phase ϕ_1, and the second atom is found to be in state $|g>$ with the phase of the $\pi/2$ pulse being ϕ_2 for its case.

Our purpose here is to calculate the Bell sum using a realistic model for the micromaser and see how it behaves as a function of the cavity parameters and decay rates. For doing so we consider the standard atom-cavity field interactions with their reservoirs and solve the equation

$$\dot{\rho} = \dot{\rho}|_{atom-reservoir} + \dot{\rho}|_{field-reservoir} + \dot{\rho}|_{atom-field} \tag{3}$$

The dynamics for an intial steady-state cavity density operator has been solved[2]. This solution can be used to calculate the probabilities in the Bell sum treating cavity dissipation and atomic decay in a realistic fashion and the formalism adopted has been used to describe the microlaser too with a suitable choice of parameters[4]. Here we calculate the Bell sum and focus on the effects of decoherence on it.

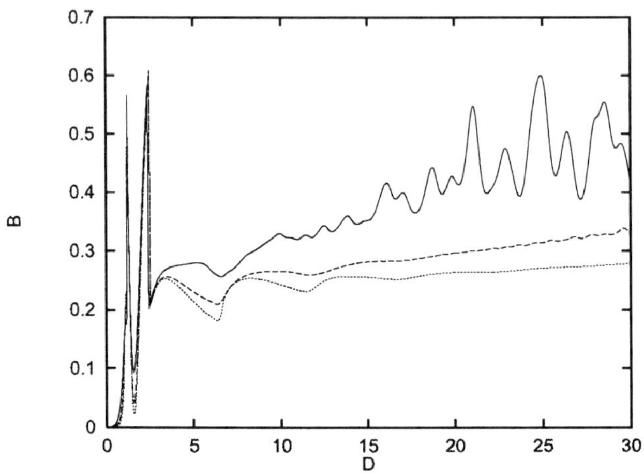

Fig. 1. Violation of Bell's inequality in a micromaser. The pump rate N, the number of individual atoms that pass through the cavity in a photon lifetime, = 20 (full), 50 (broken), and 100 (dotted).

The Bell sum B is plotted versus the pump parameter $D = \phi\sqrt{N}$ where the Rabi angle $\phi = g\tau$ in Figure 1. The cavity dissipation parameter chosen here is $\kappa/g = 0.1 \times 10^{-6}$. B decreases with the increase of pump rate N for a large range of interaction times τ. Dissipative effects creep into the dynamics through two parameters N and τ[2]. For shorter values of single atom interaction times we find that the correlations build up sharply with τ, and the peak value of B signifying maximum violation of BI is larger for higher values of N, (as a magnification of Figure 1 reveals). The maximum violation (the value of B at its second peak) is 0.5812 for $N = 20$ (full line), 0.6079 for $N = 50$ (broken) and 0.6241 for $N = 100$ (dotted) This feature is a curious example of multiparticle induced nonlocality[5].For short interaction times, naturally the effects of decoherence are too small to affect the correlations. To conclude, our study serves to highlight the fact that systems such as the micromaser[3] can be effectively used to study fundamental problems of quantum theory like entanglement and decoherence.

References and links

1. M. Loffler, B.-G. Englert and H. Walther, Appl. Phys. B **63**, 511 (1996).
2. N. Nayak, Opt. Commun. **118**, 114 (1995).
3. G. Rampe, F. Schmidt-Kaler and H. Walther, Phys. Rev. Lett. **64**, 2783 (1990).
4. A. S. Majumdar and N. Nayak, quant-ph/0005123.
5. N. D. Mermin, Phys. Rev. Lett. **65**, 1838 (1990); S. M. Roy and V. Singh, Phys. Rev. Lett. **67**, 2761 (1991); D. Home and A. S. Majumdar, Phys. Rev. A**52**, 4959 (1995).

Fiber generated quantum correlations for quantum-optical communications

M. Fiorentino, J.E. Sharping, A. Coker, and P. Kumar

Center for Photonic Communication and Computing
ECE Dept. Northwestern University, 2145 Sheridan Rd.
Evanston, Il 60208, USA
+1-847-467-3223, +1-847-491-4455, mfiore@ece.northwestern.edu

One of the main challenges facing the development of Quantum Information Technology is the difficulty associated with transmission of quantum information over lossy and noisy channels. Our approach to mitigating this problem relies on the use of fiber-based devices generating entanglement to be used as a shared quantum resource in the quantum communication network. This approach offers the advantage of technological synergy between the generation and propagation components of the overall quantum network. In this paper we present our initial work that is focused on the realization of pulsed, quantum-correlated systems that can be used as building blocks for fiber-based sources of quadrature and polarization entanglement. In this preliminary phase we evaluated several different schemes in order to determine which one can satisfy the demanding specifications of the quantum network.

To create quadrature correlations in pulses with wavelengths in the standard fiber-optic communication band around 1550 nm we used a polarization Mach-Zehnder interferometer. This interferometer can be configured as a highly-asymmetric [figure 1-(a)] or symmetric [figure 1-(b)] Sagnac loop leading to amplitude-squeezed and vacuum-squeezed light, respectively. In this experiment we use self-phase modulation of 180-fs pulses to generate squeezing in the fiber. The Mach-Zehnder geometry gives us the ability to manipulate the output pulses in order to efficiently detect the noise reduction. So far 4.4 dB of amplitude squeezing and 1.3 dB of vacuum squeezing have been measured in our apparatus [1]. The results have also been compared with numerical models of noise propagation in optical fiber.

To create polarization correlations in pulses with wavelengths around 1550 nm we have set up a non-degenerate four-wave mixing experiment in a dispersion-shifted fiber Sagnac loop [figure 1-(c)]. In a four-wave mixing experiment, a strong pump is injected into the fiber and the $\chi^{(3)}$ interaction provides parametric gain to a weak signal accompanied by the production of an an idler. For the interaction to be efficient, phase-matching requirements have to be met, which can be done when the pump is close to the zero-dispersion wavelength of the fiber. From a quantum point of view, the signal and idler beams are "twins" in the sense that their amplitude fluctuations show correlations below the shot-noise level. In a simplified photon picture of the process, two pump photons are destroyed and one signal and one idler photon are created. We have observed 1.1 dB of twin-beam quantum correlations when an injected signal is amplified through the loop [2]. To demonstrate the quantum nature of the four-wave scattering process at the single photon level, we are now in the process of setting up an all-optical-fiber version of the Franson experiment [3], which will also allow us, by measuring the degree of violation of Bell's inequalities, to quantify the purity of our source of quantum correlations. Using polarization multiplexing this source of correlated photons can then be used to build a source of polarization-entangled photons pairs.

We also plan to reproduce this experiment at a wavelength close to 800 nm, where promising quantum memory experiments are currently being performed [4]. This will

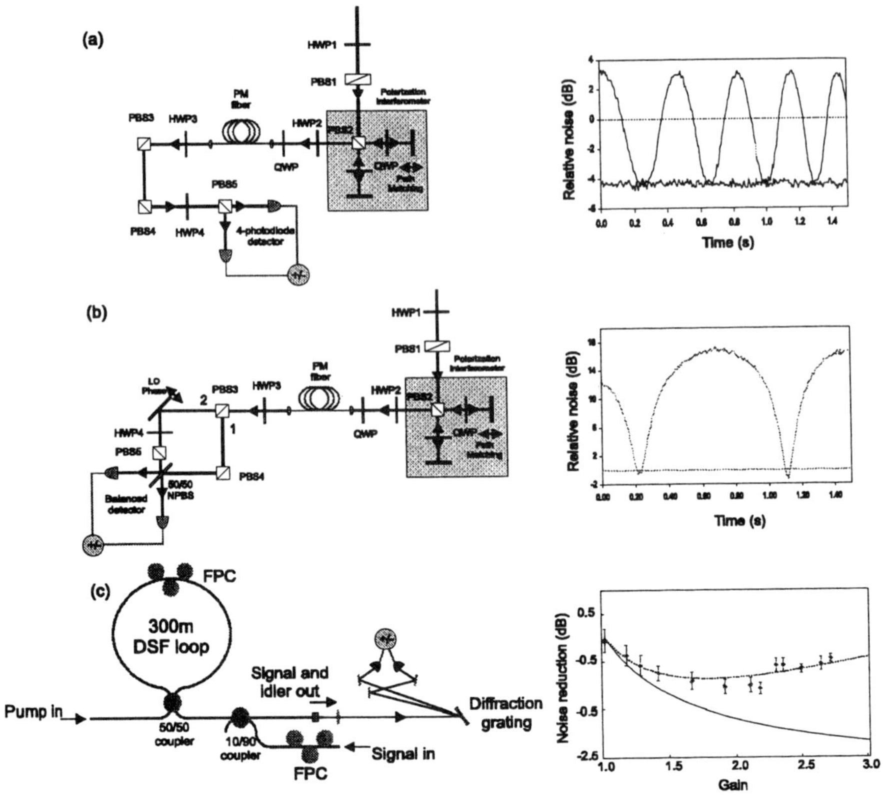

Fig. 1. Schematic of the experimental setups with noise reduction data sets: (a) Asymmetric Mach-Zehnder interferometer with sub-Poissonian light detection. (b) Symmetric Mach-Zehnder interferometer with vacuum squeezing detection. (c) Fiber Sagnac interferometer with twin-beams noise detection. HWP, half-wave plate; QWP, quarter-wave plate; PBS, polarizing beam-splitter; FPC, fiber polarization controller; DSF, dispersion shifted fiber.

allow us to integrate our quantum communication schemes with viable quantum memories, thus providing a first building block for a quantum network. Our group has recently observed, to our knowledge for the first time, four-wave mixing in microstructure (holey) fibers [5] at frequencies around 800 nm.

In conclusion, we present an array of experiments aimed at developing new sources of entanglement in optical fibers. Our experiments are part of a collaborative effort exploring the possibility of building an integrated quantum communication network. The authors acknowledge support from the U.S. Army Research Office through MURI grant DAAD19-00-1-0177.

References and links

1. M. Fiorentino, J.E. Sharping, P. Kumar, M. Vasiliev, and D. Levandovsky, "Soliton squeezing in a Mach-Zehnder interferometer", *submitted to Phys. Rev. A*
2. J.E. Sharping, M. Fiorentino, and P. Kumar, "Observation of twin-beams type quantum correlations in optical fiber", Opt. Lett. **26**, 367 (2001)
3. J.D. Franson "Bell inequality for position and time", Phys. Rev. Lett. **62**, 2205 (1989)
4. S.M. Shahriar, S. Lloyd, and P.R. Hemmer, "Long distance, unconditional teleportation of atomic states via complete Bell state measurements", *submitted to Phys. Rev. Letters.*
5. J.E. Sharping, A. Coker, M. Fiorentino, P. Kumar, and R.S. Wiendler, "Four-wave mixing in microstructure fiber", to appear in Opt. Lett.

Bloch-Siegert oscillation for detection and quantum teleportation of the phase of an oscillating field

M.S. Shahriar
Research Laboratory of Electronics, Massachusetts Institute of Technology, Cambridge, MA 02139
smshahri@mit.edu

When considering the resonant or near-resonant excitation of an atomic dipole by an oscillating field, it is customary to make the so-called rotating wave approximation (RWA), which amounts to ignoring the effect of the energy non-conserving counter-rotating term (CRT) in the Hamiltonian. It is well known that the CRT leads to the so-called Bloch-Siegert shift. Here, we show that the CRT can also lead to observable high-frequency oscillations in the amplitudes of the ground state, for example. Detection of this oscillation can be used to measure the absolute phase of the oscillating field at the beginning or the end of the excitation period. This may open up the possibility of mapping the phase distribution of a field with ultra-high resolution. Furthermore, this phase information can be transmitted, without measuring, to a remote observer via quantum teleportation with high fidelity, given the presence of a pair of atoms whose degenerate internal states are entangled. We have shown elsewhere [1,2] how such an entanglement can be produced in a robust manner over a distance of up to 50 km. The ability to teleport this phase information directly may be useful for the purpose of relative phase-locking of remote oscillators [3,4] under proper contexts, as well for devising novel methods of quantum cryptography.

To see the basic premise of this effect, consider first the reason why usually, under the RWA, an atom acts as a square-law detector. The amplitude of an atomic state is necessarily complex; whenever a measurement is made, the square of the absolute value of the amplitude is what we can observe. The (semi-classical) amplitude of an oscillating field, on the other hand, is real, composed of the sum of two complex components. This difference is side-stepped under the RWA, where only one of the two complex components is kept. As a result, generally an atom interacting with a field enables one to measure only the intensity --- and not the phase--- of the field. Of course, one can measure the phase of the field via heterodyne detection, an example of which occurs in experiments involving quadrature squeezing. In such an experiment, the weak squeezed field is multiplied by a strong field (local oscillator: LO). An atom, acting still under the *square-law* limit, can detect this *multiplied* signal, which varies with the phase difference between the weak field and the LO. Here, we show how the effect of the CRT also leads to an heterodyning effect, by making use of the beat between the co- and counter-rotating parts of the field.

We assume a field of the form $B=B_o Cos(\omega t+\phi)$, where the choice of the origin of the time variable, t, determi es value of ϕ. Consider, for example, the resonant excitation of a dipolar transition coupling le els $|1\rangle$ to $|2\rangle$. As shown in detail in ref. 4, the coefficients of the two states, under the rotati wave transformation, but without the RWA, can be written as:

$$|\tilde{\xi}(t)\rangle = \begin{bmatrix} \tilde{C}_1 \\ \tilde{C}_2 \end{bmatrix} = \sum_{n=-\infty}^{\infty} |\xi_n\rangle \beta^n, \quad |\xi_n\rangle \equiv \begin{bmatrix} a_n \\ b_n \end{bmatrix} \tag{1}$$

Inserting eqn. 1 into the effective Schroedinger eqn., and equating coefficients with same frequencies, we get, for all n:

$$\dot{a}_n = i2n\omega a_n + ig_o(b_n + b_{n-1})/2\,,\quad \dot{b}_n = i2n\omega b_n + ig_o(a_n + a_{n+1})/2 \tag{2}$$

Figure 1 shows a pictorial representation of these equations. Here, the coupling between a_o and b_o is the conventional one present when the RWA is made. The coupling to additional levels results from virtual

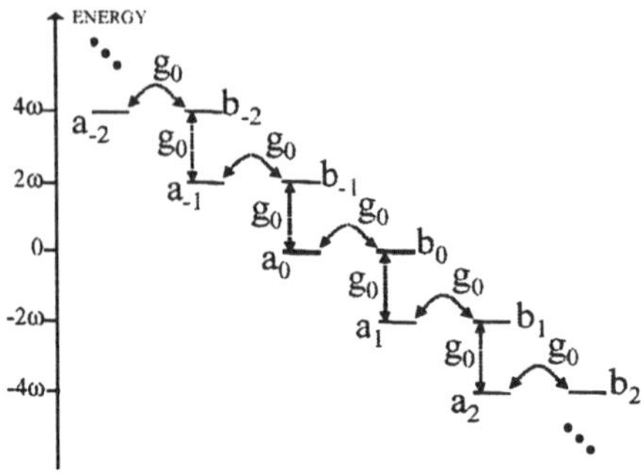

Figure 1. Schematic illustration of the multiple orders of interaction when the rotating wave approximation is not made. The strengths of the first higher order interaction, for example, is weaker than the zeroth order interaction by the ratio of the Rabi frequency, g_o, and the effective detuning, 2ω.

multi-photon processes in the absence of the RWA. The couplings to the nearest neighbors, $a_{\pm l}$ and $b_{\pm l}$ are detuned by an amount 2ω, and so on. The same set of equations also result from considering the field to be a coherent state, applying the Jaynes-Cummings interaction, and taking the semi-classical limit where the mean value of the photon number is much larger than unity. To the lowest order in (g_o/ω), we can ignore terms with $/n/>1$, thus yielding a truncated set of six eqns. In order to solve these equations, one may employ the method of adiabatic elimination valid to first order in $\sigma\equiv(g_o/4\omega)$. As derived in ref. 3, we get the following expressions for the amplitudes of the states:

$$C_1(t) = Cos(g_o t/2) - 2\sigma\Sigma\cdot Sin(g_o t/2)\,;$$
$$C_2(t) = ie^{-i(\omega t+\phi)}[Sin(g_o t/2) + 2\sigma\Sigma^*\cdot Cos(g_o t/2)] \tag{3}$$

where we have defined $\Sigma \equiv (i/2)\exp[-i(2\omega t + 2\phi)]$, and assumed that the system starts in level 1. We have verified this solution via numerical methods as well [4]. To lowest order in σ this solution is normalized at all times. In the limit of $\sigma\rightarrow 0$, we recover the result under RWA. In that case, measurement of the population of level 1 or level 2 only yields information about the interval of the excitation process, and the amplitude of the field. On the other hand, when σ is not ignored, measurement of the amplitude of either state yields the additional information about the phase of the field at the instant of the measurement. As shown in ref. 4, this process can also be used to transmit the phase information, without measuring, to a remote oberver via quantum teleportation. This work was supported by AFOSR grant # F49620-98-1-0313, ARO grant #s DAAG55-98-1-0375 and DAAD19-001-0177, and NRO grant # NRO-000-00-C-0158.

References:

1. J.H. Shapiro, in *Proc. Fifth International Conf. On Quantum Communication, Measurement, and Computing*, Capri 2000, edited by O. Hirota and P. Tombesi.
2. S.Lloyd, M.S. Shahriar, and P.R. Hemmer, , *http://xxx.lanl.gov/pdf/quant-ph /0003147.*
3. R. Jozsa, D.S. Abrams, J.P. Dowling, and C.P. Williams, *Phys. Rev. Letts. 85, 2010(2000).*
4. M.S. Shahriar, *http://xxx.lanl.gov/pdf quant-ph/0010007*

Two-photon diffraction and quantum lithography

Milena D'Angelo, Maria V. Chekhova and Yanhua Shih

Department of Physics, University of Maryland, Baltimore County, Baltimore, Maryland 21250

dmilena1@umbc.edu

Abstract: We report a proof-of-principle experimental demonstration of quantum lithography. The experimental results, obtained using the entangled nature of a two-photon state, beat the classical diffraction limit by a factor of two.

Classical optical lithography technology is facing its limit due to the diffraction effect of light. However, this classical limit can be surpassed, surprisingly, by utilizing the quantum nature of entangled multi-photon states. In an idealized experimental situation, the minimum width of the entangled N-photon diffraction pattern can be N times narrower than the width of a classical diffraction pattern. The working principle of the effect has been discussed theoretically by Boto et al [1].

We wish to report a proof-of-principle quantum lithography experiment. By using entangled photon pairs in Young's two-slit experiment, we found that *under certain experimental conditions*, the two-photon interference-diffraction pattern has spatial interference modulation period and diffraction pattern width smaller than the classical case by a factory of two.

Classically, the diffraction limit for a well collimated laser beam passing through a slit is given by $\mathrm{sinc}^2(\beta)$, where the parameter $\beta = (\pi a/\lambda)\sin\theta \simeq (\pi a/\lambda)\theta$, a is the width of the slit, and θ is the observation angle with respect to the incidence direction. Our scheme is to utilize the entangled nature of an N-particle system to surpass the diffraction limit. Consider entangled photon pairs generated in a source in such a way that the photons belonging to the same pair can only propagate "almost" horizontally, as shown in Fig. 1(a). In this case the two-photon counterpart of the classical intensity, the joint detection counting rate, is $\mathrm{sinc}^2(2\beta)$. Hence, the distribution is narrower than the classical pattern by a factor of two. If one replaces the single slit with a double-slit, the combined interference-diffraction coincidence counting rate, obtained taking into account the quantum nature of the two-photon state, is given by:

$$R_c(x) \propto \mathrm{sinc}^2[(2\pi a/\lambda)\theta)\cos^2[(2\pi d/\lambda)\theta], \tag{1}$$

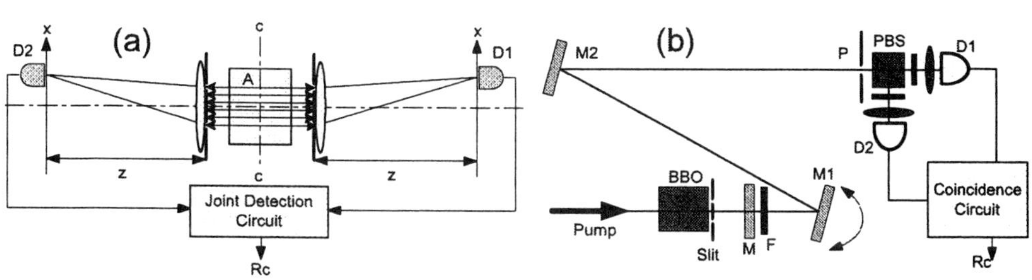

Fig. 1. ((a) Schematic of a two-photon diffraction-interference *gedankenexperiment*. The left and right sides of the picture represent signal and idler photons of an entangled pair. Detectors D1, D2 perform the coincidence measurement. (b) Schematic of the experimental setup.

where a is the width of each slit, b is the distance between the two slits and λ is the wavelength of the entangled photons. This represents the product of a standard Young's double-slit interference pattern and a standard single-slit diffraction pattern, except it has half of the classical modulation period and half of the classical pattern width, as if the wavelength of the light were reduced by one-half.

The heart of the *gedankenexperiment* is a special two-photon source: the entangled pair has to be generated in such a way as described above. The two-photon state generated via spontaneous parametric down conversion satisfies the above requirement.

The schematic setup of the experiment is illustrated in Fig. 1(b). Entangled photon pairs are produced in a $5mm$ BBO crystal, which is cut for degenerate collinear type-II phase matching to produce pair of orthogonally polarized photons(916 nm). The signal-idler beam passes through a double-slit and is reflected by two mirrors, M_1 and M_2, onto a pinhole P followed by a polarizer beam splitter PBS. The signal and idler photons are separated by PBS and are detected by the photon counting detectors D_1 and D_2, respectively. The output pulses of each detector are sent to a coincidence counting circuit. To "scan" the spatial interference-diffraction pattern, we rotate the mirror M_1 relative to the detectors. To emphasize the characteristic behavior of the entangled two-photon source, we also measured the first-order interference-diffraction pattern of a classical light with $916nm$ wavelength by the same double-slit in the same experimental setup.

One important point to be emphasized is that the double-slit must be placed *as close as possible* to the output surface of the BBO crystal, in such a way that the signal and idler photons always go through the same slit and never go through different slits. Only in this case, the observed diffraction pattern can be narrower than in the classical case by a factor of 2.

Figure 2 reports the experimental results of the distribution of coincidences (Figure 2(a)) and of the first-order interference-diffraction pattern (Figure 2(b)), versus the rotation angle θ of mirror M_1. With respect to the the classical case, the two-photon pattern has a faster spatial interference modulation and a narrower diffraction pattern width, by a factor of 2. The quantum mechanical predictions are therefore confirmed by the experimental observations.

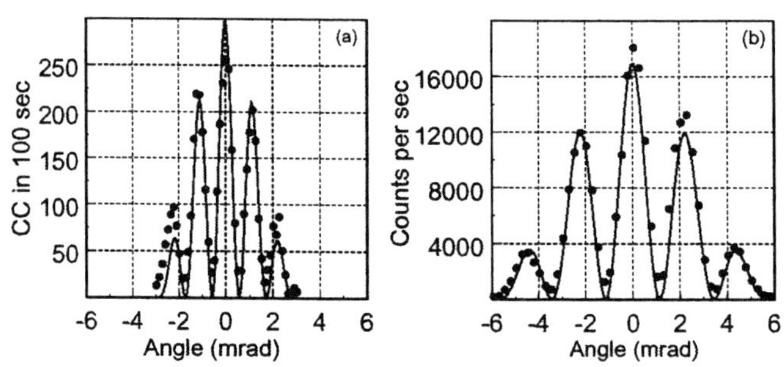

Fig. 2. Experimental measurement of the double-slit interference-diffraction pattern: (a) for the two-photon coincidences; (b) for classical light.

In conclusion, we have demonstrated the possibility of quantum lithography by using an entangled two-photon state generated from a specially designed spontaneous parametric down conversion.

References and links

1. A.N. Boto et al, Phys. Rev. Lett., **85**, 2733 (2000).

Suppressing decoherence and heating with quantum "bang-bang" controls

D. Vitali, P. Tombesi

Dip. Matematica e Fisica and Unità INFM, Università di Camerino via Madonna delle Carceri, 62032 Camerino, Italy

tel. +39 0737 402540; fax +39 0737 632525; david@camcat.unicam.it

Decoherence is the main limiting factor for quantum information processing. We can store information in qubits, which can become entangled each other, but decoherence can destroy any quantum superposition, reducing the system to a mixture of states, and the stored information is lost. For this reason decoherence control is now becoming a rapidly expanding field of investigation. In a series of previous papers [1, 2] we have faced the control of decoherence by actively modifying the system's dynamics through a feedback loop. This procedure turned out to be very effective [2], to slow down the decoherence of the linear superposition of two coherent states of a mode in a microwave cavity [3]. The main limiting aspect of this procedure is connected with the need of a measurement. In order to do the feedback in the appropriate way, one has first to perform a measurement and then the result of this measurement can be used to operate the feedback. However, any physical measurement is subject to the limitation associated with a non-unit detection efficiency.

We have shown [2] that with detection efficiency approaching unity the quantum superposition of states stored in a cavity can be protected against decoherence for many decoherence times. We wish now to face the problem of eliminating the measurement in controlling the decoherence. We show, here, how it is possible to inhibit decoherence through the application of suitable open-loop control techniques to the system of interest, that is, by using appropriately shaped time-varying control fields. To be more specific, decoherence can be inhibited by subjecting a system to a sequence of very frequent *parity kicks*, i.e., pulses designed in such a way that their effect is equivalent to the application of the parity operator on the system. As we have proved in Ref. [4], in the ideal limit of infinitely frequent pulses, this parity kicks method completely decouples the system from its environment whenever the interaction Hamiltonian anticommutes with the parity operator. Subsequent papers [5] have considered a generalization of this parity kicks method and have showed that a complete decoupling between system and environment can be obtained for a generic system if one considers an appropriate sequence of infinitely frequent kicks, realizing a symmetrization of the evolution with respect to a given group. The present parity kicks method is in fact equivalent to symmetrize the time evolution of the open system with respect to the group $\mathcal{Z}_2$, composed by the identity and the system parity operator P. Here we focus only on this case because the experimental realization of parity kicks is quite easy in many cases and moreover the applicability condition of the present method, that is the fact that the interaction with the thermal bath anticommutes with the parity operator, is satisfied by many interesting physical systems.

The continuous kicks limit is only a mathematical idealization of no practical interest. However it shows that there is in principle no limitation in the decoherence suppression one can achieve using parity kicks. Moreover it clearly shows that one has to use the most possible frequent pulses in order to achieve a significative inhibition of decoherence. Therefore the relevant question from a physical point of view is to determine at which values of the kicking period one begins to have a significant decoherence suppression. In Ref. [4] we have shown that one has a sort of phase transition at kicking rates of the order of the frequency cutoff of the thermal environment: one has a significant decoherence suppression only when the kicks becomes as fast as the characteristic timescale of the environment. In Ref. [4] the case of an harmonic oscillator coupled to a zero-temperature bath was solved in detail. In this particular case the decoherence is a consequence of dissipation and the two processes are strongly related, even in the presence of parity kicks. In this work we extend the treatment on Ref. [4] to the generic finite temperature case, in which the environment not only absorbs but also provides energy to the system of interest. In the high temperature limit, this thermal environment is essentially responsible for system's heating and we show that, when the kicking rate becomes as fast as the frequency cutoff of the bath, this quantum bang-bang control is able to significantly diminish the heating. This open-loop technique can be applied for the center-of-mass motion of a linear chain of trapped ions. In this case the free oscillation frequency is of the order of $\omega_0/2\pi \simeq 10$ Mhz and a frequent sequence of parity kicks could be obtained in principle by appropriately pulsing at about 10 Mhz the static potential applied to the end segments of the rods confining the ions along the z-axis [6]. The duration τ_0 and the intensity E_0 of the pulses have to be tailored so to have the parity kick condition $E_0\tau_0 = \pi\hbar$ and this implies having a very good control of the pulse area. Controlling the decoherence of the center-of-mass mode is crucial for the realization of quantum information processing with trapped ions. In fact, even though only the higher frequency vibrational modes will be used for quantum gate transitions involving the ions motion, suppressing decoherence of the center-of-mass motion is still important because the heating of the center-of-mass motion partially couples also with the other vibrational stretching modes [6]. The parity kicks method could be used even in the case in which the center-of-mass mode itself is used as quantum bus for the realization of quantum gates. However in this case, the use of parity kicks cannot be applied to protect against decoherence all kinds of quantum gates. In fact the parity kicks tends to average to zero any term in the Hamiltonian which is an odd function of the bosonic operators a and $a^\dagger$, and this means that the parity kicks could average to zero just the system dynamics we want to protect from decoherence, as for example, some gate operations in an ion-trap quantum computer. For example, it is easy to see that all the gate operations involving *first* red or blue sideband transitions, which implies having *linear* terms in a and $a^\dagger$ in the system Hamiltonian, as for example the C-NOT gate experimentally realized at NIST by Monroe *et al.* [7], tend to be averaged to zero by frequent parity kicks. To overcome this problem, it is however sufficient to restrict to quantum gates based on carrier transitions, which involve functions of $a^\dagger a$ only, and are not therefore affected by parity kicks, as for example the one-pulse gate proposed by Monroe *et al* [8].

References and links

1. P. Tombesi and D. Vitali, Phys. Rev. A **51**, 4913 (1995); Appl. Phys. B **60**, S69 (1995); P. Goetsch, P. Tombesi and D. Vitali, Phys. Rev. A **54**, 4519 (1996).
2. D. Vitali, P. Tombesi and G.J. Milburn, Phys. Rev. Lett. **79**, 2442 (1997); Phys. Rev. A **57**, 4930 (1998).
3. M. Brune *et al.*, Phys. Rev. Lett. **77**, 4887 (1996).
4. D. Vitali, and P. Tombesi, Phys. Rev. A **59**, 4178 (1999).
5. L. Viola, E. Knill and S. Lloyd, Phys. Rev. Lett. **82**, 2417 (1999); L. Viola, S. Lloyd, and E. Knill, Phys. Rev. Lett. **83**, 4888 (1999); P. Zanardi, Phys. Lett. A **258**, 77 (1999).
6. D.J. Wineland *et al.*, J. Res. Natl. Inst. Stand. Technol. **103**, 259 (1998).
7. C. Monroe *et al.*, Phys. Rev. Lett. **75**, 4714 (1995).
8. C. Monroe *et al.*, Phys. Rev. A. **55**, R2489 (1997).

Entanglement disentangled by spacetime vortices

J. E. Carroll

Engineering Department, University of Cambridge, Cambridge CB2 1PZ, UK
Fax 44–1223–332616 e-mail jec@eng.cam.ac.uk

1. Photons as spacetime vortices

A new view of Maxwell's equations as vorticity in 3d-space balancing vorticity in 3d-time sheds light on quantization and entanglement [1]. A '6d' spacetime algebra [2, 3] explores

$$\mathrm{curl}_{space}\,\underline{\mathbf{E}} = [\mathrm{curl}_{time}\,\underline{\mathbf{B}}^{tr}]^{tr}; \quad \mathrm{curl}_{space}\,\underline{\mathbf{B}} = [\mathrm{curl}_{time}\,\underline{\mathbf{E}}^{tr}]^{tr}; \qquad (c = 1 \text{ units}) \tag{1}$$

Components of $\underline{\mathbf{E}} = \{E_{pk}\}$ (p = 1, 2, 3; $k = x, y, z$) form spatial 'vectors' $\mathbf{E}_p = \{E_{pk},\}$ associated with temporal axes Ot_p and *temporal* vectors $\mathbf{E}_k = \{E_{pk}\}^{tr}$ associated with spatial axes Ok; similarly for $\underline{\mathbf{B}}$. The temporal equivalent of curl_{space} is curl_{time}. All fields satisfy a '6d' wave equation. With a preferred temporal axis $Ot_3 = Ot$, the Klein-Gordon equation implies that $(\partial t_1^{\ 2} + \partial t_2^{\ 2}) = m_o^{\ 2}$. Zero rest mass m_o means $\partial t_1 = \partial t_2 = 0$ with $\mathbf{E}_3 = 0 = \mathbf{B}_3$. The real spatial fields associated with the directions of 'transverse' time can then be combined using i to give:

$$\mathbf{E}_{classic} = \mathbf{E}_1 + i\,\mathbf{E}_2\,; \qquad \mathbf{B}_{classic} = -i\,(\mathbf{B}_1 + i\,\mathbf{B}_2)\,; \tag{2}$$

$\mathbf{E}_{classic}$ ($\mathbf{B}_{classic}$) is a complex spatial vector (pseudo-vector) recovering Maxwell's equations:

$$\mathrm{curl}(\mathbf{E}_{classic}) = -\,\partial_t(\mathbf{B}_{classic})\,; \qquad \mathrm{curl}(\mathbf{B}_{classic}) = \partial_t(\mathbf{E}_{classic})\,. \tag{3}$$

Source 'fields' with mass [3] are not discussed here. Methods of ref. 4, give normal modes for (2) and (3) (given a k-vector $\mathbf{k}$) as complex *analytic* vector functions of t (= t_3) :

$$\alpha(\mathbf{k},t) = \mathbf{a}(\mathbf{k})\exp[i(\mathbf{k}.\mathrm{r} - kt)]\,; \quad \beta(\mathbf{k},t) = \mathbf{b}(\mathbf{k})\exp[i(\mathbf{k}.\mathrm{r} + kt)]\,; \tag{4}$$

where $k > 0$; normally $|\alpha(\mathbf{k}, t)| \neq |\beta(\mathbf{k}, t)|$; i implies rotation through 90° in transverse time.

Poynting's theorem gives an energy density U and energy transfer vector P invariant to orientation of α and β in transverse time (averaging over volume V shown as < >) from :-

$$\mathrm{U}(\mathbf{k}) = \langle \alpha(\mathbf{k}, t)^*.\,\alpha(\mathbf{k}, t) + \beta(-\mathbf{k}, t)^*.\,\beta(-\mathbf{k}, t) \rangle \tag{5}$$

$$\mathrm{P}(\mathbf{k}) = \langle \alpha(\mathbf{k}, t)^*.\,\alpha(\mathbf{k}, t) - \beta(-\mathbf{k}, t)^*.\,\beta(-\mathbf{k}, t) \rangle \tag{6}$$

Symmetry requires U(− **k**) = U(**k**). With a 'single' k–vector, the direction of energy flow is undetermined. Energy can be said to circulate between ± **k** with frequencies ± k. This is the 'STV' concept: a *non–local* circulation of energy. Vortex chirality determines the spatial polarization through the Stokes parameters for normal mode fields. Polarization is a non-local property of the whole vortex and it is inappropriate to say that polarization propagates.

2. Promotion/Demotion

Modulation, of 'α' modes, by $\mathbf{a}^\dagger = A\exp[i(\kappa.\mathrm{r} - \kappa t)]$ promotes $\alpha(\mathbf{k}_0)$ to $\mathbf{a}^\dagger\,\alpha(\mathbf{k}_0) = \alpha(\mathbf{k}_0+\kappa)$: $\kappa \| \mathbf{k}_0$; $(\mathrm{k}_0+\kappa) > 0$; t variation implicit. Inverse operator **a** demotes α. Annihilation by demotion is a consequence of analytic function theory forcing $\mathbf{a}\alpha = 0$ if $(\mathrm{k}_0 - \kappa) < 0$. STVs promoted from a ground state have frequencies $(k_0 + \mathrm{N}\kappa) > 0$ (integer N). Modes $\alpha(\mathbf{k}_{N+1})$ and $\beta(-\mathbf{k}_N)$ move forward in space as time increases but have positive and negative frequencies respectively. With a single **k**, spacetime symmetry makes it impossible to tell the spatial direction of energy flow with time. This violation of causality is removed by using wave packets with adjacent k–vectors: $\mathbf{k}_{N+1}$ and $\mathbf{k}_N$. Equations (5) and (6) are then rewritten with $\alpha \to \alpha(\mathbf{k}_{N+1})$ [$\beta \to \beta(-\mathbf{k}_N)$] for forward waves; $\alpha \to \alpha(-\mathbf{k}_{N-1})$ [$\beta \to \beta(\mathbf{k}_N)$] for reverse waves.

Coherence and Quantum Optics VIII
Edited by Bigelow et al., Kluwer Academic/Plenum Publishers, 2003

3. Quantization without Quantum Theory

Replace α and β as above; use k-symmetry; require zero energy transfer in reverse direction:

$$< \alpha(\mathbf{k}_{N-1})^*.\alpha_+(\mathbf{k}_{N-1}) > - < \beta(-\mathbf{k}_N)^*.\beta(-\mathbf{k}_N) > = 0 \qquad (7)$$

The co-operative net forward energy may be re-written in terms of positive frequencies as:

$$U_N = < \alpha(\mathbf{k}_{N+1})^*.\alpha(\mathbf{k}_{N+1}) + \beta(-\mathbf{k}_N)^*.\beta(-\mathbf{k}_N) > \; = < \alpha(\mathbf{k}_N)^*[\mathbf{a}\,\mathbf{a}^\dagger + \mathbf{a}^\dagger\,\mathbf{a}\,]\alpha(\mathbf{k}_N) > \qquad (8)$$

The co-operative forward energy transfer can also be determined from positive frequencies:

$$P_N = U_0 = < \alpha(\mathbf{k}_N)^*[\mathbf{a}\,\mathbf{a}^\dagger - \mathbf{a}^\dagger\,\mathbf{a}\,]\alpha(\mathbf{k}_N) > \qquad (9)$$

Given U_0 as the unit of unidirectional energy transfer associated with two adjacent k–vectors, unidirectionality forces 'quantum' formalisms without initially postulating quantum theory. Energy propagates with the group velocity recovering causality. Operator algebra with field normalization gives $U_N = (N+½)U_0$; $k_0 = ½\,\kappa$ and recovers standard quantised e.m. theory. Further work shows how STVs behave statistically like photons [1].

4. Application to entanglement

Understanding entanglement is vital for future quantum information processing [5]. Consider entangled photons 'L' and 'R' propagating left and right, exciting a ground state (Fig.1) into a coherent STV web. '3+3 spacetime' gives enough degrees of freedom to Stokes parameters so that the polarizations for 'L' and 'R' need not be specified *individually* at the launch, but vortex chiralities correlate 'L' and 'R' polarizations. Localized creation/detection of vortex energy in 'L' and 'R' requires interfering k–vectors (eg $\mathbf{k}_n$, $\mathbf{k}_{n+1}$) determining direction of the *group* velocity and consistency with 'causality'. Removal of energy in, say, 'R' by a detector completes the measurement of 'R' with the polarization now specified. There is an *immediate* reorganization, without energy change, of the interconnecting web of STVs leaving the polarization of 'L' correlated [6] with that of 'R', consistent with quantum theory.

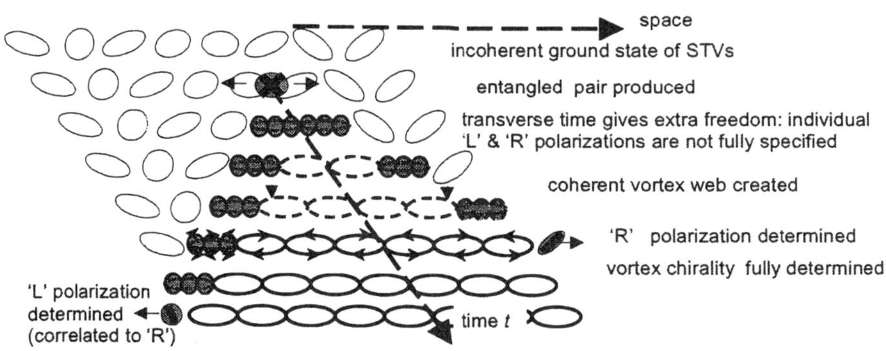

Fig.1. Disentangling Entanglement (schematic)

5. References

1 J.E.Carroll, "Spacetime Vortices," http://www2.eng.cam.ac.uk/~jec/spacetimevortices.pdf (2000).

2 D Hestanes, *Space-Time Algebra* (Gordon and Breach 1966).

3 J.E.Carroll, "Spacetime Vortices: A Geometric Algebra Approach," http://www2.eng.cam.ac.uk/~jec/spacetimevortices2.pdf (2001).

4 C. Cohen-Tannoudji, J. Dupont-Roc, and G. Grynberg *Photons and Atoms* (Wiley 1989).

5 D.Bouwmeester, A.Ekert and A.Zeilinger, *The Physics of Quantum Information*. (Springer 2000).

6 N.D.Mermin, "What is quantum mechanics trying to tell us," Am.J.Phys. **66,** 753-767 (1998).

Atomic inversion for an atom coupled to a squeezed vacuum in a coherently driven cavity

Daniel Erenso, Reeta Vyas, and Surendra Singh

Physics Department, University of Arkansas, Fayetteville, AR 72701

rvyas@comp.uark.edu

With the generation of squeezed light in many laboratories around the world, quantum optical systems interacting with squeezed light have received considerable attention. One of the simplest system among these is a single two level atom interacting with a squeezed vacuum. Studies of a single two-level atom in free space driven by coherent light and decaying into a squeezed vacuum have indicated that the fluorescent spectrum shows several modifications over the spectrum of normal resonance fluorescent [1, 2]. Here, we have studied the atomic inversion for a single two-level atom coupled to a broad band squeezed vacuum centered about the atomic resonance frequency. The atom is placed inside a coherently driven cavity. In particular we investigate the effects of atomic and cavity detunings on inversion. The time evolution for the density operator of this system is governed by

$$\begin{aligned}\frac{\partial \hat{\rho}}{\partial t} &= -i\Delta_c[\hat{a}^\dagger \hat{a}, \hat{\rho}] - i\Delta_a[\hat{\sigma}_z, \hat{\rho}] + g[\hat{a}^\dagger\hat{\sigma}_- - \hat{a}\hat{\sigma}_+, \hat{\rho}] + \epsilon[\hat{a}^\dagger - \hat{a}, \hat{\rho}] \\ &\quad + \frac{\gamma}{2}(N+1)(2\hat{\sigma}_-\hat{\rho}\hat{\sigma}_+ - \hat{\sigma}_+\hat{\sigma}_-\hat{\rho} - \hat{\rho}\hat{\sigma}_+\hat{\sigma}_-) + \frac{\gamma}{2}N(2\hat{\sigma}_+\hat{\rho}\hat{\sigma}_- - \hat{\sigma}_-\hat{\sigma}_+\hat{\rho} - \hat{\rho}\hat{\sigma}_-\hat{\sigma}_+) \\ &\quad - \gamma(M\hat{\sigma}_+\hat{\rho}\hat{\sigma}_+ + M^*\hat{\sigma}_-\hat{\rho}\hat{\sigma}_-) + \kappa(2\hat{a}\hat{\rho}\hat{a}^\dagger - \hat{a}^\dagger\hat{a}\hat{\rho} - \hat{\rho}\hat{a}^\dagger\hat{a}) \qquad (1)\end{aligned}$$

where $\Delta_c = \omega_c - \omega_0$ and $\Delta_a = \omega_a - \omega_0$ are cavity and atomic detunings with respect to the driving field, γ is the atomic decay rate for spontaneous emission into an ordinary vacuum, 2κ is the rate at which the cavity is losing photons, ϵ is the driving field amplitude, g is atom field coupling constant, and N and M are parameters that describe the squeezed reservoir coupled to the atom. They are related to the squeezing parameter r by $N = \sinh^2(r)$ and $|M| = \sinh(r)\cosh(r)$.

We obtain a generalized second-order Fokker-Planck equation for the density matrix in Eq. (1) without using system size expansion or any truncation by following an approach due to Wang and Vyas [3]. By considering the bad-cavity limit we eliminate the field variables adiabatically and obtain a Fokker-Planck equation for the atomic variables only. From this equation we obtain the steady-state expectation value of the atomic variables analytically.

In Fig. 1 the steady state atomic inversion is shown as a function of atomic detuning parameter δ_a $(= 2\Delta_a/\gamma)$ when the squeezed quadrature is out of phase $(\theta = \pi)$ with the coherent excitation. For $r = 0$ we see a single peak at zero atomic detuning. For large detunings the inversion approaches -0.5 indicating that atom is most likely to be in the ground state. This is in agreement with the result for an the atom decaying to ordinary vacuum [3]. Far away from resonance, the overall effect of increasing the squeezing parameter r is to increase the atomic inversion. This is because, even when the driving field is detuned from the atom, the atom can still sense the broad band squeezed vacuum. Consequently, as r increases the number of quanta in the squeezed vacuum increases, resulting in an increase in atomic inversion.

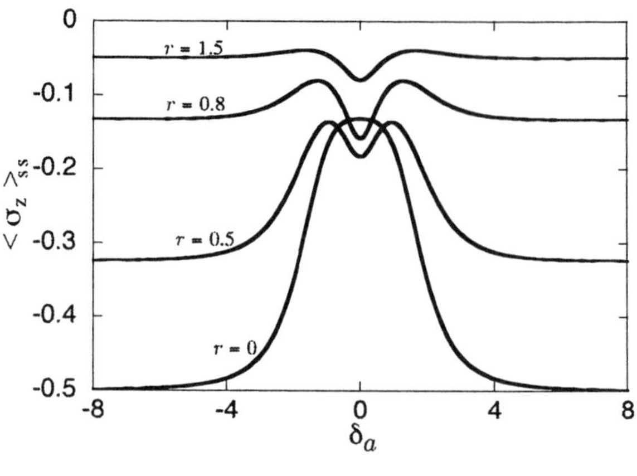

Fig. 1. Atomic inversion $\langle\sigma_z\rangle_{ss}$ as a function of the atomic detuning parameter δ_a for several different values of squeezing parameter r. Other parameters are: pump parameter $Y = 2\sqrt{2}g\epsilon/\kappa\gamma = 3$, atomic cooperativity $C = g^2/\gamma\kappa = 0.4$, cavity detuning parameter $\delta_c = \Delta_c/\kappa = \delta_a$, and $\theta = \pi$.

Atomic inversion shows interesting features near resonance. The on-resonance atomic inversion can fall below its value for the ordinary vacuum ($r = 0$). The on-resonance inversion initially decreases as r increases. With further increase in r, it rises above its value for the ordinary vacuum. Another interesting feature is that for small value of r, atomic inversion shows two peaks located symmetrically at nonzero atomic detunings. This behavior near resonance is similar to vacuum Rabi splitting. The presence of squeezed light effectively reduces the threshold value of C for seeing the vacuum Rabi splitting. With increase in r the two peaks broaden and eventually disappear. As a function of detuning the inversion shows a dip for large value of r. The peaks seen in atomic inversion in figure 1 are phase sensitive as they are not seen when the squeezed quadrature is in phase ($\theta = 0$) with the coherent excitation. These peaks are also absent when the atom decays into a non-zero temperature thermal reservoir.

This work was supported in part by the Office of Naval Research. D. Erenso acknowledges support from the ICSC World Laboratory.

References and links

1. H. J. Carmicheal, A.S. Lane, and D. F. Walls, "Resonance fluorescence from an atom in a Squeezed Vacuum," Phys. Rev. Lett. **24**, 2539 (1987); "Resonance fluorescence in a Squeezed Vacuum," J. Mod. Opt. **34**, 821 (1987).
2. Perry R. Rice and Leno M. Pedrotti, "Fluorescent spectrum of a single atom in a cavity with injected squeezed vacuum," J. Opt. Soc. Am. B 9, 2008 (1992).
3. C.Wang and R.Vyas, "Fokker-Planck equation for a single two-level atom: Application in the bad-cavity limit," Phys. Rev. A **51**, 2516 (1995); "Cavity modified Maxwell-Bloch equations for the vacuum Rabi splitting," Phys. Rev. A **55**, 823 (1997).

QUANTIZATION OF THE FIELD IN DIELECTRICS

Michael E. Crenshaw and Charles M. Bowden

AMSAM-RD-WS-ST, U. S. Army Aviation and Missile Command
Redstone Arsenal, AL 35898

Since the 1940s, the macroscopic Maxwell equations of continuum electrodynamics have been employed as the basis for quantization of the electromagnetic field in a dielectric. [1,2] Equations of motion for two-level atoms embedded in a dielectric, as well as the spontaneous emission rate, can be derived from the macroscopic quantization. Recently, we derived equations of motion for embedded atoms using the microscopic Dirac quantization of the field in the vacuum with discrete oscillators. [3] There are substantial qualitative differences in the results of the two methods that are due to the difference between the continuum and atomistic models of electrodynamics in matter.

In this proceedings, we outline the derivation of equations of motion for two-level atoms embedded in a dielectric utilizing *i*) macroscopic quantization of the Maxwell equations in dielectric media [2] and *ii*) microscopic quantization of the Maxwell equations in vacuum with enumerated discrete oscillators. [3] We invoke the correspondence principle to discriminate between the qualitatively different results and prove that the macroscopic quantization of the field in a dielectric is forbidden in this application. [4]

We begin by developing equations of motion for two-level atoms embedded in a dielectric using the two different Hamiltonians for a dielectric medium. One Hamiltonian was developed using the macroscopic quantization procedure in which the medium is lumped with the vacuum electromagnetic field. The second Hamiltonian was developed using the microscopic quantization procedure in which the medium is treated as quantum harmonic oscillators embedded in the vacuum. Starting, fundamentally, from the Hamiltonians, we derive Langevin–Bloch equations of motion for the two-level atoms from the Heisenberg equations of motion by adiabatically eliminating the field operator variables and evaluating the resulting integrals and sums in the Markov approximation.

Using the macroscopic quantization of a dispersionless and absorptionless linear dielectric, the Hamiltonian for the medium with embedded two-level atoms is

$$H = \hbar \sum_{l,\lambda} \omega_l a_l^\dagger a_l + \frac{\hbar\omega_a}{2}\sigma_3^j - \frac{i\hbar}{n}\sum_j \sum_{l,\lambda}\left(g_l^j a_l \sigma_+^j e^{i\mathbf{k}_l\cdot\mathbf{r}_j} - g_l^{j*} a_l^\dagger \sigma_-^j e^{-i\mathbf{k}_l\cdot\mathbf{r}_j}\right). \tag{1}$$

Here, a_l and $a_l^\dagger$ are the field mode operators, ω_l is the frequency of the field in mode l, σ_3^j and $\sigma_\pm^j$ are atomic operators for the jth atom, g_l^j is the coupling between an atom and the field, ω_a is the atomic resonance frequency, and n is the index of refraction of the host dielectric. Writing Heisenberg equations of motion and adiabatically eliminating the field mode operators, we obtain equations of motion for the atoms: [4]

$$\frac{d\sigma_-^j}{dt} = i\Delta_a\sigma_-^j + \frac{\mu_a}{\hbar}\sigma_3^j\left(n^2 f^+ + \frac{n\mathcal{E}}{2}\right) - in\epsilon_a\sigma_3^j\bar{\sigma}_- - \frac{n\gamma_a}{2}\sigma_-^j \tag{2}$$

$$\frac{d\sigma_3^j}{dt} = -2\left[\frac{\mu_a}{\hbar}\sigma_+^j\left(n^2 f^+ + \frac{n\mathcal{E}}{2}\right) - in\epsilon_a\sigma_+^j\bar{\sigma}_- + \frac{n\gamma_a}{4}(\sigma_3^j + 1) + H.c.\right]. \tag{3}$$

Here, $\Delta_a = \omega - \omega_a$ is the detuning of the field from resonance, f^+ is a Langevin force operator, $\gamma_a = 4\omega_a^3|\mu_a|^2/(3c^3\hbar)$ is the fundamental population decay rate for the atoms,

μ_a is the dipole moment, $\bar{\sigma}_{-}$ is a spatially averaged operator, $\epsilon_a = 4\pi N_a \mu_a^2/(3\hbar)$ is the near-dipole–dipole interaction strength, and N_a is the number density of atoms. We have taken a near-resonant field to be in a coherent state and $\mathcal{E}$ is the field amplitude.

Next, we develop equations of motion for the atoms using the microscopic quantization of the dielectric. The Hamiltonian for the medium with embedded atoms is

$$H = \hbar \sum_{l,\lambda} \omega_l a_l^\dagger a_l + \sum_n \hbar\omega_b b^{\dagger n} b^n - i\hbar \sum_n \sum_{l,\lambda} \left(h_l^n a_l b^{\dagger n} e^{i\mathbf{k}_l \cdot \mathbf{r}_n} - h_l^{n*} a_l^\dagger b^n e^{-i\mathbf{k}_l \cdot \mathbf{r}_n} \right)$$

$$+ \frac{\hbar\omega_a}{2}\sigma_3^j - i\hbar \sum_j \sum_{l,\lambda} \left(g_l^j a_l \sigma_+^j e^{i\mathbf{k}_l \cdot \mathbf{r}_j} - g_l^{j*} a_l^\dagger \sigma_-^j e^{-i\mathbf{k}_l \cdot \mathbf{r}_j} \right) \tag{4}$$

Here, b^n and $b^{\dagger n}$ are the operators for the nth oscillator at $\mathbf{r}_n$, h_l^n is the coupling between the oscillator at $\vec{r}_n$ and the quantized radiation field, and ω_b is the oscillator resonance frequency. Writing Heisenberg equations of motion, adiabatically eliminating the field operator variables and the harmonic oscillator variables, we obtain [3]

$$\frac{d\sigma_-^j}{dt} = i\Delta_a \sigma_-^j + \frac{\mu_a}{\hbar}\sigma_3^j \left(\ell f^+ + \frac{\ell\mathcal{E}}{2} \right) - i\ell\epsilon_a \sigma_3^j \bar{\sigma}_- - \frac{\ell\gamma_a}{2}\sigma_-^j \tag{5}$$

$$\frac{d\sigma_3^j}{dt} = -2\left[\frac{\mu_a}{\hbar}\sigma_+^j \left(\ell f^+ + \frac{\ell\mathcal{E}}{2} \right) - i\ell\epsilon_a \sigma_+^j \bar{\sigma}_- + \frac{\ell\gamma_a}{4}(\sigma_3^j + 1) + H.c. \right]. \tag{6}$$

The dielectric local-field enhancement factor ℓ is equivalent to the usual phenomenological local-field enhancement factor $\ell = (n^2+3)/3$, where n is the complex index of refraction in the Drude–Lorentz model.

Classical harmonic oscillators embedded in a dielectric obey an oscillator equation

$$\frac{dx_j^2}{dt^2} + \omega_a^2 x_j = \frac{e}{m_a} E_L. \tag{7}$$

For near-resonance conditions, time varying quantities are written conveniently in terms of envelope functions. Let $E_L = (\mathcal{E}_L e^{-i\omega t} + c.c)/2$ and $x_j = x_0(\rho_j e^{-i\omega t} + c.c.)$, where x_0 is a characteristic amplitude. The classical local field is $\mathcal{E}_L = \mathcal{E} + (4\pi/3)\mathcal{P} = \mathcal{E} + (4\pi/3)N_a e\bar{\rho} + (4\pi/3)\alpha_b N_b \mathcal{E}_L$, where N_a is the number density of harmonic oscillators, N_b is the number density of linearly polarizable particles, and α_b is the linear polarizability. Applying the Clausius–Mossotti–Lorentz–Lorenz relation, Eq. (7) becomes

$$\frac{d\rho_j}{dt} = i\Delta_a \rho_j + \frac{ie}{2x_0\omega m_a}\frac{\ell\mathcal{E}}{2} + i\ell\frac{4\pi}{3}\frac{N_a e^2}{2\omega m_a}\bar{\rho} \tag{8}$$

in the slowly varying envelope and rotating-wave approximations.

A valid quantum theory must be consistent with classical physics in the limit of large quantum numbers. Taking the expectation value of Eqs. (2a) and (4a) in the linear limit $\sigma_3 \to -1$, it is clear that the application of the macroscopic field quantization to the dynamics of dopants is forbidden by the correspondence principle of quantum mechanics while the microscopic quantization is consistent with the correspondence principle. [4]

[1] V. L. Ginzburg, *J. Phys. USSR* **2,** 441 (1940); J. M. Jauch and K. M. Watson, *Phys. Rev.* **74,** 950 (1948).

[2] R. H. Pantell and H. E. Puthoff *Fundamentals of Quantum Electronics* (Wiley, NY, 1969); A. Yariv, *Quantum Electronics*, 2nd. ed. (Wiley, New York, 1975).

[3] M. E. Crenshaw and C. M. Bowden, *Phys. Rev. Lett.* **85,** 1851 (2000).

[4] M. E. Crenshaw and C. M. Bowden, submitted to *Phys. Rev. A.*

Quantum optical effects across a metallic film

C R Bennett and M. Babiker
Department of Physics, University of York, Heslington, York YO10 5DD, England
crb5@york.ac.uk

J. B. Kirk
Department of Physics, University of Essex, Colchester, Essex CO4 3SQ, England

Abstract: The electromagnetic fields localised on the plane of a thin metallic film provide a decay channel for electric dipoles positioned in the vicinity of the film. We point out screening effects, unusual pair cooperative effects and explore variations with metallic density and dipole orientation.

OCIS codes: (240.0310) Thin films; (270.6630) Superradiance, superfluorescence

There is currently considerable interest in the quantum optics near thin film structures. This is fuelled partly by recent advances in the production of films, which can be as thin as one monolayer. Metallic films, in particular, have the advantage of providing enhancement effects as, for example, in evanescent mode atomic mirrors [1]. Remarkably, however, the decay characteristics for electric dipoles in the vicinity of a metallic film have not been explored before, nor the correlation effects between pairs of electric dipoles located on both sides or opposite sides of the film. Pair cooperative phenomena are of interest currently, especially in the context of quantum logic. We outline the theory involving a single metallic film and describe briefly the unusual effects associated with the spontaneous emission in such an environment.

The system in question is an infinitesimally thin film characterised by its areal electron density n_s occupying the x-y plane and so separates two identical dielectric half spaces of dielectric constant ε. The electromagnetic fields which are localised on the metallic film and which decay with distance on both sides of the film can be easily shown to follow a dispersion relation of the form $\omega^2 = k_z\left(n_s e^2/2m^*\varepsilon_0\varepsilon\right)$ where $k_z^2 = k_{\|}^2 - \varepsilon\omega^2/c^2$ with $k_{\|}$ the in-plane wavevector. The decay rate of an excited electric dipole of frequency ω_0 located at a distance z from the film follows using Green functions. The results for a dipole moment vector $\boldsymbol{\mu}$ oriented parallel ($\Gamma_{\|}$) or perpendicular (Γ_z) to the film plane are given by

$$\Gamma_{\|}(z) = \frac{\mu^2 k_z^3}{4\varepsilon_0\varepsilon\hbar} e^{-2k_z|z|}, \qquad \Gamma_z(z) = \frac{\mu^2 k_z k_{\|}^2}{2\varepsilon_0\varepsilon\hbar} e^{-2k_z|z|}, \tag{1}$$

with $\omega = \omega_0$. Figure 1 exhibits the variations of $\Gamma_{\|}(z)$ with dipole frequency ω_0 and sheet areal density n_s. In all cases $\varepsilon = 1$ and $\Gamma_z(z)$ is approximately twice the results of $\Gamma_{\|}(z)$. The important points to note here are: (i) clear enhancement relative to free space; (ii) that two dipoles at the same distance from the same film, but oscillating at different frequencies can have the same decay rate; and (iii) two dipoles of the same frequency and at equal distances from two different metallic films can decay at the same rate. These aspects can only be explained in terms of screening effects due to the conductivity of the metallic film.

The simplest Dicke-type cooperative effects realisable near the film are those for a dipole-pair. There are two types of two-dipole decay rates, one corresponding to the symmetric state (+) and the second to the antisymmetric state (-) and each depends on the dipole orientations, with the dipoles positioned at different points on the vertical axis. We quote here only the cases where both dipoles are parallel or both perpendicular to the film plane. We have

$$\Gamma_{xx}^{\pm}(z_1,z_2) = \Gamma_{yy}^{\pm}(z_1,z_2) = \left[e^{-2k_z|z_1|} + e^{-2k_z|z_2|} \pm 2e^{-k_z(|z_1|+|z_2|)}\right]\mu^2 k_z^3/8\varepsilon_0\varepsilon\hbar, \tag{2}$$

$$\Gamma_{zz}^{\pm}(z_1,z_2)=\left[e^{-2k_z|z_1|}+e^{-2k_z|z_2|}\pm 2\,\mathrm{sgn}(z_1)\mathrm{sgn}(z_2)e^{-k_z(|z_1|+|z_2|)}\right]\mu^2 k_z k_{\|}^2/4\varepsilon_0\varepsilon\hbar. \quad (3)$$

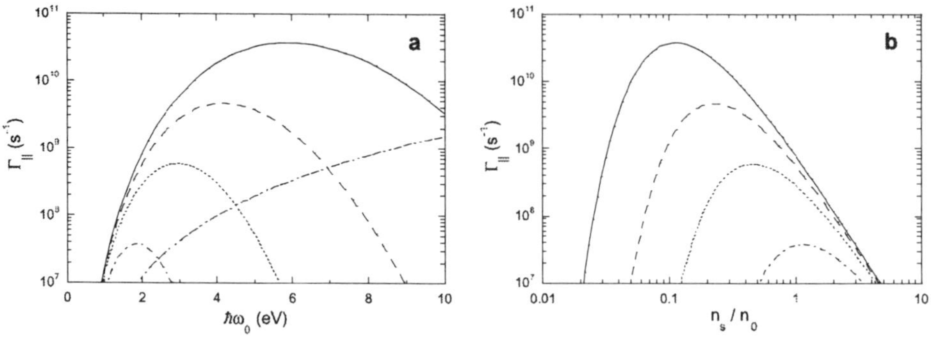

Fig. 1: The variations of the spontaneous emission with (a) the dipole frequency ω_0 where $n_s = n_0 = 1.7\times10^{20}\mathrm{m}^{-2}$ and (b) the areal density of the layer n_s where $\hbar\omega_0 = 2.0\mathrm{eV}$ for a dipole parallel to the layer and $z = 50$Å (full curve), $z = 100$Å (dashed curve), $z = 200$Å (dotted curve) and $z = 500$Å (dot-dash curve). The dot-dot-dash-dash curve is the free-space rate.

Figure 2(a) shows variations of $\Gamma_{xx}^{\pm}(z_1,-200\text{Å})$ with z_1 the position of dipole 1. The dotted curve corresponds to $\Gamma_{xx}^{-}(z_1,-200\text{Å})$ and the dashed curve shows that of the symmetric state $\Gamma_{xx}^{+}(z_1,-200\text{Å})$. The full curve is $\Gamma_{\|}(z_1)$ for a single dipole in the absence of dipole 2. The important points to note here are the superradiance and subradiance effects at image positions $z_1 = -z_2$, as well as close proximity $z_1 = z_2$. Figure 2(b) shows the variations of $\Gamma_{zz}^{+}(z_1,-200\text{Å})$ with z_1. The full curve in Fig. 2(b) shows the variations of the perpendicular decay rate $\Gamma_z(z_1)$ of a single dipole 1 (in the absence of dipole 2). The result for the antisymmetric state $\Gamma_{zz}^{-}(z_1,-200\text{Å})$ is the mirror image of $\Gamma_{zz}^{+}(z_1,-200\text{Å})$ in Fig. 2(b). The important points to note here are, first the marked change in behaviour compared with the case in Fig.2(a). Although superradiance occurs at close proximity $z_1 = z_2$, the symmetric state unexpectedly shows subradiance behaviour at the image position $z_1 = -z_2$. A second feature is the discontinuity at the film position $z = 0$ which can be traced to the discontinuity of the normal component of the electric field. Further details will be reported elsewhere.

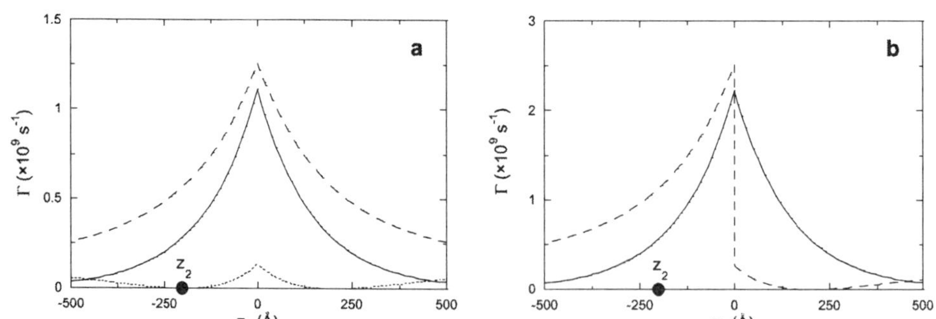

Fig. 2: The cooperative decay rate of two dipoles (a) parallel and (b) perpendicular to the layer with dipole 2 fixed at $z_2 = -200$Å and $\hbar\omega_0 = 2.0\mathrm{eV}$, $n_s = 1.7\times10^{20}\mathrm{m}^{-2}$. The solid curve shows the variations of the single dipole decay rate, with changing position of dipole 1 without dipole 2 present, while the dashed (dotted) curve is shows the results for the symmetric (antisymmetric) initial state.

1. C R Bennett, J B Kirk and M Babiker, "Theory of evanescent mode atomic mirrors with a metallic layer", Phys. Rev. A **63**, 033405 (2001).

Coupled interferometer scheme for generating maximally entangled photonic states

Christopher C. Gerry
Department of Physics and Astronomy, Lehman College, CUNY, Bronx, NY 10468-1589

Richard A. Campos
4 Perry Street, Cortlandt Manor, NY 10567

Abstract: We present a scheme for the generation of maximally entangled photonic states by the use of interferometers coupled through a Kerr medium. The device operates as a conditional beam splitter where one of its components is an optical Fredkin gate.

Currently, there is considerable interest in maximally entangled photonic states (MES) of a two-mode quantized field. Denoting the two modes by the letters a and b, the MES for N photons in one mode and the vacuum in the other have the form

$$\frac{1}{\sqrt{2}}\left(|N\rangle_a|0\rangle_b + e^{i\xi}|0\rangle_a|N\rangle_b\right). \tag{1}$$

These states potentially have at least two applications: 1) in high resolution interferometry where it is possible to obtain sensitivity at the Heisenberg limit [1] and 2) in beating the diffraction limit in photo-lithography [2]. But generating such states is a nontrivial task. For example, ordinary beam splitters cannot generate states of the form of Eq.(1) for $N > 2$. Here we describe what amounts to a conditional beam splitter (CBS) consisting of two coupled Mach-Zehnder interferometers (MZIs) coupled through a Kerr medium realizing a cross-Kerr interaction. A sketch of the device is given in Fig. 1 below.

Referring to the figure, the lower interferometer, with just the c mode of the upper MZI passing through Kerr medium, can be shown to act as a optical realization of an quantum optical Fredkin gate [3]. The cross-Kerr interaction has the form $\propto a^\dagger a b^\dagger b$. A Fredkin gate effects a controlled swap. In the present context it works as follows: Suppose mode a contains N photons initially while mode b is in the vacuum state. If mode c is also in the vacuum state then output state of the a-b modes is the same as the input. But if the c mode contains a single photon, the output states are reversed; there is a vacuum in the mode a ouput and N photons in the b. It can be shown [4] that the unitary operator for the Fredkin gate is of the form

$$U_F = \exp\left(i\chi c^\dagger c J_0\right)\exp\left(i\chi c^\dagger c J_2\right) \tag{2}$$

where $J_0 = \left(a^\dagger a + b^\dagger b\right)/2$ and $J_2 = \left(a^\dagger b - a b^\dagger\right)/2i$ are elements of the Schwinger realization of the angular momentum algebra, well know to be of great facility in describing beam splitters and interferometers [5]. The first term merely contributes a phase while the second is a conditional "rotation". For the above operator to truly describe a Fredkin gate we must have the condition $\chi = \pi$, where χ is a parameter proportional to the Kerr nonlinear susceptibility. Henceforth we assume this condition holds.

The purpose of embedding the lower MZI into the upper is to create superpositions involving four modes which, upon state reduction on the two output modes of the upper MZI, generate states of the form of Eq.(1). If the c mode initially contains a single photon state obtained, say, from a down-conversion process, and mode d is in the vacuum, then the output of the coupled interferometer device will be

$$\frac{1}{2}\left\{\left[|N\rangle_a|0\rangle_b + e^{i\xi_N}|0\rangle_a|N\rangle_b\right]|1\rangle_c|0\rangle_d + \left[|N\rangle_a|0\rangle_b - e^{i\xi_N}|0\rangle_a|N\rangle_b\right]|0\rangle_c|1\rangle_d\right\} \tag{4}$$

where $\xi_N = N\pi/2$. Evidently, state reduction on the c and d modes (detecting the presence or not of photons in these modes) reduces the output of the a-b modes to the MES

$$\frac{1}{\sqrt{2}}\left(|N\rangle_a|0\rangle_b \pm e^{i\xi_N}|0\rangle_a|N\rangle_b\right). \tag{5}$$

The signs $\pm$ are correlated with the detection of the states $|1\rangle_c|0\rangle_d$ or $|0\rangle_c|1\rangle_d$ respectively.

If the input of the a mode is initially a coherent state $|\alpha\rangle_a$, then it can be shown that the outputs of the device will be the entangled coherent states

$$\frac{1}{\sqrt{2}}\left(1+e^{-|\alpha|^2}\right)^{-1/2}\left(|\alpha\rangle_a|0\rangle_b \pm |0\rangle_a|i\alpha\rangle_b\right). \tag{6}$$

Details of the proposal, including the application to interferometry and lithography, will be published elsewhere.

References

[1]. J.J Bollinger, W. M. Itano, D. J. Wineland, and D. J. Heinzen, "Optimal frequency measurements with maximally correlated states", Phys. Rev. A **54**, R4649-R4652 (1996).

[2]. A. N. Boto, P. Kok, D. S. Abrams, S. Braunstein, C. P. Williams, and J. P. Dowling., "Quantum interferometric optical lithography: exploiting entanglement tp beat the diffraction limit" , Phys. Rev. Lett. **85**, 2733-2736 (2000).

[3]. G. J. Milburn, "Quantum optical Fredkin gate", Phys. Rev. Lett. **62**, 2124-2127 (1989).

[4]. C. C. Gerry and R. A. Campos, unpublished.

[5]. R. A. Campos, B. E. A. Saleh, and M. C. Teich, "Quantum-mechanical loss-less beam splitter: SU(2) symmetry and photon statistics", Phys. Rev. A **40**, 1371-1384 (1989).

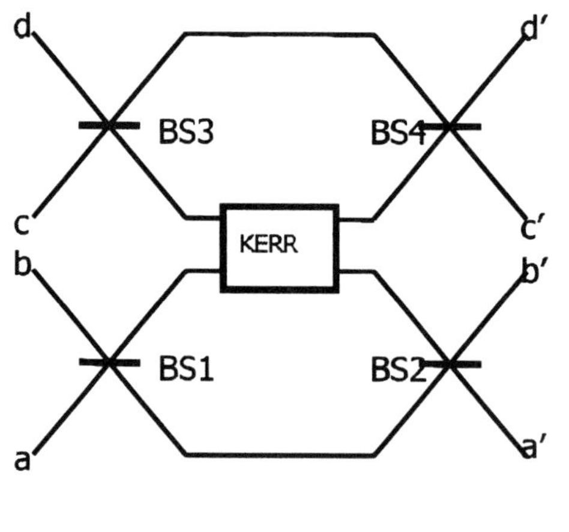

Fig. 1. Two Mach-Zehnder interferometers coupled via a cross-Kerr interaction.

Enhanced Second-Harmonic Generation due to a Local-Field Effect

I.V. Mel'nikov,[a] P.G. Kazansky,[b] and C.M. Bowden[c]

a) *Center for Research in Engineering and Applied Sciences, Morelos State Autonomous University, 1001 Av. Universidad, col. Chamilpa, 62210 Cuernavaca, Morelos, México*

b) *Optoelectronics Research Centre, University of Southampton, Southampton SO17 1BJ, Hampshire, United Kingdom*

c) *Weapons Science Directorate, US Army Aviation and Missile Command, Redstone Arsenal, Alabama 35898-5248*

Abstract: We study local-field effects as possible tool of enhancement of quadratic nonlinearity in doped optical fibers. For realistic dopant, we show that second-harmonic generation may be fundamentally altered by the change of fundamental wave dynamics due to the local-field effect. Novel effects that are discovered include a local-field enhancement of quadratic nonlinearity and depletion compensation of the fundamental wave.

In the recent years, there has been a great deal of interest in local-field correction in the dynamical response of three-level systems. Local-field corrections stem from the interaction of neighboring dipoles in a dense medium, quoted as near dipole-dipole (NDD) interaction. The inclusion of NDD effects leads to such interesting phenomena as intrinsic optical bistability [1], self-phase modulation in self-induced transparency [2], linear and nonlinear phase shift [3], enhanced gain and absorptionless index of refraction [4]. While much of the experimental and theoretical research in NDD phenomenology has been so far performed for atomic vapors, an effective and compact optical device has to be implemented in an all-solid-state format. The possibility of use of a multi-wavelength light wave in such a device is also of great practical interest.

On the other hand, there are many examples that can be quoted in the list of ions embedded in a dielectric with quadratic nonlinearity. We can mention, e.g., self-doubling crystals [5] and especially rare-earth-doped quasi-phase-matched structures, such as Er-doped periodically-poled lithium niobate [6]. In this Report, we investigate the local-field effects for optically nonlinear material in which the fundamental wave propagates along with its second harmonic. We find that the equations of motion for the process of second-harmonic generation are fundamentally altered by the presence of three-level dopant quasi-resonant with the fundamental frequency. Novel effects, which we have discovered, include quadratic nonlinearity enhancement and pump depletion compensation.

We consider an optical material as being composed of a large number of three-level ions embedded into a quadratic nonlinear material. The field of fundamental frequency ω induces the dipole polarization $\boldsymbol{p}e^{-i\omega t}$ in the ion. Then, inside the sphere of the radius $|r| \leq \lambda$ with λ being the fundamental wavelength, the local field is the sum of the macroscopic field $\boldsymbol{E}$ and the dipole field:

$$E_l = E + \frac{3(p \cdot r)r - r^2 p}{r^5}, \tag{1}$$

If $\boldsymbol{p} \propto \chi_{res}\boldsymbol{E}$, where χ_{res} is the resonant susceptibility of the dopant, then $\boldsymbol{E}_{dipole}/\boldsymbol{E} \sim \chi_{res}/r^3$, and at moderate r the dipole field may considerably exceed the macroscopic one. Since the polarization at the harmonic wavelength depends quadratically on $\boldsymbol{E}_l$, the contribution due to the local field does not disappear upon averaging over the volume. That is, at $\boldsymbol{p} = \boldsymbol{e}_z p$ and $\chi_{zzz} \neq 0$, the local field causes the macroscopic polarization at the harmonic frequency as follows:

$$P_z^{2\omega} = \chi_{zzz} p^2 N / \rho^3, \tag{2}$$

where N is the density of the resonant centers, and the averaging is made over all of the nonlinear matrix volume with exception of small spheres of the radius ρ around the centers.

In the case of three-level centers, the oscillating dipole polarization $\boldsymbol{p}$ can be of considerable magnitude where there is an induced coherence at the longest wavelength transition in the system. But then, it is exactly the system where both the enhanced lasing without inversion and absorptionless index of refraction can be reached. This means that the power losses of the fundamental wave, which are due to the harmonic generation, can be compensated by the inversionless gain. What is more, that is the magnitude of the quadratic susceptibility may be enhanced too, as it depends on the refractive index at the fundamental frequency in a periodically poled material. This fact is of extreme importance for possible implementation in the converter based on an optical fiber where the quadratic effects suffer from the low level of refraction. Our numerical modeling based on the full set of coupled Maxwell-Bloch equations confirms our speculations. They also show the pump pulse propagates faster than that of the harmonic, and there is no backward conversion from the harmonic into the pump.

In conclusion, we derive self-consistent equation of motion for quadratic nonlinear media high doped with resonant centers. We show that novel nonlinear effects may occur owing to the coupling of polarization components in the context of the near dipole-dipole interaction.

References:

[1] Y. Ben-Aryeh, C.M. Bowden, and J. Englund, Phys. Rev. **A 34**, 3917 (1986).
[2] C.M. Bowden, A. Postan, and R. Inguva, JOSA **B 8**, 1081 (1991).
[3] J.J. Maki *et al.*, Phys. Rev. Lett. **67**, 972 (1991).
[4] J.P. Dowling and C.M. Bowden, Phys. Rev. Lett. **70**, 1421 (1993).
[5] J. Bartschke *et al.*, IEEE J. Quant. Electron. **QE-33**, 2295 (1997).
[6] I.V. Mel'nikov, Opt. Commun. **132**, 479 (1996)

Non classical behaviours of optical field in Photonics band gaps structures

S.Severini, C. Sibilia and M. Bertolotti

INFM at Dipartimento di Energetica, Università di Roma "La Sapienza", Via A. Scarpa, 16, I-00161 Rome, Italy.

M. Scalora and C. Bowden

Weapon Science Directorate, AM SMI- RD-WS, Redstone Arsenal, AL 35898 5000 USA

Abstract: A theoretical analysis of the quantum behaviours of radiation field's propagation in Photonic Band Gaps structures is performed. We use quantum scattering approach to obtain some interesting bounds for propagation coefficients.

OCIS codes: 270.0270, 270.2500

1. Introduction

Our goal is to evaluate the quantum bounds imposed on the propagation coefficients, like Bogolubov coefficients, transmission and reflection coefficients, by the propagation of electromagnetic field inside a Photonic Band Gaps structures [1].

The study of this kind of structures is a fertile field of research. These structures have a number of interesting classical properties, linked to the manipulation of the electromagnetic dispersion [2] or easier establishment of phase matching condition respect to the homogeneous media [3] in the case of non-linear optical effects.

From a quantum point of view, some interesting classical results, as the frequency dependence of the transmission coefficient, in 1D problem, could be viewed as quantum bounds imposed on the frequencies permitted. The phenomenon is very similar to the concept of forbidden bands and allowed bands that we find in conductor structures as metals [4].

The purpose of the present study is to analyze the PBG structure from a quantum standpoint. Starting from quantum scattering problem [5], applied to these structures, we find a down-limit to the transmission coefficient, an upper-limit to the reflection coefficient and same bounds to Bogolubov coefficients, of the amplitude field propagation inside the structure. We are interested on the bounds imposed by the Schroedinger dynamics in the scalar 1D case only [6].

This study is the foundation of more detailed studies on the quantum properties of these versatile structures.

References:

[1]E. Yablonovitch, K.M. Leug, Nature, 351-278 (1991).

[2] M. Centini, C. Sibilia, M. Scalora, G. D'Aguanno, M. Bertolotti, Phys. Rev. E 60, 4891 (1999).

[3] M. Scalora, J.P. Dowling, M.J. Bloemer, A.S. Manka, J.W. Hauss, C.M. Bowden, Phys. Rev. A 56, 31166 (1997).

[4] Cohen-Tannoudji "Quantum Mechanics", Jhon wiley & sons, 1977.

[5] K Chadan and P.C. Sabatier, "Inverse problems in quantum scattering theory", Springer Verlag, New York, 1989.

[6] Matt Visser, Physical Review A 59 (1999), 427-438.

The Photon Model and Non-Hermitean Modes

S. A. Brown [(a)] **and B. J. Dalton** [(a), (b)]

(a) Department of Physics, University of Queensland, St Lucia, Queensland 4072, Australia
(b) SCOAP, CPES, University of Sussex, Brighton BN1 9QH, United Kingdom

Abstract: Field quantization in unstable optical systems is treated by expanding the vector potential in terms of non-Hermitean (Fox-Li) modes satisfying biorthogonality conditions. Spontaneous atomic decay is treated giving a rate enhanced by the Petermann factor.

The photon model of the quantum electromagnetic (EM) field has recently been developed further for the areas of cavity QED and quantum optics in dielectric media. Recent work has demonstrated its validity in microscopic theories [1], where EM field quantization is the same as in free space, and in macroscopic theories, where the classical optical device or material medium is treated as a spatially dependent permittivity. Canonical quantization via expanding the vector potential in terms of the true (or universe) modes for the system (obtained from a Helmholtz equation) shows that the EM field is again equivalent to uncoupled quantum harmonic oscillators (QHO), one for each true mode [2-5]. If idealised or quasi modes are used instead of true modes - as obtained from an approximate optical system permittivity, coupled QHO are obtained, one for each quasi mode [3, 6] and photon exchange processes between quasi modes occur. Phenomena such as beam splitter effects [7], light energy loss from cavities [8], reflection and refraction [9] can be explained via such processes using quasi mode theory. Both localised quasi mode (as in cavity or external regions for the standard cavity QED model [8]) and non-localised quasi mode (as in the beam splitter [7]) cases occur. Quasi mode theory has been extended [10] to cases [9] where two sets of quasi modes are needed.

The photon model obtained by quantization based on normal mode (free space, true or quasi mode) expansions involves mode functions determined from Hermitean eigenvalue equations and which satisfy standard orthogonality conditions. However, for unstable optical resonators or gain-guided lasers, the natural EM field modes to use (Fox-Li modes) are eigenfunctions of a non-Hermitian operator. Biorthogonality conditions apply for these non-Hermitean modes (NHM) and the related set of Hermitean adjoint mode functions (HAM). The appropriateness of the photon model [11] has been questioned for such systems, where phenomena such as linewidth enhancement in gain-guided lasers occur. The so-called excess noise is described by the Petermann factor [12], and several semiclassical studies (for example [13-15]) have successfully accounted for experimentally observed Peterman factors using the normalisation integrals for the HAM functions (the original NHM are normalised to unity). A number of papers [16-21] have been published with the aim of accounting for the success of the semiclassical approach in terms of a fully quantum treatment of the field.

The present paper is a further quantum treatment involving a standard canonical quantization proceedure based on expanding the vector potential in the unstable cavity region via the non-Hermitean (Fox-Li) modes and their Hermitean adjoint modes. Our approach is similar to those of [19, 21], although here we treat three dimensional systems, using the paraxial and monochromaticity approximations [19] for the cavity NHM. As in [19, 21] we consider all field modes, rather than a small number as in the simplified model treated in [18]. Our NHM and HAM are also obtained from the properties of the unstable optical system, rather than from treating atom-field interactions, as in [20]. As in [21], the external region is treated and the field described via further NHM and HAM. Similarly to [19, 21], we find that the field is equivalent to a set of QHO's, associated now with non-Hermitean modes rather than true modes and thus confirming the validity of the photon model for the case of unstable optical systems. Similarly to [19, 21], the annihilation, creation operator pairs for each QHO are not Hermitean adjoints. However, our results differ significantly in detail from those in [19, 21]. Our canonical quantization procedure involving both right and left travelling modes leads to a doubling of the annihilation, creation operator pairs compared to [19, 21]. We show that the total number of true mode QHO's equals the total number of QHO's associated with *either* with the cavity *or* the external region NHM, one oscillator for each annihilation, creation operator pair. As in [21], the final form of our quantum Hamiltonian for the

EM field is the sum of non-commuting cavity and external region contributions, but is simpler as there are no off-diagonal terms. To a good approximation, both the cavity and external region Hamiltonians can be expressed as the sum of independent QHO Hamiltonians for each NHM, but the external region Hamiltonian also includes a coupling term responsible for external NHM photon exchange processes. The two independent QHO Hamiltonians are alternative choices for an unperturbed Hamiltonian. Analogous to [19] for the cavity region Hamiltonian, we obtain left and right eigenstates for each of these unperturbed Hamiltonians and the energy is given by the usual QHO result. As in [21], certain cavity NHM and external region NHM annihilation, creation operators do not commute, leading to cavity energy gain and loss processes, though again the details differ. We find a simple description of the resultant cavity-external region light coupling in terms of surface integrals involving products of cavity and external NHM functions. Atom-field interactions are treated in the electric dipole and rotating wave approximations. We show that atomic transitions leading to cavity NHM photon absorption are associated with a different coupling constant to that for atomic transitions leading to photon emission. This feature is directly related to the treatment being based on NHM functions and leads to enhanced emission rates. Using the essential states approach, we consider spontaneous decay of a two level atom located in the cavity. The coupling term in the field Hamiltonian is neglected, assuming that atomic decay into the cavity is much faster than cavity decay. Coupled equations are obtained for atom-field states amplitudes involving the excited atom and no photons or the ground state atom with one photon in a cavity NHM. Markovian decay occurs under certain conditions, the decay rate being enhanced by the Petermann factor.

References

[1] B. Huttner and S. M. Barnett, Phys. Rev. **A46**, 4306 (1992)
[2] L. Knoll, W. Vogel and D.-G. Welsch, Phys. Rev. **A36**, 3803 (1987)
[3] R. J. Glauber and M. Lewenstein, Phys. Rev. **A43**, 467 (1991)
[4] B. J. Dalton, E. S. Guerra and P. L. Knight, Phys. Rev. **A54**, 2292 (1996)
[5] B. J. Dalton and M. Babiker, Phys. Rev. **A56**, 905 (1997)
[6] B. J. Dalton, S. M. Barnett and P. L. Knight, J. Mod. Opt. **46**, 1315, 1495 (1999)
[7] B. J. Dalton, S. M. Barnett and P. L. Knight, J. Mod. Opt. **46**, 1559 (1999)
[8] B. J. Dalton and P. L. Knight, J. Mod. Opt. **46**, 1817, 1839 (1999)
[9] S. Brown and B. J. Dalton, J. Mod. Opt. **48**, 639 (2001)
[10] S. Brown and B. J. Dalton, J. Mod. Opt. **48**, 597 (2001)
[11] A. E. Siegmann, App. Phys. **B60**, 247 (1995)
[12] K. Petermann. IEEE J. Quantum Electron. **15**, 566 (1979)
[13] A. E. Siegman, Phys. Rev. **A39**, 1253, 1264 (1989)
[14] W. A. Hamel and J. P. Woerdman, Phys. Rev. **A40**, 2785 (1989)
[15] M. A. Rippin and G. H. C. New, J. Mod. Opt. **43**, 993 (1996)
[16] I. H. Deutsch, J. C. Garrison and E. M. Wright, J. Opt. Soc. Amer. **B6**, 1244 (1991)
[17] P. Goldberg, P. W. Milonni and B. Sundaram, Phys. Rev. **A44**, 1969 (1991)
[18] P. Grangier and J.-P. Poizat, Eur. Phys. **D1**, 97 (1998)
[19] C. Lamprecht and H. Ritsch, Phys. Rev. Lett. **82**, 3787 (1999)
[20] P. J. Bardroff and S. Stenholm, Phys. Rev. **A60**, 2529 (1999)
[21] S. M. Dutra and G. Nienhuis, Phys. Rev. **A62**, 063805 (2000)

Induced coherence without induced emission in parametric down-conversion

A. Lukš and V. Peřinová

Laboratory of Quantum Optics, Faculty of Natural Sciences, Palacký University, Třída Svobody 26, 77146 Olomouc, Czech Republic
Tel.: 00420 068 5634263, Fax: 00420 68 5225246
E-mail: perinova@risc.upol.cz

Abstract:
We study quantum statistics of the multimode radiation from two coherently pumped down-conversion crystals with aligned idler beams and confirm that induced coherence need not be accompanied by induced emission.

OCIS codes: (270.0270) Quantum optics;(270.1670) Coherent optical effects

The experimentally demonstrated possibility to induce coherence in down-conversion without inducing emission [1] has been analyzed in the Schrödinger picture and recently the Heisenberg approach has been adopted [2]. The experimental situation is considered in which two similar nonlinear crystals are optically pumped by two mutually coherent pump waves of complex amplitudes $V_1 = V_2 = V$. Parametric down-conversion occurs at both crystals and the paths of the idler photons are aligned. The quantum field amplitudes of the incoming signal and idler beams can be expanded as usual

$$\hat{E}^{(+)}_{m_j 0}(\mathbf{r},t) = i \sum_{[\mathbf{k}]_{m_j}} \sqrt{\frac{\hbar\omega_{\mathbf{k}}}{L^3}} e^{i\mathbf{k}\cdot\mathbf{r}} e^{-i\omega_{\mathbf{k}} t}\, \hat{a}_{j0\mathbf{k}}(0), \quad m_j = s_j, i_j, \quad j = 1,2,$$

where L^3 is the normalization volume, $\hat{a}_{j0\mathbf{k}}(0)$ is the annihilation operator for a photon whose wave vector is $\mathbf{k}$, $\omega_{\mathbf{k}} = c|\mathbf{k}|$, and the square brackets in the summation symbol indicate that the sum is restricted to the set of $\mathbf{k}$ pertaining to beam m_j. The quantum field amplitudes of the outgoing signal and idler beams are defined as follows

$$\hat{E}^{(+)}_{m_j}(\mathbf{0}_j,t) = \overrightarrow{\mathcal{T}} \exp\left[\frac{i}{\hbar}\int_{-(\Delta t)_j}^{0} \hat{H}^{(\mathrm{slow})}_{j\,\mathrm{int}}(t')\,dt'\right] \hat{E}^{(+)}_{m_j 0}(\mathbf{0}_j,t)\, \overleftarrow{\mathcal{T}} \exp\left[-\frac{i}{\hbar}\int_{-(\Delta t)_j}^{0} \hat{H}^{(\mathrm{slow})}_{j\,\mathrm{int}}(t')\,dt'\right],$$

$$m_j = s_j, i_j, \quad j = 1,2,$$

where $\mathbf{0}_j$ is the point where the center of the jth crystal is placed, $(\Delta t)_j$ is the time taken for the radiation to cross the jth crystal, $\overleftarrow{\mathcal{T}}$ ($\overrightarrow{\mathcal{T}}$) is the usual (dual) time-ordering prescription, and $\hat{H}^{(\mathrm{slow})}_{j\,\mathrm{int}}(t')$ are the interaction Hamiltonians,

$$\hat{H}^{(\mathrm{slow})}_{j\,\mathrm{int}}(t') = i\hbar g' V \sum_{\mathbf{k},\mathbf{k}'} f(\mathbf{k},\mathbf{k}') e^{-i(\omega_0-\omega_{\mathbf{k}}-\omega_{\mathbf{k}'})t'}\, \hat{a}^{\dagger}_{j0\mathbf{k}}(0)\hat{a}^{\dagger}_{j0\mathbf{k}'}(0) + \mathrm{H.\ c.}, \quad j = 1,2,$$

with g' being a coupling constant and $f(\mathbf{k},\mathbf{k}')$ being a symmetrical function. The operators $\hat{a}_{20\mathbf{k}}(0)$ are such that $\hat{E}^{(+)}_{i_2 0}(\mathbf{0}_2,t) = \hat{E}^{(+)}_{i_1}(\mathbf{0}_2,t)$. We consider the partial alignment realized by a beam splitter inserted between the two crystals. We study the dependence of the visibility of interference on the filter transmissivity $\mathtt{t}$ and the improvement of visibility achieved in the coincidence-counting arrangement. The resulting coincidence detection probability is of the form

$$P_{ab}(t_a, t_b) = \frac{\eta^2 |\mathrm{t}|^2 g^2 |V|^2}{8\hbar^2 \omega_s \omega_i} \int_0^w \int_0^w \left\{ \left| \nu\left(\tau - \tau' + \frac{(l+f-d)}{c}\right)\right|^2 + \left|\mathrm{t}^* \nu\left(\tau - \tau' + \frac{(l-h)}{c}\right)\right|^2 \right.$$

$$\left. -2\mathrm{Im}\left[\mathrm{t}\nu\left(\tau - \tau' + \frac{(l+f-d)}{c}\right)\nu^*\left(\tau - \tau' + \frac{(l-h)}{c}\right)\exp\left\{i\left[\omega_s \frac{(d-h)}{c} + \omega_i \frac{f}{c}\right]\right\}\right]\right\} d\tau d\tau',$$

which, as expected, shows an intensity–intensity interference. Here η is the efficiency of the detectors, g is an effective coupling constant, w is a detection window. The function $\nu(\tau)$ defines a coherence time between signal and idler, d, f, h, l are distances, c is the speed of light. We take into account that the time intervals $\frac{t_a}{\omega_s}$ and $\frac{t_b}{\omega_i}$ are small in comparison with the coherence time $\nu(\tau)$.

We use the Wigner representation according to [2] to the detailed analysis of the induced coherence without induced emission. The experimentally demonstrated possibility to induce coherence in down-conversion without inducing emission conformed with the multimode formalism introduced by Ou et al. [3]. The application in [1] was successful, because the first-order processes dominated in the experiment. Essential features of the effect can be derived also in the framework of the usual modal formalism [4]. We modify the theory by Casado et al. [2], which has been developed to the second order in the coupling constant, to model also the absence of induced emission.

This work was supported by the Ministry of Education of Czech Republic (project no. LN00A015).

References and links

1. X. Y. Zou, L. J. Wang, and L. Mandel, "Induced coherence and indistinguishability in optical interference," Phys. Rev. Lett. **67**, 318–321 (1991).
2. A. Casado, A. Fernández-Rueda, T. Marshall, R. Risco-Delgado, and E. Santos, "Fourth-order interference in the Wigner representation for parametric down-conversion experiments," Phys. Rev. A **55**, 3879–3890 (1997).
3. X. Y. Ou, L. J. Wang, and L. Mandel, "Vacuum effects on interference in two-photon down conversion," Phys. Rev. A **40**, 1428–1435 (1989).
4. H. M. Wiseman and K. Mølmer, "Induced coherence with and without induced emission," Phys. Lett. A **270**, 245–248 (2000).

Correlation functions, measurement operators, and conditioned atomic states in resonance fluorescence with spectral resolution

V. N. Shatokhin and S.Ya.Kilin

B. I. Stepanov Institute of Physics, National Acad. Sci. of Belarus, Minsk, Belarus, email: shatok@ifanbel.bas-net.by

Resonance fluorescence (RF) conditioned by spectral detection has a long history. In recent years, the interest to this field has been renewed [1, 2, 3, 4] motivated by needs of the quantum measurement theory. The quantities of interest in the quantum measurement theory – the probabilities for different outcomes of the measurement process and the state of a quantum system straight after the measurement outcome – can be clearly recognized in RF with spectral resolution as the probability for detection of the spectrally filtered photon and the state of the atom following the detection.

Recently, we have considered a setup [4] in which a single atom resonance fluorescence is divided in such a way that one half of the intensity of the scattered light is directed toward a broadband detector, whereas another half is directed toward the spectral filter followed by a detector monitoring transmitted or reflected photons. Using a special type of correlation measurement of RF, in which detection of the filtered photon is followed by detection of an unfiltered photon, we derived a general expression for the conditioned atomic state following detection of the transmitted and reflected photons. This expression is related to the normalized stationary second-order intensity correlation function of a photocount at the detector D_i that monitors transmitted ($i = 1$) or reflected ($i = 2$) photons followed by a photocount at the detector D_0 (monitoring unfiltered photons) after delay τ, through the equation

$$g_{i0}^{(2)}(\tau) = \frac{\rho_{22}(\tau)\,|_{\rho(0)=\rho_i^{\mathrm{c}}}}{\rho_{22}^{\mathrm{ss}}}, \tag{1}$$

where ρ_{22}^{ss} is the stationary excited state population, and ρ_i^{c} is the stationary conditioned atomic state prepared from the detection of the photon by the detector D_i.

The conditioned atomic state ρ_i^{c} can be described in terms of the qeneralized measurement theory [5, 6]. In the limit of well separated spectral lines measurement operators A_1 and A_2 are associated with the detection of the transmitted and reflected photons, respectively, that is, the stationary conditioned atomic state ρ_i^{c} can be represented in the form $\rho_i^{\mathrm{c}} = A_i\rho^{\mathrm{ss}}A_i^{\dagger}/\mathrm{tr}[A_i^{\dagger}A_i\rho^{\mathrm{ss}}]$, $\quad (i = 1, 2)$. At resonance excitation

$$A_1 = \frac{\Gamma}{2i\Omega}\left\{\frac{1}{1-\xi}\left|+\right\rangle\left\langle-\right| + \frac{1}{1+\xi}\left|-\right\rangle\left\langle+\right| + \frac{1}{\xi}(\left|+\right\rangle\left\langle+\right| - \left|-\right\rangle\left\langle-\right|)\right\}, \tag{2}$$

where Γ is the filter bandwidth, Ω is the Rabi frequency, $\xi = (\delta - i\Gamma)/\Omega$ with δ being the filter-laser detuning, and $\left|+\right\rangle$, $\left|-\right\rangle$ are the dressed atomic states. Knowledge of A_1 is sufficient for defining A_2, due to the relation $A_2 = A_1 - \sigma_-$. Measurement operators A_1 and A_2 determine mappings of the initial atomic state $\rho^{\mathrm{ss}} = \frac{1}{2}(\left|+\right\rangle\left\langle+\right| + \left|-\right\rangle\left\langle-\right|)$ into the conditioned states ρ_1^{c} and ρ_2^{c}, respectively. Examples of the mappings realized on the Bloch sphere by the operator A_1 are illustrated by Fig. 1. It can be seen from equation (2) that if $|\delta| \sim \Omega/2$ then the terms $(1 \pm \xi)^{-1}$ and ξ^{-1} are on the same order, and

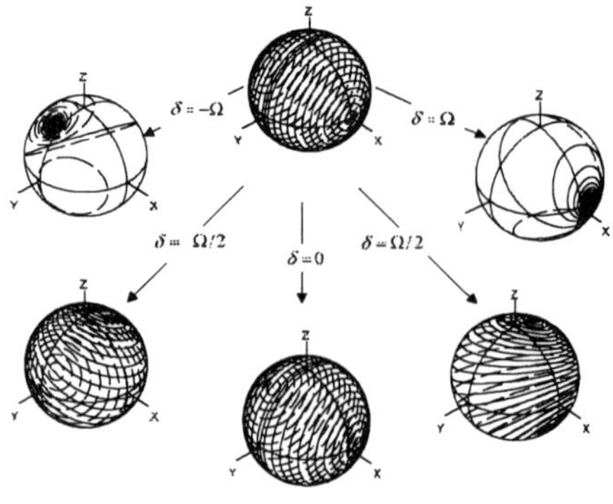

Fig. 1. Mappings $A_1(\delta)|\vartheta,\varphi\rangle / \|A_1(\delta)|\vartheta,\varphi\rangle\|$. Initial set (top center) of the equally displaced states $|\vartheta_n,\varphi\rangle$, $\vartheta_n = \pi n/25$, $n = 0, 1, ...25$ uniformly covering the Bloch sphere is transformed into the non-uniform cover of the sphere for different detunings δ. When the filter transmits photons from the high- (low-) frequency sideband, that is, $\delta = \Omega$ ($\delta = -\Omega$), the initial set of states is mapped close to the point $|+\rangle$ ($|-\rangle$). The initial set of states is not changed markedly under the transformation A_1 ($\delta = 0$), when the filter is tuned to the central line of the RF triplet. When the filter is tuned in between the RF lines ($\delta = \pm\Omega/2$), the initial set of states is transformed so that the states located near state $|+\rangle$ ($|-\rangle$) are mapped close to the upper state $|2\rangle$ of the bare atom, which is reflected in the population inversion of the conditioned state ρ^c. The filter bandwidth $\Gamma = 20\gamma$, and the Rabi frequency $\Omega = 200\gamma$.

therefore all the operators in the r.h.s. of equation (2) describing transitions between the dressed atomic states affect the properties of the operator A_1 leading to quantum interference of the probability amplitudes of the conditioned atomic state eigenvectors. It is this quantum interference that is the reason for the atomic inversion of the conditioned atomic state which can be observed experimentally.

In conclusion, we have considered the problem of the conditioned atomic state following the detection of the photon passed through or reflected from the interference filter. We have found the measurement operators corresponding to these detections in the limit of well separated spectral lines of the RF spectrum, and studied their properties using the Bloch state representation. The stationary conditioned atomic state following the detection of the transmitted photon can be found inverted if the filter is tuned to the frequency domain lying in between the central line and one of the sidebands. Atomic inversion here is the consequence of quantum interference between the atomic radiative processes that contribute to the detected photon but cannot be distinguished. As for the conditioned atomic state following the detection of the reflected photon, it is close to the ground atomic state unless the filter is tuned to any of the separate components of the spectrum. In the latter case quantum interference between the processes giving rise to the rest two components leads to maximums in the excited state population.

References and links

1. M. Holland, Phys. Rev. Lett. **81**, 5117 (1998); D. B. Horoshko and S. Ya. Kilin, *ibid.* **83**, 2469 (1999); M. Holland, *ibid.* **83**, 2470 (1999).
2. M. W. Jack, M. J. Collett, and D. F. Walls, Phys. Rev. A **59**, 2306 (1999).
3. H. M. Wiseman and G. E. Toombes, Phys. Rev. A **60**, 2474 (1999).
4. V. N. Shatokhin and S. Ya. Kilin, Opt. Commun. **174**, 157 (2000); Phys. Rev. **A** 63, 023803 (2001); Opt. Spectrosc. (in the press).
5. K. Kraus, *States, Effects, and Operations* (Springer-Verlag, Berlin, 1983).
6. C. W. Gardiner, *Quantum Noise* (Springer-Verlag, Berlin, 1991).

Quantum Algebras for Quantum Optics

S. M. Chumakov[1] **and A. Ballesteros**[2]

[1]*Depto. de Física, Universidad de Guadalajara, Revolución 1500, 44420, Guadalajara, Jalisco, México.*
[2]*Depto. de Física, Universidad de Burgos, 09001-Burgos, Spain*
sergei@cencar.udg.mx, angelb@ubu.es

Many models in Quantum Optics, such as two- and three-wave mixing and interaction of two-level atoms with a single-mode radiation field (Dicke model), can be described by interaction Hamiltonians of the form

$$H = \begin{pmatrix} 0 & A_l & 0 & \dots & 0 \\ A_l & 0 & A_{l-1} & \dots & 0 \\ \dots & \dots & \dots & \dots & \dots \\ 0 & \dots & A_{-l+2} & 0 & A_{-l+1} \\ 0 & \dots & 0 & A_{-l+1} & 0 \end{pmatrix} . \tag{1}$$

The dimension $d = (2l + 1)$ of this matrix is by no means small (for instance, in the Dicke model, $N = 2l$ is the number of atoms considered). Therefore, the analytical diagonalization of H is essential in order to solve the dynamics of the model. In some cases, the spectrum of the matrix H is nearly equidistant and the slight nonlinearity can be taken into account by perturbation theory with the zeroth-order Hamiltonian proportional to the generator J_x of the d–dimensional irreducible representation of the $SU(2)$ group (approximate $su(2)$-dynamical symmetry). For instance, this happens in the Dicke model, when the initial mean number of photons is much smaller or much larger than the number of atoms. This fact has been successfully used to reveal many interesting dynamical features (see *e.g.*, [1,2]). However, for the collective spontaneous emission or for the second harmonic generation, the spectrum of H is strongly nonlinear and the $su(2)$-dynamical symmetry cannot be used. Note, that the semiclassical parametric approximation gives reasonable results only for very short times (comparing to the natural oscillation period in the system). Thus, new analytic tools are needed.

Recently, it has been shown [4] that the quantum (deformed) algebra $su_q(2)$ (see *e.g.* [3]) can be used to define an exactly solvable Hamiltonian, $h(q)$, of the type (1). $h(q)$ has a nonlinear spectrum that depends on the deformation parameter q. We show here, that one can use $h(q)$ as a zeroth-order Hamiltonian to construct a perturbation theory for optical models of the type (1) in the highly nonlinear domain.

The most important property of the common $su(2)$ algebra is that tensor products of generators of two representations of $su(2)$, $J_{x,+,-} \otimes 1 + 1 \otimes J_{x,+,-}$, obey the same commutation relations as the initial generators. (Here we write operators form the first and second irreducible representations of $su(2)$ on the left and on the right, correspondingly.) The quantum deformation $su_q(2)$ of the common $su(2)$ generalizes the structure of this tensor product. The operators J_z , $J_\pm(q)$ generate the quantum algebra $su_q(2)$ if we define the generalized tensor product (*coproduct*) as follows,

$$\Delta(J_\pm) = J_\pm \otimes q^{-J_z} + q^{J_z} \otimes J_\pm\,, \quad \Delta(J_z) = J_z \otimes 1 + 1 \otimes J_z\,.$$

Here, the deformation parameter q is an arbitrary real number. The deformed commutation rules, consistent with the above coproduct are well-known. Unitary irreducible representations of $su_q(2)$ have the dimensions $2l + 1$, $l = 0, 1/2, 1, \ldots$. In the basis of eigenvectors of J_z, we have, $J_z|l,m\rangle = m|l,m\rangle$ and $q^{2J_z}|l,m\rangle = q^{2m}|l,m\rangle$.

Coherence and Quantum Optics VIII
Edited by Bigelow et al., Kluwer Academic/Plenum Publishers, 2003

Our starting point [4] is the fact that the operator $h(q) = q^{J_z/2}\,(J_+ + J_-)\,q^{J_z/2}$, for a given l has the spectrum $\lambda_m = (q^{2m} - q^{-2m})/(q - q^{-1})$, $m = -l, \ldots, l$. It can be proven by induction using the coproduct to construct higher-dimensional representations from the three-dimensional one. Eigenvectors are given in Ref. [4]. The $q \to 1$ limit of $h(q)$ is just $2\,J_x$. In the $(2\,l + 1)$-dimensional representation, $h(q)$ takes the form (1) with

$$A_m = q^{m-1/2}\sqrt{(q^{l+m} - q^{-l-m})(q^{l-m+1} - q^{-l+m-1})/(q - q^{-1})}\,. \tag{2}$$

Now, let us consider the Dicke model, which describes the interaction of a system of $N = 2l$ two-level atoms with the quantum radiation field in an ideal cavity. This model is mathematically equivalent to the three-photon Hamiltonian describing three-wave mixing. The Hamiltonian can be written in the matrix form (1) with the matrix elements

$$B_m = \sqrt{(l+m)(l+1-m)(s+1-m)}, \tag{3}$$

where $s \geq l$ is a parameter ($2s$ is the excitation number which is conserved in this model). Here we restrict ourselves with the case of the highest nonlinearity, $s = l$. It corresponds to the spontaneous emission in the Dicke model and to the second harmonics generation. We are specially interested in the limit of large l (high photon numbers). There is a strong resemblance between the two functions, A_m from Eq. (2) and B_m from Eq. (3), which allows us to take the operator $H_0 = \Omega\, h(q)$, as a zeroth-order Hamiltonian and find the spectrum by the perturbation theory. The values of q and Ω are to be chosen. The simplest way to fit them is to provide the coincidence of the points where the matrix elements B_m of the three-wave Hamiltonian and the matrix elements ΩA_m of H_0 take their maximum values. It gives (for $s = l$)

$$N \log q = \frac{3}{2}\log\frac{\sqrt{5}-1}{2} \approx -0.7218\,,$$

We chose Ω to make equal the values of B_m and ΩA_m in their maxima, $m_0 = -(l-1)/3$. Now, we find the approximation for the three-wave Hamiltonian in the form: $B_m \approx \Omega A_m \phi(m - m_0)$, where $\phi(m - m_0)$ is a polynomial whose coefficients ϕ_j are found comparing the Taylor expansions for $B(m)$ and $A(m)$ at the point m_0. We restrict ourselves to polynomial of the third degree. In the matrix form: $H \approx \Omega\,[h(q)f(J_z) + f(J_z)h(q)]$. The new function $f(J_z)$ is also a polynomial of degree three, whose coefficients are easily found. It is straightforward now to find the approximate eigenvalues and eigenvectors to the Hamiltonian (1).

Numerical tests show that the spectrum and eigenvectors are reproduced with high accuracy and have the correct asymptotic behavior, when $N \to \infty$. The complete dynamical analysis of the three-wave processes by means of the present approach will be given elsewhere.

References and links

1. S. M. Chumakov and M. Kozierowski, "Dicke model: quantum nonlinear dynamics and collective phenomena," Quantum Semiclass. Opt. **8**, 775-803 (1996).
2. S. M. Chumakov, A. B. Klimov and J. J. Sánchez-Mondragón, "General properties of quantum optical systems in a strong field limit," Phys. Rev. A **49**, 4972–4978 (1994).
3. V. Chari and A. Pressley, *A Guide to Quantum Groups* (Cambridge University Press, 1994).
4. A. Ballesteros and S. M. Chumakov, "On the spectrum of a Hamiltonian defined on $su_q(2)$ and quantum optical models," J. Phys. A: Math. Gen. **32**, 6261-6269 (1999).

A Delayed-Choice Quantum Eraser Experiment

Yoon-Ho Kim, Rong Yu, Sergei P. Kulik, and Yanhua Shih
Department of Physics, University of Maryland, Baltimore County, Baltimore, MD 21250 USA
Tel: (410)455-1932, Fax: (410)455-1072, Email: yokim@umbc.edu

Marlan O. Scully
Department of Physics, Texas A&M University, College Station, TX77842, USA

Abstract: We report a "Quantum Eraser" experiment proposed by Scully and Drühl in 1982. The experimental results demonstrated the possibility of simultaneously observing both particle- and wave-like behavior of a quantum via quantum mechanical entanglement. The which-path information of a quantum can be erased or marked by its entangled twin even after the registration of the quantum.
OCIS codes: (000.1600) Classical and quantum physics; (270.0270) Quantum optics

We say that the quantum observables "position" and "momentum" are "complementary" because the precise knowledge of the position (momentum) implies that all possible outcomes of measuring the momentum (position) are equally probable. Bohr illustrated complementarity with "wave-like" and "particle-like" attributes of a quantum mechanical object[1]. Since then, complementarity has often been superficially identified with "wave-particle duality of matter".

Over the years two-slit interference experiment has been emphasized as a good example of the enforcement of complementarity. In two-slit experiment, the common understanding is that the position-momentum uncertainty relation makes it impossible to determine which slit the photon (or electron) passes through without at the same time disturbing the object enough to destroy the interference pattern. In 1982, Scully et al suggested a way around this position-momentum uncertainty obstacle and proposed a quantum eraser to obtain which-path or particle-like information without scattering or otherwise introducing large uncontrolled phase factors to disturb the interference[2]. Since 1982, quantum eraser has been reported in several experiments; however, the original scheme has not been fully demonstrated.

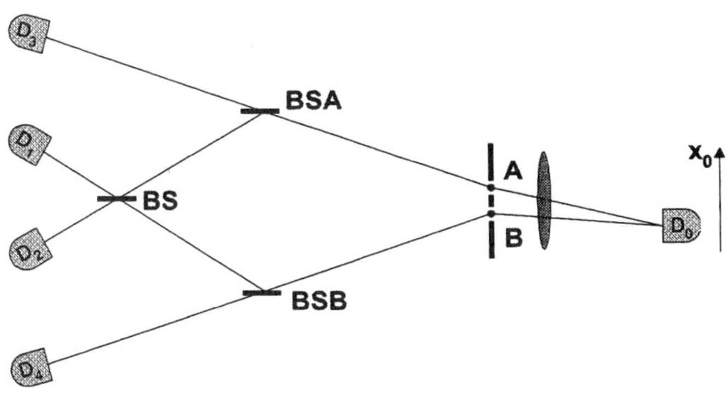

Fig. 1. Proposed quantum eraser experiment. A pair of entangled photons is emitted from either atom A or atom B. BSA, BSB, and BS are all 50/50 beamsplitters.

A quantum eraser experiment very close to the 1982 proposal (and our present experiment) is illustrated in Fig. 1. Two atoms, labeled by A and B, are excited by a laser. A pair of entangled photon, photon 1 and photon 2, is then emitted from either atom A or atom B by atomic cascade decay. Photon 1, propagates to the right, is detected by D_0, which can be scanned by a step motor along its x-axis for the observation of interference fringes. Photon 2, propagates to the left, is injected into an interferometer. If the pair is generated in atom A (B), photon 2 will follow A (B) path to meet BSA (BSB) with 50% chance to be reflected or to be transmitted. The registration of D_3 or D_4 provides which-path information of photon 2 and in turn provides which-path information of photon 1 due to the entanglement. The triggering of D_1 or D_2 erases the which-path information of photon 2 and in turn erases the which-path information of photon 1. The absence or the restoration of interference can be arranged by an appropriately contrived photon correlation study.

The experiment is designed in such a way that L_0, the optical distance between atoms A, B and the detector D_0, is much shorter than L_i, the optical distance between atoms A, B and detectors D_i (i=1~4) so that D_0 will be triggered much earlier by photon 1. After the detection of photon 1, we look at these "delayed" detection events of photon 2 which have constant time delay relative to the triggering time of D_0.

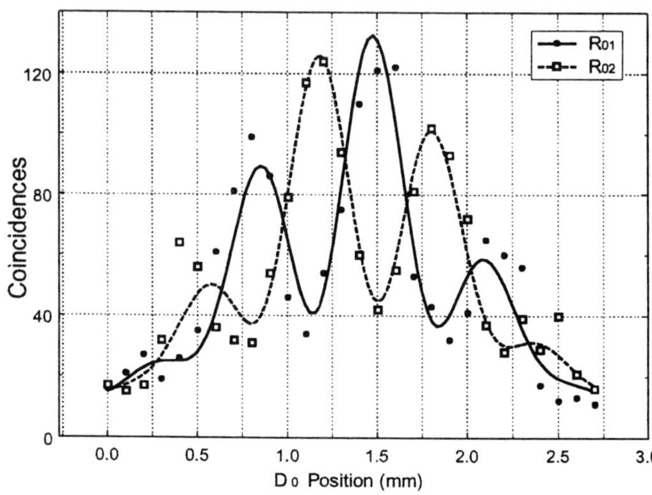

Fig. 2. R01 and R02 against the x coordinate of detector D_0. Standard Young's double slit interference patterns are observed. Note the shift between R01 and R02.

Instead of atomic cascade decay, spontaneous parametric down conversion (SPDC) is used to prepare the entangled two-photon state. A double-slit is placed in front of a BBO crystal to create regions A and B and orthogonal polarized signal-idler photon pair is generated either A or B region. Fig. 2 shows the coincidence counting between D_0-D_1 and D_0-D_2. Between D_0 and D_3, D_4 no interference is observed as we have expected (although not shown here). These experimental results well agree with standard quantum mechanical calculation using two-photon state of SPDC.

In conclusion, we have realized a quantum eraser experiment of the type proposed in Ref.[2]. The experimental results demonstrate the possibility of determining particle-like and wavelike behaviors of a photon via quantum entanglement. The which-path or both-path information can be erased or marked by its entangled twin even after the registration of the quantum itself.

---

[1] N. Bohr, Naturwissenschaften, **16**, 245 (1928)
[2] M.O. Scully and K. Drühl, Phys. Rev. A **25**, 2208 (1982)

Delayed single-photon self-interference - A double slit experiment in the time domain

Nicklas Ohlsson and Stefan Kröll
Dept. of Physics. Lund Institute of Technology (LTH), Box 118, S-221 00 Lund, Sweden
Nicklas.Ohlsson@fysik.lth.se

Serguei A. Moiseev
Zavoiskii Physicotechnical Institute, Kazan Scientific Center, Russian Academy of Sciences, Kazan, Tatarstan, 420029 Russia

If a single photon is split into two wave-packets using a beam splitter and the two possible photon paths are made to overlap, an interference pattern due to interference between the two wave-packets can arise. If a classical detector, e.g. a photographic plate, is used for detecting the interference, the interference pattern will only be visible if the difference in length between the two possible photon paths is less than the coherence length of the photon wave-packet, i.e. the two wave-packets must overlap in both space and time. If however, instead of the normal detector, a material that can record and remember the phase of the electro-magnetic field is placed in the region where the overlap takes place, it has been proposed that there can be interference even if the path difference is much larger than the coherence length of the photon [1], see Fig. 1 for a schematic experimental set-up. We denote this as 'delayed single-photon self-interference'. In this case, the path difference still allowing interference to take place would only be limited by the phase memory time of the material. The interference of the photon wave-packets will be manifested as a frequency dependent population modulation in the material. The periodicity of this modulation is related to the time differences between the two possible photon paths. The frequency dependent modulation of the population is the time domain analogue to a spatial two-slit interference pattern. The fact that two non-overlapping wave packets can produce an interference pattern could be detected by diffracting a light pulse on the frequency modulated population.

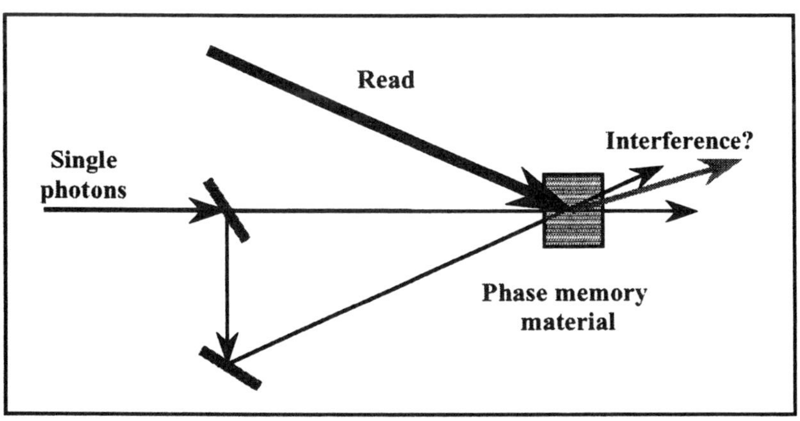

Fig. 1. Experimental set-up for the interference of single photons in a material with a long phase memory.

The delayed single-photon self-interference experiment has several unique features. It is a four-wave mixing experiment where the two first pulses share a total of only one photon. Thus interaction between these two pulses and the material, normally considered as a nonlinear multiphoton process, is here carried out by a single photon. Detailed calculations have shown [2] that if several single photon events are accumulated before the readout, the obtainable signal strength should be large enough to make the experiment feasible. We have performed experiments with multiphoton excitation pulses and extrapolated the data to the single photon level. These experiments support the conclusion from the theoretical calculations that delayed single-photon self-interference could be observed.

We have started experiments with weak pulses and observed interference between pulse pairs with a total of less than 700 photons per pair. We will gradually decrease this number of photons to get to (and below) the single photon level.

Parallel to the experimental work, a number of alternative schemes for the delayed single-photon self-interference experiment are currently being analyzed. If we e.g. consider a three level atomic system in cascade configuration and let the split single photon be resonant with the transition between the two lower lying levels, the interference could be detected at the transition between the two upper levels. This would give an advantage in that only atoms contributing to the interference pattern would be affected by the detection pulse, which would yield a significantly reduced noise level.

In the delayed single-photon self-interference, the absorption of the single photon is spread out over a large number of atoms in a material of macroscopic size, but without the normal loss of coherence. We therefore believe that the single-photon self-interference experiment is a new and exciting tool for studying microscopic and macroscopic quantum states and think that it could be used for gaining a better understanding of the quantum mechanical measurement process.

A.R. Kessel and S.A. Moiseev, "Delayed self-interference of a photon", *JETP Lett*, **58**, 80, (1993)

R. Krishna Mohan, B Luo, S. Kröll and A. Mair, "Delayed single-photon self-interference", *Phys.Rev. A*, **58**, 4348, (1998)

Retrodiction with Two-Level Atoms: Atomic Previvals and Micromaser Field Measurements

John Jeffers[1], Stephen M. Barnett[1] and David T. Pegg[2]

(1) Department of Physics, The University of Strathclyde, 107 Rottenrow, Glasgow, G4 0NG, U.K.
(2) Faculty of Science, Griffith University, Brisbane, Queensland, Australia.
john@phys.strath.ac.uk

Quantum retrodiction is quantum mechanics based on measurement results. If the state prepared by a preparaton device is not completely certain then the retrodictive state, which is the state measured by a detector, can give information about the state at earlier times. If it is used to calculate a set of retrodictive conditional probabilities, then these must be related to the predictive probabilities by Bayes' Theorem [1].

To date we have derived the master equation for the retrodiction of quantum states in open systems [2]. We have shown that the action of a predictive attenuator on a prepared state is equivalent to the action of a retrodictive amplifier on a measured state [3]. Furthermore we have applied retrodiction to a two-level atom interacting with a mode of the electromagnetic field. If the atom is driven the inversion shows retrodictive Rabi oscillations which decay to the value 1/2 [4].

Here we again study retrodiction with two-level atoms. We show, given a coherent initial field state, an unmeasured final state and a measured state of the two-level atom, that the retrodictive Rabi oscillations undergo collapses and previvals. We also retrodict micromaser field states given a measurement of the atom after te micromaser interacton. Repeated atomic detections can place the initial state of the field in one of definite photon number.

References and links

1. Stephen M. Barnett, David, T. Pegg and John Jeffers, "Bayes' Theorem and Quantum Retrodiction," J. Mod. Opt. **47**, 1779 (2000).

2. Stephen M. Barnett, David, T. Pegg, John Jeffers and Ottavia Jedrkiewicz, "Master Equation for Retrodiction of Quantum Communicaton Signals," J. Opt. Soc. Am. A **14**, 741-755 (1997).

3. Stephen M. Barnett, David, T. Pegg, John Jeffers and Ottavia Jedrkiewicz, "Retrodiction for quantum optical communications," Phys.rev A **62**, 022313 (2000).

4. Stephen M. Barnett, David, T. Pegg, John Jeffers and Ottavia Jedrkiewicz, "Atomic Retrodiction," J. Phys.B, At. Mol. Opt. Phys. **33**, 3047 (2000).

Measurements of the geometric phase of first-order optical Gaussian beams

E.J. Galvez, H.I. Sztul and P.J. Haglin
Department of Physics and Astronomy, Colgate University, Hamilton NY 13346

Geometric phase is ubiquitous in systems that undergo cyclic transformations in either parameter or state space. Manifestations of this phase have been found in many physical systems. In optics, geometric phase has had an important effect, enhancing the way optical systems are analyzed, and stimulating the discovery of interesting effects and new applications. Two well-investigated manifestations of geometric phase in optics are the *coiled optics* phase, produced by sending an optical beam in a three-dimensional (coiled) path; and *Pancharatnam* phase, produced by changes in the state of polarization. A third phase, produced by transformations in the phase structure or mode of a Gaussian beam, has just begun to be explored [1]. High-order Gaussian beams have received much attention in the last few years due to their ability to carry orbital angular momentum [2]. Previous experimental investigations of this phase have been done in the microwave regime [3,4]. Here we present the first measurements of this geometric phase in the optical regime.

The space of modes for first-order Gaussian beams is the Poincare sphere [5]. This space is similar to the space of states of polarization: the north and south poles are the two eigenstates of the paraxial wave equation with cylindrical symmetry: the Laguerre-Gauss functions LG_0^1 and LG_0^{-1}, respectively, while points in the equator represent different orientations of the eigenstates with rectangular symmetry: the Hermite-Gauss functions HG_{10} or HG_{01}. This mode-space has recently been shown to have a SU(2) structure [6]. Transformations between LG and HG modes are effected by π- and π/2-converters [7,8]. These consist pairs of cylindrical lenses that alter the phase structure of the beam via the Gouy phase. A π-converter transforms a LG_0^1 mode into a LG_0^{-1} mode and vice versa. Thus when a beam in a LG_0^1 mode is sent through two consecutive π-converters the beam returns to its initial mode possessing a geometric phase that depends on the path followed in mode space [5]. When the transverse axes of the converters form an angle α this path is given by two great semicircles that form an angle 2α in the Poincare sphere, as shown by the solid line in Fig. 1a. If the initial mode were a LG_0^{-1}, the path would be the one shown by the dashed line in Fig. 1a. The geometric phases for the LG_0^1 and LG_0^{-1} beams will be $+2\alpha$ and -2α, which are respectively half of the solid angles enclosed by each path.

We performed two type of experiments to measure this geometric phase, which manifests itself as a phase shift in the wave. In one experiment a LG_0^1 beam was produced by a TEM_{00} beam diffracting off a charge-1 binary grating. It was sent through one of the arms of a Mach-Zender interferometer and interfered with the zero-order beam going through the other arm of the interferometer, as shown in Fig 1b. The LG_0^1 beam passed through two consecutive π-converters. One of them was in a custom mount attached to a commercial rotation stage, and thus was free to rotate about the beam axis. A CCD camera and imaging software recorded the resulting interference pattern. When one of the π-converters was rotated about the beam axis the interference fringes shifted in the interference pattern.

A second experiment eliminated the dynamical phases completely. This was done without the interferometer by passing the interfering beams collinearly through two consecutive π-converters. Any dynamical phases were introduced to both beams and thus the interference produced by them was insensitive to misalignments. The change in the phase difference between the interfering beams was then exclusively due to the geometric phase. We performed several experiments with collinear beams. One case is presented below.

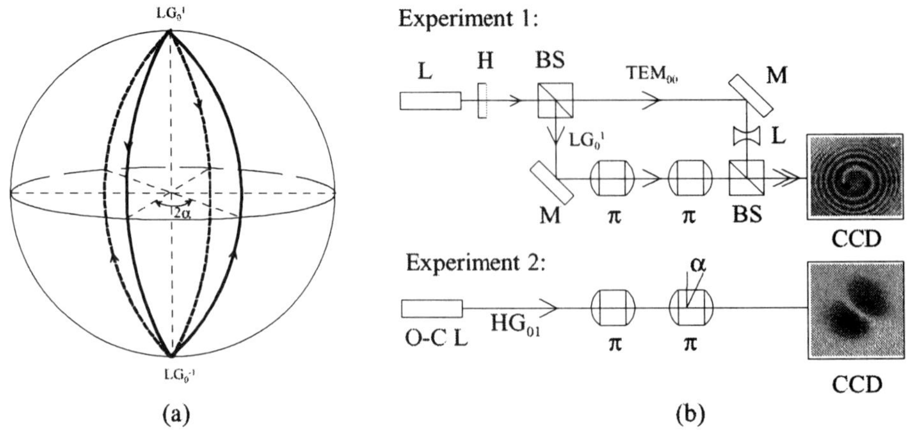

Fig.1. (a) Poincare sphere for mode transformations showing the paths that a beam in LG_0^1 (solid) and LG_0^{-1} (dashed) modes follow when going through two consecutive π-converters forming an angle α; (b) Experimental layouts for geometric phase measurements (see text) with examples of the recorded images.

A first-order Hermite-Gaussian mode rotated an angle θ relative to some fixed frame with transverse (x,y) coordinates, $HG_{10}(x',y')$, can be expressed as the superposition of the LG_0^1 and LG_0^{-1} modes with a phase difference of 2θ between them:

$$HG_{10}(x',y') = 2^{-1/2}\, LG_0^1\,(r,\phi)\, \exp[i\,\theta] + 2^{-1/2}\, LG_0^{-1}\,(r,\phi)\, \exp[-\,i\,\theta], \qquad (1)$$

where $x' = x \cos\theta + y \sin\theta$, $y' = y \cos\theta - x \sin\theta$, with $x = r \cos\phi$ and $y = r \sin\phi$. If we send a HG_{10} beam through two consecutive π-converters whose transverse axes form an angle α, the geometric-phase difference between the two component LG modes will be 4α. This will result in a rotation of the HG_{10} mode by $\theta = 2\alpha$, as given by Eq. 1. We tested this in a second type of experiment (see Fig. 1b), where we produced a HG_{10} beam with an open-frame HeNe laser forced to oscillate in the HG_{10} mode. The rotation of the HG_{10} mode by an angle θ was recorded and measured as a function of the rotation angle α of the rotating π-converter. A fit to data shows excellent agreement with the theoretical prediction: $\theta = (1.998 \pm 0.004)\,\alpha$.

This research was supported by an award from Research Corporation and a Colgate University Schlichting Fellowship. We thank R. Williams for the construction of mounts.

[1] S.J. van Enk, "Geometric phase, transformations of gaussian light beams and angular momentum transfer," Opt. Comm. **102**, 59-64 (1993).

[2] L. Allen, M.J. Padgett and M. Babiker, "The orbital angular momentum of light" in *Progress in Optics XXXIX*, E. Wolf, ed. (Elsevier Science B.V. 1999).

[3] J. Courtial, K. Dholakia, D.A. Robertson, L. Allen and M. Padgett, "Measurement of the rotational frequency shift imparted to a rotating light beam possessing orbital angular momentum," Phys. Rev. Lett. **80**, 3217-3219 (1998).

[4] G.F. Brand, "A new millimeter wave geometric phase demonstration," Int. J. Infrared Millimeter Waves **21**, 505-518 (2000).

[5] M.J. Padgett and J. Courtial, "Poincare-sphere equivalent for light beams containing orbital angular momentum," Opt. Lett. **24**, 430-432 (1999).

[6] G.S. Agarwal, "SU(2) structure of the Poincare sphere for light beams with orbital angular momentum," J. Opt. Soc. Am. A **16**, 2914-2916 (1999).

[7] L. Allen, M.W. Beijersbergen, R.J.C. Sprew and J.P. Woerdman, "Orbital angular momentum of light and the transformation of Laguerre-Gaussian laser modes," Phys. Rev. E **45**, 8185-8189 (1992).

[8] M.W. Beijersbergen, L. Allen, H.E.L.O. van der Veen, and J.P. Woerdman, "Astigmatic laser mode converters and transfer of orbital angular momentum," Opt. Comm. **96**, 123-132 (1993).

Fresnel diffraction of circular apertures illuminated with partially-coherent light: cross spectral density propagation

J. G. Suárez-Romero
Centro Nacional de Metrologia, km 4,5 Los Cués, Querétaro 76241, MÉXICO
telephone (4) 211 05 00 ext. 3328, fax (4) 211 00 53, jsuarez@cenam.mx

E. Tepichin-Rodríguez
Instituto Nacional de Astrofísica, Óptica y Electrónica, L. E. Erro 1,Tonantzintla, Pue. 72840, MÉXICO
telephone (2) 247 22 19, fax (2) 247 29 40, tepichin@haro.inaoep.mx

Abstract

In this work we present the calculation of the cross-spectral density diffracted by a circular aperture in the Fresnel approximation. Our proposal is a generalization to partially coherent illumination of the problem of near field diffraction due to a circular aperture. We show that our generalization contains as particular cases the ones already reported in the literature. We present numerical and experimental results that verify our proposal. This work is a contribution in the treatment of diffraction by multiple apertures.

1. Introduction

In many practical situations of radiant flux measurements it is recommended to use two or more successive apertures [1, 2]. For such cases, so far up to our knowledge, there has not been presented evaluations of the corresponding diffraction effects, mainly because does not exist a practical formulation that can allow us to evaluate them. In this direction, we are interested in applying the theory of partially coherence light to solve this problem. Previous authors have calculated the partially-coherent diffraction pattern of a single aperture of any shape in the Fresnel approximation [3, 4] but since only the irradiance was calculated the result cannot be applied to two apertures. Therefore the aim of this work is to calculate the irradiance distribution in the plane of the detector taking into account two successive apertures, applying the theory of partially coherence light. We discuss in section 2 the problem in terms of the partial coherence theory and in section 3 we show a comparison between numerical calculation and experimental results obtained in our laboratory.

2. Basic Theory

In a system with two successive apertures, the first one can be thought as a partially coherent source with a cross spectral density denoted as $W(P_1, P_2, \omega)$, which can be obtained with the help of the van Cittert-Zernike theorem. We are interested in evaluating the cross spectral density $W'(U_1, U_2, \omega)$ propagated through the second aperture. We show in this work that the near field propagation is given by

$$W'(U_1,U_2,\omega)=\frac{1}{(2\pi)^4}\frac{z_2^2}{z_1^2}e^{i\frac{z_1+z_2}{2R^2k}\left(v_1^2-v_2^2\right)}$$

$$\int_{P_2'}\int_{P_1'}W_0(P_1',P_2',\omega)\,e^{i\frac{z_1}{2R^2k}\left(L_1^2-L_2^2\right)}\,e^{i\frac{z_1}{R^2k}\left(L_2v_2\cos(\Psi_2-\beta_2)-L_1v_1\cos(\Psi_1-\beta_1)\right)}$$

$$\pi^2\left[C(u,L_2)-iS(u,L_2)\right]\left[C(u,L_1)+iS(u,L_1)\right]$$

$$L_1d\Psi_1L_2d\Psi_2dL_1dL_2\ , \tag{1}$$

where $C(u,L)$-$iS(u,L)$ is the diffraction field due to a point source and is given in terms of the Lommel Functions [5]. We have in Eq. (1) the cross spectral density propagated through a circular aperture (Fresnel approximation). The irradiance is obtained making $U_2=U_1$. The particular cases of diffraction that correspond to a point source and to an extended circular incoherent source are contained in Eq. (1). In the next section we evaluate it.

3. Numerical and experimental results

We considered a circular incoherent source of radio $a_0 = 4{,}7$ mm, the radii of the apertures are 1,5 mm and 1,0 mm respectively. We specifically designed our optical arrangement in such a way that the first aperture is always completely fulfilled by the incoherent source when it is observed from the detector. The incoherent source is separated from the first aperture by a distance of 0,5 m. The distance between the two apertures is 1 m, and the distance between de second aperture and the detector is 0,5 m. The wavelength of the monochromatic source is 600 nm. We show in Fig. 1 a comparison between theoretical and experimental results, it is clear that the theoretical calculation reproduces very well the experimental results.

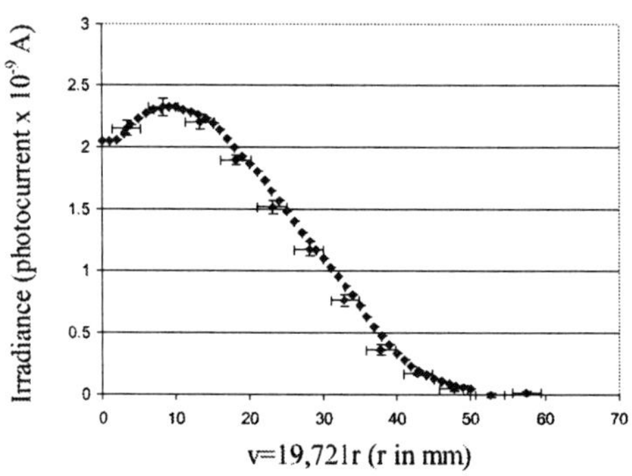

Fig. 1. Comparison between theoretical and experimental results.

4. Conclusions

We presented the calculation of the cross spectral density propagated through a circular aperture in the Fresnel approximation. Such calculation is used in reproducing the diffraction pattern of a two-successive aperture system. The numerical and experimental results coincide perfectly.

5. References

[1] L. P. Boivin, "Diffraction losses associated with tungsten lamps in absolute radiometry", Applied Optics, **14**, 197-200 (1975).

[2] E. L. Shirley, "Revised formulas for diffraction effects with point and extended sources", Applied Optics, **37**, 6581-6590 (1998).

[3] K. A. Nugent, "A generalization of Schell's theorem" Opt. Comm, **79**, 267-269 (1990).

[4] K. A. Nugent, "Partially coherent diffraction patterns and coherence measurements", JOSA A, **8**, 1574-1579 (1991).

[5] M. Born, E. Wolf, *Principles of Optics, 6th Edition* (Oxford, Pergamon Press, 1993) 517p.

Radiation forces produced by non-paraxial beam trapping

Qiang Lin, Zhaoying Wang

Department of Physics, Zhejiang University, Tian-Mu-Shan Road, Hangzhou 310028, China

Tel/Fax: +86-571-8273 086, E-mail: qlin@mail.hz.zj.cn

Abstract: Radiation forces on a micro dielectric sphere produced by standing-wave trapping of non-paraxial beams are analyzed theoretically and calculated numerically. The results show that Non-paraxial standing-wave trapping has some advantages in controlling minute Rayleigh particles.

Laser beam trapping is a powerful tool to capture and control the motion of neutral dielectric particles. When the diameter of a particle is very small, it is necessary to focus the beam very tightly to match the particle size. If the beam waist size is smaller than wavelength, the paraxial beam theory will be inapplicable, and the rigorous Maxwell's equations must be used. Our analyses are based on the rigorous solutions of Maxwell's equations for the field produced by an electric dipole, which is called E_{01} mode. Let us assume the standing-wave trapping formed by superposition of the E_{01} mode and its reflected beam through a reflector. The two E_{01} modes can be expressed as follows [1]:

$$\bar{E}_m(\bar{r},t) = E_0\left\{\left[g(kR_m) + [f(kR_m) - g(kR_m)]\frac{x^2}{R_m^{\,2}}\right]\bar{i} + [f(kR_m) - g(kR_m)]\frac{xy}{R_m^{\,2}}\bar{j}\right.$$

$$\left. + [f(kR_m) - g(kR_m)]\frac{x(z_m - iz_0)}{R_m^{\,2}}\bar{k}\right\}e^{i\omega t}, \qquad (m = 1,2). \tag{1}$$

There are two different kinds of radiation forces on the dielectric sphere. One is scattering force $\bar{F}_{scat}$ produced due to scattering of light by dielectric sphere [2,3]:

$$\bar{F}_{scat} = \frac{8n_2}{3c}\pi(ka)^4 a^2\left(\frac{n_{12}^{\,2} - 1}{n_{12}^{\,2} + 2}\right)^2 \bar{S}(\bar{r}). \tag{2}$$

Another kind of radiation force is Lorenz force produced by electro-magnetic field on the dielectric sphere, which is called gradient force $\bar{F}_{grad}$ [4]:

$$\bar{F}_{grad} = \pi n_2^2 \varepsilon_0 a^3 \left(\frac{n_{12}^2 - 1}{n_{12}^2 + 2} \right) \nabla \left| \bar{E}(\bar{r}) \right|^2 . \tag{3}$$

Substituting the corresponding electric field of the superimposed beam formed by two E_{01} modes into eq.(2) and eq.(3) respectively, we can obtain the analytical expressions of the scattering force and gradient force of the standing-wave trapping. Fig.1 shows the radiation forces profile along three coordinates for non-paraxial beam trapping ($w_0 = 0.2\lambda$).

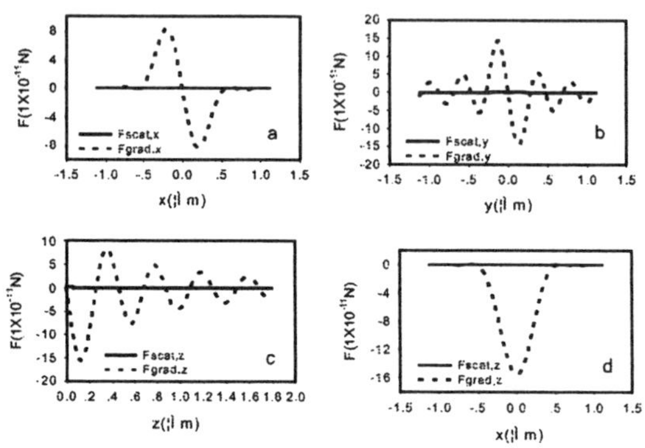

Fig.1. Radiation forces produced by superimposed non-paraxial Gaussian beams ($w_0 = 0.2\lambda$)

By comparison, we find that the non-paraxial standing-wave trapping has some advantages over paraxial standing-wave trapping in several respects. At first, it provides much larger and sharper radiation forces under the same power. Therefore, less beam power is needed to provide the same radiation force by using non-paraxial beam. Secondly, non-paraxial standing-wave trapping provides multiple equilibrium positions in both lateral and longitudinal directions. This is important in capturing micro particles. Furthermore, the beam size is matched to the particle size.

References:

[1] C. J. R. Sheppard and S. Saghafi, "Electric and magnetic dipole beam modes beyond the paraxial approximation," Optik **110**, 487-491 (1999).

[2] A. Ashkin, J. M. Dziedzic, J. E. Bjorkholm, et al., "Observation of a single-beam gradient force optical trap for dielectric particles," Opt. Lett. **11**, 288-290 (1986).

[3] M. kerker, *The scattering of light and other electromagnetic radiation* (Academic, New York, 1969), Chap.3.

[4] J. A. Stratton, *Electromagnetic theory* (McGraw-Hill, New York, 1941), Chap.3.

Measurement of the polarization-induced asymmetry in focal spots

G. Leuchs, S. Quabis, R. Dorn, O. Glöckl, and M. Reichle

Zentrum for Modern Optics, University of Erlangen, Germany

leuchs@physik.uni-erlangen.de

Abstract: Polarization effects may have a strong influence on the shape of the focal spot. Measurements and theoretical calculations for various input fields are presented. We present the first experimental evidence for an asymmetry in the case of linear polarization and for a strong longitudinal field component on the axis in the case of radial polarization.

OCIS codes: 110.0110 Imaging Systems, 110.0180 Microscopy, 260.0260 Physical optics, 260.5430 Polarization

For strong focusing with high numerical apertures, the scalar diffraction theory is no longer valid and the full vectorial character of the electromagnetic field has to be taken into account. Fundamental calculations, which are based on the Debye–approximation, were performed by Richards and Wolf [1]. They found that for a linearly polarized beam the focal spot is asymmetrically deformed along the direction of linear polarization. This asymmetry can be revoked when an input field with a doughnut-shaped intensity distribution is used, which is radially or azimuthally polarized. Here we present measurements of intensity distributions in the focal region for input fields with different polarization states and show the experimental verification of the predicted asymmetry for linear polarization.

To measure the focal intensity distribution with high lateral resolution we have developed an experimental setup which is based on the knife-edge method combined with tomographic reconstruction [2]. The incoming beam is focused by a microscope objective onto a specially designed photodiode where parts of the active area are covered with sharp edged opaque structures that are used as a knife edge. The measurement of the photocurrent as a function of the displacement of the diode allows for the calculation of the projection of the intensity distribution onto the translation axis. To reconstruct the intensity distribution using the Radon transformation several projections have to be measured by moving the edge through the focus under different angles. When this process is repeated for different distances between the diode and the focal plane, a three dimensional intensity distribution can be obtained.

In the experiment the beam of a linearly polarized Helium–Neon laser (λ=633nm) was expanded to allow for a homogeneous illumination of the entrance pupil of the focusing microscope objective (NA=0.9, 100x, pol.). Intensity profiles were measured for various orientations of the polarization with respect to the diode and for different defocus positions with a relative distance of 100nm.

Fig. 1a and 1b show the measured projections of the intensity distribution onto a plane parallel to the optical axis and the direction of displacement of the edge for perpendicular and parallel polarized input beams with respect to the edge. It is remarkable that in contrast to the theoretical predictions (fig. 1c and 1d) the experimentally obtained intensity distribution does not show forward–backward symmetry with respect to the focal plane (z=0) which we yet do not understand. In the area between lens and focal plane one finds a qualitatively good agreement and a significant difference between the two polarizations. A projection of the intensity distribution in the focal plane for the two orthogonal polarizations is depicted in fig. 2.

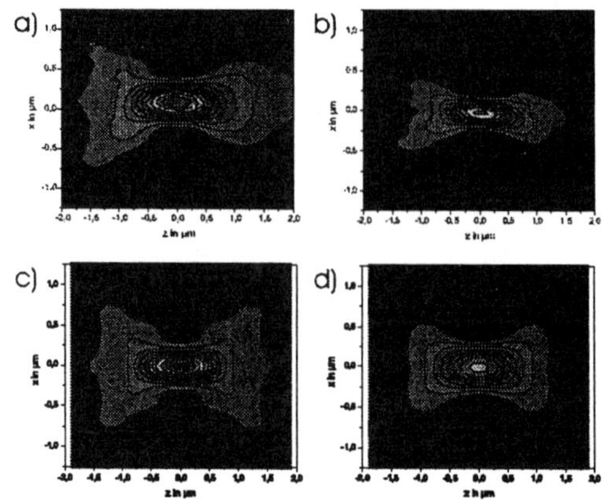

Fig. 1. Projection of the intensity distribution onto a plane containing the optical axis and the displacement axis along which the knife edge was moved. a) and c): measured and theoretical distribution with direction of polarization parallel to the edge, b) and d): perpendicular to the edge. The direction of propagation of the beam is from the left to the right.

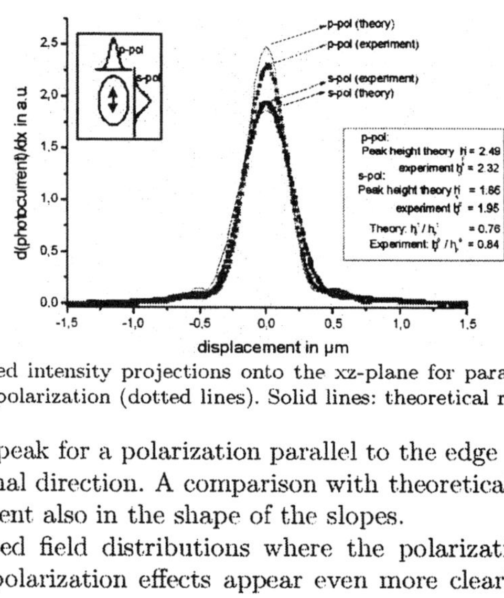

Fig. 2. Measured intensity projections onto the xz-plane for parallel (p) and perpendicular (s) polarization (dotted lines). Solid lines: theoretical results

As expected, the peak for a polarization parallel to the edge is higher and narrower than for the orthogonal direction. A comparison with theoretical curves shows a qualitatively good agreement also in the shape of the slopes.

For annular shaped field distributions where the polarization distribution shows cylinder symmetry, polarization effects appear even more clearly. For an azimuthally polarized field the dark spot in the center of the beam is maintained in the focus. In contrast, for a radially polarized field an additional longitudinal field component on the optical axis was also observed experimentally which shows a very narrow intensity distribution.

References and links

1. B. Richards, E. Wolf, "Electromagnetic diffraction in optical systems II", Proc. Roy. Soc. A **253**, 359 (1959)
2. S. Quabis, R. Dorn, M. Eberler, O. Glöckl, and G. Leuchs, "The focus of light - theoretical calculation and experimental tomographic reconstruction", Appl. Phys. B **72**, 109-113 (2001)

Partially coherent imaging through turbulence with an all-optical adaptive-optics system

Tomohiro Shirai
Photonics Research Institute, National Institute of Advanced Industrial Science and Technology (AIST), 1-2 Namiki, Tsukuba 305-8564, Japan
t.shirai@aist.go.jp

T. H. Barnes
Physics Department, University of Auckland, Private Bag 92019, Auckland, New Zealand

Abstract: A liquid-crystal adaptive-optics system using all-optical feedback interferometry is applied to partially coherent imaging through turbulence. We show both theoretically and experimentally that a blurred image due to the phase disturbance is restored in real-time.

1. Introduction

An early objective of adaptive optics was to compensate for blurring of astronomical images [1]. Nowadays the use of adaptive optics has increasingly expanded beyond astronomy and become important in a wide range of applications such as visual science [2].

In conventional applications almost all adaptive-optics (AO) systems use segmented or deformable mirrors as wavefront correctors. However, it has recently been shown elsewhere that liquid-crystal (LC) spatial light modulators (SLMs) have many advantages over mirror devices because of their compactness, low cost, low power consumption and potential for high-resolution correction [3]. In this connection we have recently proposed a novel AO system based on all-optical feedback interferometry where, using an optically addressed phase-only LC SLM, the wavefront correction is achievable simultaneously with wavefront sensing without any electronic calculations [4]. We have applied this to a system of coherent imaging through phase disturbance and shown by experiments that a blurred image was successfully restored immediately after the feedback loop was closed [5].

In this paper we demonstrate both theoretically and experimentally that the blurred image due to phase disturbance can be restored in our setup in real-time, irrespective of the state of coherence of the light illuminating the object.

2. Optical setup for partially coherent imaging with adaptive optics

Figure 1 (a) shows our all-optical AO system [4]. This setup is basically an all-optical feedback interferometer consisting of a Mach-Zehnder interferometer with large radial shear and a Hamamatsu PAL-SLM (a high resolution, non-pixelized, optically addressed, phase-only, parallel-aligned nematic LC SLM). The aberrated wavefronts to be corrected are linearly polarized and incident onto the phase-modulating side (R side) of the PAL-SLM through beam splitter BS1. It is shown previously [5] that, when the system works properly, the phase distribution across the R side of the PAL-SLM becomes nearly half of the conjugate of the input aberrated wavefronts and thus the input aberrated wavefronts are compensated for on reflection at the PAL-SLM.

Figure 1 (b) shows a setup for imaging through phase disturbance [5]. This is combined with the AO system described above by BS1 and the write light to the PAL-SLM. The object, illuminated by light (PW1) of arbitrary state of coherence, is imaged onto a CCD camera (CCD3) by a four-lens afocal system with the same focal length f, via wedge plates WP3 and WP1 acting as beam splitters. An aberration plate (AP) acting as phase disturbance is located anywhere between WP3 and lens L11, and the R side of the PAL-SLM is arranged at the image plane of AP by two lenses L11 and L12. To drive the AO system, a linearly polarized plane wave (PW2) is incident upon AP via WP3 and then imaged on the R-side of the PAL-SLM. The reflected light from the PAL-SLM is redirected at BS1 and goes to lens L1. Each optical element in our setup was so arranged that the light for the imaging system (PW1) had no influence on the AO system, and vice versa.

To find the imaging characteristics, we examined the propagation of the cross-spectral density of the incident light through the imaging system. We found that the cross-spectral density of light just behind the object is perfectly reproduced in the observation (CCD3) plane when the AO system works properly. In other words, the blurred image resulting from phase disturbance is successfully restored in our adaptive-optical imaging system, irrespective of the state of coherence of light illuminating the object. Furthermore, this performance holds true, irrespective of the location of phase disturbance (AP), insofar as the R side of the PAL-SLM is exactly arranged at the image plane of AP by L11 and L12.

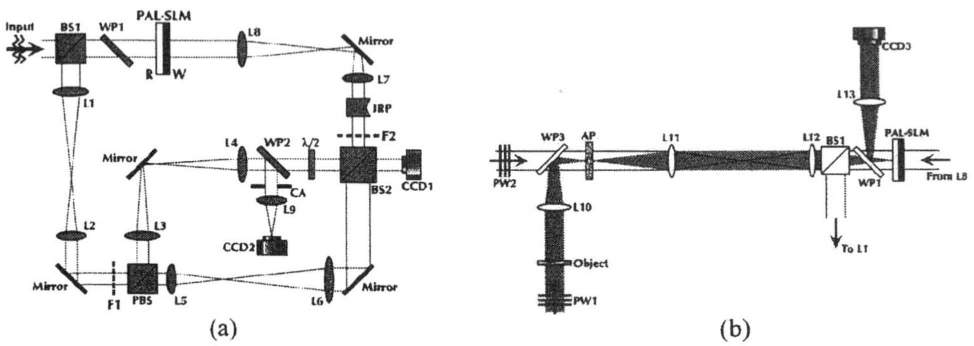

Fig. 1. Experimental setups for (a) adaptive optics and for (b) imaging through phase disturbance.

3. Experimental results

To verify the theoretical findings, we performed adaptive-optical imaging experiments for two cases when the object to be imaged is illuminated by spatially coherent light originating from a He-Ne laser and spatially incoherent white light originating from a halogen lamp. As a result, we observed in both cases that the images blurred by the phase disturbance were successfully restored immediately after the AO system was activated, though the quality of the restored images was different from each other. Typical results are shown in Fig.2.

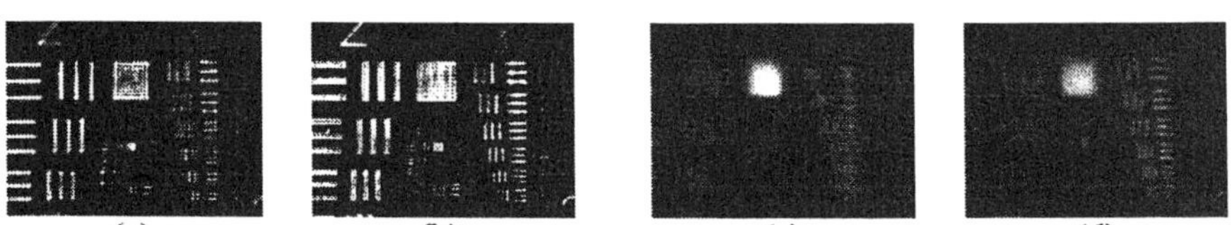

(a) (b) (c) (d)

Fig. 2. Images of a coherently-illuminated object (a) without and (b) with adaptive optics, and those of an incoherently-illuminated object (c) without and (d) with adaptive optics.

Acknowledgement

This research was supported by Industrial Technology Research Grant Program in '00 from the New Energy and Industrial Technology Development Organization (NEDO) of Japan.

References

1. For example, R. K. Tyson, *Principles of Adaptive Optics*, 2nd ed. (Academic Press, New York, 1998).
2. D. R. Williams, J. Liang, D. T. Miller, and A. Roorda, "Wavefront Sensing and Compensation for the Human Eye," in *Adaptive Optics Engineering Handbook*, R. K. Tyson, ed. (Marcel Dekker, New York, 2000), pp.287-310.
3. For example, G. D. Love, "Liquid Crystal Adaptive Optics," in *Adaptive Optics Engineering Handbook*, R. K. Tyson, ed. (Marcel Dekker, New York, 2000), pp.273-285.
4. T. Shirai, T. H. Barnes, and T. G. Haskell, "Adaptive wave-front correction by means of all-optical feedback interferometry," Opt. Lett. **25**, 773-775 (2000).
5. T. Shirai, T. H. Barnes, and T. G. Haskell, "Real-time restoration of a blurred image with a liquid-crystal adaptive-optics system based on all-optical feedback interferometry," Opt. Commun. **188**, 275-282 (2001).

Interrelation between spatial and temporal reciprocity inequalities of partially-coherent beams

M. U. Karelin and A. M. Lazaruk (late)

B. I. Stepanov Institute of Physics, National Academy of Sciences, F. Scaryna Ave., 68, Minsk, 220072, Belarus

karelin@dragon.bas-net.by

As it was shown in recent papers [1, 2, 3] both spatial and temporal structure of a beam pulse are subject to generalized reciprocity inequalities:[1] a product of characteristic pulse duration on its frequency bandwidth grows proportionally to effective number of modes in this pulse

$$\Delta\omega\,\Delta t \propto N_{eff}. \tag{1a}$$

The relation between width of angular spectrum of a beam and its cross-section width behaves in just the same manner

$$(\Delta\theta\,\Delta r)^2 \propto N_{eff}. \tag{1b}$$

Light fields, which achieve the minimum of reciprocity inequality, is of special interest because of their optimal structure and absence of extra noise or additional divergence. The effective number of modes (or number of degrees of freedom) N_{eff} in the field is a characteristic of the global coherence of the field, i. e. the ability to create long-time interference effects, such as speckle-structure and so on.

The main task of our investigation was to find out if it is possible to create *optimal* fields, which simultaneously realize minima of reciprocity relations both in temporal and spatial domains. According to two variants of modal representation (deterministic and statistical [4]) of partially coherent pulses, there are two approaches to this problem.

In the first case we investigate the structure of individual partially coherent light pulse with desired properties. As it shown in [1] at minimum of reciprocity inequality, eigenfunctions of field decomposition are Hermite functions $\psi_m(x)$:

$$E(\mathbf{r},t) = \sum_{n,m_1,m_2} a_{n,m_1,m_2}\,\psi_n(t)\,\psi_{m_1}(x)\,\psi_{m_2}(y),$$

($\mathbf{r} = (x,y)$), all values in inequalities (1) can be expressed via a_{n,m_1,m_2}. And it can be shown, that, in parametric form (N is a limit of summation on n),

$$\min(\Delta\omega\,\Delta t) = (1/3)(2N+1), \quad N_{eff} = (3/2)\big(N\,(N+1)\big)/(2\,N+1). \tag{2}$$

In the spatial domain decomposition coefficients would be degenerate and depend only on $m_+ = m_1 + m_2$ (here m_+ varies from 0 to M), And reciprocity inequality takes the parametric form:

$$\min(k\,\Delta\theta\,\Delta r) = M+1, \quad N_{eff} = (1/3)M\,(M+2). \tag{3}$$

[1] By analogy with quantum mechanics such inequalities are often referred to as "uncertainty relations". Indeed, it is necessary to mention that our results can, with suitable change of terminology, be readily applied to analysis of the density matrix of, for example, a system in entangled state.

Structures of modal eigenvalues in minimum of inequalities (1) strongly depend on subspace dimensionality. In temporal domain with dimension 1 modal weights are nondegenerate, but for beam cross-section spatial structure (with dimension equal to 2) modes have to be degenerate.

Strictly speaking, it is impossible to create the field, realizing minima for **both** inequalities (1a) and (1b). One pulse can just be very close to this optimum, what can be achieved by special fitting of spatial and temporal modes. For instance, one can choose modal weight a_{n,m_1,m_2} accordingly to temporal inequality (1a) minimum, and then spatial structure of a beam will have some difference from optimum. Our calculations show, that in most cases a divergence from absolute minimum does not exceed 10^{-2}, and discrepancy between ideal and really possible structure of light beam is negligible. The main problem is that the proposed procedure of modal fitting is not quite suitable for real experiment

From practical point of view it is more important to find out, how close to optimum one can approach, using real light sources and procedures of spatial and angle filtration. Naturally, such approach deals with a typical pulse from some ensemble — that is we need to use second variant of modal representation [4]. For an ensemble with Gaussian statistics $1/\overline{N_{\mathit{eff}}} = 1/N_T + 1/N_S$, where $\overline{N_{\mathit{eff}}}$ — average number of modes for the ensemble, N_T and N_S are effective number of terms in decomposition of ensemble averaged temporal and spatial correlation functions, respectively. These functions one should use also for evaluation of variables in left-hand side of reciprocity inequality. Now, with the help of formalism similar to used above, we can show that ensemble, which is subject to condition

$$(\overline{\Delta\omega}\,\overline{\Delta t})/\overline{N_{\mathit{eff}}} \propto 1$$

is possible only in limiting case $N_S \to \infty$ of infinite number of modes in adjacent subspace. Thus in spatial domain the ensemble structure will be **very** far from optimum. A similar situation takes place for minimum of relation between width of angle spectrum of a beam and size of its cross-section. So, there is a problem of finding of some reasonable compromise between spatial and temporal structure of an ensemble.

To this end we propose a concept of balanced spatio-temporal structure of the ensemble, with minimum of product of spatial and temporal phase volumes of the field. The main condition for such a criterion is (approximate) equality of numbers of spatial and temporal degrees of freedom in the ensemble. Then it follows $\overline{N_{\mathit{eff}}} \propto N_S/2 \propto N_T/2$, and in either domain the absolute reciprocity inequality minimum will be exceeded (approximately, in order of magnitude 2). From author's point of view, this is an optimal limit, which should be aimed for specific sources of partially coherent irradiation for optical information processing schemes, working in extremal conditions.

The work is partially supported by Belarusian Fund for Fundamental Research under grants F97-253 and F99-211.

References and links

1. M. J. Bastiaans, "Uncertainty principle for partially coherent light", J. Opt. Soc. Am., **73**, 251-255 (1983).
2. A. T. Friberg, T. D. Visser, and E. Wolf, "A reciprocity inequality for Gaussian Schell-model beams and some of its consequences", Opt. Lett., **25**, 366-368 (2000).
3. M. U. Karelin, A. M. Lazaruk, "Structure of the density matrix providing the minimum of generalized uncertainty relation for mixed states", J. Phys. A, **33**, 6807-6816 (2000).
4. A. M. Lazaruk, M. U. Karelin, "Application of modal theory of coherence for nonstationary partially coherent optical fields", submitted on 8th Rochester Conference on Coherent and Quantum Optics (see accompanuing paper).

Ultra-sensitive autocorrelation of 1.5 μm light with a photon-counting silicon avalanche photodiode

Jeffrey M. Roth, Chris Xu†, Wayne H. Knox†, Keren Bergman

Princeton University, Department of Electrical Engineering, Princeton, NJ 08544 USA
†Bell Laboratories, Lucent Technologies, Holmdel, NJ 07733 USA
jroth@princeton.edu

Abstract: We report unprecedented sensitivity to 1.5 μm two-photon absorption events in a photon-counting silicon avalanche photodiode. Autocorrelation of a 10 GHz, 1.67 ps pulsetrain using this device demonstrates record peak-power times average-power sensitivity of 1.5×10^{-3} $(\text{mW})^2$ without lock-in detection.

OCIS codes: 190.5970) Semiconductor nonlinear optics including MQW; (270.4180) Multiphoton processes; (320.7100) Ultrafast measurements

Experimentalists in optics rely widely on nonlinear autocorrelation techniques for picosecond and femtosecond pulsewidth measurements because of their straightforward and well-studied nature. The nonlinear process enables one to distinguish the pulse from coherent noise, and it most commonly entails second-harmonic generation in a crystal followed by electrical detection of the upconverted light. Recently two-photon absorption (TPA) autocorrelation has attracted considerable attention because it achieves quadratic nonlinearity using simple direct electrical detection in a semiconductor material[1, 2] that obviates the complexities of phase-matching in a nonlinear crystal. The functional behavior of TPA depends upon peak power and average power. The remedy for low power situations typically involves boosting the power in an optical amplifier to perform the measurement, even though this degrades signal to noise ratio and distorts the pulseshape. This investigation reports on a novel TPA autocorrelation technique that uses a silicon avalanche photodiode (APD) to achieve enhanced sensitivity over other TPA techniques. The sensitivity of the APD, together with the lack of single-photon background events makes this method ideally suited for characterization of low power pulses.

Figure 1 (center) shows the experimental setup we have used to demonstrate the silicon APD autocorrelator. Picosecond pulses from a 1.555 μm modelocked fiber laser with a 10 GHz repetition rate are launched into an interferometric free-space Michelson autocorrelator and the output of the autocorrelator is focused by a 10X/0.26NA objective lens (Melles Griot) to about a 10 μm spot size on the photon-counting silicon APD module (EG&G, SPCM-AQ-16). Each two-photon absorption event produces a TTL signal that is recorded by the photon counter as a function of time delay to form the autocorrelation trace.

The lower right of figure 1 shows the autocorrelation trace obtained with 7.2 μW of average power incident onto the APD with the pulses not overlapping in time; this corresponds to only 220 μW of peak power and note that under these conditions neither lock-in detection nor synchronous photon counting were used. The measured autocorrelation FWHM (full-width, half-maximum) is 2.82 ps. If we assume a transform-limited

$sech^2(\tau)$ pulse shape, the corresponding 1.67 ps pulse width is consistent with the measured 1.6 nm spectral width shown in Figure 1. We have also measured two-photon generated photon-counts as a function incident power using a single mode CW 1.575 μm laser as the excitation source. A single mode CW laser provides the advantage that pulse width and the exact pulse shape will not affect the TPA response measurement. Figure 1 includes the power dependence curve by varying the incidence power over about two decades. At the lowest power of our measurement the slope in the log-log plot is 1.4, indicating some one-photon contribution possibly due to material defects or band tail. The slope increases to >1.95 at power levels above 100 μW indicating negligible one-photon contribution.

The ultra-high sensitivity in measuring two-photon absorption is provided by the combination of large internal gain of the silicon APD ($\sim 10^6$), the quantum-limited photon-counting detection scheme, and the large TPA response inherent to silicon. The autocorrelation data shown exhibits predominantly quantum shot noise, so longer counting intervals could extend the measurement capability to lower powers. The results could be further enhanced by focusing to a smaller spot size on the APD and by counting photons synchronously with optical chopping.

In summary, we have shown that two-photon absorption in a photon-counting silicon APD provides unprecedented sensitivity in nonlinear autocorrelation measurements using direct detection from the output of the photon counting module. A high signal to noise ratio trace was obtained at an average, peak power product of 1.5×10^{-3} (mW)^2, which to our knowledge is a record for two-photon absorption autocorrelation performed without the use of a lock-in amplifier. CW data in Figure 1 reveals that one-photon absorption should remain negligable for average-power levels of the 10 GHz pulsetrain down to 2 μW.

The authors acknowledge Dr. Manyalibo J. Matthews for providing the APD.

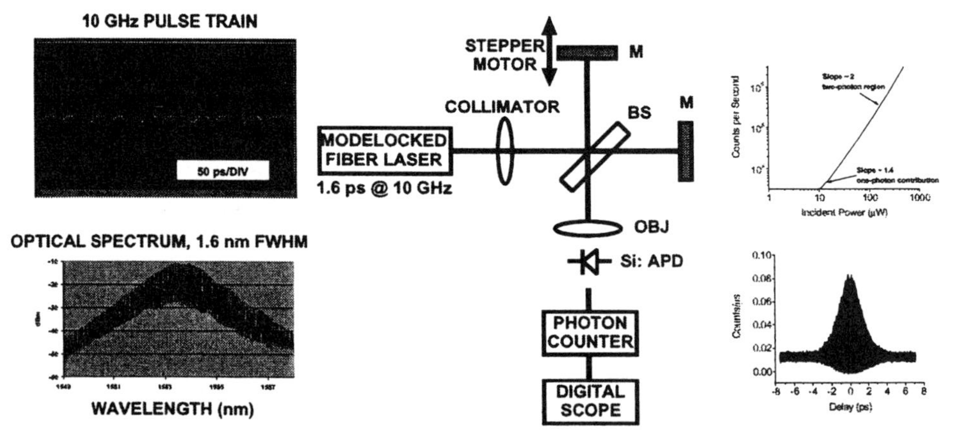

Fig. 1. Autocorrelator schematic (center); input pulse train and optical spectrum (left). CW TPA response (top right) and 7.2 μW autocorrelation trace (lower left).

References and links

1. L.P. Barry, B.C. Thomsen, J.M. Dudley, J.D. Harvey, "Autocorrelation and ultrafast optical thresholding at 1.5 um using a commercial InGaAsP 1.3 um laser diode", Electronics Letters **34**, 358-360 (1997).
2. F.R. Laughton, J.H. Marsh, D.A. Barrow, E.L. Portnoi, "The Two-Photon Absorption Semiconductor Waveguide Autocorrelator," IEEE J. Quant. Elec. **30**, 838-845 (1994).

A New Phase Space Product for Light Beams of any State of Coherence

Sergey A. Ponomarenko* **and Emil Wolf**

Department of Physics and Astronomy,
University of Rochester, Rochester, NY 14627

** email: ponos@pas.rochester.edu*

To date, many papers have been published which deal with the second-order moments of the spatial and angular intensity distributions of partially coherent beams [see, for example, 1-12,14]. A possible choice for the appropriate phase-space product of a beam, which involves the second-order moments of the Wigner distribution function, has been introduced and measured [2]. Later, the term M^2 quality factor for this product was coined by Siegman [3]. However, the beam cross-section in the source plane has been generally introduced in terms of a second-order moment of the intensity distribution [1-4,8-11]. Defined in this way, the cross-section does *not* depend on the coherence properties of the source. In order to examine the influence of the state of coherence of the source on the value of the phase-space product of the beam generated by the source, one has to define the effective cross-section and the effective angular spread so that these quantities contain information about both the intensity of the source *and* its coherence properties.

Let us consider the field generated by a planar, secondary, statistically stationary source of any state of coherence, situated in the plane $z = 0$ and radiating into the half-space $z > 0$. Let $W^{(0)}(\boldsymbol{\rho}_1, \boldsymbol{\rho}_2, \omega)$ be the cross-spectral density of the light at frequency ω [13, sec. 4.3.2] at a pair of points with position vectors $\boldsymbol{\rho}_1$ and $\boldsymbol{\rho}_2$ in the source plane. It represents the field correlations at frequency ω which exist at pairs of source points. We define the normalized width δ_W of $W^{(0)}$ by the expression

$$\delta_W^2 = \frac{\int d^2\rho_1 \int d^2\rho_2 \, (\boldsymbol{\rho}_1 - \boldsymbol{\rho}_2)^2 \, |W^{(0)}(\boldsymbol{\rho}_1, \boldsymbol{\rho}_2, \omega)|^2}{\int d^2\rho_1 \int d^2\rho_2 \, |W^{(0)}(\boldsymbol{\rho}_1, \boldsymbol{\rho}_2, \omega)|^2}, \tag{1}$$

where the integrations extend over the source. Next, we introduce the angular correlation function $\mathcal{A}(\mathbf{s}_1, \mathbf{s}_2, \omega)$ [see, for example, 13, sec. 5.6.3], which represents the correlation between pairs of plane-wave modes propagating in the directions specified by unit vectors $\mathbf{s}_1$ and $\mathbf{s}_2$. We define the normalized width $\delta_{\mathcal{A}}$ of the angular correlation function by the formula

$$\delta_{\mathcal{A}}^2 = \frac{\int d^2 s_{1\perp} \int d^2 s_{2\perp} \, (\mathbf{s}_{1\perp} - \mathbf{s}_{2\perp})^2 \, |\mathcal{A}(\mathbf{s}_1, \mathbf{s}_2, \omega)|^2}{\int d^2 s_{1\perp} \int d^2 s_{2\perp} \, |\mathcal{A}(\mathbf{s}_1, \mathbf{s}_2, \omega)|^2}. \tag{2}$$

In this expression, $\mathbf{s}_{1\perp}$ and $\mathbf{s}_{2\perp}$ are the projections, considered as two-dimensional vectors, of the unit vectors $\mathbf{s}_1$ and $\mathbf{s}_2$ onto the source plane $z = 0$. The integrations on the right-hand side of Eq. (2) are formally taken over the entire $\mathbf{s}_{1\perp}$ and $\mathbf{s}_{2\perp}$-planes though, in practice, only those $\mathbf{s}$-directions will contribute in which the far field is appreciable.

We now define a new phase-space product for beams of any state of coherence by the expression

$$Q = \left(\frac{\pi}{\lambda}\right) \delta_W \, \delta_{\mathcal{A}}, \tag{3}$$

where λ is the wavelength corresponding to the frequency ω. We then show that in the coherent limit, the new phase space product reduces to

$$Q_{coh} = \left(\frac{2\pi}{\lambda}\right) \sigma_I \sigma_J, \tag{4}$$

Coherence and Quantum Optics VIII
Edited by Bigelow et al., Kluwer Academic/Plenum Publishers, 2003

which is just the M^2 factor of Siegman [3]. In Eq. (4), σ_I and σ_J denote the variances of the intensity and of the radiant intensity, respectively.

A detailed analysis indicates that the novel phase space product is bounded from below by unity. We also demonstrate that the lowest value of Q is attained when the cross-spectral density of the beam is given by the expression

$$W_{min}^{(0)}(\boldsymbol{\rho}_1, \boldsymbol{\rho}_2) = B \exp\left[-\frac{(\rho_1^2+\rho_2^2)}{4\sigma_S^2}\right] \exp\left[-\frac{(\boldsymbol{\rho}_1-\boldsymbol{\rho}_2)^2}{2\sigma_g^2}\right]. \quad (5)$$

Equation (5) will be recognized as the cross-spectral density of a Gaussian Schell-model source [13, sec. 5.3.2], with the width of intensity distribution being σ_S and with the spatial coherence length being σ_g.

Finally, let us briefly compare the present approach with some others discussed in the literature. We have defined the effective size and the effective angular width in terms of the appropriate *correlations,* thus taking into account coherence properties of the source at a *local* level, via the spectral degree of coherence. Bastiaans [5-7], on the other hand, took account of the coherence properties of the source by using quite a different approach. He utilized the same definitions of the source size and the beam spread as those used in the coherent case, but has introduced an overall degree of coherence of the source, which amounts to incorporating its coherence properties at a *global* level. He then derived a series of inequalities involving the phase-space product as well as the overall degree of coherence. However, there appears to be no physical principle indicating which is the most appropriate measure of the overall degree of coherence. In fact, different choices of such measure have led Bastiaans to different results for the beams with minimum phase-space products [6, 7].

A fuller account of this work is given in Ref. [14].

References and links

1. M. J. Bastiaans, J. Opt. Soc. Am. **69**, 1710 (1979).
2. S. Lavi, R. Prochaska, and E. Keren, Appl. Opt. **27**, 3696 (1988).
3. A. E. Siegman, Proc. SPIE **1224**, 2-14 (1990).
4. R. Simon, N. Mukunda, and E. C. G. Sudarshan, Opt. Commun. **65**, 322 (1988).
5. M. J. Bastiaans, J. Opt. Soc. Am. **73**, 251 (1983).
6. M. J. Bastiaans, J. Opt. Soc. Am. A **1**, 711 (1984).
7. M. J. Bastiaans, J. Opt. Soc. Am. A **3**, 1243 (1986).
8. J. Serna, R. Martinez-Herrero, and P. M. Mejias, J. Opt. Soc. Am. A **8**, 1094 (1991).
9. G. Nemes and A. E. Siegman, J. Opt. Soc. Am. A **11**, 2257 (1994).
10. M. Santarsiero, F. Gori, B. Borghi, G. Cincotti and P. Vahimaa, J. Opt. Soc. Am. A **16**, 106 (1999).
11. S. A. Ponomarenko and E. Wolf, Opt. Lett. **25**, 663 (2000).
12. S. R. Seshadri, J. Opt. Soc. Am. A **17**, 780 (2000).
13. L. Mandel and E. Wolf, *Optical Coherence and Quantum Optics* (Cambridge University Press, Cambridge, 1995).
14. S. A. Ponomarenko and E. Wolf, Opt. Lett., **26**, 122 (2001).

Spectral Changes in Focused Partially Coherent Wave Fields

T.D. Visser[1] **and E. Wolf**[2]

[1]*Dept. of Physics and Astronomy, Free University*
De Boelelaan 1081, 1081 HV Amsterdam, The Netherlands
[2]*Dept. of Physics and Astronomy, University of Rochester*
River Campus Station, Rochester, NY 14627-0171, USA

E. Wolf: ewlupus@pas.rochester.edu

When light is focused, there are several mechanisms which may cause spectral changes. Among them are diffraction, the state of coherence of the light [1], and propagation through a linear time-invariant system [2]. The focusing configuration is depicted in Fig. 1, it being assumed that $f \gg a \gg \lambda$. The cross-spectral density function of a field $U^{(0)}$ of arbitrary state of coherence on a reference sphere [3] is given by the expression [4, Sec.4.3.2]

$$W^{(0)}(Q_1, Q_2, \nu) = W^{(0)}(\boldsymbol{\rho}_1, \boldsymbol{\rho}_2, \nu) = \langle U^{(0)*}(\boldsymbol{\rho}_1, \nu) U^{(0)}(\boldsymbol{\rho}_2, \nu) \rangle. \tag{1}$$

Here $\boldsymbol{\rho}_i$ $(i = 1, 2)$ is the projection onto the $\xi\eta$-plane of the position vector of the point Q_i on S. The cross-spectral density of the focused field is given by the formula

$$W(\boldsymbol{r}_1, \boldsymbol{r}_2, \nu) = \langle U^*(\boldsymbol{r}_1, \nu) U(\boldsymbol{r}_2, \nu) \rangle = \left(\frac{k}{2\pi}\right)^2 \int_S \int_S W^{(0)}(\boldsymbol{\rho}_1, \boldsymbol{\rho}_2, \nu) \frac{e^{-ik(s_1 - s_2)}}{s_1 s_2} \mathrm{d}^2\rho_1 \mathrm{d}^2\rho_2, \tag{2}$$

with $\boldsymbol{r}_i$ being points on the z-axis. We assume that the field on the reference sphere is of a Gaussian Schell-model form [4, Sec. 5.3.2]. Its cross-spectral density is then given by

$$W^{(0)}(\boldsymbol{\rho}_1, \boldsymbol{\rho}_2, \nu) = \frac{1}{f^2} \sqrt{I(\boldsymbol{\rho}_1, \nu) I(\boldsymbol{\rho}_2, \nu)} e^{-(\rho_2 - \rho_1)^2 / 2\sigma_g^2}, \tag{3}$$

where σ_g is a positive constant. The spectral intensity of the focused field at an axial point $\boldsymbol{r}$ is given by the formula

$$I(\boldsymbol{r}, \nu) = W(\boldsymbol{r}, \boldsymbol{r}, \nu). \tag{4}$$

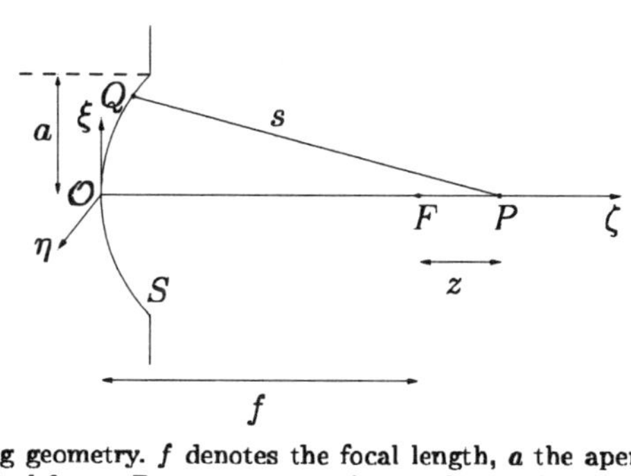

Fig. 1. Focusing geometry. f denotes the focal length, a the aperture radius, and F the geometrical focus. P represents a point on the z-axis, and Q a point on the Gaussian reference sphere S. s is the distance QP.

Coherence and Quantum Optics VIII
Edited by Bigelow et al., Kluwer Academic/Plenum Publishers, 2003

We assume that on the reference sphere the spectrum of the field is independent of position, i.e.

$$\sqrt{I(\rho_1,\nu)I(\rho_2,\nu)}/f^2 = \mathcal{S}(\nu). \tag{5}$$

After a long calculation one finds that

$$I(z,\nu) = \mathcal{S}(\nu)\mathcal{M}(z,\nu), \tag{6}$$

where the *"spectral modifier function"* $\mathcal{M}$ is given by the formula

$$\begin{aligned}\mathcal{M}(z,\nu) &= k^2 \int_0^a \int_0^a e^{-(\rho_2^2+\rho_1^2)/2\sigma_g^2} I_0\left(\frac{\rho_1\rho_2}{\sigma_g^2}\right)\cos[k[s_1(\rho_1)-s_2(\rho_2)]] \\ &\quad \times \frac{\rho_1\rho_2}{s_1(\rho_1)s_2(\rho_2)}\,\mathrm{d}\phi_1\mathrm{d}\rho_1\,\mathrm{d}\phi_2\mathrm{d}\rho_2, \end{aligned} \tag{7}$$

I_0 being a modified Bessel function. As an example, let us choose the spectrum of the light in the aperture a be a Lorentzian line. The results are shown in Fig. 2. The spectrum of the light in the aperture consists of a single line, whereas the observed spectrum consists of two lines of equal intensity. For other observation points on the axis the spectrum can also be red-shifted or blue-shifted.

We will also present results relating to spectral changes when the coherence parameter σ_g is a simple function of the frequency.

Acknowledgements
This research was supported by the U.S. Air Force Office of Scientific Research under grant No. F49260-96-1-0400, and by the Engineering Research Program of the Office of Basic Energy Sciences at the U.S. Department of Energy under grant No. DE-FG02-90 ER 14119, and by the Dutch Technology Foundation (STW).

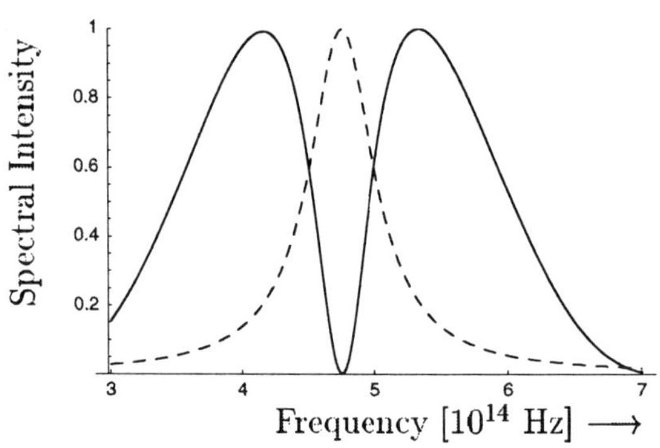

Fig. 2. The normalized spectral intensity in the aperture (dashed line), and the observed spectrum (solid line) of the focal field at a selected axial point.

References and links

1. E. Wolf and D.F.V. James, Rep. Prog. Phys. **59**, pp. 771–818 (1996).
2. E. Wolf and J.R. Fienup, Opt. Commun. **82**, pp. 209–212 (1991).
3. M. Born and E. Wolf, ***Principles of Optics***, 7th (expanded) ed. (Cambridge University Press, Cambridge, 1999). See Sec. 5.1.
4. L. Mandel and E. Wolf, ***Optical Coherence and Quantum Optics*** (Cambridge University Press, Cambridge, 1995).

Nonachromaticity and reversals of topological phase as a function of wavelength

Rajendra Bhandari

Raman Research Institute, Bangalore-560 080, India

91-80-3340122, 91-80-3340492, bhandari@rri.ernet.in

Achromaticity of the topological phase arising from polarization transformations of light [1] has been the basis of several applications [2]. Achromatic retarders based on Pancharatnam's work are routinely used in Astronomical polarimetry. Using the standard example of the "QHQ retarder", where Q and H are quarterwave and halfwave retarders, the topological origin of achromaticity was explained in ref.[3] where it was also shown that at wavelengths far removed from the design wavelength λ_0 of the retarder it can show the opposite behaviour i.e. sharp changes and reversals of phase. The new result I wish to report here is that such sharp changes and reversals can in fact be made to occur at wavelengths arbitrarily close to λ_0. This would happen if the retardation of Q and H were equal to $(2n + 1/2)\pi$ and $(4n + 1)\pi$ respectively, where n is an integer which could be made large.

A simple way to understand the effect is to imagine a standard two-slit interference experiment in which fringes are formed on a screen placed some distance away from the slits. A QHQ retarder is placed in the path of one of the two beams. Let the eigenstates of the retarder at λ_0 be $\mid X >$, i.e. linear polarization along the x-direction. The intensity on the screen is given by

$$I = I_1 + I_2 + 2\sqrt{(I_1 I_2)}Re(< X \mid U \mid X > e^{i\phi}), \tag{1}$$

where U is the 2x2 unitary polarization transformation matrix representing the QHQ device and ϕ is a path-dependent phase varying along the screen. U is a function of two parameters, (i) the retardation 2δ of Q (hence 4δ of H) at the wavelength λ, assuming a dependence $\delta = (\lambda_0/\lambda)(\pi/4)$, (ii) ρ, the orientation of the fast axis of H with respect to the X-direction. The modulus and phase of the complex visibility function $\gamma =< X \mid U \mid X >$ determine respectively the fringe contrast and the phase shift caused by the QHQ device. The assumed λ-dependence of the retardation results when the two refractive indices of Q and H are independent of wavelength. It was shown in ref.[3] that (1) when $2\delta = (2m+1)(\pi/4)$, which happens when $\lambda = (2\lambda_0)/(2m+1)$, phase singularities occur at the HWP orientations $\rho = m\pi$; m being an integer. At these values, the modulus of γ goes to zero and its phase undergoes a sharp jump of magnitude π. In a closed circuit around any one of these singularities, the phase changes by $\pm 2\pi$. Such phase changes have been verified experimentally in interference experiments [4]

(2) When $2\delta = 2m\pi - 3\pi/2$, which happens when $\lambda = \lambda_0/(4m - 3)$, the visibility is constant in magnitude and its phase increases linearly with ρ, the device acting as a pure phase retarder. (3) When $2\delta = 2m\pi - \pi/2$, which happens when $\lambda = \lambda_0/(4m - 1)$, the visibility is constant in magnitude and its phase decreases linearly with ρ, the device acting as a pure phase retarder with the opposite sign.

The key point of this paper is that if the retardation of Q and H were chosen to be $2n\pi + \pi/2$ and $4n\pi + \pi$ respectively (multi-order waveplates), the spacing in λ at which singularities and hence the phase reversals occur can be made arbitrarily small. If we assume again a $1/\lambda$ dependence of retardation of Q and H, the condition for phase reversal closest to the design wavelength for such a retarder is given by, $(2n\pi + \pi/2)(\lambda_0/\lambda) = (2n\pi + 3\pi/2)$. This gives, for $n = 1$, $\lambda = .71\lambda_0$; for $n = 2$, $\lambda = .82\lambda_0$; for $n = 3$, $\lambda = .87\lambda_0$ and so on. It can be concluded therefore that while topological phases *can* be made achromatic which can be useful, they are not necessary so.

It may also be pointed out that there are other contexts in interferometry wherein π phase jumps accompanying zero crossings of a real visibility function can be seen as manifestations of a phase singularity of a complex visibility function in a higher dimensional space. In such an enlarged description, the π phase shift acquires a sign that is measurable. We wish to cite one such example.

In an interesting extension of Pancharatnam's phase to the physics of interference of mixed states, Sjöqvist et.al. [5] have shown that if a beam of particles with internal degrees of freedom , in a mixed state with a density matrix ρ_0, is split in a Mach-Zhender interferometer and a unitary transformation U_i acts on the space of N internal states in one of the two paths, a phase difference given by arg $\text{Tr}(U_i\rho_0)$ is introduced between the two beams. Here "phase difference" is defined as the shift in the maximum of the interference pattern. This quantity, which is measurable, reduces to the phase shift $\arg(<\psi_0|\ U_i\ |\psi_0>)$, defined by Pancharatnam [1] for pure states $|\ \psi_0>$. We wish to point out that the mixed state phase as defined in ref.[5] becomes indeterminate at points in the parameter space where $|\ Tr(U_i\rho_0)\ |=0$. Eqn.(8) in the paper shows that at such points the interference pattern is uniform, with no fringes. Such points constitute singularities of phase involving discontinuous phase jumps and, depending upon the context, may occur as singular lines or surfaces. The mixed state phase involves singularities in new kinds of parameter spaces which include variables representing decoherence of quantum states or depolarization in case of polarization states of light. One can therefore have measurable topological phases resulting from cycles in parameter spaces consisting of degree of polarization and parameters of unitary transformation of polarization states.

References and links

1. S. Pancharatnam, "Generalized theory of interference and its applications", Proc. Indian. Acad. Sci. **A 44**, 247 (1956).
2. R. Bhandari, "Polarization of light and topological phases", Phys. Rep. **281**, 1 (1997).
3. R. Bhandari, "Phase jumps in a QHQ phase-shifter - some consequences", Phys. Lett. **A 204**, 188 (1995).
4. R. Bhandari, "Observation of Dirac singularities with polarization", Phys. Lett.**A 171**, 262, 267 (1992).
5. E. Sjöqvist et.al., "Geometric phases for mixed states in interferometry", Phys. Rev. Lett. **85**, 2845 (2000).

The law of conservation of radiance is a consequence of the partial coherence of the optical field

J. G. Suárez-Romero
Centro Nacional de Metrologia, km 4,5 Los Cués, Querétaro 76241, MÉXICO
telephone (4) 211 05 00 ext. 3328, fax (4) 211 00 53, jsuarez@cenam.mx

E. Tepichin-Rodríguez
Instituto Nacional de Astrofísica, Óptica y Electrónica, L. E. Erro 1, Tonantzintla, Pue. 72840, MÉXICO
telephone (2) 247 22 19, fax (2) 247 29 40, tepichin@haro.inaoep.mx

Abstract

The radiance and the law of conservation of radiance are two concepts created in an effort to explain experimental observations with extended incoherent sources. In this work we show that such observations are an effect of the partial coherence of the fields and that they can be explained by calculating the irradiance of the diffraction pattern of apertures with partial coherent illumination.

1. Introduction

The radiance is a radiometric magnitude that characterizes to sources. The concept of radiance is a development based on experimental observations, where the presence of an invariant (the measured radiant flux) lead to think that a certain feature of sources (the radiance) produces the invariant [1]. Additionally, the concept of ray that transports energy helps in demonstrating the so call radiance conservation law. This law explains the invariance of the measured radiant flux within the approximation of geometric optics. In this work we show that the observed invariance in the measured radiant flux is a consequence of a diffraction pattern that keeps practically constant its irradiance. This feature of the diffraction pattern can be explained with the theory of partial coherence.

2. Basic theory

A typical radiometric system used in the measurement of radiance is depicted in Fig. 1. There an incoherent source illuminates two successive apertures and finally an optical detector measures the radiant flux.

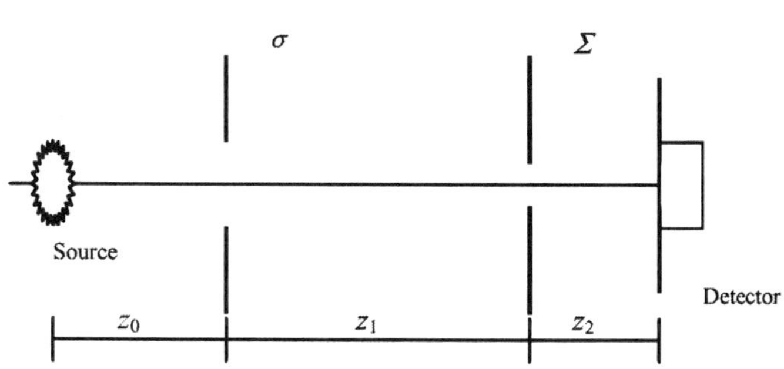

Fig. 1. Typical radiometric bench used in Radiometry

It is known that for this system, holding the apertures fixed, the measured radiant flux does not change when the source is moved farther away from, or closer to, the apertures. In order to explain such effect let us calculated the irradiance in the plain of the detector. Irradiance is given by [2]

$$W'(U_1,U_2,\omega)=\frac{1}{(2\pi)^4}\frac{z_2^2}{z_1^2}e^{i\frac{z_1+z_2}{2R^2k}\left(v_1^2-v_2^2\right)}$$

$$\int\limits_{P_2'}\int\limits_{P_1'}W_0(P_1',P_2',\omega)\,e^{i\frac{z_1}{2R^2k}\left(L_1^2-L_2^2\right)}e^{i\frac{z_1}{R^2k}\left(L_2v_2\cos(\Psi_2-\beta_2)-L_1v_1\cos(\Psi_1-\beta_1)\right)}$$

$$\pi^2\left[C(u,L_2)-iS(u,L_2)\right]\left[C(u,L_1)+iS(u,L_1)\right]$$

$$L_1d\Psi_1L_2d\Psi_2dL_1dL_2\ , \tag{1}$$

where $W(P'_1, P'_2, \omega)$ is the cross spectral density in the aperture σ, and it is given by the theorem of van Cittert-Zernike.

3. Numerical calculations

Calculations with the source to z_0=1,0 m and z_0=0,5 m of distance from the apertures are depicted in Fig2.

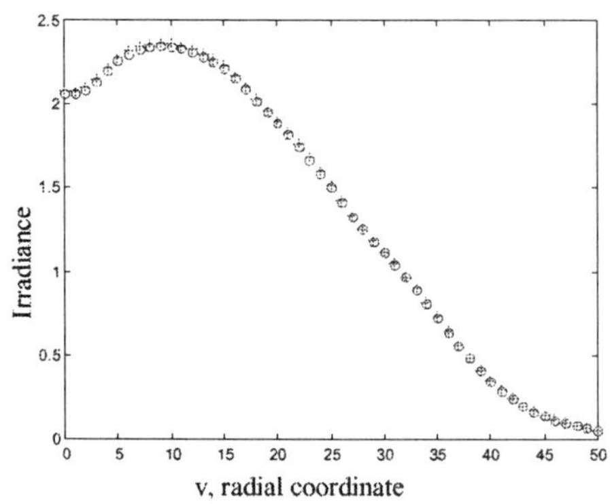

Fig 2. Numerical calculation of irradiance; 'o' for 0,5 m and '+' for 1,0 m.

The two graphics practically coincides with each other. In other words, we observe that the absolute irradiance of the diffraction pattern is practically invariant for the present parameters. In consequence the total flux that passes through the second aperture is nearly a constant.

A first explanation of why the irradiance pattern is constant can be as follows: When we separate the source from the apertures the irradiance in the aperture σ decreases as the inverse square of the distance, but its partial coherence increases. This results in a cancellation of the respective effects, leaving the irradiance at the plane of the detector as a constant.

4. Conclusions

We show in this work that the law of conservation of radiance can be explained with the theory of partial coherence. The invariance of the total radiant flux is a consequence of the partial coherence of the optical field.

7. References

[1] *Self-Study Manual on Optical Radiation Measurements: Part I—Concepts, Chapters 1 to 3*, Fred E. Nicodemus, Editor. (Gaithersburg, Md., National Bureau of Standards, 1976).

[2] J. G. Suárez-Romero, E. Tepichin-Rodríguez, "New irradiance approach for evaluation of diffraction errors", in Aplications of Photonic Technology 4, Roger A Lessard, George A. Lampropoulos, Editors, Proceedings of SPIE 4087, 1310-1314.

Application of modal theory of coherence for nonstationary partially coherent optical fields

A. M. Lazaruk (late) and M. U. Karelin

B. I. Stepanov Institute of Physics, National Academy of Sciences, F. Scaryna ave, 68, Minsk, 220072, Belarus

karelin@dragon.bas-net.by

Modal description of light coherence [1] in frequency domain proved to be very convenient tool for analysis of light from various stationary sources. In this work we present a review of some consequences from another variant of modal decomposition of light fields, based on biorthogonal expansion (a variant of Karhunen-Loéve or Schmidt decomposition). Such an approach is useful for description of nonstationary waves, especially in various nonlinear processes [2, 3, 4], where the partial coherence just means a high complexity of interacting waves. Due to mathematical similarity, two modal treatments have a lot in common and can successfully use results of each other.

Each partially coherent field can be expressed as a superposition of factorized orthonormal components $E(\mathbf{r},t) = \sum_n \sqrt{u_n}\mathcal{E}_n(\mathbf{r})\, e_n(t)$. The decomposition basic functions and modal energies u_n are eigenvectors and eigenvalues of Fredholm's integral equation

$$u_n\, \mathcal{E}_n(\mathbf{r}) = \int d^2r'\, \Gamma_S(\mathbf{r},\mathbf{r}')\, \mathcal{E}_n(\mathbf{r}'), \tag{1}$$

which kernel is a time averaged transverse correlation function of a beam $\Gamma_S(\mathbf{r},\mathbf{r}')$, and temporal envelopes $e_n(t)$ can be calculated via projection procedure $\sqrt{u_n}\, e_n(t) = \int d^2r\, E(\mathbf{r},t)\, \mathcal{E}_n^*(\mathbf{r})$.

For physically admissible light pulses of finite total energy U the projections $e_n(t)$ are also orthonormal, which leads to absence of any interference between different terms of spatial basis.

Another approach to evaluation of temporal basis lies in use of dual integral equation

$$u_n\, e_n(t) = \int dt'\, \Gamma_T(t,t')\, e_n(t'), \tag{2}$$

where $\Gamma_T(t,t') = \int d^2r\, E(\mathbf{r},t)\, E^*(\mathbf{r},t') = \sum_n u_n\, e_n(t)\, e_n^*(t')$ is a global temporal correlation function of the field, which expresses the overall correlation the wavefront patterns.

One can easily see that equations (1) and (2) make up two *equivalent* variants for evaluation the decomposition. Them both have identical spectra of eigenvalues and, in case of no degeneracy, for complete description of modal structure of a field one needs to know only one set of basic functions $\mathcal{E}_n(\mathbf{r})$ or $e_n(t)$.

The most important characteristic of field structure here is an effective number of terms in the decomposition [1, 2, 5, 6], which is expressed through 2nd order moment of modal weights u_n, $N_{\mathit{eff}} = \left(\sum_n u_n\right)^2 / \sum_n u_n^2$, and for evaluation of N_{eff} one does not need to know the exact basic functions $e_n(t)$, $\mathcal{E}_n(\mathbf{r})$:

$$N_{\mathit{eff}} = U^2 \Big/ \int d^2r \int d^2r'\, \left|\Gamma_S(\mathbf{r},\mathbf{r}')\right|^2 = U^2 \Big/ \int dt \int dt'\, \left|\Gamma_T(t,t')\right|^2. \tag{3}$$

Two equivalent forms of (3) reflect real interconnection between spatial and temporal parameters of a beam (equality for spatial and temporal degrees of freedom of wavefield).

The value of N_{eff} specifies the ability of the total field to produce interference effects between two arbitrary separate points of the beam cross-section [2, 5] and changes from unity for coherent wave to infinity for completely incoherent field. Large values of N_{eff} is necessary for suppression of speckle-noise in optical registration and lithographic applications [1, 6, 7]. The value of N_{eff} is a very convenient measure of a field overall degree of (in)coherence and ties together geometrical (or temporal) and spectral characteristics of the beam in reciprocity inequalities (uncertainty relations) [8]. Another area, where concept of the effective number of modes may become useful, is the evaluation of modal structure of an optical beam (see [9] for reference).

Till this point the basic formalism has dealt with a wave-field $E(\mathbf{r}, t)$ as with the deterministic one. In a case of statistical ensemble of partially coherent pulses the duality between two integral equations (1) and (2) disappears, and one needs to take into account two *different and complementary* expansions that describing coherence of the ensemble (but not for a particular wave). For each decomposition one can introduce its own number of degrees of freedom (3) — spatial N_S and temporal N_T (angle brackets stays for statistical averaging)

$$N_S = \langle U^2 \rangle \Big/ \left(\int d^2r \int d^2r' \, |\langle \Gamma_S(\mathbf{r}, \mathbf{r}') \rangle|^2 \right) = \lim_{N_T, T \to \infty} N_{\mathit{eff}}, \tag{4a}$$

$$N_T = \langle U^2 \rangle \Big/ \left(\int dt \int dt' \, |\langle \Gamma_T(t, t') \rangle|^2 \right) = \lim_{N_S, S \to \infty} N_{\mathit{eff}}. \tag{4b}$$

In order to preserve spatio-temporal duality of the formalism, the ensemble averaging should be applied at the later stages of consideration. As an example of such approach let us estimate number of terms in the modal decomposition of a mean light pulse from the ensemble. The simplest and the most popular type of field statistics is Gaussian, which leads to a very simple formula

$$1/\overline{N_{\mathit{eff}}} = 1/N_S + 1/N_T, \tag{5}$$

which, as an asymptotic, can be applied to rather wide class of fields [5, 6, 7].

References and links

1. L. Mandel, E. Wolf, "Optical Coherence and Quantum Optics" (Cambridge University Press, Cambridge, 1995).
2. G. A. Pasmanik, V. G. Sidorovich, "Interrelation between coherent properties of light beams and space-time structure", Radiophys. & Quantum Electron., **23**, 809-814 (1980).
3. B. Ya. Zel'dovich, N. F. Pilipetsky, V. G. Shkunov, "Principles of phase conjugation" (Springer-Verlag, Berlin, 1985)
4. N. V. Karelin, A. M. Lazaruk, "Transformation of Spatial Coherence and Coherence and Spectral Broadening in Self-Phase Modulation of Light Pulses", Optics and Spectroscopy, **83**, 377-382 (1997).
5. A. A. Leshchev, "Method for measuring the number of spatially coherent modes in optical radiation", Optics & Spectroscopy (USA), **55**, 599-600 (1983).
6. M. V. Vasil'ev, V. Yu. Venedictov, A. A. Leshchev, P. M. Semenov, V. G. Sidorovich, O. V. Solodyankin, "Reduction of speckle-structure contrast in image under laser illumination", Optics & Spectroscopy (USA), **70**, pp. 6-7 (1991).
7. E. V. Ivakin, A. I. Kitsak, N. V. Karelin, A. M. Lazaruk, A. S. Rubanov, "Approaches to coherence destruction of short laser pulses", tp be published
8. M. U. Karelin, A. M. Lazaruk, "Interrelation between spatial and temporal reciprocity inequalities of partially-coherent beams", sent on 8th Rochester Conference on Coherent and Quantum Optics (see accompanuing paper).
9. X. Xue, H. Wei, A. G. Kirk, "Intensity-based modal decomposition of optical beams in terms of Hermite-Gaussian functions", J. Opt. Soc. Am. A, **17**, 1086-1091 (2000).

Photonic electro-magnetic resonance

Boris V. Gisin

Department of Electrical Engineering - Physical Electronics,
Faculty of Engineering, Tel Aviv University,
Tel Aviv 69978, Israel

gisin@eng.tau.ac.il

It is well known that the Maxwell equation in the slowly varied amplitude approximation may be written in the form of the Schrödinger equation. Moreover, for light propagating along the optical axis of the three-fold electro-optical crystals under the action of the electric field the Maxwell equation may be written in the form of the Pauli equation [1].

The analogy is important as measurements of the magnetic moment based on the Pauli equation are most precise in physics [2]. In the analogy the motion of electron in a spatially rotating magnetic field corresponds to the modulation of a optical pulse by a circularly polarized wave traveling with the pulse group velocity. On the other hand the optical pulse propagation in a spatially rotating electric field corresponds to the motion of electron in a time rotating magnetic field. This analogy, however, is not complete as in electro-optics such a construction can not have a matrix with imaginary elements whereas one from the Pauli matrices has such elements.

In the present paper we consider both the electro-optical and Faraday effect in cubic crystals. In this case a full analogy with the Pauli equation exists and an experiment similar to the magnetic resonance is possible.

Let us assume that the plane light wave propagates along three-fold axis of the crystal possessing the Faraday effect. Using the slowly varying magnitude method [3] we may obtain from the Maxwell equation

$$i\frac{\partial}{\partial z}\chi + \frac{k_2}{2}\frac{\partial^2}{\partial \eta^2}\chi + \gamma\left(\sigma_1 V_1 + \sigma_2 V_2\right)\chi + \gamma_m \sigma_3 H \chi = 0, \tag{1}$$

where V_1, V_2 are the transverse components of the applied electric field, $\eta = t - z/u$, t is time, z is the coordinate along the three-fold axis, u is the group velocity, the parameter k_2 characterizes the dispersion of the group velocity in the first approximation [3], $\gamma = \omega d/(nc\sqrt{6})$, ω is the frequency, $d = n^4 r$, n is the refractive index, r is the electro-optical coefficient of the crystal,H is the magnetic field along the three-fold axis, $\gamma_m = g\omega/(2nc)$, g is the magneto-gyration coefficient of the crystal, c is the speed of light.

The Eq. (1) has the form of the Pauli equation where instead t, z, the variable z, η are used, the role of the transverse magnetic field, the mass and the magnetic moment is played by the applied electric field, the parameters $1/k_2$ and γ, respectively, finally the longitudinal magnetic fields in the Pauli equation and in Eq. (1) play the same role. The matrices σ_k are non other than the Pauli matrices σ_k^p : $\sigma_1 = \sigma_1^p$, $\sigma_2 = \sigma_3^p$, $\sigma_3 = -\sigma_2^p$. The change to the Pauli matrices in Eq. (1), is realized by the transformation $\psi \to (1/\sqrt{2})(1 + i\sigma_1)\chi$ [1]. This transformation turns out a circular polarization into a linear one.

For a plane light wave modulated by a electric field spatially rotating around the three-fold axis at a period L and an amplitude V : $V_1 = V \sin\beta_L z$, $V_2 = V\cos\beta_L z$, where $\beta_L = 2\pi/L$, and a constant magnetic field applied along the axis the solution of Eq. (1) is similar to that of the Pauli equation for the motion of spin in a magnetic field [4]. The condition of the photonic electro-magnetic resonance is

$$\beta_L = 2\gamma_m H \tag{2}$$

The photonic electro-magnetic resonance may be a very important tool for investigations of properties of three-fold crystals.

The resonance opens way for performing an experiment in optics like the electron magnetic resonance. Such an experiment may give a new look on the problem of anomalous magnetic moment and the problem of the frequency superposition by transition to rotating frame [5]. Indeed, consider a plane circularly polarized light wave moving through an electro-optical crystal with the rotating optical indicatrix. The light and the indicatrix, for definiteness, are assumed to rotate in the opposite direction with frequencies ω and $\nu/2$ respectively. Inside the crystal the optical frequency is $(\omega+\nu/2)$ relative to the rotating indicatrix. Without the interaction between the crystal and the light the optical frequency outside the crystal will be initial one, i.e., ω. If the interaction between the light and the crystal results in a reverse rotation relative to the indicatrix, then at output the light frequency will be $-(\omega+\nu)$. It occurs when the amplitude of the applied electric field equals a half-wave value. In the approach the linear superposition frequencies is assumed *a priori*. But what is the superposition? In Refs. [1, 5] it was shown that nonlinear superposition causes an extra energy shift in the magnetic as well as photonic resonance. This shift may be associated with an anomalous magnetic moment. The corresponding parameter in the photonic resonance in this case also must be anomalous.

References and links

1. B. V. Gisin, "Analogy between the propagation of light in a single-sideband modulator and the motion spin in a magnetic field", Phys. Lett. **A 209**, 285-289 (1995).
2. "The Physical Review - the first hundred years: a selection of seminal papers and commentaries", edited by H. Henry Stroke (AIP Press, Woodbury, NY, 1995).
3. A. Hasegawa and Y. Kodama, "Solitons in Optical Communications", (Oxford University Press: Oxford, 1995).
4. L. D. Landau, and E. M. Lifshiz, "Quatum mechanics", (Pergamon press, Oxford, 1965).
5. B. V. Gisin, "Light propagation in a medium with rotating index ellipsoid", Phys. Rev., **A 61**, 053808-1 - 053808-5 (2000).

Partially coherent volume diffraction

George Barbastathis[a] **and David J. Brady**[b]

[a]*Massachusetts Institute of Technology, Cambridge, MA 02139 USA*
[b]*Duke University, Durham, NC 27708 USA*

gbarb@mit.edu, dbrady@ee.duke.edu

Abstract: The response of a volume hologram to a partially coherent probe field is calculated under the 1st-order Born and paraxial approximations. The special case of a Gauss-Schell model probe beam is carried out and results are presented in the limits of quasi-coherent and quasi-incoherent fields.

OCIS codes: (030.1640) Coherence; (050.7330) Volume holographic gratings

The design of volume holograms as optimal imaging filters [1] is facilitated by *a priori* known properties of the objects, such as the statistical nature of the light that the objects emit or scatter. In most imaging applications, the condition of the associated inverse problem is improved if the coherence of the observed field is reduced and field correlations are measured instead of intensities, e.g. allowing 4D (3 spatial dimensions plus wavelength spectrum) object reconstruction.

In this paper we compute the response of a volume hologram under partially coherent illumination. This is the first time, to our knowledge, that this calculation has been carried out in the literature. In addition to the obvious benefit in analyzing and designing volume holographic imaging systems, our results will also influence volume holographic storage and processing (where incoherent readout using LEDs reduces hardware costs.)

Let $\Delta\epsilon(\mathbf{r})$, $\mathbf{r} \in \mathcal{V}_{\mathrm{H}} = \{-\infty < x, y < +\infty, -L/2 < z < L/2\}$ denote the dielectric modulation of a volume diffractive element. Let E_{p} denote the random process describing the amplitude of a quasi-monochromatic (wavelength λ) partially coherent probe field illuminating the hologram, and let E_{d} denote the random process of the resulting diffracted field. Using the scalar 1st-order Born and paraxial approximations [2, sec. 2.1,2] and standard notation [3], we find the diffracted mutual intensity

$$W_{\mathrm{d}}\left(\mathbf{r}_1', \mathbf{r}_2'\right) = \iint_{\mathcal{V}_{\mathrm{H}}} W_{\mathrm{p}}\left(\mathbf{r}_1, \mathbf{r}_1\right) \Delta\epsilon(\mathbf{r}_1) \Delta\epsilon^*(\mathbf{r}_2)\, \mathrm{G}\left(\mathbf{r}_1 - \mathbf{r}_1'\right) \mathrm{G}^*\left(\mathbf{r}_2 - \mathbf{r}_2'\right) \mathrm{d}^3(\mathbf{r}_1) \mathrm{d}^3(\mathbf{r}_2), \quad (1)$$

where W_{p} is the mutual intensity of E_{p} and $\mathrm{G}(\mathbf{r}) = \mathrm{e}^{i2\pi|\mathbf{r}|/\lambda} / |\mathbf{r}|$ (Green's function).

As example, consider a volume hologram recorded in the transmission geometry by the interference of a coherent Gaussian reference beam with waist w_0 located at $\mathbf{r}_{\mathrm{f}} = (0, 0, -z_0)$, $z_0 \gg L/2$, and a plane wave signal beam propagating at direction (u_0, v_0), $|u_0|, |v_0| \ll 1$ radian, with respect to the $\hat{\mathbf{z}}$-axis. The Gaussian beam and plane wave are mutually coherent. The dieletric modulation $\Delta\epsilon$ is given by the product of the conjugated Gaussian beam field [3, eq. 5.6-34,35] and the plane wave field. The hologram is probed by a Gauss-Schell model beam with correlation domain σ_{g} and source extent σ_{S} originating at $\mathbf{r}_{\mathrm{p}} = (x_{\mathrm{p}}, y_{\mathrm{p}}, z_{\mathrm{p}})$ in the vicinity of the Gaussian reference beam. Using [3, sec. 5.6.4] and (1), we find the far-field angular mutual intensity of the diffracted field at observation directions (u_1, v_1), (u_2, v_2)

$$W_{\mathrm{d}}^{(\infty)}(u_1, v_1; u_2, v_2) = W_{\mathrm{d},0}^{(\infty)} \iint_{-\infty}^{\infty} \frac{\mathrm{e}^{K_{\mathrm{p}}(z_1, z_2) + iD(z_1, z_2)}}{S_{\mathrm{p}}(z_1, z_2)} \operatorname{rect}\left(\frac{z_1}{L}\right) \operatorname{rect}\left(\frac{z_2}{L}\right) \mathrm{d}z_1 \mathrm{d}z_2, \quad (2)$$

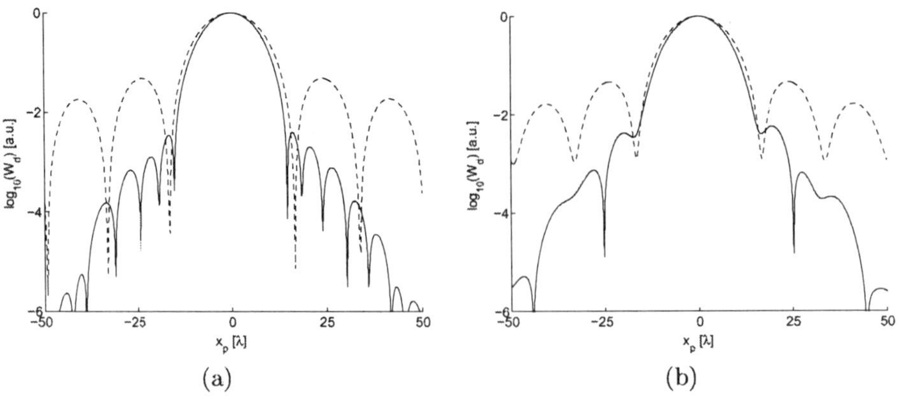

Fig. 1. Normalized output intensity (see text) for partially-coherent probe with waist matched to the reference beam (*i.e.*, $w_0 = \sigma_S$), and correlation distance *(a)* $\sigma_g = 100\lambda$ (coherent limit), *(b)* $\sigma_g = 0.1\lambda$ (incoherent limit). Solid line: $w_0 = 6.4\lambda \Leftrightarrow (\mathrm{NA}) \approx 5\times10^{-2}$, dashed line: $w_0 = 64\lambda \Leftrightarrow (\mathrm{NA}) \approx 5\times10^{-3}$.

where $W_{\mathrm{d},0}^{(\infty)}$ is a normalizing constant,

$$K_{\mathrm{p}}(z_1,z_2) = \frac{\kappa_1\left(c_2^2+d_2^2\right)+\kappa_2\left(c_1^2+d_1^2\right)-2\zeta\left(c_1c_2+d_1d_2\right)}{4\left(\kappa_1\kappa_2-\zeta^2\right)} + \frac{\left(\gamma_1+\gamma_2-2\beta\right)\left(x_{\mathrm{p}}^2+y_{\mathrm{p}}^2\right)}{4\left(\gamma_1\gamma_2-\beta^2\right)},$$

$$D(z_1,z_2) = \pi\left[\left(u_1^2+v_1^2\right)z_1-\left(u_2^2+v_2^2\right)z_2\right]-\left[\psi(z_1-z_0)-\psi(z_2-z_0)\right],$$

$$S_{\mathrm{p}}(z_1,z_2) = \left(\gamma_1\gamma_2-\beta^2\right)\left(\kappa_1\kappa_2-\zeta^2\right)w(z_1-z_0)\,w(z_2-z_0),$$

$$c_j = -\frac{(\gamma_{3-j}-\beta)\,x_{\mathrm{p}}}{4\left(\gamma_1\gamma_2-\beta^2\right)}+\frac{u_0-u_j}{\lambda},\qquad d_j = -\frac{(\gamma_{3-j}-\beta)\,y_{\mathrm{p}}}{4\left(\gamma_1\gamma_2-\beta^2\right)}+\frac{v_0-v_j}{\lambda},\qquad j=1,2$$

$$\kappa_j = \frac{\gamma_{3-j}}{4\left(\gamma_1\gamma_2-\beta^2\right)}-\frac{1}{w^2(z_j-z_0)}-\frac{i\pi}{\lambda R(z_j-z_0)},\quad j=1,2;\quad \zeta=\frac{\beta}{4\left(\gamma_1\gamma_2-\beta^2\right)},$$

$$\gamma_j = \alpha - i\frac{\lambda(z_j-z_0)}{2\pi},\quad j=1,2;\qquad \text{and}\quad \alpha,\beta\quad \text{are as in [3, eq. 5.6-66,71].}$$

Fig. 1 shows the maximal far-field intensity $W_{\mathrm{d}}^{(\infty)}(u_0 - x_{\mathrm{p}}/z_0, v_0; u_0 - x_{\mathrm{p}}/z_0, v_0)$ [2, ex. 2, eq. 59] resulting from a volume hologram of thickness $L = 3\times10^3\lambda$, recorded with $(u_0, v_0) = (0.2, 0)$, $z_0 = 10^4\lambda$, and probed from $(x_{\mathrm{p}}, 0, -z_0)$, calculated numerically. Notice the decrease in Bragg selectivity, blunting of the Bragg nulls and growth of the side-lobes as the numerical aperture and/or the degree of probe beam coherence decrease. Future work includes the 3-D point-spread function (PSF) calculation for this and more general geometries, the inverse problem of designing the volume hologram according to the object coherence modes for imaging applications, and experimental demonstrations.

G. B. gratefully acknowledges the CAREER Award from the National Science Foundation for supporting this work.

References and links

1. G. Barbastathis, M. Balberg, and D. J. Brady, "Confocal microscopy with a volume holographic filter", *Opt. Lett.*, 24(12):811–813, 1999. G. Barbastathis and D. J. Brady, "Multidimensional tomographic imaging using volume holography," *Proc. IEEE*, 87(12):2098–2120, 1999;
2. G. Barbastathis and D. Psaltis, "Volume holographic multiplexing methods," in *Holographic data storage*, H. Coufal, D. Psaltis, and G. Sincerbox, editors, Springer, 2000.
3. L. Mandel and E. Wolf, *Optical coherence and quantum optics*, Cambridge Univ. Press, 1995.

Inverse diffractive optics: asymptotic technique for light of any state of coherence

Ari T. Friberg, Anna Thaning

Royal Institute of Technology (KTH), Department of Microelectronics and Information Technology, SE-100 44 Stockholm, Sweden

ari.friberg@optics.kth.se

Abstract: A method to design diffractive axicons for partially coherent light is derived from the asymptotic wave theory.

Diffractive optics involves micro and nanotechnology to take advantage of light's wave nature. State-of-the-art design procedures yield structures, such as array generators and polarization-sensitive devices, for ideal laser light [1]. The methods often employ geometrical optics, e.g. diffractive ray tracing, or are iterative. Despite increasing use of diode lasers and partially coherent sources, very little work has been done on diffractive elements in fluctuating light and, in particular, on partially coherent inverse diffractive optics, i.e., on synthesizing diffractive elements that take into account the reduced state of coherence. We introduce a general method for the design of diffractive axicons [2], based on scalar wave theory in the asymptotic limit of large wave numbers.

The Fresnel diffraction integral [3] applicable to the cross-spectral density of a stationary, partially coherent radiation at frequency $\omega = k/c$ is evaluated by the method of stationary phase and the leading contribution in $1/k$ is retained. The on-axis intensity at some distance behind a symmetric element takes on the form

$$I_{\rm sp}(0,z) = \frac{k}{2\pi z^2} I_{\rm in}(\rho_c)\tilde{g}_{\rm in}(\rho_c)\rho_c^2 \frac{1}{\psi''(\rho_c,z)}, \tag{1}$$

where $I_{\rm in}(\rho)$ and $\tilde{g}_{\rm in}(\rho)$ are the illumination intensity and (angularly averaged) degree of spatial coherence, respectively. Moreover, $\psi''(\rho_c, z)$ is the second derivative of $\psi(\rho, z) = \rho^2/2z + \phi(\rho)$, where $\phi(\rho)$ is the axicon phase retardation function, evaluated at the stationary point ρ_c. On making use of the results

$$\phi'(\rho_c) = -\rho_c/z, \qquad \psi''(\rho_c, z) = 1/z + \phi''(\rho_c), \tag{2}$$

and on interpreting Eq. (1) as a differential equation for phase $\phi(\rho)$, we obtain

$$F(z)[\phi''(\rho) - \frac{\phi'(\rho)}{\rho}] = G(\rho)\,[\phi'(\rho)]^2, \tag{3}$$

where we have introduced $F(z) = I_{\rm sp}(0,z)$ and $G(\rho) = (k/2\pi)I_{\rm in}(\rho)\tilde{g}_{\rm in}(\rho)$. Since $z(\rho) = -\rho/\phi'(\rho)$, we find at once from Eq. (3) that $\rho G(\rho) = F(z)z'(\rho)$, which can be integrated to yield the axicon design criterion in the form

$$\int_r^\rho \rho' G(\rho')\mathrm{d}\rho' = \int_d^z F(z')\mathrm{d}z'. \tag{4}$$

Here r denotes the inner rim radius of an annular-aperture axicon that is normally used to enhance the image uniformity, and the axicon line image begins at d.

We note, firstly, that Eq. (4) formally resembles the phenomenological design principle based on geometrical optics and energy conservation [4]. However, Eq. (4) extends the method to illumination of any state of coherence. Although the right-hand side may be difficult to interpret physically, all quantities in Eq. (4) are mathematically unambiguously defined. Secondly, Eq. (4) [or differential equation (3)] can in principle be solved for axicon phase $\phi(\rho)$ or its derivative $\phi'(\rho)$, when the desired on-axis intensity profile $F(z)$ and the illumination specified by $G(\rho)$ are known. Explicit results exist for several special cases. These include uniform-intensity annular axicons [$F(z)$ = constant] in the limiting cases of fully coherent Gaussian beam (fundamental Hermite-Gaussian mode) and uniform intensity of Gaussian coherence [5]. And thirdly, for a given illumination and desired image profile Eq. (4) can be solved numerically first for $z = z(\rho)$, from which the phase function is obtained via relation $\phi'(\rho) = -\rho/z(\rho)$. Such a novel brute-force approach renders analytical manipulations unnecessary.

As a demonstration we consider Gaussian Schell-model (GSM) beams representing multi-mode lasers [6]. The phase function $\phi(\rho)$ designed for GSM beam illumination is computed from Eq. (4), after which the axicon line image can be assessed by numerical simulations using partially coherent Fresnel diffraction. Figure 1 shows calculated axial intensity profiles for an optimized, annular-aperture, uniform-intensity axicon. The GSM beam intensity width equals the element's radius (5 mm), and the state of coherence varies from complete coherence (a) to global incoherence (e). The axicon is apodized at both (radial) aperture edges.

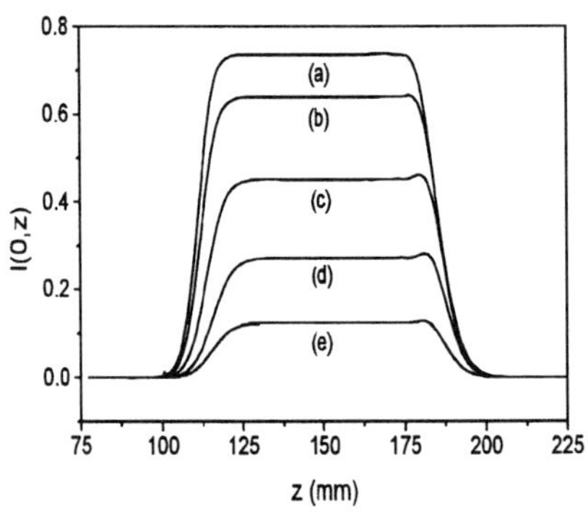

Fig. 1. Wave-theoretic on-axis image intensity profiles for GSM beam illumination. Coherence width is (a) ∞, (b) 10 mm, (c) 5 mm, (d) 3 mm, and (e) 1.5 mm.

Our tehnique replaces geometrical-optics rays with stationary points and extends inverse diffractive optics, with new ways of solution, to the regime of partial coherence. The authors acknowledge useful discussions with S.Yu. Popov and Z. Jaroszewicz.

References and links

1. See, for example, J. Turunen and F. Wyrowski, eds., *Diffractive Optics for Industrial and Commercial Applications* (Akademie Verlag, Berlin, 1997).
2. L.M. Soroko, *Meso-Optics — Foundations and Applications* (World Scientific, Singapore, 1996); Z. Jaroszewicz, *Axicons: Design and Propagation Properties*, Research & Development Treatises, Vol. 5 (SPIE Polish Chapter, Warsaw, 1997).
3. A.T. Friberg and S.Yu. Popov, "Effects of partial spatial coherence with uniform-intensity diffractive axicons," J. Opt. Soc. Am. A **16**, 1049–1058 (1999).
4. J. Sochacki et al., "Nonparaxial design of generalized axicons," Appl. Opt. **31**, 5326–5330 (1992).
5. J. Pu et al., "Annular-aperture diffractive axicons illuminated by Gaussian beams," J. Opt. A **1**, 730–734 (1999); S.Yu. Popov and A.T. Friberg, "Design of diffractive axicons for partially coherent light," Opt. Lett. **23**, 1639–1641 (1998).
6. L. Mandel and E. Wolf, *Optical Coherence and Quantum Optics* (Cambridge University Press, Cambridge, 1995), Sec. 5.6.4.

Demonstration of a coherent thermal source.

J.J. Greffet, R. Carminati, K. Joulain, J.P. Mulet
Laboratoire EM2C, Ecole Centrale Paris, 92295 Châtenay-Malabry Cedex, France
Greffet@em2c.ecp.fr

Y. Chen
Laboratoire de Microstructures et de Microélectronique, 92220 Bagneux, France

S. Mainguy
CEA CESTA, 33114 Le Barp, France

OCIS : 020.1670; 030.1640; 030.5620; 030.1670.

Abstract: We report experimental measurements of the emission of thermal light by a lamellar grating ruled on SiC. This sample is a partially spatially coherent source that emits light in a narrow angle as an antenna. In addition, this surface emits different spectra in different directions displaying the Wolf effect.

1. Introduction

Coherence properties of thermal sources can be drastically modified when looking at them in the near-field. It has recently been shown [1] that the thermal excitation of surface phonon-polaritons yields an enhancement of four orders of magnitude of the density of states at a distance of 100 nm to the surface. This results in a density of energy almost monochomatic and enhanced by four orders of magnitude. It has also been shown that the coherence length of the electromagnetic field emitted by such an interface can be as large as 20 λ[2]. This property suggests that directional antennas can be produced if one is able to convert them into propagating waves. This is possible in principle by using a grating as a coupler[3].

2. A thermal source which is partially spatially coherent : an infrared antenna.

In order to optimize the conversion, we have used a code to design the best grating. It has been realized on a substrate of SiC. This *particular thermal source behaves as a coherent antenna* displaying a narrow lobe for a well defined angle. Thus we have done an infrared coherent antenna. This is possible because we take advantage of the coherence of the surface wave.

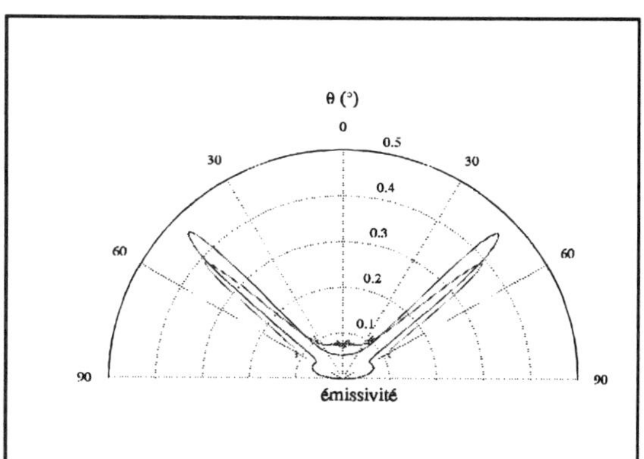

Fig. 1. Predicted and measured angular dependence of the emissivity ε of a SiC grating at $\lambda=11.36\mu m$.

In other words, this device allows to look directly at the surface phonon-polariton. Indeed when changing the wavelength, the angle of emission changes following the dispersion relation of the phonon-polariton. This dispersion relation can be derived from the measurements as it was done in ref. [3]. Note in Fig. 1 the enhancement of the emissivity in the peak by more than an order of magnitude due to the coupling of the surface modes with the propagating modes.

3. A natural thermal source that produces the Wolf effect.

Finally, we note that this source has a spectrum that has a strong dependence on the angle of emission. To our knowledge, this is the first example of a natural source displaying the Wolf effect in emission.

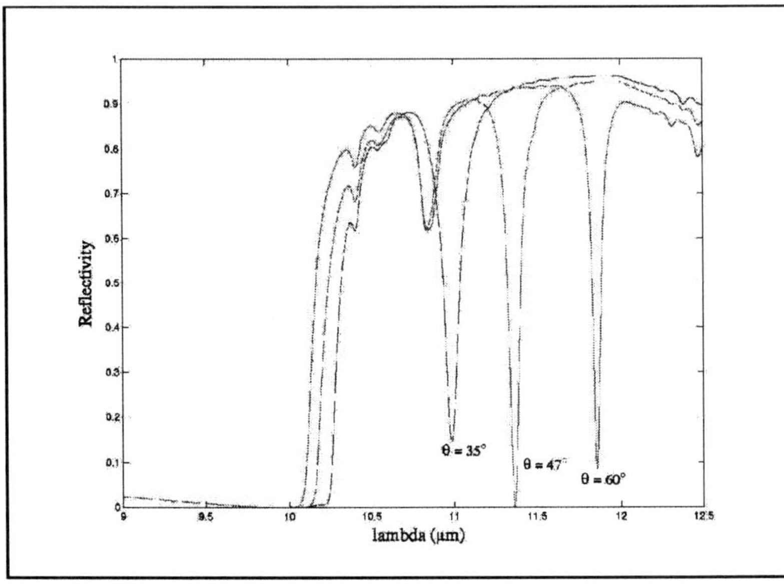

Fig.2. Experimental measurements of the spectral reflectivity R of the sample for different angles. Since the grating supports only one order, we have $\varepsilon=1-R$.

A discussion of the properties of this source will be given. Comparison between experimental measurements and theory will be also reported.

4. Conclusion

We have demonstrated that a thermal source can be partially spatially coherent. This unusual behaviour is due to the underlining spatial coherence of collective surface excitations.

5. References

1. A. Shchegrov , K. Joulain, R. Carminati and J.J. Greffet, "Near-field spectral effects due to electromagnetic surface excitations", Phys. Rev. Lett. **85**, 1548 (2000).
2. R. Carminati and J.J. Greffet, "Near-field effects in spatial coherence of thermal sources" Phys. Rev. Lett. **82**, 1660 (1999).
3. J.Le Gall, M. Olivier and J.J. Greffet, "Experimental and theoretical study of reflection and coherent thermal emission by a SiC grating supporting a surface-phonon polariton" Phys. Rev. B **55**, 10105 (1997).

Image reconstruction in scalar wave optics

M. Ježek, Z. Hradil

Department of Optics, Palacký University, 17. listopadu 50, 772 00 Olomouc, Czech Republic

tel: 00420 68 563 4255, fax: 00420 68 522 5246 , e-mail: nathan@thunder.upol.cz
http://thunder.upol.cz/~nathan/physics/reconstruction/

Optics was established as a science to acquire information through observation and measurement. This effort is, however, always affected by errors of various natures. There are tight analogies between scalar wave optics and nonrelativistic quantum mechanics. Paraxial Fresnel approximation of Helmholtz wave equation in wave optics is nothing else than Schrödinger equation. Field may be described by coherent scalar amplitude, or more generally, by a second order correlation function (mutual intensity) in the case of partially coherent fields,

$$\Gamma(x,x',z) = \langle \psi^*(x',z)\psi(x,z)\rangle, \qquad \Gamma(x,x',z) = \Gamma^*(x',x,z), \qquad \Gamma(x,x,z) \geq 0.$$

The brackets denote the averaging over complex amplitudes and z being the longitudinal coordinate of the optical setup. For the sake of simplicity a one dimensional transverse space will be assumed (coordinate x). Quantum counterparts may be easily established in the Dirac notation. Complete knowledge may be represented using a pure state, whereas any randomness in the ensemble of identically prepared particles corresponds to the incoherent mixture of pure states – a density operator

$$\widehat{\rho} = \sum_k \lambda_k |\varphi_k\rangle\langle\varphi_k|, \qquad \widehat{\rho}^+ = \widehat{\rho}, \qquad \langle\psi|\widehat{\rho}|\psi\rangle \geq 0.$$

The second order correlation function may be interpreted as x-representation of density operator $\mathrm{Tr}[\widehat{\rho}|x'\rangle\langle x|] = \Gamma(x,x',z)$. Transformation of complex amplitude in wave optics is given by integral transformation. This corresponds to the coordinate representation of a generic transformation of quantum state,

$$|\psi\rangle = \widehat{T}|\psi_0\rangle, \qquad \langle x|\psi\rangle = \int \mathrm{d}x_0 \langle x|\widehat{T}|x_0\rangle\langle x_0|\psi_0\rangle.$$

The states $|x_0\rangle$ are related to the position in the object plane, whereas $|x\rangle$ to the position in the image plane. The kernel of the integral transformation is the propagator

$$\mathcal{K}(x,x_0) = \langle x|\widehat{T}|x_0\rangle$$

corresponding to the response function (point spread function) in scalar wave optics. Loosely speaking it relates the point source in the object plane ($z=0$) with its image in the image plane. This mapping is fuzzy in realistic image processing due to the effect of diffraction. The explicit form of response function is fully specified by the geometrical parameters of the optical device. In the case of realistic observations with finite aperture, operator $\widehat{T}$ is not unitary and observations are not orthogonal. Detection represents the last step necessary for establishing the full analogy. Intensity of partially coherent wave in the position (x,z) is given by diagonal elements of mutual intensity $I(x,z) = \Gamma(x,x,z)$. This quantity is measurable and may be sampled using relative frequencies f_i. It plays the role of probability in quantum theory $p(x) = I(x)$.

The analogy between scalar wave and quantum theories will be used for the formulation of the following inverse problem. Assuming an unknown signal propagating through optical elements, the output field may be detected. Provided that properties of optical system and properties of detection process are known, the input signal should be inferred from the measurements performed on the output. This scheme has already been considered in analogy with quantum tomography [1]. However, adopting this procedure for realistic noisy data one always runs into nonphysical results, since the inversion is an ill posed problem. To overcome this difficulty the maximum likelihood estimation was proposed for various quantum reconstruction schemes [2, 3, 4]. It has all the features of genuine quantum measurement. This technique will be adopted here for reconstruction of correlation function which yields with the highest likelihood the detected data. Ultimate limitation of image processing beyond the standard resolution will be investigated [5].

The numerical simulation shows the feasibility of the method. The object state created by double slit illuminated by coherent light is measured for twenty discrete transverse displacements of measuring apparatus. This is an ideal thin lens with finite aperture in imaging arrangement and intensity detector consisting from forty discrete pixels in image plane. The correlation function of input state is reconstructed as hermitian matrix 100×100. Real and imaginary parts of the estimated mutual intensity are shown in the Figs. 1 and 2. The separation of the two slits is two times below the Rayleigh limit producing intensity in image plane as plotted in the Fig. 3. Due to the anisoplanasy of the setup the peaks could be distinguished provided that transverse shift is induced. Reconstructed intensity in the object plane exhibits two distinguishable peaks as shown in the Fig. 4.

Fig. 1

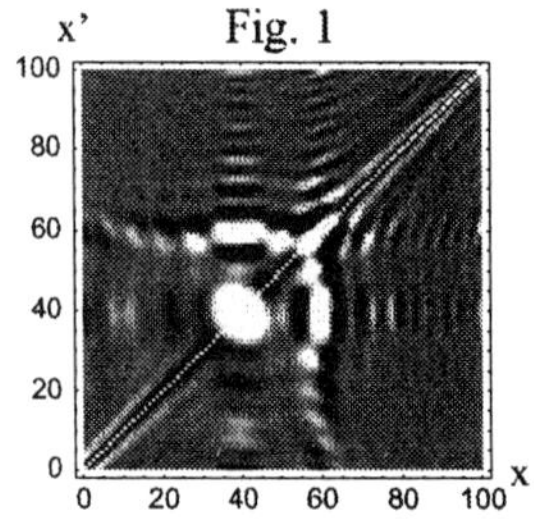

Fig. 2

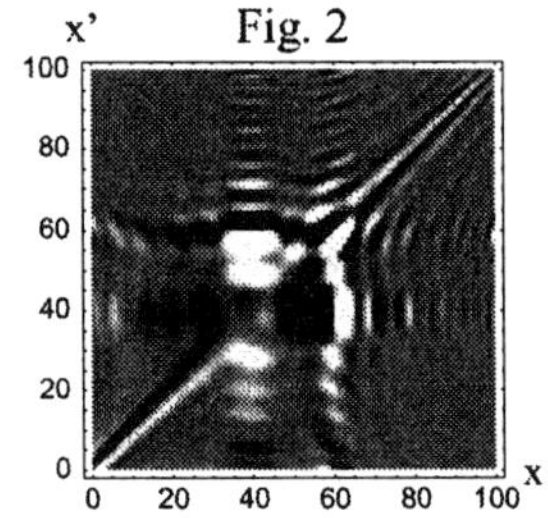

Fig. 3

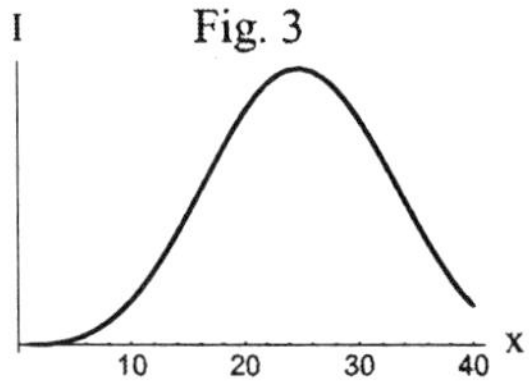

Fig. 4

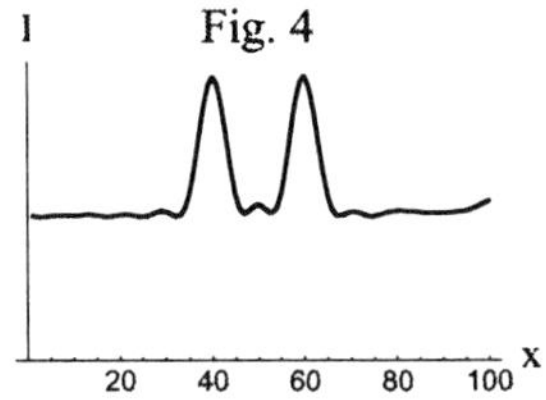

References and links

1. M. G. Raymer, M. Beck, and D. F. McAlister, "Complex wave–field reconstruction using phase–space tomography," Phys. Rev. Lett. **72**, 1137–1140 (1994).
2. Z. Hradil, "Quantum–state estimation," Phys. Rev. A **55**, R1561 (1997).
3. Z. Hradil, J. Summhammer, and H. Rauch, "Quantum tomography as normalization of incompatible observations," Phys. Lett. A **261**, 20 (1999).
4. J. Řeháček, Z. Hradil, and M. Ježek, "Iterative algorithm for reconstruction of entangled states," Phys. Rev. A **63**, 040303(R) (2001).
5. B. R. Frieden, "Probability, Statistical Optics, and Data Testing," Springer, Berlin 1983.

High depth of field microscopy with an interferometric imaging system

P. Potuluri, M. R. Fetterman and D. J. Brady

1 Introduction

In conventional imaging systems there is a trade-off between numerical aperture and depth of field. We have previously discussed the elimination of this trade-off in coherence imaging systems[1, 2]. In coherence imaging systems the relevant trade off is between source complexity and numerical aperture. Coherence systems have infinite angular depth of field, but the reconstructed image SNR degrades as a function of the number of independent source channels. One can reduce the effective number of source channels by decreasing the numerical aperture.

Microscopic applications are particularly attractive for coherence imaging because the spatial complexity of microscopic sources is often relatively low while simultaneous demand for high numerical aperture, high resolution and high depth of field is strong.

Researchers have demonstrated various techniques to increase the depth of field of a microscope. One approach is to reduce the aperture of the exit pupil slightly, thereby increasing the depth of field, but at the expense of low throughput and low resolution. A more interesting technique is to use a cubic phase plate [3].

We describe a microscope, hereafter termed the mRSI, formed by replacing the eyepiece of a light microscope with an RSI (rotational shear interferometer). The advantage of the RSI over the eye or a camera is that one obtains infinite depth of field in the reconstructed microscopic image without sacrificing numerical aperture.

2 Theory and Results

For a quasi-monochromatic and spatially incoherent source, described by the radiant density $I_s(r_s)$, the mutual intensity [4] between two sampling points at $(\frac{\Delta x}{2}, \frac{\Delta y}{2}, 0)$ and $(\frac{-\Delta x}{2}, \frac{-\Delta y}{2}, 0)$ under the paraxial approximation is given by [1],

$$J(\Delta x, \Delta y) = \int \frac{I_s(r_s)}{z_s^2} \exp\left[-j\frac{2\pi}{\lambda z_s}(x_s\Delta x + y_s\Delta y)\right] d^3 r_s \tag{1}$$

This integral when inverted with respect to Δx and Δy yields the 3D source intensity.

$$S(u, v) = \int \frac{I_s(r_s)}{z_s^2} \delta\left(u - \frac{x_s}{\lambda z_s}, v - \frac{y_s}{\lambda z_s}\right) d^3 r_s \tag{2}$$

It can be seen that $S(u, v)$ is a line integral through $\frac{I_s(r_s)}{z_s^2}$ along a ray passing through the points $(x_s = \lambda z_s u, y_s = \lambda z_s v)$. If a system can measure the mutual intensity, then Eq. 2 can be used to obtain the source intensity. We use an RSI to measure the mutual intensity function J.

Fig 1 shows a schematic diagram of our optical system. The object is placed under a Zeiss Axioplan transmission-light microscope. Just as the purpose of the objective/eyepiece combination in a conventional microscope is to match the numerical aperture of the source to the numerical aperture of the eye, the purpose of the objective/RSI combination in the mRSI is to match the numerical aperture of the source to the numerical aperture of the RSI. This match is achieved with a microscope objective and relay lenses.

Coherence and Quantum Optics VIII
Edited by Bigelow et al., Kluwer Academic/Plenum Publishers, 2003

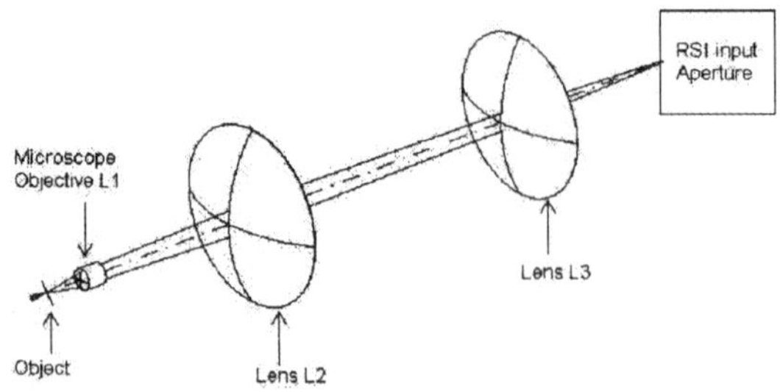

Fig. 1. Schematic diagram of an mRSI

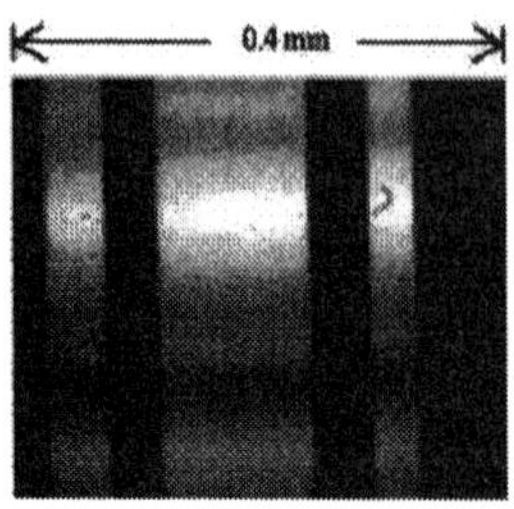

(a)

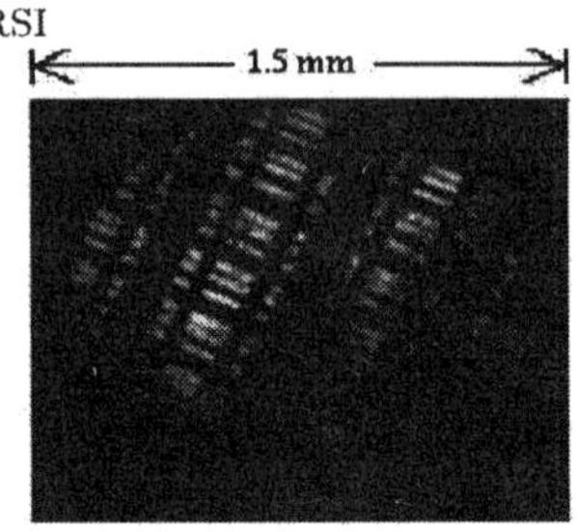

(b)

Fig. 2. Images of the two diffraction masks obtained from (a) the standard microscope (b) the mRSI

To illustrate the high depth of field imaging with the mRSI, we have taken two diffraction masks (one containing horizontal lines and the other containing vertical lines) and we placed one on top of the other. The spatial frequency of the mask is roughly 10 lines/mm. Fig. 2(a) shows the image obtained from a video camera placed at the output of just the microscope. We could focus to either the horizontal lines or to the vertical lines but both lines could not be focussed at a time. Fig. 2(b) shows the image of the same masks obtained by the mRSI. It can be seen clearly that both the lines are in focus.

3 Conclusion

The mRSI is capable of imaging objects through a microscopic objective with an infinite depth of field. This should have applications to areas such as biology, fluid mechanics, and material science. In this work, we have shown the design and analysis of the mRSI, along with experimental results.

The authors are grateful to Daniel Marks, Jason Gallicchio, Ronald Stack and Remy Tumbar for their comments. This work was supported by the Defense Advanced Research Projects Agency through Army Research Office grant DAAG55-98-1-0039.

References and links

1. Daniel L. Marks, Ronald A. Stack, David J. Brady, David C. Munson and Rachael B. Brady, "Visible Cone-Beam tomography with a lensless interferometric camera," Science **284**, 2164-2166 1999.
2. Daniel L. Marks, Ronald A. Stack and David J. Brady " Three-dimensional coherence imaging in the Fresnel domain" Appl. Opt. **8**, 38, 1332-1342 1999.
3. Sara Bradburn, Thomas W. Cathey and Edward R. Dowski, Jr. "Realizations of focus invariance in opyical-digital systems with wave-front coding" Appl. Opt. **35**, 36, 9157-9166 1997.
4. L. Mandel and E. Wolf, *Optical Coherence and Quantum Optics*. Cambridge, UK: Cambridge University Press, 1995.

On the nonlinear orbital motion of the driven two-photon Jaynes-Cummings model

Amitabh Joshi

Laser and Plasma Technology Division, BARC, Mumbai 400 085, India

ajoshi@apsara.barc.ernet.in

One of the simple yet very important quantum optical model describing the interaction of a single mode quantized electromagnetic field with a two-level atom or an effective two-level atom, better known as one-photon or two-photon Jaynes-Cummings model(JCM) in the literature[1], play very significant roles both in the theoretical predictions as well as in interpretation of experiments in cavity quantum electrodynamics. Both these models have a common feature in the sense that they exhibit collapse and revival phenomenon of Rabi oscillations (but on different time scales) when the initial field is in superposition of Fock states[1].

In a recent generalization the usual one-photon JCM together with external driving field has been considered[2] and it has been shown that the mean photon number oscillates like the orbital motion of a bound mechanical system and it also exhibits collapse and revival phenomenon on a time scale much larger than the time scale of the usual collapse-revival phenomenon of Rabi oscillations in the atomic inversion. Here we consider a usual two-photon JCM driven by an external field in order to study: (1) the dynamic evolution of mean photon number in the presence of external driving field, (2) to compare the nonlinear models of driven two-photon JCM with that of driven one-photon JCM and (3) to examine the role of dynamic Stark shift on the orbital oscillations.

We consider an effective two-level atom with atomic transition frequency ω_0, undergoing a non-resonant two-photon transition in the single mode radiation field of frequency ω such that $2\omega = \omega_0$ in an ideal and closed cavity so that the field damping and the radiative damping can be ignored. For such a system the two-photon JCM hamiltonian which includes Stark shift (due to virtual transitions to the intermediate levels) and driven by an external undepleted field can be given in rotating wave approximation (RWA) and in interaction picture as

$$H = \hbar(a^\dagger + \frac{\varepsilon_{ext}}{\sqrt{g}})(a + \frac{\varepsilon_{ext}}{\sqrt{g}})(\beta_1|e><e| + \beta_2|g><g|) + i\hbar g(\sigma_+(a + \frac{\varepsilon_{ext}}{\sqrt{g}})^2 - \sigma_-(a^\dagger + \frac{\varepsilon_{ext}}{\sqrt{g})})^2), \tag{1}$$

in which β_1 and β_2 are the measures of Stark splittings, g is the two-photon coupling constant ε_{ext} is the real amplitude of the external driving field, $a(a^\dagger)$ is the annihilation (creation) operator, $\sigma_\pm$ are pseudo spin operators and $|e>$ ($|g>$) is the excited (ground) state of the atom. This kind of two-photon JC model is directly applicable to experiments describing the process of two-photon transition between $^{44}S_{1/2} \longleftrightarrow ^{43} S_{1/2}$ levels in Rydberg ^{85}Rb atoms through the relay level $^{39}P_{3/2}$. Also, a trapped and laser irradiated ion exhibits a Jaynes-Cummings dynamics in appropriate limits. Recently, it has been proposed that the multiquantum JCM (similar to Eq.(1) above) could be realized for a trapped and laser irradiated ion, far from Lamb-Dicke regime under appropriate resonance condition.

The time dependent Schrödinger equation for the hamiltonian $\tilde{H} = D(\alpha)HD^\dagger(\alpha)$ is

$$i\hbar\frac{\partial}{\partial t}|\tilde{\psi}>= \tilde{H}|\tilde{\psi}> . \tag{2}$$

The wave function $|\tilde{\psi}>$ is related to $|\psi>$ of hamiltonian (1) as follows:

$$|\tilde{\psi}(t)>= D(\alpha)|\psi(t)>, \qquad |\tilde{\psi}(0)>= D(\alpha)|\psi(0)>= |\alpha>|e> . \tag{3}$$

We consider our effective two state system to be consisting of the excited state $|e>$ and the ground state $|g>$ with initial state of total system of atom and electromagnetic field to be $|\psi(0)>= |0>|e>$ and $|\alpha>= D(\alpha)|0>$, a coherent state with $D(\alpha) = exp[\alpha(a^\dagger - a)]$.

It is very straightforward to calculate mean photon number:

$$< a^\dagger a >=< \psi|a^\dagger a|\psi > \quad =< \tilde{\psi}|(a^\dagger-\alpha)(a-\alpha)|\tilde{\psi} > \quad = 2\alpha(\alpha- < \tilde{\psi}|a|\tilde{\psi} >)+2(1-P_e), \tag{4}$$

where $P_e =< \tilde{\psi}|e >< e|\tilde{\psi} >$ is occupation probability of the excited state $|e >$.

It is observed that the mean photon number $< a^\dagger a >$ gives rise to orbital oscillations. The time period of these oscillations is $2\pi/g$ and $2\pi/\sqrt{g^2 + (\beta_1 - \beta_2)^2}$ respectively in the absence and in the presence of Stark shifts. Due to non-zero Stark shifts there is modulation in the quantity $< a^\dagger a >$. However, the dynamical evolution of $< a^\dagger a >$ or the orbital motion also exhibits collapse and revival phenomenon of Rabi oscillations over a very long period as compared to revival time of Rabi oscillations in the atomic inversion. The revival time in atomic inversion is given by π/g whereas the revival time of orbital oscillations is given by $\sim 8\pi\alpha^6/g$ (where $\alpha = \frac{\varepsilon_{ext}}{\sqrt{g}}$) in the absence of Stark shifts and by $\sim 8\pi\alpha^6(1 + (\beta_1 - \beta_2)^2/g^2)^{3/2}/(g|1 - 8(\beta_1 - \beta_2)^2/g^2|)$ when the Stark shift is non-zero.

The most important similarity between the driven two-photon JCM[3] and the driven one-photon JCM [2] is synchronization of orbital oscillations and the collapse and revival phenomenon of Rabi oscillations. The Stark shift in the two-photon JCM influences both the orbital oscillations as well as the collapse and revival phenomenon in the excitation probability. At short times the Stark shift introduces modulation in the orbital oscillations in the two-photon JCM. The orbital oscillations exhibit a kind of collapse and revival phenomenon over a very long time period in both one and two-photon driven JCMs. However, the revival times for the long time collapse and revival phenomenon are $8\pi\alpha^6/g$ for two-photon JCM and $8\pi\alpha^3/g$ for one-photon JCM respectively. Further the revival-time period for the driven two-photon JCM can elongate with the inclusion of Stark shift.

References and links

1. B. W. Shore and P. L. Knight, "The Jaynes-Cummings Model," J. Mod. Opt. **40**, 1195-1238 (1993); A. Joshi, "Cavity Quantum electrodynamics of two-level atoms: some theoretical studies," Ph. D. thesis (1991), University of Bombay (unpublished).
2. Y. -T. Chough and H. J. Carmichael, "Nonlinear oscillator behaviour in the Jaynes-Cummings model," Phys. Rev. A **54**, 1709-1714 (1996).
3. A. Joshi, "Nonlinear dynamical evolution of the driven two-photon Jaynes-Cummings model," Phys. Rev. A **62**, 043812-6 (2000).

Jaynes-Cummings model with moving phased atoms

Amitabh Joshi

Laser and Plasma Technology Division, BARC,
Mumbai 400 085, India

ajoshi@apsara.barc.ernet.in

Among many generalization and extensions of Jaynes-Cummings model (JCM)[1], the very important one is to include the atomic motion in the JCM so that the spatial field structure of cavity mode can be incorporated in this model[2]. By including the cavity mode structure in the JCM, some interesting effects similar to self induced transparency (SIT) and adiabatic following (AF) in the atomic inversion can be observed[2-4]. We have examined these effects when a phased atom is sent through the cavity and have predicted that for a particular choice of atomic coherence, cavity cooling can be observed[5].

The two-level atom with ground state $|g>$ (energy $E_g = \hbar\omega_g$) and excited state $|e>$ (energy $E_e = \hbar\omega_e$) is interacting with cavity field of frequency ω. The one-photon transition between $|e>$ and $|g>$ can be described by the JCM hamiltonian [2]:

$$H = \hbar\omega_g|g><g| + \hbar\omega_e|e><e| + \hbar\omega a^\dagger a + \hbar g f(z)(S_- a^\dagger + S_+ a). \tag{1}$$

Here, $S_+ = |e><g|$, $S_- = |g><e|$, $a_i(a_i^\dagger)$ is the cavity field annihilation (creation) operator of the cavity field mode and g represents atom-field coupling. We restrict our studies for atomic motion along the z axis so that only the z dependence of the mode function $f(z)$ would be necessary to consider. The atomic motion would be incorporated as $f(z) \to f(vt)$, where v denotes the atomic velocity. Hence we can define our cavity mode TE_{mnp} : $f(vt) = sin(p\pi vt/L)$, in which p stands for the number of half wavelengths of the mode inside a cavity of length L.

Since the atom absorbs and emits a single photon in an ideal cavity so the basis vectors are $|n, e>$ and $|n+1, g>$, and the wavefunction of the above system can be written in the following way:

$$\psi(t) = \sum_n (a_n(t)|n, e> + b_n(t)|n+1, g>). \tag{2}$$

Clearly, the dynamical evolution of the system for each value of n is determined by two coupled equations for the complex amplitudes $a_n(t)$ and $b_n(t)$. But, we can also define three real quantities as follows[2]: $u_n = 2Re(a_n b_n^*)$, $v_n = -2Im(a_n b_n^*)$, $w_n = |a_n|^2 - |b_n|^2$ and thus obtain a discrete set of Bloch-like equations of motion of atom-field system, due to the discretized photon distribution of the electromagnetic field in the cavity[2]

$$\begin{aligned} \frac{d}{dt}w_n &= 2g\sqrt{(n+1)}sin(p\pi vt/L)v_n, \\ \frac{d}{dt}v_n &= \Delta u_n - 2g\sqrt{(n+1)}sin(p\pi vt/L)w_n, \\ \frac{d}{dt}u_n &= -\Delta v_n. \end{aligned} \tag{3}$$

Here, detuning $\Delta = \omega_e - \omega_g - \omega$. Let the atom enters the cavity at $t = 0$, in a coherent superposition of its state, i.e., $\psi_{atom}(0) = \frac{1}{\sqrt{1+|\mu|^2}}(|e> + \mu|g>)$, in which μ is a complex number: $\mu = |\mu|e^{i\theta}$. Here we assume $|\mu| = 1$ and $\theta = \pi/2$ so the initial conditions become $w_n(0) = 0$, $v_n(0) = 1$, $u_n(0) = 0$ and the measurement of the population inversion be carried out at the transit time $t_T = L/v$ when the atom leaves the cavity.

For $p = 1$ or the odd modes of cavity, atomic dynamics is not much interesting. However, for $p = 2$ or the even modes, we get something very interesting. Under this condition at $\Delta = 0$, the 'pulse area' (see Ref.[2-4]) vanishes over the 'p' half wavelengths of mode structure and the atomic inversion and coherence are not affected by the passage of atom thru the cavity. This is different from the self-induced transparency (SIT). With $p = 2$ mode the field driving the atom undergoes a phase reversal half-way through the evolution, this will reverse the direction of Rabi oscillations returning the configuration to the initial state where as the SIT works through nonlinearly without the phase reversal business. For very large detuning (Δ), atom undergoes adibatic following(AF).

The very interesting situation arises for $\Delta = \pm 3$ where we find that the atomic inversion at the exit of the cavity is 0.5 (we have started with atomic inversion to be 0 at the cavity entrance). So, there is gain in the atomic inversion. This kind of gain is not observed with the usual JCM with the same initial conditions. Also, we do not find such a gain without the input of phased atoms as studied in Ref.[2].

So, it is only the initial atomic coherence, which can lead to gain in the population inversion at a particular value of atomic detuning (symmetric). Consequently, this will result into cooling of the resonator cavity as the quantity $< S_z + a^\dagger a >$ is a constant of motion for this model. Since the atomic inversion at the cavity exit is more than that at cavity entrance, there will be depletion of cavity field as a result of the above constant of motion. The mean photon number of the thermal field is given by the Planck's function $\bar{n} = (e^{\beta \hbar \omega} - 1)$ and $\beta = k_B T$ (where k_B is the Boltzmann constant). If the incoming atom is in its upper state $|e>$ at the entrance then it can deposit the photon inside the cavity provided it satisfy the proper conditions and hence mean number of photon inside the cavity increases and effectively the temperature changes as per the Planck's function. This phenomenon will be very sensitive to how good the thermal isolation of cavity from the environment and the field damping in the cavity. In our case the outcoming atom is removing photons at the two symmetric values of detuning meaning it is decreasing the mean number of photons inside the cavity at these values of detuning[5] provided cavity is kept isolated with respect to the environment (which implies that no heat exchange is allowed once the cavity acquires a temperature of $T^0 K$ prior to the start of the experiment). In other words the rate of injection of phased atoms is such that the removal of photons from the cavity is much larger than what the cavity is gaining from the environment as a result of thermalization. The thermal photon distribution inside the cavity is narrowed down in such situation[5].

References and links

1. B. W. Shore and P. L. Knight, "The Jaynes-Cummings Model," J. Mod. Opt. **40**, 1195-1238 (1993); A. Joshi, "Cavity Quantum electrodynamics of two-level atoms: some theoretical studies," Ph. D. thesis (1991), University of Bombay (unpublished).
2. R. R. Schlicher, " Jaynes-Cummings model with atomic motion," Opt. Comm. **70**, 97-102 (1989).
3. A. Joshi, "Effect of atomic motion on Rydberg atoms undergoing two-photon transition in a lossless cavity," Phys. Rev. A **42**, 1752-1756 (1990).
4. A. Joshi, "Two-mode two-photon Jaynes-Cummings model with atomic motion," Phys. Rev. A **58**, 4662-4667 (1998).
5. A. Joshi, "Effect of atomic coherence on Jaynes-Cummings model with atomic motion," Phys. Lett. A (to be published).

Dynamics of two atoms coupled to a cavity field

A. Vidiella-Barranco and J.A. Roversi

Instituto de Física "Gleb Wataghin", Universidade Estadual de Campinas, 13083-970 Campinas SP Brazil

vidiella@ifi.unicamp.br, roversi@ifi.unicamp.br

H. Moya-Cessa

INAOE, Coordinación de Optica, Apdo. Postal 51 y 216, 72000 Puebla, Pue., Mexico

hmmc@inaoep.mx

The problem of the interaction of two-level atoms with a cavity field has been already addressed in the literature [1, 2]. This constitutes an interesting quantum optical system, given that it involves three fully quantized subsystems, and also admits an exact solution. The investigation of tripartite systems is particularly attractive specially due to the interest in three-system entanglement with implications on quantum information, for instance [3]. Nevertheless, it seems that such a physical situation has been not fully explored, and deserves further investigation. In most treatments, the atoms are supposed to be either both resonant with the field [1] or both nonresonant [2]. We have two subsystems in principle able to *store* quantum information and one subsystem (field) able to *transmit* quantum information, and one may ask about what kind of correlations could arise between the atomic subsystems in virtue of their interaction with the field. Here we consider a different (asymmetric) situation; one atom resonant with the cavity field and the other very far from resonance.

We consider that two (two-level) atoms are coupled to the electromagnetic field inside a high-Q cavity, so that in the present approach losses will be neglected. As a first approach we are not considering direct coupling (of a dipole type, for instance) between the atoms in the present work. The corresponding hamiltonian may be written as

$$\hat{H} = \hbar \left[\omega \hat{n} + \omega \sigma_z^{(1)} + \omega_2 \sigma_z^{(2)} + \lambda_1 \left(\hat{a}^\dagger \sigma_-^{(1)} + \sigma_+^{(1)} \hat{a}\right) + \lambda_2 \hat{n} \sigma_z^{(2)}\right]. \tag{1}$$

Atom 1 is in resonance with the cavity field (of frequency ω), having a coupling constant λ_1 while atom 2 is out of resonance (atomic frequency ω_2), and coupling constant $\lambda_2 = \lambda_1^2/\delta$ (with $\delta = \omega_2 - \omega$). We may therefore treat the atom 2-field interaction in the so-called dispersive limit [4], which is valid provided $|\delta| \gg \sqrt{n+1}\,\lambda_1$. This results in an effective interaction hamiltonian $\hat{H}_i' = \lambda_2 \hat{n} \sigma_z^{(2)}$. The hamiltonian $\hat{H}$ above may be further transformed into the following interaction picture hamiltonian

$$\hat{H}_I = \hat{U}_0 \hat{H} \hat{U}_0^\dagger = (\delta + \lambda_2 \hat{n}) \sigma_z^{(2)} + \lambda_1 \left(\hat{a}^\dagger \sigma_-^{(1)} + \sigma_+^{(1)} \hat{a}\right), \tag{2}$$

where

$$\hat{U}_0(t) = \exp[-i\omega t(\hat{n} + \sigma_z^{(1)} + \sigma_z^{(2)})]. \tag{3}$$

We may assume that initially the atoms are decoupled from the field, so that we have an initial product state for the whole system

$$|\Psi(0)\rangle = |\varphi\rangle \otimes (a_1|g_1\rangle + b_1|e_1\rangle) \otimes (a_2|g_2\rangle + b_2|e_2\rangle), \tag{4}$$

i.e., both atoms prepared in superpositions of their energy eigenstates, and with $|\varphi\rangle = \sum A_n|n\rangle$ being the initial arbitrary (pure) state of the field. After solving Schrödinger's equation, we obtain the following time dependent state vector

$$|\Psi(t)\rangle = \sum_{n=0}^{\infty} \{C_{1;n}e^{-i[\delta+\lambda_2(n-1/2)]t}|n-1\rangle|e_1\rangle|e_2\rangle + C_{2;n}e^{-i[\delta+\lambda_2(n-1/2)]t}|n\rangle|g_1\rangle|e_2\rangle \quad (5)$$
$$+C_{3;n}e^{-i(\delta_n+\lambda_2/2)t}|n\rangle|e_1\rangle|g_2\rangle + C_{4;n}e^{-i[\delta+\lambda_2(n+1/2)]t}|n+1\rangle|g_1\rangle|g_2\rangle\},$$

with

$$\begin{aligned} C_{1;n} &= \left(\cos\Delta_n^{1/2}t + i\frac{\lambda_2}{2}\sin\frac{\Delta_n^{1/2}t}{\Delta_n^{1/2}}\right)A_{n-1}b_1b_2 - i\lambda_1\sqrt{n}\sin\frac{\Delta_n^{1/2}t}{\Delta_n^{1/2}}A_na_1b_2 \quad (6)\\ C_{2;n} &= i\lambda_1\sqrt{n}\sin\frac{\Delta_n^{1/2}t}{\Delta_n^{1/2}}A_{n-1}b_1b_2 + \left(\cos\Delta_n^{1/2}t - i\frac{\lambda_2}{2}\sin\frac{\Delta_n^{1/2}t}{\Delta_n^{1/2}}A_na_1b_2\right)\\ C_{3;n} &= \left(\cos\Delta_{n+1}^{1/2}t - i\frac{\lambda_2}{2}\sin\frac{\Delta_{n+1}^{1/2}t}{\Delta_{n+1}^{1/2}}\right)A_nb_1a_2 - i\lambda_1\sqrt{n+1}\sin\frac{\Delta_{n+1}^{1/2}t}{\Delta_{n+1}^{1/2}}A_{n+1}a_1a_2\\ C_{4;n} &= -i\lambda_1\sqrt{n+1}\sin\frac{\Delta_{n+1}^{1/2}t}{\Delta_{n+1}^{1/2}}A_nb_1a_2 + \left(\cos\Delta_{n+1}^{1/2}t + i\frac{\lambda_2}{2}\sin\frac{\Delta_{n+1}^{1/2}t}{\Delta_{n+1}^{1/2}}A_{n+1}a_1a_2\right) \end{aligned}$$

where $\Delta_n = \lambda_1^2 n + \frac{\lambda_2^2}{4}$ is a modified Rabi frequency.

As an example we may consider the simple initial conditional for the global atoms-field state

$$|\Psi(0)\rangle = |0\rangle \otimes |e_1\rangle) \otimes (a_2|g_2\rangle + b_2|e_2\rangle), \quad (7)$$

or the field in the vacuum state, atom 1 in the excited state and atom 2 in a superposition state. After tracing over the atom 2 as well as the field variables, we obtain the following reduced density operator for atom 1

$$\hat{\rho}_1(t) = \left(\cos^2\Delta_1^{1/2}t + \frac{\lambda_2^2}{4}\frac{\sin^2\Delta_1^{1/2}}{\Delta_1}\right)|e_1\rangle\langle e_1| + \lambda_1^2\frac{\sin^2\Delta_1^{1/2}t}{\Delta_1}|g_1\rangle\langle g_1|, \quad (8)$$

with $\Delta_1 = \frac{\lambda_2^2}{4} + \lambda_1^2$. We note that the effect of the non-resonant atom 2 on the resonant one (atom 1) is in fact a frequency shift equivalent to a atom-cavity detuning, although atom 1 is actually in resonance with the cavity field.

We present an investigation of features such as entanglement among subsystems of interest under different initial conditions for the atoms and the field.

References and links

1. "Atom-atom correlations induced by resonant coupling with a laser field", Y. Gontier, Phys. Rev. A **59**, 4747-4755 (1999).
2. S.-B. Zheng and G.-G. Guo, "Efficient Scheme for Two-Atom Entanglement and Quantum Information Processing in Cavity QED", Phys. Rev. Lett. **85**, 2392-2395 (2000).
3. A.K. Pati, "Existence of Schmidt decomposition for tripartite systems", Phys. Lett. A **278**, 118-122 (2000).
4. M. Brune, S. Haroche, J.M. Raimond, L. Davidovich, and N. Zagury, "Manipulation of photons in a cavity by dispersive atom-field coupling - quantum non-demolition measurements and generation of Schrödinger cat states", Phys. Rev. A **45**, 5193-5214 (1992).

Pattern Formation, Localization and Coherent Structures in Nonlinear Wave Dynamics

Antonina N. Fedorova and Michael G. Zeitlin

Russian Academy of Sciences,
Institute of Problems of Mechanical Engineering,
V.O., Bolshoj pr., 61, 199178, St. Petersburg, Russia
anton@math.ipme.ru
http://www.ipme.ru/zeitlin.html
http://www.ipme.nw.ru/zeitlin.html

Abstract: We apply variational-wavelet approach for constructing multiscale high-localized eigenmodes expansions in different models of nonlinear waves. We demonstrate appearance of coherent localized structures and stable pattern formation in different collective dynamics models.

OCIS codes: (030.0030) Coherence and statistical optics; (000.3860) Mathematical methods in physics

We consider the applications of a new numerical-analytical technique which is based on the methods of local nonlinear harmonic analysis or wavelet analysis to the nonlinear wave dynamics. Such approach may be useful in all models in which it is possible and reasonable to reduce all complicated problems related with statistical/stochastic distributions to the problems described by systems of nonlinear ordinary/partial differential equations with or without some (functional) constraints (nonlinear Schroedinger, Vlasov-Maxwell -Poisson, etc and especially for the modelling of optical gap solitons, coherent matter waves in optical lattices). Wavelet analysis gives us the possibility to work with well-localized bases in functional spaces and gives for the general type of operators (differential, integral, pseudodifferential) in such bases the maximum sparse forms. In our variational-wavelet approach [1]–[3] we can consider polynomial and rational

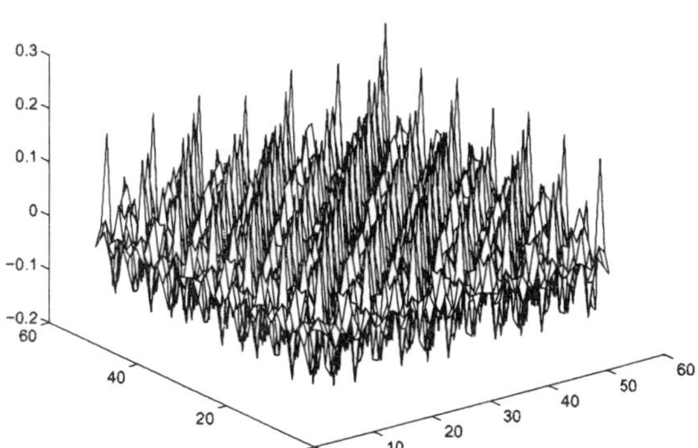

Fig. 1. Appearance of coherent structure

Coherence and Quantum Optics VIII
Edited by Bigelow et al., Kluwer Academic/Plenum Publishers, 2003

type of nonlinearities. The solution has the multiscale/multiresolution decomposition via nonlinear high-localized eigenmodes which corresponds to the full multiresolution expansion in all time/space scales.This gives us expansion into the slow part and fast oscillating parts. So, we may move from coarse scales of resolution to the finest one for obtaining more detailed information about our dynamical process. The first term corresponds on the global level of function space decomposition to resolution space and the second one to detail space. In this way we give contribution to our full solution from each scale of resolution or each time/space scale or from each nonlinear eigenmode. The same is correct for the contribution to power spectral density (energy spectrum): we can take into account contributions from each level/scale of resolution. Our functional space

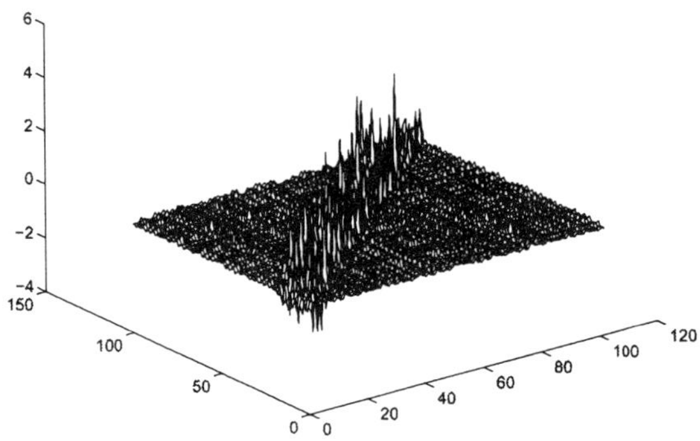

Fig. 2. Stable pattern

decomposition corresponds to exact nonlinear eigenmodes decompositions.It should be noted that such representations give the best possible localization properties in the corresponding (phase)space/time coordinates. In contrast with different approaches we do not use perturbation technique or linearization procedures and represent dynamics via generalized nonlinear localized eigenmodes expansion with the best possible convergence properties. So, by using wavelet bases with their good (phase)space/time localization properties we can modelling the dynamics of high-localized coherent structures in spatially-extended stochastic/turbulent systems with collective behaviour. In all these models numerical modelling demonstrates the appearence of coherent high-localized structures (Fig.1) and stable patterns formation (Fig.2).

References and links

1. A.N. Fedorova, M.G.Zeitlin, "Variational Approach in Wavelet Framework to Polynomial Approximations of Nonlinear Accelerator Problems", American Institute of Physics , Conf. Proc., **468**, Nonlinear and Collective Phenomena in Beam Physics, 48-68, (1999).
2. A.N. Fedorova, M.G.Zeitlin, "Variational-Wavelet Approach to RMS Envelope Equations", Proc. 2nd Advanced Accelerator Workshop on The Physics of High Brightness Beams pp.235–254, World Scientific, (2000).
3. A.N. Fedorova, M.G.Zeitlin, "Localized Coherent Structures and Patterns Formation in Collective Models of Beam Motion", Quantum Aspects of Beam Physics, World Scientific,(2001); Los Alamos preprint,physics/0101007, http://arXiv.org/abs/physics/0101007

Stochastic simulations and discreteness of light

R. Filip

Department of Optics, Palacký University,
17. listopadu 50, 772 07 Olomouc,
Czech Republic

email:filip@optnw.upol.cz, tel:+420-68-5631572, fax:+420-68-5224246

The stochastic simulations of nonlinear interactions are frequently employed method in quantum optics [1, 2]. The simulations use the continuous stochastic variables and one can put a question, whether it is possible to generate the effects of discreteness of light in the nonlinear systems. In addition, it is well known that for large variety of the nonlinear systems, the stochastic simulations exhibit an instability and generate invalid solutions [3] if the boundary terms in the phase-space become non-negligible. On the other hand, it will be presented here on an illustrative example that this instability is closely related to the omitting of a quantum discreteness in the continuous stochastic simulations. Because, if the discrete character is incorporated in a proper way, the stochastic simulations offer the correct results in agreement with the exact solutions. In summary, we suggest that the continuous stochastic simulations are not able to correctly describe a discreteness of light in the quantum nonlinear systems [4]. To illustrate this, we use the nonlinear photonic interaction (anharmonic oscillator) described by the interaction Hamiltonian $\hat{H}_I = \frac{\hbar}{2}\kappa\hat{A}^{+2}\hat{A}^2$, where κ is a real nonlinear constant and $\hat{A}$ is the annihilation operator of the light mode. An evolution of mean-value of the quadrature operator $\hat{Q} = \frac{1}{2}(\hat{A} + \hat{A}^+)$ for initial coherent state $|\alpha\rangle$ exhibits well-known periodical behavior with revival period $\Delta t = \kappa/2\pi$, as can be seen in Fig. 1(b), which is an exhibition of the interference between the discrete eigenstates of the photon-number operator [5].

The stochastic simulation method describes the system by means of a stochastic mapping for particular ordering. For the normal ordering a stochastic mapping for the Euler-method of simulations can be found in the form [3]

$$\alpha_{j+1} = \alpha_j - i\kappa\Delta t\alpha_{j*}\alpha_j^2 + \alpha_j\sqrt{-i\kappa\Delta t}\eta_j, \quad \alpha_{j+1*} = \alpha_{j*} + i\kappa\Delta t\alpha_{j*}^2\alpha_j + \alpha_{j*}\sqrt{i\kappa\Delta t}\eta_{j*}, \quad (1)$$

where α_j and α_{j*} are the stochastic variables in the double dimensional phase-space, Δt is a small time step of iteration and η_j, η_{j*} are Gaussian stochastic forces with the following correlation properties $\langle\eta_j\rangle = \langle\eta_{j*}\rangle = 0$, $\langle\eta_j\eta_k\rangle = \langle\eta_{j*}\eta_{k*}\rangle = \delta_{jk}$, $\langle\eta_j\eta_{k*}\rangle = 0$. The double dimensional phase-space must be introduced to correctly incorporate the quantum noise effects. However, this step introduces an instability in the system (1) and trajectories in the phase space are unbounded. Thus, the stochastic simulations diverge if we directly iterate the stochastic mapping and we can use only a short-time interval of solution, as it can be seen in Fig. 1(a). The number of divergent trajectories increases seriously as the system is going into the revival region. What must be considered in order to avoid the divergent trajectories and to restore a correct behavior of the simulations? One can use the symmetrically ordered simulations $\alpha_{j+1} = \alpha_j + i\kappa\Delta t(|\alpha_j|^2 - 1)\alpha_j$ which do not require the extension to double dimensional phase-space. This maping can simply be simulated, starting from the Wigner-function of the coherent state, but the revival effect is not obtained, as it can be seen in Fig. 1(c).

For this interaction, a different construction of stochastic simulations will be used, directly from the evolution of wave function in the Fock state basis for the initial coherent

Coherence and Quantum Optics VIII
Edited by Bigelow et al., Kluwer Academic/Plenum Publishers, 2003

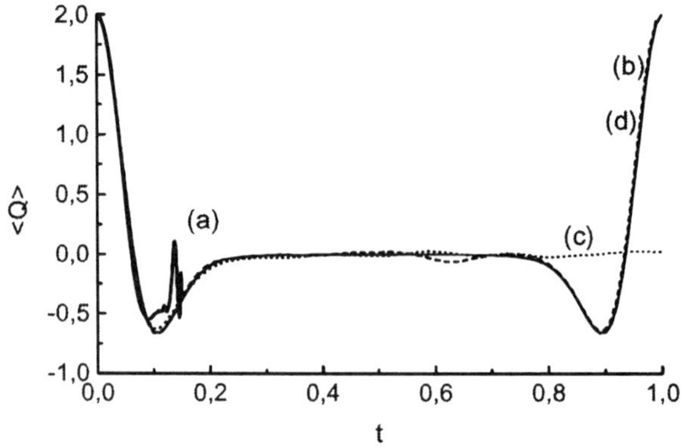

Fig. 1. Dynamics of the mean-value of the quadrature of light: (a) instable the continuous stochastic simulations, (b) exact solution, (c) semiclassical continuous stochastic simulations, (d) discrete stochastic simulations.

state. Then, the correct first and second normally ordered moments can be reconstructed from the following stochastic mapping

$$\alpha_{j+1} = \alpha_j - i\kappa\Delta t n\alpha_j + \alpha_j\sqrt{-i\kappa\Delta t}\eta_j, \quad \alpha_{j+1*} = \alpha_{j*} + i\kappa\Delta t n\alpha_{j*} + \alpha_{j*}\sqrt{i\kappa\Delta t}\eta_{j*}, \quad (2)$$

including a discrete photon-number n, which is generated with the probability given by the Poissonian distribution $p(n) = \exp(-|\alpha|^2)|\alpha|^{2n}/n!$, where $|\alpha|$ is the initial amplitude of the coherent state. The additional noise source introduced into the map (2) leads to the more time-consuming simulations with a very small time step and in addition, about 10^6 trajectories must be averaged to obtain the excepted mean values. If we perform the symmetrical and exact stochastic simulations to obtain mean value of quadrature $\langle\hat{Q}\rangle$ and to compare them with exact solution, we simply find that the use of only continuous stochastic variable cannot describe effects connected with discrete character of eigenvalues of the photon-number operator. On the other hand, if we introduce the discrete stochastic parameter n into the continuous simulations in a proper way, the agreement with exact solutions is apparent, as it can be seen in Fig. 1(d). The divergent trajectories are removed from Eqs. (2) by the discretization procedure. Comparison with the continuous mapping (1) implies: an inability of the continuous simulation method to correctly describe the light discreteness and an inconsistency of the simulations with solution of the Fokker-Planck equation [5]. Even for large amplitude $|\alpha|$ of the initial coherent state, the revival effects occur in the time evolution and the continuous simulations are not still appropriate. However, if we consider a decoherence during the nonlinear interaction, the interference of photon-number eigenstates can be suppressed and the stochastic simulations can provide a correct description of the quantum system. This work was supported by the project LN00A015 by the Ministry of Education of Czech Republic.

References and links

1. D.F. Walls and G.J. Milburn, *Quantum Optics* (Springer-Verlag, Berlin 1994).
2. C.W. Gardiner, *Quantum Noise* (Springer-Verlag, Berlin 1991).
3. A. Gilchrist, C.W. Gardiner, and P.D. Drummond, "Positive P representation: Application and validity," Phys. Rev. A **55**, 3014 (1997).
4. R. Filip, "Stochastic simulations: continuous approximations of quantum noise" J. Opt. B: Quantum Semiclass. Opt. **1**, 16 (2001).
5. G.J. Milburn, "Quantum and classical Liouville mechanics of the anharmonic-oscillator," Phys. Rev. A **33**, 674 (1986).

From the ordinary threshold regime to ideal thresholdless lasing

S. M. Dutra, J. Visser, G. Nienhuis, and J. P. Woerdman

Huygens Laboratory, University of Leiden, P.O. Box 9504, 2300 RA Leiden, The Netherlands

phone: +3171 5275931, fax: +3171 5275819, e-mail: dutra@molphys.leidenuniv.nl

Abstract: We show how by varying the speed of an atomic beam crossing a high-Q cavity, a regime with a well-defined laser threshold is replaced by another where the photon statistics is always Poissonian.

OCIS codes: (270.3430) Laser theory; (270.2500) Fluctuations, relaxations, and noise

Even though there has been experimental evidence for thresholdless lasing, the so-called ideal thresholdless laser [1] has never been realized in the laboratory. An ideal thresholdless laser is a thresholdless laser where the photon statistics is Poissonian for any pump power. Rice and Carmichael determined the conditions for this in the adiabatic limit where the decay rate of the atomic polarization is much larger than all the other rates in the laser [1]. Here we present an experimentally feasible scheme where not only the ideal thresholdless laser conditions can be studied outside the adiabatic limit, but where a convenient control parameter allows us to go from the ordinary laser regime to the ideal thresholdless regime.

We consider a single-mode high-Q cavity and a beam of excited two-level atoms going through it, one at a time, as in a micromaser [2]. We assume that the vacuum Rabi frequency, g, is much larger than both the cavity and the atomic decay rates, κ and $\gamma_{\|}$. But, unlike a micromaser, we do not assume strong coupling because we allow for decay of the atomic polarization (via Stark broadening as discussed below) at a rate $\gamma_{\perp} > g$. Spontaneous emission is irreversible as in an ordinary laser, with the spontaneous emission rate given by $4g^2/\gamma_{\perp}$ in the limit of $\gamma_{\perp} \gg g$. On the other hand, unlike an ordinary laser, we assume that $(g/\gamma_{\|})g \gg \gamma_{\perp}$, which together with $g \gg \gamma_{\|}$ and $g \gg \kappa$, guarantees that virtually all spontaneous emission photons will go into the cavity mode *when the atom is inside the cavity.* Thus β, the fraction of spontaneous emission into the lasing mode, would always be very close to one here, were it not for an extra loss channel: unlike atoms of the gain medium of ordinary lasers, these atoms leave the cavity. Even though virtually all spontaneous emission photons would end up in the cavity mode if the atoms never left the cavity, some spontaneous emission photons will be emitted into free-space modes by atoms that fail to emit while they are crossing the cavity. This provides us with a neat way of controlling the effective β of the laser simply by varying the speed at which the atoms cross the cavity: at speeds slow enough for the crossing time t_{int} to be much longer than the inverse of the spontaneous emission rate, β will be very close to one, whilst at high speeds β will be much smaller than one.

A general theory incorporating incoherent atomic and cavity decay into the Jaynes-Cummings dynamics has been worked out by Briegel and Englert [3] using damping bases. The case considered here, where there is only polarization decay, is much simpler. Assuming ordinary Poissonian pumping and a negligible average number of thermal photons, we find that the probability of having n photons in the cavity mode in the

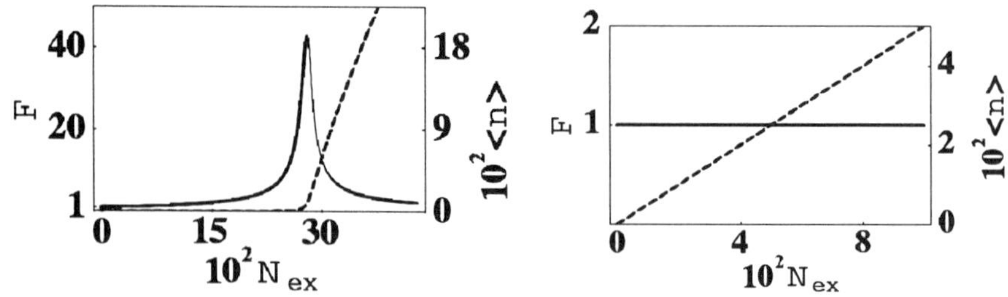

Fig. 1. The Fano parameter (full line) and the average photon number (dotted line) as functions of N_{ex} for $\Gamma = 2$ and $W = 0.04$ on the left, and for $\Gamma = 2$ and $W = 44$ on the right.

"steady state" is given by

$$p_n = p_0 \prod_{k=1}^{n} \frac{N_{ex}}{2k}(1 - G_k), \tag{1}$$

where N_{ex} is the average number of atoms that cross the cavity within a photon lifetime, the counterpart of the number of excited atoms in the gain medium of an ordinary laser (i.e. a measure of the pumping power), p_0 is the probability of finding no photons in the cavity mode (obtained as a normalization constant from the requirement $\sum p_n = 1$), and G_k, which describes the combined effects of the Jaynes-Cummings evolution and dephasing, is given by

$$G_k = e^{-\Gamma W}\left\{\frac{\Gamma}{\sqrt{\Gamma^2 - k}}\sinh(\sqrt{\Gamma^2 - k}W) + \cosh(\sqrt{\Gamma^2 - k}W)\right\}, \tag{2}$$

with $\Gamma \equiv \gamma_\perp/(4g)$ and $W \equiv gt_{\text{int}}$. From (1), we see that for weak pumping where $N_{ex} \ll \Gamma/W$, $p_n \approx [N_{ex}W/(4\Gamma)]^n p_0$ and the photon statistics is thermal. For strong pumping where $N_{ex} \gg \Gamma/W$ the photon statistics is Poissonian. This is just like an ordinary laser, with the laser threshold given by $N_{ex} = 4\Gamma/W$. Now, if $W/\Gamma \gg 1$, the device becomes an ideal thresholdless laser with a Poissonian photon statistics for all pumping powers given by $p_n = \exp(-N_{ex}/2)(N_{ex}/2)^n/n!$. Figure 1 shows the result of a numerical simulation using the exact eq. (1) both in the ordinary laser regime and in the ideal thresholdless regime.

At present, in the *optical* regime g can be made only slightly larger than κ and $\gamma_\parallel$. However, the current state of the art in cavity QED *microwave* experiments readily provides $g \gg \gamma_\parallel, \kappa$ and at the lowest temperature achieved in such experiments (0.15K in [2]) the average number of thermal photons will be negligible. The atomic polarization decay required can be produced by introducing random fluctuations in the electric field that is applied across the superconducting mirrors of the open high-Q cavity used in many of these experiments [4], thus allowing an experimental test of our theory.

References and links

1. P. R. Rice and H. J. Carmichael, "Photon statistics of a cavity-QED laser: A comment on the laser-phase-transition analogy," Phys. Rev. A **50**, 4318-4329 (1994).
2. G. Rempe, F. Schmidt-Kaler, and H. Walther, "Observation of sub-Poissonian photon statistics in a micromaser," Phys. Rev. Lett. **64**, 2783-2786 (1990).
3. H.-J. Briegel and B.-G. Englert, "Quantum optical master equations: The use of damping bases," Phys. Rev. A **47**, 3311-3329 (1993).
4. M. Brune, F. Schmidt-Kaler, A. Maali, J. Dreyer, E. Hagley, J. M. Raimond, and S. Haroche, "Quantum Rabi oscillation: A direct test of field quantization in a cavity," Phys. Rev. Lett. **76**, 1800-1803 (1996).

Criticality and dynamics of quantum fluctuations in an OPO

Peter D. Drummond and Kaled Dechoum

Department of Physics, University of Queensland, Brisbane, Queensland, Australia 4067

A system consisting of a nonlinear crystal inside a Fabry-Perot interferometer, that couples subharmonic intra-cavity modes to a second-harmonic pump, is known as an optical parametric oscillator (OPO). This simple quantum system is rich enough to show non-equilibrium critical phenomena, as occurs in other fields of physics. It is natural to apply some of the theory of finite size scaling and the renormalization group to get an insight into the critical quantum dynamics.

The OPO system can be described by a set of coupled classical stochastic equations, mapped from the the quantum density matrix, using operator representation theory. The following stochastic equations are obtained for any driven field for the optical parametric oscillator with the presence of transverse modes using the positive P-representation

$$\begin{aligned}\frac{\partial A_0}{\partial t}(\vec{x},t) &= -\gamma_0 A_0 + \mathcal{E} - \frac{g}{2}A_1^2 + \frac{ic^2}{2\omega_0}\nabla^2 A_0 \\ \frac{\partial A_1}{\partial t}(\vec{x},t) &= -\gamma_1 A_1 + gA_1^+ A_0 + \frac{ic^2}{2\omega_1}\nabla^2 A_1 + \sqrt{gA_0}\xi(\vec{x},t)\ ,\end{aligned}$$

Here we write the two dimensional Laplacian as $\nabla^2 = \partial^2/\partial x^2 + \partial^2/\partial y^2$, and omit the conjugate equations for A_1^+, which has a similar equation with an uncorrelated noise term. The stochastic correlations, $\langle\xi(\vec{x},t)\rangle = 0$ and $\langle\xi(\vec{x},t)\xi(\vec{x}',t')\rangle = \delta^{(2)}(\vec{x}-\vec{x}')\delta(t-t')$, means that $\xi(\vec{x},t)$ and $\xi^+(\vec{x},t)$ represent two real Gaussian and uncorrelated stochastic processes.

If we drop the transverse modes of the above equations we return to the well known single mode OPO. This system has a quantum critical point - a phase transition in the infinite volume limit, where the stochastic terms become small. In a recent analysis [1] of this problem near the critical point, going beyond the linear theory, we obtained a scaling law for the squeezing quadrature spectrum near threshold, and the parameters for the optimum squeezing. We found that the optimum noise-reduction or quantum squeezing is sensitive to the relative pump-field damping, and improves in the limit of $\gamma_0/\gamma_1 \to 0$. This prediction is confirmed by a full numerical simulation of the equations, and gives substantial improvements over the previous Plimak-Walls scaling [2] - which we agree with in the regime of $\gamma_0 \simeq \gamma_1$.

If we include the transverses modes, after adiabatic elimination of the pump mode, we get a reduced equation for the enhanced signal quadrature $X_1 = A_1^+ + A_1$, which has the large and slow critical fluctuations:

$$\frac{\partial X_1}{\partial t} = -\gamma_1\left(1 - \frac{g\mathcal{E}}{\gamma_0\gamma_1}\right)X_1 - \frac{g^2}{8\gamma_0}X_1^3 - \frac{1}{\gamma_1}\left(\frac{c^2}{2\omega_1}\right)^2\nabla^4 X_1 + \sqrt{\frac{g\mathcal{E}}{\gamma_0}}\left[\xi(\vec{x},t) + \xi^+(\vec{x},t)\right] \tag{1}$$

It is possible to write a Fokker Planck equation for the probability density $P(X,t)$, and look for the equilibrium distribution in the form $P(X) = N exp[-V(X)]$, where $V(X)$ is a potential function. The solution for the distribution of X is given by

$$P(X) = N\exp\left[-\int\left(\frac{\eta_1}{2}X^2(x) + \frac{\eta_2}{4}X^4(x) + \frac{\eta_3}{2}[\nabla^2 X]^2\right)d^2x\right] \tag{2}$$

Coherence and Quantum Optics VIII
Edited by Bigelow et al., Kluwer Academic/Plenum Publishers, 2003

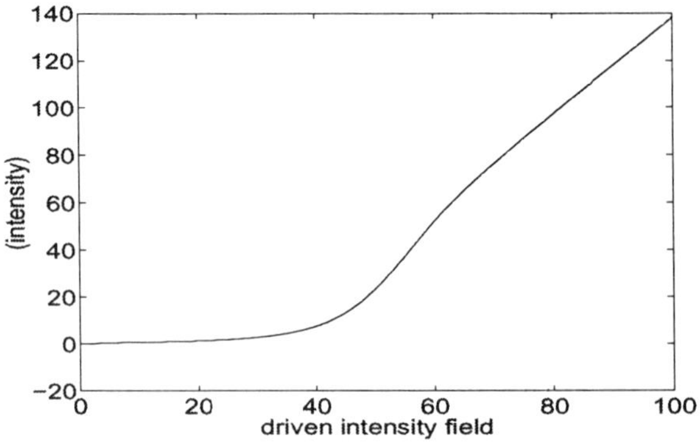

Fig. 1. Amplitude of the sub-harmonic field as a function of the driving field

One can looks at this expression and recognize the Landau-Ginzburg free energy in the exponent without first neighbor interactions and including next neighbor interactions. Here the quantum noise term plays the role of a finite temperature. That is, we have been able to map this problem into a known magnetic phase-transition with a Lifshitz point. We therefore expect two ordered phase, one of them a spatially modulated phase associated with a pattern formation. Exactly this behaviour is known to occur in an OPO, from previous analyses [3, 4]. However, it is also known from condensed matter physics that the Lifshitz point is only a true phase-transition at finite temperature, in dimension greater than two [5]. As our system has two dimensions with transverse modes, we expect that this system does not have a true phase transition in the infinite volume limit, but rather just a continuous transition without threshold.

A complete numerical simulation for sub-harmonic generation with a continuously scanned input shows that, in fact, due to the presence of the transverse term, the subharmonic field is a continuously varying function of the input, and apparently doesn't have a true critical point in two transverse dimensions.

In summary, we have analysed the OPO critical dynamics both with and without transverse terms. Our main results are that squeezing is optimised with a very low loss pump, and that the OPO with transverse modes has the same universality class as a magnetic Lifshitz point in two dimensions.

References and links

1. S. Chaturvedi, K. Dechoum, and P. D. Drummond, and P. D. Drummond, K. Dechoum and S. Chaturvedi, submitted to Phys. Rev. A.
2. L. I. Plimak and D.F. Walls, Phys. Rev. A **50**, 2627 (1994).
3. L. A. Lugiato, M. Brambilla, and A. Gatti, Advances in Atomic, Molecular, and Optical Physics, edited by B. Bederson and H. Walthers (Academic Press, New York), **40**, 229, (1998).
4. M. Vaupel, A. Maître, and C. Fabre, Phys. Rev. Lett. **83**, 5278 (1999).
5. T. A. Kaplan, Phys. Rev. Lett. **44**, 760 (1980).

Quantum deflection of ultracold atoms by ideal mirrors

V. V. Dodonov and M. A. Andreata

Departamento de Física, Universidade Federal de São Carlos, Via Washington Luiz, km 235, 13565-905 São Carlos, SP, Brasil

vdodonov@df.ufscar.br

Suppose that one throws a particle in the direction parallel to the surface of an impenetrable wall. A classical particle would not "feel" the presence of the wall. However, the quantum particle is associated with some "wave packet". This packet rapidly spreads out, and eventually all its plane-wave components will be reflected back from the boundary. As a consequence, the particle will be *deflected* from its initial direction of motion.

To give a quantitative analysis, consider the evolution of the mean position $\langle\widehat{x}\rangle \equiv \int \psi^*\widehat{x}\psi\,dx$ and the mean momentum $\langle\widehat{p}\rangle \equiv -i\hbar\int\psi^*(\partial\psi/\partial x)\,dx$ in the direction x perpendicular to the wall, remembering that the motions in the x and z (along the surface) directions are independent (therefore the parallel component p_z does not change). Calculating the time derivatives of these quantities with the account of the Schrödinger equation $\partial\psi/\partial t = (i\hbar/2m)\partial^2\psi/\partial x^2$ (for $x > 0$) and the boundary condition $\psi(0,t) \equiv 0$, we obtain the modified Ehrenfest equations in the presence of the ideal impenetrable wall: $d\langle x\rangle/dt = \langle p\rangle/m$, $d\langle p\rangle/dt = (\hbar^2/2m)\left|\partial\psi/\partial x\right|^2_{x=0} \equiv F_r(t)$, where $F_r(t)$ can be interpreted as an effective "quantum repulsive force". This force is always positive, so it causes the "center of mass" of the quantum packet to move far from the boundary. The equations of motion for the variances, $\sigma_x \equiv \langle\widehat{x}^2\rangle - \langle\widehat{x}\rangle^2$, $\sigma_p \equiv \langle\widehat{p}^2\rangle - \langle\widehat{p}\rangle^2$, and $\sigma_{px} = \sigma_{xp} \equiv \frac{1}{2}\langle\widehat{x}\widehat{p} + \widehat{p}\widehat{x}\rangle - \langle\widehat{x}\rangle\langle\widehat{p}\rangle$, are as follows, $\dot{\sigma}_x = 2\sigma_{xp}/m$, $\dot{\sigma}_{xp} = \sigma_p/m - F_r(t)\langle x\rangle$, $\dot{\sigma}_p = -2F_r(t)\langle p\rangle$. If $\langle p(0)\rangle = 0$, the quantities $F_r(t)$, $\langle x\rangle$ and $\langle p\rangle$ are always nonnegative, and all three (co)variances turn out less than they would be in the case of the free space evolution (for the same initial conditions). This means that the "nonlocal quantum interaction" with the wall makes the packet narrower, both with respect to the momentum and coordinate. Besides, the final mean value of the momentum component in the direction perpendicular to the surface, $\langle p(\infty)\rangle$, is different from its initial zero value. The increase of the kinetic energy of the center of mass of the particle, $E_c = \langle p\rangle^2/(2m)$, is achieved at the expense of the *energy of quantum fluctuations* of the packet, $E_{fluc} = \sigma_p/(2m)$, since the total energy is conserved.

Consider an exact solution to the Schrödinger equation in the half-space [1]

$$\psi_\beta(x,t) = \left[\frac{16m}{\pi\hbar\varepsilon^2(t)}\right]^{1/4} \exp\left[\frac{im\dot{\varepsilon}}{2\hbar\varepsilon}x^2 - \frac{\mathrm{Re}\,\varepsilon}{\varepsilon}\beta^2\right] \sinh\left(\sqrt{\frac{2m}{\hbar}}\frac{\beta x}{\varepsilon}\right), \quad \varepsilon(t) = b^{-1} + ibt,$$

where positive coefficient b is responsible for the initial width of the wave packet, and positive coefficient β determines the initial position of its center. If $\beta \gg 1$, the coordinate distribution $\mathcal{P}(x,t) = |\psi(x,t)|^2$ is very close to the Gaussian at the initial moment: $\mathcal{P}(x,0) \approx \left(\pi s^2\right)^{-1/2}\exp\left[-(x-x_c)^2/s^2\right]$, where $s = \sqrt{\hbar/(mb^2)}$, $x_c/s = \beta\sqrt{2} \gg 1$. The wave packet involved possesses a nonzero *correlation coefficient* $\mathcal{R} \equiv \sigma_{px}/\sqrt{\sigma_p\sigma_x}$. Therefore, it belongs to the family of *correlated states* introduced in [2] and studied under the name "contractive states" in [3].

It is convenient to use dimensionless variables $\tilde{x} = x/x_c$, $\tilde{p} = ps/\hbar$, $\tau = b^2t/\beta$. The initial mean value of the transverse momentum is zero. However, for $\tau \gg 1$, $\langle\tilde{p}(\tau)\rangle$ tends to the asymptotical value [4] $\langle\tilde{p}(\infty)\rangle = \pi^{-1/2}$, and $\langle\widehat{p}(\infty)\rangle^2/(2mE) = 2/\pi$. Therefore, more than half of the energy of quantum fluctuations is transformed finally into the kinetic energy of the center of the wave packet. The initial smooth Gaussian momentum distribution $\widetilde{\mathcal{P}}(\tilde{p},0) = \pi^{-1/2}\exp\left(-\tilde{p}^2\right)$ eventually transforms to the asymptotical asymmetrical distribution, which equals zero for $\tilde{p} < 0$ and strongly oscillates for $\tilde{p} > 0$: $\widetilde{\mathcal{P}}_\infty(\tilde{p}) = 4\pi^{-1/2}\exp\left(-\tilde{p}^2\right)\sin^2(\sqrt{2}\,\beta\tilde{p})$. The coordinate probability density also strongly oscillates as $\exp(-\tilde{x}^2/\tau^2)\sin^2(2\beta\tilde{x}/\tau)$, and the *invariant uncertainty product* [5] $\widetilde{\Delta} \equiv \left[\sigma_x\sigma_p - \sigma_{xp}^2\right]/\hbar^2$ rapidly increases from $\widetilde{\Delta}(0) \approx 1/4$ to $\widetilde{\Delta}_\infty = \beta^2(1-2/\pi) \gg 1/4$.

We have studied also other forms of wave packets, in particular, packets with nonzero initial mean momentum and packets corresponding to exact solutions in the potential gx^{-2}, as well as packets describing initial localized *mixed* (thermal squeezed) quantum states. Besides, we have analyzed the influence of the nonperfect reflectivity of the mirror, using different model dependences of the complex reflection coefficient $r(k)$.

A significant deflection can occur if the initial mean velocity $\langle v_z\rangle$ has the same order of magnitude as the asymptotical transverse velocity $v_\infty \approx \hbar/(2ms)$. Taking $s = 10^{-6}$ cm, we obtain $v_\infty \sim 2$ cm/s for Cs atoms and $v_\infty \sim 2$ m/s for hydrogen atoms. The mean velocity assumes the asymptotical value v_∞ for $\tau > 5$. Consequently, the "deflection time" can be evaluated as $t_d \approx 5\beta/b^2 \approx 4msx_c/\hbar$. For the initial distance from the boundary $x_c = 1$ cm we have $t_d \sim 1$ s for Cs atoms and $t_d \sim 0.01$ s for hydrogen atoms. During this time the atom will pass about $2x_c$ along the surface and about x_c in the perpendicular direction. So, the 'quantum deflection' can be quite observable. The effect is rather impressive from the classical point of view: a particle passes in 1 cm from the wall, nonetheless, in the absence of any visible force, it 'feels' the presence of the boundary and changes the direction of motion by 45° (or even more, if the initial parallel velocity is less than v_∞). For light atoms, from H to Be, one can increase the initial distance even to 10 cm. In a wide sense, the effect of "quantum deflection" can be considered as an analog of the Aharonov-Bohm effect, since both phenomena have the same origin: quantum non-locality and the existence of "wave properties" of quantum objects.

Since the ideal wall is the limiting case of repulsive potentials, one may suppose that the initial form of the wave packet related to the "external" [6] degrees of freedom could be important in the collision processes, when absolute values of their initial mean momenta are less than the momenta dispersions. As a result of a distortion of the shapes of the particle wave packets during the collision, the absolute values of the *mean momenta* will not be conserved. In this sense, the collisions between ultraslow particles turn out inelastic, as far as the initial states are not idealized plane waves, but wave packets of finite spatial extensions. This new effect could be important for the physics of ultracold atoms.

References and links

1. V. V. Dodonov, I. A. Malkin, and V. I. Man'ko, "Even and odd coherent states and excitations of a singular oscillator," Physica **72**, 597-615 (1974).
2. V. V. Dodonov, E. V. Kurmyshev, and V. I. Man'ko, "Generalized uncertainty relation and correlated coherent states," Phys. Lett. A **79**, 150-152 (1980).
3. H. P. Yuen, "Contractive states and the standard quantum limit for monitoring free-mass positions," Phys. Rev. Lett. **51**, 719-722 (1983); P. Storey, T. Sleator, M. Collett, and D. Walls, "Contractive states of a free atom," Phys. Rev. A **49**, 2322-2328 (1994).
4. V. V. Dodonov and M. A. Andreata, "Deflection of quantum particles by impenetrable boundary," Phys. Lett. A **275**, 173-181 (2000).
5. V. V. Dodonov and O. V. Man'ko, "Universal invariants of quantum-mechanical and optical systems," J. Opt. Soc. Am. A **17**, 2403-2410 (2000).
6. N. P. Bigelow, W. Chalupczak, R. Ejnisman, H. Pu, P. Rudy, and J. Shaffer, "Quantum control of motional states of neutral atoms: Exploiting the external degrees of freedom," Acta Phys. Polon. A **93**, 11-29 (1998).

Spontaneous emission near metal-coated dielectric cylindrical structures

G. D. Banyard, C. R. Bennett and M. Babiker
Department of Physics, University of York, Heslington. York YO10 5DD, England
gdb104@york.ac.uk

Abstract: Electromagnetic field distributions are determined for a system comprising a dielectric cylinder immersed in another dielectric, with a thin metallic coating at the interface. The decay properties of a dipole emitter are then investigated. Enhancement effects arising from the metallic coating are explored.

OCIS codes: (240.0310) Thin films; (020.5580) Quantum electrodynamics

Small scale cylindrical structures have featured prominently in recent years, not just as microguides for electromagnetic fields, but also as atom guides where the guiding mechanism is mainly controlled by an excited cavity mode [1]. The response of quantum systems to the vacuum fields supported by structures of this kind is also of considerable interest currently [2]. In the case of metal-coated dielectric cylinders, dipole emission needs to be assessed before the structure is explored for its merits as an atom guide.

To this end we have studied a general problem in which a dielectric cylinder of dielectric constant ε_1 is embedded in a different material of dielectric constant ε_2. A thin film metallic coating is introduced at the cylindrical interface between the two dielectrics, with the main purpose being the enhancement of the evanescent modes arising under certain conditions. The situation is reminiscent of planar atomic mirrors in which a thin metallic overlayer deposited on a planar dielectric surface has been shown, both experimentally and theoretically, to lead to pronounced enhancement of the repulsive potential arising from the modified evanescent mode [3]. The role of the metallic film in aiding the guiding action in metal-coated guides is currently under investigation. The purpose here is to report the results of a study exploring the main characteristics of the spontaneous emission in such a structure.

The system under investigation is a dipole emitter of moment vector $\boldsymbol{\mu}$ and oscillation frequency ω_0 situated at some point inside the cylinder ($r < a$) or outside it ($r > a$), where a is the cylinder radius. Once the electromagnetic fields have been determined and normalised, the spontaneous emission rate emerges by application of the Fermi golden rule

$$\Gamma(\mathbf{r}) = \frac{2\pi}{\hbar^2}\sum_{\mathbf{q},n}\left|\langle e;\{0\}| -\boldsymbol{\mu}.\mathbf{E}(\mathbf{r})|g;\{\mathbf{q},n\}\rangle\right|^2 \delta(\omega-\omega_0), \tag{1}$$

where $\mathbf{E}(\mathbf{r})$ is the electric field, evaluated here using the photon tunnelling technique. The notation is such that $|e\rangle$ and $|g\rangle$ are the two quantum states representing the dipole system; $|\{0\}\rangle$ stands for the vacuum field state; $|\{\mathbf{q},n\}\rangle$ is a single-quantum field state of frequency $\omega(\mathbf{q},n)$, with $\mathbf{q}$ either the bulk wavevector of an incoming wave or a one-dimensional wavevector along the cylinder axis for confined modes; n is an integer. Briefly, the photon tunnelling technique involves individual incident plane waves, each reflected and transmitted at the cylinder surface, producing appropriate linear combinations of Bessel functions in a manner satisfying electromagnetic boundary conditions. The wave can continue to propagate through the cylinder or decay evanescently (if $\varepsilon_2 > \varepsilon_1$). Confined (if $\varepsilon_2 < \varepsilon_1$) and surface modes may also exist and will satisfy a dispersion relation and separate normalisation conditions. Besides the continuity of the tangential components of the electric field vector at the cylinder interface, a second set of boundary conditions must ensure that all tangential magnetic field components experience a jump due to the conductivity $\sigma = in_se^2/m^*(\omega+i\gamma)$ arising from the metallic coating, where $\gamma \ll \omega$ represents a small plasma loss term. Here n_s is the two dimensional electron density of the coating, and e and m^* are the electron charge and effective mass, respectively. The resulting field distribution inside and outside emerges in terms of Bessel functions, reflecting the cylindrical symmetry of the structure. The dipole decay rate can be explored with varying cylinder radius; dipole position inside and outside the cylinder and with varying metallic coating density. The results, including the local field correction [4], are shown in Figs. 1 and 2. Figure 1 exhibits the changes in the rate with dipole frequency ω_0 and metallic areal density n_s. We note the maxima in the rate at certain frequencies and areal densities. Figure 2 shows the variation of the rate with position r for a fixed radius a. These results clearly indicate that the metallic coating leads to an enhancement of the spontaneous rate, particularly close to the metal, due to the surface mode. The results also reproduce various appropriate asymptotic limits, most notably: the high n_s limit, corresponding to the perfect conductor case; the zero n_s limit, corresponding to the bare dielectric cylinder case and the case of small (large) radius, leading to the bulk results of material 1 (2). The study of atom dynamics in such a structure acting as an atom guide will be described in a future report.

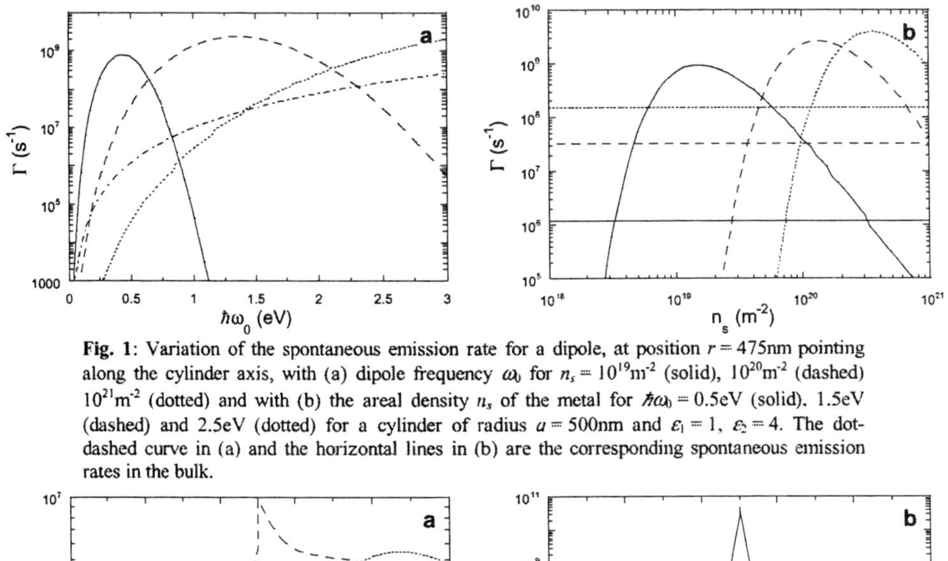

Fig. 1: Variation of the spontaneous emission rate for a dipole, at position $r = 475$nm pointing along the cylinder axis, with (a) dipole frequency ω_0 for $n_s = 10^{19}\text{m}^{-2}$ (solid), 10^{20}m^{-2} (dashed) 10^{21}m^{-2} (dotted) and with (b) the areal density n_s of the metal for $\hbar\omega_0 = 0.5$eV (solid), 1.5eV (dashed) and 2.5eV (dotted) for a cylinder of radius $a = 500$nm and $\varepsilon_1 = 1$, $\varepsilon_2 = 4$. The dot-dashed curve in (a) and the horizontal lines in (b) are the corresponding spontaneous emission rates in the bulk.

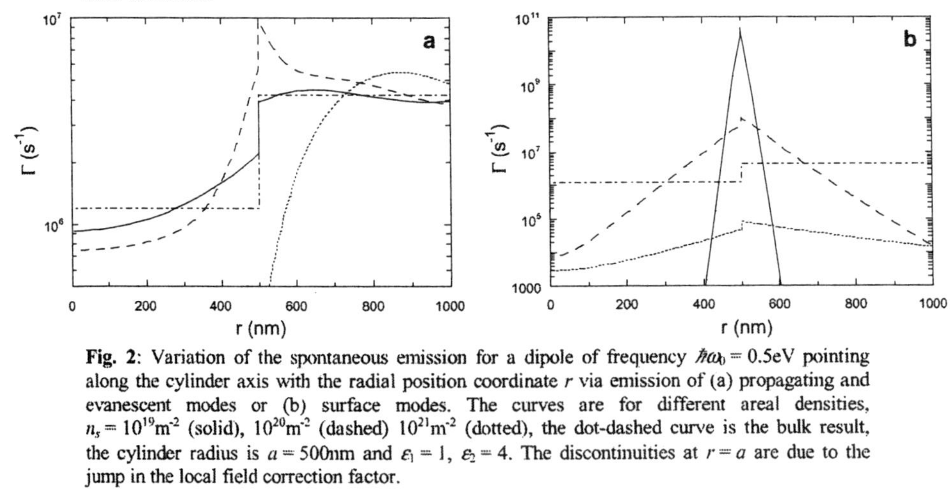

Fig. 2: Variation of the spontaneous emission for a dipole of frequency $\hbar\omega_0 = 0.5$eV pointing along the cylinder axis with the radial position coordinate r via emission of (a) propagating and evanescent modes or (b) surface modes. The curves are for different areal densities, $n_s = 10^{19}\text{m}^{-2}$ (solid), 10^{20}m^{-2} (dashed) 10^{21}m^{-2} (dotted), the dot-dashed curve is the bulk result, the cylinder radius is $a = 500$nm and $\varepsilon_1 = 1$, $\varepsilon_2 = 4$. The discontinuities at $r = a$ are due to the jump in the local field correction factor.

1. J P Dowling and J Gea-Banacloche, "Evanescent light-wave atom mirrors, resonators, waveguides and traps", Adv. Atom Mol. Opt. Phys **37** 1-94 (1997).
2. M. A. Rippin and P. L. Knight, "Modified spontaneous emission in cylindrical microcavities: Waveguiding and distributed Bragg reflecting structures", J. Mod. Opt. **47** 807-853 (1996).
3. C R Bennett, J B Kirk and M Babiker, "Theory of evanescent mode atomic mirrors with a metallic layer", Phys. Rev. A **63** 033405 (2001); and references therein.
4. R. J. Glauber and M. Lewenstein, "Quantum optics of dielectric media", Phys. Rev. A. **43** 467-491 (1991).

Time-frequency Wigner function of ultrashort light pulses

Wictor C. Magno
Departamento de Física, Universidade Federal de Pernambuco, 50670-901 Recife, PE, Brazil
+55 081 3271-8450, fax +55 081 3271-0359 : wcm@npd.ufpe.br

The Wigner function is a powerful tool for analyzing ultrashort light pulses. Starting from the original paper of Paye [1], some works [2-3] have been published involving this function for the complete determination of the electric field of optical pulses. The chronocyclic Wigner function $W(\omega, t)$ of an optical pulse is a Fourier transform of the two-time correlation function $C(t,t')$ of the electric field $\mathcal{E}(t)$:

$$W(\omega,t)=\int dt' C(t+t'/2\,,\,t-t'/2)\exp(i\omega t')=\int dt' \mathcal{E}(t+t'/2)\,\mathcal{E}^*(t-t'/2)\exp(i\omega t')\ . \quad (1)$$

For a normalized Gaussian pulse unchirped the Wigner function is [1]:

$$W_G(\omega,t)=2\exp\left(\frac{-t^2}{2\Delta t^2}\right)\cdot\exp\left(\frac{-\omega^2}{2\Delta\omega^2}\right)\quad , \quad (2)$$

where Δt and $\Delta\omega$ are respectively the rms time and spectral widths of the pulse, with $\Delta t\Delta\omega = ½$ (Fourier limited). This function is a bell-shaped surface in the phase space and is positively defined.

An interesting pulse shape is the hyperbolic secant function because it is a solution of the nonlinear wave equation which propagates without alteration and with time width τ_S (τ_P=1.75τ_S FWHM): $\mathcal{E}(t)$ = sech(t/τ_S) exp($-i\omega_o t$), where ω_o is the carrier frequency of the laser. Some colliding-pulse mode-locked (CPM) dye lasers present this time profile. The Wigner function of this pulse cannot be factorized in the phase space and can be written as:

$$W_S(\omega,t)=4\pi\,\tau_S sin[2\,t\,(\omega-\omega_o)]\cdot\csc h\left(\frac{2t}{\tau_S}\right)\cdot\csc h[\,\pi\,\tau_S(\omega-\omega_o)]\quad . \quad (3)$$

The sinusoidal and hyperbolic functions indicates that this distribution is not positive defined and cannot be associated with a distribution of probability. In fact, the possibility of existence of interference fringes in this function is a property of its nonclassical characteristic. We have analyzed the effect of the propagation of this distribution in a dispersive medium such as an optical fiber. The electric field of the pulse after propagating a length z along the fiber can be written as: $\mathcal{E}(\omega,z)$ = $\mathcal{E}_0(\omega, 0)$ exp[i k(ω)z]. We can expand the wave vector k(ω) around the central frequency ω_o as $k(\omega) \approx k_o + k_o'(\omega - \omega_o) + ½\, k_o''(\omega - \omega_o)^2$, with $k_o = \omega_o n_o/c$ and n_o is the refractive index at the central frequency. After propagate a distance z along the fiber, the output Wigner function assumes the following form

$$W_S^{\rm out}(\omega,t)=W_S^{\rm in}(A\omega-Ct,Dt-B\omega)=\frac{4\pi\,\tau_S\cdot sin\left\{2\left[(t-k_o'z)-k_o''(\omega-\omega_o)z\right](\omega-\omega_o)\right\}}{sinh\left\{\dfrac{2\left[(t-k_o'z)-k_o''(\omega-\omega_o)z\right]}{\tau_S}\right\}\ sinh[\pi\tau_S(\omega-\omega_o)]}\quad , \quad (4)$$

where W_S^{in} and W_S^{out} are the Wigner function before and after the fiber. The parameters A=1, B=$k_o''z$, C=0, D=1, are elements of the ABCD matrix of the dispersive medium [3]. We have included the propagation time ($k_o'z$) at the group velocity $(k_o')^{-1}$. In Fig.1(a) we show the propagation of the Wigner function (4) for three values of the fiber length: z = 0, 10 and 20 (dispersion length). The time unit used here is the duration τ_P of the pulse and the angular frequency unit is radian per time unit. The parameters of the fiber are: $k_o=1$, $k_o'=1$, $k_o''=0.18$. This last parameter induces frequency chirp through the group-velocity dispersion (GVD) and results in changes in the time width of the pulse during the propagation. In Fig.1(b) we show the contour plots of the Wigner function.

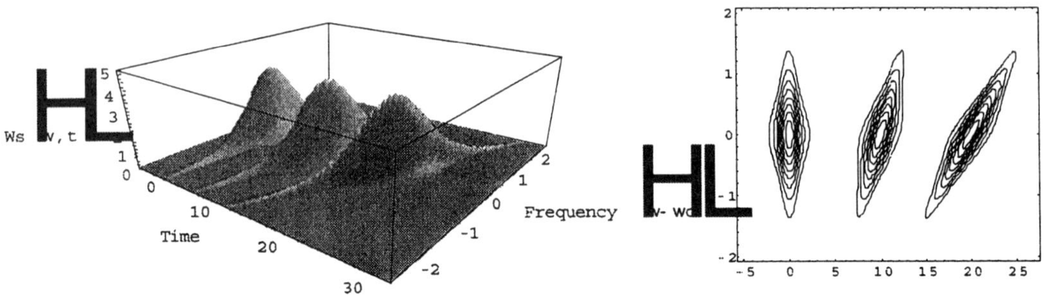

Fig.1 (a) Propagation of the Wigner function for three different fiber length: z=0, 10 and 20. (b) Contour plots of these functions.

An optical device well-known as chirp modulator introduces a quadratic phase modulation into the pulse profile. Its ABCD elements are: A=1, B=0, $C=-(k_o''f)^{-1}$, D=1, where k_o'' characterizes the dispersion of the modulator, and f determines the strength of the chirp [3]. The effect of this device on the Wigner function is shown in Fig.2 (a) and (b), with $k_o''f = 1/3$, for the parameters of the modulator.

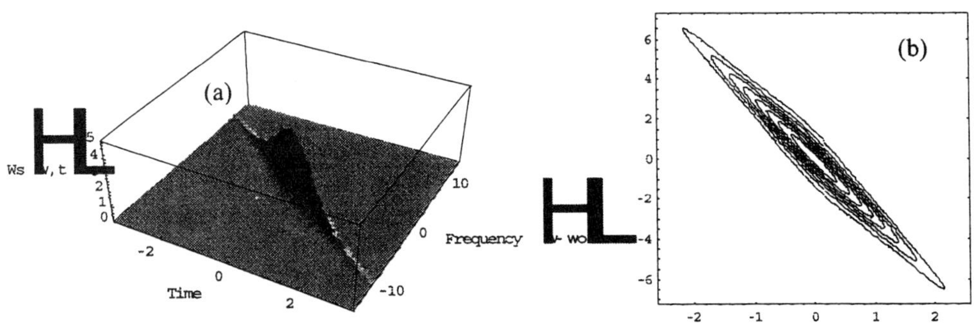

Fig.2 (a) Propagation of the Wigner function in a chirp modulator and (b) contour plot of this function.

In summary, we have calculated the Wigner function of a hyperbolic secant ultrashort pulse. The simple analytic result obtained simplifies the propagation analysis, which we did for a dispersive medium and in a chirp modulator. These results will be discussed.

[1] J. Paye, "The Chronocyclic Representation of Ultrashort Light Pulses", IEEE Journal of Quantum Electronics **28**, 2262-2272 (1991).

[2] M. Beck, M. G. Raymer, I. A. Walmsley and V. Wong, "Chronocyclic tomography for measuring the amplitude and phase structure of optical pulses", Optics letters **18**, 2041-2043 (1993).

[3] D. F. V. James and G. S. Agarwal, "Generalized Radon transform for tomographic measurement of short pulses", J. Opt. Soc. Am. B **12**, 704-708 (1995).

Rotation angle of the Wigner distribution function of the partially coherent fractional Fourier field

Hiroyuki Yoshimura and Koichi Ito
Faculty of Engineering, Chiba University, 1-33, Yayoi-cho, Ikage-ku, Chiba, 263-8522, Japan
Tel: +81-43-290-3932, Fax: +81-43-290-3933, E-mail: yosimura@tu.chiba-u.ac.jp

1. Introduction

The Fractional Fourier transform (FRT) is the generalization of the conventional Fourier transform. Recently, the application of the FRT to optical information processing has been widely studied [1]. We have been expanding the idea of the FRT to the partially coherent light field. In [2], we first proposed the idea of fractional cross-spectral density (FRCSD), and the characteristics of partially coherent light beam in the fractional Fourier plane produced by the GSM (Gaussian Schell-model) source were investigated in the viewpoint of the extent of the beam, the wavefront-curvature radius of the beam and coherence induced spectral changes. In addition, the effects of lens aperture to the GSM source size on the resultant partially coherent fractional Fourier field were analyzed numerically [3]. Furthermore, coherence properties of the partially coherent fractional Fourier fields with different orders were analyzed [4].

In this study, we analyze the rotation angle of the Wigner distribution function (WDF) of the partially coherent fractional Fourier field produced by the GSM source. We discuss the behavior of the rotation angle by changing the parameters associated with the GSM source and the optical system realizing the FRT.

2. Equi-WDF ellipse of the partially coherent fractional Fourier field

The conventional CSD of a monochromatic GSM source is given by

$$W_0(\xi,\Delta\xi) = A\exp(-2\xi^2/w^2)\exp[-(1+1/\sigma^2)\Delta\xi^2/2w^2], \tag{1}$$

where A, w and σ denote a positive constant, the GSM source size and the scaled spatial coherence length of the GSM source, respectively. For simplicity, we consider the one-dimensional case in this study. Then, the WDF of the partially coherent light field in the p-order Fourier plane becomes

$$WDF_p(x,\nu) = Aw\sqrt{2\pi/(1+1/\sigma^2)}\exp(-2x^2/w_p^2)\exp[-2(\pi w_p)^2(\nu + x/\lambda R_p)^2/(1+1/\sigma^2)], \tag{2}$$

where x and ν are the spatial coordinate and the spatial frequency, respectively. Moreover, w_p and R_p are the extent and wavefront-curvature radius of the beam in the p-order Fourier plane, respectively [2]. By defining the equi-WDF ellipse as the trajectory of the WDF in space of x and ν, which it takes the value of exp(-2) of the maximum value, we can expressed it as

$$\left\{\frac{1}{w_p^2}+\frac{(\pi w_p)^2}{(\lambda R_p)^2(1+1/\sigma^2)}\right\}x^2+\frac{2(\pi w_p)^2}{\lambda R_p(1+1/\sigma^2)}x\nu+\frac{(\pi w_p)^2}{(1+1/\sigma^2)}\nu^2=1. \tag{3}$$

3. Results

In Fig. 1, dependence of the rotation angle of the equi-WDF ellipse θ on the order of the Fourier transform p is shown. As an example, Figs.1(a) and 1(b) correspond to the results for σ=0.1 and 10, respectively. In each figure, the Fresnel number associated with the GSM source size, N_w, takes a value of 0.1, 1.0 and 10. The positive value of the rotation angle θ means that the equi-WDF ellipse rotates clockwise.

From these figures, we can find that the behavior of the rotation angle θ by variation of the order p is completely different for the different values of N_w. In particular, when N_w =0.1 and 10, the behavior of the rotation angle θ does not correspond to the Lohmann's definition of the FRT based on the WDF, where the rotation angle θ changes linearly by variation of the order p [5]. When N_w=1.0, however, the behavior agrees well with the Lohmann's definition and is independent of the value of σ.

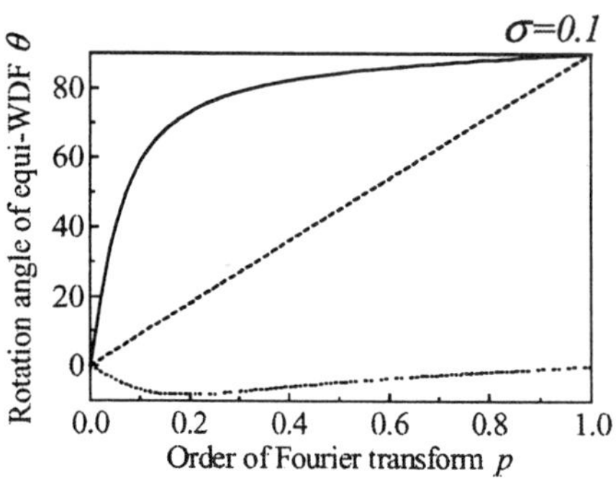

(a)

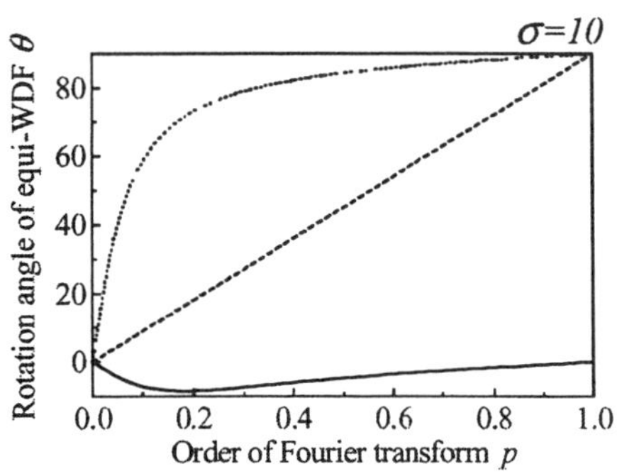

(b)

Fig.1. Dependence of the Rotation angle of the equi-WDF θ on the order p. The solid, dashed, dotted curves are the results for N_w =0.1, 1.0 and 10, respectively.

4. Conclusions

In this paper, we have investigated the WDF of the partially coherent fractional Fourier field produced by the GSM source. In particular, the dependence of the rotation angle of the WDF-ellipse on the order of the Fourier transform, the spatial coherence length of the GSM source and the Fresnel number associated with the GSM source size has been analyzed. As a result, it is found that the rotation angle changes almost linearly for variation of the order when the Fresnel number takes the value of 1.0. This fact shows that the condition of the optical system realizing the FRT exists.

5. References

[1] H. M. Ozaktas, Z. Zalevsky and M. A. Kutay, *The fractional Fourier transform* (John Wiley & Sons, 2001).
[2] H. Yoshimura and T. Iwai: "Properties of the Gaussian Schell-model source field in a fractional Fourier plane," J. Opt. Soc. Am. A **14**, 3388-3393 (1997).
[3] H Yoshimura and T. Iwai: "Effects of lens aperture on the average intensity in a fractional Fourier plane," Pure Appl. Opt. **7**, 1133-1141 (1998).
[4] H. Yoshimura and K. Ito: "Coherence properties of fractional Fourier fields with different orders," Conf. Prog. OSA Ann. Meet. Exh. 2000, 62 (2000).
[5] A. W. Lohmann: "Image rotation, Wigner rotation, and the fractional Fourier transform," J. Opt. Soc. Am. A **10**, 2181-2186 (1993).

Quasiclassical Calculations of Wigner Functions in Nonlinear Dynamics via Wavelets

Antonina N. Fedorova and Michael G. Zeitlin

Russian Academy of Sciences,
Institute of Problems of MechanicalEngineering, Mathematical Methods in Mechanics Group
V.O., Bolshoj pr., 61, 199178, St. Petersburg, Russia

zeitlin@math.ipme.ru
http://www.ipme.ru/zeitlin.html
http://www.ipme.nw.ru/zeitlin.html

Abstract: We present the application of variational-wavelet analysis to numerical/analytical calculations of Wigner functions in (nonlinear) quasiclassical dynamics problems. (Naive) deformation quantization and multiresolution representations are the key points.

OCIS codes: (270.0270) Quantum optics; (000.3860) Mathematical methods in physics

We present the application of variational-wavelet analysis to numerical calculations of Wigner functions in (nonlinear) quasiclassical dynamical problems. (Naive) deformation quantization and multiresolution representations are the key points. We construct the representation via multiscale expansions in generalized coherent states or high-localized nonlinear eigenmodes in the base of compactly supported wavelets and wavelet packets. Our approach is based on extension of our variational-wavelet approach [1]-[2]. Wavelet

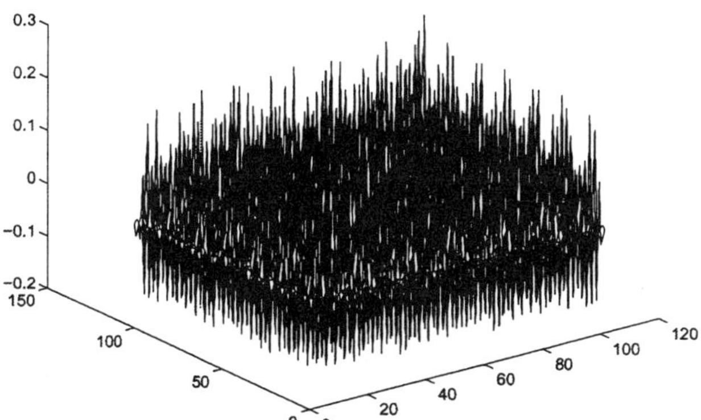

Fig. 1. Six-eigenmodes decomposition

analysis is some set of mathematical methods, which gives us the possibility to work with well-localized bases in functional spaces and gives maximum sparse forms for the general type of operators (differential, integral, pseudodifferential) in such bases. These bases are natural generalization of standard coherent, squeezed, thermal squeezed states, which correspond to quadratical systems (pure linear dynamics) with Gaussian Wigner functions. So, we try to calculate quantum corrections to classical dynamics described by arbitrary polynomial nonlinear Hamiltonians from papers [1]-[3]. The common point for classical/quantum calculations is that any solution which comes from full multiresolution expansion in all space/time (or phase space) scales represents expansion into a slow part and fast oscillating parts. So, we may move from the coarse scales of resolution to the finest one for obtaining more detailed information about our dynamical classical/quantum process. In this way we give contribution to our full solution from each scale of resolution. The same is correct for the contribution to power spectral density (energy spectrum): we can take into account contributions from each level/scale of resolution (Fig.1). Our (nonlinear) eigenmodes are more realistic for the modelling of nonlinear classical/quantum dynamical process than the corresponding linear gaussian-like coherent states. We mention only the best convergence properties of expansions based on wavelet packets, which realize the so called minimal Shannon entropy property, save CPU time and HDD space. We present numerical calculations of Wigner functions as solutions of (non)stationary Wigner equations for a few models of nonlinear underlying classical dynamics, which explicitly demonstrates quantum behaviour properties and the appearance of stable pattern formation (Fig.2).

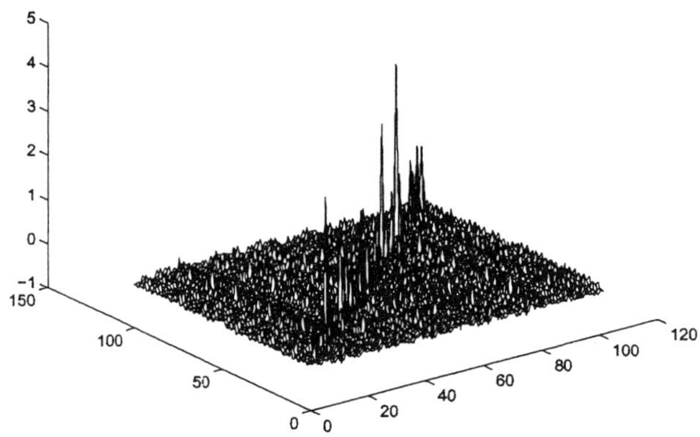

Fig. 2. Stable pattern

References and links

1. A.N. Fedorova, M.G.Zeitlin, "Symmetry, Hamiltonian Problems and Wavelets in Accelerator Physics", American Institute of Physics , Conf. Proc., **468**, Nonlinear and Collective Phenomena in Beam Physics, 69-93, (1999).
2. A.N. Fedorova, M.G.Zeitlin, "Variational-Wavelet Approach to RMS Envelope Equations", Proc. 2nd Advanced Accelerator Workshop on The Physics of High Brightness Beams pp.235–254, World Scientific, (2000).
3. A.N. Fedorova, M.G.Zeitlin, "Quasiclassical Calculations for Wigner Functions via Multiresolution, Quantum Aspects of Beam Physics", World Scientific, (2001); Los Alamos preprint, physics/0101006, http://arXiv.org/abs/physics/0101006

Quantum correlations in a passive nonlinear optical fiber resonator

Sara Schneider[(1)], **Edouard Brainis**[(2)], **Marc Haelterman**[(2)]

[(1)]*Ecole Polytechnique, CP 165/56, Av. F.D. Roosevelt 50, Université Libre de Bruxelles, B-1050 Bruxelles, Belgium*
[(2)]*Service d'Optique Théorique et Appliquée, CP 194/5, Av. F.D. Roosevelt 50, Université Libre de Bruxelles, B-1050 Bruxelles, Belgium*

Tel.: + 32 2 650 2956; Fax: + 32 2 650 2941; E-mail: sara.schneider@ulb.ac.be

In recent experiments [1] a passive nonlinear optical fiber oscillator was stabilised and operated in several regimes showing bistability, modulational instability and period doubling. As well as showing these classical nonlinear phenomenons, nonlinear optical cavities can, in principle, generate quantum states of light such as two-mode squeezed states [2, 3]. However, these quantum signatures in fibers are rather fragile and subject to noise from various sources such as coupling of the field modes to accoustical waves of the fiber (GAWBS) [4].

We investigate the quantum version of the non-linear Schrödinger equation

$$\frac{\partial E(t,\tau)}{\partial t} = \left[-1 + i\left(|E(t,\tau)|^2 - \Delta\right) - i\eta\frac{\partial^2}{\partial\tau^2}\right] E(t,\tau) + S(\tau) \tag{1}$$

for the field envelope $E(t,\tau)$ taking into account dispersion (see e.g. [5]) and derive a Fokker-Planck equation for the field inside the cavity. The corresponding Langevin equations for the field mode operators give expressions for the quantum fluctuations taking into account the effects of noise in the various operating schemes of the cavity.

Possible experimental modifications to the existing system to diminish the noise sources as well as detector requirements will be discussed briefly.

References and links

1. S. Coen and M. Haelterman, Phys. Rev. Lett. **79**, 4139 (1997); S. Coen, M. Tlidi, P. Emplit, and M. Haelterman, Phys. Rev. Lett. **83**, 2328 (1999).
2. G.J. Milburn *et al.*, J. Opt. Soc. Am. B **4**, 1476 (1987).
3. R.E. Slusher *et al.*, J. Opt. Soc. Am. B **4**, 1453 (1987).
4. R.M.Shelby *et al.*, Phys. Rev. B **31**, 5244 (1985); R.M. Shelby *et al.*, Phys. Rev. Lett. **54**, 939 (1985).
5. M. Brambilla *et al.*, Opt. Commun. **83**, 367 (1991); L.A. Lugiato and F. Castelli, Phys. Rev. Lett. **68**, 3284 (1992).

Time Asymmetry in Measurements based on Photoelectric Detection

A. Denisov and H. J. Carmichael

Department of Physics, University of Oregon, Eugene, OR 97403-1274

tel: (541) 346-5863, denisov@darkwing.uoregon.edu

The most notable feature of the conditional homodyne detection (CHD) technique of Foster et al. [1, 2] is that it makes a *conditional* measurement, detecting the amplitude fluctuations of an optical field by collecting data from a balanced homodyne detector only on the cue of a conditioning photon count. The average fluctuation of the field is recovered as a conditional evolution over time, both before and after the conditioning photoelectric event.

In this paper we investigate the evolution of the measured correlation function, $h(\tau)$, contrasting the time dependence in the positive ($\tau \geq 0$) and negative ($\tau \leq 0$) directions of time. The time evolution may be asymmetric. In this regard $h(\tau)$ is quite unlike the correlation functions considered previously in quantum optics, such as the intensity correlation function, $g^{(2)}(\tau)$, or the autocorrelation function for quadrature amplitudes that underlies the spectrum of squeezing.

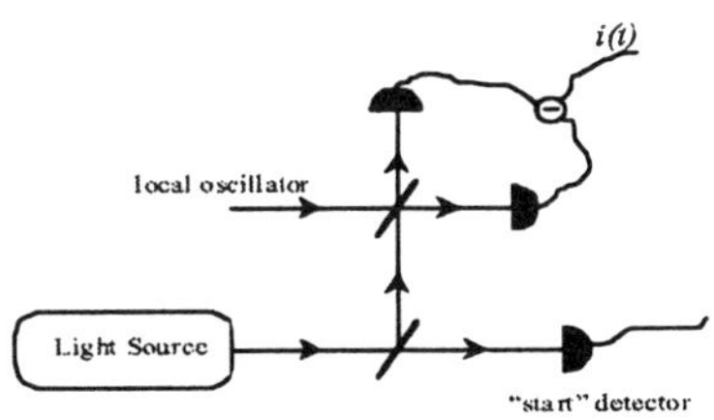

Fig. 1. CHD scheme for measuring the correlation function $h(\tau)$.

The correlator used to measure $h(\tau)$ is illustrated in Fig. 1. It is based on the "start"/"stop" scheme employed in a measurement of $g^{(2)}(\tau)$. The main difference is the balanced homodyne detector (BHD) which replaces the photon multiplier in the "stop" channel. The origin of time for the measurement is referenced to each "start", where, over an interval of a few correlation times before and after the "start", the BHD current, $i(t)$, is digitized, recorded, and used to update a cumulative average. Averaging many such samples to reduce the shot noise, the recorded signal is a conditional average of the fluctuations of the quadrature amplitude in phase with the local oscillator (LO). Expressing field operators in terms of the source quasimode annihilation and creation operators, $\hat{a}$ and $\hat{a}^\dagger$, the setup measures the normal-odered, time-ordered correlation function

$$h(\tau) = \frac{\langle : (\hat{a}^\dagger \hat{a})(t) \hat{A}_\theta(t+\tau) : \rangle}{\langle \hat{a}^\dagger \hat{a} \rangle_{\rm ss} \langle \hat{A}_\theta \rangle_{\rm ss}}, \tag{1}$$

where $\hat{A}_\theta \equiv \frac{1}{2}(\hat{a}e^{-i\theta} + \hat{a}^\dagger e^{i\theta})$; θ is the LO phase. The choice of normalization assumes that the steady-state mean $\langle \hat{A}_\theta \rangle_{\rm ss}$ is different from zero.

For a classical field we would replace the operators with complex random variables, α_t and α_t^*, and evaluate the average as an integration over a two-time probability distribution for α_t and $\alpha_{t+\tau}$. Assuming stationarity and microreversibility, we can show in this case that $h_{\rm class}(-\tau) = h_{\rm class}(\tau)$. On the other hand, for the quantum field operators, the product $(\hat{a}^\dagger \hat{a})(t)\hat{X}_\theta(t+\tau)$ is to be explicitly time ordered because the measurement is realized through photon absorption (photoelectric detection); but the time-ordered form of $(\hat{a}^\dagger \hat{a})(t)\hat{X}_\theta(t-\tau)$ is not what one reaches from $(\hat{a}^\dagger \hat{a})(t)\hat{X}_\theta(t+\tau)$ using microreversibility. It is thus possible to find $h(-\tau) \neq h(\tau)$. Such a time asymmetry is a direct manifestation of the time ordering in photoelectric detection.

More specifically, the asymmetry may be seen to arise from interchanging the time order in a sequence of two meaurement of noncommuting observables: for positive τ, the photon number is measured and *then* the field amplitude, while for negative τ the field amplitude is measurement before the photon number. We have analyzed the two measurement sequences forwards in time using the method of quantum trajectories [3]. In this approach the different time orders are traced through a different state reduction and conditional evolution [4].

If we assume Gaussian fuctuations about a mean field amplitude, $\langle \hat{a} \rangle_{ss} = |\alpha| e^{i\phi}$, the correlation function (1) becomes [2]

$$h(\tau) = 1 + 2\frac{|\alpha|\langle : \Delta\hat{A}_\phi(t)\Delta\hat{A}_\theta(t+\tau) : \rangle}{\langle \hat{a}^\dagger \hat{a} \rangle_{ss} \langle \hat{A}_\theta \rangle_{ss}}, \tag{2}$$

where $\Delta\hat{A}_\phi$ and $\Delta\hat{A}_\theta$ are quadrature amplitude operators for the field fluctuation. In this case $h(\tau)$ is time symmetric; hence, non-Gaussian fluctuations are necessary for there to be asymmetry. The presence of non-Gaussian fluctuations is not sufficient, however. We demonstrate this in Fig. 2(a) with the example of resonance fluorescence, where, although the fluctuations are non-Gaussian, $h(\tau)$ is time symmetric; moreover, one may prove that the correlation function is symmetric for all driving field strengths. We have obtained asymmetric correlation functions, however, in cavity QED. The example shown in Fig. 2(b) is for a cavity containing one resonant two-level atom, driven on resonance by a coherent external field. The correlation function is measured—as in [1]—on the light transmitted by the cavity. Making both analytical and numerical calculations, we investigate this time asymmetry as a function of the system parameters.

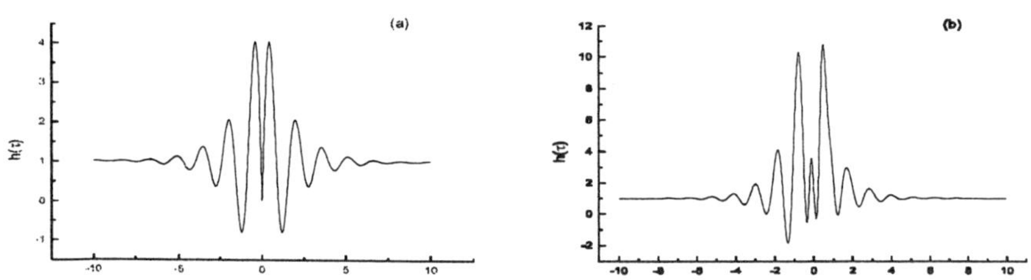

Fig. 2. Numerical simulations of conditional homodyne detection for (a) resonance fluorescence, with atomic decay rate $\gamma = 1$ and Rabi frequency $\Omega = 4$, (b) single-atom cavity QED, with dipole coupling constant $g = 6$, atomic decay rate $\gamma = 1$, and driving field strength $\mathcal{E} = 1$.

References and links

1. G. T. Foster, L. A. Orozco, H. M. Castro-Beltran, and H. J. Carmichael, Phys. Rev. Lett **85**, 3149-52 (2000).
2. H. J. Carmichael, H. M. Castro-Beltran. G. T. Foster, and L. A. Orozco, Phys. Rev. Lett. **85**, 1855-8 (2000).
3. H. J. Carmichael, *An Open Systems Approach to Quantum Optics*, Lecture Notes in Physics, Vol. m18 (Springer-Verlag, Berlin, 1993).
4. H. J. Carmichael, R. J. Brecha, and P. R. Rice Opt. Commun. **82**, 73-9 (1991).

Photon correlations in the homodyne detection of squeezed light

Daniel Erenso, Hua Deng, Reeta Vyas, and Surendra Singh

Physics Department, University of Arkansas, Fayetteville, AR 72701

(501)575-6569, (501)575-4580, rvyas@comp.uark.edu

When the squeezed and highly bunched light from the DPO interferes with the coherent field of a local oscillator (LO), the superposed field can exhibit a rich variety of nonclassical photon correlations [1]. We show that these nonclassical correlations are a result of conditioned interference between the light from the DPO and the LO.

The DPO and LO fields are combined (homodyned) at a beamsplitter of reflectivity $\mathcal{R}$ and transmittivity $\mathcal{T}$ $(\mathcal{R}+\mathcal{T}=1)$ to produce the superposed field. We will refer to the superposed field as the HDPO field. The annihilation operator for the field at one of the output ports of the beam splitter can be written as $\hat{a}=\sqrt{\mathcal{R}}\,\hat{a}_\ell+\sqrt{\mathcal{T}}\,\hat{a}_d$, where $\hat{a}_\ell$ and $\hat{a}_d$ are annihilation operators for the LO and the DPO. Then the second-order intensity correlation function which is proportional to the probability of detecting a photon at time $t=T$ conditioned on the detection of a first photon at time $t=0$ in the steady-state can be written as

$$g^{(2)}(T)=\frac{\langle:\hat{n}(T)\hat{n}(0):\rangle_{ss}}{\langle\hat{n}\rangle_{ss}^2}=\frac{\langle\hat{n}(T)\rangle_c}{\langle\hat{n}\rangle_{ss}}, \tag{1}$$

where $\hat{n}=\hat{a}^\dagger\hat{a}$ is the photon number operator for the HDPO field, and the colons stand for normal and time ordering of the operators between them. Thus $g^{(2)}(T)$ is equal to the conditional mean photon number $\langle\hat{n}(T)\rangle_c$ normalized to the steady-state value $\langle\hat{n}\rangle_{ss}$. The conditional mean amplitude and photon number for the HDPO can be written as

$$\begin{aligned}\langle\hat{a}(T)\rangle_c &= \sqrt{\mathcal{R}}\alpha_\ell+\sqrt{\mathcal{T}}\langle\hat{a}_d(T)\rangle_c, &(2)\\ \langle\hat{n}(T)\rangle_c &= \mathcal{R}\bar{n}_\ell+\mathcal{T}\langle\hat{n}_d(T)\rangle_c+\sqrt{\mathcal{R}\mathcal{T}\bar{n}_\ell}\left[e^{i\phi}\langle\hat{a}_d^\dagger(T)\rangle_c+e^{-i\phi}\langle\hat{a}_d(T)\rangle_c\right], &(3)\end{aligned}$$

where $\alpha_\ell=\sqrt{\bar{n}_\ell}\,e^{i\phi}$ is the coherent amplitude for the LO field, $\bar{n}_\ell$ is the mean photon number and ϕ is its phase referred to the DPO pump. The term inside the square brackets represents interference between the conditional DPO field and the LO field. In the limit of the DPO mean photon number $\bar{n}_d<<1$, the mean values for the DPO field conditioned on a first detection are found to be [2, 3]

$$\langle\hat{a}_d(T)\rangle_c\approx\frac{\sqrt{2\bar{n}_d\mathcal{R}\mathcal{T}}}{2\langle\hat{n}\rangle_{ss}}e^{-\gamma T}\alpha_\ell^*=\frac{\sqrt{2\bar{n}_d\bar{n}_\ell\mathcal{R}\mathcal{T}}}{2\langle\hat{n}\rangle_{ss}}e^{-\gamma T}e^{-i\phi},\quad\langle\hat{n}_d(T)\rangle_c\approx\bar{n}_d+\frac{\mathcal{T}\bar{n}_d}{2\langle\hat{n}\rangle_{ss}}e^{-2\gamma T}. \tag{4}$$

Here 2γ is the DPO cavity decay rate. Using Eq. (4) in (2) and (3) we find, with the help of Eq. (1)

$$\begin{aligned}\langle\hat{a}(T)\rangle_c &\approx \sqrt{\mathcal{R}}\alpha_\ell+\frac{\mathcal{T}\sqrt{2\bar{n}_d\mathcal{R}}}{2\bar{n}_{ss}}e^{-\gamma T}\alpha_\ell^*=\sqrt{\mathcal{R}\bar{n}_\ell}\left[e^{i\phi}+\frac{\sqrt{2\bar{n}_d}\,\mathcal{T}}{2\bar{n}_{ss}}e^{-\gamma T}e^{-i\phi}\right], &(5)\\ g^{(2)}(T) &\approx 1+\frac{\bar{n}_d\mathcal{T}^2}{2\bar{n}_{ss}^2}e^{-2\gamma T}+\frac{\mathcal{T}\sqrt{2\bar{n}_d}}{\bar{n}_{ss}}e^{-\gamma T}\cos(2\phi)=\frac{|\langle\hat{a}(T)\rangle_c|^2}{|\langle\hat{a}\rangle_{ss}|^2}, &(6)\end{aligned}$$

In the limit $T\to\infty$, Eqs (2)-(4) lead to the steady-state values

$$\langle\hat{a}_d\rangle_{ss}=0,\quad\langle\hat{a}\rangle_{ss}=\sqrt{\mathcal{R}}\alpha_\ell,\quad\langle\hat{n}_d\rangle_{ss}=\bar{n}_d,\quad\langle\hat{n}\rangle_{ss}=\mathcal{R}\bar{n}_\ell+\mathcal{T}\bar{n}_d. \tag{7}$$

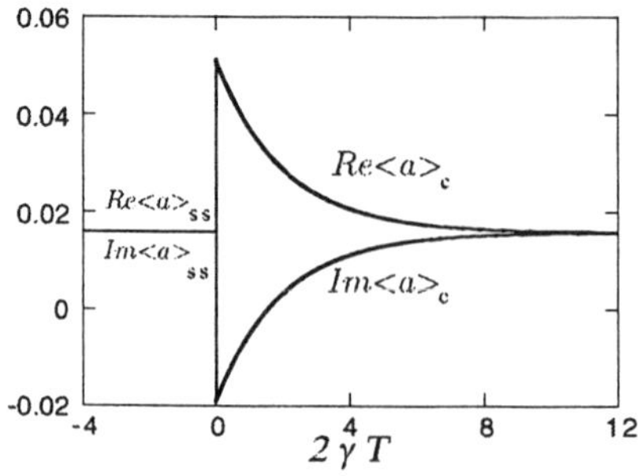

Fig. 1. Time evolution of the real and imaginary parts of the HDPO field amplitude as a function of $2\gamma T$ following a first detection at $T = 0$ in the steady-state for $\bar{n}_\ell = 10^{-3}$, $\bar{n}_d = 10^{-5}$, $\phi = 45^o$, and $\mathcal{T} = 0.5 = \mathcal{R}$. Note that as the amplitude relaxes toward the steady-state, its imaginary part passes through zero but the real part never vanishes.

It is clear from this equation that the interference term in Eq. (3) vanishes in the steady-state indicating that the steady-state DPO field is incoherent with respect to the LO field. After a first detection, however, the DPO radiates [Eq. (4)] in phase with the phase-conjugate of the LO field. Moreover, for a short period of the order of $(2\gamma)^{-1}$ following the detection, the amplitude of the field radiated by the DPO can exceed the LO field amplitude. Figure (1) shows the time-evolution of the HDPO field amplitude for $\phi = 45^o$. Following the measurement at $T = 0$, the real part of the amplitude exceeds its steady-state value significantly and then relaxes back to it. The imaginary part, on the other hand, changes sign after the measurement and then relaxes back to its steady-state value. If we choose phase $\phi = \pi/2$, the LO and the DPO fields are pure imaginary and out of phase as can be seen from Eq. (5). In this case the HDPO field can vanish as it relaxes toward its steady-state. It follows from the Eq. (6) that whenever the HDPO field vanishes $g^{(2)}(T)$ will also vanish leading to violations of classical inequalities satisfied by $g^{(2)}(T)$ [1, 4]. Nonclassical behavior of $g^{(2)}(T)$ is thus a direct consequence of post-detection coherence and enhanced emission from the DPO following a measurement. The equality of Eq. (6), although excellent for $\bar{n}_d << 1$, is not exact. As $\bar{n}_d$ increases these violations become less pronounced.

This work was supported in part by the Office of Naval Research.

References and links

1. A. B. Dodson and Reeta Vyas, "Homodyne photon statistics of the subthreshold degenerate parametric oscillator," Phys. Rev. A **47**, 3396-3412 (1993).
2. R. Vyas and S. Singh, "Quantum statistics of broadband squeezed light," Opt. Lett. **14**, 1110-1112 (1989); "Photon-counting statistics of the degenerate parametric oscillator," Phys. Rev. A **40**, 5147-5159 (1989).
3. H. Deng, D. Erenso, R. Vyas, and S. Singh, "Entanglement, interference, and measurement in a degenerate parametric oscillator," Phys. Rev. Lett. **86**, 2770 (2001).
4. P. R. Rice and H. J. Carmichael, "Single-atom cavity-enhanced absorption. I. Photon statistics in the bad-cavity limit," IEEE J. Quantum Electronics **24**, 1351 (1988).

Rigorous analysis of homodyne detection

Tomáš Tyc[1,2] and Barry C. Sanders[1]

[1] *Department of Physics, Macquarie University, Sydney, New South Wales 2109, Australia*
[2] *Institute of Theoretical Physics, Masaryk University, 61137 Brno, Czech Republic*

OCIS codes: (270.5290) Photon statistics

Abstract: A rigorous analysis of four– and eight–port homodyne detection, employing asymptotic methods for su(2) Wigner functions, connects the operational formulation of photon counting with the usual quadrature and coherent state projection descriptions.

We present a rigorous analysis of four– and eight–port homodyne detection (4PHD and 8PHD, respectively) used in quantum optics for quadrature phase (QP) measurements and simultaneous joint QP measurements, respectively [1]. Beginning with the operational positive operator–valued measure (POVM) [2] for photon counting at the interferometer output ports, we employ asymptotic methods for su(2) Wigner functions [3] to obtain the POVMs for QP and joint QP measurements.

Balanced homodyne detection scheme (BHD), depicted in Fig. 1(a), corresponds to 4PHD with a 50/50 beam splitter (BS): the state $|\psi\rangle$ to be measured is mixed at this BS with a strong local oscillator (LO) assumed to be in a coherent state $|\alpha\rangle = |a\mathrm{e}^{\mathrm{i}\varphi}\rangle$, and the photon number difference $2m \in \mathbb{Z}$ from the two BS output ports is measured. The photon number sum $2j \in \mathbb{Z}$ can also be measured but in practice is not. In a strong LO limit, it is generally assumed that the number–difference distribution $\mathcal{P}(m)$ converges to the field QP distribution $P(x)$ with $x = \sqrt{2}\,m/a$ [4]. We establish the correspondence $\mathcal{P}(m) \to P(x)$ rigorously via asymptotic methods; this correspondence is crucial to connect the operational photon counting POVM to the simplified POVM approach employed in ususal analyses of BHD.

The probability amplitude for $j \pm m$ photons emerging from each of the two interferometer output ports is $M^j_m(a,\varphi) = {}_{12}\langle jm|\hat{B}|\psi\rangle_1 |a\mathrm{e}^{\mathrm{i}\varphi}\rangle_2$ with $\hat{B} = \exp[\pi(\hat{a}_2^\dagger \hat{a}_1 - \hat{a}_1^\dagger \hat{a}_2)/4]$ the unitary BS operator and ${}_{12}\langle jm|$ the joint eigenstate of number operators $\hat{a}^\dagger_{1,2}\hat{a}_{1,2}$ with eigenvalues $j \pm m$, respectively. The asymptotic formulæ for Wigner rotation matrices [3] and Stirling's formula (neglecting terms of order n^2/a^2 and higher with n the average photon number in the state $|\psi\rangle$) yield

$$M^j_m(a,\varphi) = \pi^{-1/4} a^{-1} \exp\left[-\frac{(2j-a^2)^2}{4a^2}\right] \mathrm{e}^{2\mathrm{i}\varphi j} \langle x_\varphi = m/\sqrt{j}|\psi\rangle. \qquad (1)$$

Here $\langle x_\varphi = m/\sqrt{j}|$ denotes the eigenstate of the field quadrature $\hat{x}_\varphi$ with the eigenvalue $m/\sqrt{j}$. The probability $P(x)\,\mathrm{d}x$ for the variable $x_\varphi = m/\sqrt{j}$ to be from the interval $(x, x+\mathrm{d}x)$ is the sum of $|M^j_m(a,\varphi)|^2$ for m, j such that $m/\sqrt{j} \in (x, x+\mathrm{d}x)$, $2j \in \mathbb{Z}$, and $m \in \{-j, -j+1, \ldots, j\}$. In the strong LO limit, we obtain (neglecting terms of order $1/a^2$) the standard result $P(x)\,\mathrm{d}x = |\langle x|\psi\rangle|^2\,\mathrm{d}x$, thus demonstrating that BHD indeed measures the QP $\hat{x}_\varphi$ with corresponding POVM $|x\rangle\langle x|$. In practice, j is not usually measured but substituted by $a^2/2$, which blurs the measured x via convolution with a Gaussian distribution that has a variable width of $x/2a$.

For balanced 8PHD (Fig. 1(b)) the state $|\psi\rangle$ is mixed at a 50/50 BS with another state $|\xi\rangle$ (mixed states can easily replace the pure states in this analysis) generally assumed to be the vacuum state. The two resulting beams are mixed with strong LOs

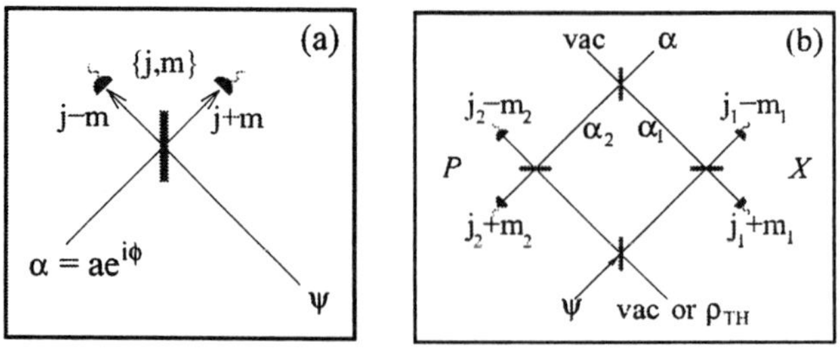

Fig. 1. Four-port (a) and eight–port (b) homodyne detection schemes. In 8PHD, the second input can be either vacuum or thermal state.

with a mutual phase difference $\pi/2$ to perform joint x and p measurements on $|\psi\rangle$. The distribution $\mathcal{P}(m_1, m_2)$ converges to the Q-function for the state $|\psi\rangle$ [5].

To obtain this result from the operational photon number measurement analysis, we apply the su(2) transformation approach as for BHD. In the strong LO limit and for $|\xi\rangle$ the vacuum state, we obtain the limiting form for the joint QP probability distribution (for $x = m_1/\sqrt{j_1}, p = m_2/\sqrt{j_2}$) to be $P(x,p)\,\mathrm{d}x\,\mathrm{d}p = P(\beta)\,\mathrm{d}^2\beta = |\langle\beta|\psi\rangle|^2\,\mathrm{d}^2\beta/\pi$ with $\beta = x + ip$. In this way is is shown that 8PHD indeed measures the Q-function of the input state. The corresponding POVM is $\hat{Q}(\beta) = |\beta\rangle\langle\beta|/\pi$ with measure $\mathrm{d}^2\beta$.

In real experiments, $\mathcal{P}(m_1, m_2)$ cannot be measured exactly due to a finite detector efficiency η. Taking this into account and rescaling x, p each by $\eta^{-1/2}$, we find that the observed distribution $P_\eta(\beta)$ is a convolution of the actual distribution $P(\beta)$ and the Gaussian distribution with variance $\sigma_\eta^2 = (\eta^{-1} - 1)/2$. The corresponding POVM is then $\hat{Q}_\eta(\beta) = (2\pi^2\sigma_\eta^2)^{-1} \int \mathrm{e}^{-|\beta-\beta'|^2/(2\sigma_\eta^2)}\,|\beta'\rangle\langle\beta'|\,\mathrm{d}^2\beta'$.

The vacuum state should be replaced by a thermal state $\rho_{\rm TH}$ with temperature T and average photon number $N = [\exp(\hbar\omega/kT) - 1]^{-1}$ for realistic situations as well as for the quantum teleportation [6] with one component of a two-mode squeezed vacuum state entering the second input port of 8PHD: this component behaves locally as a thermal state. The probability distribution $P(\beta)$ then corresponds to the convolution of Wigner distributions of the two input states ψ and $\rho_{\rm TH}$ [5], or, equivalently, to the convolution of the Q-function of the state ψ and the Glauber-Sudarshan P-function of the state $\rho_{\rm TH}$. As the latter is $P_{\rm TH}(\beta) = \exp(-|\beta|^2/N)/(\pi N)$, we find that measurement with thermal noise corresponds to a POVM with $\hat{Q}_{\rm TH}(\beta) = (\pi^2 N)^{-1} \int \exp(-|\beta - \beta'|^2/N)\,|\beta'\rangle\langle\beta'|\,\mathrm{d}^2\beta'$. If both the effects of thermal noise and $\eta < 1$ are present, the resulting distribution $P'(x,p)$ will be the convolution of the Q-function and the Gaussian distribution with variance $\sigma^2 = \sigma_\eta^2 + \sigma_{\rm TH}^2 = (\eta^{-1} - 1 + N)/2$.

To conclude, we have described BHD and 8PHD in terms of the POVM for photon counting and employed asymptotic methods for su(2) Wigner functions to obtain rigorously the QP and joint QP POVMs, respectively. Finite detector efficiency and thermal fields have been included. This rigorous analysis is essential for future studies of quantum information in terms of operational photon counting POVMs.

We appreciate the hospitality of the Erwin Schrödinger Institute in Vienna, valuable discussions with V. Bužek, T. Rudolph and Ch. Simon and support by Macquarie University Research Grants and an Australian Research Council Large Grant.

References and links

1. N. G. Walker, J. Mod. Optics **34**, 15 (1987).
2. K. Kraus, *States, Effects and Operations* (Springer, Berlin, 1983).
3. D. J. Rowe, B. C. Sanders, H. de Guise, J. Math. Phys. (in press).
4. B. C. Sanders, K. S. Lee and M. S. Kim, Phys. Rev. A **52**, 735 (1995).
5. U. Leonhardt, Phys. Rev. A **48**, 3265 (1993).
6. S. L. Braunstein and H. J. Kimble, Phys. Rev. Lett. **80**, 869 (1998).

Joint quantum measurement using unbalanced array detection

M. Beck[†], C. Dorrer and I.A. Walmsley
The Institute of Optics, University of Rochester, Rochester, New York 14627
[†] Permanent address: Department of Physics, Whitman College. Walla Walla, Washington 99362
Ph: 509-527-5260, Fax: 509-527-5904, beckmk@whitman.edu

Here we describe an experiment that measures the joint statistics of a multitemporal mode field. This experiment is based on a recent proposal [1], however, rather than using balanced heterodyne detection with array detectors, here we perform unbalanced detection. This allows a significant reduction in experimental complexity, as it simplifies alignment and avoids the need to obtain proper registration between corresponding pixels whose outputs need to be subtracted. Balancing is typically used to greatly reduce the degradation caused by classical noise on the local oscillator (LO) [2]. In our experiments we operate at the shot-noise level (SNL) without balancing. We are able to do this because we use an array detector, which demonstrates one more advantage to using array detectors [1,3,4]. The mode statistics are expressed in terms of the joint Q-function of the modes $Q(x_p,y_p)$ [1,5].

Pulses from a mode-locked Ti:sapphire laser enter a Michelson interferometer. The LO beam emerges from one arm of the Michelson, while the other arm delays the signal pulse by approximately 3.5 ps and contains 1.5 cm of glass which adds dispersion to the signal pulse. The signal pulse is thus stretched and chirped. After leaving the interferometer, the pulses are frequency doubled to a wavelength of 400 nm in a nonlinear crystal. The resulting blue LO pulses are approximately 100 fs in duration. We then focus the beam into a spectrometer, and detect the spectrum with a CCD array, obtaining a 512 point spectrum. We use an exposure time of 300ms, so each "shot" is actually composed of millions of pulses.

The operator that we measure in our experiments is the Fourier transform of the number operator for the individual pixels of the detector $\hat{n}_j$,

$$\hat{K}_p \equiv \sum_j \exp[-i2\pi pj/512]\hat{n}_j \ . \tag{1}$$

The index p corresponds to the temporal delay between the LO and signal pulses, where $t=p\delta t$; for our experimental parameters, δt= 29.8 fs. The real and imaginary parts of K_p correspond to the quadrature amplitudes x_p & y_p of the p'th temporal mode of the signal field, once we have normalized by the amplitude of the LO. We histogram our measurements of x_p & y_p into a 32x32 array of bins to obtain the Q-function. We are able to obtain SNL operation without balancing because the classical LO noise is localized near the p=0 modes, while our signal is placed in modes with much larger time delays.

Figure 1 shows measured Q-functions corresponding to two different temporal modes of this pulse; Fig. 1(a) is for a mode near the peak of the pulse, while Fig. 1(b) is for a mode in the trailing edge. We see that these Q-functions are circular, indicating that our signal field has had its phase randomized to a large extent. This is because of phase drift between the two arms of the interferometer during the 8 hours it took to acquire the over 14,000 shots used to construct these Q-functions. The average number of photons in the mode corresponding to Fig. 1(a) is 120, while there are 6 photons in the mode of Fig. 1(b); consequently, the radius of the Q-function is larger in Fig. 1(a) than it is in Fig. 1(b).

Because the 4-dimensional joint Q-function of two modes is difficult to display, we instead display simply the correlations between x-quadratures of two modes in terms of joint distributions $Q'(x_p, x_{p'})$. Figure 2(a) shows correlations between two modes that are both near the peak of the pulse, and consequently have nearly the same amplitude and phase. This

Coherence and Quantum Optics VIII
Edited by Bigelow et al., Kluwer Academic/Plenum Publishers, 2003

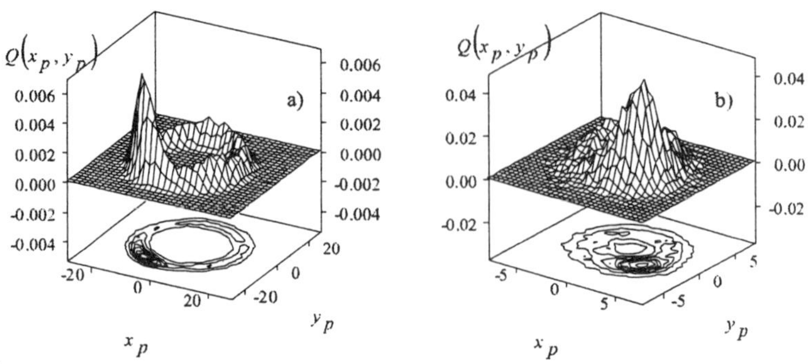

Fig. 1 The measured Q-functions of two different temporal modes of a chirped signal pulse. The mode delay time is t=3.457 ps for a), while it is t=3.785 ps in b)

joint distribution lies mainly along a line whose slope is 1, indicating strong positive correlation between the x-quadratures. This is what we would expect for two modes whose phases are nearly the same. In Fig. 2(b) we show correlations between the same modes displayed in Fig. 1. The average phase difference between these modes is nearly $\pi/2$, and this produces a joint distribution that is very different.

We have demonstrated a new technique for performing joint quantum measurements of multitemporal mode optical fields, which results in the determination of the joint Q-function of the modes. We have shown that by using array detection we are able to effectively eliminate the classical noise on the local oscillator field without the need to perform balanced detection, which is a considerable experimental simplification.

[1] C. Iaconis, E. Mukamel, and I.A. Walmsley, "Temporal heterodyne detector ...," J. Opt. B: Quantum Semiclass. Opt. **2**, 510-516 (2000).

[2] H.P. Yuen and V.W.S. Chan, "Noise in homodyne and heterodyne detection," Opt. Lett. **8**, 177-179 (1983).

[3] M. Beck, "Quantum state tomography with array detectors," Phys. Rev. Lett. **84**, 5748-5751 (2000); A.M. Dawes and M. Beck, "Simultaneous quantum-state measurements using array detection," Phys. Rev. A **63**, 040101(R) (2001).

[4] M.G. Raymer, J. Cooper, and M. Beck, "Many-port homodyne detection of an optical phase," Phys. Rev. A **48**, 4617-4628 (1993).

[5] J.H. Shapiro and S.S. Wagner, "Phase and amplitude uncertainties...," IEEE J. Quantum Electron. **20**, 803-813 (1984).

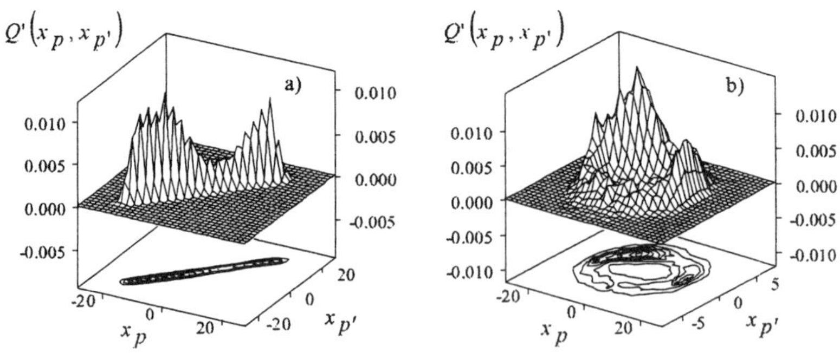

Fig. 2 The measured joint distributions for two different pairs of temporal modes of a chirped signal pulse. The mode delay times in a) are t=3.457 ps and t'=3.576 ps, while in b) they are t=3.457 ps and t'=3.785 ps.

A new view of Bell's inequalities

Louis Sica
Code 5630, Naval Research Laboratory, Washington, D.C. 20375
Ph: 202-767-9466, Fax: 202-767-9203, sica@ccf.nrl.navy.mil

Abstract:

Bell's inequality has been derived from a single assumption in a new proof. An experimental procedure is herein described to obtain data consistent with this assumption. The correlations to be expected are computed from quantum mechanics.

1. Bell's Theorem

In Refs. [1,2] it has been shown that a Bell's inequality is identically satisfied by four lists of numbers of length N, each number restricted to ± 1, and independently of N. The 4N numbers enter into the inequality in the form of estimates of four cross correlations among the four lists of data. If limits exist as N -> ∞, then the correlations must satisfy the inequality as well, yielding the standard Bell's inequality constraint.

Bell's inequality is not an identity in the correlation functions, however, unless they are consistent with this limiting procedure. If the inequality is violated by an arbitrary set of correlation functions, then four data sets having those cross-correlations cannot exist, since any data sets whatsoever of any length identically satisfy the inequality. Bell's inequality is a constraint that correlations must satisfy in order that they represent any data sets that could possibly exist.

The form that Bell's inequality takes depends explicitly on a factoring step that follows from the fact that four cross-correlations are obtained from only four lists of data. If consistency with the factoring step is not respected when cross-correlations are computed among data lists obtained from experiments, the logical link with the derivation is broken. If the conditions leading to the constraint do not apply to the data, then the correlations among the data may or may not satisfy the constraint.

2. Experimental Issues

In the historic experiments of Aspect [3] and in more recent experiments using parametric down conversion [4,5], the four correlations have been computed among eight data lists rather then four as assumed in derivations of Bell's inequalities. This has allowed violation of the inequality. However, a modification of the usual data processing procedure derives four data sets from measurements of Bell-pairs in a way consistent with the factoring condition used in the derivation. The procedure is to measure $A(\theta_A)$ and $B(\theta_B)$, the outputs of a Bell experiment for a series of Bell pairs, followed by $A'(\theta_{A'})$ and $B(\theta_B)$, in a second experiment. A third experiment may be employed to obtain a series of readings at $A(\theta_A)$ with $B'(\theta_{B'})$. Consistency with the derivation of Bell's inequality implies reordering the pairs in the second experiment so that values for the B-list are the same as in the first run. Similarly, pairs A-B' are reordered so that the A-values are the same as in the first experiment. In the limit N -> ∞, the fraction of data items with values equal to + 1 and - 1 are equal, allowing the rearrangements to be carried out, and there are no changes in the computed values of the correlations <AB>, <A'B>, and <AB'>. The data sequences now conform to the factoring condition, and <A'B'> may be computed from the data in hand in accord with this condition.

3. Correlations Obtained

Probabilities for the procedure described, which depends on taking the data in repeated trials, can be computed from probability theory, since conditional probabilities p(A'|B), p(B'|A), and joint probability p(A,B) are available from quantum mechanics. The correlations obtained from the procedure described are

$$\langle AB\rangle = -\cos(\theta_A - \theta_B) \tag{1a}$$

$$\langle A'B\rangle = -\cos(\theta_{A'} - \theta_B) \tag{1b}$$

$$\langle AB'\rangle = -\cos(\theta_A - \theta_{B'}) \tag{1c}$$

$$\langle A'B'\rangle = -\cos(\theta_A - \theta_B)\cos(\theta_{A'} - \theta_B)\cos(\theta_A - \theta_{B'}) \tag{1d}$$

That the correlation <A'B'> is different from the other three follows from the fact that it does not arise from an actual physical interaction. It is also consistent with the fact that the Bell pair correlation cannot be represented as a wide sense stationary process in angle coordinates [2].

4. Interpretation of Counterfactuals

If measurements can be obtained from only one of a mutually exclusive set of apparatus settings, the others that would have been obtained had they been accessible are termed counterfactuals. The procedure described above uses a common statistical approach for the comparison of results contingent on such mutually exclusive alternative measurements. It is less strict, however, than that invoked in Bell's statement of his theorem, since repeated trials are used. In the original derivation of Bell's inequality, Bell demanded that alternative mutually exclusive measurement possibilities be considered for the same trial under the assumption that quantum mechanical variables were supplemented with hypothetical hidden variables whose specification was sufficient to predict quantum mechanical outcomes. The correlation of these measurement outcomes implies repetitive experiments in which all initial conditions are held constant for each different apparatus setting. A different sample space from that applicable to quantum mechanics experiments with Bell-pairs is thereby defined along with the corresponding definition of a joint probability density for all observables under consideration, some of which are noncommuting. Quantum mechanics indicates that noncommuting variables are not statistically equivalent to commuting variables. Bell's theorem is consistent with the quantum mechanical view. When correlations for all alternative measurements in a single trial are assigned the same quantum mechanical correlation function, Bell's inequality is violated, indicating that no data sets exist having these hypothetical correlations.

5. References

1. L. Sica, "Bell's inequalities I: An explanation for their experimental violation", Opt. Commun. 170, 55-60 (1999).
2. L. Sica, "Bell's inequalities II: Logical loophole in their interpretation", Opt. Commun. 170, 61-66. (1999).
3. A. Aspect, P. Grangier, G. Roger, "Experimental realization of Einstein-Podolsky-Rosen-Bohm gedankenexperiment: A new violation of Bell's inequalities", Phys. Rev. Lett. 49, 1804-1807 (1982).
4. Z. Y. Ou and L. Mandel, "Violation of Bell's Inequality and Classical Probability in a Two-Photon Correlation Experiment", Phys. Rev. Lett. 61, 50-53 (1988).
5. P. T. Kwiat, K. Mattle, H. Weinfurter, and A. Zeilinger, "New High-Intensity Source of Polarization-Entangled Photon Pairs", Phys. Rev. Lett. 75, 4337-4341 (1995).

An experiment for testing de Broglie – Bohm theory against standard quantum mechanics

G. Brida, M.Genovese and C.Novero
Istituto Elettrotecnico Nazionale Galileo Ferraris
Str. delle Cacce 91
I-10135 Torino, Italy

G.Introzzi
Dipartimento di Fisica Nucleare e Teorica
Via Bassi 6, I-27100 Pavia, Italy

P. Ghose and A.S. Majumdar
S.N. Bose National Centre for Basic Sciences
Block JD, Sector III, Salt Lake, Calcutta 700098, India

In a recent paper P. Ghose [1] has proposed an experiment addressed to compare de Broglie -Bohm theory and standard quantum mechanics, based on different predictions for coincidence counts in a double slit experiment.

In this talk we discuss the theoretical motivations and the realisation of this proposal based on pairs of photons produced by parametric down conversion. We describe the implementation of the experimental set-up under realisation in Torino.

Preliminary results will be discussed.

1) P. Ghose, quant-ph/0001024.

Bell's inequalities for multiphoton states

Arvind

Department of Physics, Guru Nanak Dev University, Amritsar 142005, India

telephon +911834508898, fax +91258820, e-mail arvind@physics.iisc.ernet.in

Abstract: The connection between quantum optical nonclassicality and the violation of Bell's inequalities is explored. Bell type inequalities are formulated for general states of quantised radiation and their violation is connected to other nonclassical properties of the field. For the family of centered Gaussian states the direct connection between violation of Bell-type inequalities and squeezing is established.

OCIS codes: (99.9999) foundations of quantum mechanics

We develop the machinery for analysing the violation of Bell type inequalities for a general state of the 4-mode radiation field in a setup of the type shown in the diagram. For the direction $\boldsymbol{k}$, the two orthogonal polarisation modes are described by the annihilation operators a_1 and a_2, with a_3 and a_4 being similarly chosen for the direction $\boldsymbol{k'}$. Without any loss of generality we choose $\boldsymbol{k}$ and $\boldsymbol{k'}$ to be in the plane of the paper. This allows a simple choice for the directions x, x' to be in the same plane while y and y' point out of this plane. The passive, total photon number conserving, canonical transformations (which will play an important role in our analysis) amount to replacing the a_j's by their complex linear combinations $a'_j = U_{jk}a_k$, with U being a unitary matrix belonging to $U(4)$. P_1 and P_2 are polarisers placed at angles θ_1 and θ_2 with respect to the x and x' axes while D_1 and D_2 are photon detectors.

Usually states with strictly one photon in each direction are considered for violation of Bell-type inequalities; a general state however could have an *arbitrary number of photons*, and could even be a *mixed state*. To handle such states one needs to generalise the concept of coincidence counts, stipulate the polariser action on general quantum states and identify precisely the hermitian operators for which a hidden variable description is being assumed. As a result of this generalisation we will show that a classical state in the quantum optical sense always obeys these inequalities while a nonclassical state may violate them, possibly after a passive $U(4)$ transformation. Starting with a general nonclassical state, we subject it to a general unitary evolution corresponding to passive canonical transformations $U(4)$ before we look for the violation of Bell-type inequalities.

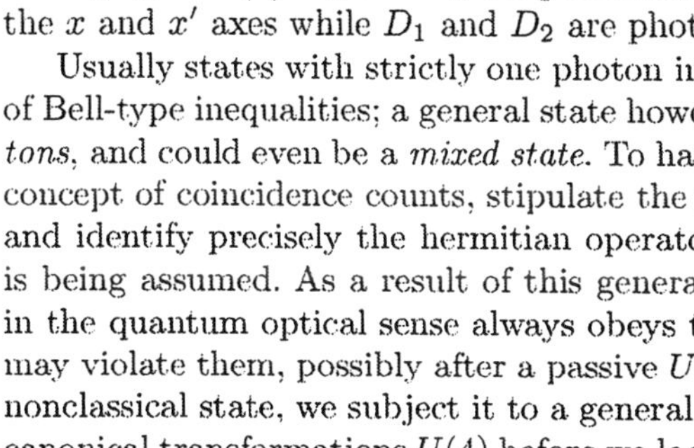

Coincidence Counts

$P(\theta_1, \theta_2)$	:	P_1 at θ_1 and P_2 at θ_2.
$P(\theta_1, \ )$	:	P_1 at θ_1 and P_2 removed.
$P(\ , \theta_2)$	:	P_1 removed and P_2 at θ_2.
$P(\ , \ )$	:	Both P_1 and P_2 removed.

A coincidence is defined to occur when both the detectors D_1 and D_2 click simultaneously *i.e., one or more photons* are detected by each.

With a proper way of quantum mechanically implementing the action of the polarizer in mind we define the following hermitian operators, with eigen values 0 and 1

$$\widehat{A}_1 = (I_{2\times 2} - |00\rangle\langle 00|)_{\mathbf{k}} \qquad \widehat{A}_2 = (I_{2\times 2} - |00\rangle\langle 00|)_{\mathbf{k}'}$$
$$\widehat{A}_1(\theta_1) = (I_{\theta_1} - |0\rangle_{\theta_1\,\theta_1}\langle 0|)\, I_{\theta_1+\frac{\pi}{2}} \qquad \widehat{A}_2(\theta_2) = (I_{\theta_2} - |0\rangle_{\theta_2\,\theta_2}\langle 0|)\, I_{\theta_2+\frac{\pi}{2}} \tag{1}$$

The subscripts θ_1 and θ_2 in the last two equations refer to the settings of P_1 and P_2. $I_{2\times 2}$ is the two-mode unit operator while $I_{\theta_{1(2)}}$ and $I_{\theta_{1(2)}+\frac{\pi}{2}}$ are one-mode unit operators for the relevant polarisation modes along the propagation directions $\mathbf{k}$ or $\mathbf{k}'$. The expectation values of these operators are the probabilities of detecting *at least one photon* of the appropriate kind(for example $\langle\widehat{A}_1\rangle$ is the probability of detecting at least one photon at D_1 with P_1 removed, and $\langle\widehat{A}_1(\theta_1)\rangle$ that at D_1 with P_1 set at θ_1).

The quantum mechanical predictions for various coincidence count rates are the expectation values of the products of pairs of these operators:

$$P(\theta_1,\theta_2) = \langle\widehat{A}_1(\theta_1)\,\widehat{A}_2(\theta_2)\rangle \qquad P(\theta_1,\ \ ) = \langle\widehat{A}_1(\theta_1)\,\widehat{A}_2\rangle$$
$$P(\ \ ,\theta_2) = \langle\widehat{A}_1\,\widehat{A}_2(\theta_2)\rangle \qquad P(\ \ ,\ \ ) = \langle\widehat{A}_1\,\widehat{A}_2\rangle \tag{2}$$

We note here that due to the definitions (1) the polariser action is automatically implemented!

In the case when a hidden variable theory is assumed the value of the hidden variable along with the state vector will give us the actual outcomes of the individual measurements for the dynamical variables $A_1, A_2, A_1(\theta_1)$ and $A_2(\theta_2)$. The locality condition of "no action at a distance" can then be readily used to calculate the coincidence count rates. Further, these rates are constrained by the following inequality due to Clauser and Horne [6]

$$-P(\ \ ,\ \ ) \leq P(\theta_1,\theta_2) - P(\theta_1,\theta_2') + P(\theta_1',\theta_2) + P(\theta_1',\theta_2') - P(\theta_1',\ \ ) - P(\ \ ,\theta_2) \leq 0$$

This is the required generalised Bell-type inequality relevant for arbitrary multi-photon states. We emphasise here that the coincidence count rates for general states have a different meaning as opposed to two-photon states. For two-photon states, extensively studied in the literature(for example the state $\frac{1}{2}(a_1^\dagger - a_3^\dagger)(a_4^\dagger - a_2^\dagger)|0\rangle_1|0\rangle_2|0\rangle_3|0\rangle_4$), our formalism reduces to the usual one. More explicitly, the single hermitian operator $\widehat{A} = I - |0\rangle\langle 0|$ giving the probability for finding one or more photons reduces effectively to $a^\dagger a$. The simplifying relations $P(\theta_1,\ \ ) = P(\theta_1,\theta_2) + P(\theta_1,\theta_2+\frac{\pi}{2})$ etc. are also obtained from the reduction of $\widehat{A}$'s and are not valid for general states.

We now turn to the analysis of interesting multiphoton states. The calculations for some examples of such states give the results sumarized in the following table

State	φ(z)	Bell type inequality
Coherent States	Classical	No violation
Classical States	Classical	No volation
Two photon States	Nonclassical	Violation
Squeezed States	Nonclasical	Violation
Squeezed Thermal States	Nonclassical	Violation

References and links

1. J. S. Bell, *Physics*, **1** 195(1964).
2. M. D. Reid and D. F. Walls. *Phys. Rev.*, **A 34**, 1260(1986).
3. M. S. Chubarov and E. P. Nikolayev. *Phys. Lett.*, **A 110**, 199(1985).
4. C. Brif, A. Mann, M. Revzen, *Phys.Rev.* **A 57** 742(1998).
5. R. A. Campos, B. E. A. Saleh, and M. C. Teich. *Phys. Rev. Lett.*, **A 40**, 1371(1989).
6. J. F. Clauser and M. A. Horne. *Phys. Rev.*, **D 10**, 526(1974).

Testing Bell Inequalities in Photonic Crystals

D.G. Angelakis and P.L. Knight

Optics Section, Blackett Laboratory, Imperial College, London SW7 2BW, England

telephone: +44-207-5947642, fax: +44-207-5947714, e-mail:d.angelakis@ic.ac.uk

During the past twenty years, a number of interesting experiments for testing the Bell Inequalities have been proposed and carried out[1]. However in all these photon experiments low detection efficiency combined with low analyzing speed of the results prevented the complete closure of the detection and/or the communication loophole. We propose a scheme to prepare entangled atomic pairs in a photonic crystal. We anticipate that the potential of using optically separated transitions along with the stability of the entangled state to spontaneous emission could potentially close both the communication and the detection loophole appearing in experiments so far.

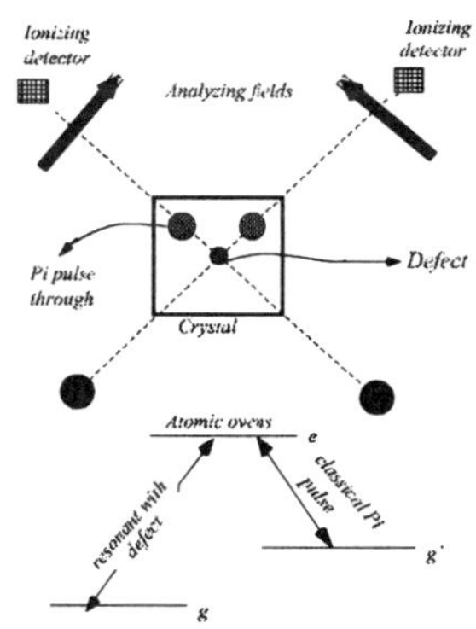

Fig. 1. The proposed scheme and the atomic system consideration

In photonic crystals the regions in which photon localization occurs can be viewed as spherical "bubbles"[2]. These can be arranged in lines through which an excited atom can be fired. Our system consists of two three level atoms, the first of which is initially prepared in the upper of two optically separated states, denoted by $|e_1\rangle$ and the second in the lower one $|g_2\rangle$. The two atoms propagate sequentially in orthogonal directions through the defect region of the crystal(see Fig. 1). The defect mode is initially prepared in the vacuum state $|0\rangle$ and it is on resonance with the atomic transition $|e_i\rangle \rightarrow |g_i\rangle$. In Fig. 2 we show contour plots of the strength of the correlations versus the velocities. The values of the parameters correspond to the case of Rb atoms traveling through a defect in an optical photonic crystal (e.g., GaP) at thermally-accesible velocities ($v \sim 150 - 400 m/s$). The extend of the defect mode is assumed to be equal to the lattice constant a and $a = 0.8\lambda$. The coupling constant has maximum value of $1.1 \times 10^{10} rad/sec$ at the center of the defect. The dashed lines in the plot represent to $|S| = 2.2, 2.4, 2.6, 2.8$, whereas the solid line corresponds to $S = 2$, i.e., the maximum value allowed by local realistic theories. As the transitions here are optically separated, decoherence due to spontaneous emission would potentially be a problem during the flight from the crystal to the analyzers/detectors(inside the crystal spontaneous emission is competely inhibited[2]). To tackle this problem through a properly engineered line defect we inject a coherent light field into the crystal which crosses the atom's path.

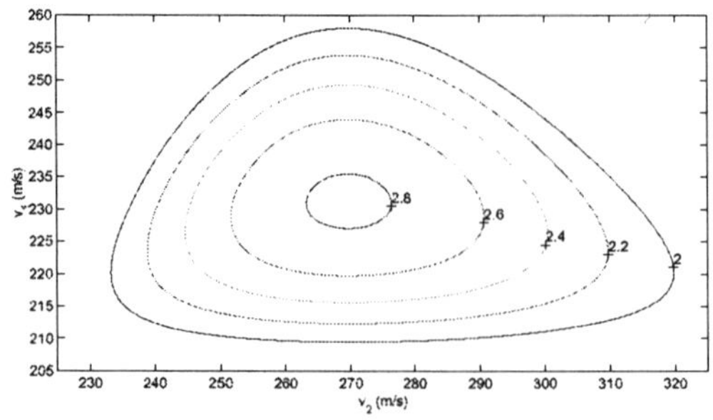

Fig. 2. Contour plot of Bell's parameter $\mathcal{S}$ as function of the atomic velocities v_1, v_2

This field is resonant with the $|e\rangle \rightarrow |g'\rangle$ transition. By adjusting the field intensity properly we can apply a Π pulse between $|e\rangle$ and $|g'\rangle$ transferring any population from $|e\rangle$ to $|g'\rangle$(see Fig. 1). The transition $|g'\rangle \rightarrow |g\rangle$ is dipole non-allowed and thus can be considered stable for our purposes(lifetime$\sim$1 sec)
We also calculate the robustness of our scheme as a function of the spread in the atomic velocities as a function of Δv for the case of $\bar{v}_1 = 231\text{m/s}$ and $\bar{v}_2 = 270\text{m/s}$. As we show the violation is quite strong even in the case of ± 25m/ which is easily within the limits of curent velocity selection technology. Detection efficiency (we detect optically

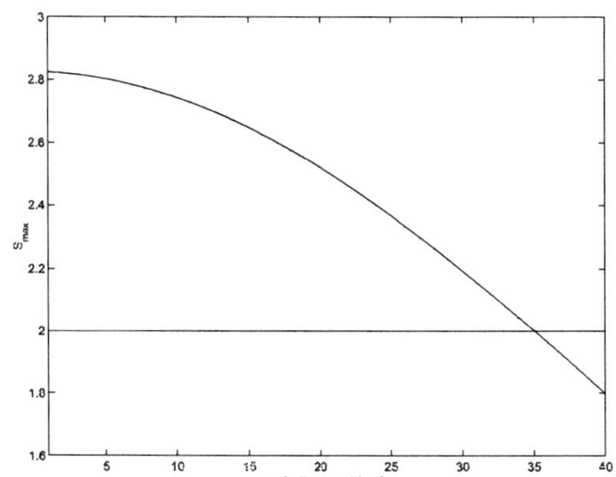

Fig. 3. The sensitivity of the scheme as a function of the spread in the atomic velocities. The vertical axis is the value the quantum mechanical sum $\bar{\mathcal{S}}$ and the horizontal the atomic velocity spread Δv . A violation of Bell's Inequality is predicted as long as $\Delta v \leq 34\text{m/s}$.

separated transitions) can be as high as 95%. This would close the detection loophole. The communication loophole can be also closed, as our entangled state stays alive for seconds which as we show is enough to prevent any kind of subluminal communication between the analyzers/detectors.

References and links

1. A. Aspect et al., Phys. Rev. Lett. **47**, 460 (1981); Z.Y.Ou and L. Mandel, Phys. Rev. Lett. **61**, 50 (1998); A. Kuzmich et al., *ibid.* **85**, 1349;E.S. Fry, Th. Walther and S. Li, Phys. Rev. A **52**, 4381 (1995); G. Weihs et al., Phys. Rev. Lett. **81**, 5039 (1998).;M. A. Rowe et al., Nature 409, 791 (2001); D. G. Angelakis et al., submitted.
2. D. Angelakis et al., Phys. Rev. A **61**, 055802;A. Blanco et al., Nature **405**, 437 (2000); E. Yablonovitch et al., Phys. Rev. Lett **58**, 2059 (1987);

On the detection of single mode quantum coherence at optical frequencies

R. W. Spekkens and J. E. Sipe

Department of Physics, University of Toronto,
60 St. George Street, Toronto, Ontario, Canada M5S 1A7

OCIS codes: (270.0270) Quantum Optics

Protocols discussed in the context of quantum information theory often require the ability to implement preparation procedures associated with one of several non-orthogonal density operators, and measurement procedures associated with one of several non-orthogonal bases. A single mode of the electromagnetic field at optical frequencies might at first appear to be a candidate system for implementing such tasks. The quantum state of a laser above threshold is typically assumed to be well approximated by a coherent state, and any two coherent states are strictly non-orthogonal. Moreover, optical homodyne tomography would appear to provide a way of measuring (at least upon repeated preparations) an arbitrary quadrature of the field. However, we shall argue that conventional preparation and measurement procedures on single mode optical fields are not appropriate for quantum information processing.

We begin by considering conventional preparations. The crucial result here is due to Mølmer[1]: conventional optical sources are associated with density operators diagonal in the number basis, that is, they do not generate quantum coherence (*i.e.*, a coherent superposition of number states). The argument is as follows. Since the interaction between the degrees of freedom of the gain medium and the light field is energy conserving, and since the gain medium begins in a thermal distribution, the interaction cannot lead to any coherence between number states of the light field.

We now demonstrate a similar result for measurement procedures: conventional detection schemes - schemes that make use of the standard optical toolkit: lasers, photodetectors, beam splitters, phase shifters and the like - only measure the number basis, that is, they cannot detect quantum coherence. Our demonstration begins with the observation that the most general combination of such devices takes the form of a $2N$-port interferometer, where one of the input ports corresponds to the signal field(the object of the measurement), while the other $N-1$ correspond to probe fields, and at each of the N output ports is a photodetector. The body of the interferometer may involve any combination of linear optical elements.

Photodetection theory tells us that standard photodetectors are represented by a POVM of the form $\{\Pi_k\}_{k=0}^{\infty}$, with

$$\Pi_k =: \frac{\left(\eta a^\dagger a\right)^k e^{-\eta a^\dagger a}}{k!} :$$

where $:\ :$ indicates normal ordering, and η is the efficiency. The probability of outcome k is of the form $Tr\left(\rho\Pi_k\right) = \sum_n \lambda_n^k \left\langle n\right| \rho \left|n\right\rangle$. Since this does not depend on the off-diagonal elements of ρ, it is clear that a photodetector cannot, on its own, detect quantum coherence. The basic reason for the appearance of the number basis is the fact that the interaction between the charges in the photodetector and the field is energy-conserving and the charges are measured in their energy eigenstates.

In considering the probe fields, all that is important is that they are described by density operators that are diagonal in the number basis. This follows from Mølmer's argument and the fact that we have assumed that these probe fields come from conventional optical sources. Next, we use the fact that the most general combination of linear optical elements is associated with a unitary transformation on the tensor product of Hilbert spaces for the modes of all the input ports. Any loss mechanisms act in the same way as a coupling into an output mode. Combining these results it follows that the effective POVM on the signal field is one whose elements are diagonal in the number basis. Thus, a $2N$-port interferometer, no matter how complex, can do nothing more than implement an error-prone measurement of the number basis.

A simplified version of the argument proceeds as follows. If the photodetectors were ideal, so that together they constituted a measurement of the total number of photons in all modes; and if the optical elements were lossless, so that they conserved the total number of photons; and if the probe fields were described by number states; then one could immediately infer the number of photons in the signal mode. Such an idealized interferometer would therefore constitute a measurement of the number basis. Error-prone photodetectors, absorption, and uncertainty in probe field photon number simply imply that the interferometer implements a measurement of the number basis that is error-prone.

A consequence of this result is that the technique of optical homodyne tomography, as conventionally implemented, can only determine the number distribution of a signal field. However, *if* one had a probe field in a coherent state then one *could* reconstruct any quadrature of the signal field. Thus, the ability to generate optical coherence implies the ability to detect it.

One might well ask: if conventional optical sources are properly represented by mixtures of number states, why is it that calculations wherein these sources are represented by coherent states have agreed so well with experiments? Our result provides the answer. Experiments that involve conventional detection schemes are completely insensitive to the values of the off-diagonal elements of the density matrix. As such, one can assume *any values at all* for these elements without affecting the result of such calculations. In particular, one can replace a mixed state of the form $\rho = \sum_n p_n |n\rangle \langle n|$ with any pure state of the form $\sum_n \sqrt{p_n} e^{i\theta_n} |n\rangle$ where the θ_n are arbitrary phases. If one could measure something besides the number basis, then the two states above would predict different results, and (presumably) only the former would yield results that were empirically correct.

It should be noted that these results do *not* establish the impossibility of generating or detecting quantum coherence, but only that these feats cannot be accomplished with conventional optical sources and detection schemes. Our analysis provides some insight into what sorts of schemes might, however, succeed in these tasks.

References and links

1. K. Mølmer, PRA **55**, 3195 (1997)

Soliton interaction control in coherently excited three-level system embedded in a nonlinear waveguide

E. J. S. Fonseca, J. F. Liberato, S. B. Cavalcanti and J. M. Hickmann
Departamento de Física, Universidade Federal de Alagoas
Cidade Universitária, 57072-970, Maceió, AL, Brazil
phone: +558221414 24. fax: +55822141645, e-mail: hickmann@loqnl.ufal.br

Abstract: We investigate the properties of the interaction between a pair of solitons that are originated from the break up dynamics of a high order optical soliton propagating through a cylindrical waveguide in the presence of three-level resonances associated with a dopant.

OCIS codes: (270.1670) Coherent optical effects; (270.5530) Pulse propagation and solitons; (190.0190) Nonlinear optics

The coexistence of a self-induced transparency (SIT) soliton with a soliton solution of the nonlinear Schrödinger (NLS) equation, henceforth called a SIT/NLS-soliton, has been suggested [1] and demonstrated [2]. The cloning of such mixed state has been reported using two resonant optical fields propagating through a doped optical nonlinear waveguide [3]. Recently, numerical simulations using a pair of nonlinear Schrödinger equations coupled with Bloch equations, exhibited the cloning of a pair of solitons originated from a mixed high order SIT/NLS-soliton, i.e. a soliton with area 4π and $N = 2$. The fields were configurated in a Λ-scheme, with the mixed high-order soliton at the pump frequency and a small signal at the Stokes frequency. The pump pulse broke up into a pair of solitons which were subsequently cloned to the signal frequency [4]. In this work we study the interaction's nature of this cloned soliton pair in function of the initial pump and signal intensitiy ratio and time overlap.

Now we shall analyze the interaction of a three-level system diffused in a nonlinear cylindrical waveguide and a pair of pulses propagating in the waveguide's z-direction. We will consider a state $|3\rangle$ as the fundamental state of our three-level system, and $|1\rangle$ and $|2\rangle$ as two excited states with different energies, being the latter a metastable state. The pump pulse with an amplitude $A_{13}(z,t)$ and a frequency ω_{13} is resonant with the transition between the states $|1\rangle$ and $|3\rangle$. The signal field $A_{12}(z,t)$ at the Stokes frequency ω_{12} connects the states $|1\rangle$ and $|2\rangle$. The fields are assumed to have the same linear polarization, and can be described by the following coupled generalized NLS equations:

$$\frac{\partial A_{12}}{\partial z} = -\frac{i\beta_{12}}{2}\frac{\partial^2 A_{12}}{\partial T^2} + i\gamma_{12}\left[|A_{12}|^2 + 2|A_{13}|^2\right]A_{12} + \frac{i\omega_{12}n_a}{2\varepsilon_0 c}c_2^* c_1 \mu_{12} \tag{1}$$

$$\frac{\partial A_{13}}{\partial z} = -\frac{i\beta_{13}}{2}\frac{\partial^2 A_{13}}{\partial T^2} + i\gamma_{13}\left[|A_{13}|^2 + 2|A_{12}|^2\right]A_{13} + \frac{i\omega_{13}n_a}{2\varepsilon_0 c}c_3^* c_1 \mu_{13} \tag{2}$$

where the probability amplitudes $c_j(z,t)$ of the atomic levels $|j\rangle$ within the rotating wave approximation are obtained through the Bloch equations,

$$\frac{\partial c_1}{\partial t} = \frac{i}{\hbar}[c_2\mu_{12}A_{12} + c_3\mu_{13}A_{13}]; \quad \frac{\partial c_2}{\partial t} = \frac{i}{\hbar}[c_1\mu_{12}A_{12}^*]; \quad \frac{\partial c_3}{\partial t} = \frac{i}{\hbar}[c_1\mu_{13}A_{13}^*] \tag{3}$$

Here μ_{12} and μ_{13} are the electrical dipole moments related to the associated transitions and we assumed $\mu_{12} = \mu_{13}$ for simplicity. Furthermore, we have introduced the nonlinear parameter $\gamma_{ij} = n_a k_{ij}$ with k_{ij} as the wave vector and $ij = 12$ (pump) or $ij = 13$ (signal). The time in the equations refers to a Galilean reference frame moving at an averaged group velocity.

The above set of coupled equations is solved numerically considering as pump pulse a high order SIT/NLS-soliton with $N = 2$ and area 4π and a signal pulse with the pump shape but lower intensity. When the population trapping is established, the pump is depleted and $c_1 \to 0$, $c_2 \to 0$, $c_3 \to 1$. At this stage one is left with a single NLS, and the initial condition is given by a two-soliton solution whose distance and relative phase depends completely on the previous coherent process. Fig. 1 shows this asymptotic behavior for $\Delta = 0.05$ (Fig. 1*a*) and $\Delta = 0.5$ (Fig. 1*b*). Here, Δ represents the squared relative intensity ratio between pump and signal at the input, i.e. $\Delta = \frac{|A_{12}|^2}{|A_{13}|^2} = r^2$. These graphs show clearly that by increasing the Δ ratio, the interaction between the two soliton pulses is strengthened up to the point that the cloning process is so accelerated that the breakup process occurs only after the cloning process.

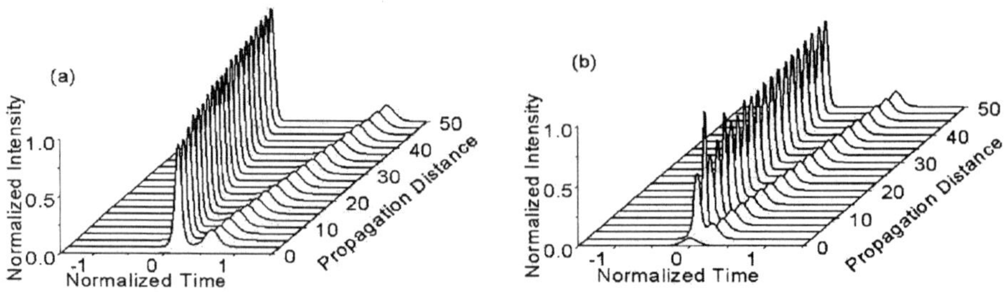

Figure 1: Signal evolution for Δ ratios equal to (a) 0.05 and (b) 0.5

Another parameter studied that influences the asymptotic soliton interaction was the initial overlap between A_{12} and A_{13}. In Fig. 2, we plot the asymptotic behavior of the soliton pair for the same squared relative intensity ratio $\Delta = 0.3$, but for different initial overlaps between pump and signal. Comparing the evolution for an initial complete overlap of pump and signal shown in Fig. 2*a* with an initial partial overlap (see inset in Fig.2*b*) shown in Fig. 2*b*, we notice that as the overlap decreases the interaction is modified from the repulsive behavior. In fact we see solitons crossing one another in a kind of intermingled state in the second case.

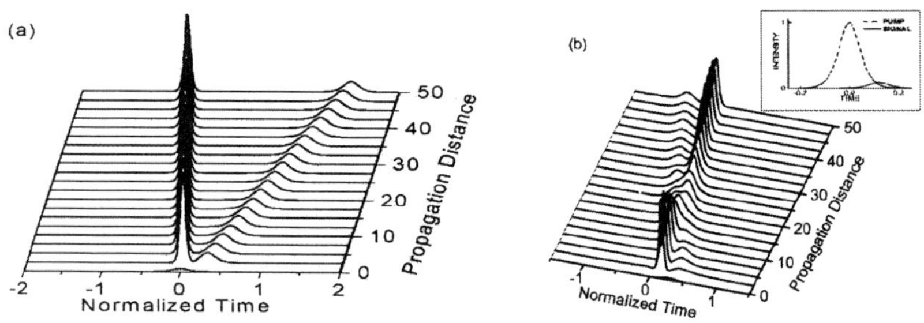

Figure 2: Signal evolution for a pump and signal initial overlap: (a) complete or (b) partial (see insert)

We have studied the simultaneous propagation of a pair of optical solitons, generated by the break up of a mixed high order SIT/NLS-soliton, through a cylindrical waveguide, taking into account the resonances provided by impurities modelled as a three-level system. The interaction between the soliton cloned pair was shown to depend on their relative phase which, due to the Kerr effect, depends on the relative intensities of pump and signal pulses. Furthermore, we showed that using an initial overlap between pump and signal it is possible to avoid the initial repulsion between the generated soliton pair. In resume, we confirmed that population trapping states offers many interesting effects on the manipulation of optical pulses through matter such as cloning, breakup, and even soliton interaction control.

References

1. A. I. Maimistov and É. A. Manykin, Zh Eksp. Teor. Fiz **85**, 685 (1983).
2. M. Nakazawa, E. Yamada and H. Kubota, Phys. Rev. Lett. **66**, 2625 (1991).
3. D. P. Caetano, S. B. Cavalcanti and J. M. Hickmann in Nonlinear Optics: Materials, Fundamentals and Applications, OSA Technical Digest (Optical Society of America, Washington DC, 1998) paper# MC32/119.
4. D. P. Caetano, S. B. Cavalcanti and J. M. Hickmann, Nonlinear Optics: Materials, Fundamentals and Applications, OSA Technical Digest (Optical Society of America, Washington DC, 2000) paper# TuB8-1/77.

Nonexponential decoherence of coherent state superposition in degenerate optical parametric amplifier

R. Filip and J. Peřina

Department of Optics, Palacký University,
17. listopadu 50, 772 07 Olomouc,
Czech Republic

email:filip@optnw.upol.cz, tel:+420-68-5631572, fax:+420-68-5224246

Schrödinger-cat states (coherent state superpositions with sufficiently large amplitudes and opposite phase) are hard to be produced experimentally. They are more rapidly exponentially destroyed in the atom-amplifier with squeezed reservoir in the strong amplification limit. On the other hand, in this contribution, a slower non-exponential decay of coherence is obtained if the one-mode amplifier exhibits the squeezing and the reservoir is not squeezed. We consider the classical pumped degenerate second-order nonlinear parametric process with amplification constant κ and with losses arising in the nonlinear medium described by the damping constant γ and reservoir thermal fluctuations with mean number $\langle n \rangle$. The mode is consider initially in the superposition of two coherent states $|\psi(0)\rangle = \frac{1}{\sqrt{N}}\left(|\xi(0)\rangle + |-\xi(0)\rangle\right)$, $N = 2\left(1+\exp(-2\xi(0)^2)\right)$, with $\xi(0) \geq 2$. For positive κ, the superposition is orientated in the phase-space in the same direction as the maximal amplification occurs, whereas for negative κ in direction, where the squeezing is occurring. The Fokker-Planck equation for the Wigner function can be analytically solved [1] and the marginal distribution for P-quadrature for the Wigner function ($s=0$) and Q function ($s=-1$) can be found

$$p(P) = \frac{2}{N\sqrt{\pi b_s}} \exp\left(-\frac{P^2}{b_s}\right)\left[1 + R_s \cos\left(2\frac{\eta(t)P}{b_s}\right)\right], \tag{1}$$

where $R_s = e^{-2\xi^2(0)} \exp\left(\frac{\eta^2(t)}{b_s}\right)$, $\eta(t) = G\xi(0)$, $G = \exp\left(-(\kappa+\gamma)t\right)$, $b_s = B - C - \frac{1+s}{2}$

$$\begin{aligned} B(t) &= c_1 - c_2 \exp(-2\gamma t)\cosh(2\kappa t) - c_3 \exp(-2\gamma t)\sinh(2\kappa t), \\ C(t) &= -c_2 \exp(-2\gamma t)\sinh(2\kappa t) + c_3\left(1 - \exp(-2\gamma t)\cosh(2\kappa t)\right), \\ c_1 &= \frac{2\gamma^2(1+\langle n\rangle) - \kappa^2}{2(\gamma^2-\kappa^2)}, \quad c_2 = \frac{2\gamma^2\langle n\rangle + \kappa^2}{2(\gamma^2-\kappa^2)}, \quad c_3 = \frac{\gamma\kappa(1+2\langle n\rangle)}{2(\gamma^2-\kappa^2)}. \end{aligned} \tag{2}$$

The ratio R_s controls the relative visibility of interference fringes in the marginal distribution $p(P)$. The parameter can be determined directly from the homodyne ($s=0$) or heterodyne ($s=-1$) detection of marginal distribution $p(P)$ for state superposition $p_{CAT}(0)$ and corresponding mix of coherent states $p_{MIX}(0)$ by the following relation: $R_0 = (p_{CAT}(0) - p_{MIX}(0))/p_{MIX}(0)$.

The amplifier can exhibit non-exponential decay for positive κ and for sufficiently negative κ, the relative visibility R_0 can be preserved [1]. This is unexpected result as compared with exponential decay in the standard treatment of decoherence. If the limits $|\kappa| \gg \gamma$ and $\frac{\gamma}{\kappa}\langle n\rangle \ll 1$ are assumed, R_s can be approximately found in the following form for the DOPA and atom amplifier

$$R_s(t) \approx \exp(-2\xi^2(0))\exp[\frac{\xi(0)^2}{\frac{1}{2} + (\frac{\gamma}{\kappa}(\langle n\rangle + \frac{1}{2}) - \frac{s}{2})\exp(2\kappa t)}], \tag{3}$$

Coherence and Quantum Optics VIII
Edited by Bigelow et al., Kluwer Academic/Plenum Publishers, 2003

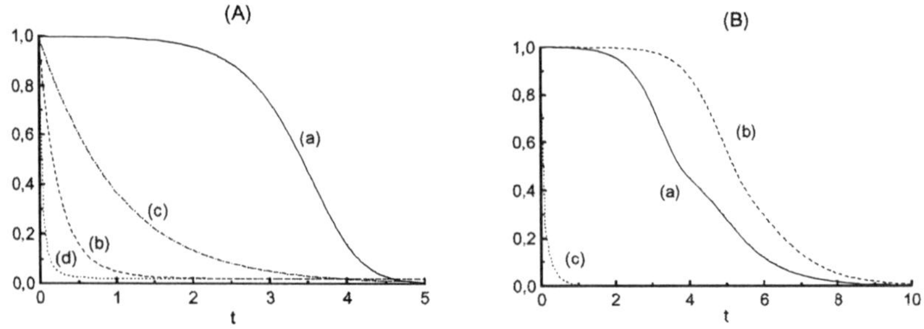

Fig. 1. (A) Time evolution of peak-to-peak ratio R_s normalized to initial value for (a) Wigner function, (b) Q-function compared with (c) squeezing of noise in P-quadrature and (d) Wigner function for unselective amplifier, with $\gamma = 0.0001$, $\kappa = 1$, $\langle n \rangle = 0$ and $\xi(0) = 2$. (B) the purity parameter for (a) $\kappa = 1$, (b) $\kappa = -1$ and (c) for unselective amplifier with $\kappa = 1$, with $\gamma = 0.0001$, $\langle n \rangle = 0$ and $\xi(0) = 2$.

in comparison with the unselective amplifier exhibiting the exponential decay, which is depicted in the Fig. 4(A). In Eq. (3), it can be found simple reason for non-exponential decay, because initially the interference is preserved, but for long-time it is exponentially-like vanishing. In the small-time limit, the $R_0(t)$ has approximately constant values whereas for long-time it exhibits an exponential decreasing. For the atom amplifier, the ratio decreases in the small-time limit in contrast to Eq. (3) and consequently the preserving of interference cannot be observed. It can be found that the decay (3) changes into exponential behavior as the ratio $\frac{\gamma}{\kappa}(\langle n \rangle + \frac{1}{2})$ increasees. Thus this aspect of decoherence can be observed only for initially not too much separated superposition of coherent states, in the small thermal noise limit. On the other hand, the peak-to-peak ration R for the Q-function exhibits rapid exponential decay comparing with the non-exponential decay of the Wigner function, as can be seen in Fig. 1(A). The purity parameter P of the state can be expressed by means of the Wigner function and can be treated as another measure of decoherence. The DOPA purity parameter exhibits the non-exponential character of decay independently of the particular representation, as can be seen in Fig. 1(B), in contrast with the exponential decay for the unselective atom-amplifier. The off-set of purity parameter is pronounced for negative κ, but in the long-time limit, the state has a strongly mixed character for both positive and negative κ. On other hand, the visibility of interference, which is described by the peak-to-peak ratio R_0, is preserved for κ negative. Thus the visibility of interference can be preserved for Schrödinger-cat state, even if the purity of the state considerably disappears. The experimental realization can be proposed for confirming non-exponential decay of quantum superposition: for small damping constant $\gamma \approx 10^5 s^{-1}$ and strong amplification $\kappa \approx 10^{10} s^{-1}$, $\xi(0) = 2$, $\langle n \rangle \ll 1$, for light velocity in the nonlinear crystal $c \approx 1.6 \times 10^8 ms^{-1}$ and the gain approximately $G \approx 25$, the length of suppression of peak-to-peak ratio to value $R_0 = 0.95$ is approximately $l \approx 0.05m$, which is corresponding to the length of the nonlinear crystal.

This work was supported by the project LN00A015 by the Ministry of Education of Czech Republic.

References and links

1. R. Filip, "Amplification of Schrödinger-cat state in degenerate optical parametric amplifier," J. Opt. B: Quantum and Semiclass. Opt. **3**, S1 (2001).

Decoherence Control by Homodyne-Mediated Feedback

Jin Wang[1], H.M. Wiseman[2,1], G. J. Milburn [3], Stefano Mancini [4]

[1] *Centre for Laser Science, Department of Physics, The University of Queensland, Brisbane, Queensland 4072, Australia*
[2] *School of Science, Griffith University, Brisbane, Queensland 4111, Australia*
[3] *Centre for Quantum Computer Technology , The University of Queensland, Brisbane, Queensland 4072, Australia*
[4] *INFM - Research Unit at Milano University, Via Celoria 16, 20133 Milano, Italy*
Tel:(617) 3365-1361, Fax:(617) 3365-1242, wang@physics.uq.edu.au

Abstract: We present an improved theory using Homodyne-mediated feedback to fight decoherence in a two-level atom to control the dynamics of the system and to quantify the effectiveness of feedback in both Markovian and non-Markovian processes.

OCIS codes: (270.0270) Quantum Optics; (999.9999) Quantum Feedback

Although classical models of feedback schemes have been used for a long time to control dynamical noise, an analogous quantum theory of feedback has been developed only in the last fifteen years [1, 2, 3, 4, 5, 6, 7]. Recently there has been considerable interest in quantum feedback as a way to fight decoherence in isolated quantum systems, using the approach of Refs. [4, 5]. For example, it has been suggested as a way to create optical squeezed states, to create micromaser number states, to correct errors in quantum bits, and to protect optical and microwave Schrodinger cat states against dissipation.

Decoherence in quantum systems can be loosely defined as loss of purity. Therefore the ultimate success in using feedback to fight decoherence would be to create an arbitrary stable pure state in the presence of dissipation. This goal was realized by Hoffman, Mahler and Hess (HMH) [8] for a very simple system: a resonantly driven two-level atom. They showed that by making part of the coherent driving proportional to the homodyne photocurrent can stabilize the state to any point on the bottom half of the sphere.

Here we reanalyze their proposal by using the technique of stochastic master equations, allowing their results to be generalized in four ways. The diagram of our proposed experimental apparatus is described in Fig.1. On the left of Fig.2, we present the stationary states we can reach by driving the atom but without feedback. The feedback can be described by the Hamiltonian

$$H_{\rm fb} = \lambda \sigma_y I(t-\tau). \tag{1}$$

Where λ is the feedback amplitude, $I(t-\tau)$ is the continuous homodyne current from a detection scheme with efficiency η,

$$I(t-\tau) = \sqrt{\gamma}\,\langle \sigma_x \rangle_{\rm c}(t-\tau) + \xi(t-\tau)/\sqrt{\eta}. \tag{2}$$

and τ is the feedback delay time for non-Markovian process while it is negligible in Markovian process.

First, we revisit the question of stability in Markovian process and find that, contrary to the claims of HMH, the states in the upper half of the Bloch sphere can be stabilized as well as those in the lower half (this is what was better than they realized) under unit-efficiency detection. The only states which cannot be stabilized, are those on the equator of the Bloch sphere. We present this result in Fig. 2a as a full circle (minus the points on the equator) in the Block sphere.

Second, we consider how effective feedback is with non-unit-efficiency detection in Markovian process. In this case it is almost impossible to stabilize the atom at any pure state, except the ground state. Instead, we aim to produce a steady state which is as close as possible to a given pure state. Again we find that states near the equator cannot be well-protected against decoherence. We also find an echo of the distinction HMH found between the upper and lower halves of the Bloch sphere, in that states in the upper half sphere are affected much more by loss of detection efficiency that those in the lower half. These features are illustrated in Fig.2c.

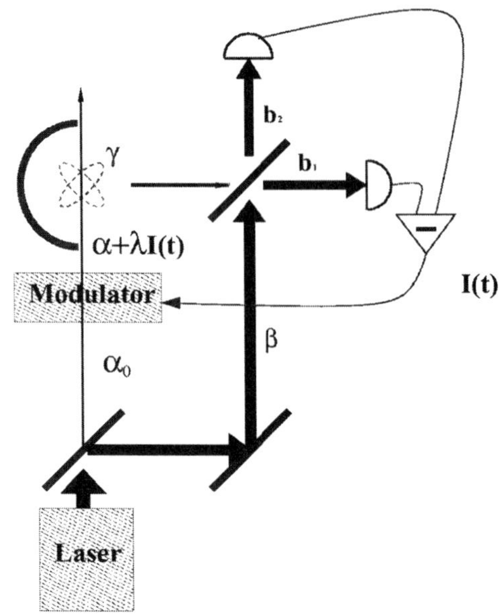

Fig. 1. Diagram of the experimental apparatus. The laser beam is split to produce both the local oscillator β and the field α_0 which is modulated using the current $I(t)$. The modulated beam, with amplitude proportional to $\alpha + \lambda I(t)$, drives an atom at the centre of the parabolic mirror, where α and λ are driving and feedback amplitude, respectively. The fluorescence thus collected is subject to homodyne detection using the local oscillator, and gives rise to the homodyne photocurrent $I(t)$.

Third,we consider the feedback stabilization in the presence of both dissipative and non-dissipative decoherence, assuming unit-detection efficiency. The population decay is a dissipative decoherence process while the polarization decay is non-dissipative. Polarization decay is also called phase decay, arising from elastic collisions in an atomic vapor, or elastic photon scattering in a solid [9]. We find that the purity of the most possible stable states in the upper half of the Bloch sphere is affected symmetrically by phase decay rate as those in the lower half, this is different from the situation of non-unit efficiency detection, in which the top half shrinks more quickly than the bottom half.

Fourth, we consider how effective feedback is in non-Markovian process, that is feedback with time delay. From Fig. 3a, we find that the time delay will result in more randomness in the steady state quantum trajectory. The longer the feedback delay time is, the more dramatic the fluctuations are. From Fig. 3b, we can see that the purity of a single trajectory is not be affected by the non-Markovian process while the purity of the average state is. Rather, the purity of the average state reflects the stability of the individual trajectory.

References

1. H. A. Haus and Y. Yamamoto, Phys. Rev. A **34**, 270 (1986).
2. Y. Yamamoto, N. Imoto and S. Machida, Phys. Rev. A **33**, 3243 (1986).
3. J. M. Shapiro *et al*, J. Opt. Soc. Am. B **4**, 1604 (1987).
4. H. M. Wiseman and G. J. Milburn, Phys. Rev. Lett. **70**, 548 (1993).
5. H. M. Wiseman, Phys. Rev. A **49**, 2133 (1994).
6. L. Plimak, Phys. Rev. A **50**, 2120 (1994).
7. A.C. Doherty and K. Jacobs, Phys. Rev. A **60**, 2700 (1999).
8. H. F. Hofmann, O. Hess, and G. Mahler, Optics Express **2**, 339 (1998).
9. D.F. Walls and G.J. Milburn, *Quantum Optics*, (Springer, Berlin, 1994).

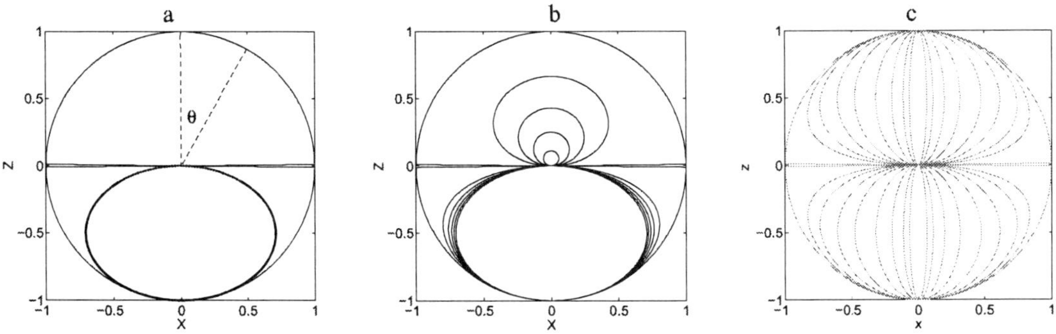

Fig. 2. In Fi2a, locus of the solutions to the Bloch equations. The ellipse in the lower half plane is the locus for the equations with driving but without feedback. The full circle (minus the points on the equator) is the locus for the equations with optimal driving and feedback. In fig.2b, locus of the solutions to the Bloch equations with optimal feedback for different values of detector efficiency η. From the outside in, we have $\eta = 1, 0.8, 0.6, 0.4, 0.2, 0$. Note: Our results are in the x–z plane on the Bloch sphere, since the effect of our driving is to rotate the atom around the y-axis in Bloch space. In Fig2c, Locus of the solutions to the Bloch equations with optimal feedback for different values of dissipative decay rate Γ. from the outside in, we have $\Gamma = 0, 0.01, 0.03, 0.1, 0.3, 1, 3, 10, 30$.

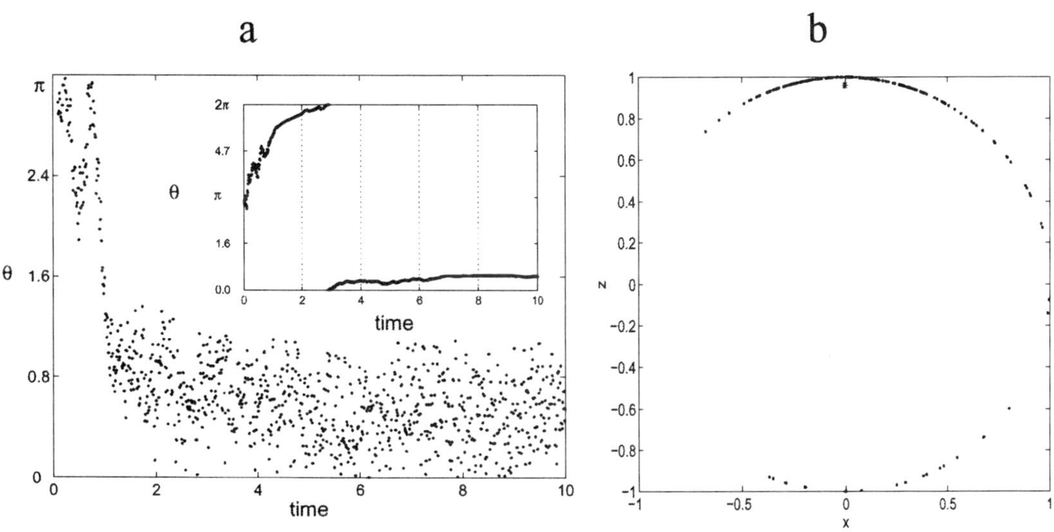

Fig. 3. Fig.3a: Typical single quantum trajectories in Bloch space for $t \in [0, 10\gamma^{-1}]$ for $\theta_0 = \pi/6$ (θ_0 is the desired state), starting at the ground state, where γ is the decay rate of the system. Note that the single trajectory can go either via the ground state or the excited state to get to the state $\theta_0 = \pi/6$. The main plot with a more dramatic fluctuations is for feedback with a large time $\tau = 0.02\gamma^{-1}$ and the trajectory is via the ground state, while the one with an inset into this figure is that for feedback with no time delay and the trajectory is via the excited state; Fig.3b: Typical single quantum trajectories in Bloch space for $t \in [0, 100\gamma^{-1}]$ with $\tau = 0.02\gamma^{-1}$ and $\theta_0 = 0$, starting at the ground state. Note that the single star dot underneath the single trajectory path indicates the loss of purity of the average state, while the single trajectory evolves stochastically along the outside full circle which is the locus of pure states.

Quantum control of population inversion in the presence of spontaneous emission

S. G. Schirmer, Andrew D. Greentree and A. V. Durrant

Quantum Processes Group, The Open University, Milton Keynes, MK7 6AA, United Kingdom

Tel: +44-1908-652326, Fax: +44-1908-652140, S.G.Schirmer@open.ac.uk

Many approaches to quantum control explicitly assume idealized cases. Starting with an idealized control scheme derived using Lie group decompositions, we employ computer models to simulate the effect of spontaneous emission that occurs in a realistic system, and attempt to optimize the performance of the control scheme. For illustration, we use a four-state Rubidium atom with transitions as indicated in Fig. 1, where ω_n are the frequencies and f_n are the absorption oscillator strengths of the transitions. Assuming that the atom is initially in the ground state $|1\rangle = |5S_{1/2}\rangle$, our aim is to maximise the population inversion between states $|1\rangle$ and $|4\rangle = |6P_{3/2}\rangle$ by applying control pulses derived from resonant, intense laser beams. For simplicity, we assume that the pulse envelopes of our lasers are Gaussian and off-resonant effects can be neglected.

If the finite lifetimes of the excited states are ignored then it can easily be shown, e.g., using Lie group decompositions [1, 2], that a complete population transfer from the ground state to the excited state $|4\rangle$ can be achieved by applying a sequence of three control pulses $F_n(t) = A_n(t)e^{i\omega_n t+\phi_n} + c.c.$, $n = 1, 2, 3$, resonant with the nth transition frequency ω_n, with pulse area $\int_{\Delta t_n} 2d_n A_n(t)\,dt = \pi$, where d_n is the dipole moment of the transition $|n\rangle \to |n+1\rangle$ and Δt_n is the length of the nth pulse. The length and shape of the pulses do not influence the outcome of the control process as long as the pulse areas remain fixed. For a realistic system, however, the decay of the excited states by spontaneous emission constantly reduces the populations of the excited states and ultimately leads to the repopulation of the ground state, which is detrimental to achieving population inversion. Moreover, the performance of the control scheme depends significantly on the shape and length of the control pulses employed.

Fig. 2 shows the difference between the populations of states $|1\rangle$ and $|4\rangle$, $\rho_{44} - \rho_{11}$, at the conclusion of the control process as a function of the total control time T_f for various ratios of the pulse lengths, as obtained from a computer simulation of the control process. For very short control times the final yield is around 90% of the theoretical maximum value of one. As the total control time increases, the final yield $\rho_{44} - \rho_{11}$ decreases since dissipative effects accumulate. However, the computer simulations also suggest that the final yield depends considerably on the ratio of the pulse lengths, not just the total control time T_f. For our four-level Rubidium system, the naive choice of equal pulse lengths, i.e., pulse ratios 1 : 1 : 1, consistently results in the worst yield, independent of the total control time T_f. Another obvious choice of the pulse lengths according to the ratio of the lifetimes of the excited states results only in a marginal improvement. However, the final yield and the threshold value of T_f for population inversion increase considerably as the ratio of the length of the first two pulses versus that of the last pulse increases. These results can be partly explained by the very different lifetimes of the excited states of Rubidium. The lifetime of state $|2\rangle$ is less than 1/3 of that of $|3\rangle$ and almost 1/4 of that of $|4\rangle$. Hence, state $|2\rangle$ is the weakest link in the chain and it is imperative to minimize the time the system spends in state $|2\rangle$ in order to reduce dissipative losses. Since the amount of time the system spends in state $|n\rangle$ in the control scheme under consideration is roughly proportional to $(\Delta t_{n-1} + \Delta t_n)/2$, not

Coherence and Quantum Optics VIII
Edited by Bigelow et al., Kluwer Academic/Plenum Publishers, 2003

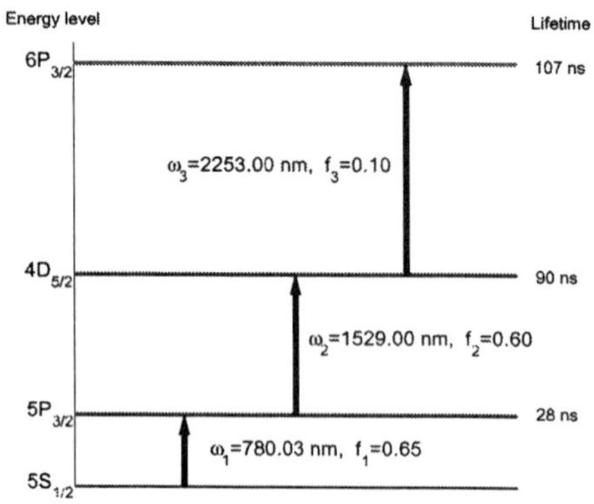

Fig. 1. Energy-level diagram

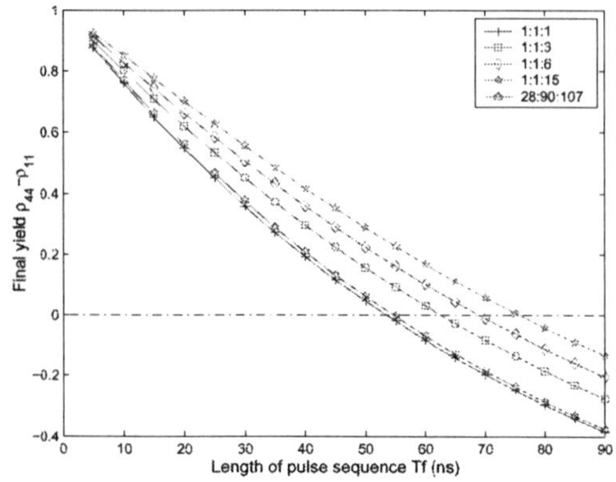

Fig. 2. Final value of $\rho_{44} - \rho_{11}$ as a function of T_f

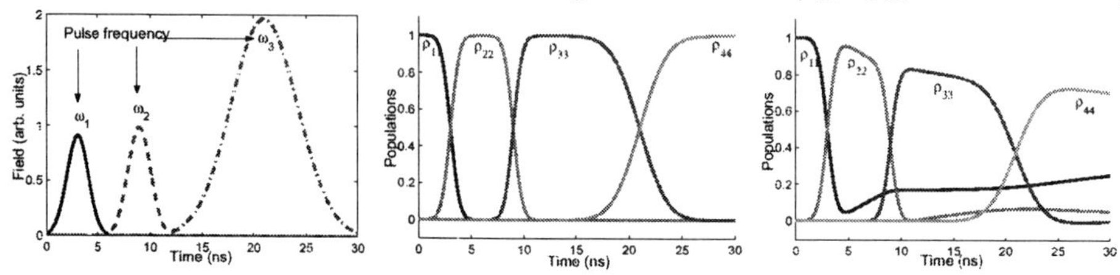

Fig. 3. Control pulse sequence (left) and corresponding evolution of the energy-level populations for an ideal non-dissipative system (center) and a realistic dissipative system (right)

simply Δt_n, we see that it is advantageous to choose the pulse lengths Δt_n, $n = 1, 2, 3$, such that $\Delta t_2 + \Delta t_3 > 3(\Delta t_1 + \Delta t_2)$, which explains why increasing the ratio of the length of the third pulse compared to those of the first two pulses tends to reduce the repopulation of the ground state due to spontaneous emission and thus improves the performance of the control scheme. Fig. 3 shows the (envelopes of the) control pulses and the corresponding evolution of the energy-level populations if dissipative effects due to spontaneous emission are neglected (center) and if realistic lifetimes of the excited states are assumed (right). The total pulse length in the example is 30 ns, where the first and second pulse are both 6 ns long while the third pulse is 18 ns. The cumulative effect of spontaneous emission is clearly noticeable by comparing the graphs.

The example studied in this paper suggests that it is possible to achieve substantial population inversions between the ground state and a highly excited state of an atom (or possibly ion or molecule) by applying a sequence of short control pulses, even if the finite lifetimes of the intermediate excited states are taken into account. In spite of inevitable population losses due to spontaneous emission, pulse lengths and ratios can be adjusted to minimize dissipative effects and achieve substantial yields. We believe such a scheme could be used to calculate the pulse configurations required to build short wavelength pulse lasers in the UV or even X-ray regime by applying a sequence of short optical-frequency laser pulses to a cloud of trapped atoms or ions to achieve population inversion between the ground state and a highly excited state.

References and links

1. V. Ramakrishna *et.al.*, Phys. Rev. A **61**, 032106 (2000)
2. S. G. Schirmer, A. D. Greentree, V. Ramakrishna and H. Rabitz, quant-ph/0105155

Decoherence of a trapped ion: The combined effect of the micromotion and the stray fields on the fidelity of motional states

S. Brouard and J. Plata

Departamento de Física Fundamental II, Universidad de La Laguna, La Laguna E38204, Tenerife, Spain.

sbrouard@ull.es, jmplata@ull.es

Different physical mechanisms can be relevant to the decoherence of an ion in a Paul trap. In particular, stray electric fields present in the trap can play an important role in *heating* the ion, i.e., in driving it out of the initially prepared motional ground state [1, 2]. For this system, the secular or pseudopotential approximation [1], valid only in restricted ranges of parameters, is usually applied. Consequently, the trapped ion is described as an undriven oscillator with an effective frequency, the residual effect of the driving, known as *micromotion*, being neglected. Here, we report on the cooperative action of the micromotion and random classical fields on the heating dynamics. The relevance of this depopulation mechanism to the *stability* of the system can be advanced from purely classical considerations: the effect of the driving in frequency is expected to get stronger as the "equilibrium position" of the ion is changed by the external field. In order to have a more complete view of the problem, we have made a nonperturbative quantum treatment that has allowed us to discuss the heating process on analytical grounds, tracing back its characteristics to both the properties of the deterministic dynamics and the statistics of the fields [3].

The system is modeled as a harmonic oscillator parametrically driven in frequency and interacting with a classical field. To solve the dynamics, we have taken the following steps. First, we have obtained the quasieigenstates in the absence of the classical field through a sequence of unitary transformations, set up from the solutions to the Mathieu's equation that describes the classical dynamics [4]. Second, by means of an additional change to a rotating frame with an oscillating frequency, our model is shown to be formally equivalent to that considered in previous studies to analyze the heating by a random field when the micromotion can be neglected [2]. Third, a methodology based on the use of the displacement operator has been applied to find the time propagator for each realization of the field [2]. Finally, the *fidelity* has been obtained by projecting the time-evolved state into the initial one and performing a statistical average over the field realizations.

Our analytic treatment establishes a connection between the quantum time evolution for a single realization of the field and its classical counterpart and therefore gives the clues to controlling the deterministic dynamics. Additionally, it provides a suitable framework for incorporating the diverse statistical properties that can be assumed for the random fields. In a previous study, a Gaussian statistics with an exponential degree of correlation was considered for the stray fields, and partial revivals of the fidelity of the ground state were predicted for low amplitudes and long coherence times [2]. We have generalized those results. We have found that the simultaneous action of the random field and the parametric driving in frequency leads to a modulation of the fidelity that has a multiplicative effect on the reduction of the heating time. For the mean excitation number, we have detected [3], apart from the damped oscillations with the secular frequency ω_s and the asymptotic linear increase with time, both features found in the absence of micromotion, a complex series of decreasing terms oscillating with $\omega_s + n\omega_{rf}$, ω_{rf} being the frequency of the parametric driving. Hence, we conclude that the presence of quasiharmonic noise centered on one of these sideband frequencies would resonantly enhance the effect of the corresponding contribution.

A simplified picture of the heating processes emerges in the white-noise limit [3]. We have assumed that the initial phase of the parametric driving is uniformly distributed, which corresponds to a different time origin for each experimental run, and have observed in this limit the following features. First, there is a linear increase with time of the effective energy, the heating rate depending on the noise strength and on the characteristics of the deterministic dynamics. Second, the micromotion gives rise to a series of oscillating terms in the mean energy, which, after the average in phase, appear as an additive constant. Finally, our results corroborate the standard assumption that, for classical motion near the periodic regime, the secular approximation accounts for the main characteristics of the heating dynamics.

References and links

1. D. J. Wineland, C. Monroe, W. M. Itano, D. Leibfried, B. E. King and D. M. Meekhof, "Experimental issues in coherent quantum-state manipulation of trapped atomic ions," J. Res. Natl. Inst. Stand. Technol. **103**, 259 (1998).
2. D. F. V. James, Phys. Rev. Lett. "Theory of heating of the quantum ground state of trapped ions," **81**, 317 (1998).
3. L. S. Brown, "Quantum motion in a Paul trap," Phys. Rev. Lett. **66**, 527 (1991).
4. S. Brouard and J. Plata, "Heating of a trapped ion by random fields: The influence of the micromotion," Phys. Rev. A **63**, 043402 (2001).

Coherent control of the excitation routes in four-wave mixing with Rydberg levels

W. C. Magno and S. S. Vianna
Departamento de Física, Universidade Federal de Pernambuco, 50670-901 Recife, PE - Brazil
R. B. Prandini and P. Nussenzveig
Instituto de Física, Universidade de São Paulo, Caixa Postal 66318, 05315-970 São Paulo, SP - Brazil

In this paper, we extend our previous experimental and theoretical results [1] by investigating the polarization dependence of the Four-Wave Mixing (FWM) signal, using rubidium as a nonlinear medium. Interference in this system results from the existence of two or more indistinguishable quantum paths for the excitation of the atomic levels involved in the nonlinear process. Four levels are explicitly considered here: the atomic ground state, two Rydberg states almost resonant with a two-photon transition from the ground state, and an intermediate state resonant with a three-photon transition from the ground state via one of the Rydberg levels. Because theses states have internal structures, there are multiple quantum paths for each step of the nonlinear process. The probability amplitudes for each of these paths depend on Clebsch-Gordan coefficients which are intimately related to the input lasers' polarizations. Hence one can manipulate the contribution of interference terms to the FWM signal by controlling these polarizations.

The experiments are carried out in a vapor - cell containing rubidium vapor. The cell is heated up to a temperature of 180°C and crossed by two copropagating and linearly polarized laser beams. One of the beams, of frequency ω_1, is produced by a pulsed dye laser ($\lambda_1 \approx 602$ nm) corresponding to two-photon transitions from the ground state 5S to 16D and 18S levels. The other beam has a fixed frequency, ω_2, corresponding to the fundamental mode (IR) of a Nd:YAG laser. This laser is quasi-resonant with the transition from the Rydberg levels back to the $6P_{3/2}$ level. The two beams are focused in the middle of the cell in order to have good phase matching. The blue generated signal at frequency $\omega_3 = 2\omega_1 - \omega_2$ ($\lambda_3 \approx 420$nm) is analyzed in a monochromator and detected by a photomultiplier tube.

Typical FWM spectra are presented in Fig.1, as a function of the dye laser detuning, $\Delta_1 = 2\,\omega_1 - \omega_{16D}$, with respect to the two-photon transition 5S→16D. The polarization of the IR laser is linear and kept fixed along the x-axis (we choose the z-axis as the propagation direction). The dye laser polarization is always linear but its orientation can be rotated by an angle θ relative to the IR beam polarization. The generated signals with parallel ($I_{//}$) or perpendicular ($I_{\perp}$) polarization relative to the IR beam polarization are shown, respectively, in Fig.1a and Fig.1b. Analyzing the parallel component of the generated signal we note that the 16D line presents great angular variation. In particular, for $\theta = 55^{\circ}$ the 16D peak disappears completely and for $\theta=0^{\circ}$ a maximum intensity is observed. The $6P_{3/2}$ line also presents dependence with the incident polarizations, while the 18S line is almost constant. For signal polarization perpendicular to the IR polarization (Fig.1b), the 18S line vanishes completely, while the other lines present dependence with the incident polarizations.

The intensity of the 16D line as a function of the θ is shown in Fig. 2 for parallel and perpendicular signal polarizations. We can immediately recognize interference fringes for both polarizations. Fringes result from the fact that there are multiple quantum paths for the intermediate state considered, and the Clebsch-Gordan coefficient for each quantum path depends on the polarization relation between the incident beams. So, the observed behavior of the generated signal with the polarization can be understood as an interference between the different quantum pathways due to the several atomic sublevels m_l and the different beam polarizations. The solid lines in Fig. 2 represent the theoretical results obtained using the density matrix formalism applied to a four-level system. The small discrepancy observed between experiment and theory for parallel signal polarization, although not yet fully understood, may be related to a nonlinear absorption in the medium. The two components of the FWM signal, parallel ($I_{//}$) and perpendicular ($I_{\perp}$) to the IR beam's polarization, are given by [2]:

$$I_{//} \propto \left|\frac{N}{\Delta_2 - i\gamma_{ab}}\right|^2 \left|\frac{1}{9}\cdot\frac{P_0(\theta)\langle 5P|r|18S\rangle\langle 18S|r|6P\rangle}{(\Delta_1-\alpha)-i\gamma_{ad}} + \frac{4}{45}\cdot\frac{P_2(\theta)\langle 5P|r|16D\rangle\langle 16D|r|6P\rangle}{\Delta_1 - i\gamma_{ac}}\right|^2 , \quad (1)$$

$$I_{\perp} \propto \left|\frac{N}{\Delta_2 - i\gamma_{ab}}\right|^2 \left|0 + \frac{2}{15}\cdot\frac{g(\theta)\langle 5P|r|16D\rangle\langle 16D|r|6P\rangle}{\Delta_1 - i\gamma_{ac}}\right|^2 , \quad (2)$$

where the subscripts a, b, c and d represent, respectively, the states 5S, $6P_{3/2}$, 16D and 18S of the rubidium. We have defined the detuning $\Delta_2 = (2\omega_1 - \omega_2) - \omega_{ab}$, and introduced as well the relaxation rates; α is the separation between the 16D and 18S Rydberg levels and the atomic density is given by N. The functions that describe the angular dependence with θ are: $P_0(\theta) = 1$, $g(\theta) = P_1(\theta)[1 - P_1^2(\theta)]^{1/2} = \cos\theta \sin\theta$ and $P_2(\theta) = \frac{1}{2}[3\cos^2\theta - 1]$ (Legendre polynomials).

relaxation rates, α is the separation between the 16D and 18S Rydberg levels and the atomic density is given by N. The functions that describe the angular dependence with θ are: $P_0(\theta) = 1$, $g(\theta) = P_1(\theta)[1 - P_1^2(\theta)]^{1/2} = \cos\theta \sin\theta$ and $P_2(\theta) = \frac{1}{2}[3\cos^2\theta - 1]$ (Legendre polynomials).

Cold collision control in Cavity QED

Ji il Kim, Roberto B. B. Santos, and Paulo Nussenzveig

Instituto de Física, Universidade de São Paulo,
Caixa Postal 66318, CEP 05315-970, São Paulo, SP, Brazil

nussen@if.usp.br

Abstract: We show how the modified electromagnetic vacuum of a resonant optical cavity can be used to manipulate the collisional dynamics between cold atoms by a modification of spontaneous emission times. Multiparticle entanglement and a higher spectral density of modes enhance spontaneous emission and alter the dynamics of cold collisions. A specific situation is considered and we show this effect can be observed as a large suppression in trap-loss rates.

Cold and ultracold atomic collisions [1] are quite different from collisions at higher temperatures. In a cold collision, the atoms move so slowly that the interaction time can be larger than the typical spontaneous emission time. Consequently, the atoms can change their internal states during a collision in the presence of light, and the dynamics of a cold collision can be manipulated with light as seen in optical shielding experiments [2]. Remarkably, it is possible to control the fate of a cold collision even in the complete absence of light by manipulating the spontaneous emission time.

Cavity quantum electrodynamics [3] provides us with very reliable and efficient ways to alter the spontaneous emission time of an excited state by means of a spatially and spectrally modified vacuum. However, the spontaneous emission rate for single atoms or molecules in the optical domain is usually not significantly enhanced owing to the relatively small solid angle encompassed by a centimeter-sized Fabry-Perot cavity. It is, nevertheless, possible to achieve a large enhancement of spontaneous emission when we consider a sample of many identical quasimolecules (weakly bound molecular states [1]) coupled to the same cavity mode. The excited quasimolecules are indistinguishable when interacting with the cavity field. Quantum interference will be important in the process of spontaneous emission into the cavity in a way analogous to the effect of superradiance [4]. Since for most experiments in cold collisions the typical separation between different quasimolecules is greater than an optical wavelength, this interference will only be constructive if the quasimolecules are excited into a multiparticle entangled [5] state by a laser matching the cavity mode.

To explore the effect of Cavity QED on cold collisions, we refer to a particular type of excited state collision between neutral alkali atoms, the so-called radiative escape collision. A more thorough discussion can be found in [6]. A quasimolecule is formed when one atom of a colliding pair is excited by a laser of frequency ω_L at an internuclear separation R_C, and it dissociates emitting a photon with energy $\hbar\omega_\gamma$. If the emission takes place with an internuclear separation smaller than R_e, the dissociated pair acquires enough kinetic energy to escape from the trap. Reducing the spontaneous emission time increases the probability for emission to occur at a separation $R > R_e$. The consequent trap-loss rate reduction is a signature of a Cavity QED effect on cold collisions.

Consider a sample of cold ^{85}Rb atoms trapped in a far-off resonance trap (FORT) [7] or in a magnetic trap [8] near the center of a quasi-confocal optical resonator. The number N of quasimolecules depends on the detuning $\delta = \omega_L - \omega_A$ ($\hbar\omega_A$ is the atomic transition energy), the number N_A of atoms in the sample and their density n_A. Initially, all quasimolecules are in the ground state $|g\rangle$ and there is just one photon in the cavity mode. This state couples to the collective excited state $|E,0\rangle = \sum_i \frac{\Omega_i}{\Omega}|i,0\rangle$ where

$|1,0\rangle \equiv |eg\cdots g\rangle|0\rangle$, $|2,0\rangle \equiv |ge\cdots g\rangle|0\rangle$ and so forth, and Ω_i and $\tilde{\Omega}$ are the single and the collective quasimolecule Rabi frequency [6]. The cavity mode has a finite width γ_c, and it can be arranged to resonate with the transition $|e\rangle \to |g\rangle$ for any quasimolecule whose internuclear separation lies in the range $R_e < R < R_C$ and not to resonate for $R < R_e$. The spontaneous decay rate Γ_c of $|E,0\rangle$ on resonance with the cavity mode depends on the distribution and orientation of the quasimolecules along the cavity mode and its ratio to its free-space counterpart Γ is given by

$$\frac{\Gamma_c}{\Gamma} \approx \left(1 + \frac{3}{2}\frac{\Delta\Omega_{\rm eff}}{4\pi} N\,\Lambda_{00}\frac{\omega_L^3}{\omega_A^3}\right), \tag{1}$$

where $\Delta\Omega_{\rm eff}$ is an effective solid angle accounting for emission into several cavity modes, and Λ_{00} is the line-shape function for the TEM_{00} mode.

In the state $|E,0\rangle$, the region $R_e < R < R_C$ (where the pairs are subject to a cavity enhanced decay rate) is covered in a time interval t_c, and the excited pairs can survive to cover the cavity-free region $0 < R < R_e$ in a time t_e. The atoms of each quasimolecule can execute multiple passages through the point R_e, oscillating back and forth until dissociation. The probability $\mathcal{L}_c$ that the trap will loose a pair of atoms is

$$\mathcal{L}_c = \frac{\sinh{(t_0 - t_c)\Gamma}}{\sinh{[t_0 + (\Gamma_c/\Gamma - 1)t_c]\Gamma}} \equiv p\,\mathcal{L}_0\,, \tag{2}$$

where $t_0 = t_e + t_c$ is the total time interval and $\mathcal{L}_0$ is the cavity-free loss-probability. Comparing the losses with and without cavity is possible provided that the quasimolecules are excited either via a cavity mode or by a laser beam propagating orthogonally to the cavity axis. In order to measure $p = \mathcal{L}_c/\mathcal{L}_0$, it would be necessary to adjust the laser intensities so as to have the same fraction of excited state quasimolecules in both situations; in this case the excitation probability factors out.

To a typical detuning $\delta = -2\pi \times 100$ MHz corresponds the total time $t_0 = 3.0 \times 10^{-8}$ s. Assuming $N_A \sim 10^6$ atoms at a density $n_A \sim 10^{12}$ cm^{-3}, we have $N \simeq 45$ quasimolecules involved in a multiparticle entangled state.

We consider a (quasi-)confocal cavity with mirrors of diameter 1.0 cm and reflectivities 0.97, separated by 2.9 cm. The excitation field, with circular polarization, matches a TEM_{00} mode. The cavity spectral width is $\gamma_c/2\pi \approx 200$ MHz. A numerical simulation gives an effective solid angle $\Delta\Omega_{\rm eff}/4\pi \approx 7.4 \times 10^{-4}$ and $\Gamma_c \approx 4.3\Gamma$, leading to a impressive reduction $p = \mathcal{L}_c/\mathcal{L}_0 \sim 3 \times 10^{-3}$ in the trap-loss probability with a 5 mK trap depth. These numbers are certainly within the reach of current experimental possibilities.

acknowledgement

This work was supported by FAPESP and CNPq (brazilian agencies).

References and links

1. See, e.g., J. Weiner *et al.*, Rev. Mod. Phys. **71**, 1 (1999).
2. L. Marcassa *et al.*, Phys. Rev. Lett. **73**, 1911 (1994); S. Bali *et al.*, Europhys. Lett. **27** 273 (1994); V. Sanchez-Villicana *et al.*, Phys. Rev. Lett. **74**, 4619 (1995).
3. S. Haroche, in *Les Houches, Session XXXVIII, 1982, Vol. I*, ed. by G. Grynberg and R. Stora (North-Holland, Amsterdam 1984), pp. 193–309
4. R.H. Dicke, Phys. Rev. **93**, 99 (1954); R. Bonifacio, P. Schwendimann, and F. Haake, Phys. Rev. A **4**, 302 (1971); *ibid.* **4**, 854 (1971); M. Gross and S. Haroche, Phys. Rep. **93**, 301 (1982).
5. C.A. Sackett *et al.*, Nature (London) **404**, 256 (2000); D. Bouwmeester *et al.*, Phys. Rev. Lett. **82**, 1345 (1999); A. Rauschenbeutel *et al.*, Science **288**, 2024 (2000); and references therein.
6. J. I. Kim, R. B. B. Santos, and P. Nussenzveig, Phys. Rev. Lett. **86**, 1474 (2001).
7. S. Chu *et al.*, Phys. Rev. Lett. **57**, 314 (1986).
8. V. Bagnato *et al.*, Phys. Rev. Lett. **58**, 2194 (1987).

Wave packet control in the Fourier time basis

Ashok Muthukrishnan and C. R. Stroud, Jr.

The Institute of Optics, University of Rochester, Rochester, NY 14627

amuthuk@optics.rochester.edu

Abstract: We introduce a discrete Fourier basis in time for Rydberg atomic wave packets in analogy to modelocked laser pulses in a cavity. The wave packet basis is related to the basis of excited energy levels by a quantum Fourier transform, and serves as a natural basis for describing multilevel quantum control with pulsed fields. We apply this paradigm to the control of radial wave packets using optical pulses shorter than a Kepler period. Recent results in coherent wave packet control using optical and terahertz pulses are elucidated in this context.

A coherent laser field in a cavity occupying several frequency modes can be represented in the time domain as a series of modelocked laser pulses [1]. An analogous situation occurs in the time evolution of an atomic wave packet, where the phases of the energy levels evolve in time in coherent superposition. By appropriately discretizing the time evolution of the wave packet, we can replace the basis of excited energy levels by a Fourier basis in time consisting of wave packets in cyclic motion about the nucleus.

We consider a uniform superposition of d Rydberg states $|n\rangle_\nu = |\bar{n}+n, 1, 0\rangle$, where $-d/2+1 \leq n \leq d/2$. Introducing a Fourier phase kernel in the superposition, we define the Fourier basis for these energy levels as

$$|m\rangle_\tau = \frac{1}{\sqrt{d}} \sum_n \exp\left(i2\pi nm/d\right) |n\rangle_\nu, \tag{1}$$

where $-d/2+1 \leq m \leq d/2$. For large quantum numbers $\bar{n}$, the time evolution of the energy level phases is nearly harmonic for short times, $\exp(-i\omega_n t) \approx \exp(i2\pi nt/T_K)$, where $T_K = 2\pi\bar{n}^3$ is the mean Kepler period. Thus the wave packet states $|m\rangle_\tau$ correspond to d discrete times $t_m = mT_K/d$ in one Kepler orbit. Each wave packet state is spatially localized in the radial dimension and oscillates between the inner and outer classical turning points [2].

The wave packet basis is a natural basis for describing the control of the Rydberg wave function using pulsed fields [3]. In particular, the $m=0$ radial wave packet state is localized near the inner turning point of the classical orbit and marks the region of strongest interaction of the atom with an applied field (see Fig. 1(a)). The Rydberg wave function can be written as

$$|\psi(t)\rangle = \sum_n c_n(t)|n\rangle_\nu = \sum_m \tilde{c}_m(t)|m\rangle_\tau, \tag{2}$$

where the wave packet amplitudes $\tilde{c}_m$ and the energy level amplitudes $\tilde{c}_n$ are related by a discrete Fourier transform. Coherent control of the wave packet amplitudes can be achieved using a series of short optical pulses, as outlined in Fig. 1(b).

In the weak field excitation regime, it is well known that the amplitude distribution of Rydberg states excited is directly proportional to the pulse spectrum [2]. Analytical extensions of this control problem to the strong-field regime have been carried out [4] relating the pulse envelope to the scaled Fourier amplitudes in the Rydberg basis.

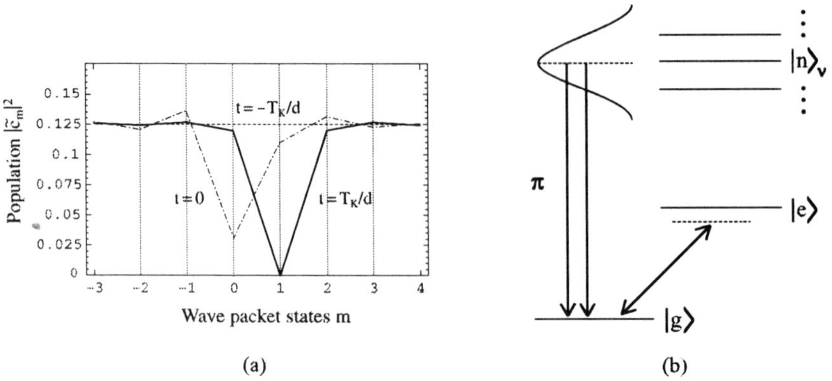

Fig. 1. (a) Pulsed inversion of the innermost wave packet state at $\bar{n} = 180$ and $d = 8$. Wave packet populations $|\tilde{c}_m|^2$ are shown before ($t = -T_K/d$), during ($t = 0$) and after ($t = T_K/d$) the pulse. Spectral width of the pulse is approximately $1.3d/T_K$. (b) Wave packet control in the Fourier basis. Broadband π pulses invert two wave packet states in the basis to low lying energy levels g and e, where their amplitudes are controlled near resonance.

These results show that in atomic wave packet control problems, the Fourier time basis in Eq. (1) is a natural basis for describing the atom-field interaction.

Recently, the strong interaction of the radial wave packet with a terahertz (THz) half-cycle field near the atomic core has been studied as a means for quantum information processing [5]. Phase information is stored in the delocalized part of the Rydberg wave function ($m \neq 0$), which is amplified by the THz pulse at the expense of the localized wave packet near the core ($m = 0$).

Precise control of individual wave packet amplitudes offers a useful paradigm for multilevel quantum computing [3]. Since the wave packet basis is defined in terms of a quantum Fourier transform, switching between the energy level and wave packet bases provides a natural and convenient method for taking Fourier transforms in the atom, a key element of many quantum algorithms [6].

Acknowledgements

This work was supported by the Army Research Office through the MURI Center for Quantum Information.

References and links

1. P. W. Milonni and J. H. Eberly, *Lasers*, (John Wiley & Sons, New York, 1988), Chap. 12.
2. Jonathan Parker and C. R. Stroud, Jr., "Coherence and decay of Rydberg wave packets," Phys. Rev. Lett. **56**, 716-9 (1986).
3. Ashok Muthukrishnan and C. R. Stroud, Jr., "Atomic wave packet basis for quantum information," (submitted for publication).
4. Luís E. E. de Araujo, Ian A. Walmsley and C. R. Stroud, Jr., "Analytic solution for strong-field quantum control of atomic wave packets," Phys. Rev. Lett. **81**, 955-8 (1998).
5. J. Ahn, D. N. Hutchinson, C. Rangan and P. H. Bucksbaum, "Quantum phase retrieval of a Rydberg wave packet using a half-cycle pulse," Phys. Rev. Lett. **86**, 1179-82 (2001).
6. R. Josza, "Quantum algorithms and the Fourier transform," Proc. R. Soc. Lond. A **454**, 323-337 (1998).

Vacuum noise in presence of atoms

Alexander S. Shumovsky
Physics Department, Bilkent University, Bilkent, Ankara, 06533, Turkey
Phone: (90 312) 290 1975, FAX: (90 312) 266 4579, e-mail: shumo@fen.bilkent.edu.tr

Abstract: It is shown that the vacuum fluctuations of electromagnetic field are concentrated near atoms. This effect makes worse the quantum limit of precision of measurements in atomic systems in the intermediate zone.
OCIS codes: (270.25000) Fluctuations, relaxations, and noise; (270.5580) Quantum electrodynamics

It is well known that the quantum nature of electromagnetic radiation manifests itself in the zero-point oscillations (ZPO) of the field strengths. In particular, ZPO are responsible for a number of effects such as the spontaneous emission, Lamb shift, Casimir-Polder force, quantum beats etc. As a source of quantum noise, ZPO define the quantum limit of precision of measurements.

In the usual treatment, ZPO are calculated as though the field is represented by the plane waves of photons. The vacuum noise, obtained in this way, is distributed uniformly over the space. This simplified picture overlooks the fact that the atomic transitions emit the multipole radiation represented by the spherical waves of photons [1]. The purpose of this note is to show that the vacuum noise in presence of atoms is concentrated in some vicinity of atoms and exceeds the level predicted by the model of plane waves of photons.

Let us note that the plane and multipole photons are specified by different number of quantum degrees of freedom. In the former case, the state of the field is specified by the energy (wave number) and two polarizations. In the latter case, the photon state is described by the energy, parity (type of radiation either electric or magnetic), angular momentum j=1,2,..., and projection m=-j,...,j of the angular momentum. Thus, even in the simplest case of a monochromatic dipole radiation (j=1) of a given type, the multipole photons are specified by three degrees of freedom instead of the two polarizations of plane photons. The increase of the number of degrees of freedom leads to an increase of ZPO. In fact, it is easily seen that the

$$H^0_{(plane)} = \sum_k \hbar\,\omega_k, \quad H^0_{(multi)} = \sum_k \sum_{j=0}^{\infty} (2j+1)\,\hbar\,\omega_k.$$

energies of ZPO of the plane and multipole photons are connected as follows
At first sight, the two expressions are equivalent because both give the infinite energy of the vacuum state. In fact, this infinity is inessential because of the following reason. The contribution of ZPO can be recognized only through a measurement that implies an averaging of physical quantities over a finite exposition time of detector. In other words, any real measurement involves a filtration, leading to a separation of finite transmission frequency band (TFB). For a given j, the ratio of the two vacuum energies at the same TFE is

$$H^0_{(multi)} / H^0_{(plane)} = (2j+1)/2.$$

Thus, the level of ZPO of the multipole field always exceeds that of plane waves.

The above relations correspond to the vacuum state energy in the whole volume of quantization. Much more interesting and important result can be obtained from the consideration of spatial properties of the field. It is well known that the mode function of the plane waves of photons is characterized by the exponential position dependence exp(ik•r). Therefore, the vacuum energy density of the plane waves of photons is independent of **r**. In turn, the mode function V(**r**) of the multipole radiation is represented by a combination of spherical Bessel functions, describing the radial dependence, and spherical harmonics. The vacuum energy density then has the form

$$H^{0}_{(multi)}(r) = \sum_{k} \sum_{\lambda jm\mu} \left| V_{k\lambda jm\mu}(r) \right|^2 ,$$

where λ denotes the parity and μ is the index of the so-called helicity basis centered at the atom location. The explicit form of V can be found in [2,3].

A more detailed analysis [4] shows that the vacuum energy density of multipole radiation does not depend on the spherical angles and decreases with the distance from the atom asymptotically attaining the level predicted by the model of plane waves. At short distances $r \leq \Lambda/2$ (where Λ is the wavelength), the level of ZPO of multipole field strongly exceeds that of the plane waves. To stress the importance of obtained result, we now note that, in a number of modern experiments on engineered entanglement in the system of trapped Ridberg atoms, the interatomic distances are of the same order or even less. Since, in this case, we have more than one atom, consider as an illustrative example the system of two identical atoms separated by distance d. Assume that one of the atoms (source) is initially in the excited state, while the other (detector) is in the ground state. The emission of a photon by the source atom and successive absorption by the detecting atom can be interpreted as a Hertz-type measurement. Definitely, this experiment should be described in terms of spherical waves of photons as a superposition of outgoing and incoming waves focused on the source and detector respectively. It then follows from the above result that both atoms "condense" ZPO around. If $d \gg \Lambda$, the detection process is influenced mainly by the vacuum noise due to the detecting atom. In the opposite case of the short distances $d \leq \Lambda/2$, there is an overlap of ZPO concentrated near the source and detector, which makes worse the quantum limit of precision of measurement.

The above results can also be applied to the calculation of Casimir-Polder forces in the atomic systems and for atoms near a surface.

References

[1]. C. Cohen-Tannouji, J. Dupont-Roc, and G. Grinberg, "Atom-photon interaction" (Wiley, New York, 1992).

[2]. W. Heitler, "The quantum theory of radiation" (Oxford University Press, New York, 1954).

[3]. A.S. Davydov, "Quantum Mechanics" (Pergamon, Oxford, 1976).

[4]. A.S. Shumovsky, "The quantum multipole radiation," in *Contemporary optics and electrodynamics*, M. Evens, ed. (Wiley, New York, 2001).

Quantum interference effect in molecular system

Jin Wang, H.M. Wiseman, Z. Ficek and G. J. Milburn

Centre for Laser Science, Department of Physics, The University of Queensland, Brisbane, Queensland 4072, Australia

wang@physics.uq.edu.au

There have been a large number of theoretical studies on the effects of quantum interference in atomic and molecular system. Quantum interference can lead to many interesting effects such as amplification without population inversion, electromagnetically-induced transparency, phase dependent spectra and population inversions, and ultranarrow spectral lines.

In 1996, Xia *et al.* [1] carried out the first experimental investigation of constructive and destructive interference effects in spontaneous emission. In the experiment they used sodium dimers, which can be modeled as five-level molecular systems with a single ground level, two intermediate and two upper levels, driven by a two-photon process from the ground level to the upper doublet. By monitoring the fluorescence from the upper levels they observed that the total fluorescent intensity, as a function of two-photon detuning, is composed of two peaks on transitions with parallel and three peaks on transitions with anti-parallel dipole moments. The observed variation of the number of peaks with the mutual polarization of the dipole moments gives compelling evidence for quantum interference in spontaneous emission.

Here we present a theory using master equation technique to explain the observed variation of the number of peaks with the mutual polarization of the molecular transition dipole moments.

The five-level molecular system we are considering is shown in Fig.1. In the experiment, a destructive interference was observed between two transitions, $|a_1\rangle \rightarrow |b\rangle$ and $|a_2\rangle \rightarrow |b\rangle$, with parallel dipole moments, and a constructive interference was observed between transitions $|a_1\rangle \rightarrow |d\rangle$ and $|a_2\rangle \rightarrow |d\rangle$ with anti-parallel dipole moments.

First, we show in Fig.2 that in the experiment, if the dye laser was only coupled to the two-photon transition $|c\rangle \rightarrow |a_1\rangle, |a_2\rangle$, quantum interference can be observed but the fluorescent intensity exhibits two peaks (as a function of laser detuning) regardless of the mutual orientation of the transition dipole moments.

Second, it is stated in the experimental paper [1] that the two-photon transition in sodium dimers was enhanced by a near-resonant intermediate level, indicating that the laser could also couple the ground state $|c\rangle$ to the upper states $|a_1\rangle, |a_2\rangle$ via cascaded one-photon transitions. If we include this one-photon coupling into consideration, we can observe three-peak structure in the UV region with anti-parallel dipole moments ($p_u = -1$) and two-peak structure in the visible region with parallel dipole moments ($p_v = 1$) as illustrated in Fig.2. The strong dependence of the amplitude of the central peak on the mutual orientation of the molecular dipole moments is precisely the quantum interference effect.

References and links

1. H.R. Xia, C.Y. Ye and S.-Y. Zhu, Phys. Rev. Lett.**77**,1032 (1996).

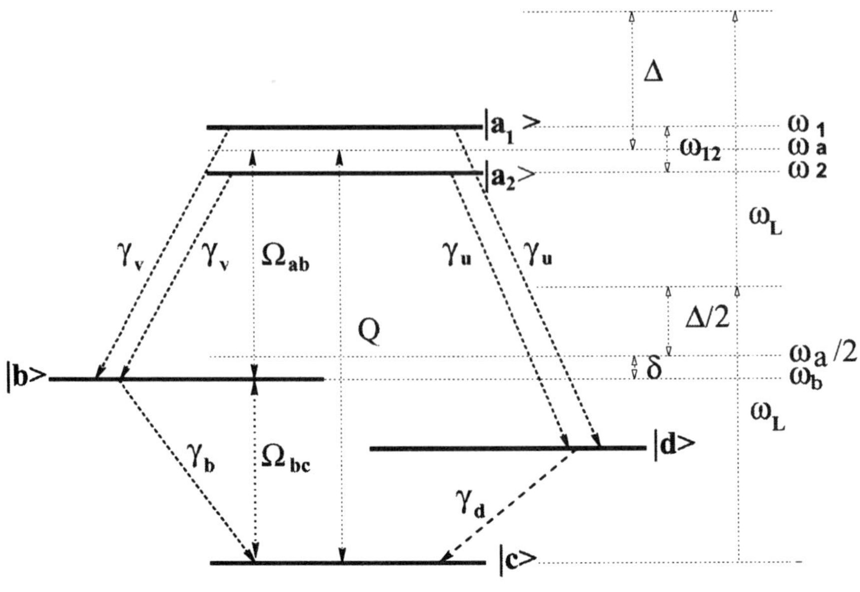

Fig. 1. Energy level structure and couplings of the molecular system.

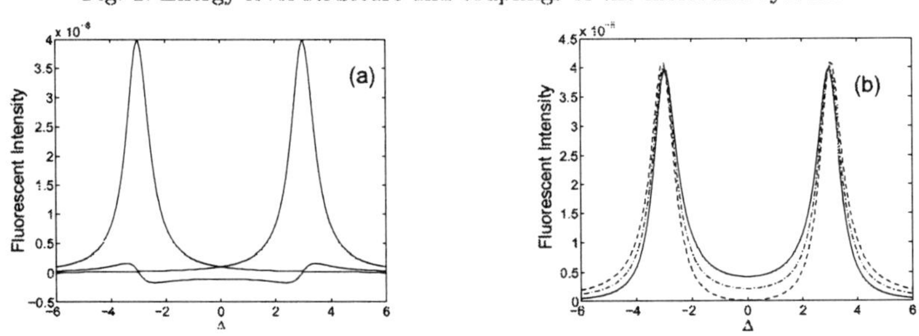

Fig. 2. Analytical results for 2-photon coupling only. (a) shows populations (ρ_{11} peaked to the left, ρ_{22} peaked to the right)and coherence ($\Re\rho_{12}$, below the axis) of the excited states. (b) shows the total fluorescent intensity in units of the spontaneous emission rate.The solid line shows the intensity on the ultraviolet transition ($p_u = -1$), the dashed line the intensity on the visible transition ($p_v = 1$), and the dash-dot line the hypothetical intensity for a transition with orthogonal dipole moments ($p = 0$).

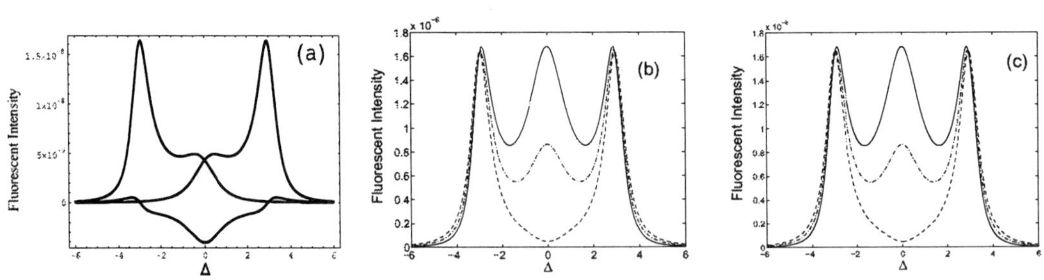

Fig. 3. Weak field assumption results for one- and two-photon driving. (a) shows analytical populations and coherences of the the excited states, as in Fig.1(a). (b) shows the analytical results of fluorescence intensity. (c) we plot the numerical fluorescent intensity for a strong driving field. It is seen that the fluorescent intensity exhibits the same behavior as that for the weak driving field, shown in Fig.2(a) and (b), despite the fact that the solutions have been derived in different regimes. The real experiment is carried out under strong field driving.

Lighting a dark state by shaping of light-induced potentials

Vladimir S. Malinovsky and Jeffrey L. Krause

University of Florida, Quantum Theory Project, Gainesville, FL 32611-8435

malinovs@qtp.ufl.edu

Abstract: We show that careful shaping of light-induced potentials offers a general method for selective population transfer, even when the desired final state is a dark state. Results are presented for a three-state model of the sodium dimer in which the objective is to populate the outer well of the $^1\Sigma_u(4s)$ double-well potential.

The method used in this work involves population transfer via precise shaping of light-induced potentials (LIPs) [1]. One version of this method is known as APLIP (adiabatic passage on light-induced potentials) [2]. In general, the APLIP procedure uses a pair of intense laser pulses to create and then adiabatically deform a set of light-induced potentials. By carefully controlling the detuning, intensity, duration and timing of the pulses, the temporal and spatial adiabaticity of the process leads to selective, efficient population transfer on a single LIP.

As an example of the type of problem we envision, consider the schematic in Fig. 1. The $^1\Sigma_u(4s)$ state of the sodium dimer, Na_2, has a double-well structure with the outer well displaced by a large distance from the Franck-Condon region of the ground state. The Franck-Condon factor for the $v = 0 \rightarrow v'' = 0$ transition is zero. For this reason, any population created in low-lying vibrational levels of the outer well would be expected to have an extremely long lifetime. Clearly, populating the vibrational states of the outer well cannot be accomplished with weak-response techniques. Nor can it be accomplished with any single shaped pulse, of any intensity.

In this work we consider, for convenience, simple pulses and simple pulse sequences. We use a pair of Gaussian pulses, and vary the Rabi frequencies, detunings and delay times between the pulses. The first pulse, conventionally called the pump pulse, connects the initial state $^1\Sigma_u(3p)$ with the intermediate state $^1\Pi_g(3p)$. The second pulse, called the Stokes pulse, connects the intermediate state with the final state, $^1\Sigma_u(4s)$. The pulses can be applied in either an intuitive ordering (I), in which the pump pulse arrives before the Stokes pulse, or a counterintuitive ordering (C), in which the pump pulse

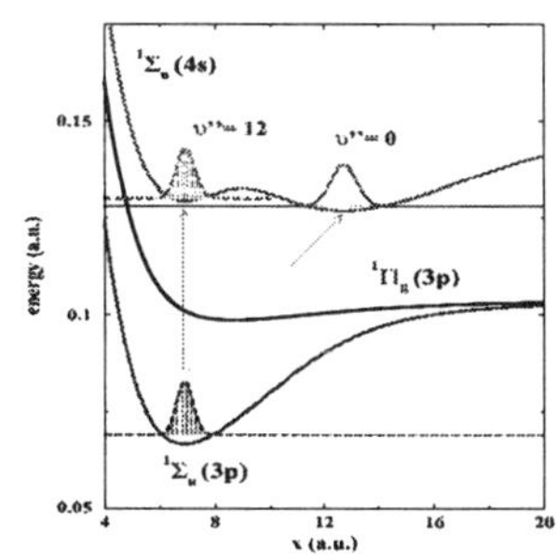

Fig. 1. Schematic for population transfer in Na_2. The initial state is a single vibrational eigenstate on the $^1\Sigma_u(3p)$ state, and the objective is to create population in the outer well of the $^1\Sigma_u(4s)$ state. The potential surfaces for Na_2 are adapted from Ref. [3].

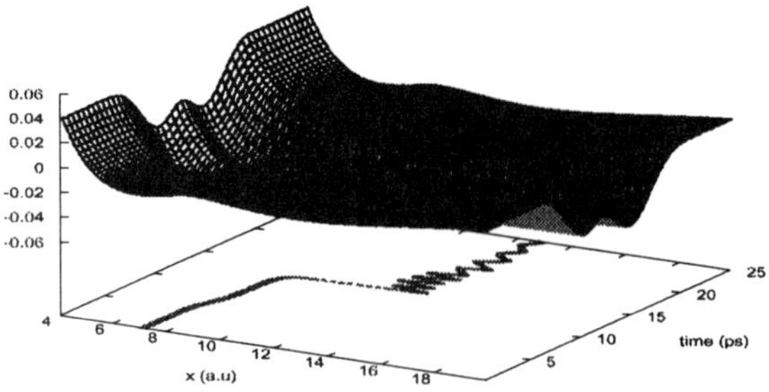

Fig. 2. Light-induced potential for the (I,+) case.

arrives after the Stokes pulse. The detuning can be either positive (+) or negative (−), which is defined with respect to first transition.

Figure 2 shows the LIP and a trajectory $\langle x(t) \rangle$ of the centroid of the wave packet for the (I,+) combination. The initial state is the $v = 6$ eigenstate of the $^1\Sigma_u(3p)$. As can be seen by the oscillations of the centroid at later times, the population transfer is not selective. Rather than producing a pure $v'' = 6$ eigenstate as the product, a wave packet is created consisting of several eigenstates near $v'' = 6$. Nevertheless, these states are all well below the barrier to the inner well, and have vanishing Franck-Condon factors with the ground state. As a result, this wave packet should be very long-lived, and detectable experimentally. Once again, the dynamics can be understood by analyzing the time-dependent LIPs. The initial state has sufficient energy to flow over the barrier to the outer well, which lowers at just the right time to allow the wave packet to cross, and then raises as the pulses turn off, trapping the population in the outer well.

We demonstrate in this work that carefully chosen pairs of intense laser pulses can be used to populate a state with zero Franck-Condon overlap with the initial state. That is, we can light a dark state and create matter in a state that cannot be created via conventional, perturbative means. The topology of this system is such that once the dark state (composed of vibrational levels in the outer well of the double-well potential) is populated, it should have a very long lifetime, since it has no direct radiative path to the ground state.

The mechanism for the LIP-shaping invoked here relies on the ability to use laser pulses to selectively deform adiabatic potentials. This is, we suspect, a rather general procedure. It augments the paradigm of weak response spectroscopy, in which Nature decides which transitions are possible, and which are not. Interestingly, in the case considered here, the double well topology breaks the symmetry of the (C,±), (I,∓) cases observed previously, in which both orderings of the pulses can be used to produce the desired goal. In the Na_2 double-well potential, one pulse order populates the inner well, and the other populates the outer well. These observations are a hint that the richness of this method has not yet been fully explored.

References and links

1. I. R. Solá, B. Y. Chang, J. Santamaría, V. S. Malinovsky, and J. L. Krause, Phys. Rev. Lett. **85**, 4241 (2000).
2. B. M. Garraway and K. A. Suominen, Phys. Rev. Lett. **80**, 932 (1998).
3. I. Schmidt, *Ph.D thesis*, Kaiserslautern University, Germany, 1987.

Dynamical Creation of Entanglement by Feedback

Jin Wang, H.M. Wiseman, and G. J. Milburn

Centre for Quantum Computer Technology , The University of Queensland, Brisbane, Queensland 4072, Australia

wang@physics.uq.edu.au

One of the increasing interests in quantum information theory is to find ways to generate as much entanglement as possible for a given state. In this proposal we are trying to answer a question: For a two-qubit system coupled to a single-mode cavity which is driven and heavily damped, what is the maximum amount of entanglement that can be generated by using homodyne-mediated feedback control [1]? The feedback experimental apparatus is shown in Fig.1.

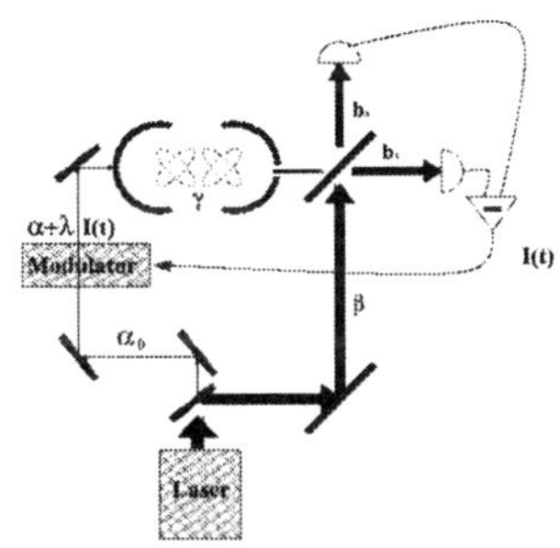

Fig. 1. Diagram of the experimental apparatus. The laser beam is split to produce both the local oscillator β and the field α_0 which is modulated using the homodyne current $I(t)$ derived from the damped cavity.

By performing adiabatic elimination of the cavity mode, we get the feedback-modified master equation.

$$d\rho_c = dt\gamma\mathcal{D}[J^-]\rho_c - idt[H_\alpha,\rho_c] - idt[F, -iJ^-\rho_c + i\rho_c J^+] + dt\frac{1}{\gamma}\mathcal{D}[F]\rho_c. \quad (1)$$

where the feedback is described by the Hamiltonian $H_{fb} = FI(t) = \lambda J_x I(t)$. Here $I(t)$ is the homodyne current.

We choose the so-called concurrence as a measure for entanglement and $(Tr\left[\rho^2\right] - \frac{1}{3})\frac{3}{2}$ for the degree of purity . As shown in Fig.2, after including feedback onto the amplitude of the driving on the atom, we see that the steady state concurrence has been encreased from 0.11 to 0.31.

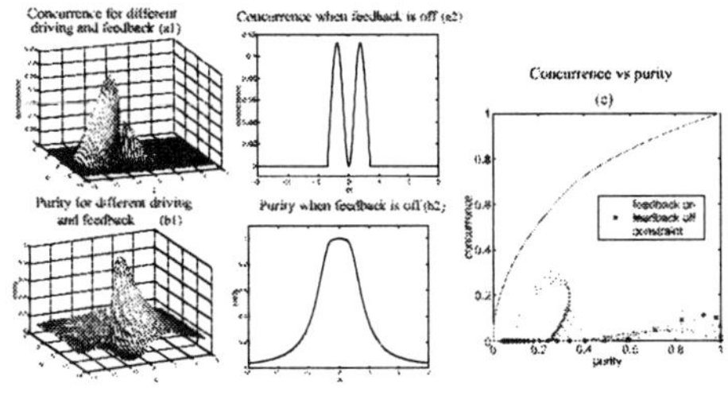

Fig. 2. Plot of concurrence and purity when feedback is on and off.

Coherence and Quantum Optics VIII
Edited by Bigelow et al., Kluwer Academic/Plenum Publishers, 2003

To gain further knowledge about the position of the steady state when it is maximally entangled, we look at the modulous of the matrix elements and Q function as shown in Fig.3. Q function is defined as

$$\begin{aligned} Q(\theta,\phi) &= \langle j,\theta,\phi|\rho|j,\theta,\phi\rangle \\ |j,\theta,\phi\rangle &= \sqrt{\frac{(2j)!}{(j+m)!(j-m)!}}\sum_{m=-j}^{j}\left[\cos^{j+m}(\frac{\theta}{2})\sin^{j-m}(\frac{\theta}{2})e^{-im\phi}\right]|j,m\rangle \end{aligned} \quad (2)$$

There are a couple of points need explanation. First, when there is only driving laser shining on the system, the maximally entangled steady-state is mainly confined to the bottom half of the Q function box and the upper state is almost unpoupulated; However, when the driving amplitude is modified by the feedback, the upper state becomes very well populated and the entanglement is greatly encreased. Second, the off-diagonal elements of the density matrix represent coherence and the presence of coherence is a necessary condition for the creation of entanglement. To illustrate this, we plot three bell states which are the maximally entangled state, in which the off-diagonal terms are presented. In contrast, we also plot three non-entangled states, they are identical, ground and excited state, where the none of the off-diagonal elements appears. We also provide a semi-classical explaination to the increased entanglement.

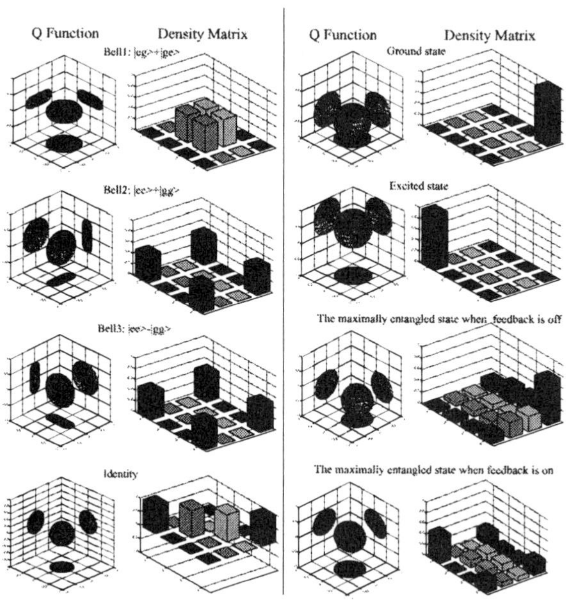

Fig. 3. Fig.3 are the plots of Q function and density matrix for eight different states as indicated in the figure. The Q function is shown as three projectives on three faces. The density matrix is shown by plotting the absolute value of density matrix elements. The diagonal elements of the density matrix plotting are shown as $|ee\rangle\langle gg|$, $|eg\rangle\langle ge|$, $|ge\rangle\langle eg|$, $|gg\rangle\langle gg|$ from left to right, respectively.

References and links

1. H. M. Wiseman and G. J. Milburn, Phys. Rev. A **47**, 1652 (1993).

State-selective phase control of molecular wave packets in two electronic states

Hans U. Stauffer[a,1], Joshua B. Ballard[a], Zohar Amitay[a], and Stephen R. Leone[a,2]
[a]*JILA, National Institute of Standards and Technology and University of Colorado, Department of Chemistry and Biochemistry, and Department of Physics, University of Colorado, Boulder, CO 80309-0440, USA*
[1]*National Research Council Postdoctoral Research Fellow, National Institute of Standards and Technology*
[2]*Staff member, Quantum Physics Division, National Institute of Standards and Technology*

Abstract: State-selective phase control over rotational wave packets in two Li_2 electronic states using tailored femtosecond pulses is demonstrated. A shaped preparation pulse simultaneously creates wave packets in the *A* and *E* electronic states, which are probed using a second time-delayed pulse. This allows independent phase control of selected rotational wave packet components in both electronic states.

The manipulation of molecular dynamics simultaneously involving multiple electronic surfaces using ultrafast laser pulses has provided a natural extension beyond coherent control on a single electronic surface. Recent experiments have taken advantage of various control parameters, such as optical pulse amplitude and phase manipulation, as well as the relative timing of pulses to control the transfer of population to vibrational wave packets on two different electronic surfaces [1,2]. An avenue of molecular coherent control that has not to date been exploited, however, is the *state-selective* manipulation of the phases of rovibrational states composing wave packets in multiple electronic states. This state-selectivity is also important in the context of quantum computation applications. Prior work in our group has demonstrated phase control of two-state rotational and ten-state rovibrational wave packets prepared in the $E(^1\Sigma_g^+)$ electronic state of lithium dimer using phase-tailored femtosecond pulses [3,4]. Here, we extend this state-selective phase control to Li_2 superpositions prepared, using a single ultrafast pulse, in two electronic states. Wave packets are prepared in the *E* state, via a one-photon absorption process, and in the $A(^1\Sigma_u^+)$ state, via a resonant impulsive stimulated Raman scattering process [5] involving purely rotational Raman transitions.

The experimental setup has been described in detail previously [3,4]. A cw laser is used to populate the $A(v = 15, J = 24)$ rovibrational state of Li_2 in a static cell held at 1023 K. From this launch state, a wave packet consisting of molecular eigenstates in both the *A* and *E* electronic states is prepared using a single linearly polarized phase- and amplitude-tailored ultrafast pulse ("preparation" pulse). A time-delayed, unshaped ultrafast probe pulse, with polarization parallel to that of the preparation pulse, allows detection of the coherent rotational dynamics in both the *E* and *A* states by, respectively, one and two photon ionization. Both pulses are centered at 12,435 cm^{-1} (FWHM ~125 cm^{-1}). Phase and amplitude manipulation of different spectral components of the preparation pulse is carried out using a 128 pixel liquid crystal spatial light modulator [6]. Rotational selection rules for one ($\Delta J = \pm 1$) and two photon ($\Delta J = 0, \pm 2$) transitions dictate that the "pump/dump" process couples the $E(v = 18, J = 23, 25)$ and $A(v = 15, J = 22, 24$, and $26)$ eigenstates. This is shown schematically in Fig. 1 with the relevant "pump" and "dump" resonant transitions. The phases, ϕ_{P*} and ϕ_{D*}, of the electric field at two of these resonant frequencies are manipulated

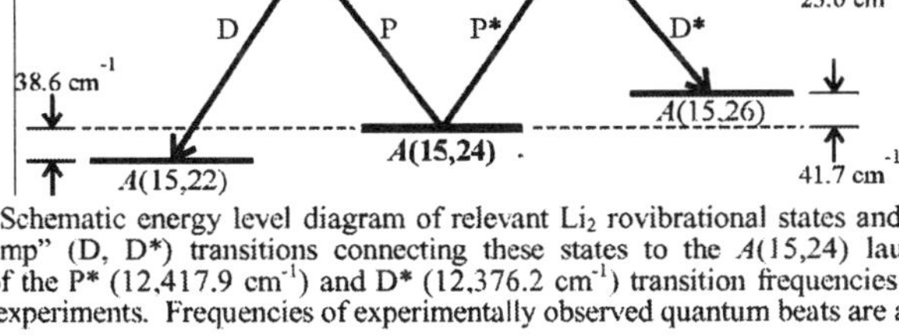

Fig. 1. Schematic energy level diagram of relevant Li_2 rovibrational states and "pump" (P, P*) and "dump" (D, D*) transitions connecting these states to the *A*(15,24) launch state. The phases of the P* (12,417.9 cm^{-1}) and D* (12,376.2 cm^{-1}) transition frequencies are manipulated in these experiments. Frequencies of experimentally observed quantum beats are also shown.

Coherence and Quantum Optics VIII
Edited by Bigelow et al., Kluwer Academic/Plenum Publishers, 2003

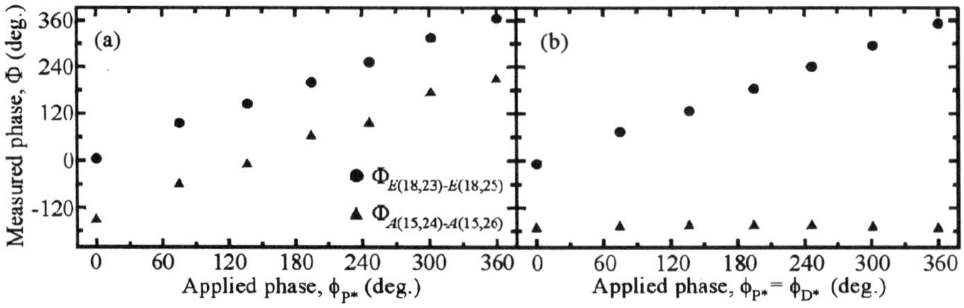

Fig. 2. Measured phases of rotational beats when (a) phase ϕ_{P*} is applied to the P* spectral component of the electric field, and (b) phase $\phi_{P*} = \phi_{D*}$ is applied to both the P* and D* transition frequencies. All other spectral components are held at zero relative phase.

independently to facilitate phase control of individual rotational states, as described below. Amplitude shaping of the preparation pulse is used to block the excitation frequencies to other E state vibrational levels (v = 17 and 19).

The time-dependent ionization signal, obtained by scanning the delay time between the preparation and probe pulses, is analyzed to extract the frequencies and phases of individual quantum beats (*i.e.*, between pairs of rotational states in each of the respective electronic states). Only two A state beats are observed experimentally, corresponding to the two newly populated rotational states $A(v = 15, J = 22$ and $26)$ beating with the $A(15,24)$ launch state. This behavior is expected since most of the wave packet amplitude remains in the launch state under intermediate field conditions. The measured phase, Φ, of a given quantum beat with respect to zero delay time between the preparation and probe pulses reflects the difference in phases acquired by the two rotational states after the pulse is over. Thus, the measured A beat phases directly reflect the phases of the newly populated rotational states relative to the phase of the launch state. Without any phase manipulation, the $E(18,23)$-$E(18,25)$ beat is found to reach a maximum at time zero ($\Phi_{E(18,23)-E(18,25)} = 0° \pm 10°$), whereas the A state beats are found to be near a minimum at time zero ($\Phi_{A(15,22)-A(15,24)} = -193° \pm 15°$, $\Phi_{A(15,24)-A(15,26)} = -156° \pm 15°$), consistent with predictions of second order time dependent perturbation theory [7].

Two different state-selective phase control cases are illustrated in Fig. 2. In Fig. 2(a), ϕ_{P*} is varied while maintaining the relative phases of all other optical frequencies at zero. As expected, the phase applied to this pump transition directly affects $E(18,25)$ and therefore $\Phi_{E(18,23)-E(18,25)}$. This pump step phase manipulation also controls the phase of the $A(15,26)$ rotational state, which is populated to a large degree via the resonant transitions P* and D*. Note that the phase acquired by the $A(15,26)$ state in Fig. 2(a) does not exactly parallel that acquired by $E(18,25)$; instead, slight deviations exist due to contributions from non-resonant transitions in populating $A(15,26)$ [7]. The case shown in Fig. 2(b), in which ϕ_{P*} and ϕ_{D*} are manipulated simultaneously while the rest of the pulse is held at zero relative phase, demonstrates independent phase control of rotational states in both electronic states. Here, the phase applied to the pulse at the D* transition frequency compensates for the phase assigned to the $E(18,25)$ rotational state; thus the phase of $A(15,26)$ can be held constant while directly manipulating the phase of a rotational state in E. Not included in Fig. 2 for purposes of clarity is $\Phi_{A(15,22)-A(15,24)}$, which remains essentially constant at $-193° \pm 22°$ in both cases. Thus, non-resonant contributions from the phase manipulated pulse frequencies corresponding to the P* and D* transitions do not significantly alter the phase of $A(15,22)$. These two examples thus demonstrate a high degree of experimental ability to simultaneously and independently manipulate rotational state phases in the E and A electronic states.

References

1. T. Hornung, R. Meier, and M. Motzkus, Chem. Phys. Lett. **326**, 445 (2000).
2. R. Pausch, M. Heid, T. Chen, H. Schwoerer, and W. Kiefer, J. Raman Spectrosc. **31**, 7 (2000).
3. R. Uberna, M. Khalil, R. M. Williams, J. M. Papanikolas, and S. R. Leone, J. Chem. Phys. **108**, 9259 (1998).
4. R. Uberna, Z. Amitay, R. A. Loomis, and S. R. Leone, Faraday Discuss., **113**, 385 (1999).
5. U. Banin, A. Bartana, S. Ruhman, and R. Kosloff, J. Chem. Phys. **101**, 8461(1994).
6. A. M. Weiner, Rev. Sci. Instrum. **71**, 1929 (2000).
7. H. U. Stauffer, J. B. Ballard, Z. Amitay, and S. R. Leone, in preparation.

Quasimodes and pseudomodes in structured reservoirs

B. J. Dalton[1,2], **Stephen M. Barnett**[3], **and B. M. Garraway**[1]

[1]*Sussex Centre for Optical and Atomic Physics, University of Sussex, Brighton BN1 9QH, United Kingdom.*
[2]*Department of Physics, University of Queensland, St Lucia, Queensland 4072, Australia.*
[3]*Department of Physics and Applied Physics, University of Strathclyde, Glasgow G4 0NG, United Kingdom.*

Abstract: The quasimode description of a quantum system coupled to a structured reservoir, such as a high-Q cavity, is compared to the pseudomode description. Master equations are found for the non-Markovian dynamics.

Quasimodes are a well established feature of cavity QED and quantum optics [1] which have been applied to several different systems such as high-Q cavities [1], Fabry-Perot cavities [2], dielectric systems [3], and beam splitters [4]. Quasimodes are a convenient alternative to true, or 'universe', modes [5] (which are the eigenstates of the radiation part of the Hamiltonian used as the starting point for deriving pseudomodes). For a cavity the quasimodes divide into two types: internal and external quasimodes (see Fig.1. below). The internal quasimodes represent cavity modes and possess a discrete

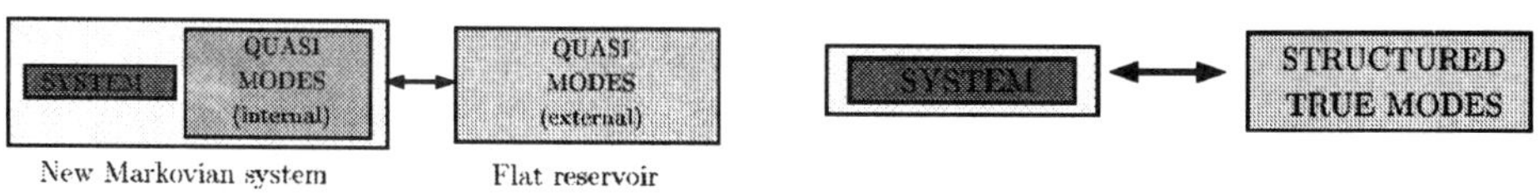

Fig. 1. Quasimode picture (left) and true mode picture (right).

set of frequencies. The external quasimodes represent modes localised outside the cavity and form a continuum. The system, e.g. an atom, couples to the internal quasimodes, whilst the reservoir structure arises from the coupling between the internal and external quasimodes. The external quasimodes can be used as a flat reservoir for the system and internal quasimodes; indeed the coupling of internal quasimodes to external quasimodes leads to cavity loss. When we include the system interaction with the internal quasimodes we can derive a master equation for the reduced density operator of the system and internal quasimodes. This is done by following a standard procedure, involving Born and Markoff approximations for the external quasimodes. The non-Markovian dynamics of the system would then be seen after tracing out the internal quasimodes from the reduced density operator. This gives a complete account of the system behaviour, but a Fano diagonalization [6] connects the quasimode model to the true mode picture and yields results for the *atom-true mode* field couplings. These couplings are used as a basis for the pseudomode model [7], from which *exact* master equations can be derived in certain circumstances.

The *pseudomode* approach, in its present form, does not rely on perturbation theory at all [7]. In fact the pseudomode represents part of the structure of the reservoir and thus the size of the Hilbert space for the problem is dramatically reduced. The method has the advantage of giving the exact equations for the dynamics of a quantum system in a form directly comparable to the master equations derived from the usual perturbation theory [8]. One of the results of this description has been the ability to describe non-Markovian dynamics with master equations usually thought of as appropriate for Markovian systems. This approach is possible because the addition of new variables, represented by the pseudomodes, makes it possible to change the non-Markovian equations to apparently Markovian ones. As a result conventional Markovian techniques can be used for further study of the non-Markovian system. The pseudomode process results in a master equation in a standard Lindblad form—as though the strongly coupled system were a Markovian one. Of course, it is not Markovian—what has happened is that the original non-Markovian system has been made Markovian by the introduction of the pseudomode (essentially an auxiliary variable). However, the key to the pseudomode technique is the choice of variable for the pseudomodes; so far only some very simple models have been treated involving single quantum excitations of the reservoir.

The pseudomode model depends directly on the analytic structure of the reservoir structure function $D(\omega) \propto \rho(\omega)|g(\omega)|^2$ where $\rho(\omega)$ is the density of states and $g(\omega)$ is the true mode system-field coupling [7]. It is clear that only this combination of $\rho(\omega)$ and $g(\omega)$ is of significance from the start of the analysis. The forms of the reservoir structure function that are tractable are ratios of polynomials; the poles of the structure function are associated with the pseudomodes and the residues at the poles determine couplings to the pseudomodes. The investigation of multiple excitation of the reservoir by the Fano diagonalization process [6] has shown that quasimodes can be associated with pseudomodes in certain situations (specifically high Q cavities). In the quasimode model the internal (discrete) quasimodes are coupled to the continuum quasimodes: if *this* coupling is flat enough we can recover simple results for the true mode coupling constant which can be linked to the pseudomode model. The *association* then identifies each quasimode with a pseudomode by examining the reservoir structure function. However, if the internal-external quasimode coupling is not flat enough the result is a very non-Markovian system where there can be more pseudomodes than quasimodes. This is true even for the single excitation case where the pseudomode model presently applies. Flatness in the internal-external quasimode coupling is a requirement for a quasimode model Markovian master equation, but is not necessarily a requirement for a pseudomode model Markovian master equation [7]. This is because the pseudomode model depends only on the reservoir structure function.

From the results outlined here we are able to see, firstly, how the quasimode picture relates to the true modes, and how this impacts on the true mode coupling constants. Then starting from a true mode picture we can see how the pseudomode model describes the same quantum system (at least for single excitations). The quasimodes can then be related to the pseudomodes. Finally, we can show how master equations can be derived for both pseudomodes [7] and quasimodes [6]. In the case of quasimodes we can treat multiple quantum reservoir excitations when the reservoir structure can be attributed to discrete quasimodes.

References

1. S.M. Barnett and P.M. Radmore, Opt. Comm. **68**, 364 (1988); B.J. Dalton, S.M. Barnett, and P. L. Knight, J. Mod. Opt. **46**, 1315 (1999), *ibid* 1495 (1999).
2. B.J. Dalton and P.L. Knight, J. Mod. Opt. **46**, 1817 (1999), *ibid.* 1839 (1999).
3. S.A. Brown and B.J. Dalton, J. Mod. Opt. **48**, 597 (2001), *ibid.* 639 (2001).
4. B.J. Dalton, S.M. Barnett, and P.L. Knight, J. Mod. Opt. **46**, 1559 (1999).
5. see, e.g., R. Lang, M.O. Scully, and W.E. Lamb Jr., Phys. Rev. A **7**, 1788 (1973); B.J. Dalton, E.S. Guerra, and P.L. Knight, Phys. Rev. A **54**, 2292 (1996).
6. B.J. Dalton, S.M. Barnett, and B.M. Garraway, e-print quant-ph/0102142.
7. B.M. Garraway, Phys. Rev. **A 55**, 2290 (1997); *ibid.* 4636 (1997).
8. see, e.g., G.S. Agarwal, *Quantum statistical theories of spontaneous emission and their relation to other approaches*, *Springer Tracts in Modern Physics*, Vol. **70**, eds. G. Höhler et al., (Springer, New York, 1974).

Pulse generation in vibrating cavities

V. V. Dodonov and M. A. Andreata

Departamento de Física, Universidade Federal de São Carlos,
Via Washington Luiz, km 235, 13565-905 São Carlos, SP, Brasil
vdodonov@df.ufscar.br

Classical and quantum phenomena in cavities with moving boundaries attracted attention of many researchers for a long time, being known now under the names *nonstationary Casimir effect*, *dynamical Casimir effect*, or *mirror (motion) induced radiation* [1, 2]. One of theoretical results obtained in the last years was the prediction of the exponential growth of the energy of the field under the resonance conditions [2, 3]. We address the problem of the *energy density distribution* inside the cavity with oscillating boundaries. It was indicated in [3] that the main part of the energy is concentrated in several sharp peaks which move from one boundary to another, becoming more and more narrow with the course of time. However, in that papers the evolution of these packets was studied in the framework of numerical calculations only. Recently the analytical solutions to the problem of a one-dimensional ideal cavity with *resonantly vibrating* boundaries were found [4]. Here we use these solutions to provide explicit formulae for the shape, height and width of the energy packets formed in the vibrating cavity.

We consider the model of a massless scalar field in a 1D cavity formed by two infinite ideal plates whose positions are given by $x_l \equiv 0$ and $x_r = L(t)$, where $L(t) = 1 + \varepsilon \sin[p\pi(1+\delta)t]$, $p = 1, 2, \ldots$, $c = \hbar = 1$. The field operator in the Heisenberg representation can be written in the form $\hat{A}(x,t) = \sum_{n=1}^{\infty}(2/\sqrt{n})\left[\hat{b}_n\psi^{(n)}(x,t) + \text{h.c.}\right]$, with $\psi^{(n)}(x,t) = [L(t)]^{-1/2}\sum_{k=1}^{\infty}\left[\rho_k^{(n)}e^{-ik\pi(1+\delta)t} - \rho_{-k}^{(n)}e^{ik\pi(1+\delta)t}\right]\sin[\pi kx/L(t)]$ and $[\hat{b}_n, \hat{b}_k^\dagger] = \delta_{nk}$. Slowly varying in time coefficients $\rho_k^{(n)}$ obey (under the resonance condition $|\delta| \ll 1$ and for small ampliude of vibrations $|\varepsilon| \ll 1$) the equations ($\tau \equiv \frac{1}{2}\varepsilon\omega_1 t$)

$$\frac{\mathrm{d}}{\mathrm{d}\tau}\rho_k^{(n)} = \sigma\left[(k+p)\rho_{k+p}^{(n)} - (k-p)\rho_{k-p}^{(n)}\right] + 2i\gamma k\rho_k^{(n)}, \quad \gamma \equiv \delta/\varepsilon, \quad \sigma \equiv (-1)^p. \tag{1}$$

Using exact solutions of (1) found in [4], we have calculated the mean value of the energy density operator in one space dimension $\hat{W}(x,t) = \left[(\partial\hat{A}/\partial t)^2 + (\partial\hat{A}/\partial x)^2\right]/(8\pi)$ at $t \geq T$, when the wall comes back to its initial position $L_0 = 1$ and does not move anymore. After extracting the regularized vacuum Casimir's term (in one dimension) $\tilde{W}^{(Cas)} = -\pi/24$, we have expressed the non-divergent part of the energy density as follows: $W(x,t) = \frac{\pi}{2}[F(u;\tau) + F(v;\tau)]$, where $u = t' + x$, $v = t' - x$, $t' \equiv t + \delta T$,

$$F(u;\tau) = -\mathcal{B} + f(u;\tau)\Big\{\mathcal{B} + \sum_{n,k=1}^{\infty}\sqrt{nk}\,\mathrm{Re}\left[\langle\hat{b}_n\hat{b}_k\rangle\Lambda^{\frac{n+k}{p}} + \langle\hat{b}_n^\dagger\hat{b}_k\rangle\Lambda^{\frac{k-n}{p}}\right]\Big\}, \tag{2}$$

$$f(u;\xi) = \frac{\left(1-\xi^2\right)^2}{[1+\xi^2+2\sigma\xi\cos(p\pi u-\varphi)]^2}, \quad \Lambda(u;\xi) = e^{i(2\varphi-\pi pu)}\frac{1+\sigma\xi\exp[i(\pi pu-\varphi)]}{1+\sigma\xi\exp[i(\varphi-\pi pu)]},$$

$$\xi(\tau) = \frac{\sinh(ap\tau)}{\left[a^2+\sinh^2(ap\tau)\right]^{1/2}}, \quad e^{i\varphi} = \sqrt{1-\gamma^2\xi^2} + i\gamma\xi, \quad a = \sqrt{1-\gamma^2}, \quad \mathcal{B} \equiv \frac{p^2-1}{24}.$$

For the initial vacuum state, function $W(x,t)$ has p peaks in the interval $0 \leq x \leq 1$. All the peaks have equal heights $W_{max}^{(vac)} = 2\pi\mathcal{B}\xi/(\xi-1)^2 = (\pi\mathcal{B}/2)\left(e^{4p\tau} - 1\right)$ (formulas containing τ are given for $\gamma = 0$), excepting some distinguished instants of time when two peaks moving in the opposite directions merge. If $\xi \to 1$ (i.e. $\tau > 1$ and $\gamma < 1$), the energy density can be approximated in the vicinity of each peak by the Lorentz-like distribution

$$W^{(vac)}(\delta x) = W^{(vac)}_{max}\left[1 + \left(2\delta x/\Delta_{1/4}\right)^2\right]^{-2}, \quad \Delta_{1/4} = \frac{2}{p\pi}\frac{1-\xi}{\sqrt{\xi}} \approx \frac{4}{p\pi}e^{-2p\tau}. \qquad (3)$$

Excepting narrow regions of the length $\Delta_+ \approx (\pi p)^{-1}\sqrt{1-\xi} \approx \sqrt{2}(\pi p)^{-1}e^{-p\tau}$ nearby the peaks, the 'dynamical' energy density is less than its initial vacuum value. The minimum values of W far off the peaks are given by $W_{min} = -4\pi\mathcal{B}\xi(\xi+1)^{-2} = \pi\mathcal{B}\left(e^{-4p\tau} - 1\right)$. The total asymptotical minimum value equals $\tilde{W}^{(as)}_{min} = -\pi p^2/24$, being independent of the initial state of the field. The positions of peaks also do not depend on the initial state. For the initial states with diagonal density matrices in the Fock basis, $\langle\hat{b}^\dagger_n\hat{b}_k\rangle = \nu_n\delta_{nk}$, $\langle\hat{b}_n\hat{b}_k\rangle = 0$ (e.g., for the thermal or Fock states), all the peaks still have equal heights, increased by the quantity (compared with the vacuum case) $\Delta W_{max} = \frac{1}{2}\mathcal{E}_0(1+\xi)^2/(1-\xi)^2$, where $\mathcal{E}_0 = \pi\sum_{n=1}^{\infty} n\nu_n$. If the initial density matrix in the Fock basis has nonzero off-diagonal elements (e.g., for the coherent states), different terms in the sum (2) can interfere, and the peaks acquire some kind of 'fine structure'. If the detuning γ is different from zero, then some deformations of the form of peaks are observed, although the maximal heights are still of the same order of magnitude as in the case $\gamma = 0$, as far as $\gamma \le 1$. But if the detuning exceeds the critical value $\gamma = 1$, the energy becomes an oscillating function of the 'slow time' τ, the amplitude of oscillations being proportional approximately to $\left(\gamma^2 - 1\right)^{-1}$ [4]. The peaks become rather wide and low, since the parameter ξ is limited by the inequality $\xi \le \gamma^{-1}$ if $\gamma > 1$.

If the walls possess some small transmission coefficient, then a small part of radiation accumulated inside the cavity can leave it. In this case one could observe sharp pulses of radiation outside the cavity [5], whose amplitudes must be proportional to the heights of the peaks inside the cavity multiplied by the small transmission coefficient. The intensity of these pulses can be significantly increased, if the initial state is different from vacuum and possesses sufficient energy, like thermal states [5, 6] or coherent states.

References and links

1. V. V. Dodonov, A. B. Klimov, and V. I. Man'ko, "Nonstationary Casimir effect and oscillator energy level shift," Phys. Lett. A **142**, 511-513 (1989); J. Schwinger, "Casimir light: a glimpse," Proc. Nat. Acad. Sci. USA **90**, 958-959 (1993); G. Barton and C. Eberlein, "On quantum radiation from a moving body with finite refractive index," Ann. Phys. (NY) **227**, 222-274 (1993).
2. V. V. Dodonov and A. B. Klimov, "Generation and detection of photons in a cavity with a resonantly oscillating boundary," Phys. Rev. A **53**, 2664-2682 (1996); A. Lambrecht, M.-T. Jaekel and S. Reynaud, "Motion induced radiation from a vibrating cavity," Phys. Rev. Lett. **77**, 615-618 (1996).
3. C. K. Law, "Resonance response of the quantum vacuum to an oscillating boundary," Phys. Rev. Lett. **73**, 1931-1934 (1994); C. K. Cole and W. C. Schieve, "Radiation modes of a cavity with a moving boundary," Phys. Rev. A **52**, 4405-4415 (1995); D. A. R. Dalvit and F. D. Mazzitelli, "Creation of photons in an oscillating cavity with two moving mirrors," Phys. Rev. A **59**, 3049-3059 (1999).
4. V. V. Dodonov, "Resonance photon generation in a vibrating cavity," J. Phys. A **31**, 9835-9854 (1998); V. V. Dodonov and M. A. Andreata, "Squeezing and photon distribution in a vibrating cavity," J. Phys. A **32**, 6711-6726 (1999); M. A. Andreata and V. V. Dodonov, "Energy density and packet formation in a vibrating cavity," J. Phys. A **33**, 3209-3223 (2000).
5. A. Lambrecht, M.-T. Jaekel and S. Reynaud, "Generating photon pulses with an oscillating cavity," Europhys. Lett. **43**, 147-152 (1998).
6. G. Plunien, R. Schützhold and G. Soff, "Dynamical Casimir effect at finite temperature," Phys. Rev. Lett. **84**, 1882-1885 (2000).

New ultra-bright source of entangled photons

G. Bitton, W.P. Grice, L. Zhang, J. Moreau, J. Barhen and V. Protopopescu
Center for Engineering Science Advanced Research
Oak Ridge National Laboratory, Oak Ridge, Tennessee 37830
bittong@ornl.gov

Abstract: A new ultra-bright entangled photon source has been realized using two cascaded type-II crystals. Entanglement of photons originating from any point on the output cones is observed provided that suitable optical delay lines are used.

1. Introduction

Entangled photon states generated by spontaneous parametric down-conversion (SPDC) in non-linear crystals have been extensively used to test basic concepts in quantum mechanics as well as in an increasing number of applications. Substantial effort is currently invested in improving the quality of those sources, both in terms of the total number of entangled states produced by the source and the degree of the entanglement of those states. In this paper, we report on a new ultra-bright source of entangled photons realized using type-II SPDC in two crystals arranged in cascade. The optical axes of the crystals are at angles θ and -θ with respect to the pump beam direction (see Fig. 1). In such an architecture, the extraordinarily (ordinarily) polarized cone of the first crystal and the ordinarily (extraordinarily) polarized cone of the second crystal overlap. This overlapping removes the association between the photons' polarizations and their spatial location over the entire area of the output cones. Hence, one can use any part of the output cones as a source for entangled states, provided that suitable delay lines are installed.

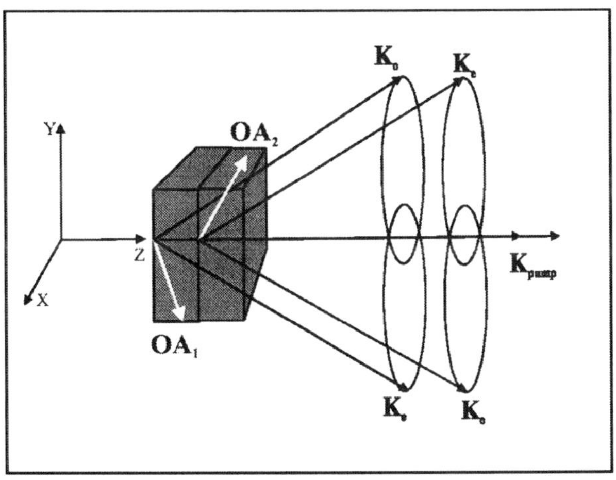

Fig. 1. The output cones from the two-cascaded crystals

A delay line is used to erase the information contained in the arrival times of the photons. Due to the birefringent nature of the crystals, photons with different polarizations travel at different speeds inside the crystals. Hence, the differences in the arrival times of the photons contain some information about their polarizations and about the crystal in which they were created. Failure to erase that information will result in a low degree of polarization entanglement. A rather simple delay-line may be used to increase the visibility over a large range of the output cones. Because the ordinary polarized photon from the first crystal (1o) travels along the same path as the extraordinarily polarized photon from the second crystal (2e), a simple birefringent delay is sufficient to equalize their average arrival times. A similar delay may be used in the path taken by the (1e) and the (2o) photons. As a result, the polarization of the photons cannot be distinguished by their arrival time.

2. Experiment and Results

The frequency doubled output of an ultra-fast Ti:Sapphire laser at 795 nm was used to pump two adjacent thin BBO crystals (1 mm each) with identically cut angles of θ=43.9° suitable for type-II down-conversion. Two variable circular irises located 7.5 cm after the crystals are used to select two specific spatial directions. Quartz wedges with continuously variable thickness were used as delay lines. The detection package consisted of polarizing beam splitters preceded by half wave plates along with single photon detectors and interference filters (IF).

Polarization interference was measured with one of the polarizers at 45°, while the angle of the other polarizer was varied. With no delay lines, poor visibility of 12% is observed, while an increase to 60% visibility is observed with the delay line in place. The bandwidth of the IF in this measurement was 20 nm. The visibility was found to increase when the bandwidth of the IF and the spectral-width of the pump beam were decreased. Further increases are expected following fine-tuning of the delay lines.

This source also exhibits space-time interference, as can be seen in Fig. 2. In this plot the coincidence counts are shown as a function of the thickness of the quartz wedge inserted into one of the optical paths, while the delay line in the other optical path was kept fixed. A fast oscillation of the coincidence counts as function of the thickness of the delaying material was observed along with a slower change in the visibility. The fringe spacing in Fig. 2. is related to a change of 88.2 μm of quartz in the delay line, which is consistent with the fringe spacing expected for 795 nm light.

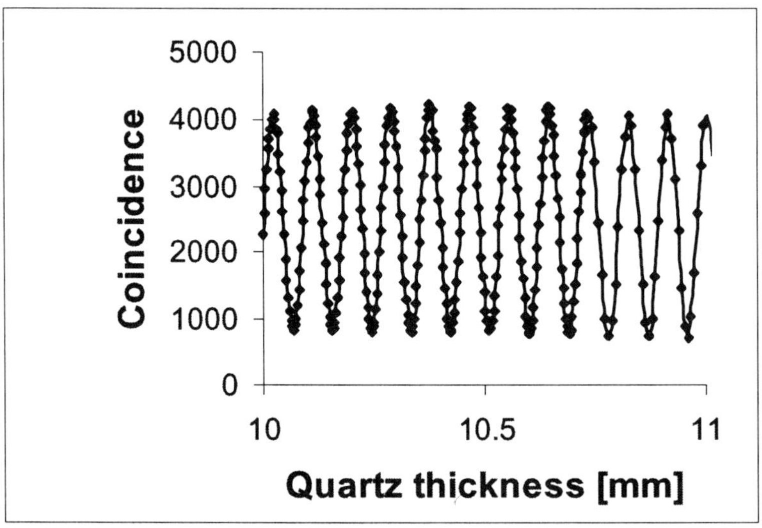

Fig. 2. Coincidence counts as function of the quartz thickness used in the delay line.

In summary, we have realized and characterized a new entangled photon source. Using two crystals in cascade with optical axes at specific directions, we were able to produce entangled photon states from any two conjugated points on the output cones. We have found that simple delay lines in the two output paths can be used to increase the visibility over a large range of the output cones.

This research was supported by the US Department of Energy, Office of Basic Energy Sciences under contract No. DE-AC05-00OR22725 with UT-Battelle, LLC.

Source of photon-number-squeezed light with adjustable level of squeezing

Ondřej Haderka[a,b], **Jan Peřina, Jr.**[a,b], **Jan Soubusta**[a,b]

[a] *Joint Laboratory of Optics of Palacký University and Institute of Physics of Academy of Sciences of the Czech Republic, 17. listopadu 50A, 772 00 Olomouc, Czech Republic*
[b] *Department of Optics, Faculty of Natural Sciences, Palacký University, 17. listopadu 50, 772 00 Olomouc, Czech Republic*

haderka@sloup.upol.cz, perina_j@sloup.upol.cz

Abstract: A source of photon-number-squeezed light based on postselection from entangled photon pairs produced by parametric frequency down-conversion is suggested. Postselection device consists of $1 \times N$ coupler and N detectors. All possible sources of noise are taken into account. The dependence of the Fano factor of generated light on various parameters is examined.

OCIS codes: (270.5290) Photon statistics;(190.2620) Frequency conversion

Entangled photon pairs produced by spontaneous parametric frequency down-conversion have recently been widely used in experimental quantum physics. They have been successfully applied in the research of fundamental problems of quantum theory [1]. Among others, their properties were conveniently applied in quantum cryptography.

Employing the entanglement of photons in pairs, we suggest to construct a single-photon source as follows. Perform a photon-number measurement in the signal beam and select only those cases in which a single photon has been detected. Then there is a single photon in the idler beam with a high probability. Thus, the contribution of the vacuum state as well as multiphoton states in the state of the idler beam is suppressed.

We assume that pairs of entangled photons are generated at the output of a nonlinear crystal with Poissonian statistics [2, 3]. The signal and idler beams are then selected by frequency filters and pinholes. The filtering is in general imperfect, i.e., sometimes only one of the members of a pair reaches a detector. We model this brokening of photon pairs by beamsplitters in both the signal and idler beams. Photons that lost their twins then contribute to noisy fields at the detectors. It can be shown [3] that their statistics are given by geometrical distributions. A $1 \times N$ coupler followed by N detectors is used as a photon-number measuring device. Detectors are assumed to have limited quantum efficiencies and they are not able to resolve the photon number. In addition, there are noise detections coming both from dark counts of the detectors and from stray light in the laboratory. The influence of noises is described quantum consistently (for details, see [3]).

We assume that postselection occurs if an arbitrary detector out of N detectors beyond the $1 \times N$ coupler detects a photon and the rest of $N-1$ detectors do not register a photon. The photon source operates in the ideal case as follows. Perfect entanglement between the signal and idler fields together with the postselection procedure eliminates the vacuum state in the idler field. On the other hand, a high number of ideal detectors beyond the $1 \times N$ coupler in the signal field suppress the Fock states with the photon number greater than one in the idler field. The idler field is then in a state being close to the one-photon Fock state. This state exhibits photon-number squeezing. For

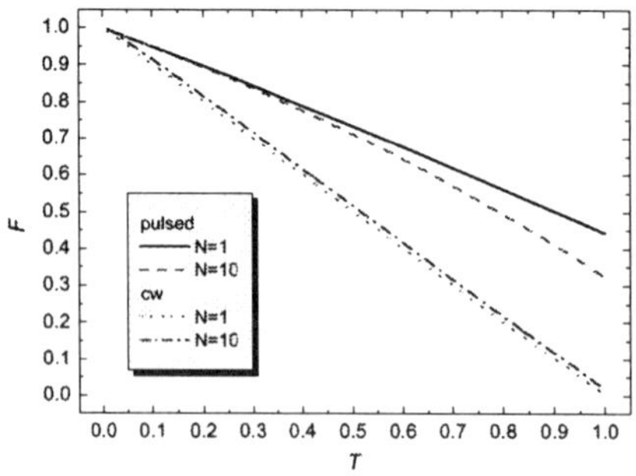

Fig. 1. Fano factor F as a function of the coupling coefficient T for typical values of the mean number of pairs μ in cw regime ($\mu = 10^{-4}$) and in pulsed regime ($\mu = 1$); for details, see [3].

a finite number of detectors, the more detectors are used, the better the exclusion of multiphoton states.

A lower quantum efficiency of real detectors results in a lower suppression of multiphoton states in the idler field. Noise coming from dark counts of the detectors lead in principle to the occurrence of the vacuum state in the idler field.

Photons in the signal and idler fields are not perfectly entangled because photon pairs may be broken as they propagate towards detectors. Photons that lost their twins then contribute to noise both in the signal and idler beams. The coincidence-count rate describing the number of photon pairs in the detectors is then lower and almost all advantages of the many-detector postselecting device are lost for weak coupling between the signal and idler fields. The weight of the vacuum state in the idler field is great owing to triggers by photons that lost their twins. Also the exclusion of multiphoton states is very inefficient.

Achievable values of the Fano factor F depend on how well the contributions of the vacuum state and multiphoton states are eliminated. This means that the mean number of generated photon pairs cannot be both too small (greater vacuum-state contribution) and too great (greater contribution of multiphoton-states). The influence of coupling between the signal and idler photons on the values of the Fano factor F is crucial (see Fig. 1). As is shown in Fig. 1, we can get good approximations of single-photon Fock states ($F \to 0$) in the cw regime and for good coupling ($T \to 1$). This is because the vacuum-state probability can be made very small and the contribution of multiphoton states is small due to low values of the mean number of pairs μ. The use of more detectors in the postselecting device does not lead to lower values of F. On the other hand, more detectors lead to lower values of F in the pulsed regime (greater values of μ), because they suppress a greater contribution of multiphoton states. The cw regime is better for the generation of photon-number squeezed light compared to the pulsed regime.

This work was supported by the Ministry of Education of the Czech Republic (projects no. LN00A015 and no. 19982003012).

References and links

1. J. Peřina, Z. Hradil, and B. Jurčo, *Quantum Optics and Fundamentals of Physics* (Kluwer, Dordrecht, 1994), Chap. 8.
2. O. Haderka and J. Peřina, Jr., "Approximations of single-photon states by postselection from correlated pairs," Proc. NATO ARW on Decoherence and its Implications in Quantum Computation and Information Transfer, (Mykonos, Greece, 2000) (in print).
3. J. Peřina, Jr., O. Haderka, and J. Soubusta, "Quantum cryptography using photon source based on postselection from entangled two-photon states," (submitted to Phys. Rev. A).

High-visibility two-photon interference in femtosecond pulse pumped type-II SPDC

Yoon-Ho Kim, Maria V. Chekhova,* Sergei P. Kulik,*
Morton H. Rubin and Yanhua Shih

Department of Physics
University of Maryland, Baltimore County
Baltimore, Maryland 21250

Tel: (410)455-1932, Fax: (410)455-1072, Email: yokim@umbc.edu

By far the most efficient source of two-particle entangled state is spontaneous parametric down-conversion (SPDC). Especially, cw laser pumped type-II SPDC has been widely treated as a good source of entangled photon pairs, since one can easily prepare a polarization Bell state [1]. However, femtosecond pulse pumped type-II SPDC shows poor quantum interference visibility when a thick crystal is used. To restore the visibility, one has to use narrow-band spectral filters at the expense of available photon flux.

In this paper, we demonstrate a scheme for high-visibility two-photon interference with thick (3.4mm) type-II BBO crystals in femtosecond pulse pumped SPDC. Polarization Bell states can be prepared using this method without spectral and amplitude post-selection.

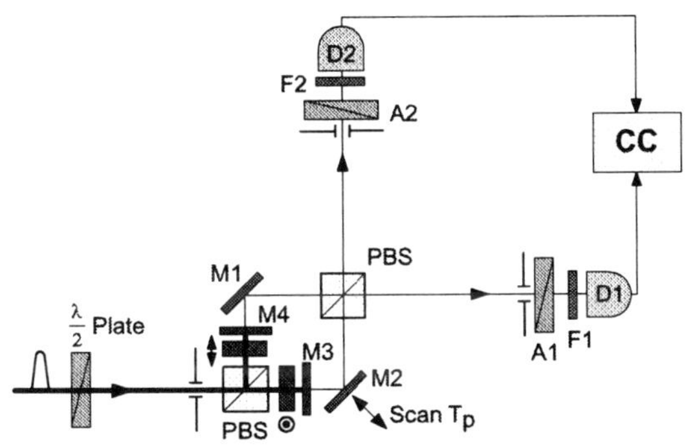

Fig. 1. Bell state preparation scheme using two type-II BBO in a Mach-Zehnder interferometer. M_3 and M_4 are the pump reflecting mirrors. T_p is the delay between the biphoton amplitudes from two different crystals. Therefore, the relative phase between the two amplitudes is $\varphi = \Omega_p T_p$. Note that the optic axes of the BBO crystals are oriented orthogonally, one vertically (↕) and the other horizontally (⊙).

Consider the experimental setup shown in Fig.1. A type-II BBO crystal is placed in each arm of the Mach-Zehnder interferometer (MZI). The pump pulse is polarized at 45° by using a $\lambda/2$ plate. There are only two amplitudes in this process, one from each crystal, since a polarizing beam splitter is used to split the signal and the idler photon pairs. Therefore, the quantum state, when the MZI is properly aligned, can be written as

$$|\psi\rangle = |H_1, V_2\rangle + e^{i\varphi}|V_1, H_2\rangle.$$

By varying $\varphi = \Omega_p T_p$, modulation in the coincidence counting rate is observed at the pump central frequency (400nm). Therefore a true Bell state can be prepared without any post-selection methods when φ is properly chosen. Note also that e-polarized pho-

Coherence and Quantum Optics VIII
Edited by Bigelow et al., Kluwer Academic/Plenum Publishers, 2003

tons are always detected by D_2 and o-polarized photons are always detected by D_1. This is of great importance when femtosecond laser is used as a pump.

Theoretically, the space-time interference pattern at $\theta_1 = \theta_2 = 45°$ shows

$$R_c = 1 + V(T_p)\cos(\Omega_p T_p), \tag{1}$$

where $V(T_p)$ is the envelope function of the space-time interference pattern. It is important to note that the envelope function $V(T_p)$ is exactly equal to that of the self-convolution of the pump pulse when no spectral filters are used: $V(T_p)$ does not involve any crystal parameters at all. This means that the measured FWHM of the quantum interference pattern should be equal to the measured FWHM of the self-convolution of the pump pulse. This is in contrast to the case reported in Ref.[2] in which type-I BBO crystals are used in the setup similar to Fig.1: the measured FWHM of the quantum interference envelope is bigger than the self-convolution of the pump pulse and is strongly dependent on the crystal thickness and the group velocity difference between the pump and the SPDC photons.

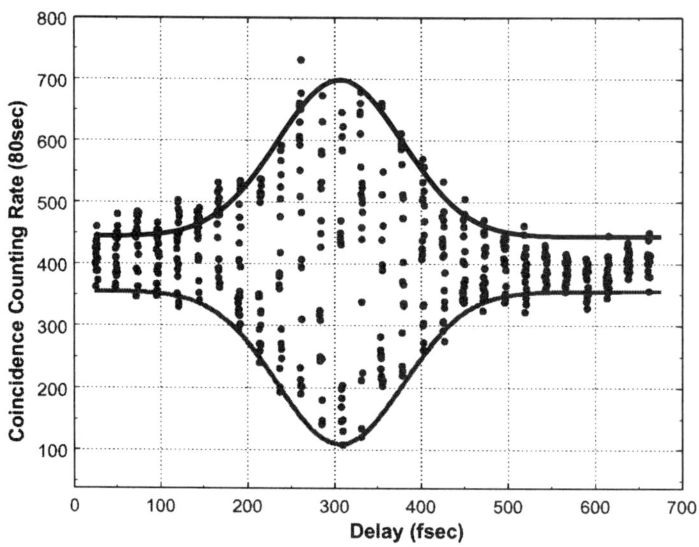

Fig. 2. Quantum interference observed in coincidence counts. No modulation is observed in single counting rates. Each column of data actually represent modulations of several wavelengths. 40nm FWHM filters are used. The measured FWHM of the envelope is 170fsec which is equal to the FWHM envelope of the pump pulse itself.

Fig.2 shows the measured space-time quantum interference pattern by scanning T_p. The envelope of the interference pattern is equal to that of the pump pulse self-convolution which confirms Eq.1. The measured visibility is, to the best of our knowledge, the highest (76%) in femtosecond pulse pumped type-II SPDC without spectral and amplitude post-selection. The majority of visibility loss is due to the imperfections in the alignment of the MZI. More careful alignment and improvement of the pump beam quality should improve the visibility.

References and links

1. Y.H. Shih and A.V. Sergienko, Phys. Lett. A **186**, 29 (1994); Phys. Lett. A **191**, 201 (1994); P.G. Kwiat *et al.* Phys. Rev. Lett. **75**, 4337 (1995).
2. Y.-H. Kim, S.P. Kulik, and Y.H. Shih, Phys. Rev. A **62**, 011802(R) (2000).

Four Photon Entanglement from Parametric Down Conversion

M. Eibl[2], S. Gaertner[1], Ch. Kurtsiefer[1], M. Żukowski[3], and H. Weinfurter[1,2]

[1] *University of Munich, Munich, Germany*
[2] *Max-Planck Institute for Quantum Optics, Garching, Germany*
[3] *Uniwersytet Gdański, Gdańsk, Poland*

harald.weinfurter@physik.uni-muenchen.de

Abstract:
The emission of two photon pairs from type-II parametric down conversion results in a highly entangled 4-photon state. We describe the features of this new state and its application for the first 4-particle test of local realism.

OCIS codes: (270.4180) Multiphoton processes, (030.1670) Coherent optical effects

Parametric down conversion has proven to be the best source of entangled photon pairs in an ever increasing number of experiments on the foundations of quantum mechanics [1] and in the new field of quantum communication. With increasing pump power, the generation of more than only one photon pair at a time can no longer be neglected. Yet, due to bosonic interference, these multi photon states are not just a product of many pair-states, but also exhibit high degree of n-photon entanglement [2]. Here we concentrate on the features of the spontaneously created four photon state and apply it in a four photon test of local realism based on a generalized Bell-inequality.

In type-II parametric down conversion polarization entangled photon pairs can be generated in a non-linear crystal [3] pumped with strong UV-pulses. Taking multiple pair emissions into account one obtains the state $Z \exp\left(-i\alpha(a_V^* b_H^* + a_H^* b_V^*)\right) |\, 0\rangle$, where Z is a normalization constant, α is proportional to the pulse amplitude, a_V^* is the creation operator of a photon with vertical polarization in mode a, etc., and where complete indistinguishability of the emitted pairs is assumed (Fig. 1). The interesting four photon emission is described by the quadratic term in the expansion of the above state [4].

In order to observe the non-classical features of the state, we first rotate the polarization of mode a and then pass the four photon state via two beam splitters transforming x into $\frac{1}{\sqrt{2}}(x + x')$, with $x = a, b$. Extracting terms that lead to 4 photon coincidence the state, after normalization, can be symbolically written as

$$\begin{aligned} &\sqrt{1/3}\,(|\, VVVV\rangle_{aa'bb'} + |\, HHHH\rangle_{aa'bb'} \\ &+\tfrac{1}{2}\,(|\, HVHV\rangle_{aa'bb'} + |\, HVVH\rangle_{aa'bb'} + |\, VHHV\rangle_{aa'bb'} + |\, VHVH\rangle_{aa'bb'})) \\ &= \sqrt{2/3}\,|\, GHZ\rangle_{aa'bb'} + \sqrt{1/3}\,|\, EPR\rangle_{aa'}\,|\, EPR\rangle_{bb'}\,, \end{aligned} \tag{1}$$

i.e., a sum of a 4-photon GHZ state[5] and a product of two EPR-Bohm states.

In the experiment pulsed parametric down conversion was obtained by pumping a type-II cut BBO-crystal with pulses of about 120fs duration and a center-wavelength of 780nm. Multi photon polarization analysis is performed behind polarizing beam splitters and analyzed by a 8-channel single photon detection and multicoincidence unit. It allows

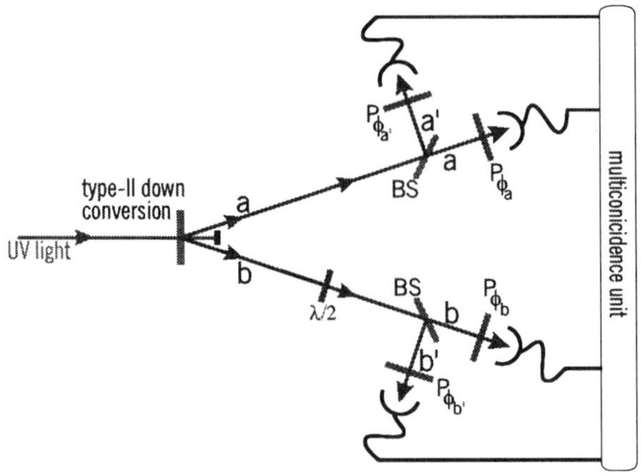

Fig. 1. Experimental set up to demonstrate the entanglement inherent in 4-photon emission from type-II paramteric down conversion. ($\lambda/2$-half wave plate to flip polarization, BS-beam splitter, and P_{ϕ_a} represents analysis of elliptic polarization with main axis of the polarization ellipse at 45° and phase angle ϕ_a, etc.)

efficient determination of all possible multi photon coincidence rates and the correlation function for a given phase setting in one single run.

When analyzing the polarization of each of the four photons in the four output modes strong correlations can be observed, which also make this state the ideal candidate for multiparty quantum cryptography and teleportation. In spite of strong correlations the state 1 does not give rise to a GHZ-contradiction, however it can be used for a four photon test of local realism according to a generalized Bell inequality [2, 6].

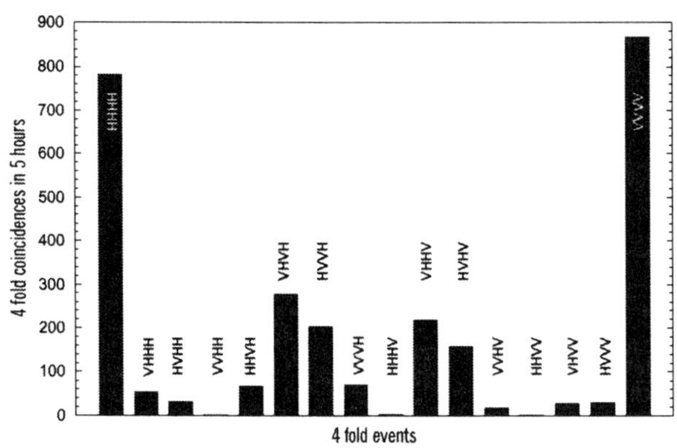

Fig. 2. Polarization analysis in the H/V-basis reveals the differences between state 1 and a product of two pairs where all allowed events would have equal probability.

This work was supported by the EU-IST-FET project QuComm.

References and links

1. L. Mandel, Rev. Mod. Phys. **71**, S274 (1999); A. Zeilinger, Rev. Mod. Phys. **71**, S288 (1999).
2. H. Weinfurter, M. Żukowski, submitted to Phys. Rev. A.
3. P.G. Kwiat, K. Mattle, H. Weinfurter, A. Zeilinger, A.V. Sergienko, and Y. Shih, Phys. Rev. Lett. **75**, 4337 (1995).
4. The emission of more than two pairs will result in entangled states of more and more photons (see also A. Lamas-Linares et al., quant-ph/0103056).
5. D.M. Greenberger, M. Horne, A. Shimony, and A. Zeilinger, Am. J. Phys., **58**, 1131 (1990).
6. M. Żukowski, C. Brukner, quant-ph/0102039.

Observation of time-dependent inhibited spontaneous emission

D. Branning and A.L. Migdall
NIST, 100 Bureau Dr., STOP 8441. Gaithersburg, MD, 20899
Telephone: (301) 975-8112 Fax: (301) 869-5700 e-mail: dbran@nist.gov

When an excited atom is placed in a small reflecting cavity, it is allowed to radiate only into the discrete set of electromagnetic modes that satisfy the boundary conditions in the cavity. If none of these modes are compatible with the spontaneous emission modes of the atom, it will not emit radiation. This phenomenon, known as "inhibited spontaneous emission," has been observed for atoms, for trapped electrons, and for parametric downconverters (PDC), in which incident "pump" photons spontaneously decay into pairs of lower-energy "signal" and "idler" photons [1].

In 1995, Fearn *et al.* proposed an interesting thought experiment [2]: suppose an excited atom is placed in an inhibiting cavity, and after some time, one mirror of the cavity is instantaneously replaced by a photon-counting detector. Will the photodetector count a photon immediately (because the counter-propagating field amplitudes have been in the cavity all along), or will there be some round-trip delay due to the necessity of the atom first "seeing" the changed mode structure of the cavity, and then radiating a photon into this new mode structure?

The thought experiment is impractical to carry out for atoms, but because large cavities can be used to inhibit a PDC, it is possible to perform a variation of the thought experiment, as shown in Fig. 1. A pump laser beam is directed into a PDC, where pairs of signal and idler photons are created. The pump, signal, and idler beams are reflected back into the PDC, whereupon another downconversion event is possible. When the relative phases are adjusted for destructive interference, the emission of the PDC will be inhibited, and no counts will be recorded at detector 1.

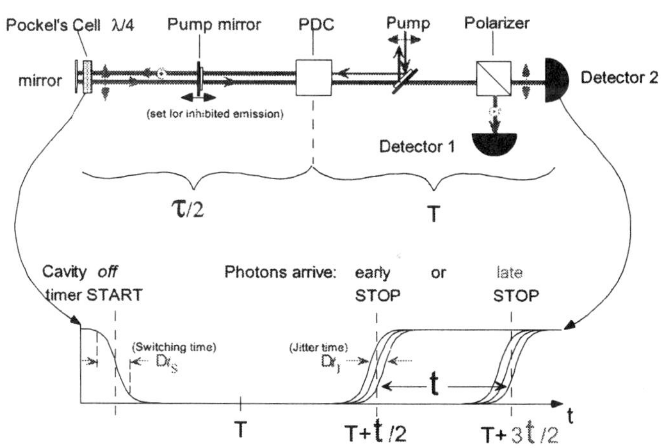

Fig. 1. Conceptual arrangement for the experiment.

A Pockels cell can then be used to quickly rotate the signal and idler polarizations by 90°, so that the cavity is effectively shut off. After this, the light in the cavity will be transmitted through the PDC, and directed by the polarizer to photodetector 2. The question is, how soon after the cavity is switched off will photons be registered? The two plausible scenarios in the thought experiment lead to two different answers: if photons were present in the cavity all along, then the delay between the Pockels cell activation and the counting of photons at the lower detector should be $\tau/2$ (half the round-trip time of the cavity) plus T (the propagation time from the PDC to the detector). On the other hand, if the PDC had to wait to emit light until the mode structure was changed, this would incur an extra round-trip delay of τ, so that the total delay would be $3\tau/2 + T$. These two possibilities can be experimentally distinguished so long as the Pockels cell switching time and the "jitter," or time resolution of the photodetection electronics, is not greater than τ, the round-trip time of light in the cavity.

The experiment conducted at NIST uses the 351 nm light from an Ar^+ laser to pump a KDP crystal cut and aligned for collinear, frequency-degenerate type-I downconversion. The Pockels cell is driven at 8 kHz with a rise time of 7 ns. After passing through 10-nm bandpass filters, the downconverted photons are counted with actively quenched silicon avalanche photodiodes. Preliminary results, for a cavity length of 1 m, are shown in Fig. 2.

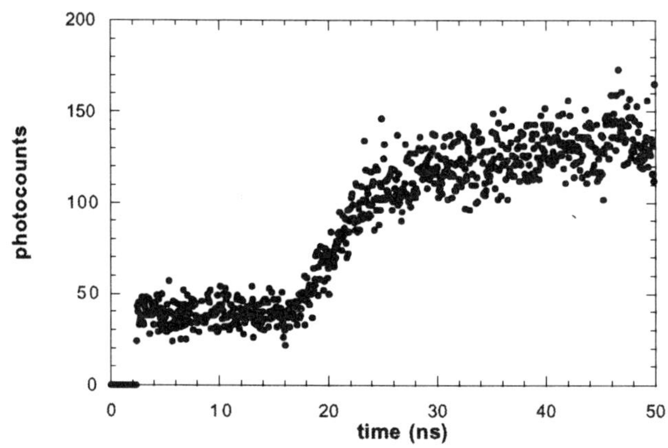

Fig. 2. Preliminary arrival-time data for downconverted photons in the cavity.

[1] T.J.Herzog *et al.*, "Frustrated Two-Photon Creation via Interference," Phys. Rev. Lett. **72**, 629-632 (1994).

[2] H. Fearn, R.J. Cook, and P. Milonni, "Sudden Replacement of a Mirror by a Detector in Cavity QED: Are Photons Counted Immediately?" Phys. Rev. Lett. **74**, 1327-1330 (1995).

[3] H. Weinfurter *et al*, "Frustrated Parametric Downconversion: Virtual or Real Photons?" Ann. New York Acad. Sci. **755**, 61-65 (1995).

Conditional coherence via phase-sensitive post-selection

K.J. Resch, J.S. Lundeen, and A.M. Steinberg
Department of Physics, University of Toronto, 60 St. George Street Toronto ON M5S 1A7 CANADA
ph: (416) 946-3146, fax: (416) 978-2537

Abstract: A single beam of light created via spontaneous parametric down-conversion has a random phase. We show it is possible to induce a well-defined *absolute* phase in one beam by overlapping the other with a weak coherent state and conditioning on detection of a photon. This post-selection fixes the phase of the down-conversion even though stimulated emission is negligibly small.

Experiment has shown that interference fringes between a local oscillator (LO) and a beam created through spontaneous parametric down-conversion (SPDC) do not occur [1], and down-conversion beams are mutually incoherent [2]. One can understand this absence of coherence in terms of which-path information. It is well known that interference cannot occur if there exists information that can identify which path a system took to reach its final state. In the case of homodyne measurement between a LO phase-reference and the SPDC idler mode, a strong number correlation between the signal and idler beams makes it possible to know whether a detected photon came from down-conversion or from the reference laser: if there is a photon in the signal mode, then the detected photon was most likely the idler, if there is no signal photon, then the photon must have come from the LO. The mere existence of which-path information, whether or not it is actually measured, precludes the possibility of interference.

In this work, we overlap a second LO beam over the signal beam from SPDC, as represented in the experimental cartoon (fig. 1.)

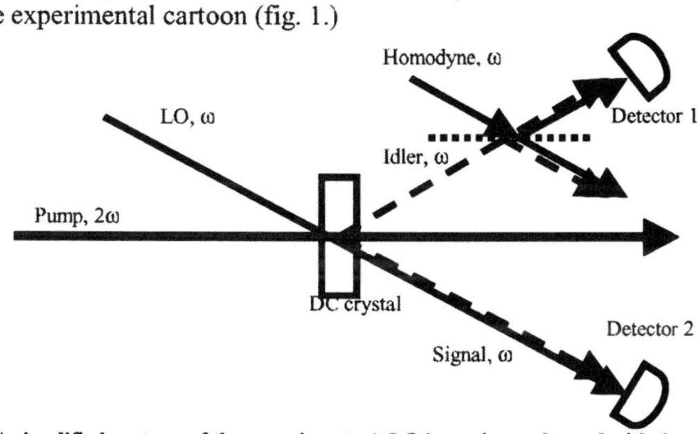

Figure 1. A simplified cartoon of the experiment. A LO beam is overlapped with the signal beam created via SPDC. The idler beam is mixed with a phase reference in an optical homodyne measurement.

Due to the number uncertainty of the LO beam, the correlations between beams 1 and 2 are thus weakened. This erases some of the which-path information; full erasure can be accomplished by conditioning on detection at D2. This detection collapses the state of the light in the idler mode to a particular superposition of the Fock states $|0\rangle$ and $|1\rangle$ that resembles a weak coherent state, $|\alpha\rangle \approx [|0\rangle + \alpha|1\rangle]$, where α depends on the experimental parameters. The optical phase of this state, $\varphi = \arg(\alpha)$, is set by the phase difference between the pump laser and the LO. In our setup, the LO, pump and homodyne laser are all phase-locked since they all originate from the same laser. The weak coherent state conditionally created in the idler mode can therefore create a stable interference pattern upon homodyne measurement. A state preparation technique of this type was considered by Hardy [3] and has more recently been generalized by Clausen et al. [4]. The interference effect studied here is closely related to studies of the phase correlations in SPDC [5]. This effect is also related to an experiment by Zou et al. [6], whereby overlapping idler modes from two different down-conversion crystals induced a well-defined *relative* phase between the two signal modes. The first-order interference effect reported in that experiment occurred without a conditional measurement. This work builds on some of our previous experiments that focus on effective nonlinearities at single photon intensities [7,8].

Light from an ultrafast Ti:Sa laser is frequency doubled to serve as the pump laser, and some of the fundamental is highly attenuated and used to create both the homodyne laser and the LO. We use type-II phase-matching for down-conversion and, after spatial and spectral filtering, send the vertically polarized beam to detector 2 and the horizontally polarized beam to the homodyne setup containing detector 1. The vertically polarized LO passes through the crystal into the signal mode and is sent, along with the signal, to detector 2. Motorized stages are placed in trombone delay arms in the pump laser path and the homodyne beam path. Fig. 2. shows the coincidence rate as a function of the delay introduced by the motor in the pump path (Fig. 2a.), and the motor in the homodyne path (Fig. 2b.).

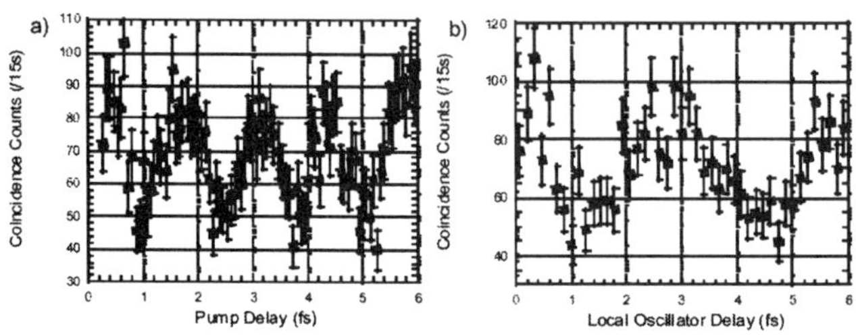

Figure 2. The conditional intensity (coincidence rate) vs time delay for a) the pump laser, and b) the homodyne laser.

This is the homodyne interference pattern conditioned on a photon in the signal mode. The fringe spacings in the two plots correspond to the pump laser frequency (Fig. 2a.) and the fundamental frequency (Fig. 2b.)

We have demonstrated that the idler beam created via SPDC can be collapsed to a state resembling a weak coherent state with a well-defined optical phase by performing a phase-sensitive measurement on the signal beam. While this setup brings to mind the classical nonlinear effect of difference frequency generation (DFG), this new effect differs in some very important respects. DFG occurs via stimulated emission and our LO is of insufficient intensity to produce a measurable amount of stimulated emission; our LO singles rate is only about 2×10^4 s^{-1}, which corresponds to 1 photon per 4000 pulses. Also, the light produced by classical DFG is first-order coherent regardless of the intensity in the signal mode; no conditional measurement would be required to observe intensity fringes. In our experiment, the detection of a photon in the signal mode is a phase-sensitive post-selection, that, upon success, collapses the idler to a weak coherent state.

[1] J.G. Rarity, P.R. Tapster, and R. Loudon, "Nonclassical interference between independent sources", quant-ph/9702032 (1997).

[2] C.K. Hong, Z.Y. Ou, and L. Mandel, "Measurement of subpicosecond time intervals between two photons by interference", Phys. Rev. Lett., **59**, 2044 (1987).

[3] L. Hardy, "Nonlocality of a single photon revisited" Phys. Rev. Lett., **73**, 2279 (1994).

[4] J. Clausen, H. Hansen, L. Knoll, J. Mlynek, and D.G. Welsch, "Conditional quantum state engineering in repeated 2-photon downconversion", quant-ph/0007050 (2000).

[5] A. Kuzmich, I.A. Walmsley, and L. Mandel, "Violation of Bell's Inequality by a Generalized Einstein-Podolsky-Rosen State Using Homodyne Detection", Phys. Rev. Lett., **85**, 1349 (2000)

[6] X.Y. Zou, L.J. Wang, and L. Mandel, "Induced coherence an indistiguishability in optical interference", Phys. Rev. Lett., 67, 318 (1991).

[7] K.J. Resch, J.S. Lundeen, and A.M. Steinberg, "Experimental observation of nonclassical effects on single-photon detection rates" Phys. Rev. A, **63**, 020102(R) (2001).

[8] K.J. Resch, J.S. Lundeen, and A.M Steinberg, "Nonlinear optics with less than one photon", quant-ph/0101020, (2000).

Quantum descriptions of light propagation in a parametric amplifier

V. Peřinová, A. Lukš

Laboratory of Quantum Optics, Faculty of Natural Sciences, Palacký University, Třída Svobody 26, 77146 Olomouc, Czech Republic

perinova@risc.upol.cz

Abstract: Quantum descriptions of light propagation in a dielectric medium have been appropriately chosen and generalized and compared in the application to a parametric amplifier.

OCIS codes: (270.5530) Pulse propagation and solitons;(270.5580) Quantum electrodynamics

Quantum analysis of light propagation in a parametric amplifier is suitable for the demonstration of a macroscopic theory of quantum electrodynamics in dielectric media. It was performed using temporal modes in [1]. The temporal mode approach differs from the ordinary spatial mode approach by the exchange of the ordinary time variable for a spatial coordinate. In [1], the field is separated into two parts: The pump field that is taken as monochromatic nondepleted classical field of frequency $\omega_{\rm p}$ and the downconverted field, which is nonmonochromatic, but centered about the frequency $\omega_{\rm c} = \omega_{\rm p}/2$. The time periodicity T of the field is assumed. The electromagnetic fields are decomposed, dependent on the refractive index $n(\omega_m)$ of the medium. The medium is assumed to be lossless and the refractive index $n(\omega_m)$ to be real. Here ω_m are discrete frequencies, $\omega_m = \frac{2m\pi}{T}$. The spatial progression is described by the momentum operator

$$\hat{G}(z) = \hat{G}_{\rm lin}(z) + \hat{G}_{\rm nonlin}(z), \tag{1}$$

$$\hat{G}_{\rm lin}(z) = \hbar \sum_m k_m \hat{a}^\dagger(z,\omega_m)\hat{a}(z,\omega_m), \tag{2}$$

$$\hat{G}_{\rm nonlin}(z) = \frac{\hbar}{4}\sum_m \lambda(\epsilon_m)\left[\hat{a}^\dagger(z,\omega_{\rm c}+\epsilon_m)\hat{a}^\dagger(z,\omega_{\rm c}-\epsilon_m)e^{ik_{\rm p}z} + \text{H. c.}\right], \tag{3}$$

where $\hat{a}(z,\omega_m)$ and their conjugates form a set of creation and annihilation operators, $\lambda(\epsilon_m)$ is a coupling constant with the property $\lambda(\epsilon_m) = \lambda(-\epsilon_m)$, $k_m \equiv k(\epsilon_m) = \frac{n(\omega_m)\omega_m}{c}$ are the wave vectors of the downcoverted fields in the medium, $\epsilon_m = \omega_m - \omega_{\rm c}$, $k_{\rm p} = \frac{n(\omega_{\rm p})\omega_{\rm p}}{c}$ is the wave vector of the pump field. In the spatial Heisenberg picture, the equations of motion are

$$\frac{\partial}{\partial z}\hat{a}(z,\omega_{\rm c}+\epsilon_m) = ik_m\hat{a}(z,\omega_{\rm c}+\epsilon_m) + \frac{i}{2}\lambda(\epsilon_m)e^{ik_{\rm p}z}\hat{a}^\dagger(z,\omega_{\rm c}-\epsilon_m) \tag{4}$$

and their solution is well known.

Deutsch and Garrison have approached degenerate parametric amplification [2] using a direct space approach, in which the usual spatial modes are not introduced and the boson quantum field $\hat{\Psi}(z,t)$ is studied instead of them. They have found that the steady state solution is a squeezed vacuum with a squeezing parameter gradually increasing in the direction of pumping beam. In [2], the envelope Hamiltonian has been considered

$$\hat{H}_{\rm env} = -i\frac{\hbar}{2}\int_{-\infty}^{\infty}\frac{c}{n(z,\omega_{\rm c})}\left[\hat{\Psi}^\dagger(z)\frac{\partial\hat{\Psi}(z)}{\partial z} - \frac{\partial\hat{\Psi}^\dagger(z)}{\partial z}\hat{\Psi}(z)\right]dz, \tag{5}$$

where $n(z, \omega_c)$ is the refractive index, which we consider to be space dependent, $\hat{\Psi} \equiv \hat{\Psi}(z,t)$. For $n(z, \omega_c) = 1$, this Hamiltonian is a generator of direct space translation

$$\exp\left\{\frac{i}{\hbar}\hat{H}_{\mathrm{env}}t\right\}\hat{\Psi}(z)\exp\left\{-\frac{i}{\hbar}\hat{H}_{\mathrm{env}}t\right\} = \hat{\Psi}(z-ct), \tag{6}$$

the field operators being taken at equal times. Here an interaction Hamiltonian is considered, which can be slightly generalized and written in the form

$$\hat{H}_{\mathrm{int}}(t) = i\frac{\hbar}{2}\int_0^L \frac{c}{n(z,\omega_c)}\left[\kappa^*(z,t)\hat{\Psi}^2(z,t) - \kappa(z,t)\hat{\Psi}^{\dagger 2}(z,t)\right]dz, \tag{7}$$

where $\kappa(z,t)$ is a coupling constant.

In addition to these macroscopic approaches the so-called microscopic approach [3] can be treated. It has the advantage of introducing the continuum of modes in the medium, but not trying to consider such a continuum in the free space. Along with the dispersion, the absorption is included. The advantage of this approach is that the refractive index can be complex, and it describes both the dispersion and absorption. In the framework of this approach forward propagating fields $\hat{c}_+(z,\omega)$ have been introduced, which satisfy equal-space commutation relations

$$[\hat{c}_+(z,\omega), \hat{c}_+^\dagger(z',\omega')] = \exp[-\mathrm{Im}\{K(\omega)\}|z-z'|]\delta(\omega-\omega')\hat{1}, \tag{8}$$

where $K(\omega)$ are complex wave vectors, $K(\omega) = \frac{\omega}{c}n(\omega)$. Choosing the Hamiltonian

$$\hat{H}_{\mathrm{nonlin}} = -\frac{\hbar c}{4}\int\int\int\Big[\lambda(\omega_c+\epsilon)g(z-z')\hat{c}_+^\dagger(z,\omega_c+\epsilon)\hat{c}_+^\dagger(z',\omega_c-\epsilon)\exp\left[ik_p\frac{(z+z')}{2}\right] + \mathrm{H.\ c.}\Big]\,d\epsilon\,dz\,dz', \tag{9}$$

with

$$g(z) = \frac{1}{2\mathrm{Im}\{K(\omega)\}}\left\{-\delta''(z) - ik_p\delta'(z) + \left[\left(\frac{k_p}{2}\right)^2 + [\mathrm{Im}\{K(\omega)\}]^2\right]\delta(z)\right\}, \tag{10}$$

we obtain the spatio-temporal equations of progression

$$\left[\frac{n(\omega_c+\epsilon)}{c}\frac{\partial}{\partial t} + \frac{\partial}{\partial z}\right]\hat{c}_+(z,\omega_c+\epsilon) = -\frac{in(\omega_c+\epsilon)}{2}e^{ik_pz}\hat{c}_+(z,\omega_c-\epsilon) + \sqrt{2\mathrm{Im}\{K(\omega)\}}\hat{f}(z,\omega_c-\epsilon), \tag{11}$$

where the Langevin operators $\hat{f}(z,\omega)$ obey the simple commutation relations

$$[\hat{f}(z,\omega), \hat{f}^\dagger(z,\omega')] = \delta(z-z')\delta(\omega-\omega')\hat{1}. \tag{12}$$

This work was supported by the Ministry of Education of Czech Republic (projects no. LN00A015 and no. CEZ 314/98).

References and links

1. B. Huttner, S. Serulnik, and Y. Ben-Aryeh, "Quantum analysis of light propagation in a parametric amplifier," Phys. Rev. A **42**, 5594–5600 (1990).
2. I. H. Deutsch and J. C. Garrison, "Quantum spatial progression of squeezed light in a degenerate parametric amplifier," Opt. Commun. **86**, 311–316 (1991).
3. E. Schmidt, J. Jeffers, S. M. Barnett, L. Knöll, and D.-G. Welsch, "Quantum theory of light in nonlinear media with dispersion and absorption," J. Mod. Opt. **45**, 377–401 (1998).

Higher order photon statistics of intracavity second harmonic generation

Daniel Erenso, Ron Adams, Hua Deng, Reeta Vyas, and Surendra Singh

Physics Department, University of Arkansas, Fayetteville, AR 72701

rvyas@comp.uark.edu

Intracavity second harmonic generation (ISHG) is known to exhibit nonclassical effects such as sub-Poissonian statistics and antibunching [1, 2]. The predicted size of these effects, however, is very small. To enhance these effects a scheme based on homodyne detection was proposed [2]. In homodyne detection the light from the ISHG is combined with a coherent local oscillator (LO) at a beam splitter. We will refer to the homodyned second harmonic field as HISHG. We calculate the factorial moments for the HISHG field. Our approach is similar to that used by Dodson and Vyas [3] to discuss homodyne detection of a degenerate parametric oscillator. We obtain analytical results for the factorial moments and show that the HISHG exhibits novel higher order nonclassical effects.

The factorial moments of the photon counting distribution can be obtained from the generating function $G(s,T)$ as

$$\left\langle m^{(l)}\right\rangle = \sum_{m=1}^{\infty} m(m-1)\cdots(m-l+1)p(m,T) = (-1)^l\left[\frac{d^l}{ds^l}G(s,T)\right]_{s=0}, \qquad (1)$$

where $p(m,T)$ is the probability of detecting m photons in the counting interval $[0-T]$, and s is an auxiliary parameter. The generating function for photoelectron counting statistics measured by a detector of efficiency η placed at the output port can be written as

$$G(s,T) = \left\langle \exp\left[-s\eta\int_0^T I(t)dt\right]\right\rangle, \qquad (2)$$

where the averaging is with respect to the positive-P function of the field. $I(t)$ represents the photon flux variable of the incident beam. For the HISHG $I(T)$ can be written as [2]

$$I(t) = 2\gamma n_o\left[\left(u_2\sqrt{\mathcal{T}} + \sqrt{\bar{n}_l\mathcal{R}}\sin\phi\right)^2 - \left(u_1\sqrt{\mathcal{T}} - i\sqrt{\bar{n}\mathcal{T}} - i\sqrt{\bar{n}_l\mathcal{R}}\cos\phi\right)^2\right]. \qquad (3)$$

where γ is the cavity linewidth, $\mathcal{T}$ is the transmittivity and $\mathcal{R}$ is the reflectivity of the beam splitter, ϕ is the phase difference between the ISHG and the LO, n_o is threshold photon number of the ISHG, $\bar{n}$ and $\bar{n}_l$ are the mean photon numbers for the ISHG and the LO in units of n_o. u_1 and u_2 are real Gaussian random variables representing the field amplitudes with zero mean and correlation function given by

$$\langle u_i(t)u_j(t')\rangle \;=\; \delta_{ij}\frac{\gamma\bar{n}}{4n_o\lambda_i}e^{-\lambda_i|t-t'|}, \qquad i=1,2. \qquad (4)$$

Here $\lambda_1 = \gamma(1+3\bar{n})$ and $\lambda_2 = \gamma(1+\bar{n})$. Using the correlation properties of u_1 and u_2 and following a procedure similar to the one outlined in Ref. [3, 4], we obtain the generating function. Factorial moments are then obtained with the help of Eq. (1).

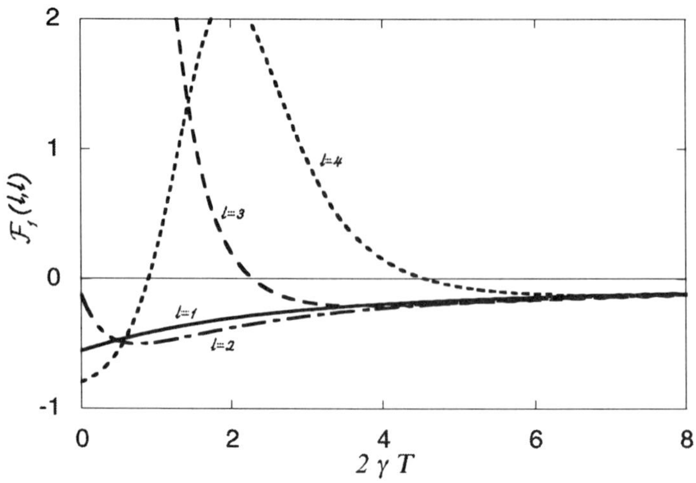

Fig. 1. Parameter $\mathcal{F}_1(\ell,\ell)$ as a function of $2\gamma T$ for $l = 1$ to 4. Parameters are $n_o = 10^6$, $\bar{n} = 0.1$, $\bar{n}_\ell = 0.0998$ $\phi = 180°$, and transmittivity $\mathcal{T}= 0.5$. $\mathcal{F}_1(\ell,\ell) < 0$ shows nonclassical effects.

It is shown in Ref. [5] that for a classical field the following parameter is always positive

$$\mathcal{F}_k(l,j) = \left[\frac{\langle m^{(l+k)}\rangle \langle m^{(j-k)}\rangle}{\langle m^{(l)}\rangle \langle m^{(j)}\rangle} - 1\right] \geq 0, \tag{5}$$

where integers $l \geq j \geq k$. However, for quantum fields this parameter can become negative [5]. Thus $\mathcal{F}_k(l,j) < 0$ reflects higher order quantum effects. Note that $\mathcal{F}_1(1,1)$ $(=< m^{(2)} > / < m >^2 -1 = Q/ < m >)$ is proportional to Q-parameter.

In Fig. 1 we have plotted $\mathcal{F}_1(\ell,\ell)$ as a function of the counting time interval $2\gamma T$ for $\ell = 1$ to 4. $\mathcal{F}_1(\ell,\ell) < 0$ shows higher order nonclassical effects. These nonclassical effects can be observed in photoelectric measurements of light and do not depend on the efficiency of detection.

This work was supported in part by the Office of Naval Research. D. Erenso acknowledges support from the ICSC World Laboratory.

References and links

1. R. Loudon, "Non-classical effects in the statistical properties of light," Rep. Prog. Phys. **43**, 913 (1980); P. D. Drumond, K. J. McNeil, and D. F. Walls, "Non-equilibrium transitions in sub/second harmonic generation. II. Quantum theory,"**28**, 211 (1981).
2. R. Vyas, C. Wang, and S. Singh, "Homodyne detection for the enhancement of antibunching," Phys. Rev. A **54**, 2391-2396 (1996).
3. A. B. Dodson and R. Vyas, "Homodyne photon statistics of the subthreshold degenerate parametric oscillator," Phys. Rev. A **47**, 3396-3412 (1993).
4. R. Vyas and S. Singh, "Photon-Counting statistics of the degenerate parametric oscillator," Phys. Rev. A **40**, 5147-5159 (1989); "Antibunching and phtoemission waiting times," J. Opt. Soc. Am. B **17**, 634-637 (2000).
5. C. T. Lee, "Higher order criteria for nonclassical effects in photon statistics," Phys. Rev. A **41**, 1721 (1990); R. Vyas and S. Singh, "Higher-order nonclassical effects in a parametric oscillator," Phys. Rev. A **62**, 033803(5) (2000).

Observation of Nonclassical Photon Statistics in Two-Photon Interference

Y. J. Lu and Z. Y. Ou
Department of Physics, Indiana University-Purdue University Indianapolis, Indianapolis, IN 46202
Tel: (317) 274-2125, FAX: (317) 274-2393, zou@IUPUI.edu

Abstract: **We report the observation of a variety of photon statistics including anti-bunching and a new type of nonclassical effect in the interference between a narrow band two-photon state and a coherent state.**

1. Introduction

Nonclassical photon statistics such as photon anti-bunching is usually observed in two-level atomic system with resonant excitation [1]. This is the first observed nonclassical effect requiring a full quantum description of light. Recently, it was suggested [2] that such an effect is also present in the homodyned field between the output of an optical parametric oscillator (OPO) and a strong coherent local oscillator. Because homodyning is involved, its origin must come from quantum interference between the two fields.

Coherent homodyne of down-converted field was first studied by Grangier *et al.* [3] to investigate nonlocal effect of two-photon system. It involves interference between coherent local fields and down-converted fields. Two-photon interference produces sinusoidal modulation of two-photon coincidence. It turns out that with a proper phase, such a two-photon interference effect will give rise to cancelled two-photon events in the mixed field leading to anti-bunching. Since the two fields involved in interference are independent of each other, observation of such an effect will require a narrow band two-photon source so that detectors can resolve them temporally.

2. Theory

Consider a two-photon state, given by $|\eta\rangle \approx |0\rangle + \eta\,|2\rangle$ where $\eta << 1$ (this is a weak squeezed state), and a coherent state $|\alpha\rangle \approx |0\rangle + \alpha|2\rangle + (\alpha^2/\sqrt{2})\,|2\rangle$ with $\alpha << 1$. Let us mix them with a 50:50 beam splitter as in Fig.1.

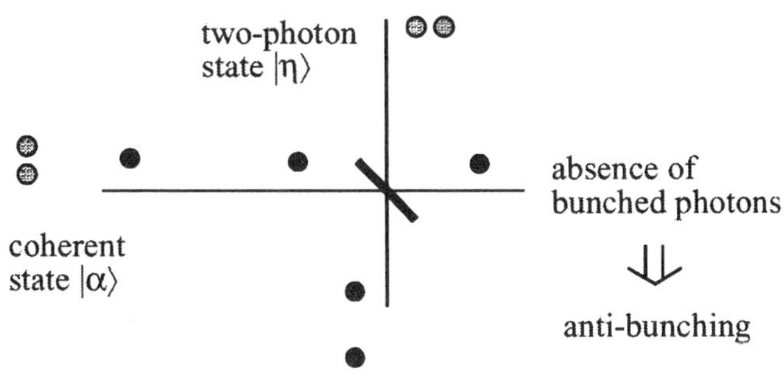

Fig. 1. Homodyne of a two-photon state with a weak coherent state for two-photon interference.

The output state can be found to be

$$|\Psi\rangle = |0\rangle + \alpha/\sqrt{2}\{|1,0\rangle+|0,1\rangle\} + \eta/\sqrt{2}\,|1,1\rangle, \tag{1}$$

when $\eta = \alpha^2/\sqrt{2}$. As seen in Eq.(1), the state $|2\rangle$ does not exists in the output state due to destructive two-photon interference. This indicates a photon anti-bunching effect.

3. Experiment

A recently produced narrow band two-photon state from an OPO operated far below threshold [4] will serve as the two-photon source and a laser of the same frequency as the coherent field. Fig.2 shows the results of Hanbury Brown-Twiss type intensity correlation measurement on the mixed field. Different types of photon statistics are obtained as the relative phase and intensity of the two fields varies. Fig.2a and b show the typical photon bunching and anti-bunching effects. Fig.2c shows a new type of nonclassical photon statistics, where the Schwartz inequality $|g^{(2)}(0) - 1| \geq |g^{(2)}(\tau) - 1|$ is violated [5]. Here $g^{(2)}(\tau)$ is the normalized intensity correlation function.

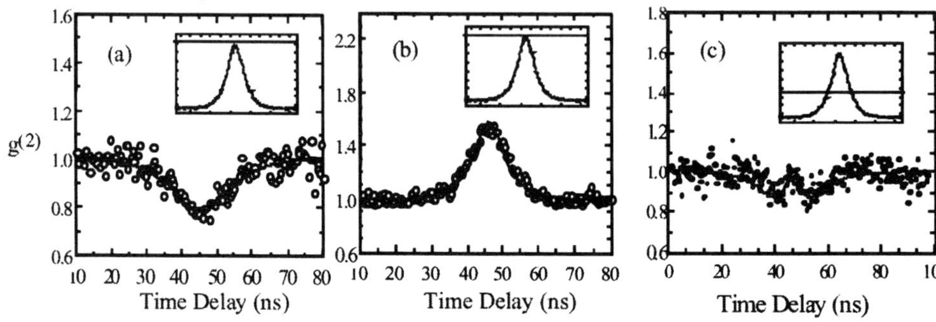

Fig. 2 Intensity correlation function for different situations in the mixing of a narrow band two-photon source and a coherent field. (a) Relative phase is for destructive interference; (b) relative phase is for constructive interference; (c) relative phase is same as (a) but relative strength is different. The inserts show the relative two-photon strength of each field. The time axis is offset and zero time delay is at about 46 ns.

Understanding of Fig.2a,b is straightforward from part 2. But the strange shape of $g^{(2)}(\tau)$ in Fig.2c needs a few words. the inserts of each figure show the relative two-photon strength of the two sources. Because of the finite correlation time for the two-photon source, the two-photon event has different strength at different time delay with highest at zero delay but coherent state gives flat dependence due to the randomness of photon arrival. Fig.2a,b correspond to the case when the two strengths are equal at zero time delay and complete destructive(constructive) interference results in less (more) two-photon events at the zero time delay. But for Fig.c, the two strengths are equal at two nonzero time delays (one before and one after) thus creating the double dip feature. The dips do not go all the way to zero because all the figures show the raw data without subtraction of background. Other factor such as phase fluctuation may also raise the level the dips.

This work is supported by the Office of Naval Research.

4. References

[1] H. J. Kimble, M. Dagenais, and L. Mandel, Phys. Rev. Lett. **39**, 691 (1977).
[2] A. B. Dodson and Reeta Vyas, Phys. Rev. A**47**, 3396 (1993).
[3] P. Grangier, M. J. Potasek, and B. Yurke, Phys. Rev. A**38**, 3182 (1988).
[4] Z. Y. Ou and Y. J. Lu, Phys. Rev. Lett. **83**, 2556 (1999); Y. J. Lu and Z. Y. Ou, Phys. Rev. A**62**, 033804 (2000).
[5] P. R. Rice and H. J. Carmichael, IEEE J. Quantum Electron. **24**, 1352 (1988).

Demonstration of the complementarity of one- and two-photon interference

A. F. Abouraddy*, B. E. A. Saleh, A. V. Sergienko, and M. C. Teich
Quantum Imaging Laboratory, Departments of Electrical & Computer Engineering and Physics, Boston University, 8 Saint Mary's Street, Boston, MA 02215
*raddy@bu.edu, http://www.bu.edu/qil

Abstract:

The visibilities of second-order (single-photon) and fourth-order (two-photon) interference have been observed in a Young's double-slit experiment using light generated by spontaneous parametric down-conversion. As the effective source size is increased, coherence is diminished while entanglement is enhanced. This is the first experimental demonstration of this complementarity in the spatial domain.

OCIS codes: (270.0270) Quantum optics, (030.0030) Coherence and statistical optics

Summary

The conventional theory of optical coherence dictates that the visibility of interference fringes equals the degree of coherence. In a Young's double-slit experiment using an incoherent light source, the degree of coherence at the slits is inversely proportional to the angular size of the source. The visibility of two-photon interference fringes in a Young's double-slit experiment is governed by the degree of entanglement. For light generated by spontaneous parametric down-conversion in a nonlinear crystal, the degree of entanglement is controlled by the size of the source (the width of the pump beam). A smaller source size corresponds to reduced entanglement, and therefore to reduced visibility of two-photon interference.

A basic complementarity between coherence and entanglement underlies the complementarity between single- and two-photon interference. We have recently shown [1] that the dependence of the second-order coherence function and the two-photon wavefunction on the source size are mathematically identical, so that there is a duality between the two systems. A source of large area generates light of low coherence, but yields highly entangled photon pairs. In contrast, a small source emits light of high spatial coherence, but generates poorly entangled photon pairs. The purpose of this paper is to report the first experimental demonstration of this complementarity.

The experimental setup is illustrated in Fig. 1. A 1-mm thick $LiIO_3$ crystal of 6×6 mm cross-section is pumped by a 35-mW Kr^+-ion laser of 406-nm wavelength to generate spontaneous parametric down-converted (SPDC) light in the degenerate collinear Type-I configuration. The down-converted beam is passed through a circular aperture of 2-mm diameter and through a double-slit aperture at a variable distance d from the circular aperture. The slits are of width 0.35 mm each and are separated by a distance a = 0.70 mm. The unconverted pump is prevented from reaching the slits by use of a combination of two Glan-Thompson prisms oriented with orthogonal polarizations before and after the crystal. The double-slit aperture is followed by a 2-f system using a C-mount Nikon lens of focal length f = 50 mm. The detector is an intensified CCD (ICCD) camera with 512×512pixels, each of size 24×24 μm^2. The digitized analog signal is transferred to a computer for subsequent analysis. The experiment was repeated 5 times at different angular sizes of the source, ranging from near field to far field (d = 5.5, 6.3, 30, 54, and 87 cm). In each case the one- and two-photon interference patterns were determined and the corresponding visibilities were estimated. The visibilities of single- and two-photon interference, V_{1m} and V_{12} respectively, for the five experiments are displayed in Fig. 2 (open circles), together with a plot of the theoretical (ideal) complementarity relation (solid curve), $V_{12}^2 + V_{1m}^2 = 1$. The ideal relationship is derived under the assumption of a thin crystal and narrow spectral SPDC bandwidth. The dashed curve in Fig. 2, which was obtained by simulating Eqs. (5.6), (5.8), and (5.10) in Ref. [1] and using an optical-system impulse response function representing free-space propagation from the crystal to the double slits and the ensuing 2-f system. This accommodates a crystal of finite thickness; however, the down-converted light was taken to be narrow with respect to the central down-conversion wavelength.

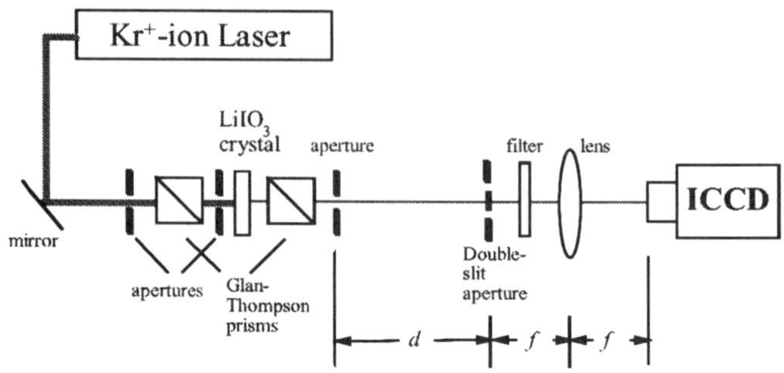

Fig. 1. Experimental arrangement.

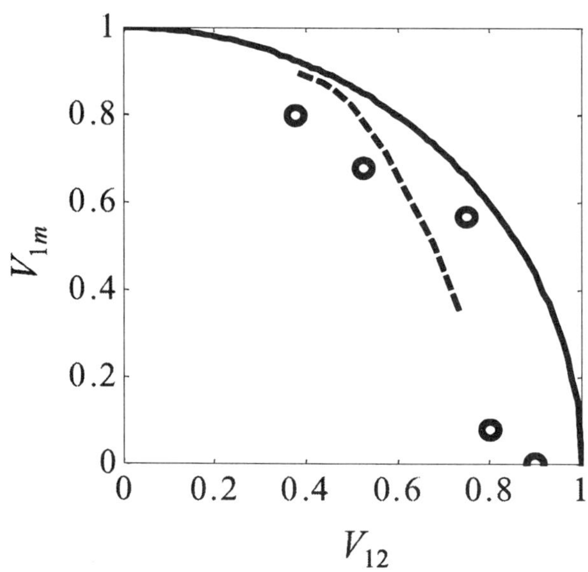

Fig. 2. The marginal one-photon visibility V_{1m} versus the two-photon visibility V_{12}. Solid curve represents the ideal complementarity relationship $V_{12}^2 + V_{1m}^2 = 1$. The experimental results are indicated by circles. Dashed curve represents the results of a simulation that takes into consideration the experimental geometry and the crystal thickness.

[1] B. E. A. Saleh, A. F. Abouraddy, A. V. Sergienko, and M. C. Teich, "Duality between partial coherence and partial entanglement," Phys. Rev. A **62**, 043816 (2000).

The role of hyperentanglement in quantum interferometry

M. Atatüre, G. Di Giuseppe, M. Shaw, A. V. Sergienko*, B. E. A. Saleh, M. C. Teich

Quantum Imaging Laboratory, Department of Electrical & Computer Engineering and Department of Physics
Boston University, 8 Saint Mary's St., Boston, MA 02215, USA
*AlexSerg@bu.edu; http://www.bu.edu/qil/

Abstract: We discuss the role of hyperentanglement in quantum-interference experiments that involve the nonlinear process of spontaneous parametric down-conversion. A multidimensional theory of parametric down-conversion is presented along with experimental verification for various interferometry arrangements.

OCIS codes: (270.0270) Quantum Optics, (190.7110) Ultrafast Nonlinear Optics

Summary

Entanglement has proven to be an important resource in generating new optical states that can be used in quantum information processing, secure communications, imaging, and precise optical measurements. The transition to experimental reality for these states is often a challenging step, however. One of the most accessible sources of entangled-photon states is a nonlinear optical process called spontaneous parametric down-conversion (SPDC), in which an optical crystal illuminated by a laser beam spontaneously generates pairs of photons. Each generated pair comprises a hyperentangled state in an infinite-dimensional Hilbert space, and is simultaneously entangled in momentum, frequency, and polarization. One can exploit the richness of this state in many powerful ways that range from probing the fundamental ideas of quantum mechanics via Bell's inequalities and GHZ tests, to practical applications including metrology, quantum imaging, and secure communications.

In particular, femtosecond parametric down-conversion serves as a source for demonstrating counterintuitive effects such as quantum teleportation and entanglement swapping, as well as for generating multi-photon entangled states. It has been shown previously that dispersion effects arising from the use of ultra-short femtosecond laser pulses in SPDC result in a dramatic degradation of entanglement. Previous attempts to describe this phenomenon have been based on separate analyses of frequency, momentum, and space-time variables and, as such, have failed to provide a comprehensive theory that accords with experimental observations.

We present a formalism that incorporates the hyperentangled nature of the quantum state generated via SPDC. The key to providing an adequate physical description of parametric down-conversion is to consider the initial entangled state in its entirety, including the interplay among the different quantum variables. To cover a broad range of experiments the theory is kept general and comprises three fundamentally distinct steps: generation, propagation, and measurement. In proceeding thusly we refrain from imposing the characteristics of a particular measurement apparatus onto the structure of the quantum state. Depending on the intended application it is now possible to design an optimal experimental system that modifies the quantum state in accordance with the desired aim of a particular experiment. The validity of our model is confirmed by its agreement with ultrafast and cw-pumped SPDC experiments carried out in our Laboratory.

Eliminating frequency and space-time entanglement in multi-photon states

W. P. Grice
Center for Engineering Science Advanced Research
Oak Ridge National Laboratory, Oak Ridge, Tennessee 37830

A. B. U'Ren and I. A. Walmsley
The Institute of Optics, University of Rochester, Rochester, NY 14627

Abstract: Visibility in interference experiments involving photon pairs from separate down-conversion events is reduced when these pairs are frequency-entangled. We present a method for eliminating the entanglement at the source, making pre-detection spectral filters unnecessary.

OCIS codes: (999.9999) Ultrafast quantum optics; (270.1670) Coherent optical effects.

Experimentalists in the field of quantum optics have performed numerous interesting experiments utilizing pairs of photons generated in the process of spontaneous parametric down-conversion. Although it was the two-photon state that was of interest in most of these experiments, a number of recent proposals require states of greater than two photons. In the few experiments that have been performed to date, these multi-photon states have been "constructed" from pairs of photons emitted in separate down-conversion events. Generating multi-photon states in this way exploits many of the properties of the constituent pairs and it simplifies matters experimentally because it reduces the number of sources required.

However, this method does present a problem: since energy must be conserved in the down-conversion process, the emitted pairs are entangled in frequency. This entanglement makes it possible to identify members of the multi-photon state as siblings, thus providing information that may render interfering pathways distinguishable. Two recent experiments[1,2], for example, relied on Bell state measurements in which it was impossible to determine the origins of the detected photons. This was accomplished by placing narrow spectral filters in front of the detectors to minimize the effect of the entanglement. However, this method of removing the distinguishing spectral information results in a greatly diminished count rate. We propose an alternate solution to this problem—one that does not require pre-detection filtering. We show that by adjusting parameters such as crystal length, crystal material, pump bandwidth, and center pump wavelength, the frequency entanglement is eliminated at the down-conversion source. With no need for spectral filters, counting rates would improve significantly.

The two-photon state produced in the process of spontaneous parametric down-conversion is $|\psi\rangle = (\frac{1}{2\pi})^2 \iint d\omega_s d\omega_i f(\omega_s, \omega_i) \hat{a}_s^\dagger(\omega_s) \hat{a}_i^\dagger(\omega_i)|vac\rangle$. In general, this state is entangled in frequency, although if the probability amplitude can be written as $f(\omega_s, \omega_i) = f_s(\omega_s) f_i(\omega_i)$, then the state becomes $|\psi\rangle = (\frac{1}{2\pi})^2 \int d\omega_s f_s(\omega_s) \hat{a}_s^\dagger(\omega_s) \int d\omega_i f_i(\omega_i) \hat{a}_i^\dagger(\omega_i)|vac\rangle$, which is a direct product state. That is, the degree of frequency entanglement is zero if the probability amplitude is factorable. In addition, the degree of *space-time* entanglement is zero when this condition is satisfied. This follows directly from the relationship between the spectral and temporal representations of the two-photon state. In the temporal domain, the state is $|\psi\rangle = \iint dt_s dt_i g(t_s, t_i) \hat{a}_s^\dagger(t_s) \hat{a}_i^\dagger(t_i)|vac\rangle$, where $\hat{a}_{s,i}(\omega)$ and $f(\omega_s, \omega_i)$ are the Fourier and double Fourier transforms of $\hat{a}_{s,i}(t)$ and $g(t_s, t_i)$, respectively. Thus, if there exist functions $f_s(\omega_s)$ and $f_i(\omega_i)$ such that $f(\omega_s, \omega_i) = f_s(\omega_s) f_i(\omega_i)$, then there also exist $g_s(t_s)$ and $g_i(t_i)$ such that $g(t_s, t_i) = g_s(t_s) g_i(t_i)$.

In order to determine how this factorability condition is related to experimental parameters, we examine the two-photon state in more detail. The general expression for the two-photon amplitude is $f(\omega_s,\omega_i)=\alpha(\omega_s+\omega_i)\phi(\omega_s,\omega_i)$, where $\alpha(\omega_s+\omega_i)$ is a function describing the envelope of the pump field and $\phi(\omega_s,\omega_i)$ is a phase-matching function[3]. A plot of $|f(\omega_s,\omega_i)|^2$ is shown in Fig. 1(a) for type-II down-conversion in Beta Barium Borate (BBO) with a 400 nm pump. It is evident in this plot that $f(\omega_s,\omega_i)$ is not factorable, since the range of idler frequencies depends somewhat upon the signal frequency.

In general, it is not possible to express the two-photon amplitude as a product of functions of the respective frequencies. However, if $\phi(\omega_s,\omega_i)$ and $\alpha(\omega_s+\omega_i)$ are both approximated by Gaussians, and if only the lowest order terms in power series expansions of the wave numbers are retained, then $f(\omega_s,\omega_i)$ is factorable (and the state is not frequency-entangled) as long as

$$\frac{1}{\sigma^2}+\gamma L^2(k_p'-k_s')(k_p'-k_i')=0\,. \tag{1}$$

In this expression, σ is the $1/e$ width of $\alpha(\omega_s+\omega_i)$, L is the crystal length, $\gamma=0.04822$, and the primes indicate derivatives with respect to frequency.

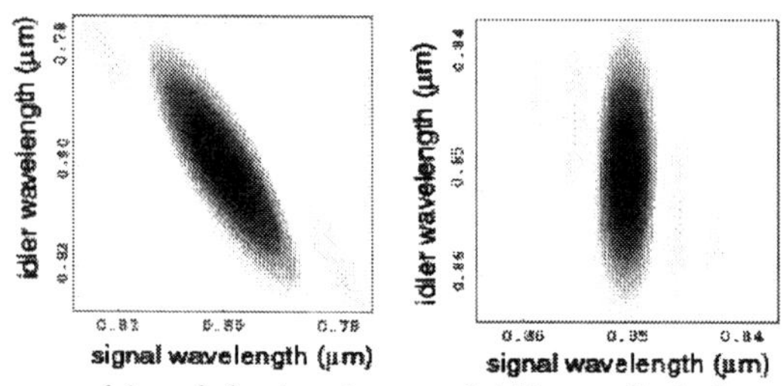

Fig. 1. The square modulus of the two-photon probability amplitude in type-II (a) BBO pumped at 400 nm; and (b) ADP pumped at 425 nm. As an aid to the reader, the axes have been converted from frequency to wavelength.

Through the parameters σ and L, the expression in Eq. (1) imposes a constraint on the relative widths of the pump envelope and phase-matching functions. The equation can not be satisfied at all, though, unless $(k_p'-k_s')(k_p'-k_i')<0$. Since the derivatives are the reciprocals of the respective group velocities, this inequality will be satisfied only if the group velocity of the pump lies between the group velocities of the signal and idler fields. This group velocity mismatch can be realized in the degenerate case only in a type-II configuration and does not occur in any of the recently reported experimental schemes. However, it can be achieved in Ammonium Dihydrogen Phosphate (ADP) as long as the down-converted fields lie within the range 0.82 µm < λ < 1.64 µm. The plot in Fig. 1(b) shows $|f(\omega_s,\omega_i)|^2$ for type-II ADP with the pump centered at 425 nm and with the pump bandwidth and crystal length chosen so as to satisfy Eq. (1). Note that the range of idler frequencies is independent of the signal frequency.

When experimental constraints (*e.g.*, available pump wavelengths) make it impossible to realize the group velocity mismatch, the two-photon state will retain some frequency-entanglement. It is possible, nonetheless, to reduce the *degree* of entanglement. That is, the degree of entanglement may be minimized with respect to the remaining parameters—crystal length and pump bandwidth. Of course, the minimum degree of entanglement attainable via this method is lower for some crystal materials than for others. This research was supported by the Engineering Research Program of the US Department of Energy, Office of Basic Energy Sciences under contract No. DE-AC05-00OR22725 with UT-Battelle, LLC.

[1] D. Bouwmeester, et al., "Experimental quantum teleportation," Nature, **390**, 575-579, (1997).
[2] J. Pan, et al., "Experimental entanglement swapping: entangling photons that never interacted," Phys. Rev. Lett., **80**, 3891-3894, (1998).
[3] W. Grice and I. Walmsley, "Spectral information and distinguishability in type-II downconversion with a broadband pump," Phys. Rev. A, **56**, 1627-1634, (1997).

OPO in transverse degenerate cavities: multimode operation and spatial quantum effects

S.Ducci, N. Treps, M. Martinelli, A. Maître, C. Fabre
Laboratoire Kastler Brossel, Université Pierre et Marie Curie, Case 74, 75252 Cedex 05 France
mmartine@spectro.jussieu.fr

Abstract:

Simultaneous OPOs oscillation on different transverse modes can be a source of spatial quantum correlation and local squeezing. We have experimentally shown such multitransverse mode operation, a preliminary condition for observation of spatial quantum effects.

Nonclassical states of light have been widely studied for the last decades. As an example of these nonclassical states, we can mention the intensity noise fluctuations reduced below the shot noise limit, the generation of intense beams with noise compressed in one quadrature, squeezed vacuum fluctuations in the output of sub-threshold OPOs, and twin beams generation with correlated intensity fluctuations. Most of time one deals with a bright beam, or vacuum fluctuations that are measured in homodyne detection with the help of a coherent TEM_{00} beam.

Generally the transverse section of the beam is completely integrated on the detection process, the light being focused on the detector to avoid loss of information that would reduce the nonclassical effect of the measurement. Blocking part of the beam or measuring the noise fluctuation in a small region of the single transverse mode squeezed beam wavefront would reduce the noise compression, just like the attenuation of the beam by a neutral density filter. For these beams, no quantum correlation between different points in the wavefront is expected.

On the other hand, multitransverse nonclassical states can present quantum correlation between different points of the beam wavefront or local squeezing. This is characterized by the noise reduction even if we sample only a small region of the beam. As shown in [1,2], the single transverse mode squeezed state cannot be used to go beyond the shot noise limit in the resolution of optical images. This problem appears also in the measurement of small displacements by light beams [3], where these multitransverse squeezed modes can improve the final resolution.

Optical Parametric Oscillators operating in cavities with degenerate transverse modes are very promising for the generation of these multitransverse nonclassical states. Many theoretical works in the subject have discussed the generation of spatial squeezing in plane cavity OPO [4], quasi-planar cavity [5], or confocal cavity pumped by a plane wave [6]. It was shown that in this configuration we can have multitransverse mode squeezed vacuum, which can be measured by any axially symmetrical local oscillator.

Another interesting possibility is the light generation of twin beams in concentric cavity operation. In this condition, a generated signal-idler pair of photons would keep their transverse momentum correlation until they leave the cavity. Output beams are then expected to present their well-known twin behavior (intensity correlation of total signal and idler beams) but also local correlation of the intensity fluctuations. More than twin, these "siamese beams" can demonstrate the spatial noise compression.

All these nonclassical multitransverse mode states require the simultaneous oscillation of many different transverse modes. We have experimentally demonstrated

Coherence and Quantum Optics VIII
Edited by Bigelow et al., Kluwer Academic/Plenum Publishers, 2003

this operation of an OPO, made of a KTP crystal in a linear cavity pumped by the second harmonic of a Nd:YAG laser. The concentric, semiconcentric and confocal cavities were studied, presenting a rich variety of patterns. In a confocal cavity, a stable operation is obtained, and the OPO can be seen to produce a multimode beam, which can be described in a Laguerre-Gauss base by a sum of 25 different modes [7]. The concentric and semiconcentric cavities are in a critical stability limit, and their operation is much more delicate. Nevertheless, patterns are also observed in these configurations.

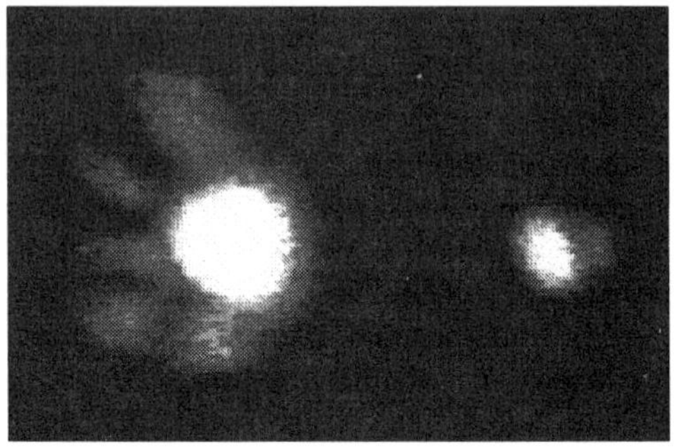

Fig. 1: Transverse patterns of the OPO. Left: Idler pattern in a confocal configuration
Right: Signal and idler structures for a concentric cavity.

We are presently working on the experimental setup for the measurement of spatial quantum effects. The measurement of spatial correlation is made by a pair of high quantum efficiency detectors with four quadrant division each. This complex eight detector system uses a dedicated demodulation circuit for the noise measurement and a computer controlled acquisition system, with specific software for data analysis. The spatial noise correlation of these transverse multimode beams can now be studied, and the noise fluctuations between each quadrant can be compared. Preliminary results of this system will be presented at the conference.

---

1 Mikhail I.Kolobov, "The spatial behavior of nonclassical light", Rev. Mod. Phys. **71**, 1539-1589 (1999).

2 Mikhail I.Kolobov, and C. Fabre, "Quantum limits in optical resolution", Phys. Rev. Lett. **85**, 3789-3792 (2000).

3 C. Fabre, J. B. Fouet, and A. Maître, "Quantum limits in the measurement of very small displacements in optical images", Opt. Lett. **25**, 76-78 (2001).

4 L. A. Lugiato, A. Gatti, H. Wiedermann, "Quantum fluctuations and nonlinear optical patterns", in *Les Houches, Session LXIII, 1995: Quantum Fluctuations* , Course 12, S. Reynaud, E. Giacobino, and J. Zinn-Justin , eds. (Elsevier Science, 1997)

5 M. Marte, H. Ritsch, K. I. Petsas, A. Gatti, L. A. Lugiato, C. Fabre, and D. Leduc, "Spatial patterns in optical parametric oscillators with spherical mirrors: classical and quantum effects", Opt. Express **3**, 71-80 (2000).

6 L. A. Lugiato, and Ph. Grangier, J. Opt. Soc. Am. **B 14**, 225 (1997).

7 M. Vaupel, A. Maître, and C. Fabre, "Observation of pattern formation in optical parametric oscillators", Phys. Rev. Lett. **83**, 5278-5281 (1999).

OPTICAL PATTERNS SUSTAINED BY QUANTUM NOISE

R. Zambrini[1], S. M. Barnett[2], M. San Miguel[1] and P. Colet[1]

[1] *Instituto Mediterráneo de Estudios Avanzados, IMEDEA (CSIC-UIB),Campus Universitat Illes Balears,07071 Palma de Mallorca,Spain.*
[2] *Department of Physics and Applied Physics,University of Strathclyde, 107 Rottenrow,Glasgow G4 0NG,Scotland,United Kingdom*

roberta@imedea.uib.es
http://www.imedea.uib.es/PhysDept/

Noise-sustained patterns have been predicted in several optical systems including cavities filled with Kerr media and Optical Parametric Oscillators[1, 2]. In OPO's, due to birefringence, walk-off between pump and signal appears. This walk-off is responsible for the existence of a convective regime, in which a transverse pattern arises as a macroscopic manifestation of amplified and spatially structured quantum noise, with magnification factors of several orders of magnitudes. These noise-sustained structures are interesting candidates for the study of quantum correlations at a macroscopic level. However, standard methods to calculate quantum fluctuations are not helpful in this situation. Linearization schemes, usually invoked to use a positive-definite Wigner distribution, can not be applied since the reference state is unstable and fluctuations are amplified. P-positive representation is also not useful because of trajectories which diverge from the reference unstable state.

We propose to study quantum correlations in the convective regime of a degenerate type-I OPO using two approximation methods. A first proposal consists in a time dependent parametric approximation, in which the intracavity pump field is taken as a classical stochastic field $\mathcal{A}_0$. In this way we have a quadratic Hamiltonian describing the quantum dynamics of the down-converted field. In the Wigner representation we obtain a stochastic equation for the c-number associated with the signal field, which is coupled to the equation defining the dynamics of $\mathcal{A}_0$:

$$\begin{aligned}\partial_t \mathcal{A}_0(\vec{x},t) &= -[(1+i\Delta_0)-i\nabla^2]\mathcal{A}_0(\vec{x},t) - \frac{1}{2}\alpha_1^2(\vec{x},t) + E_0(\vec{x}) \qquad (1)\\ \partial_t \alpha_1(\vec{x},t) &= -\left[(1+i\Delta_1)-2i\nabla^2+v\partial_y\right]\alpha_1(\vec{x},t) + \mathcal{A}_0(\vec{x},t)\alpha_1^*(\vec{x},t) + \frac{1}{\sqrt{a}}\frac{g}{\gamma}\xi_1(\vec{x},t)\end{aligned}$$

$\xi(\vec{x},t)$ is a Gaussian white noise with correlation $\langle \xi(\vec{x},t)\xi^*(\vec{x}',t')\rangle = \delta(\vec{x}-\vec{x}')\delta(t-t')$, γ is the cavity decay rate, $\Delta_{0,1}$ are the cavity detunings, E_0 is the input field and v and g are the walk-off and nonlinear coefficients.

A second method consists in the use of the Q representation: due to non-linearity diffusion is not always positive, but in the range of parameters of the convective regime we find that fluctuations of the stable pump, for which positiveness of diffusion is not fulfilled, have a negligible probability to occur. We obtain two Langevin equations, similar to the ones above but with a Gaussian white noise also in the pump field and a *phase sensitive multiplicative* noise in the signal, depending on the instantaneous value of the pump.

Results obtained with both approximations are in good agreement and reveal the importance of non-linearities in the dynamics in particular regimes, as at threshold (of signal generation) or in the convective regime.

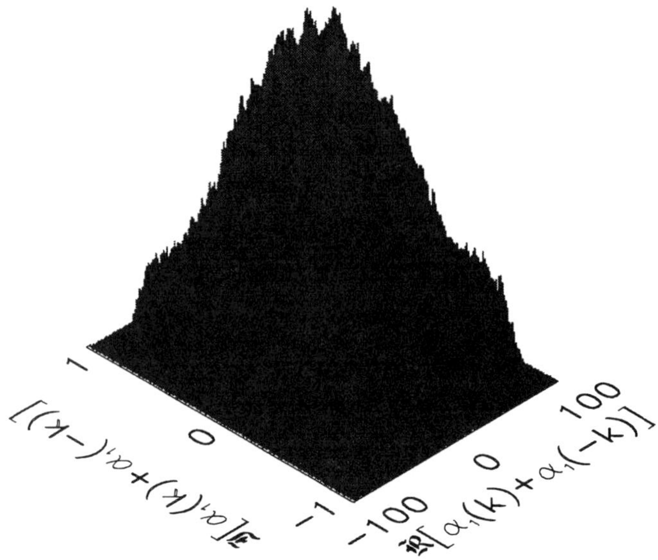

Fig. 1. Wigner distribution function associated with the superposition of signal modes $W(\alpha_1(k)+\alpha_1(-k))$, for a couple of excited modes in the convective regime.

We have studied the behavior of fluctuations in the transverse pattern appearing at threshold in the signal field, increasing the pump intensity.

The *non-linear* study through the critical point reveals continuity in the variation of correlations depending on the pump intensity, while a linear model would fail, predicting diverging amplitudes [3]. The profile of near-field correlations is modulated at the wavelength of the stripe pattern appearing. Quantum effects appear: quadratures of signal modes superposition are squeezed below the shot Noise level and statistics of intensity difference of two critical modes $\pm k_c$ is sub-Poissonian.

Increasing more the pump we reach the convective regime and we observe strong amplification of signal fluctuations and consequent importance of non-linearities. The probability distributions for signal modes are not Gaussian: in the figure we show the Wigner distribution function associated with the superposition of modes $\alpha_1(k)+\alpha_1(-k)$. Very near to the threshold fluctuations are microscopic and quantum effects are preserved; however, increasing the pump intensity, fluctuations are amplified of several orders of magnitude and the quantum effects of squeezing in quadratures and sub-Poissonian statistics in twin beams intensity difference are destroyed.

References and links

1. M. Santagiustina, P. Colet, M. San Miguel, and D.Walgraef, "Noise-sustained convective structures in nonlinear optics" Phys. Rev. Lett. **79**, 3633 (1997);
2. M. Santagiustina, P. Colet, M. San Miguel, and D.Walgraef, "Two-dimensional noise-sustained structures in optical parametric oscillators," Phys. Rev. E **58**, 3843 (1998).
3. A. Gatti, H. Wiedemann, L. A. Lugiato, I. Marzoli, G. L. Oppo, and S. M. Barnett, "Langevin treatment of quantum fluctuations and optical patterns in optical parametric oscillators below threshold," Phys. Rev. A **56**, 877 (1997).

Continuously induced coherence and quantum Zeno effect

J. Řeháček, L. Mišta Jr., J. Fiurášek and J. Peřina

Department of Optics and Joint Laboratory of Optics, Palacký University, 17. listopadu 50, 772 00, Olomouc, Czech Republic

e-mail: rehacek@alpha.inf.upol.cz

In the famous experiment by Zou *et al.* it has been shown that the coherence of signal photons originating in two downconversion processes can be induced by aligning the idler beam of the first process with the idler beam of the other and making thus signal photons indistinguishable. A natural question arises whether such interplay between quantum indistinguishability and coherence also occurs in the dynamics of two downconverters experiencing continuous rather than discrete coupling.

The setup of the thought experiment which will be discussed in the following is shown in Fig. 1. Two downconversion processes of coupling strengths Γ_1 and Γ_2 are allowed to exchange energy with rate κ via idler beams. The system is effectively described by the interaction momentum operator

$$\hat{G} = \hbar(\Gamma_1 \hat{A}^\dagger_{s_1} \hat{A}^\dagger_{i_1} + \Gamma_2 \hat{A}^\dagger_{s_2} \hat{A}^\dagger_{i_2} + \kappa \hat{A}_{i_1} \hat{A}^\dagger_{i_2} + \text{h.c.}). \tag{1}$$

A typical behavior of the normalized mutual coherence function γ well below threshold is shown in Fig. 1 (solid line). The coherence of the signal modes periodically attains its maximum allowed values ± 1 and hence the signal modes become fully coherent every now and then. This coherence is not induced by stimulated emission due to the small mean number of photons present in the system (see Fig. 1, dotted line). Like Zou *et al.* [1] one can say that for the corresponding interaction lengths the state of the output light is such that no matter what measurement is performed on the output idler modes, no information about the mode in which the signal photon has left the coupled downconverters is gained.

To see the origin of the coherence induced by the continuous coupling the original device can be conveniently replaced by a sequence of simpler devices in the spirit of [2].

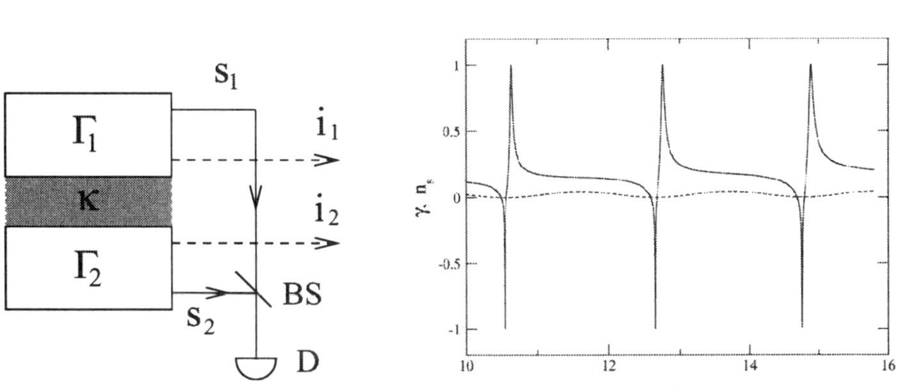

Fig. 1. Left: Experimental arrangement; gray color indicates the linear interaction region; BS – beamsplitter, D – detector. Right: Mutual coherence of signal beams (solid line); mean total number of signal photons for the same interaction length L (dotted line); $\kappa = 3$, $\Gamma_1 = 0.1$, $\Gamma_2 = 0.3$.

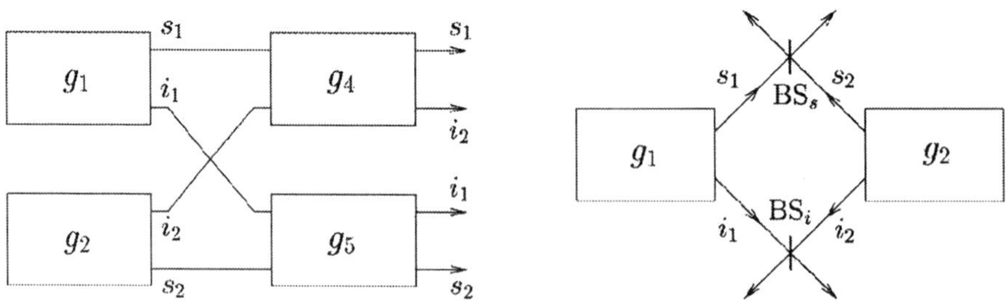

Fig. 2. Two substituting schemes for the two continuously coupled downconverters. Notice that the continuous interaction of three modes is replaced by a sequence of four simpler devices.

This corresponds to the following decomposition of the evolution operator:

$$e^{\frac{i}{\hbar}\hat{G}z} = e^{ig_5(z)\hat{X}_5} e^{ig_4(z)\hat{X}_4} e^{ig_1(z)\hat{X}_1} e^{ig_2(z)\hat{X}_2}, \tag{2}$$

where

$$\begin{aligned} \hat{X}_1 &= \hat{A}_{s_1}\hat{A}_{i_1} + \hat{A}^\dagger_{s_1}\hat{A}^\dagger_{i_1}, & \hat{X}_2 &= \hat{A}_{s_2}\hat{A}_{i_2} + \hat{A}^\dagger_{s_2}\hat{A}^\dagger_{i_2}, \\ \hat{X}_4 &= i(\hat{A}_{s_1}\hat{A}_{i_2} - \hat{A}^\dagger_{s_1}\hat{A}^\dagger_{i_2}), & \hat{X}_5 &= i(\hat{A}_{i_1}\hat{A}_{s_2} - \hat{A}^\dagger_{i_1}\hat{A}^\dagger_{s_2}), \end{aligned} \tag{3}$$

are Hermitian generators of those simpler devices, see Fig. 2 (left). Here we assumed that all modes start from vacuum. For a general input state two additional beam-splitters are needed to replace the original setup.

Another representation shown in Fig. 2 (right) is also possible. Notice that the two schemes are in fact a modified setup of Zou *et al.* and the two-photon interferometer of Ou *et al.* [3], respectively. They are convenient for quantitative analysis in terms of which-way information. One can easily show that for interaction lengths giving rise to full signal coherence no information about the path of the signal photon is accessible. On the other hand, $\gamma = 0$ implies the possibility to localize the signal photon by measuring the Stokes operator,

$$\hat{S} = e^{-i\hat{\sigma}_y\phi}\hat{S}_z e^{i\hat{\sigma}_y\phi}, \tag{4}$$

on the idler beams. The angle ϕ is determined by the parameters g_i of the substituting scheme. With increasing lengths of the original continuously coupled device the amount of which-way information continuously varies between the two extreme cases. Two continuously coupled downconverters can thus be regarded as an interesting generalization of the experimental setup of Zou *et al.* where the distinguishability of signal photons can be controlled by the length of the device itself rather than by using auxiliary optical elements.

As the coupling between the idler beams becomes stronger the points of perfect signal-idler entanglement ($\gamma = 0$) get closer and the efficiency of downconversion processes goes down, since such entanglement (observation) necessarily disturbs the phase relations between the interacting beams. In a broader sense this can be interpreted as a manifestation of the quantum Zeno effect.

This work was supported by grant LN00A015 of the Czech Ministry of Education.

References and links

1. X. Y. Zou, L. J. Wang and L. Mandel, "Induced coherence and indistinguishability in optical interference," Phys. Rev. Lett. **67**, 318-321 (1991).
2. S. L. Braunstein, "Squeezing as an irreducible resource," arXiv: quant-ph/9904002.
3. Z. Y. Ou, L. J. Wang, X. Y. Zou, and L. Mandel, "Evidence for phase memory in 2-photon down conversion through entanglement with the vacuum," Phys. Rev. A **41**, R566-568 (1990).

Towards a quantum Zeno tomography

P. Facchi, G. Krenn, S. Pascazio

Dipartimento di Fisica, Università di Bari, I-70126 Bari, Italy

Z. Hradil, J. Řeháček

Department of Optics, Palacký University, 17. listopadu 50, 772 00, Olomouc, Czech Republic

e-mail: rehacek@alpha.inf.upol.cz

Absorption tomography is an important experimental technique revealing the internal structure of material bodies. In many cases it is desirable to minimize the exposition of the sample being examinated. Unlike totally transmitting and absorbing bodies [1] two partially transmitting bodies cannot be distinguished with probability one [2]. Let us compare the efficiencies of the "Zeno" interferometric setup (Fig. 1a) and standard absorption tomography (Fig. 1b). In the Zeno setup a particle is initially injected from the left, crosses the interferometer L times and is finally detected by one of two detectors. Provided the number of loops is sufficiently large the probabilities of particle being detected in the Zeno channel, ordinary channel and being absorbed read

$$p_z(\tau) = 1 - \frac{\pi^2}{4L}\frac{1+\tau}{1-\tau} + O(L^{-2}), \quad p_o(\tau) = O(L^{-2}), \quad p_a(\tau) = \frac{\pi^2}{4L}\frac{1+\tau}{1-\tau} + O(L^{-2}), \tag{1}$$

respectively; τ is the amplitude transmission coefficient of the gray sample. Notice that the statistics of the Zeno tomography becomes binomial for large L. In the standard case the detection/absorption probabilities read

$$p'_d(\tau) = \tau^2, \quad p'_a(\tau) = 1 - \tau^2 \tag{2}$$

and the statistics is also binomial.

Let us first discuss the estimation of an unknown transmission coefficient $T = \tau^2$. The lower bound on the resolution ΔT of an unbiased estimator $\hat{T}$ of T is set by the Cramér-Rao inequality. A straightforward calculation shows that for a given number of absorbed particles, Zeno and standard optimum estimators of T are bounded by the same quantity:

$$\Delta\hat{T}^{\text{opt}}_{st,Ze} \geq \frac{\tau(1-\tau^2)}{\sqrt{N_a}}, \tag{3}$$

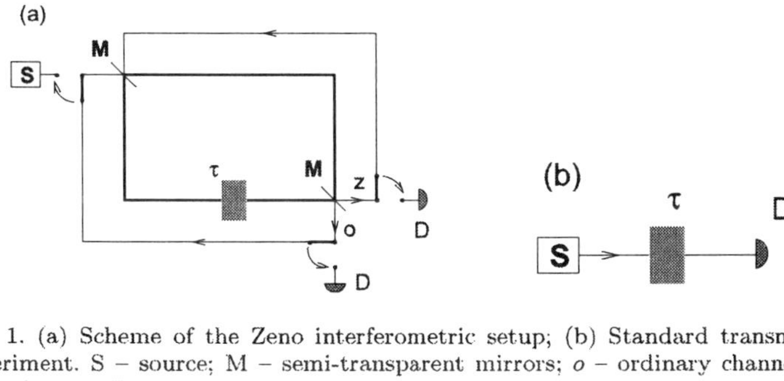

Fig. 1. (a) Scheme of the Zeno interferometric setup; (b) Standard transmission experiment. S – source; M – semi-transparent mirrors; o – ordinary channel; z – Zeno channel; D – detectors.

Coherence and Quantum Optics VIII
Edited by Bigelow et al., Kluwer Academic/Plenum Publishers, 2003

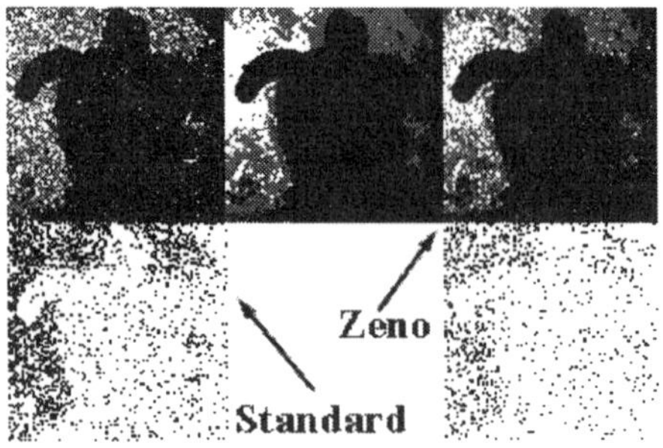

Fig. 2. Comparison of standard and Zeno tomographic techniques. The original object (a turtle, center) consits of three levels of gray: $\tau_1 = 0.99$, $\tau_2 = 0.98$ and τ_3=0.92. Reconstructed objects (above) and misinterpreted pixels (below) are shown for standard (left) and Zeno (right) tomography.

which can be attained in the standart setup. This tells us that the Zeno setup cannot provide a better resolution than standard tomography. Loosely speaking, one gains information only through the absorbed particles.

Despite this pessimistic result standard tomography can still be outperformed by the Zeno method, provided only a few gray levels are to be distinguished and the frequency of occurrence of different grays in the sample is not uniform. In the case of n gray levels τ_i occurring in the sample with frequencies f_i we proceed as follows: Each pixel is illuminated by N particles n of which are detected. The pixel is assigned the gray level that maximizes the likelihood

$$\mathcal{L} = P(n|H_i)f_i, \tag{4}$$

where $P(n|H_i)$ is the binomial detection statistics for the Zeno or standard experimental setup. The number of incident particles N can be chosen so as to make the total number of absorbed particles in the Zeno and standard apparatus approximately the same. The result of a simulation is shown in Fig. 2. The object consists of three levels of gray $\tau_1 = 0.99$ (white), $\tau_2 = 0.98$ (gray) and $\tau_3 = 0.92$ (black) with relative frequencies $f_1 = 0.1$, $f_2 = 0.2$ and $f_3 = 0.7$. For the same number of absorbed particles the Zeno tomography *halves* the number of misinterpreted pixels. It is not difficult to see why the Zeno method sometimes works better. According to (1) the probability of a particle being absorbed in the Zeno setup is an increasing function of τ. Therefore the Zeno tomographic image is a kind of "negative" of the standard tomographic image. Provided that the object contains only a small fraction of more transparent pixels surrounded by a more absorbing material, the Zeno setup reduces the exposition.

A simple interferometric setup cannot increase the resolution per absorbed particle if nothing about the sample is known. However, for a certain class of samples, the Zeno method can be advantageous. Using the interferometric setup the gray scale of the sample can be manipulated. In this way, a prior knowledge about the grayness of the sample can be utilized in the Zeno scheme for decreasing the irradiation.

This work was supported by TMR Network ERB FMRXCT 96-0057 "Perfect Crystal Neutron Optics" of the European Union and by Grant LN00A015 of the Czech Ministry of Education.

References and links

1. A. C. Elitzur, L. Vaidman, "Quantum-mechanical interaction-free measurements," Found. Phys. **23**, 987-997 (1993).
2. G. Krenn, J. Summhammer, K. Svozil, "Interferometric information gain versus interaction–free measurements," Phys. Rev. A **61**, 052102 (2000); G. Mitchinson and S. Massar, "Interaction-free discrimination between semi-transparent objects," Phys. Rev. A **63**, 032105 (2001).

Separability of Gaussian States

Krzysztof Wódkiewicz [‡,*] **and Berthold-Georg Englert**[†]

[‡] *Instytut Fizyki Teoretycznej, Uniwersytet Warszawski, Warszawa 00–681, Poland*
[*] *Department of Physics and Astronomy, University of New Mexico, Albuquerque NM 87131, USA*
[†] *Max-Planck-Institut für Quantenoptik, Hans-Kopfermannn-Straße 1, 85748 Garching, Germany*

wodkiew@fuw.edu.pl, bge@mpq.mpg.de

Abstract: Separability and nonlocality of quantum two-party Gaussian states corresponding to continuous variable systems is investigated. Such states are a natural generalization of the original Einstein Podolsky Rosen entangled state. The separability of such generally mixed Gaussian states is studied in the framework of the operator formalism for the density operator. An explicit construction of separable and nonseparable Gaussian density operators is presented.

OCIS codes: (270.6570) Squeezed states

The nonlocality of entangled states described by continuous variable system has been at the core of all the discussions associated with the original Einstein Podolsky Rosen (EPR) wave function. The EPR state describes a correlated two-particle system characterized by a Gaussian wave function. The nonlocality of this original EPR pure state has been recently exhibited in the Wigner representation [1]. Recent application of entangled two-mode squeezed states in quantum teleportation and quantum information has generated a lot of interest in the separability properties of general mixed Gaussian states in quantum optics [2]. In one approach the separabilty criterion for continuous variable systems described by a Gaussian Wigner function has been investigated using the properties of the Heisenberg uncertainty relations [3]. In a different approach, based on the criterion of positivity under partial trace transpose, the separability of such states has been exhibited [4].

In this presentation a direct operator method is used to study the separability of an arbitrary two-mode quantum Gaussian operator $\hat{G}(\hat{a}, \hat{a}^\dagger, \hat{b}, \hat{b}^\dagger)$, that has unit trace $\mathrm{Tr}\{\hat{G}\} = 1$. In this direct operator approach the specific algebraic form of the Gaussian operator will determine if $\hat{G}$ is a density operator that is separable or not. We exploit the fact than an arbitrary Gaussian operator $\hat{G}$ can be obtained using an appropriate transformation S on a product of two basic Gaussian operators:

$$\hat{G} = S^\dagger\, \lambda_a(1-\lambda_a)^{\hat{A}^\dagger \hat{A}}\, \lambda_b(1-\lambda_b)^{\hat{B}^\dagger \hat{B}}\, S\,. \tag{1}$$

In this formula λ_a and λ_a are positive and transformed boson operators are used.

The operator (1) must be positive in order to describe a physical quantum state represented by a density operator $\hat{\rho}$. This is achieved if and only if $1 \geq \lambda_a$ and $1 \geq \lambda_a$. This is the criterion for a general Gaussian operator $\hat{G}$ to be a positive operator.

From the general form given by (1) one can derive the condition for which the density operator becomes a pure Gaussian state. The density operator (1) becomes a pure state if and only if $\lambda_a = \lambda_b = 1$. In this case the density operator is a projector on a wave function:

$$\hat{\rho} = |\Psi\rangle\langle\Psi| = S^{\dagger}|0_A, 0_B\rangle\langle 0_A, 0_B|S\,. \tag{2}$$

In quantum optics applications the state $|\Psi\rangle$ from (2) is the well known two-mode squeezed state generated in the process of nondegenerate optical parametric amplification. This state represents the optical analogue of the original EPR state.
The partial transposition of the Gaussian density operator $\hat{\rho}^{T_a}$ can be investigated. From (1) it follows that such an operator is equivalent to

$$\hat{\rho}^{T_a} = \tilde{S}^{\dagger}\,\tilde{\lambda}_a(1-\tilde{\lambda}_a)^{\hat{A}^{\dagger}\hat{A}}\,\tilde{\lambda}_b(1-\tilde{\lambda}_b)^{\hat{B}^{\dagger}\hat{B}}\,\tilde{S}\,. \tag{3}$$

Following Peres criterion, the positivity of this operator leads to the conclusion that the Gaussian state is separable if and only if $1 \geq \tilde{\lambda}_a$ and$1 \geq \tilde{\lambda}_a$. Using these properties a construction of the decomposition of a separable state as concave superposition of the form

$$\hat{\rho} = \sum_i p_i(\hat{\rho}_a \otimes \hat{\rho}_b)_i \tag{4}$$

is given and its relation to the properties of the P-representation established. Explicit examples of separable and nonseparable Gaussian states will be presented and discussed.

References and links

1. K. Banaszek, and K. Wódkiewicz "Nonlocality of the Einstein-Podolsky-Rosen State in the Wigner representation," Phys. Rev. A **58**, 4345 (1998).
2. S. L. Braunstein, Ch. A. Fuchs, H. J. Kimble, and P. van Loock, "Quantum vesrsu Classical Domains for Teleportation with Continous Variables," preprint quantum-ph/0012001, http://xxx.lanl.gov/
3. Lu-Ming Duan, G. Giedke, J. I. Cirac, and P. Zoller, "Inseparability Criterion for Continous Variable Systems," Phys. Rev. Lett. **84**, 2722 (2000).
4. R. Simon, "Peres-Horodecki Separability Criterion for Continous Variable Systems," Phys. Rev. Lett. **84**, 2726 (2000).

Wigner function dynamics on the sphere

A. B. Klimov and S. M. Chumakov

Depto. de Física, Universidad de Guadalajara, Revolución 1500, 44420, Guadalajara, Jalisco, México.

klimov@cencar.udg.mx, sergei@cencar.udg.mx

Phase-space description of quantum dynamics in terms of the Wigner quasiprobability function is widely used. The common Wigner function is connected with the Heisenberg-Weyl (HW) group, and one has to write the Hamiltonian in terms of its generators (creation and annihilation operators). In many cases, the Hamiltonian can be also written in terms of spin operators $S_{x,y,z}$, making use of the Schwinger mapping. The quasiprobability distribution function on SU(2) was introduced by Stratonovich [1] in 1956 (see also [2]). Natural phase space here is a two-dimensional sphere. Knowing the dynamics of the SU(2)-Wigner function on the sphere one can find the time behavior of any observable. A similar procedure has been performed for the two- and three-wave mixing in terms of the common HW-Wigner functions in 4- and 6-dimensional phase spaces [3]. The quasiclassical Wigner function dynamics was found numerically and then used to reconstruct the mean values. For the SU(2)-Wigner function on the sphere, we have a reduction of the phase space dimension from four or six (two or three harmonic oscillators) to two (spin system), which strongly simplifies numeric procedures.

Here we analyze the dynamical equation for the $SU(2)$-Wigner function in the quasiclassical limit. This equation may be expanded in the inverse powers of the representation dimension. The principal term leads to the classical evolution of the initial distribution on the sphere (where every point moves along its classical trajectory). Other terms represent quantum correction whose contribution strongly depends on the initial state. For some initial states the quantum terms do not essentially affect the initial stage of evolution (when one can neglect the self-interference) and the dynamics may be described in terms of classical trajectories. By such a way, we arrive at the results which are drastically better then the "naive" solution of the Heisenberg equations of motion with decoupled correlators. However, the quantum phenomena which follow from the self-interference (like Shcrödinger cats) are beyond the scope of this semiclassical approximation. To illustrate this idea, we consider the simplest nontrivial example of the finite-level analog of the Kerr Medium [4],[5].

Consider a $2S+1$-dimensional irreducible representation of the SU(2) group. The $SU(2)$-Wigner function for a spin state with a density matrix ρ (the Wigner symbol of ρ) is defined [1,2] as $W_\rho(\theta,\phi) = \mathrm{Tr}\,(\rho\hat{w}(\theta,\phi))$, where $\hat{w}(\theta,\phi)$ is the Wigner operator

$$\hat{w}(\theta,\phi) = \alpha_S \sum_{L=0}^{2S} \sum_{M=-L}^{L} Y_{L,M}(\theta,\phi)^* \,\hat{T}_{L,M}^{(S)}, \quad \alpha_S = \frac{2\sqrt{\pi}}{\sqrt{2S+1}} \tag{1}$$

Here we use the spherical harmonics $Y_{L,M}(\theta,\phi)$ and the irreducible tensor operators $\hat{T}_{L,M}^{(S)}$ (*cf.* [6], Eq. 2.4(6)). The function $W_\rho(\theta,\phi)$ is covariant under rotations and provides the overlap relation $\alpha_S^{-2}\int \sin\theta\, d\theta\, d\phi W_\rho(\theta,\phi)\, W_A(\theta,\phi) = \mathrm{Tr}\,(\rho A)$, for any operators A and ρ. One can prove the following integral representation for the Wigner operator [7],

$$\hat{w}(\theta,\phi) = \int_0^{2\pi} d\omega e^{-i\omega\,\vec{n}\cdot\vec{S}} f(\omega). \tag{2}$$

Here $e^{-i\omega\,\vec{n}\cdot\vec{S}}$ is a group rotation, $\vec{n} = (\cos\phi\sin\theta, \sin\phi\sin\theta, \cos\theta)$ is a unit vector and the weight function $f(\omega)$ is given in Ref. [7]. In the limit of large representation dimensions, $S \gg 1$, it takes the following asymptotic form [7],

$$f(\omega) \to (-1)^S \left[\delta(\omega - \pi) - \frac{i}{S}\frac{\partial}{\partial\omega}\delta(\omega - \pi)\right], \quad S \to \infty, \tag{3}$$

where the limit is understood in a weak sense. The last formula leads to convenient approximations for Wigner functions of coherent states. For instance, for the eigenstate of S_x one obtains, $W_{CS}(\theta, \phi) = (\sin\theta\cos\phi)^{2S-1}[1 + \sin\theta\cos\phi]$. Eq. (3) also provides the way to reconstruct the HW-Wigner function from the SU(2)-Wigner function, considering contraction of the SU(2) to the Heisenberg-Weyl group.

Eqs. (2),(3) allow us to write the dynamical equation for the SU(2)-Wigner function in the following form:

$$\frac{\partial W_\rho}{\partial t} = \frac{1}{\sqrt{S(S+1)}}\{W_H, W_\rho\} + Q.$$

Here W_H stands for the Wigner symbol (1) of the Hamiltonian, and Q is the quantum correction, whose contribution depends on the initial state, $\{.,.\}$ is the classical Poisson bracket on the sphere.

Let us consider the simplest example, the "finite Kerr Hamiltonian" $H = \chi S_z^2$, which leads to generation of squeezed atomic states [4] and atomic Schrödinger cats [5]. (Here S_z is the atomic inversion for a collection of $2S$ two-level atoms.) Neglecting quantum corrections, we arrive at the following quasiclassical form of the Wigner function for the finite Kerr Hamiltonian:

$$W_\rho(\theta, \phi|t) = W_\rho\left(\theta, \phi - \frac{\chi t}{2S+1}\middle| t = 0\right).$$

Let us stress, that for some initial states quantum corrections may be of principal importance. However, it can be shown that quantum corrections are ~ 1 (while the classical term is $\sim S$) when the dynamics starts from the coherent states located close to the equator of the sphere. In this case the evolution is essentially semiclassical, and the initial distribution is deformed in such a way that every point on the sphere moves along its classical trajectory. Note that the quasiclassical Wigner function evolution does not correspond to the solution of classical equations of motion for the average values of spin operators (in the sense that we do not arrive at equations with decoupled correlators).

References and links

1. R. L. Stratonovich, "On distributions in representation space," Sov. Phys. JETP **4**, 891-898 (1957).
2. G. S. Agarwal, "Relation between atomic coherent-state representation, state multipoles, and generalized phase-space distributions," Phys. Rev. A **24**, 2889-2896 (1981).
3. G. Drobný, A. Bandilla and I. Jex, "Quantum description of nonlinearly interacting oscillators via classical trajectories," Phys. Rev. A **55**, 78-93 (1997).
4. M. Kitagawa and M. Ueda, "Squeezed spin states" Phys. Rev. A **47**, 5138-5143 (1993).
5. S. M. Chumakov, A. Frank and K. B. Wolf, "Finite Kerr medium: Macroscopic quantum superposition states and Wigner functions on the sphere," Phys. Rev. A **60**, 1817-1822 (1999).
6. D. A. Varshalovich, A. N. Moskalev and V. K. Khersonskii, *Quantum Theory of Angular Momentum,* (World Scientific, Singapore, 1988).
7. A. B. Klimov and S. M. Chumakov, "Quasi-probability distributions for the simplest dynamical groups", J. Opt. Soc. Am. A **17**, 2315-2318 (2000).

An experiment on the stimulated emission of the evanescent photon

Hyun Ho Choi, Hyungju Kim, Jaewoo Noh,
Department of Physics, InhaUuniversity, Inchon 402-751 South Korea
Phone:82-32-860-7662, Fax: 83-32-872-7562, email: jnoh@inha.ac.kr
Wonho Jhe
Department of Physics, Center for Near-field Optics and Technology, Seoul National University, Seoul, South Korea
phone: 82-2-880-6600, email: whjhe@phsa.snu.ac.kr

Abstract:

We studied experimentally the amplification of light beam reflected from a plane boundary, where a gain medium is placed in the transmitted beam region. It is shown that the reflected light beam is amplified due to the stimulated emission of photons from the dye molecules coupled to the evanescent wave. The experimental result is compared with the quantum theory developed recently, and the meaning of photon is reexamined.

1. Introduction

It has been known that an atom near the planar boundary can absorb and emit photons via the coupling with the evanescent wave [1]. Recently we presented a full quantum theory for the atom and the evanescent wave interaction [2]. According to the theory, an atom present in the evanescent wave region can amplify a light beam making a total internal reflection at a planar boundary, and the amplification process was explained by the stimulated emission of a triplet mode photon. The evanescent wave is treated as a part of the triplet mode, which is made of the incident, reflected, and the transmitted wave. Because the evanescent wave is present in the transmitted region when the light incident angle is close to the critical angle, it is expected that the rate of the spontaneous and the stimulated emission of photon is higher near the critical angle. It is also expected that the reflected light beam can be amplified even when the light incident angle is slightly less that the critical angle. Comparing the situation with the well-known beam splitter case where a gain medium is placed in the transmitted wave region, we analyze the meaning of the photon and reexamine the role of vacuum field that acts as a decorrelator in quantum optics theory.

2. Theory

The interaction of an atom and a light field is explained by an interaction Hamiltonian, which involves the electric field and the atom dipole moment. Because there exists the electric field for the non-propagating evanescent wave, the interaction of the atom and the light field is possible, and the full quantum theory can be used to calculated the rate of spontaneous and stimulated emission of photon. We used the triplet mode for the treatment of the evanescent wave, and considered the photon as the elementary excitation of electromagnetic field mode. The stimulated emission of the triplet mode photon is always possible when there exists the near field component in the triplet mode. The emission rate for the spontaneous and stimulated processes were calculated and the result show interesting dependence of the atom position and the light incident angle.

3. Experiment

A solution of DCM dye was prepared as a gain medium and a chopped CW DPSS Laser pumps the dye cell. To observe the angular spectrum of the light emission, a hemisphere lens is attached to the dye cell. A HeNe laser is used as a probe and the amplification of the reflected light beam is detected with a low noise photodiode. A Lock-in amplifier is used for the phase sensitive detection of the light signal. Both the spontaneous emission and the stimulated emission of light are detected with the same detector, and the result shows interesting property of the angular spectrum. Due to the different symmetry associated with the emission, the location of the peak and the shape of the spectrum for the two emission processes are different.

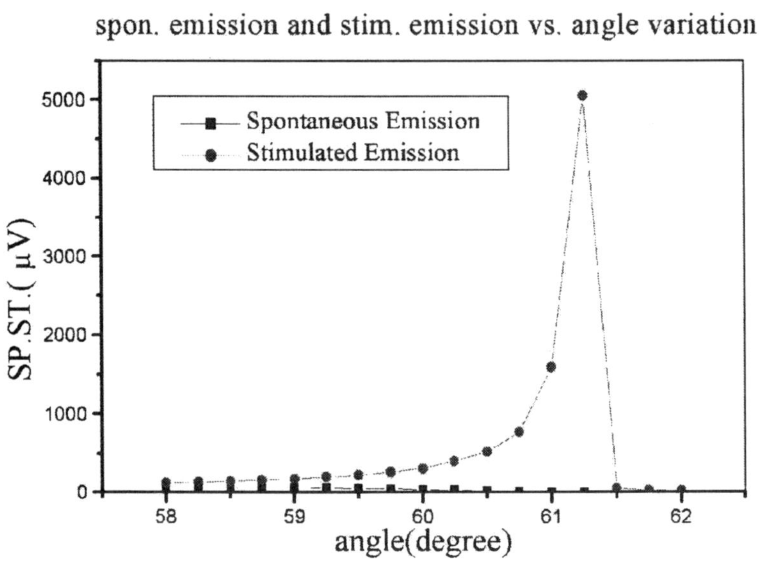

References

[1] C. K. Carniglia, L. Mandel, and K. H. Drexhage, "Absorption and Emission of Evanescent Photons," J. Opt. Soc. Am. **62**, 479-486 (1972).

[2] Chang-Woo Lee, Kyungbum Kim, Jaewoo Noh, and Wonho Jhe, "Quantum theory of amplified total internal reflection due to evanescent-mode coupling", Phys. Rev. A **62**, 053805 (2000).

Nonclassical Measure of Multi-mode Radiation Field

Youngchul Kim, Kisik Kim, and Dae-Yoon Park

Department of Physics, Inha University, Inchon 402-751, Korea

The concept of the nonclassical light has drawn a great attention in modern quantum optics. In a wide sense, the nonclassical light is a light whose properties cannot be described in the classical wave theory. One practical definition for nonclassical lights can be made through the coherent state representation for density operators of light fields: [1]

$$\hat{\rho} = \int \phi(v)|v\rangle\langle v|d^2v, \tag{1}$$

where $|v\rangle$ is the coherent state and the integral is to be taken over the entire complex v-plane. $\phi(v)$ is called the P-function and plays the role of a classical probability density function, however, it sometimes becomes highly singular or takes on negative values. This behavior of $\phi(v)$ can be used to distinguish nonclassical lights from classical ones. [2] When $\phi(v)$ does not become a classical probability density function, the corresponding light is referred to as nonclassical.

The nonclassical light has been analyzed more quantitatively by introducing the nonclassical measure. For this purpose, the R-function is defined as a parametrized convolution [3]

$$R_\tau(v) = \frac{1}{\pi\tau}\int e^{-|v-v'|^2/\tau}\phi(v')d^2v'. \tag{2}$$

The smallest τ, denoted by τ_m, which makes $R_\tau(v)$ a true probability density function is defined as the nonclasscial measure. Since $R_1(v) = \langle v|\hat{\rho}|v\rangle$, the relevant domain of τ is $0 \leq \tau \leq 1$. The nonclassical measure of the nonclassical property has also been defined and the various nonclassical properties have been compared quantitatively. [4]

In this paper, we extend the nonclassical measure of a single mode field to a multi-mode field. When we denote the multi-mode P-function by $\phi(\{v\})$, the corresponding R-function is defined as

$$R_\tau(\{v\}) = \int\cdots\int\prod_i\left(\frac{1}{\pi\tau}e^{-|v_i-v_i'|^2/\tau}\right)\phi(\{v'\})\prod_i(d^2v_i'). \tag{3}$$

We apply the multi-mode nonclassical measure to the output fields from a beam splitter. When we send a single photon toward the beam splitter, the output fields become entangled and the corresponding R-function is obtained as

$$R_\tau(v_1,v_2) = \left(1-\frac{1}{\tau}+\frac{1}{\tau^2}|tv_1+rv_2|^2\right)\frac{1}{\tau}e^{-|v_1|^2/\tau}\frac{1}{\tau}e^{-|v_2|^2/\tau}, \tag{4}$$

where t and r are the transmission and the reflection coefficients. From this expression, we immediately obtain $\tau_m = 1$ and know the nonclassical measure is preserved. On the other hand, the single mode nonclassical measure of the transmitted field is calculated to be

$$R_\tau(v_1) = \int R_\tau(v_1,v_2)d^2v_2 = \frac{1}{\tau^2}e^{-|v_1|^2/\tau}\left(\tau-|t|^2+\frac{|t|^2}{\tau}|v_1|^2\right), \tag{5}$$

and $\tau_m = |t|^2$ is obtained. This reduction is due to the partial trace over the reflected field.

References

[1] R. J. Glauber, Phys. Rev. **130**, 2529 (1963).
[2] L. Mandel and E. Wolf, *Optical Coherence and Quantum Optics* (Cambridge University Press, Cambridge, 1995), Sec. 11.8.
[3] C. T. Lee, Phys. Rev. A **44**, R2775 (1991); **45**, 6586 (1992).
[4] Y. Kim, J. Noh, and K. Kim, J. Kor. Phys. Soc. **29**, 598 (1996).

Generation of states with holes in the photon distribution at controlled positions

J. M. C. Malbouisson
Instituto de Física, Universidade Federal da Bahia, 40210-340, Salvador, BA, Brazil
jmalboui@ufba.br

B. Baseia
Instituto de Física, Universidade Federal de Goiás, 70401-970, Goiânia, GO, Brazil

Abstract: We discuss the generation of states of a field mode in a high-Q cavity that possess holes in their photon-number distributions located at controlled positions.

Dispersive atom-field interaction, combined with Ramsey interferometry, has been used to generate superpositions of field states in a cavity, a remarkable example being Schrödinger cat states [1]. In this scheme, a Rydberg atom with selected velocity crosses the system consisting of a Ramsey zone with controlable intensity (where it is prepared in a specific superposition $c_g|g\rangle + c_e|e\rangle$), a high-Q cavity and a second Ramsey (producing a $\pi/2$ rotation in the atomic space), before being detected. The atom-field dispersive interaction in the cavity is given by the effective Hamiltonian $\hat{H} = 2\hbar d^2\delta^{-1}(|i\rangle\langle i| - |e\rangle\langle e|)\hat{a}^+\hat{a}$, where **d** is the atomic dipole moment associated with the transition $|g\rangle \to |e\rangle$ (frequency ω_0), $\delta = \omega - \omega_0$ is the detuning parameter and $\hat{a}$ is the photon annihilation operator of the field mode with frequency ω. If the field mode in the cavity is initially in a coherent state $|\alpha\rangle = |re^{i\theta}\rangle$, this dispersive interaction entangles the atomic state components with the coherent state and a phase shifted partner $|\alpha'\rangle$; the atom-field state is then rearranged after the atom passes the second Ramsey zone. Now, by detecting the atom in its ground (excited) state, the field state in the cavity is projected into a superposition of $|\alpha\rangle$ and $|\alpha'\rangle$ having the same coefficients as those of the atomic superposition before the atoms enters the cavity (detection in $|e\rangle$ introduces an additional phase difference of π between c_g and c_e), $|\psi\rangle = M(c_g|\alpha\rangle \pm c_e|\alpha'\rangle)$. The coherent state $|\alpha'\rangle$ has $\alpha' = \alpha e^{i\phi}$ where $\phi = 2\hbar d^2\delta^{-1}\tau$, τ being the time the atom takes to cross the cavity. Thus, by controlling the intensity of the first Ramsey zone and the detuning, one can manipulate the field-state superposition created in the cavity; even and odd coherent states are obtained adjusting the parameters to have $c_g = c_e$ and $\phi = \pi$. If one considers successive atoms, properly prepared, crossing the system and being detected in a specific sequence of atomic states then superpositions of Schrödinger cat states can be produced [2].

The process briefly described above can be used to generate states possessing holes in the photon-number distribution (PND) located at desired positions. In fact, sending a first atom to the system and controlling the atomic transit time, the detuning and the intensity of the first Ramsey zone to match the conditions $\phi_1 = \pi/N$, $c_g^{(1)} = 1/\sqrt{2}$ and $c_e^{(1)} = \exp[i(1 - N_1/N)\pi]/\sqrt{2}$, where N and N_1 are integers, the detection of the atom in the state $|g\rangle$ generates a field state in the cavity having a PND, $P(n) = |\langle n|\psi\rangle|^2$, which is null for the values of n belonging to the set $\{N_1, N_1+2N, N_1+4N, \ldots\}$. If N is taken large compared with r^2, only the hole at $n=N_1$ is significant since all the others lie far way from the peak of the PND where it effectively vanishes; one thus can create states having one hole in the PND located in a desired position [3,4].

The production of states possessing more than one hole in the PND can be implemented by performing sequential selective detection of atoms which cross the system after have been appropriately prepared. Owing to the form of the dispersive Hamiltonian, an unoccupied photon-number state remains unoccupied after interaction so that the passage of an atom does not affect the holes already burned. Therefore, if K atoms pass through the cavity containing initially the coherent state $|r\exp(i\theta)\rangle$, prepared such that $\phi_j = \pi/N$, $c_g^{(j)} = 1/\sqrt{2}$ and $c_e^{(j)} = \exp[i(1-N_j/N)\pi]/\sqrt{2}$ for $j=1,\cdots,K$, and all of them detected in $|g\rangle$, one generates a state with number-basis representation given by

$$|\Psi\rangle = M_K\, e^{-r^2/2} \sum_{n=0}^{\infty} \frac{r^n e^{in\theta}}{\sqrt{2^K n!}} \prod_{j=1}^{K} \left(1 + e^{i[1+(n-N_j)/N]\pi}\right)|n\rangle ,$$

where M_K is the normalization constant. Such a state has holes in its PND for $n = N_j + 2lN$ with $j=1,\cdots,K$, l being any nonnegative integer.. Thus, by fixing the atomic transit time through the cavity and the dispersive interaction coupling to have all $\phi_j = \pi/N$ with N sufficiently greater than r^2, the control of the intensity of the first Ramsey zone leads to a state possessing K holes in the PND for $n \in \{N_1, N_2, \cdots, N_K\}$ if the largest N_j is sufficiently small (less than $2r^2$ approximately) to be within the region where $P(n)$ is significant.

Roughly speaking, the probability of sequential detection of K atoms in the state $|g\rangle$ decreases like 2^{-K} and so the production of states possessing a large number of holes in their PND is not favored due to decoherence effects. However, for a moderate intensity of the initial field state in the cavity, many distinct states with a few number of holes can be generated. Although the parameter N does not interfere in the positions of the holes, it influences the height of the hills between the holes in the PND; as N is increased, the PND tends to a shape independent of N. The normalized standard deviation of the photon number can be easily calculated and indicates that states having the same number of holes (and with the same r) does not necessarily have the same kind of statistics. Having states of a field mode possessing holes in the PND at controlled positions may be of interest if one is able to construct a readout system identifying these holes. Such a system can be an atomic deflection setup as, for example, the one suggested in reference [5].

This work was partially supported by CNPq, Brazil.

[1] S. Haroche and J. M. Raimond, "Manipulation of nonclassical field states in a cavity by atom interferometry" in *Cavity quantum electrodynamics*, P. Berman, ed. (Academic Press, New York, 1994).

[2] J. M. C. Malbouisson and B. Baseia, "Higher-generation Schrödinger cat states in cavity QED," J. Mod. Opt. **46**, 2015-2041 (1999).

[3] B. Baseia, M. H. Y. Moussa and V. S. Bagnato, "Hole burning in Fock space," Phys. Lett. A**240**, 277-281 (1998).

[4] B. Baseia and C. M. A. Dantas, "A note on hole burning in the Fock space," Phys. Lett. A**253**, 123-124 (1999).

[5] M. Freyberger and A. M. Herkommer, "Probing a quantum state via atomic deflection," Phys. Rev. Lett. **72**, 1952-1955 (1994).

The spin-orbit decomposition of the angular momentum of free electromagnetic fields is artificial.

H. H. Arnaut
Departamento de Física, Universidade Federal de Minas Gerais, Belo Horizonte 30161-970, MG, Brazil

G. A. Barbosa
Northwestern University, Center for Photonic Communication and Computing ECE Department, 2145 N. Sheridan Rd., Evanston, Illinois, 60208-3118, USA

In the standard decomposition of the total angular momentum (AM) $\mathbf{J}(\mathbf{r_0}) = \mathbf{L}(\mathbf{r_0}) + \mathbf{S}$ of a free electromagnetic field the vector $\mathbf{S}$ represents an intrinsic angular momentum also refered as the photon *spin*. In 1936 Beth [1] performed a famous experiment in which a intense beam of circularly polarized light reaching a quartz half-wave plate emerged on the opposite side with the inverse polarization. The mechanical torque on the plate could be measured in their apparatus and it coincides in sign and magnitude with what would be expected if one consider a variation of $\pm 2\hbar$ per photon due the polarization inversion. That experiment is sometimes referred to as the measurement of the spin AM of the photon. In atomic theory is very well know that a absorption or emission of a photon circularly polarized changes the z component of the atomic AM of $\pm\hbar$ in perfect concordance with the image of a intrinsic AM carried for photons. The uncountable number of experiments carrying similar interpretations turned the concept of photon spin undisputable.

Nevertheless the meaning of $\mathbf{S}$ is far from clear. It has already been shown [2] that in quantum theory neither $\widehat{\mathbf{L}}$ or $\widehat{\mathbf{S}}$ is a true AM:

$$\left[\widehat{S}_i, \widehat{S}_j\right] = 0 \quad , \quad \left[\widehat{L}_i, \widehat{L}_j\right] = i\hbar\epsilon_{ijk}\left(\widehat{L}_k + \widehat{S}_k\right) \quad , \quad \left[\widehat{L}_i, \widehat{S}_j\right] = i\hbar\epsilon_{ijk}\widehat{S}_k \; . \tag{1}$$

The last Eq. (1) implies that the $\widehat{\mathbf{J}}$ eigenspace may not be factorized as a tensor product of $\widehat{\mathbf{L}}$ and $\widehat{\mathbf{S}}$ eingenspaces. It has also been shown [3] that the eigenvalues of $\widehat{S}_i$ lies in a continuum and changes in the intrinsic part of the angular momentum are followed by opposite changes in the orbital part and for each Cartesian component:

$$\widehat{S}_z\left|\psi\right\rangle = s\cos\phi_0\hbar\left|\psi\right\rangle \quad , \quad \widehat{L}_z\left|\psi\right\rangle = \left(l - s\cos\phi_0\right)\hbar\left|\psi\right\rangle \text{and} \quad \widehat{J}_z\left|\psi\right\rangle = l\hbar\left|\psi\right\rangle \; , \tag{2}$$

where $|\psi\rangle = |\psi(s = \pm 1, \phi_0)\rangle$ designates the light state for an arbitrary single-photon state and indexes $s = \pm 1$ (usually called "spin" values) indicate circular polarized light states , ϕ_0 is the polar angle. Actually, those classical experiments can be interpreted differently. Starting from $\mathbf{L} = \epsilon_0 \int d^3r\, E_i(\mathbf{r},t)\,(\mathbf{r}\times\nabla_\mathbf{r})\,A_i(\mathbf{r},t)$ and $\mathbf{S} = \epsilon_0 \int d^3r\, \mathbf{E}(\mathbf{r},t)\times\mathbf{A}(\mathbf{r},t)$ we may use the Coulomb gauge and the plane wave expansion of the electromagnetic potential $\mathbf{A}(\mathbf{r},t) = \sum_s \int d^3k \left[u(\mathbf{k},s)\,\hat{\varepsilon}_\mathbf{s}(\mathbf{k})\,e^{i(\mathbf{k}\cdot\mathbf{r}-\omega t)} + c.c.\right] / \left[(2\pi)^{\frac{3}{2}}\sqrt{2\epsilon_0}\right]$ and $\mathbf{E}(\mathbf{r},t) = -\partial\mathbf{A}(\mathbf{r},t)/\partial t$. Without any loss of generality we may consider the field modes as circularly polarized with a polarization basis vectors $\left(\hat{\varepsilon}_1(\mathbf{k}), \hat{\varepsilon}_2(\mathbf{k}), \hat{\mathbf{k}}\right)$ be a right-handed real orthonormal basis, thus $\hat{\varepsilon}_s(\mathbf{k}) = \left(\hat{\varepsilon}_1(\mathbf{k}) + is\hat{\varepsilon}_2(\mathbf{k})\right)/\sqrt{2}$. The vector $\varepsilon_\sigma(\mathbf{k})$ is well defined but for a global phase factor $e^{i\varphi_s(\mathbf{k})}$ which reflects the freedom of the choice of the real vectors $\varepsilon_i(\mathbf{k})$ over a rotation around $\mathbf{k}$. Choosing $\widehat{\varepsilon}_1(\mathbf{k}) = (\cos\phi\cos\theta, \cos\phi\sin\theta, -\sin\phi)$ and $\widehat{\varepsilon}_2(\mathbf{k}) = (-\sin\theta, \cos\theta, 0)$ where θ and ϕ are the azimutal and polar angles of $\mathbf{k}$, one can obtain

$$\mathbf{L} = \sum_s \left(\int d^3k\,\omega\,|u(\mathbf{k},\mathbf{s})|^2\,\mathbf{k}\times\nabla_\mathbf{k}\Phi(\mathbf{k},\mathbf{s}) + \int d^3k\,\omega\,|u(\mathbf{k},s)|^2\,\frac{s(\cos\theta,\sin\theta,0)}{\sin\phi} - \int d^3k\,\omega\,|u(\mathbf{k},\mathbf{s})|^2\,s\hat{\mathbf{k}} \right) \quad \text{and} \quad \mathbf{J} = \mathbf{L} + \int d^3k\,\omega\,|u(\mathbf{k},\mathbf{s})|^2\,s\hat{\mathbf{k}} \tag{3}$$

The z component of the total AM is then shown to be provenient of the orbital part and the intrinsic component does not depend on $\Phi(\mathbf{k},\mathbf{s})$.

The Beth experiment – This experiment can be revisited if one looks closely at Eq. (3). We see that unless $\lim_{\phi\to 0,\pi} |u(\mathbf{k},\mathbf{s})|^2 / \sin\phi = 0$ some restriction apply to $\Phi(\mathbf{k},\mathbf{s})$. Specifically, $\Phi(\mathbf{k},\mathbf{s}) = s\theta + g_s(\mathbf{k})$ where $g_s(\mathbf{k})$ is a 2π periodic function in θ such that $\frac{\partial}{\partial\theta}\lim_{\phi\to 0,\pi} g_s(\mathbf{k}) = 0$. If we consider a half wave plate lying in xy plane then, for quasi-normal incidence, it can be shown that $\Phi(\mathbf{k},\mathbf{s})$ is transformed as $\Phi(\mathbf{k},\mathbf{s}) \to \Phi(\mathbf{k},\mathbf{s}) - 2s\theta$. This leads to a total z AM variation of $\Delta J_z = -2\sum_s s\int d^3k\omega\, |u(\mathbf{k},\mathbf{s})|^2$ which explain the results obtained by Beth without using the concept of photon spin.

Quarter-waveplate – A look at a quarter-waveplate transforming linearly polarized light to circularly polarized shows, in the paraxial approximation, that this transformation gives $(\hat{\mathbf{x}} + \hat{\mathbf{y}})/\sqrt{2} \to (\hat{\mathbf{x}} + is\hat{\mathbf{y}})/\sqrt{2} \simeq e^{is\theta}\left(\hat{\theta} + is\hat{\phi}\right)/\sqrt{2} \to u_{out} = u(\mathbf{k})\, e^{i(\mathbf{k}\cdot\mathbf{r}-\omega t)} e^{is\theta}\hat{\varepsilon}_{\mathbf{s}}(\mathbf{k})$. The original intensity profile is recovered for the output beam. Of course, this is also in agreement with Eqs. (2) in the limit $\phi_0 \to 0$ and the beam shows no intensity minimum at the center because $l' \equiv (l - s\cos\phi_0) \to 0$.

Quantized field – In dealing with the AM of the quantized field we could include the phase factor in the polarization vectors $\hat{\varepsilon}(\mathbf{k},\mathbf{s}) = e^{i\varphi(\mathbf{k},\mathbf{s})}(\hat{\varepsilon}_1(\mathbf{k}) + is\hat{\varepsilon}_2(\mathbf{k}))/\sqrt{2}$. The quantum operator for the intrinsic part of the AM of a free electromagnetic field reads $\widehat{\mathbf{S}} = \hbar\sum_s s\int d^3k\, \widehat{n}(\mathbf{k},s)\frac{\mathbf{k}}{k}$. The quantum operators are

$$\widehat{\mathbf{L}} = \sum_s \left(\left(\int d^3k \left[\widehat{a}^\dagger(\mathbf{k},s)(\mathbf{k}\times\nabla_{\mathbf{k}})\widehat{a}\left(\mathbf{k},s'\right) - hc\right] + \hbar\int d^3k\,\widehat{n}(\mathbf{k},s)\frac{s(\cos\theta,\sin\theta,0)}{\sin\phi} \right.\right.$$

$$\left.\left. -\hbar s\int d^3k\,\widehat{n}(\mathbf{k},s)\frac{\mathbf{k}}{k} + \hbar\int d^3k\,\widehat{n}(\mathbf{k},s)\left[\mathbf{k}\times\nabla_{\mathbf{k}}\varphi(\mathbf{k},s)\right]\right)\times\left(\frac{-i\hbar}{2}\right)\right), \text{ and } \widehat{\mathbf{J}} = \widehat{\mathbf{L}} + \widehat{\mathbf{S}} \qquad (4)$$

The eigenvectors of $\widehat{S}_z$ and $\widehat{J}_z$ for the one-photon field give

$$|\xi(\phi_0,s)\rangle = \int d^3k\, g(\mathbf{k},s)\,\delta(\phi-\phi_0)\,\widehat{a}^\dagger(\mathbf{k},s)|0\rangle \ , \ \widehat{S}_z|\xi(\phi_0,s)\rangle = s\hbar\cos\phi_0|\xi(\phi_0,s)\rangle \qquad (5)$$

$$|\zeta_l(s)\rangle = \int d^3k\, h(\phi,k,s)\, e^{-[i(l\theta-\varphi(\mathbf{k},s))]}\,\widehat{a}^\dagger(\mathbf{k},\mathbf{s})|0\rangle \ , \ \widehat{J}_z|\zeta_l(s)\rangle = \hbar l|\zeta_l(s)\rangle \qquad (6)$$

where h and g are arbitrary functions. All contributions for the z component of the total AM are contained in the orbital part.

Atomic transition – In an atomic transition where the emitted field is symmetrical around the z axis right handed (left handed) circularly polarized photons carries an AM of $+\hbar$ $(-\hbar)$ in the z direction. The continuity of the fields operator requires that $\varphi(\mathbf{k},s) = s\theta$ leading to $|\zeta_{l=1}(s)\rangle = \int d^3k\, h(\phi,k,s=1)\, a^\dagger(\mathbf{k},s=1)|0\rangle$ and $|\zeta_{l=-1}(s)\rangle = \int d^3k\, h(\phi,k,s=-1)\, a^\dagger(\mathbf{k},s=-1)|0\rangle$, and only the total AM eigenfunctions were used without the need of $|\xi\rangle$.

Acknowledgments– H. H. Arnaut has a graduate fellowship from CNPq , Brazil. G. A. Barbosa had partial support from the U. S. Office of Naval Research.

---

[1] R. A. Beth, Phys. Rev. **48**, 471 (1935); **50**, 115-125 (1936).
[2] S. J. Van Enk and G. Nienhuis, J. Mod. Opt. **41**, 963-977 (1994).
[3] H. H. Arnaut and G. A. Barbosa, Phys. Rev. Letters **85**, 286-289 (2000)

Heisenberg equation-of-motion methods applied to non-Markovian open systems

James D. Cresser

Department of Physics, Macquarie University, NSW 2109, Australia

cresser@physics.mq.edu.au

Abstract: A Heisenberg equation-of-motion method is develpoed for deriving the stochastic Schrödinger equations and master equations for non-Markovian open systems. Exact results are obtained for a damped harmonic oscillator, and a damped two level atom, and a perturbative method developed for the driven two level atom.

OCIS codes: (270.2500) Fluctuations, relaxations, and noise, (030.1640) Coherence

It is well-known that a master equation of Lindblad form

$$\dot{\rho} = i\hbar^{-1}[H_S, \rho] - \tfrac{1}{2}\left[L^\dagger L\rho + \rho L^\dagger L - 2L\rho L^\dagger\right] \tag{1}$$

where the Hamiltonian H_S and the operator L are those of a system coupled to a Markovian reservoir can be *unravelled* into a ensemble of stochastically evolving pure states or quantum trajectories of two different kinds – smooth evolution interrupted by 'quantum jumps' or by continuous diffusive evolution [1, 2]. In the latter case, the stochastic equation for the unnormalized state vector takes the form:

$$\frac{d}{dt}|\psi_{[\xi(t)]}\rangle = -i\hbar^{-1}H_S|\psi_{[\xi(t)]}(t)\rangle + \xi(t)L|\psi_{[\xi(t)]}(t)\rangle|\psi_{[\xi(t)]}(t)\rangle - \tfrac{1}{2}L^\dagger L|\psi_{[\xi(t)]}(t)\rangle \tag{2}$$

where $\xi(t)$ is a zero-mean, complex Gaussian white noise i.e. $\langle\langle\xi^*(t)\xi(s)\rangle\rangle = \delta(t-s)$.

If the system-reservoir coupling has a non-negligible memory time, the master equation is no longer of Lindblad form. It can be written either with an explicit memory kernel or else, by use of the time-convolutionless method of [3], as a local-in-time master equation with, however, time dependent coefficients. But given the value of the stochastic Schrödinger equation in obtaining numerical solutions of the master equation, as well as the possibility of a measurement interpretation [2], it would be useful to be able to derive stochastic Schrödinger equations for non-Markovian systems. Results have been obtained along these lines [4, 5, 6, 7] by a variety of means for systems coupled to a bosonic reservoir. A further approach, based on a method originally used by Mollow [8], has been proposed in [9]. It makes use of the Heisenberg equations of motion for the system and reservoir operators to eliminate all virtual processes (i.e. photon reabsorption) that give rise to damping and energy level shifts, from the Schrödinger equation of motion for the system plus reservoir, without making the Markov approximation. The result is a Schrödinger equation that includes these (time dependent) energy level widths and shifts, and involves emission processes only. For a bosonic reservoir, these emission processes are shown to be equivalent to a classical colored noise driving term. The result is a diffusive stochastic Schrödinger equation for the system, the general Diósi-Strunz stochastic Schrödinger equation first derived in [5]:

$$i\hbar\frac{d}{dt}|\psi_{[\xi(t)]}(t)\rangle = \left[H_S - i\hbar L^\dagger\int_0^t ds\,\mathcal{G}(t-s)\overline{L}(s,t,[\xi(t)]) + \hbar\xi(t)L\right]|\psi_{[\xi(t)]}(t)\rangle. \tag{3}$$

where $\xi(t)$ is a zero-mean, Gaussian, colored noise driving term for which $\langle\langle\xi^*(t)\xi(s)\rangle\rangle = \mathcal{G}(t-s)$ and where $\mathcal{G}$ is a memory kernel obtained from the system-reservoir coupling. The principal difficulty lies in determining the form of the operator $\overline{L}(s,t,[\xi(t)])$, and it is the way in which this is done that further distinguishes the method developed here from that of [5, 6, 7].

Exact results for the stochastic Schrödinger equation can be obtained in a number of cases, including an excited two level system, a driven damped harmonic oscillator, and quantum Brownian motion as described by the Calderia-Leggett model [10]. All result in a stochastic Schrödinger equation with time dependent coefficients tha satisfy linear integro-differential equations. For example it is found, for the damped two level atom (raising and lowering operators $a^\dagger$ and a respectively) in the RWA:

$$i\frac{d}{dt}|\psi_{[\xi(t)]}(t)\rangle = \left[(\omega_0 + \Delta(t) - i\Gamma(t))a^\dagger a + a\xi(t)\right]|\psi_{[\xi(t)]}(t)\rangle \tag{4}$$

where the time dependent energy level width $\Gamma(t)$, and shift $\Delta(t)$ are given in terms of a quantity α

$$\frac{\dot{\alpha}}{\alpha} = -i(\omega_0 + \Delta(t)) - \Gamma(t). \tag{5}$$

that satisfies

$$\dot{\alpha}(t) = -i\omega_0\alpha(t) - \int_0^t ds\,\mathcal{G}(t-s)\alpha(s); \qquad \alpha(0) = 1, \tag{6}$$

a result first obtained in [11].

For other cases of interest, such as a damped, driven two level atom, or even a two level atom coupled to a bosonic reservoir with counter-rotating terms included, no exact result seems to be possible. However, the method developed here offers a perturbative approach to obtaining approximate stochastic equations based on obtaining perturbative solutions to the Heisenberg equations of motion that are used to eliminate the virtual processes from the Schrödinger equation. The stochastic equations so obtained, and the corresponding master equations are compared with those found by the alternate perturbative treatments of [6, 7].

References and links

1. J. Dalibard, Y. Castin, and K. Mølmer, "Wave-function approach to dissipative processes in quantum optics", Phys. Rev. Lett. **68**, 580-583, (1992).
2. H. J. Carmichael, *An Open Systems Approach to Quantum Optics*, (Springer, Berlin, 1993)
3. S. Chaturvedi and F. Shibata, "Time-convolutionless projection operator formalism for elimination of fast variables. Applications to Brownian motion", Z. Phys. **B35**, 297-308 (1979).
4. H.-P. Breuer, B. Kappler, and F. Petruccione,"Stochastic wave-function method for non-Markovian quantum master equations", Phys. Rev. A **59**, 1633-1643, (1999).
5. L. Diósi and W. T. Strunz, "The non-Markovian stochastic Schrdinger equation for open systems", Phys. Lett. A. **235**, 569-573, (1997).
6. L. Diósi, N. Gisin, and W. T. Strunz, "Non-Markovian quantum state diffusion", Phys. Rev. A**58**, 1699-1712, (1998).
7. W. T. Strunz, L. Diósi, and N. Gisin, "Open System Dynamics with Non-Markovian Quantum Trajectories", Phys. Rev. Lett.**82**, 1801-1805, (1999).
8. B. R. Mollow, "Pure-state analysis of resonant light scattering: Radiative damping, saturation, and multiphoton effects", Phys. Rev. A **12**, 1919-1943, (1975).
9. J. D. Cresser, "A Heisenberg Equation-of-Motion Derivation of Stochastic Schrödinger Equations for Non-Markov Open Systems", Laser Phys. **10**, 337-347, (2000).
10. A. O. Caldeira and A. J. Leggett, "Path integral approach to quantum brownian-motion", Physica A**121**, 587-616, (1983).
11. B. M. Garraway, "Decay of an atom coupled strongly to a reservoir ", Phys. Rev. A **55**, 4636-4639, (1997)

A phase operator from discretized complete set of coherent states

D.M. Davidović

The institute of Nuclear Sciences "Vinča", PO Box 522, 11001 Belgrade, Yugoslavia

D. Arsenović

Institute of Physics, P. O. Box 57, Belgrade, Yugoslavia

Many various ways to give a consistent and useful definition of the phase operator for a quantum oscillator have been suggested over the years [0]. Physically appealing idea, based on the canonical quantization of the classical relation leads to the following expression

$$\hat{\phi}_L = \frac{1}{2i}(\ln \hat{a}^+ - \ln \hat{a}) \tag{1}$$

where $\hat{a}^+$ and $\hat{a}$ are the creation and annihilation operators. However the expression 1 faces serious mathematical difficulties. These are due to the simultaneous appearance of the non-entire function $\ln \alpha$ and the overcomplete resolution of the identity on $L_2(\mathbf{R})$: $I = \int |\alpha\rangle\langle\alpha| d\alpha$, given by the coherent states $|\alpha\rangle$.

In order to provide a well defined object with an intuitively correct interpretation we utilized a subset $\{|\alpha_{m,n}\rangle\}$ of the oscillator's coherent states, which is complete, but not overcomplete [0]. The states $\{|\alpha_{m,n}\rangle\}$ and its biorthogonal states $\{|w_{m,n}\rangle\}$ are parameterized by points in a discrete lattice in the complex plain. Furthermore, periods of the lattice are chosen such that the surface of the cell is equal to π. The latest condition, provided that one and only one node from this lattice is excluded, is necessary and sufficient for the set of states to be complete and not overcomplete.

Vectors from the biorthogonal basis are characterized by [0]:

$$\langle\alpha|w_{kl}\rangle = (-1)^{kl+k+l} e^{-\frac{|\alpha|^2}{2}} \alpha^*_{kl} \frac{\sigma(\alpha^*)}{\alpha^*(\alpha^* - \alpha^*_{kl})} \tag{2}$$

where $\sigma(\alpha)$ is Weierstrass sigma function:

$$\sigma(\alpha) = \alpha \prod_{\substack{m,n \\ (m,n)\neq(0,0)}} \left(1 - \frac{\alpha}{\beta_{mn}}\right) e^{\frac{\alpha}{\beta_{mn}} + \frac{1}{2}\frac{\alpha^2}{\beta^2_{mn}}} \tag{3}$$

and $\beta_{mn} = m\omega_1^* + n\omega_2^*$. Complex numbers ω_1 and ω_2 determine lattice. We choose $\omega_1 = \sqrt{\pi}$, $\omega_2 = i\sqrt{\pi}$ for convenience.

Using the sets $\{|\alpha_{m,n}\rangle\}$ and $\{|w_{m,n}\rangle\}$ we define the following expression:

$$\hat{\phi} = \frac{1}{2i} \sum_{m,n} \ln \alpha_{m,n} |\alpha_{m,n}\rangle\langle w_{m,n}| - \ln \alpha^*_{m,n} |w_{m,n}\rangle\langle\alpha_{m,n}| \tag{4}$$

to represent the phase $\hat{\phi}$ when acting on the dense set of states represented as convergent sums of $|\alpha_{m,n}\rangle$.

The domain of $\hat{\phi}$ evidently contains the dense set $\{|\alpha_{m,n}\rangle\}$, and in addition all the vectors represented as convergent sums such that the action of $\hat{\phi}$ is also represented as a convergent sum. Although $\ln a$, $a \in \mathbf{C}$ is not an entire function, the representation (1)

Coherence and Quantum Optics VIII
Edited by Bigelow et al., Kluwer Academic/Plenum Publishers, 2003

enables one to calculate matrix elements of $\hat{\phi}$ and its commutation relations with $\hat{N}$ on a dense set where they are defined. These results can be used to approximate the mean values of the phase operator in other physically relevant states which do not belong to its domain. For example, numerical calculations show that:

$$\langle\alpha|\hat{\phi}|\alpha\rangle \approx \phi \tag{5}$$

and

$$\langle\alpha|[\hat{\phi}, \hat{H}]|\alpha\rangle \approx i, \tag{6}$$

where ϕ is the phase of the coherent state $|\alpha\rangle$, and $\hat{H}$ the hamiltonian of the harmonic oscillator. The errors in the previous approximate equations are indeed quite small. Moreover, eq. 5 is exact for coherent states from the nodes of the lattice. Also, eq. 6 is satisfied with the great accuracy for all states which are linear combinations of $|\alpha_k\rangle$ where k lies in any half-plane and is enough far from the origin.

On physical grounds independently of whatever definition of phase operator one would expect that average value of phase for the highly excited coherent states $|\alpha\rangle$ equals $\arg\alpha$, and also that for the states $|n\rangle$ phase is distributed randomly what leads to corresponding average value. The proposed operator satisfies both of these conditions.

Lynch R 1995 *Phys. Rep.* **256**, 367

Perelomov A 1986 *Generalized Coherent States and Their Applications* (Springer, Berlin)

Dynamics of a two-photon laser

Daniel J. Gauthier, Olivier Pfister,[†] William J. Brown,[‡] and Michael D. Stenner

Duke Univesity, Department of Physics, Durham, NC 27708

The high degree of temporal coherence of laser light arises from a complex interplay between the fundamental light-matter interactions of absorption, spontaneous emission, and stimulated emission. The coherence properties of the generated light can be altered significantly, and often in a surprising manner, by modifying the type of light-matter interaction on which the laser is based. As an example, the two-photon laser [1, 2] is based on the higher-order two-photon stimulated emission process, whereby two incident photons stimulate an excited atom to a lower energy state and four photons are scattered.

We have recently achieved degenerate two-photon optical lasing [3] using laser-driven potassium atoms [4] contained in a very high finesse optical resonator. This new type of quantum oscillator displays the expected turn-on behavior whereby it can only turn on when an injected coherent field brings the intercavity light above a critical value. Once the laser is above threshold, we observe polarization instabilities and chaos even though the total intensity remains nearly constant. This observation is not encompassed within the framework of current theoretical models of two-photon lasers and results from the atomic Zeeman degeneracy of the states participating in the two-photon stimulated emission process. We suggest that such a two-photon laser could emit polarization-entangled twin beams of light if this degeneracy is lifted, which may have quantum information applications.

To achieve two-photon optical lasing, we use a two-photon gain medium consists of a dense, collimated, effusive beam of laser-driven ^{39}K atoms that passes through the center of a high-finesse optical resonator. We use the D_1 transition of ^{39}K whose atomic states are denoted by $|\alpha F_\alpha M_\alpha\rangle$, with $\alpha = g$ for $4^2S_{1/2}$ and $\alpha = e$ for $4^2\mathrm{P}_{1/2}$ levels and with a ground-state hyperfine splitting of $\Delta_g = 462$ MHz. To create a population inversion for multiphoton transitions starting from the state $|g22\rangle$, the atoms are continuously optically pumped into this state by two $\hat{\sigma}_+$-polarized fields whose frequencies are set close to ω_{21} and ω_{22} (Bohr frequencies are denoted by $\omega_{F_eF_g}$). The two-photon laser resonator is in a linear (standing-wave) transverse-mode-degenerate sub-confocal configuration with a cavity finesse of 15, 140 and contains a maximum of 7×10^6 atoms.

Two-photon stimulated emission into the laser cavity mode (frequency ω_l) occurs when the optically pumped atoms are illuminated by an intense $\hat{\sigma}_-$-polarized Raman pump beam of frequency ω_d, and ω_l is adjusted to the two-photon transition frequency ($\simeq \omega_d + \Delta_g/2$). Photons are added two at a time to the cavity field via stimulated emission when two photons are annihilated from the $\hat{\sigma}_-$-polarized Raman pump field and the atom undergoes a multi-photon transition from the state $|g22\rangle$ to any one of the states $|g1M_g\rangle$. Note that such a hyper-Raman process can be mapped onto the idealized two-photon stimulated emission process by working in a dressed-state basis. The atomic, Raman pump beam, and optical cavity axes are mutually orthogonal, thereby suppressing competing phase-matched processes that might give rise to parametric oscillation.

Conclusive evidence that we have achieved two-photon lasing is indicated by the threshold behavior of the laser. Due to the nonlinear gain, lasing will only commence when a critical atomic inversion density and a critical cavity photon number is present. To observe the threshold behavior, we measure the power of the beam emitted from

Coherence and Quantum Optics VIII
Edited by Bigelow et al., Kluwer Academic/Plenum Publishers, 2003

the two-photon laser resonator and inject pulses of light into the cavity to trigger the laser to the 'on' state and shown in Fig. 1(a) for $\sim 3.3\times10^5$ injected photons. The duration of the injected pulse is indicated by the dashed vertical lines. It is seen that the power emitted by the cavity rises to approximately 0.2 μW and remains at this value, corresponding to an average cavity photon number of 2.2×10^6 and an intracavity intensity of 7.3 W/cm^2 at the cavity waist.

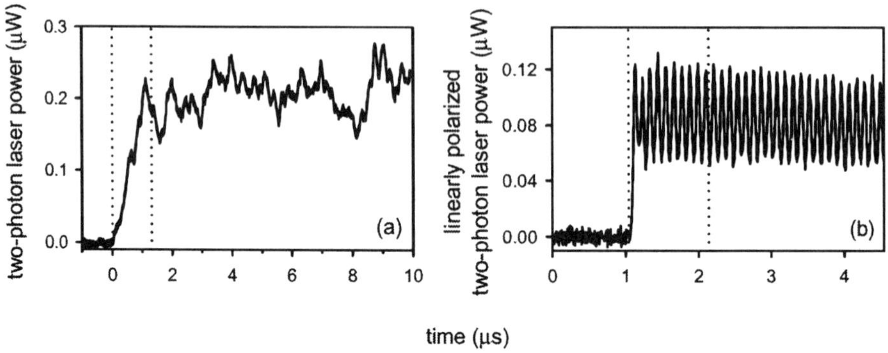

Fig. 1. (a) Turn on behavior of the two-photon laser. (b) Polarization dynamics of the two-photon laser.

While the total emitted power is seen to remain nearly constant, we find that the two-photon laser displays polarization instabilities. By placing a linear polarization before the detector, it is seen in Fig. 1(b) that the state of polarization undergoes very regular oscillations of period 0.11 μs, with a 50% depth of modulation. Similar dynamical behavior is observed for all polarizer orientations, suggesting that the state of polarization is elliptical with an ellipticity of 0.5 and a rotating major axis. We find that the laser dynamics is quite sensitive to the applied magnetic field: an increase by as little as 0.5 Gauss is sufficient to generate a much more complicated pattern, turning into a possibly chaotic pattern for a field strength of 2 Gauss. We believe that it may be possible to suppress the instabilities using a stronger magnetic field to lift the degeneracy of the different quantum pathways.

Lifting the degeneracy between the pathways also has the promise of yielding polarization entanglement. Consider, for simplicity, a single emission event. When the Zeeman levels are degenerate, the superposition of all possible amplitudes gives a photon pair (1,2) in the state $[\alpha_1|\hat{z}\rangle_1+\beta_1|\hat{x}\rangle_1][\alpha_2|\hat{z}\rangle_2+\beta_2|\hat{x}\rangle_2]$ with $|\alpha_i^2|+|\beta_i^2|=1$, where $|\hat{\varepsilon}\rangle_i$ ($\varepsilon=x,y$) is the polarization-labeled Fock state of the ith photon of the emitted pair. In a strong magnetic field, the $\hat{x}\hat{z}$ and $\hat{z}\hat{x}$ pairs become off-resonant and should stop lasing, creating the polarization-entangled state $\alpha_1\alpha_2|\hat{z}\rangle_1|\hat{z}\rangle_2+\beta_1\beta_2|\hat{x}\rangle_1|\hat{x}\rangle_2$.

This work is supported by NSF grant PHY-9876988.

†University of Virginia, Dept. of Physics, 382 McCormick Rd, Charlottesville, VA 22904.
‡Corvis Corp., 7015 Albert Einstein Drive, Columbia, MD 21046.

References and links

1. A.M. Prokhorov, Science **10**, 828 (1965); P.P. Sorokin and N. Braslau, IBM J. Res. Dev. **8**, 177 (1964); R.Z. Garwin, *ibid.* **8**, 338 (1964).
2. D.J. Gauthier, Q. Wu, S.E. Morin, and T.W. Mossberg, Phys. Rev. Lett. **68**, 464 (1992).
3. O. Pfister, W.J. Brown, M.D. Stenner, and D.J. Gauthier, "Polarization instabilities and in a two-photon laser," to appear in Phys. Rev. Lett. (2001).
4. O. Pfister, W.J. Brown, M.D. Stenner, and D.J. Gauthier, Phys. Rev. A **60**, R4249 (1999).

The principles of active spectroscopy of elastic light scattering

Anatoly Sukhodolsky,
General Physics Institute, Vavilov 38, Moscow Russia
http://www.gpi.ru/~sukhodol/

Abstract: The paper is to introduce active spectroscopy of elastic light scattering. The new principal approach is based on the application of an innovative vector optical heterodyning technique for active signal processing, which appears to be available by beating several scattered vector waves within a square-low photo-detector by their registration [1]. While the classical spectroscopy of optical beating deals with the problem to study a scalar optical signal [2] that is due to elastic scattering of a single narrow-band laser beam, the active spectroscopy of elastic scattering deals with two vector waves simultaneously undergo simultaneous scattering by a scatterer. In case one of the vector waves (heterodyne) has strong periodical modulation of direction of polarization, Coherent Light Beating Scattering (CLBS) occurs [1], which units elastic Light Scattering and Coherent non-linear Beating (mixing) within the square-low photo-detector. As a result, the vector optical signal provides with the information about both size and velocity of a scatterer, which is not available for well-known spectroscopy of scalar optical beating [2].

Vector optical heterodyning

The paper has twofold objective: a) to make an introduction into theory of vector optical signal and its application for CLBS by consideration of both a linear apertures and a particle. In other words, the first goal is to find out how such a well-known classical diffraction problems, solved for scalar waves, can be studied within CLBS, b) to find out the basic features of CLBS that can improve the performance of optical instrumentation, where standard "passive" light scattering is being used. In other words, the second objective is to show the concrete problems of optical measurements where CLBS has a visible advantage with regards to usual elastic light scattering using the algorithms of traditional scalar optical signal.

a) The wave theory of CLBS is based on a suggestion to separately apply Huygen's principle for different Jones component of a vector wave in propagation. As a result, theory of CLBS comes to the scalar Kirchhoff's integral solution for different elements of a CLBS-matrix responsible to vector optical signal. While the scalar optical signal is due to the product wave amplitude and its complex conjunction, the vector optical signal by coherent heterodyning is due to vector multiplication a Jones vector and its Hermitian conjugation.

b) The several practical problems where CLBS can find application are being considered. The problems of simultaneous sizing&velocimetry of macroscopic particles and the longitudinal and transverse CLBS-Doppler monitoring of mechanical vibrations are in scope of calculations. The problems of signal and noise will take visible part in consideration of CLBS-applications for practice.

The non-linear laser spectroscopy with application of tunable laser, such as for example Coherent Antistokes Raman Scattering (CARS) [3], states the problem if there is a way to apply "active" approaches developed in non-linear laser spectroscopy within the linear spectroscopy of nondestructive elastic light scattering. The idea of CLBS as a solution to the problem was firstly experimentally tested by a quite specific problem of motility of microorganisms [1]. A special problem that is solved within the paper is the more general problem of active spectroscopy in elastic light scattering. This problem is supposed to consider by introduction into theory of CLBS, which supported by several experiments. The problem to find velocity of a remote object by tuning of frequency of CLBS – is solved for both transverse and longitudinal Doppler velocities together with simultaneous k-sizing by variation of vector wave difference. The talk will be accompanied by animated solutions (please see one of examples at site www.come.to/vectorwaves) for several concrete problems to better understand the common scope of proposed approach and to compare "active" concept in elastic light scattering with classical "passive" scalar light scattering. Besides, the problem to have simultaneous w-spectroscopy of velocities and k-spectroscopy of sizes is solved by CLBS at the fundamental level that can be useful to be promoted into other spheres of spectroscopy, such as, for example, spectroscopy of molecular Rayleigh scattering.

Jones vector&matrix calculation and the principle of superposition in linear optics and scalar method of Kirchhoff integral used for concept of Scalar and Vector Optical Signal within a square-law photo-detector as a sole nonlinear element in elastic light scattering of vector waves.

The principles of CLBS sizing and velocimetry

The classic linear aperture and a single particle are considered to introduce the principles of CLBS sizing and velocimetry. CLBS matrix for studies of a particle distribution is derived. The direct vector solution of Maxwell's equations for diffraction of polarized waves becomes very difficult to describe the elliptical states of polarization specific for CLBS. Therefore, the principal solutions for CLBS are based on scalar Kirchhoff integral methods to apply for different components of Jones vectors separately. Such suggestion is based on the principle of linear superposition of electromagnetic waves in linear optics. As far as this is promotion of the popular Jones calculations into new sphere of wave optics to consider diffraction of polarized waves, the method can be interested in solving other problems of vector waves optics beyond CLBS. Actually, CLBS as a new active spectroscopy of elastic scattering can be attractive in applied optics and optical signal processing. The principal possibility to measure longitudinal and transverse Doppler velocities, by involving methods of vector optical heterodyning, can help with several basic formulas for vector signal processing in optical engineering. Several new innovative solutions for CLBS particle sizing and CLBS Doppler velocimetry can help to build new optical instruments.

References

1. Sukhodolsky A.T., "Coherent light beating scattering for size&velocity control", Internet lecture for International Workshop and Fall School for Young Scientists and Students "Light Scattering Technologies in Mechanics, Biomedicine and Material Science" October 6 - 9, 1998 Saratov, Russia (or click http://www.gpi.ru/~sukhodol/AS/)

2. H.Z.Cummins,E.R.Plike ed., Photon correlation and light beating spectroscopy, Plenum press, New York and London, 1974

3. Akhmanow A.A., Koroteev N.I. Methods of nonlinear optics in Raman scattering, Nauka (in Russian), Moscow, 1981

4. R.M.A.Arram, N.M.Bashar Ellipsometry and polarized light, North-Holland publishing, Amsterdam-New York-Oxford, 1977

Off-diagonal radiative-decay-coupled optical transitions in double quantum wells

Danhong Huang, and D. A. Cardimona
Air Force Research Lab, 3550 Absentee Avenue S. E., Kirtland AFB, NM 87117
Phone: (505)-846-5788, FAX: (505)-846-6098, email: danhong.huang@kirtland.af.mil

1. Introduction

When a laser is applied to resonantly excite a multi-level atomic system, there exists *quantum interference* in the absorption and emission spectra. A proper theory [1] for the radiative decay of excited electrons in an atomic system requires a quantum electrodynamic treatment of photons and electrons in second quantization. In addition to the usual diagonal radiative-decay time, there is an *off-diagonal radiative-decay coupling* (**ODRDC**) that becomes important when two or more electron transition energies are very close. [2] The effect of **ODRDC** describes a nearly resonant absorption of a spontaneously emitted photon from one downward electron transition by another upward electron transition. With the existence of **ODRDC**, any two nearly-degenerate electron transitions which are excited by a pump laser and share a common initial state will be coupled and interfere with each other, giving rise to a zero absorption [2-4] in the region between the two overlapping absorption peaks. In this paper, we study the effect of **ODRDC** on optical transitions of electrons in double-quantum wells (**DQW's**) which provides two nearly-degenerate excited states by electron tunneling. [5]

2. Results

By using the *rotating-wave approximation*, we can simplify the density-matrix equations by keeping only the nearly-resonant terms in the steady-state. In our density-matrix equations for the pump laser, the terms $\Omega^R{}_{12}(k_\|)$ $\mathrm{Im}[\rho_{12}(k_\|, \omega_C)]$ and $\Omega^R{}_{13}(k_\|)$ $\mathrm{Im}[\rho_{13}(k_\|, \omega_C)]$ represent the stimulated electron transitions between the lower state and the upper doublet, where ω_C is the frequency of the pump laser, $\Omega^R{}_{12}(k_\|)$ and $\Omega^R{}_{13}(k_\|)$ are the Rabi frequencies associated with the pump laser. Moreover, the terms $\beta_{21,13}[\Omega_{31}(k_\|)]$ $\rho_{13}(k_\|, \omega_C)$ and $\beta_{31,12}[\Omega_{21}(k_\|)]$ $\rho_{12}(k_\|, \omega_C)$ stand for the ODRDC which is responsible for the quantum interference between the nearly-resonant electron transitions from $|1\mathbf{k}_\|>$ to $|2\mathbf{k}_\|>$ and $|1\mathbf{k}_\|>$ to $|3\mathbf{k}_\|>$, where $\Omega_{31}(k_\|)$ and $\Omega_{21}(k_\|)$ are related to the energy separations between the lower state and the upper doublet. Finally, the terms $[i\Omega^R{}_{13}(k_\|)/2]$ $\rho^*{}_{23}(k_\|, \omega_C)$ and $[i\Omega^R{}_{12}(k_\|)/2]$ $\rho_{23}(k_\|, \omega_C)$ come from the induced quantum coherence, *i.e.*, non-zero induced $\rho_{23}(k_\|, \omega_C)$.

Similarly, we can get a pair of coupled equations for the density-matrix elements $\rho'_{12}(k_\|, \omega_P)$ and $\rho'_{13}(k_\|, \omega_P)$ associated with the weak probe field, where ω_P is the frequency of the probe field. In these two coupled equations, the difference of occupation probabilities $\rho_{jj}(k_\|, \omega_C)$-$\rho_{11}(k_\|, \omega_C)$ for j=2, 3 contains the effect of *transition blocking*, namely non-zero $\rho_{22}(k_\|, \omega_C)$ and $\rho_{33}(k_\|, \omega_C)$ due to laser excitation. The terms containing $\rho_{23}(k_\|, \omega_C)$ and $\rho^*{}_{23}(k_\|, \omega_C)$ come from induced quantum coherence due to laser pumping, which directly modifies the difference of occupation probabilities. The coupling terms $\beta_{21,13}[\Omega_{31}(k_\|)]$ $\rho'_{13}((k_\|, \omega_P)$ and $\beta_{31,12}[\Omega_{21}(k_\|)]$ $\rho'_{12}(k_\|, \omega_P)$ correspond to the ODRDC between the two nearly-resonant optical transitions from $|1\mathbf{k}_\|>$ to $|2\mathbf{k}_\|>$ and $|1\mathbf{k}_\|>$ to $|3\mathbf{k}_\|>$.

The nonlinear laser-field absorption coefficient at zero temperature can be found from the Lorentz ratio function, [6] and the Lorentz ratio function can be calculated from the *quantum-statistical average* of the laser-field Rabi frequency given by

$$< \Omega_L{}^R > = \Sigma_{j<j'} \Sigma_{\mathbf{k}\|} \rho_{jj'}(k_\|, \omega_C) \, \Omega^R{}_{jj'}(k_\|) \,,$$

where the phase cancellation between the two Rabi frequencies reflects the quantum interference between the two optical transitions of electrons in **DQW's**.

In our numerical calculation, the laser-field absorption coefficient and refractive-index function for asymmetric **DQW's** with intersubband pumping have been studied for different laser-field strengths, momentum-dependent energy-level separations, electron tunneling, and degrees of off-resonance. The zero absorption in the overlapping region within a small tunneling gap due to quantum interference between a pair of nearly-degenerate electron transitions has been found for the pump laser and explained. The quantum interference has been clearly demonstrated through the phase cancellation between the two statistically-averaged transition dipole moments. The laser frequency ω_C for the zero absorption is tunable within the tunneling gap if a small *dc* bias field is applied in the growth direction. The momentum-dependent energy-level separation has been found to be the most predominant in destroying the quantum interference in the system. For a weak probe field, we have investigated its linear absorption spectrum and refractive-index function for the same system selectively coupled by a laser field with different intensities. The optical gain of the probe field has been observed as a hole in the weak absorption peak, which has been demonstrated as a result of the partial inversion of the electron occupation probabilities in momentum-space after the system is selectively pumped by a laser. The optical gain of the probe field has been found to increase with the pump-laser strength. The influences of the induced quantum coherence and **ODRDC** on the probe-field absorption have been quantitatively analyzed. The relevance of the **ODRDC** in the current system with two nearly-degenerate energy levels to the coherent laser-coupled systems proposed for electro-magnetically induced transparency [7] and lasing without inversion [8] has been demonstrated by directly comparing the density-matrix equations in these cases.

3. Conclusions

In conclusion, we have derived the density-matrix equations in the presence of a pump-laser field for electrons in **DQW's** by working in second quantization and including **ODRDC** between a pair of nearly-degenerate electron transitions. These fully quantized density-matrix equations allow us to calculate simultaneously the optical response of electrons and radiative decay of excited-state electrons quantum-electrodynamically. On the basis of the derived density-matrix equations, the nonlinear optical absorption coefficient for the pump laser has been calculated. When another weak probe field is applied, its linear absorption spectrum has also been calculated by including the interactions between the probe field and electrons in the **DQW's** and between the probe and pump-laser fields.

4. References

[1] P. W. Milonni,"Semiclassical and quantum-electrodynamical approaches in nonrelativistic radiation theory", Phys. Rep. C **25**, 1-81 (1976).

[2] D. A. Cardimona, M. G. Raymer, and C. R. Stroud Jr., "Steady-state quantum interference in resonance fluorescence", J. Phys. B **15**, 55-64 (1982).

[3] D. A. Cardimona, M. P. Sharma, and M. A. Ortega, "Effect of three-level atomic state coherence on collapse and revival dynamics", J. Phys. B **22**, 4029-4039 (1989).

[4] D. A. Cardimona, "Effect of atomic-state coherence and spontaneous emission on three-level dynamics", Phys. Rev. A **41**, 5016-5025 (1990).

[5] A. Imamoglu and R. J. Ram, "Semiconductor lasers without population inversion", Opt. Lett. **19**, 1744-1746 (1994).

[6] D. H. Huang and M. O. Manasreh, "Intersubband transitions in strained $In_{0.07}Ga_{0.93}As/Al_{0.40}Ga_{0.60}As$ multiple quantum wells and their application to a two-colors photodetector", Phys. Rev. B **54**, 5620-5628 (1996).

[7] Imamoglu and Harris, "Laser without inversion: interference of dressed lifetime-broadened states", Opt. Lett. **14**, 1344-1346 (1989).

[8] M. O. Scully, "Enhancement of the index of refraction via quantum coherence", Phy. Rev. Lett. **67**, 1855-1858 (1991).

Forward four-wave mixing mediated by a Kerr nonlinearity

Jandir M. Hickmann†, Colin McCormick, Martino Trassinelli and Raymond Y. Chiao

Department of Physics, University of California
Berkeley, California 94720-7300
phone: 15106427166, fax: 15106438497, e-mail: chiao@physics.berkeley.edu

OCIS codes: (190.4380) Nonlinear optics, four-wave mixing; (270.4180) Multiphoton processes

Abstract: We investigate a forward scattering process which result from the four-wave interactions between two beams crossing a nonlinear media where the phase matching is mediated by the Kerr nonlinearity. We also discuss a possible analogy to this process in the time domain.

In a recent experiment it was observed that a photon-photon scattering process is possible in a rubidium vapor cell. In this experiment, two counter-propagating beams "collide" in the nonlinear sample and photons at 90-degree from the line of the colliding beams are shown to be correlated by means of coincidence detection, suggesting a photon-photon scattering mediated by the Kerr nonlinearity[1]. Here we intend to discuss a related process, where the two beams are in a degenerated four-wave mixing configuration usually utilized in the self-diffraction dephasing time measurements, with a rather small angle between them. Normally this configuration generates high orders diffracted beams, used for the dephasing time measurements. However a different kind of four-wave mixing process should be also possible, where two photons of noncollinear pump beams are absorbed and two collinear signal photons are generated in the direction of a bisector line between the pump beams. We will demonstrate that through a phenomenon similar to the *weak-wave retardation* [2][3], the phase matching is possible even if all four waves have basically the same frequency.

In order to understand this idea, we must look at the nonlinear polarization component P_4 and for simplicity we will assume that we have a stimulated process putting a small seed E_3 in the direction of the bisector line (see Fig. 1). In this case we may write the polarization for the stimulated field as:

$$\begin{aligned} P_4 &= \frac{3\varepsilon_0}{4}\chi^{(3)}_{xxxx}\{|E_4|^2 + 2(|E_1|^2 + |E_2|^2 + |E_3|^2) \\ &\quad +2E_1E_2E_3^* \exp\{i[(k_1 + k_2 - k_3 - k_4)\, z - (\omega_1 + \omega_2 - \omega_3 - \omega_4)t]\}\} \end{aligned}$$

where $\chi^{(3)}_{xxxx}$ is a component of the third order nonlinear susceptibility tensor, and E_j are the fields associated to the pump ($j = 1, 2$) and signal ($j = 3, 4$) beams. In this equation, the first term accounts for the self-phase modulation, the second for the cross-phase modulation and the third for the four wave mixing. Usually for a weak signal, the self-phase modulation will not be important as well the cross-phase modulation from the seed field and these terms may be neglected. The cross-phase modulation terms from the pump beams are large and in fact will play an important part in the four wave mixing process. The exponential term in the four-wave mixing term shows us that in order for the process be efficient, we need energy conservation and a phase-matching between the different wave-vectors associated with each field, and the degenerated case where all the frequencies are the same should be a very improbable process.

However, note that the signal waves have a decrease in the refractive index, generated by the cross-phase modulation term associated to the pump beams, if we assume that the nonlinear Kerr coefficient is negative, i.e., the sign corresponding to a self-defocusing nonlinearity. This difference in the refractive index is given by:

$$\Delta n = -\frac{3}{4n_0}|\,\mathrm{Re}[\chi^{(3)}_{xxxx}]|(|E_1|^2 + |E_2|^2)$$

where n_0 is the index of refraction. Consequently, there exists an subtractive shortening of the weak-wave vector by $\Delta k = -|\Delta n|\omega_0/c$ which we will call *weak-wave advancement*, in close analogy with an effect called "weak-wave retardation" in the case of the self-focussing sign of the Kerr nonlinearity[3]. In this way a phase matching between the wave-vectors is guaranteed, and we believe that this process must be almost as efficient as its counterpart, which was even suggested to be used to measure the nonlinear index[2].

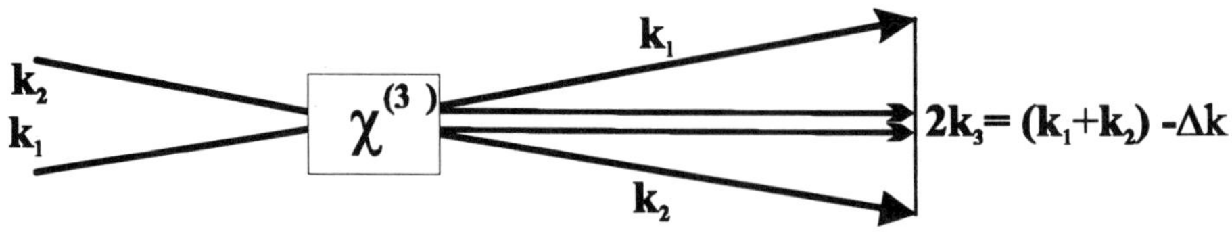

Fig. 1. The wave-vector configuration for the weak-wave advancement.

Now, we remember that the weak-wave retardation effect in the degenerated four-wave mixing is closely related to the nonlinear process of modulational instability associated with the propagation of light in an optical fiber in the anomalous dispersion regime. In this effect, a CW beam breakups into a periodic pulse train with a simultaneous appearance of associated spectral side bands[5]. If we extend this analogy in order to incorporate the effect discussed here, we expect that the inverted process of modulational instability should also be possible in optical fibers in the normal dispersion regime. In this time-domain version of the four-wave mixing mediated by the Kerr nonlinearity, we have as a input a oscillating field modulated at the frequency 2Ω. In order to generate this oscillating field we use spectral components with optical frequencies components equal to $\omega_0 \pm \Omega$, in such way that we will create a beating at 2Ω. These two optical components correspond to the noncollinear pump beams in spatial case. As a result, we obtain a stationary signal at the frequency ω_0, corresponding to the signal beams in the spatial case. This behavior is in complete agreement with the idea that the modulational instability may be thought as a four-wave mixing process where the fiber nonlinearity provides the phase-matching between the different fields involved[6].

In conclusion, we have discussed a four-wave mixing process mediated by the Kerr nonlinearity and related this process to a different, inverse kind of modulational instability. We believe that this nonlinear scattering process is at the heart of a experimental realization of a photonic Bose-Einstein condensate[4].

This work was supported in part by the ONR. Martino Trassinelli is on leave from Ecole Normale, Paris, France.

†On leave from the Department of Physics, Universidade Federal de Alagoas, Maceió, AL, Brazil with financial support from CNPq.

References

1. M. W. Mitchell, C. J. Hancox, and R. Y. Chiao, Phys. Rev. A **62**, 043819 (2000).
2. R. W. Boyd,*Nonlinear optics*, (Academic Press, Boston, 1992), and the references therein.
3. R. Y. Chiao, P. L. Kelley, and E. Garmire, Phys. Rev. Lett. **17**, 1158 (1966).
4. R. Y. Chiao, Opt. Comm. **179**, 157 (2000).
5. K. Tai. A. Hasegawa, and A. Tomita, Phys. Rev. Lett. **56**, 135 (1986).
6. G. P. Agrawal, *Nonlinear Fiber Optics*, 2nd ed. (Academic, San Diego, 1995).

Honeycomb Pattern Formation by Laser-Beam Filamentation in Atomic Sodium Vapor

R. S. Bennink, V. Wong, D. L. Aronstein, R. W. Boyd, S. G. Lukishova, A. M. Marino, and C. R. Stroud, Jr.
Institute of Optics, University of Rochester, Rochester, NY 14627
phone: (716) 275-5030, FAX: (716) 244-4936
bennink@optics.rochester.edu

Spontaneous pattern formation remains an intriguing problem in the field of nonlinear optics.. Such phenomena have been studied previously in a variety of material systems [1], including atomic vapors, liquid crystals, photorefractive crystals, organic liquids, glasses, semiconductors, and bacteriorodopsin. Most previous observations of regular pattern formation were obtained in systems in which optical feedback plays an important role. This feedback can be produced by the use of optical resonators and/or by the use of counterpropagating beams. Here we report a striking form of pattern formation that occurs when *single* laser beam (i.e. with no feedback mechanism) passes through an atomic sodium vapor. Pattern formation of this type appears to have been previously unreported, although pattern formation has been observed in sodium vapor under somewhat different experimental conditions [2], as well as in studies of small-scale self-focusing with a single beam in Kerr (non-saturable) media, e.g. [3].

In our experiment, we inject a 150 mW collimated laser beam with a diameter of 220 μm into a 6 cm-long sodium vapor cell containing 8×10^{12} atoms/cm^3. The laser is tuned 1.5 GHz to the blue side of the D2 resonance line. The beam is found to break into multiple components in passing through the cell as a consequence of self-focusing effects. Under certain input conditions, the beam is found to break up into three components of comparable power whose positions form an equilateral triangle. An example of such an arrangement is shown in the central panel of Fig. 1. The far field pattern of this emission has the symmetric honeycomb pattern shown in the right-most panel of the figure. We find both the near- and far-field patterns to be stable for tens of minutes, suggesting that filamentation of the light within the sodium vapor cell has produced a high-order soliton-like structure, perhaps similar to structures that have been predicted and observed recently [4]. Solitons of single [5] and multiple [6] orders are well known for non-saturable media, as are various self-trapping solutions for saturable media [7], Moreover, the highly structured beam of the sort we have observed may constitute a system in which to study quantum statistical effects [8].

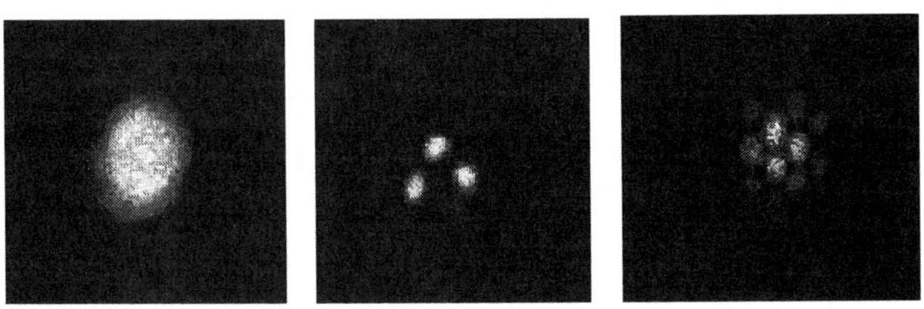

Fig. 1. Experimental observation of spontaneous pattern formation in atomic sodium vapor, showing the incident laser beam (left), the beam at the cell output (center beam in the far field (right).

Coherence and Quantum Optics VIII
Edited by Bigelow et al., Kluwer Academic/Plenum Publishers, 2003

We have also developed a numerical model that explains our experimental observations. The propagation of a laser beam through a medium with a saturable nonlinearity can be described by the equation

$$\left[\frac{\partial}{\partial z}+\frac{1}{2ik}\nabla_T^2\right]A=i\frac{n_0\omega}{2\pi}\frac{n_2}{1+\gamma|A|^2}A$$

where ∇_T^2 is the transverse Laplacian, ω is the frequency of the laser, and γ is a parameter related to the saturation intensity. The term involving n_2 describes self-focusing effects. We have performed a numerical integration of this equation for the conditions of our experiment, and we find that under appropriate conditions the beam does in fact break into three components as it propagates through the cell. These results are shown in Fig. 2.

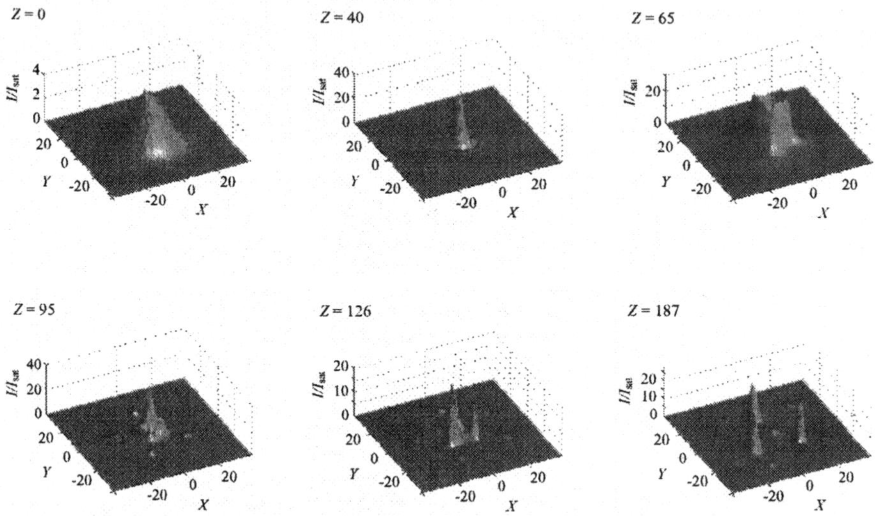

Fig. 2. Numerical simulation of the break-up of a laser beam into three components in passing through an atomic sodium vapor.

In summary, we have observed a new sort of pattern formation that occurs without external feedback and is observed when a single laser beam passes through an atomic vapor cell. Our observations are well modeled by a simple theoretical model.

References:

1. "Tranverse effects in nonlinear-optical systems," Special issue, J. Opt.Soc.Am. B **7**, No. 6 (1990).
2. J.W. Grantham, H.M. Gibbs, G. Khitrova, J.F. Valley, and Xu Jiajin, "Kaleidoscopic spatial instability: bifurcations of optical transverse solitary waves," Phys. Rev. Lett., **66**, 1422-1425 (1991).
3. A.J. Campillo, S.L. Shapiro, and B.R. Suydam, "Periodic breakup of optical beams due to self-focusing," Appl. Phys. Lett. **23**, 628-630 (1973).
4. W. Krolikowski, E. Ostrovskaya, C. Weilnau, M. Geisser, G. McCarthy, Y. S. Kivshar, C. Denz, and B. Luther-Davies, "Observation of dipole-mode vector solitons," Phys. Rev. Lett. **85**, 1424-1427 (2000).
5. R.Y. Chiao, E. Garmire, and C.H. Townes, "Self-trapping of optical beams," Phys. Rev. Lett. **13**, 479-482 (1964).
6. H.A. Haus, "Higher order trapped light beam solutions", Appl. Phys. Lett. **8**, 128-129 (1966).
7. M.L. Dowell, B.D. Paul, A. Gallagher, and J. Cooper, "Self-focused light propagation in a fully saturable medium: Theory," Phys. Rev. A **52**, 3244-3253 (1995).
8. A. Gatti,, E. Brambilla, L.A. Lugiato, and M.I. Kolobov, "Quantum entangled images", Phys. Rev. Lett. **83**, 1763-1766 (1999).

Initiation of superfluorescence: Collective and noncollective emission

J. P. Clemens and H. J. Carmichael

Department of Physics, University of Oregon, Eugene, OR 97403

tel: (541) 346-5863, jclemens@darkwing.uoregon.edu

The collective interaction of atoms with light is a topic of fundamental interest. In particular, the phenomena of superradiance and superfluorescence have received much attention since Dicke published his original paper on the topic [1]. Most work has focused either on the overall geometry dependent coupling between the atoms and the field to establish a preferred direction of emission, or it has assumed a preferred direction and then focused on the time evolution of the emission, often using a semiclassical approximation which neglects fluctuations in the atomic state (see references in [2]).

In this work we extend previous results to include the effects of competition between the collective emission in the preferred direction and noncollective emission into other directions during the early evolution from the inverted state. We begin with the master equation for the reduced density operator of the atoms derived by Bonifacio, et al. [3], though we do not assume that the collective emission dominates over ordinary spontaneous emission. The quantum trajectory method provides a simple numerical approach to this problem. Among other things, we calculate the photon statistics of the emitted light and demonstrate a correlation between pulse intensity and pulse delay time.

As a model, we consider N two level atoms inside an optical cavity. The atoms are assumed to be located at antinodes of the cavity mode so that they couple to the field with equal strengths and can be described by the collective operators $J_\pm = \sum_{i=1}^{N} \sigma_\pm$ (alternatively consider a ring cavity). The cavity decay rate is κ and the atoms spontaneously emit out of the cavity at the rate γ. In the bad cavity limit (κ larger than the collective emission rate), the field mode will adiabatically follow the atomic polarization. The master equation in the Born-Markov and rotating wave approximations is then

$$\dot{\rho} = \frac{\Gamma}{2}\left(2J_-\rho J_+ - J_+J_-\rho - \rho J_+J_-\right) + \frac{\gamma}{2}\sum_{i=1}^{N}\left(2\sigma_{i-}\rho\sigma_{i+} - \sigma_{i+}\sigma_{i-}\rho - \rho\sigma_{i+}\sigma_{i-}\right), \quad (1)$$

where the first term describes collective emission, and the second, the noncollective emission. The effects of noncollective emission are not fully included in earlier work.

We simulate superfluorescent pulses using quantum trajectories, where the system is described by a pure state conditioned on a sequence of emission events. The time evolution is generated by the non-Hermitian effective Hamiltonian

$$H_{\text{eff}} = -\Gamma J_+J_- - \gamma\sum_{i=1}^{N}\sigma_{i+}\sigma_{i-}, \quad (2)$$

which acts between emissions, punctuated by collective emissions, $|\psi\rangle \rightarrow J_-|\psi\rangle$ (plus normalization), at the rate $\Gamma\langle\psi|J_+J_-|\psi\rangle$ and noncollective emissions, $|\psi\rangle \rightarrow \sigma_{i-}|\psi\rangle$, at the rates $\gamma\langle\psi|\sigma_{i+}\sigma_{i-}|\psi\rangle$, $i = 1, \ldots, N$.

We are interested in the case of superfluorescence, where the system is initially inverted. Using the Dicke states, $|j, m\rangle$, $j = (0, \frac{1}{2}), \ldots, N/2$, $m = -j, \ldots, j$, greatly simplifies the trajectories. The initial state is $|N/2, N/2\rangle$. The collapse operators act on a Dicke state to give $J_-|j,m\rangle \propto |j, m-1\rangle$ and $\sigma_{i-}|j,m\rangle \propto |j-\frac{1}{2}, m-\frac{1}{2}\rangle \otimes |-\rangle_i$.

Coherence and Quantum Optics VIII
Edited by Bigelow et al., Kluwer Academic/Plenum Publishers, 2003

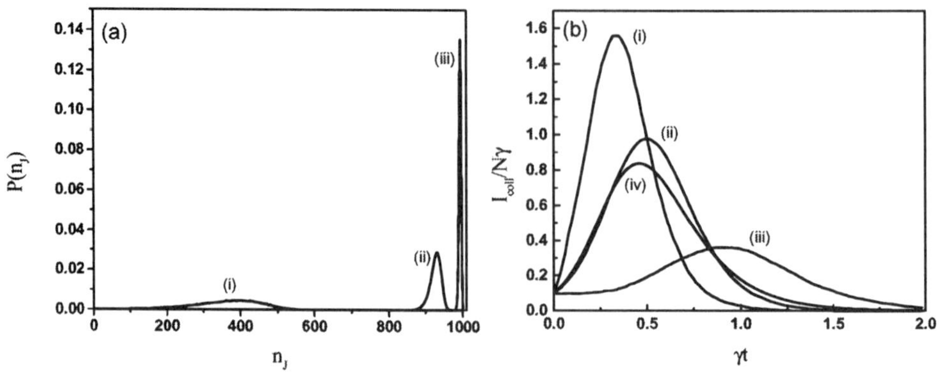

Fig. 1. (a) Photon statistics of the collective emission: $N = 1000$ and $\Gamma/\gamma = 0.01$ (i), 0.1 (ii), 1 (iii). (b) Average collective pulse intensity as a function of time: $N = 100$ and $\Gamma/\gamma = 0.1$, given a total pulse photon number of $n_J = 75$ (i), 57 (ii), 31 (iii). Curve (iv) is the average over n_J and the most probable photon number is $n_J = 57$. $P(n_J)$ is relatively broad for the parameters used in (b), similar to curve (i) in (a).

Thus, the noncollective emission puts one atom in the ground state and leaves the remaining atoms in a Dicke state. Furthermore, the Dicke states are eigenstates of the Hamiltonian. Hence, when started in a Dicke state, the system remains in a Dicke state under both kinds of emission as well as under the continuous evolution. In fact, since $[J_-, \sigma_{i-}] = 0$, the conditioned state of the system at any time depends only on the numbers of emissions of each kind, n_J and n_σ, that have occurred prior to that time; for given n_J and n_σ the state is $|(N - n_\sigma)/2, (N - n_\sigma)/2 - n_J\rangle$.

The trajectory formalism allows us to gain detailed statistical information about the emitted light in a straightforward manner. For example we have access to the complete photon number distribution of the collective pulse; previously, this was only calculated approximately for times such that $n_J, n_\sigma \ll N$ [4]. For purposes of illustration we plot in Fig. 1(a) the photon number distribution for $N = 1000$ and various ratios Γ/γ. Here $P(n_J)$ is the probability that n_J photons are emitted collectively and $N - n_J$ noncollectively. Apparently a significant fraction of the photons may be emitted collectively even when Γ/γ is quite small, but when there are only $N = 1000$ atoms available, as here, the distribution for small Γ/γ is quite broad.

We can go beyond photon statistics to study correlations between the peak intensity and the pulse delay time. We show in Fig. 1(b) the average intensity as a function of time, given a definite number of photons in the emitted pulse. We see that the pulse intensity and delay time are correlated, with higher intensity giving shorter delays. Thus, for parameters where the photon number distribution is broad, we find there are delay time fluctuations that have not been calculated before.

We have numerically simulated the emission of superfluorescent pulses, extending the one-mode model of superradiance to include spontaneous emission into other modes. New results include the photon statistics of the collective pulse and correlations between the peak intensity and the pulse delay time. We anticipate that an extension of our approach will allow us to go beyond the simple model used here to include the full angular dependence of the emission for a collection of atoms distributed in space.

References and links

1. R. H. Dicke, Phys. Rev. **93**, 99-110 (1954).
2. M. Gross and S. Haroche, Phys. Rep. **93**, 301-396 (1982).
3. R. Bonifacio, P. Schwendimann, and F. Haake, Phys. Rev. A **4**, 302-313 (1971).
4. C. T. Lee, Phys. Rev. A **14**, 1926-1928 (1976).

Frequency offset locking of 35 mW 657nm visible laser diodes

Shintaro Hisatake, Masahiro Fukushima,
Yoshihiro Kurokawa, and Wakao Sasaki
Department of Electronics, Doshisha University, Kyotanabe City, Kyoto 610-0321 JAPAN
Fax: +81-774-65-6783 E-mail: etal101@mail4.doshisha.ac.jp

Abstract: Frequency of a 35 mW visible laser diode has been offset locked to another 35mW frequency stabilized laser diode by using delay-line microwave discriminator. The linewidth of the heterodyne beat spectrum has been estimated to be about 1 MHz under the offset locking condition.

1. Introduction

Accurate frequency control of a narrow-linewidth high output power visible laser diode (LD) is important for not only various scientific experiments but also industrial fields. Although various attempts have been made for linewidth reduction and frequency control of visible LD's, present narrow-linewidth visible laser light sources are still expensive and have rather low output power to be applied in the extensive fields of industry.

In this work, we have composed a narrow-linewidth visible laser light source, as a master laser, using a commercially available 35mW 657nm LD based on simple frequency stabilization system. The frequency of a similar type 35 mW visible LD (slave laser) has been offset locked to the frequency of the master laser by negative electrical feedback to the injection current of the slave laser.

2. Experimental set up and results

Fig.1. shows an experimental set up for offset locking. Two samples of 35 mW 657nm type LD were used as master and slave lasers. The frequency of the master laser was stabilized to a confocal Fabry-Perot cavity by negative electrical feedback to the injection current based on Pound-Drever technique. The achieved minimum square root of the Allan variance of the master laser was 3.9×10^{-11} at $\tau = 0.1$ msec , and the obtained linewidth was less than 400 kHz [1]. A Littman type external cavity laser was used as a slave laser. The front facets of the slave laser have no anti-reflection coating. Coarse tuning of the slave laser wavelength was achieved by changing the laser temperature and the external cavity length so that both lasers might operate at almost the same wavelength. Accurate tuning of the slave laser frequency was carried out by adjusting the injection current of the slave laser. In the experiments, the injection currents and temperatures of the master and slave lasers were set at 80.09 mA, 15.56 °C and 95.17mA, 23.00 °C, respectively. As a frequency discriminator, we used a delay-line microwave discriminator. The frequency error signal from the discriminator was fed back to the injection current of the slave laser through PI servo amplifier to stabilize heterodyne beat frequency.

As a result, Fig.2. shows the achieved heterodyne beat spectrum under the frequency offset locking condition. The resolution of the spectrum analyzer was 300 kHz.

Coherence and Quantum Optics VIII
Edited by Bigelow et al., Kluwer Academic/Plenum Publishers, 2003

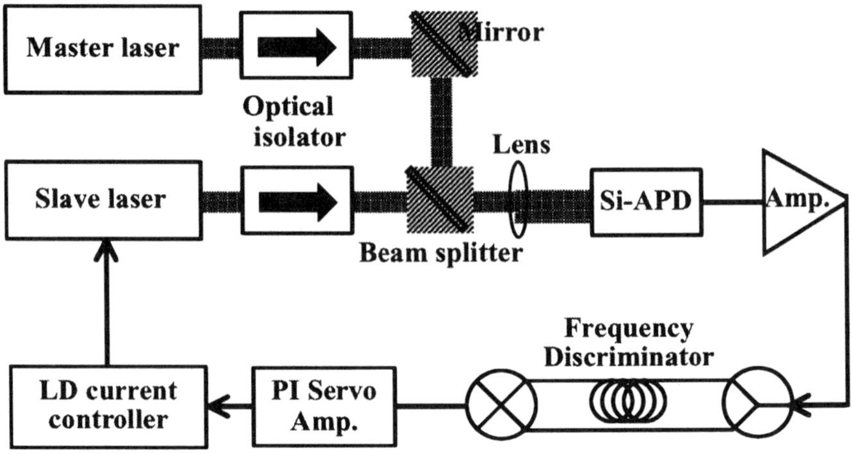

Fig. 1. Experimental set up for frequency offset locking.

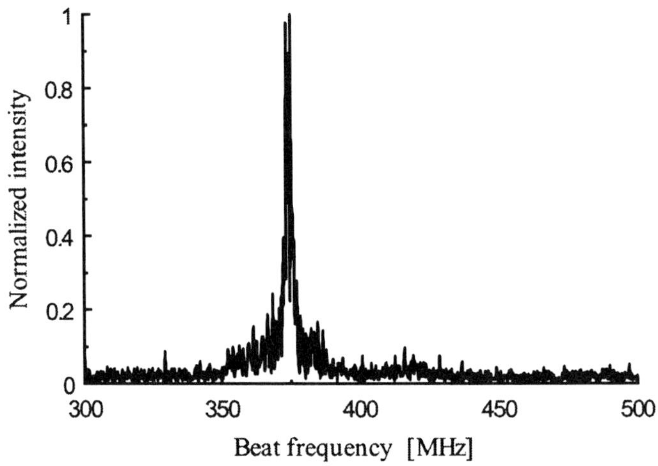

Fig.2. Heterodyne beat spectrum under the offset locking condition.

3. Conclusion

The frequency of a 35 mW 657 nm visible LD has been offset locked to the frequency of the narrow-linewidth master LD. The beat frequency has been stabilized at about 375 MHz and the linewidth of the beat signal has been estimated to be about 1 MHz.

4. Reference

[1] S. Hisatake, et al., SPIE Proc. 45th Annual Meeting, Vol. 4285, San Jose, California, Jan. 2001

Collective interaction between radiation and atoms: phase transition, optical gratings and the influence of the modeling hypotheses

J. Javaloyes, M. Perrin, G.L. Lippi

Institut Non Linéaire de Nice, UMR 6618 CNRS–UNSA
1361 Route des Lucioles, F-06560 Valbonne, France
Tel. +33 (0)4 92 96 73 04; Fax +33 (0)4 93 65 25 17
perrin@inln.cnrs.fr

A.Politi

Istituto Nazionale di Ottica Applicata
L.go E. Fermi 6, I-50125 Firenze, Italy
politi@ino.it

Abstract: We present results on pump-probe interaction in a sample including self-consistently a feedback loop between atoms and field. A phase transition, different optical gratings and atomic momentum distributions are analyzed. Various modeling choices are discussed.

OCIS codes: (999.9999) Collective atom-field effects; (350.2770) Gratings

The role of momentum conservation in the atom-field interaction has been considered in detail by a model that allows for a global coupling between atoms and field (and therefore for a feedback between these components) [1]. As a consequence of the dipolar interaction, atoms adopt preferential positions in space. Such a self-generated grating has been credited as the physical origin (Bragg mirror) of gain in a counterpropagating (weak) probe beam.

Experiments devoted at testing the predictions [2] have provided results that could be interpreted within the framework of such an instability but have failed to provide direct evidence for an atomic density grating. An alternative interpretation has therefore been proposed, based on a polarization grating [3], but tests conducted on one of the setups [4] do not confirm the validity of this hypothesis.

The main discrepancy between the original model [1] and the experiments [2] lies in the difference between the hypotheses (monocinetic, collimated beam) and reality (thermal sample in a cell). To improve the numerical predictions and give a better interpretation of the experimental results, we have introduced collisions with a thermally distributed buffer gas [5].

In order to account for the interaction with a thermal bath, we have followed the traditional approach adopted in molecular dynamics: as a result of collisions (occurring at random, exponentially distributed times) the atomic momentum is redistributed according to the temperature of the heat bath. Simultaneously, the *phase* of the atomic polarization (we consider only polarization-preserving shocks) is randomly reset to a value uniformly distributed between 0 and 2π.

As a result, the numerical simulation of the dynamical equations (up to 4096 atoms) shows that the average variables and output intensity reach a steady state. The output intensity approaches a sharp transition as the number of atoms increases. Below a critical pump value, the gain (ratio between output and input intensity) on steady state is negligible. Only above this value, we observe a net gain.

An important difference between our approach and the one of [1] is that the transition is *not* accompanied by the generation of a density grating. Indeed, in our simulations the atomic bunching is always nearly zero. At the same time, we see that the grating in the modulus of atomic polarization is very weak (and noisy), a fact that is not incompatible with the choice of circularly polarized light – with the same helicity – for both pump and probe. Hence, in this configuration, which reflects the experimental choices [2, 4], one can also exclude strong effects coming from the grating of the modulus of the atomic polarization.

Conversely, we observe a strong grating in the phase of the polarization, and that the *coherence parameter*:

$$c = \frac{1}{N} \sum_{j=1}^{N} S_j e^{-i\theta_j},$$

reaches a non-zero value (S_j represents the atomic (complex) polarization of the jth atom and θ_j its position along the pump beam axis –normalized to $2k$, k being the pump's wavenumber). The fact that c differs from zero indicates that the atomic polarization adapts its *phase* (adiabatically) in space to take values that are compatible with the e.m. field distribution. This appears to be the actual source term for the instability. In addition, we also observe the presence of a fairly strong grating in the inversion of population. This is understandable because of the existence of a modulated field structure.

The distribution of atomic momenta in the sample turns out to be strongly affected by the interaction. Above threshold, the initially gaussian-distributed velocities rearrange themselves to give rise to a double-peaked structure which is compatible with one group of atoms at rest in the modulated intensity wave, and another (wider) moving with a most probable velocity relative to it.

Some further insight about the atomic collective behaviour can be gained by using an "effective" macroscopic model for collisions, where we neglect fluctuations. This can be done by just adding a damping term to the dynamical equation for the momentum [6]. Using this hypothesis, we have investigated the possibility of further simplifying the model equations, adiabatically eliminating the internal atomic degrees of freedom. As a result, we find that under certain approximations, the model can be reduced to a system of globally coupled oscillators and that this explains the clustering behaviour which we observe numerically in specific parameter ranges.

On the other hand, by adopting a Langevin-type of approach (i.e. adding a suitable stochastic force to the damping term) we find a qualitative agreement between its predictions and those obtained from the complete model which includes collisions (in molecular-dynamics fashion).

References and links

1. R. Bonifacio and L. De Salvo, Nucl. Instr. Methods Phys. Res. Sec. A **341**, 360 (1994); R. Bonifacio, L. De Salvo, L.M. Narducci, and E.J. D'Angelo, Phys. Rev. A **50**, 1716 (1994).
2. G.L. Lippi, G.P. Barozzi, S. Barbay, and J.R. Tredicce, Phys. Rev. Lett. **76**, 2452 (1996); P.R. Hemmer , N.P. Bigelow, D.P. Katz, M.S. Shariar, L. De Salvo, and R. Bonifacio, Phys. Rev. Lett. **77**, 1468 (1996).
3. W.J. Brown, J.R. Gardner, D.J. Gauthier, and R. Vilaseca, Phys. Rev. A **55**, R1601 (1997).
4. S. Barbay, G. Fabre, and G.L. Lippi, Opt. Commun. **165**, 119 (1999).
5. M. Perrin, G.L. Lippi, and A. Politi, submitted to Physical Review Letters; M. Perrin, G.L. Lippi, and A. Politi, submitted to Journal of Modern Optics.
6. Bonifacio, Verkerk, Opt. Commun. **124**, 469 (1996).

Local field effects on the spatiotemporal dynamics of broad-area two-level lasers

V. Ahufinger, J. Mompart, R. Corbalán
Departament de Física. Universitat Autònoma de Barcelona, E-08193, Bellaterra, Spain

J. Garcia-Ojalvo, M. C. Torrent, R. Vilaseca
Departament de Fisica i Enginyeria Nuclear. Universitat Politècnica de Catalunya, Colom 11, E-08222, Terrassa, Spain
Telf: (34) 93 581 16 53, Fax: (34) 93 581 21 55, e-mail address: Ramon.Corbalan@uab.es

Abstract:

We show that local field effects on the spatiotemporal dynamics of broad-area two-level lasers change the width of the neutral stability curve around its minimum increasing the range of transverse wavevectors that participate in pattern formation.

Summary

In the last few years there has been an intense activity in the study of transverse instabilities in fundamental nonlinear optical systems such as homogeneously broadened two and three-level laser systems [1,2], feedback Kerr media, photorefractive oscillators, atomic vapors, and optical parametric oscillators, among others [3]. In the case of homogeneously broadened two-level systems, just above threshold, there is a different behavior for positive and negative cavity detunings [1]. Whereas for cavities tuned above resonance with respect to the two level transition, the emission is in the form of a uniform plane wave propagating along the cavity axis, for cavities tuned below resonance the optical wavevector develops a small component in the transverse plane, in such a way that laser emission becomes off-axis, and the field projection in the transverse plane defines a travelling wave or "phase roll". This off-axis emission can be interpreted as a way of increasing the oscillation frequency of the electromagnetic field and making it energetically resonant with the two level transition without losing the resonance condition with the cavity.

In this paper we investigate the effects on the spatiotemporal dynamics of broad-area two-level systems when a dense gain medium is considered. It is well known that in dense media the dipole-dipole interactions become non-negligible and the effective field acting on an atom/molecule is no longer the macroscopic field but the so-called local field, which is due to all external sources and to all dipole moments of the atoms/molecules within the sample except the one under consideration [4]. In quantum optics, the inclusion of the local field effects gives rise to a large number of interesting phenomena, such as spectral red shifts [5], mirrorless optical bistability [6], "piezophotonic" effects [7], and enhancement of inversionless gain and absorptionless index of refraction in lasing without inversion [8].

Through a linear stability analysis (LSA) of the trivial non-lasing solution, we show that the inclusion of the local field corrections in the density-matrix equations of the system does not change the pumping threshold at which transverse instabilities develop, but strongly broadens the neutral stability curve around its minimum. In particular, for a certain range of cavity detunings, the neutral stability curve is almost flat in a wide domain, which implies that, at a given pumping, the domain of wavenumbers participating in the pattern formation strongly increases. This makes the transverse dynamics more sensitive to noise and entails the appearance of spatio-temporal regimes at low pumping rates. It is also worth noticing that due to the high value of the density and therefore of the gain and cavity losses, there is an important pulling effect that leads to values of the cavity detunings much larger than in the previous studies with a medium of low density [2]. Numerical simulations of the full nonlinear model confirm the results of the LSA and add further insight into the transverse dynamics of the system.

[1] P. K. Jakobsen, J. Lega, Q. Feng, M. Staley, J. V. Moloney and A. C. Newell, "Nonlinear transverse modes of large-aspect-ratio homogeneously broadened lasers: I. Analysis and numerical simulations", Phys. Rev. A **49**, 4189 (1994).
[2] J. Garcia-Ojalvo, M. C. Torrent and R. Vilaseca, "Spatiotemporal dynamics of dressed broad-area lasers", Quantum Semiclass. Opt. **10**, 809 (1998).
[3] G. J. de Valcarcel, E. Roldan and R. Vilaseca, "Special issue on Paterns in Nonlinear Optical Systems", J. Optics B **1**, n1 (1999).
[5] R. W. Boyd, "Nonlinear Optics", Academic Press. (San Diego, 1992).
[6] R. Friedberg, S. R. Hartmann, J. T. Manassah, "Effect of local field correction on a strongly pumped resonance", Phys. Rev. A **40**, 2446 (1989).
[7] F. A. Hopf, C. M. Bowden, W. H. Louisell, "Mirrorless optical bistability with the use of LFC", Phys. Rev. A **29**, 2591 (1984). Y. Ben-Aryeh, C. M. Bowden, and J. C. Englund, "Intrinsic optical bistability in collections of spatially distributed two-level atoms", Phys. Rev. A **34**, 3917 (1986).
[8] A. S. Manka, J. P. Dowling, C. M. Bowden and M. Fleischhauer, "Piezophotonic switching due to local field effects in a coherently prepared medium of three-level atoms", Phys. Rev. Lett. **73**, 1789 (1994).
[9] J. P. Dowling, C. M. Bowden, "Near dipole-dipole effects in lasing without inversion: An enhancement of gain and absorptionless index of refraction", Phys. Rev. Lett. **70**, 1421 (1993).

Dynamics of multiple-collapse with ultrashort laser pulses

K. D. Moll and Alexander L. Gaeta
School of Applied and Engineering Physics, Cornell University, 212 Clark Hall, Ithaca, NY 14853
kdm14@cornell.edu, a.gaeta@cornell.edu

Abstract: We investigate the self-focusing dynamics of ultrashort pulses in dispersive media. In the anomalous dispersion regime, the pulse may undergo multiple collapse at locations separated by several diffraction lengths.

The three mechanisms that primarily influence the propagation of ultrashort laser pulses in bulk media are diffraction, dispersion, and nonlinearity. For example, in the normal dispersion regime it is found that at powers above the critical power for self-focusing the pulse can exhibit pulse-splitting dynamics rather than undergo spatio-temporal collapse [1,2]. We have experimentally investigated the interplay between these effects in both the normal and anomalous group velocity dispersion (GVD) regimes and find significant differences in the self-focusing dynamics under conditions in which multiple collapse events occur.

For the experiments operating in the normal dispersion regime, we used 50-fs pulses from an amplified Ti:sapphire system operating at a wavelength of 800 nm. For the anomalous dispersion regime, we utilized 50-fs pulses from an optical parametric amplifier (OPA) at a wavelength of 1550 nm. The laser beam was focused into various transparent materials and allowed to self-focus (see Fig. 1). The collapsing beam was observed through a microscope and imaged by a digital camera.

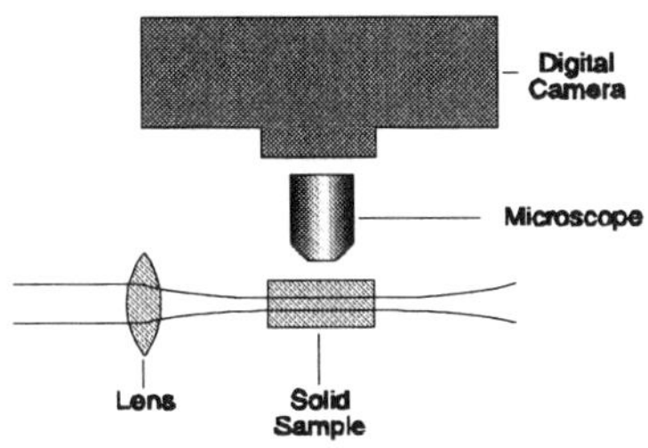

Fig. 1. Schematic of the experimental setup. The digital camera images the scattered light of the focused beam as it collapses due to self-focusing.

Images are taken as the pulse energy is increased. Figure 2 shows a series of photographs taken of the beam as it is focused into sapphire (normal dispersion at 800 nm) with a 7.5-cm lens. As the energy is increased, the pulse undergoes self-focusing which causes the location of the collapse point to move towards the input face. However as the energy is increased further, a second collapse point is observed. Further increases in the power result in the formation of additional collapse points and in the movement of the collapse point towards the front surface. In the presence of normal GVD, the locations of the collapse points occur between the input face and the geometric focal point of the lens. Similar behavior is observed in fused-silica and BK7 glasses and with lenses of 5, 17.5, and 50 cm focal lengths.

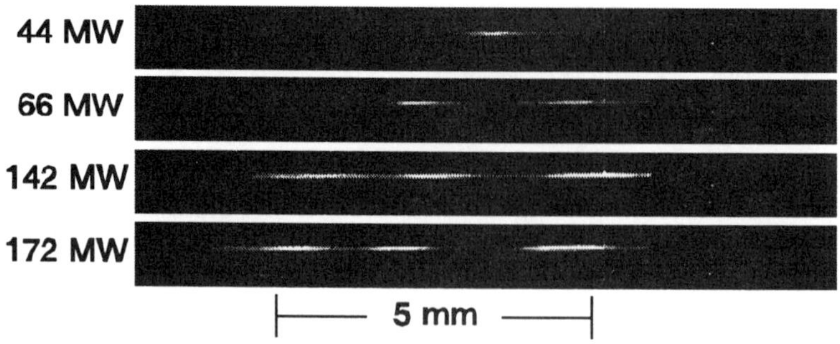

Fig. 2. Photographs of multiple collapse events in the normal dispersion regime as the peak input power (shown on the left) of the pulse is increased. The incident beam is propagating from left to right.

Similar experiments are repeated for the anomalous GVD regime with the pulses from the OPA, which are focused into a 30-cm long block of BK7-glass. As in the case of normal dispersion, multiple collapse events are observed, however the location of these points is not confined to the interval between the input face and the geometrical focus. We find, for example, that the location of the additional collapse points could be separated from the position of the initial collapse by as much as 20 cm, which is a phenomenon unseen in the case of normal dispersion.

We propose the following mechanism for these observations. At powers just above the point of collapse, the beam self-focuses, and the collapse is halted by plasma formation that partially absorbs the pulse and defocuses the beam. As the incident pulse energy is increased, enough energy remains in the pulse after the interaction with the plasma such that self focusing can overcome diffraction and the beam can collapse again. This mechanism applies to both the normal and anomalous GVD regimes. However, the ability for the beam that experiences anomalous dispersion to refocus at much larger distances than its normal dispersion counterpart is a result of to the manner in which dispersion and nonlinearity chirp the pulse. In the case of normal (anomalous) dispersion, the pulse is chirped in such a way that self-phase modulation cannot (can) compensate the phase across the pulse. As the Ti:sapphire pulse self-focuses, dispersion and nonlinearity both chirp the pulse in the same sense, leading to a temporal broadening and decrease in the peak power. As the pulse continues to propagate, dispersion continues to chirp the pulse and leads to a further reduction in the peak power such that self-focusing becomes negligible. However in the anomalous GVD regime, as the pulse self-focuses, a large chirp is introduced from the nonlinearity near the focus in the opposite sense as that produced by anomalous dispersion. Hence, as the pulse propagates past the collapse point, the ensuing dispersion as it propagates leads to compression of the pulse and an increase in the peak power above the point needed for self-focusing to overcome diffraction (i.e., the threshold for collapse).

In conclusion, we have observed that the self-focusing dynamics can differ significantly in the normal or anomalous GVD regimes. We have seen the formation of multiple collapse events separated by distances far larger than the confocal parameter of the initial beam demonstrating that our observation represents a true defocusing-focusing cycle and not the occurrence of a moving focus.

[1] J. E. Rothenberg, *Opt. Lett.* **17**, 583 (1992).
[2] J. K. Ranka, R. W. Schirmer, and A. L. Gaeta, *Phys. Rev. Lett.* **77**, 3783 (1996); S. A. Diddams, H. K. Eaton, A. A. Zozulya, and T. S. Clement, *Opt. Lett.* **23**, 379 (1998).

Spectroscopic Methods for Subwavelength Atom Localization

M. Suhail Zubairy[1,2], Sajid Qamar[2], and Shi-Yao Zhu[3]

[1]*Department of Physics, Texas A&M University, College Station, TX 77843*

[2]*Department of Electronics, Quaid-i-Azam University, Islamabad, Pakistan*

[2]*Department of Physics, Hong Kong Baptist University, Kowloon Tong, Hong Kong*

zubairy@physics.tamu.edu

Abstract: We propose a simple spectroscopic scheme of atom localization for an atom passing through a standing wave field. The Rabi frequency is position dependent and therefore the spontaneously emitted photon carries the information of the atomic center-of-mass motion. This leads to atom localization even during the flight through the standing wave field.

OCIS codes: (999.9999)

In recent years, several schemes have been considered for the localization of an atom using the standing optical light field. These schemes are based on the possibilities of measuring the phase shift of either the standing wave or the atomic dipole caused by the passage of the atom through the field. We can also correlate a spatially varying potential with an atomic resonance frequency and the position distribution of the atoms. In these methods the position information is available only after the atom has passed through the field. However for many potential applications, it is desirable to obtain position information of the atom during its passage through the standing wave.

In this talk, we suggest a simple scheme to localize an atom inside the standing wave during its motion. This scheme utilizes the idea that the frequency of the spontaneously emitted photon carries the information about the position of the atom due to its dependence on the position-dependent Rabi frequency of the driving field. Therefore an atom is localized as soon as the spontaneously emitted photon is detected. This scheme presents a simple method for the localization of an atom using simple two-level atom interacting with the classical standing-wave field. In the presence of the driving field, dynamic Stark splitting of the atomic levels takes place and we get a three-peak resonance fluorescence spectrum. The splitting is directly proportional to the position dependent Rabi frequency. Our scheme exploits this fact and by measuring the frequency of the spontaneously emitted photon we can localize the atom during its motion through the standing field.

We consider a two-level atom A with energy levels $|a\rangle$ and $|b\rangle$ and transition frequency ω_{ab} that is described by a center-of-mass wavefunction $f(x)$. The atom is moving along z-axis and interacts with a resonant standing-wave light field of wave-vector $\kappa = \omega_{ab}/c$ aligned along the x direction. The velocity component of the atom along z-axis is considered large enough so that the motion in this direction is treated classically. The driven atom radiates spontaneously and one of the modes of the scattered light interacts with the detector atom B, initially in its ground state. The detector atom consists of the ground level $|\beta\rangle$ and a set of excited levels $|\alpha_{\mathbf{k}}\rangle$. We assume that the scattered light of wave-vector $\mathbf{k}_0$ is absorbed by the detector atom and is excited to an appropriate energy level $|\alpha_{\mathbf{k}_0}\rangle$. Our aim is to find the conditional position distribution of the atom A, i.e., the conditional probability $W(x;t \mid \alpha_{\mathbf{k}_0})$ of finding the atom A at position x at time t when the detector atom B is excited to the level $|\alpha_{\mathbf{k}_0}\rangle$.

We assume that the center-of-mass momentum of the atom A along x-axis does not change appreciably during its passage through the standing-wave. We can then neglect the kinetic energy term for the atom in the Raman-Nath approximation.

Our scheme utilizes the fact that the frequency of the spontaneously emitted photon is directly related to the x dependent Rabi frequency of the driving field. We now see how the spectrum of the spontaneously emitted photons or scattered light mimic the position probability of the center-of-mass motion of an atom. The conditional probability $W(x;t \mid \alpha_{\mathbf{k}_0})$ of finding the atom A at position x at time t when the detector atom excites to the level $|\alpha_{\mathbf{k}_0}\rangle$ is

$$W(x;t \mid \alpha_{\mathbf{k}_0}) = \sum_{n_{\mathbf{q}}} \left|\langle x| \psi_{b,0_{\mathbf{k}_0},n_{\mathbf{q}},\alpha_{\mathbf{k}_0}}\rangle\right|^2. \tag{1}$$

A detailed calculation yields the following expression for the conditional position probability

$$W\left(x;t \mid \alpha_{_{\mathbf{k}_0}}\right) \equiv W(x) = |f(x)|^2 \, P(\omega,x,t), \tag{2}$$

where

$$P(\omega,x,\infty) = \frac{I_0(r)}{8\pi}\left[\frac{3\Gamma/4}{(\Delta+2g(x))^2+(3\Gamma/4)^2} + \frac{\Gamma}{\Delta^2+(\Gamma/2)^2} + \frac{3\Gamma/4}{(\Delta-2g(x))^2+(3\Gamma/4)^2}\right]. \tag{3}$$

This is the well-known three-peak Mollow spectrum, the only difference being the position dependence of the Rabi frequency.

In the resonance fluorescence spectrum we have three peaks centered at $\Delta = 0$ and $\Delta = \pm 2g$. In our scheme of localization of an atom we replace the Rabi frequency g with the position-dependent Rabi frequency $g(x) = G\sin(\kappa x)$. The peaks are now x dependent and are located at $\Delta = 0$ and $\Delta = \pm 2G\sin(\kappa x)$. Atom now undergoes different Rabi oscillation at different position in a standing wave and we get maxima in the position distribution corresponding to these Rabi frequencies. for an initially broad wave packet as a function of the normalized position κx and detuning Δ. We note that for zero detuning there is a uniform position probability distribution over the wavelength domain of the standing wave. This is due to the fact that the atom exhibits a peak at $\Delta = 0$ for any value of Rabi frequency, and hence for all values of κx. The heights of the peaks for all values of position are the same and we therefore obtain a uniform position distribution. Thus the conditional position distribution provides no information about the atom localization for $\Delta = 0$. An increase in detuning corresponds to the localization of the atom at different positions inside the standing wave depending on the value of the position-dependent Rabi frequency $g(x)$. We obtain four maxima of same heights and widths in the region $0 \le \kappa x \le 2\pi$ located at $\kappa x = \pm\sin^{-1}(\Delta/2G) \pm n\pi$ $(n = 0, \pm 1)$. For small values of Δ, these maxima are located near the nodes of the standing wave. However with the increased detuning, these peaks move towards the antinodes of the standing wave. For $\Delta = \pm 2G$, four maxima merge into two and lie on the antinodes of the standing wave. There are no resonances for $|\Delta| > 2G$ and we obtain a flat position distribution over the wavelength domain.

These results indicate a strong correlation between the detuning of the scattered light and the position of the atom. The measurement of a particular frequency corresponds to the localization of the atom in a subwavelength domain of the standing wave.

Laser Accelerating of Neutral Atoms: Vecksler-Macmillan Phase Stability

I.V. Mel'nikov[a),b)] and P.G. Kazansky[c)]

a) *Centro de Investigaciones en Ingeniería y Ciencias Aplicadas, Universidad Autónoma del Estado de Morelos, Ave. Universidad 1001, col. Chamilpa, Cuernavaca 62210, Morelos, México*
Voice/Fax: +52.73.29.70.84, E-mail: Igor.Melnikov@ciicap.uaem.mx

b) *General Physics Institute of the Russian Academy of Sciences, ul. Vavilova 38, Moscow 117942, Russian Federation*

c) *Optoelectronics Research Centre, University of Southampton, Highfield, Southampton SO17 1BJ, United Kingdom*

Abstract: We report a neutral atom accelerator that exploits Vecksler-Macmillan principle of phase stability originally suggested for charged-particle beams and may be implemented in a hollow or photonic fibre.The Wigner density-matrix formalism is used to derive closed expressions which supply us with condition of trapping atoms into an acceleration mode and parameters of the accelerated atom bunch.

The recent high level of interest in atom optics is due largely to dramatic achievements in fabrication of optical structures for guiding neutral atoms and even Bose-Einstein condensate. This has gained understanding of fundamental physics behind the manipulation of the macroscopic state of an atom ensemble by means of laser light. On such a base, it has been proposed theoretically [1] and implemented experimentally [2] an acceleration of neutral atoms by Ti:sapphire laser.

This Report suggests such a modification of the atom guiding scheme that may allow accelerate a bunch of neutral atoms up to a keV energy. This can be implemented in the two ways: (i) scanning of the wavelength of the driving cw laser by means of, e.g., a scanning Fabry-Perot interferometer inserted into the laser cavity, or (ii) applying the magnetic field to the guiding structure such that it changes the wavelength of the transition between two Zeeman sublevels gradually along the acceleration trajectory.

The acceleration of an ensemble of neutral atoms can be modelled by Wigner density matrix. This, being a function of quantum-mechanical co-ordinate and momentum, provides a self-consistent transformation to the description of the motion in terms of classical variables. In the framework of this Report, this enables us to make use of the atom distribution function that is the sum of diagonal elements of Wigner matrix. We derive a corresponding equation for the distribution function and use it in our analysis of the spatio-temporal dynamics and stability of its solution that is, of the atom ensemble. We find that the atom can be trapped into an acceleration mode within a certain gap window of the rate of the wavelength scanning. The upper bound is due to the fact the maximum rate is not allowed to be larger than the Doppler shift of the absorption line of accelerated atoms otherwise the wavelength is taken away from the contour and subsequently there is no interaction between the laser field and ensemble. At small values of the scanning rate, the line broadens because of the spontaneous decay and holds the same width of the Doppler contour.

We show that it might be an atom-optics analogue of Vecksler-Macmillan phase stability for a charged-particle beam. That is scanning of the wavelength drives the area of the interaction along the longitudinal velocity axis. The synchronously moving atom acquires a momentum in the direction of the light propagation and hence experiences an acceleration w which is due to the photon absorption. Let its velocity v_1 is on the leading edge of the driving force where $\partial w/\partial v < 0$, where v is the atom velocity. The atom with $v > v_1$ experiences a smaller acceleration whereas that with $v < v_1$ moves with larger one. This means that the stable bunch to be formed around v_1 from an initially wide atom ensemble what is an atom-optic analogue of Vecksler-Macmillan phase stability. Our numerical results obtained for a 10-W 780-nm cw laser confirm our speculations. And what is more, we show that if the guiding structure is made on the based on a up-chirped Bragg grating, than it is possible to drive atoms with a gap soliton which will gather all the atoms on its leading edge into an extremely narrow bunch which is then accelerated as a singe particle due to the spontaneous decay suppression.

References:

[1] M.K. Oberthalet *et al.*, Phys. Rev. Lett. **83**, 4447 (1999)
[2] R.M. Godun *et al.*, Phys. Rev. A **62**, 013411 (2000)

Self-Pulsing in an Optical Ring Cavity Coupled with Three-Level Atoms

Hai Wang, D. J. Goorskey, and Min Xiao
Department of Physics. University of Arkansas, Fayetteville. AR 72701
Tel.: (501)575-6568, Fax: (501)575-4580
E-mail: mxiao@mail.uark.edu

One of the reasons for the recent rising interest in electromagnetically induced transparency (EIT) in multilevel atomic systems [1] is its potential application in enhancing nonlinear optical processes. Several experiments were carried out recently to show that the atomic coherence effect in multi-level atomic systems could reduce absorption and enhance frequency conversion and four-wave mixing processes [2,3]. On the other hand, when two-level atoms are put into an optical cavity, many nonlinear processes, such as optical bistability and instability, occur. Recently, several theoretical papers proposed switching effects in three-level and four-level atomic systems with and without an optical cavity [4,5]. These nonlinear optical effects are caused by enhanced Kerr-nonlinearity and dramatic changes in the dispersion slope due to atomic coherence in such multi-level atomic systems. Recently, we have measured the enhanced Kerr-nonlinear index of refraction n_2 of three-level Λ-type atoms by using an optical ring cavity and found a large modification of n_2 as functions of the frequency detunings of the probe and coupling lasers. We also observed self-pulsation of cavity transmission when the coupling frequency is tuned to atomic resonance and the probe frequency is detuned from resonance by a few MHz. In this paper, we report such observation and discuss its physical mechanism.

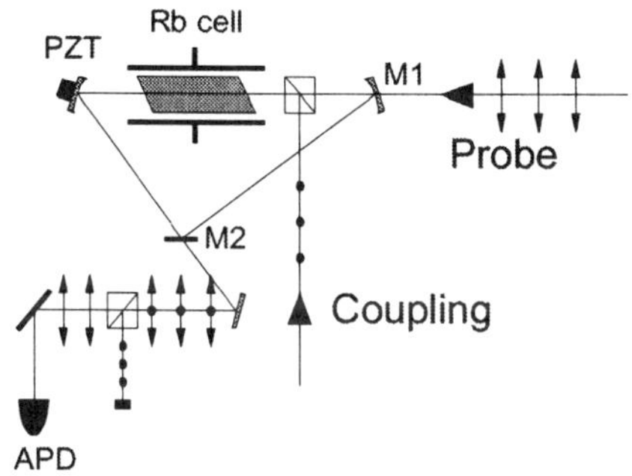

Fig. 1. Experimental setup: PZT is a mirror mounted on a piezoelectric transducer for adjusting the cavity length. M1 and M2 are mirrors. APD is an avalanche photodiode detector.

The basic experimental system is shown in Fig. 1 with an optical ring cavity containing rubidium atoms in a 5 cm long, Brewster-cut vapor cell. We consider the three-level Λ-type system in the D_1 lines of ^{87}Rb. The two fine structure sub-levels of $5S_{1/2}$ serve as the two lower states and the $5P_{1/2}$, F′=2 level serves as the upper state to form the Λ-type system. The probe beam interacts with the $5S_{1/2}$, F=1 to $5P_{1/2}$, F′=2 transition and the coupling beam couples the levels $5S_{1/2}$, F=2 and $5P_{1/2}$, F′=2. The atomic vapor cell is heated to 64°C. The Kerr nonlinearity n_2 for the probe beam is enhanced due to atomic coherence in the three-level medium and is measured to be about +/- 6.5×10^{-6} cm^2/W at a probe detuning of -/+ 7 MHz with a coupling detuning of zero. The experimental set-up is very similar to our earlier experiment in measuring the cavity-linewidth narrowing effect due to EIT [6]. The probe beam is focused by a lens, through mirror M1, to the center of the vapor cell and is mode-matched with the optical cavity. The coupling beam is introduced through a cubic polarizing beam-splitter and intersects the probe laser at a 2° angle in the center of the Rb vapor cell. This misalignment is enough to keep the coupling beam from circulating within the cavity. The two beams are orthogonally polarized and co-propagate in the atomic medium to eliminate the first-order Doppler effect [7]. The coupling beam power is about 23 mW and the probe beam power is about 140 μW. The probe and coupling beam radii at the center of the vapor cell are about 80 μm and 700 μm, respectively. The frequencies of the coupling and probe beams are locked to atomic transition lines by saturation absorption spectroscopy in another rubidium-filled vapor cell and their detunings are monitored by a Fabry-Perot cavity. The cavity transmission of the probe beam (cavity field) is detected by an APD and the coupling beam is rejected by another cubic polarizing beam-splitter.

With the probe and coupling lasers frequency-locked, we scan the cavity length from below resonance towards on resonance. When the cavity length reaches the cavity resonant condition, the circulating intensity of the probe beam reaches a maximum. Due to the enhanced Kerr nonlinearity in the atomic vapor due to atomic coherence, the cavity transmission intensity quickly drops off when the cavity length increases slightly past resonance. As the cavity length is scanned further, the cavity transmission intensity starts to oscillate, as shown in Fig. 2. The oscillating period is about a few μs and is very regular. Next, we slow down the cavity length scanning rate and record the oscillation. The oscillation appears persistent for a long time with an oscillation period of 1–2 μs, as shown in Fig. 3. Such self-pulsation only exists when both the coupling and probe beams are on and only happens for a small range of coupling and probe detunings where the Kerr nonlinearity is greatly enhanced.

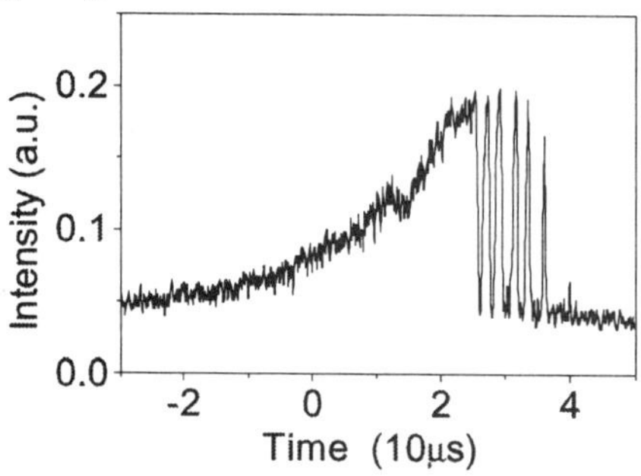

Fig.2. Cavity transmission as cavity is scanned through resonance showing self-pulsing immediately after the intensity reaches the resonant maximum.

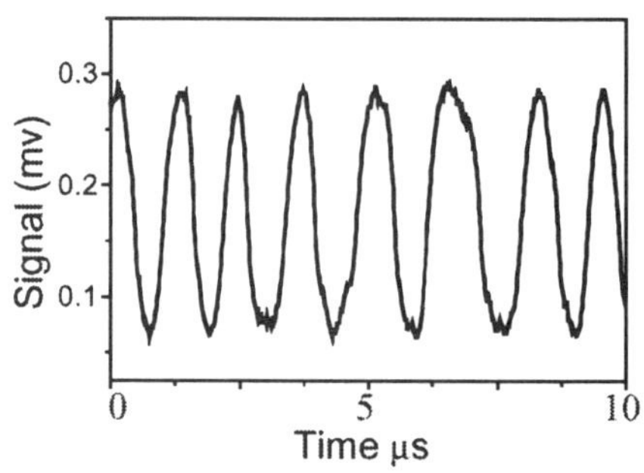

Fig. 3. Signal output from optical ring cavity showing self-pulsing. Probe detuning is at +7 MHz and coupling frequency is locked on resonance.

We are currently measuring the parametric dependence of this self-pulsing phenomenon and are working on the theoretical explanation. We expect this effect to have applications in optical switching with low light levels and in quantum information processing.

This work is supported by the Office of Naval Research and the National Science Foundation.

References:

[1] S. E. Harris, "Electromagnetically Induced Transparency", Physics Today, **50**(7), 36 (1997).

[2] A.J. Merriam, S.J. Sharpe, M. Shverdin, D. Manuszak, G.Y. Yin, and S.E. Harris, "Efficient Nonlinear Frequency Conversion in an All-resonant Double-Λ System", Phys. Rev. Lett., **84**, 5308 (2000).

[3] B. Lü, W.H. Burkett, and M. Xiao, "Nondegenerate four-wave mixing in a double-Λ system under the influence of coherent population trappings", Opt. Lett., **23**, 804 (1998).

[4] A. Imamoğlu, H. Schmidt, G. Woods, and M. Deutch, "Strongly Interacting Photons in a Nonlinear Cavity", Phys. Rev. Lett., **79**, 1467 (1997).

[5] S.E. Harris and Y. Yamamoto, "Photon Switching by Quantum Interference", Phys. Rev. Lett., **81**, 3611 (1998).

[6] H. Wang, D.J. Goorksey, W.H. Burkett, and M. Xiao, "Cavity-linewidth narrowing by means of electromagnetically induced transparency", Opt. Lett., **25**, 1732 (2000).

[7] J. Gea-Banacloche, Y. Li, S. Jin, and M. Xiao, "Electromagnetically induced transparency in ladder-type inhomogeneously broadened media: Theory and experiment", Phys. Rev. A, **51**, 576 (1995) ; M. Xiao, Y. Li, S. Jin, and J. Gea-Banacloche, "Measurement of Dispersive Properties of Electromagnetically Induced Transparency in Rubidium Atoms", Phys. Rev. Lett., **74**, 666 (1995).

Matter-Wave Mixing by Optical Cavity Fields

Hong Y. Ling
Department of Chemistry and Physics, Rowan University, Glassboro, NJ 08028

Recently, there has been a growing interest in matter-wave mixing involving Bose-Einstein condensates (BEC) [1-4]. We consider a model [1] consisting of a cavity field (probe), an opposing non-depletion pump field, and a matter wave traveling along x normal to the propagation dimension z of the optical fields, which are tuned far away from the atomic transition. Our model is a matter wave version of the collective atomic recoil laser (CARL) [5-7], which shares the same gain mechanism with recoil-induced resonance (RIR) [7,8]. Here, virtual two-photon transitions couple the |0> (injected atomic mode) to the |±ℏK> z-momentum modes, where ℏK is the two-photon momentum. If the scattering of the atoms from |±ℏK> to higher order modes is ignored, the equations of motion are reduced into

$$\left[\frac{\partial}{\partial t}+v_{0x}\frac{\partial}{\partial x}\right]\Phi_0(x,t)=iA^*\Phi_{+1}(x,t)+iA\Phi_{-1}(x,t) \tag{1a}$$

$$\left[\frac{\partial}{\partial t}+v_{0x}\frac{\partial}{\partial x}\right]\Phi_{-1}(x,t)=-i\omega_{2r}\Phi_{-1}(x,t)+iA^*\Phi_0(x,t) \tag{1b}$$

$$\left[\frac{\partial}{\partial t}+v_{0x}\frac{\partial}{\partial x}\right]\Phi_{+1}(x,t)=-i\omega_{2r}\Phi_{+1}(x,t)+iA\Phi_0(x,t) \tag{1c}$$

$$\frac{dA(t)}{dt}=i\delta_C A(t)+i\alpha\frac{1}{L_x}\int_0^{L_x}dx\left[\Phi_0^*(x,t)\Phi_{+1}(x,t)+\Phi_{-1}^*(x,t)\Phi_0(x,t)\right]-\kappa A(t) \tag{2}$$

where $\Phi_0(x,t)$ and $\Phi_\pm(x,t)$ are, respectively, the |0> and |±ℏK> components of the matter wave, ω_{2r} the two-photon recoil frequency shift, v_{0x} the speed of the injected wave, A(t) the scaled cavity field amplitude, δ_C the pump frequency shift relative to the cavity mode, α the gain coefficient, κ the decay rate of the cavity, and L_x the effective interaction distance along x dimension. In addition to the probe field, two momentum side modes will be made into

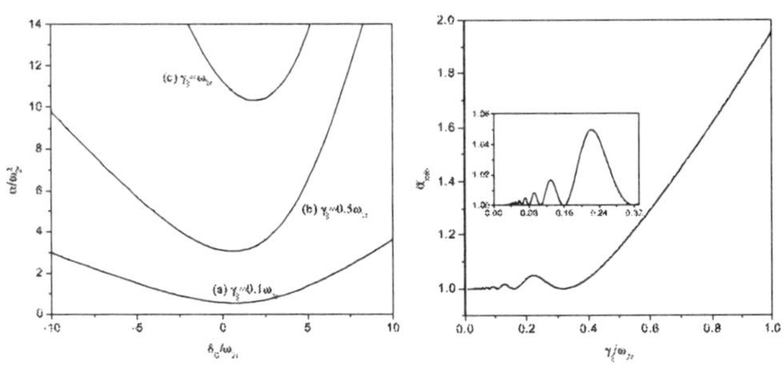

Fig. 1. The figure on the left side represents the threshold α as a function of δ_C with κ=2.63 ω_{2r}. The figure on the right side shows how $\overline{\alpha}_{\min}$ changes with $\gamma_{\parallel}$.

"lasing" as a result of nonlinear coupling between the optical and atomic fields. For this reason, we treat this problem as a multimode "laser" operation, where "lasing" can mean both optical and matter waves. The requirement that the beat frequencies from both optical and atomic oscillations must be mode-locked at the steady state means the following steady-state ansatz:

$$A(t)=|A_L|e^{-i\delta_L t-i\theta_L},\Phi_0(x,t)=\phi_0(x),$$

$$\Phi_{+1}(x,t) = \phi_{+}(x)e^{-i\delta_L t - i\theta_L}, \Phi_{-1}(x,t) = \phi_{-}(x)e^{i\delta_L t + i\theta_L}, \tag{3}$$

where δ_L and θ_L are, respectively, the frequency shift and arbitrary phase of the cavity field relative to the pump field. This ansatz transforms Eqs. (1) and (2) into the steady-state equations involving $|A_L|$, δ_L, and ϕ_α (α=0,-,+), that are subject to the boundary condition: ϕ_0=1, $\phi_\pm$=0. We first derive the threshold lasing condition from the steady sate equations in a perturbative manner, and discuss it in analogy to the four-wave parametric generation of Stokes and anti-Stokes signals in nonlinear optics. The left side of Fig. 1 illustrates how the threshold α increases with both $|\delta_C|$ and $\gamma_{\parallel}(=v_{0x}/L_x)$. Let $\bar{\alpha} = \alpha/2\gamma_{\parallel}\kappa$. The smaller the $\bar{\alpha}$ the easier the system is to produce lasing. The minimum α required for lasing as a function of $\gamma_{\parallel}$

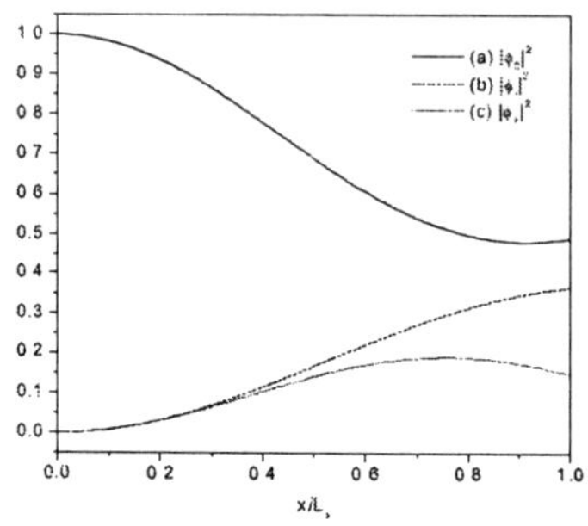

Fig. 2. Spatial distribution of atomic field modes under the condition that $\gamma_{\parallel}$=0.4ω_{2r}, κ=0.1ω_{2r},α=0.3 ω_{2r}^2, and δ_C=0. The calculated $|A_L|$=0.36ω_{2r} and frequency shift δ_L= -0.31ω_{2r}.

is depicted on the right side of Fig. 1. It shows that $\bar{\alpha}_{min}$ =1 can be realized in the locations determined by $\gamma_{\parallel}$= $\omega_{2r}/n\pi$ (n=1,2,3,...), at which the gain of ϕ_- is maximum while the absorption of ϕ_+ is zero. To calculate the steady state, we take the advantage of the collective nature of the field variable, and switch the base from ($|0\rangle$,$|\pm\hbar K\rangle$) to a new one by a unitary transformation composed of the eigenvectors of a symmetric matrix. With this transformation, the spatial degree of freedom in the steady state formalism can be integrated analytically, leading to a set of steady sate equations much easier to solve numerically. Figure 2 gives an example of steady-state spatial distributions of atomic modes, indicating how the two momentum side modes grow with distance at the expense of the injected mode. A linear stability analysis procedure is also developed based on the same unitary transformation and a new operation called the pseudo-complex conjugate operation. Numerical study of the instability threshold reveals a dramatic reduction in the instability threshold value α at small $\gamma_{\parallel}$, suggesting that BEC may hold a great promise for nonlinear instability studies because of its unprecedented long interaction time with photons.

References:

1. M. G. Moore and P. Meystre, "Optical control and entanglement of atomic Schrodinger fields", Phys. Rev. A. 59, R1754 (1999).
2. L. Deng *et al.*, "Four-wave mixing with matter waves", Nature(London) 398, 218(1999).
3. S. Inonye *et al.*, "Superradiant rayleigh scattering from a Bose-Einstein condensate", Science 285, 571(1999).
4. M. Trippenbach *et al.*, "Theory of four-wave mixing of matter waves from a Bose-Einstein condensate", Phys. Rev. A 62, 023608(2000).
5. R. Bonifacio and L. DeSalvo, "Collective atomic recoil laser (CARL) optical gain without inversion by collective atomic recoil and self-bunching of two-level atoms", Nucl. Instrum. Methods Phys. Res. A 341, 360 (1994).
6. M. G. Moore and P. Meystre, "Effects of atomic diffraction on the collective atomic recoil laser", Phys. Rev. A 58, 3248(1998).
7. H. Y. Ling *et al.*, "Theory of a collective atomic recoil laser", Phys. Rev. A 63, 053810(2001).
8. J. Guo *et al.*, "Recoil-induced resonance in nonlinear spectroscopy", Phys. Rev. A 46, 1426(1992).

Saturation law and the resonance fluorescence spectrum of a forced two-level atom described by modified optical Bloch equations

Gregory Surdutovich and Alexandre Ghiner*
Departamento de Engenharia Eletrica, UFPR, Cx.P.19011, Curitiba,PR, Brazil
Tel..55(41)361-3221, fax 55(41)361-3228, e-mail **gregory@eletr.ufpr.br**
*Departamento de Fisica, UFMA, CEP 65080-040, Sao Luis, MA, Brazil
Tel.55(98)217-8221, fax 55(98)217-8201 e-mail **ghiner@ufma.br**

1.Introduction

An increasing interest to nonlinear modifications of optical Bloch equations (OBE) raises a problem about proper account of the spontaneous relaxation in these equations. Usually in order to calculate the influence of the nonlinear terms to the resonance fluorescence line shape function the relaxation terms were putting into the generalized Bloch equations *ad hoc* [1]. Making use of only energetic considerations we found earlier the relaxation mechanism in the form of additional terms in the OBE which, on one hand, guarantees the fulfillment of the balance of the fluxes of energy between atomic and field subsystems and, on the other hand, allows to include arbitrary collision-like relaxation mechanism without violation of this balance [2]. In a remarkable manner these additional terms exactly correspond to renormalization of the external field with allowance of the classical radiation damping (RD) effect

$$\vec{E}^{(r)} \rightarrow \vec{E}^{(r)} + \frac{2}{3c^3}\dddot{\vec{d}}^{(r)} \equiv \tilde{\vec{E}}^{(r)} \tag{1}$$

II. Modified OBE

A system of the modified optical Bloch equations acquires the form

$$\ddot{\vec{d}}^{(r)} + 2\Gamma\dot{\vec{d}}^{(r)} + \omega_0^2\vec{d}^{(r)} = -2fN^{(r)}\vec{E}^{(r)} - \frac{4}{3}\frac{f}{c^3}N^{(r)}\dddot{\vec{d}}^{(r)} \tag{2a}$$

$$\dot{N}^{(r)} + \gamma\left(N^{(r)} - N^{e}\right) = \frac{2}{\hbar\omega_0}\vec{E}^{(r)}\cdot\dot{\vec{d}}^{(r)} + \frac{4}{3\hbar\omega_0 c^3}\dot{\vec{d}}^{(r)}\cdot\dddot{\vec{d}}^{(r)} \tag{2b}$$

Where $f \equiv \frac{\omega_0}{\hbar}\frac{\left|d_{eg}\right|^2}{3} = \frac{e^2}{m}F$, $F = \frac{m}{e^2}\frac{\omega_0}{\hbar}\frac{\left|d_{eg}\right|^2}{3}$ and the rest notations are conventional.

To find the resonance fluorescence spectrum we apply to this system the ordinary procedure of calculation of the complete nonsaturating probe absorption coefficient of the waves ω_1 and ω_2 in presence of strong field ω_3.

$$\vec{E}^{(r)}(t) = \frac{1}{2}\left(\vec{E}_1 e^{-i\omega_1 t} + \vec{E}_2 e^{-i\omega_2 t} + \vec{E}_3 e^{-i\omega_3 t} + c.c.\right) \tag{3a}$$

$$\vec{d}^{(r)}(t) = \frac{1}{2}\left(\vec{d}_1 e^{-i\omega_1 t} + \vec{d}_2 e^{-i\omega_2 t} + \vec{d}_3 e^{-i\omega_3 t} + c.c.\right) \tag{3b}$$

$$N^{(r)}(t) = N + \frac{1}{2}\left(N_\Delta e^{-i\Delta t} + N_{2\Delta}e^{-i2\Delta t} + c.c.\right) \tag{3c}$$

$$\Delta \equiv \omega_2 - \omega_1 = \omega_3 - \omega_2 \qquad \Delta_2 \equiv \omega_2 - \omega_0$$

Under $|E_2| >> |E_2|, |E_3|$. In the slow –amplitude approximation one obtains instead of (2) the following system of equations

$$\left(\Delta_2 + i\tilde{\Gamma}\right)\vec{d}_2 = \frac{f}{\omega_0} N \vec{E}_2 \,, \tag{4a}$$

$$\gamma\left(N - N_e\right) = \frac{i}{2\hbar}\left(\vec{E}_2 \cdot \vec{d}_2^{\,*} - c.c\right) - \frac{2k^3}{3\hbar}\left|\vec{d}_2\right|^2 \,, \tag{4b}$$

$$\left(\Delta_1 - \Delta + i\tilde{\Gamma}\right)\vec{d}_1 = \frac{f}{\omega_0} N \vec{E}_1 + \frac{f}{2\omega_0} N_\Delta^{\,*}\left(\vec{E}_2 + \frac{2}{3} ik^3 \vec{d}_2\right) \,, \tag{4c}$$

$$\left(\Delta_3 - \Delta + i\tilde{\Gamma}\right)\vec{d}_3 = \frac{f}{\omega_0} N \vec{E}_3 + \frac{f}{2\omega_0} N_\Delta\left(\vec{E}_2 + \frac{2}{3} ik^3 \vec{d}_2\right) \,, \tag{4d}$$

$$\left(\gamma - i\Delta\right)N_\Delta = \frac{i}{\hbar}\left(\vec{E}_2 \cdot \vec{d}_1^{\,*} - \vec{E}_1^{\,*} \cdot \vec{d}_2 + \vec{E}_3 \cdot \vec{d}_2^{\,*} - \vec{E}_2^{\,*} \cdot \vec{d}_3\right) - \frac{4k^3}{3\hbar}\left(\vec{d}_1^{\,*} \cdot \vec{d}_2 + \vec{d}_2^{\,*} \cdot \vec{d}_3\right) \tag{4e}$$

Where $\tilde{\Gamma} \equiv \Gamma - \Gamma_d N$, $\qquad \Gamma_d \equiv \frac{2k^2}{3c} f$. (5)

As a result, the saturation law takes a form of the cubic equation

$$N\left[1 + \frac{\left|\frac{E_2}{E_s}\right|^2}{\left(\frac{\Delta_2}{\Gamma}\right)^2 + \left(1 - \frac{\Gamma_d}{\Gamma} N\right)^2}\right] = N_e \,, \tag{6}$$

Whereas in a general case for the dipole moment one finds

$$\vec{d}_1 = \frac{f}{\omega_0}\frac{N}{\Delta_2 - \Delta + i\tilde{\Gamma}}\left\{\vec{E}_1 - \frac{F_\Delta^{\,*}}{2E_s^2}\gamma\Gamma\frac{\Delta - 2i\Gamma}{\Delta - i\gamma}\left[\frac{\vec{E}_2\left(\vec{E}_2^{\,*} \cdot \vec{E}_1\right)}{\left(\Delta_2 - i\tilde{\Gamma}\right)\left(\Delta_2 - \Delta + i\tilde{\Gamma}\right)} + \frac{\vec{E}_2\left(\vec{E}_2 \cdot \vec{E}_3^{\,*}\right)}{\left(\Delta_2 + i\tilde{\Gamma}\right)\left(\Delta_2 + \Delta - i\tilde{\Gamma}\right)}\right]\right\}$$

In the limiting case of exact resonance, $\Delta_2 = 0$, and weak saturation $\left|\frac{E_2}{E_s}\right|^2 << 1$ for the absorption coefficient α_1'' it follows

$$\alpha_1'' = \frac{fN_e}{\omega_0(1+\delta^2)}\left[-1 + \frac{\gamma' - 2\delta^2 + [\gamma' - (2+\gamma')\delta^2]\frac{\Gamma_d}{\tilde{\Gamma}}N_e}{(1+\delta^2)(\gamma^2+\delta^2)}\tilde{I}_2\right], \tag{8}$$

where

$$\delta = \frac{\Delta}{\tilde{\Gamma}}, \quad \gamma' = \frac{\gamma}{\tilde{\Gamma}}, \quad \tilde{I}_2 = \frac{\gamma\Gamma^2}{\tilde{\Gamma}^3}\left|\frac{E_2}{E_s}\right|^2$$

The deviation from the Lorentzian profile of the response function of the two level system even in a weak external field, re-scaling of the integrated intensities and positions of the central peak and sidebands of the Mollow triplet [3] for sufficiently strong driving should be interpreted as signatures of the universal radiation damping effects .

1. K.Wodkiewicz and M.O.Scully, Phys.Rev, A42 (1990),511.
2. G.I.Surdutovich and A.V.Ghiner, Physica A, 283 (2000), 212.
3. P.Meystre and M.O.Sargent III, Elements of Quantum Optics, Springer, Berlin, 1993.

Efficient 1-D Velocity Selection in Laser-Cooled Atom Clouds

S.H. Myrskog, J.K. Fox, A.M. Jofre, L.R. Segal, S.R. Mishra and A.M. Steinberg

Dept. of Physics, University of Toronto, Canada

In this paper we demonstrate a "velocity-selection" technique which selects the lowest energy atoms from a laser-cooled cloud. The technique selects in one dimension, and confines the colder, selected atoms in a small auxiliary trap, spatially separated from hotter atoms. For one dimensional experiments, velocity selection can have an efficiency (defined as the fraction of the initial cloud which remains after selection) that varies as $\sqrt{U/k_BT}$, where U defines a maximum energy and T is a cloud temperature, making it more efficient than 3D cooling methods such as evaporative cooling, which typically retains less than 1% of the atoms when cooling to 1% of the initial temperature[1]. The process of selection creates non-thermal distributions with a sharp cutoff in velocity, which can be important for experiments on effects such as tunneling[2] and quantum collisions[3].

We begin the selection process by placing an atom cloud in a quadrupole magnetic potential, combined with a sharply peaked, dipole-force potential which can act as a barrier. The atoms accelerate toward the minimum of the quadrupole field(see Figure 1). As they begin to move, they encounter the barrier, which is located between the initial position of the atoms and the minimum of the quadrupole potential. The barrier is 13 microns wide, while even at temperatures of 1 μK, the atoms have de Broglie wavelengths on the order of 0.5 μm. Therefore the collision is essentially classical. (In future experiments, narrow barriers and colder clouds will be used for the study of quantum collisions.) If an atom has enough kinetic energy then it will be transmitted through the barrier while the lower-energy atoms will be reflected by the barrier. Reflected atoms are trapped in the local potential minimum produced by the combination of the potential gradient and the barrier, and therefore have velocities $v \leq v_{\max} = \sqrt{U/2m}$, where m is the mass of the atom. For low barriers and negligible potential energy, this efficiency should scale with $v_{\max} : \eta \simeq \sqrt{\frac{2U}{\pi k_B T}} \approx v_{\max}/v_{rms}$, where k_B is the Boltzmann constant and T is the initial temperature of the cloud. For steeper gradients the region over which selection can occur becomes limited to a maximum distance $x_m = U/g$ due to excess potential energy (see Figure 1a.) The efficiency then varies as $\left(\frac{U}{T}\right)^{3/2}$, decreasing the overall efficiency. The resulting clouds have non-thermal velocity distributions with mean kinetic energies varying from $U/6$ for steep gradients to $U/3$ for minimal potential energy. With very small gradients, the technique selects atoms with an efficiency that varies linearly with velocity, therefore being more efficient (in 1-D) than evaporative cooling, which varies roughly as the square of the velocity.

We perform the selection process by placing laser-cooled ^{85}Rb atoms, optically pumped to the F=3, m_F=3 state, in a magnetic gradient of order 3 G/cm. The potential barrier is produced by focussing a Ti:Sapph laser into a sheet of light 13 microns wide and 5 mm tall which intersects the potential gradient (see Figure 1a). The laser is detuned 210 GHz to the blue of the D2 resonance for Rubidium, creating a repulsive potential which can be varied in height up to 100 microKelvin. The selection occurs over a period of about 20 ms during which time the atoms encounter the barrier and either get reflected or transmitted. The temperature of the resulting clouds is measured by a time of flight method. The temperature measured has a mean value of approximately

Coherence and Quantum Optics VIII
Edited by Bigelow et al., Kluwer Academic/Plenum Publishers, 2003

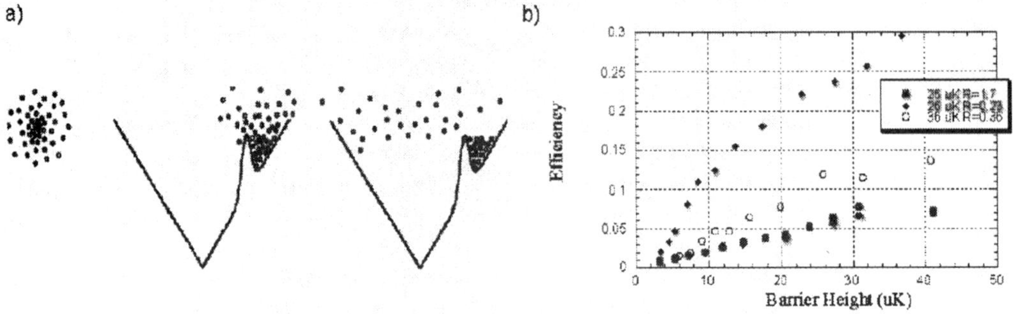

Fig. 1. a)Velocity Selection procedure. Atom cloud starting in free space Displaced quadrupole potential and barrier are turned on, followed by the system at the end of the selection process. b) Selection efficiency versus well depth.

1/4 of the well depth.

The efficiency is determined by measuring the number of atoms in the selected peak and comparing it to the total number of atoms 0.5 milliseconds after the selection process is complete. Figure 1b shows the efficiency of selection as a function of barrier height for different ratio of potential to kinetic energy. The diamonds refer to a experimental parameters where the ratio of potential to kinetic energy is 0.23. The efficiency is shown to vary in a nearly square-root fashion with barrier height, growing as $U^{0.58}$. The data sets with higher potential energy (ratios of 0.36 and 1.7 for the hollow and solid circles respectively) show a marked decrease in efficiency, and vary approximately linearly with the barrier height.

In order to observe a square-root dependence of efficiency on barrier height, it is necessary to start the barrier in the region of highest density, in the center of the cloud, although this halves the number of trappable atoms. We observed a selection efficiency of 2.1% (4.2% of the total capturable atoms) at a temperature of 750 nK which is 1/35th of the initial temperature of 26 uK. For very small barrier heights the potential energy always becomes larger than the kinetic energy, and the efficiency then varies as $\left(\frac{U}{T}\right)^{3/2}$, which then shows up as a displacement from the origin.

We have demonstrated a new technique to select low-temperature atoms from a cloud of laser-cooled atoms. With an efficiency of 2.1%, we have selected clouds as much as 1/35th of the initial cloud temperature, although the technique has no fundamental limit in temperature. In the limit of negligible potential energy the technique can be as efficient as $\sqrt{\frac{2U}{\pi k_B T}}$. The technique imparts a sharp cut in the velocity distribution, making it useful for experiment on effects such as tunneling and collisions. For extremely small, narrow barriers it is possible to create wells which contain only a single bound state, allowing for the creation of custom wavefunctions[4].

References and links

1. W. Ketterle and N.J. Van Druten, " Evaporative cooling of trapped atoms", in *Adv. Atomic, Molecular and Optical Physics*, **37**,edited B. Bederson and H. Walther, (Academic Press, Cambridge,1996)
2. A. M. Steinberg, S. Myrskog, Han Seb Moon, Hyun Ah Kim, Jalani Fox, Jung Bog Kim, Ann. Phys. (Leipzig) **7**, 593 (1998)
3. S. Brouard and J.G. Muga, Phys. Rev. Lett. **81**, 2621 (1998)
4. J.K.Fox, S.R. Mishra, S.H. Myrskog, A.M. Jofre, L.R. Segal and A.M. Steinberg, To be Published

Integrability, chaos and mixed phase-space in atom optics billiards

Nir Friedman, Ariel Kaplan, Mikkel Andersen, and Nir Davidson
Department of Physics of Complex Systems, Weizmann Institute of Science, Rehovot 76100, Israel
fenir@wise.weizmann.ac.il

1. Introduction

The dynamics of particles inside billiards has been extensively investigated both classically and quantum mechanically, since this very simple system yet exhibits a rich variety of non-linear dynamics phenomena [1]. Hence, billiards where used to explore the connections between classical and quantum chaos, and also to shed light on the origins of statistical mechanics. Recently, we realized an "atom-optics billiard", where ultra-cold atoms are trapped inside a dark optical trap, whose shape can be varied to create different atom dynamics [2]. In this work, we present numerical investigation and experimental realization of atom-optics billiards with ordered, chaotic or mixed phase-space dynamics.

To investigate the dynamics of the trapped atoms we measured their decay through a small hole on the trap's boundary. For systems with chaotic dynamics the decay is exponential, with a decay time $\tau_c = \pi A/vL$ (where v is the atomic velocity, A the billiard's area, and L the hole's length [4]). The decay from integrable systems is slower, with an algebraic tail at long times [3]. For systems with mixed phase-space, the decay depends on the position of the hole – if the hole does not include trajectories which belong to a stable island, these stable trajectories are trapped for long times, and a strong deviation from exponential decay is observed. On the other hand, if the hole contains trajectories from the island, these will decay fast and the rest of the atoms will decay exponentially, since they belong to the chaotic part of phase-space.

The atom-optics billiards are realized using a tightly focused laser beam which is detuned far above the atomic resonance and is rapidly scanning along the desired trap shape, hence generating a time-averaged potential wall [4], with trap shapes such as an ellipse, a circle and a Bunimovich stadium.

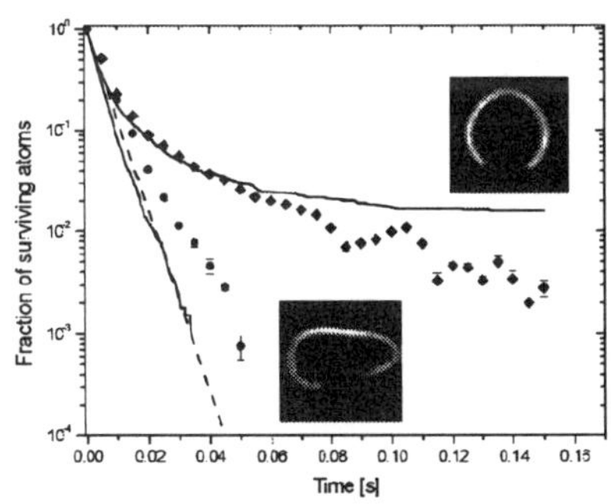

Fig. 1. Decay of the number of trapped atoms from a circular (♦) and a tilted-stadium (•) atom-optics billiard. The solid lines are the results of numerical simulations, the dashed line is an exponential decay with a decay rate τ_c calculated for the experimental parameters. The insets show measured CCD images (300X300 μm) of the billiards cross section at the beam focus.

2. Circle and stadium atom-optics billiards: ordered and chaotic motion

We first demonstrate the decay from a circular billiard (integrable) and a tilted Bunimovich stadium (two half circles with different radii, connected by two non-parallel straight lines, in which the motion is chaotic). The mean kinetic energy of the atoms (~ 400 times E_{recoil}, the recoil energy) is much higher than the average gravitational energy, so the last can be neglected. The measured decay curves are shown in Fig. 1. It can be seen that the decay from the stadium is almost exponential as expected for chaotic dynamics, whereas the decay from the circle is much slower, indicating the integrable dynamics of the atoms.

Another integrable system that we studied is the elliptical billiard. Here, the effect of random scattering was investigated by exposing the atoms to a series short pulses of optical molasses which randomize the atoms direction of motion. When the rate of these randomizing pulses was comparable to the average escape time, the stability of the atomic motion was significantly reduced.

3. Wall softness: mixed phase-space

An island of stability can be induced inside a chaotic phase space, for example, by changing the softness of the trapping potential. For some particular billiard shapes, a theory which predicts where these islands will appear, and what would be their size was developed [5]. For the stadium case, however, we are not aware of such a theoretical treatment. Nevertheless, our numerical simulations indicate the clear appearance of islands around the singular points (where the half circle meets the straight line, see inset to Fig. 2).

In Fig. 2(A) we show part of a phase-space diagram, for a softness parameter $w_0 = 30$ μm ($1/e^2$ radius of the trapping laser beam). For this soft wall an island appears, as can be seen also from the stable trajectory in the upper inset. For a harder wall ($w_0 = 8$ μm) no island appears and the motion is ergodic over the whole phase-space. Experimental results for the decay from a tilted-stadium, for a soft and hard walls ($w_0 = 14$ and $w_0 = 24$ μm) are shown in Fig. 2(B).. It can be seen that the decay becomes slower as the wall becomes softer, indicating an island whose size depends on the softness parameter. Similar measurements for a circular stadium show that the decay is almost independent of w_0.

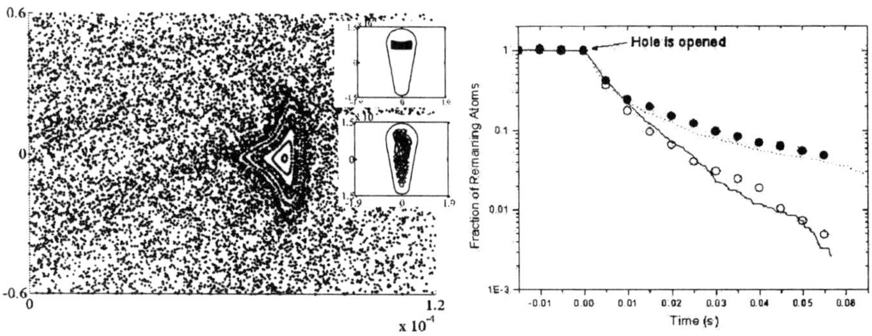

Fig. 2. The effect of wall softness. A) part of a phase-space diagram for the tilted stadium, with a soft wall ($w_0 = 30$ μm), where a big island appears. The insets shows a typical trajectory in the island (upper) and outside the island (lower). B) Measured decay curves for a tilted stadium, with a soft wall (full circles, $w_0 = 24$ μm), and a hard wall (open circles, $w_0 = 14$ μm). The dashed and full lines are results of full numerical simulations, for the soft- and hard-wall cases, respectively.

4. References

[1] M. C. Gutzwiller, *Chaos in calssical and quantum mechnics* (Springer-Verlag, New York, 1990).

[2] N. Friedmam, A. Kaplan, D. Carasso and N. Davidson, "Observation of chaotic and regular dynamics in atom-optics billiards", Phys. Rev. Lett. **86**, 1518-1521 (2001).

[3] W. Bauer and G. F. Bertsch, "Decay of ordered and chaotic systems", Phys. Rev. Lett. **65**, 2213-2216 (1990).

[4] N. Friedman, L. Khaykovich, R. Ozeri and N. Davidson, "Compression of cold atoms to very high densities in a novel rotating-beam blue-detuned optical dipole trap", Phys. Rev. A **61**, 031403(R) (2000).

[5] V. Rom-Kedar and D. Turaev, "Big islands in dispersing billiard-like potentials", Physica D **130**, 187-210 (1999).

Atom lasers, coherent states, and coherence

H. M. Wiseman[1] and John A. Vaccaro[2]

[1]*School of Science, Griffith University, Brisbane 4111 Australia.*
[2]*Division of Physics and Astronomy, University of Hertfordshire, Hatfield AL10 9AB, UK.*

+61 7 3875 7271, +61 7 3875 7685, H.Wiseman@gu.edu.au

In elementary presentations of quantum optics it is more or less an axiom that lasers produce coherent light. However, theoretical models for typical laser processes imply that the state of the cavity mode for a laser has no mean coherent amplitude at all. Rather, far above threshold, it is a Poissonian mixture of number states, or, equivalently, a mixture of coherent states of all phases [1, 2]

$$\rho_{\rm ss} = \sum_{n=0}^{\infty} e^{-\mu}\frac{\mu^n}{n!}|n\rangle\langle n| = \int \frac{d\phi}{2\pi}|\sqrt{\mu}\,e^{i\phi}\rangle\langle\sqrt{\mu}\,e^{i\phi}|. \tag{1}$$

The existence of these two representations warns against believing that the laser really is in a coherent state $|\sqrt{\mu}\,e^{i\phi}\rangle$ of definite but unknown phase ϕ.

The "unknown coherent state" description and the "unknown number state" description are *mathematically equivalent* representations of the stationary state matrix $\rho_{\rm ss}$. However, in the physical context that $\rho_{\rm ss}$ is the stationary state of an open quantum system in *dynamical equilibrium*, the two representations are *not physically equivalent.* This idea is at the heart of this paper, as we investigate whether these, and other pure state ensembles, are *physically realizable* [3]. We will show that under some circumstances, the "unknown coherent state" description is not physically realizable, in contrast to the "unknown number state" description which always is.

Before proceeding further, it is necessary to clarify what we mean by "physically realizable" (PR). A stationary pure state ensemble of a given system is PR if it is possible, *without altering the dynamics of the system*, to know that its state at equilibrium is definitely one of the pure states in the ensemble. Of course *which* pure state cannot be predicted beforehand. It may seem contradictory to say that the system at equilibrium is mixed, but that nevertheless we can know it to be in a pure state. The resolution is that, by monitoring the system's environment, the system state can, under suitable circumstances, be collapsed over time into a pure state. Being simply an example of a quantum measurement, this process, called an *unraveling* [4], will be stochastic. *On average*, the system evolution is not changed and the ensemble of pure states produced by the unraveling is guaranteed to be equivalent to the equilibrium mixed state.

The question of whether an ensemble is PR cannot be determined from the stationary mixed state $\rho_{\rm ss}$. Rather, it depends upon the dynamics (reversible and irreversible) which produced that stationary state. The fact that different dynamics can lead to the same stationary mixed state is easy to see for the case of a laser. Any process which commutes with boson number will not alter the stationary laser state $\rho_{\rm ss}$, since its eigenstates are the number states, as shown by Eq. (1). An example of an irreversible process which commutes with boson number is phase diffusion. This is relevant to all real lasers, which have some phase diffusion in excess of the fundamental limit. There are also reversible processes which commute with boson number, such as that generated by degenerate four-wave mixing. While this dynamics is unimportant in most optical lasers, it is expected to be very significant in *atom lasers* [5].

An atom laser is a device which would produce an output beam of bosonic atoms analogous to an optical laser's beam of photons [5]. The idea for an atom laser was published independently by a number of authors [6, 7, 8, 9], shortly after the first achievement of Bose-Einstein condensation (BEC) of weakly interacting atoms [10, 11, 12]. There have since been some important advances in the coherent release of pulses [13, 14] and beams [15, 16] of atoms from a condensate. Because the condensate is not replenished in these experiments, the output coupling cannot continue indefinitely, so these devices cannot be considered true lasers. Nevertheless they represent the first steps towards achieving a continuously operating atom laser.

Even though the atoms in most BEC experiments are gaseous, elastic collisions may dominate the dynamics of the condensate. This self-interaction does not alter the number of atoms in the condensate, and is analogous to a $\chi^{(3)}$ nonlinearity in optics. It is convenient to define a dimensionless self-interaction strength $\chi = (8\mu\pi\hbar a_s/\kappa m)\int d^3\mathbf{r}|\psi(\mathbf{r})|^4$, where a_s is the atomic s-wave scattering length, κ is the exponential loss rate of the condensate, m is the atomic mass, and $\psi(\mathbf{r})$ is the condensate wavefunction. This parameter has has an enormous influence on what ensembles are PR. It also determines the laser linewidth, and hence the coherence of the laser output [5].

We find that for any nonzero self-energy χ of the bosons in the laser mode, the coherent state ensemble is not PR; the closest one can come to it is an ensemble of squeezed states. These squeezed states approach coherent states only for $\chi \ll 1$, which is a much stronger condition than the $\chi \ll \mu^{3/2}$ needed for the laser output to be coherent (Bose degenerate) [5]. Current experiments related to atom lasers [13, 14, 15, 16] have taken place in the opposite regime, $\chi \gg 1$. By contrast with the coherent state ensemble, the number state ensemble is always PR.

As the self-energy χ increases, the states in the PR ensemble closest to the coherent state ensemble become increasingly amplitude squeezed. For $\chi \gg \mu^2$ these states are so amplitude squeezed that, like number states, they have no mean field. Nevertheless, there is a regime ($\mu^{3/2} < \chi < \mu^2$) for which states with well-defined coherent amplitudes are PR even though the atom laser is not coherent (in the sense of having a Bose-degenerate output). We interpret this anomaly in terms of a new concept, namely conditional coherence (and hence conditional Bose degeneracy). That is to say, knowing which member of physically realizable ensemble one has at a given point in time allows the coherence to be demonstrated, where it could not be in the absence of that knowledge.

References and links

1. W.H. Louisell, *Quantum Statistical Properties of Radiation* (John Wiley & Sons, New York, 1973).
2. M. Sargent, M.O. Scully, and W.E. Lamb, *Laser Physics* (Addison-Wesley, Reading Mass., 1974).
3. H.M. Wiseman and J.A. Vaccaro, Phys. Lett. A **250**, 241 (1998).
4. H.J. Carmichael, *An Open Systems Approach to Quantum Optics* (Springer-Verlag, Berlin, 1993).
5. H.M. Wiseman, Phys. Rev. A **56**, 2068 (1997).
6. H.M. Wiseman and M.J. Collett, Phys. Lett. A **202**, 246 (1995).
7. R.J.C. Spreeuw, T. Pfau, U. Janicke, and M. Wilkens, Europhys. Lett. **32**, 469 (1995).
8. M. Olshanii, Y. Castin, and J. Dalibard, in *Proc. XII Conference on Lasers Spectroscopy*, edited by M. Inguscio, M. Allegrini and A. Sasso (World Scientific, 1995).
9. M. Holland *et al*, Phys. Rev. A **54**, R1757 (1996).
10. M.H. Anderson *et al*, Science **269**, 198 (1995).
11. C.C Bradley, C.A. Sackett, J.J. Tollett, and R.G. Hulet, Phys. Rev. Lett. **75**, 1687 (1995).
12. K.B. Davis *et al.*, Phys. Rev. Lett. **75**, 3969 (1995).
13. M.-O. Mewes, M. R. Andrews, D. M. Kurn, D. S. Durfee, C. G. Townsend, and W. Ketterle, Phys. Rev. Lett. **78**, 582 (1997)
14. B.P. Anderson, M.A.Kasevich, Science **282**, 1686 (1998).
15. E.W. Hagley, L. Deng, M, Kozuma, J. Wen, K. Helmerson, S.L. Rolston, W.D. Phillips, Science **283**, 1706 (1999);
16. I. Bloch, T.W. Hänsch, and T. Esslinger Phys. Rev. Lett. **82**, 3008 (1999).

Atom laser dynamics

Nicholas Robins and Elena A. Ostrovskaya

Optical Sciences Center, The Australian National University, Canberra ACT 0200, Australia

Craig Savage

Department of Physics and Theoretical Physics, The Australian National University, Canberra ACT 0200, Australia

craig.savage@anu.edu.au

http://www.anu.edu.au/Physics/Savage

Fundamentally, it is the coherence properties, such as linewidth and intensity correlations, that best capture the unique physics of lasers [1, 2]. Schawlow and Townes showed that optical lasers' first order coherence, or linewidth, is ultimately limited by spontaneous emission, and also decreases in inverse proportion to the laser power. Analogous results have been derived for atom lasers [3]. In practice, however, the Schawlow-Townes limit is not achieved, because the linewidth is limited by technical noise and by dynamical effects rather than by quantum noise. For example relaxation oscillations, due to the nonlinear interaction between the inversion and the light, are the primary determinant of the low frequency noise spectrum in many optical lasers.

We focus on one aspect of the nonlinear dynamics of atom lasers: the dynamics of the atom laser output beam density. In particular, we investigate how the spectrum depends on the pumping rate. We choose to focus on the density because it can be measured experimentally using sensitive optical techniques such as phase modulation spectroscopy [4]. A conclusion of our work is that three-body recombination plays a major role in the high frequency dynamics. As the pumping rate is increased the spectral peaks weaken and move to higher frequencies.

Our atom laser model is schematically illustrated on the left of Fig. 1. It consists of a dilute gas Bose-Einstein condensate trapped by an effectively one-dimensional potential. This is pumped from a reservoir of non-condensed trapped atoms. The output atom laser beam is generated by a Raman transition to an untrapped atomic state. Our model builds on that of Kneer *et al.* [5] by adding three-body recombination and a propagating output beam. In the following we will refer to the trapped Bose-Einstein condensate as the "condensate", and to the output beam as the "beam". We make the mean-field approximation in which the condensate and beam are described by the classical fields $\Psi_a(x,t)$ and $\Psi_b(x,t)$ respectively, which obey Gross-Pitaevskii type equations [6]. Our model is more complete than previous mean-field treatments since we include pumping, output coupling, and three-body recombination, and also explicitly model the dynamics of the pump reservoir, the condensate, and the beam. To describe the pumping we use a phenomenological model, which mimics the pumping of an optical laser [5]. The condensate atoms are coupled, by a reversible Raman transition, to an untrapped electronic state forming the atom laser beam. The beam evolves under the influences of gravity and atom-atom interactions, which are dominated by those due to the condensate.

On the right of Fig. 1 we show the output beam density $|\Psi_b(x_0,t)|^2$ as a function of time at a fixed position far outside the trapped condensate. The time span has been chosen to be longer than that of the transient dynamics associated with the growth of the condensate from a seed value, and short enough to be experimentally accessible.

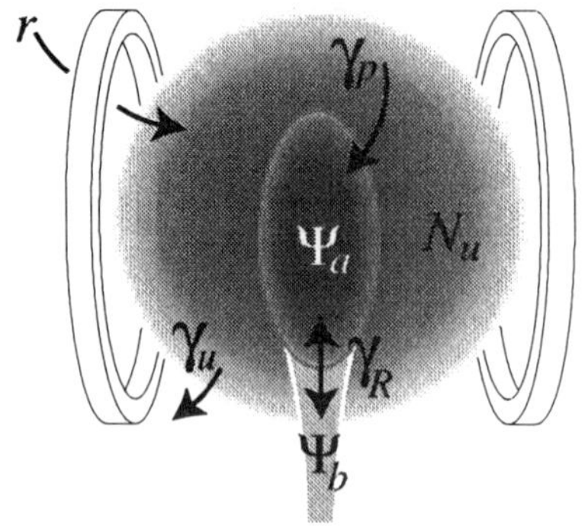

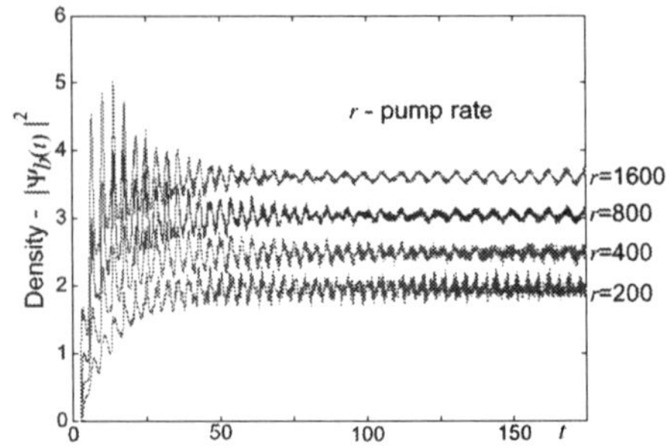

Fig. 1. (Left) Schematic representation of our atom laser model. The rings represent coils producing the trapping potential. The symbols are defined in our paper [7]. (Right) The output beam density, $|\Psi_b(x_0, t)|^2$, as a function of time at a position far outside the condensate, for different pump rates: from bottom to top $r = 200$, 400, 800, and 1600.

Quasi-stationary, nearly periodic dynamics develop, becoming approximately harmonic for large pump rates r. This fundamental oscillation of the beam, at higher pump rates, arises from the well known Kohn mode [6], corresponding to a rigid "sloshing" back and forth of the condensate at the trap frequency.

The peaks in the frequency spectra of the density time series are well approximated by the eigen-frequencies of an uncoupled 1D condensate [5]. As the pumping rate r increases, the slow beam oscillations change from the frequency of the condensate breathing mode to that of the sloshing mode. This is because the condensate itself changes from breathing to sloshing at higher pumping rates. As Kneer *et al.* [5] noted, the sloshing mode must be excited by a spatially asymmetric perturbation, such as the output beam. The spectra also contain a high frequency group of peaks. As the pumping rate increases the frequency of this group increases and its total power decreases.

We have investigated the role of three-body recombination in a simplified pumped atom laser model similar to that of Kneer *et al.* [5]. That is, with a spatially uniform phenomenological loss rather than an explicitly modelled output beam. Without three-body recombination the condensate density power spectrum simply rolls off at high frequencies. Adding three-body recombination produces a high frequency spectral group. Hence we conclude that three-body recombination has an important effect on the spectrum of the output atom laser beam.

References and links

1. L. Mandel and E. Wolf. *Optical coherence and quantum optics* (Cambridge University Press, 1995).
2. R.J. Ballagh and C.M. Savage, "The theory of atom lasers" in *Bose-Einstein Condensation: from atomic physics to quantum fluids. Proceedings of the 13th Physics Summer School*, C.M. Savage and M. Das, ed. (World Scientific, Singapore, 2001).
3. R. Graham, "Decoherence of Bose-Einstein condensates in traps at finite temperature", Phys. Rev. Lett. **81**, 5262-5265 (1998).
4. J.E. Lye, B.D. Cuthbertson, H.-A. Bachor, J.D. Close, "Phase modulation spectroscopy: a non-destructive probe of Bose-Einstein condensates" J. Opt. B: Quant. Semiclass. Opt. **1**, 402-407 (1999).
5. B. Kneer, T. Wong, K. Vogel, W. P. Shleich, and D. F. Walls, "Generic model of an atom laser", Phys. Rev. A **58**, 4841-4853 (1998).
6. F. Dalfovo, S. Giorgini, L. P. Pitaevskii and S. Stringari, "Theory of Bose-Einstein condensation in trapped gases", Rev. Mod. Phys. **71**, 463-512 (1999).
7. N. Robins, C.M. Savage, and E. A. Ostrovskaya, "Atom laser dynamics", cond-mat/0103597.

Atom-Laser Line-Narrowing by Feedback

H. M. Wiseman and L.K. Thomsen

School of Science, Griffith University, Brisbane, 4111 Australia.

+61 7 3875 7271, +61 7 3875 7685, H.Wiseman@gu.edu.au

An atom laser is a continuous source of coherent atom waves, analogous to an ordinary laser which is a continuous source of coherent photon waves (light) [1]. Ideas for creating an atom laser were published independently by a number of authors [2, 3, 4, 5], shortly after the first achievement of Bose-Einstein condensation (BEC) of gaseous atoms [6, 7, 8]. There have since been some important advances in the coherent release of pulses [9, 10] and beams [11, 12] of atoms from a condensate. Since the condensate is not replenished in these experiments, the output coupling cannot continue indefinitely. Nevertheless they represent major steps towards achieving a continuously operating atom laser.

One defining feature of a laser is that it has a relatively narrow linewidth [1]. The dominant contributer to the linewidth of an atom laser will be the collisional interaction of atoms, which is negligible for photons. In this work [13] we show that this contribution can be reduced by many orders of magnitude using a feedback mechanism. For a single-mode condensate the dominant effect of collisions is to turn atom number fluctuations in the condensate into fluctuations in the energy, which are equivalent to frequency fluctuations. By monitoring the number fluctuations, it is possible using feedback to largely compensate for the linewidth caused by these frequency fluctuations. The key practical points are that the measurement does not rely upon any external condensate phase reference, and that the control requires only the ability to change the energy of the atoms, which could be done with a spatially uniform optical or magnetic field.

We begin with the standard linewidth, which can be derived using the usual single-mode model of the laser [14, 15]. Far above threshold, its dynamics are modeled by the completely positive master equation [16, 17]

$$\dot{\rho} = \kappa\mu\mathcal{D}[a^\dagger]\mathcal{A}[a^\dagger]^{-1}\rho + \kappa\mathcal{D}[a]\rho. \tag{1}$$

Here $\mathcal{D}[A]B \equiv ABA^\dagger - \mathcal{A}[A]B$ where $\mathcal{A}[A]B \equiv (A^\dagger AB + BA^\dagger A)/2$, κ is the loss rate, $\mu \gg 1$ the stationary mean boson number, and a the annihilation operator for the laser mode. This yields the standard linewidth $\ell = \ell_{\rm st} = \kappa/2\mu$, as expected.

As a model for an atom laser we take the standard master equation (1) and add

$$\dot{\rho} = -iC[a^\dagger a^\dagger aa, \rho]\ ;\ \ C = (2\pi\hbar a_s/m)\int d^3\mathbf{r}|\psi(\mathbf{r})|^4 \equiv \kappa\chi/4\mu. \tag{2}$$

This represents the self-energy of the atoms due to collisions, where $\psi(\mathbf{r})$ is the wavefunction for the condensate mode, and a_s is the s-wave scattering length. This Hamiltonian term causes shearing in phase space, turning number fluctuations into frequency fluctuations. The linewidth can be evaluated analytically in two limits:

$$\ell = \kappa(1+\chi^2)/2\mu \text{ for } \chi \ll \sqrt{\mu}\ ;\ \ \ell = 2\kappa\chi/\sqrt{2\pi\mu} \text{ for } \chi \gg \sqrt{\mu}. \tag{3}$$

Using the preliminary atom laser experiments [9, 10, 11, 12] as a guide to realistic parameter values, the dimensionless interaction strength χ is found to always satisfy $\chi \gg 1$. This implies a linewidth for the atom laser far above the standard limit.

Since number fluctuations cause the linewidth broadening, by measuring the number we should be able to compensate for them. QND atom number measurements can be performed on the condensate *in situ* using dispersive imaging techniques [18]. We consider a far-detuned probe laser beam of power P, frequency ω_p and cross-sectional area A (assumed larger than the condensate) which passes through the condensate. This adds the quantum back-action term

$$\dot{\rho} = N\mathcal{D}[a^\dagger a]\rho\ ; \quad N = (P/\hbar\omega_p)\left(\hbar\omega_p\Gamma^2/8A\Delta I_{\rm sat}\right)^2 \tag{4}$$

to the master equation, causing extra phase diffusion (and hence linewidth broadening).

By CW homodyne measurement, the phase shift of the probe beam can be measured. In order to control the phase dynamics of the condensate, we wish to use the homodyne current to modulate the phase. This can be done, for example, by applying a uniform magnetic field or far-detuned light field across the whole condensate. This adds a term proportional to $a^\dagger a$ to the condensate Hamiltonian. For Markovian feedback [19] of strength parameterized by F, this yields the total master equation

$$\dot{\rho} = \kappa\mu\mathcal{D}[a^\dagger]\mathcal{A}[a^\dagger]^{-1}\rho + \kappa\mathcal{D}[a]\rho - i(C-F)[a^\dagger a^\dagger aa, \rho] + +(N + F^2/N)\mathcal{D}[a^\dagger a]\rho. \tag{5}$$

For $\chi \gg 1$ (that is, $C \gg \ell_{\rm st}$), the optimal regime is $F = C = N$, so the interaction energy term is completely canceled, but the measurement-induced phase diffusion is doubled. This is due to the shot noise in the homodyne photocurrent being fed-back. Then the linewidth is given simply by

$$\ell = \kappa/2\mu + 2C = \kappa(1+\chi)/2\mu, \tag{6}$$

which is smaller than (3) by a factor of up to $\mu^{-1/2}$, perhaps 10^{-3}.

We have found good agreement between the above analytical expressions, and numerical calculations of the linewidth from the master equations [13]. We have also checked that a QND measurement of suitable strength ($N \simeq C$) can be performed, without causing significant atom loss due to spontaneous emission by atoms excited by the detuned probe beam. In summary, there appear to be no fundamental reasons that this proposal could not be put to good use when a continuous atom laser is realized.

References and links

1. H.M. Wiseman, Phys. Rev. A **56**, 2068 (1997).
2. H.M. Wiseman and M.J. Collett, Phys. Lett. A **202**, 246 (1995).
3. R.J.C. Spreeuw, T. Pfau, U. Janicke, and M. Wilkens, Europhys. Lett. **32**, 469 (1995).
4. M. Olshanii, Y. Castin, and J. Dalibard, in *Proc. XII Conference on Lasers Spectroscopy*, edited by M. Inguscio, M. Allegrini and A. Sasso (World Scientific, 1995).
5. M. Holland *et al*, Phys. Rev. A **54**, R1757 (1996).
6. M.H. Anderson *et al*, Science **269**, 198 (1995).
7. C.C Bradley, C.A. Sackett, J.J. Tollett, and R.G. Hulet, Phys. Rev. Lett. **75**, 1687 (1995).
8. K.B. Davis *et al.*, Phys. Rev. Lett. **75**, 3969 (1995).
9. M.-O. Mewes, M. R. Andrews, D. M. Kurn, D. S. Durfee, C. G. Townsend, and W. Ketterle, Phys. Rev. Lett. **78**, 582 (1997)
10. B.P. Anderson, M.A.Kasevich, Science **282**, 1686 (1998).
11. E.W. Hagley, L. Deng, M, Kozuma, J. Wen, K. Helmerson, S.L. Rolston, W.D. Phillips, Science **283**, 1706 (1999);
12. I. Bloch, T.W. Hänsch, and T. Esslinger Phys. Rev. Lett. **82**, 3008 (1999).
13. H.M. Wiseman and L.K.W. Thomsen, Phys. Rev. Lett. **86**, 1143 (2001); L.K.W. Thomsen and H.M. Wiseman, in preparation.
14. W.H. Louisell, *Quantum Statistical Properties of Radiation* (John Wiley & Sons, New York, 1973).
15. M. Sargent, M.O. Scully, and W.E. Lamb, *Laser Physics* (Addison-Wesley, Reading Mass., 1974).
16. H.M. Wiseman, Phys. Rev. A **47**, 5180 (1993).
17. H.M. Wiseman, Phys. Rev. A **60**, 4083 (1999).
18. M.R. Andrews, M.-O. Mewes, N.J. van Druten, D.S. Durfee, D.M. Kurn, and W. Ketterle, Science **273**, 84 (1996).
19. H.M. Wiseman, Phys. Rev. A **49**, 2133 (1994); Errata *ibid.*, **49** 5159 (1994) and *ibid.* **50**, 4428 (1994).

Coherence in an Optically Trapped Fermi Gas

K. M. O'Hara, M. E. Gehm, S. R. Granade, M.-S. Chang and J. E. Thomas

Physics Department, Duke University, Durham, NC 27708-0305

jet@phy.duke.edu

Abstract: We demonstrate continuous evaporative cooling of lithium fermions in a stable optical trap. This system offers exciting prospects for attaining extremely long hyperfine coherence lifetimes at high density, and for studying multiple-particle coherence in Fermi superfluids.

OCIS codes: (020.0020), (020.7010)

Trapping and cooling of neutral fermionic atoms offers exciting prospects for fundamental precision studies of single- and multi-particle coherence in Fermi gases for which the interaction strength, density, and temperature can be experimentally controlled. At low temperatures, where s-wave scattering is dominant, the Pauli exclusion principle forbids scattering in a single component Fermi gas, so that coherence degrading two- and three-body collisions are suppressed [1]. Hence, a dense one-component sample in a pure superposition state can exhibit extremely long coherence lifetimes.

In two-component Fermi gases with strongly attractive interactions, i.e., large and negative scattering lengths, a transition to a superfluid is predicted to occur at experimentally accessible temperatures [2]. The superfluid exhibits multi-particle coherence as a result of Cooper pairing. Certain stable two-state mixtures of ^{6}Li and ^{40}K which have very strong, magnetically tunable, attractive interactions are promising candidates for observing this transition.

Recently, evaporation to degeneracy has been accomplished in fermionic ^{40}K by magnetically trapping two different spin components [3]. Sympathetic cooling of fermionic ^{6}Li to degeneracy also has been accomplished by using bosonic ^{7}Li as a refrigerator for ^{6}Li contained in the same magnetic trap [4]. These experiments have been used to dramatically illustrate the effects of Fermi statistics in a degenerate one-component Fermi gas. Unfortunately, magnetic traps are not well suited to studying long hyperfine coherence lifetimes. Further, none of the stable attractive mixtures required for superfluidity can be confined in a magnetic trap, since the required spin states are repelled. Hence, optical traps are essential for studying fundamental features of coherence in Fermi gases.

To enable trapping of arbitrary mixtures of fermions, we have developed an all-optical trap based on an ultrastable CO_2 laser. This trap has been used to confine ^{6}Li atoms in a nearly spin-state-independent potential with a lifetime of 370 seconds with negligible heating [5]. The extreme detuning from resonance leads to optical scattering rates of only one photon per hour, and very little perturbation of the ground state hyperfine transition frequency. However, the small difference in the confining potentials of the upper and lower hyperfine states leads to a spread in the hyperfine transition frequency which is proportional to the spread in vibrational energy. This causes the coherence to dephase unless the atoms are very cold. To minimize the dephasing rate, an ultracold gas of ^{6}Li can be confined in a shallow CO_2 laser trap, where the minimum trap depth is limited only by the gravitational force. In this case, hyperfine coherence lifetimes beyond a hundred seconds should be achievable.

Ultracold temperatures can be attained by employing continuous evaporative cooling by adiabatically lowering the depth of the CO_2 laser trap [6]. To enable evaporative cooling, a mixture of the two lowest spin states of ^{6}Li is directly loaded into the trap from a ^{6}Li MOT. This stable mixture has a negligible scattering length at magnetic fields up to a few gauss, but the scattering length increases to more than -300 a_0 for magnetic fields of a few hundred gauss and exhibits a Feshbach resonance for magnetic fields around 800 gauss. Another stable mixture can be created from the first by using a pair of optical fields to induce a Raman π-pulse which transfers one spin component to a higher lying state. The stable two-component gas so produced exhibits a scattering length of up to -1600 a_0 at low magnetic field and enables rapid evaporative cooling [7].

We will discuss the physics of continuous evaporative cooling of a two-state mixture of ^{6}Li fermions in a time-dependent optical trap, our progress toward achieving very low temperatures, and the prospects for observing superfluidity in this system.

References and links

1. K. Gibble and B. J. Verhaar, "Eliminating Cold-Collision Frequency Shifts," Phys. Rev. A **52**, 3370 (1995).
2. H. T. C. Stoof, M. Houbiers, C. A. Sackett, and R. G. Hulet,Phys. Rev. Lett. **76**, 10 (1996); See also, M. Houbiers, R. Ferwerda, H. T. C. Stoof, W. I. McAlexander, C. A. Sackett, and R. G. Hulet, Phys. Rev. A **56**, 4864 (1998).
3. B. DeMarco and D. S. Jin, *Science* **285**, 1703 (1999).
4. A. G. Truscott, K. E. Strecker, W. I. McAlexander, G. B. Patridge, and R. G. Hulet, to appear in *Science*, 1 March, 2001.
5. K. M. O'Hara, S. R. Granade, M. E. Gehm, T. A. Savard, S Bali, C. Freed and J. E. Thomas, Phys. Rev. Lett. **82**, 4204 (1999).
6. For a discussion of evaporative cooling in an optical trap, see K. M. O'Hara, Ph.D. Thesis, (Duke University, 2000), unpublished.
7. K. M. O'Hara, M. E. Gehm, S. R. Granade, S. Bali, and J. E. Thomas, Phys. Rev. Lett. **85**, 2092 (2000).

Phonon excitations and stability of a Bose-Fermi mixture quantum gas

H. Pu, W. Zhang and P. Meystre

Optical Sciences Center, University of Arizona, Tucson, AZ 85721, USA

hp@yoda.optics.arizona.edu

Abstract: We consider a mixture of Bose-Einstein condensate and degenerate fermi gas at zero temperature. The phonon spectrum of the condensate is calculated and the stability regime is identified.

OCIS codes: (999.9999) Bose-Einstein condensate, degenerate Fermi gas

New developments in the field of ultracold atoms are now creating opportunities to study the quantum degenerate fermi gases[1, 2]. Due to the Pauli exclusion principle, the direct evaporative cooling technique does not work for spin-polarized fermionic atoms. To circumvent this obstacle, one must cool the fermi gas either simutaneously using two different spin states or mixing the fermi gas with a gas of bosonic atoms and then sympathetically cool both in a mixture. The experimental progress along this latter direction has stimulated a great interest in studying the Bose-Fermi mixtures in the quantum degenerate regime[3, 4, 5]. Here we report our work on the elementary excitations of such a mixture at zero temperature. Using an example of homogeneous system, we are able to identify the stable and unstable regime of the mixture and calculate the sound velocity of bosonic phonons under the influence of fermions. We restrict our discussions to the mixture with positive scattering lengths.

We start with two coupled equations derived from the system Hamiltonian, which describe respectively the equations of motion for the boson fluctuation operator $\hat{\xi}_B$ and fermion field operator $\hat{\psi}_F$:

$$i\hbar\frac{\partial\hat{\xi}_B}{\partial t} = \left(\hat{T}_B + V_B - \mu_B + 2gn_B + fn_F\right)\hat{\xi}_B + g\phi_B^2\hat{\xi}_B^+ + f\phi_B\delta\hat{\rho}_F \qquad (1)$$

$$i\hbar\frac{\partial\hat{\psi}_F}{\partial t} = \left(\hat{T}_F + V_F - \mu_F + fn_B\right)\hat{\psi}_F + f(\phi_B\hat{\xi}_B^+ + \phi_B^*\hat{\xi}_B)\hat{\psi}_F \qquad (2)$$

where the subscripts B and F represent 'boson' and 'fermion', respectively, $\hat{T}$ is the kinetic energy operator, V the trapping potential, μ the chemical potential, $n_B = |\phi_B|^2$ and $n_F = \langle\hat{\psi}_F^+\hat{\psi}_F\rangle$ are steady state density distributions, while $\delta\hat{\rho}_F = \hat{\psi}_F^+\hat{\psi}_F - n_F$ is the density fluctuation operator for fermions, and g and f represent the boson-boson and boson-fermion interaction strengths, respectively. They are related to the scattering lengths as $g = 4\pi\hbar^2 a_B/m_B, f = 2\pi\hbar^2 a_{BF}/m_r$ with m_r being the reduced mass. As we are considering a spin-polarized fermion gas, the fermion-fermion interaction is neglected.

We first solve the fermion equation (2) using first-order Born approximation. This allows us to write $\delta\hat{\rho}_F$ in terms of boson fluctuation operators $\hat{\xi}_B^+$, $\hat{\xi}_B$ and the mean-field wavefunctions. Putting this back to the bosonic equatiop (1), we are able to calculate the excitation spectrum of the condensate.

As an example, let us consider a homogeneous mixture of Bose-Fermi gas. In this case, we can solve the coupled equations (1) and (2) analytically. We find that the phonon spectrum can be written as:

$$\hbar^2\omega^2 = \frac{\hbar^2 k^2}{2m_B}\left[\frac{\hbar^2 k^2}{2m_B} + 2gn_B + \frac{2f^2 n_B}{(2\pi)^2\hbar}\ell_\omega(k)\right] \tag{3}$$

where

$$\ell_\omega(k) = \int_0^{k_F} k_m^2 dk_m \int_{-1}^{1} dx \left(\frac{\Theta(|\mathbf{k_m}+\mathbf{k}|-k_F)}{\omega - \frac{\hbar k^2}{2m_F} - \frac{\hbar k k_m}{m_F}x - i0^+} - \frac{\Theta(|\mathbf{k_m}+\mathbf{k}|-k_F)}{\omega + \frac{\hbar k^2}{2m_F} + \frac{\hbar k k_m}{m_F}x - i0^+}\right)$$

with k_F being the Fermi wavenumber.

To have a stable system, the imaginary part of $\ell_\omega(k)$ has to vanish. Otherwise, the system will possess modulational instability and will break into distinct phases. The condition for $\mathrm{Im}[\ell_\omega(k)] = 0$ determines the stability regime which depends on both the scattering lengths and the densities. Figure 1 shows the phonon spectrum for various values of a_{BF}. For sufficiently large a_{BF}, the spectrum breaks into two branches with the lower one falls into the unstable regime. In the stable regime, we can show that the sound velocity are given approximately as: $c \approx c_0[1 + 4a_{BF}^2/((2+c_0)^2\pi a_B)]$ where c_0 is the sound velocity in a pure condensate.

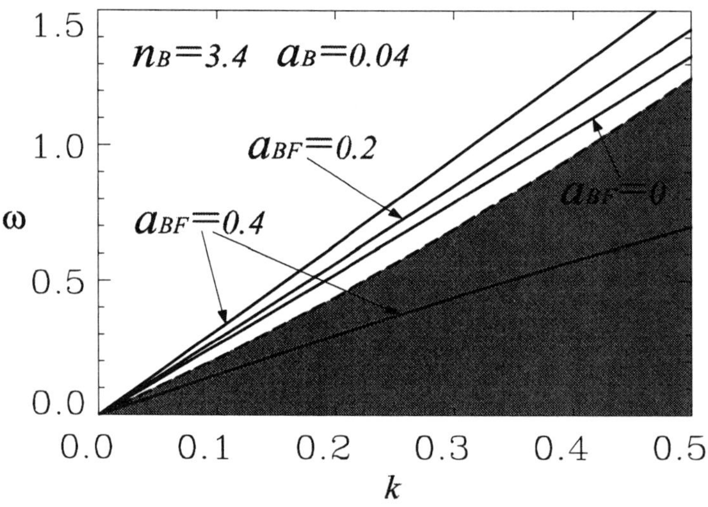

Fig. 1. Phonon spectrum of a boson-fermion mixture. The gray area below the dashed line represents the unstable regime. We assume $m_F = m_B = m$. The units for frequency, length, and wavenumber are $\hbar k_F^2/2m$, $1/k_F$, and k_F, respectively.

References and links

1. B. DeMarco and D. S. Jin, "Onset of Fermi degeneracy in a trapped atomic gas", Science **285**, 1703 (1999).
2. A. G. Truscott *et al.*, "Observation of Fermi pressure in a gas of trapped atoms", Science Express Reports, March 1, 2001.
3. K. Mølmer, "Bose condensates and Fermi gases at zero temperature", Phys. Rev. Lett. **80**, 1804 (1998).
4. M. Amoruso *et al.*, "Temperature-dependent density profiles of trapped boson-fermion mixtures", Eur. Phys. J. D **4**, 261 (1998).
5. L. Viverit *et al.*, "Zero-temperature phase diagram of binary boson-fermion mixtures", Phys. Rev. A **61**, 053605 (2000).

Slow Polaritons in Atomic Bose-Einstein Condensates

G. Juzeliūnas [1,2] and H. J. Carmichael [2]

[1] *Institute of Theoretical Physics and Astronomy, A. Goštauto 12, Vilnius 2600, Lithuania*
[2] *Department of Physics, University of Oregon, Eugene, OR 97403-1274, USA*

juzeliun@oregon.uoregon.edu, hjcarm@oregon.uoregon.edu

Recently electromagnetically induced transparency (EIT) was shown to slow down dramatically [1], or even to stop completely, laser pulses in atomic gases [2, 3]. Such behavior can be understood in terms of a branch of slow polaritons appearing between two close atomic resonances, as shown in Fig. 1 (adapted from Fig. 2b of ref.[4]). Here we present a general theory of polaritons in atomic BEC, followed by the analysis of the EIT polaritons. The possibility of reverting a stopped polariton is demonstrated.

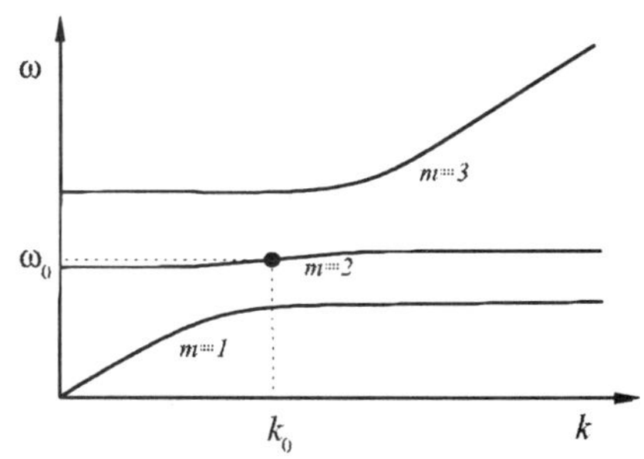

Fig. 1. Polariton dispersion branches in a medium with two resonance frequencies. The EIT resonance corresponds to a point where $\omega = \omega_0 = ck_0$ and hence $n = 1$.

Consider a gas of Bose atoms containing two hyperfine ground states, $|g\rangle$ and $|q\rangle$, and an electronic excited state $|e\rangle$. The atomic fields are described by the Bose-operators $\widehat{\psi}_j(\mathbf{r}) = V_0^{-1/2}\sum_{\mathbf{k}} b_{j,\mathbf{k}} e^{-i\mathbf{k}.\mathbf{r}}$, where $b_{\mathbf{k},j}$ annihilates an atom with internal state j ($j = g, q, e$) and wave-vector $\mathbf{k}$, and V_0 is the quantization volume. Suppose N_0 atoms, in internal state $|g\rangle$, populate the $\mathbf{k} = \mathbf{0}$ translational level to form a homogeneous BEC, for which $b_{g,\mathbf{0}} \approx \sqrt{N_0}$. (The homogeneous model may also be suitable for the analysis of a trapped BEC [5].) Suppose a control laser with wave-vector $\mathbf{k}_c$ and angular frequency ω_c couples the states $|q\rangle$ and $|e\rangle$. The Hamiltonian describing the probe field in such a medium, reads in the rotating frame: $H = H_{rad} + H_{at} + H_{rad-at} + H_{cont}$, where

$$H_{rad} = \sum_{\mathbf{k}} \hbar ck \left(a_{\mathbf{k}}^{\dagger} a_{\mathbf{k}} + 1/2\right), \tag{1}$$

$$H_{at} = \hbar \sum_{\mathbf{k}} \left[\tilde{\omega}_{e,\mathbf{k}} b_{e,\mathbf{k}}^{\dagger} b_{e,\mathbf{k}} + \tilde{\omega}_{q,\mathbf{k}} b_{q,\mathbf{k}}^{\dagger} b_{q,\mathbf{k}} + \Omega_c b_{q,\mathbf{k}-\mathbf{k}_c}^{\dagger} b_{e,\mathbf{k}} + \Omega_c b_{e,\mathbf{k}}^{\dagger} b_{q,\mathbf{k}-\mathbf{k}_c}\right] \tag{2}$$

$$H_{rad-at} = -\frac{i}{2}\sum_{\mathbf{k}} \mu \left(\frac{\hbar ck\rho}{2\varepsilon_0}\right)^{1/2} \left(a_{-\mathbf{k}} - a_{\mathbf{k}}^{\dagger}\right)\left(b_{e,\mathbf{k}} + b_{e,-\mathbf{k}}^{\dagger}\right) + h.c., \tag{3}$$

and

$$H_{cont} = (x/2\varepsilon_0)\int \mathbf{p}(\mathbf{r})\mathbf{p}(\mathbf{r})d^3\mathbf{r} = x\sum_{\mathbf{k}} \frac{\mu^2\rho}{2\varepsilon_0}\left(b_{e,-\mathbf{k}} + b_{e,\mathbf{k}}^{\dagger}\right)\left(b_{e,\mathbf{k}} + b_{e,-\mathbf{k}}^{\dagger}\right) \tag{4}$$

Coherence and Quantum Optics VIII
Edited by Bigelow et al., Kluwer Academic/Plenum Publishers, 2003

with $x = 2/3$, $\rho = N_0/V_0$, $\tilde{\omega}_{e,\mathbf{k}} = \omega_{e,\mathbf{k}} - \omega_{g,\mathbf{0}}$, and $\tilde{\omega}_{q,\mathbf{k}-\mathbf{k}_c} = \omega_{q,\mathbf{k}-\mathbf{k}_c} + \omega_c - \omega_{g,\mathbf{0}}$. Here $\omega_{j,\mathbf{k}} = \omega_j + \hbar k^2/2M$ are the atomic frequencies containing electronic and translational contributions, Ω_c is the Rabi frequency of the control field, $a_{\mathbf{k}}$ is the annihilation operator for a probe photon polarized along the $|g\rangle \to |e\rangle$ transition, and the summation is over the wave-vectors $\mathbf{k} \equiv \mathbf{k}_p$ involved in a wave-packet (probe pulse) formation. The operator H_{at-rad} describes the coupling between the atoms and the probe field, $\mathbf{p}(\mathbf{r})$ is the condensate part of the polarization field, and the operator H_{cont} represents the contact interaction which appears in the multipolar formulation of QED [6].

It is noteworthy that we have chosen $x = 2/3$ in H_{cont}, rather than the commonly used $x = 1$ [6]. In fact, by taking $x = 1$, one implicitly assumes that the electrostatic interaction between atoms is given by $(2\varepsilon_0)^{-1} \int \mathbf{p}^{\|}(\mathbf{r})\mathbf{p}^{\|}(\mathbf{r})d^3\mathbf{r}$, which contains the Dirac delta singularity at the origin characteristic of the dipole-dipole coupling between point atoms. Such a singularity is an artifact because atoms, which are not point objects, cannot overlap in reality. The singularity can be eliminated by subtracting the term $(6\varepsilon_0)^{-1} \int \mathbf{p}(\mathbf{r})\mathbf{p}(\mathbf{r})d^3\mathbf{r}$, leading to our $x = 2/3$. This choice does indeed yield the proper local field corrections, featured in the refractive index (5).

The Hamiltonian H has been diagonalized using the Bogoliubov transformation to arrive at a set of polariton operators $P_{m,\mathbf{k}}$ ($m = 1, 2, 3$), each a linear combination of the operators $a_{\mathbf{k}}$, $b_{e,\mathbf{k}}$ and $b_{q,\mathbf{k}-\mathbf{k}_c}$, as well as the appropriate creation operators. The eigenfrequencies are determined by the equation:

$$\omega = ck/n(\omega), \quad \text{with} \quad n^2 = \frac{1 + x\alpha\rho/\varepsilon_0}{1 - (1-x)\,\alpha\rho/\varepsilon_0}, \quad \text{where} \quad \alpha = \frac{1}{\hbar}\sum_{\pm} \frac{2\,|\mu_{\pm}|^2\,\omega_{k,\pm}}{\omega_{k,\pm}^2 - \omega^2} \tag{5}$$

is the atomic polarizability; $\omega_{k,\pm}$ is an eigenfrequency of the atomic Hamiltonian H_{at} and $\mu_{\pm}$ is the corresponding transition dipole. Consequently one obtains three dispersion branches, the center one describing slow polaritons, as depicted in Fig. 1.

The polaritons are generally not stable quasi-particles due to their mutual interaction involving phonons (to be analyzed elsewhere). However, they appear to be almost completely decoupled from phonons in the vicinity of the EIT resonance at $\omega = \tilde{\omega}_{q,\mathbf{k}_0-\mathbf{k}_c} \equiv \omega_0$. In such a situation, the susceptibility $\alpha\rho/\varepsilon_0 \approx (\omega - \omega_0)\,\mu^2\rho/\hbar\varepsilon_0\Omega_c^2$ is close to zero, giving the dispersion $\omega \approx \omega_0 + v_g(k - k_0)$, where $v_g = c(1 + \omega_0\mu^2\rho/2\Omega_c^2\hbar\varepsilon_0)^{-1} \approx 2\Omega_c^2\hbar\varepsilon_0 c/\omega_0\mu^2\rho$ is the familiar group velocity [1] and $k_0 = \omega_0/c$.

A slow polariton is a mixture of a magnetic excitation and a photon. The share of the latter is v_g/c. Therefore, by switching off the control beam, one maps the probe pulse onto the collective spin excitations, as suggested in [7]. In our case this corresponds to the transition $P_{2,\mathbf{k}_p} \to b_{q,\mathbf{k}_p-\mathbf{k}_c}$, making the $m = 2$ polariton branch flat. Applying subsequently a control pulse with wave-vector $\mathbf{k}'_p$, one reaches the opposite transformation: $b_{q,\mathbf{k}_p-\mathbf{k}_c} \to P_{2,\mathbf{k}'_p}$, where $\mathbf{k}'_p = \mathbf{k}_p - \mathbf{k}_c + \mathbf{k}'_c$. With co-propagating control and probe beams, one can change the direction of the probe pulse by changing the direction of the control pulse. For instance, if $\mathbf{k}'_c = -\mathbf{k}_c$, then $\mathbf{k}'_p = -\mathbf{k}_p - 2\,(\mathbf{k}_c - \mathbf{k}_p)$, i. e. the regenerated polariton experiences also a small frequency shift $2v_g\,(k_c - k_p)$.

Work by G. J. carried out as a visiting Fulbright Scholar at the University of Oregon.

References and links

1. L.V. Hau, S.E. Harris, Z. Dutton, and C.H. Behroozi, Nature **397**, 594 (1999).
2. C. Liu, Z. Dutton, C.H. Behroozi and L.V. Hau, Nature **409**, 490 (2001).
3. D. P. Phillips *et al.*, Phys. Rev. Lett. **86**, 783 (2001).
4. G. Juzeliūnas, Phys. Rev.**A 53**, 3543 (1996).
5. G. Morigi and G. S. Agarwal, Phys. Rev.**A 62**, 013801 (2000).
6. M. Lewenstein, L. You, J. Cooper and K. Burnett, Phys. Rev. A **50**, 2207 (1994).
7. M. Fleishhauer and M.D. Lukin, Phys. Rev. Lett. **84**, 5094 (2000).

Master Equation Approach for an Ideal Bose Gas

K. T. Kapale[1,2], M. O. Scully[1,2], and M. S. Zubairy[1,3]

[1]*Department of Physics, Texas A&M University, College Station, TX 77843*
[2]*Max-Planck Institut fur Quantenoptik, 85748 Garching, Germany*
[3]*Department of Electronics, Quaid-i-Azam University, Islamabad, Pakistan*
k-kapale@physics.tamu.edu

Abstract: Though experimentally more relevant, treatment of a confined ideal Bose gas through canonical ensemble is a formidable task due to the particle number conservation condition. We show here that the recently developed non-equilibrium master equation approach gives an equilibrium solution which is faithful to the canonical ensemble result.

Recently, we [1, 2, 3] have developed a non-equilibrium master equation approach on the lines of the quantum theory of laser for an ideal Bose gas of N atoms in an arbitrary trapping potential in contact with a thermal reservoir at temperature T. The condensate statistics obtained from the detailed balance equilibrium solution of the master equation compares very nicely with the exact numerical result even in the transition region. We study the success of this approach more closely and show that the equilibrium solution to the master equation essentially replicates the canonical ensemble distribution with an added ease of dealing with the troublesome particle number conservation condition. The setails can be seen in our recent paper [4]

Starting with an appropriate interaction Hamiltonian and designing the density matrix equation of motion one obtains the Bose gas master equation for the probability of certain distribution of atoms among the single particle energy levels of the trap as

$$\begin{aligned}
&\dot{p}_{n_0,\{n_k\}_{n_0}} = \\
&-\kappa \sum_{k>l>0} (\eta_{kl}+1) \left[(n_l+1)\, n_k\, p_{n_0,\{n_k\}_{n_0}} - n_l\,(n_k+1)\, p_{n_0,\{\ldots,n_l-1,\ldots,n_k+1,\ldots\}_{n_0}}\right] \\
&-\kappa \sum_{k>l>0} \eta_{kl} \left[n_l\,(n_k+1)\, p_{n_0,\{n_k\}_{n_0}} - (n_l+1)\, n_k\, p_{\{\ldots,n_l+1,\ldots,n_k-1,\ldots\}_{n_0}}\right] \\
&-\kappa \sum_{k>0} (\eta_{k0}+1) \left[(n_0+1)\, n_k\, p_{n_0,\{n_k\}_{n_0}} - n_0\,(n_k+1)\, p_{n_0-1,\{\ldots,n_k+1,\ldots\}_{n_0-1}}\right] \\
&-\kappa \sum_{k>0} \eta_{k0} \left[n_0\,(n_k+1)\, p_{n_0,\{\ldots,n_k,\ldots\}_{n_0}} - (n_0+1)\, n_k\, p_{n_0+1,\{\ldots,n_k-1,\ldots\}_{n_0+1}}\right]. \qquad (1)
\end{aligned}$$

Here the terms with $k > l > 0$ and $k > 0, l = 0$ are separated to delineate the transitions among the excited trap levels and the transitions involving the ground state. Looking at the terms with like population coefficients one arrives at the detailed balance conditions

$$p_{n_0,\{n_k\}_{n_0}} = e^{-\Omega_{kl}}\, p_{n_0,\{\ldots,n_l+1,\ldots,n_k-1,\ldots\}_{n_0+1}}, \qquad (2)$$

$$p_{n_0,\{n_k\}_{n_0}} = e^{-\Omega_{k0}}\, p_{n_0+1,\{\ldots,n_k-1,\ldots\}_{n_0}}, \qquad (3)$$

pertaining to the transitions among two of the excited states and between ground state and one of the excited states respectively. Just to note, here we have also used

$$\frac{\eta_{kl}}{\eta_{kl}+1} = \frac{1}{e^{\Omega_{kl}}-1}\frac{1}{\left[\frac{1}{e^{\Omega_{kl}}-1}+1\right]} = e^{-\Omega_{kl}}, \tag{4}$$

with $\Omega_{kl} = \beta(\varepsilon_k - \varepsilon_l)$ being the parameter signifying the energy difference between levels k and l and $\beta = 1/k_BT$. It can be easily shown that both the groups of detailed balance conditions given by Eq. (2) and (3) form a complete set independently, in the sense that the other group can be generated easily from either one. For mathematical simplicity we deal with the set signifying transitions involving the ground state explicitly, the final conclusion being independent of the set of the detailed balance conditions employed.

With a repetitive application of Eq. (3), we can sweep all the particles in a generic state k to the condensate, finally accumulating all of $N = n_0 + \sum_{k>0} n_k$ particles in the ground state. In other words

$$\begin{aligned} p_{n_0,\{n_k\}_{n_0}} &= e^{-\Omega_{10}} p_{n_0+1,\{n_1-1,\ldots,n_k,\ldots\}_{n_0+1}} \\ &= e^{-n_1\Omega_{10}} p_{n_0+n_1,\{0,\ldots,n_k,\ldots\}_{n_0+n_1}} \\ &= e^{-n_1\Omega_{10}} e^{-\Omega_{20}} p_{n_0+n_1+1,\{0,n_2-1,\ldots,n_k,\ldots\}_{n_0+n_1+1}} \\ &= e^{-n_1\Omega_{10}} e^{-n_2\Omega_{20}} p_{n_0+n_1+n_2,\{0,0,\ldots,n_k,\ldots\}_{n_0+n_1+n_2}} \\ &= p_{N,\{0\}_N} \prod_{k>0} e^{-n_k\Omega_{k0}}, \end{aligned} \tag{5}$$

where we have utilized the canonical condition $N = n_0 + \sum_{k>0} n_k$. A little observation of Eq. (5) should convince us that it is indeed equivalent to canonical ensemble result, which can be rearranged in a more transparent form as

$$p_{n_0,\{n_k\}_{n_0}} = p_{N,\{0\}_N} \prod_{k>0} e^{-n_k\beta(\varepsilon_k-\varepsilon_0)} = \mathcal{C} e^{-n_0\beta\varepsilon_0} \left(\prod_{k>0} e^{-n_k\beta\varepsilon_k}\right) \delta_{N,n_0+\sum_{k>0} n_k}, \tag{6}$$

where $\mathcal{C} = p_{N,\{0\}} e^{N\beta\varepsilon_0}$ is the normalization constant, which is related to the canonical partition function. The appearance of particle number conservation condition is signified by the Dirac delta function.

To summarize, we have shown that the equilibrium detailed balance solution of the ideal Bose gas master equation gives a result which is an exact replica of canonical ensemble probability of a certain distribution of atoms among the trap levels. Furthermore, these detailed balance conditions when adapted to the condensate with appropriate recipe allow us to determine the condensate occupation probability [4]. The conclusion being that the condensate master equation faithfully replicates the canonical ensemble result. Also, it is fairly easy to incorporate the particle number conservation condition in the master equation approach as opposed to the conventional canonical ensemble approach. Thus, we have a powerful and easy to handle tool, to deal with the problem of condensate statistics of trapped Bose gases.

References and links

1. M. O. Scully, Phys. Rev. Lett. **82**, 3927 (1999).
2. M. Scully and K. Kapale, in *Proceedings of the R. Arnowitt Fest: Relativity, Particle Physics and Cosmology*, edited by R. E. Allen (World Scientific, Singapore, 1999), pp. 181-189.
3. V. V. Kocharovsky, M. O. Scully, S. Y. Zhu, and M. S. Zubairy, Phys. Rev. A **61**, 023609 (2000).
4. K. Kapale, and M. S. Zubairy, Opt. Comm. **191**, 299 (2001).

Monopole charge distributions of rotating condensates

L. G. Boussiakou, C. R. Bennett and M. Babiker
Department of Physics, University of York, Heslington. York YO10 5DD, England
lb135@york.ac.uk

Abstract: We show that a vortex state Bose-Einstein condensate of atoms carrying electric (magnetic) dipoles gives rise to magnetic (electric) charge and field distributions. The theory is completely symmetric concerning the electric and magnetic aspects and preserves overall charge neutrality.

OCIS codes: (270.5580) Quantum electrodynamics; (000.6800) Theoretical physics

We have explored the electromagnetic properties of rotating Bose-Einstein condensates (BECs) and examined the two cases in which the constituent atoms are characterised by (i) an electric dipole moment (electric BEC) [1] and (ii) a magnetic dipole moment (magnetic BEC). In an electric BEC the rotational motion gives rise to a Röntgen interaction, while in the magnetic BEC it involves the Aharonov-Casher interaction. Our theory requires a step by step derivation of the Lagrangian [2] for a model constituent, a neutral atom composed of two oppositely charged particles expressed in terms of their centre of mass and relative coordinates **R** and **q**

$$L = M\dot{R}^2/2+\overline{m}\dot{q}^2/2+\int d^3r\{\varepsilon_0\left[E^2(\mathbf{r})+c^2B^2(\mathbf{r})\right]/2+\mathbf{P}(\mathbf{r}).\mathbf{E}(\mathbf{r})$$
$$+\mathbf{M}(\mathbf{r}).\mathbf{B}(\mathbf{r})-\dot{\mathbf{R}}.[\mathbf{P}(\mathbf{r})\times\mathbf{B}(\mathbf{r})]+\dot{\mathbf{R}}.[\mathbf{M}(\mathbf{r})\times\mathbf{E}(\mathbf{r})]\}, \qquad (1)$$

where M is the total mass and $\overline{m}$ is the reduced mass of the two particles; $\mathbf{p}$ and $\mathbf{m}$ are the electric polarisation and magnetisation fields. Our Lagrangian has the significant feature of being exactly symmetric between the electric and magnetic aspects of the problem. The constitutive relations linking the matter fields to the electromagnetic fields follow from the Lagrangian as [2]

$$\mathbf{H}(\mathbf{r})=\mathbf{B}(\mathbf{r})/\mu_0-\mathbf{M}(\mathbf{r})+\dot{\mathbf{R}}\times\mathbf{P}(\mathbf{r}), \qquad \mathbf{D}(\mathbf{r})=\varepsilon_0\mathbf{E}(\mathbf{r})+\mathbf{P}(\mathbf{r})+\dot{\mathbf{R}}\times\mathbf{M}(\mathbf{r})/c^2, \qquad (2)$$

where **D** is the electric displacement and **H** is the magnetic field intensity.

The polarisation and magnetisation of the condensate will depend on its wavefunction, so it is necessary to solve the Gross-Pitaevskii equation, arising from the corresponding Hamiltonian in a many-body treatment. For a condensate with a total of N atoms each with either an electric dipole $\mathbf{d}$ or a magnetic dipole $\boldsymbol{\mu}$ we write $\mathbf{p}(\mathbf{R}) = N\mathbf{d}|\psi(\mathbf{R})|^2$ and $\mathbf{m}(\mathbf{R}) = N\boldsymbol{\mu}|\psi(\mathbf{R})|^2$ where ψ is the condensate wavefunction. We assume that the BEC is confined within a cylinder of radius R_0 occupying the region $|z| < z_0$ of the z-axis. We perform a Madelung transformation and, assuming that the wavefunction is independent of z, write $\psi(\mathbf{R})=|\psi(r)|e^{in\phi}/\sqrt{4\pi z_0 R_0^2}$, where (r,ϕ,z) are the cylindrical coordinates of **R**. We thus obtain (ignoring field effects) the velocity profile of the BEC as $\dot{\mathbf{R}} = \hat{\boldsymbol{\phi}}\hbar n/Mr$, and also the radial wavefunction satisfies the equation [3]

$$\left[d^2/d\xi^2+1/\xi\, d/d\xi-n^2/\xi^2+2ME/\hbar^2-4n_{1d}a|\psi(\xi)|^2\right]\psi(\xi)=0, \qquad (3)$$

with $\xi = r/R_0$, $n_{1d} = N/2z_0$ is the one-dimensional atomic density parameter and we have set $\psi(R_0) = 0$. In the absence interactions the wavefunction would be proportional to the Bessel function J_n.

For an electric BEC we have in the static limit [3]

$$\nabla.\mathbf{H} = (\hbar n/Mr)[\delta(r) + d/dr]n_{1d}d|\psi(r)|^2/2\pi R_0^2 = \rho_m(r)/\mu_0 \,, \tag{4}$$

where we have identified the right hand side of Eq. (4) as a magnetic monopole charge density produced by the rotating BEC. Note that this charge density is zero outside the condensate and overall charge neutrality is preserved. This is in contrast to [1]. Thus, we may write $\mathbf{H} = -\nabla\Phi$ and seek a solution for a BEC confined to a cylinder of a finite height $2z_0$. Using Green functions we obtain

$$\Phi(\xi, z) = \Phi_0^m \int_0^1 d\xi' \int_0^{2\pi} d\phi' \frac{\xi' - \xi\cos(\phi')}{h^2}\left[\frac{1-z}{\sqrt{h^2 f^2 + (1-z)^2}} + \frac{1+z}{\sqrt{h^2 f^2 + (1+z)^2}}\right], \tag{5}$$

where z is in units of z_0, $\Phi_0^m = \hbar n n_{1d} d/8\pi^2 M R_0^2$, $h^2 = \xi^2 + \xi'^2 - 2\xi\xi'\cos(\phi')$ and $f = R_0/z_0$. Figure 1 shows a contour plot of Φ on a symmetry plane through the cylinder axis.

For a magnetic BEC we have the analogue of Eq. (4) [3]

$$\varepsilon_0 \nabla.\mathbf{E} = -(\hbar n/c^2 Mr)[\delta(r) + d/dr]n_{1d}\mu|\psi(r)|^2/2\pi R_0^2 = \rho_e(r), \tag{6}$$

where ρ_e is an electric charge density distribution. We can see that the above equation for $\mathbf{E}$ becomes identical to that for $\mathbf{H}$ when d is replaced by $-\mu_0\mu$. The equation for the electrostatic potential, $\mathbf{E} = -\nabla\Phi$, is given by Eq. (5) with Φ_0^m replaced by $\Phi_0^e = -\hbar n n_{1d}\mu_0\mu/8\pi^2 M R_0^2$. It can then be seen that Fig. 1 also presents a contour plot of the electric potential.

In conclusion we have shown how, even if no condensate atoms occupy the BEC vortex core, a monopole charge distribution, either electric or magnetic, arises which is globally neutral. The problem becomes similar to that of finding the potential in a cylindrical capacitor. Unfortunately order of magnitude estimates show that the effects are small when calculated on the basis of current estimates, their experimental detection will have to await the advent of denser BECs than are currently available.

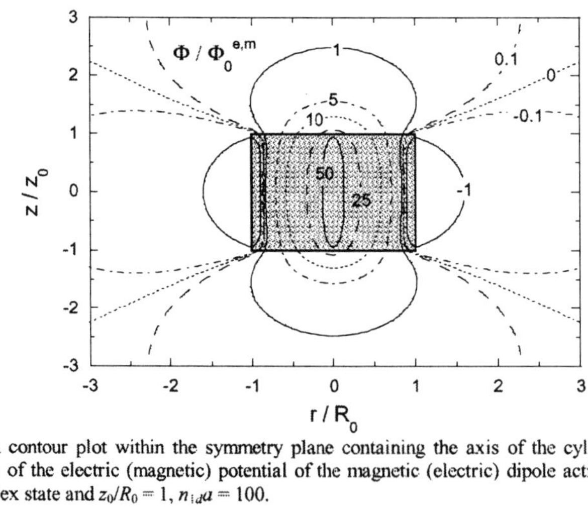

Fig. 1: A contour plot within the symmetry plane containing the axis of the cylinder showing variations of the electric (magnetic) potential of the magnetic (electric) dipole active BEC for a $n = 1$ vortex state and $z_0/R_0 = 1$, $n_{1d}a = 100$.

1. U. Leonardt and P. Piwnicki, "Quantized Röntgen Effect in Bose-Einstein Condensates", Phys. Rev. Lett. **82** 2426-2429 (1999).
2. L. G. Boussiakou, C. R. Bennett and M. Babiker, "Unified theory of electric and magnetic monopoles due to vortices in Bose-Einstein condensates", in preparation.
3. C. R. Bennett, L. G. Boussiakou and M. Babiker, "Quantized electric and magnetic monopoles in rotating Bose-Einstein condensates", submitted to Phys. Rev. Lett.

Observation of vortex phase singularities in stirred Bose-Einstein condensates

S. Inouye, A.P. Chikkatur, A. Görlitz, S. Gupta, T.L. Gustavson, A.E. Leanhardt, D.E. Pritchard, T.P. Rosenband, and W. Ketterle

Department of Physics and Research Laboratory of Electronics, Massachusetts Institute of Technology, Cambridge, MA 02139, USA
sinouye@mit.edu

Abstract:
We have observed phase singularities due to vortex excitation in Bose-Einstein condensates. This was done by exciting vortices by stirring a condensate with a laser beam and then interfering it with an unperturbed condensate.

OCIS codes: (000.1600) Classical and quantum physics; (020.0020) Atomic and molecular physics

Quantized vortices play an essential role in understanding superfluidity. In liquid helium, the critical velocity for the motion of a macroscopic object is determined by the excitation of vortices. Vortices are considered to be the source of mutual friction between the normal fluid and the superfluid. If we rotate a bucket filled with superfluid, the fluid starts to rotate only when the vortex state is energetically more favorable.

The motion of a quantized vortex and the resulting flow field is, however, far from simple and has been the subject of intensive study for a long time [1]. Vortices can tilt, bend, develop oscillations and form rings resulting in a complicated flow field. If there is more than one such excitation present, they can form a lattice or vortex tangle, depending on their geometry and charge.

Until recently, studies of vortices in superfluid were limited to liquid helium. Recent advances in dilute gas Bose-Einstein condensates [2] offer a unique opportunity for observing more details of superfluidity, and several outstanding experiments have already been performed to show quantized circulation [3, 4]. Since the interaction is weak, the quantum depletion is negligible and all atoms share a single-variable macroscopic wave function. Since the flow field is the gradient of the phase part of the wave function, it can be visualized by interferometric techniques. We extracted the phase information by overlapping an excited condensate with an unperturbed condensate, which served as a local oscillator.

If we overlap two Bose-Einstein condensates moving relative to each other with velocity v, the density profile shows sinusoidal modulation of periodicity h/mv. If one of the condensates has a vortex excitation in it, the density modulation should show characteristic "forking" at the position of the vortex [5]. This can be understood as a consequence of different fringe spacing on both sides of the vortex, since the additional flow due to a vortex increases or decreases the relative velocity between the two fluids on either side of it. The difference in the number of fringes on the two sides of the fork should not only give the direction of the flow but also the multiplicity of the vortex.

Our setup is a combination of two experiments conducted in our lab in the past [6, 7]. Briefly, a double-well potential to hold two condensates was prepared by producing a potential hill at the center of a cigar-shaped magnetic trap. For this purpose, blue-detuned far-off resonant laser light was focused into a light sheet and aligned to the center of the magnetic trap. Laser cooled sodium atoms were loaded into the double-well potential and further cooled by rf-induced evaporation. After two condensates each containing $\sim 1 \times 10^6$ atoms in $F = 1, m_F = -1$ state formed at the bottoms of the double-well potential, we slash through one of the condensate with another blue-detuned laser beam, while the other remains intact in its trap. After slashing the beam once across the "sample" condensate, all the confining potentials were switched off and the two condensates were overlapped during 41 ms time-of-flight.

Figure 1 shows typical interference fringes with(a) and without(b) vortices. Clear forking of the fringes is observed at the center of (a), while all the fringes are straight in (b). The cigar axis is vertical and the light sheet is horizontal in the figure, forming two condensates which are displaced vertically. Since vortices were excited in the condensate on top, the direction of the vortex flow in (a) should be clockwise. This is a strong

Fig. 1. Observation of phase singularities due to vortex formation.

technique for observing nucleation of vortex pairs, recombination, and break up of multiply-charged vortices. By changing the timing of the release of atoms from the trap after the slashing, it is possible to determine the time it takes for the perturbed condensate to establish a uniform phase.

We have observed vortex excitation in Bose-Einstein condensate by observing the phase profile across the condensate. The vortex was excited by moving the focus of a laser beam across the condensate. Our results will contribute to the study of superfluid turbulence, where detailed comparison with theory is necessary.

References

1. R. J. Donnelly, Quantized Vortices in Helium II (Cambridge University Press, Cambridge, England, 1991).
2. M. H. Anderson *et.al.*, Science **269**, 198 (1995); K. B. Davis *et.al.*, Phys. Rev. Lett. **75**, 3969 (1995); C. C. Bradley *et.al.*, Phys. Rev. Lett. **78**, 985 (1997); D. Fried *et al.*, Phys. Rev. Lett. **81**, 3811 (1998).
3. M. R. Matthews *et.al.* Phys. Rev. Lett. **83**, 2498 (1999).
4. K. W. Madison, F. Chevy, W. Wohlleben, and J. Dalibard Phys. Rev. Lett. **84**, 806 (2000).
5. E.L. Bolda and D. F. Walls Phys. Rev. Lett. **81**, 5477 (1998).
6. C. Raman *et. al.* Phys. Rev. Lett. **83** 2502 (1999).
7. M.R. Andrews *et.al.* Science **275**, 637 (1997).

Inner-shell x-ray lasing without population inversion

J. Mompart and R. Corbalán
Departament de Física, Universitat Autònoma de Barcelona, E-08193, Bellaterra, Spain
Telf: (34) 93 581 1653; Fax: (34) 93 581 2155; email address: Ramon.Corbalan@uab.es

L. Roso
Departamento de Física Aplicada, Universidad de Salamanca, E-37008, Salamanca, Spain
Telf: (34) 923 29 4678; Fax: (34) 923 29 4584; email address: roso@gugu.usal.es

Abstract:
A novel scheme to achieve cw x-ray lasing without population inversion involving inner-shell ionization is analyzed. In the small pumping regime, the absorption of the most abundant atoms/ions without inner vacancies is cancelled through quantum interference while the rest of the atoms/ions contribute to x-ray laser emission. Different difficulties inherent to x-ray lasing, e.g., the recoil effect and Doppler broadening, are addressed and possible solutions are discussed.

Summary

In the last years, lasing without population inversion (LWI) has attracted much attention for its potential to generate continuous-wave (cw) laser emission in the short-wavelength domain where conventional methods based on population inversion are not available or difficult to implement [1-4]. In the most common approach to LWI, a so-called driving laser couples an electronic transition with the aim of generating a laser field in an adjacent non-inverted transition. Up to now several experiments have reported lasing without inversion demonstrating the feasibility of the LWI idea. Unfortunately, no one of these experiments was performed in the so-called frequency up-conversion regime since the drive laser had a frequency equal or larger than that of the generated laser field (for a detailed review of the state of the art in LWI see [4]). All these experiments were performed in infrared or optical transitions in gas media involving external valence electrons and the main difficulties were the implementation of an incoherent pumping mechanism, needed to bring some population to the upper laser level, and Doppler broadening. To sum up, LWI has been demonstrated although frequency up-conversion LWI is still a very interesting and challenging problem.

Here we propose a novel approach to short-wavelength lasing: To ionize an inner-shell electron and then use this empty level as the lower level of a laser transition in a Λ-type three-level configuration (see Fig. 1). X-ray lasing by inner-shell ionization is not a new idea in the literature [5-7] and indeed it has been intensively investigated in the pulse regime since only transient population inversion can be achieved in these schemes. The new aspect of the present model consists of using the ideas of LWI to propose cw x-ray lasing even with only a few percent of the atoms with an inner-shell vacancy.

In our scheme, a drive laser couples the optical transition $|2\rangle$-$|3\rangle$ and x-ray lasing takes place between levels $|1\rangle$-$|2\rangle$. Typically, x-ray pumping sources, e.g., a relativistic electron beam, create only a few percent of atoms/ions with a removed inner-shell electron due to the fast recombination rate. Under standard conditions, the most abundant atoms are those without an inner-shell vacancy that, consequently, absorb the x-ray laser field and avoid x-ray laser emission. Nevertheless, we demonstrate here that it is possible to reduce and even to cancel the x-ray absorption of these atoms/ions in the presence of an optical laser field in an adjacent transition, thus allowing x-ray lasing.

We have developed a theoretical model consisting of two different species. Specie 1: atoms/ions without inner-shell vacancy; and, Specie 2: atoms/ions with an inner-shell vacancy. The population inversion condition in this model is defined straightforwardly as population of specie 2 being larger than population of specie 1. In the following, we restrict ourselves to the non-inversion case.

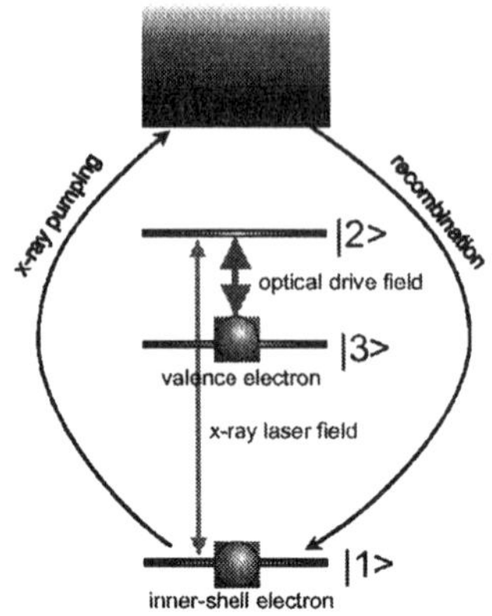

Fig. 1.- Λ-type three-level scheme under investigation.

There is an appealing description of the mechanism for lasing in our scheme. In the presence of the drive field, level |2> splits into the so-called Rabi sidebands or dressed states (AC-Stark splitting) and, by selecting appropriate parameter values, quantum interference between these dressed states, i.e., electromagnetically induced transparency, gives rise to cancellation of x-ray absorption for the atoms/ions of specie 1. On the other hand, x-ray pumping is continuously creating atom/ions of the specie 2. Thus, once an atom/ion of specie 2 is created it can amplify the x-ray laser field through two different processes: (i) direct one-photon stimulated emission from |2> to |1>; and (ii) by a two-photon or Raman process with stimulated emission of one x-ray photon simultaneously to absorption of one drive photon, i.e., a |3>→|1> process. By a convenient selection of the parameter setting, we demonstrate that x-ray lasing is possible with more than 80% of the atoms/ions as specie 1, i.e., without an inner-shell vacancy.

In the x-ray domain several problem arise as, for instance, the recoil effect and Doppler broadening. The most natural way to avoid these problems is to place the active atoms/ions in a host crystal. In particular, the recoil effect can be drastically reduced due to the large number of atoms among which the photon momentum is distributed. In this case, a convenient selection of the atoms/ions and the host matrix has to be done in such a way that the host-matrix do not absorb the x-ray emitted by the active atoms/ions. As an example of the proposed x-ray LWI scheme, we have studied erbium in different host matrices as a suitable x-ray lasing candidate.

[1] O. Kocharovskaya, Amplification and lasing without inversion, Phys. Rep. **219**, 175 (1992).
[2] M. O. Scully, From lasers and masers to phaseonium and phasers, Phys. Rep. **219**, 191 (1992).
[3] P. Mandel, Lasing without inversion: a useful concept?, Contemp. Phys. **34**, 235 (1993).
[4] J. Mompart and R. Corbalán, Lasing without inversion, J. Opt. B: Quantum Semiclass. Opt. **2**, R7-R24 (2000).
[5] S. J. Moon and D. C. Eder, Theoretical investigation of an ultrashort-pulse coherent x-ray source at 45 Å, Phys. Rev. A **57**, 1391 (1998).
[6] K. Moribayashi, A. Sasaki, and T. Tajima, X-ray emission by ultrafast inner-shell ionization from vapors of Na, Mg, and Al, Phys. Rev. A **59**, 2732 (1999).
[7] D. Kim, S. H. Son, J. H. Kim, C. Tóth, and C. P. J. Barty, Gain characteristics of inner-shell photoionization-pumped $L_{23}M_1$ transition in Ca, Phys. Rev. A **63**, 023806 (2001).

Spatiotemporal dynamics of three-level inversionless lasers

J. Mompart, V. Ahufinger, R. Corbalán
Departament de Física, Universitat Autònoma de Barcelona, E-08193, Bellaterra, Spain

J. Garcia-Ojalvo, M. C. Torrent, R. Vilaseca
Departament de Física i Enginyeria Nuclear, Universitat Politècnica de Catalunya, Colom 11, E-08222, Terrassa, Spain

Telf: (34) 93 581 16 53, Fax: (34) 93 581 21 55, e-mail address: Ramon.Corbalan@uab.es

Abstract:

We investigate the spatiotemporal dynamics of broad-area three-level inversionless lasers in continuous wave and self-pulsing regimes. We show analytically and numerically that the scenario for pattern formation completely differs from that occurring in inverted three-level systems.

Summary

The spatiotemporal dynamics of broad-area two- and three-level optical systems has been extensively studied during the last years [1-5]. In inverted two-level laser systems, the interplay between nonlinearities and diffraction results, depending on the sign of the cavity detuning, in that laser emission takes place with a small wavevector component transverse to the cavity-axis which, in turn, results in the formation of a transverse traveling wave (TW). In three-level systems with population inversion in the lasing transition, the presence of the amplifying Rabi sidebands gives rise to a completely different scenario for pattern formation. Some characteristics of this scenario are, for instance, cohexistence of transverse wavevector values near lasing threshold and the suppression of the Autler-Townes doublet [6].

We study here the spatiotemporal dynamics of inversionless lasing in coherently driven broad-area three-level systems. In these schemes, a resonant external laser field drives one transition while laser emission takes place in an adjacent non inverted transition. Due to the fact that there is no population inversion, the Rabi sidebands or dressed states built up by the driving field contribute to absorption although, depending on the parameter values, gain can appear at line center or at two symmetrically located sidebands outside the dressed states resonances. In this last case, it has been shown recently that the simultaneous excitation of these two sidebands can give rise to inversionless lasing in a self-pulsing regime [7].

Through a linear stability analysis (LSA) of the trivial non-lasing solution we determine the conditions for lasing in a continuous wave (cw) or self-pulsing regime in the presence of diffraction. We show that the scenario for pattern formation in lasing without inversion (LWI) with gain between the Rabi sidebands is quite similar to that in two-level systems, i.e., for negative cavity tunings cw laser emission takes place with a transverse TW. On the other hand, LWI based on gain at the two sidebands outside the dressed states resonances gives rise to a completely new scenario for pattern formation. The trivial non-lasing solution can be destabilized through a pitchfork bifurcation with one or two possible transverse wavevectors or through a Hopf bifurcation giving rise in some cases to a stable self-pulsing regime. For a small unsaturated gain parameter the lower threshold corresponds to the pitchfork bifurcation although as we increase this parameter the threshold values for both type of bifurcations become quite similar. Finally, a numerical simulation of the full set of equations confirms the results coming from the LSA and allows us to study the interplay between the transverse spatial dynamics and the longitudinal temporal dynamics.

[1] F. Prati, M. Brambilla and L. A. Lugiato, "Pattern formation in lasers", Riv. Nuovo Cimento **17**, 1 (1994).

[2] P. K. Jakobsen, J. Lega, Q. Feng, M. Staley, J. V. Moloney and A. C. Newell, "Nonlinear transverse modes of large-aspect-ratio homogeneously broadened lasers: I. Analysis and numerical simulations", Phys. Rev. A **49**, 4189 (1994).

[3] G. Huyet, M. C. Martinoni, J. R. Tredicce and S. Rica, "Spatiotemporal dynamics of lasers with a large Fresnel number", Phys. Rev. Lett. **75**, 4027 (1995).

[4] J. Garcia-Ojalvo, M. C. Torrent and R. Vilaseca, "Spatiotemporal dynamics of dressed broad-area lasers", Quantum Semiclass. Opt. **10**, 809 (1998).

[5] G. J. de Valcarcel, E. Roldan and R. Vilaseca, "Special issue on Paterns in Nonlinear Optical Systems", J. Optics B **1**, n1 (1999).

[6] J. Garcia-Ojalvo, R. Vilaseca, and M. C. Torrent, "Suppression of Autler-Townes gain splitting in lasers with a planar resonator" Europhys. Lett., **43**, 261 (1998).

[7] J. Mompart, C. Peters, and R. Corbalán, "Lasing without inversion: self-pulsing in cascade schemes", Phys. Rev. A **57**, 2163 (1998).

Spectral width of electromagnetically induced transparency in hot atomic gases

Ali Javan[1], Olga Kocharovskaya[2,3], Hwang Lee[4], Yuri Rostovtsev[2], Marlan O. Scully[2,5], and Alexander Zibrov[2]

[1]*Department of Physics, Massachusetts Institute of Technology, Cambridge, MA 02139*
[2]*Institute for Quantum Studies and Department of Physics, Texas A & M University, College Station, TX 77843*
[3]*Institute of Applied Physics,Nizhny Novgorod 603120, Russia*
[4]*Jet Propulsion Laboratory, California Institute of Technology, Pasadena, CA 91109*
[5]*Max-Planck-Institut fur Quantenoptik,D-85748 Garching, Germany*

1. Introduction

Experiments in Electromagnetically Induced Transparency (EIT) have produced ultraslow light propagation [1-4] with spatial compression, and have made it possible to enhance nonlinear optical processes by orders of magnitude [5,6]. Similar techniques have been utilized to store light pulses both in an ultracold atomic medium [7] and in a hot gas [8]. Small group velocities are accomplished by the large index changes near the EIT resonance, and the steepness of the dispersion function is directly related to the width of the transparency window. Therefore, in addition to achieving EIT, the spectral width of the EIT resonance is an important aspect in the study of slow light as well as precision measurements [9-12]. In hot gas experiments, not like in the case of ultracold atoms, Doppler broadening becomes crucial as it is the dominant dephasing mechanism. In this paper, we present an analytic expression of the spectral width of EIT resonance in a Doppler broadened.

2. Doppler broadening and spectral width of EIT resonance

We consider a simple three-level Λ scheme where one of the two lower-levels (c) is coupled to the upper level a by a coherent drive laser with its Rabi frequency Ω, and the transition a to b is probed by a weak coherent field. There are three different time scales involved in this simple model. They are the upper level decay rate (2γ), coherence decay rate between the two lower levels (γ_{bc}), and the inhomogeneous decay rate ($2W$).

Due to the Doppler effect the atoms in a gas see the radiation field with shifted frequency. Hence the macroscopic polarization representing medium's response to the radiation, needs to be averaged over the frequency distribution determined by velocity distribution of the atoms. Doppler broadening is usually modeled by convolution of a given function over a Maxwell-Boltzmann velocity distribution. In order to obtain a simple expression of the linewidth, we approximate the usual Gaussian distribution with a Lorentzian function; this leads to rather simple form of inhomogeneously broadened susceptibility with which detailed analysis is possible [13]. Under the assumption the inhomogeneous width $2W$ is bigger than any other parameters, the linewidth of EIT resonance then can be found as

$$\Gamma_{EIT} = \gamma_0\sqrt{1+\frac{\Omega^2}{\Omega_s^2}},$$

where $\gamma_0 = \sqrt{2\gamma_{bc}/\gamma}\ \Omega$ and $\Omega_s^2 = 2\gamma_{bc}W^2/\gamma$ with given Doppler width $2W$. Two extreme cases are noticeable: For $\Omega >> \Omega_s$, the linewidth of EIT resonance is proportional to the intensity of drive laser Ω^2 ($\Gamma_{EIT} \approx \Omega^2/W$). The inhomogeneous width ($2W$) essentially replaces the homogeneous width (2γ) when we recall $\Gamma_{EIT} \approx \Omega^2/\gamma$ for a homogeneously broadened system.

For $\Omega << \Omega_s$, it is proportional to the Rabi frequency of drive laser and furthermore it does not depend on the width of Doppler broadening. This recovers the result obtained by Feld and Javan many years ago. They predicted the spectral width of the narrow line to be independent on Doppler width and linearly proportional to the driving field Rabi frequency, which has an essential resemblance with EIT experiments in gaseous media except that it dealt with a gain medium [14].

2. The role of optical pumping

Let us suppose we the system is homogeneously broadened. The optical pumping rate from the level c to b is Ω^2/γ for the resonant driving field. In order to have a complete optical pumping within the homogeneous linewidth, this rate should be much greater than the pumping from level b to c, i.e. $\Omega^2/\gamma >> \gamma_{bc}$. This means that the driving field should be sufficiently strong to achieve EIT.

In the case of inhomogeneously broadened system the optical pumping for a certain velocity group of atoms are decreased due to the relative detuning of the resonant radiation field, and on the average it becomes $\Omega^2 2\gamma/(4\gamma^2+4W^2) \approx \Omega^2\gamma/2W^2$. For it to be larger than γ_{bc}, we obtain the condition $\Omega >> \Omega_s$. Let us write the width of EIT as $\Gamma_{EIT} = \Omega^2/\Delta$, which is equivalent to the linewidth expression for the homogeneously broadened case with Δ as an effective width given by the number of atoms participating EIT process. If $\Omega >> \Omega_s$, $\Delta = W$. If $\Omega << \Omega_s$, Δ can be estimated by $\Omega^2(\gamma/\Delta^2) = \gamma_{bc}$, which leads to $\Gamma_{EIT} = \gamma_0$.

5. References

[1] L.V. Hau, *et. al.*, "Light spped reduction to 17 metres per second in an ultracold atomic gas" Nature **397**, 594-598 (1999).

[2] M.M. Kash *et al.*, "Ultraslow Group Velocity and Enhanced Nonlinear Optical Effects in a Coherently Driven Hot Atomic Gas" Phys. Rev. Lett. **82**, 5229-5232 (1999).

[3] D. Budker *et al.*, "Nonlinear Magneto-optics and Reduced Group Velocity of Light in Atomic Vapor with Slow Ground State Relaxation" Phys. Rev. Lett. **83**, 1767-1770 (2000).

[4] O. Kocharovskaya, Y. Rostovtsev, and M.O. Scully, "Stopping light via hot atoms" Phys. Rev. Lett. **86**, 628-631 (2001).

[5] S.E. Harris, "Electromagnetically induced transparency" Physics Today **50** (7) 36-42 (1997).

[6] S.E. Harris and L.V. Hau, "Nonlinear optics at low light levels" Phys. Rev. Lett. **82**, 4611-4614 (1999).

[7] L.V. Hau, *et al.*, , "Observation of coherent optical information in an atomic medium using halted light pulses" Nature **409**, 490-493 (2001).

[8] D.F. Phillips, *et al.*, "Storage of light in atomic vapor" Phys. Rev. Lett. **86**, 783-786 (2001).

[9] M.O. Scully and M. Fleischhauer, "High-sensitivity magnetometer based on index-enhanced media", Phys. Rev. Lett. **69**, 1360-1363 (1992).

[10] M.D. Lukin *et al.*, "Spectroscopy in dense coherent media : line narrowing and interference effects" Phys. Rev. Lett. **79**, 2959-2962 (1997).

[11] A. Nagel *et al.*, "Experimental realization of coherent dark-state magnetometers", Europhys. Lett. **44**, 31-36 (1998).

[12] D. Budker, V. Yashchuk, and M. Zolotorev, "Nonlinear magneto-optic effects with ultranarrow widths," Phys. Rev. Lett. **81**, 5788-5791 (1998).

[13] A. Javan, O. Kocharovskaya, H. Lee, and M.O. Scully "From laser induced line narrowing to electromagnetically induced transparency" (to be published).

[14] M.S. Feld and A. Javan, "Laser induced line-narrowing effects in coupled Doppler-broadened transitions", Phys. Rev. **177**, 540-562 (1969).

Absorptive photon switching by quantum interference

Min Yan, Edward Rickey, and Yifu Zhu

Florida International University
Miami, Florida 33199

Electromagnetically induced transparency (EIT) in an absorbing medium refers to the suppression of linear absorption of a probe laser at the line center by destructive quantum interference. The quantum interference may be destroyed by adding a switching laser coupled to a separate state in a four-state system. The four-state system exhibits unusually large nonlinear probe absorption at the EIT center and may be used to realize absorptive photon switching as proposed by Harris and Yamamoto [1].

We report an experimental study of enhanced third-order nonlinear absorption and absorptive photon switching in ^{87}Rb atoms cooled and confined in a magneto-optical trap. In our experimental scheme, a coupling laser drives D_1 F=2 – F'=1 transition and a weak probe laser couples the D_1 F=1 – F'=1 transition. This forms a Λ-type EIT system. A switching laser couples the D_2 F=2 – F'=3 transition and thus forms a four-state Rb EIT system that exhibits the large third-order nonlinear absorption for the probe laser. The experiment is run sequentially with a repetition rate of 10 Hz. All lasers are turned on and off by Acousto-Optic Modulators. For each period of 100 ms, cooling and trapping of the Rb atoms takes ~ 99 ms during which only the trap laser and the repump laser are on. Data collection takes about 1 ms during which the trap laser and the repump laser are turned off while the coupling laser, the probe laser, and the switching laser are turned on. We recorded the nonlinear absorption spectrum of the probe laser by scanning the probe frequency across the atomic transition. We observed the greatly enhanced, pure nonlinear absorption with amplitude comparable to that of single-photon absorption in a simple two-state system. Switching of the probe absorption is demonstrated when the probe frequency is fixed at the EIT resonance center and the switching laser pulse is turned on and off with a pulse duration smaller than 1 ms. We observed switching of the probe absorption with both a continuous-wave probe laser and a pulsed probe laser, the duration of which matches that of the switching pulse.

Fig. 1(a) plots the calculated Im (ρ_{13}) (proportional to the probe absorption) for the Rb four-level EIT system versus time when a switching pulse simulated from the experiment (the top, positive curve) is turned on at $t=t_2=0$. The rise time of the switching pulse is 150 ns. Fig. 1(b) shows the measured switching pulse (the top curve in arbitrary units) and the probe transmission (the lower curves) versus time. The probe absorption is turned on by the switching pulse and the percentile nonlinear absorption of ~30% is observed. The switching rise time of the probe absorption follows that of the switching pulse, which is ~150 ns. The fall time delay of the absorption switching caused by the smaller EIT width is seen at $\Omega \approx 4.5$ MHz (the solid curve). We varied the pulse width of the third pump laser from 1 μs to 300 μs, and found that the time duration of the probe absorption follows exactly that of the third pump laser pulse. With a pulsed probe laser, we also observed attenuation of the pulse amplitude ~ 30% when the switching pulse overlaps with the probe pulse. Our experimental measurements agree well with theoretical calculations based on the simple four-state model system.

Coherence and Quantum Optics VIII
Edited by Bigelow et al., Kluwer Academic/Plenum Publishers, 2003

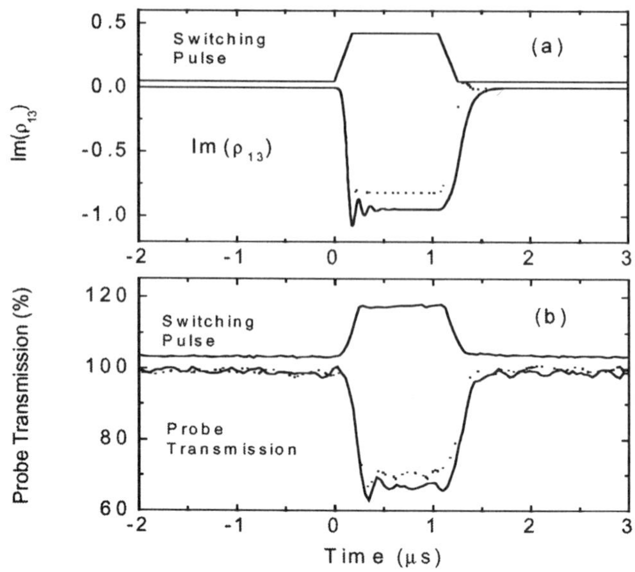

Fig. 1. (a) Calculated Im (ρ_{13}) for the Rb four-level EIT system versus time when a switching pulse simulated from the experiment (the top, positive curve) is turned on at $t=t_2=0$. The rise time of the switching pulse is 150 ns. The Rabi frequencies are $\Omega'=1.5\Gamma_3$, $g=0.1\Gamma_3$, and $\Omega=0.5\Gamma_3$ (dotted line) and Γ_3 (solid line) for the switching laser, the probe laser, and the coupling laser, respectively (Γ_3 =5.3 MHz is the natural linewidth of the upper state). (b) The measured switching pulse (the top curve in arbitrary units) and the probe transmission (the lower curves) versus time. For the dotted curve, $2\Omega\approx9$ MHz while for the solid curve, $2\Omega\approx4.5$ MHz. The switching rise time of the probe absorption follows that of the switching pulse, which is ~150 ns. The fall time delay of the absorption switching caused by the smaller EIT width is seen at $\Omega\approx4.5$ MHz (the solid curve).

1. S. Harris and Y. Yamamoto, Phys. Rev. Lett. **81**, 3611 (1998).

Coherent VUV and XUV production enhanced by EIT

Marcel Anscombe, Rebeca de Nalda, Jon Marangos
The Blackett Laboratory, Imperial College of Science Technology and Medicine, London. UK.
Tel: +44 (0)20 594 7889, Fax: +44 (0)20 7594 7777, email: m.anscombe@ic.ac.uk

Abstract: We report experimental investigations of four-wave mixing schemes in krypton enhanced by electromagnetically induced transparency (EIT). High conversion efficiency was demonstrated and interpreted for vacuum ultraviolet (VUV) generation. Work in progress is described which aims to produce bright extreme-ultraviolet (XUV) light at 87nm using enhancement from an autoionising state.

Bench-top sources of coherent VUV and XUV light will be of use in many fields of science and technology. Gases can by used for sum-frequency mixing to the XUV/VUV from easily produced visible wavelengths [1]. The main obstacle to efficient production of these wavelengths is that at resonance where the non-linearity is highest the absorption is also large. Harris [2] proposed that the efficiency would be improved by orders of magnitude by exploiting the atomic coherence effect known as electromagnetically induced transparency (EIT). In this process, the absorption is cancelled due to the opposite signs of absorption amplitudes in the laser-dressed medium, whilst the amplitudes responsible for the non-linear response experience constructive interference. VUV sum-frequency mixing enhanced by EIT was first demonstrated in hydrogen by Hakuta, Stoicheff and co-workers [3].

For high conversion efficiencies it is necessary to use a dense Doppler-broadened medium. The Rabi frequency of the field producing the dressed state must be large enough to overcome the inhomogeneous broadening and for this reason pulsed lasers are required. However, if this Rabi frequency is too big then the non-linearity will be reduced due to the large Autler-Townes splitting. Krypton was chosen in our experiments due to a relatively small (1GHz) Doppler width at room temperature, at least an order of magnitude smaller than the earlier hydrogen experiments [3]. This enables larger conversion efficiencies and large density–length (*NL*) products to be studied.

A sum-difference frequency mixing scheme resonant with the bound-states of krypton was investigated. A weakly focused coupling beam at 759nm was employed in a 1cm krypton-filled cell along with a focused UV beam at 212nm with which it mixed to generate 123nm VUV light [4,5]. These experiments showed that the EIT enhancement was optimized at an *NL* of 5×10^{16} cm^{-2}, with a conversion efficiency approaching 1%. The reduction of VUV signal as the density increased was well modelled qualitatively by a simple background susceptibility parameter. This background susceptibility was due to the rest of the atomic levels not included in the 3-level theoretical model [6]. The main conclusion is that when working at high NL this phase mismatch becomes a problem even though the transparency remains good and thus limits the brightness of the source.

EIT has also been demonstrated for an autoionising state [7]. This suggests that reduction of absorption due to EIT in the proximity of an autoionising state that provides a large non-linearity could be used to enhance XUV production. Work is in progress to investigate this type of scheme using a narrow autoionising state in Kr.We are producing coherent radiation at 87nm (XUV) in this frequency mixing scheme (Fig. 1).

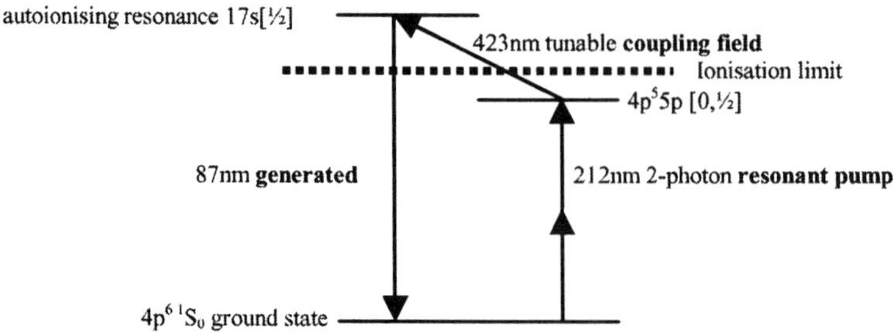

Fig 1. The krypton energy levels used in the sum-frequency mixing experiment using an autoionising state. Light is generated at 87nm.

The single-mode field at 423nm was provided by Type I doubling in a 6mm long BBO crystal. The fundamental at 846nm was from a Continuum Mirage 800 optical parametric oscillator (OPO) pumped by the second harmonic of an injection-seeded Nd:YAG. Typically 0.5mJ pulses of 2ns duration were produced. The two-photon pump at 212nm was produced from a dye laser pumped by a XeCl Excimer. The dye laser output was not single mode (since this is not required to satisfy the conditions for EIT) but the dye oscillator had an intracavity etalon to reduce the bandwidth (nominally 1.2GHz). Typically 200μJ pulses of 16ns duration were produced after doubling in BBO. In order to detect any possible enhancement due to the autoionising resonance, the fundamental was scanned through continuum in single longitudinal mode from 846.2nm to 847.4nm.

In the precursor experiment generating VUV at 123nm (see above) a magnesium fluoride window was used as the cell's exit window. For XUV no transparent window material exists. This necessitated using a cell with a small (100μm) exit hole at the vacuum end. The pump and coupling field were brought to a focus in front of the exit hole on one arm of a vacuum spectrometer. The first order of the generated XUV was measured using an electron multiplier. Pressures of up to 7mbar could be used before the limit is reached where the electron multiplier undergoes breakdown ($>10^{-4}$ mbar). At these pressures the XUV sees a optically deep medium due to the large photoionisation associated with the autoionising state, therefore we expect significant improvements in the XUV generation due to EIT under these conditions. Preliminary results showed good XUV generation and further work is in progress to detect any resonant enhancement effects.

[1] J. F. Reintjes, Nonlinear Optical Processes in Liquids and Gases, (Quantum electronics – principles and applications), Academic Press, (1984)

[2] S.E. Harris, Nonlinear optical processes using electromagnetically induced transparency, Phys. Rev. Let, **64**, 1107-1108 (1990)

[3] G. Z. Zhang, K. Hakuta, B. P. Stoicheff, Nonlinear optical generation using electromagnetially induced transparency in atomic hydrogen, Phys. Rev. Lett. **71**, 3099 (1993)

[4] C. Dorman, I. Kucukkara, J. P. Marangos, Density dependent effects due to residual susceptibilities in a four-wave mixing scheme with electromagnetically induced transparency, Opt. Comms. **180**, 263-270 (2000).

[5] C. Dorman, I. Kucukkara, J. P. Marangos, Measurement of high conversion efficiency to 123.6nm radiation in a four-wave-mixing scheme enhanced by electromagnetically induced transparency, Phys. Rev. A, **61**, 013802

[6] J. C. Petch, C. H. Keitel, P.L. Knight, J. P. Marangos, Role of electromagnetically induced transparency in resonant four-wave mixing schemes, Phys. Rev. A, **53**, 543-561 (1996)

[7] K.J. Boller, A. Imamoglu, S. E. Harris, Observation of electromagnetically induced transparency, Phys. Rev. Lett., **66**, 2593-2596 (1991)

[8] S. E. Harris, Z.-F Luo, Preparation energy for electromagnetically induced transparency, Phys. Rev. A, **52**, R928-R931

Laser Cooling of an EIT material inside a double cavity

F.A. Narducci

EO Sensors Division , Naval Air Warfare Center, 22347 Cedar Point Rd., Patuxent River, Md. 20670

NarducciFA@navair.navy.mil

The work presented here is the first attempt (to this author's knowledge) to investigate the response of a three level atom in the lambda configuration to a double cavity with two injected fields. The details of the calculations are extremely long and will be presented elsewhere. In this paper, we outline only the starting point and the main results.

The effect of an optical cavity on the radiative properties of an atom has long been the focus of the field known as cavity quantum electro-dynamics. Of particular relevance to this work is the recent article by Vuletic and Chu [1] that suggests the ability to cool complex systems with the aid of an optical cavity. This paper and the papers of reference [2] form the basis for a recent paper which takes a detailed look at the action of an injected field on a collection of two level atoms with attention to the boundary conditions imposed by the optical cavity [3]. A main conclusion from these works [2-4] is that, for large atomic detuning, the sign of the detuning of the injected field with respect to the cavity resonance controls whether a collection of two level atoms undergoes heating or cooling.

In recent years, a lot of attention has been paid to the case of a three level atom driven by two external fields [4]. Under the appropriate conditions, one of the fields (usually weak) undergoes no absorption, but an enhanced dispersion. This condition is known as electro-magnetically-induced transparency (EIT) and is responsible for the recent work in ultra-slow light [5] and is related to the work on trapped light [6].

In this work, we consider the situation depicted in Fig. 1.

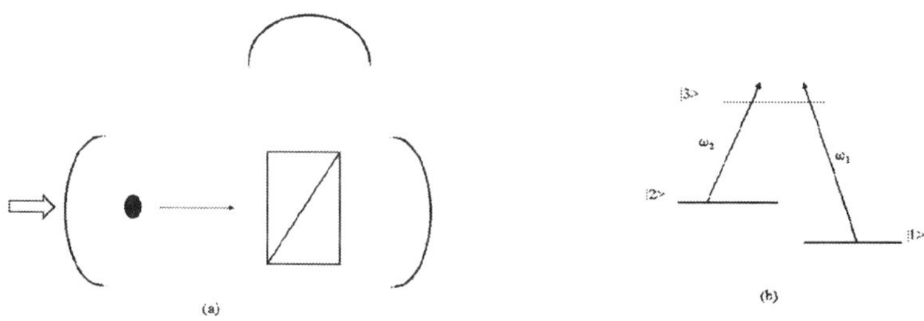

Figure 1: (a) Conceptual picture of a three level atom moving in a double cavity. (b) Level structure of the three level atom.

A three level atom is placed inside the double resonator. A classical field is injected with two orthogonally polarized components oscillating at ω_1 and ω_2. The orthogonal polarization components allow independent control of the two cavity resonances with respect to the field components. We begin with the Hamiltonian

$$H = \hbar\Delta_{21} \mid 2 \rangle\langle 2 \mid + \hbar\omega_{31} \mid 3 \rangle\langle 3 \mid - \hat{\mu}\hat{E}_{total}(z,t) + \frac{\hat{p}^2}{2m} \tag{1}$$

where

$$\hat{E}_{total}(z,t) = \hat{E}_1(z,t)e^{-i\omega_1 t} + \hat{E}_2(z,t)e^{-i\omega_2 t} + h.c. \tag{2}$$

After careful accounting of the boundary conditions imposed by the two pairs of mirrors, and after making the semi-classical approximation and taking the uniform field limit, we arrive at the final set of equations:

$$
\begin{aligned}
\frac{d}{d\tau}X_1 &= -\tilde{\kappa}_1(1+i\theta_1)X_1 + \tilde{\kappa}_1 Y_1 + 4C_1\tilde{\kappa}_1\cos(k\tilde{z})\sigma_1 \\
\frac{d}{d\tau}X_2 &= -\tilde{\kappa}_2(1+i\theta_2)X_2 + \tilde{\kappa}_2 Y_2 + 4C_2\tilde{\kappa}_2\cos(\tilde{z})\sigma_2 \\
\frac{d}{d\tau}\tilde{z} &= \tilde{\omega}_r\tilde{p} \\
\frac{d}{d\tau}\tilde{p} &= \frac{i}{2}k\tilde{W}_{31}[\sigma_1 X_1^- - \sigma_1^+ X_1]\sin(k\tilde{z}) + \frac{i}{2}\tilde{W}_{32}[\sigma_2 X_2^- - \sigma_2^+ X_2]\sin(\tilde{z}) \\
\frac{d}{d\tau}\sigma_1 &= -(\tilde{\gamma}_1 - i\tilde{\delta}_1)\sigma_1 + 2\tilde{\gamma}_1[2R_{z3} + R_{z2} - 1]X_1\cos(kz) - \frac{1}{2}\tilde{W}_{32}\sigma_{12}X_2\cos(\tilde{z}) \\
\frac{d}{d\tau}\sigma_2 &= -(\tilde{\gamma}_2 - i\tilde{\delta}_2)\sigma_2 + 2\tilde{\gamma}_2[R_{z3} - R_{z2}]X_2\cos(\tilde{z}) - \frac{1}{2}\tilde{W}_{31}\sigma_{12}^+ X_1\cos(k\tilde{z}) \\
\frac{d}{d\tau}\sigma_{12} &= -[\tilde{\gamma}_{12} - i(\tilde{\delta}_1 - \tilde{\delta}_2)]\sigma_{12} + 2\gamma_1\sigma_2^+ X_1\cos(k\tilde{z}) + 2\gamma_2\sigma_1 X_2\cos(\tilde{z}) \\
\frac{d}{d\tau}R_{z2} &= -\tilde{W}_{21}R_{z2} + \tilde{W}_{32}R_{z3} + \frac{1}{2}\tilde{W}_{32}(\sigma_2 X_2^- + \sigma_2^+ X_2)\cos(\tilde{z}) \\
\frac{d}{d\tau}R_{z3} &= -(\tilde{W}_{31} + \tilde{W}_{32})R_{z3} - \frac{1}{2}\tilde{W}_{31}(\sigma_1 X_1^- + \sigma_1^+ X_1)\cos(k\tilde{z}) - \frac{1}{2}\tilde{W}_{32}(\sigma_2 X_2^- + \sigma_2^+ X_2)\cos(\tilde{z})
\end{aligned}
\tag{3}
$$

where X_i is the ith field, σ's are polarizations, R_{zi} the population in the ith level, κ_i the ith cavity decay rate, ω_r the recoil induced Doppler shift, θ_i the ith cavity detuning, δ_i the detuning of the ith transition, Y_i the amplitude of the injected field, C_i the ith co-operativity parameter and W_{ij} the decay rate from level i to j. Here, $\tau = \gamma t$, $\tilde{z} = k_2 z$, $\tilde{p} = p/(\hbar k_2)$, and $k=k_1/k_2$. A tilde denotes scaling to an arbitrary optical frequency γ.

This model allows for the complete study of cooling in an EIT medium. Of particular interest are the cases in which the fields are near the two-photon resonance but have large single photon detuning, which results in no absorption (and little slowing of the atom) and the case in which the detuning of the two cavities are set to allow for competing heating and cooling effects. Numerical studies are in progress and will be presented.

This work has been supported by a grant from the Office of Naval Research and the Naval Air Warfare Center.

---

[1] V. Vuletic and S. Chu, "Laser Cooling of Atoms, Ions or Molecules by Coherent Scattering," Phys. Rev. Lett., **84**, 3787-3790 (2000)

[2] (a) P. Horak, G. Heckenblaikner, K.M. Gheri, H. Slecher, and H. Ritsch, "Cavity Induced Atom Cooling in the Strong Coupling Regime," Phys. Rev. Lett., **79**, 4974-4977 (1997); (b) G. Heckenblaikner, M. Gangl, P. Horak, H. Ritsch, "Cooling an atom in a weakly driven high-Q cavity," Phys. Rev. A., **58**, 3030-3042, (1998).

[3] F.A. Narducci, Z. Ye, L.M. Narducci and H-Y Ling, "Laser Cooling in a High Finesse Cavity," submitted to Phys. Rev. A.

[4] (a) see for example S.E. Harris, "Electromagnetically induced Transparency," Physics Today, **50**, 36-42 (1997) (b) M.O. Scully and S. Zubairy, *Quantum Optics*, Cambridge University Press (1997).

[5] (a) L.V. Hau, S.E. Harris, Z. Dutton, and C.H. Behroozi, "Light Speed Reduction to 17 metres per second in an ultracold atomic gas," Nature, **397**,594-598 (1999); (b) M.M. Kasj, V.A. Sautenkov, A.S. Zibrov, L. Hollberg, G.R. Welch, M.D. Lukin, Y. Rostovstev, E.D. Fry and M.O. Scully, "Ultraslow Group Velocity and Enhanced Nonlinear Optical Effects in a Coherently Driven Atomic Gas," Phys. Rev. Lett., **82**, 5229-5232 (1999).

[6] (a)C. Liu, Z. Dutton, C.H. Behroozi, and L.V. Hau, "Observation of coherent optical information storage in an atomic medium using halted light pulses," Nature, **409**, 490-493 (2001) (b) D. F. Phillips, A. Fleishhauer, A. Mair, R. L. Walsworth, M.D. Lukin, "Storage of Light in Atomic Vapor," Phys. Rev. Lett., **86**, 783-786 (2001).

Lambda propagation laws

V. V. Kozlov[(1,2)] **and J. H. Eberly**[(2)]

[(1)]*Abteilung für Quantenphysik, Universität Ulm, D-89069, Germany;*
[(2)]*Department of Physics and Astronomy and Rochester Theory Center for Optical Science and Engineering, University of Rochester, Rochester, New York 14627*

Abstract: We report new equations and solutions that govern the propagation of two unmatched non-steady state arbitrary-shaped pulses in an inhomogeneously broadened three-level lambda medium.

OCIS codes: (270.0270) Quantum Optics; (270.5530) Pulse propagation and solitons.

A wide variety of quantum optical coherence effects, as well as experimental techniques, are based on double resonance. An important theoretical challenge arises however in cases where propagation self-consistently couples the dynamics of the fields and the matter. This is where the Area Theorem discovered by McCall and Hahn [1] is so useful in analysing propagation in two-level media. The Area Theorem shows that a key parameter in temporal evolution, the tipping angle of the Bloch vector, has a role to play in spatial evolution as well. However, for double-resonance interactions there has been no general guide to self-consistent propagation. To fill the gap, we report new propagation laws suggesting a three-level area theorem [2].

Evolution equations for a generic "lambda" configuration can be written using the standard travelling-wave coordinates $\zeta = z$, and $\tau = t - z/c$, the local time in a reference frame moving with the velocity of light in the medium:

$$\frac{\partial}{\partial \zeta}\Omega_g = i\mu \langle a g^* \rangle , \qquad \frac{\partial}{\partial \zeta}\Omega_{g'} = i\mu' \langle a g'^* \rangle , \tag{1}$$

$$i\frac{\partial}{\partial \tau}\, g = -\frac{1}{2}\Omega_g^*\, a, \quad i\frac{\partial}{\partial \tau}\, g' = -\frac{1}{2}\Omega_{g'}^*\, a, \quad i\frac{\partial}{\partial \tau}\, a = \Delta a - \frac{1}{2}(\Omega_{g'}\, g' - \Omega_g\, g) . \tag{2}$$

The notation is conventional [see Fig. 1(a)]. We assume the fields are initially and remain unchirped. The angular brackets in (1) refer to Doppler averages. We do not require $\mu = \mu'$ but we treat only that case here.

The usual dressed basis follows from a rotation in g-g' space:

$$\begin{bmatrix} B \\ D \end{bmatrix} = \begin{bmatrix} \cos\theta/2 & \sin\theta/2 \\ -\sin\theta/2 & \cos\theta/2 \end{bmatrix} \begin{bmatrix} g \\ g' \end{bmatrix}, \tag{3}$$

where the dressing angle θ is defined (up to $\pm n\pi$) by $\cos\theta/2 \equiv \Omega_g/\Omega_B$ and $\sin\theta/2 \equiv \Omega_g'/\Omega_B$, and the "bright field" $\Omega_B \equiv \sqrt{\Omega_g^2 + \Omega_{g'}^2}$, and the bright-dark coupling scheme is indicated in Fig. 1(b).

Our theory asserts that the dressing angle θ is the primary dynamical variable for the system. It is already known that $\dot{\theta}$ separates non-adiabatic from adiabatic regimes, but θ has not previously been exploited as an evolution variable. Probably, the most surprising consequence is that it brings a wide range of three-level propagation dynamics under direct and simple analytical control. We introduce the "dark Rabi frequency" via the Fleischhauer-Manka relation [3]: $\Omega_D = -i(\dot{\Omega}_g\Omega_{g'} - \dot{\Omega}_{g'}\Omega_g)/\Omega_B^2$, which supplies an unambiguous formula for θ: $\theta(\zeta,\tau) = i\int_{-\infty}^{\tau} d\tau'\, \Omega_D(\zeta,\tau')$.

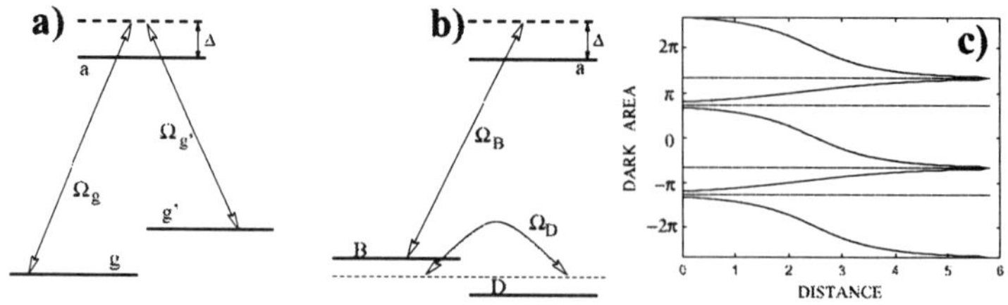

Fig. 1. Sketches of the a) bare-state and b) bright/dark-state pictures of a typical atom in a Λ-type medium. c) Dark area branches for $\alpha' = 4\alpha$, a more general case [2] than Eq. (4) for $\theta(\zeta, T)$. Distance is measured in $\alpha\zeta$ and $\phi = \pi/3$.

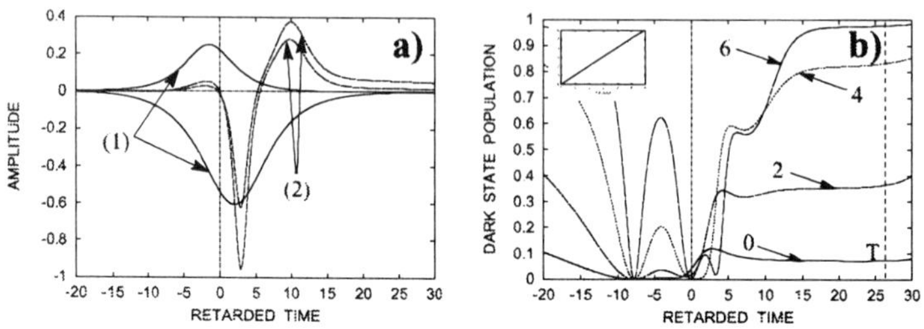

Fig. 2. a) Curves (1) show initially mismatched input pulses with areas 1.5π and 4.8π. Curves (2) show output pulses after propagating 14 Beer's lengths. b) Dark state $|D(\tau)|^2$ for $\Delta = 0$ at $\alpha\zeta = 0, 2, 4, 6$. Inset shows Eq. (5) for $\Delta = 0$ at the trailing edge ($T = 26$). Retarded time units are the same in a) and b) and $\alpha = \alpha'$.

After a "long" time T, following the passage of the pulses, the total area of the dark field obeys the nonlinear propagation law:

$$\frac{\partial \theta(\zeta, T)}{\partial \zeta} = -\frac{\alpha}{2} \sin(\theta + \phi), \tag{4}$$

where ϕ is determined by the ground-state coherence of the medium, and α is the normal inverse Beer's length. Eqn. (4) has the form of an area theorem [2]. There are infinitely many branches for the dark area θ, as indicated in Fig. 1(c). We expect shapes to change as the pulses evolve to reach a stable area with matched forms, as is seen in Fig. 2(a). Our theory also provides a new equation and solution for the ***spatial evolution*** of the dark state population on resonance:

$$\frac{\partial D^2}{\partial \zeta} = \alpha D^2(1 - D^2); \qquad \text{or} \qquad \ln \frac{1 - D^2(\zeta)}{1 - D^2(0)} = \alpha\zeta\,. \tag{5}$$

To check this prediction, we have numerically integrated equations (1) - (2) and in Fig. 2(b) the inset shows the logarithm in Eq. (5) vs. $\alpha\zeta$ on the trailing edge of the pulses. As predicted, we obtain straight line behavior.

[1] S. L. McCall and E. L. Hahn, Phys. Rev. **183**, 457 (1969).
[2] V.V. Kozlov and J.H. Eberly, Phys. Rev. Lett. (submitted).
[3] M. Fleischhauer and A. Manka, Phys. Rev. A **54**, 794 (1996).

Signal velocity and quantum fluctuation in superluminal light pulse propagation

A. Kuzmich, A. Dogariu, and L. J. Wang

NEC Research Institute, 4 Independence Way, Princeton, NJ 08540, USA

Lwan@research.nj.nec.com

P. W. Milonni

Los Alamos National Laboratory, Los Alamos, NM 87545, USA

R. Y. Chiao

Physics Department, University of California, Berkeley, CA 94720, USA

Pulse propagation with a superluminal group velocity is a rather counterintuitive phenomenon that at first might appear to violate the special theory of relativity's pillar statement that no signal can be transmitted faster than c. However, it has been argued by Sommerfeld and Brillouin [1] and discussed later in a series of works by Chiao et al. ([2] and references therein) that the faster-than-c propagation is perfectly in accord with special relativity. The reason, they noted, is that the signal velocity should be properly defined as the velocity of a step-function-like signal, not that of a smooth pulse. While this argument is certainly a correct one, it is based on the classical wave theory. On the other hand, the phenomenon of gain-assisted superluminal pulse propagation [3] can be studied in the situations when a pulse consists of only a few photons, or when the number of photons is large, but the sensitivity of the light detection system is at or better than the shot photon noise level. In both situations, quantum mechanical treatment of both the light and the medium is necessary to provide a meaningful description of the system's dynamics. Here we report the results of such a study [4].

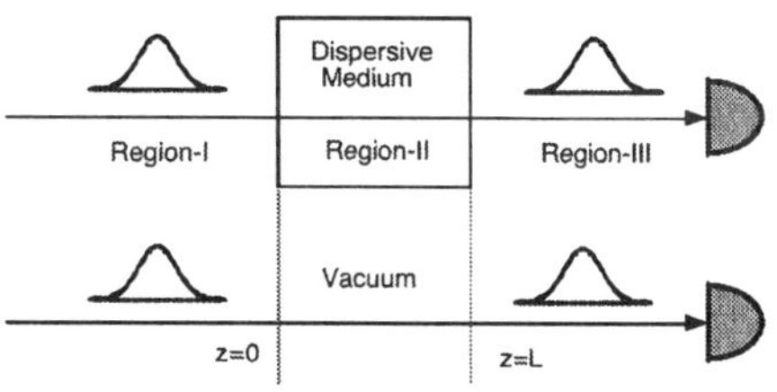

Fig.1, Pulse propagation through a transparent anomalous dispersion medium of a length *L. Pulse propagation through the same length in vacuum is also shown for comparison.*

Fig.2 Signal-to-noise ratios for light pulses propagating through the gain-assisted anomalous dispersion medium SNRmed(L,t), and through the same distance in a vacuum SNRvac(L,t).

We consider the experimental situation illustrated in Fig.1. We suggest here an operational definition of the signal velocity and apply it to the recently observed superluminal propagation of a light pulse in a gain medium [3]. We assign a time window T centered about a pre-arranged time t0 at the detector and monitor the photocurrent produced by the detector. The observable that we base our definition on is the integrated photon number

$$S(L,t)=\int_{-T/2}^{t} E^{(+)}(L,t_1)E^{(-)}(L,t_1)dt_1 .$$

E(+) and E(-) are respectively the positive- and negative-frequency parts of the reduced electric field operator at the exit point (z=L) of the medium and T/2 is half the time window assigned to the pulse, typically a few times the pulse width. The second-order variance of the integrated photon number gives the noise power due to quantum fluctuations. Hence we can define an optical signal-to-noise ratio SNR(L,t) following Ref.[4,5]. We define the arrival time ts of a signal as the time at which SNR(L,t) reaches a prescribed threshold level.

Coherence and Quantum Optics VIII

Edited by Bigelow et al., Kluwer Academic/Plenum Publishers, 2003

Using parameters appropriate to the experiments [3], we compute SNRmed(L,t) for the pulse that propagates through the anomalously dispersive medium [4]. In Fig.2 we plot the results of such computations for SNRmed(L,t) as a function of time on the output signal. For reference we also show SNR for the identical pulse propagating over the same length in vacuum. It is evident from the results shown that the pulse propagating in vacuum always maintains a higher SNR, so that the actual arrival time of the signal is delayed, even though the pulse itself is advanced compared with propagation over the same distance in vacuum.

By requiring that at some time t' the SNR of the output pulse be equal to that of the input pulse at a time t, we obtain a time difference t=t'-t that marks the propagation time of the light signal, and L/ t gives the signal velocity. In Fig.3 we plot t as a function of gain for (t-Tc)/ = -3 and -1. This corresponds to cases where the signal point is set at 3 and 1 times the pulse width on the leading edge of the pulse. We also plot for reference the pulse advance L/vg. It is evident that the retardation in the SNR far exceeds the pulse advancement.

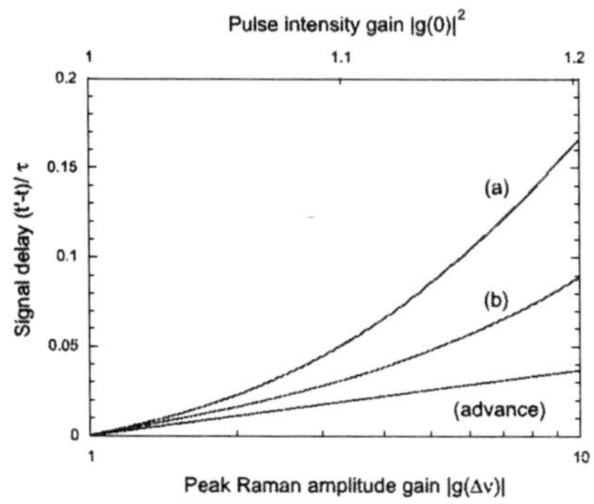

Fig.3, Delay $\delta t = t' - t$ due to reduced signal-to-noise ratio as a function of the gain coefficient. Curves (a) and (b) are for $t/\tau = -3$ and -1, respectively. Curve (a) is delayed more because at the early stage of the pulse, amplified spontaneous emission (ASE)-self beat noise produces noise much greater than the shot noise level.

References

[1] L. Brillouin, *Wave Propagation and Group Velocity*, (Academic, New York, 1960).

[2] R. Y. Chiao, *in Amazing Light: A Volume Dedicated to Charles Hard Townes on His 80th Birthday*, edited by R. Y. Chiao
(Springer-Verlag, New York, 1996), p. 91.

[3] L. J. Wang, A. Kuzmich, and A. Dogariu, Nature **406**, 277-279 (2000).

[4] A. Kuzmich, et al. Phys. Rev. Lett. (to be published).

[5]E. Desurvire, *Erbium-Doped Fiber Amplifiers: Principles and Applications*, (Wiley, New York, 1994), Chapter 2.

Slow light in a solid without using electromagnetically induced transparency

Matthew J. Sellars and Neil B Manson
Laser Physics Centre, Research School of Physical Sciences and Engineering, Australian National University, Canberra, A. C. T., 0200, Australia.
Tel: 61 261254204, Fax: 61 261250029, e-mail: Neil.Manson@anu.edu.au

Abstract:
A spectral hole is burnt in a sample of Eu^{3+} doped Y_2SiO_5 and measured using a short light pulse. For a 4 mm thick sample the transmission is reduced from 80 % to 10% over a bandwidth of 2 kHz and the associated dispersion corresponded to a light speed of 60 m/s.

Introduction:
To observe steep normal dispersion and slow group velocities it is necessary to generate a narrow spectral region of low absorption within a broad absorption line with high optical density. In the recent demonstrations of slow light the narrow spectral feature has been prepared by electromagnetically induced transparency (EIT) [1, 2]. By employing EIT they have been able to generate spectral features significantly narrower than the natural line width of the optical transition. In solids the situation can be different as the line widths of the transmission feature obtained using EIT are determined by inhomogeneous effects and in the case of rare earth ions these can be larger than the homogeneous line widths of the optical transitions. However, spectral hole burning in solids can give transmission features of the order of the homogeneous line width and result in slower group velocities than could be achieved with EIT. This is illustrated in the present work where a narrow spectral transmission is obtained with spectral hole burning and measurement shows it gives rise to the slowest group velocity reported for a solid.

Experiment:
In this work a crystal of 0.5% Eu^{3+} doped Y_2SiO_5 was used. It has been shown that for a site in the crystal at low temperature the Eu^{3+} $^7F_0 \Leftrightarrow {}^5D_0$ transition at 579.9 nm has a dephasing time of 2.6 ms [3]. This corresponds to a transition with a line width of 100 Hz whereas the transition has an inhomogeneous linewidth of 10 GHz. The isotopes of europium have a nuclear spin of 5/2 and second order hyperfine and electric quadrupole interactions result in splitting of both the ground and excited state into three doubly degenerate levels with splittings in the range of 20 – 200 MHz. Optical excitation at 579.9 nm redistributes the ground state hyperfine populations and gives a spectral hole which last for several hours. The optical density at line centre was 0.8 for a sample 4 mm thick. The minimum width of the spectral hole would be 200 Hz (twice the homogeneous line width) and if 100% deep corresponds to an achievable group velocity of 0.6 m/s.

The spectral holes were burnt using a modified Coherent 699-21 ring dye laser with a frequency stability of the order of 100 Hz over time periods of 100 ms. Measurements involved a Mach-Zehnder interferometer. The reference beam was shifted in frequency by 10 MHz and the heterodyne signal was detected in two orthogonal phases to enable simultaneous determination of absorption and dispersion [4]. The spectral hole was prepared by exposing the sample to a 3 nW beam for 100 ms. 10 ms after the hole was excited with a 20 μs long 50 nW probe pulse with an area of $\pi/100$. The complex amplitude of the transmission of the probe was recorded and spectrum generated by taking the Fourier transform (figure 1). The spectral hole has a width of 2 kHz and the transmission reduced from 80% to 10%.. The peak to peak change in refractive index across the hole is 2.0×10^{-5}

Coherence and Quantum Optics VIII
Edited by Bigelow et al., Kluwer Academic/Plenum Publishers, 2003

and the dispersion at the centre was $dn/d\nu = 1.0 \times 10^{-8}\ Hz^{-1}$. This dispersion corresponds to a group velocity of 60 ± 5 m/s.

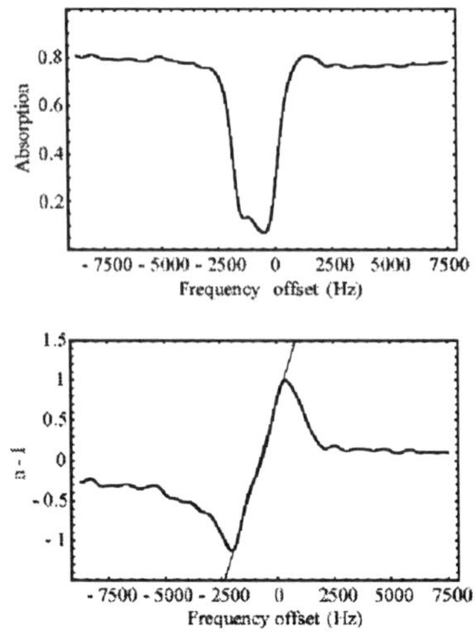

Figure 1. Spectrum of hole generated by taking Fourier transform of transmitted pulse's complex amplitude

By reducing the burn pulse duration it was possible to reduce the width of the hole at the expense of the hole depth. For example, with a 8 ms long pulse the hole width was reduced to 800 Hz and the hole depth reduced to 4 %. The larger width with the longer burn pulse can be attributed to instabilities in the laser frequency accompanied by saturation in the hole burning. Thus, slower group velocities may be achieved with improvement of laser stability. However, it may not be possible to attain the theoretical limit for the present sample as the effects giving the 800 Hz line width remains to be established. Another way of increasing the dispersion across the hole is to increase the optical density of the sample. However, Eu^{3+} doped Y_2SiO_5 have the property that further increase of the concentration above 0.1% increases the inhomogeneous line width without an increase in the peak optical density. Therefore, reduction in the group velocity cannot be achieved by using higher concentration.

Discussion:

EIT has not been demonstrated in Eu^{3+} doped Y_2SiO_5 but the linewidth will be given by the inhomogeneous linewidth of the ground state hyperfine transitions and in similar materials this is of the order of 10's kHz. The hole depth will be restricted due to the inhomogeneous broadening of the hyperfine transition and the inhomogeneous broadening of the optical transition. Thus, EIT is unlikely to give a reduced group velocity as significant as that obtained here using spectral holeburning. It is recognized that the narrow features generated in this way do not have the same applications as those obtained by EIT but there are complimentary optical processing applications such as time-domain optical memories [5] which use spectral hole burning.

References:

[1]. L. V. Hau, S. E. Harris, Z. Dutton, and C. H. Behroozi, *"Light speed reduction to 17 metres per second in an ultracold atomic gas"*. Nature **397** 594-598 (1999).

[2]. M. M. Kash, V. A. Sautenkov, A. S. Zibrov, L. Hollerg, M. Scully, *"Ultraslow group veleocity and enhanced nonlinear optical effects in a coherently driven hot atomic gas"*. Phys. Rev. Lett. **82** 5229-5232 (1999).

[3]. R.W. Equal, Y. Sun, R. L. Cone and R. M. Macfarlane, *"Ultraslow optical dephasing in Eu^{3+} : Y_2SiO_5"*, Phys. Rev. Lett., **72** 2179 – 2182 (1994).

[4]. T. R. Dyke, M. J. Sellars, G. Pryde, N. B. Manson, U. Elman, S. Kroll, *"Experimental demonstration of data erasure for time domain optical memories"*, J of Opt. Soc. Am. B **16** 805-811 (1999).

[5]. T. W. Mossberg, *"Time-domain frequency-selective optical data storage"*, Opt. Lett. 7 77-79 (1982).

Electromagnetically induced transparency with standing-wave drive in the frequency up-conversion regime

F. Silva
Departament d'Òptica, Universitat de València, E-46100 Burjassot, Spain

J. Mompart, V. Ahufinger and R. Corbalán
Departament de Física, Universitat Autònoma de Barcelona, E-08193, Bellaterra, Spain
Telf: (34) 93 581 16 53, Fax: (34) 93 581 21 55, e-mail address: Ramon.Corbalan@uab.es

Abstract:
We study Electromagnetically Induced Transparency in Doppler-broadened three-level media for probe to drive field frequency ratio R>1 and moderate drive intensities. We show that in this regime a standing-wave drive is more efficient than a travelling-wave one, specially when R is close to half integer values.

Summary

Linear and nonlinear optical properties of an atomic medium can be controlled and modified by applying a strong "drive" laser field that induces atomic coherence. Atomic coherence manifests itself in a rich variety of phenomena that are a subject of current research interest, such as Electromagnetically Induced Transparency (EIT); amplification and lasing without inversion; steep dispersion and either ultraslow group velocity or gain-assisted superluminal light propagation; and enhanced nonlinear optical susceptibility (for reviews see, for instance [1-5]). Here we consider EIT, which is among the most basic of these phenomena and has many useful applications, ranging from laser cooling of free atoms to subrecoil temperatures [6] to storage of light [7].

In EIT, an otherwise absorbing medium is made transparent to a coherent field on resonance with a certain atomic transition by applying an intense coherent drive field to an adjacent transition. Up to now, most papers dealing with EIT in gas media have considered only the case of a traveling-wave (TW) drive field. In this contribution we study EIT in Doppler-broadened three-level media with a drive field composed by two equal-frequency counterpropagating TW fields. When these two components have equal intensities one has a standing wave (SW) drive. We also consider the most general case with different intensities for both TW components, which we call for simplicity asymmetric SW drive. Our analysis is focused on the advantages of using a (asymmetric) SW drive configuration instead of a TW one in the so-called frequency up-conversion regime where the probe wavelength is substantially shorter than that of the drive laser.

By using standard semiclassical density matrix formalism and the dressed-atom approach we show that with SW driving electromagnetically induced probe transparency windows appear close to half integer values of the probe to drive field frequency ratio R. Many real three-level systems exist with $R \approx n+1/2$ (n=1,2,3,...). Restricting ourselves to the most interesting case of moderate drive intensities (the Rabi frequencies involved are smaller than the Doppler width of the driven transition) we show that for R>1 a SW drive field is appreciably more efficient than a TW drive in inducing probe transparency, specially when R is close to half integer values. In particular, we have considered parameters for a real closed cascade scheme in Barium atoms with R=1.48 obtaining probe transmission resonances with values above 50% when the transmission is almost negligible either without drive field or with only one of the TW components of the SW drive. We have shown that an asymmetric SW drive field with two counterpropagating components of different intensity is even more efficient than a SW one in inducing probe transparency. In ladder schemes, to achieve optimum probe transmission with a given total drive intensity, the drive component counterpropagating with the probe should be stronger. In addition to the Ba case, we have considered another system in Beryllium atoms with R=3.51, obtaining transmission resonances above 80%.

A short account of some of the results to be presented at the Conference was published recently [8].

[1] M. O. Scully and M. S. Zubairy, Quantum Optics, (Cambridge Univ. Press., 1997) Chap. 7, pp. 220-245.
[2] E. Arimondo, Coherent population trapping in laser spectroscopy, in Progress in Optics XXXV, Ed. E. Wolf (Elsevier, Amsterdam 1996) Chap. V, pp. 258-354.
[3] S. E. Harris, Electromagnetically Induced Transparency, Phys. Today **50** (7), 36 (1997).
[4] J. P. Marangos, Topical review: Electromagnetically Induced Transparency, J. Mod. Opt. **45**, 471 (1998)
[5] J. Mompart and R. Corbalán, Lasing without inversion, J. Opt. B: Quantum Semiclass. Opt. **2**, R7-R24 (2000).
[6] A. Aspect, E. Arimondo, R. Kaiser, N. Vansteenkiste, and C. Cohen–Tannoudji, Laser cooling below the one-photon recoil by velocity-selective coherent population trapping , Phys. Rev. Lett. **61**, 826 (1988).
[7] Ch. Liu, Z. Dutton, C.H. Behroozi, and L.V. Hau, Observation of coherent optical information storage in an atomic medium using halted light pulses, Nature, **409**, 490 (2001); D.F. Phillips, A. Fleischhauer, A. Mair, R.L. Walsworth, and M.D. Lukin, Storage of light in atomic vapor, Phys. Rev. Lett. **86**, 783 (2001).
[8] F. Silva, J. Mompart, V. Ahufinger and R. Corbalán, Electromagnetically induced transparency in Doppler broadened three-level systems with resonant standing wave drive, Europhys. Lett. **51**, 286 (2000).

Pulse propagation in a high-gain bichromatically-driven Raman amplifier

Daniel J. Gauthier and Michael D. Stenner

Duke University, Department of Physics and The Fitzpatrick Center for Photonics and Communication Systems, Durham, NC 27708

Using intense electromagnetic fields, it is now possible to tailor the absorption, amplification, and dispersion properties of multi-level atoms [1]. Dramatic examples include the reduction of the group velocity of a pulse of light to 17 m/s [2] and increasing the group velocity v_g to values greater than c or even to negative values [3]. Very recently, pulses of light have been effectively stopped in an atomic medium and the information initially contained in the pulse is encoded reversibly in the atomic coherence [4].

In the recent experiment on 'fast' light [3] (defined as $v_g > c$ or $v_g < 0$), an alkali atom is driven by an intense bichromatic optical field that creates a gain doublet via the stimulated Raman process. Pulses of light are injected into the vapor where the spectrum of the pulse coincides with the spectral region of large anomalous dispersion between the doublet and where the size of the optical amplification is small [5]. Using a simple model, we find that the advancement of the pulse t_{adv} in comparison to its propagation time through vacuum is given by

$$\frac{t_{adv}}{\tau_p} \approx 0.13gL, \tag{1}$$

where τ_p is the pulse width, g is the line-center unsaturated gain coefficient, and L is the length of the medium. To obtain Eq. 1, we adjust the separation between the gain peaks and the bandwidth of the pulse to their optimum values. In the experiments of Wang *et al.* [3], gL was small, resulting in a small relative pulse advancement.

We describe our recent experimental observations of pulse propagation through a bichromatically driven potassium vapor in a regime where gL is large [6]. We observe several new effects that occur in this high-gain regime. One observation is that t_{adv}/τ_p is limited by a convective instability initiated by quantum fluctuations. When $gL \approx 15$ [7], quantum fluctuations at the input face of the atomic vapor cell lead to the generation of a macroscopic field that obscures the transmitted pulse and eventually depletes the pump beam power in the very high gain limit. This process of superfluorescence limits $t_{adv}/\tau_p \approx 2$.

Another observation is that the behavior of a bichromatically-driven Raman amplifier differs dramatically from its monochromatic counterpart. We observe that the bichromatic driving does more that just create a Raman gain doublet, as assumed in Ref. [3]. In addition to the Raman gain process, the bichromatic field drives four-wave mixing processes that occur even for low pulse intensities, but whose effects become most apparent when gL is large. These four-wave mixing processes lead to a dramatic broadening of the spectrum of the pulse as it propagates through the dispersion-tailored medium. The spectrum of the transmitted pulse consists of a comb of frequencies centered on pulse carrier frequency and a spacing equal to the frequency splitting of the bichromatic driving field [8]. Our observations can be motivated by an analysis of the interaction where the atom is dressed by the bichromatic field [9].

Surprisingly, we observe that the optical pulse still experiences advancement even in the presence of these complicating effects as it propagates through the high-gain amplifier. The figure shows our experimental observations for the case when $gL = 9.5$ and the splitting of the bichromatic driving field is 24 MHz. The dashed line corresponds to the case when the pulse propagates through vacuum; it is nearly gaussian-shaped with a duration of 184 ns. The solid line shows the case when the atoms are present. It is seen that the pulse envelope is advanced and it is modulated at 24 MHz due to the creation of new frequencies via the four-wave mixing processes.

We will discuss the mechanisms responsible for the four-wave mixing and superfluorescence processes and pulse propagation through high-gain, bichromatically-driven Raman amplifiers.

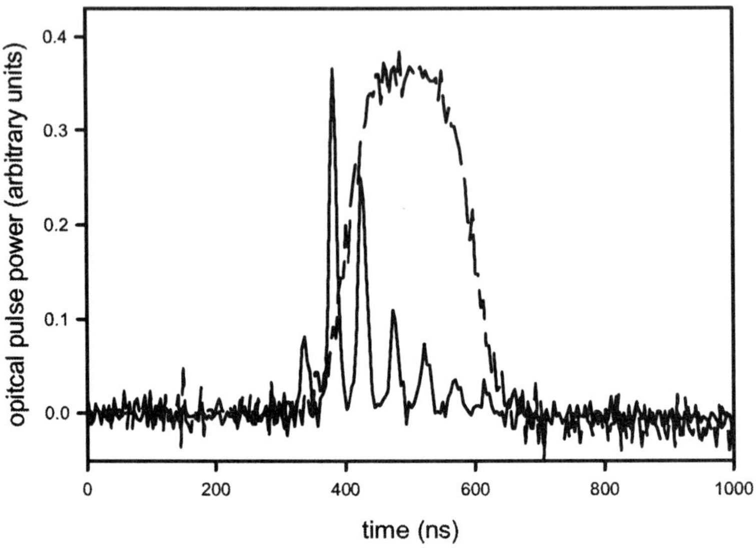

Fig. 1. Pulse propagation through vacuum (solid line) and through a high-gain bichromatically-driven Raman amplifier.

We gratefully acknowledge the financial support of the National Science Foundation through grant PHY-9876988.

References and links

1. S.E. Harris, Phys. Today **50**, 36 (1997).
2. L.V. Hau, S.E. Harris, Z. Dutton, and C.H. Behroozi, Nature **397**, 594 (1999).
3. L.J. Wang, A. Kuzmich, and A. Dogariu, Nature **406**, 277 (2000).
4. C. Liu, Z. Dutton, C.H. Behroozi, and L.V. Hau, Nature **409**, 490 (2001); and D.F. Phillips, A. Fleischhauer, A. Mair, R.L. Walsworth, and M.D. Lukin, Phys. Rev. Lett. **86** 783 (2001).
5. A.M. Steinberg and R.Y. Chiao, Phys. Rev. A **49**, 2071 (1994).
6. H.M. Concannon, W.J. Brown, J.R. Gardner, and D.J. Gauthier, Phys. Rev. A **56**, 1519 (1997).
7. R.W Boyd, *Nonlinear Optics* (Academic Press, Boston, 1992), Chs. 8 and 9.
8. A.V. Sokolov, D.R. Walker, D.D. Yavuz, G.Y. Yin, and S.E. Harris, Phys. Rev. Lett. **85**, 562 (2000).
9. Y. Zhu, Q. Wu, A Lezama, D.J. Gauthier, and T.W. Mossberg, Phys. Rev. A **41**, 6574 (1990).

Superluminal nonlinear anti-echos

J. M. Hickmann† and Raymond Y. Chiao

Department of Physics, University of California
Berkeley, California 94720-7300
phone: 15106427166, fax: 15106438497, e-mail: chiao@physics.berkeley.edu

OCIS codes: (270.1670) Coherent optical effects; (190.4380) Nonlinear optics, four-wave mixing; (270.4180) Multiphoton processes

Abstract: We investigate the possibility of a superluminal photon anti-echo in a degenerate four-wave mixing configuration when two pulses with different carrier frequencies collide in a nonlinear sample.

An experiment recently performed at NEC Princeton by Wang *et al.* [1] has shown that laser pulses can travel through a cesium vapor cell with group velocities so much faster than the vacuum speed of light c that they become *negative*. The meaning of negative group velocities is that the peak of the output pulse leaves the exit face of the cell *before* the peak of the input pulse arrives at the entrance face[2]. These results do not imply that the theory of special relativity has been overthrown, but only that special relativity does not imply that group velocities can never exceed c, contrary to common belief.

In an earlier *nonlinear* optics experiment[3], it was observed that a sequence of two strong laser pulses, peaking at times $t_1 < t_2$, generate, in the presence of a Kerr nonlinearity and spatial dispersion, not only a photon-echo-like pulse at the *later* time $2t_2 - t_1$, which *follows* the two strong pulses, but also a photon "anti-echo" pulse at the *earlier* time $2t_1 - t_2$, which *precedes* the two strong pulses. It was noted that the "anti-echo" pulse was just as intense as the echo pulse. The experiment employed a complicated method in which the temporal pulses were first converted into spatially separated frequency components by means of a grating. Four-wave mixing in a Kerr medium then generated new spatial sidebands, and these spatial sidebands were converted back to temporal sidebands by means of a second grating. However, it was pointed out that the Kramers-Kronig relations could not be applied to this nonlinear situation, and they therefore called causality into question.

We want to show that causality is not violated, analyzing a simplified version of their experiment in which two laser pulses directly propagate in a dispersive and nonlinear medium. These pulses directly generate echoes and anti-echoes from a temporal index grating created by the collision of the two pulses inside the nonlinear sample. Thus our idea has no need of any temporal-to-spatial conversions and back-conversions by means of gratings.

The physics of the generation of an "anti-echo" in addition to an "echo" in four-wave mixing can be most easily understood by viewing four-wave mixing as a phase-conjugation process [4]. Recall that via a fast $\chi^{(3)}$ nonlinearity of a medium pumped by two strong pump waves E_{pump1} and E_{pump2}, a weak probe wave E_{probe} produces a polarization which is the *complex conjugate* of the injected probe wave

$$P_{conjugate} = \chi^{(3)} E_{pump1} E_{pump2} E^*_{probe} \tag{1}$$

where, assuming phase matching, there results a radiated conjugate field $E_{conjugate}$ produced by this polarization, which is proportional to the *complex conjugated* probe field E^*_{probe}. This conjugate wave is produced in a different (i.e., conjugate) direction from the probe wave, in the case of noncollinear, monochromatic, phase-matched plane waves. Under complex conjugation, time reverses sign, so that $t \rightarrow -t$. Thus the complex-conjugated field represents a *time-reversed*, or *wavefront-reversed* copy of the injected probe field (see Fig. 1). Now consider a more general situation in which the outgoing monochromatic plane waves are modulated by a slowly-varying envelope, which can be represented in the frequency domain by a Fourier integral over frequency-dependent phase factors $\exp(i\phi(\omega_i))$, where the index i labels the probe and conjugate fields produced by the four-wave mixing process. In this way, one can describe the envelopes of the outgoing time-varying pulses. It follows from Eq. (2) that under complex conjugation, and hence phase conjugation through the four-wave mixing process,

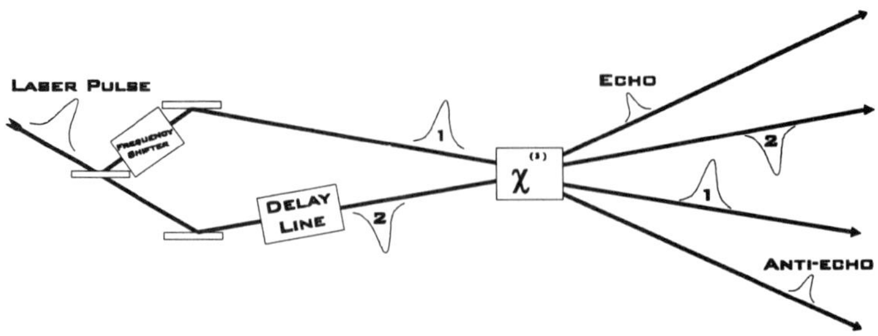

Fig. 1. Schematic setup

$$\exp(i\phi(\omega_{conjugate})) \rightarrow \exp(-i\phi(\omega_{probe})), \tag{2}$$

so that $\phi(\omega_{conjugate}) \rightarrow -\phi(\omega_{probe})$ in four-wave mixing. However, in the stationary phase approximation, the peak of any outgoing scattered pulse is given by the Wigner formula for the "group delay" [5], viz.,

$$\tau_{group} = \left[\frac{d\phi(\omega)}{d\omega}\right]_{\omega_{carrier}}. \tag{3}$$

In four-wave mixing, the group delay for the conjugate beam is reversed in sign relative to the group delay of probe beam, so that a group *delay* for the probe pulse becomes a group *advance* for the conjugate pulse

$$\tau_{group}(\omega_{\text{conjugate}}) = \left[\frac{d\phi(\omega)}{d\omega}\right]_{\text{conjugate}} = -\left[\frac{d\phi(\omega)}{d\omega}\right]_{\text{probe}} = -\tau_{group}(\omega_{\text{probe}}). \tag{4}$$

Therefore, an injected probe wave, which produces an "echo" wave with a *positive* group delay, also produces an "anti-echo" pulse corresponding to a conjugate wave with a *negative* group delay. For two equally strong pump pulses and a fast Kerr nonlinearity, the probe "echo" pulse is equal in amplitude to the conjugate "anti-echo" pulse.

However, it is clear that if there is a front for the earlier of the two strong pump pulses, before which the electric field is strictly zero, which we shall call a "Sommerfeld front", then by inspection of Eq. (1), even if one assumes an *instantaneous* response for $\chi^{(3)}$, the polarization of the conjugate wave will also be strictly zero. Therefore we see that although the Kramers-Kronig relations may not be valid for nonlinear responses, causality cannot be violated, contrary to what was previously alleged[3]. We believe that it is possible to demonstrate this experimentally by imposing a sharp front onto the earlier of the two strong pulses used in a degenerate four-wave mixing setup, and thus observe that the limit set on the "anti-echo" advancement is given by the Sommerfeld front.

†On leave from the Department of Physics, Universidade Federal de Alagoas, Maceio - AL - Brazil with financial support from CNPq.

References

1. L. J. Wang, A. Kuzmich, and A. Dogariu, Nature **406**, 277 (2000).
2. R. Y. Chiao, Phys. Rev. A **48**, R34 (1993);"Population Inversion and Superluminality," R. Y. Chiao in *Amazing Light: A Volume Dedicated to Charles Hard Townes on his 80th Birthday*, edited by R. Y. Chiao (Springer Verlag, New York, 1996), p. 91.
3. V. L. Da Silva, Y. Silberberg, J. P. Heritage, Optics Letters, **18**, 580 (1993).
4. R. A. Fisher, *ed.*, *Optical Phase Conjugation* (Academic Press, New York, 1983).
5. E. P. Wigner, Phys. Rev. **98**, 145 (1955).

On the collisional cooling of co-trapped atomic and molecular ions by ultracold atoms: Ca^+ + Na and $Na_2^+(v^*,J^*)$ + Na.

W.W. Smith, E. Babenko, R. Cote' and H.H. Michels
Department of Physics, University of Connecticut, Storrs, Connecticut 06269-3046
winthrop.smith@uconn.edu

Abstract:

Calculated potentials for collisions of one-electron Ca^+ ions with co-trapped Na atoms suggest the excited (triplet sigma +) state of (NaCa)+ is metastable with respect to charge transfer. Measurements of the elastic cross sections for cooling of Ca+ and of internally hot Na molecular ions by ultracold Na are planned.

New experiments are being initiated to investigate the regime of cold atomic and molecular ion-atom collision processes in a special hybrid neutral-atom--ion trap under high vacuum collisions [1]. The idea is to use the techniques of laser cooling, trapping and manipulation of cold neutral atoms to make a refrigerator for the sympathetic cooling of an overlapping sample of atomic or molecular ions to very low temperatures. We intend to probe fundamental collision processes and cooling kinetics between ultracold trapped neutral atoms and an overlapping sample of atomic or molecular ions in as low a temperature range as possible. An example of such processes is resonant and near-resonant charge transfer, for which the semiclassical cross section has been predicted to diverge at low velocities (although the quantal cross section for s-wave scattering will be finite). The hybrid trap we are building consists of a conventional magneto-optic neutral atom trap (MOT) combined with a linear Paul radiofrequency quadrupole ion trap. The linear Paul trap, with longitudinal electrostatic confinement, is chosen so as to suppress most of the micromotion and r.f. heating associated with conventional Paul traps [2]. Normal modes of excitation of laser-cooled Wigner crystal ion strings have recently been observed in such traps[3]. Studies of the type we are undertaking may find use, for example in estimating decoherence due to background gas collisions in trapped-ion quantum computing schemes as well as trapped ion frequency standards.

Recently after discussions with one of us (WWS), Coté and Dalgarno undertook some theoretical studies of ultracold atom-ion collisions, for the particular case of Na+ + Na collisions [4]. They found the elastic and charge transfer cross sections to be very large - ~106 atomic units (ao2) at millikelvin temperatures, orders of magnitude larger than ultracold neutral-neutral elastic cross sections as the same temperature because of the long range (e.g. R-4) polarization potential. Their calculations span the transition from the semiclassical Langevin regime to the fully quantal regime, dominated by s-wave scattering at the lowest energies, not accessible at present. For example, the elastic scattering cross section for Na+ + Na collisions was found to follow an E-1/3 law in the semiclassical regime where numerous partial waves contribute, given by 4171E-1/3 in atomic units. These cross sections are large enough so that collisions and cooling can be expected to occur even in the high vacuum of a hybrid trap (neutral density ~109-1010 atoms/cm3).

For the experiments, it is desirable that the trapped ions can be initially laser cooled,

Coherence and Quantum Optics VIII
Edited by Bigelow et al., Kluwer Academic/Plenum Publishers, 2003

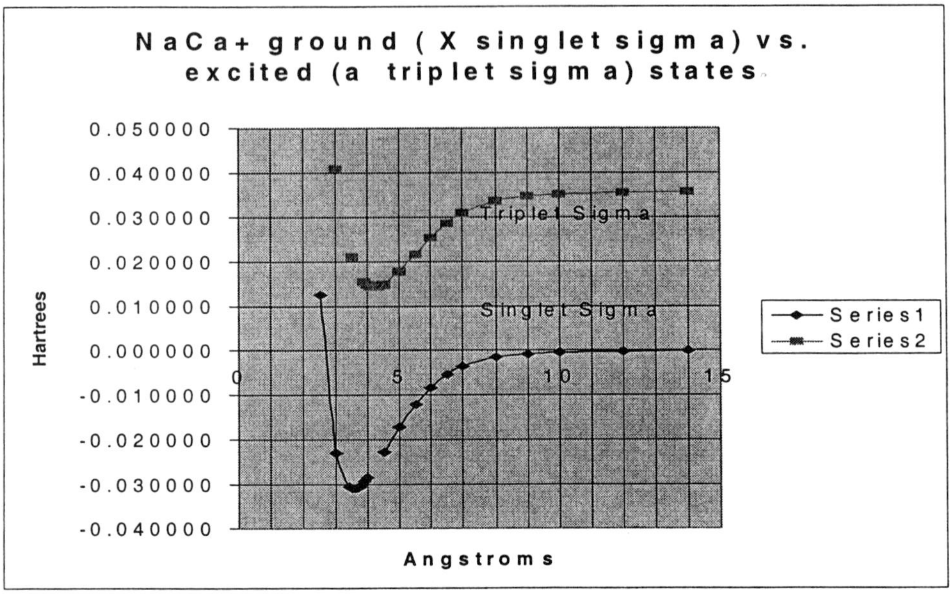

Fig. 1. *Ab initio* Second-order Mollar-Plesset perturbation calculation by H. H. Michels of the adiabatic potential curves for the selected states of the $NaCa^+$ molecular system. Basis set employed is a triple zeta including diffuse plus d & f polarization functions.

which is not really feasible for closed-shell Na^+ ions. For this reason, we are gearing up to study the collisional cooling of laser pre-cooled Ca^+ ions by ultracold Na atoms. To calculate the elastic cross section for cold Ca^+ - Na collisions we have performed *ab initio* calculations of the adiabatic potential curves for ($NaCa^+$) molecular system in the X $^1\Sigma^+$, a $^3\Sigma^+$ and the A $^1\Sigma$ states, which go asymptotically to the ground (Na^+ + Ca) and excited (Ca^+ + Na(3s)) states of the ion-atom quasimolecular system. These calculations imply a similarly large value of the elastic scattering cross section, again ~$E^{-1/3/}$ as in the Na^+ + Na case [4], demonstrating the feasibility of collisional cooling of co-trapped Ca^+ ions by ultracold Na. One might wonder if the excited state of the Ca^+ + Na system would relax to the ground state (Na^+ + Ca), which is almost 1 eV lower. Our calculations have demonstrated, however (see Fig. 1) that the X (spin singlet) and a (spin triplet) states *do not mix* for low energy collisions; furthermore the system should be *metastable* against radiation since a spin-flip is forbidden for electric dipole radiation. Thus, we expect only *elastic scattering* and metastability vs. charge exchange.

A second planned experiment, for which we have no theoretical calculations at present, is to produce translationally cold but vibrorotationally hot (by ~0.7 eV) Na_2^+ *molecular* ions by photoassociative ionization via Na(3p) + Na(3d) collisions [5]. These are *off resonance* for Franck-Condon photodissociation and we plan to study the *internal state cooling* of the dimer ions by collisions with co-trapped, ultracold Na atoms, which can be detected because the cooling will bring them into resonance for dissociation by the trapping 589 nm light.

References

[1] W. W. Smith, "Cold Atomic and Molecular Ion-Atom Collisions in a Hybrid Trap", proposal 9/15/99 to the National Science Foundation.

[2] See, for example, M. G. Raizen, J. M. Gilligan, J. C. Bergquist, W. M. Itano, and D. J. Wineland, "Ionic crystals in a linear Paul trap," Phys. Rev. A **45**, 6493 (1992).

[3] H. C. Naegerl, D. Leibfried, F. Schmidt-Kaler, J. Eschner and Rainer Blatt, "Coherent excitation of normal modes in a string of Ca+ ions," Optics Express **3**, 89 (1988); see also references therein.

[4] R. Coté and A. Dalgarno, "Ultracold atom-ions collisions," Phys. Rev. A **62**, 012709 (2000).

[5] E. Babenko, C. Tapalian and W. W. Smith, "Associative ionization in laser-excited sodium 3p + 3d collisions," Chem. Physics Letters **244**, 121 (1995). See also C. Tapalian and W. W. Smith, "Resonant collisional dissociation of Na_2^+ by Na(3p) in an effusive beam," Phys. Rev. A **49**, 921 (1994).

Mandel's Q-parameters for the outcoming atom statistics of a micromaser*

R.K. Bullough[(3)], S. Arun Kumar[(1)], and R.R. Puri[(2)]

[(1)] *Formerly at the Institute of Mathematical Sciences, C.I.T. Campus, Tasamani, Chennai- 600 113, India.*

[(2)] *Theoretical Physics Division, Central Complex, Bhabha Atomic Research Centre, Mumbai -400 085, India.*

[(3)] *Dept. of Mathematics, University of Manchester Institute of Science and Technology, PO Box 88, Manchester M60 1QD*

tel. +44-161-200-3672; fax. +44-161-200-3669; email. robin.bullough@umist.ac.uk

*Paper dedicated to Leonard Mandel.

We have studied the statistics of the atoms emerging from the cavity of a micromaser in a dynamical discrete-time 'stroboscopic' description which takes into account the measurements made, in general with imperfect efficiencies $\eta < 1$, on the states of the outcoming atoms. Inverted atoms enter stochastically, in general with a binomial distribution in discrete time; but we also consider the continuous time limit of this input statistics which is Poissonian. We calculate probability distributions for the states of the atoms emerging from the cavity and use them to calculate the variances of these distributions taken specifically in Mandel's Q-parameter form [1]. Likewise we calculate the variance Q_f for the cavity field.

In order to extract as much information as it may be possible to obtain about the state of the cavity field from observations made only on the states of the outcoming atoms we envisage two alternative experimental procedures: one of these is to consider a fixed number N of atoms pumped into the cavity and subsequently leaving it to undergo state detection; the other is to consider input of the excited atoms and their subsequent detection and collection in a fixed time $t \geq T_c$. Here T_c is the cavity damping time. We consider in particular steady state behaviours achieved in the two limits $N \to \infty$ and $t \to \infty$, as well as the approaches to these two limits. Although these limits are the same for the photon statistics of the cavity field and its corresponding variances Q_f, they are not the same, in general, for the observable outcoming atom statistics. We study in particular $Q_e(Q_g)$ for outcoming atoms detected in their excited states (ground states), for both $N \to \infty$ and $t \to \infty$, as functions of $N_{ex} = RT_c$: R is the mean rate of entry for the incoming atoms.

We find the smooth behaviours, to be displayed in examples, as functions of N_{ex} for the *trapping state fields* in equilibrium at zero temperature. But these smooth behaviours are generally different for the two cases $N \to \infty$ and $t \to \infty$, and it is only Q_e, for fixed N and $N \to \infty$, which follows Q_f as a function of N_{ex} relatively closely. As we will also show these smooth, featureless, behaviours as functions of N_{ex} disappear at small finite temperatures. Although these thresholds (in Q_f) have long since been familiar at the larger temperatures ~ 0.5 K no very simple connection, even at strictly *qualitative* level, seems to exist between any of the four atomic Q_s and Q_f at these larger temperatures and this is so despite actual experimental attempts to infer Q_f from $\tilde{Q}_g$, the Q_g measureed at fixed t as $t \to \infty$.

In this particular analysis the field is supposed already to have reached its equilibrium state, and this state is not disturbed by the atomic state detections. *In contrast* $Q_e(Q_g)$ at fixed N evolves with N to its steady state value as $N \to \infty$; and $\tilde{Q}_e(\tilde{Q}_g)$ at fixed t evolves similarly with t as $t \to \infty$. Even so the detector efficiencies η prove to be simple factors of each of the atomic Q_s and η does not appear at all in Q_f, while the explicit effects of measurement on the atoms vanish from the analysis via conventional ensemble averages which simply give equal weight to each realization of a sequence of atomic inputs combined with any sequence of subsequent atomic state measurements.

We conclude that there is no simple formula connecting *any* of $Q_e, Q_g, \tilde{Q}_e, \tilde{Q}_g$ as measured with the Q_f which it was hoped could be determined from them; and that only by combining *all possible* measurements on the atoms can we hope to infer any *qualitative*, but by no means quantitative, behaviour of Q_f. Note that our work for a fixed collection time t compares directly with e.g. the results of [2] in the continuous time (i.e. Poissonian) limit for the incoming atoms although [2] adopts a fixed rate R for the incoming atoms sometimes called "coarse graining": the work for a collection of a fixed number N of atoms provides potentially *new* information for determining Q_f as explained.

References and links

1. L. Mandel, "Sub-Poissonian photon statistics in resonance fluorescence", Opt. Lett. **4**, 205 (1974).
2. H.-J. Briegel, B.-G. Englert, N. Sterpi and H. Walther, "One-atom maser: Statistics of detector clicks", Phys. Rev. A **49**, 2962 (1994).
3. See also R.R. Puri, S. Arun Kumar, and R.K. Bullough, "Stroboscopic theory of atomic statistics in the micromaser", submitted to Josa B, August 22, 2000.
4. For the reference [1] (1. above) and 'photon anti-bunching' see also R.K. Bullough and S.S. Hassan, "Theory of resonance fluorescence in mixed coherent/chaotic fields" in Royal Society Discussion Meeting "Light scattering in physics, chemistry and biology", Phil. Trans. Roy. Soc. London, A **293**, 232-236 (1979) and following pages.
5. Also see for the reference [4] (4. above) R.K. Bullough and S.S. Hassan, "Atoms in coherent and incoherent fields: I, Microscopic theory of Einstein's rate equation", Physica A **151**, 397-424 (1988); and "II, Resonance fluorescence in mixed coherent-chaotic fields", Physica A **151**, 425-446

Finite temperature correlations in the trapped Bose-Einstein gas

N.M. Bogoliubov*, R.K. Bullough†, V.S. Kapitonov**, C. Malyshev* and J Timonen††

*Steklov Institute of Mathematics at St. Petersburg, Fontanka 27, 191011 St. Petersburg, Russia

**St Petersburg Technological Institute, Moskovsky 26, 196013 St. Petersburg, Russia

††Department of Physics, University of Jyväskylä, PO Box 35, FIN-40351, Jyväskylä, Finland

†Dept. of Mathematics, University of Manchester Institute of Science and Technology, PO Box 88, Manchester M60 1QD

tel. +44-161-200-3672; fax. +44-161-200-3669; email. robin.bullough@umist.ac.uk

There is a large literature (cf. eg. [1, 2]) which, under conditions of translational invariance, has used functional integral methods to calculate, *ab initio*, the equilibrium finite temperature 2-point correlation functions (Green's functions) $G(\mathbf{r},\mathbf{r}') \equiv \langle\hat{\psi}(\mathbf{r},\tau)\hat{\psi}^\dagger(\mathbf{r}',\tau')\rangle$ for a Bose gas in each of $d=3$, $d=2$, $d=1$ space dimensions: $\langle\ldots\rangle$ means thermal average and τ,τ' are 'thermal times' for which $0<\tau,\tau'<\beta$ and $\beta^{-1}=k_BT$, T the temperature. These functional integral methods [1, 2] solve the many-body problem, to good and controlled approximations, for a weakly coupled Bose gas typically modelled by the quantum repulsive Nonlinear Schrödinger (NLS) equation in $d+1$ dimensions – essentially as was first used by N.N. Bogoliubov (1947) for $d=3$ and as is now described by eg. Huang [3]: in G $\hat{\psi}$ (adjoint $\hat{\psi}^\dagger$) are Bose field operators, $[\hat{\psi}(\mathbf{r}),\hat{\psi}^\dagger(\mathbf{r}')] = \delta(\mathbf{r}-\mathbf{r}')$ for $\hbar=1$ and $\mathbf{r},\mathbf{r}'$ are two spatial points in d dimensions: translational invariance means $G(\mathbf{r},\mathbf{r}')=G(\mathbf{r}-\mathbf{r}')$ depending only on $\mathbf{r}-\mathbf{r}'\equiv\mathbf{R}$. Results show a critical temperature T_c such that for $T<T_c$ there is a long range constant valued condensate density ρ_0 for, and only for, $d=3$; and for $d=3$ G is asymptotic to constant $\times\rho_0$ for large enough $R=|\mathbf{R}|$. For $d=2,1$, G falls off with R ultimately to zero, and eg. for $d=1$ exact calculations [4] show an *exponential* decay with R.

Functional integral calculations of this kind have been the stamping ground of high-powered theoretical and mathematical physicists working with little relevance to eg. laser spectroscopy and apparently with little bearing on experiments. However, recent experiments [5] on Bose condensed ^{87}Rb vapour in an elongated paraboloidal trap ($\omega_\perp = 2\pi\times 140Hz$, $\omega_{ax}=2\pi\times 13$Hz) have successfully *measured* $G(\mathbf{r},\mathbf{r}')$ viewed as a function $\rho(z,z')$ depending on two axial coordinates z,z', a distance $\Delta z=|z-z'|$ $(=R)$ apart: the Figure 4 in [5] (for example) shows existence of a condensate for and only for $T<T_c$ – this seen as a constant finite off-set in $V\equiv\rho(z,z')/\rho(z,z)$ for large values of R $(=\Delta z)$: no such off-sets occur for $T>T_c$ where correlations observed through V are Gaussians in Δz in character. The Figure 4 also shows for smaller values of R $(=\Delta z)$ and $T<T_c$ a fall-off in V towards these finite off-set values *particularly* well fitted by a curve $\propto R^{-1}\equiv|z-z'|^{-1}$ although Gaussian fits with off-sets are also considered (see curve for 310 nK in the Figure 4). The reference here, in [5], is to Huang's *ideal* Bose gas

Coherence and Quantum Optics VIII
Edited by Bigelow et al., Kluwer Academic/Plenum Publishers, 2003

result where R^{-1} behaviour is modified by an exponential decay ([3], p.304). However N.N. Bogoliubov [6] predicted k^{-2} behaviour in momentum space $\mathbf{k}$ ($k = |\mathbf{k}|$) if, and only if, $d = 3$. In $\mathbf{R}$- space, for $d = 3$, k^{-2} behaviour becomes *exactly* R^{-1} behaviour, so that this fit at 310 nK is evidence for the quantum repulsive NLS model and *three* space dimensions (*not* $d = 1$ coordinatised by z alone).

Under conditions of translational invariance calculations in $\mathbf{k}$-space are particularly convenient: they are used in [1, 2, 6] eg. where it is shown [1] that in the condensate the system is in a Glauber type coherent state $|\alpha\rangle : a_{\mathbf{k}}|\alpha\rangle = \alpha|\alpha\rangle$ for, and only for, the $\mathbf{k} = \mathbf{0}$ *mode*; $|\alpha|^2$ is then essentially the constant condensate density, ρ_0. In the experiments [5] translational invariance is broken by the trap, which cannot be one dimensional as noted, while it is not at all plain whether the trap introduces modifications and eg. more so transversely than axially – although the Gaussian behaviours for $T > T_c$ is already evidence of the presence of the trap. In order to put numerical values onto these modifications we have re-calculated $G(\mathbf{r}, \mathbf{r}')$ by comparable (but technically more difficult) functional integral methods; and we have calculated the order parameters $\langle\psi(\mathbf{r})\rangle$, $\langle\psi^\dagger(\mathbf{r}')\rangle$ all for each of $d = 3, 2, 1$ dimensions in the presence of a paraboloidal trap potential. Our main results for $T << T_c$ can be summarized by the *analytical* formulae for the first order coherence function $G^{(1)}(\mathbf{r}, \mathbf{r}') \equiv G(\mathbf{r}, \mathbf{r}')/\langle\psi(\mathbf{r})\rangle\langle\psi^\dagger(\mathbf{r}')\rangle$

$$G^{(1)}(\mathbf{r}, \mathbf{r}') \sim \exp\left(\frac{\nu}{2\pi\beta}\frac{1}{\rho_0(\mathbf{S})R}\right) \quad (d = 3) \tag{1}$$

$$\sim (R/\lambda_T)^{-\nu}/\pi\beta\rho_0(\mathbf{S}) \quad (d = 2) \tag{2}$$

$$\sim \exp\left(\frac{-\nu}{\beta\rho_0(\mathbf{S})}R\right) \quad (d = 1); \tag{3}$$

for $T \leq T_c$ these expressions are modified by additional Gaussian corrections from thermal excitations, and for $T > T_c$ only these Gaussian corrections survive (in agreement with [5]). In these expressions (1)-(3) $\nu \equiv m/2\hbar^2$, m the Boson mass, $\rho_0(\mathbf{r})$ is the condensate density roughly the negative of the trap potential cut-off at vectors $\mathbf{r} = \mathbf{R}_c$, $\lambda_T = \hbar\beta v$ with $v = \sqrt{g\rho_0(0)/m}$ the sound velocity at the centre of the trap $\mathbf{r} = \mathbf{0}$ $g(> 0)$ being the coupling constant of the quantum NLS. The vector $\mathbf{S} \equiv \frac{1}{2}(\mathbf{r} + \mathbf{r}')$. Thus each expression depends on $R = |\mathbf{R}|$ *and* on $\mathbf{S}$, a 'centre of mass coordinate', and so on $\mathbf{r}$ and $\mathbf{r}'$ separately. For $d = 3$ we find $\langle\psi(r)\rangle \simeq \sqrt{\rho_0(\mathbf{r})}e^{i\delta}$, $\langle\psi^\dagger(\mathbf{r}')\rangle \simeq \sqrt{\rho_0(\mathbf{r}')}e^{-i\delta}$, δ an undetermined phase and it is easy to see that

$$G(\mathbf{r}, \mathbf{r}') = \sqrt{\rho_0(\mathbf{r})\rho_0(\mathbf{r}')}\left[1 + \frac{\nu}{2\pi\beta}\frac{1}{\rho_0(\mathbf{S})R}\right], \tag{4}$$

up to order R^{-1} for $R >> 2\pi\beta\rho_0(\mathbf{S})/\nu$. There is therefore long range order and coherence in the trap with $G \sim \sqrt{\rho_0(\mathbf{r})\rho_0(\mathbf{r}')}$, $G^{(1)} \sim 1$ for a coherent state, for such large R. Thus G is no longer asymptotic to a constant ρ_0 but depends on each of $\mathbf{r}$, $\mathbf{r}'$ inside the condensate. The fall-off of $G^{(1)}$ to its asymptotic, large R, value remains as R^{-1} (as in the translationally invariant problem) except that this is modified by the correlation length $\beta\rho_0(\mathbf{S})/\nu$ depending on $\mathbf{S}$ ie. again depending on both of $\mathbf{r}, \mathbf{r}'$. Unfortunately although [5] confirms for $T = 310$ nK in the Figure 4 the R^{-1} fall-off predicted by equation (4) the trap geometry used is not yet able to clearly show up the slowly varying behaviour due to the $\rho_0(\mathbf{S})$ modifying the R^{-1} fall-off in (4). From our *exact* translationally invariant results in [4] for $d = 1$ we expect additional oscillatory behaviour of G for $d = 3$ for which there is already some experimental evidence in the data of [5]. Recently Shlyapnikov *et al.* have reported results for G based on cumulant methods for both of $d = 2$ [7] and $d = 1$ [8] which include effects of the magnetic trap. Our result (3) applies for $S << R << R_c$ as in [5]: more generally our result for $d = 1$ compares directly with [8] with however a very different exponent for thermal fluctuations: other thermal effects add to these.

Our major conclusion is that functional integral methods are ideal methods for calculating finite but very low temperature behaviours of Bose gases capable of achieving *quantitative* agreement with experiment (subject to the validity of the quantum NLS model adopted). Variants on the formulae (1)-(3) were first reported at ICOLS XIV but were apparently little understood then. A summary is now published in [9], the Appendix A. In our present paper we will NOT concern ourselves with the intricacies of the mathematical physics of functional integration: our main concern is to compare the new theoretical results like those above obtained for the trap present with those for the trap absent, and to compare each of these with the results of experiments obtained so far, *and* to make further theoretical predictions for future experiments.

References and links

1. V.N. Popov, *Functional Integrals in Quantum Field Theory and Statistical Physics* (Reidel, Dordrecht, 1983) and refs.
2. V.N. Popov, *Functional integrals and collective excitations* (CUP, Cambridge, 1990) (ISBN 0521-407-8777) and refs.
3. K. Huang, *Statistical Mechanics*, 2nd Edn. (Wiley, New York, 1987).
4. N.M. Bogoliubov, R.K. Bullough and J. Timonen, *Critical Behaviour for Correlated Strongly Coupled Boson Systems in 1+1 Dimensions*, Phys. Rev. Lett. **72**, 3933 (1994).
5. I. Bloch, T.W. Hänsch and T. Esslinger, *Measurement of the spatial coherence of a trapped Bose gas at the phase transition*, Nature **403**, 166 (2000).
6. N.N. Bogoliubov, *Quasi-expectation values in problems of statistical mechanics*, JINR preprint D-781 (1963) (in Russian).
7. D.S. Petrov, M. Holzman and G.V. Shlyapnikov, *Bose-Einstein Condensation in Quasi-2D Trapped Gases*, Phys. Rev. Lett. **84**, 2551 (2000).
8. D.S. Petrov, G.V. Shlyapnikov and J.T.M. Walraven, *Regimes of Quantum Degeneracy in Trapped 1D Gases*, Phys. Rev. Lett., **85**, 3745 (2000).
9. R.K. Bullough, *The optical solitons of QE1 are the BEC of QE14: has the quantum soliton arrived? Special Foundation Lecture: a personal view.* Proc. 14th National Quantum Electronics and Photonics Meeting, 08 September, 1999, Manchester UK, J. Mod. Optics **47**, No.11, 2029-2065, (2000) (Erratum: J. Mod. Optics **48**, No. 4, 15 March 2001).

Quantum solitons as possible qubits

R.K. Bullough* and Miki Wadati†

**Department of Mathematics, University of Manchester Institute of Science and Technology, PO Box 88, Manchester M60 1QD*

†*Department of Physics, Graduate School of Science, University of Tokyo, Bunkyo-ku, Tokyo 113-0033, Japan*

In a recent article [1] Izo Abram, of the Centre National d'Etudes des Télécommunications (CNET), Bagneux, France suggested that *quantum solitons* in optical fibres could both be 'squeezed' below the short noise level, thus *reducing* quantum noise, and that they could be used as natural 'qubits' for quantum information, quantum computing, logic gates, etc. The point here is that *multi*-quantum solitons are more easily realised experimentally (in an optical fibre) than are the current realisations of qubits using spin-$\frac{1}{2}$ systems or as in [2] using 2-level atoms and/or single photons in the context of cavity q.e.d. The 'quantum soliton' as described by Abram [1] is the 1-soliton solution of the well known classical one-dimensional attractive Nonlinear Schrödinger (NLS) equation which solution can be put in the form [3]

$$\psi^s(x,t) = 2|c|^{-\frac{1}{2}}\eta(\exp i\Phi(x,t))\text{sech}[2\eta(x-x_0)+8\xi\eta t] \tag{1}$$

where the oscillatory phase $\exp i\Phi(x,t)$ is determined by $\Phi \equiv -4(\xi^2-\eta^2)t - 2\xi x + \theta_0$ – where ξ, η, θ_0 are all real valued parameters: $c < 0$ is the coupling constant of the NLS equation. Quantum features on this classical soliton are then introduced ([1, 4, 5] and see [6]) by imposing momentum-position uncertainly or number phase uncertainly, the latter in particular satisfying $\Delta n \Delta\phi \geq \frac{1}{2}$, a Heisenberg bound for the fluctuations in particle number n and phase ϕ. For large enough n (a 'photon number') $[\hat{N}, \hat{\phi}] = -i$ for operators $\hat{N}, \hat{\phi}$ hence the bound: $\hat{N}$ has eigenvalues n. Because of the well-known stability of classical NLS solitons in optical fibres for very long distance optical communication (especially in the presence of amplifiers at distance separations $\sim$ 100 km), and because such classical 'bits' must in fact be *quantum* 'bits', and therefore 'qubits', the quantum soliton solutions of the *quantum* attractive NLS equation seem worth investigating as such and in these terms – and this is what we shall do in this paper.

For general reference the quantum (attractive) NLS equation in one dimension can be put in the form, with $c < 0$,

$$-i\hbar\frac{\partial\hat{\psi}}{\partial t} = \frac{\hbar^2}{2m}\frac{\partial^2\hat{\psi}}{\partial x^2} - 2c\hat{\psi}^\dagger\hat{\psi}\hat{\psi} \tag{2}$$

where the quantum fields $\hat{\psi}(x,t)$, adjoint $\hat{\psi}^\dagger(x,t)$, satisfy the equal time Bose field commutators like $[\hat{\psi}(x,t), \hat{\psi}^\dagger(x,t)] = \hbar\delta(x-x')$. Conveniently we set $\hbar = 1$ and the particle mass $m = \frac{1}{2}$. One thinks of these particles as 'photons' and they are massive ($m > 0$) because the one dimensional fibre acts as a dielectric with a (nonlinear) refractive index $\neq 1$. Equation (2) derives from a Hamiltonian $\hat{H}$ [3] and there are a number operator $\hat{N} \equiv \int\psi^\dagger\psi dx$ with eigenvalues n (the photon numbers) and a momentum operator $\hat{P} = -i\int\psi^\dagger(\partial\psi/\partial x)dx$, with eigenvalues P: $\hat{H}, \hat{N}, \hat{P}$ all mutually commute as operators (and there are actually infinitely many more such mutually commuting operators – inducing *exact solutions* of equation (2)). For soliton solutions we need $c < 0$ in (2) and it is $|c|$ which figures in the classical 1-soliton expression equation (1). Indeed in classical limit $\hat{\psi} \to \psi$, $\hat{\psi}^\dagger \to \psi^*$ so that $\{\psi, \psi^*\} = i\delta(x-x')$ and $\{...\}$ is the Poisson bracket. It is then easy to see that (1) solves the classical attractive NLS equation which is analogous to (3).

One then discovers [3] that the classical soliton (1) derives *exactly* from a quantum mechanical *matrix element* for photon numbers $n \to \infty$ which is

$$\lim_{n\to\infty}\langle n, X', t|\hat{\psi}(x-vt)|n+1, X, t\rangle \tag{3}$$

where $\hat{\psi}(x-vt)$ with v a velocity, solves (2) (with $m = \frac{1}{2}$, $\hbar = 1$, $c < 0$). The states $|n+1, X, t\rangle$ derive as wave packets (in a manner which will be explained) from simultaneous eigenstates $|n+1, P\rangle$ of $\hat{H}, \hat{N}, \hat{P}$ etc. In general the eigenstates parametrised by wave-vectors $k_1, ..., k_n$ satisfy

$$\hat{N}|k_1, ..., k_n\rangle = n|k_1, ..., k_n\rangle$$

Coherence and Quantum Optics VIII
Edited by Bigelow et al., Kluwer Academic/Plenum Publishers, 2003

$$\hat{P}|k_1, ..., k_n\rangle = \left(\sum_{j=1}^{n} k_j\right) |k_1, ..., k_n\rangle, \tag{4}$$

and the corresponding eigenvalue of $\hat{H}$ is $\left(\sum_{j=1}^{n} k_j^2\right)$; but for a quantum soliton these states are *bound states* forming an 'n-string' such that $k_j = p + \frac{1}{2}\{n - (2j-1)\ ic\}$ for $j = 1, 2, ..., n$ (and $c < 0$) in the complex k- plane ($i = \sqrt{-1}$; p is a real momentum and indeed $P = \left(\sum_{j=1}^{n} k_j\right) = np$ while n is the eigen number (photon number) deriving from $\hat{N}$. Thus, as n is fixed, despite the ideas of ref. [1] these states which are eigenstates of $\hat{N}$ are infinitely squeezed in n and cannot be squeezed any more! Indeed it is relevant that the *semi-classical* quantisation of the classical NLS model leads to exactly the same conclusion [7]).

On the other hand experiments on laser light (coherent state light) traversing the two different loops of a Sagnac interferometer demonstrate *pronounced* squeezing of the photon number [1, 8, 9] and the conclusion must be that the true quantum soliton (of the quantum NLS equation) will only manifest itself after the traversal of a significant length of fibre (for intially coherent light sources). In [1], and with reference to [7], it is suggested that two quantum solitons (two qubits) will exhibit quantum entanglement and it is certainly well known [7, 10] that the two soliton solutions of the classical attractive NLS equation exhibit mutual 'phase shifts' in which the reference point x_0 of the 1-soliton equation (1) is shifted by amount Δ through interaction with a second soliton (which will have a reference point $x_0' \neq x_0$) while x_0' is shifted by $-\Delta$ (conservation of phase shift). To check this suggestion of [1] for the true quantum solitons being described in this paper we evidently need to work out matrix elements like (3) for two different n-string solutions like (4) with the two different reference points x_0, x_0' and examine the *quantum* phase shifts involved: Ref. [11] and its references is already concerned with classical *and* quantum phase shifts. This analysis is currently in hand and results will be reported at the meeting as a part of the more general analysis of N-quantum soliton solutions here to be viewed as N qubits for quantum information purposes.

It is noteworthy that these quantum solitons might be created as an attractive BEC in a *strictly one dimensional* magnetic trap containing ^{7}Li vapour or tuned ^{85}Rb vapour for example: the instability evident in $d = 3$ space dimensions [11] may not be exhibited for $d = 1$ and the quantum solitons created there. Indeed since multi-quantum soliton solutions are *bound* states of $\hat{H}$ these should be the natural condensates for $d = 1$ of the quantum attractive NLS model.

Results in these several directions will be reported at the meeting.

References

1. I. Abram 'Quantum Soitons', Physics World, February 1999, pp.21-25.
2. S. Haroche, G. Nogues, A. Rauschenbeutel, S. Onaghi, M. Brune, and J.M. Raimond, *Quantum Knitting in Cavity QED* in 'Laser Spectroscopy: XIV International Conference', eds. Rainer Blatt *et al.* (World Scientific: Singapore, 1999) pp. 140-149.
3. M. Wadachi and M. Sakagami, *Classical Soliton as Limit of the Quantum Field Theory*, J. Phys. Soc. Japan **53**, 1933-1928 (1984).
4. P.D. Drummond, R.M. Shelby, S.R. Friberg and Y. Yamamoto, *Quantum solitons in optical fibres*, Nature **365**, 307 (1993).
5. M. Rosenblatt and R.M. Shelby, *Squeezed Optical Solitons*, Phys. Rev. Letts. **66**, 153 (1991).
6. S.R. Friberg, S. Machida and Y. Yamamoto, *Quantum non-demolition measurement of the photon number of an optical soliton*, Phys. Rev. Lett. **69**, 3165 (1992).
7. R.K. Bullough, N.M. Bogoliubov and J.T. Timonen, *Solitons of the Nonlinear Schrödinger Equations and Their Quantum Noise*, paper QThJ4 IQEC, 10-15 Sept. 2000, Nice Acropolis, Nice, France, Advance Programme p.93.
8. S. Schmitt *et al.*, *Photon number squeezed solitons from an asymmetric fibre-optic Sagnac interferometer*, Phys. Rev. Lett. **81**, 2446 (1998).
9. S. Spätter *et al.*, *Observation of multimode quantum correlations in fibre optical solitons*, Phys. Rev. Lett. **81**, 786 (1998).
10. R.K. Bullough and J. Timonen, *Quantum and Classical Integrable Models and Statistical Mechanics*, in Statistical Mechanics and Field Theory, eds. V.V. Bazhanov and C.J. Burden (World Scientific, Singapore, 1995) pp. 336-414 and references.
11. S.L. Cornish *et al.*, *Stable* ^{85}*Rb Bose-Einstein Condensates with Widely Turnable Interactions*, Phys. Rev. Lett. **85**, 1795 (2000); A. Rybin, C.G. Varzugin, M. Lindberg, J. Timonen and R.K. Bullough, *Similarity solutions and collapse in the attractive Gross-Pitaevskii equation*, Phys. Rev. E **62**, 6224 (2000).

Manifestation of classical phase in a single spontaneously emitted photon

M. Ligare and D. R. Taylor

Physics Department, Bucknell University, Lewisburg, PA 17837 USA

mligare@bucknell.edu

Abstract: We note a manifestation of the classical phase of a traveling wave in a multimode quantized field. We study spontaneous emission, scattering, and re-absorption by two-level atoms in a one-dimensional optical cavity, and observe classical phase information in the complex quantum amplitude for re-absorption of the scattered field.

The concept of phase for a quantum radiation field has been investigated from the early days of quantum mechanics. (For a review, see reference [1].) Most previous work has focussed on defining an appropriate phase operator for states of a single-mode field. In this paper we investigate a one-photon field that is the result of spontaneous emission; this field state is a linear combination of many singly-occupied field modes. The classical phase and amplitude information does not appear in any quantity directly related to a field operator, but rather it appears in the complex quantum amplitude for transitions to states in which the spontaneously emitted photon has been subsequently re-absorbed.

We study a fully quantized analog of the the simple classical treatment of scattering given by Feynman [2]. We consider various sets of two-level atoms at fixed positions within a one-dimensional multimode optical cavity of length L. The specific results presented here are for two atoms: atom 1 located at $x_1 = L/2$ and atom 2 located at $x_2 = 3L/4$. Atom 1 begins in the excited state and spontaneously emits a photon, which subsequently scatters from atom 2. The quantum amplitude for the re-excitation of atom 1 by the scattered photon exhibits the same phase and magnitude information contained in a classical field scattered by a harmonic oscillator dipole.

Our Hamiltonian assumes both the electric dipole and rotating-wave approximations [3]; both of the atoms have space-dependent couplings to all of the quantized standing-wave cavity modes. The state of the system is written as

$$|\psi(t)\rangle = c_1(t)|e,g;\emptyset\rangle + c_2(t)|g,e;\emptyset\rangle + \sum_k d_k(t)|g,g;k\rangle, \tag{1}$$

where $|e,g;\emptyset\rangle$ represents a state with atom 1 excited, atom 2 in the ground state, and no photons in the field, $|g,e;\emptyset\rangle$ represents a similar state, except that it is atom 2 which is excited, and $|g,g;k\rangle$ represents a state with both atoms in the ground state and a single photon in the kth field mode. The atoms have resonance frequencies $\omega^{(1)}_{atom}$ and $\omega^{(2)}_{atom}$; the central frequency of the field emitted by atom 1 is detuned from the frequency of atom 2 by $\omega^{(1)}_{atom} - \omega^{(2)}_{atom}$. We determine the time evolution of the system by calculating eigenstates and energies of the *total* Hamiltonian, and then projecting the initial state of the system onto the eigenstates, whose evolution is known [4]. Our results were calculated using 299 field modes, resulting in a 301×301 Hamiltonian matrix.

Fig. 1 illustrates the magnitude of the amplitude for atom 1 to be found in the excited state, $|c_1(t)|$, for several values of the detuning of the atoms. This amplitude decays exponentially from its initial value of 1. (The atom-field couplings were chosen

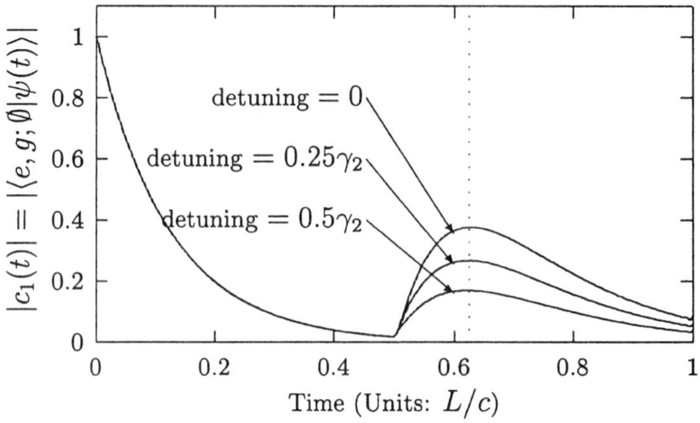

Fig. 1. Amplitude for initially excited atom to be found in the excited state.

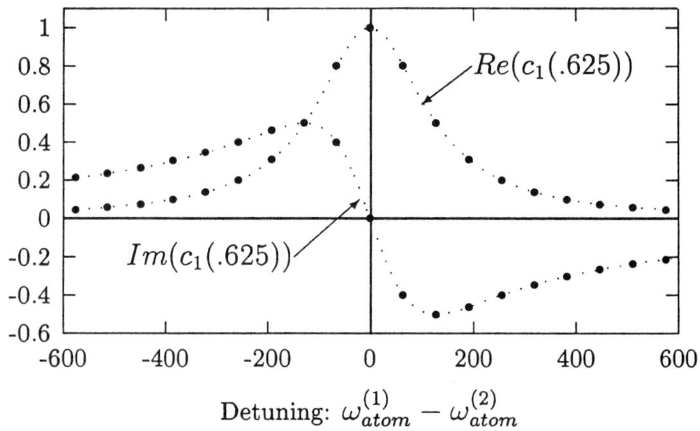

Fig. 2. Real and imaginary parts of the complex amplitude for the initially excited atom to be in the excited state ($c_1 = \langle e, g; \emptyset|\psi\rangle$) at time $t = 0.625$. (The data have been normalized to the zero-detuning amplitude.)

so that the atoms have probability decay constants $\gamma_1 = 16$ and $\gamma_2 = 256$ in the units of the figures.) At time $t = 0.5L/c$ the field reradiated by atom 2 first returns to atom 1, causing an abrupt increase in $|c_1(t)|$. As the relative detuning between the two atoms increases, the effect of the scattered photon on atom 1 decreases. The phase of the complex number c_1 also changes. Fig. 2 illustrates the real and imaginary parts of c_1 at time $t = .625$ (when the effect of the scattered field from atom 2 is the largest) as a function of the relative detuning of the atoms. The points are the results of our calculations, and the dotted lines give the well-known theoretical prediction for the real and imaginary parts of the complex amplitude of the *classical* field scattered from a harmonic oscillator dipole having the same energy decay constant as atom 2.

References and links

1. M. Orszag, *Quantum Optics*, (Springer-Verlag, Berlin, 2000), Chapter 15.
2. R. P. Feynman, R. B. Leighton, and M. Sands, *The Feynman Lectures on Physics*, (Addison-Wesley, Reading, MA, 1963), vol. I, Chapters 31 and 32.
3. P. Meystre and M. Sargent III, *Elements of Quantum Optics*, (Springer, Berlin, 1999)
4. M. Ligare and S. Becker, Am. J. Phys., **63**, 788-796 (1995)

Dynamics of single-atom, single-photon trapping in cavity QED

T. W. Lynn, C. J. Hood, A. C. Doherty, K. Birnbaum, and H. J. Kimble
Norman Bridge Laboratory of Physics 12-33, California Institute of Technology, Pasadena, CA 91125
tel: (626) 395-8343, fax: (626) 793-9506, e-mail: theresa@its.caltech.edu

Abstract: Single atoms are trapped via strong coupling to single-photon fields in optical cavity QED. Properties of the atom-cavity interaction are explored through experimental observation of trapped-atom dynamics and lifetimes as trap parameters are systematically varied.

Much recent work has focused on the developing ability to observe and control individual quantum systems. We pursue this goal in the context of optical cavity QED, where the physical system of interest is a single atom strongly coupled to the electromagnetic field of a high-Q optical cavity. The atom-cavity coupling is characterized by the rate of coherent interaction g_0 (one-half the single-photon Rabi frequency), which for strong coupling must dominate the cavity decay rate κ and the atomic spontaneous emission γ. In this situation we may have critical atom and photon numbers orders of magnitude less than one [1]. These conditions have recently been integrated with the additional requirement that strength of the atom-cavity coupling exceed the atomic kinetic energy, thus realizing strong coupling for both *internal* and *external* degrees of freedom. These tools allow the real-time detection and manipulation of the atom-cavity system, as illustrated by the trapping of single atoms with intracavity fields at the single-photon level [2,3]. We observe the dynamics of trapped atoms over a range of driving parameters to elucidate the trapping mechanism and to enable more detailed control of the atom-cavity system by real-time variation of these parameters.

Our experimental apparatus is shown schematically in Figure 1, and is discussed in greater detail in Ref [3]. Laser-cooled Cesium atoms are dropped from a magneto-optical trap (MOT) approximately 3mm above the center of the optical cavity, and fall one at a time through the cavity mode. The cavity length is set such that the cavity resonance frequency ω_{cavity} is nearly equal to the atomic transition frequency ω_{atom}, with $\omega_{atom} - \omega_{cavity} = \Delta_{ac}$. The system response is now described by the Jaynes-Cummings ladder of eigenstates of coupled atom-cavity excitation. The cavity is driven by a weak probe laser of frequency $\omega_{probe} = \omega_{atom} - \Delta_p$, and the transmission of this probe beam is monitored to provide real-time information about the atom-cavity dynamics.

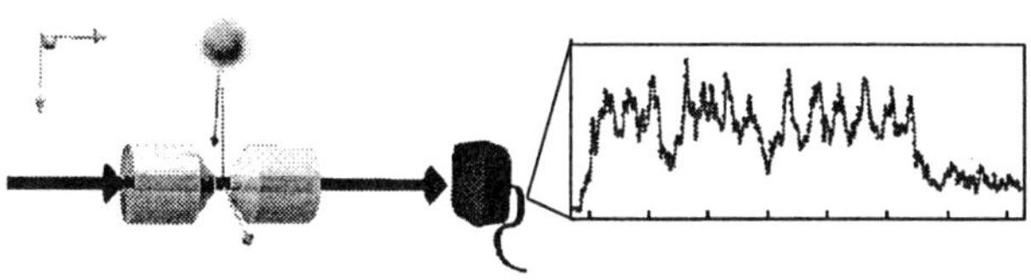

Fig. 1. Schematic of experimental apparatus. An atom passing through the optical cavity is detected and trapped by an intracavity field at the level of a single photon. Atomic position information is encoded in the measured probe transmission through the cavity. For this experiment, the atom-cavity system has rates $(g_0, \kappa, \gamma)/2\pi = (110, 14.2, 2.6) MHz$.

An example of an atom transit signal is shown in Figure 1. Before an atom falls into the cavity, the probe beam is tuned near the lower dressed state of the coupled atom-cavity system to provide small (0.05 intracavity photons), off-resonant excitation of the empty cavity. As an atom enters the cavity mode, the probe transmission increases; when this transmission increase is detected, we switch the probe power up to a higher driving level n_d to excite the system into the lower dressed state and create the accompanying effective potential to trap the atom within the cavity mode. Once the atom is bound by the cavity field, the measured cavity transmission reflects excursions of the atom towards (high transmission) and away from (low transmission) regions of strongest coupling. The atom eventually heats out of the trap due to momentum diffusion associated with repeated decays and re-excitations.

From a series of trapping events we can extract trap times for a given set of driving parameters (n_d,Δ_{ac},Δ_p). Trap-time histograms for three different values of atom-cavity detuning Δ_{ac} are shown in Figure 2. For these data sets, the drive strength n_d is held constant, while the probe detuning Δ_p is adjusted so that the probe is always tuned to excite the lower dressed state for each value of Δ_{ac}. In all three cases shown, a trapping potential associated with the lower dressed state is predicted. However, the potential depth and heating rate vary with Δ_{ac}, leading to the observed variation in trap lifetime.

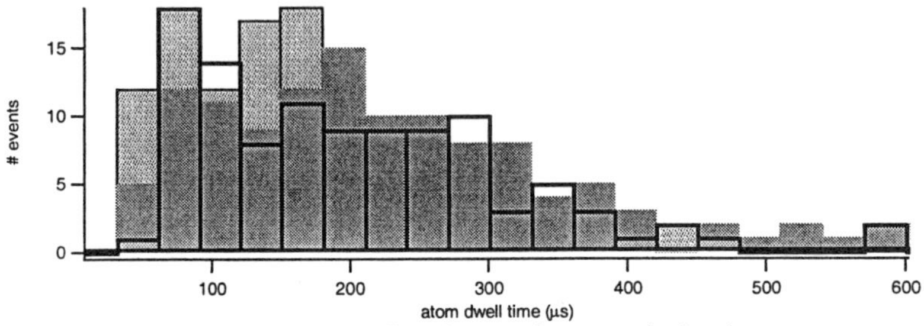

Fig. 2. Trap-time histograms showing effect of changes in atom-cavity detuning: $\omega_{cavity} < \omega_{atom}$ (solid), $\omega_{cavity} = \omega_{atom}$ (shaded), and $\omega_{cavity} > \omega_{atom}$ (hollow). For comparison, the time for a freely falling atom to transit the cavity mode is approximately 75µs.

We can also use the information in individual trapped-atom transit signals to study the dynamics of atomic motion in the spatially varying trapping potential created by the cavity mode. In our experimental regime, motion along the cavity axis is strongly suppressed, and the observed oscillations in the cavity transmission signal can be attributed to atomic motion in the cylindrically symmetric radial potential [2,4]. Thus such oscillations should display a characteristic frequency associated with motion in that potential. In fact, this effective potential is significantly anharmonic for amplitudes of motion comparable to the Gaussian cavity waist, and so rather than a single frequency we observe a well-defined relationship between oscillation period and amplitude of atomic motion, as encoded in the depth of oscillations in transmission. We plot period vs. oscillation depth for several values of cavity driving strength, with detunings held constant. At each drive strength, the data closely follow the predictions based on a calculated effective potential. Comparisons between different drive strengths show differences arising from changes in effective potential as well as saturation of the atom-cavity response.

In conclusion, we present experimentally observed behavior of the bound single-atom, single-photon system as different driving parameters are independently varied. These results are qualitatively explained by the fully quantum-mechanical treatment of the atom-cavity system, as discussed above. Quantitative comparison also demonstrates close correspondence between data and the quantum theory [4]. This understanding of the trap dynamics and measured transmission signals should enable further refinement of real-time control in this system, leading to quantum feedback algorithms for the atomic position in the cavity.

H. J. Kimble, "Strong interaction of single atoms and photons in cavity QED," Phys. Scr. **T76**, 127-137 (1998).

C. J. Hood et al., "The atom-cavity microscope: single atoms bound in orbit by single photons," Science **287**, 1447-1453 (2000).

P. W. H. Pinkse et al., "Trapping an atom with single photons," Nature **404**, 365-368 (2000).

A. C. Doherty et al., "Trapping of single atoms with single photons in cavity QED," PRA **63**, art. 013401 (2000).

Entanglement in cascaded-crystal parametric down-conversion

G. Di Giuseppe, M. Atatüre, M. Shaw, A. V. Sergienko*, B. E. A. Saleh, M. C. Teich

Quantum Imaging Laboratory, Department of Electrical & Computer Engineering and Department of Physics, Boston University, 8 Saint Mary's St., Boston, MA 02215, USA
*AlexSerg@bu.edu; http://www.bu.edu/qil/

Abstract: Quantum interference experiments have been carried out to investigate the entanglement of two-photon states generated by two nonlinear optical crystals separated in space. We demonstrate the joint role of position-direction and frequency-time indistinguishability in defining the generated state.

OCIS codes: (270.0270) Quantum Optics

Summary

The generation of photon pairs via optical spontaneous parametric down-conversion (SPDC) has been in the spotlight for quantum-optical experiments such as quantum teleportation, metrology, imaging, cryptography, and computing. A number of experimental configurations based on the use of SPDC have been designed over the last decade with the aim of producing various types of entangled states. The manipulation of the probability amplitudes of photon pairs originating from two independent regions of nonlinear crystal separated in space-time provides a route to the intelligent engineering of entangled states with increased flexibility.

The experiment we report utilizes two separate nonlinear crystals illuminated by a highly coherent, monochromatic continuous-wave laser. The crystals are cut for type-II down-conversion, producing degenerate collinear entangled photon pairs with orthogonal polarizations. The power of the pump laser is sufficiently low to ensure, with high probability, that only one photon pair is produced in a given time window. A photon pair produced in either of the crystals travels through the same polarization-intensity interferometer, which consists of a birefringent quartz plate (which introduces relative temporal delay between the photons of a pair), a nonpolarizing beam splitter (which directs them to both interferometer arms), and a Glan-Thompson polarizer aligned at 45° with respect to the horizontal axis at each output of the beam splitter (which symmetrizes the response of each detector to the two polarizations) . Coincidence-count measurements are performed by two single-photon-counting avalanche photodiodes (APDs) placed at the two output ports of the interferometer. If only a single crystal is used in such an apparatus, it would yield a single triangular dip in the coincidence-count rate as a function of the relative temporal delay between the two photons of a pair. However, when a cascaded-crystal source of entangled photons is used, more complex interference patterns are observed.

In particular, we experimentally observed two contiguous distinct triangular dips, corresponding to independent contributions of the two nonlinear regions, but only for a particular spacing between the crystals. Changing this spacing has enabled us to produce a variety of unusual interference patterns, corresponding to novel entangled states. In previous treatments of fourth-order quantum interference, transverse and longitudinal effects were always considered separately and assumed to have no effect on each other. In contrast to the single-crystal configuration, we show that a full spatial-temporal analysis is required to successfully explain and predict dual crystal experimental results. We also demonstrate that the unique structure of such entangled states is irreversibly altered by spatial and spectral filtering.

Absorption spectra of the four level N system with two strong fields, in laser-cooled rubidium

S.R. de Echaniz, Andrew D Greentree, A.V. Durrant

Quantum Processes Group, Department of Physics and Astronomy, The Open University, Walton Hall, Milton Keynes MK7 6AA, UK

tel: +44 1908 652123, fax: +44 1908 654192, email: s.r.echaniz@open.ac.uk

D.M. Segal, J.P. Marangos

Laser Optics and Spectroscopy Group, Blackett Laboratory, Imperial College of Science, Technology and Medicine, London SW7 2BZ UK,

J.A. Vaccaro

Department of Physics and Astronomy, University of Hertfordshire, Hatfield, AL10 9AB, UK

Abstract: Spectra illustrating the doubly driven N system are presented with accompanying dressed state analysis. The spectra show a three-photon absorption peak in addition to the normal EIT profile, due to the second coupling field.

OCIS codes: 020.1670, 270.1670

There has recently been much interest in the properties of coherently prepared atomic systems. Of particular interest is the study of electromagnetically induced transparency and associated optical non-linearities. As part of our ongoing investigations into the properties of coherently driven four level systems we present the spectra associated with the doubly driven system presented in figure 1. The system has 2 strong coupling fields in a three-level V configuration which is probed to a fourth level by a weak probe. This scheme is an absorptive classical precursor to realising true single photon quantum interference schemes [1,2].

Our experiments were performed in a standard rubidium magneto-optical trap, the details of which can be found elsewhere [3]. The laser fields were derived from external cavity grating controlled laser diodes with the linewidths of the lasers being less than 1MHz.

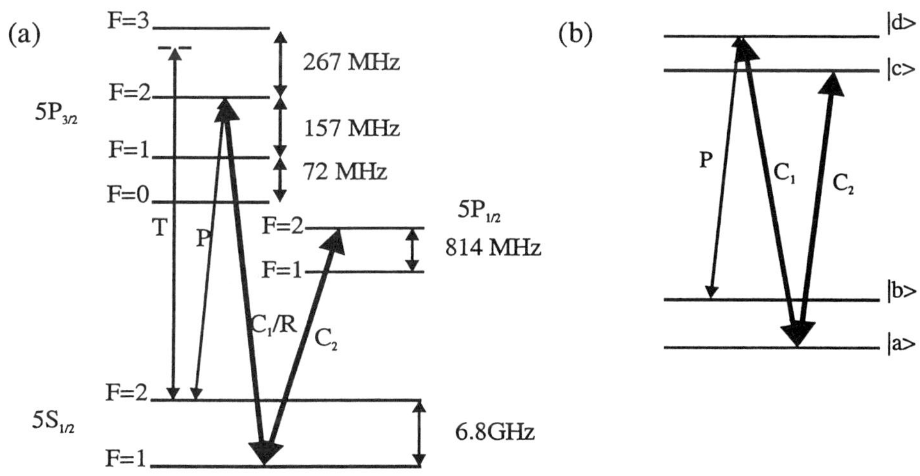

Fig.1: (a) Field configuration in the rubidium MOT. Fields T and R are the trapping and repumping fields. (b) The model four state system. The probe field, P is scanned across the |b>-|d> transition whilst the coupling fields C_1 and C_2 were kept at fixed detunings.

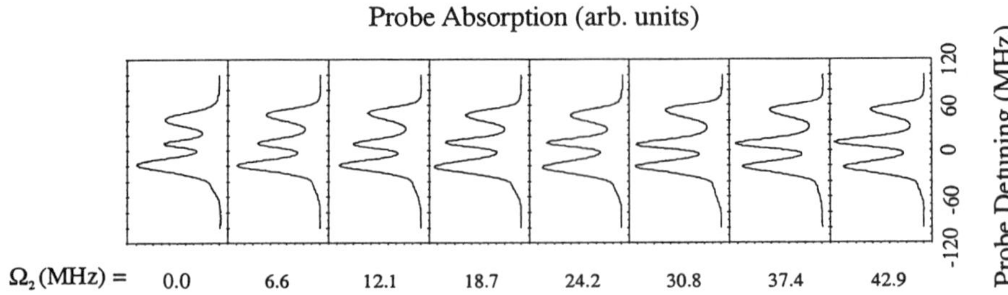

Fig. 2: Probe absorption spectra for increasing second coupling field Rabi frequency, Ω_2, and constant first coupling field Rabi frequency, Ω_1. Note the increase in height of the central peak with increasing Ω_2 indicating the presence of a three-photon absorption resonance for $\Omega_2 > 0$. The asymmetry in the spectra is due to the inclusion of a small mutual detuning of the coupling fields.

Experimental spectra showing the probe absorption for various values of the second coupling field Rabi frequency, Ω_2, are shown in figure 2. The central peaks contain a constant contribution from uncoupled absorptions as well as a three photon absorption peak. Further spectra showing the results with varying coupling field detunings have also been performed [3].

A simple doubly dressed analysis is found to adequately describe all the phenomena and yields a different, intuitive picture for the understanding of this system from the standard density matrix approach. In figure 3 we show the comparison between experiment and theory for the case where C_1 is fixed on resonance and the detuning C_2 is stepped through the three photon resonance, highlighting the migration of the doubly dressed states.

The authors would like to acknowledge the financial support of the EPSRC.

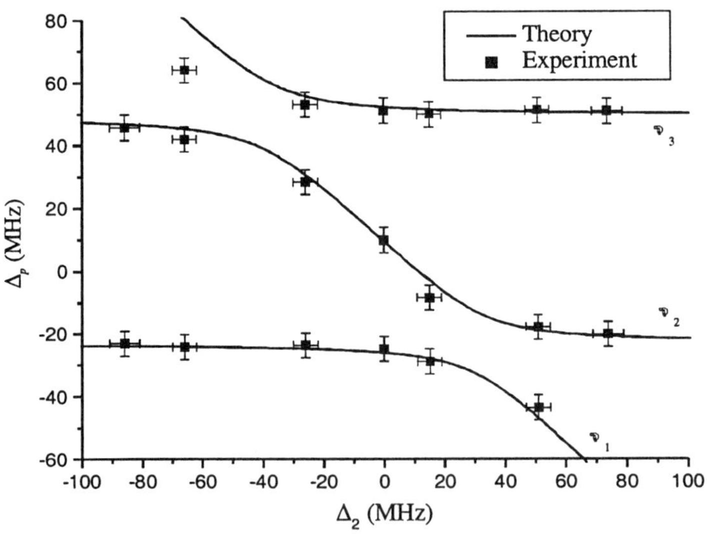

Fig. 3: Migration of the doubly dressed states. Experimental points showing the positions of the three absorption peaks as a function of the detuning of C_2 (Δ_2) give good agreement with the results of the doubly dressed analysis.

References:

[1] A. Imamoğlu, H. Schmidt, G. Woods and M. Deutsch, "Strongly interacting photons in a non-linear cavity", Phys. Rev. Lett. **79**, 1467 (1997)

[2] A.D. Greentree *et al.*, "Prospects for photon blockade in four-level systems in the N configuration with more than one atom", J. Opt. B: Quantum Semiclass. Opt. **2**, 252 (2000)

[3] S.R. de Echaniz *et al.*, "Observations of a doubly driven V system probed to a fourth level in laser-cooled rubidium", accepted for publication in Phys. Rev. A and at quant-ph/0102098

From local to non local entanglement

A. Napoli, A. Messina and G. Compagno
INFM, MURST and Dipartimento di Scienze Fisiche ed Astronomiche, via Archirafi 36, 90123 Palermo, Italy

May 31, 2001

Abstract

A simple experimental scheme for entangling two spatially separated cavities is presented. We exploit the possibility of trasferring the entanglement initially stored in a single resonator, into entanglement between two modes of the separated cavities.

The preparation of prefixed entangled states is the subject of an intense theoretical as well as experimental activity. For example it has very recently been generated in laboratory a maximally entangled state of two orthogonal modes of a high-Q cavity using the interaction with a single two-level Rydberg atom only [1]. In this paper we present a very simple procedure aimed at entangling two modes of spatially separated resonators having different frequencies. The implemention of the scheme we are going to describe requires a two-level atom and two cavities, the first one sustaining two quantized modes of frequency ω_1 and ω_2 respectively, and the second one having its fundamental frequency equal to ω_1. The first step of our procedure consists in preparing the bimodal cavity in the entangled state

$$\frac{1}{\sqrt{2}}[|0_{(1)}\rangle_A|1_{(2)}\rangle_A + e^{i\chi}|1_{(1)}\rangle_A|0_{(2)}\rangle_A] \tag{1}$$

exploiting, for example, the scheme reported in ref. [1]. The states $|k_{(i)}\rangle_A$ appearing in eq. (1), denote the Fock states $|k\rangle$ $(k = 0,1)$ of the mode of frequency ω_i $(i = 1,2)$ of the first cavity A whereas χ is an assigned phase factor. Suppose now to inject an effective two-level atom, initially prepared in its ground state, into the bimodal resonator A where it interacts with the mode of frequency ω_1 by a JC coupling whose strenght is measured by the real constant g_A. Let's indicate by t_{int}^A the interaction time between the atom and the cavity A. It is possible to prove that, manipulating the atomic velocity in such a way that the condition $g_A t_{int}^A = \frac{\pi}{2}$ is satisfied, when the atom leaves the bimodal cavity the state of the system under scrutiny can be written down as

$$|\psi(t_{int}^A)\rangle = \frac{1}{\sqrt{2}}|0_{(1)}\rangle_A\left[|1_{(2)}\rangle_A|-\rangle - ie^{i\chi}|0_{(2)}\rangle_A|+\rangle\right] \tag{2}$$

Coherence and Quantum Optics VIII
Edited by Bigelow et al., Kluwer Academic/Plenum Publishers, 2003

In other words the supposed JC interaction between the two-level atom and the mode of frequency ω_1 produces a disentanglement between the two modes in A entangling, at the same time, the atom with the mode of frequency ω_2. Suppose now that the atom leaving the cavity A enters the second single-mode cavity B previously prepared in its vacuum state $|0\rangle_B$. We assume, once again, that the interaction between the two-level atom and the cavity B is well described by a JC model characterized by a real coupling constant g_B. If the interaction time t_{int}^B between the atomic system and the second cavity B is such that $g_B t_{int}^B = \frac{\pi}{2}$, then, when the atom leaves the single-mode resonator, the state of the coupled system becomes

$$|\psi\rangle = \frac{1}{\sqrt{2}}|0_{(1)}\rangle_A \left[|1_{(2)}\rangle_A|0\rangle_B - e^{i\chi}|0_{(2)}\rangle_A|1\rangle_B\right]|-\rangle \tag{3}$$

The result obtained allows us to state that starting from the entangled state (1) of the two modes of the bimodal cavity, we succeed in establishing an entanglement between two modes of the two spatially separated cavities A and B having different frequencies. Very recently, other papers treating the generation of maximally entangled states of two spatially separated cavities, have been presented [2]. In ref. [2] M. Ikram proposes a scheme for generating a two-cavity entangled field state for fixed number of photons inside the two resonators.In particular, injecting a single two-level excited atom through two cavities, initially prepared in their respective vacuum states, the maximally entangled state $\frac{1}{\sqrt{2}}(|0,1\rangle + |1,0\rangle)$ of the two resonators, can be successfully generated. In spite of the fact that the proposal of Ikram et al is simpler than our procedure, expecially when the respective initial conditions are compared, it should however be appreciated that, with the help of our method, it is possible to built up a wider class of entangled target states. The procedure reported in ref. [2] can be indeed successfully exploited for entangling two modes of separated cavities having almost the same frequency. In our scheme, on the contrary, the two entangled modes can also have very different frequencies. In addition, whilst the relative phase between the two superposed two-mode components in the entangled state proposed in ref. [2] is fixed, in our case such a phase may be varied appropriately controlling the entangled state initially stored in the cavity A.

Acknowledgments

This research was supported by Comitato Regionale di Ricerche Nucleari e di Struttura della Materia. One of the authors (A.N.) is indebted to MURST and FSE for partially supporting this work in the context of Programma Operativo Multiregionale 1994/99 940023I1.

References

[1] A. Rauschenbeutel et al, arXiv:quant-ph/010562 14 May 2001

[2] Ikram, M., 2000, Optics Comm. 184, 417.

LEONARD MANDEL SYMPOSIUM

Immediately following the Eighth Rochester Conference on Coherence and Quantum Optics, a symposium was held at the University of Rochester on Saturday the 16th of June 2001 in celebration of the life and career of Leonard Mandel. In the symposium, presentations were made by several of Professor Mandel's former Ph. D. students. Here are the manuscripts that followed from these talks.

Symposium Coordinators

H. J. Kimble and F. Davidson

LEONARD MANDEL SYMPOSIUM

16 June 2001
Rochester, NY

Leonard Mandel 1927-2001

The Eighth Rochester Conference on Coherence and Quantum Optics was dedicated to the career of Leonard Mandel. This last section of the conference Proceedings contains detailed and personal appreciations of his work. They were coordinated and contributed by five students representing all 39 recipients of the Ph.D. degree under his guidance.

Leonard Mandel was one of the founding fathers of the field of quantum optics. His genius was revealed in a series of uniquely deep and beautiful studies of the nature of light. He pursued this theme through insightful theoretical analyses and pioneering experiments that have became landmarks in the field. Not since the early days of quantum mechanics has one individual so intimately investigated and so dramatically advanced our understanding of the quantum aspects of light.

Professor Mandel developed and undertook more and more refined observations of higher than second-order interference. In this way Mandel was a recognized leader in the quest to understand states of light with manifestly nonclassical characteristics. His group made the first observation of photon anti-bunching and later made the first measurements of sub-Poissonian photon statistics and single photon phase.

His experiments were almost always simple and elegant, and they provided repeated and convincing demonstrations of the inconsistency of quantum mechanics with local realistic theories, which is to say that they neatly contradicted conventional common sense. With pairs of photons Mandel and his students demonstrated remarkable effects: quantum spatial beating, violations of local realism, and phase memory due to quantum entanglement with the vacuum. His group showed that Dirac's well-known statement about photon interference must be expanded to add that a pair of photons interferes only with the pair itself. The famous Hong-Ou-Mandel interferometer forms the basis of Bell-state measurements. It allows a test of non-classicality based on fourth-order interference at a beam splitter, now used in quantum-communications protocols such as teleportation. Mandel himself proposed several schemes for secure communications based on photon entanglement in the early 1980s, well before current methods came to prominence.

Mandel and his students reported a remarkable experiment involving interference from the superposition of fields from two different but coherent parametric processes. They showed that the mere possibility of making a measurement destroyed an interference pattern, even if this measurement was not made. Later Mandel's group, in an experiment with photon pairs, demonstrated a central tenet of quantum mechanics: that there is no physical reality for elemental quantum processes in the absence of a measurement.

Mandel received many prestigious awards during his long career. He won both the Frederick Ives Medal and Max Born award of the Optical Society of America (OSA), the Italian National Research Council's Marconi Medal, and the Thomas Young Medal of the Institute of Physics. He was a fellow of the American Physical Society, the OSA and the American Academy of Arts and Science, and was elected posthumously to the National Academy of Sciences. At the University of Rochester he was known for a light touch in the classroom and for careful attention to his students. He was appointed the first Lee DuBridge professor of Physics and Optics and received the university's Graduate Teaching Award.

Born in Berlin on May 9, 1927, Mandel moved to England with his family as a young boy. He attended Birkbeck College, University of London, where he received a B.Sc. in Mathematics and Physics in 1947, a B.Sc. in Physics in 1948 and a Ph.D. in Nuclear Physics under E. Paul George in 1951. He worked as a Technical Officer at Imperial Chemical Industries for four years and then, in 1955, was appointed Lecturer and later Senior Lecturer at Imperial College, University of London. He remained there until 1964 when he left for the U.S.A. to join the University of Rochester as a Professor of Physics.

A Thumb-Nail Sketch of the Career of Professor Leonard Mandel

H. J. Kimble
Norman Bridge Laboratory of Physics 12-33
California Institute of Technology
Pasadena, CA 91125, USA

As part of the Memorial Symposium at the Eight Rochester Conference on Coherence and Quantum Optics, this paper offers an overview of the remarkable career of Professor Leonard Mandel. The presentation upon which this paper is based can be accessed at *http://www.its.caltech.edu/~qoptics/*.

1. Introduction

The Eighth Rochester Conference on Coherence and Quantum Optics was dedicated to Professor Leonard Mandel, who died on February 9, 2001. At the time of his death at age 73, Professor Mandel was the Lee DuBridge Professor Emeritus of Physics and Optics at the University of Rochester, having become emeritus only a few months before he died. Over the period 1966 – 1995, Professor Mandel was one of the main organizers of Rochester Conferences on Coherence and Quantum Optics (CQO), which since its inception has been the premier international conference for the field. In view of his close association with CQO and beyond the dedication of the conference in his honor, the organizers of CQO-VIII agreed that a special symposium in Professor Mandel's honor would be most appropriate, and that the format of the symposium would be an overview of his remarkable career. The program that emerged was to have this story told by his former graduate students, with seven of Professor Mandel's thirty nine graduate students making presentations that spanned more or less chronologically his research activities over roughly thirty-six years at the University of Rochester. The schedule of speakers and topics for the symposium was as follows:

Fig. 1. Professor Leonard Mandel in his office in the Bausch and Lomb Building at the University of Rochester.

In Memorium -
Professor Leonard Mandel as Recalled by His Students

- H. Jeff Kimble (PhD 1977) - *Introduction and Overview of the Career of Professor Leonard Mandel*
- Fred Davidson (PhD 1969) - *Photon Counting - The Early Days of Intensity Correlation Measurements*
- Mario Dagenais (PhD 1978) - *Photon Antibunching and Resonance Fluorescence Experiments*
- Paul Lett (PhD 1986) - *Dye and Ring Laser Experiments*
- Jeff Ou (PhD 1990) - *Parametric Down Conversion - The Twins Are Born for a Life of Entanglement*
- Jaewoo Noh (PhD 1993) - *The Quantum Phase - A New Theory Meets Experiment*
- David Branning (PhD 1998) - *Parametric Down Conversion - The Twins Grow Up to Become Even Bizarre*

Certainly, over the course of Saturday afternoon, June 16, 2001, the speakers could provide only the briefest of surveys of Professor Mandel's many scientific achievements. But perhaps more to the point, these speakers could and did offer personal accounts of daily interactions with Professor Mandel while engaged in research that in many cases became landmark scientific advances. They thereby provided the broader community with invaluable insight into the career of one of the great scientists of the latter half of the 20th century. Each of the speakers at the symposium has contributed a written version of his presentation in the articles that follow.

The symposium opened with an overview of Professor Mandel's career, which is summarized in the following. Note that reproductions of the viewgraphs upon which this paper is based can be accessed at *http://www.its.caltech.edu/˜qoptics/*.

2. Early Career

Leonard Mandel attended Birkbeck College of the University of London, where he received a B.Sc. degree in Mathematics and Physics in 1947 and a Ph.D. Degree in Nuclear Physics in 1951. His Ph.D. thesis "Interactions of Non-Ionizing Cosmic Ray Particles" involved research in a laboratory high in the Swiss Alps, with two photographs from this era shown in Figure 2. After graduate school, he took up an appointment as Technical Officer at Imperial Chemical Industries Ltd, where he published his first paper "Accuracy Limitations of β-Ray Thickness Measurements" in the British Journal of Applied Physics **5**, 287 (1954). Even in this earliest paper it is perhaps possible to glimpse the origins of what would become a lifelong interest in stochastic processes in the paper's discussion of limitations to measurement accuracy due to noise from Geiger tubes.

From Imperial Chemical, Mandel moved in 1955 to Imperial College, University of London, to become a Lecturer and later Senior Lecturer. While at Imperial College, Mandel's interest in optical coherence was stimulated by the Hanbury Brown and Twiss experiment in which the correlations between photocurrents from independent detectors revealed information about the spatial coherence of light from an extended source. Mandel's entrance into the debate about the appropriate theoretical description of this experiment was a classic paper in 1958 in which the so-called "Mandel formula" for photoelectric counting made its first appearance. This work has become a mainstay of photon counting measurements, including for the earliest studies of lasers and modern photon correlation spectroscopy.

Mandel remained at Imperial College until 1964, when he left for USA, to become Professor of Physics at the University of Rochester. By the time Mandel arrived in Rochester his research interests had largely shifted from nuclear physics to optical physics, and it was hoped that his appointment at the University of Rochester would strengthen a new research group in this rapidly developing field. These hopes were fully realized, with Leonard Mandel being widely credited as being one of the founding fathers of the field of Quantum Optics. Although he made seminal contributions across all of Quantum Optics, a central theme of his research was a continuing quest to explore and elucidate the quantum character of light by way of insightful theoretical analyses and a set of pioneering experiments that have become landmarks in the field.

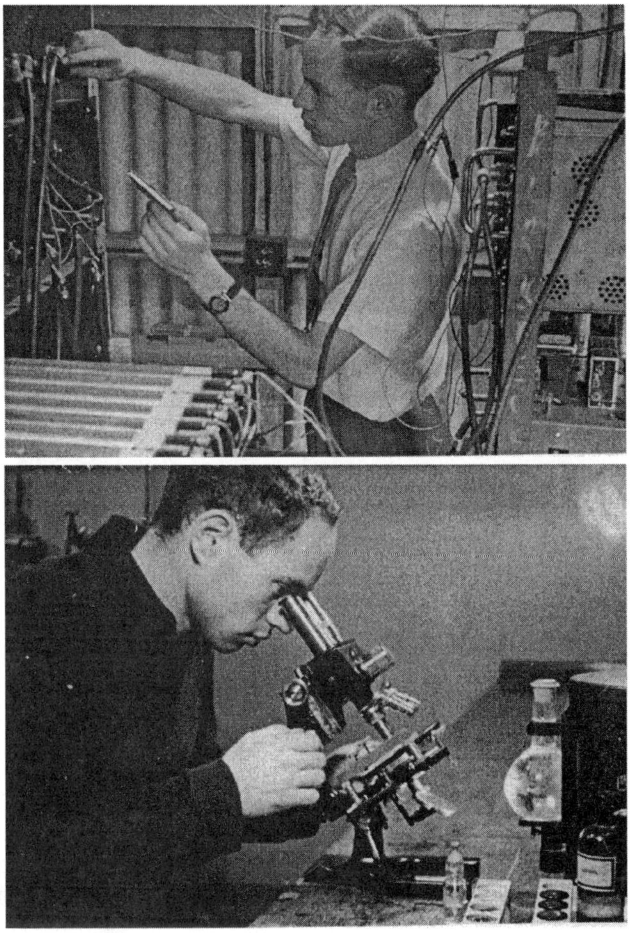

Fig. 2. Leonard Mandel at work as a graduate student in cosmic ray physics atop the Swiss Alps at the Jungfrau, circa 1950. Photographs courtesy of Jeanne Mandel.

3. Interference between Independent Lasers at the Single-Photon Level –

As detailed in the contribution by Fred Davidson in this volume, Mandel had already become quite interested in the possibility of interference between independent lasers while at Imperial College. This early work laid the foundation for a now famous experiment involving interference at the single photon level carried out by Mandel and his student R. Pfleegor soon after his move to the University of Rochester. More precisely, two beams derived from independent lasers were attenuated to a level such that the average interval between photon emissions was very long compared to the transit time through the apparatus. Even though each photoelectric detection event was recorded long before the next emission by one of the sources, interference fringes were nonetheless recorded. As in the Hanbury Brown and Twiss experiment, Mandel employed a detection scheme exploiting interference of fourth-order in the amplitudes of the relevant fields, as opposed to the more familiar second-order interference (e.g., as in a Michelson interferometer), but now with an ingenious extension. Over the next three decades, Mandel continued to refine fourth-order interferometry into an extremely powerful tool for investigations of the quantum character of light, so much so that it is now a standard technique employed by groups worldwide.

Of course the growth of Quantum Optics was triggered and sustained by the invention of the laser. With the hindsight of history, it may be surprising that there was considerable controversy about the quantum statistical properties of laser and other light sources, in which Mandel and his colleagues G. Sudarshan and E. Wolf were actively involved. Perhaps spurred by this debate, Mandel and his students F. Davidson, D. Meltzer, and others were among the first to carry out basic experiments to elucidate the coherence properties of the laser in the transition from incoherent to coherent emission around threshold. Carried forward by several successive generations of graduate students, these early studies of the light from a single-mode laser grew over the ensuing decades into diverse investigations of more complex laser behavior, including mode correlation and competition in ring lasers, and optical bistability and first-order phase transitions in dye lasers. These experiments on laser photon statistics were of remarkable breadth and are reviewed in the contribution by Paul Lett.

4. Nonclassical Light – Resonance Fluorescence from Single Atoms

Although quantum theories are sufficient to understand the laser and general optical interactions, it is nonetheless the case that historically the statistical properties of optical fields could be described almost without exception in terms of so-called semiclassical theories in which the material system is quantized but the field is treated classically. Since the earliest days of Quantum Optics, much attention has thus been directed to finding physical processes that generate manifestly quantum or nonclassical fields, and to identifying unique experimental signatures. Through diverse theoretical analyses, Mandel played a vital role in this effort, including the prediction that the fluorescent light from a single atom would exhibit "photon antibunching," which is a manifestly quantum effect derived independently by H. J. Carmichael and D. F. Walls. Mandel and his students H. J. Kimble and M. Dagenais carried out an experiment to investigate this effect, and in 1977 succeeded in making the first observation of photon antibunching. The apparatus employed for this work was constructed by Kimble and graduate student J. A. Abate, who are shown in the photograph in Figure 3. A more detailed discussion of the actual antibunching experiments is presented in the contribution by Mario Dagenais. Subsequently, R. Short and Mandel made the first measurements of another nonclassical effect, namely sub-Poissonian photon statistics, again in the setting of single-atom resonance fluorescence.

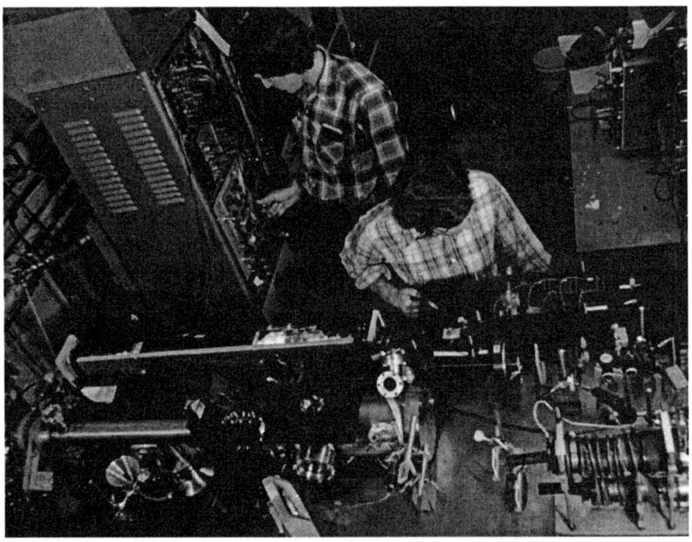

Fig. 3. H. J. Kimble (upper) and J. A. Abate (lower) with atomic beam apparatus and associated dye laser that eventually lead to the first observations of photon antibunching, circa 1975. Photograph courtesy of Jeanne Mandel.

The observation of photon antibunching by Mandel's group is widely regarded as a landmark experiment that initiated the modern era in Quantum Optics involving investigations of nonclassical states of the electromagnetic field. Beyond being simply a matter of semantics, nonclassical fields such as were first observed by Mandel and his students are required to achieve measurement precision beyond the standard quantum limits, to explore the quantum-classical interface, and more generally to advance the new field of Quantum Information Science as it relates to quantum computation and communication.

5. Photon Twins – Parametric Down Conversion

To investigate yet further nonclassical light, Mandel's group in the early 1980s undertook what would become a landmark set of experiments involving pairs of photons produced by the fission of a pump photon in the process of parametric down conversion. Following earlier work in 1970 by Burnham and Weinberg on the correlation time between the photon pairs, Mandel and his students S. R. Friberg and C. K. Hong first demonstrated the nonclassical character of the down conversion process, and were thereby able to achieve a localized one-photon state. Many years earlier, Mandel had helped to lay the foundations for these and similar measurements with his classic 1966 paper on configuration-space photon number operators. Building upon the pioneering work of J. Clauser and others from the 1970s who had observed the quantum nature of photon pairs produced in atomic cascades, Mandel recognized and soon exploited several unique advantages of parametric down conversion for investigations of nonclassical and nonlocal effects in Quantum Optics.

Returning to the theme of fourth-order interference but now with pairs of photons, in the late 1980s Mandel and his students produced a dazzling set of experimental advances, including demonstrations of quantum spatial beating, of violations of local realism, and of phase memory due to quantum entanglement with the vacuum. Led by Jeff Ou (whose review in this volume provides an expanded discussion), Mandel's group showed that Dirac's well known statement about single-photon interference must be modified to read instead - *In fourth order interference, a pair of photons interferes only with the pair itself.* They introduced the so-called Hong-Ou-Mandel interferometer, which has become a workhorse of modern research in Quantum Optics, including for Bell-state detection to enable diverse tasks such as quantum teleportation.

Beyond what had by then become "the standard model" of manifestly quantum effects, Mandel's genius for probing the quantum realm led to yet further startling advances throughout the 1990s that challenged and ultimately redefined our understanding of the quantum character of light and, more generally, of quantum measurement. For example, in 1991 Mandel and his students L. J. Wang and X. Y. Zou reported a remarkable experiment involving interference from the superposition of fields from two different but coherent parametric processes. In their experiment the mere possibility of making a measurement was shown to destroy an interference pattern, even if this possibility went unrealized and absent any external "disturbance." In yet another experiment involving photon pairs from parametric down conversion in 1995, Mandel and his students D. Branning and J. Torgerson and research associate C. Monken demonstrated with brilliant clarity a central tenet of quantum mechanics – namely that there is no physical reality for elemental quantum processes in the absence of a measurement. David Branning's article provides a discussion of this last era in Mandel's career.

6. A Prodigious Career –

Regrettably it is not possible to offer more than these few glimpses into the set of topics and collaborators encompassed over the remarkable career of Professor Leonard Mandel. In addition to the study of nonclassical fields, he was also deeply interested in the question of quantum phase for the electromagnetic field, and together with his students A. Fougeres and J. W. Noh developed an operationally based theory and carried out extensive experiments to test this and other competing theories. Jaewoo Noh describes this research in his contribution to these proceedings. Within the setting of optical coherence, Mandel's ability to identify critical issues and make their consequences manifest are evidenced in many important contributions, including his work on cross-spectral purity, concerning a criterion ensuring that the spectrum of light remains unchanged on superposition in a two-slit interference experiment.

Mandel supervised the thesis research of some thirty-nine students, many of whom subsequently became leading figures in diverse areas of science and technology. He earned the life-long respect of his students for his detailed and conscientious stewardship of their scientific development. Research problems were chosen to fit a particular student's capabilities with an eye toward improvement with each experience. Most often twice per day, he made the rounds of his laboratories to interact with each one of his students, with help ranging from optical alignment to electronic design to theoretical calculations. Not wanting to interrupt a student at work, he would sometimes quietly approach in his soft-soled shoes to view the activity at hand, leading to more than a few startled heart skips. In response to a student's query "Can you see a signal yet?" as some component of an apparatus was being aligned, would come his distinctively British reply "Not a sausage!" Once when the need arose to regulate the water cooling for an oil diffusion pump, he arrived the next morning with a flow switch commandeered from a washing machine at home, which did indeed solve the problem.

Notes with calculations or designs given to him were promptly returned with extensive annotations (corrections of physics as well as of grammar), most often the following morning. At least one student carried these exchanges over the Queen's English forward into later life by sending Mandel a letter with a dangling participle hidden in the text only to receive a prompt reply with the postscript "I found it!".

In addition to his work at the frontiers of physics, Mandel established a reputation as an excellent teacher at all levels of the university curriculum. He had a special interest in neophyte science students and developed the first course at the University of Rochester for non-science majors, which he taught regularly for 20 years. In 1992 he received the university's Faculty Award for Graduate Teaching.

Mandel published about 300 papers and was the co-editor of several conference proceedings. In 1980, a paper on coherence properties of optical fields, which he coauthored with Emil Wolf in 1965, was designated a Citation Classic by the Institute of Scientific Publications and in 1988 the paper was listed as one of the 100 most cited articles published in the Review of Modern Physics since 1955. He was the coauthor with Wolf of a comprehensive book entitled *Optical Coherence and Quantum Optics*, published by Cambridge University Press in 1995.

Mandel received many top awards, including the Frederic Ives Medal and the Max Born Medal of the Optical Society of America. He was the recipient of the Marconi Medal of the Italian Research Council and of the Thomas Young Award of the British Institute of Physics. In 1994 and 1996 Mandel was elected member of the New York Academy of Sciences and of the American Academy of Arts and Sciences, respectively. As already mentioned, during the period 1966 – 1995 he was one of the main organizers of the Rochester Conferences on Coherence and Quantum Optics. In April, 2001 he was elected posthumously to membership in the National Academy of Sciences.

7. A Great Man –

Beyond being a great scientist, Professor Leonard Mandel was also a great man. He had a warm personality, humility, and compassion and a delightful sense of humor. Interactions with him enriched the lives of colleagues and friends in an international community, including significantly his faculty associates at the University of Rochester of whom Professors Nick Bigelow, Joe Eberly, Carlos Stroud, Ian Walmsley, and Emil Wolf are well known to the Quantum Optics Community. Indeed, the duo of Mandel and Wolf pictured in Figure 4 has had a profound impact on the course of physics over the past five decades, having been close friends and partners in a variety of scientific endeavors. In addition to his immediate students and close colleagues, Mandel had many enduring friendships and a lasting influence on individuals the world over. His legacy is by now pervasive in physics, and will undoubtedly become all the more so with the passing of time.

Mandel enjoyed a happy family life with his wife, Jeanne, a ballet teacher, whom he met when they were students at the University of London, and with whom he shared the love for ballet; and with their two children, Barry and Karen. He adored his four grandchildren, and was actively involved in their lives and interests.

Those of us who were fortunate to have been associated with Len Mandel will always remember him with deep affection and utmost respect. His example as mentor, colleague, and friend will vividly continue to shape our own lives. His career provides a standard for accomplishment and integrity that exemplifies the best of the scientific profession. We will greatly miss him.

8. Acknowledgements –

Some parts of this article are based on an obituary which appeared in the August, 2001, issue of *Physics Today* and which was coauthored by the present author and by Professor Emil Wolf. The contributions of Mrs. Jeanne Mandel are also gratefully acknowledged.

Fig. 4. Professors Leonard Mandel and Emil Wolf, whose collaboration and friendship spanned five decades.

Dye and ring laser experiments or: Photon statistics of lasers and atoms (and a whole lot of other stuff)

Paul D. Lett
National Institute of Standards and Technology, 100Bureau Dr. Mail Stop 8424, Gaithersburg, MD 20899-8424
paul.lett@nist.gov

1. Introduction

Prof. Leonard Mandel not only led his students through a number of rather widely recognized experiments over the course of his career, but also through an extremely thorough investigation of lasers and other light sources. Leonard Mandel was not a self-promoter, but so many of his experiments became in some sense "famous" (at least within the quantum optics community) that some of us (who worked on the "other" experiments) might have felt left behind. We were, instead, inspired. We were inspired by the man, by his ideas, and by his care and regard for his students. In the following few pages I will attempt to give a brief sketch of some of the work that Leonard Mandel became involved in around the time that I had the privilege of being one of his graduate students. Although some of this research might not be well-known outside of a very narrow community, it all fits into the overall picture of what interested Leonard Mandel professionally and what he became known for over the course of his career.

I, personally, learned so much from working with him, and from working alongside the other students who were also inspired by him, that I could never have felt neglected. He always had time for his students. If you were willing to go find him, he had time for you - otherwise he gave you pretty free reign over how you scheduled your day. Pretty much the ideal, from a graduate student's point of view!

2. The Person

His door was always open...literally. If he was in the office, the door was always open unless he was with someone or preparing for a class coming up in a few minutes. It usually was only partly open - maybe 8 or 10 inches. Often I would come up and peek in - to find him writing away at his desk. I would softly knock on his door and inquire if he had a minute. I'm sure he must have said no occasionally, but I can only recall once when he was just leaving for a class and I had a rather involved question, that he asked me to come back later. In fact, the thing that amazed me was how easily he took interruption. He would often be in the middle of working on some theoretical problem and writing out some involved equation and he would say, "just a moment, please." He would take no more than a moment to finish writing down the equation, and then he would take as much time with me as I needed - be it signing a purchase request, or having an hour discussion about some detail of my experiment. When I finished I would walk out of the office and again close the door most of the way behind me, as it was when I came in. By this time he was often already back at his desk, with pen in hand, writing again. While such an interruption would set a mere mortal like myself back for quite some time to try to figure out where I was in the problem again, I don't think he missed a beat - he had an amazing ability to pick up right where he had left off.

Leonard Mandel was the man who, while I was around (and before his back began to give him real problems), ran up stairs. His students all knew that and joked about it. If you were to engage him in conversation between his office (on the third floor of the physics building, next to that of his close friend, Emil Wolf) and the labs (on the fourth floor) you had to brace yourself for taking on the staircase. He seemed to feel obliged to attack the stairs at full speed - often leaving you, mid-sentence, most of a flight of steps behind him. Embarrassing when you were a student. It also usually left you rather breathless, finishing your sentence or asking him to repeat what you missed.

He went for a walk outdoors most days around noontime. He made the time to take a break. I am not certain that he did this for any reason other than to take a break or to get a little fresh air, but I have always admired it, especially as I moved on and my own schedule became more demanding than that of the typical graduate student. Professor Mandel was always a wonderful role model for this kind of balance in one's life. He put an emphasis on his family that is not always apparent in the successful academic. He did not travel to conferences particularly often (as his students we sometimes wished he would leave town more - so that we could as well!) and he did not seek out invited talks or publicity for his work. He always gave us the impression that good work speaks for itself. He inspired a loyalty in his students who then wanted to do well in order to make him proud of us. As a beginning graduate student, not knowing anything of his reputation but finding his work with lasers attractive, I can remember getting this feeling that I was joining the group of a remarkable man. It was apparent just from the respect that the senior students like Rajarshi Roy and Surendra Singh expressed for him.

Sometimes we were amazed at the man who, although seemingly himself a theorist in his day-to-day life, could often diagnose your experimental problems from sitting in his office. He liked to suggest the emitter-follower circuits that he was familiar with instead of op-amps, but allowed that we could probably do the job with the more complicated devices if we wanted. Although rather formal in day-to-day dealings, he would occasionally surprise you. He always wore short-sleeved white shirts. On the other hand, at the group picnics that he had at his home we discovered that he was a killer table-tennis player. The only student that I ever saw score more than a few points against him was Jeff Ou. This was sort of an embarrassment to the rest of us who were all half his age but couldn't score a point on him! (His wife Jeanne Mandel, as well as Emil Wolf were also rumored to be quite the table tennis players - although we never convinced them to play while I was around.)

While I know of many students who complained of their thesis advisors, (a species who are the world-over known for their rather poor record for responding to thesis chapters that are given to them for comments) this was not the case for Leonard Mandel. We all know of professors who held onto thesis chapters for months without comment (perhaps for the purpose of getting cheap labor out of the poor grad student for even longer, perhaps for less evil reasons having to do with laziness). Quite the opposite with Leonard Mandel. I can recall staying up very late one night, typing a thesis chapter into what was perhaps only describable as an “adequate” word processor. I was closing-in on finishing it and decided to hang around really late just to be done with a chapter. I handed the chapter to Professor Mandel at around 9 am the next morning - having had little sleep but quite happy that I could finally move on to the next chapter. That afternoon, around maybe 3, he appeared at my office with the marked-up copy of the chapter that I had just given him that morning. I had not yet started to write on the next chapter and already I had revisions to do on the last one! That was the last time I handed him a chapter first thing in the morning!

3. The science

Photodetection, coherence and all the complications that went with different sources of light were amongst the topics that interested Leonard Mandel. He investigated thermal sources, lasers, parametric down conversion, resonance fluorescence, and so on. If it made photons, or if it was made up of photons, he was interested in it!

A unifying way in which to examine these light sources was by investigating the intensity correlations or photon statistics of the light that they gave out. In this rather superficial survey of one segment of his work I will try to give a flavor for what problems he pursued, and who he pursued them with. There are many places to learn the physics, in particular in the original references, so I will not try to go into detail, but just to sketch the story.

The following definitions of the correlation functions of the light will be useful:

$$\lambda(\tau)=\frac{\langle T:\Delta I(r,t)\Delta I(r,t+\tau):\rangle}{\langle I(r,t)\rangle\langle I(r,t+\tau)\rangle}=\frac{\langle I(t)I(t+\tau)\rangle}{\langle I\rangle^2}-1 \tag{1}$$

$$\lambda(\tau)=\frac{\langle T:\Delta I_1(r,t)\Delta I_2(r,t+\tau):\rangle}{\langle I_1(r,t)\rangle\langle I_2(r,t+\tau)\rangle}=\frac{\langle I_1(t)I_2(t+\tau)\rangle}{\langle I_1\rangle\langle I_2\rangle}-1 \tag{2}$$

They express the time-ordered and normally-ordered expectation values of the auto- and cross-correlation functions of one or two light beams. $\lambda(t) > 0$ for most sources for short times τ, and for long times the correlation function approaches zero. Truly coherent light also has $\lambda(t) = 0$ for all t. Leonard Mandel was not just interested in "most sources," however, and spent much of his time using these correlations to characterize quantum, that is truly non-classical light sources. Of course, the famous formula that connects $\langle I \rangle$ with P(n), the probability of detecting n photons in an interval, derived by Professor Mandel in the late 50's, lies at the basc of any photon-counting implementation of these formulas.

I am somewhat sloppy with my notation here - something that Professor Mandel would not have done! He always included a footnote or parenthetic comment in his papers to the effect that "Hilbert space operators are indicated by the caret." I have omitted them here. I will simply note that, in my experience, there are some things that, try as you might, your students will refuse to learn!

Leonard Mandel had a bit of a dry sense of humor about it all, however. Tom Chyba reminded me of one scientific meeting that we attended (it was a few years ago and laser pointers were just becoming the thing to have) where Professor Mandel gave a talk on quantum states of light, or photon statistics of a laser, or some such topic. He was given a laser pointer for the presentation and he was having trouble getting it to turn on. As a conference worker was bringing him another pointer he made a quiet aside remark to the audience to the effect that, “I never did understand lasers.” This brought out a roar of laughter from the audience. The conference aide then handed him a “conventional” pointer that projected a small arrow, using a light bulb and a lens in a rather large handheld device. This pointer worked immediately and he remarked, “Ahh... a thermal source.” Again, to a roar of laughter from the audience.

Professor Mandel had an early and lasting interest in the photon statistics of laser sources. The early laser theory of, for example, Scully and Lamb [1], and Risken and Vollmer [2], and the early experiments of Arecchi [3] and others were the comparisons of the time. Fred Davidson and Mandel [4] performed correlation measurements of the intensity fluctuations of a HeNe laser near threshold. He measured the intensity fluctuations ($\lambda(0)$) and the correlation times of these fluctuations. Later David Meltzer and Mandel [5] looked at the photon counting statistics, $\lambda(\tau)$ for times after a Q-switch was activated in a HeNe laser. Even in these measurements, although the early lasers were harder to control then, the theoretical basis of the tools for understanding them were remarkably the same as they are some 20 or 30 years later. The intensity correlations as measured by equations (1) and (2) were what we all relied on. Prof. Mandel went on to study laser photon statistics in systems varying from single-mode HeNe’s and dye lasers, to HeNe ring laser gyroscopes, to multi-mode ring dye lasers.

4. HeNe ring laser experiments

One of the systems to receive some measure of Leonard Mandel's attention was the HeNe ring laser. To quote from one of his papers, “Although the ring laser has been the subject of a great deal of both theoretical and experimental work over the last decade or so, the emphasis has generally been on applications, and the fundamental properties of the ring laser have received much less attention.” (M-Tehrani and Mandel '78) [6]. This was always a theme with him - the fundamentals. How do we understand the very basics? In particular, this was true for light, it’s detection, and its sources. And, especially with the introduction of the parametric down conversion sources, this extended to an understanding of quantum mechanics itself. I have known very few others who have approached the immediate grasp of quantum mechanics and its consequences that seemed to come naturally to Leonard Mandel.

The gain in a HeNe laser is due to an inhomogeneously broadened medium, and this has some interesting consequences. In particular, in the ring laser gyroscope where two counter-propagating modes in the ring compete for gain, the competition is affected by the degree of coupling that is present, determined by the detuning of the laser from line center. A convenient description of the laser is given by the Langevin equations of motion for the field amplitudes of a two-mode laser:

$$(d/dt)E_1(t) = [a - |E_1(t)|^2 - \xi\,|E_2(t)|^2\,]\,E_1(t) + q_1(t) \tag{3}$$

$$(d/dt)E_2(t) = [a - |E_2(t)|^2 - \xi\,|E_1(t)|^2\,]\,E_2(t) + q_2(t) \tag{4}$$

Where $E_{1,2}(t)$ are the (complex) laser field amplitudes, a is the pump parameter, ξ is the cross-saturation parameter, and $q_{1,2}(t)$ are (complex) noise terms representing spontaneous emission in the system. This "white" noise is delta-correlated in time in the model,

$$\langle q(t)q^*(t+\tau)\rangle = \delta(\tau), \tag{5}$$

and for the HeNe laser on resonance the saturation parameter $\xi = 1$. A number of Professor Mandel's students came to be on intimate terms with these equations. Mohammad M-Tehrani and Mandel [7] examined the mode competition in the ring laser, measuring the fluctuations and correlations of the light intensities, as well as the cross-correlations between the two beams. The measurements show how, as the pump parameter is increased, both modes begin to turn on but finally the intensity of one of the beams (directions) in the ring "wins-out" over the other. The relative fluctuations in the beams at first start to be reduced as the pump parameter is increased and then as one mode "wins" its fluctuations are reduced further (the light becomes coherent), while the other returns to a thermal state.

Surendra Singh was also intimately familiar with the HeNe ring laser and its theory. He measured the intensity correlation functions in a ring laser gyro but also contributed much to the theory of the system. Studies were made of the mode competition, intensity fluctuations and correlations, and the frequency dependence of the fluctuations by Singh and Mandel [8, 9] and, along with Professor F. Hioe, an associate from St. John Fisher College, by Hioe, Singh, and Mandel [10].

The operation of a ring laser with backscattering was also an interesting problem. This was a particularly important problem in the HeNe ring laser, which was used as the basis of ring laser gyros before fiber gyros were introduced. The backscattering couples the two modes and locks them together in frequency, eliminating their sensitivity to small rotation rates. This problem was studied in a number of papers by Bill Christian and Mandel [11, 12] and Christian, Ed Gage, and Mandel [13]. The backscattering problem was also studied in a dye ring laser by Rajarshi Roy and Mandel [14]. This system has a homogeneously broadened gain medium, leading to even stronger coupling between the directional modes, as discussed below.

5. Dye lasers

The first dye lasers were developed around 1970 [15]. Mandel and his students almost immediately took up these tremendously tunable and useful systems, both as a subject of study, and as a means of exciting atomic sodium for resonance fluorescence experiments. The Langevin equations that describe a single mode dye laser are:

$$(d/dt)E(t) = [\,a - |E(t)|^2\,]\,E(t) + q(t), \tag{6}$$

where again, E(t) is the (complex) laser field, a is the pump parameter, and q(t) is the delta-correlated spontaneous emission noise term with correlations defined by $\langle q(t)q^*(t+\tau)\rangle = \delta(\tau)$.

The first lasers studied were not, in fact, single-mode lasers, but rather multi-mode standing-wave systems. The photon statistics of such a dye laser were first investigated in the group by Joe Abate, Jeff Kimble and Mandel [16]. The behavior of this laser approached that expected for single-mode behavior when the laser was operated well above the threshold for lasing, but showed much larger fluctuations in the region of threshold. It was noted that the argon ion laser used to pump the dye laser had intensity fluctuations that seemed to correlate with the intensity fluctuations seen in the dye laser output. At this point, accounting for the pump noise became an interesting problem. A correction for the pump noise was introduced by subtracting-out the longer-time-scale correlations and reasonable agreement was then achieved for $\lambda(0)$ in the region of threshold.

In fact, the effects of a whole variety of different perturbations in lasers were studied over the careers of various graduate students. Almost anything that could affect the behavior of the laser would affect the photon statistics, and thus became interesting. The variation of the photon statistics with the cavity length in a HeNe laser was studied, with a visiting Professor, Les Allen, by Chour-Yih Huang and Mandel [17, 18]. They found that the fluctuations of the laser became progressively larger as the cavity length was decreased, consistent with the idea that fewer and fewer photons could be stored in the cavity. The effect of the flow velocity of the liquid dye stream on the photon statistics of the dye laser was measured by Ralph Short, Rajarshi Roy and Mandel [19]. Although this might seem to be a rather mundane consideration, its effect is significant and the problem was revisited a number of times. It was investigated in the guise of chaos studies, as discussed below, and again a few years later by Samir Bali, Jae Woo Noh, Andre Fougeres, Fat Chi Cheng, and Mandel, [20].

Nearly all of Professor Mandel's students got involved in some measurements of intensity correlations, even those whose thesis did not center on such measurements. The frequency dependence of the intensity fluctuations of a laser detuned off-resonance was examined by Surendra Singh, Steve Friberg, and Mandel [21]. (Steve Friberg spent most of his time setting up the new parametric down conversion lab.) The ring dye laser with controlled asymmetry was studied by Ed Gage and Mandel [22]. Random π phase jumps in a HeNe ring laser with backscattering were studied by Bill Christian, Tom Chyba, Ed Gage and Mandel [23]. A ring dye laser with injected signal was studied by Fat Chi Cheng and Mandel [24]. Even from such an abbreviated list of the topics studied, it is clear that we were all "brought up" learning how to understand photon statistics.

The intensity correlation functions of dye lasers were studied again, this time using a single-mode laser. Such measurements were made by Ralph Short, Rajarshi Roy and Mandel [25] and, with another skilled visitor, Professor Katsuji Kaminishi, more detailed measurements were reported by Kamanishi, Roy, Short and Mandel [26]. The measurements were again found to be at variance with the theory near threshold. Graham, Hohnerbach and Schenzle [27] then published a model in which the pumping fluctuations were introduced directly into Eq. (6). They dropped the additive noise term q(t), and added a (multiplicative) noise term to the pump parameter so that a became a(t). They suggested that the dye laser behavior was, in fact, dominated by pump fluctuations. The problem began to be one of noise in nonlinear dynamical systems, which was becoming an extremely active field at the time.

Experiments by Short, Mandel and Roy [28] showed that this delta-correlated (infinite bandwidth) fluctuating-pump model did not account for the laser behavior under conditions other than those originally reported. Colored noise (finite bandwidth) was then suggested for the pumping fluctuations by Dixit and Sahni [29]. This improvement, plus the retention of the additive spontaneous emission noise of the original model, was finally shown to work by Lett, Short and Mandel [30]. Finally, the single mode dye laser with pump noise can be modeled as:

$$(d/dt)E(t) = [\, a + \eta(t) - |E(t)|^2 \,]\, E(t) + q(t) \tag{7}$$

where E(t) is the (complex) laser field, a is now the average pump parameter. q(t) represents spontaneous emission as a delta-correlated additive noise $\langle q(t)q^*(t+\tau)\rangle = \delta(\tau)$, and $\eta(t)$ is a colored noise term, multiplying the field, representing pump noise and having the correlation property $<\eta(t)\eta^*(t+\tau)>= Qe^{-t/\Gamma}$.

6. Sub-Poissonian Statistics

One of the better-known contributions of Leonard Mandel to the foundations of quantum optics was his discussion and measurements of sub-Poissonian photon statistics. There are at least two separate indications of a non-classical field: $\lambda(0) < \lambda(\tau)$ and $\lambda(\tau) < 0$ separately violate classical inequalities and indicate that the optical field is truly quantum mechanical and has no classical-field explanation. The first of these is related to photon antibunching and the second to sub-Poissonian photon statistics. The experiments of Jeff Kimble, Mario Dagenais and Mandel [31] demonstrated photon antibunching in resonance fluorescence from sodium atoms, but $\lambda(0) < \lambda(\tau)$ does not necessarily lead to sub-Poissonian photon statistics. While resonance fluorescence should produce both effects, atom number fluctuations in the field of view, for example, can remove the sub-Poissonian nature of the light, making $\lambda(\tau) > 0$ for all τ.

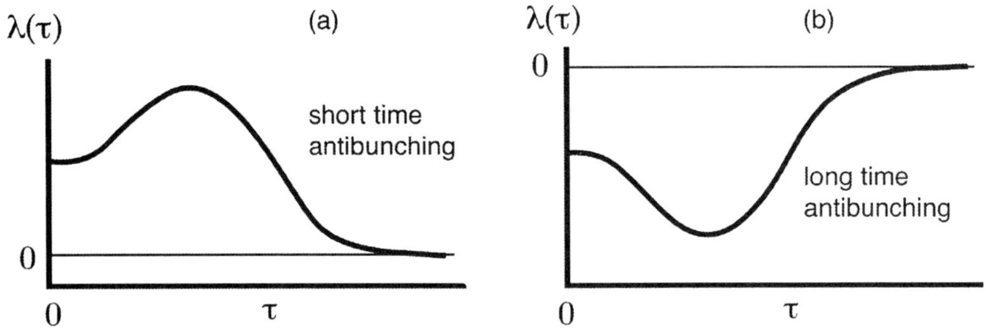

Figure 1. Correlation functions representing photon antibunching at (a) short-times, and (b) long times. Case (b) leads to sub-Poissonian photon statistics, whereas (a) does not.

The departure from Poisson statistics can be measured by what has come to be known as "Mandel's Q – parameter," which is negative for sub-Poissonian statistics:

$$Q = \frac{\langle(\Delta n)^2\rangle - \langle n\rangle}{\langle n\rangle}, \tag{8}$$

where $<(\Delta n)^2>$ is the variance of the number of the number of photon counts and is equal to the mean for Poisson statistics. $\lambda(0) < \lambda(t)$ or short-time antibunching can occur even if $\lambda(t) > 0$ for all t (see Fig. 1), while only if $\lambda(0) < 0$ does one obtain sub-Poissonian photon statistics.

The observation of sub-Poissionian photon statistics in resonance fluorescence was made by Ralph Short and Mandel [32]. As in the observation of photon antibunching, a very weak atomic beam was excited by a laser and the resonance fluorescence was collected and counted. The photon detection period in the experiment was triggered on the presence of fluorescence in an upstream region to reduce the atom number fluctuations in the field of view during the counting period. The results are shown in Table 1.

Table 1. Photon statistics of resonance fluorescence. Total counts, N, and the number of 0's, 1's, etc., for the observed and corrected data and Poisson distributions with equal counts. (From [33])

	Observed	Poisson	Corrected	Poisson
N	24 725 000	24 725 000	24 505 918	24 505 918
N(0)	24 565 295	24 564 898	24 355 000	24 355 167
N(1)	159 307	159 065	150 620	150 287

N(2)	397	515	297	464
N(3)	1	1	1	1
<n>	$(6.475 \pm 0.016) \times 10^{-3}$	6.475×10^{-3}	$(6.171 \pm 0.16) \times 10^{-3}$	6.171×10^{-3}
Q	$(-1.48 \pm 0.25) \times 10^{-3}$	0	$(-2.20 \pm 0.23) \times 10^{-3}$	0

Corrections were made to account for effects like background counts, detector deadtimes, and residual atom number fluctuations. It is, by all accounts, not exactly a large effect, but it is a many-standard-deviation effect. There have, of course, been many experiments showing sub-Poissonian statistics since, including many showing larger deviations from Poisson statistics. A nice review from a few years ago was given by Teich and Saleh [34].

As an aside here I would like to note that many of Leonard Mandel's theoretical treatments of this and other problems were published in single-author papers. In fact in the period 1975-2000 (which was easy to search electronically!) he published 36 single-author papers. In addition, of course, he had many other theoretical collaborations, with visitors like Professor Daan Lenstra while I was there, and others as well.

7. The two-mode dye laser

The two-mode dye laser was also a favorite in Mandel's lab. The Langevin equations for it are just as in Eqs. (3) and (4), however, for a homogeneously broadened system like a dye laser $\xi = 2$. Such a strong coupling between the modes leads to rather dramatic behavior where well above threshold the two directions cannot lase simultaneously. In fact a spontaneous direction switching occurs in a ring dye laser above threshold and this behavior received a good deal of study in Mandel's labs as well.

First passage time distributions for the intensity of a single mode of such a laser, under the influence of the quantum fluctuations of spontaneous emission in a dye ring laser was investigated by Rajarshi Roy, Ralph Short, Jim Durnin and Mandel [35] (Jim Durnin later left the group and turned into a theorist...). Similar measurements under the influence of an injected signal were also made by Fat Chi Cheng and Mandel [24].

I also did a thesis on this kind of switching behavior. With Ralph Short, simulations of the ring dye laser with pumping fluctuations were performed that looked remarkably like the real physical behavior. I performed two-time intensity cross-correlation measurements and studied the effects of the finite correlation time of the colored noise pumping fluctuations [36]. Ed Gage looked at the dye ring laser with a controlled asymmetry between the losses in the two counter-propagating modes [22] as well as other behaviors.

Another aspect of laser theory that Leonard Mandel was interested in was bistability and phase transition analogies in laser systems. Rajarshi Roy looked at bistability and a first order phase transition analogy in a dye ring laser system [37]. “Potentials” describing the state of the system could be derived by converting the first-order Langevin equations of motion (with white noise) to second-order differential equations of the Fokker-Planck type. The system is then described by drift and diffusion on the potential surface and a probability distribution can be written down. This can be used to show, for example, the variation of an order parameter like the most probable intensity, with the pump parameter, a, or the coupling parameter, ξ.

Finally, we found dynamical chaos in the lab while I was there (in addition to the more usual kinds of chaos...). The question of noise in nonlinear dynamical systems and the whole topic of dynamical chaos was just beginning to be explored in detail and gas lasers were commonly used in such investigations. For our experiments we recorded a series of intensity measurements of the laser. Tom Chyba, who had experience with the gas laser experiments before joining the group, wrote up an analysis code using a procedure due to Grassberger and Procaccia [38]. This was used to extract the dimensionality of the phase space of the system from the time series of intensity measurements. If the dimensionality of the phase space visited by the system is equal to the number of variables in the system, it is deterministic. If it has a integer dimension less than the number of variables, then there are one or more constants of the motion. If it has a fractional dimension less than the dimension of the system, then dynamical chaos is present. If the dimensionality is very large, perhaps infinite, the system is completely noise-driven.

We found that there was chaos present, with a dimensionality of around 6.5, in measurements of the light emerging from the dye laser. The measurements showed that a dye laser that used an open jet stream, rather than a flowing-dye cell, did not show this behavior. Chaos was also not found in the noise on the pump laser itself, however, in measurements of the pump laser light remaining after having passed through the dye cell, it did appear. Thus the chaotic noise found on the light was probably due to turbulent flow (known to commonly exhibit dynamical chaos) in the flowing-dye cell of our dye laser system. The work was published by Chyba, Gage, Rupa Ghosh, Lett, Mandel, and Ian McMackin (a student of Mike Raymer) [39]. One additional conclusion of this work is that using the dye cell instead of a jet for a "more stable" laser, as we had often claimed, was probably not such a good idea.

8. Final thoughts

One result of Leonard Mandel's great research activity was that his book with Emil Wolf, "Optical Coherence and Quantum Optics" [40] had to be continuously revised. This meant that his secretary of many years, Ruth Andrus needed to retype the book a number of times. I watched as the book evolved from one form to another, and even one word processor to another. One problem of writing a book on an evolving and expanding field is that, the longer you take to finish the job, the longer the book becomes. Leonard Mandel and Ruth Andrus both seemed to maintain good humor about it all. Although much of the book was written in some form before I arrived in Rochester, it was not finished until well after I left.

Degrees were eventually received, and those of us who worked with Leonard Mandel went on with our lives after graduate school. Nonetheless, he left an indelible mark on all of our lives. Shaping the way in which we view physics, and the way in which we have done our science, and even lived our lives. I went off to the National Bureau of Standards - now the National Institute of Standards and Technology, where I work in the group of Bill Phillips. Bill has always thought very highly of Leonard Mandel and has often referred to a certain parametric down-conversion experiment of his [41] as, "the amazing Mandel experiment". I think that it is not unintentional that there is some ambiguity there - and that we are allowed to take that modifier "amazing" and apply it to either the person or the experiment.

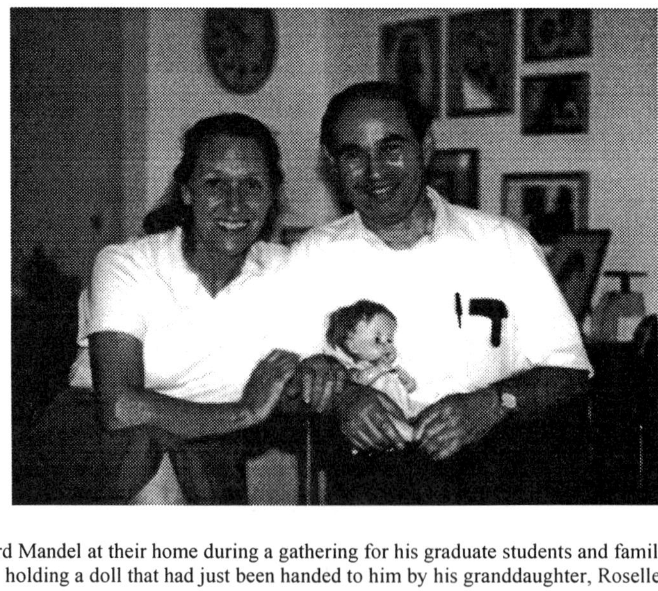

Jeanne and Leonard Mandel at their home during a gathering for his graduate students and family (circa 1985); he is holding a doll that had just been handed to him by his granddaughter, Roselle.

9. Acknowledgements

I would like to thank Joe Eberly, Fred Davidson and Jeff Kimble for putting together this symposium and giving me the chance to reminisce and review my time at Rochester. I would also like to thank those other graduate students with whom I shared an advisor, and so much else, with. In particular I thank those that I overlapped with in time: Rajarshi Roy, Surendra Singh, Ralph Short, Steve Friberg, Bill Christian, Rupa Ghosh, Tom Chyba, Chung Ki Hong, Ed Gage and Jeff Ou. I learned much from each of them. Thanks also to Ian Walmsley, who survived Professor Mandel's Quantum Optics II class with me and has been a friend and collaborator since. A particular part of my experience at Rochester associated with Leonard Mandel was his long-time secretary, Mrs. Ruth Andrus, and I especially thank her for fond memories of my time in the group. I am also grateful to Jeanne Mandel and the rest of the Mandel family for all of their efforts in making this memorial session a special event for us, and for making us welcome over the years. Finally, I would like to thank Leonard Mandel for all he has done for me.

References

1. M. Scully and W. Lamb, Phys. Rev. Lett. **16**, 853 (1966).
2. H. Risken and H. Vollmer, Z. Phys. **204**, 240 (1967).
3. F. Arecchi , V. Degiorgio and B. Querzola, Phys. Rev. Lett. **19**, 1168 (1967).
4. F. Davidson and L. Mandel, Phys. Lett. **25A**, 700 (1967).
5. D. Meltzer and L. Mandel, Phys. Rev. Lett. **25**, 1151 (1970).
6. M. M-Tehrani and L. Mandel, Phys. Rev. A **17**, 694 (1978).
7. M. M-Tehrani and L. Mandel, Opt. Lett. **1**, 196 (1977).
8. S. Singh and L. Mandel, Phys. Rev. A **20**, 2459 (1979).
9. S. Singh and L. Mandel, Opt. Comm. **40**, 139 (1981).
10. F. T. Hioe , S. Singh and L. Mandel, Phys. Rev. A **19**, 2036 (1979).
11. W. R. Christian and L. Mandel, Phys. Rev. A **34**, 3932 (1986).
12. W. R. Christian and L. Mandel, J. Opt. Soc. Am. B **5**, 1406 (1988).
13. W. R. Christian , E. C. Gage and L. Mandel, Opt. Lett. **12**, 328 (1987).
14. R. Roy and L. Mandel, Opt. Comm. **35**, 247 (1980).
15. O. Peterson , S. Tuccio and B. Snavely, Appl. Phys. Lett. **17**, 245 (1970).
16. J. A. Abate , H. J. Kimble and L. Mandel, Phys. Rev. A **14**, 788 (1976).
17. L. Allen , C. Y. Huang and L. Mandel, Opt. Comm. **22**, 251 (1977).
18. C. Y. Huang and L. Mandel, Opt. Comm. **32**, 345 (1980).
19. R. Short , R. Roy and L. Mandel, Appl. Phys. Lett. **37**, 973 (1980).
20. S. Bali, J. Noh, A. Fougeres, F. C. Cheng, and L. Mandel, Opt. Comm. **78**, 377 (1990).
21. S. Singh , S. Friberg and L. Mandel, Phys. Rev. A **27**, 381 (1983).
22. E. C. Gage and L. Mandel, Phys. Rev. A **38**, 5166 (1988).
23. W. R. Christian, T. H. Chyba, E. C. Gage, and L. Mandel, Opt. Comm. **66**, 238 (1988).
24. F. C. Cheng and L. Mandel, J. Opt. Soc. Am. B **8**, 1681 (1991).
25. R. Short , R. Roy and L. Mandel, Appl. Phys. Lett. **37**, 973 (1980).
26. K. Kaminishi, R. Roy, R. Short, and L. Mandel, Phys. Rev. A **24**, 370 (1981).
27. R. Graham , M. Hohnerbach and A. Schenzle, Phys. Rev. Lett. **48**, 1396 (1982).
28. R. Short , L. Mandel and R. Roy, Phys. Rev. Lett. **49**, 647 (1982).
29. S. N. Dixit and S. Sahni, Phys. Rev. Lett. **50**, 1273 (1983).
30. P. Lett , R. Short and L. Mandel, Phys. Rev. Lett. **52**, 341 (1984).
31. H. J. Kimble , M. Dagenais and L. Mandel, Phys. Rev. Lett. **39**, 691 (1977).
32. R. Short and L. Mandel, Phys. Rev. Lett. **51**, 384 (1983).
33. R. Short and L. Mandel, in *Coherence and Quantum Optics V*, edited by L. Mandel and E. Wolf (Plenum, New York, 1984), p. 671.
34. M. Teich and B. Saleh, in *Progress in Optics*, edited by E. Wolf (North Holland, Amsterdam, 1988), Vol. 26, p. 1.
35. R. Roy, R. Short, J. Durnin, and L. Mandel, Phys. Rev. Lett. **45**, 1486 (1980).
36. P. Lett and L. Mandel, J. Opt. Soc. Am. B **2**, 1615 (1985).

37. R. Roy and L. Mandel, Opt. Comm. **34**, 133 (1980).
38. P. Grassberger , I. Procaccia and A. Ben Mizrachi, Phys. Rev. A **29**, 975 (1984).
39. T. Chyba, et al., Opt. Lett. **12**, 422 (1987).
40. L. Mandel and E. Wolf, *Optical Coherence and Quantum Optics* (Cambridge University Press, Cambridge, 1995).
41. X. Y. Zou , L. J. Wang and L. Mandel, Phys. Rev. Lett. **67**, 318 (1991).

Parametric Down-Conversion: The Twins are born for a Life of Entanglement

Z. Y. Ou
Department of Physics, Indiana University-Purdue University Indianapolis, Indianapolis, IN 46202
(317)274-2125, FAX: (317) 274-2393, Email: zou@iupui.edu

1. Introduction

The existence of quantum entanglement is what makes quantum physics different from classical physics. First introduced by Schrödinger in his discussion of the famous cat paradox [1], the amazing effects of quantum entanglement are very counter-intuitive and sometimes mind-boggling. As a pioneer and an expert in both classical and quantum coherence theory, Mandel is the right person to use light for the exploration of quantum entanglement. Among many of Mandel's contributions, the pioneering work and subsequent development on parametric down-conversion (PDC) opened up a whole field of research to experimentally observe the phenomena of quantum entanglement. Today, in the newly established field of quantum information science, quantum entanglement has become an essential ingredient in basically all the discussions. Schrödinger may have been the first person to mention the word of "Entanglement", but it was Mandel who first demonstrated it experimentally and created further a variety of different entangled states, which include entangled states of polarization, of frequency, and of phase variables. Besides quantum entanglement, Mandel also exploited PDC in the study of fundamental physics as well as practical applications in optical communication.

2. The early work on PDC

Throughout Mandel's scientific career, there was one basic scheme, that is, the demonstration of quantum mechanical behavior of light. Mandel's work on parametric down-conversion followed his monumental work on resonant fluorescence for the demonstration of photon anti-bunching [2] and sub-Poissonian photon statistics [3]. Recognizing the exceedingly large coincidence in the early work on parametric down-conversion (PDC) by Burnham and Weinerg [4], Mandel realized that this process is a natural nonclassical source that is easier to handle than resonant fluorescence.

2.1 Correlation time between the twin photons

The process of parametric frequency down-conversion is a process in which an incoming photon, after interacting with a nonlinear crystal, is split into two photons of lower energy. Due to historic reason, the two down-converted fields are labeled as signal and idler fields. Although there were numerous theoretical investigations on PDC [5], there was only one experimental study on the photon correlation in PDC, which is the one from Burnham and Weinberg in 1973 [4]. In this work, Burnham and Weinberg measured correlation time between the arrival of the twin photons and found an upper bound of 4 ns limited by their electronics response time. Equipped with an ultra fast detector made of micro-channel plates with about 50 ps response time, Mandel and his students Steve Friberg and Chung Ki Hong made a similar measurement as Burnham and Weinberg on the correlation time and set the upper limit to 150 ps [6]. However, this was not a simple repeat of Burnham-Weinberg experiment. They made a full measurement of the time delay distribution and found that the distribution is basically independent of the pump intensity and bandwidth and the nonlinear medium's length. This measurement answered a number of questions raised in the theoretical investigations on the correlation time. Furthermore, Mandel and Hong devised a new theory about down-conversion based on a more realistic model of continuous mode structure for the twin photons [7]. Although this theory predicted a far smaller correlation time (about 1 ps) than what was measured, it produces all other features demonstrated in the experiment. This theory turned out to be the first theory to correctly describe spontaneous parametric down-conversion with traveling modes for all three waves and later became the starting point for all the treatments on two-photon interference.

2.2 Applications of PDC

Immediately after the initial study on PDC was done, Mandel and his students started demonstrating the nonclassical property of this new unique source. They found the exceedingly large coincidence recorded between the signal and idler beams violates the classical Schwartz inequality: $\langle I_1 I_2\rangle^2 \leq \langle I_1^2\rangle\langle I_2^2\rangle$ [8].

The strong correlation between the signal and idler beams was also used to realize an optical communication cryptographic protocol to have information hidden in the background of each field but revealed in the coincidence measurement [9]. This is one of the early attempts of application of quantum source in communication.

Realization of a single-photon Fock state is another straightforward application of PDC. Ideally, the twin photons from PDC are 100% correlated. In other words, the detection of one photon will indicate the existence of the other one. So even though the arrival of the twin is completely random, we can still guarantee that there is always one photon in the signal beam so long as a photon is detected in the idler beam. This was demonstrated in a photon counting experiment by Hong and Mandel [10].

3. Two-photon interference

3.1 Initial work on two-photon interference

Mandel's interest in photon interference traces back to his landmark experiment of demonstrating interference between two independent lasers [11]. This experiment seems to pose a serious challenge to Dirac's statement that "a photon only interferes with itself, interference between two photons never occurs". To see if the interference between two independent fields still exists even at single photon level for testing Dirac's statement, Mandel and his student Pfleegor designed an ingenious experiment that reveals the unstable interference pattern with two detectors [12]. The principle behind this experiment is that if interference pattern exists, there tends to be a negative correlation for the two detectors spaced at half the fringe spacing but a positive correlation at full fringe spacing even for an unstable moving fringe pattern. This is the very first two-photon interference experiment.

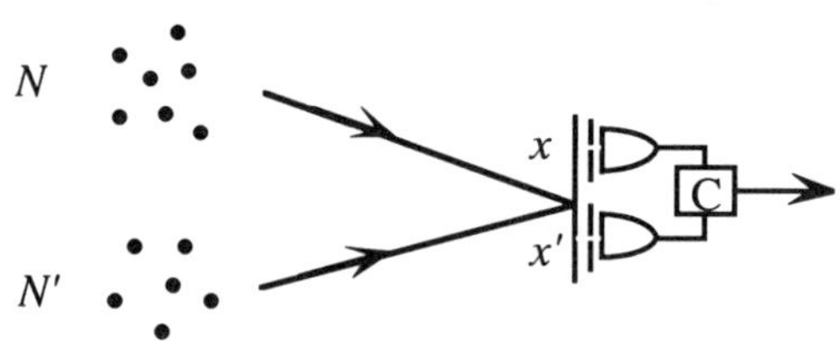

Fig. 1. Scheme for nonclassical two-photon interference with two groups of excited atoms.

In the next decade that follows, Mandel's attention turned away from interference to the more interesting phenomena of resonant fluorescence, which exhibits nonclassical properties such as photon anti-bunching and sub-Poissonian photon statistics. From that point on, Mandel started to devote much of his later career to looking for the subtle differences between classical and quantum description of light. His intuition in this brought him back to the problem of interference between two independent sources. But this time the fields are from quantum sources of atomic emitters and the phenomena are nonclassical. In a theoretical paper in 1983 [13], Mandel investigated the two-photon correlation in the interference between light fields emitted from two groups of two-level atoms (Fig.1). He found a typical interference pattern in the two-photon correlation as expressed in the form of

$$P_2(x_1,x_2) = 1 + V\cos 2\pi(x_1 - x_2)/L \tag{1}$$

where x_1 and x_2 are the positions of the two detectors and L is the fringe spacing. V is the visibility ranging from 0 to 1. Mandel went on to show that there is an upper bound of 50% for V if the fields are classical fields. For the atomic system, V depends on the number of atoms N, N' in each group and in particular, $V = 100\%$ when $N = N' = 1$. Mandel was the first to point out that an interference effect for which classical wave theory is best to describe has to rely on quantum theory for its explanation. This work laid the foundation for future work on two-photon interference and two-photon entanglement. The special situation for which the visibility reaches 100% corresponds to the initial state of $|1\rangle_A|1\rangle_B$, i.e., one photon for each field. This sets up the role for parametric down-conversion [14].

3.2 Two-photon interference in parametric down-conversion

In non-collinear parametric down-conversion, the output fields consist of two separable signal and idler fields which each have one photon making it a perfect source for two-photon interference. Even the wide band nature of this source does not prevent it from showing a visibility of 100%. So after repeated unsuccessful tries with atomic system, which is the proposed system in the initial paper [13], Mandel and his student R. Ghosh turned their effort to PDC and tried to superpose the two down-converted fields for interference. They met their success immediately. This is first ever observed nonclassical two-photon interference experiment [15].

In the meanwhile, another project in Mandel's group involving students C. K. Hong and Z. Y. Ou was under way to utilize the two photons in PDC to create an EPR-Bohm singlet polarization state [16], following a suggestion from Chubarov and Nikolayev [17]. Such a state can be obtained by superposing the two photons with orthogonal polarization at a beam splitter (see below, Section 4.2). While playing the newly discovered formula for deriving output state from a given input state in a beam splitter [18], the group found an interesting effect: when the two input photons have the same polarization, i.e., $|1\rangle_a|1\rangle_b$, the output state becomes

$$|\Phi\rangle_{out} = \frac{1}{\sqrt{2}}\left(|2\rangle_A|0\rangle_B + |0\rangle_A|2\rangle_B\right), \tag{2}$$

with one possible term $|1\rangle_A|1\rangle_B$ clearly missing (a, b are two input modes and A, B are two output modes). The intuition led Mandel to suggest that it is the destructive interference between the two possible paths (either both transmit or both are reflected, see Fig.2) that makes the $|1,1\rangle$ state missing in Eq.(2). Subsequent experimental demonstration by Hong, Ou and Mandel [19] gave birth to the Hong-Ou-Mandel (HOM) interferometer that has many applications ranging from measurement of the width of a weak femto-second pulse [20] to the measurement of photon tunneling time in a barrier [21].

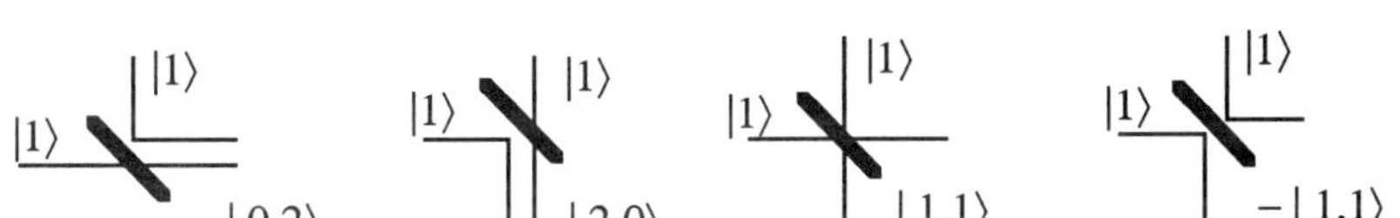

Fig. 2. Four possible output states for HOM interferometer. Destructive interference occurs for last two possibilities.

4. Demonstration of quantum entanglement

Two-photon interference discussed above has a fringe pattern that does not depend on the phases of the superposing fields. This was originally thought to be a unique property of fourth-order correlation. But it was later shown by Horne et al [22] and by Grangier et al [23] that phase dependent intensity correlation is a result of quantum entanglement. And the experiments discussed above are a special case of the quantum entanglement – the two paths for interference are overlapping and there is a fixed phase difference of 180° for a beam splitter in Hong-Ou-Mandel interferometer. Since the first experiment on quantum two-photon interference, Mandel and his group had demonstrated quantum entanglement of various kinds involving many degrees of freedom such as polarization, frequency, space-time and phase.

4.1 Entanglement with vacuum

The very first demonstration of phase dependent two-photon interference pattern was performed with two parametric down-conversion processes as shown in Fig.3 [24, 25]. This experiment was initially intended to show the phase memory via entanglement with vacuum in parametric down-conversion process: the quantum state of parametric down-conversion is of the form (single mode) of

$$|\Phi\rangle = |0\rangle + \eta V_p |1_s, 1_i\rangle, \tag{3}$$

where V_p is the amplitude of the pump field and carries the phase of pump field into the two-photon state. With two such processes, an entangled state is obtained:

$$|\Phi_1\rangle \otimes |\Phi_2\rangle = |0\rangle + \eta_1 V_{p1} |1_{s1}, 1_{i1}\rangle + \eta_2 V_{p2} |1_{s2}, 1_{i2}\rangle, \tag{4}$$

where we omit the higher order terms. The two-photon correlation at A and B will depend on the relative phase of V_{p1} and V_{p2} [25]. This is the entanglement in direction of propagation of photons.

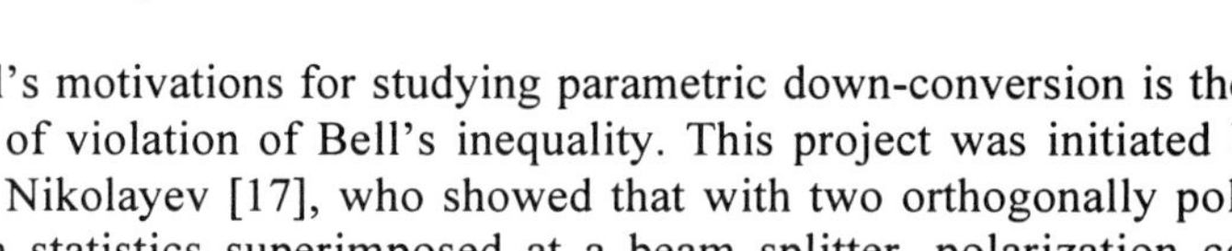

Fig. 3. Two-photon interference between two parametric down-converters.

4.2 Polarization entanglement

One of Mandel's motivations for studying parametric down-conversion is the possibility of a demonstration of violation of Bell's inequality. This project was initiated by a paper from Chubarov and Nikolayev [17], who showed that with two orthogonally polarized beams of sub-Poissonian statistics superimposed at a beam splitter, polarization correlation at the outputs will violate Bell's inequality. Such fields are exactly the ones from PDC. Actually, when the input state for a beam splitter is $|1\rangle_x |1\rangle_y$ the output state is [18]

$$|\Phi\rangle_{out} = \frac{1}{2}\left(|1_{x1}, 1_{y2}\rangle - |1_{y1}, 1_{x2}\rangle + i|1_{x1}, 1_{y1}\rangle + i|1_{x2}, 1_{y2}\rangle\right). \tag{5}$$

If we ignore the last two terms that make no contribution to correlation measurement, the state is the Bohm-type polarization entangled state and violation of Bell's inequality is expected. Violation of 6 standard deviations of Bell's inequality was demonstrated in 1988 [26]. Shih and Alley also independently performed a similar experiment in 1986 [27].

4.3 Frequency entanglement

Almost in the same time when polarization entanglement was demonstrated, Mandel's group realized that parametric down-conversion process produces a state of superposition of different frequency components as shown in the following state:

$$|\Phi\rangle = \int d\Omega \Phi(\Omega)\,|\,\omega_0 + \Omega\rangle_s\,|\,\omega_0 - \Omega\rangle_i \tag{6}$$

Here Ω runs from $-\Delta\Omega/2$ to $\Delta\Omega/2$ with $\Delta\Omega$ the bandwidth of the down-conversion. If two frequency components are picked, a simple frequency entangled state is obtained:

$$|\Phi\rangle = \frac{1}{\sqrt{2}}\left(|\,\omega_1\rangle_s\,|\,\omega_2\rangle_i + |\,\omega_2\rangle_s\,|\,\omega_1\rangle_i\right) \tag{7}$$

and we should expect to see beat signal in time correlation measurement. However, since the spectrum of down-conversion is so wide that it is impossible to observe directly such a frequency entanglement in the form of temporal beating. Fortunately, HOM interferometer converts time into space and makes the observation of spatial beating possible. To ensure the entanglement, filters cannot be placed before the beam splitter but after the beam splitter (Fig.4). What happens here is that when D1 sees a ω_1 photon and D2 a ω_2 photon, there are two possible contributions: ω_1 from signal while ω_2 from idler or vice versa. Such a scheme was also used to observe spatial beating that is normally to fast to see in time domain [28].

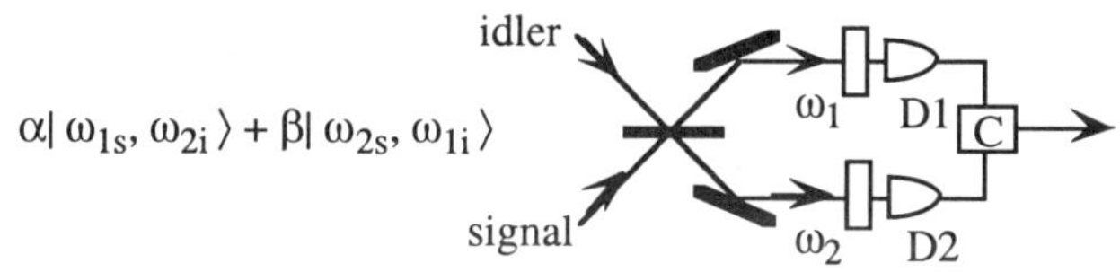

Fig. 4. Scheme for demonstrating frequency entanglement.

4.4 Franson Interferometer

Almost in the same time when Mandel's group demonstrated phase dependent two-photon interference, Franson, while working in a different context, proposed a scheme for demonstrating locality violation of time and energy [29]. After some discussion with Franson during the CQO6 in Rochester, Mandel realized that down-conversion system is best to implement such a proposal. The idea of Franson is that the photon pairs emitted at different time cannot be distinguished by the detectors. This results in two-photon interference. This is

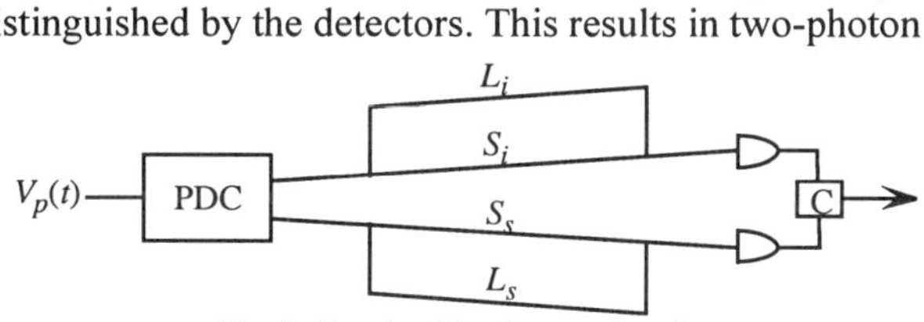

Fig. 5. Sketch of the Franson interferometer.

achieved by two unbalanced Mach-Zehnder interferometers shown in Fig.5, where the two photons arriving in the two detectors for coincidence can go through either the short paths or the long ones. The entangled state here has the form of

$$|\Phi\rangle_{SYS} = \eta V_p(t)\,|\,S_s, S_i\rangle + \eta V_p(t + \Delta L/c)\,|\,L_s, L_i\rangle, \tag{8}$$

where $\Delta L/c$ is the time delay between the long and short paths. Therefore, the visibility is directly related to the coherence of the pump field, $V_p(t)$. This was first analyzed and demonstrated by Mandel's group [30].

4.5 Two-photon interferometer and beyond

In the entangled two-photon states discussed above, the two photons involved are separable or non-degenerate. A degenerate two-photon superposition state of the form can be obtained in the outputs of HOM interferometer. Notice here that the two photons are nonseparable – they come out of the beam splitter all together either from port 1 or port 2. If the two output fields are brought together for interference, a sinusoidal modulation in two-photon coincidence measurement should occur. This was demonstrated by Mandel's group in 1990 [31] and independently by Rarity et al [32]. Such a two-photon interferometer is very much similar to a single photon interferometer. The only difference is that the two photons as a whole instead of a single photon travel in the interferometer. An immediate generalization would be an *N*-photon interferometer. Such an interferometer can increase the accuracy and ultimately reach its fundamental limit in phase measurement. Recently, it was suggested and demonstrated that this *N*-photon interferometer can be used to improve optical lithography [33].

At the dawn of the new millennium, we find ourselves in a new information age, an age that is characterized by the involvement of quantum mechanical principles in information processing. The concept of quantum entanglement, an essential element in quantum mechanics, together with its many nonclassical consequences will gradually become common knowledge to the general public, just like Copernicus' heliocentric theory in fourteenth century. Mandel's pioneering work and many contributions in understanding quantum entanglement will be forever remembered.

5. References

[1] E. Schrödinger, Naturwissenschaften **23**, 807-812; 823-828; 844-849 (1935).
[2] H. J. Kimble, M. Dagenais, and L. Mandel, Phys. Rev. Lett. **39**, 691 (1977).
[3] R. Short and L. Mandel, Phys. Rev. Lett. **51**, 384 (1983).
[4] D. C. Burnham and D. L. Weinberg, Phys. Rev. Lett. **25**, 84 (1970).
[5] See references in Ref.[7].
[6] S. R. Friberg, C. K. Hong, and L. Mandel, Phys. Rev. Lett. **54**, 2011 (1984).
[7] C. K. Hong and L. Mandel, Phys. Rev. A**31**, 2409 (1985).
[8] S. R. Friberg, C. K. Hong, and L. Mandel, Opt. Comm. **54**, 311 (1985).
[9] C. K. Hong, S. R. Friberg, and L. Mandel, Appl. Opt. **24**, 3877 (1985).
[10] C. K. Hong and L. Mandel, Phys. Rev. Lett. **56**, 58 (1986).
[11] G. Magyar and L. Mandel, Nature (London) **198**, 255 (1963).
[12] R. L. Pfleegor and L. Mandel, Phys. Rev. **159**, 1084 (1967).
[13] L. Mandel, Phys. Rev. A**28**, 929 (1983).
[14] R. Ghosh, C. K. Hong, Z. Y. Ou, and L. Mandel, Phys. Rev. A**34**, 3962 (1986).
[15] R. Ghosh and L. Mandel, Phys. Rev. Lett. **59**, 1903 (1987).
[16] Z. Y. Ou, C. K. Hong, and L. Mandel, Phys. Lett. **122**A, 11 (1987).
[17] M. S. Chubarov and E. P. Nikolayev, Phys. Lett. **110**A, 199 (1985).
[18] Z. Y. Ou, C. K. Hong, and L. Mandel, Opt. Comm. **63**, 118 (1987).
[19] C. K. Hong, Z. Y. Ou, and L. Mandel, Phys. Rev. Lett. **59**, 2044 (1987).
[20] Y. Miyamoto, T. Kuga, M. Baba, and M. Matsuoka, Opt. Lett., **18**, 900 (1993).
[21] A. M. Steinberg, P. G. Kwiat, and R. Y. Chiao, Phys. Rev. Lett. **71**, 708 (1993)
[22] M. A. Horne, A. Shimony, and A. Zeilinger, Phys. Rev. Lett. **62**, 2209 (1989).
[23] P. Grangier, M. J. Potasek, and B. Yurke, Phys. Rev. A**38**, 3132 (1988).
[24] Z. Y. Ou, L. J. Wang, and L. Mandel, Phys. Rev. A**40**, 1428 (1989).
[25] Z. Y. Ou, L. J. Wang, X. Y. Zou, and L. Mandel, Phys. Rev. A**41**, 566 (1990).
[26] Z. Y. Ou, and L. Mandel, Phys. Rev. Lett. **61**, 50 (1988).
[27] Y. H. Shih and C. O. Alley, Phys. Rev. Lett. **61**, 2921 (1988).
[28] Z. Y. Ou, and L. Mandel, Phys. Rev. Lett. **61**, 54 (1988).
[29] J. D. Franson, Phys. Rev. Lett. **62**, 2205 (1989).
[30] Z. Y. Ou, X. Y. Zou, L. J. Wang, and L. Mandel, Phys. Rev. Lett. **65**, 321 (1990).
[31] Z. Y. Ou, X. Y. Zou, L. J. Wang, and L. Mandel, Phys. Rev. A**42**, 2957 (1990).
[32] J. G. Rarity et al, Phys. Rev. Lett. **65**, 1348 (1990).
[33] A. N. Boto et al, Phys. Rev. Lett. **85**, 2733 (2000); M. D'Angelo et al, Phys. Rev. Lett. **87,** 013602 (2001).

The Quantum Phase – A New theory Meets Experiment

Jaewoo Noh
Inha University,402-751, Inchon, South Korea
Tel)82-32-860-7662, Fax)82-32-872-7562, e-mail)jnoh@inha.ac.kr

Abstract:
The operational approach to the phase problem, studied by Dr. Mandel and his students during the period 1990-1995, is reviewed. The theory of the operationally defined phase difference between two optical fields and the experiments performed to measure the phase difference using photon counting method are explained, and the importance of the research is re-examined.

1. Introduction

The old question of how to represent the optical phase in quantum mechanics has been the subject of discussion and controversy for a long time [1-8]. Because it is one of the basic assumption of the quantum mechanics that a physical observable can be represented by a Hermitian operator, and the phase difference between the two optical fields are measured in common experimental situations, many scientists tried to find a proper Hermitian operator for the optical phase to make the quantum theory of light more complete. However, most theories developed for the quantum phase operators were considered to be unsatisfactory for one reason or another, and there had been few experiments to test the validity of the theories.

Dr. Mandel and his students including myself approached the phase problem in a different, more operational way, by analyzing what is actually measured in a typical homodyne experiments to determine the phase difference between two fields, both in classical and in quantum optics [9-10]. We also performed experiments to measure the phase difference between two fields with 8-port homodyne detection scheme and reported the result [11]. It turned out that the phase difference of two fields could be measured by counting photons, and the phase probability distribution of a given field could be obtained [12]. Careful analysis of the experimental result showed that the phase could be associated with the field in a single photon level.

The purpose of this paper is to review the former works of Dr. Mandel and re-examine the importance and the impact of works to the scientific research.

2. Theory

In a typical simple phase measurement scheme, two optical fields can be combined using a beam splitter and the resultant output field intensities are measured by two photo-detectors. The output intensities have either sine or cosine modulations that depend on the phase difference between the two input fields so that one can obtain the sine or cosine of the phase difference. The measurement scheme has one drawback. One cannot measure the sine and cosine of the phase difference simultaneously, so that one cannot determine the value of the phase difference from one measurement. Inserting a $\lambda/4$ plate in one input field can change the output intensity modulation from sine to cosine or vice versa, but it is not guaranteed that one can obtain the correct value of the phase difference because the quantum state changes when one makes an operation to the field. Another measurement scheme is to use the 8-port homodyne scheme, shown in the fig. 1. Because the $\lambda/4$ plate is always in the one arm of the interferometer, two outputs from the interferometer have sine modulations and the other two outputs have cosine modulations. One can obtain the phase difference of the two input fields in one measurement, and the repeated measurement will show the distribution of the phase difference values for the input fields in certain states.

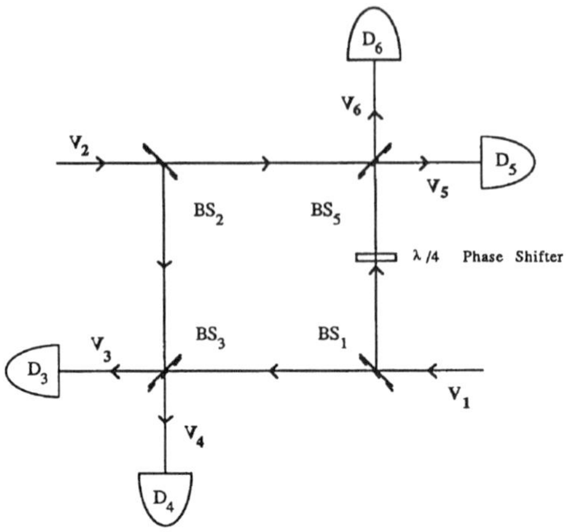

Fig. 1.8-port homodyne detection scheme to measure the phase difference between two input fields [].,

We defined inferred and measured phase difference operators which are made of the photon number operators for the output fields [9], as below.

$$C \,..\, n_4 - n_3, \qquad S \,..\, n_6 - n_5, \qquad [C,S] = 0 .$$

$$C_M \,..\, C\,?(C^2 + S^2)^{-1/2}, \quad S_M \,..\, S\,?(C^2 + S^2)^{-1/2}$$

The output photon number operators n_3, n_4 ... include the vacuum field coming from the unused input ports. The C_M and S_M operators are called measured phase operators. The relation between the theoretical phase difference operators and the measured output port photon numbers became clear with the measured phase operators, because what we measure in the photon counting experiment are the number of output photons. The inferred phase operators are normally ordered operators, related to the classical phase information contained in the two input fields, and they are the phase difference value without the vacuum field contribution.

3. Experiments

We started experiment with the light in the coherent state. A beam splitter split a frequency stabilized HeNe laser output and the two resulting fields were directed into the 8-port homodyne interferometer. Four PMT's were used to measure the output photon numbers. Using the computer interface, we kept the recorded number of photons for many repeated measurements. The phase difference measurement result for various input field intensity is shown in fig. 2.

The average value of the measure sine and cosine of the phase difference, and the dispersion of the sine and cosine values, which are considered as the fluctuation of the phase difference, are plotted. The experimental results agree with the theoretical calculations, shown in the fig. 2 as solid lines, very well. In the case of two very weak field, we still get the information of the phase difference, because we discarded all-null results from the measured data set and renormalized the measured values.

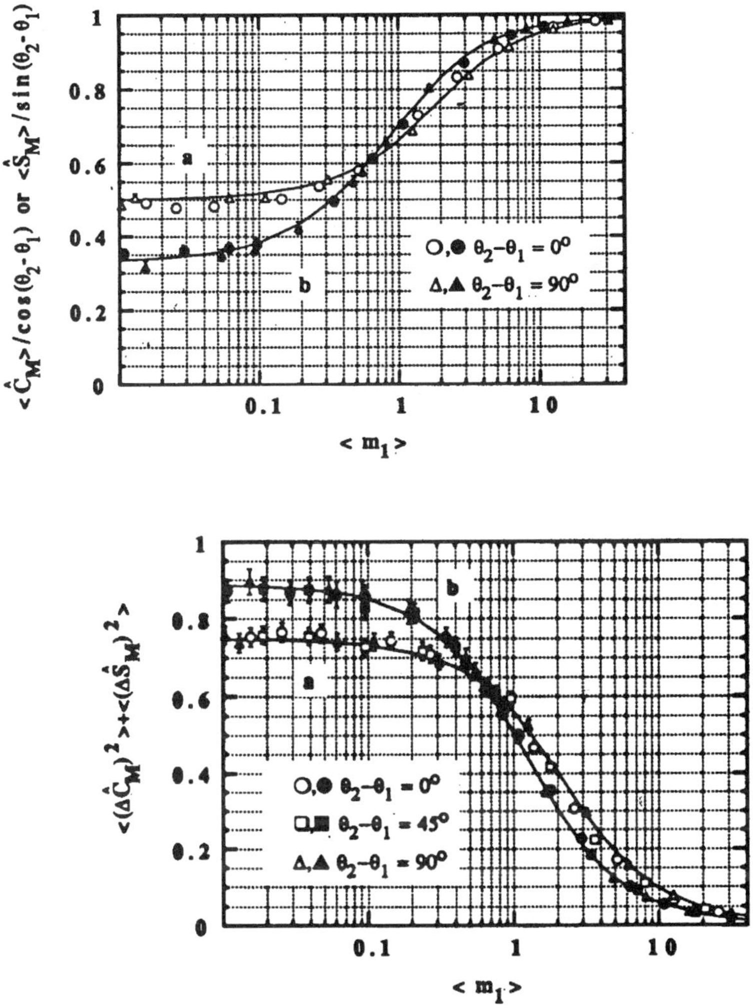

Fig. 2. Experimental data for the coherent state light.

It turned out that, if we discard the all-null results from our experiment with a very weak intensity coherent state light, we were effectively measuring the phase difference of the split single photon state. We also found that we could get the probability distribution of the phase difference by changing the phase shift introduced in the one arm of the interferometer. The experimental results for the coherent state and the spilt single photon are shown in the fig. 3 and fig. 4 respectively. In the case of fig. 3, one input field is rather strong, thus acts like a local oscillator or a reference field, and the other input field has average photon number of 1.0.

The case of split photon state, mathematically represented as $|0\rangle|1\rangle + |1\rangle|0\rangle$, is shown in the fig. 4. In this case, the effective output photon number is always 1, detected by one of the four PMT's installed in the experiment set-up. The probability of detecting the photon in one of the 4 detectors changes as the phase shift value in the interferometer is changed.

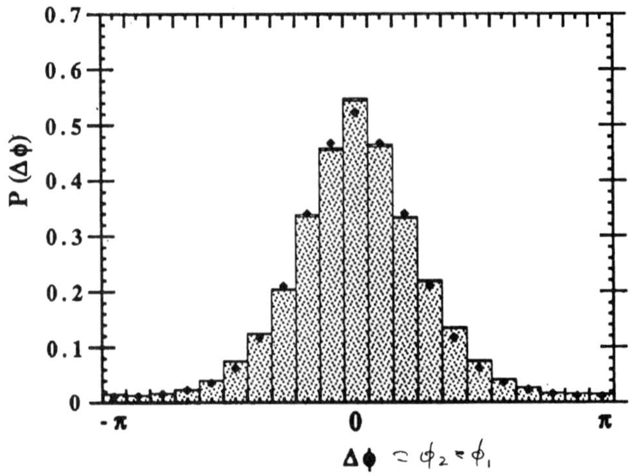

Fig. 3. Experimental result for the probability distribution of the coherent state.

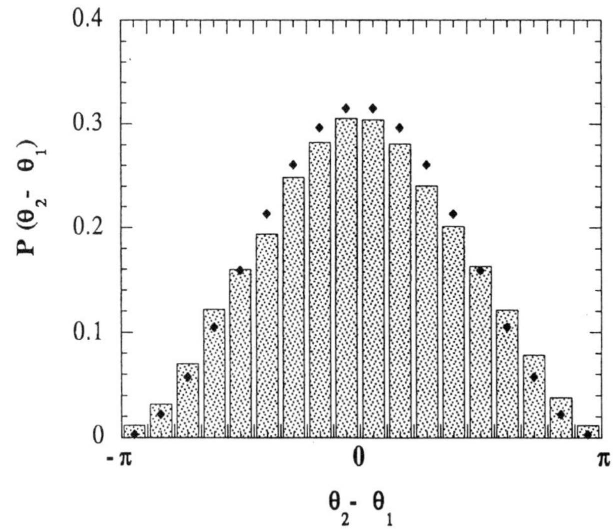

Fig. 4. Experimental data for the split single photon state.

The fact that we could measure the phase probability distribution of the split single photon state tells us that the phase difference is a physical quantity related to the state of the field. In other words, the photon has probability amplitude to reach any of the four detectors and the probability is determined by the wave nature of light.

We also did another experiment to reveal the wave property of light. We made a partially coherent light by putting a rotating ground glass in the laser output. The degree of coherence was changed as we change the speed of the rotating ground glass, and the probability distribution of the phase difference was measured for several values of the degree of coherence. The result is shown in fig. 5. Again, we found excellent agreement between the theory and the experiment.

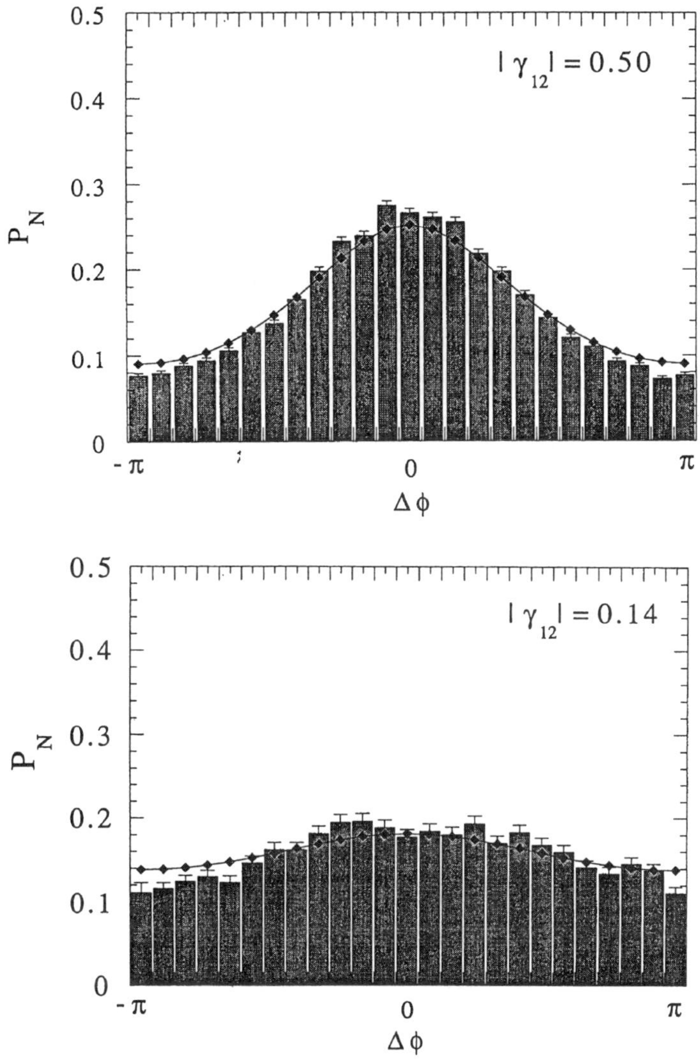

Fig. 4. Experimental data for the partially coherent light

4. Summary

From the operational theory and the experiment we learned that the concept of the optical phase is closely related to the classical optics, where the optical field is treated as the electromagnetic wave. The phase measurement results tell us that what we really measure is the phase difference between the two input fields. The experiment with split single photon shows that we can measure the phase in the single photon level, and what we measure is the phase of the probability amplitude, which is just the electromagnetic wave mode associated with the photon. Each photon detected by the photo-detectors carries the information of the phase in that sense. The photon is a quantized electromagnetic field mode, and it interferes with itself.

We wish that the research work performed by us under the supervision of Dr. Mandel helped people in this research field to get better understanding of the nature of photon and the concept of optical phase.

References

[1] P.A.M. Dirac, Proc. R. Soc. London A **114**, 243 (1927).
[2] W.H.Louisell, Phys. Lett. 7, 60 (1963).
[3] L. Susskind and J. Glogower, Physics (Long Island City, N.Y.) **1**, 49 (1964).
[4] P. Carruthers and M.M. Nieto, Phys. Rev. Lett. **14**, 387 (1965).
[5] C.M. Caves, Phys. Rev. D **26**, 1817 (1982).
[6] J.H. Shapiro and S.S. Wagner, IEEE J. Quant. Elect. **20**, 803 (1984).
[7] S.M. Barnett and D.T. Pegg, J. Phys. A **39**, 1665 (1989).
[8] D.T. Pegg and S.M. Barnett, Europhys. Lett. **6**, 483 (1988).
[9] J.W. Noh, A. Fougeres, and L. Mandel, Phys. Rev. A **45**, 424 (1992).
[10] J.W. Noh, A. Fougeres, and L. Mandel, Phys. Rev. A **46**, 2840 (1992).
[11] J.W. Noh, A. Fougeres, and L. Mandel, Phys. Rev. Lett. 67, 1426 (1992).
[12] J.W. Noh, A. Fougeres, and L. Mandel, Phys. Rev. A **71**, 2579 (1993).

Parametric Downconversion II – The Twins Grow Up

David Branning
Department of Physics, University of Illinois at Urbana-Champaign, 1110 W. Green St., Urbana, IL 61801
Tel: (217) 265-5457 Fax: (217) 244-7559 e-mail: branning@uiuc.edu

Abstract: I present a personal account of the research and activities of Leonard Mandel and his students in the 1990's. In that final decade, Professor Mandel's group expanded the use of parametric downconversion to explore the foundations of quantum mechanics, building upon the techniques which he had pioneered in the 1980's. The investigations of the operational quantum phase reached their conclusion, and experiments continued in atomic resonance fluorescence. At the end of the decade, new collaborations led to experiments in spin squeezing, homodyne detection, and pulsed downconversion.

Leonard Mandel in his office at the University of Rochester, circa 1994.

Introduction

"The first thing I noticed when I entered Professor Mandel's office in the spring of 1992 was a bookshelf stacked to the ceiling with numbered reams of publication reprints. Upon closer inspection, I realized two things: that every article on the shelves bore his name in the list of authors, and that there were some 220 of them. Although I had never heard of Professor Mandel before that day, standing in front of that wall of paper stretching back to 1954, I began to feel somewhat awed. I was shown in to the second part of his office, where, in front of another wall of reprints, Professor Mandel kindly explained to me some of the recent work his research group had carried out. As our discussion continued, my intimidation gave way to excitement, and I realized that this was the kind of work I wanted to do." [1]

Thus began my relationship with Leonard Mandel. As his 38th student, I met him at a time when he had already made a vast array of contributions to quantum optics. I came to him in search of a direction for my own research and growth as a physicist; what he showed me on that first day captured my imagination.

A Mind-Boggling Experiment

As Dr. Mandel explained, there is a special kind of light source, called a *parametric downconverter*, that creates photons two at a time. It requires an input beam of light, called the *pump*, which is effectively split into two lower-frequency output beams called the *signal* and the *idler* (see Figure 1.)

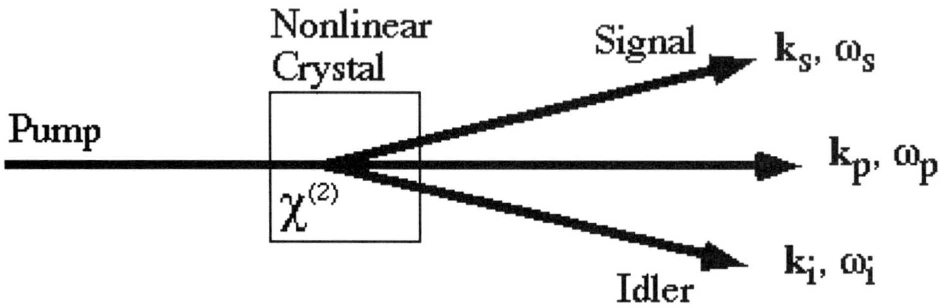

Figure 1: Spontaneous parametric downconversion.

The process occurs in particular crystals that have a nonlinear dielectric response, so that the induced polarization within them depends on the square of the pump's electric field. The nonlinearity is very small, so that downconversion is an inefficient process; even with a very intense laser supplying the pump beam, relatively few signal and idler photons are produced. But whenever one is created, the other must be created with it in order to conserve the total energy and momentum of the system.

Now it is possible to arrange for the signal beams from two identical downconverters to be superposed, so that interference between them is possible. This can be done by shining the signal beams from each downconverter onto opposite sides of a half-silvered mirror (or *beamsplitter*) which reflects half of the incident light from each side, and transmits the other half. If we look at one of the beams leaving the beamsplitter, it contains a superposition of reflected light from one downconverter, and transmitted light from the other. If we were to measure the intensity of this combined beam, say, by counting the arrival of photons at a detector in a given time interval, we might expect to see interference – a modulation of the intensity – as the relative phase between the two input signal beams is varied. The phase could be varied by making small changes to the path lengths from each downconverter to the beamsplitter.

In fact, we do not see this interference in general. We therefore say that the signal beams from the two downconverters are not *coherent* with one another – that is, they do not possess the definite phase relationship between them that is required for interference to occur. This is true even if each downconverter is pumped by the same laser beam, as shown in Figure 2.

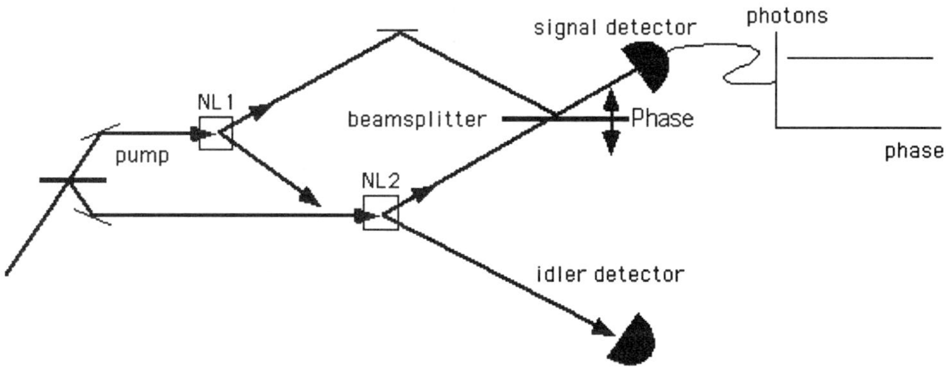

Figure 2: Signal beams from two downconverters are combined but do not interfere.

However, there is a special circumstance in which interference *can* be seen in the superposed signal beam. If the *idler* beams from the two crystals – traveling in a totally different region of space – are overlapped, as in Figure 3, then it turns out that the *signals* become coherent, and display interference!

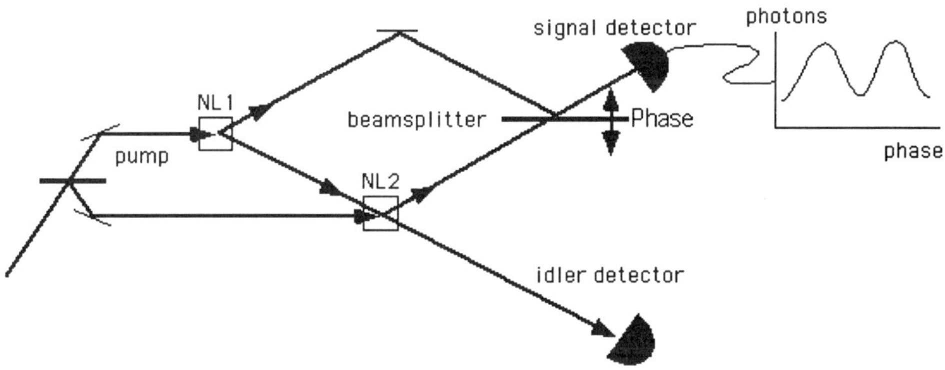

Figure 3: When the idlers are overlapped, the signals interfere.

Why should this be so? One might guess that the photons in the first idler beam are stimulating the downconversion process in the second crystal, and that this produces coherence in the same way that stimulated emission produces coherence in laser light. But here is the remarkable fact: the coherence persists even when the downconversion process is so weak that only *one* signal photon is in the apparatus at a time! Moreover, the counting rates are unaffected by the alignment of the idlers. So there cannot be any stimulated emission.

For this reason, the phenomenon is called *induced coherence without induced emission.* But how does it work? In this experiment, there are two processes that lead to our counting a signal photon. Either the signal photon came from the first crystal and was reflected to the detector, or it came from the second crystal, and was transmitted to the detector. The rules of quantum mechanics say the following: if these two processes are *distinguishable*, then the total counting rate is just given by adding the rate for each process, and no interference can occur. If, on the other hand, the two processes are *indistinguishable*, then we must add the *amplitudes* for each process to obtain the total rate, and interference is possible.

In light of this rule, it is easy to see why there was initially no interference, when the idlers were not aligned (Figure 2). Suppose we had placed an extra, perfectly efficient, detector into the idler beam from the first crystal. Then for each photon that arrived at the signal detector, we could have determined its origin simply by checking to see whether our idler detector also fired. If it did so, then we know that this particular signal photon must have come from the first crystal; if not, then we know it came from the second. Because the two paths for the signal photons were distinguishable in this way, it was impossible for them to interfere.

However, when we align the idler beams in space (Figure 3), we make it impossible to perform such a test. Adding a detector that counts the arrival of idler photons will do us no good in this situation, because no matter which crystal produces the idler, our added detector will fire! And now, because there is no test that can tell us which crystal produced a given signal (or idler) photon, the signal paths become indistinguishable, so the signal photon counting rate must display interference. The interference fringe observed in Dr. Mandel's lab is shown in Figure 4, in contrast to the non-interference shown when the idler connection was blocked by an absorber (equivalent to misaligning the idlers, because again the idlers from crystal 2 could be used to test the origin of each pair.)

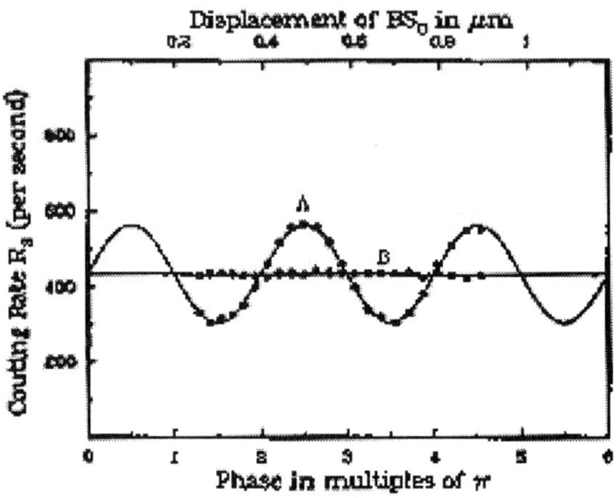

Figure 4: Interference of the signals (A), and non-interference when the idlers are blocked (B) [2].

This simple experiment showed that we can make interference come and go for a set of photons, by manipulating a completely different set of photons, in a different region of space, that determine what information is available about them in the outside world. And perhaps the most mind-boggling aspect of this is the following: the interference comes and goes at the signal detector *whether or not we detect the idlers.* In other words, to destroy interference, it is not even necessary to record the distinguishing information – it is merely necessary that the distinguishing information *exists*, somewhere in the universe. Conversely, to create interference, we must arrange it so that this distinguishing information does not exist, anywhere in the universe.[1] It does not matter whether anyone actually bothers to record this information, or indeed, whether we are able to record it in practice. This led Dr. Mandel to conclude that the quantum state, which predicts whether or not interference will occur, "reflects not only what is known, but to an extent also what is knowable, in principle, about the photon."[4]

---

[1] This also implies that if we do not see interference between two beams of light, it means that there is some method, in principle, to tell from which beam each photon originated, even if we have not discovered what that method is. The relationship between indistinguishability and interference was succinctly derived by Professor Mandel in ref. [3], and has a direct analogy with a modern-day realization about any quantum system: namely, if the state of a quantum system is mixed (i.e., not capable of displaying interference), this can always be viewed as the result of ignoring (or "tracing over") some other degrees of freedom that are correlated to the system.

Variations on the Mind-Boggling Experiment

After this conversation, I left Dr. Mandel's office very excited, sure that I had seen something profound. I decided to go to Rochester for graduate school, hoping to join his group and continue this kind of work. One year later, in the spring of 1993, my hopes were realized as I was accepted into his research group. At that time the group contained seven students working in three major areas: Lijun Wang, Xingaou Zou, and Tim Grayson worked on downconversion; Samir Bali and Frank Narducci carried on with atomic resonance fluorescence; and Jaewoo Noh and André Fougères investigated the quantum phase operator. Joining concurrently with me was Justin Torgerson, bringing the total number of students to nine.

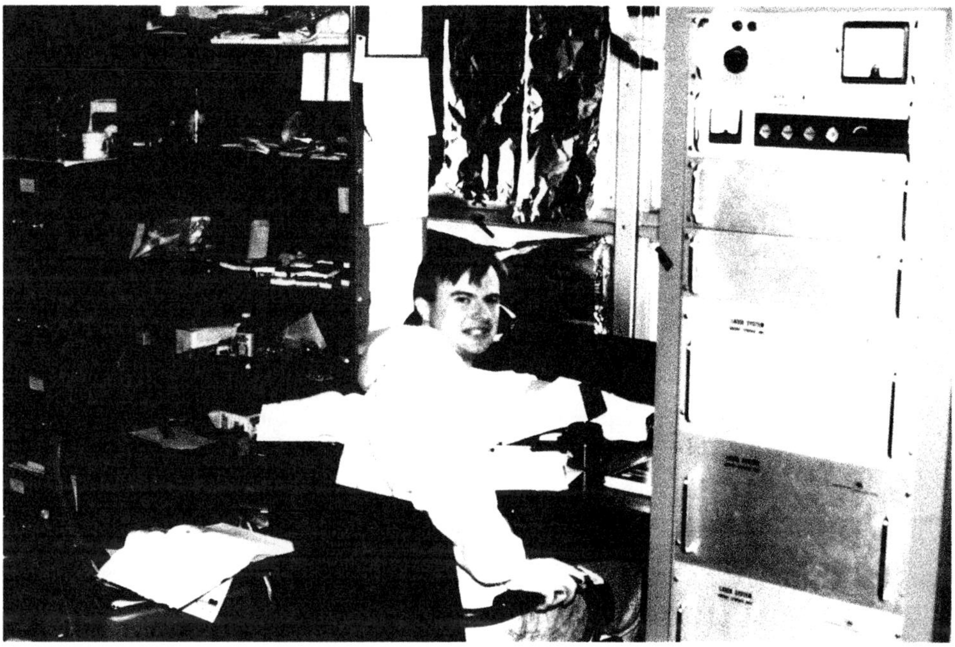

The author, shortly after joining Dr. Mandel's group.

Initially, Lijun and Zou occupied the downconversion lab, where the induced coherence experiment had been performed. They were continuing the work begun by Jeff Ou and C.K. Hong[2]. Their work had already garnered some attention in *Scientific American*, with the apt title "Quantum Philosophy" [5]. Shortly after Justin and I joined the group, Lijun graduated, and Zou began writing his thesis, leaving Tim in charge of the downconversion lab.

---

2 see "Parametric Downconversion I – The Twins Are Born," in this volume.

Tim Grayson in the computer room.

Tim was constructing a variation on the induced coherence experiment, in which mirrors were placed around the idler connection so that the idler photons bounced back and forth before possibly being ejected by a beamsplitter in the middle (see Figure 5).

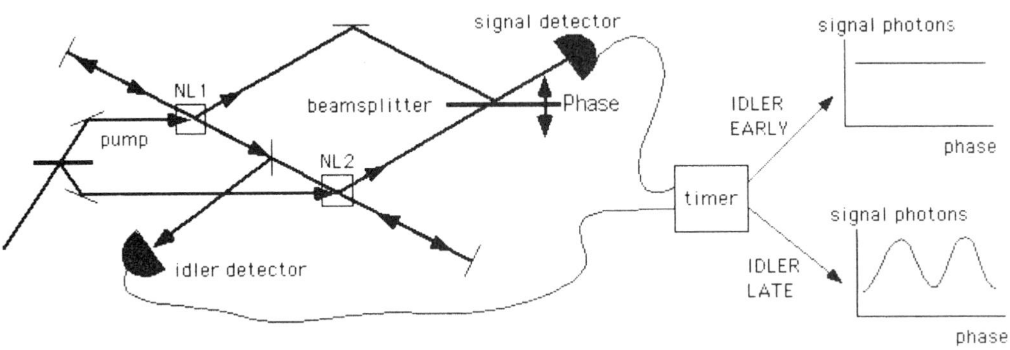

Figure 5: Induced coherence with a cavity around the idler connection.

If the idler was ejected right away, it would reveal that the corresponding signal photon came from the first downconverter; otherwise, if the idler made one or more cavity round trips before being ejected, it could have come from either crystal, and therefore the signal origins once again became indistinguishable. So by correlating the detection of the signals with the arrival of "early" or "late"-ejected idler photons, one could see the signal photons "choosing" to be indistinguishable (and displaying interference) or "choosing" to be distinguishable (and not displaying interference.)

Justin and I were both eager to work in the downconversion lab. Tim had us machine some parts for him, and then, one afternoon he brought us into the darkened lab with him as he took data. We adjusted various controls during the experiment, and then, without thinking much of it, went back to our studies as first-year graduate students. Shortly afterwards, a draft of the paper to be submitted [6] was in our mailboxes, and we were shocked to see ourselves listed as authors! We asked Professor Mandel about it, not sure that our limited participation really merited co-authorship, but it was his feeling that our contributions should be acknowledged. We were thrilled, and of course Dr. Mandel must have known that this taste of authorship would provide even more motivation for Justin and me, the "rookies," to carry out new experiments.

The next variation on the induced coherence experiment was to see what the effect of a phase shift on the idler connection would be for the signal interference. The phase shift in this case was induced by a set of waveplates that sent the polarization of the idlers around a closed loop, with a net geometric phase accrued [7].

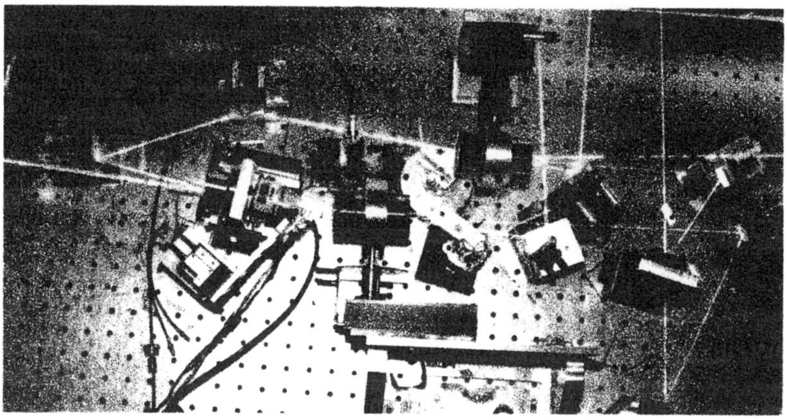

The geometric phase experiment.

Geraldo Barbosa was visiting from Universidade Federal de Minas Gerais, in Bello Horizonte, Brazil, and was also working on the geometric phase experiment. While this work was carried out by Justin, Tim, and Geraldo, I moved three doors down the hall to help Frank Narducci with his experiments on resonance fluorescence.

Frank Narducci in the resonance fluorescence lab.

Frank had come into the group under Samir, and the two of them had been working with light emitted by sodium atoms, while these atoms themselves were driven on or near resonance with a laser. This condition, called "resonance fluorescence," was a long-time interest of Dr. Mandel's[3]. Frank and Samir were working in the same lab where the pioneering observations of antibunching had been made by Jeff Kimble and Mario Dagenais . In fact, Frank and Samir used much of the same equipment, and when things weren't working, they would often comb through the lab notebooks and Ph.D. theses left behind by Kimble and Dagenais, looking for some clue as to how they might have solved a similar problem twenty years earlier.

However, once Samir began writing his thesis, Frank was left with nobody to help out. He knew that I was interested in quantum interference, and so he cleverly lured me into his lab by explaining a very interesting experiment.

Quantum Interference with Atomic Emission

Suppose we have a single, two-level atom, undergoing resonance fluorescence – emitting photons at random times, even as it is being made to oscillate regularly between its ground and excited states. If the photons emitted from such an atom could be superposed on top of the driving field at a detector, would interference occur? It turns out that the answer is "sometimes yes, sometimes no" (see Figure 6). For according to the distinguishability

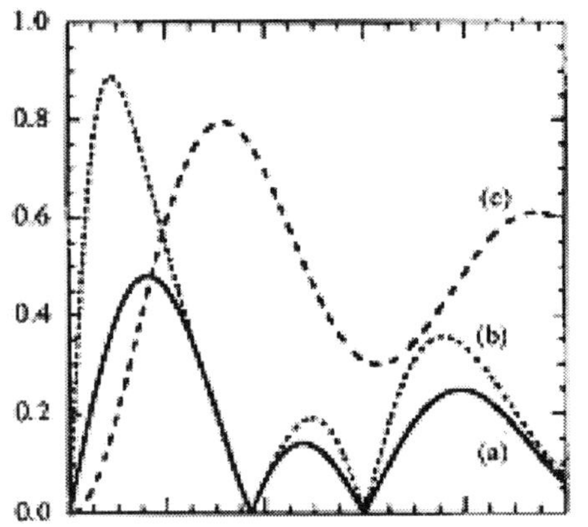

Figure 6: Evolution of the fluorescent field (a), interference visibility (b), and atomic excitation (c) [8].

arguments above, the interference should only occur for those detected photons whose origin (the atom, or the driving field) could not be determined, in principle, by any other measurements. This was not always the case, because at certain times during its oscillation, the state of the atom would reveal the source of the photon! For example, if the atom were in the ground state, it could not have emitted the photon. This means the detected photon could only have come from the classical driving field. Similarly, at other times the atom would be driven into its excited state, where it would be possible to measure its energy without disturbing it. If the photon happened to be emitted from this energy eigenstate of the atom, we

[3] See "Photon Antibunching and Resonance Fluorescence Experiments," in this volume.

would again be able to identify its origin[4]. Either way, interference would not be possible. On the other hand, for those times when the atom was not in a definite energy state, the origin of the photon would be uncertain, and interference would occur [8].

The experiment had first been proposed by Hong, Ou, and Mandel in 1988 [9], but it was Frank and Samir who had begun the work of actually trying to observe this time-dependent interference between the nonclassical light from the atom and the classical light from the driving field. The difficulty with actually trying to observe the interference was caused by the relative motion between the atoms in the beam and the driving field; such motions amount to an uncontrolled phase disturbance that will destroy any possible interference[5]. To get around this problem, Dr. Mandel had suggested that they use the elastic scattering from the atom as the classical light source, thereby canceling the effects of the atomic motion [11]. The elastic scattering might be separated from the fluorescence using a Fabry-Perot filter, as shown in Figure 7.

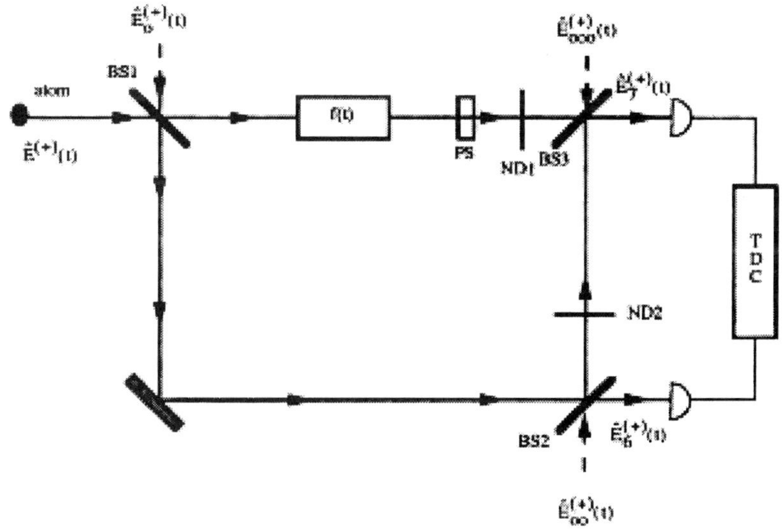

Figure 7: Scheme to interfere resonance fluorescence with classical light [8].

The atom would start a clock when it emitted a photon, indicating that it was beginning its oscillation from the ground state. Then one could record interference at various times throughout the oscillation cycle, and see that the visibility, or contrast, of the interference would vary depending on how far the atom had gone through this cycle.

The coffee machine was in Frank's lab, and one of the first events of the day would be that Tim and André would come over for coffee to discuss whatever results or difficulties they had encountered the previous night. Frank, Tim, and André called themselves the "Dad Squad," and would also share stories related to their toddlers at home. These informal meetings between the three labs were almost a daily ritual. The closeness with which the "Dad Squad" worked and advised each other was something that truly inspired me about Dr. Mandel's group. There was an atmosphere of intellectual honesty and mutual respect, which brought forth the best work from all of us, and was clearly part of a long tradition. It was in this atmosphere that the idea to combine Tim's induced coherence apparatus with André's phase operator experiments was born.

4 Another way of putting it is that a photon emitted from a fully excited atom cannot have a definite phase.

5 See, for example, a recent related interference experiment with single trapped ions [10].

Downconversion and the Quantum Phase Operator

A very old question in quantum mechanics had been, "is there an operator for the phase of a photon, and if so, how is it constructed?" Dr. Mandel had taken up this problem with Jaewoo and André. They had constructed an 8-port homodyne apparatus that determined the relative phase between two classical light beams that were put into two of the ports, by subtracting and dividing various combinations of the intensities recorded at the output ports. To determine the quantum phase between two photons, they reasoned, we should simply replace the classical (coherent) light beams with single-photon input states, and replace the intensity measurements at the detectors with photon-counting operators. This was the "operational approach" to the quantum phase problem[6].

The operational approach had made correct predictions for weak, coherent beams of laser light, and for partially coherent fields as well with the aid of Tim [12] and Justin [13]. But now the true test would come from the almost perfect single-photon states emitted from parametric downconverters. Thus Tim and André began conspiring to use the two signal beams from the induced coherence experiment as inputs to the quantum phase apparatus (see Figure 8), with the bonus that the coherence between the signals could be directly modified by blocking or unblocking the idler beams [14].

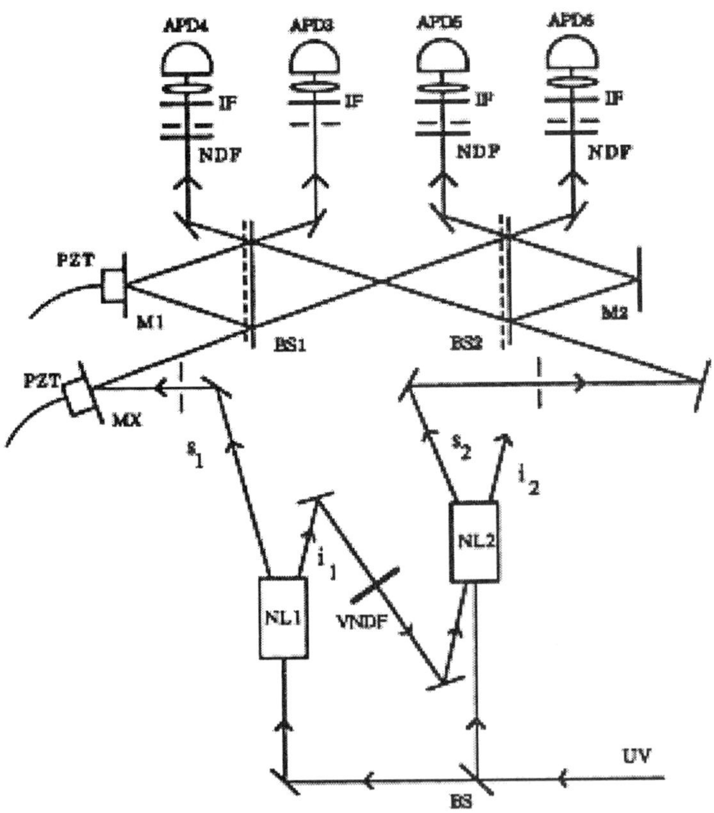

Figure 8: Diagram of the operational phase experiment with "induced coherence" inputs [15].

[6] see "The Quantum Phase: A New Theory Meets Experiment," in this volume.

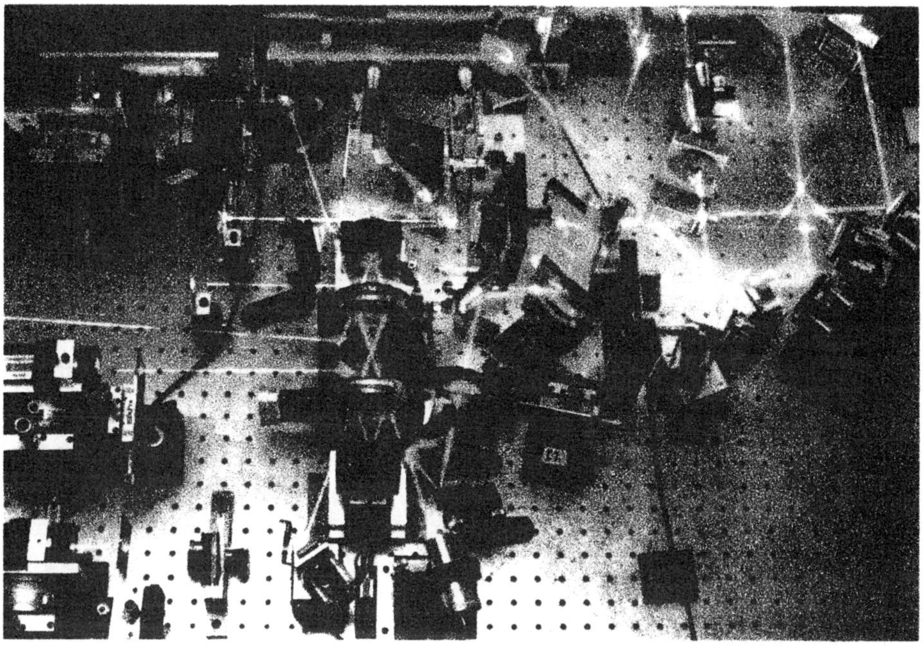

Measurement of the operational quantum phase between two signal photons under induced coherence.

During these early days, Justin and I were keen to have lunch with the other members of the group, particularly the Dad Squad. But this often posed a difficult three-body problem. André would come down the hall to Frank's lab and announce, with a single word in his thick French Canadian accent, "lunch-time!" Frank would reply, "give me five minutes." By the time Frank was ready, André or Tim would have gotten interested in something else, and request their own five-minute delays. Not wanting to waste the five minutes, Frank would go back to work, and the cycle would continue. Justin and I were left to ask each other, "which do you think is harder, quantum optics, or getting these guys to go to lunch?"

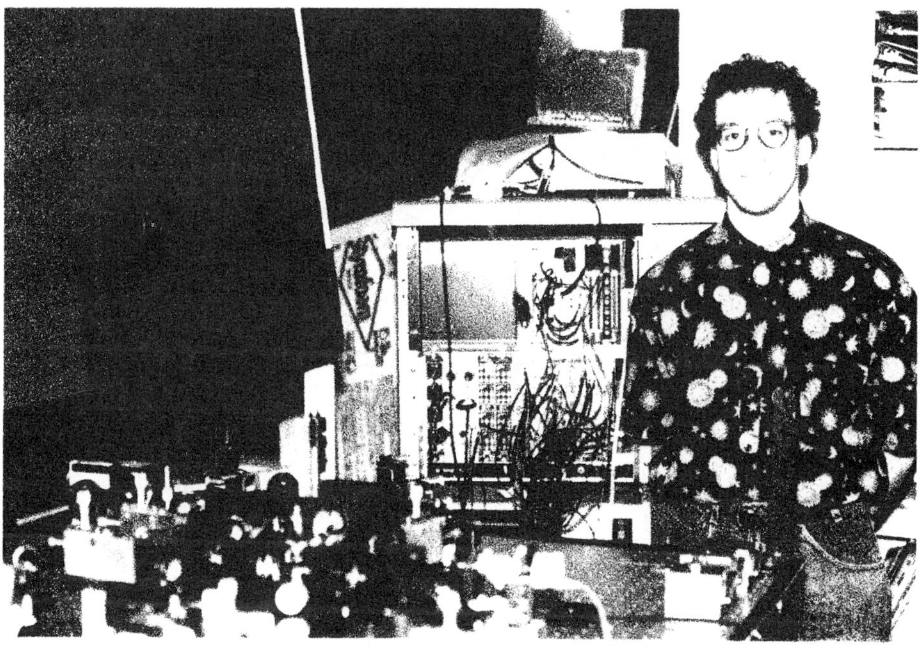

André Fougères in the downconversion lab.

New Visitors

Tim went on with Geraldo to study some effects relating to the spatial overlap of the downconversion beams [16], and then began writing his thesis. But André was not left entirely alone with the operational phase problem, for we were joined in the spring of 1994 by Carlos Monken, a postdoctoral researcher from Geraldo's institution in Brazil. We were also joined by Augusto Garuccio, a visiting professor from the University of Bari, Italy.

Carlos Monken and Augusto Garuccio.

Augusto and Dr. Mandel had long been interested in problems arising from quantum *entanglement* – the situation that arises when two (or more) parts of a quantum system are considered separately, each behaving in a way that is random, yet somehow correlated with the other. For example, a pair of photons might be entangled in their polarizations. This would mean that a measurement of each photon's polarization alone would appear to give random results, but that when the results for the two photons were compared, they would be identical in every trial. The question of how the outcomes of these measurements could be perfectly *random* (as demanded by quantum mechanics) and yet perfectly *correlated* (also as demanded by quantum mechanics) dated back almost to the birth of quantum theory itself, and was the basis of Einstein's most jarring criticism of quantum mechanics [17].

Einstein and his co-authors had used this property of entanglement to argue that the randomness of quantum-mechanical outcomes could not be the end of the story, and that some deeper theory must be developed to complete the quantum description of nature – to explain the appearance of "random" results. To satisfy Einstein, such a theory would have to embrace two very appealing ideas: *locality*, which demands that two separate systems cannot influence each other unless they can be connected by a light signal, and *realism*, which insists that properties like polarization must have definite values even when they are not being measured.

Augusto was, and is, one of those modern-day physicists seeking clues about what such "local realistic" theories should look like. Meanwhile, Dr. Mandel and his students had already come up with two ways to use parametric downconverters to test some alternatives to quantum mechanics. The first had been to use the photon pairs to test whether nature obeyed a mathematical constraint on local realistic theories discovered by J.S. Bell in 1964. The "Bell inequality" had been tested in earlier, more famous experiments, but never with such ease and simplicity as with the downconverter [18]. The second had been a test of an alternative "Pilot wave" theory developed by Bohm and deBroglie; Dr. Mandel and his students had shown that at least one version of this theory did not correctly predict the behavior of the downconverted photons in an ingeniously designed experiment [19].

This second demonstration had drawn some criticism from Augusto [20], and it was clear to me from the first day I met him that he disagreed strongly with Dr. Mandel's traditional interpretation of quantum mechanics. But it also impressed me that, despite their disagreements, Dr. Mandel would nevertheless invite Augusto to visit, and that Augusto would accept. Dr. Mandel clearly had a desire to learn from an opposing point of view, while Augusto just as clearly held Dr. Mandel in the highest regard. This was beneficial for all of us, as we learned much from Augusto during his visit.

The Hardy-Jordan Experiment

Dr. Mandel and Augusto had become interested in a proposal for a new way to test whether local-realistic theories could be true. The proposal was made by Lucien Hardy [21], and brought to Dr. Mandel's attention by Thomas Jordan [22]. The general idea was to construct a system of entangled quantum particles, and to make four sets of measurements on them that, if considered from a local-realistic point of view, would generate a paradox, and show that nature could not be operating in a local-realistic fashion. Dr. Mandel translated this proposal into a design [23] for a downconversion experiment (see Figure 9), and so it began.

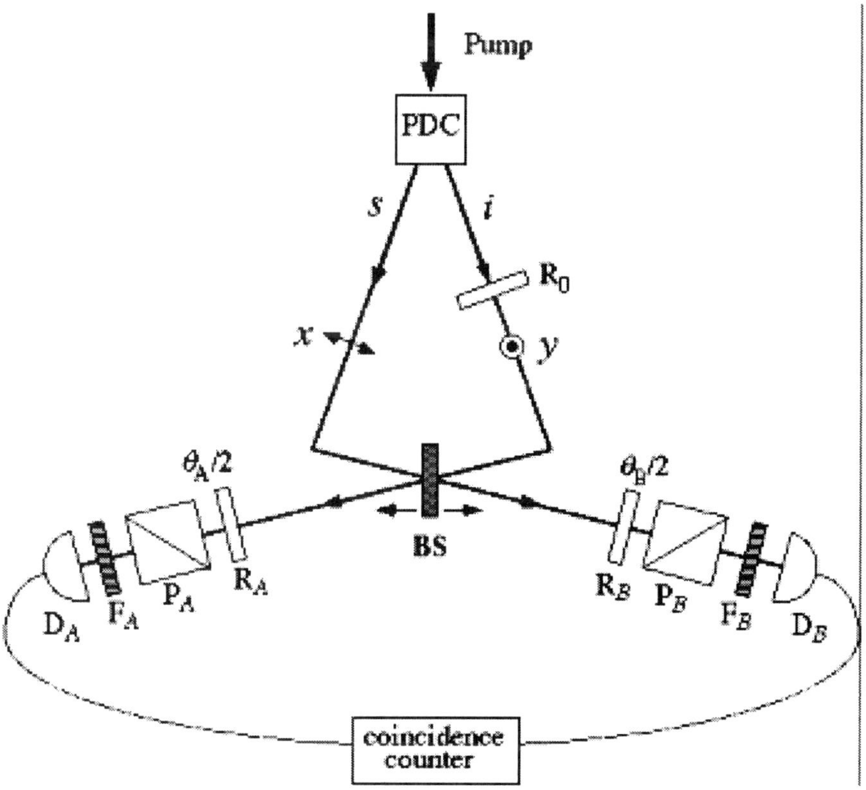

Figure 9: The Hardy-Jordan experiment.

When talk of the Hardy-Jordan experiment began, I realized that it was time to leave Frank's lab and go back to downconversion. After all, as I recently joked with Frank, I had turned two knobs for Tim and gotten my name on a paper, whereas after eight months in Frank's lab, no paper had appeared! Frank saw the humor in this, as well as the harsh reality: his interference experiment had turned out to be extremely difficult (and would later be shown to be impossible, with his equipment[7]). Dr. Mandel did not object to my switching over, nor did Frank, even though the situation was terribly unbalanced – Augusto, Justin, and myself working on downconversion, Carlos and André on phase measurements, and Frank left alone to struggle with the atomic beam.

Justin and Augusto computing polarizations.

As we carried out the Hardy-Jordan experiment, Augusto taught us the history of the Einstein criticism of quantum mechanics, and of the experimental tests of Bell's inequality. Meanwhile, André finished his phase measurements and began writing his thesis, leaving Carlos free to help Frank and the rest of us. Carlos was enormously helpful in the lab, and taught us many "tricks of the trade" as we aligned optics on into the night. When the Hardy-Jordan data was finally complete, there was a flurry of excitement. Drafts of the paper circulated about the group, with many rounds of corrections and debates. Augusto was actively involved, but in the end chose not to be an author [24]. After the paper was published, the work received some attention in *Science* [25] and in *Optics & Photonics News* [26]. The editors of *American Scientist* also solicited an article from us [27].

Next, a small modification of the Hardy-Jordan apparatus allowed us to carry out a test of Bell's inequality, but this time measuring circular, rather than linear, polarizations. As

---

[7] In Frank's own words, "the effect Samir and I were looking for is a 'strong' single atom effect, in the sense that two atoms will not enhance but rather kill the effect. The suggestion that Dr. Mandel made about using the elastic scattering was a beautiful suggestion. But, the problem was that, to get any kind of signal for a single atom, we needed a microscope objective to collect as much light as possible. This meant, however, that both the fluorescence and the elastic scattering were Doppler broadened, in such as way as to make them indistinguishable from each other by making use of the Fabry-Perot filter."

Augusto had informed us, this had been tried before, but the violation of the inequality demanded by quantum mechanics had never been observed. We were able to succeed with downconversion, and to cast some light on why the previous attempts may have failed [28]. A further modification then allowed us to refute yet another alternative to quantum mechanics – a local-realistic theory of photodetection proposed by Franco Selleri [29].

During this period, *Optical Coherence and Quantum Optics*, the book co-written by Dr. Mandel and Professor Emil Wolf, appeared [30]. It was the book many of us wished had existed when we had taken Dr. Mandel's course in quantum optics, codifying as it did all of our carefully recorded class notes and problem sets. Also during this period, Dr. Mandel was named to the Lee DuBridge Chair in the physics department.

After a final paper on "anticoherence," to which the entire group contributed [31], Augusto and Carlos returned to their respective home institutions. Frank investigated the squeezing in resonance fluorescence (subsequently observed by Samir and co-workers [32]). After measuring the transition from single atom anti-bunching to bunching from many atoms, Frank began writing his thesis. Meanwhile Justin and Dr. Mandel returned to the quantum phase problem, performing new experiments [33] and eventually concluding that it could not be a unique quantity. That is, there is a different operationally defined phase operator corresponding to each apparatus one might construct to measure it [34]. Soon after this, Justin began writing his thesis.

New Collaborations

In 1996, I was invited by Professor Ian Walmsley, in the nearby Department of Optics, to give a talk on the "quantum eraser." This was an experiment based on a proposal by Scully and Druhl [35] that Carlos and I had performed with two downconverters, similar to the "mind-boggling" experiment of 1991: it showed that one could make interference come and go in a system by first creating, and then eliminating, distinguishing information [36]. Dr. Mandel had not been enthusiastic about the quantum eraser when we first proposed it, perhaps because he felt it was covering old ground [37], but he had eventually supported it, saying, "well, I suppose there is still some interest in people seeing these kinds of results."

Ian was certainly among those who were interested. His group worked with incredibly brief pulses of light, only tens of femtoseconds in duration. Ian and one of his students, Warren Grice, were studying what happened when such "ultrafast" or "ultra-short" pulses were used as the pump for a parametric downconverter. In particular, what were the effects on the frequency entanglement between the signal and idler photons? They had come up with an experiment that resembled a quantum eraser, to be performed with an ultrafast-pumped downconverter [38]. Once again, I was hooked by someone else's description of a quantum interference experiment, and asked Dr. Mandel if it would be possible for our groups to collaborate so that I could join their effort. He agreed, and I spent the final two years of graduate school working in Ian's group on several experiments with ultrafast-pumped downconversion. This happy arrangement was possible only through Dr. Mandel's generosity.

The downconversion experiments undertaken by Ian's group were based on the two-photon interference effect discovered by Hong, Ou, and Mandel in 1987 [39]. Since its discovery, this effect – that two indistinguishable photons, incident from opposite sides of a beamsplitter, must exit together, on the *same* side – has played an ever-increasing role in modern quantum optics. The "Hong-Ou-Mandel Interferometer" has been used in quantum teleportation [40,41], in entanglement swapping [42], and in determining the relative flight times for photons through dispersive media [43-47] and tunneling barriers [48,49]. New applications continue to appear as optical quantum information science moves forward [50].

A few months after I began working with Ian and Warren, Dr. Mandel's group was joined by a new student, Alex Kuzmich. From the time Justin and I had joined, only a few younger students had sought to join the group, and none had remained for longer than a year. Alex came in as a very advanced student, with a mastery of quantum mechanics and his name on several papers already. He and Dr. Mandel quickly performed an experiment on sub-shot-noise polarization measurements with downconverters [51], and he also provided crucial input and data for a paper on interference between multiple photons at a beamsplitter [52].The latter paper was also a result of our collaboration with Ian, and soon Alex began to work more closely with Ian as well. This ultimately led to a new experimental violation of Bell's inequality with a downconverter, using homodyne detection [53].

In addition to all of this, Alex and Dr. Mandel pursued another line of research: the creation of "spin squeezing" in the collective quantum state of a group of cesium atoms [54,55]. This was achieved in collaboration with Professor Nick Bigelow in the physics department. The latter of these papers appeared in August of 2000, the final article on a list of over 300 authored or co-authored by Dr. Mandel. Alex graduated as the 39th and last of Dr. Mandel's Ph.D. students.

Final Thoughts

The picture of Dr. Mandel's research that has emerged here from my recollections is certainly not a complete one, and of course neither is the picture of the man himself. André had a special term of endearment for him: we often called him "The Boss." This was accurate, for he guided the course of our research, taught us how to perform calculations, and was unfailingly available for consultation in his office, but he was not "in the trenches" aligning mirrors with us on a daily basis. We understood that this was not his role, but ours. Though he was The Boss, we students were encouraged to participate in every aspect of the work, from the conception of experiments and preliminary calculations, to their design, construction, operation, and ultimate summation in journal articles. If we were denied anything at all, it was perhaps only the chance to stand shoulder-to-shoulder with him in the lab, and get to know him better in the familiar ways we all came to know each other – as friends and comrades.

But in talking to various members of the group over the years, there is one note that consistently sounds among us as we reflect on those times: a feeling of gratitude. Each of us was given a tremendous opportunity by Dr. Mandel, to pursue directions in physics that were fascinating and exciting to us. For this, and for his guidance, his kindness, and his example, we cannot thank him enough.

Left to right: Alex Kuzmich, Ron Jodoin, Paul Lett, Samir Bali, Joe Abate, Jeff Ou, Chuck Carniglia, Fred Davidson, Jeanne Mandel, Ruth Andrus, Jeff Kimble, Jennifer VanRemmen, Dave Meltzer, Xingaou Zou, Surrendra Singh, David Branning, Lijun Wang, Mario Dagenais, Ed Martin gathered at the Mandel home after the Eighth Rochester Conference on Coherence and Quantum Optics. (Photograph courtesy of Ron Jodoin.)

(For a review of Leonard Mandel's work with parametric downconversion in his own words, the reader is referred to [56]. This was a comprehensive article commissioned for the special centenary issue of *Reviews of Modern Physics*, and reprinted in [57], where Dr. Mandel is found in the company of many of the most prominent figures in 20th century physics. Several of his experiments are also explained in [58], and in less technical terms in [59].)

References

[1] D. Branning, Ph.D. thesis, University of Rochester, 1998 (unpublished).

[2] T. P. Grayson, Ph.D. thesis, University of Rochester, 1994 (unpublished).

[3] L. Mandel, "Coherence and indistinguishability," Optics Letters **16**, 1882 (1991).

[4] X. Y. Zou, L. J. Wang, and L. Mandel, "Induced Coherence and Indistinguishability in Optical Interference," Physical Review Letters **67**, 318 (1991).

[5] J. Horgan, "Quantum Philosophy," *Scientific American* (1992).

[6] T. P. Grayson, X. Y. Zou, D. Branning, J. R. Torgerson, and L. Mandel, "Interference and indistinguishability governed by time delays in a low-Q cavity," Physical Review A **48**, 4793 (1993).

[7] T. P. Grayson, J. R. Torgerson, and G. A. Barbosa, "Observation of a nonlocal Pancharatnam phase shift in the process of induced coherence without induced emission," Physical Review A **49**, 626 (1994).

[8] F. Narducci, Ph.D. thesis, University of Rochester, 1996 (unpublished).

[9] C. K. Hong, Z. Y. Ou, and L. Mandel, "Interference between a fluorescent photon and a classical field: An example of nonclassical interference," Physical Review A **37**, 3006 (1988).

[10] J. Eschner, C. Raab, R. Schmidt-Kaler, and R. Blatt, "Light interference from single atoms and their mirror images," Nature (London) (to be published).

[11] S. Bali, F. A. Narducci, and L. Mandel, "Coherence and elastic scattering in resonance fluorescence," Physical Review A **47**, 5056 (1993).

[12] A. Fougères, J. W. Noh, T. P. Grayson, and L. Mandel, "Measurement of phase differences between two partially coherent fields," Physical Review A **49**, 530 (1994).

[13] A. Fougères, J. R. Torgerson, and L. Mandel, "New investigations of the phase distribution with variable coherence," Optics Communications **105**, 199 (1994).

[14] A. Fougères, C. H. Monken, and L. Mandel, "Measurements of the probability distribution of the phase difference between two quantum fields," Optics Letters **19**, 1771 (1994).

[15] A. Fougères, Ph.D. thesis, University of Rochester, 1994 (unpublished).

[16] T. P. Grayson and G. A. Barbosa, "Spatial properties of spontaneous parametric down-conversion and their effect on induced coherence without induced emission," Physical Review A **49**, 2948 (1994).

[17] A. Einstein, B. Podolsky, and N. Rosen, "Can Quantum-Mechanical Description of Physical Reality Be Considered Complete?," Physical Review **47**, 777 (1935).

[18] Z. Y. Ou and L. Mandel, "Violation of Bell's Inequality and Classical Probability in a Two-Photon Correlation Experiment," Physical Review Letters **61**, 50 (1988).

[19] L. J. Wang, X. Y. Zou, and L. Mandel, "Experimental Test of the de Broglie Guided-Wave Theory for Photons," Physical Review Letters **66**, 1111 (1991).

[20] J. R. Croca, A. Garuccio, V. L. Lepore, and R. N. Moreira, "Comment on "Experimental Test of the de Broglie Guided-Wave Theory for Photons"," Physical Review Letters **68**, 3813 (1992).

[21] L. Hardy, "Nonlocality for Two Particles without Inequalities for Almost All Entangled States," Physical Review Letters **71**, 1665 (1993).

[22] T. F. Jordan, "Testing Einstein-Podolsky-Rosen assumptions without inequalities with two photons or particles with spin 1/2," Physical Review A **50**, 62 (1994).

[23] J. R. Torgerson, D. Branning, and L. Mandel, "A method of demonstrating violation of local realism with a two-photon downconverter without use of Bell inequalities," Applied Physics B **60**, 267 (1994).

[24] J. R. Torgerson, D. Branning, C. H. Monken, and L. Mandel, "Experimental demonstration of the violation of local realism without Bell inequalities," Physics Letters A **204**, 323 (1995).

[25] J. Glanz, "Measurements Are the Only Reality, Say Quantum Tests," Science **279**, 1439 (1995).

[26] E. Kreifeldt, "Optical experiment defies classical physics," *Optics & Photonics News* **7** (3), 10 (1996).

[27] D. Branning, "Does Nature Violate Local Realism?," American Scientist **85**, 160 (1997).

[28] J. R. Torgerson, D. Branning, C. H. Monken, and L. Mandel, "Violations of locality in polarization-correlation measurements with phase shifters," Physical Review A **51**, 4400 (1995).

[29] A. Garuccio, J. R. Torgerson, C. H. Monken, D. Branning, and L. Mandel, "Experimental test of Selleri's photodetection model," Physical Review A **53**, 2944 (1996).

[30] L. Mandel and E. Wolf, *Optical Coherence and Quantum Optics* (Cambridge University Press, Cambridge, 1995).

[31] C. H. Monken, A. Garuccio, D. Branning, J. R. Torgerson, F. Narducci, and L. Mandel, "Generating mutual coherence from incoherence with the help of a phase-conjugate mirror," Physical Review A **53**, 1782 (1996).

[32] Z. H. Lu, S. Bali, and J. E. Thomas, "Observation of Squeezing in the Phase-Dependent Fluorescence Spectra of Two-Level Atoms," Physical Review Letters **81**, 3635 (1998).

[33] J. R. Torgerson and L. Mandel, "Sharpening the measured quantum phase distribution by discarding data," Optics Communications **133**, 153 (1997).

[34] J. R. Torgerson and L. Mandel, "Is There a Unique Operator for the Phase Difference of Two Quantum Fields?," Physical Review Letters **76**, 3939 (1996).

[35] M. O. Scully and K. Druhl, "Quantum eraser: a proposed photon correlation experiment concerning observation and 'delayed choice' in quantum mechanics," Physical Review A **25**, 2208 (1982).

[36] C. H. Monken, D. Branning, and L. Mandel, in *Coherence and Quantum Optics VII*, edited by J. H. Eberly, L. Mandel and E. Wolf (Plenum Press, 1995), p. 701.

[37] A. G. Zajonc, L. J. Wang, X. Y. Zou, and L. Mandel, "Quantum eraser," Nature (London) **353**, 507 (1991).

[38] W. P. Grice and I. A. Walmsley, "Spectral information and distinguishability in type-II down-conversion with a broadband pump," Physical Review A **56**, 1627 (1997).

[39] C. K. Hong, Z. Y. Ou, and L. Mandel, "Measurement of subpicosecond time intervals between two photons by interference," Physical Review Letters **59**, 2044 (1987).

[40] S. L. B. Braunstein, S.L. and A. Mann, "Measurement of the Bell operator and quantum teleportation," Physical Review A **51**, R1727 (1995).

[41] D. Bouwmeester, J. W. Pan, K. Mattle, M. Eibl, H. Weinfurter, and A. Zeilinger, "Experimental quantum teleportation," Nature **390**, 575 (1997).

[42] J. W. Pan, D. Bouwmeester, H. Weinfurter, and A. Zeilinger, "Experimental Entanglement Swapping: Entangling Photons That Never Interacted," Physical Review Letters **80**, 3891 (1998).

[43] A. M. Steinberg, P. G. Kwiat, and R. Y. Chiao, "Dispersion Cancellation in a Measurement of the Single-Photon Propagation Velocity in Glass," Physical Review Letters **68**, 2421 (1992).

[44] A. M. Steinberg, P. G. Kwiat, and R. Y. Chiao, "Disperision cancellation and high-resolution time measurements in a fourth-order optical interferometer," Physical Review A **45**, 6659 (1992).

[45] A. V. Sergienko, Y. H. Shih, and M. H. Rubin, "Experimental evaluation of a two-photon wave packet in type-II parametric downconversion," Journal of the Optical Society of America B **12**, 859 (1995).

[46] E. Dauler, G. Jaeger, A. Muller, A. Migdall, and A. V. Sergienko, "Tests of a Two-Photon Technique for Measuring Polarization Mode Dispersion With Subfemtosecond Precision," Journal of Research of the National Institute of Standards and Technology **104**, 1 (1999).

[47] D. Branning, A. Migdall, and A. V. Sergienko, "Simultaneous measurement of group and phase delay between two photons," Physical Review A **62**, 063808 (2000).

[48] A. M. Steinberg, P. G. Kwiat, and R. Y. Chiao, "Measurement of the Single-Photon Tunneling Time," Physical Review Letters **71**, 708 (1993).

[49] R. Y. Chiao, P. G. Kwiat, and A. M. Steinberg, "Faster than Light?," *Scientific American*, 52 (1993).

[50] D. Bouwmeester, A. K. Eckert, and A. Zeilinger, *The Physics of Quantum Information* (Springer-Verlag, Heidelberg, 2000).

[51] A. Kuzmich and L. Mandel, "Sub-shot-noise interferometric measurements with two-photon states," Journal of Optics B: Quantum and Semiclassical Optics **10**, 493 (1998).

[52] A. Kuzmich, D. Branning, L. Mandel, and I. A. Walmsley, "Multiphoton interference effects at a beamsplitter," Journal of Modern Optics **45**, 2233 (1998).

[53] A. Kuzmich, I. A. Walmsley, and L. Mandel, "Violation of Bell's Inequality by a Generalized Einstein-Podolsky-Rosen State Using Homodyne Detection," Physical Review Letters **85**, 1349 (2000).

[54] A. Kuzmich, N. P. Bigelow, and L. Mandel, "Atomic quantum non-demolition measurements and squeezing," Europhysics Letters **42**, 481 (1998).

[55] A. Kuzmich, L. Mandel, and N. P. Bigelow, "Generation of Spin Squeezing via Continuous Quantum Nondemolition Measurement," Physical Review Letters **85**, 1594 (2000).

[56] L. Mandel, "Quantum effects in one-photon and two-photon interference," Reviews of Modern Physics **71**, S274 (1999).

[57] L. Mandel, "Quantum Effects in One-Photon and Two-Photon Interference," in *More things in heaven and earth: A celebration of physics at the millennium*, edited by B. Bederson (Springer-Verlag, New York, 1999), p. 460.

[58] J. Perina, Z. Hradil, and B. Jurco, *Quantum Optics and Fundamentals of Physics* (Kluwer, Norwell, MA, 1994).

[59] G. Greenstein and A. G. Zajonc, *The Quantum Challenge* (Jones and Bartlett, Boston, 1997).

AUTHOR INDEX

INDEX